The cover art for this edition, as for three of the four previous editions, was rendered by William B. Westwood, a noted medical illustrator. Bill has a talent for capturing the dynamic nature of physiological processes in a way that excites the imagination and delights the eye. As an author with an appreciation for art but little artistic ability, I've been enormously gratified to see my vague suggestions transformed into artwork so engaging. I wish I could say "Yes, that's what I had in mind!" But it wouldn't be true, because my mental images always pale before Bill's artistic realization.

The montage Bill created depicts different, but interacting and interdependent, physiological processes. The prominently displayed heart is receiving blood from some vessels (veins) and pumping blood into others (arteries). The purpose of this activity is to keep blood moving through the many capillaries of the body, one of which is shown large and open in the foreground. The central position of the capillary is justified because capillaries are the "business end" of the circulation. The blood in capillaries carries oxygen (in the numerous red blood cells seen in the vessel); antibodies (the Y-shaped structures); white blood cells for defense (the cells squeezing out of the capillary); and many other substances needed by the body. The oxygen carried by the blood is obtained from air sacs (alveoli) of the lungs, shown in the middle right of the picture.

The large cell in the left foreground is an endocrine cell of the adrenal medulla, which is secreting the hormone epinephrine into the blood capillary. Nerve axons are shown innervating the adrenal cell to emphasize the interaction of the nervous and endocrine systems in regulating other body systems. Just visible above the adrenal cell is a mast cell. Mast cells release histamine and are important in immune function and allergy. Also important in immune function is the macrophage, here seen engulfing another cell near the middle of the picture. Nerve axons releasing chemical neurotransmitters and neurohormones are also evident.

A look at the cover art awakens the viewer to the vast complexity and beauty of human physiology. It is my hope that the knowledge gained through study of the text will heighten this initial sense of wonder and further stimulate the imagination.

Human Physiology

fifth edition

STUART IRA FOX

PIERCE COLLEGE

WCB **Wm. C. Brown Publishers**

Dubuque, IA Bogota Boston Buenos Aires Caracas Chicago
Guilford, CT London Madrid Mexico City Sydney Toronto

Book Team

Editor   *Colin H. Wheatley*
Developmental Editor   *Kristine Noel*
Production Editor   *Julie L. Wilde*
Designer   *Christopher E. Reese*
Art Editor   *Mary E. Powers*
Photo Editor   *John C. Leland*
Permissions Coordinator   *Vicki Krug*

WCB **Wm. C. Brown Publishers**

President and Chief Executive Officer   *Beverly Kolz*
Vice President, Publisher   *Kevin Kane*
Vice President, Director of Sales and Marketing   *Virginia S. Moffat*
Vice President, Director of Production   *Colleen A. Yonda*
National Sales Manager   *Douglas J. DiNardo*
Marketing Manager   *Craig S. Marty*
Advertising Manager   *Janelle Keeffer*
Production Editorial Manager   *Renée Menne*
Publishing Services Manager   *Karen J. Slaght*
Royalty/Permissions Manager   *Connie Allendorf*

Copyedited by Ann Mirels

Cover Illustration by William B. Westwood

The credits section for this book begins on page 685 and
is considered an extension of the copyright page.

612/70X

y wife, Ellen

PUMP UP YOUR GRADES!

TRIM YOUR STUDY TIME

Student Study Guide for *Human Physiology*
Laurence G. Thouin, Jr.
ISBN 0-697-20987-3

For each chapter in this text, there is a corresponding study guide chapter—including questions for review, objective questions, and essay questions—that reviews major chapter concepts.

Coloring Guide to Anatomy and Physiology
by Judith Stone and Robert Stone
ISBN 0-697-17109-4

This helpful manual serves as a framework for learning human anatomy and physiology. By labeling and coloring each drawing, you will easily learn key anatomical and physiological structures and functions.

Atlas of the Skeletal Muscles
by Judith Stone and Robert Stone
ISBN 0-697-10618-7

This inexpensive atlas depicts all of the human skeletal muscles in precise drawings that show their origin, insertion, action, and innervation. Each muscle is presented on a separate page, and the spinal cord level of the nerve fibers is included in parentheses. These features will help you visually understand the action of muscles.

Study Cards for Human Anatomy and Physiology
by Kent Van De Graaff et al.
ISBN 0-697-26447-5

Make studying a breeze with this boxed set of 300, two-sided study cards. Each card provides a complete description of terms, clearly labeled drawings, pronunciation guides, and clinical information on diseases.

 Wm. C. Brown Publishers

Brief Contents

Expanded Contents

Preface

At one time or another, all of us wonder about how our body works. The human body—its functions and mechanisms—is the focus of human physiology. Not only is body function a topic of general interest, it is also a required course of study for many college students. Human physiology provides the scientific foundation for the field of medicine and all other technologies related to human health and physical performance. The scope of topics included in a human physiology course is therefore wide-ranging, yet each topic must be covered in sufficient detail to provide a firm basis for future expansion and application. The rigor of the course, however, need not diminish the student's initial fascination with how the body works. On the contrary, a basic understanding of physiological mechanisms can instill a deeper appreciation for the awesome complexity and beauty of the human body and motivate the student to learn still more.

This text is designed to serve the needs of students enrolled in an undergraduate physiology course. The beginning chapters introduce basic chemical and biological concepts to provide these students—many of whom do not have extensive science backgrounds—with the framework they need to comprehend physiological principles. In the chapters that follow, the material is presented in such a way as to promote conceptual understanding rather than rote memorization of facts. Every effort has been made to help students integrate related concepts and to understand the relationships between anatomical structures and their functions.

Abundant summary flowcharts and tables serve as aids for review. Beautifully rendered figures, with a functional use of color, are designed to enhance learning. Health applications are discussed often to heighten interest, deepen understanding of physiological concepts, and help students relate the material they have learned to their individual career goals. In addition, various other pedagogical devices are used extensively (but not intrusively) to add to the value of the text as a comprehensive learning tool. These devices are discussed in detail under the heading "Learning Aids—A Guide to the Student."

Shaping the Fifth Edition

Before I began writing this new edition, Wm. C. Brown Publishers requested users of the fourth edition to send in their suggestions and comments, focusing on a chapter of particular interest. The response was enthusiastic, warm, encouraging, insightful—in a word, it was wonderful! Because so many people sent in one-chapter reviews, every chapter in the book was reviewed several times over. The fifth edition benefited enormously by this input, as evidenced by the reorganization of chapters, the expanded coverage of many topics, and the introduction of new material.

Examples of organizational changes include the grouping of the nervous system chapters to follow each other (chapter 10 in the fourth edition is now chapter 9) and the shifting of the immune system chapter (formerly chapter 19) to follow the chapters on the cardiovascular system (as chapter 15). An introduction to blood acid-base balance, previously distributed between two chapters (15 and 16), is now consolidated in chapter 13. Expanded discussions include those on neurotransmitter release and action, mechanisms of hormone action, development of atherosclerosis, mechanism of muscle contraction, paracrine regulation, regeneration of damaged axons, and T cell function, to name but a few.

Our knowledge about a number of physiological processes has grown since the last edition of this text was written. The physiological roles of nitric oxide, for example, are just now becoming clear. Associated topics, including long-term potentiation in neurons and paracrine regulation of blood vessels, are thus covered in more depth than was previously possible. Physiological and clinical advances based on breakthroughs in molecular genetics are emphasized in accordance with their current prominence. This emphasis is apparent in discussions on cystic fibrosis, muscular dystrophy, ALS, and Laron dwarfism, among other topics. New material on oncogenes, apoptosis, brown fat, oral rehydration therapy, G-proteins, olfactory receptor genetics, acute mountain sickness, neurotrophins, and the mechanism of ADH action further illustrate how the text has been extensively updated in response to new research findings.

The fifth edition features many new and revised figures to accompany the new topics and updated discussions. These figures are beautifully rendered and stylistically consistent with the previous edition. The new pedagogical devices introduced in the last edition have been retained in this edition. These include the Clinical Investigations and the appendix on exercise physiology. The latter has been expanded in accordance with the increased coverage of exercise physiology throughout the text.

Supplementary Materials

The supplementary materials that accompany the fifth edition of *Human Physiology* are designed to help students in their learning activities and assist instructors in planning coursework and presentations. These supplementary materials include

1. An *Instructor's Manual* prepared by Jeffrey and Karianne Price. This manual can be used as an aid to planning lessons. It also includes an extensive bank of questions for constructing examinations.

2. A *Laboratory Guide to Human Physiology: Concepts and Clinical Applications,* seventh edition, written by Stuart I. Fox. This laboratory manual is self-contained, so that students can prepare for the laboratory exercises and quizzes without having to bring the textbook to the laboratory. The manual provides exercises that reinforce many of the topics covered in a human physiology course. These exercises have been classroom tested for a number of years.

3. An *Instructor's Manual* for the laboratory guide. This manual is designed to help instructors set up the laboratory. It also provides answers to the questions in the laboratory reports.

4. A *Student Study Guide to Accompany Human Physiology* written by Dr. Lawrence G. Thouin, Jr. This guide

for students provides excellent sample objective questions, tips for answering essay questions, flowchart activities, crossword puzzles, and other devices to promote effective studying.

5. A set of *150 acetate transparencies*. These transparencies were made from selected illustrations in the text and were chosen for their value in reinforcing lecture presentations. They are available to instructors who adopt the text.

6. A *Student Study Art Notebook* containing all the illustrations from the transparency set is available to students. With this notebook at their desks, students no longer have to worry about whether they will be able to see leader lines and labels in a large lecture hall.

7. *Microtest*. This computerized system enables the instructor to create customized exams quickly and easily. Test questions are available from the test item file that appears in the Instructor's Manual.

8. *QuickStudy*. This student self-testing program operates on an IBM/PC or Macintosh computer. The questions have been extensively revised and expanded by Dr. L. G. Thouin, Jr.

Multimedia

To provide an even more comprehensive supplement package, the fifth edition of *Human Physiology* introduces two new and exciting multimedia learning tools for the student.

1. *Explorations in Human Anatomy and Physiology*. This set of fifteen interactive animations on CD-ROM covers key topics in human physiology. Explorations in Human Anatomy and Physiology consists of hands-on interactive activities that can be used by an instructor in lecture and/or placed in a resource center for student use. This software is available for use with Macintosh and IBM Windows computers. Appropriate modules are listed in many of the text chapters, following the chapter summaries.

2. *WCB Life Science Animations*. A series of five videotapes contains more than fifty animations of physiological processes integral to the study of human anatomy and physiology. Some of the topics are better learned when movement and the changes it produces are visualized. The videotapes can be used in class as a lecture tool, or students can visit a resource center to view them on their own. The five videotapes are
   #1 Chemistry, The Cell, and Energetics
   #2 Cell Division, Heredity, Genetics, Reproduction, and Development
   #3 Animal Biology I
   #4 Animal Biology II
   #5 Plant Biology, Evolution, and Ecology

3. *WCB Anatomy and Physiology Videodisc*. This two-disc, four-sided videodisc contains thirty animations integral to the study of physiology. Along with the animations are illustrations from *Human Physiology*, histology slides, clinical slides, and radiographic slides. The videodisc can be used in class to supplement lectures.

Learning Aids—A Guide to the Student

The clarity of organization and writing style in the fifth edition will help you to integrate and synthesize information, rather than merely memorize facts. Think of the major sections in the text as packets of instructions on how to assemble a device or appliance. Take each concept a step at a time, reread when necessary, scrutinize the figures, and actually write out answers to the study questions that appear at the end of these sections. Don't be intimidated by a long explanation—after all, a detailed set of instructions is easier to follow than one that is too brief. The more actively you interact with your text, the better will be your understanding of physiology, and the more enjoyable the study will become.

To help you gain the most from your textbook, consider the following information about how to use the learning aids.

Chapter Outline and Objectives

Look over the chapter outline before reading a chapter to get a feel for the topics to be covered, and use the outline later to help you look up topics and integrate concepts with those covered later in the text. Check off the objectives as you complete each major section to see if you are getting what is required from the text.

Perspectives

Immediately following each major section heading is a short paragraph, double-spaced and bold-faced to set it apart. It is a concise statement of the section's central concepts, or organizing themes, that are illustrated in detail in the text that follows. Read these statements carefully. As their name implies, they will help you place a section in perspective as you begin it, before getting involved with the specifics. Read them again after you have completed a section to help you maintain your conceptual focus.

Boxed Information

Following a discussion of a basic concept, you may find a block of text printed on a colored background with an accompanying icon. These "boxes" feature clinical or fitness applications of the information just covered. You will find it enjoyable, as well as instructive, to see how your newly acquired basic knowledge is being applied to practical problems.

Cross-References

As you read about a particular physiological mechanism, you may come across a reference to a concept that was discussed earlier in the text. If this concept is no longer clear, look it up in the referenced chapter. This is a good way to review, and it will help you to more completely integrate related physiological concepts. You may also see a reference to a related topic that is covered in more detail in a later chapter. Go ahead—take a peek at what's ahead. You may not be responsible for the detail now, but you will be better prepared for the detail now, but you will be better prepared to integrate this information with previous knowledge when you reach this chapter later on in the course.

Study Activities

Each major chapter section ends with a list of study activities. These may be essay questions to answer, or perhaps diagrams or flowcharts to draw. Don't just think about how you might respond. Use a pencil and paper to write or draw.

Illustrations and Tables

The text includes abundant tables and illustrations to support the concepts presented. Careful study of the tables will increase your understanding of the text, and the summary tables will be useful when you review for examinations. Although many of the figures can be admired for their beauty alone, they were created with one primary purpose—to illustrate concepts presented in the text. Therefore, refer to the figures and analyze them as you read. Each one has been placed as close as possible to its text reference to spare you from flipping through pages.

Chapter Summaries

At the end of each chapter, the material is summarized for you in outline form. This outline summary is organized by major section headings followed by the key points in the section. Read the summary after studying the chapter to be sure that you have not missed any points, and use the chapter summaries to help you review for examinations.

Clinical Investigations

Following each chapter summary is a Clinical Investigation. Think of each one as a puzzle, and use the clues provided to solve it. After writing out your solution, see how closely it agrees with the solution given in Appendix B.

Review Activities

A section called Review Activities follows the Clinical Investigations. These activities include *objective questions* (with the answers in Appendix C) and *essay questions*. The first essay question in each chapter is answered in the Student Study Guide. Be sure to take these self-quizzes in a "closed book" fashion after studying the chapters, and then correct your answers using the appendix. Be sure to review the information relating to any questions you might have missed. These practice exams will help you to anticipate the kinds of questions that are likely to appear on real exams. They will also provide you with feedback as to the depth of your learning and understanding.

Selected Readings and Multimedia

Each chapter closes with a list of books and articles. Interested students can use these lists as sources of additional information on topics covered in the chapter. Many of the articles are from popular journals, such as *Scientific American*, and are written for audiences with limited science backgrounds.

Appendices

Appendix A is entitled "Exercise Physiology: Summary and Text References." It is designed to help you integrate the physiology of different organs and systems into a comprehensive view of how the body works, using exercise as a theme. Appendix B contains solutions to the Clinical Investigations. Refer to this appendix only after you have attempted to solve an investigation yourself. Appendix C consists of answers to the objective questions included in the Review Activities sections.

Glossary

The glossary of terms at the end of the book is particularly noteworthy for its comprehensiveness. The definitions for almost all of the terms are accompanied by pronunciation keys, and synonyms are indicated as appropriate. Whenever you encounter an unfamiliar term or would like additional information about a term, look it up in the glossary. Also, look to the glossary as you study for exams, to check your understanding of the technical terminology.

Student Study Guide

Written by Dr. Lawrence G. Thouin Jr., this optional book can help you to derive more benefit from the text. The answer to the first essay question in the Review Activities at the end of each chapter is provided here, along with helpful hints on how to answer essay questions on physiology. In addition, the study guide provides additional objective questions (with answers), fill-in-the-blank questions, crossword puzzles, and other devices to help you use your textbook more effectively.

Acknowledgments

As mentioned previously, many users of the fourth edition contributed individual chapter reviews. I am extremely grateful to all of them and have endeavored to incorporate their suggestions wherever possible. In addition to the many professors who contributed chapter reviews of the fourth edition, several colleagues reviewed the entire fifth edition in manuscript form. I am grateful to them for taking on this arduous task, and I assure them that their efforts resulted in a much improved final draft. I would also like to thank H. A. Pershadsingh, Ph.D., M.D. (Kern County Medical Center) for his review of the entire fourth edition textbook and his expert clinical and research input. His suggestions made the fifth edition significantly more current. As in the past, my colleagues at Pierce College have been very supportive and helpful. In particular, I would like to thank Dr. Lawrence G. Thouin Jr., Dr. James Rikel, and Mr. Edmont Katz.

Fifth-Edition Manuscript Reviews:

Joseph K. Allamong
West Virginia University
Laren Barker
Southwest State University
Rose Ann Bast
Mount Mary College
Barbara W. Birge
University of Kentucky

Don R. Boyer
Washburn University of Topeka
Thomas A. Burns
Northwestern State University
Melvin C. Ching
James Madison University
John A. Chisler
Glenville State College
Peter Claussen
South Dakota State University
Paul V. Cupp, Jr.
Eastern Kentucky University
Dwayne H. Curtis
California State University, Chico
Joseph A. De Guzman
Merritt College
Kamiab Delfanian
Worthington Community College
Carolyn A. Dennehy
University of Northern Colorado
Kathryn A. Elias
IVY Tech State College–Indianapolis
Robert E. Farrell, Jr.
Penn State University–York
Daniel S. Fertig
East Los Angeles College
Sheldon R. Gordon
Oakland University
John C. Grew
St. Francis College
Janice J. Halsne
Luther College
Vicki J. Harber
University of Alberta
John P. Harley
Eastern Kentucky University
L. Mark Harrison
University of Indianapolis
Ceil Ann Herman
New Mexico State University
Sandra Hsu
Merritt College
David H. Jones
Grove City College
Roderick P. Kernan
University College Dublin
T. Daniel Kimbrough
Virginia Commonwealth University
Loren W. Kline
University of Alberta
Jeanne Kowalczyk
University of South Carolina
David S. Mallory
Marshall University
Elden W. Martin
Bowling Green State University

Donald M. McKinstry
Penn State Erie, Behrend College
Jacqueline Shea McLaughlin
*Pennsylvania State University
–Allentown Campus*
Steven D. Mercurio
Mankato State University
Glenn W. Merrick
Duluth Community College
Gail L. Miller
York College
Lowell D. Neudeck
Northern Michigan University
Olalekan E. Odeleye
National Institute of Health
Eileen S. O'Neill
*University of Massachusetts
Dartmouth*
Virginia A. Pascoe
Mt. San Antonio College
H. A. Pershadsingh
Kern Medical Center
Scott K. Powers
University of Florida
Nancy Rauch
Merritt College
Larry A. Reichard
Maple Woods Community College
Roscoe B. Root
Lansing Community College
Albert J. Roy
University of Rhode Island
Leland S. Shapiro
Los Angeles Pierce College
Kevin Sinchak
Lansing Community College
Lloyd C. Stavick
*University of South Dakota
Lake Area Vo-Tech Institute*
David E. Taylor
*University of South Carolina at
Spartanburg*
Kenneth A. Thomas
University of Rhode Island
Richard L. Walker
University of Calgary
Danny Wann
Carl Albert State College
DeLoris Wenzel
University of Georgia
Gary L. Whitson
University of Tennessee
Karin E. Winnard
Sonoma State University
David A. Woodman
University of Nebraska–Lincoln

Human Physiology

1

The Study of Body Function

1. describe, in a general way, the topics studied in physiology and explain the importance of physiology in modern medicine.

2. describe the characteristics of the scientific method.

3. define *homeostasis* and explain how this concept is used in physiology and medicine.

4. describe the nature of negative feedback loops and explain how these mechanisms act to maintain homeostasis.

5. explain how antagonistic effectors help to maintain homeostasis.

6. describe the nature of positive feedback loops and explain how these mechanisms function in the body.

7. distinguish between intrinsic and extrinsic regulation and describe, in a general way, the roles of the nervous and endocrine systems in body regulation.

8. explain how negative feedback inhibition helps to regulate the secretion of hormones, using insulin as an example.

9. list the four primary tissues and their subtypes and describe the distinguishing features of each primary tissue.

10. relate the structure of each primary tissue to its functions.

11. describe how the primary tissues are organized into organs, using the skin as an example.

12. describe the nature of the extracellular and intracellular compartments of the body and explain the significance of this compartmentalization.

OUTLINE

Introduction to Physiology

Human physiology is the study of how the human body functions, with emphasis on specific cause-and-effect mechanisms. Knowledge of these mechanisms has been obtained experimentally through applications of the scientific method.

Physiology is the study of biological function—of how the body works, from cell to tissue, tissue to organ, organ to system, and of how the organism as a whole accomplishes particular tasks essential for life. In the study of physiology, the emphasis is on *mechanisms*—with questions that begin with the word *how* and answers that involve cause-and-effect sequences. These sequences can be woven into larger and larger stories that include descriptions of the structures involved (anatomy) and that overlap with the sciences of chemistry and physics.

The separate facts and relationships of these cause-and-effect sequences are derived empirically from experimental evidence. Explanations that seem logical are not necessarily true; they are only as valid as the data on which they are based, and they can change as new techniques are developed and further experiments are performed. The ultimate objective of physiological research is to understand the normal functioning of cells, organs, and systems. A related science—pathophysiology—is concerned with how physiological processes are altered in disease or injury.

Pathophysiology and the study of normal physiology complement one another. For example, a standard technique for investigating the functioning of an organ is to observe what happens when it is surgically removed from an experimental animal or when its function is altered in a specific way. This study is often aided by "experiments of nature"—diseases—that involve specific damage to the functioning of an organ. The study of disease processes has thus aided our understanding of normal functioning, and the study of normal physiology has provided much of the scientific basis of modern medicine. This relationship is recognized by the Nobel Prize committee, whose members award prizes in the category "Physiology or Medicine."

The physiology of invertebrates and of different vertebrate groups is studied in the science of comparative physiology. Much of the knowledge gained from comparative physiology has benefited the study of human physiology. This is because animals, including humans, are more alike than they are different. This is especially true when comparing humans with other mammals. The small differences in physiology between humans and other mammals can be of crucial importance in the development of pharmaceutical drugs (discussed later in this section), but these differences are relatively slight in the overall study of physiology.

Scientific Method

All of the information in this text has been gained by application of the scientific method. Although many different techniques are involved in the scientific method, all share three attributes: (1) confidence that the natural world, including ourselves, is ultimately explainable in terms we can understand; (2) descriptions and explanations of the natural world that are honestly based on observations and that could be modified or refuted by other observations; and (3) humility, or the willingness to accept the fact that we could be wrong. If further study should yield conclusions that refuted all or part of an idea, the idea would have to be modified accordingly. In short, the scientific method is based on a confidence in our rational ability, honesty, and humility. Practicing scientists may not always display these attributes, but the validity of the large body of scientific knowledge that has been accumulated—as shown by the technological applications and the predictive value of scientific hypotheses—are ample testimony to the fact that the scientific method works.

The scientific method involves specific steps. In the first step, a **hypothesis** is formulated. In order for this hypothesis to be scientific, it must be capable of being refuted by experiments or other observations of the natural world. For example, one might hypothesize that people who exercise regularly have a lower resting pulse rate than other people. Experiments are conducted, or other observations are made, and the results are analyzed. Conclusions are then drawn as to whether the new data either refute or support the hypothesis. If the hypothesis survives such testing, it might be incorporated into a more general **theory.** Scientific theories are statements about the natural world that incorporate a number of proven hypotheses. They serve as a logical framework by which these hypotheses can be interrelated and provide the basis for predictions that may as yet be untested.

The hypothesis in the preceding example is scientific because it is *testable;* the pulse rates of 100 athletes and 100 sedentary people could be measured, for example, to see if there were statistically significant differences. If there were, the statement that athletes, on the average, have lower resting pulse rates than sedentary people would be justified *based on these data.* One must still be open to the fact that this conclusion could be wrong. Before the discovery could become generally accepted as fact, other scientists would have to consistently replicate the results. Scientific theories are based on *reproducible* data.

It is quite possible that when others attempt to replicate the experiment their results will be slightly different. They may then construct scientific hypotheses that the differences in resting pulse rate also depend on factors such as the nature of the exercise performed, or on other variables. When scientists attempt to test these hypotheses, they will likely encounter new problems, requiring new explanatory hypotheses, which then must be tested by additional experiments.

In this way, a large body of highly specialized information is gradually accumulated, and a more generalized explanation (a scientific theory) can be formulated. This explanation will almost always be different from preconceived notions. People who follow the scientific method will then appropriately modify their concepts, realizing that their new ideas will probably have to be changed again in the future as additional experiments are performed.

Development of Pharmaceutical Drugs

The development of new pharmaceutical drugs can serve as an example of how the scientific method is used in physiology and its health applications. The process usually starts with scientists conducting basic physiological research, often at cellular and molecular levels. Perhaps a new family of drugs is developed using cells in tissue culture (in vitro, or outside the body). For example, cell physiologists, studying membrane transport, may discover that a particular family of compounds blocks membrane channels for Ca^{++}. Because of their knowledge of physiology, other scientists may predict that a drug of this nature might be useful in the treatment of hypertension (high blood pressure). This drug may then be tried in experimental animals.

If a drug is effective at extremely low concentrations in vitro, there is a chance that it may work in vivo (in the body) at concentrations low enough not to be toxic (poisonous). This must be thoroughly tested utilizing experimental animals, primarily rats and mice. More than 90% of drugs tested in experimental animals are too toxic for further development. Only in those rare cases when the toxicity is low enough may the drug be moved to human/clinical trials.

In **phase I clinical trials,** the drug is tested on healthy human volunteers. This is done to test its toxicity in humans and to study how the drug is "handled" by the body: how it is metabolized, how rapidly it is removed from the blood by the liver and kidneys, how it can be most effectively administered, and so on. If no toxic effects are observed, the drug can proceed to the next stage. In **phase II clinical trials,** the drug is tested on the target human population (for example, those with hypertension). Only in those exceptional cases where the drug seems to be effective but has minimal toxicity does testing move to the next phase. **Phase III trials** occur in many research centers across the country to maximize the number of test participants. At this point, the test population must include a sufficient number of subjects of both sexes, as well as people of different ethnic groups. In addition, people are tested who have other health problems besides the one that the drug is intended to benefit. For example, those who have diabetes in addition to hypertension would be included in this phase. If the drug passes phase III trials, it goes to the Federal Drug Administration (FDA) for approval. **Phase IV trials** test other potential uses of the drug.

The percentage of drugs that make it all the way through these trials to eventually become approved and marketed is very low. Notice the crucial role of basic research, using experimental animals, in this process. Virtually every prescription drug on the market owes its existence to such research.

Homeostasis and Feedback Control

The regulatory mechanisms of the body can be understood in terms of a single, shared function: that of maintaining constancy of the internal environment. A state of relative constancy of the internal environment is known as homeostasis, and it is maintained by effectors that are regulated by sensory information from the internal environment.

The Greek philosopher Aristotle (384–322 B.C.) speculated on the function of the human body, but another ancient Greek, Erasistratus (304–250? B.C.), is considered the father of physiology because he attempted to apply physical laws to the study of human function. Galen (A.D. 130–201) wrote widely on the subject and was held as the supreme authority until the Renaissance. Physiology became a fully experimental science with the revolutionary work of the English physician William Harvey (1578–1657), who demonstrated the function of the heart and proved the circulation of the blood.

However, the father of modern physiology is the French physiologist Claude Bernard (1813–1878), who observed that the *milieu interieur* ("internal environment") remains remarkably constant despite changing conditions in the external environment. In a book entitled *The Wisdom of the Body,* published in 1932, the American physiologist Walter Cannon (1871–1945) coined the term **homeostasis** to describe this internal constancy. Cannon further suggested that the many mechanisms of physiological regulation have but one purpose—the maintenance of internal constancy.

The concept of homeostasis has been of immense value in the study of physiology because it allows diverse regulatory mechanisms to be understood in terms of their "why" as well as their "how." The concept of homeostasis also provides a major foundation for medical diagnostic procedures. When a particular measurement of the internal environment, such as a blood measurement (table 1.1), deviates significantly from the normal range of values, it can be concluded that homeostasis is not being maintained and that the person is sick. A number of such measurements, combined with clinical observations, may allow the particular defective mechanism to be identified.

Negative Feedback Loops

In order for internal constancy to be maintained, the body must have sensors that are able to detect deviations from a **set point.** The set point is analogous to the temperature set on a house thermostat. In a similar manner, there is a set point for body temperature, blood glucose concentration, the tension on a tendon, and so on. When a sensor detects a deviation from a particular set point, it must relay this information to an **integrating**

Table 1.1	Approximate Normal Ranges for Measurements of Some Fasting Blood Values
Measurement	**Normal Range**
Arterial pH	7.35–7.45
Bicarbonate	24–28 mEq/L
Sodium	135–145 mEq/L
Calcium	4.5–5.5 mEq/L
Oxygen content	17.2–22.0 ml/100 ml
Urea	12–35 mg/100 ml
Amino acids	3.3–5.1 mg/100 ml
Protein	6.5–8.0 g/100 ml
Total lipids	400–800 mg/100 ml
Glucose	75–110 mg/100 ml

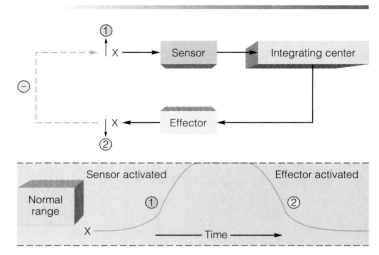

Figure 1.1

A rise in some factor of the internal environment ($\uparrow$X) is detected by a sensor. This information is relayed to an integrating center, which causes an effector to produce a change in the opposite direction ($\downarrow$X). The initial deviation is thus reversed, completing a negative feedback loop (shown by the dashed arrow and negative sign). The numbers indicate the sequence of changes.

center, which usually receives information from many different sensors. The integrating center is often a particular region of the brain or spinal cord, but in some cases it can also be cells of endocrine glands. The relative strengths of different sensory inputs are weighed in the integrating center, and, in response, the integrating center either increases or decreases the activity of particular effectors, which are generally muscles or glands.

The thermostat of a house can serve as a simple example. Suppose you set the thermostat at a set point of 70°F. If the temperature of the house rises sufficiently above the set point, a sensor within the thermostat will detect the deviation. This will then act, via the thermostat's equivalent of an integrating center, to activate the effector. The effector in this case may be an air conditioner, which acts to reverse the deviation from the set point.

If the body temperature exceeds the set point of 37°C, sensors in a part of the brain detect this deviation and, acting via an integrating center (also in the brain), stimulate activities of effectors (including sweat glands) that lower the temperature. If, as another example, the blood glucose concentration falls below normal, the effectors act to increase the blood glucose. One can think of the effectors as "defending" the set points against deviations. Since the activity of the effectors is influenced by the effects they produce, and since this regulation is in a negative, or reverse, direction, this type of control system is known as a **negative feedback loop** (fig. 1.1). (Notice that in fig. 1.1 and in subsequent figures, negative feedback is illustrated with a dashed line and a negative sign.)

The nature of the negative feedback loop can be understood by again referring to the analogy of the thermostat and air conditioner. After the air conditioner has been on for some time, the room temperature may fall significantly below the set point of the thermostat. When this occurs, the air conditioner will be turned off. The effector (air conditioner) is turned on by a high temperature and, when activated, produces a negative change (lowering of the temperature) that ultimately causes the effector to be turned off. In this way, constancy is maintained.

It is important to realize that these negative feedback loops are continuous, ongoing processes. Thus, a particular nerve fiber that is part of an effector mechanism may always display some activity, and a particular hormone, which is part of another effector mechanism, may always be present in the blood. The nerve activity and hormone concentration may decrease in response to deviations of the internal environment in one direction (fig. 1.1), or they may increase in response to deviations in the opposite direction (fig. 1.2). Changes from the normal range in either direction are thus compensated for by reverse changes in effector activity.

Since negative feedback loops respond after deviations from the set point have stimulated sensors, the internal environment is never absolutely constant. Homeostasis is best conceived as a state of **dynamic constancy,** in which conditions are stabilized above and below the set point. These conditions can be measured quantitatively, in degrees Celsius for body temperature, for example, or in milligrams per deciliter (one-tenth of a liter) for blood glucose. The set point can be taken as the average value within the normal range of measurements (fig. 1.3).

Antagonistic Effectors

Most factors in the internal environment are controlled by several effectors, which often have antagonistic actions. Control by antagonistic effectors is sometimes described as "push-pull," where the increasing activity of one effector is accompanied by decreasing activity of an antagonistic effector. This affords a finer degree of control than could be achieved by simply switching one effector on and off.

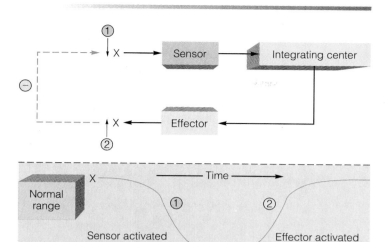

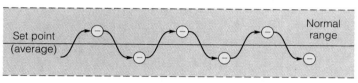

Figure 1.2

A negative feedback loop that compensates for a fall in some factor of the internal environment (↓X). (Compare this figure with fig. 1.1.)

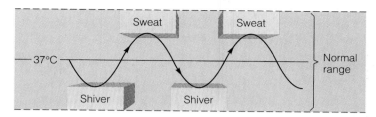

Figure 1.3

Negative feedback loops (indicated by negative signs) maintain a state of dynamic constancy within the internal environment.

Figure 1.4

A simplified scheme by which body temperature is maintained within the normal range (with a set point of 37°C) by two antagonistic mechanisms—shivering and sweating. Shivering is induced when the body temperature falls too low, and it gradually subsides as the temperature rises. Sweating occurs when the body temperature is too high, and it diminishes as the temperature falls. Most aspects of the internal environment are regulated by the antagonistic actions of different effector mechanisms.

Room temperature can be maintained for example, by simply turning an air conditioner on and off, or by just turning a heater on and off. A much more stable temperature, however, can be achieved if the air conditioner and heater are both controlled by a thermostat. Then the heater is turned on when the air conditioner is turned off, and vice versa. Normal body temperature is maintained about a set point of 37°C by the antagonistic effects of sweating, shivering, and other mechanisms (fig. 1.4).

The blood concentrations of glucose, calcium, and other substances are regulated by negative feedback loops involving hormones that promote opposite effects. While insulin, for example, lowers blood glucose, other hormones raise the blood glucose concentration. The heart rate, similarly, is controlled by nerve fibers that produce opposite effects: stimulation of one group of nerve fibers increases heart rate; stimulation of another group slows the heart rate.

Quantitative Measurements

Normal ranges and deviations from the set point must be known quantitatively in order to study physiological mechanisms. For these and other reasons, quantitative measurements are basic to the science of physiology. One example of this, and of the actions of antagonistic mechanisms in maintaining

homeostasis, is shown in figure 1.5. Blood glucose concentrations were measured in five healthy people before and after an injection of insulin, a hormone that acts to lower the blood glucose concentration. A graph of the data reveals that the blood glucose concentration decreased rapidly but was brought back up to normal levels within 80 minutes after the injection. This demonstrates that negative feedback mechanisms acted to restore homeostasis in this experiment. These mechanisms involve the action of hormones that act antagonistically to insulin and promote the secretion of glucose from the liver (chapter 18).

Positive Feedback

Constancy of the internal environment is maintained by effectors that act to compensate for the change that served as the stimulus for their activation; in short, by negative feedback loops. A thermostat, for example, maintains a constant temperature by increasing heat production when it is cold and decreasing heat production when it is warm. The opposite occurs during **positive feedback**—in this case, the action of effectors *amplifies* those changes that stimulated the effectors. A thermostat that works by positive feedback, for example, would increase heat production in response to a rise in temperature.

It is clear that homeostasis must ultimately be maintained by negative rather than by positive feedback mechanisms. The effectiveness of some negative feedback loops, however, is increased by positive feedback mechanisms that amplify the actions of a negative feedback response. Blood clotting, for example, occurs as a result of a sequential activation of clotting factors; the activation of one clotting factor results in activation of many in a positive feedback, avalanche-like, manner. In this way, a single change is amplified to produce a blood clot. Formation of the clot, however, can prevent further loss of blood, and thus represents the completion of a negative feedback loop.

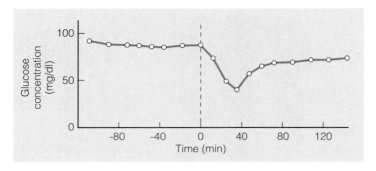

Figure 1.5

Average blood glucose concentrations of five healthy individuals before and after a rapid intravenous injection of insulin. The "0" indicates the time of the injection.

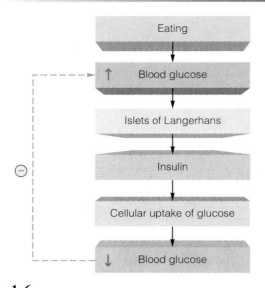

Figure 1.6

Negative feedback control of insulin secretion and blood glucose concentration. Mechanisms such as this maintain homeostasis.

Neural and Endocrine Regulation

The effectors of most negative feedback loops include the actions of nerves and hormones. In both neural and endocrine regulation, particular chemical regulators released by nerve fibers or endocrine glands stimulate target cells by interacting with specific receptor proteins in these cells. The mechanisms by which this regulation is achieved will be described in later chapters.

Homeostasis is maintained by two general categories of regulatory mechanisms: (1) those that are *intrinsic*, or "built-in," to the organs being regulated and (2) those that are *extrinsic*, as in regulation of an organ by the nervous and endocrine systems.

The endocrine system functions closely with the nervous system in regulating and integrating body processes and maintaining homeostasis. The nervous system controls the secretion of many endocrine glands, and some hormones in turn affect the function of the nervous system. Together, the nervous and endocrine systems regulate the activities of most of the other systems of the body.

Regulation by the endocrine system is achieved by the secretion of chemical regulators called **hormones** into the blood. Since hormones are secreted into the blood, they are carried by the blood to all organs in the body. Only specific organs can respond to a particular hormone, however; these are known as the *target organs* of that hormone.

Nerve fibers are said to *innervate* the organs that they regulate. When stimulated, these fibers produce electrochemical nerve impulses that are conducted from the origin of the fiber to its end point in the target organ innervated by the fiber. These target organs can be muscles or glands that may function as effectors in the maintenance of homeostasis.

Feedback Control of Hormone Secretion

The details of the nature of the endocrine glands, the interaction of the nervous and endocrine systems, and the actions of hormones will be explained in later chapters. For now, it is sufficient to describe the regulation of hormone secretion very broadly, since it so superbly illustrates the principles of homeostasis and negative feedback regulation.

Hormones are secreted in response to specific chemical stimuli. A rise in the plasma glucose concentration, for example, stimulates insulin secretion from structures known as the islets of Langerhans in the pancreas. Hormones are also secreted in response to nerve stimulation and to stimulation by other hormones.

The secretion of a hormone can be inhibited by its own effects, in a negative feedback manner. Insulin, as previously described, produces a lowering of blood glucose. Since a rise in blood glucose stimulates insulin secretion, a lowering of blood glucose caused by insulin's action inhibits further insulin secretion. This closed-loop control system is called **negative feedback inhibition** (fig. 1.6).

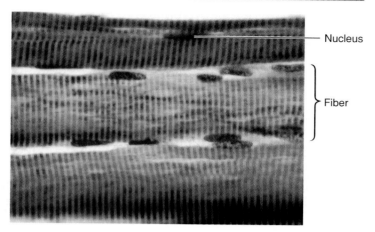

Figure 1.7

Three skeletal muscle fibers showing the characteristic cross striations.

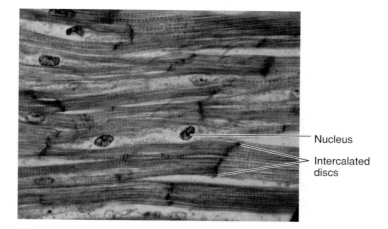

Figure 1.8

Human cardiac muscle. Notice the striated appearance and dark-staining intercalated discs.

1. Define *homeostasis* and describe how this concept can be used to explain physiological control mechanisms.
2. Define the term *negative feedback* and explain how it contributes to homeostasis. Illustrate this concept by drawing a negative feedback loop.
3. Describe positive feedback and explain how this process functions in the body.
4. Explain how the secretion of a hormone is controlled by negative feedback inhibition. Use the control of insulin secretion as an example.

The Primary Tissues

The organs of the body are composed of four different primary tissues, each of which has its own characteristic structure and function. The activities and interactions of these tissues determine the physiology of the organs.

Although physiology is the study of function, it is difficult to properly understand the function of the body without some knowledge of its anatomy, particularly at a microscopic level. Microscopic anatomy constitutes a field of study known as *histology*. The anatomy and histology of specific organs will be discussed together with their functions in later chapters. In this section, the common "fabric" of all organs is described.

Cells that have similar functions are grouped into categories called *tissues*. The entire body is composed of only four types of tissues. These **primary tissues** include (1) muscle, (2) nervous, (3) epithelial, and (4) connective tissues. Groupings of these four primary tissues into anatomical and functional units are called *organs*. Organs, in turn, may be grouped together by common functions into *systems*. The systems of the body act in a coordinated fashion to maintain the entire organism.

Muscle Tissue

Muscle tissue is specialized for contraction. There are three types of muscle tissue: **skeletal, cardiac,** and **smooth.** Skeletal muscle is often called *voluntary muscle* because its contraction is consciously controlled. Both skeletal and cardiac muscles are **striated;** they have striations, or stripes, that extend across the width of the muscle cell (figs. 1.7 and 1.8). These striations are produced by a characteristic arrangement of contractile proteins, and for this reason skeletal and cardiac muscle have similar mechanisms of contraction. Smooth muscle (fig. 1.9) lacks these cross striations and has a different mechanism of contraction.

Skeletal Muscle

Skeletal muscles are generally attached to bones at both ends by means of tendons; hence, contraction produces movements of the skeleton. There are exceptions to this pattern, however. The tongue, superior portion of the esophagus, anal sphincter, and diaphragm are also composed of skeletal muscle, but they do not cause movements of the skeleton.

Figure 1.9

A photomicrograph of smooth muscle cells. Notice that these cells contain single, centrally located nuclei (N) and lack striations.

Since skeletal muscle cells are long and thin, they are called **fibers,** or **myofibers** (from the Greek *myos,* meaning "muscle"). Despite their specialized structure and function, each myofiber contains structures common to all cells (nuclei, mitochondria, and other organelles, described in chapter 3).

The muscle fibers within a skeletal muscle are arranged in bundles, and within these bundles the fibers extend in parallel from one end to the other of the bundle. The parallel arrangement of muscle fibers (shown in fig. 1.7) allows each fiber to be controlled individually: one can thus contract fewer or more muscle fibers and, in this way, vary the strength of the whole muscle's contraction. The ability to vary, or "grade," the strength of skeletal muscle contraction is obviously needed for proper control of skeletal movements.

Cardiac Muscle

Although cardiac muscle is striated, it differs markedly from skeletal muscle in appearance. Cardiac muscle is found only in the heart, where the **myocardial cells** are short, branched, and intimately interconnected to form a continuous fabric. Special areas of contact between adjacent cells stain darkly to show *intercalated discs* (fig. 1.8), which are characteristic of heart muscle.

The intercalated discs couple myocardial cells together mechanically and electrically. Unlike skeletal muscles, therefore, the heart cannot produce a graded contraction by varying the number of cells stimulated to contract. Because of the way it is constructed, the stimulation of one myocardial cell results in the stimulation of all other cells in the mass and a "wholehearted" contraction.

Smooth Muscle

As implied by the name, smooth muscle cells (fig. 1.9) do not have the cross striations characteristic of skeletal and cardiac muscle. Smooth muscle is found in the digestive tract, blood vessels, bronchioles (small air passages in the lungs), and in the ducts of the urinary and reproductive systems. Circular

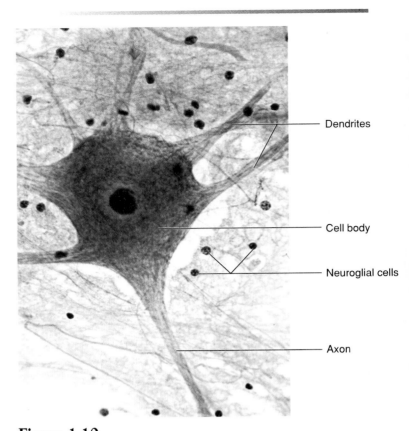

Figure 1.10

A photomicrograph of nerve tissue showing a single neuron and numerous smaller neuroglial cells.

arrangements of smooth muscle in these organs produce constriction of the *lumen* (cavity) when the muscle cells contract. The digestive tract also contains longitudinally arranged layers of smooth muscle. Rhythmic contractions of circular and longitudinal layers of muscle produce *peristalsis*—a process that pushes food from one end of the digestive tract to the other.

Nervous Tissue

Nervous tissue consists of nerve cells, or **neurons,** which are specialized for the generation and conduction of electrical events, and of **neuroglia,** which provide the neurons with anatomical and functional support.

Each neuron consists of three parts: (1) a *cell body,* (2) *dendrites,* and (3) an *axon* (fig. 1.10). The cell body contains the nucleus and serves as the metabolic center of the cell. The dendrites (literally, "branches") are highly branched cytoplasmic extensions of the cell body that receive input from other neurons or from receptor cells. The axon is a single cytoplasmic extension of the cell body that can be quite long (up to a few feet in length). It is specialized for conducting nerve impulses from the cell body to another neuron or to an effector (muscle or gland) cell.

Table 1.2 Summary of Membranous Epithelial Tissues

Type	Structure and Function	Location
Simple Epithelia	Single layer of cells; diffusion and filtration	Covering visceral organs, linings of lumina and body cavities
Simple squamous epithelium	Single layer of flattened, tightly bound cells; diffusion and filtration	Capillary walls, alveoli of lungs, covering visceral organs, linings of body cavities
Simple cuboidal epithelium	Single layer of cube-shaped cells; excretion, secretion, or absorption	Surface of ovaries; linings of kidney tubules, salivary ducts, and pancreatic ducts
Simple columnar epithelium	Single layer of nonciliated, tall, columnar-shaped cells; protection, secretion, and absorption	Lining of most of digestive tract
Simple ciliated columnar epithelium	Single, ciliated layer of columnar-shaped cells; transportive role through ciliary motion	Lining lumina of the uterine tubes
Pseudostratified ciliated columnar epithelium	Single layer of ciliated, irregularly shaped cells, many goblet cells; protection, secretion, ciliary movement	Lining of respiratory passageways
Stratified Epithelia	Two or more layers of cells; protection, strengthening, or distension	Epidermal layer of skin; linings of body openings, ducts, and urinary bladder
Stratified squamous epithelium (keratinized)	Numerous layers containing keratin, outer layers flattened and dead; protection	Epidermis of skin
Stratified squamous epithelium (nonkeratinized)	Numerous layers lacking keratin, outer layers moistened and alive; protection and pliability	Linings of oral and nasal cavities, vagina, and anal canal
Stratified cuboidal epithelium	Usually two layers of cube-shaped cells; strengthening of luminal walls	Larger ducts of sweat glands, salivary glands, and pancreas
Transitional epithelium	Numerous layers of rounded, nonkeratinized cells; distension	Luminal walls of ureters and urinary bladder

Source: From Kent M. Van De Graaff, *Human Anatomy,* 4th ed. Copyright © 1995 Wm. C. Brown Communications, Inc., Dubuque, Iowa. Reprinted by permission of Times Mirror Higher Education Group, Inc., Dubuque, Iowa. All Rights Reserved.

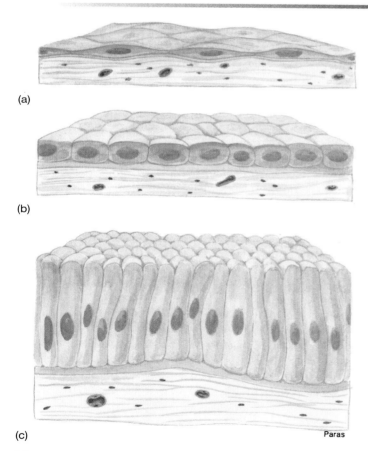

(a)

(b)

(c)

Paras

Figure 1.11

(*a*) Simple squamous, (*b*) simple cuboidal, and (*c*) simple columnar epithelial membranes. The tissue beneath each membrane is connective tissue.

The neuroglia, composed of *neuroglial cells,* do not conduct impulses but instead serve to bind neurons together, modify the extracellular environment of the nervous system, and influence the nourishment and electrical activity of neurons. Neuroglial cells are about five times more abundant than neurons in the nervous system and, unlike neurons, maintain a limited ability to divide by mitosis throughout life.

Epithelial Tissue

Epithelial tissue consists of cells that form **membranes,** which cover and line the body surfaces, and of **glands** that are derived from these membranes. There are two categories of glands. *Exocrine glands* (*exo* = outside) secrete chemicals through a duct that leads to the outside of a membrane and thus to the outside of a body surface. *Endocrine glands* (from the Greek *endon* = within) secrete chemicals called *hormones* into the blood.

Epithelial Membranes

Epithelial membranes are classified according to the number of their layers and the shape of the cells in the upper layer (table 1.2). Epithelial cells that are flattened in shape are *squamous;* those that are taller than they are wide are *columnar;* and those that are as wide as they are tall are *cuboidal* (fig. 1.11). Those epithelial membranes that are only one cell layer thick are known as *simple membranes;* those that are composed of a number of layers are *stratified membranes.*

A simple squamous membrane is adapted for diffusion and filtration. Such a membrane lines all blood vessels, where it is known as an *endothelium.* A simple cuboidal epithelium lines the ducts of exocrine glands and part of the tubules of the kidney. A simple columnar epithelium lines the lumen of the

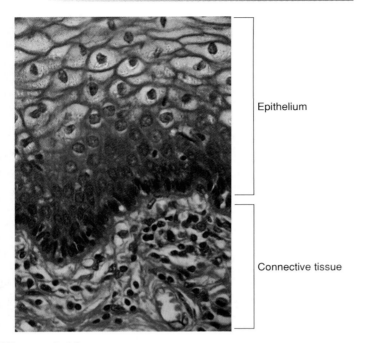

Epithelium

Connective tissue

Figure 1.12

The stratified squamous nonkeratinized epithelial membrane of the vagina.

stomach and intestine. Dispersed among the columnar epithelial cells are specialized unicellular glands called *goblet cells* that secrete mucus. The columnar epithelial cells in the uterine (fallopian) tubes of females and in the respiratory passages contain numerous *cilia* (hairlike structures, described in chapter 3) that can move in a coordinated fashion and aid the functions of these organs.

The epithelial lining of the esophagus and vagina that provides protection for these organs is a stratified squamous epithelium (fig. 1.12). This is a *nonkeratinized* membrane, and all layers consist of living cells. The *epidermis* of the skin, by contrast, is *keratinized*, or *cornified* (fig. 1.13). Since the epidermis is dry and exposed to the potentially desiccating effects of the air, the surface is covered with dead cells that are filled with a water-resistant protein known as *keratin*. This protective layer is constantly flaked off from the surface of the skin and therefore must be constantly replaced by the division of cells in the deeper layers of the epidermis.

The constant loss and renewal of cells is characteristic of epithelial membranes. The entire epidermis is completely replaced every 2 weeks; the stomach lining is renewed every 2 to 3 days. Examination of the cells that are lost, or "exfoliated," from the outer layer of epithelium lining the female reproductive tract is a common procedure in gynecology (as in the Pap smear).

In order to form a strong membrane that is effective as a barrier at the body surfaces, epithelial cells are very closely packed and are joined together by structures collectively called **junctional complexes.** There is no room for blood vessels between adjacent epithelial cells. The epithelium must therefore receive nourishment from the tissue beneath, which has large intercellular spaces that can accommodate blood vessels and nerves. This underlying tissue is called *connective tissue*. Epithelial membranes are attached to the underlying connective tissue by a layer of proteins and polysaccharides known as the **basement lamina.** This layer can be observed only under the microscope using specialized staining techniques.

Exocrine Glands

Exocrine glands are derived from cells of epithelial membranes. The secretions of these cells are expressed to the outside of the epithelial membranes (and hence to the surfaces of the body) through *ducts*. This is in contrast to endocrine glands, which lack ducts and which therefore secrete into capillaries within the body (fig. 1.14). The structure of endocrine glands will be described in chapter 11.

The secretory units of exocrine glands may be simple tubes, or they may be modified to form clusters of units around branched ducts (fig. 1.15). These clusters, or **acini,** are often surrounded by tentacle-like extensions of *myoepithelial cells,* which contract and squeeze the secretions through the ducts. The rate of secretion and the action of myoepithelial cells are influenced by hormones and by the autonomic nervous system, as will be described in later chapters.

Examples of exocrine glands in the skin include the lacrimal (tear) glands, sebaceous glands (which secrete oily sebum into hair follicles), and sweat glands. There are two types of sweat glands. The more numerous, the *eccrine* (or *merocrine*) *sweat glands,* secrete a dilute salt solution that serves in thermoregulation (evaporation cools the skin). The *apocrine sweat glands,* located in the axillae (underarms) and pubic region, secrete a protein-rich fluid. This provides nourishment for bacteria that produce the characteristic odor of this type of sweat.

All of the glands that secrete into the digestive tract are also exocrine. This is because the lumen of the digestive tract is a part of the external environment, and secretions of these glands go to the outside of the membrane that lines this tract. Mucous glands are located throughout the length of the digestive tract. Other relatively simple glands of the tract include salivary glands, gastric glands, and simple tubular glands in the intestine.

The *liver* and *pancreas* are exocrine (as well as endocrine) glands, derived embryologically from the digestive tract. The exocrine secretion of the pancreas—pancreatic juice—contains digestive enzymes and bicarbonate and is secreted into the

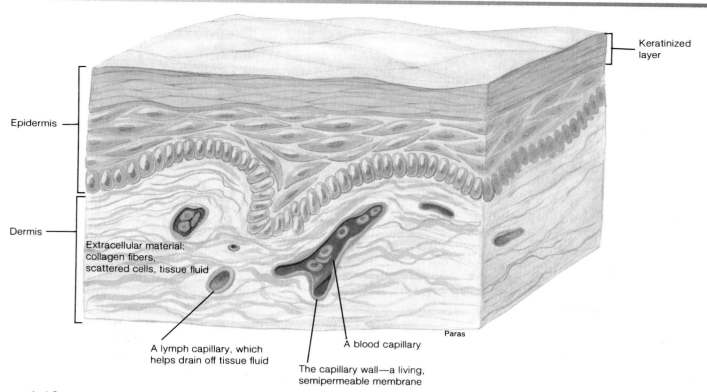

Keratinized layer

Epidermis

Dermis

Extracellular material: collagen fibers, scattered cells, tissue fluid

A lymph capillary, which helps drain off tissue fluid

A blood capillary

The capillary wall—a living, semipermeable membrane

Paras

Figure 1.13

A section of skin showing the loose connective tissue dermis beneath the cornified epidermis. Loose connective tissue contains scattered collagen fibers in a matrix of protein-rich fluid. The intercellular spaces also contain cells and blood vessels.

small intestine via the pancreatic duct. The liver produces and secretes bile (an emulsifier of fat) into the small intestine via the gallbladder and bile duct.

Exocrine glands are also prominent in the reproductive system. The female reproductive tract contains numerous mucus-secreting exocrine glands. The male accessory sex organs—the *prostate* and *seminal vesicles*—are exocrine glands that contribute to semen. The testes and ovaries (the gonads) are both endocrine and exocrine glands. They are endocrine because they secrete sex steroid hormones into the blood; they are exocrine because they release gametes (ova and sperm) into the reproductive tracts.

Connective Tissue

Connective tissue is characterized by large amounts of extracellular material in the spaces between the connective tissue cells. This extracellular material may be of various types and arrangements and, on this basis, several types of connective tissues are recognized: (1) connective tissue proper, (2) cartilage, (3) bone,

and (4) blood. Blood is usually classified as connective tissue because about half its volume is composed of an extracellular fluid known as *plasma*.

Connective tissue proper includes a variety of subtypes. An example of *loose connective tissue* (or *areolar tissue*) is the dermis of the skin (see fig. 1.13). This connective tissue consists of scattered fibrous proteins, called *collagen*, and tissue fluid, which provides abundant space for the entry of blood and lymphatic vessels and nerve fibers. Another type of connective tissue proper is *dense fibrous connective tissue*, which contains densely packed fibers of collagen that may be in an irregular or a regular arrangement. Dense irregular connective tissue (fig. 1.16) contains a meshwork of randomly oriented collagen fibers and forms the tough capsules and sheaths around organs. Tendons, which connect muscle to bone, and ligaments, which connect bones together at joints, are examples of dense regular connective tissue. The collagen fibers of this tissue are oriented in the same direction (fig. 1.17). The characteristics of these and other types of connective tissue proper are summarized in table 1.3.

The Study of Body Function

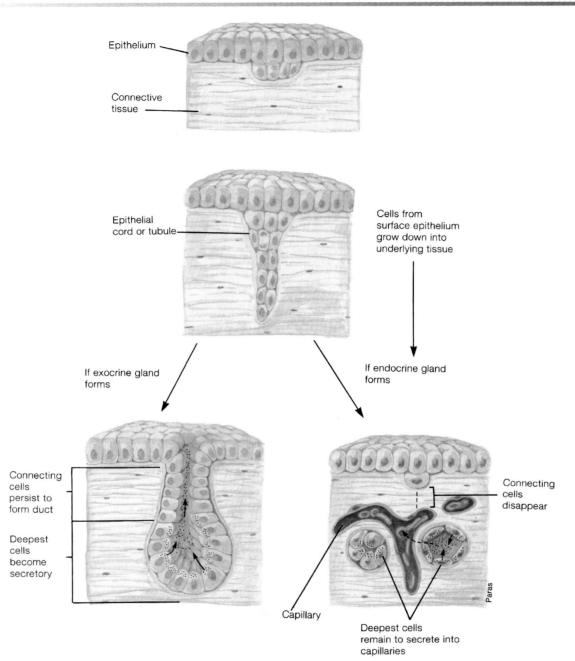

Figure 1.14

The formation of exocrine and endocrine glands from epithelial membranes.

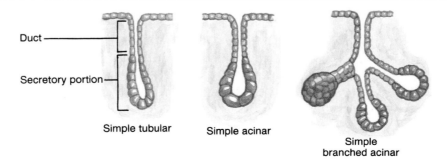

Figure 1.15

The structure of exocrine glands.

Table 1.3 Summary of Connective Tissue Proper

Type	Structure and Function	Location
Loose connective (areolar) tissue	Predominantly fibroblast cells with lesser amounts of collagen and elastin cells; binds organs, holds tissue fluids, diffusion	Surrounding nerves and vessels, between muscles, beneath the skin
Dense fibrous connective tissue	Densely packed collagen fibers; provides strong, flexible support	Tendons, ligaments, sclera of eye, deep skin layers
Elastic connective tissue	Predominantly irregularly arranged elastic fibers; supports, provides framework	Large arteries, lower respiratory tract, between vertebrae
Reticular connective tissue	Reticular fibers forming supportive network; phagocytic	Lymph nodes, liver, spleen, thymus, bone marrow
Adipose connective tissue	Adipose cells; protects, stores fat, insulates	Hypodermis of skin, surface of heart, omentum, around kidneys, back of eyeball, surrounding joints

Source: From Kent M. Van De Graaff, *Human Anatomy,* 4th ed. Copyright © 1995 Wm. C. Brown Communications, Inc., Dubuque, Iowa. Reprinted by permission of Times Mirror Higher Education Group, Inc., Dubuque, Iowa. All Rights Reserved.

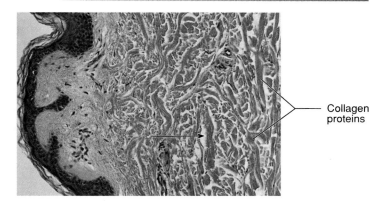

Collagen proteins

Figure 1.16

A photomicrograph of dense irregular connective tissue. Notice the tightly packed, irregularly arranged collagen proteins.

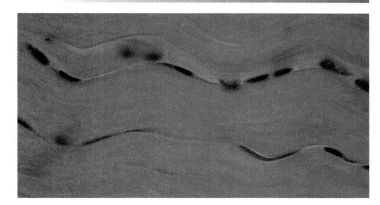

Figure 1.17

A photomicrograph of a tendon showing a dense regular arrangement of collagen fibers.

Cartilage consists of cells, called *chondrocytes,* surrounded by a semisolid ground substance that imparts elastic properties to the tissue. Cartilage is a type of supportive and protective tissue commonly called "gristle." It forms the precursor to many bones that develop in the fetus and persists at the articular (joint) surfaces on the bones at all movable joints in adults.

Bone is produced as concentric layers, or *lamellae,* of calcified material laid around blood vessels. The bone-forming cells, or *osteoblasts,* surrounded by their calcified products, become trapped within cavities (called *lacunae*). The trapped cells, which are now called *osteocytes,* remain alive because they are nourished by "lifelines" of cytoplasm that extend from the cells to the blood vessels in *canaliculi* (little canals). The blood vessels lie within central canals, surrounded by concentric rings of bone lamellae with their trapped osteocytes. These units of bone structure are called *haversian systems* (fig. 1.18).

The *dentin* of a tooth (fig. 1.19) is similar in composition to bone, but the cells that form this calcified tissue are located in the pulp (composed of loose connective tissue). These cells send cytoplasmic extensions, called *dentinal tubules,* into the dentin. Dentin, like bone, is thus a living tissue that can be remodeled in response to stresses. The cells that form the outer *enamel* of a tooth, by contrast, are lost as the tooth erupts. Enamel is a highly calcified material, harder than bone or dentin, that cannot be regenerated; artificial "fillings" are therefore required to patch holes in the enamel.

1. List the four primary tissues and describe the distinguishing features of each type.
2. Compare and contrast the three types of muscle tissue.
3. Describe the different types of epithelial membranes and state their locations in the body.
4. Explain why exocrine and endocrine glands are considered to be epithelial tissues and distinguish between these two types of glands.
5. Describe the different types of connective tissues and explain how they differ from one another in their content of extracellular material.

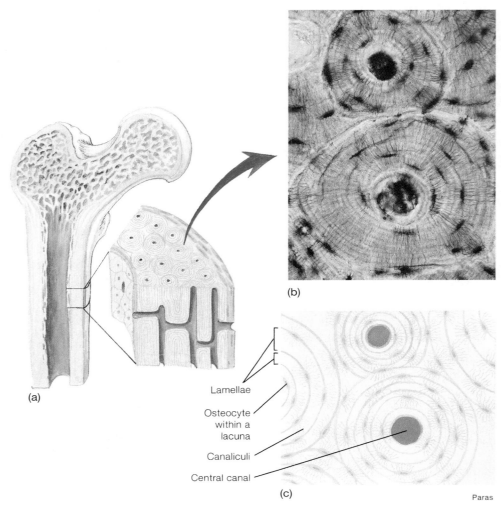

Lamellae

Osteocyte within a lacuna

Canaliculi

Central canal

(a)

(b)

(c)

Paras

Figure 1.18

(*a*) The structure of bone tissue. (*b*) A photomicrograph showing haversian systems, which are illustrated in (*c*).

Organs and Systems

Organs are composed of two or more primary tissues, which serve the different functions of the organ. The skin is an organ that has numerous functions provided by its constituent tissues.

An organ is a structure composed of at least two, and usually all four, primary tissues. The largest organ in the body, in terms of surface area, is the skin. In this section, the numerous functions of the skin serve to illustrate how primary tissues cooperate in the service of organ physiology.

An Example of an Organ: The Skin

The cornified *epidermis* protects the skin (fig. 1.20) against water loss and against invasion by disease-causing organisms. Invaginations of the epithelium into the underlying connective tissue *dermis* create the exocrine glands of the skin. These include hair follicles (which produce the hair), sweat glands, and sebaceous glands. The secretion of sweat glands cools the body by evaporation and produces odors that, at least in lower animals, serve as sexual attractants. Sebaceous glands secrete oily sebum into hair follicles, which transport the sebum to the surface of the skin. Sebum lubricates the cornified surface of the skin, helping to prevent it from drying and cracking.

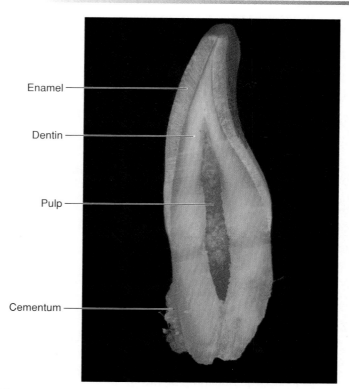

Figure 1.19

A cross section of a tooth showing pulp, dentin, and enamel. The root of the tooth is covered by a calcified connective tissue called cementum, which helps to anchor the tooth in its bony socket.

The skin is nourished by blood vessels within the dermis. In addition to blood vessels, the dermis contains wandering white blood cells and other types of cells that protect against invading disease-causing organisms. It also contains nerve fibers and fat cells; however, most of the fat cells are grouped together to form the *hypodermis* (a layer beneath the dermis). Although fat cells are a type of connective tissue, masses of fat deposits throughout the body—such as subcutaneous fat—are referred to as **adipose tissue.**

Sensory nerve endings within the dermis mediate the cutaneous sensations of touch, pressure, heat, cold, and pain. Some of these sensory stimuli directly affect the sensory nerve endings. Others act via sensory structures derived from non-neural primary tissues. The Pacinian corpuscles in the dermis of the skin (fig. 1.21), for example, monitor sensations of pressure. Motor nerve fibers in the skin stimulate effector organs, resulting in, for example, the secretions of exocrine glands and contractions of the arrector pili muscles, which attach to hair follicles and surrounding connective tissue (producing goose bumps). The degree of constriction or dilation of cutaneous blood vessels—and therefore the rate of blood flow—is also regulated by motor nerve fibers.

The epidermis itself is a dynamic structure that can respond to environmental stimuli. The rate of its cell division—and consequently the thickness of the cornified layer—increases under the stimulus of constant abrasion. This produces calluses. The skin also protects itself against the dangers of ultraviolet light by increasing its production of *melanin* pigment, which absorbs ultraviolet light while producing a tan. In addition, the skin is an endocrine gland that produces and secretes vitamin D (derived from cholesterol under the influence of ultraviolet light), which functions as a hormone.

The architecture of most organs is similar to that of the skin. Most are covered by an epithelium that lies immediately over a connective tissue layer. The connective tissue contains blood vessels, nerve endings, scattered cells for fighting infection, and possibly glandular tissue as well. If the organ is hollow—as with the digestive tract or blood vessels—the lumen is also lined with an epithelium overlying a connective tissue layer. The presence, type, and distribution of muscular and nervous tissue vary in different organs.

Systems

Organs that are located in different regions of the body and that perform related functions are grouped into **systems.** These include the nervous system, endocrine system, cardiovascular system, respiratory system, urinary system, musculoskeletal system, integumentary system, reproductive system, digestive system, and immune system (table 1.4). By means of numerous regulatory mechanisms, these systems work together to maintain the life and health of the entire organism.

Body-Fluid Compartments

Tissues, organs, and systems can all be divided into two major parts, or compartments. The **intracellular compartment** is that part inside the cells; the **extracellular compartment** is that part outside the cells. Both compartments consist primarily of water—they are said to be aqueous. The two compartments are separated by the cell membrane surrounding each cell (chapter 3).

The extracellular compartment is subdivided into two parts. One part is the *blood plasma,* the fluid portion of the blood. The other is the fluid that bathes the cells within the organs of the body. This is called *tissue fluid,* or *interstitial fluid.* Blood plasma and tissue fluid communicate freely with each other through blood capillaries in most parts of the body. The kidneys regulate the volume and composition of the blood plasma, and thus, indirectly, the fluid volume and composition of the entire extracellular compartment.

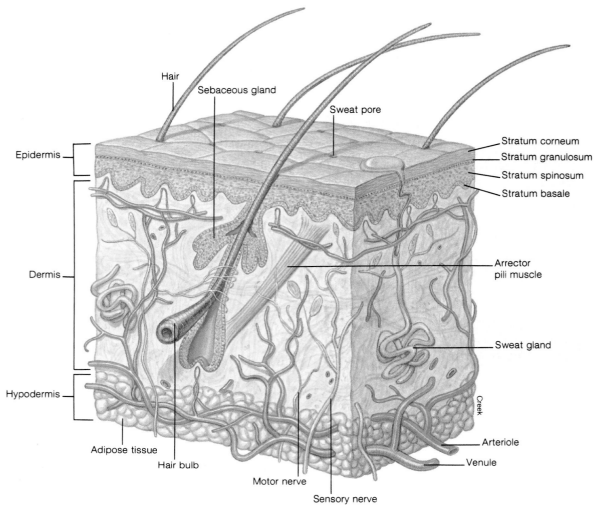

Hair

Sebaceous gland

Sweat pore

Epidermis

Stratum corneum

Stratum granulosum

Stratum spinosum

Stratum basale

Arrector
pili muscle

Dermis

Sweat gland

Hypodermis

Creek

Adipose tissue

Arteriole

Venule

Hair bulb

Motor nerve

Sensory nerve

Figure 1.20

A diagram of the skin. The skin is an organ that contains all four types of primary tissues.

There is also selective communication between the intracellular and extracellular compartments through the movement of molecules and ions through the cell membrane (chapter 6). This is how cells obtain the molecules they need for life and how they eliminate waste products.

1. State the location of each type of primary tissue in the skin.

2. Describe the functions of nerve, muscle, and connective tissue in the skin.

3. Describe the functions of the epidermis and explain why this tissue is called "dynamic."

4. Distinguish between the intracellular and extracellular compartments and explain their significance.

Sensory neuron —————

Figure 1.21

A diagram of a Pacinian corpuscle. This receptor for deep pressure consists of epithelial cells and connective tissue proteins that form concentric layers around the ending of a sensory neuron.

Table 1.4   Organ Systems of the Body

System	Major Organs	Primary Functions
Circulatory	Heart, blood vessels, lymphatic vessels	Movement of blood and lymph
Respiratory	Lungs, airways	Gas exchange
Urinary	Kidneys, ureters, urethra	Regulation of blood volume and composition
Immune	Bone marrow, lymphoid organs	Defense of the body against invading pathogens
Integumentary	Skin, hair, nails	Protection, thermoregulation
Skeletal	Bones, cartilages	Movement and support
Endocrine	Hormone-secreting glands, such as the pituitary, thyroid, and adrenals	Secretion of regulatory molecules called hormones
Nervous	Brain, spinal cord, nerves	Regulation of other body systems
Muscular	Skeletal muscles	Movements of the skeleton
Digestive	Mouth, stomach, intestine, liver, gallbladder, pancreas	Breakdown of food into molecules that enter the body
Reproductive	Gonads, external genitalia, associated glands and ducts	Continuation of the human species

Summary

Introduction to Physiology   p. 4

I. Physiology is the study of how cells, tissues, and organs function.

  A. In the study of physiology, cause-and-effect sequences are emphasized.

  B. Knowledge of physiological mechanisms is deduced from data obtained experimentally.

II. The science of physiology overlaps with chemistry and physics and shares knowledge with the related sciences of pathophysiology and comparative physiology.

  A. Pathophysiology is concerned with the functions of diseased or injured body systems and is based on knowledge of how normal systems function, which is the focus of physiology.

  B. Comparative physiology is concerned with the physiology of animals other than humans and shares much information with human physiology.

III. All of the information in this book has been gained by applications of the scientific method. This method has three essential characteristics.

  A. It is assumed that the subject under study can ultimately be explained in terms we can understand.

  B. Descriptions and explanations are honestly based on observations of the natural world and can be changed as warranted by new observations.

  C. Humility is an important characteristic of the scientific method; the scientist must be willing to change his or her theories when warranted by the weight of the evidence.

Homeostasis and Feedback Control   p. 5

I. Homeostasis refers to the dynamic constancy of the internal environment.

  A. Homeostasis is maintained by mechanisms that act through negative feedback loops.

    1. A negative feedback loop requires (1) a sensor that can detect a change in the internal environment and (2) an effector that can be activated by the sensor.

    2. In a negative feedback loop, the effector acts to cause changes in the internal environment that compensate for the initial deviations that are detected by the sensor.

  B. Positive feedback loops serve to amplify changes and may be part of the action of an overall negative feedback mechanism.

  C. The nervous and endocrine systems provide extrinsic regulation of other body systems and act to maintain homeostasis.

D. The secretion of hormones is stimulated by specific chemicals and is inhibited by negative feedback mechanisms.

II. Effectors act antagonistically to defend the set point against deviations in any direction.

The Primary Tissues, Organs, and Systems p. 9

I. The body is composed of four primary tissues: muscular, nervous, epithelial, and connective tissues.

A. There are three types of muscle tissue: skeletal, cardiac, and smooth muscle.
  1. Skeletal and cardiac muscle are striated.
  2. Smooth muscle is found in the walls of the internal organs.

B. Nervous tissue is composed of neurons and neuroglial cells.
  1. Neurons are specialized for the generation and conduction of electrical impulses.
  2. Neuroglial cells provide the neurons with anatomical and functional support.

C. Epithelial tissue includes membranes and glands.
  1. Epithelial membranes cover and line the body surfaces and their cells are tightly joined by junctional complexes.
  2. Epithelial membranes may be simple or stratified and their cells may be squamous, cuboidal, or columnar.
  3. Exocrine glands, which secrete into ducts, and endocrine glands, which lack ducts and secrete hormones into the blood, are derived from epithelial membranes.

D. Connective tissue is characterized by large intercellular spaces that contain extracellular material.
  1. Connective tissue proper is categorized into subtypes, including loose, dense fibrous, adipose, and others.
  2. Cartilage, bone, and blood are classified as connective tissues because their cells are widely spaced with abundant extracellular material between them.

II. Organs are units of structure and function that are composed of at least two, and usually all four, primary tissues.

A. The skin is a good example of an organ.
  1. The epidermis is a stratified squamous keratinized epithelium that protects underlying structures and produces vitamin D.
  2. The dermis is an example of loose connective tissue.
  3. Hair follicles, sweat glands, and sebaceous glands are exocrine glands found within the dermis.
  4. Sensory and motor nerve fibers enter the spaces within the dermis to innervate sensory organs and smooth muscles.
  5. The arrector pili muscles that attach to the hair follicles are composed of smooth muscle.

B. Organs that are located in different regions of the body and that perform related functions are grouped into systems. These include among others, the cardiovascular system, digestive system, and endocrine system.

III. The fluids of the body are divided into two major compartments.

A. The intracellular compartment refers to the fluid within cells.

B. The extracellular compartment refers to the fluid outside of cells; extracellular fluid is subdivided into plasma (the fluid portion of the blood) and tissue fluid.

Review Activities

Objective Questions

Match the following:
1. Glands are derived from
2. Cells are joined closely together in
3. Cells are separated by large extracellular spaces in
4. Blood vessels and nerves are usually located within

a. nervous tissue
b. connective tissue
c. muscular tissue
d. epithelial tissue

Multiple Choice

5. Most organs are composed of
   a. epithelial tissue.
   b. muscle tissue.
   c. connective tissue.
   d. all of the above.

6. Sweat is secreted by exocrine glands. This means that
   a. it is produced by epithelial cells.
   b. it is a hormone.
   c. it is secreted into a duct.
   d. it is produced outside the body.

7. Which of the following statements about homeostasis is *true?*
   a. The internal environment is maintained absolutely constant.
   b. Negative feedback mechanisms act to correct deviations from a normal range within the internal environment.

c. Homeostasis is maintained by switching effector actions on and off.
   d. All of the above are true.

8. In a negative feedback loop, the effector organ produces changes that are
   a. in the same direction as the change produced by the initial stimulus.
   b. opposite in direction to the change produced by the initial stimulus.
   c. unrelated to the initial stimulus.

9. A hormone called *parathyroid hormone* acts to help raise the blood calcium concentration. According to the principles of negative feedback, an effective stimulus for parathyroid hormone secretion would be
   a. a fall in blood calcium.
   b. a rise in blood calcium.

10. Which of the following consists of dense, parallel arrangements of collagen fibers?
    a. skeletal muscle tissue
    b. nervous tissue
    c. tendons
    d. dermis of the skin

11. The act of breathing raises the blood oxygen level, lowers the blood carbon dioxide concentration, and raises the blood pH. According to the principles of negative feedback, sensors that regulate breathing should respond to
    a. a rise in blood oxygen.
    b. a rise in blood pH.
    c. a rise in blood carbon dioxide concentration.
    d. all of the above.

Essay Questions

1. Describe the structure of the various epithelial membranes and explain how their structures relate to their functions.[1]

2. Compare bone, blood, and the dermis of the skin in terms of their similarities. What are the major structural differences between these tissues?

3. Describe the role of antagonistic negative feedback processes in the maintenance of homeostasis.

4. Explain, using insulin as an example, how the secretion of a hormone is controlled by the effects of that hormone's actions.

5. Describe the steps in the development of pharmaceutical drugs and evaluate the role of animal research in this process.

6. Why is Claude Bernard considered to be the father of modern physiology? Why is the concept he introduced so important in physiology and medicine?

Selected Readings

Adolph, E. F. 1968. *Origins of Physiological Regulations*. New York: Academic Press.

Benison, S., A. C. Barger, and E. L. Wolfe. 1987. *Walter B. Cannon: The Life and Times of a Young Scientist*. Cambridge, MA: Harvard University Press.

Bloom, W. B., and D. W. Fawcett. 1975. *A Textbook of Histology*. 10th ed. Philadelphia: W. B. Saunders Co.

Cannon, W. B. 1932. *The Wisdom of the Body*. New York: Norton and Company.

Di Fiore, M. S. H. 1974. *An Atlas of Histology*. 4th ed. Philadelphia: Lea and Febiger.

Fye, B. 1987. *The Development of American Physiology: Scientific Medicine in the Nineteenth Century*. Baltimore: Johns Hopkins University Press.

Jones, R. W. 1973. *Principles of Biological Regulation: An Introduction to Feedback Systems*. New York: Academic Press.

Kessel, R. G., and R. H. Kardon. 1979. *Tissues and Organs: A Text-Atlas of Scanning Electron Microscopy*. San Francisco: W. H. Freeman.

Pappenheimer. J. R. 1987. A silver spoon. *Annual Review of Physiology* 49:1.

Prosser, C. L. 1986. The making of a comparative physiologist. *Annual Review of Physiology* 48:1.

Rapport, Samuel, and Helen Wright, eds. 1964. *Science: Method and Meaning*. New York: New York University Press.

Schmidt-Nielsen, Knut. 1990. *Animal Physiology: Adaptations and Environment*. 4th ed. New York: Cambridge University Press.

Schmidt-Nielsen, Knut. 1994. How are control systems controlled? *American Scientist* 82:38.

Soderberg, U. 1964. Neurophysiological aspects of homeostasis. *Annual Review of Physiology* 26:271.

[1]*Note:* This question is answered on page 9 of the Student Study Guide.

Chemical Composition of the Body

OBJECTIVES   *After studying this chapter, you should be able to . . .*

1. describe the structure of an atom and explain the meaning of the terms *atomic mass* and *atomic number*.

2. explain how covalent bonds are formed and distinguish between nonpolar and polar covalent bonds.

3. describe the structure of an ion and explain how ionic bonds are formed.

4. describe the nature of hydrogen bonds and explain their significance.

5. describe the structure of a water molecule and explain why some compounds are hydrophilic and others are hydrophobic.

6. define the terms *acid* and *base* and explain what is meant by the pH scale.

7. explain how the pH of the blood is stabilized by bicarbonate buffer and define the terms *acidosis* and *alkalosis*.

8. describe the various types of carbohydrates and give examples of each type.

9. describe the mechanisms of dehydration synthesis and hydrolysis reactions and explain their significance.

10. state the common characteristic of lipids and describe the different categories of lipids.

11. describe how peptide bonds are formed and discuss the different orders of protein structure.

12. list some of the functions of proteins and explain why proteins can provide the specificity required to perform these functions.

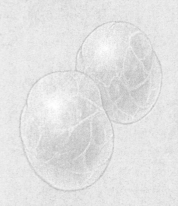

OUTLINE

Atoms, Ions, and Chemical Bonds

The study of physiology requires some familiarity with the basic concepts and terminology of chemistry. A knowledge of atomic and molecular structure, the nature of chemical bonds, and the nature of pH and associated concepts provides the foundation for much of physiology.

The structures and physiological processes of the body are based, to a large degree, on the properties and interactions of atoms, ions, and molecules. Water is the major solvent in the body and accounts for 65% to 75% of the total weight of an average adult. Of this amount, two-thirds is contained within the body cells (in the *intracellular compartment*); the remainder is contained in the *extracellular compartment*, including the blood and tissue fluids. Dissolved in this water are many organic molecules (carbon-containing molecules such as carbohydrates, lipids, proteins, and nucleic acids) and inorganic molecules and ions (atoms with a net charge). Before describing the structure and function of organic molecules within the body, it would be useful to consider some basic chemical concepts, terminology, and symbols.

Atoms

Atoms are much too small to be seen individually, even with the most powerful electron microscope. Through the efforts of many scientists, however, the structure of an atom is now well understood. At the center of an atom is its *nucleus*. The nucleus contains two types of particles—*protons,* which bear a positive charge, and *neutrons,* which carry no charge (are neutral). The mass of a proton is approximately equal to the mass of a neutron, and the sum of the protons and neutrons in an atom is equal to the **atomic mass** of the atom. For example, an atom of carbon, which contains six protons and six neutrons, has an atomic mass of 12 (table 2.1).

The number of protons in an atom is given as its **atomic number.** Carbon has six protons and thus has an atomic number of 6. Outside the positively charged nucleus are negatively charged **electrons.** Since the number of electrons in an atom is equal to the number of protons, atoms have a net charge of zero.

Although it is often convenient to think of electrons as orbiting the nucleus like planets orbiting the sun, this older view of atomic structure is no longer believed to be correct. A given electron can occupy any position in a certain volume of space called the *orbital* of the electron. The orbital is like an energy "shell," or barrier, beyond which the electron usually does not pass.

There are potentially several such orbitals around a nucleus, with each successive orbital being farther from the nucleus. The first orbital, closest to the nucleus, can contain only two electrons. If an atom has more than two electrons (as do all atoms except hydrogen and helium), the additional electrons must occupy orbitals that are more distant from the nucleus. The second orbital can contain a maximum of eight electrons; the third can also contain a maximum of eight, and the fourth can contain a maximum of eighteen. The orbitals are filled from the innermost outward. Carbon, with six electrons, has two electrons in its first orbital and four electrons in its second orbital (fig. 2.1).

It is always the electrons in the outermost orbital, if this orbital is incomplete, that participate in chemical reactions and form chemical bonds. These outermost electrons are known as the *valence electrons* of the atom.

Isotopes

A particular atom with a given number of protons in its nucleus may exist in several forms that differ from one another in their number of neutrons. The atomic number of these forms is thus the same, but their atomic mass is different. These different forms are called **isotopes.** All of the isotopic forms of a given atom are included in the term **chemical element.** The element hydrogen, for example, has three isotopes. The most common of these has a nucleus consisting of only one proton. Another isotope of hydrogen (called *deuterium*) has one proton and one neutron in the nucleus, whereas the third isotope (*tritium*) has one proton and two neutrons. Tritium is a radioactive isotope that is commonly used in physiological research and in many clinical laboratory procedures.

Table 2.1 Atoms Commonly Present in Organic Molecules

Atom	Symbol	Atomic Number	Atomic Mass	Orbital 1	Orbital 2	Orbital 3	Number of Chemical Bonds
Hydrogen	H	1	1	1	0	0	1
Carbon	C	6	12	2	4	0	4
Nitrogen	N	7	14	2	5	0	3
Oxygen	O	8	16	2	6	0	2
Sulfur	S	16	32	2	8	6	2

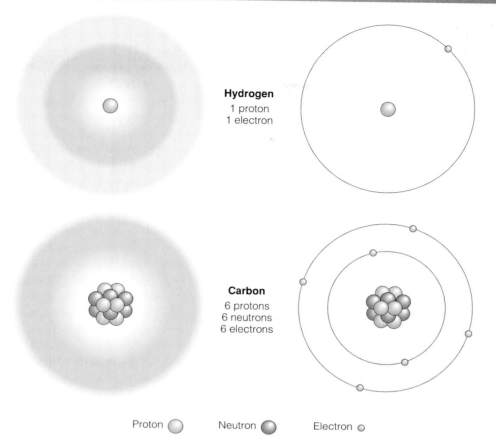

Hydrogen
1 proton
1 electron

Carbon
6 protons
6 neutrons
6 electrons

Proton ○ Neutron ● Electron ○

Figure 2.1

Diagrams of the hydrogen and carbon atoms. The electron orbitals on the left are represented by shaded spheres indicating probable positions of the electrons. The orbitals on the right are represented by concentric circles.

Chemical Bonds, Molecules, and Ionic Compounds

Molecules are formed through interaction of the valence electrons between two or more atoms. These interactions, such as the sharing of electrons, produce **chemical bonds** (fig. 2.2). The number of bonds that each atom can have is determined by the number of electrons needed to complete the outermost orbital. Hydrogen, for example, must obtain only one more electron—and can thus form only one chemical bond—to complete the first orbital of two electrons. Carbon, by contrast, must obtain four more electrons—and can thus form four chemical bonds—to complete the second orbital of eight electrons (fig. 2.3, *left*).

Covalent Bonds

Covalent bonds result when atoms share electrons. Covalent bonds that are formed between identical atoms, as in oxygen gas (O_2) and hydrogen gas (H_2), are the strongest because their electrons are equally shared. Since the electrons are equally distributed between the two atoms, these molecules are said to be

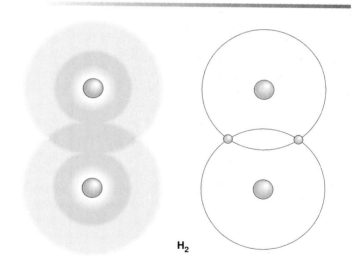

H_2

Figure 2.2

A hydrogen molecule showing the covalent bonds between hydrogen atoms formed by the equal sharing of electrons.

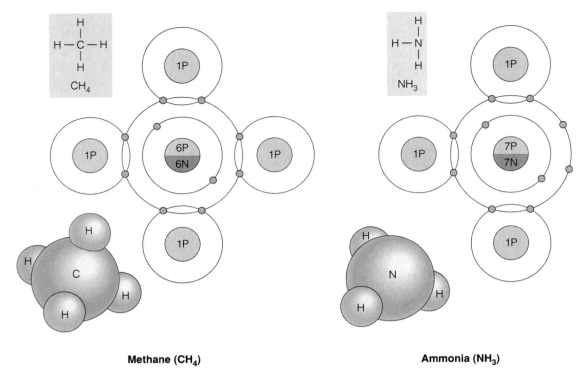

Methane (CH₄) — Methane (CH$_4$)

Ammonia (NH₃) — Ammonia (NH$_3$)

Figure 2.3

The molecules methane and ammonia represented in three different ways. Notice that a bond between two atoms consists of a pair of shared electrons (the electrons from the outer orbital of each atom).

nonpolar and are bonded by nonpolar covalent bonds. When co-valent bonds are formed between two different atoms, however, the electrons may be pulled more toward one atom than the other. The end of the molecule toward which the electrons are pulled is electrically negative in comparison to the other end. Such a molecule is said to be *polar* (has a positive and negative "pole"). Atoms of oxygen, nitrogen, and phosphorus have a par-ticularly strong tendency to pull electrons toward themselves when they bond with other atoms.

Water is the most abundant molecule in the body and serves as the solvent of body fluids. Water is a good solvent because it is polar; the oxygen atom pulls electrons from the two hydrogens toward its side of the water molecule, so that the oxygen side is more negatively charged than the hydrogen side of the molecule (fig. 2.4). The significance of the polar nature of water in its function as a solvent is discussed in the next section.

Ionic Bonds

Ionic bonds result when one or more valence electrons from one atom are completely transferred to a second atom. Thus, the electrons are not shared at all. The first atom loses elec-trons, so that its number of electrons becomes smaller than its

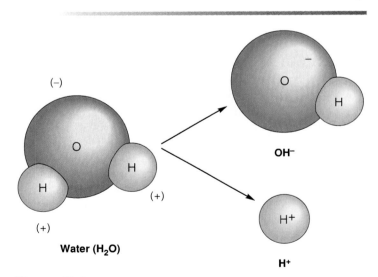

Water (H₂O) — Water (H$_2$O)

Figure 2.4

A model of a water molecule showing its polar nature. Notice that the oxygen side of the molecule is negative, whereas the hydrogen side is positive. Polar covalent bonds are weaker than nonpolar covalent bonds. As a result, some water molecules ionize to form a hydroxyl ion (OH⁻) and a hydrogen ion (H⁺).

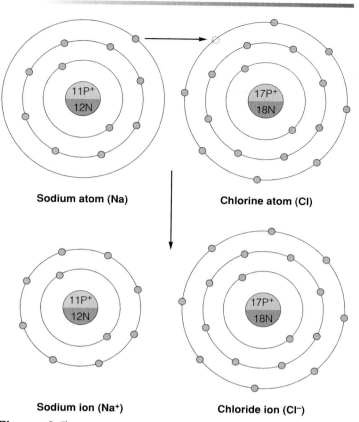

Sodium atom (Na)   **Chlorine atom (Cl)**

Sodium ion (Na⁺)   **Chloride ion (Cl⁻)**

Figure 2.5

The dissociation of sodium and chlorine to produce sodium and chloride ions. The positive sodium and negative chloride ions attract each other to produce the ionic compound sodium chloride (NaCl).

number of protons; it becomes a positively charged **ion.** Positively charged ions are called *cations* because they move toward the negative pole, or cathode, in an electric field. The second atom now has more electrons than it has protons and becomes a negatively charged ion, or *anion* (so called because it moves toward the positive pole, or anode, in an electric field). The cation and anion attract each other to form an **ionic compound.**

Common table salt, sodium chloride (NaCl), is an example of an ionic compound. Sodium, with a total of eleven electrons, has two in its first orbital, eight in its second orbital, and only one in its third orbital. Chlorine, conversely, is one electron short of completing its outer orbital of eight electrons. The lone electron in sodium's outer orbital is attracted to chlorine's outer orbital. This creates a chloride ion (represented as Cl^-) and a sodium ion (Na^+). Although table salt is shown as NaCl, it is actually composed of Na^+Cl^- (fig. 2.5).

Ionic bonds are weaker than polar covalent bonds, and therefore ionic compounds easily dissociate when dissolved in water to yield their separate ions. Dissociation of NaCl, for example, yields Na^+ and Cl^-. Each of these ions attracts polar water molecules; the negative ends of water molecules are attracted to the Na^+, and the positive ends of water molecules are attracted to the Cl^- (fig. 2.6). The water molecules that surround these ions in turn attract other molecules of water to form *hydration spheres* around each ion.

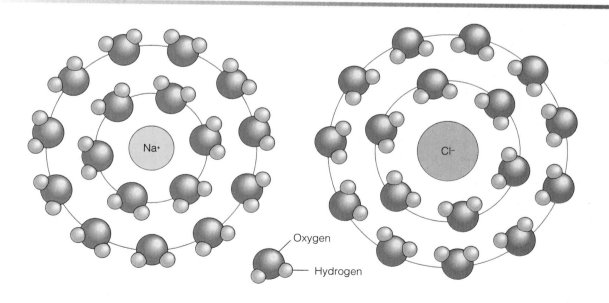

Oxygen

Hydrogen

Water molecule

Figure 2.6

The negatively charged oxygen-ends of water molecules are attracted to the positively charged Na^+, whereas the positively charged hydrogen-ends of water molecules are attracted to the negatively charged Cl^-. Other water molecules are attracted to this first concentric layer of water, forming hydration spheres around the sodium and chloride ions.

Chemical Composition of the Body

The formation of hydration spheres makes an ion or a molecule soluble in water. Glucose, amino acids, and many other organic molecules are water-soluble because hydration spheres can form around atoms of oxygen, nitrogen, and phosphorus, which are joined by polar covalent bonds to other atoms in the molecule. Such molecules are said to be **hydrophilic.** By contrast, molecules composed primarily of nonpolar covalent bonds, such as the hydrocarbon chains of fat molecules, have few charges and thus cannot form hydration spheres. They are insoluble in water, and in fact actually avoid it. For this reason, nonpolar molecules are said to be **hydrophobic.**

Hydrogen Bonds

Hydrogen bonds are very weak bonds that help to stabilize the delicate folding and bending of long organic molecules such as proteins. When hydrogen forms a polar covalent bond with an atom of oxygen or nitrogen, the hydrogen gains a slight positive charge as the electron is pulled toward the other atom. This other atom is thus described as being *electronegative*. Since the hydrogen has a slight positive charge, it will have a weak attraction for a second electronegative atom (oxygen or nitrogen) that may be located near it. This weak attraction is called a **hydrogen bond.** Hydrogen bonds are usually shown with dashed or dotted lines (fig. 2.7) to distinguish them from strong covalent bonds, which are shown with solid lines.

Hydrogen bonds can be formed within the folds of a protein (discussed in a later section), and between the chains of DNA (chapter 3). They can also be formed between adjacent water molecules (fig. 2.7). The hydrogen bonding between water molecules is responsible for many of the physical properties of water, including its *surface tension* and its ability to be pulled as a column through narrow channels in a process called *capillary action*.

Acids, Bases, and the pH Scale

The bonds in water molecules joining hydrogen and oxygen atoms together are, as previously discussed, polar covalent bonds. Although these bonds are strong, a small proportion of them break as the electron from the hydrogen atom is completely transferred to oxygen. When this occurs, the water molecule ionizes to form a *hydroxyl ion* (OH^-) and a *hydrogen ion* (H^+), which is simply a free proton (see fig. 2.4). A proton released in this way does not remain free for long, however, because it is attracted to the electrons of oxygen atoms in water molecules. This forms a *hydronium ion*, shown by the formula H_3O^+. For the sake of clarity in the following discussion, however, H^+ will be used to represent the ion resulting from the ionization of water.

Ionization of water molecules produces equal amounts of OH^- and H^+. Since only a small proportion of water molecules ionize, the concentration of H^+ and OH^- ions are each equal to only 10^{-7} molar (the term *molar* is a unit of concentration described in chapter 6; for hydrogen, one molar equals one gram per liter). A solution with 10^{-7} molar hydrogen ion, which is produced by the ionization of water molecules in which the H^+ and OH^- concentrations are equal, is said to be **neutral.**

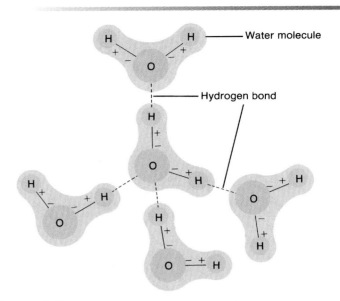

Figure 2.7

The oxygen atoms of water molecules are weakly joined together by the attraction of the electronegative oxygen for the positively charged hydrogen. These weak bonds are called *hydrogen bonds*.

A solution that has a higher H^+ concentration than that of water is called *acidic*; one with a lower H^+ concentration is called *basic*. An **acid** is defined as a molecule that can release protons (H^+) to a solution; it is a "proton donor." A **base** is a negatively charged ion (anion), or a molecule that ionizes to produce the anion, which can combine with H^+ and thus remove the H^+ from solution; it is a "proton acceptor." Most strong bases release OH^- into a solution, which combines with H^+ to form water and which thus lowers the H^+ concentration. Examples of common acids and bases are shown in table 2.2.

pH

The H^+ concentration of a solution is usually indicated in pH units on a pH scale that runs from 0 to 14. The pH number is equal to the logarithm of 1 over the H^+ concentration:

$$pH = \log \frac{1}{[H^+]}$$

where $[H^+]$ = molar H^+ concentration.

Pure water has a H^+ concentration of 10^{-7} molar at 25°C, and thus has a pH of 7 (neutral). Because of the logarithmic relationship, a solution with 10 times the hydrogen ion concentration (10^{-6} M) has a pH of 6, whereas a solution with one-tenth the H^+ concentration (10^{-8} M) has a pH of 8. The pH number is easier to write than the molar H^+ concentration, but it is admittedly confusing because it is *inversely related* to the H^+ concentration: a solution with a higher H^+ concentration has a lower pH number; one with a lower H^+ concentration has a higher pH number. A strong acid with a high H^+ concentration of 10^{-2} molar, for example, has a pH

Table 2.2 Common Acids and Bases			
Acid	**Symbol**	**Base**	**Symbol**
Hydrochloric acid	HCl	Sodium hydroxide	NaOH
Phosphoric acid	H_3PO_4	Potassium hydroxide	KOH
Nitric acid	HNO_3	Calcium hydroxide	$Ca(OH)_2$
Sulfuric acid	H_2SO_4	Ammonium hydroxide	NH_4OH
Carbonic acid	H_2CO_3		

Table 2.3 The pH Scale

	H^+ Concentration (Molar)	pH	OH^- Concentration (Molar)
Acids	1.0	0	10^{-14}
	0.1	1	10^{-13}
	0.01	2	10^{-12}
	0.001	3	10^{-11}
	0.0001	4	10^{-10}
	10^{-5}	5	10^{-9}
	10^{-6}	6	10^{-8}
Neutral	10^{-7}	7	10^{-7}
Bases	10^{-8}	8	10^{-6}
	10^{-9}	9	10^{-5}
	10^{-10}	10	0.0001
	10^{-11}	11	0.001
	10^{-12}	12	0.01
	10^{-13}	13	0.1
	10^{-14}	14	1.0

of 2, whereas a solution with only 10^{-10} molar H^+ has a pH of 10. **Acidic solutions,** therefore, have a pH of less than 7 (that of pure water), whereas **basic solutions** have a pH between 7 and 14 (table 2.3).

Buffers

A *buffer* is a system of molecules and ions that acts to prevent changes in H^+ concentration and thus serves to stabilize the pH of a solution. In blood plasma, for example, the pH is stabilized by the following reversible reaction involving the bicarbonate ion (HCO_3^-) and carbonic acid (H_2CO_3):

$$HCO_3^- + H^+ \rightleftharpoons H_2CO_3$$

The double arrows indicate that the reaction could go either to the right or to the left; the net direction depends on the concentration of molecules and ions on each side. If an acid (such as lactic acid) should release H^+ into the solution, for example, the increased concentration of H^+ would drive the equilibrium to the right and the following reaction would be promoted:

$$HCO_3^- + H^+ \longrightarrow H_2CO_3$$

Blood pH

Lactic acid and other organic acids are produced by the cells of the body and secreted into the blood. Despite the release of H^+ by these acids, the arterial blood pH normally does not decrease but remains remarkably constant at pH 7.40 ± 0.05. This constancy is achieved, in part, by the buffering action of bicarbonate shown in the preceding equation. Bicarbonate serves as the major buffer of the blood.

Certain conditions could cause an opposite change in pH. For example, excessive vomiting that results in loss of gastric acid could cause the concentration of free H^+ in the blood to fall and the blood pH to rise. In this case, the reaction previously described could be reversed:

$$H_2CO_3 \longrightarrow H^+ + HCO_3^-$$

The dissociation of carbonic acid yields free H^+, which helps to prevent an increase in pH. Bicarbonate ions and carbonic acid thus act as a *buffer pair* to prevent either decreases or increases in pH, respectively. This buffering action normally maintains the blood pH within the narrow range of 7.35 to 7.45.

If the arterial blood pH falls below 7.35, the condition is called acidosis. A blood pH of 7.20, for example, represents significant acidosis. Notice that acidotic blood need not be acidic. An increase in blood pH above 7.45, conversely, is known as alkalosis. Acidosis and alkalosis are normally prevented by the action of the bicarbonate/carbonic acid buffer pair and by the functions of the lungs and kidneys. Regulation of blood pH is discussed in more detail in chapter 13.

Organic Molecules

Organic molecules are those molecules that contain the atoms carbon and hydrogen. Since the carbon atom has four electrons in its outer orbital, it must share four additional electrons by covalent bonding with other atoms to fill its outer orbital with eight electrons. The unique bonding requirements of carbon enable it to join with other carbon atoms to form chains and rings, while still allowing the carbon atoms to bond with hydrogen and other atoms.

Most organic molecules in the body contain hydrocarbon chains and rings, as well as other atoms bonded to carbon. Two adjacent carbon atoms in a chain or ring may share one or two pairs of electrons. If the two carbon atoms share one pair of electrons, they are said to have a *single covalent bond*; this leaves each carbon atom free to bond with as many as three other atoms. If the two carbon atoms share two pairs of electrons, they have a *double covalent bond*, and each carbon atom can bond with a maximum of only two additional atoms (fig. 2.8).

The ends of some hydrocarbons are joined together to form rings. In the shorthand structural formulas of these molecules, the carbon atoms are not shown but are understood to be located at the corners of the ring. Some of these cyclic molecules have a double bond between two adjacent carbon atoms. Benzene and related molecules are shown as a six-sided ring with alternating double bonds. Such compounds are called *aromatic*. Since all of the carbons in an aromatic ring are equivalent, double bonds can be shown between any two adjacent carbons in the ring (fig. 2.9), or even as a circle within the hexagonal structure of carbons.

The hydrocarbon chain or ring of many organic molecules provides a relatively inactive molecular "backbone" to which more reactive groups of atoms are attached. Known as

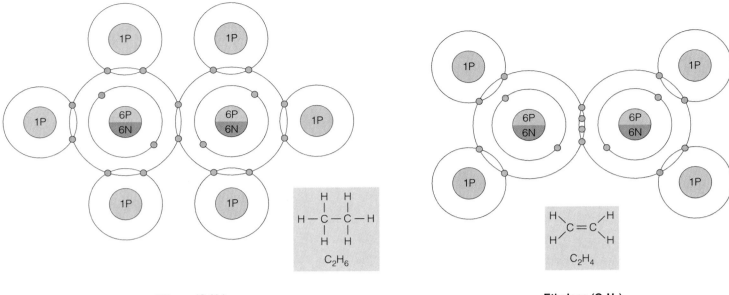

Ethane (C_2H_6)

Ethylene (C_2H_4)

Figure 2.8

Two carbon atoms joined by a single covalent bond (*left*) and by a double covalent bond (*right*). In both cases, each carbon atom shares four pairs of electrons (has four bonds) to complete the eight electrons required to fill its outer orbital.

functional groups of the molecule, these reactive groups usually contain atoms of oxygen, nitrogen, phosphorus, or sulfur. They are largely responsible for the unique chemical properties of the molecule (fig. 2.10).

Classes of organic molecules can be named according to their functional groups. *Ketones*, for example, have a carbonyl group within the carbon chain. An organic molecule is an *alcohol* if it has a hydroxyl group bound to a hydrocarbon chain. All *organic acids* (acetic acid, citric acids, lactic acid, and others) have a *carboxyl group* (fig. 2.11).

A carboxyl group can be abbreviated COOH. This group is an acid because it can donate its proton (H⁺) to the solution. Ionization of the OH part of COOH forms COO⁻ and H⁺ (fig. 2.12). The ionized organic acid is designated with the suffix -*ate*. For example, when the carboxyl group of lactic acid ionizes, the molecule is called *lactate*. Since both ionized and un-ionized forms of the molecule exist together in a solution (the proportion of each depends on the pH of the solution), one can correctly refer to the molecule as either lactic acid or lactate.

Stereoisomers

Two molecules may have exactly the same atoms arranged in exactly the same sequence yet differ with respect to the spatial orientation of a key functional group. Such molecules are called **stereoisomers** of each other. Depending upon the direction in which the key functional group is oriented with respect to the molecules, stereoisomers are called either D-isomers (for *dextro*, or right-handed) or L-isomers (for *levo*, or left-handed). Their relationship is similar to that between a right and left glove—if the palms are both pointing forward, the two cannot be superimposed.

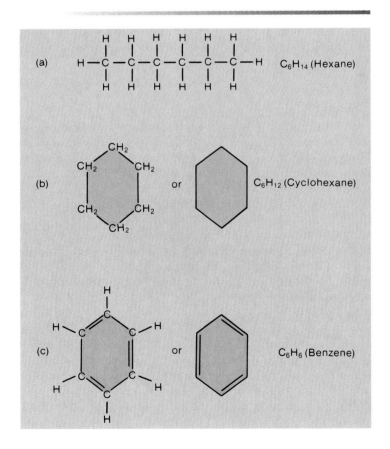

Figure 2.9

Hydrocarbons that are (*a*) linear, (*b*) cyclic, and (*c*) aromatic rings.

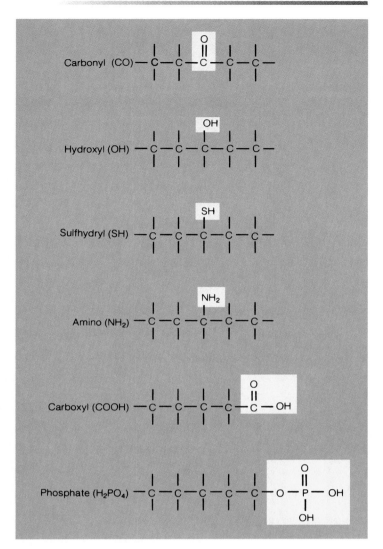

Figure 2.10

Various functional groups of organic molecules.

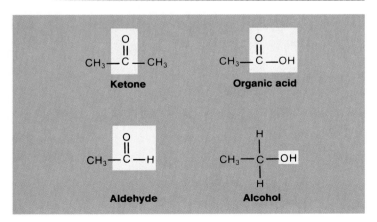

Figure 2.11

Categories of organic molecules based on functional groups.

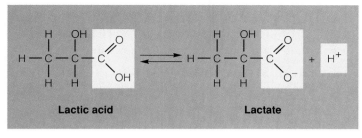

Figure 2.12

The carboxyl group of an organic acid, such as that of lactic acid, can ionize to yield a free proton, or hydrogen ion (H^+). The reaction is reversible, as indicated by the double arrows.

Birth defects resulted when pregnant women used a tranquilizer drug known as *thalidomide* in the early 1960s. The drug available at the time contained a mixture of both right-handed (D) and left-handed (L) forms. This tragic circumstance emphasizes the clinical importance of stereoisomers. It has since been learned that the L-stereoisomer is a potent tranquilizer, but the right-handed version causes disruption of fetal development and the resulting birth defects.

These subtle differences in structure are extremely important biologically. They ensure that enzymes—which interact with such molecules in a stereo-specific way in chemical reactions—cannot combine with the "wrong" stereoisomer. The enzymes of all cells (human and others) can combine only with L-amino acids and D-sugars, for example. The opposite stereoisomers (D-amino acids and L-sugars) cannot be absorbed into the body from the intestine or be used by any enzyme in metabolism.

1. Describe the structure of an atom and define the terms *atomic weight* and *atomic number*. Explain why different atoms are able to form characteristic numbers of chemical bonds.
2. Describe the nature of nonpolar and polar covalent bonds, ionic bonds, and hydrogen bonds. Explain why ions and polar molecules are soluble in water.
3. Define the terms *acidic, basic, acid,* and *base*. Define *pH* and describe the relationship between pH and the H^+ concentration of a solution.
4. Explain how carbon atoms can bond with each other and with atoms of hydrogen, oxygen, and nitrogen.

Carbohydrates and Lipids

Carbohydrates are a class of organic molecules that includes monosaccharides, disaccharides, and polysaccharides. All of these molecules are based on a characteristic ratio of carbon, hydrogen, and oxygen atoms. Lipids are a category of diverse organic molecules that share the physical property of being nonpolar and thus insoluble in water.

Carbohydrates and lipids are similar in many ways. Both groups of molecules consist primarily of the atoms carbon, hydrogen, and oxygen, and both serve as major sources of energy in the body (accounting for most of the calories consumed in food). Carbohydrates and lipids differ, however, in some important aspects of their chemical structures and physical properties. Such differences significantly affect the functions of these molecules in the body.

Carbohydrates

Carbohydrates are organic molecules that contain carbon, hydrogen, and oxygen in the ratio described by their name—*carbo* (carbon) and *hydrate* (water, H_2O). The general formula of a carbohydrate molecule is thus CH_2O; the molecule contains twice as many hydrogen atoms as carbon or oxygen atoms.

Monosaccharides, Disaccharides, and Polysaccharides

Carbohydrates include simple sugars, or **monosaccharides,** and longer molecules that contain a number of monosaccharides joined together. The suffix *-ose* denotes a sugar molecule; the term *hexose*, for example, refers to a six-carbon monosaccharide with the formula $C_6H_{12}O_6$. This formula is adequate for some purposes, but it does not distinguish between related hexose sugars, which are *structural isomers* of each other. The structural isomers glucose, fructose, and galactose, for example, are monosaccharides that have the same ratio of atoms arranged in slightly different ways (fig. 2.13).

Two monosaccharides can be joined covalently to form a **disaccharide,** or double sugar. Common disaccharides include table sugar, or *sucrose* (composed of glucose and fructose), milk sugar, or *lactose* (composed of glucose and galactose), and malt sugar, or *maltose* (composed of two glucose molecules). When numerous monosaccharides are joined together, the resulting molecule is called a **polysaccharide.** *Starch*, for example, is a polysaccharide found in many plants that is formed by the bonding together of thousands of glucose subunits. **Glycogen** (animal starch), found in the liver and muscles, likewise consists of repeating glucose molecules but differs from plant starch in that glycogen is more highly branched (fig. 2.14).

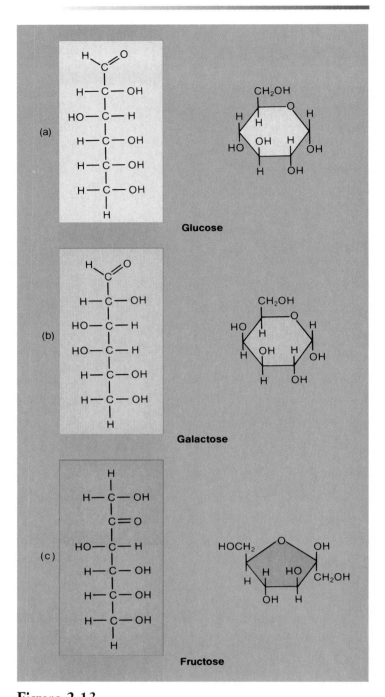

Figure 2.13

The structural formulas of three hexose sugars: (*a*) glucose, (*b*) galactose, and (*c*) fructose. All three have the same ratio of atoms — $C_6H_{12}O_6$.

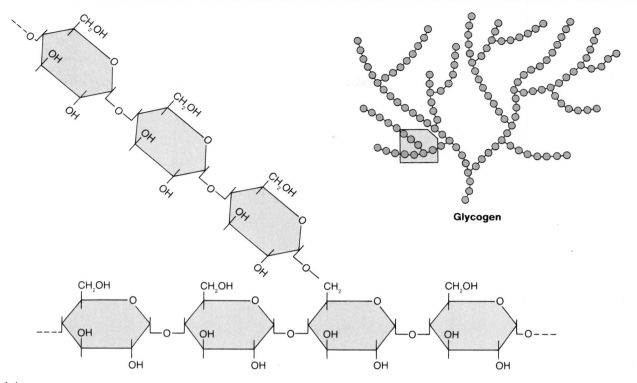

Figure 2.14

Glycogen is a polysaccharide composed of glucose subunits joined together to form a large, highly branched molecule.

Many cells store carbohydrates for use as an energy source, as described in chapter 5. If a cell were to store many thousands of separate monosaccharide molecules, however, their high concentration would draw an excessive amount of water into the cell, damaging or even killing it. The net movement of water through membranes is called osmosis, and is discussed in chapter 6. Cells that store carbohydrates for energy minimize this osmotic damage by instead joining the glucose molecules together to form the polysaccharides starch or glycogen. Since there are fewer of these larger molecules, less water is drawn into the cell by osmosis (see chapter 6).

Dehydration Synthesis and Hydrolysis

In the formation of disaccharides and polysaccharides, the separate subunits (monosaccharides) are bonded together covalently by a type of reaction called **dehydration synthesis,** or **condensation.** In this reaction, which requires the participation of specific enzymes (chapter 4), a hydrogen atom is removed from one monosaccharide and a hydroxyl group (OH) is removed from another. As a covalent bond is formed between the two monosaccharides, water (H_2O) is produced. Dehydration synthesis reactions are illustrated in figure 2.15.

When a person eats disaccharides or polysaccharides, or when the stored glycogen in the liver and muscles is to be used by tissue cells, the covalent bonds that join monosaccharides into disaccharides and polysaccharides must be broken. These *digestion reactions* occur by means of **hydrolysis.** Hydrolysis (from the Greek *hydro*=water; *lysis*=break) is the reverse of dehydration synthesis. A water molecule is split, and the resulting hydrogen atom is added to one of the free glucose molecules as the hydroxyl group is added to the other (fig. 2.16).

When a potato is eaten, the starch within it is hydrolyzed into separate glucose molecules within the intestine. This glucose is absorbed into the blood and carried to the tissues. Some tissue cells may use this glucose for energy. Liver and muscles, however, can store excess glucose in the form of glycogen by dehydration synthesis reactions in these cells. During fasting or prolonged exercise, the liver can add glucose to the blood through hydrolysis of its stored glycogen.

Dehydration synthesis and hydrolysis reactions do not occur spontaneously; they require the action of specific enzymes. Similar reactions, in the presence of other enzymes, build and break down lipids, proteins, and nucleic acids. In general, therefore, hydrolysis reactions digest molecules into their subunits, and dehydration synthesis reactions build larger molecules by the bonding together of their subunits.

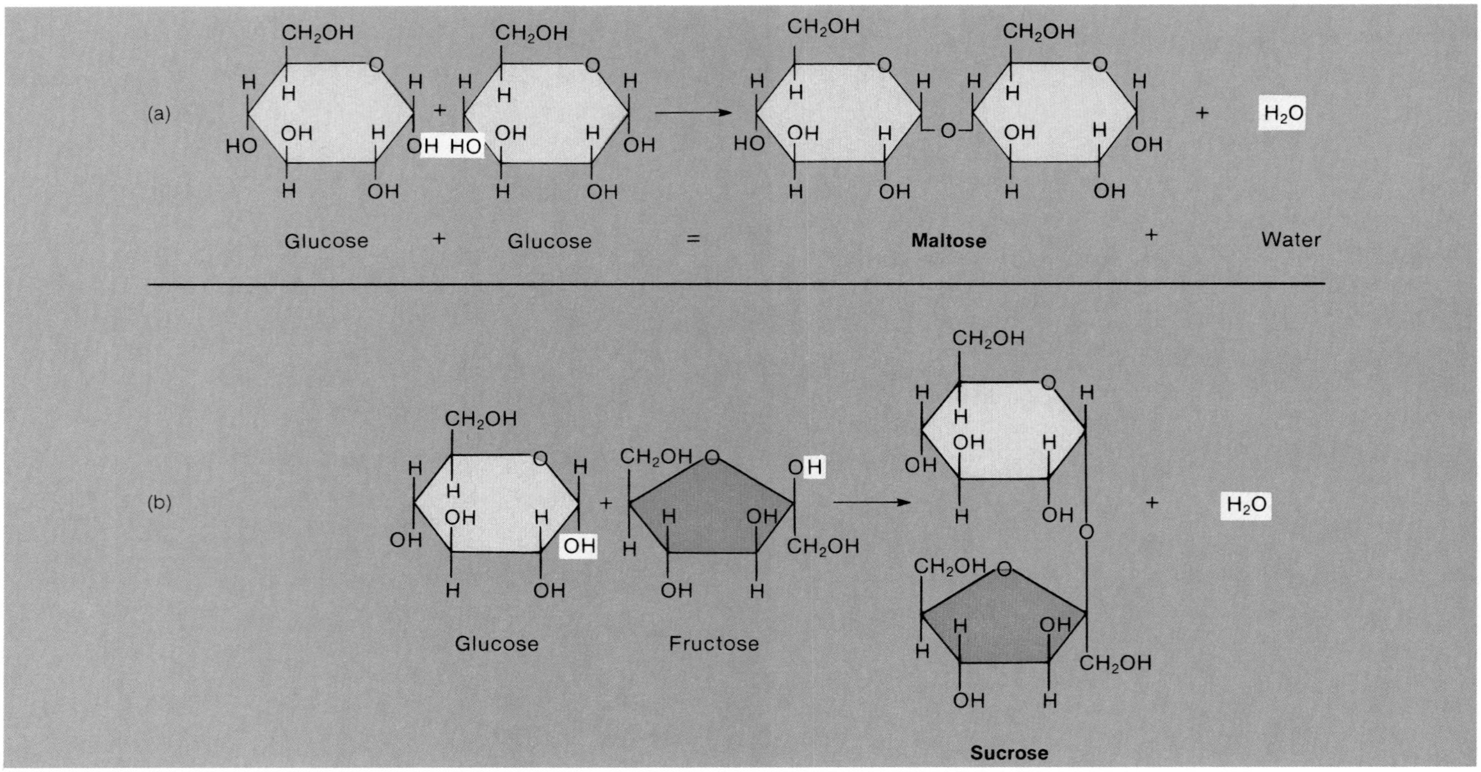

Figure 2.15

Dehydration synthesis of two disaccharides: (*a*) maltose and (*b*) sucrose. Notice that as the disaccharides are formed, a molecule of water is produced.

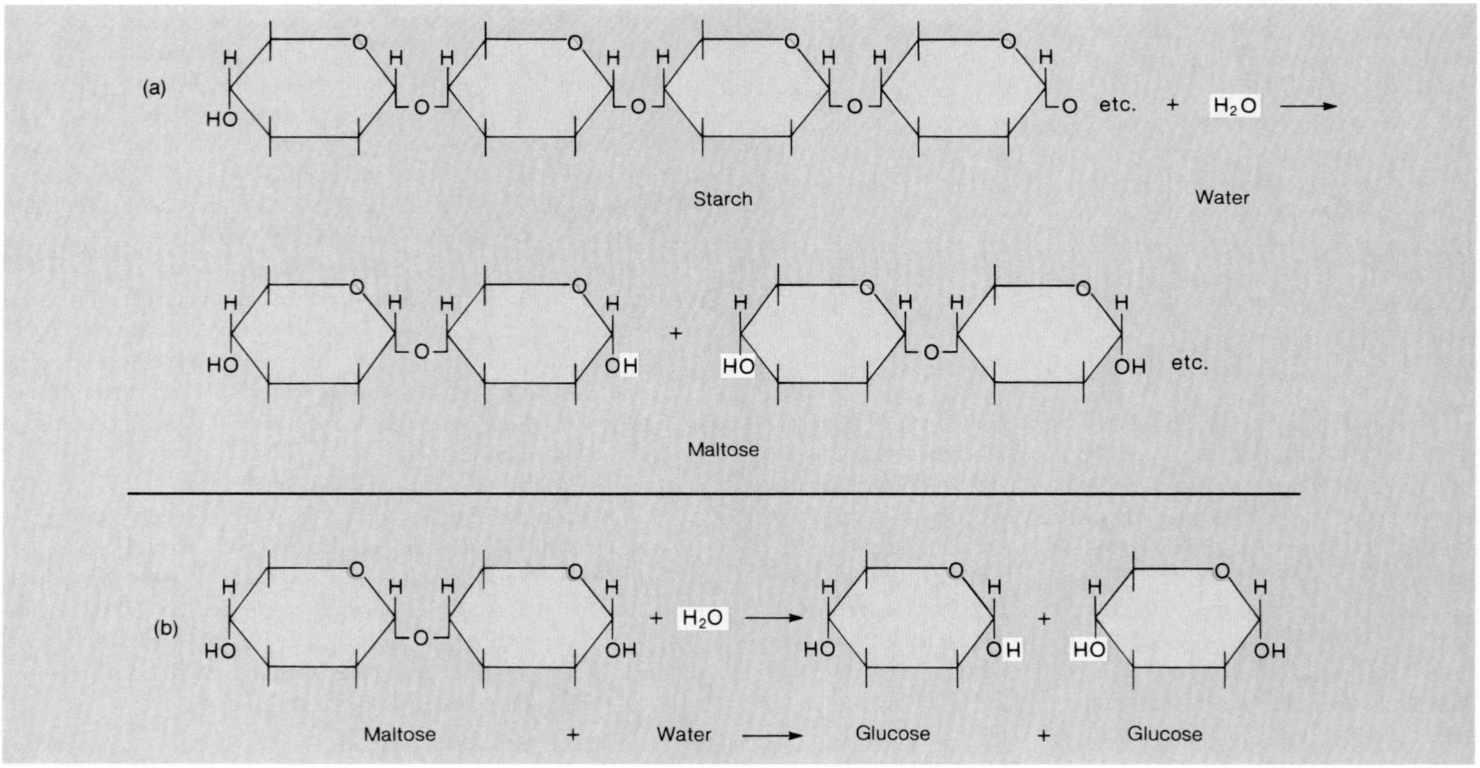

Figure 2.16

The hydrolysis of starch (*a*) into disaccharides (maltose) and (*b*) into monosaccharides (glucose). Notice that as the covalent bond between the subunits breaks, a molecule of water is split. In this way, the hydrogen atom and hydroxyl group from the water are added to the ends of the released subunits.

Lipids

The category of molecules known as lipids includes several types of molecules that differ greatly in chemical structure. These diverse molecules are all in the lipid category by virtue of a common physical property—they are all *insoluble in polar solvents* such as water. This is because lipids consist primarily of hydrocarbon chains and rings, which are nonpolar, and therefore hydrophobic. Although lipids are insoluble in water, they can be dissolved in nonpolar solvents such as ether, benzene, and related compounds.

Triglycerides

Triglycerides are the subcategory of lipids that includes fat and oil. These molecules are formed by the condensation of one molecule of *glycerol* (a three-carbon alcohol) with three molecules of *fatty acids*. Each fatty acid molecule consists of a nonpolar hydrocarbon chain with a carboxylic acid group (abbreviated COOH) on one end. If the carbon atoms within the hydrocarbon chain are joined by single covalent bonds so that each carbon atom can also bond to two hydrogen atoms, the fatty acid is said to be *saturated*. If there are a number of double covalent bonds within the hydrocarbon chain so that each carbon atom can bond to only one hydrogen atom, the fatty acid is said to be *unsaturated*. Triglycerides that contain saturated fatty acids are called **saturated fats;** those that contain unsaturated fatty acids are **unsaturated fats** (fig. 2.17).

Within the adipose cells of the body, triglycerides are formed as the carboxylic acid ends of fatty acid molecules condense with the hydroxyl groups of a glycerol molecule (fig. 2.18). Since the hydrogen atoms from the carboxyl ends of fatty acids form water molecules during dehydration synthesis, fatty acids that are combined with glycerol can no longer release H^+ and function as acids. For this reason, triglycerides are described as *neutral fats*.

 The saturated fat content (expressed as a percentage of total fat) for some food items is as follows: canola, or rapeseed, oil (6%), olive oil (14%), margarine (17%), chicken fat (31%), palm oil (51%), beef fat (52%), butter fat (66%), and coconut oil (77%). Health authorities recommend that a person's total fat intake not exceed 30% of the total energy intake per day, and that saturated fat contribute less than 10%. This is because saturated fat in the diet may contribute to high blood cholesterol, which is a significant risk factor in heart disease and stroke (see chapter 13). Animal fats, which are solid at room temperature, are generally more saturated than vegetable oils because the hardness of the triglyceride is determined partly by the degree of saturation. Palm and coconut oil, however, are notable exceptions. Though very saturated, they nonetheless remain liquid at room temperature due to the fact that they have short fatty acid chains.

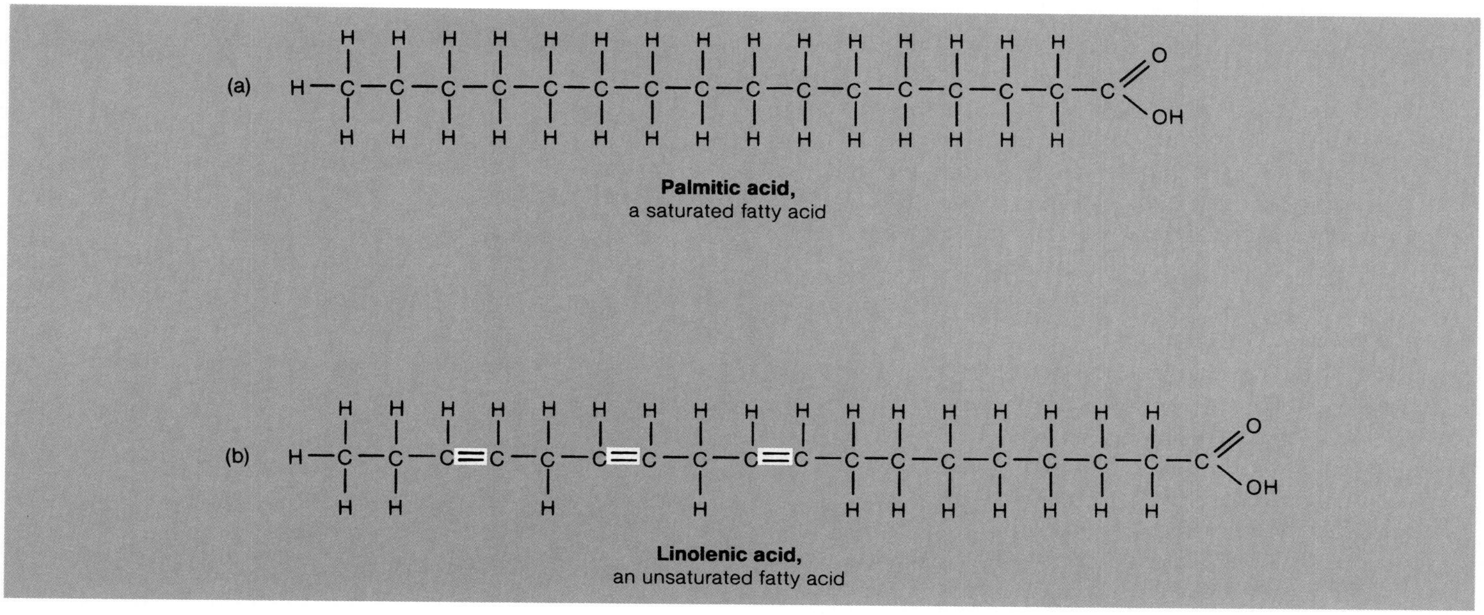

Palmitic acid,
a saturated fatty acid

Linolenic acid,
an unsaturated fatty acid

Figure 2.17

Structural formulas (*a*) for saturated fatty acids and (*b*) for unsaturated fatty acids.

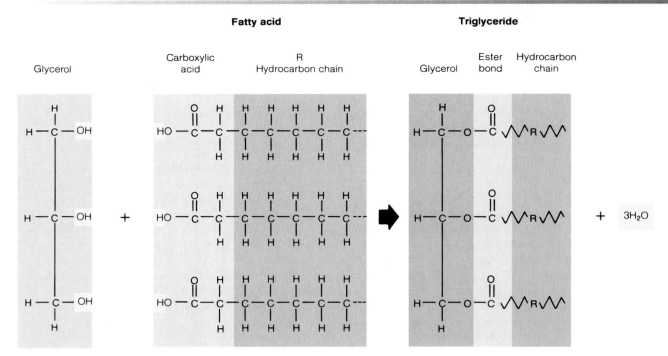

Fatty acid   **Triglyceride**

Glycerol   Carboxylic acid   R Hydrocarbon chain   Glycerol   Ester bond   Hydrocarbon chain

Figure 2.18

Dehydration synthesis of a triglyceride molecule from a glycerol and three fatty acids. A molecule of water is produced as an ester bond forms between each fatty acid and the glycerol. Sawtooth lines represent hydrocarbon chains, which are symbolized by an R.

Ketone Bodies

Hydrolysis of triglycerides within adipose tissue releases *free fatty acids* into the blood. Free fatty acids can be used as an immediate source of energy by many organs; they can also be converted by the liver into derivatives called *ketone bodies*. These include four-carbon-long acidic molecules (acetoacetic acid and β-hydroxybutyric acid) and acetone (the solvent in nail-polish remover). A rapid breakdown of fat, as occurs during dieting and in uncontrolled diabetes mellitus, results in elevated levels of ketone bodies in the blood. This is a condition called **ketosis.** If there are sufficient amounts of ketone bodies in the blood to lower the blood pH, the condition is called **ketoacidosis.** Severe ketoacidosis, which may occur in diabetes mellitus, can lead to coma and death.

Phospholipids

The class of lipids known as *phospholipids* includes a number of different categories of lipids, all of which contain a phosphate group. The most common type of phospholipid molecule is one in which the three-carbon alcohol molecule glycerol is attached to two fatty acid molecules; the third carbon atom of the glycerol molecule is attached to a phosphate group, and the phosphate group in turn is bonded to other molecules. If the phosphate group is attached to a nitrogen-containing choline molecule, the phospholipid molecule thus formed is known as *lecithin*. Figure 2.19 shows a simple way of illustrating the structure of a phospholipid—the parts of the molecule capable of ionizing (and thus becoming charged) are shown as a circle, whereas the nonpolar parts of the molecule are represented by wavy lines.

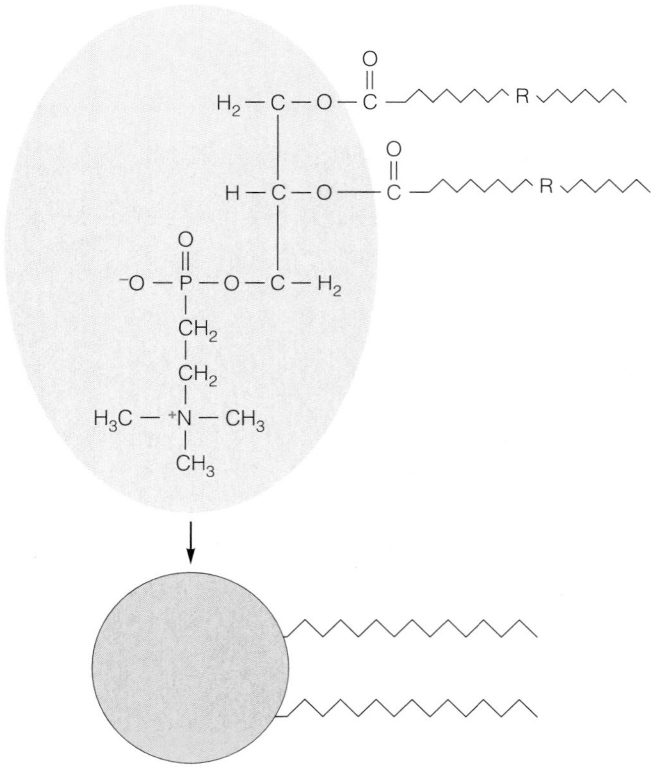

Figure 2.19

The structure of lecithin, a typical phospholipid (*top*), and its more simplified representation (*bottom*).

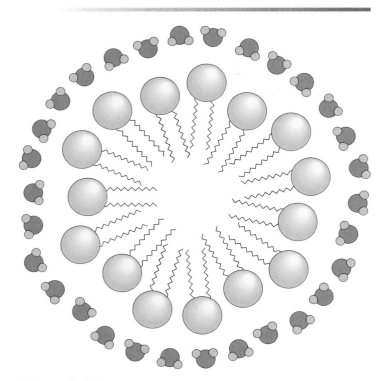

Figure 2.20

The formation of a micelle structure by phospholipids such as lecithin.

Since the nonpolar ends of phospholipids are hydrophobic, they tend to group together when mixed in water. This allows the hydrophilic parts (which are polar) to face the surrounding water molecules (fig. 2.20). Such aggregates of molecules are called **micelles.** The dual nature of phospholipid molecules (part polar, part nonpolar) allows them to alter the interaction of water molecules and thus decrease the surface tension of water. This function of phospholipids makes them *surfactants* (surface-active agents). The surfactant effect of phospholipids prevents the lungs from collapsing due to surface tension forces (chapter 15). Phospholipids are also the major component of cell membranes, as will be described in chapter 3.

Steroids

The structure of steroid molecules is quite different from that of triglycerides or phospholipids, and yet steroids are still included in the lipid category of molecules because they are nonpolar and insoluble in water. All steroid molecules have the same basic structure: three six-carbon rings are joined to one five-carbon ring (fig. 2.21). However, different kinds of steroids have different functional groups attached to this basic structure, and they vary in the number and position of the double covalent bonds between the carbon atoms in the rings.

Cholesterol is an important molecule in the body because it serves as the precursor (parent molecule) for the steroid hormones produced by the gonads and adrenal cortex. The testes and ovaries (collectively called the *gonads*) secrete **sex steroids,** which include estradiol and progesterone from the

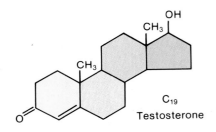

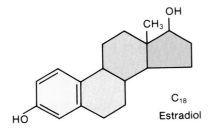

Figure 2.21

Cholesterol and some of the steroid hormones derived from cholesterol.

ovaries and testosterone from the testes. The adrenal cortex secretes the **corticosteroids,** including hydrocortisone and aldosterone, among others.

Prostaglandins

Prostaglandins are a type of fatty acid with a cyclic hydrocarbon group. Although their name is derived from the fact that they were originally noted in the semen as a secretion of the prostate, it has since been shown that they are produced by and are active in almost all organs, where they serve a variety of regulatory functions. Prostaglandins are implicated in the

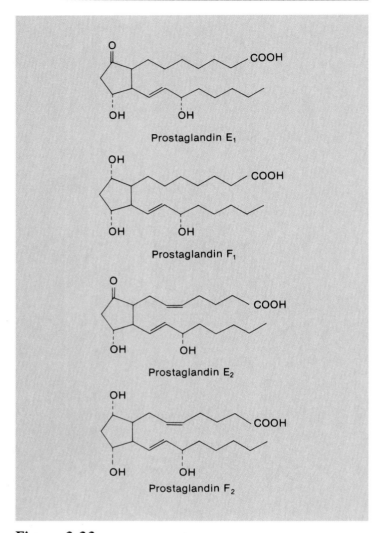

Figure 2.22
Structural formulas of various prostaglandins.

regulation of blood vessel diameter, ovulation, uterine contraction during labor, inflammation reactions, blood clotting, and many other functions. Structural formulas of different types of prostaglandins are shown in figure 2.22.

1. Describe the common structure of all carbohydrates and distinguish between monosaccharides, disaccharides, and polysaccharides.
2. Using dehydration synthesis and hydrolysis reactions, explain how disaccharides and monosaccharides can be interconverted and how triglycerides can be formed and broken down.
3. Describe the characteristics of a lipid and discuss the different subcategories of lipids.
4. Relate the functions of phospholipids to their structure and explain the significance of the prostaglandins.

Proteins

Proteins are large molecules composed of amino acid subunits. Since there are twenty different types of amino acids that can be used in constructing a given protein, the variety of protein structures is immense. This variety allows each type of protein to perform very specific functions.

The enormous diversity of protein structure results from the fact that there are twenty different building blocks—the amino acids—that can be used to form a protein. These amino acids, as will be described in the next section, are joined together to form a chain that can twist and fold in a specific manner due to chemical interactions between the amino acids. The specific sequence of amino acids in a protein, and thus the specific structure of the protein, is determined by genetic information. This genetic information for protein synthesis is contained in another category of organic molecules, the nucleic acids. The structure of nucleic acids, and the mechanisms by which the genetic information they encode directs protein synthesis, is described in chapter 3.

Structure of Proteins

Proteins consist of long chains of subunits called **amino acids.** As the name implies, each amino acid contains an *amino group* (NH_2) on one end of the molecule and a *carboxylic acid group* (COOH) on another end. There are twenty different amino acids, with different structures and chemical properties, that are used to build proteins. The differences between the amino acids are due to differences in their *functional groups.* "R" is the abbreviation for *functional group* in the general formula for an amino acid (fig. 2.23). The R symbol actually stands for the word *residue,* but it can be thought of as indicating the "rest of the molecule."

When amino acids are joined together by dehydration synthesis, the hydrogen from the amino end of one amino acid combines with the hydroxyl group of the carboxylic acid end of another amino acid. As a covalent bond is formed between the two amino acids, water is produced (fig. 2.24). The bond between adjacent amino acids is called a **peptide bond,** and the compound formed is called a *peptide.* When a number of amino acids are joined in this way, a chain of amino acids, or a **polypeptide,** is produced.

The lengths of polypeptide chains vary widely. A hormone called *thyrotropin-releasing hormone,* for example, is only three amino acids long, whereas *myosin,* a muscle protein, contains about 4,500 amino acids. When the length of a polypeptide chain becomes very long (greater than about 100 amino acids), the molecule is called a **protein.**

The structure of a protein can be described at four different levels. At the first level, the sequence of amino acids in the protein, called the **primary structure** of the protein, is

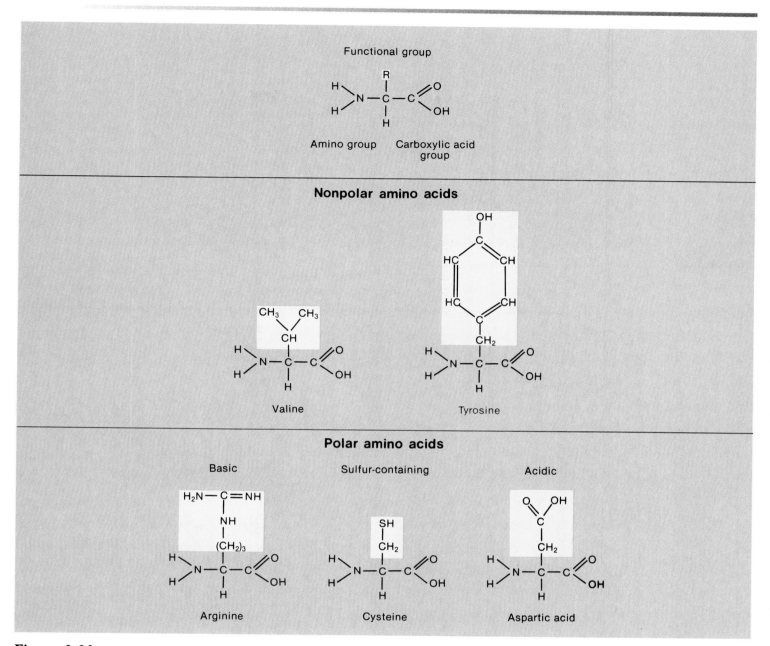

Figure 2.23

Representative amino acids, showing different types of functional (R) groups.

described. Each type of protein has a different primary structure. All of the billions of *copies* of a given type of protein in the body have the same structure, however, because the structure of a given protein is coded by the genes. The primary structure of a protein is illustrated in figure 2.25a.

Weak hydrogen bonds may form between the hydrogen atom of an amino group and an oxygen atom from a different amino acid nearby. These weak bonds cause the polypeptide chain to twist into a *helix*. The extent and location of the helical structure is different for each protein because of differences in amino acid composition. A description of the helical structure of a protein is termed its **secondary structure** (fig. 2.25b).

Most polypeptide chains bend and fold upon themselves to produce complex three-dimensional shapes, called the **tertiary structure** of the proteins. Each type of protein has its own characteristic tertiary structure. This is because the folding and bending of the polypeptide chain is produced by chemical interactions between particular amino acids that are located in different regions of the chain.

Most of the tertiary structure of proteins is formed and stabilized by weak chemical bonds (such as hydrogen bonds) between the functional groups of widely spaced amino acids. Since most of the tertiary structure is stabilized by weak bonds, this structure can easily be disrupted by high temperature or by

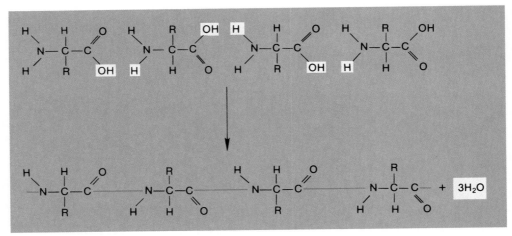

Figure 2.24

The formation of peptide bonds by dehydration synthesis reactions between amino acids.

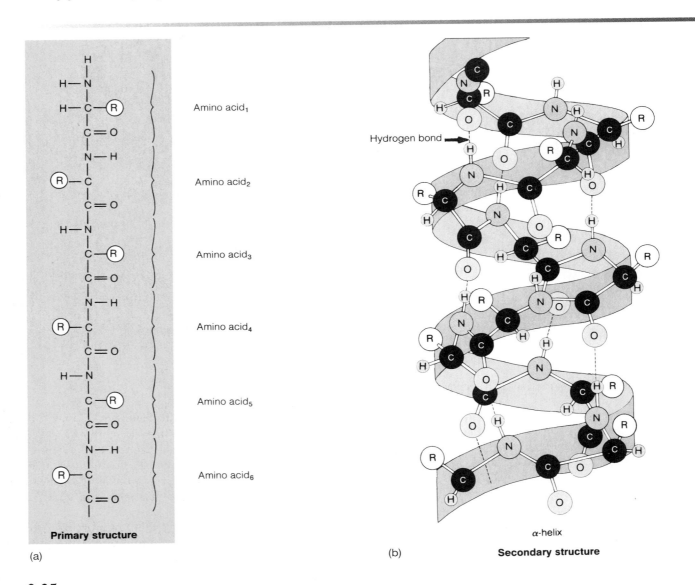

Amino acid₁

Amino acid₂

Amino acid₃

Amino acid₄

Amino acid₅

Amino acid₆

Primary structure

(a)

Hydrogen bond

α-helix

Secondary structure

(b)

Figure 2.25

A polypeptide chain showing (*a*) its primary structure and (*b*) its secondary structure.

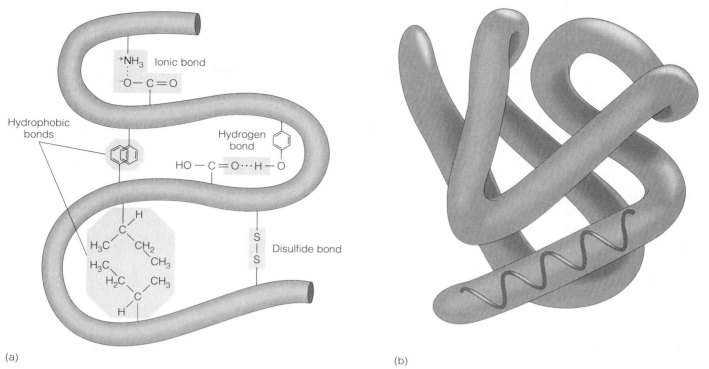

(a)

(b)

Figure 2.26

The tertiary structure of a protein. (*a*) Interactions between functional (R) groups of amino acids result in (*b*) the formation of proteins with complex three-dimensional shapes.

Table 2.4   Composition of Selected Proteins Found in the Body

Protein	Number of Polypeptide Chains	Nonprotein Component	Function
Hemoglobin	4	Heme pigment	Carries oxygen in the blood
Myoglobin	1	Heme pigment	Stores oxygen in muscle
Insulin	2	None	Hormonal regulation of metabolism
Luteinizing hormone	1	Carbohydrate	Hormonal stimulation of the gonads
Fibrinogen	1	Carbohydrate	Involved in blood clotting
Mucin	1	Carbohydrate	Forms mucus
Blood group proteins	1	Carbohydrate	Produces blood types
Lipoproteins	1	Lipids	Transports lipids in blood

changes in pH. Irreversible changes in the tertiary structure of proteins that occur by these means are referred to as *denaturation* of the proteins. The tertiary structure of some proteins, however, is made more stable by strong covalent bonds between sulfur atoms (called *disulfide bonds* and abbreviated S — S) in the functional group of an amino acid known as cysteine (fig. 2.26).

Denatured proteins retain their primary structure (the peptide bonds are not broken) but have altered chemical properties. Cooking a pot roast, for example, alters the texture of the meat proteins—it doesn't result in an amino acid soup. Denaturation is most dramatically demonstrated by frying an egg. Egg albumin proteins are soluble in their native state, in which they form the clear, viscous fluid of a raw egg. When denatured by cooking, these proteins change shape, cross-bond with each other, and by this means form an insoluble white precipitate—the egg white.

Some proteins (such as hemoglobin and insulin) are composed of a number of polypeptide chains covalently bonded together. This is the **quaternary structure** of these proteins. Insulin, for example, is composed of two polypeptide chains—one that is twenty-one amino acids long, the other that is thirty amino acids long. Hemoglobin (the protein in red blood cells that carries oxygen) is composed of four separate polypeptide chains (see chapter 16, fig. 16.32). The composition of various body proteins is shown in table 2.4.

Many proteins in the body are normally found combined, or *conjugated*, with other types of molecules. **Glycoproteins** are proteins conjugated with carbohydrates. Examples of such molecules include certain hormones and some proteins found in the cell membrane. **Lipoproteins** are proteins conjugated with lipids. These are found in cell membranes and in the plasma (the fluid portion of the blood). Proteins may also be conjugated with pigment molecules. These include hemoglobin, which transports oxygen in red blood cells, and the cytochromes, which are needed for oxygen utilization and energy production within cells.

Functions of Proteins

Because of their tremendous structural diversity, proteins can serve a wider variety of functions than any other type of molecule in the body. Many proteins, for example, contribute significantly to the structure of different tissues and in this way play a passive role in the functions of these tissues. Examples of such *structural proteins* include collagen (fig. 2.27) and keratin. Collagen is a fibrous protein that provides tensile strength to connective tissues, such as tendons and ligaments. Keratin is found in the outer layer of dead cells in the epidermis, where it prevents water loss through the skin.

Many proteins play a more active role in the body, where specificity of structure and function is required. *Enzymes* and *antibodies*, for example, are proteins—no other type of molecule could provide the vast array of different structures needed for their tremendously varied functions. Other examples of specificity include proteins in cell membranes, which may serve as *receptors* for specific regulator molecules (such as hormones) and as *carriers* for transport of specific molecules across the membrane. Proteins provide the diversity of shape and chemical properties required by these functions.

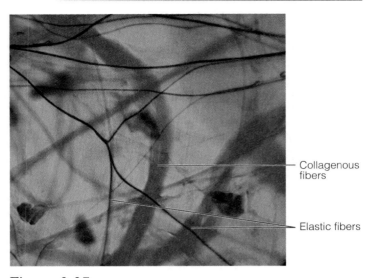

Collagenous fibers

Elastic fibers

Figure 2.27

A photomicrograph of collagen fibers within connective tissue.

---

1. Write the general formula for an amino acid and describe how amino acids differ from each other.
2. Describe the different levels of protein structure and explain how they are produced.
3. Describe the different categories of protein function in the body and explain why proteins can serve such diverse functions.

Summary

Atoms, Ions, and Chemical Bonds   p. 24

I. Covalent bonds are formed by atoms that share electrons. They are the strongest type of chemical bonds.
- A. Electrons are equally shared in nonpolar covalent bonds and unequally shared in polar covalent bonds.
- B. Atoms of oxygen, nitrogen, and phosphorus strongly attract electrons and become electrically negative compared to the other atoms sharing electrons with them.

II. Ionic bonds are formed by atoms that transfer electrons. These weak bonds join atoms together in an ionic compound.
- A. If one atom in this compound takes an electron from another atom, it gains a net negative charge and the other becomes positively charged.
- B. Ionic bonds easily break when the ionic compound is dissolved in water. Dissociation of the ionic compound yields charged atoms called ions.

III. When hydrogen bonds with an electronegative atom, it gains a slight positive charge and is weakly attracted to another electronegative atom. This weak attraction is a hydrogen bond.

IV. Acids donate hydrogen ions to solution, whereas bases lower the hydrogen ion concentration of a solution.
- A. The pH scale is a negative function of the logarithm of the hydrogen ion concentration.

B. In a neutral solution, the concentration of H$^+$ is equal to the concentration of OH$^-$, and the pH is 7.

C. Acids raise the H$^+$ concentration and thus lower the pH below 7; bases lower the H$^+$ concentration and thus raise the pH above 7.

V. Organic molecules contain atoms of carbon joined together by covalent bonds. Atoms of nitrogen, oxygen, phosphorus, or sulfur may be present as specific functional groups in the organic molecule.

Carbohydrates and Lipids p. 32

I. Carbohydrates contain carbon, hydrogen, and oxygen, usually in a ratio of 1:2:1.

A. Carbohydrates consist of simple sugars (monosaccharides), disaccharides, and polysaccharides (such as glycogen).

B. Covalent bonds between monosaccharides are formed by dehydration synthesis, or condensation. Bonds are broken by hydrolysis reactions.

II. Lipids are organic molecules that are insoluble in polar solvents such as water.

A. Triglycerides (fat and oil) consist of three fatty acid molecules joined to a molecule of glycerol.

B. Ketone bodies are smaller derivatives of fatty acids.

C. Phospholipids (such as lecithin) are phosphate-containing lipids with a polar group that is hydrophilic; the rest of the molecule is hydrophobic.

D. Steroids (including the hormones of the adrenal cortex and gonads) are lipids with a characteristic five-ring structure.

E. Prostaglandins are a family of cyclic fatty acids that serve a variety of regulatory functions.

Proteins p. 38

I. Proteins are composed of long chains of amino acids bonded together by covalent peptide bonds.

A. Each amino acid contains an amino group, a carboxyl group, and a functional group that is different for each of the more than twenty different amino acids.

B. The polypeptide chain may be twisted into a helix (secondary structure) and bent and folded to form the tertiary structure of the protein.

C. Proteins that are composed of two or more polypeptide chains are said to have a quaternary structure.

D. Proteins may be combined with carbohydrates, lipids, or other molecules.

E. Because they are so diverse structurally, proteins serve a wider variety of specific functions than any other type of molecule.

Clinical Investigation

A student, feeling it is immoral to eat plants or animals, decides to eat only artificial food. After raiding a chemistry laboratory, he places himself on a diet consisting only of the D-amino acids and L-sugars he obtained in his raid. After a couple of weeks, he begins to feel weak and seeks medical attention. Laboratory analysis of his urine reveals very high concentrations of ketone bodies (ketonuria). What might be the cause of his weakness and ketonuria?

Clues

See the description of stereoisomers in the section on organic molecules and the description of ketone bodies in the section on lipids.

Review Activities

Objective Questions

1. Which of the following statements about atoms is *true?*
   a. They have more protons than electrons.
   b. They have more electrons than protons.
   c. They are electrically neutral.
   d. They have as many neutrons as they have electrons.

2. The bond between oxygen and hydrogen in a water molecule is
   a. a hydrogen bond.
   b. a polar covalent bond.
   c. a nonpolar covalent bond.
   d. an ionic bond.

3. Which of the following is a nonpolar covalent bond?
   a. The bond between two carbons.
   b. The bond between sodium and chloride.
   c. The bond between two water molecules.
   d. The bond between nitrogen and hydrogen.

4. Solution A has a pH of 2, and solution B has a pH of 10. Which of the following statements about these solutions is *true?*
   a. Solution A has a higher H$^+$ concentration than solution B.
   b. Solution B is basic.
   c. Solution A is acidic.
   d. All of the above are true.

5. Glucose is
   a. a disaccharide.
   b. a polysaccharide.
   c. a monosaccharide.
   d. a phospholipid.

6. Digestion reactions occur by means of
   a. dehydration synthesis.
   b. hydrolysis.
7. Carbohydrates are stored in the liver and muscles in the form of
   a. glucose.
   b. triglycerides.
   c. glycogen.
   d. cholesterol.
8. Lecithin is
   a. a carbohydrate.
   b. a protein.
   c. a steroid.
   d. a phospholipid.

9. Which of the following lipids have regulatory roles in the body?
   a. steroids
   b. prostaglandins
   c. triglycerides
   d. both *a* and *b*
   e. both *b* and *c*
10. The tertiary structure of a protein is *directly* determined by
   a. the genes.
   b. the primary structure of the protein.
   c. enzymes that "mold" the shape of the protein.
   d. the position of peptide bonds.

11. The type of bond formed between two molecules of water is
   a. a hydrolytic bond.
   b. a polar covalent bond.
   c. a nonpolar covalent bond.
   d. a hydrogen bond.
12. The carbon-to-nitrogen bond that joins amino acids together is called
   a. a glycosidic bond.
   b. a peptide bond.
   c. a hydrogen bond.
   d. a double bond.

Essay Questions

1. Compare and contrast nonpolar covalent bonds, polar covalent bonds, and ionic bonds.[1]
2. Define *acid* and *base* and explain how acids and bases influence the pH of a solution.
3. Using dehydration synthesis and hydrolysis reactions, explain the relationships between starch in an ingested potato, liver glycogen, and blood glucose.
4. "All fats are lipids, but not all lipids are fats." Explain why this is an accurate statement.
5. Explain the relationship between the primary structure of a protein and its secondary and tertiary structures. What do you think would happen to the tertiary structure if some amino acids were substituted for others in the primary structure? What physiological significance might this have?
6. What are the similarities and differences between a fat and an oil? What is the physiological and clinical significance of the degree of saturation of fatty acid chains?

Selected Readings

Demers, L. M. September 1984. The effects of prostaglandins. *Diagnostic Medicine*, p. 37.
Doolittle, R. F. October 1985. Proteins. *Scientific American*.
Hakomori, Sen-itiroh. May 1986. Glycosphingolipids. *Scientific American*.
Jackson, R. W., and A. M. Gotto. 1974. Phospholipids in biology and medicine. *New England Journal of Medicine* 290:24.

Kuehl, F. A. Jr., and R. W. Egan. 1980. Prostaglandins, arachidonic acid, and inflammation. *Science* 210:978.
Prockop, D. J., and N. A. Guzman. 1977. Collagen disease and the biosynthesis of collagen. *Hospital Practice* 12:61.
Sharon, N. May 1974. Glycoproteins. *Scientific American*.
Sharon, N. November 1980. Carbohydrates. *Scientific American*.

Sharon, N., and H. Lis. January 1993. Carbohydrates in cell recognition. *Scientific American*.
Weinberg, R. A. October 1985. The molecules of life. *Scientific American*.
Whitney, E., C. Cataldo, and S. Rolfes. 1991. *Understanding Normal and Clinical Nutrition*. New York: West.

Life Science Animation

The animation that relates to chapter 2 is #1 Formation of an Ionic Bond.

---

[1]Note: This question is answered on page 20 of the Student Study Guide.

Cell Structure and Genetic Control

OBJECTIVES   *After studying this chapter, you should be able to . . .*

1. describe the structure of the cell membrane and explain its functional importance.

2. describe how cells move by amoeboid motion and state which cells in the human body move in this manner.

3. describe the structure of cilia and flagella and state some of their functions.

4. describe the processes of phagocytosis, pinocytosis, receptor-mediated endocytosis, and exocytosis.

5. state the functions of the cytoskeleton, lysosomes, mitochondria, and the endoplasmic reticulum.

6. describe the structure of the cell nucleus and explain its significance.

7. describe the structure of nucleotides and distinguish between the nucleotides of DNA and RNA.

8. explain how DNA is constructed and what is meant by the law of complementary base pairing.

9. explain how RNA is produced according to the genetic information in DNA and distinguish between the different types of RNA.

10. describe how proteins are produced according to the information contained in messenger RNA.

11. describe the structure of the rough endoplasmic reticulum and Golgi apparatus and discuss their functions in the secretion of proteins.

12. explain what is meant by the semiconservative mechanism of DNA replication.

13. describe the different stages of the cell cycle and the events that occur in the different phases of mitosis.

14. define the terms *hypertrophy* and *hyperplasia* and explain their physiological importance.

15. describe the events that occur in meiosis, compare them to those that occur in mitosis, and discuss the function of meiotic cell division in human physiology.

OUTLINE

Cell Membrane and Associated Structures

The cell is the basic unit of structure and function in the body. Many of the functions of cells are performed by particular subcellular structures known as organelles. The cell membrane is an extremely important structure that allows selective communication between the intracellular and extracellular compartments. The dynamic nature of the cell membrane also aids cellular movement.

The cell looks so small and simple when viewed with the ordinary (light) microscope that it is difficult to conceive of each cell as a living entity unto itself. Equally amazing is the fact that the physiology of our organs and systems derives from the complex functions of the cells of which they are composed. Complexity of function demands complexity of structure even at the subcellular level.

As the basic functional unit of the body, each cell is a highly organized molecular factory. Cells come in a wide variety of shapes and sizes. This great diversity, which is also apparent in the subcellular structures within different cells, reflects the diversity of function of different cells in the body. All cells, however, share certain characteristics; for example, they are all surrounded by a cell membrane, and most of them possess the structures listed in table 3.1. Thus, although no single cell can be considered "typical," the general structure of cells can be indicated by a single illustration (fig. 3.1).

For descriptive purposes, a cell can be divided into three principal parts:

1. **Cell (plasma) membrane.** The selectively permeable cell membrane surrounds the cell, gives it form, and separates the cell's internal structures from the extracellular environment.
2. **Cytoplasm and organelles.** The cytoplasm is the aqueous content of a cell between the nucleus and the cell membrane. Organelles are subcellular structures within the cytoplasm of a cell that perform specific functions. The term *cytosol* is frequently used to describe the soluble portion of the cytoplasm; that is, the part that cannot be removed by centrifugation.
3. **Nucleus.** The nucleus is a large, generally spheroid body within a cell. It contains the DNA, or genetic material, of the cell and thus directs the cell's activities.

Structure of the Cell Membrane

Because both the intracellular and extracellular environments (or "compartments") are aqueous, a barrier must be present to prevent the loss of cellular molecules such as enzymes, nucleotides, and others that are water-soluble. Since this barrier cannot itself be composed of water-soluble molecules, it makes sense that the cell membrane should be composed of lipids.

Table 3.1	Structure and Function of Cellular Components	
Component	**Structure**	**Function**
Cell (plasma) membrane	Membrane composed of phospholipid and protein molecules	Gives form to cell and controls passage of materials in and out of cell
Cytoplasm	Fluid, jellylike substance in which organelles are suspended	Serves as matrix substance in which chemical reactions occur
Endoplasmic reticulum	System of interconnected membrane-forming canals and tubules	Smooth endoplasmic reticulum metabolizes nonpolar compounds and stores Ca^{++} in striated muscle cells; rough endoplasmic reticulum assists in protein synthesis
Ribosomes	Granular particles composed of protein and RNA	Synthesize proteins
Golgi apparatus	Cluster of flattened, membranous sacs	Synthesizes carbohydrates and packages molecules for secretion; secretes lipids and glycoproteins
Mitochondria	Membranous sacs with folded inner partitions	Release energy from food molecules and transform energy into usable ATP
Lysosomes	Membranous sacs	Digest foreign molecules and worn and damaged organelles
Peroxisomes	Spherical membranous vesicles	Contain enzymes that produce hydrogen peroxide and use this for oxidation reactions
Centrosome	Nonmembranous mass of two rodlike centrioles	Helps to organize spindle fibers and distribute chromosomes during mitosis
Vacuoles	Membranous sacs	Store and excrete various substances within the cytoplasm
Fibrils and microtubules	Thin, hollow tubes	Support cytoplasm and transport materials within the cytoplasm
Cilia and flagella	Minute cytoplasmic extensions from cell	Move particles along surface of cell or move cell
Nuclear membrane	Membrane surrounding nucleus, composed of protein and lipid molecules	Supports nucleus and controls passage of materials between nucleus and cytoplasm
Nucleolus	Dense, nonmembranous mass composed of protein and RNA molecules	Produces ribosomal RNA for ribosomes
Chromatin	Fibrous strands composed of protein and DNA	DNA comprises the genetic material of cells and therefore controls cellular activities

Source: From Kent M. Van De Graaff, *Human Anatomy*, 4th ed. Copyright © 1995 Wm. C. Brown Communications, Inc., Dubuque, Iowa. Reprinted by permission of Times Mirror Higher Education Group, Inc., Dubuque, Iowa. All Rights Reserved.

Chapter Three

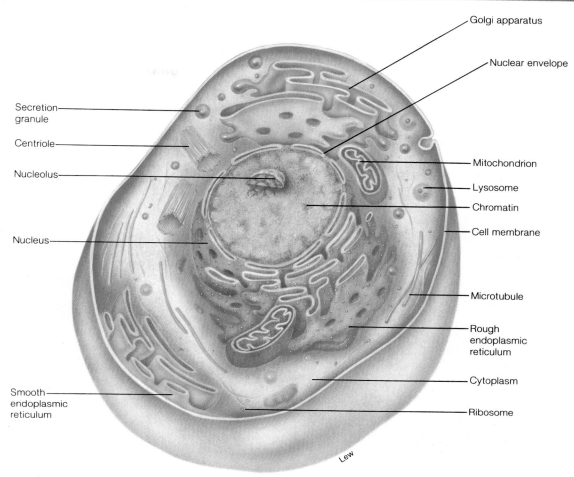

Golgi apparatus

Nuclear envelope

Secretion granule

Centriole

Nucleolus

Nucleus

Mitochondrion

Lysosome

Chromatin

Cell membrane

Microtubule

Rough endoplasmic reticulum

Cytoplasm

Smooth endoplasmic reticulum

Ribosome

Lew

Figure 3.1

A generalized cell and the principal organelles.

The **cell membrane** (also called the **plasma membrane,** or **plasmalemma**), and indeed all of the membranes surrounding organelles within the cell, are composed primarily of phospholipids and proteins. Phospholipids, as described in chapter 2, are polar on the end that contains the phosphate group and nonpolar (and hydrophobic) throughout the rest of the molecule. Since there is an aqueous environment on each side of the membrane, the hydrophobic parts of the molecules "huddle together" in the center of the membrane, leaving the polar ends exposed to water on both surfaces. This results in the formation of a double layer of phospholipids in the cell membrane.

The hydrophobic core of the membrane restricts the passage of water and water-soluble molecules and ions. Certain of these polar compounds, however, do pass through the membrane. The specialized functions and selective transport properties of the membrane are believed to be due to its protein content. Some proteins are found partially submerged on each side of the membrane; other proteins span the membrane completely from one side to the other. Since the membrane is not solid—phospholipids and proteins are free to move laterally—the proteins within the phospholipid "sea"

The cell membranes of all higher organisms contain cholesterol. The cells in the body with the highest content of cholesterol are the Schwann cells, which form insulating layers around certain nerve fibers (see chapter 7). Their high cholesterol content is believed to be important in this insulating function. The ratio of cholesterol to phospholipids also helps to determine the flexibility of a cell membrane. When there is an inherited defect in this ratio, the flexibility of the cell may be reduced. This could result, for example, in the inability of red blood cells to flex at the middle when passing through narrow blood channels and thereby cause occlusion of these small vessels.

are not uniformly distributed, but rather present a constantly changing mosaic pattern. This structure is known as the **fluid-mosaic model** of membrane structure (fig. 3.2).

The proteins found in the cell membrane serve a variety of functions, including structural support, transport of molecules

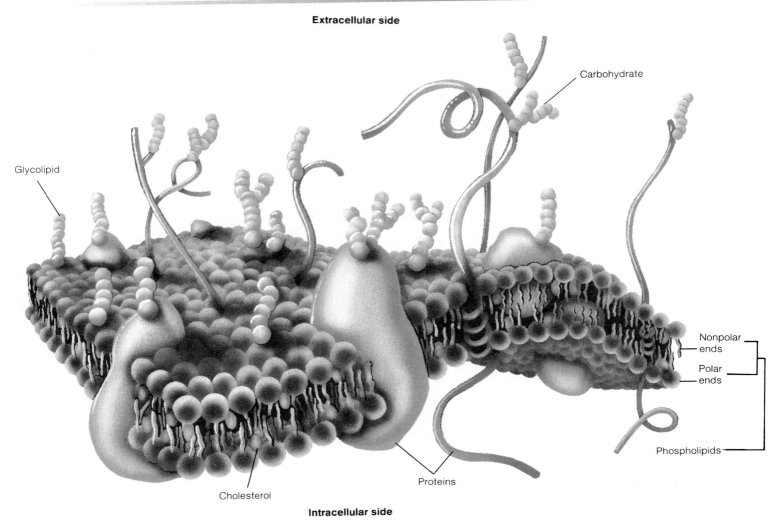

Extracellular side

Carbohydrate

Glycolipid

Nonpolar ends

Polar ends

Phospholipids

Proteins

Cholesterol

Intracellular side

Figure 3.2

The fluid-mosaic model of the cell membrane. The membrane consists of a double layer of phospholipids, with the polar phosphates (shown by spheres) oriented outward and the nonpolar hydrocarbons (wavy lines) oriented toward the center. Proteins may completely or partially span the membrane. Carbohydrates are attached to the outer surface.

across the membrane, and enzymatic control of chemical reactions at the cell surface. Some proteins function as receptors for hormones and other regulatory molecules that arrive at the outer surface of the membrane or as cellular "markers" (antigens), which identify the blood and tissue type.

In addition to lipids and proteins, the cell membrane also contains carbohydrates, which are primarily attached to the outer surface of the membrane as glycoproteins and glycolipids. These surface carbohydrates have many negative charges and, as a result, affect the interaction of regulatory molecules with the membrane. The negative charges at the surface also affect interactions between cells—they help keep red blood cells apart, for example. Stripping the carbohydrates from the outer red blood cell surface results in their more rapid destruction by the liver, spleen, and bone marrow.

Endocytosis and Exocytosis

Most of the movement of molecules and ions between the intracellular and extracellular environment involves passage through the cell membrane. These membrane transport processes are discussed in chapter 6. If molecules are too large to pass through the cell membrane, however, they may be taken into a cell by a different process. In this case, a portion of the cell membrane forms an invagination, or pouch, that contains some of the extracellular fluid and other material. This invagination eventually pinches off to form a new, membrane-enclosed organelle within the cell cytoplasm that contains the engulfed extracellular material. The general name for this process is **endocytosis.** There are three subtypes of endocytosis: phagocytosis, pinocytosis, and receptor-mediated endocytosis.

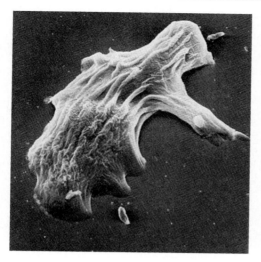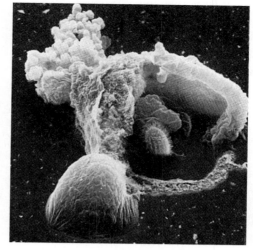

Figure 3.3

Scanning electron micrographs of phagocytosis, showing the formation of pseudopods and the entrapment of the prey within a food vacuole.

Amoeboid Movement and Phagocytosis

Some body cells—including certain white blood cells and macrophages in connective tissues—are able to move like an amoeba (a single-celled animal). This **amoeboid movement** is performed by the extension of parts of the cytoplasm to form *pseudopods*, which attach to a substrate and pull the cell along.

Cells that move by amoeboid motion—as well as certain liver cells, which are not mobile—use pseudopods to surround and engulf particles of organic matter (such as bacteria). This process is a type of cellular "eating" called **phagocytosis.** It serves to protect the body from invading microorganisms and to remove extracellular debris.

Phagocytic cells surround their victim with pseudopods, which join together and fuse (fig. 3.3). After the inner membrane of the pseudopods has become a continuous membrane around the ingested particle, it pinches off from the cell membrane. The ingested particle is now contained in an organelle called a *food vacuole* within the cell. The particle will subsequently be digested by enzymes contained in a different organelle (the lysosome), described in a later section.

Pinocytosis

Pinocytosis—a process related to phagocytosis—is performed by many cells. Instead of forming pseudopods, the cell membrane invaginates to produce a deep, narrow furrow. The membrane near the surface of this furrow then fuses, and a small vacuole containing the extracellular fluid is pinched off and enters the cell. In this way a cell can take in large molecules, such as proteins, which may be present in the extracellular fluid.

Receptor-Mediated Endocytosis

Another type of endocytosis involves the smallest area of cell membrane, and it occurs only in response to specific molecules in the extracellular environment. Since the extracellular molecules must bind to very specific *receptor proteins* in the cell membrane, this process is known as *receptor-mediated endocytosis*.

In receptor-mediated endocytosis, the interaction of specific molecules in the extracellular fluid with specific membrane receptor proteins causes the membrane to invaginate, fuse, and pinch off to form a vesicle (fig. 3.4). Vesicles formed in this way contain extracellular fluid and molecules that could not have passed by other means into the cell. Cholesterol attached to specific proteins, for example, is taken up into artery cells by receptor-mediated endocytosis. (This is in part responsible for atherosclerosis, as described in chapter 13.)

Exocytosis

Exocytosis is the reverse of endocytosis and serves as a mechanism for the secretion of cellular products into the extracellular environment. Proteins and other molecules produced within the cell that are destined for export (secretion) are packaged inside of the cell within vesicles by an organelle known as the Golgi apparatus. In the process of exocytosis, these secretory vesicles fuse with the cell membrane and release their contents into the extracellular environment (see fig. 3.27). Nerve endings release their chemical neurotransmitters in this manner, for example (chapter 7).

When the vesicle containing the secretory products of the cell fuse with the cell membrane during exocytosis, the total amount of cell membrane is increased. This process replaces material that was lost from the cell membrane during endocytosis.

Cilia and Flagella

Cilia are tiny hairlike structures that protrude from the cell and, like the coordinated action of rowers in a boat, stroke in unison. Cilia in the human body are found on the apical surface (the surface facing the lumen, or cavity) of stationary epithelial cells in the respiratory and female reproductive tracts. In the respiratory system, the cilia transport strands of mucus to the throat (pharynx), where the mucus can either be swallowed or expectorated. In the female reproductive tract, ciliary movements in the

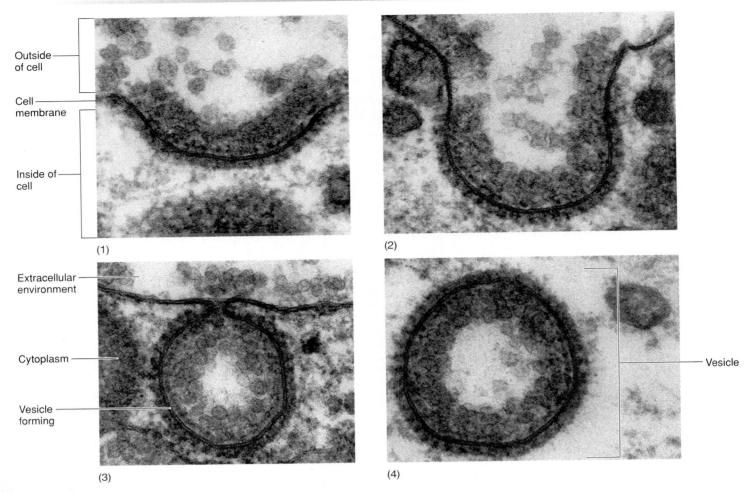

Outside of cell

Cell membrane

Inside of cell

(1)

(2)

Extracellular environment

Cytoplasm

Vesicle forming

(3)

Vesicle

(4)

Figure 3.4

Stages 1 through 4 of receptor-mediated endocytosis, during which specific bonding of extracellular particles to membrane receptor proteins is believed to occur.

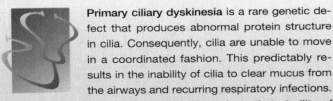 **Primary ciliary dyskinesia** is a rare genetic defect that produces abnormal protein structure in cilia. Consequently, cilia are unable to move in a coordinated fashion. This predictably results in the inability of cilia to clear mucus from the airways and recurring respiratory infections. The structure of flagella is also affected, so that sterility of males occurs due to defective flagellar movement of sperm.

Microvilli

In areas of the body that are specialized for rapid diffusion, the surface area of the cell membranes may be increased by numerous folds. The rapid passage of the products of digestion across the epithelial membranes in the intestine, for example, is aided by such structural adaptations. The surface area of the apical membranes (the part facing the lumen) in the intestine is increased by numerous tiny folds that form fingerlike projections called **microvilli** (fig. 3.6). Similar microvilli are also found in the kidney tubule epithelium, which must reabsorb various molecules that are filtered out of the blood.

epithelial lining of the uterine tube draw the ovum (egg) into the tube and move it toward the uterus.

Sperm are the only cells in the human body that have **flagella.** The flagellum is a single, whiplike structure that propels the sperm cell through its environment. Both cilia and flagella are composed of *microtubules* (formed from proteins) arranged in a characteristic way. One pair of microtubules in the center of a cilium or flagellum is surrounded by nine other pairs of microtubules, to produce what is often described as a "9 + 2" arrangement (fig. 3.5).

1. Draw the fluid-mosaic model of the cell membrane and describe the structure of the membrane.
2. Describe the structure of cilia and flagella and list some of their functions.
3. Draw a figure showing phagocytosis and pinocytosis and explain the significance of these processes.
4. Describe the structure and function of microvilli.

Chapter Three

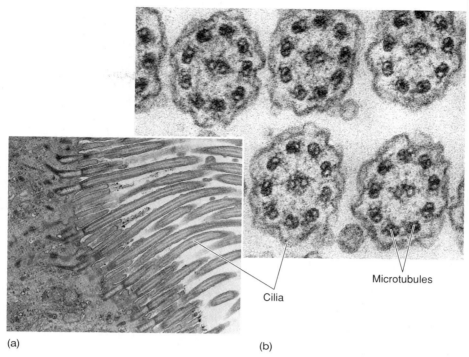

(a)

Cilia

Microtubules

(b)

Figure 3.5

Electron micrographs of cilia, showing (a) longitudinal and (b) cross sections. Notice the characteristic "9 + 2" arrangement of microtubules in the cross sections.

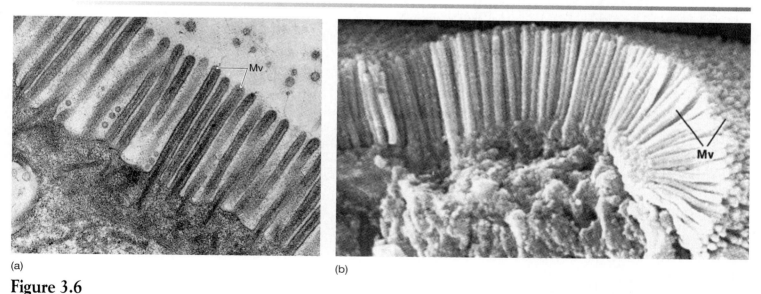

Mv

(a)

Mv

(b)

Figure 3.6

Microvilli (Mv) in the small intestine, as seen with (a) the transmission and (b) the scanning electron microscope.

Reproduced from R. G. Kessel and R. H. Kardon, Tissues and Organs: A Text Atlas of Scanning Electron Microscopy, W. H. Freeman and Co., 1979.

Cytoplasm and Its Organelles

Many of the functions of a cell that are performed in the cytoplasmic compartment result from the activity of specific structures called organelles. Among these are the lysosomes, which contain digestive enzymes, and the mitochondria, where most of the cellular energy is produced. Other organelles participate in the synthesis and secretion of cellular products.

Cytoplasm and Cytoskeleton

The jellylike matrix within a cell (exclusive of that within the nucleus) is known as **cytoplasm.** When viewed in a microscope without special techniques, the cytoplasm appears to be uniform and unstructured. According to recent evidence, however, the cytoplasm is not a homogenous solution; it is, rather, a highly organized structure in which protein fibers—in the form of *microtubules* and *microfilaments*—are arranged in a complex latticework. Using fluorescence microscopy, these structures can be visualized with the aid of antibodies against their protein components (fig. 3.7). The interconnected microfilaments and microtubules are believed to provide structural organization for cytoplasmic enzymes and support for various organelles.

The latticework of microfilaments and microtubules is said to function as a **cytoskeleton** (fig. 3.8). The structure of this "skeleton" is not rigid; it is capable of quite rapid movement and reorganization. Contractile proteins—including actin and myosin, which are responsible for muscle contraction—are microfilaments found in most cells. Such microfilaments aid in amoeboid motion, for example, so that the cytoskeleton is also

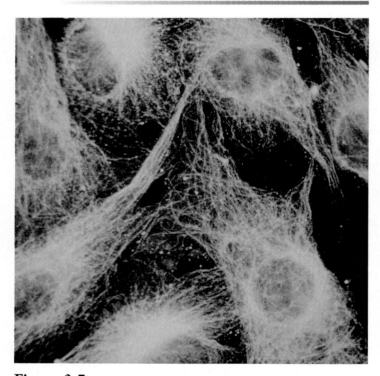

Figure 3.7

An immunofluorescence photograph of microtubules forming the cytoskeleton of a cell. Microtubules are visualized with the aid of antibodies against tubulin, the major protein component of the microtubules.

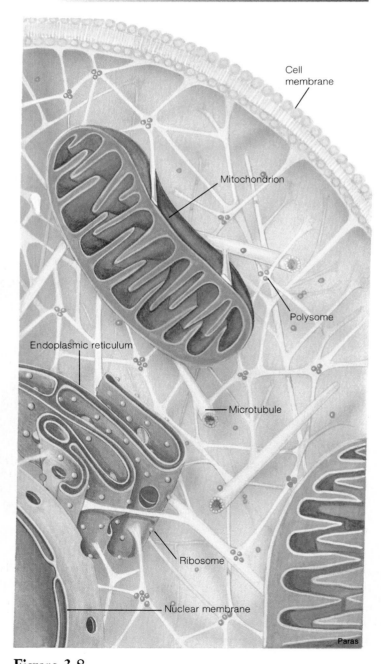

Figure 3.8

A diagram showing how microtubules form the cytoskeleton.

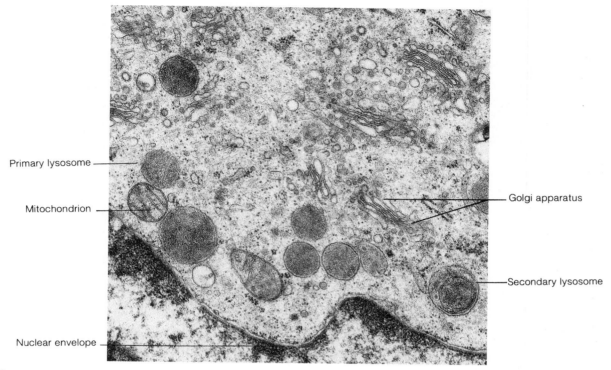

Figure 3.9

An electron micrograph showing primary and secondary lysosomes, mitochondria, and the Golgi apparatus.

the cell's "musculature." Microtubules, as another example, form the *spindle apparatus* that pulls chromosomes away from each other in cell division. Microtubules also form the central parts of cilia and flagella and contribute to the structure and movements of these structures.

Lysosomes

After a phagocytic cell has engulfed the proteins, polysaccharides, and lipids present in a particle of "food" (such as a bacterium), these molecules are still kept isolated from the cytoplasm by the membranes surrounding the food vacuole. The large molecules of proteins, polysaccharides, and lipids must first be digested into their smaller subunits (amino acids, monosaccharides, and so on) before they can cross the vacuole membrane and enter the cytoplasm.

The digestive enzymes of a cell are isolated from the cytoplasm and concentrated within membrane-bound organelles called **lysosomes** (fig. 3.9). A *primary lysosome* is one that contains only digestive enzymes (about forty different types) within an environment that is considerably more acidic than the surrounding cytoplasm. A primary lysosome may fuse with a food vacuole (or with another cellular organelle) to form a *secondary lysosome* in which worn-out organelles and the products of phagocytosis can be digested. Thus, a secondary lysosome contains partially digested remnants of other organelles and ingested organic material. A lysosome that contains undigested wastes is called a *residual body*. Residual

 Most, if not all, molecules in the cell have a limited life span. They are continuously destroyed and must be continuously replaced. Glycogen and some complex lipids in the brain, for example, are digested normally at a particular rate by lysosomes. If a person, because of some genetic defect, does not have the proper amount of these lysosomal enzymes, the resulting abnormal accumulation of glycogen and lipids could destroy the tissues. Examples of such defects include *Tay–Sach's disease* and *Gaucher's disease.*

bodies may eliminate their wastes by exocytosis, or the wastes may accumulate within the cell as the cell ages.

Partly digested membranes of various organelles and other cellular debris are often observed within secondary lysosomes. This is a result of **autophagy,** a process that destroys worn-out organelles so that they can be continuously replaced. Lysosomes are thus aptly described as constituting the "digestive system" of the cell.

Lysosomes have also been called "suicide bags" because a break in their membranes would release their digestive enzymes and thus destroy the cell. This happens normally in *programmed cell death* (or *apoptosis*), described in a later section. An example is the destruction of tissues that occurs during the remodeling processes in embryological development.

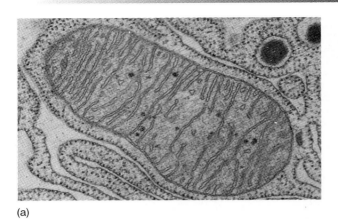

(a)

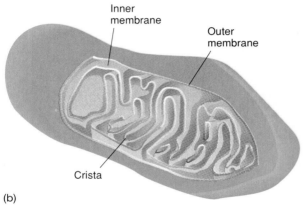

Inner
membrane

Outer
membrane

Crista

(b)

Figure 3.10

(*a*) An electron micrograph of a mitochondrion. The outer membrane and the infoldings of the inner membrane—the cristae—are clearly seen. The fluid in the center is the matrix. (*b*) A diagram of the structure of a mitochondrion.

Mitochondria

All cells in the body, with the exception of mature red blood cells, have from a hundred to a few thousand organelles called **mitochondria.** Mitochondria serve as sites for the production of most of the energy of cells (chapter 5). For this reason, mitochondria are sometimes called the "powerhouses" of the cell.

Mitochondria vary in size and shape, but all have the same basic structure (fig. 3.10). Each is surrounded by an *outer membrane* that is separated by a narrow space from an *inner membrane*. The inner membrane is characterized by many folds, called *cristae*, which extend into the central area (or *matrix*) of the mitochondrion. The cristae and the matrix compartmentalize the space within the mitochondrion and have different roles in the generation of cellular energy. The structure and functions of mitochondria will be described in more detail in the context of cellular metabolism in chapter 5.

Mitochondria are able to migrate through the cytoplasm of a cell, and it is believed that they are able to reproduce themselves. Indeed, mitochondria contain their own DNA! This is a more primitive form of DNA than that found within the cell nucleus. For this and other reasons, many scientists be-

An ovum (egg cell) contains mitochondria; the head of a sperm cell contains none. Therefore, all of the mitochondria in a fertilized egg are derived from the mother. The mitochondrial DNA replicates itself and the mitochondria divide, so that all of the mitochondria in the fertilized ovum and the cells derived from it during embryonic and fetal development are genetically identical to those in the original ovum. This provides a unique form of inheritance that is passed only from mother to child. A rare cause of blindness known as *Leber's hereditary optic neuropathy,* as well as several other disorders, are inherited only along the maternal lineage and are known to be caused by defective mitochondrial DNA.

lieve that mitochondria evolved from separate organisms, related to bacteria, that invaded the ancestors of animal cells and remained in a state of symbiosis.

Endoplasmic Reticulum

Most cells contain a system of membranes known as the endoplasmic reticulum, of which there are two types: (1) a **rough,** or **granular, endoplasmic reticulum** and (2) a **smooth endoplasmic reticulum** (fig. 3.11). A rough endoplasmic reticulum bears ribosomes (discussed in a later section) on its surface, whereas a smooth endoplasmic reticulum does not. The smooth endoplasmic reticulum serves a variety of purposes in different cells; it provides a site for enzyme reactions in steroid hormone production and inactivation, for example, and a site for the storage of Ca^{++} in striated muscle cells. The rough endoplasmic reticulum is found in cells that are active in protein synthesis and secretion, such as those of many exocrine and endocrine glands.

Details of the structure and function of the rough endoplasmic reticulum and its associated ribosomes, and of another organelle called the Golgi apparatus, will be described in a later section on protein synthesis. The structure of centrioles and the spindle apparatus, which are involved in DNA replication and cell division, will also be described in a separate section.

The smooth endoplasmic reticulum in liver cells contains enzymes used for the inactivation of steroid hormones and many drugs. This inactivation is generally achieved by reactions that convert these compounds to more water-soluble and less active forms, which can be more easily excreted by the kidneys. When people take certain drugs (such as alcohol and phenobarbital) for a long period of time, increasingly large doses of these compounds are required to produce a given effect. This phenomenon, called *tolerance,* is accompanied by growth of the smooth endoplasmic reticulum and thus an increase in the amount of enzymes charged with inactivation of these drugs.

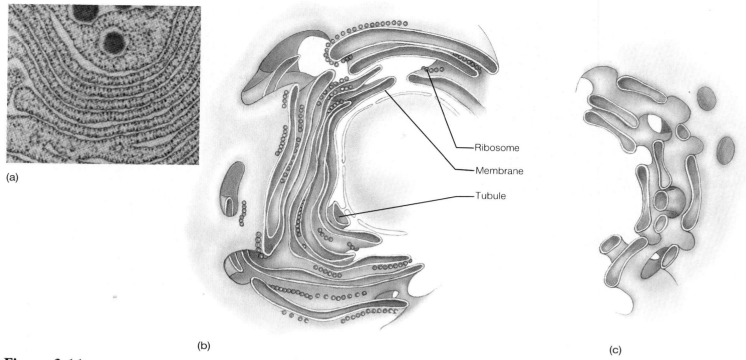

Figure 3.11

(a) An electron micrograph of endoplasmic reticulum (about 100,000×). Rough endoplasmic reticulum (b) has ribosomes attached to its surface, whereas smooth endoplasmic reticulum (c) lacks ribosomes.

1. Explain why microtubules and microfilaments can be thought of as the skeleton and musculature of a cell.

2. Describe the contents of lysosomes and explain the significance of autophagy.

3. Describe the structure and functions of mitochondria.

4. Explain how mitochondria can provide a genetic inheritance that is derived only from the mother.

5. Distinguish between a rough and a smooth endoplasmic reticulum in terms of their structure and function.

Cell Nucleus and Nucleic Acids

The genetic code is based on the structure of DNA. DNA and RNA are composed of subunits called nucleotides, and together these molecules are known as nucleic acids. The sequence of DNA nucleotides is the basis of the genetic code and serves to direct the synthesis of RNA molecules. It is through the synthesis of RNA, and eventually of protein, that the genetic code is expressed.

Most cells in the body have a single **nucleus,** although some—such as skeletal muscle cells—are multinucleate. The nucleus is surrounded by a *nuclear envelope* (fig. 3.12), which is composed of an inner membrane and an outer membrane. The outer membrane is continuous with the endoplasmic reticulum in the cytoplasm. At various points, the inner and outer membranes of the nuclear envelope are fused together by structures called *nuclear pore complexes.* These structures function as rivets, holding the nuclear envelope together. Each nuclear pore complex has a central opening, the *nuclear pore* (fig. 3.13), surrounded by interconnected rings and columns of proteins. Small molecules may pass through the complexes by diffusion, but movement of protein and RNA through the nuclear pores is a selective, energy-requiring process.

Transport of specific proteins from the cytoplasm, through the nuclear pores, and into the nucleus may serve a variety of functions, including regulation of gene expression by hormones (see chapter 11). Transport of RNA out of the nucleus, where it is formed, is required for gene expression. As described in this section, genes are regions of the DNA within the nucleus. Each gene contains the code for the production of a particular type of RNA called messenger RNA (mRNA). As an mRNA is transported through the nuclear pore, it becomes associated with ribosomes that are either free in the cytoplasm or associated with the rough endoplasmic reticulum. The mRNA then provides the code for the production of a specific type of protein.

The primary structure of the protein (its amino acid sequence) is determined by the sequence of bases in mRNA. The

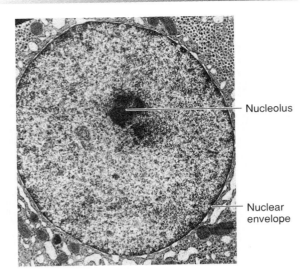

Figure 3.12

The nucleus of a liver cell showing the nuclear envelope and nucleolus.

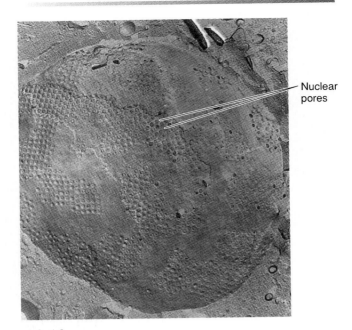

Figure 3.13

An electron micrograph of a freeze-fractured nuclear envelope showing the nuclear pores.

base sequence of mRNA has been previously determined by the sequence of bases in the region of the DNA (the gene) that codes for the mRNA. Genetic expression therefore occurs in two stages: first **genetic transcription** (synthesis of RNA) and then **genetic translation** (synthesis of protein).

Each nucleus contains one or more dark areas (see fig. 3.12). These regions, which are not surrounded by membranes, are called **nucleoli.** The DNA within the nucleoli contains the genes that code for the production of ribosomal RNA (rRNA), an essential component of ribosomes.

Nucleic Acids

Nucleic acids include the macromolecules of **DNA** and **RNA,** which are critically important in genetic regulation, and the subunits from which these molecules are formed. These subunits are known as *nucleotides*.

Nucleotides are used as subunits in the formation of long polynucleotide chains. Each nucleotide, however, is composed of three smaller subunits: a five-carbon sugar, a phosphate group attached to one end of the sugar, and a *nitrogenous base* attached to the other end of the sugar (fig. 3.14). The nitrogenous bases are cyclic nitrogen-containing molecules of two kinds: pyrimidines and purines. The *pyrimidines* contain a single ring of carbon and nitrogen, whereas the *purines* have two such rings (fig. 3.15).

Deoxyribonucleic Acid

The structure of DNA serves as the basis for the genetic code. For this reason, it might seem logical that DNA should have an extremely complex structure. DNA is indeed larger than any other molecule in the cell, but its structure is actually simpler than that of most proteins. This simplicity of structure deceived some of the early scientists into believing that the protein content of chromosomes, rather than their DNA content, provided the basis for the genetic code.

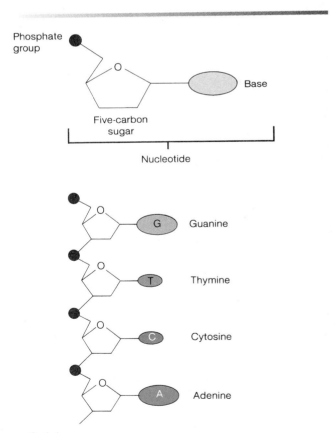

Figure 3.14

The general structure of a nucleotide and the formation of sugar-phosphate bonds between nucleotides to form a polymer.

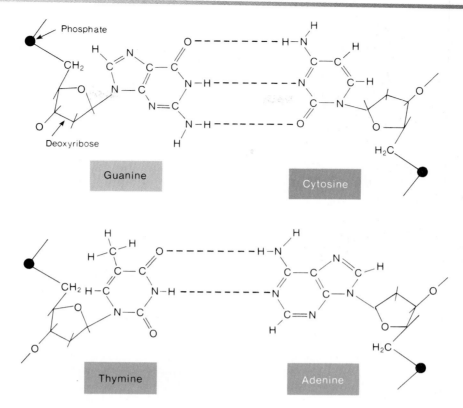

Figure 3.15

The four nitrogenous bases in deoxyribonucleic acid (DNA). Notice that hydrogen bonds can form between guanine and cytosine and between thymine and adenine.

Sugar molecules in the nucleotides of DNA are a type of pentose (five-carbon) sugar called **deoxyribose.** Each deoxyribose can be covalently bonded to one of four possible bases. These bases include the two purines (*adenine* and *guanine*) and the two pyrimidines (*cytosine* and *thymine*). There are thus four different types of nucleotides that can be used to produce the long DNA chains.

When nucleotides combine to form a chain, the phosphate group of one condenses with the deoxyribose sugar of another nucleotide. This forms a sugar-phosphate chain as water is removed in dehydration synthesis. Since the nitrogenous bases are attached to the sugar molecules, the sugar-phosphate chain looks like a "backbone" from which the bases project. Each of these bases can form hydrogen bonds with other bases, which are in turn joined to a different chain of nucleotides. Such hydrogen bonding between bases thus produces a *double-stranded* DNA molecule; the two strands are like a staircase, with the paired bases as steps (fig. 3.15).

Actually, the two chains of DNA twist about each other to form a **double helix,** so that the molecule resembles a spiral staircase (fig. 3.16). It has been shown that the number of purine bases in DNA is equal to the number of pyrimidine bases. The reason for this is explained by the **law of complementary base pairing:** adenine can pair only with thymine (through two hydrogen bonds), whereas guanine can pair only with cytosine (through three hydrogen bonds). Knowing this rule, we could predict the base sequence of one DNA strand if we knew the sequence of bases in the complementary strand.

Although we can be certain of which base is opposite a given base in DNA, we cannot predict which bases will be above or below that particular pair within a single polynucleotide chain. Although there are only four bases, the number of possible base sequences along a stretch of several thousand nucleotides (the length of a gene) is almost infinite. Yet, even with this amazing variety of possible base sequences, almost all of the billions of copies of a particular gene in a person are identical. The mechanisms by which this is achieved will be described in a later section.

Chromatin

The DNA within the cell nucleus is combined with protein to form **chromatin,** the threadlike material that makes up the chromosomes. Much of the protein content of the chromatin is of a type known as *histones*. Histone proteins (there are several different forms) are positively charged and organized to form spools about which the negatively charged strands of DNA are wound. Such spooling creates particles known as **nucleosomes** (fig. 3.17).

Some genes are permanently repressed in all cells; others are repressed much of the time but are activated under particular conditions. Recent evidence suggests that histones act to repress genes. According to one model, a different kind of protein that functions as a gene activator may work by causing histones

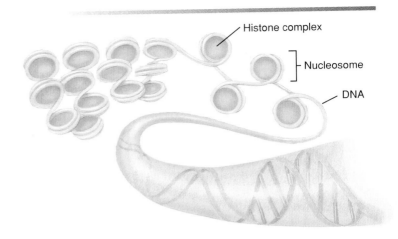

Figure 3.17

The structure of chromatin. Part of the DNA is wound around complexes of histone proteins, forming particles known as *nucleosomes*.

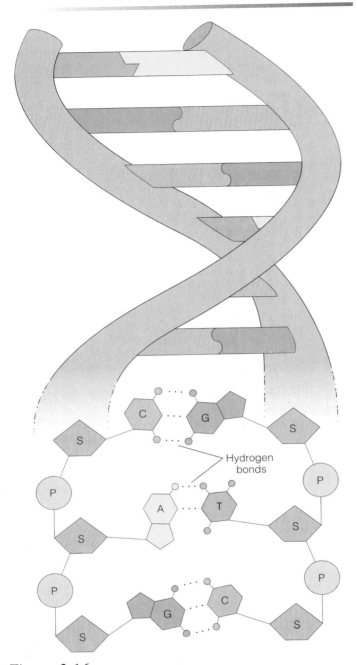

Figure 3.16

The double-helix structure of DNA.

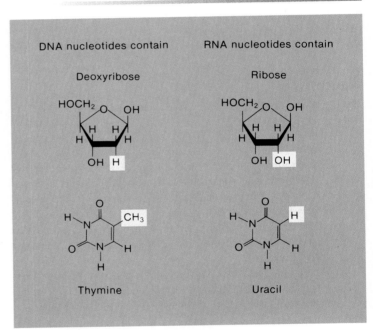

Figure 3.18

Differences between the nucleotides and sugars in DNA and RNA.

to dissociate from the specific region of DNA that promotes the gene's activation. Removal of the repressive effects of the histones, then, would activate the gene. Active regions of chromatin, called *euchromatin*, have a threadlike appearance in the electron microscope, whereas inactive chromatin appears as blotches called *heterochromatin*.

Ribonucleic Acid

The genetic information contained in DNA functions to direct the activities of the cell through its production of another type of nucleic acid—RNA (*ribonucleic acid*). Like DNA, RNA consists of long chains of nucleotides joined together by sugar-phosphate bonds. Nucleotides in RNA, however, differ from those in DNA (fig. 3.18) in three ways: (1) a **ribonucleotide** contains the sugar *ribose* (instead of deoxyribose), (2) the base *uracil* is found in place of thymine, and (3) RNA is composed of a single polynucleotide strand (it is not double-stranded like DNA).

There are three types of RNA molecules that function in the cytoplasm of cells: *messenger RNA* (mRNA), *transfer RNA* (tRNA), and *ribosomal RNA* (rRNA). All three types are made within the cell nucleus by using information contained in DNA as a guide.

RNA Synthesis

One gene codes for one polypeptide chain. Each gene is a stretch of DNA that is several thousand nucleotide pairs long. The DNA in a human cell contains 3 billion to 4 billion base pairs—enough to code for at least 3 million proteins. Since the average human cell contains less than this amount (30,000 to 150,000 different proteins), it follows that only a fraction of the DNA in each cell is used to code for proteins. The remainder of the DNA may be inactive, may be redundant, or it may serve to regulate those regions that do code for proteins.

In order for the genetic code to be translated into the synthesis of specific proteins, the DNA code first must be transcribed into an RNA code. This is accomplished by DNA-directed RNA synthesis, or **genetic transcription.**

In RNA synthesis, the enzyme *RNA polymerase* breaks the weak hydrogen bonds between paired DNA bases. This does not occur throughout the length of DNA, but only in the regions that are to be transcribed (there are base sequences that code for "start" and "stop"). Double-stranded DNA, therefore, separates in these regions so that the freed bases can pair with the complementary RNA nucleotide bases, which are freely available in the nucleoplasm.

This pairing of bases, like that which occurs in DNA replication (described in a later section), follows the law of complementary base pairing: guanine bonds with cytosine (and vice versa), and adenine bonds with uracil (because uracil in RNA is equivalent to thymine in DNA). Unlike DNA replication, however, only *one* of the two freed strands of DNA serves as a guide for RNA synthesis (fig. 3.19). Once an RNA molecule has been produced, it detaches from the DNA strand on which it was formed. This process can continue indefinitely, producing many thousands of RNA copies of the DNA strand that is being transcribed. When the gene is no longer to be transcribed, the separated DNA strands can then go back together again.

Types of RNA

There are four types of RNA produced within the nucleus by genetic transcription: (1) **precursor messenger RNA (pre-mRNA),** which is altered within the nucleus to form mRNA; (2) **messenger RNA (mRNA),** which contains the code for the synthesis of specific proteins; (3) **transfer RNA (tRNA),** which is needed for decoding the genetic message contained in mRNA; and (4) **ribosomal RNA (rRNA),** which forms part of the structure of ribosomes. The DNA that codes for rRNA synthesis is located in the part of the nucleus called the nucleolus. The DNA that codes for pre-mRNA and tRNA synthesis is located elsewhere in the nucleus.

In bacteria, where the molecular biology of the gene is best understood, a gene that codes for one type of protein produces an mRNA molecule that begins to direct protein synthesis as soon as it is transcribed. This is not the case in higher organisms, including humans. In higher cells, a pre-mRNA is produced that must be modified within the nucleus before it can enter the cytoplasm as mRNA and direct protein synthesis.

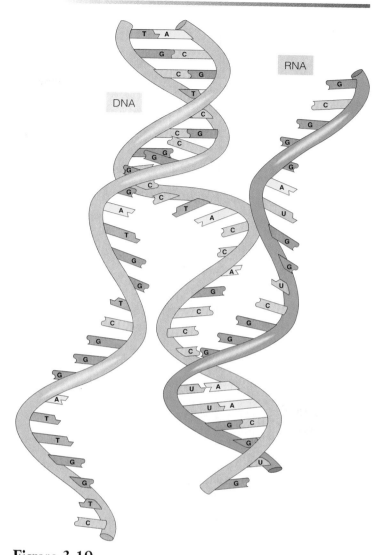

Figure 3.19

RNA synthesis (genetic transcription). Notice that only one of the two DNA strands is used to form a single-stranded molecule of RNA.

The **Human Genome Project** was launched by Congress in 1988, with the goal of completely mapping the human genome by September 30, 2005. That allowed just 17 years to determine the exact sequences of bases with which the 3 billion base pairs are arranged to form the 50,000 to 100,000 genes in the haploid human genome. (The haploid genome is the genome of a sperm cell or oocyte.) Such a detailed map will greatly aid the diagnosis and treatment of the 4,000 different genetic diseases that are directly caused by particular abnormal genes. Other diseases—the majority—have a genetic component but are caused by a variety of factors. The diagnosis and treatment of these diseases also may be greatly aided by research utilizing a complete map of the human genome.

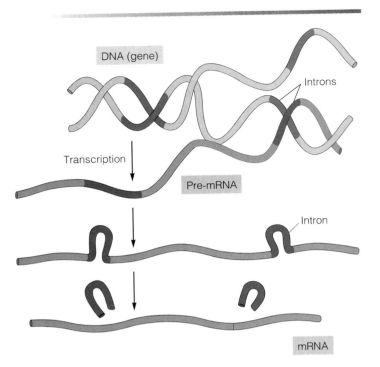

Figure 3.20

Processing of a pre-mRNA molecule into mRNA. Noncoding regions of the genes, called introns, produce excess bases within the pre-mRNA. These excess bases are removed, and the coding regions of mRNA are spliced together.

Precursor mRNA is much larger than the mRNA that it forms. Surprisingly, this large size of pre-mRNA is not due to excess bases at the ends of the molecule that must be trimmed; rather, the excess bases are *within* the pre-mRNA. The genetic code for a particular protein, in other words, is split up by stretches of base pairs that do not contribute to the code. These regions of noncoding DNA within a gene are called *introns;* the coding regions are known as *exons.* As a result, pre-mRNA must be cut and spliced to make mRNA (fig. 3.20). This cutting and splicing can be quite extensive—a single gene may contain up to 50 introns, which must be removed from the pre-mRNA in order to convert it to mRNA.

---

1. Describe the appearance and composition of chromatin and the structure of nucleosomes. Explain the significance of histone proteins.
2. Describe the structure of DNA and explain the law of complementary base pairing.
3. Describe the structure of RNA and list the different types of RNA.
4. Explain how RNA is produced within the nucleus according to the information contained in DNA.
5. Explain how precursor mRNA is modified to produce mRNA.

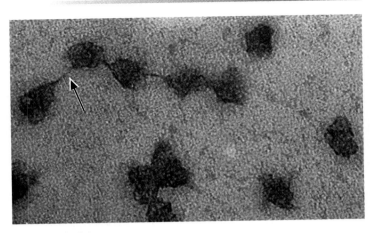

Figure 3.21

An electron micrograph of polyribosomes. An RNA strand (*arrow*) joins the ribosomes together.

Protein Synthesis and Secretion

In order for a gene to be expressed, it first must be used as a guide, or template, in the production of a complementary strand of messenger RNA. The mRNA is then used as a guide to produce a particular type of protein whose sequence of amino acids is determined by the sequence of base triplets (codons) in the mRNA.

When mRNA enters the cytoplasm, it attaches to **ribosomes,** which appear in the electron microscope as numerous small particles. A ribosome is composed of three molecules of ribosomal RNA and fifty-two proteins, arranged to form two subunits of unequal size. The mRNA passes through a number of ribosomes to form a "string-of-pearls" structure called a *polyribosome* (or *polysome,* for short), as shown in figure 3.21. The association of mRNA with ribosomes is needed for genetic translation—the production of specific proteins according to the code contained in the mRNA base sequence.

Each mRNA molecule contains several hundred or more nucleotides, arranged in the sequence determined by complementary base pairing with DNA during genetic transcription (RNA synthesis). Every three bases, or *base triplet,* is a code word—called a **codon**—for a specific amino acid. Sample codons and their amino acid "translations" are shown in table 3.2 and in figure 3.22. As mRNA moves through the ribosome, the sequence of codons is translated into a sequence of specific amino acids within a growing polypeptide chain.

Chapter Three

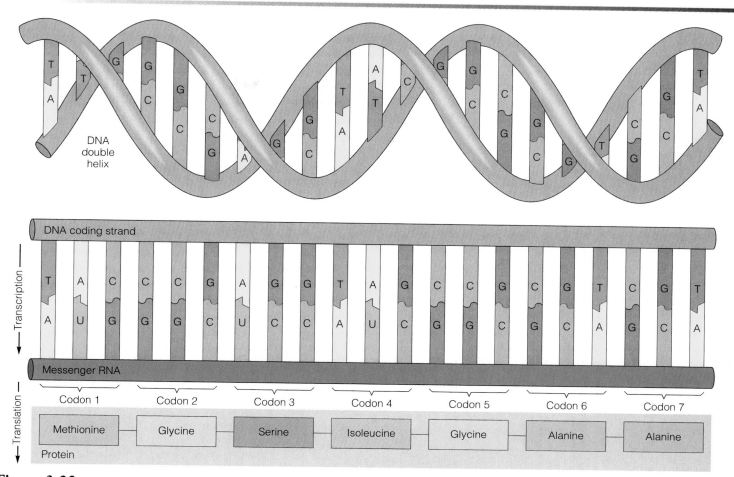

Figure 3.22

The genetic code is first transcribed into base triplets (codons) in mRNA and then translated into a specific sequence of amino acids in a protein.

Table 3.2	Selected DNA Base Triplets and mRNA Codons	
DNA Triplet	**RNA Codon**	**Amino Acid**
TAC	AUG	"Start"
ATC	UAG	"Stop"
AAA	UUU	Phenylalanine
AGG	UCC	Serine
ACA	UGU	Cysteine
GGG	CCC	Proline
GAA	CUU	Leucine
GCT	CGA	Arginine
TTT	AAA	Lysine
TGC	ACG	Tyrosine
CCG	GGC	Glucine
CTC	GAG	Aspartic acid

Transfer RNA

Translation of the codons is accomplished by tRNA and particular enzymes. Each tRNA molecule, like mRNA and rRNA, is single-stranded. Although tRNA is single-stranded, it bends in on itself to form a cloverleaf structure (fig. 3.23*a*), which is believed to be further twisted into an upside down "L" shape (fig. 3.23*b*). One end of the "L" contains the **anticodon**—three nucleotides that are complementary to a specific codon in mRNA.

Enzymes in the cell cytoplasm called *aminoacyl-tRNA synthetase* enzymes join specific amino acids to the ends of tRNA, so that a tRNA with a given anticodon can bind to only one specific amino acid. There are twenty different varieties of synthetase enzymes, one for each type of amino acid. Each synthetase must not only recognize its specific amino acid, it must be able to attach this amino acid to the particular

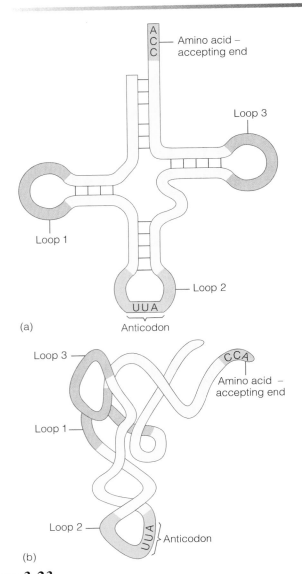

Figure 3.23

The structure of transfer RNA (tRNA). (*a*) A simplified cloverleaf representation and (*b*) the three-dimensional structure of tRNA.

tRNA that has the correct anticodon for that amino acid. The cytoplasm of a cell thus contains tRNA molecules that are each bonded to a specific amino acid, and which are each capable of bonding with their anticodon base triplet to a specific codon in mRNA.

Formation of a Polypeptide

The anticodons of tRNA bind to the codons of mRNA as the mRNA moves through the ribosome. Since each tRNA molecule carries a specific amino acid, the joining together of these amino acids by peptide bonds creates a polypeptide whose amino acid sequence has been determined by the sequence of codons in mRNA.

The first and second tRNA bring the first and second amino acids together, and a peptide bond forms between them. The first amino acid then detaches from its tRNA, so that a dipeptide is linked by the second amino acid to the second tRNA. When the third tRNA binds to the third codon, the amino acid it brings forms a peptide bond with the second amino acid (which detaches from its tRNA). A tripeptide is now attached by the third amino acid to the third tRNA. The polypeptide chain thus grows as new amino acids are added to its growing tip (fig. 3.24). This growing polypeptide chain is always attached by means of only one tRNA to the strand of mRNA, and this tRNA molecule is always the one that has added the latest amino acid to the growing polypeptide.

As the polypeptide chain grows in length, interactions between its amino acids cause the chain to twist into a helix (secondary structure) and to fold and bend upon itself (tertiary structure). At the end of this process, the new protein detaches from the tRNA as the last amino acid is added. Many proteins are further modified after they are formed; these modifications occur in the rough endoplasmic reticulum and Golgi apparatus.

Function of the Rough Endoplasmic Reticulum

Proteins that are to be used within the cell are produced in polyribosomes that are free in the cytoplasm. If the protein is a secretory product of the cell, however, it is made by mRNA-ribosome complexes located in the rough endoplasmic reticulum. The membranes of this system enclose fluid-filled spaces (cisternae), which the newly formed proteins may enter. Once in the cisternae, the structure of these proteins is modified in specific ways.

When proteins that are destined for secretion are produced, the first thirty or so amino acids are primarily hydrophobic. This *leader sequence* is attracted to the lipid component of the membranes of the endoplasmic reticulum. As the polypeptide chain elongates, it is "injected" into the cisterna within the endoplasmic reticulum. The leader sequence is, in a sense, an "address" that directs secretory proteins into the endoplasmic reticulum. Once the proteins are in the cisterna, the leader sequence is removed so the protein cannot reenter the cytoplasm (fig. 3.25).

The processing of the hormone insulin can serve as an example of the changes that occur within the endoplasmic reticulum. The original molecule enters the cisterna as a single polypeptide composed of 109 amino acids. This molecule is called preproinsulin. The first twenty-three amino acids serve as a leader sequence that allows the molecule to be injected into the cisterna within the endoplasmic reticulum. The leader sequence is then quickly removed, producing a molecule called proinsulin. The remaining chain folds within the cisterna so that the first and last amino acids in the polypeptide are brought close together. The central region is then enzymatically removed, producing two chains—one of them, twenty-one amino acids long; the other, thirty amino

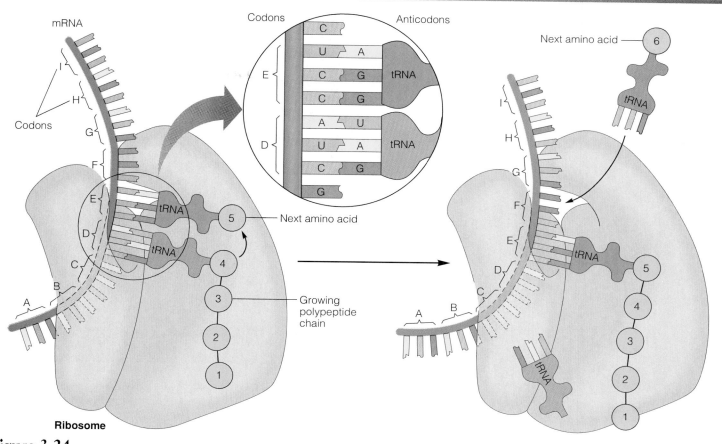

Figure 3.24

The translation of messenger RNA (mRNA). As the anticodin of each new aminoacyl-tRNA bonds with a codon on the mRNA, new amino acids are joined to the growing tip of the polypeptide chain.

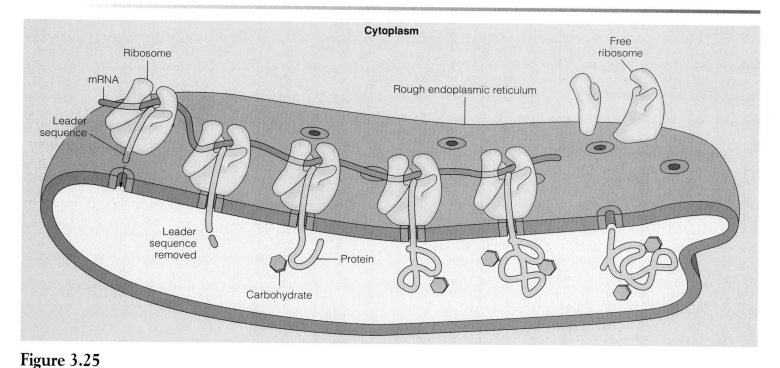

Figure 3.25

A protein destined for secretion begins with a *leader sequence* that enables it to be inserted into the endoplasmic reticulum. Once it has been inserted, the leader sequence is removed and carbohydrate is added to the protein.

Figure 3.26

The long polypeptide chain called proinsulin is converted into the active hormone insulin by enzymatic removal of a length of amino acids. The insulin molecule produced in this way consists of two polypeptide chains (*colored circles*) joined by disulfide bonds.

acids long—which are subsequently joined together by disulfide bonds (fig. 3.26). This is the form of insulin that is normally secreted from the cell.

Function of the Golgi Apparatus

Secretory proteins do not remain trapped within the rough endoplasmic reticulum; they are transported to another organelle within the cell—the **Golgi apparatus.** This organelle serves three interrelated functions: (1) further modifications of proteins (such as the addition of carbohydrates to form *glycoproteins*) occur in the Golgi apparatus; (2) different types of proteins are separated according to their function and destina-

Table 3.3	Regulation of the Processes Involved in Genetic Expression
Stage of Gene Expression	**Possible Regulatory Mechanisms**
RNA synthesis	Duplication of genes; cutting and splicing of genes to different positions on a chromosome
	Association of histone and nonhistone proteins with chromatin
Nuclear processing of RNA	Cutting and splicing of nuclear RNA Capping of messenger RNA with 7-methyl guanosine and addition of about 200 adenylate nucleotides
Translation of mRNA	Availability of specific tRNA molecules
Post-translational changes in protein structure	Cleavage of parent protein into biologically active fragments
	Chemical modification of amino acids
	Association of polypeptide chains together to form quaternary structure
	"Addressing" of protein for secretion by leader sequence and insertion into rough endoplasmic reticulum
	Addition of carbohydrate to proteins within Golgi apparatus

tion; and (3) the final products are packaged and shipped to their destinations. In the Golgi apparatus, for example, proteins that are to be secreted are separated from those that will be incorporated into the cell membrane and from those that will be introduced into lysosomes, and these different proteins are packaged into separate membrane-enclosed vesicles.

The Golgi apparatus consists of a stack of several flattened sacs. One side of the sac serves as the site of entry for cellular products that arrive in vesicles from the endoplasmic reticulum. After specialized modifications of the proteins are made within one sac, the modified proteins are passed by means of vesicles to the next sac until the finished products leave the Golgi apparatus from the opposite side of the stack (fig. 3.27). Depending on the nature of the specific product, the vesicles that leave the Golgi apparatus may become lysosomes, storage granules of secretory products, or additions to the cell membrane.

A summary of the processes involved in genetic expression and the points at which regulation may occur is provided in table 3.3.

1. Explain how mRNA, rRNA, and tRNA function during the process of protein synthesis.
2. Describe the rough endoplasmic reticulum and explain how the processing of secretory proteins differs from the processing of proteins that remain within the cell.
3. Describe the structure and functions of the Golgi apparatus.

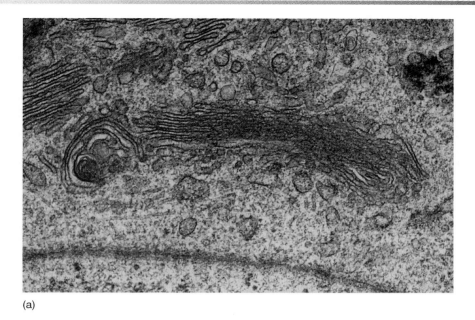

(a)

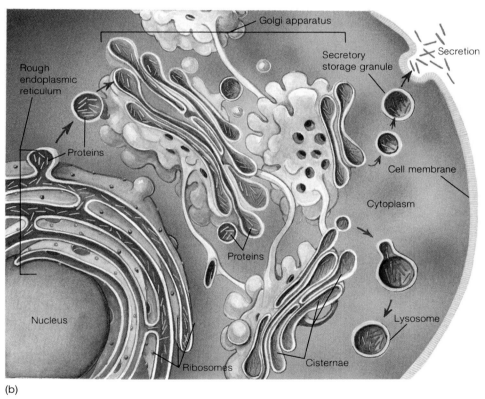

(b)

Figure 3.27

(*a*) An electron micrograph of a Golgi apparatus. Notice the formation of vesicles at the ends of some of the flattened sacs. (*b*) An illustration of the processing of proteins by the rough endoplasmic reticulum and Golgi apparatus.

DNA Synthesis and Cell Division

When a cell is going to divide, each strand of the DNA within its nucleus acts as a template for the formation of a new complementary strand. Organs grow and repair themselves through a type of cell division known as mitosis. The two daughter cells produced by mitosis contain the same genetic information as the parent cell. Gametes contain only half the number of chromosomes as their parent cell and are formed by a type of cell division called meiosis.

Genetic information is required for the life of the cell and for the ability of the cell to perform its functions in the body. Each cell obtains this genetic information from its parent cell through the process of DNA replication and cell division. DNA is the only type of molecule in the body capable of replicating itself, and mechanisms exist within the dividing cell to ensure that the duplicate copies of DNA are properly distributed to the daughter cells.

DNA Replication

When a cell is going to divide, each DNA molecule replicates itself, and each of the identical DNA copies thus produced is distributed to the two daughter cells. Replication of DNA requires the action of a specific enzyme known as *DNA polymerase*. This enzyme moves along the DNA molecule, breaking the weak hydrogen bonds between complementary bases as it travels. As a result, the bases of each of the two DNA strands become free to bond with new complementary bases (which are part of nucleotides) that are available within the surrounding environment.

According to the rules of complementary base pairing, the bases of each original strand will bond with the appropriate free nucleotides: adenine bases pair with thymine-containing nucleotides; guanine bases pair with cytosine-containing nucleotides, and so on. In this way, two new molecules of DNA, each containing two complementary strands, are formed. The DNA polymerase enzyme links the phosphate groups and deoxyribose sugar groups together to form a second polynucleotide chain in each DNA that is complementary to the first DNA strands. Thus, two new double-helix DNA molecules are produced that contain the same base sequence as the parent molecule (fig. 3.28).

When DNA replicates, therefore, each copy is composed of one new strand and one strand from the original DNA molecule. Replication is said to be **semiconservative** (half of the original DNA is "conserved" in each of the new DNA molecules). Through this mechanism, the sequence of bases in DNA—which is the basis of the genetic code—is preserved from one cell generation to the next.

Advances in the identification of human genes, methods of cloning (replicating) isolated genes, and other technologies have made gene therapy a realistic possibility. The first federally approved human genetic engineering experiment began in 1989 when patients with advanced cancer were treated with special white blood cells from their own body. These "tumor infiltrating lymphocytes" had been given an exogenous (foreign) genetic marker. The exogenous gene didn't aid the treatment but served only to test whether or not the procedure was workable and safe. The first approved gene therapy of a disease began testing in 1990. This involved the attempted correction of a genetic defect for an enzyme called adenosine deaminase (ADA) that causes failure of the immune system. Clinical human trials (see chapter 1) of gene therapy for diseases of the blood-forming cells, liver, lungs, clotting system, and other diseases are currently in progress.

The Cell Cycle

Unlike the life of an organism, which can be pictured as a linear progression from birth to death, the life of a cell follows a cyclical pattern. Each cell is produced as a part of its "parent" cell; when the daughter cell divides, it in turn becomes two new cells. In a sense, then, each cell is potentially immortal as long as its progeny can continue to divide. Some cells in the body divide frequently; the epidermis of the skin, for example, is renewed approximately every 2 weeks, and the stomach lining is renewed about every 2 or 3 days. Other cells, however, such as nerve and striated muscle cells in the adult, do not divide at all. All cells in the body, of course, live only as long as the person lives (some cells live longer than others, but eventually all cells die when vital functions cease).

The nondividing cell is in a part of its life cycle known as **interphase** (fig. 3.29), which is subdivided into G_1, S, and G_2 phases, as will be described. The chromosomes are in their extended form (as euchromatin), and their genes actively direct the synthesis of RNA. Through their direction of RNA synthesis, genes control the metabolism of the cell. During this time the cell may be growing, and this part of interphase is known as the G_1 *phase*. Although sometimes described as "resting," cells in the G_1 phase perform the physiological functions characteristic of the tissue in which they are found. The DNA of resting cells in the G_1 phase thus produces mRNA and proteins as previously described.

If a cell is going to divide, it replicates its DNA in a part of interphase known as the S *phase* (S stands for *synthesis*). Once DNA has replicated in the S phase, the chromatin condenses in the G_2 *phase* to form short, thick, rodlike structures by the end of G_2. This is the more familiar form of chromosomes because they are easily seen in the ordinary (light) microscope.

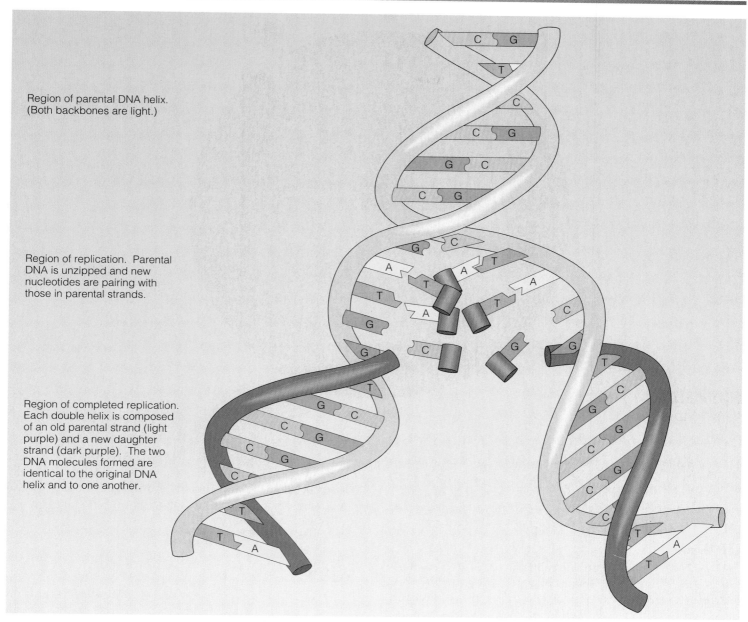

Region of parental DNA helix.
(Both backbones are light.)

Region of replication. Parental DNA is unzipped and new nucleotides are pairing with those in parental strands.

Region of completed replication. Each double helix is composed of an old parental strand (light purple) and a new daughter strand (dark purple). The two DNA molecules formed are identical to the original DNA helix and to one another.

Figure 3.28

The replication of DNA. Each new double helix is composed of one old and one new strand. The base sequence of each of the new molecules is identical to that of the parent DNA because of complementary base pairing.

Cyclins and p53

A group of proteins known as the **cyclins** promote different phases of the cell cycle. The concentration of **cyclin D** proteins within the cell, for example, rises during the G_1 phase of the cycle and acts to move the cell quickly through this phase. Therefore, overactivity of a gene that codes for a cyclin D might be predicted to cause uncontrolled cell division, as occurs in a cancer. Indeed, overexpression of the gene for cyclin D1 has been shown to occur in some cancers,

including those of the breast and esophagus. Genes that contribute to cancer are called **oncogenes.**

While oncogenes promote cancer, other genes—called **tumor suppressor genes**—inhibit its development. One very important tumor suppressor gene is known as **p53.** This name refers to the protein coded by the gene, which has a molecular weight of 53,000. The normal gene protects against cancer by indirectly blocking the ability of cyclins to stimulate cell division. For this reason, a mutated, and therefore ineffective, p53

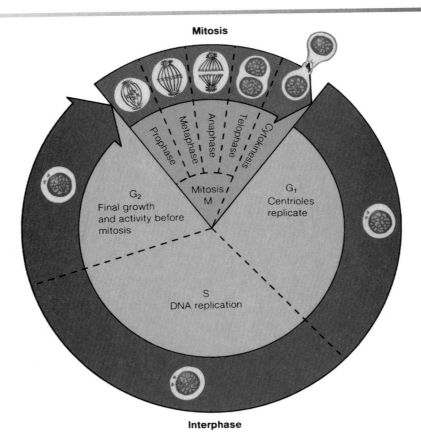

Figure 3.29

The life cycle of a cell.

gene promotes cancer. Indeed, mutated p53 genes are found in over 50% of all cancers. Mice that had their p53 genes "knocked out" through an exciting new technology in genetic engineering all developed tumors. These important discoveries have obvious relevance to cancer diagnosis and treatment.

Cell Death

Cell death occurs both pathologically and naturally. Pathologically, cells deprived of a blood supply may swell, rupture their membranes, and burst. Such cellular death is known as **necrosis.** In certain cases, however, a different pattern is observed. Instead of swelling, the cells shrink. The membranes remain intact but become bubbled, and the nuclei condense. This pattern has been named **apoptosis** (from a Greek term describing the shedding of leaves from a tree).

Apoptosis occurs normally as part of programmed cell death—a process described previously in the section on lysosomes. Programmed cell death refers to the physiological process responsible for the remodeling of tissues during embryonic development, tissue turnover in the adult body, and the functioning of the immune system. A neutrophil (a type of white blood cell), for example, is programmed to die by apoptosis 24 hours after its creation in the bone marrow. Killer T lymphocytes (another type of white blood cell) destroy targeted cells by binding to molecules in the membrane of their victim and thereby triggering apoptosis.

Using mice with their gene for p53 knocked out, scientists have learned that p53 is also needed for the apoptosis that normally occurs when a cell's DNA is damaged. If the p53 gene has mutated to an ineffective form, the cell will not be destroyed by apoptosis as it should but instead will divide to produce daughter cells with damaged DNA. This may be one mechanism responsible for the development of a cancer.

Mitosis

At the end of the G_2 phase of the cell cycle, which is generally shorter than G_1, each chromosome consists of two strands called **chromatids,** which are joined together by a *centromere* (fig 3.30). The two chromatids within a chromosome contain identical DNA base sequences because each is produced by the semiconservative replication of DNA. Each chromatid, therefore, contains a complete double-helix DNA molecule that is a copy of the single DNA molecule existing prior to replication. Each chromatid will become a separate chromosome once cell division has been completed.

The G_2 phase completes interphase. The cell next proceeds through the various stages of cell division, or **mitosis** (from the Greek *mitos* = thread). This is the M *phase* of the cell cycle. Mitosis is subdivided into four parts: *prophase, metaphase, anaphase,* and *telophase* (fig. 3.31). In metaphase of mitosis, the chromosomes line up single file along the equator of the cell. This aligning of chromosomes at the equator is believed to result from the action of **spindle fibers,** which are attached to the centromere of each chromosome (fig. 3.31).

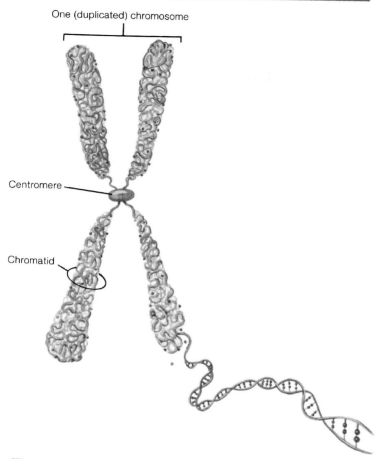

One (duplicated) chromosome

Centromere

Chromatid

Figure 3.30

The structure of a chromosome after DNA replication, at which point it consists of two identical strands, or chromatids.

Anaphase begins when the centromeres split apart and the spindle fibers shorten, pulling the two chromatids in each chromosome to opposite poles. Each pole therefore gets one copy of each of the forty-six chromosomes. Division of the cytoplasm (*cytokinesis*) during telophase results in the production of two daughter cells that are genetically identical to each other and to the original parent cell.

Role of the Centrosome

All animal cells have a **centrosome,** located near the nucleus in a nondividing cell. At the center of the centrosome are two **centrioles,** which are positioned at right angles to each other. Each centriole is composed of nine evenly spaced bundles of microtubules, with three microtubules per bundle (fig. 3.32). Surrounding the two centrioles is an amorphous mass of material called the *pericentriolar material.* Microtubules grow out of the pericentriolar material, which is believed to function as the center for the organization of microtubules in the cytoskeleton.

Through a mechanism that is still not understood, the centrosome replicates itself during interphase if a cell is going to divide. The two identical centrosomes then move away from each other during prophase of mitosis and take up positions at

opposite poles of the cell by metaphase. At this time, the centrosomes produce new microtubules. These new microtubules are very dynamic, rapidly growing and shrinking as if they were "feeling" out randomly for chromosomes. When a microtubule finally binds to the proper region of a chromosome, it becomes stabilized. In this way, the microtubules from both centrosomes form the spindle fibers that are attached to each of the replicated chromosomes at metaphase.

The spindle fibers pull the chromosomes to opposite poles of the cell during anaphase, so that, when the cell pinches inward at telophase, two identical daughter cells will be produced. This also requires the centrosomes, which somehow organize a ring of contractile filaments halfway between the two poles. These filaments are attached to the cell membrane, and when they contract, the cell is pinched in two. The filaments consist of actin and myosin proteins, the same contractile proteins present in muscle.

Hypertrophy and Hyperplasia

The growth of an individual from a fertilized egg into an adult involves an increase in cell number and an increase in cell size. Growth due to an increase in cell number results from mitotic cell division and is termed **hyperplasia.** Growth of a tissue or organ due to an increase in cell size is termed **hypertrophy.**

Most growth is due to hyperplasia. A callus on the palm of the hand, for example, involves thickening of the skin by hyperplasia due to frequent abrasion. An increase in skeletal muscle size as a result of exercise, by contrast, is produced by hypertrophy.

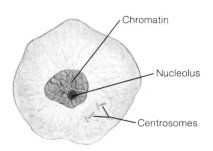

Chromatin

Nucleolus

Centrosomes

(a) Interphase

- The chromosomes are in an extended form and seen as chromatin in the electron microscope.
- The nucleus is visible.

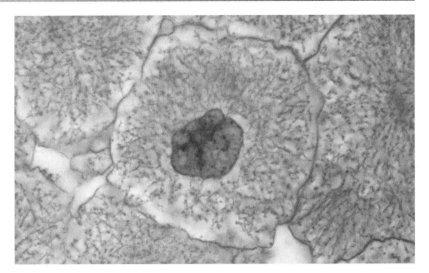

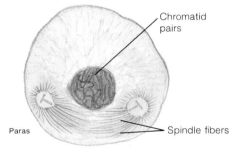

Chromatid pairs

Paras

Spindle fibers

(b) Prophase

- The chromosomes are seen to consist of two chromatids joined by a centromere.
- The centrioles move apart toward opposite poles of the cell.
- Spindle fibers are produced and extend from each centrosome.
- The nuclear membrane starts to disappear.
- The nucleolus is no longer visible.

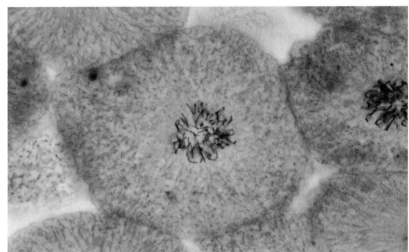

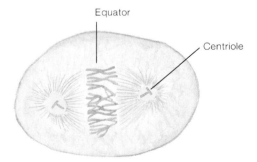

Equator

Centriole

(c) Metaphase

- The chromosomes are lined up at the equator of the cell.
- The spindle fibers from each centriole are attached to the centromeres of the chromosomes.
- The nuclear membrane has disappeared.

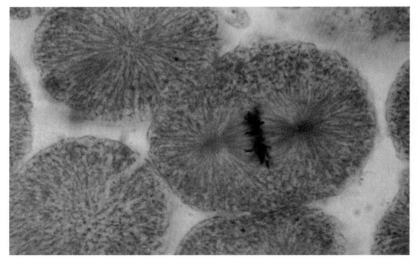

continued

Figure 3.31

The stages of mitosis.

Chapter Three

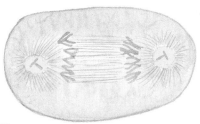

(d) Anaphase

The centromeres split, and the sister chromatids separate as each is pulled to an opposite pole.

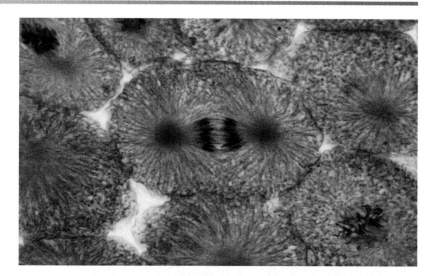

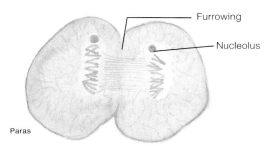

Paras

(e) Telophase

- The chromosomes become longer, thinner, and less distinct.
- New nuclear membranes form.
- The nucleolus reappears.
- Cell division is nearly complete.

Furrowing

Nucleolus

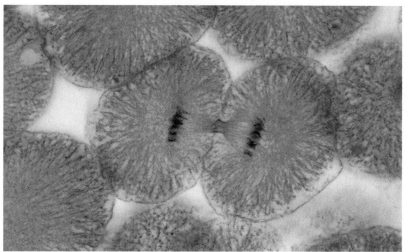

Figure 3.31

continued

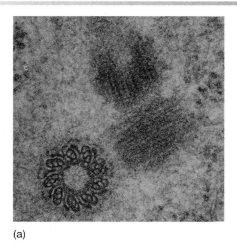

(a)

Paras

(b)

Figure 3.32

The centrioles. (*a*) A micrograph of the two centrioles in a centrosome. (*b*) A diagram showing that the centrioles are positioned at right angles to one another.

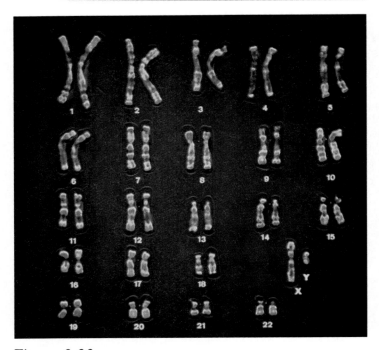

Figure 3.33

A false-color light micrograph showing the full complement of male chromosomes arranged in numbered homologous pairs.

Stage	Events
Table 3.4 Stages of Meiosis	
First Meiotic Division	
Prophase I	Chromosomes appear double-stranded.
	Each strand, called a chromatid, contains duplicate DNA joined together by a structure known as a centromere.
	Homologous chromosomes pair up side by side.
Metaphase I	Homologous chromosome pairs line up at equator.
	Spindle apparatus is complete.
Anaphase I	Homologous chromosomes separate; the two members of a homologous pair move to opposite poles.
Telophase I	Cytoplasm divides to produce two haploid cells.
Second Meiotic Division	
Prophase II	Chromosomes appear, each containing two chromatids.
Metaphase II	Chromosomes line up single file along equator as spindle formation is completed.
Anaphase II	Centromeres split and chromatids move to opposite poles.
Telophase II	Cytoplasm divides to produce two haploid cells from each of the haploid cells formed at telophase I.

Meiosis

When a cell is going to divide, either by mitosis or meiosis, the DNA is replicated (forming chromatids) and the chromosomes become shorter and thicker, as previously described. At this point the cell has forty-six chromosomes, each of which consists of two duplicate chromatids.

The short, thick chromosomes seen at the end of the G_2 phase can be matched into pairs, the members of which appear to be structurally identical. These matched pairs of chromosomes are called **homologous chromosomes.** One member of each homologous pair is derived from a chromosome inherited from the father, and the other member is a copy of one of the chromosomes inherited from the mother. Homologous chromosomes do not have identical DNA base sequences; one member of the pair may code for blue eyes, for example, and the other for brown eyes. There are twenty-two homologous pairs of *autosomal chromosomes* and one pair of *sex chromosomes*, described as X and Y. Females have two X chromosomes, whereas males have one X and one Y chromosome (fig. 3.33).

Meiosis is a special type of cell division that occurs only in the gonads (testes and ovaries) and is used only in the production of gametes (sperm and ova). In meiosis, the homologous chromosomes line up side by side, rather than single file, along the equator of the cell. The spindle fibers then pull one member of a homologous pair to one pole of the cell, and the other member of the pair to the other pole of the cell. Each of the two daughter cells thus acquires only one chromosome from each of the twenty-three homologous pairs contained in the parent. The daughter cells, in other words, contain twenty-three rather than forty-six chromosomes. For this reason, meiosis (from the Greek *meion* = less) is also known as **reduction division.**

At the end of this cell division, each daughter cell contains twenty-three chromosomes—but *each of these consists of two chromatids.* (Since the two chromatids per chromosome are identical, this does not make forty-six chromosomes; there are still only twenty-three *different* chromosomes per cell at this point.) The chromatids are separated by a second meiotic division. Each of the daughter cells from the first cell division itself divides, with the duplicate chromatids going to each of two new daughter cells. A grand total of four daughter cells can thus be produced from the meiotic cell division of one parent cell. This occurs in the testes, where one parent cell produces four sperm. In the ovaries, one parent cell also produces four daughter cells, but three of these die and only one progresses to become a mature egg cell (as will be described in chapter 20).

The stages of meiosis are subdivided according to whether they occur in the first or the second meiotic cell division. These stages are designated as prophase I, metaphase I, anaphase I, telophase I; and then prophase II, metaphase II, anaphase II, and telophase II (table 3.4 and fig. 3.34).

The reduction of the chromosome number from forty-six to twenty-three is obviously necessary for sexual reproduction, where the sex cells join and add their content of chromosomes together to produce a new individual. The significance of meiosis, however, goes beyond the reduction of chromosome number. At metaphase I, the pairs of homologous chromosomes can line up with either member facing a given pole of the cell. (Recall that each member of a homologous pair came from a different

Chapter Three

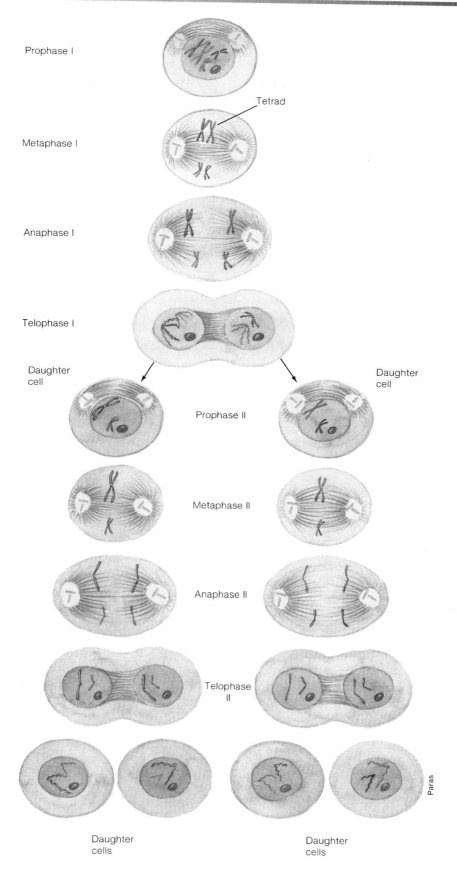

Prophase I

Metaphase I

Tetrad

Anaphase I

Telophase I

Daughter cell

Prophase II

Daughter cell

Metaphase II

Anaphase II

Telophase II

Daughter cells

Daughter cells

Paras

Figure 3.34

Meiosis, or reduction division. In the first meiotic division, the homologous chromosomes of a diploid parent cell are separated into two haploid daughter cells. Each of these chromosomes contains duplicate strands, or chromatids. In the second meiotic division, these chromosomes are distributed to two new haploid daughter cells.

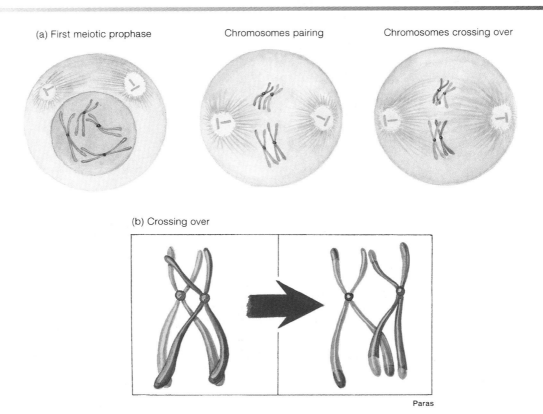

(a) First meiotic prophase   Chromosomes pairing   Chromosomes crossing over

(b) Crossing over

Paras

Figure 3.35

(*a*) Genetic variation results from the crossing over of tetrads, which occurs during the first meiotic prophase. (*b*) A diagram depicting the recombination of chromosomes that occurs as a result of crossing over.

parent.) Maternal and paternal members of homologous pairs are thus randomly shuffled. Hence, when the first meiotic division occurs, each daughter cell will obtain a complement of twenty-three chromosomes that are randomly derived from the maternal or paternal contribution to the homologous pairs of chromosomes of the parent.

In addition to this "shuffling of the deck" of chromosomes, exchanges of parts of homologous chromosomes can occur at metaphase I. That is, pieces of one chromosome of a homologous pair can be exchanged with the other homologous chromosome in a process called *crossing over* (fig. 3.35). These events together result in **genetic recombination** and ensure that the gametes produced by meiosis are genetically unique. This provides genetic diversity for organisms that reproduce sexually, and genetic diversity is needed to promote survival of species over evolutionary time.

1. Draw a simple diagram of the semiconservative replication of DNA using stick figures and two colors.

2. Describe the cell cycle using the proper symbols to indicate the different stages of the cycle.

3. List the phases of mitosis and briefly describe the events that occur in each phase.

4. Distinguish between mitosis and meiosis in terms of their final result and their functional significance.

5. Summarize the events that occur during the two meiotic cell divisions and explain the mechanisms by which genetic recombination occurs during meiosis.

Summary

Cell Membrane and Associated Structures p. 48

I. The structure of the cell (plasma) membrane is described by a fluid-mosaic model.
   A. The membrane is composed predominately of a double layer of phospholipids.
   B. The membrane also contains proteins, distributed in a mosaic pattern.
II. Some cells move by extending pseudopods; cilia and flagella protrude from the cell membrane of some specialized cells.
III. Invaginations of the cell membrane, in the process of endocytosis, allow the cells to take up molecules from the external environment.
   A. In phagocytosis, the cell extends pseudopods that eventually fuse together to create a food vacuole; pinocytosis involves the formation of a narrow furrow in the membrane, which eventually fuses.
   B. Receptor-mediated endocytosis requires the interaction of a specific molecule in the extracellular environment with a specific receptor protein in the cell membrane.
   C. Exocytosis, the reverse of endocytosis, is a process that allows the cell to secrete its products.

Cytoplasm and Its Organelles p. 54

I. Microfilaments and microtubules produce a cytoskeleton that aids movements of organelles within a cell.
II. Lysosomes contain digestive enzymes and are responsible for the elimination of structures and molecules within the cell and for digestion of the contents of phagocytic food vacuoles.
III. Mitochondria serve as the major sites for energy production within the cell. They contain an outer membrane with a smooth contour and an inner membrane with infoldings called cristae.
IV. The endoplasmic reticulum is a system of membranous tubules in the cell.
   A. The rough endoplasmic reticulum is covered with ribosomes and is involved in protein synthesis.
   B. The smooth endoplasmic reticulum is the site for many enzymatic reactions and, in skeletal muscles, serves to store Ca^{++}.

Cell Nucleus and Nucleic Acids p. 57

I. The cell nucleus is surrounded by a nuclear membrane and contains chromatin, which consists of DNA and protein.
II. Nucleic acids include DNA, RNA, and their nucleotide subunits.
   A. The DNA nucleotides contain the sugar deoxyribose, whereas the RNA nucleotides contain the sugar ribose.
   B. There are four different types of DNA nucleotides, which contain one of four possible bases: adenine, guanine, cytosine, and thymine. In RNA, the base uracil substitutes for the base thymine.
   C. DNA consists of two long polynucleotide strands twisted into a double helix. The two strands are held together by hydrogen bonds between specific bases; adenine pairs with thymine, and guanine pairs with cytosine.
   D. RNA is single-stranded. There are four types of RNA molecules: ribosomal RNA, transfer RNA, precursor messenger RNA, and messenger RNA.
III. Active euchromatin directs the synthesis of RNA in a process called genetic transcription.
   A. The enzyme RNA polymerase causes separation of the two strands of DNA along the region of the DNA that constitutes a gene.
   B. One of the two separated strands of DNA serves as a template for the production of RNA. This occurs by complementary base pairing between the DNA bases and ribonucleotide bases.

Protein Synthesis and Secretion p. 62

I. Messenger RNA leaves the nucleus and attaches to the ribosomes.
II. Each transfer RNA, with a specific base triplet in its anticodon, binds to a specific amino acid.
   A. As the mRNA moves through the ribosomes, complementary base pairing between tRNA anticodons and mRNA codons occurs.
   B. As each successive tRNA molecule binds to its complementary codon, the amino acid it carries is added to the end of a growing polypeptide chain.
III. Proteins destined for secretion are produced in ribosomes located in the rough endoplasmic reticulum and enter the cisternae of this organelle.
IV. Secretory proteins move from the rough endoplasmic reticulum to the Golgi apparatus, which consists of a stack of membranous sacs.
   A. The Golgi apparatus modifies the proteins it contains, separates different proteins, and packages them in vesicles.
   B. Secretory vesicles from the Golgi apparatus fuse with the cell membrane and release their products by exocytosis.

DNA Synthesis and Cell Division p. 68

I. Replication of DNA is semiconservative; each DNA strand serves as a template for the production of a new strand.
   A. The strands of the original DNA molecule gradually separate along their entire length and,

through complementary base pairing, form a new complementary strand.

B. In this way, each DNA molecule consists of one old and one new strand.

II. During the G_1 phase of the cell cycle, the DNA directs the synthesis of RNA, and hence that of proteins.

III. During the S phase of the cycle, DNA directs the synthesis of new DNA and replicates itself.

IV. After a brief rest (G_2), the cell begins mitosis (M stage of the cycle).

A. Mitosis consists of the following phases: interphase, prophase, metaphase, anaphase, and telophase.

B. In mitosis, the homologous chromosomes line up single file and are pulled by spindle fibers to opposite poles.

C. This results in the production of two daughter cells, each containing forty-six chromosomes, just like the parent cell.

V. Meiosis is a special type of cell division that results in the production of gametes in the gonads.

A. The homologous chromosomes line up side by side, so that only one of each pair is pulled to each pole.

B. This results in the production of two daughter cells, each containing only twenty-three chromosomes, which are duplicated.

C. The duplicate chromatids are separated into two new daughter cells during the second meiotic cell division.

Clinical Investigation

A liver biopsy is taken from a teenage boy with apparent liver disease, and different microscopic techniques for viewing the sample are performed. The biopsy reveals an unusually extensive smooth endoplasmic reticulum, and the patient has admitted to having a history of substance abuse. In addition, an abnormally large amount of glycogen granules are found, and many intact glycogen granules are seen within secondary lysosomes. Laboratory analysis reveals a lack of the enzyme that hydrolyzes glycogen. What is the relationship between these observations?

Clues

See the sections on lysosomes and the endoplasmic reticulum, paying particular attention to the boxed information in these sections.

Review Activities

Objective Questions

1. According to the fluid-mosaic model of the cell membrane,
   a. protein and phospholipids form a regular, repeating structure.
   b. the membrane is a rigid structure.
   c. phospholipids form a double layer, with the polar parts facing each other.
   d. proteins are free to move within a double layer of phospholipids.

2. After the DNA molecule has replicated itself, the duplicate strands are called
   a. homologous chromosomes.
   b. chromatids.
   c. centromeres.
   d. spindle fibers.

3. Nerve and skeletal muscle cells in the adult, which do not divide, remain in the
   a. G_1 phase.
   b. S phase.
   c. G_2 phase.
   d. M phase.

4. The phase of mitosis in which the chromosomes line up at the equator of the cell is called
   a. interphase.
   b. prophase.
   c. metaphase.
   d. anaphase.
   e. telophase.

5. The phase of mitosis in which the chromatids separate is called
   a. interphase.
   b. prophase.
   c. metaphase.
   d. anaphase.
   e. telophase.

6. The RNA nucleotide base that pairs with adenine in DNA is
   a. thymine.
   b. uracil.
   c. guanine.
   d. cytosine.

7. Which of the following statements about RNA is *true*?
   a. It is made in the nucleus.
   b. It is double-stranded.
   c. It contains the sugar deoxyribose.
   d. It is a complementary copy of the entire DNA molecule.

8. Which of the following statements about mRNA is *false*?
   a. It is produced as a larger pre-mRNA.
   b. It forms associations with ribosomes.
   c. Its base triplets are called anticodons.
   d. It codes for the synthesis of specific proteins.

9. The organelle that combines proteins with carbohydrates and packages them within vesicles for secretion is
   a. the Golgi apparatus.
   b. the rough endoplasmic reticulum.
   c. the smooth endoplasmic reticulum.
   d. the ribosome.

10. The organelle that contains digestive enzymes is
    a. the mitochondrion.
    b. the lysosome.
    c. the endoplasmic reticulum.
    d. the Golgi apparatus.

11. If four bases in one DNA strand are A (adenine), G (guanine), C (cytosine), and T (thymine), the complementary bases in the RNA strand made from this region are
    a. T,C,G,A.
    b. C,G,A,U.
    c. A,G,C,U.
    d. U,C,G,A.

12. Which of the following statements about tRNA is *true*?
    a. It is made in the nucleus.
    b. It is looped back on itself.
    c. It contains the anticodon.
    d. There are over twenty different types of tRNA.
    e. All of the above are true.

13. The step in protein synthesis during which tRNA, rRNA, and mRNA are all active is known as
    a. transcription.
    b. translation.
    c. replication.
    d. RNA polymerization.

14. The anticodons are located in
    a. tRNA.
    b. rRNA.
    c. mRNA.
    d. ribosomes.
    e. endoplasmic reticulum.

Essay Questions

1. The cell membrane is an extremely dynamic structure. Using examples, explain why this statement is true.[1]
2. Explain how one DNA molecule serves as a template for the formation of another DNA and why DNA synthesis is said to be semiconservative.
3. What is the genetic code, and how does it affect the structure and function of the body?
4. Why may tRNA be considered the "interpreter" of the genetic code?
5. Compare the processing of cellular proteins with that of proteins that are secreted by a cell.
6. Explain the interrelationship between the endoplasmic reticulum and the Golgi apparatus. What becomes of vesicles released from the Golgi apparatus?
7. Explain the functions of centrioles in nondividing and dividing cells.
8. Describe the phases of the cell cycle and explain how it may be regulated.
9. Distinguish between oncogenes and tumor suppressor genes and give examples of how such genes may function.
10. Define *apoptosis* and explain its physiological significance.

Selected Readings

Afzelius, Björn. 1986. Disorders of ciliary motility. *Hospital Practice* 21:73.

Allen, R. D. February 1987. The microtubule as an intracellular engine. *Scientific American*.

Anderson, W. F. 1992. Human gene therapy. *Science* 256:808.

Berlin, R. D., T. E. Oliver, and H. H. Yin. 1975. The cell surface. *New England Journal of Medicine* 292:515.

Bretscher, M. S. October 1985. The molecules of the cell membrane. *Scientific American*.

Brown, D. D. 1981. Gene expression in eukaryotes. *Science* 211:667.

Capaldi, R. A. March 1974. A dynamic model of cell membranes. *Scientific American*.

Capecchi, M. R. March 1994. Targeted gene replacement. *Scientific American*.

Chambon, P. May 1981. Split genes. *Scientific American*.

Cohen, J. J. 1993. Apoptosis: The physiologic pathway of cell death. *Hospital Practice* 28:35.

Crick, F. October 1962. The genetic code. *Scientific American*.

Crick, F. 1979. Split genes and RNA splicing. *Science* 204:264.

Crystal, R. G. 1995. The gene as the drug. *Nature Medicine* 1:15.

Danielli, J. F. 1973. The bilayer hypothesis of membrane structure. *Hospital Practice* 8:63.

Darnell, J. E. Jr. October 1985. RNA. *Scientific American*.

Dautry-Varsat, A., and H. F. Lodish. May 1984. How receptors bring proteins and particles into cells. *Scientific American*.

Davidson, E. H., and R. J. Britten. 1979. Regulation of gene expression: Possible role of repetitive sequences. *Science* 204:1052.

DeDuve, C. May 1963. The lysosome. *Scientific American*.

DeDuve, C. May 1983. Microbodies in the living cell. *Scientific American*.

Dingwall, C., and R. Laskey. 1992. The nuclear membrane. *Science* 258:942.

Dustin, P. August 1980. Microtubules. *Scientific American*.

Felsenfeld, G. October 1985. DNA. *Scientific American*.

Fox, C. F. February 1972. The structure of cell membranes. *Scientific American*.

Friend, S. 1994. p53: A glimpse at the puppet behind the shadow play. *Science* 265:334.

Glover, D. M., C. Gonzalez, and J. W. Raff. June 1993. The centrosome. *Scientific American*.

Grivell, L. A. March 1983. Mitochondria DNA. *Scientific American*.

Grunstein, M. October 1992. Histones as regulators of genes. *Scientific American*.

Hartwell, L. H., and M. B. Kastan. 1994. Cell cycle control and cancer. *Science* 266:1821.

Hayflick, H. January 1980. The cell biology of human aging. *Scientific American*.

Kornfeld, S., and W. S. Sly. 1985. Lysosomal storage defects. *Hospital Practice* 20:71.

Lake, J. A. August 1981. The ribosome. *Scientific American*.

---

[1]*Note:* This question is answered on page 33 of the Student Study Guide.

Lazarides, E., and J. P. Revel. May 1979. The molecular basis of cell movement. *Scientific American*.

Levine, A. J. 1995. Tumor suppressor genes. *Science & Medicine* 2:28.

Lodish, H. F., and J. E. Rothman. January 1979. The assembly of cell membranes. *Scientific American*.

Mazia, D. January 1974. The cell cycle. *Scientific American*.

McKusick, V. A. 1981. The anatomy of the human genome. *Hospital Practice* 16:82.

Miller, O. L. Jr. March 1973. The visualization of genes in action. *Scientific American*.

Murray, A. W., and M. W. Kirschner. March 1991. What controls the cell cycle. *Scientific American*.

Nomura, M. October 1969. Ribosomes. *Scientific American*.

Oakley, B. R., and C. E. Oakley. 1995. Tubulin and microtubulis. *Science and Medicine* 2:58.

Oliver, J. D. 1994. The role of p53 in cancer development. *Science and Medicine* 1:16.

Palade, G. 1975. Intracellular aspects of the process of protein synthesis. *Science* 189:347.

Porter, K. R., and J. B. Tucker. March 1981. The ground substance of the living cell. *Scientific American*.

Ptashne, M. January 1989. How gene activators work. *Scientific American*.

Racker, E. February 1968. The membrane of the mitochondrion. *Scientific American*.

Rich, A., and S. H. Kim. January 1978. The three-dimensional structure of transfer RNA. *Scientific American*.

Rothman, J. E. September 1985. The compartmental organization of the Golgi apparatus. *Scientific American*.

Rothman, J. E., and L. Orci. 1992. Molecular dissection of the secretory pathway. *Nature* 355:409.

Scott-Burden, T. 1994. Extracellular matrix: The cellular environment. *News in Physiological Sciences* 9:110.

Singer, S. J. 1973. Biological membranes. *Hospital Practice* 8:81.

Singer, S. J., and G. L. Nicolson. 1972. The fluid mosaic model of the structure of cell membranes. *Science* 175:720.

Sloboda, R. D. 1980. The role of microtubules in cell structure and cell division. *American Scientist* 68:290.

Stein, G., J. S. Stein, and L. J. Kleinsmith. February 1975. Chromosomal proteins and gene regulation. *Scientific American*.

Steitz, J. A. June 1988. "Snurps." *Scientific American*.

Steller, H. 1995. Mechanisms and genes of cellular suicide. *Science* 267:1445.

Stossel, T. P. September 1994. The machinery of cell crawling. *Scientific American*.

Thompson, C. B. 1995. Apoptosis in the pathogenesis and treatment of disease. *Science* 267:1456.

Wallace, D. C. 1986. Mitochondria genes and disease. *Hospital Practice* 21:77.

Wallace, D. C. 1992. Mitochondrial genetics: A paradigm for aging and degenerative diseases? *Science* 256:628.

Weber, K., and M. Osborn. October 1985. The molecules of the cell matrix. *Scientific American*.

White, R., and J. M. Lalouel. February 1988. Chromosome mapping with DNA markers. *Scientific American*.

Life Science Animations

The animations that relate to chapter 3 are #2 Journey into a Cell, #3 Endocytosis, #4 Cellular Secretion, #12 Mitosis, #13 Meiosis, #14 Crossing Over, #15 DNA Replication, #16 Transcription of a Gene, and #17 Protein Synthesis.

4

Enzymes And Energy

OBJECTIVES   *After studying this chapter, you should be able to . . .*

1. state the principles of catalysis and explain how enzymes function as catalysts.

2. describe how the names of enzymes are derived and explain the significance of isoenzymes.

3. describe the effects of pH and temperature on the rate of enzyme-catalyzed reactions and explain how these effects are produced.

4. describe the roles of cofactors and coenzymes in enzymatic reactions.

5. explain how the law of mass action helps to account for the direction of reversible reactions.

6. explain how enzymes work together to produce a metabolic pathway and how this pathway may be affected by end-product inhibition and inborn errors of metabolism.

7. explain how the first and second laws of thermodynamics can be used to predict whether metabolic reactions will be endergonic or exergonic.

8. describe the production of ATP and explain the significance of ATP as the universal energy carrier.

9. define the terms *oxidation*, *reduction*, *oxidizing agent*, and *reducing agent*.

10. describe the derivation of NAD and FAD in oxidation-reduction reactions and explain the functional significance of these two molecules.

OUTLINE

Enzymes as Catalysts

Enzymes are biological catalysts that function to increase the rate of chemical reactions. Most enzymes are proteins, and their catalytic action results from their complex structure. The great diversity of protein structure allows different enzymes to be specialized in their action.

The ability of yeast cells to make alcohol from glucose (a process called *fermentation*) had been known since antiquity, yet even as late as the mid-nineteenth century no scientist had been able to duplicate the trick in the absence of living yeast. Also, yeast and other living cells could perform a vast array of chemical reactions at body temperature that could not be duplicated in the chemistry laboratory without adding a substantial amount of heat energy. These observations led many mid-nineteenth-century scientists to believe that chemical reactions in living cells were aided by a "vital force" that operated beyond the laws of the physical world. This "vitalist" concept was squashed along with the yeast cells when a pioneering biochemist, Eduard Buchner, demonstrated that juice obtained from yeast could ferment glucose to alcohol. The yeast juice was not alive—evidently some chemicals in the cells were responsible for fermentation. Buchner didn't know what these chemicals were, so he simply named them **enzymes** (Greek for "in yeast").

Enzymes are proteins, as a general rule. The only known exceptions are a few special cases in which RNA demonstrates enzymatic activity. Recent experiments, for example, suggest that the RNA component of ribosomes (see chapter 3) serves as the enzyme that helps form peptide bonds within the growing polypeptide. Regardless of its chemical nature, an enzyme acts as a *biological catalyst*. A catalyst is a chemical that (1) increases the rate of a reaction, (2) is not itself changed at the end of the reaction, and (3) does not change the nature of the reaction or its final result. The same reaction would have occurred to the same degree in the absence of the catalyst, but it would have progressed at a much slower rate.

In order for a given reaction to occur, the reactants must have sufficient energy. The amount of energy required for a reaction to proceed is called the **activation energy.** By analogy, a match will not burn and release heat energy unless it is first "activated" by striking the match or by placing it in a flame.

In a large population of molecules, only a small fraction will possess sufficient energy for a reaction. Adding heat will raise the energy level of all the reactant molecules, thus increasing the percentage of the population that has the activation energy. Heat makes reactions go faster, but it also produces undesirable side effects in cells. Catalysts make the reaction go faster at lower temperatures by *lowering the activation energy required* thus ensuring that a larger percentage of the population of reactant molecules will have sufficient energy to participate in the reaction (fig. 4.1).

Since a small fraction of the reactants will have the activation energy required for the reaction even in the absence of a catalyst, the reaction could theoretically occur spontaneously at a slow rate. This rate, however, would be much too slow for the needs of a cell. So, from a biological standpoint, the presence or absence of a specific enzyme catalyst acts as a switch—the reaction will occur if the enzyme is present and will not occur if the enzyme is absent.

Mechanism of Enzyme Action

The ability of enzymes to lower the activation energy of a reaction is a result of their structure. Enzymes are large proteins with complex, highly ordered, three-dimensional shapes produced by physical and chemical interactions between their amino acids. Each type of enzyme has a characteristic shape, or *conformation*, with ridges, grooves, and pockets that are lined with specific amino acids. The particular pockets that are active in catalyzing a reaction are called the *active sites* of the enzyme.

The reactant molecules, which are the *substrates* of the enzyme, have shapes that allow them to fit into the active sites. The fit may not be perfect at first, but a perfect fit may be induced as the substrate gradually slips into the active site. This induced fit, together with temporary bonds that form between the substrate and the amino acids lining the active sites of the enzyme, weakens the existing bonds within the substrate molecules and allows them to be more easily broken. New bonds are more easily formed as substrates are brought close together in the proper orientation. The *enzyme-substrate complex*, formed temporarily in the course of the reaction, then dissociates to yield *products* and the free unaltered enzyme. This model of how enzymes work is known as the **lock-and-key model** of enzyme activity (fig. 4.2).

Naming of Enzymes

Although an international committee has established a uniform naming system for enzymes, the names that are in common use do not follow a completely consistent pattern. With the exception of some older enzyme names (such as pepsin, trypsin, and renin), all enzyme names end with the suffix *-ase* (table 4.1). Classes of enzymes are named according to their activity, or "job category." *Hydrolases*, for example, promote hydrolysis reactions. Other enzyme categories include *phosphatases*, which catalyze the removal of phosphate groups;

Table 4.1	Selected Enzymes and the Reactions They Catalyze
Enzyme	**Reaction Catalyzed**
Catalase	$2\ H_2O_2 \rightarrow 2\ H_2O + O_2$
Carbonic anhydrase	$H_2CO_3 \rightarrow H_2O + CO_2$
Amylase	starch $+ H_2O \rightarrow$ maltose
Lactate dehydrogenase	lactic acid $\rightarrow$ pyruvic acid $+ H_2$
Ribonuclease	RNA $+ H_2O \rightarrow$ ribonucleotides

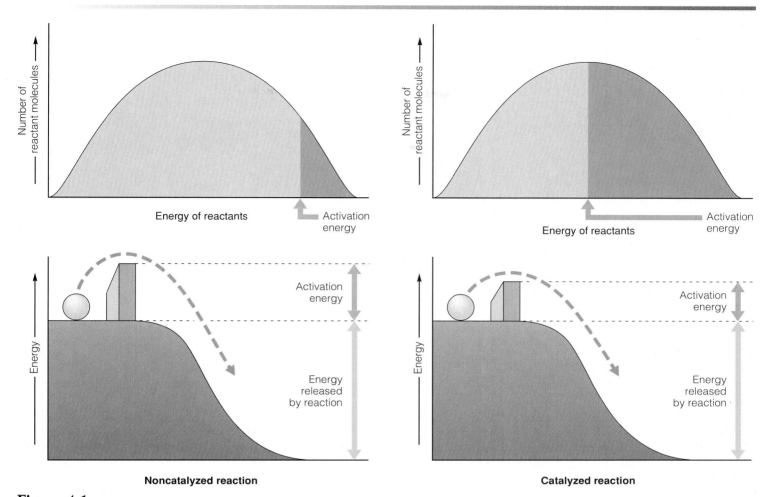

Figure 4.1

A comparison between a noncatalyzed reaction and a catalyzed reaction. The upper figures compare the proportion of reactant molecules that have sufficient activation energy to participate in the reaction (shown in green). This proportion is increased in the enzyme-catalyzed reaction because enzymes lower the activation energy required for the reaction (shown as a barrier on top of an energy "hill" in the lower figures). Reactants that can overcome this barrier are able to participate in the reaction, as shown by arrows pointing to the bottom of the energy hill.

synthetases, which catalyze dehydration synthesis reactions; and *dehydrogenases*, which remove hydrogen atoms from their substrates. Enzymes called *isomerases* rearrange atoms within their substrate molecules to form structural isomers, such as glucose and fructose.

The names of many enzymes specify both the substrate of the enzyme and the job category of the enzyme. Lactic acid dehydrogenase, for example, removes hydrogens from lactic acid. Since enzymes are very specific as to their substrates and activity, the concentration of a specific enzyme in a sample of fluid can be measured relatively easily. This is usually done by measuring the rate of conversion of the enzyme's substrates into products under specified conditions.

 When tissues become damaged due to diseases, some of the dead cells disintegrate and release their enzymes into the blood. Most of these enzymes are not normally active in the blood because of the absence of their specific substrates, but their enzymatic activity can be measured in a test tube by the addition of the appropriate substrates to samples of plasma. Such measurements are clinically useful, because abnormally high plasma concentrations of particular enzymes are characteristic of certain diseases (table 4.2).

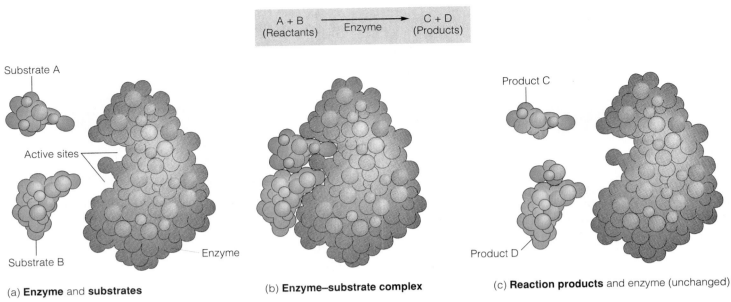

$$\boxed{\begin{array}{c} \text{A + B} \\ \text{(Reactants)} \end{array} \xrightarrow{\text{Enzyme}} \begin{array}{c} \text{C + D} \\ \text{(Products)} \end{array}}$$

Substrate A

Active sites

Substrate B

Enzyme

Product C

Product D

(a) **Enzyme** and **substrates**

(b) **Enzyme–substrate complex**

(c) **Reaction products** and enzyme (unchanged)

Figure 4.2

The lock-and-key fit model of enzyme action. (*a*) Substrates A and B fit into active sites in the enzyme, forming an enzyme-substrate complex. (*b*) This then dissociates (*c*), releasing the products of the reaction and the free enzyme.

Enzymes that do exactly the same job (that catalyze the same reaction) in different organs have the same name, since the name describes the activity of the enzyme. Different organs, however, may make slightly different "models" of the enzyme that differ in one or a few amino acids. These different models of the same enzyme are called **isoenzymes.** The differences in structure do not affect the active sites (otherwise they would not catalyze the same reaction), but they do alter the structure of the enzymes at other locations, so that the different isoenzymatic forms can be separated by standard biochemical procedures. These techniques are useful in the diagnosis of diseases.

1. Define *activation energy* and explain how it is affected by a catalyst. Explain how structure of an enzyme enables it to function as a catalyst.
2. Diagram the lock-and-key model of enzyme action, including labels, and write a generalized formula for this reaction.
3. Explain how enzymes are named.

Different organs, when they are diseased, may liberate different isoenzymatic forms of an enzyme that can be measured in a clinical laboratory. For example, the enzyme **creatine phosphokinase**, abbreviated either **CPK** or **CK**, exists in three isoenzymatic forms. These forms are identified by two letters that indicate two components of this enzyme. One form is identified as MM and is liberated from diseased skeletal muscle; the second is BB, released by a damaged brain; and the third is MB, released from a diseased heart. Newer clinical tests utilize antibodies that can bind to the M and B components to specifically measure the level of the MB form in the blood when heart disease is suspected.

Table 4.2 Examples of the Diagnostic Value of Some Enzymes Found in Plasma

Enzyme	Diseases Associated with Abnormal Plasma Enzyme Concentrations
Alkaline phosphatase	Obstructive jaundice, Paget's disease (osteitis deformans), carcinoma of bone
Acid phosphatase	Benign hypertrophy of prostate, cancer of prostate
Amylase	Pancreatitis, perforated peptic ulcer
Aldolase	Muscular dystrophy
Creatine kinase (or creatine phosphokinase-CPK)	Muscular dystrophy, myocardial infarction
Lactate dehydrogenase (LDH)	Myocardial infarction, liver disease, renal disease, pernicious anemia
Transaminases (AST and ALT)	Myocardial infarction, hepatitis, muscular dystrophy

Control of Enzyme Activity

The rate of an enzyme-catalyzed reaction depends on the concentration of the enzyme, the pH and temperature of the reaction, and on a number of other factors. Genetic control of enzyme concentration, for example, affects the rate of progress along particular metabolic pathways and thus regulates cellular metabolism.

The activity of an enzyme, as measured by the rate at which its substrates are converted to products, is influenced by a variety of factors, including (1) the temperature and pH of the solution; (2) the concentration of cofactors and coenzymes, which are needed by many enzymes as "helpers" for their catalytic activity; (3) the concentration of enzyme and substrate molecules in the solution; and (4) the stimulatory and inhibitory effects of some products of enzyme action on the activity of the enzymes that helped to form these molecules.

Effects of Temperature and pH

An increase in temperature will increase the rate of non-enzyme-catalyzed reactions. A similar relationship between temperature and reaction rate occurs in enzyme-catalyzed reactions. At a temperature of 0°C the reaction rate is immeasurably slow. As the temperature is raised above 0°C the reaction rate increases, but only up to a point. At a few degrees above body temperature (which is 37°C) the reaction rate reaches a plateau; further increases in temperature actually *decrease* the rate of the reaction (fig. 4.3). This decrease is due to the fact that the tertiary structure of enzymes becomes altered at higher temperatures.

A similar relationship is observed when the rate of an enzymatic reaction is measured at different pH values. Each enzyme characteristically has its peak activity in a very narrow pH range, which is the **pH optimum** for the enzyme. If the pH is changed so that it is no longer within the enzyme's optimum range, the reaction rate will decrease (fig. 4.4). This decreased

Although the pH of other body fluids shows less variation than that of the fluids of the digestive tract, the pH optima of different enzymes found throughout the body do show significant differences (table 4.3). Some of these differences can be exploited for diagnostic purposes. Disease of the prostate, for example, may be associated with elevated blood levels of a prostatic phosphatase with an acidic pH optimum (descriptively called acid phosphatase). Bone disease, on the other hand, may be associated with elevated blood levels of alkaline phosphatase, which has a higher pH optimum than the similar enzyme released from the diseased prostate.

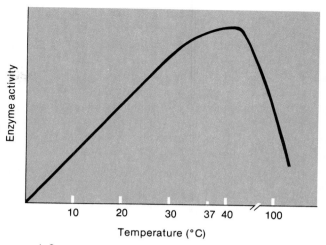

Figure 4.3

The effect of temperature on enzyme activity, as measured by the rate of the enzyme-catalyzed reaction under standardized conditions.

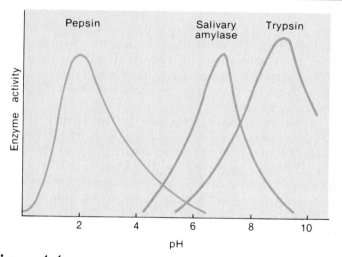

Figure 4.4

The effect of pH on the activity of three digestive enzymes.

Table 4.3   pH Optima of Selected Enzymes

Enzyme	Reaction Catalyzed	pH Optimum
Pepsin (stomach)	Digestion of protein	2.0
Acid phosphatase (prostate)	Removal of phosphate group	5.5
Salivary amylase (saliva)	Digestion of starch	6.8
Lipase (pancreatic juice)	Digestion of fat	7.0
Alkaline phosphatase (bone)	Removal of phosphate group	9.0
Trypsin (pancreatic juice)	Digestion of protein	9.5
Monoamine oxidase (nerve endings)	Removal of amine group from norepinephrine	9.8

enzyme activity is due to changes in the conformation of the enzyme and in the charges of the R groups of the amino acids lining the active sites.

The pH optimum of an enzyme usually reflects the pH of the body fluid in which the enzyme is found. The acidic pH optimum of the protein-digesting enzyme *pepsin,* for example, allows it to be active in the strong hydrochloric acid of gastric juice. Similarly, the neutral pH optimum of *salivary amylase* and the alkaline pH optimum of *trypsin* in pancreatic juice allow these enzymes to digest starch and protein, respectively, in other parts of the digestive tract.

Cofactors and Coenzymes

Many enzymes are completely inactive when they are isolated in a pure state. Evidently some of the ions and smaller organic molecules that are removed in the purification procedure play an essential role in enzyme activity. These ions and smaller organic molecules needed for the activity of specific enzymes are called **cofactors.**

Cofactors include metal ions such as Ca^{++}, Mg^{++}, Mn^{++}, Cu^{++}, Zn^{++}, and selenium. Some enzymes with a cofactor requirement do not have a properly shaped active site in the absence of the cofactor. In these enzymes, the attachment of cofactors causes a conformational change in the protein that allows it to combine with its substrate. The cofactors of other enzymes participate in the temporary bonds between the enzyme and its substrate when the enzyme-substrate complex is formed (fig. 4.5).

Coenzymes are cofactors that are organic molecules derived from niacin, riboflavin, and other water-soluble vitamins. Coenzymes participate in enzyme-catalyzed reactions by transporting hydrogen atoms and small molecules from one enzyme to another. Examples of the actions of cofactors and coenzymes in specific reactions will be given in the context of their roles in cellular metabolism later in this chapter.

Substrate Concentration and Reversible Reactions

The rate at which an enzymatic reaction converts substrates into products depends on the enzyme concentration and on the concentration of substrates. When the enzyme concentration is at a given level, the rate of product formation will increase as the substrate concentration increases. Eventually, however, a point will be reached where additional increases in substrate concentration do not result in comparable increases in reaction rate. When the relationship between substrate concentration and reaction rate reaches a plateau of maximum velocity, the enzyme is said to be *saturated.* If we think of enzymes as workers and substrates as jobs, there is 100% employment when the enzyme is saturated; further availability of jobs (substrate) cannot further increase employment (conversion of substrate to product). This concept is illustrated in figure 4.6.

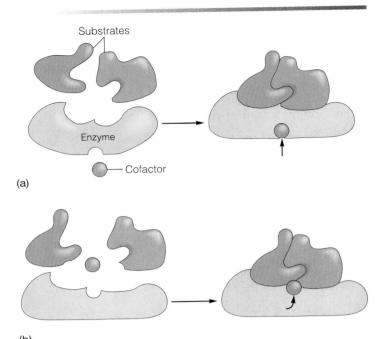

(a)

(b)

Figure 4.5

The roles of cofactors in enzyme function. In (*a*) the cofactor changes the conformation of the active site, allowing for a better fit between the enzyme and its substrates. In (*b*) the cofactor participates in the temporary bonding between the active site and the substrates.

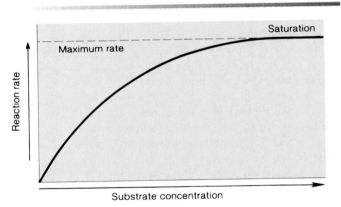

Figure 4.6

The effect of substrate concentration on the reaction rate of an enzyme-catalyzed reaction. When the reaction rate is maximum, the enzyme is said to be *saturated.*

Some enzymatic reactions within a cell are reversible, with both the forward and backward reactions catalyzed by the same enzyme. The enzyme *carbonic anhydrase,* for example, is named because it can catalyze the following reaction:

$$H_2CO_3 \rightarrow H_2O + CO_2$$

The same enzyme, however, can also catalyze the reverse reaction:

$$H_2O + CO_2 \rightarrow H_2CO_3$$

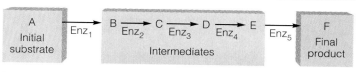

Figure 4.7

A metabolic pathway, where the product of one enzyme becomes the substrate of the next in a multienzyme system.

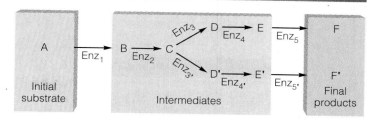

Figure 4.8

A branched metabolic pathway.

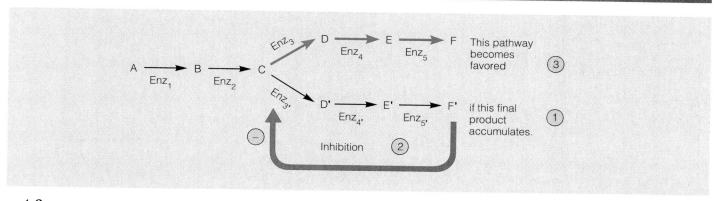

Figure 4.9

End-product inhibition in a branched metabolic pathway. Inhibition is shown by the arrow in step 2.

The two reactions can be more conveniently illustrated by a single equation:

$$H_2O + CO_2 \rightleftarrows H_2CO_3$$

The direction of the reversible reaction depends, in part, on the relative concentrations of the molecules to the left and right of the arrows. If the concentration of CO_2 is very high (as it is in the tissues), the reaction will be driven to the right. If the concentration of CO_2 is low and that of H_2CO_3 is high (as it is in the lungs), the reaction will be driven to the left. The principle that reversible reactions will be driven from the side of the equation where the concentration is higher to the side where the concentration is lower is known as the **law of mass action.**

Although some enzymatic reactions are not directly reversible, the net effects of the reactions can be reversed by the action of different enzymes. Some of the enzymes that convert glucose to pyruvic acid, for example, are different from those that reverse the pathway and produce glucose from pyruvic acid. Likewise, the formation and breakdown of glycogen (a polymer of glucose) are catalyzed by different enzymes.

Metabolic Pathways

The many thousands of different types of enzymatic reactions within a cell do not occur independently of each other. They are, rather, all linked together by intricate webs of interrelationships, the total pattern of which constitutes cellular metabolism. A part of this web that begins with an *initial substrate,* progresses through a number of *intermediates,* and ends with a *final product* is known as a **metabolic pathway.**

The enzymes in a metabolic pathway cooperate in a manner analogous to workers on an assembly line where each contributes a small part to the final product. In this process, the product of one enzyme in the line becomes the substrate of the next enzyme, and so on (fig. 4.7).

Few metabolic pathways are completely linear. Most are branched so that one intermediate at the branch point can serve as a substrate for two different enzymes. Two different products can thus be formed that serve as intermediates of two pathways (fig. 4.8).

End-Product Inhibition

The activities of enzymes at the branch points of metabolic pathways are often regulated by a process called **end-product inhibition.** In this process, one of the final products of a divergent pathway inhibits the branch-point enzyme that began the path toward the production of this inhibitor. This inhibition prevents that final product from accumulating excessively and results in a shift toward the final product of the alternate pathway (fig. 4.9).

The mechanism by which a final product inhibits an earlier enzymatic step in its pathway is known as **allosteric inhibition.** The allosteric inhibitor combines with a part of the enzyme at a location other than the active site. This causes the active site to change shape so that it can no longer combine properly with its substrate.

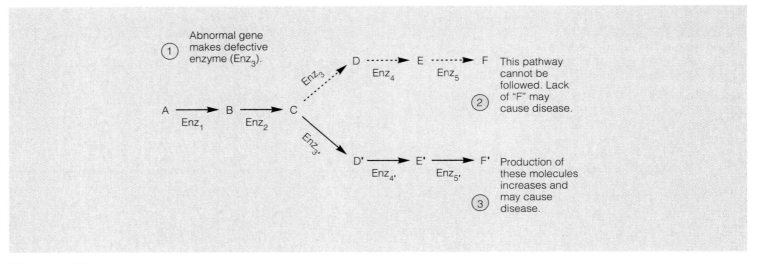

Figure 4.10

The effects of an inborn error of metabolism on a branched metabolic pathway.

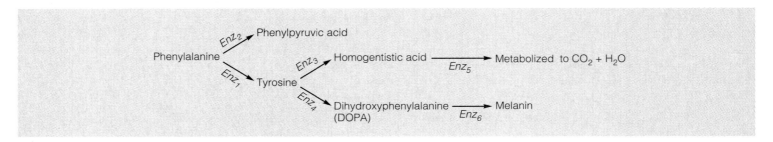

Figure 4.11

Metabolic pathways for the degradation of the amino acid phenylalanine. Defective *enzyme$_1$* produces phenylketonuria (PKU), defective *enzyme$_2$* produces alcaptonuria (not a clinically significant condition), and defective *enzyme$_3$* produces albinism.

Inborn Errors of Metabolism

Since each different polypeptide in the body is coded by a different gene (chapter 3), each enzyme protein that participates in a metabolic pathway is coded by a different gene. An inherited defect in one of these genes may result in a disease known as an "inborn error of metabolism." In this type of disease, the quantity of intermediates formed *prior to* the defective enzymatic step *increases*, and the quantity of intermediates and final products formed *after* the defective step *decreases*. Diseases may result from deficiencies of the normal end product, or from accumulation to toxic levels of intermediates formed prior to the defective step. If the defective enzyme is active at a step that follows a branch point in a pathway, the intermediates and final products of the alternate pathway will increase (fig. 4.10). Abnormally increased production of these alternate products can be the cause of some metabolic diseases.

One of the conversion products of phenylalanine is a molecule known as *DOPA*, which is an acronym for dihydroxyphenylalanine. DOPA is a precursor of the pigment molecule *melanin*. An inherited defect in the enzyme that catalyzes the

The branched metabolic pathway that begins with phenylalanine as the initial substrate is subject to a number of inborn errors of metabolism (fig. 4.11). When the enzyme that converts this amino acid to the amino acid tyrosine is defective, the final product of a divergent pathway accumulates and can be detected in the blood and urine. This disease—**phenylketonuria (PKU)**—can result in severe mental retardation and a shortened life span. PKU occurs sufficiently often (although no inborn error of metabolism is common) to warrant the testing of all newborn babies for this defect. If the disease is detected early, brain damage can be prevented by placing the child on an artificial diet low in the amino acid phenylalanine.

formation of melanin from DOPA results in albinism (fig. 4.11). Besides PKU and albinism, there are many other inborn errors of amino acid metabolism, as well as errors in carbohydrate and lipid metabolism (table 4.4).

Table 4.4 Examples of Inborn Errors in the Metabolism of Amino Acids, Carbohydrates, and Lipids

Metabolic Defect	Disease	Abnormality	Clinical Result
Amino acid metabolism	Phenylketonuria (PKU)	Increase in phenylpyruvic acid	Mental retardation, epilepsy
	Albinism	Lack of melanin	Susceptibility to skin cancer
	Maple-syrup disease	Increase in leucine, isoleucine, and valine	Degeneration of brain, early death
	Homocystinuria	Accumulation of homocystine	Mental retardation, eye problems
Carbohydrate metabolism	Lactose intolerance	Lactose not utilized	Diarrhea
	Glucose 6-phosphatase deficiency (Gierke's disease)	Accumulation of glycogen in liver	Liver enlargement, hypoglycemia
	Glycogen phosphorylase deficiency (McArdle syndrome)	Accumulation of glycogen in muscle	Muscle fatigue and pain
Lipid metabolism	Gaucher's disease	Lipid accumulation (glucocerebroside)	Liver and spleen enlargement, brain degeneration
	Tay–Sachs disease	Lipid accumulation (ganglioside G_{M2})	Brain degeneration, death by age 5
	Hypercholestremia	High blood cholesterol	Atherosclerosis of coronary and large arteries

1. Draw graphs to represent the effects of changes in temperature, pH, and enzyme and substrate concentration on the rate of enzymatic reactions. Explain the mechanisms responsible for the effects you have graphed.

2. Using arrows and letters of the alphabet, draw a flowchart of a metabolic pathway with one branch point.

3. Describe a reversible reaction and explain how the law of mass action affects this reaction.

4. Define *end-product inhibition* and use your diagram of a branched metabolic pathway to explain how this process will affect the concentrations of different intermediates.

5. Suppose, due to an inborn error of metabolism, that the enzyme that catalyzed the third reaction in your pathway (question no. 2) was defective. Describe the effects this would have on the concentrations of the intermediates in your pathway.

Bioenergetics

Living organisms require the constant expenditure of energy to maintain their complex structures and processes. Central to life processes are chemical reactions that are coupled, so that the energy released by one reaction is incorporated into the products of another reaction. The transformation of energy in living systems is largely based on reactions that produce and destroy molecules of ATP and on oxidation-reduction reactions.

Bioenergetics refers to the flow of energy in living systems. Organisms maintain their highly ordered structure and life-sustaining activities through the constant expenditure of energy obtained ultimately from the environment. The energy flow in living systems obeys the first and second laws of a branch of physics known as *thermodynamics.*

According to the **first law of thermodynamics,** energy can be transformed (changed from one form to another), but it can neither be created nor destroyed. This is sometimes called the *law of conservation of energy.* As a result of energy transformations, according to the **second law of thermodynamics,** the universe and its parts (including living systems) become increasingly disorganized. The term *entropy* is used to describe the degree of disorganization of a system. Energy transformations thus increase the amount of entropy of a system. Only energy that is in an organized state—called *free energy*—can be used to do work. Thus, since entropy increases in every energy transformation, the amount of free energy available to do work decreases. As a result of the increased entropy described by the second law, systems tend to go from states of higher free energy to states of lower free energy.

The chemical bonding of atoms into molecules obeys the laws of thermodynamics. Atoms that are organized into complex organic molecules, such as glucose, have more free energy (less entropy) than six separate molecules each of carbon dioxide and water. Therefore, in order to convert carbon dioxide and water to glucose, energy must be added. Plants perform this feat using energy from the sun in the process of *photosynthesis* (fig. 4.12).

Endergonic and Exergonic Reactions

Chemical reactions that require an input of energy are known as **endergonic reactions.** Since energy is added to make these reactions "go," the products of endergonic reactions must contain more free energy than the reactants. A portion of the

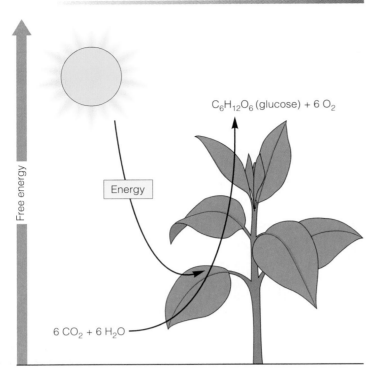

$C_6H_{12}O_6$ (glucose) + 6 O_2

Free energy

Energy

6 CO_2 + 6 H_2O

Figure 4.12

A simplified diagram of photosynthesis. Some of the sun's radiant energy is captured by plants and used to produce glucose from carbon dioxide and water. As the product of this endergonic reaction, glucose has more free energy than the initial reactants.

energy added, in other words, is contained within the product molecules. This follows from the fact that energy cannot be created or destroyed (first law of thermodynamics) and from the fact that a more-organized state of matter contains more free energy (less entropy) than a less-organized state (as described by the second law).

The fact that glucose contains more free energy than carbon dioxide and water can easily be proven by the combustion of glucose to CO_2 and H_2O. This reaction releases energy in the form of heat. Reactions that convert molecules with more free energy to molecules with less—and, therefore, that release energy as they proceed—are called **exergonic reactions.**

As illustrated in figure 4.13, the amount of energy released by an exergonic reaction is the same whether the energy is released in a single combustion reaction or in the many small, enzymatically controlled steps that occur in tissue cells. The energy that the body obtains from the consumption of particular foods can therefore be measured as the amount of heat energy released when these foods are combusted.

Heat is measured in units called *calories.* One calorie is defined as the amount of heat required to raise the temperature of one cubic centimeter of water one degree on the Celsius scale. The caloric value of food is usually indicated in kilocalories (one kilocalorie = 1000 calories), which are often called large calories, designated with a capital C.

Coupled Reactions: ATP

In order to remain alive, a cell must maintain its highly organized, low-entropy state at the expense of free energy in its environment. Accordingly, the cell contains many enzymes that catalyze exergonic reactions using substrates that come ultimately from the environment. The energy released by these exergonic reactions is used to drive the energy-requiring processes (endergonic reactions) in the cell. Since the cell cannot use heat energy to drive energy-requiring processes, chemical-bond energy that is released in the exergonic reactions must be directly transferred to chemical-bond energy in the products of endergonic reactions. Energy-liberating reactions are thus *coupled* to energy-requiring reactions. This relationship is like that of two meshed gears; the turning of one (the energy-releasing exergonic gear) causes turning of the other (the energy-requiring endergonic gear). This relationship is illustrated in figure 4.14.

The energy released by most exergonic reactions in the cell is used, either directly or indirectly, to drive one particular endergonic reaction (fig. 4.15): the formation of **adenosine triphosphate (ATP)** from adenosine diphosphate (ADP) and inorganic phosphate (abbreviated P_i).

The formation of ATP requires the input of a fairly large amount of energy. Since this energy must be conserved (first law of thermodynamics), the bond that is produced by joining P_i to ADP must contain a part of this energy. Thus, when enzymes reverse this reaction and convert ATP to ADP and P_i, a large amount of energy is released. Energy released from the breakdown of ATP is used to power the energy-requiring processes in all cells. As the **universal energy carrier,** ATP serves to more efficiently couple the energy released by the breakdown of food molecules to the energy required by the diverse endergonic processes in the cell (fig. 4.16).

Coupled Reactions: Oxidation-Reduction

When an atom or a molecule gains electrons, it is said to become **reduced;** when it loses electrons, it is said to become **oxidized.** Reduction and oxidation are always coupled reactions: an atom or a molecule cannot become oxidized unless it donates electrons to another, which therefore becomes reduced. The atom or molecule that donates electrons *to* another is a

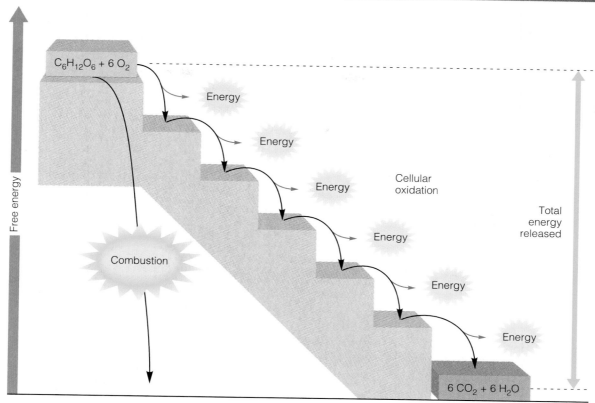

Figure 4.13

Since glucose contains more energy than six separate molecules each of carbon dioxide and water, the combustion of glucose is an exergonic reaction. The same amount of energy is released when glucose is broken down stepwise within the cell.

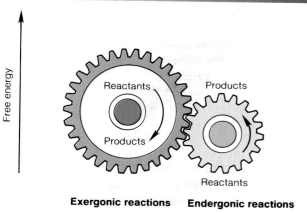

Exergonic reactions   Endergonic reactions

Figure 4.14

A model of the coupling of exergonic and endergonic reactions. The reactants of the exergonic reaction (represented by the larger gear) have more free energy than the products of the endergonic reaction because the coupling is not 100% efficient—some energy is lost as heat.

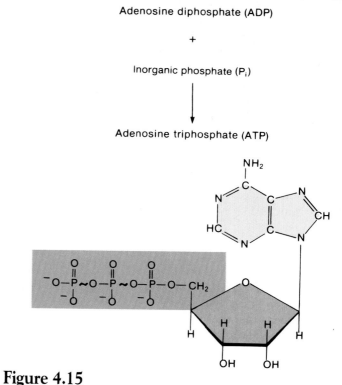

Adenosine diphosphate (ADP)

+

Inorganic phosphate (P_i)

Adenosine triphosphate (ATP)

Figure 4.15

The formation and structure of adenosine triphosphate (ATP).

Enzymes and Energy

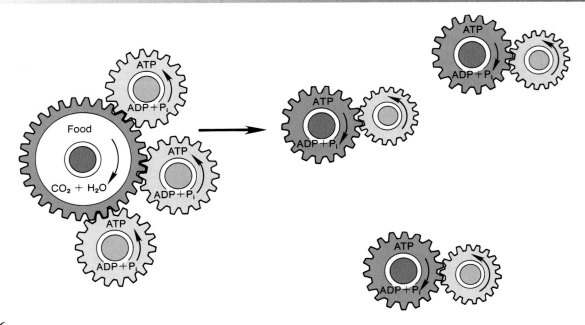

Figure 4.16

A model of ATP as the universal energy carrier of the cell. Exergonic reactions are shown as gears with arrows going down (these reactions produce a decrease in free energy); endergonic reactions are shown as gears with arrows going up (these reactions produce an increase in free energy).

reducing agent, and the one that accepts electrons *from* another is an **oxidizing agent.** It should be noted that an atom or a molecule may function as an oxidizing agent in one reaction and as a reducing agent in another reaction; it may gain electrons from one atom or molecule and pass them on to another in a series of coupled oxidation-reduction reactions—like a bucket brigade.

Notice that the term *oxidation* does not imply that oxygen participates in the reaction. This term is derived from the fact that oxygen has a great tendency to accept electrons; that is, to act as a strong oxidizing agent. This property of oxygen is exploited by cells; oxygen acts as the final electron acceptor in a chain of oxidation-reduction reactions that provides energy for ATP production.

Oxidation-reduction reactions in cells often involve the transfer of hydrogen atoms rather than of free electrons. Since a hydrogen atom contains one electron (and one proton in the nucleus), a molecule that loses hydrogen becomes oxidized, and one that gains hydrogen becomes reduced. In many oxidation-reduction reactions, pairs of electrons—either as free electrons or as a pair of hydrogen atoms—are transferred from the reducing agent to the oxidizing agent.

Two molecules that serve important roles in the transfer of hydrogens are **nicotinamide adenine dinucleotide (NAD),** which is derived from the vitamin niacin (vitamin B_3), and **flavin adenine dinucleotide (FAD),** which is derived from the vitamin riboflavin (vitamin B_2). These molecules (fig. 4.17) are coenzymes that function as *hydrogen carriers* because they accept hydrogens (becoming reduced) in one enzyme reaction and donate hydrogens (becoming oxidized) in a different enzyme reaction (fig. 4.18). The oxidized forms of these molecules may be written simply as NAD and FAD.

Each FAD can accept two electrons and can bind two protons. Therefore, the reduced form of FAD is combined with the equivalent of two hydrogen atoms and may be written as $FADH_2$. Each NAD can also accept two electrons but can bind only one proton. The reduced form of NAD is therefore indicated by $NADH + H^+$ (the H^+ represents a free proton). When the reduced forms of these two coenzymes participate in an oxidation-reduction reaction, they transfer two hydrogen atoms to the oxidizing agent (fig. 4.18).

 Production of the coenzymes NAD and FAD is the major reason that we need the vitamins niacin and riboflavin in our diet. As described in chapter 5, NAD and FAD are required to transfer hydrogen atoms in the chemical reactions that provide energy for the body. Niacin and riboflavin do not themselves provide the energy, although this is often claimed in misleading advertisements for health foods. Nor can eating extra amounts of niacin and riboflavin provide extra energy. Once the cells have obtained sufficient NAD and FAD, the excess amounts of these vitamins are simply eliminated in the urine.

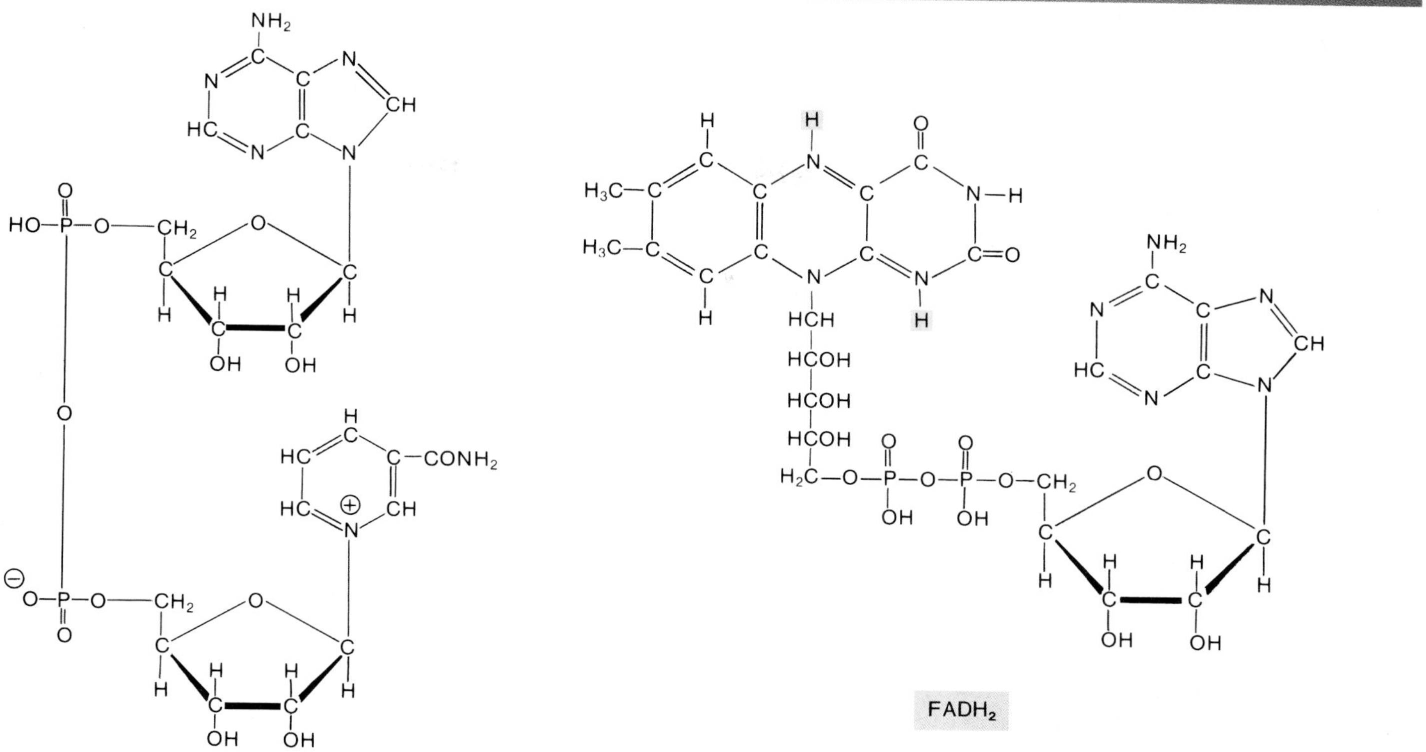

(a) NAD

(b) FADH₂

Figure 4.17

Structures (*a*) of the oxidized form of NAD (nicotinamide adenine dinucleotide) and (*b*) of the reduced form of FAD (flavin adenine dinucleotide). Notice the two additional hydrogen atoms (shown in color) that reduce FAD.

1. Describe the first and second laws of thermodynamics. Use these laws to explain why the chemical bonds in glucose represent a source of potential energy and describe the process by which cells can obtain this energy.

2. Define the terms *exergonic reaction* and *endergonic reaction*. Use these terms to describe the function of ATP in cells.

3. Using the symbols $X\text{-}H_2$ and Y, draw a coupled oxidation-reduction reaction. Designate the molecule that is reduced and the one that is oxidized and state which one is the reducing agent and which is the oxidizing agent.

4. Describe the functions of NAD, FAD, and oxygen (in terms of oxidation-reduction reactions) and explain the meaning of the symbols *NAD*, *NADH + H⁺*, *FAD*, and *FADH₂*.

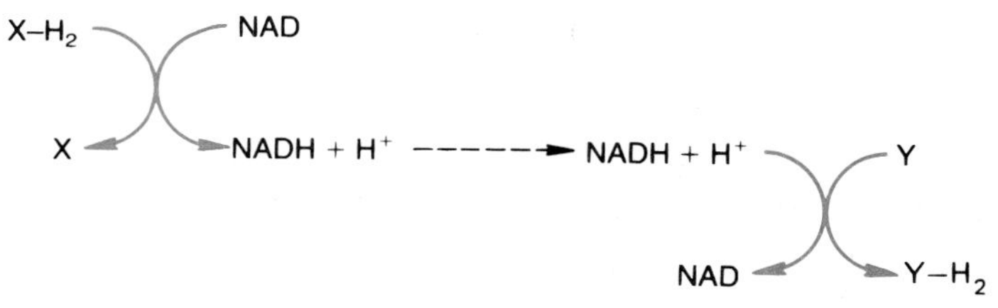

NAD is oxidizing agent

NADH is reducing agent

Figure 4.18

NAD is a coenzyme that functions to transfer pairs of hydrogen atoms from one molecule to another. In the first reaction, NAD is reduced (acts as an oxidizing agent); in the second reaction, NAD is oxidized (acts as a reducing agent).

Summary

Enzymes as Catalysts   p. 84

I. Enzymes are biological catalysts.
   A. Catalysts increase the rate of chemical reactions.
      1. A catalyst is not altered by the reaction.
      2. Catalysts do not change the final result of a reaction.
   B. Catalysts lower the activation energy of chemical reactions.
      1. The activation energy is the amount of energy needed by the reactant molecules to participate in a reaction.
      2. In the absence of a catalyst, only a small proportion of the reactants possess the activation energy to participate.
      3. By lowering the activation energy, enzymes allow a larger proportion of the reactants to participate in the reaction, thus increasing the reaction rate.
II. Most enzymes are proteins.
   A. Protein enzymes have specific three-dimensional shapes that are determined by the amino acid sequence and, ultimately, by the genes.
   B. The reactants in an enzyme-catalyzed reaction—called the substrates of the enzyme—fit into a specific pocket in the enzyme called the active site.
   C. By forming an enzyme-substrate complex, substrate molecules are brought into proper orientation and existing bonds are weakened. This allows new bonds to be more easily formed.

Control of Enzyme Activity   p. 87

I. The activity of an enzyme is affected by a variety of factors.
   A. The rate of enzyme-catalyzed reactions increases with increasing temperature, up to a maximum.
      1. This is because increasing the temperature increases the energy in the total population of reactant molecules, thus increasing the proportion of reactants that have the activation energy.
      2. At a few degrees above body temperature, however, most enzymes start to denature, and the rate of the reactions that they catalyze therefore decreases.
   B. Each enzyme has optimal activity at a characteristic pH—called the pH optimum for that enzyme.
      1. Deviations from the pH optimum will decrease the reaction rate because the pH affects the shape of the enzyme and charges within the active site.
      2. The pH optima of different enzymes can vary widely—pepsin has a pH optimum of 2, for example, while trypsin is most active at a pH of 9.
   C. Many enzymes require metal ions in order to be active—these ions are therefore said to be cofactors for the enzymes.
   D. Many enzymes require smaller organic molecules for activity. These smaller organic molecules are called coenzymes.
      1. Many coenzymes are derived from water-soluble vitamins.
      2. Coenzymes transport hydrogen atoms and small substrate molecules from one enzyme to another.
   E. The rate of enzymatic reactions increases when either the substrate concentration or the enzyme concentration is increased.
      1. If the enzyme concentration is constant, the rate of the reaction increases as the substrate concentration is raised, up to a maximum rate.
      2. When the rate of the reaction does not increase upon further addition of substrate, the enzyme is said to be saturated.
II. Metabolic pathways involve a number of enzyme-catalyzed reactions.
   A. A number of enzymes usually cooperate to convert an initial substrate to a final product by way of several intermediates.
   B. Metabolic pathways are produced by multienzyme systems in which the product of one enzyme becomes the substrate of the next.
   C. If an enzyme is defective due to an abnormal gene, the intermediates that are formed following the step catalyzed by the defective enzymes will decrease, and the intermediates that are formed prior to the defective step will accumulate.
      1. Diseases that result from defective enzymes are called inborn errors of metabolism.
      2. Accumulation of intermediates often results in damage to the organ that contains the defective enzyme.

D. Many metabolic pathways are branched, so that one intermediate can serve as the substrate for two different enzymes.
E. The activity of a particular pathway can be regulated by end-product inhibition.
   1. In end-product inhibition, one of the products of the pathway inhibits the activity of a key enzyme.
   2. This is an example of allosteric inhibition, in which the product combines with its specific site on the enzyme, changing the conformation of the active site.

Bioenergetics p. 91

I. The flow of energy in the cell is called bioenergetics.
A. According to the first law of thermodynamics, energy can neither be created nor destroyed but only transformed from one form to another.
B. According to the second law of thermodynamics, all energy transformation reactions result in an increase in entropy (disorder).
   1. As a result of the increase in entropy, there is a decrease in free (usable) energy.
   2. Atoms that are organized into large organic molecules thus contain more free energy than more disorganized, smaller molecules.

C. In order to produce glucose from carbon dioxide and water, energy must be added.
   1. Plants use energy from the sun for this conversion, in a process called photosynthesis.
   2. Reactions that require the input of energy to produce molecules with higher free energy than the reactants are called endergonic reactions.
D. The combustion of glucose to carbon dioxide and water releases energy in the form of heat.
   1. A reaction that releases energy, thus forming products that contain less free energy than the reactants, is called an exergonic reaction.
   2. The same total amount of energy is released when glucose is converted into carbon dioxide and water within cells, even though this process occurs in many small steps.
E. The exergonic reactions that convert food molecules into carbon dioxide and water in cells are coupled to endergonic reactions that form adenosine triphosphate (ATP).
   1. Some of the chemical-bond energy in glucose is therefore transferred to the "high energy" bonds of ATP.
   2. The breakdown of ATP into adenosine diphosphate (ADP) and inorganic phosphate results in the liberation of energy.
   3. The energy liberated by the breakdown of ATP is used to power all of the energy-requiring processes of the cell—ATP is thus the "universal energy carrier" of the cell.

II. Oxidation-reduction reactions are coupled and usually involve the transfer of hydrogen atoms.
A. A molecule is said to be oxidized when it loses electrons and to be reduced when it gains electrons.
B. A reducing agent is thus an electron donor; an oxidizing agent is an electron acceptor.
C. Although oxygen is the final electron acceptor in the cell, other molecules can act as oxidizing agents.
D. A single molecule can be an electron acceptor in one reaction and an electron donor in another.
   1. NAD and FAD can become reduced by accepting electrons from hydrogen atoms removed from other molecules.
   2. $NADH + H^+$, and $FADH_2$, in turn, donate these electrons to other molecules in other locations within the cells.
   3. Oxygen is the final electron acceptor (oxidizing agent) in a chain of oxidation-reduction reactions that provide energy for ATP production.

Clinical Investigation

A 77-year-old man is brought to the hospital after experiencing severe chest pains. Analysis of blood samples reveals an abnormally high plasma concentration of the MB isoform of creatine phosphokinase. During his hospital stay, the patient complains of difficulty in urination, and an additional blood test reveals a high concentration of acid phosphatase. What do these blood tests suggest?

Clues

Read the section "Naming of Enzymes" and the boxed information in this section. Examine table 4.2.

Review Activities

Objective Questions

1. Which of the following statements about enzymes is *true*?
   a. Most proteins are enzymes.
   b. Most enzymes are proteins.
   c. Enzymes are changed by the reactions they catalyze.
   d. The active sites of enzymes have little specificity for substrates.

2. Which of the following statements about enzyme-catalyzed reactions is *true*?
   a. The rate of reaction is independent of temperature.
   b. The rate of all enzyme-catalyzed reactions is decreased when the pH is lowered from 7 to 2.
   c. The rate of reaction is independent of substrate concentration.
   d. Under given conditions of substrate concentration, pH, and temperature, the rate of product formation varies directly with enzyme concentration up to a maximum, at which point the rate cannot be increased further.

3. Which of the following statements about lactate dehydrogenase is *true*?
   a. It is a protein.
   b. It oxidizes lactic acid.
   c. It reduces another molecule (pyruvic acid).
   d. All of the above are true.

4. In a metabolic pathway,
   a. the product of one enzyme becomes the substrate of the next.
   b. the substrate of one enzyme becomes the product of the next.

5. In an inborn error of metabolism,
   a. a genetic change results in the production of a defective enzyme.
   b. intermediates produced prior to the defective step accumulate.
   c. alternate pathways are taken by intermediates at branch points that precede the defective step.
   d. All of the above are true.

6. Which of the following represents an endergonic reaction?
   a. $ADP + P_i \rightarrow ATP$
   b. $ATP \rightarrow ADP + P_i$
   c. $glucose + O_2 \rightarrow CO_2 + H_2O$
   d. $CO_2 + H_2O \rightarrow glucose$
   e. both *a* and *d*
   f. both *b* and *c*

7. Which of the following statements about ATP is *true*?
   a. The bond joining ADP and the third phosphate is a high-energy bond.
   b. The formation of ATP is coupled to energy-liberating reactions.
   c. The conversion of ATP to ADP and P_i provides energy for biosynthesis, cell movement, and other cellular processes that require energy.
   d. ATP is the "universal energy carrier" of cells.
   e. All are true.

8. When oxygen is combined with two hydrogens to make water,
   a. oxygen is reduced.
   b. the molecule that donated the hydrogens becomes oxidized.
   c. oxygen acts as a reducing agent.
   d. both *a* and *b* apply.
   e. both *a* and *c* apply.

9. Enzymes increase the rate of chemical reactions by
   a. increasing the body temperature.
   b. decreasing the blood pH.
   c. increasing the affinity of reactant molecules for each other.
   d. decreasing the activation energy of the reactants.

10. According to the law of mass action, which of the following conditions will drive the reaction $A + B \rightarrow C$ to the right?
    a. an increase in the concentration of A and B
    b. a decrease in the concentration of C
    c. an increase in the concentration of enzyme
    d. both *a* and *b*
    e. both *b* and *c*

Essay Questions

1. Explain the relationship between the chemical structure and the function of an enzyme and describe how both structure and function may be altered in various ways.[1]

2. Explain how the rate of enzymatic reactions may be regulated by the relative concentrations of substrates and products.

3. Explain how end-product inhibition represents a form of negative feedback regulation.

4. Using the first and second laws of thermodynamics, explain how ATP is formed and how it serves as the universal energy carrier.

5. The coenzymes NAD and FAD can "shuttle" hydrogens from one reaction to another. Explain how this process serves to couple oxidation and reduction reactions.

6. Explain what is meant by "inborn errors of metabolism" using albinism and phenylketonuria as examples.

7. Why do we need to eat food containing niacin and riboflavin? Explain the function of these vitamins in the body.

---

[1]*Note:* This question is answered on page 42 of the Student Study Guide.

Selected Readings

Alberts, B. et al. 1989. *Molecular Biology of the Cell*. 2d ed. New York: Garland Publishing.

Baker, J. J. W., and G. E. Allen. 1982. *Matter, Energy, and Life*. 4th ed. Reading, MA: Addison-Wesley.

Berry, H. K. March 1984. The spectrum of metabolic disorders. *Diagnostic Medicine*, p. 39.

Devlin, T. M. 1992. *Textbook of Biochemistry with Clinical Correlations*. New York: Wiley.

Dyson, R. D. 1978. *Essentials of Cell Biology*. 2d ed. Boston: Allyn & Bacon.

Hinkle, P., and R. E. McCarty. March 1979. How cells make ATP. *Scientific American*.

Lehninger, A. L. September 1961. How cells transform energy. *Scientific American*.

Lehninger, A. L. 1982. *Principles of Biochemistry*. New York: Worth.

Sheeler, P., and D. E. Bianchi. 1980. *Cell Biology: Structure, Biochemistry, and Function*. New York: Wiley.

Stryker, L. 1981. *Biochemistry*, 2d ed. New York: Freeman.

Life Science Animation

The animation that relates to chapter 4 is #11 ATP as an Energy Carrier.

5

Cell Respiration and Metabolism

OBJECTIVES

After studying this chapter, you should be able to . . .

1. describe the steps of glycolysis and explain the significance of this metabolic pathway.

2. describe how lactic acid is formed and explain the physiological significance of this pathway.

3. define the term *gluconeogenesis* and describe the Cori cycle.

4. describe the pathway for the aerobic respiration of glucose through the steps of the Krebs cycle.

5. explain the functional significance of the Krebs cycle in relation to the electron-transport system.

6. describe the electron-transport system and oxidative phosphorylation.

7. describe the role of oxygen in aerobic respiration.

8. compare the lactic acid pathway and aerobic respiration in terms of initial substrates, final products, cellular locations, and the total number of ATP molecules produced per glucose respired.

9. explain how glucose and glycogen can be interconverted and how the liver can secrete free glucose derived from its stored glycogen.

10. define the terms *lipolysis* and β-*oxidation* and explain how these processes function in cellular energy production.

11. explain how ketone bodies are formed.

12. describe the processes of oxidative deamination and transamination of amino acids and explain how these processes can contribute to energy production.

13. explain, in terms of the metabolic pathways involved, how carbohydrates or protein can be converted to fat.

14. state the preferred energy sources of different organs.

OUTLINE

Glycolysis and the Lactic Acid Pathway

In cellular respiration, energy is released by the stepwise breakdown of glucose and other molecules, and some of this energy is used to produce ATP. The complete combustion of glucose requires the presence of oxygen and yields thirty-eight ATP per glucose. Some energy can be obtained, however, in the absence of oxygen by the pathway that leads to the production of lactic acid. This process results in a net gain of two ATP per glucose.

All of the reactions in the body that involve energy transformation are collectively termed **metabolism.** Metabolism may be divided into two categories: *anabolism* and *catabolism*. Catabolic reactions release energy, usually by the breakdown of larger organic molecules into smaller molecules. Anabolic reactions require the input of energy and include the synthesis of large energy-storage molecules, such as glycogen, fat, and protein.

The catabolic reactions that break down glucose, fatty acids, and amino acids serve as the primary sources of energy for the synthesis of ATP. That is, some of the chemical-bond energy in glucose, for example, is transferred to the chemical-bond energy in ATP. Since energy transfers can never be 100% efficient (according to the second law of thermodynamics, as described in chapter 4), some of the chemical-bond energy from glucose is lost as heat.

These energy transfers involve oxidation-reduction reactions. As explained in chapter 4, oxidation of a molecule occurs when the molecule loses electrons. This must be coupled to the reduction of another atom or molecule, which accepts the electrons. In the breakdown of glucose, fatty acids, and other molecules for energy, some of the electrons initially present in the molecules are first transferred to intermediate carriers of these electrons, and then to a *final electron acceptor*. If a molecule is completely broken down to carbon dioxide and water within an animal cell, the final electron acceptor will always be an atom of oxygen. Because of the involvement of oxygen, the metabolic pathway that converts molecules such as glucose or fatty acid to carbon dioxide and water (transferring some of the energy to ATP) is called **aerobic cell respiration.** The oxygen for this process is obtained from the blood. The blood, in turn, obtains oxygen from air in the lungs through the process of breathing, or ventilation, as described in chapter 16. Ventilation also serves the important function of eliminating the carbon dioxide produced by aerobic cell respiration.

Unlike the process of burning, or combustion, which quickly releases the energy content of molecules as heat (and which can be measured as kilocalories—see chapter 4), the conversion of glucose to carbon dioxide and water within the cells occurs in small, enzymatically catalyzed steps. Oxygen is used only at the last step. Since a small amount of the chemical-bond energy of glucose is released at early steps in the metabolic pathway, some tissue cells can obtain energy for ATP production in the temporary absence of oxygen. This process is described in the next two sections.

Glycolysis

The breakdown of glucose for energy involves a metabolic pathway known as **glycolysis.** This term is derived from the Greek *glykys* = sweet and *lysis* = a loosening, and it refers to the breakage of sugar. Glycolysis is the metabolic pathway by which glucose—a six-carbon (hexose) sugar—is converted into two molecules of pyruvic acid, or pyruvate. (See chapter 2 for a discussion of how these two terms are used for describing organic acids.) Even though each pyruvic acid molecule is roughly half the size of a glucose, glycolysis is *not* simply the breaking in half of glucose. Glycolysis is a metabolic pathway involving many enzymatically controlled steps.

Each pyruvic acid molecule contains three carbons, three oxygens, and four hydrogens. The number of carbon and oxygen atoms in one molecule of glucose—$C_6H_{12}O_6$—can thus be accounted for in the two pyruvic acid molecules. Since the two pyruvic acids together account for only eight hydrogens, however, it is clear that four hydrogen atoms are removed from the intermediates in glycolysis. Each pair of these hydrogen atoms is used to reduce a molecule of NAD. In this process, each pair of hydrogen atoms donates two electrons to NAD, thus reducing it. The reduced NAD binds one proton from the hydrogen atoms, leaving one proton unbound as H^+ (described in chapter 4). Starting from one glucose molecule, therefore, glycolysis results in the production of two molecules of NADH and two H^+. The H^+ will follow the NADH in subsequent reactions, so for simplicity we can refer to reduced NAD simply as NADH.

Glycolysis is exergonic, and a portion of the energy that is released is used to drive the endergonic reaction ADP + P_i → ATP. At the end of the glycolytic pathway there is a net gain of two ATP molecules per glucose molecule, as indicated in the overall equation for glycolysis:

$$Glucose + 2NAD + 2ADP + 2P_i \rightarrow$$

$$2 \text{ pyruvic acid} + 2NADH + 2ATP$$

Although the overall equation for glycolysis is exergonic, glucose must be "activated" at the beginning of the pathway before energy can be obtained. This activation requires the addition of two phosphate groups derived from two molecules of ATP. Energy from the reaction ATP → ADP + P_i is therefore consumed at the beginning of glycolysis. This is shown as an "up-staircase" in figure 5.1. Notice that the P_i is not shown in these reactions in figure 5.1; this is because the phosphate is not released but instead is added to the intermediate molecules of glycolysis. At later steps in glycolysis, four molecules of ATP are produced (and two molecules of NAD are reduced) as energy is liberated (the "down-staircase" in fig. 5.1). The two molecules of ATP used in the beginning, therefore, represent an energy investment; the net gain of two ATP and two NADH by the end of the pathway represents an energy profit.

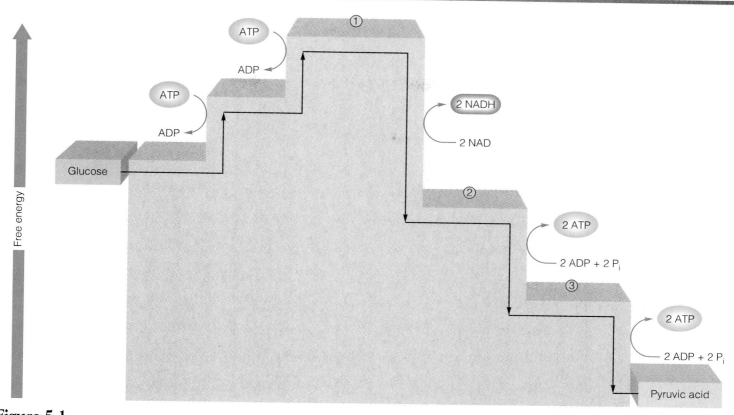

Figure 5.1

The energy expenditure and gain in glycolysis. Notice that there is a "net profit" of two ATP and two NADH molecules for every molecule of glucose that enters the glycolytic pathway. Molecules listed by number are (1) fructose 1,6-diphosphate, (2) 1,3-diphosphoglyceric acid, and (3) 3-phosphoglyceric acid (see fig. 5.2).

Table 5.1 Enzymes, Cofactors, and Coenzymes Required for Glycolysis

Step	Enzyme	Coenzyme or Cofactor	Comments
1	Hexokinase	Mg^{++}	Catalyzes the phosphorylation of glucose, fructose, or mannose
2	Hexose phosphate isomerase	—	Interconverts glucose and fructose
3	Phosphofructokinase	Mg^{++}, ATP	Allosteric enzyme that is inhibited by high ATP
4	Aldolase	—	Splits hexose sugar into two three-carbon compounds
5	Phosphoglyceraldehyde dehydrogenase	NAD	Adds inorganic phosphate and oxidizes aldehyde to acid as NAD is reduced by removal of two hydrogens
6	Phosphoglycerate kinase	Mg^{++}	Two molecules of ATP formed at this step
7	Phosphoglyceromutase	—	Phosphate group transferred to different carbon
8	Enolase	Mg^{++}, Mn^{++}	Catalyzes molecular rearrangement
9	Pyruvate kinase	Mg^{++}, K^+	Two molecules of ATP formed at this step

The overall equation for glycolysis obscures the fact that this is a metabolic pathway consisting of nine separate steps. The individual steps in this pathway are shown in figure 5.2, and the enzymes that catalyze these steps are listed in table 5.1.

Lactic Acid Pathway

In order for glycolysis to continue, there must be adequate amounts of NAD available to accept hydrogen atoms. Therefore, the NADH that is produced in glycolysis must become oxidized by donating its electrons to another molecule. (In aerobic respiration this other molecule is located in the mitochondria and ultimately passes its electrons to oxygen.)

When oxygen is *not* available in sufficient amounts, the NADH (+ H^+) produced in glycolysis is oxidized in the cytoplasm by donating its electrons to pyruvic acid. This results in the re-formation of NAD and the addition of two hydrogen atoms to pyruvic acid, which is thus reduced. This addition of two hydrogen atoms to pyruvic acid produces *lactic acid* (fig. 5.3).

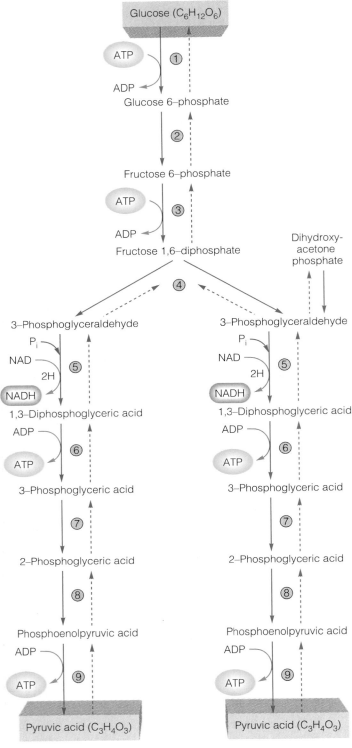

Figure 5.2

In glycolysis, one glucose molecule is converted into two pyruvic acid molecules in nine separate steps. In addition to two pyruvic acids, the products of glycolysis include two molecules of NADH and four molecules of ATP. Since two ATP molecules were used at the beginning, however, the net gain is two ATP molecules per glucose. Dashed arrows indicate reverse reactions that may occur under other conditions.

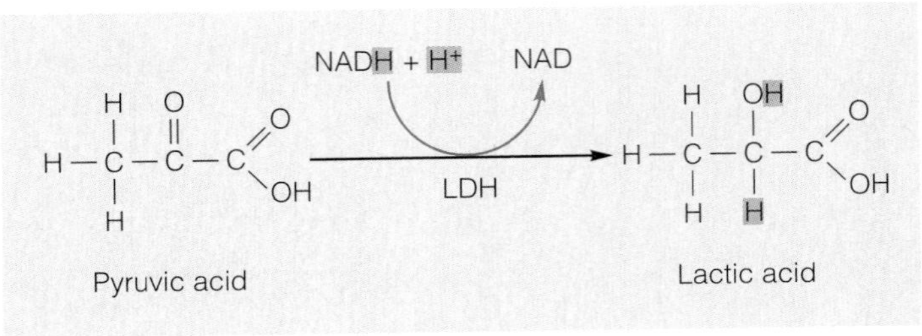

Figure 5.3

The addition of two hydrogen atoms (colored boxes) from reduced NAD to pyruvic acid produces lactic acid and oxidized NAD. This reaction is catalyzed by lactic acid dehydrogenase (LDH).

The metabolic pathway by which glucose is converted to lactic acid is frequently referred to by physiologists as **anaerobic respiration.** "Anaerobic" describes the fact that oxygen is not used in the process. However, many biologists prefer that "anaerobic respiration" be reserved for the pathways in certain bacteria that employ sulfur or iron as a final electron acceptor in place of oxygen. In this sense, "respiration" refers to the use of an inorganic atom as the final electron acceptor. If anaerobic respiration is used to describe this bacterial metabolism, then another term is needed for the production of lactic acid. In this case the term **lactic acid fermentation** may be used, since the metabolic pathway is analogous to that used by yeast cells to produce ethyl alcohol. In both the production of alcohol by yeast and of lactic acid by human cells, the electron acceptor is an organic molecule. In this text, anaerobic respiration and lactic acid fermentation will be used synonymously.

The lactic acid pathway yields a net gain of two ATP molecules (produced by glycolysis) per glucose molecule. A cell can survive without oxygen as long as it can produce sufficient energy for its needs in this way and as long as lactic acid concentrations do not become excessive. Some tissues are better adapted to anaerobic conditions than others—skeletal muscles survive longer than cardiac muscle, which in turn survives under anaerobic conditions longer than the brain.

Except for red blood cells, which can use only the lactic acid pathway (thus sparing the oxygen they carry), anaerobic respiration provides only a temporary sustenance for tissues that have energy requirements in excess of their aerobic ability. Anaerobic respiration can occur for only a limited period of time—longest for skeletal muscles, shorter for the heart, and shortest for the brain. It occurs when the *ratio of oxygen supply to oxygen need* (related to the concentration of NADH) falls below a critical level. Anaerobic respiration is, in a sense, an emergency procedure that provides some ATP until the emergency (oxygen deficiency) has passed.

It should be noted, though, that there is no real "emergency" in the case of skeletal muscles, where anaerobic respiration is a normal daily occurrence that does not harm

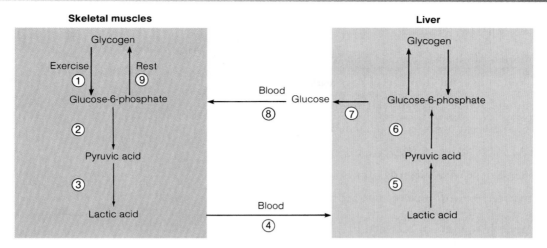

Skeletal muscles

Glycogen

Exercise ① ⑨ Rest

Glucose-6-phosphate

② Pyruvic acid

③ Lactic acid

Liver

Glycogen

Glucose-6-phosphate ⑥

Pyruvic acid ⑤

Lactic acid

Blood ⑧ Glucose ⑦

Blood ④

Figure 5.4

The Cori cycle. The sequence of steps is indicated by numbers 1 through 9.

the tissue or the individual. Lactic acid production by muscles, however, is associated with pain and muscle fatigue. (The metabolism of skeletal muscles will be discussed in chapter 12.) In contrast to skeletal muscles, the heart normally respires only aerobically. When anaerobic conditions do occur in the heart, a potentially dangerous situation may be present.

The Cori Cycle

Some of the lactic acid produced by exercising skeletal muscles is delivered by the blood to the liver. The enzyme lactic acid dehydrogenase within liver cells is then able to convert lactic acid to pyruvic acid. In the process, NAD is reduced to NADH + H⁺. Unlike most other organs, the liver contains the enzymes needed to take pyruvic acid molecules and convert them to glucose 6-phosphate, a process that is essentially the reverse of glycolysis. Glucose 6-phosphate in liver cells can then be used as an intermediate for glycogen synthesis, or it can be converted to free glucose that is secreted into the blood. The conversion of noncarbohydrate molecules (lactic acid, amino acids, and glycerol) through pyruvic acid into glucose is an extremely important process called **gluconeogenesis.** The significance of this process in exercise and starvation will be discussed later in this chapter.

During exercise, some of the lactic acid produced by skeletal muscles may be transformed through gluconeogenesis in the liver to blood glucose. This new glucose can serve as an energy source during exercise and can be used after exercise to help replenish the depleted muscle glycogen. This two-way traffic between skeletal muscles and the liver is called the **Cori cycle** (fig. 5.4). Through the Cori cycle, gluconeogenesis in the liver allows depleted skeletal muscle glycogen to be restored within 48 hours.

Ischemia refers to inadequate blood flow to an organ, such that the rate of oxygen delivery is insufficient to maintain aerobic respiration. Inadequate blood flow to the heart, or *myocardial ischemia,* may occur if the coronary blood flow is occluded by atherosclerosis, a blood clot, or by an artery spasm. People with myocardial ischemia often experience *angina pectoris*—severe pain in the chest and left (or sometimes, right) arm area. This pain is associated with increased blood levels of lactic acid, which are produced by the ischemic heart muscle. If the ischemia is maintained, the cells may die and produce an area called an infarct. The degree of ischemia and angina can be decreased by vasodilator drugs, such as nitroglycerin and amyl nitrite, which improve blood flow to the heart and also decrease the work of the heart by dilating peripheral blood vessels.

1. Define the term *glycolysis* in terms of its initial substrates and products. Explain why there is a net gain of two ATP in this process.

2. Discuss the two meanings of the term *anaerobic respiration*. As the term is used in this text, what are its initial substrates and final products?

3. Describe the physiological functions of anaerobic respiration. In which tissue(s) is anaerobic respiration normal? In which tissue is it abnormal?

4. Define the term *gluconeogenesis* and explain how this process replenishes the glycogen stores of skeletal muscles following exercise.

Aerobic Respiration

In the aerobic respiration of glucose, pyruvic acid is formed by glycolysis and then converted into acetyl coenzyme A. This begins a cyclic metabolic pathway called the Krebs cycle. As a result of these pathways, a large amount of reduced NAD and FAD (NADH and FADH$_2$) is generated. These reduced coenzymes provide electrons for an energy-generating process that drives the formation of ATP.

Aerobic respiration is equivalent to combustion in terms of its final products (CO_2 and H_2O) and in terms of the total amount of energy liberated. In aerobic respiration, however, the energy is released in small, enzymatically controlled oxidation reactions, and a portion (38% to 40%) of the energy released in this process is captured in the high-energy bonds of ATP.

The aerobic respiration of glucose begins with glycolysis. Glycolysis in both anaerobic and aerobic respiration results in the production of two molecules of pyruvic acid, two molecules of ATP, and two molecules of NADH + H$^+$ per glucose molecule. In aerobic respiration, however, the electrons in NADH are *not* donated to pyruvic acid and lactic acid is not formed, as happens in anaerobic respiration. Instead, the pyruvic acids will move to a different cellular location and undergo a different reaction; the NADH produced by glycolysis will eventually be oxidized, but that occurs later in the story.

In aerobic respiration, pyruvic acid leaves the cell cytoplasm and enters the interior (the matrix) of mitochondria. Once pyruvic acid is inside a mitochondrion, carbon dioxide is enzymatically removed from each three-carbon-long pyruvic acid to form a two-carbon-long organic acid—acetic acid. The enzyme that catalyzes this reaction combines the acetic acid with a coenzyme (derived from the vitamin pantothenic acid) called *coenzyme A*. The combination thus produced is called **acetyl coenzyme A,** abbreviated **acetyl CoA** (fig. 5.5).

Glycolysis converts one glucose molecule to two molecules of pyruvic acid. Since each pyruvic acid molecule is converted into one molecule of acetyl CoA and one CO_2, two molecules of acetyl CoA and two molecules of CO_2 are derived from each glucose. The acetyl CoA molecules serve as substrates for mitochondrial enzymes in the aerobic pathway; the carbon dioxide is a waste product in this process, and is carried by the blood to the lungs for elimination. It should be noted that the oxygen in CO_2 is derived from pyruvic acid, not from oxygen gas.

The Krebs Cycle

Once acetyl CoA has been formed, the acetic acid subunit (two carbons long) combines with oxaloacetic acid (four carbons long) to form a molecule of citric acid (six carbons long). Coenzyme A acts only as a transporter of acetic acid from one enzyme to another (similar to the transport of hydrogen by NAD). The formation of citric acid begins a cyclic metabolic pathway

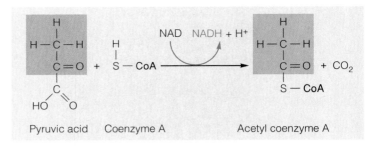

Figure 5.5

The formation of acetyl coenzyme A in aerobic respiration.

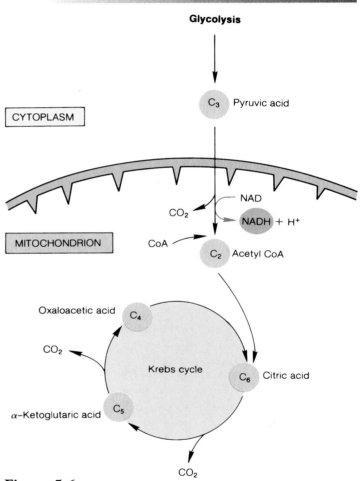

Figure 5.6

A simplified diagram of the Krebs cycle showing how the original four-carbon-long oxaloacetic acid is regenerated at the end of the cyclic pathway. Only the numbers of carbon atoms in the Krebs cycle intermediates are shown; the numbers of hydrogens and oxygens are not accounted for in this simplified scheme.

known as the **citric acid cycle,** or **TCA cycle** (for tricarboxylic acid; citric acid has three carboxylic acid groups). Most commonly, however, this cyclic pathway is called the **Krebs cycle,** after its principal discoverer, Sir Hans A. Krebs. A simplified illustration of this pathway is shown in figure 5.6.

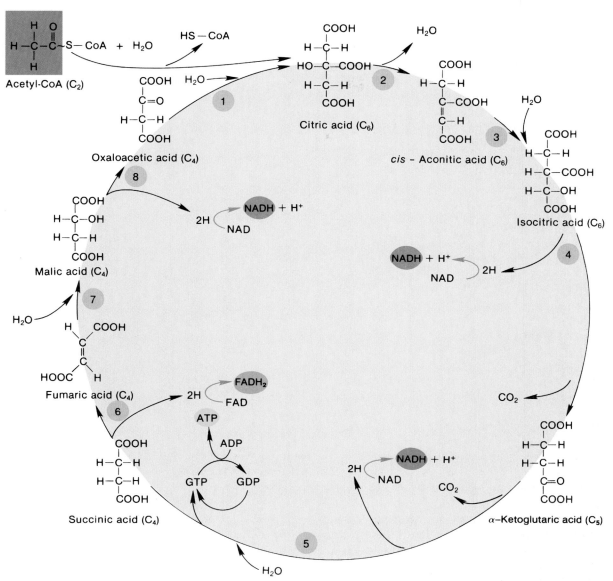

Figure 5.7

The Krebs cycle.

Through a series of reactions involving the elimination of two carbons and four oxygens (as two CO_2 molecules) and the removal of hydrogens, citric acid is eventually converted to oxaloacetic acid, which completes the cyclic metabolic pathway. In this process the following events occur: (1) one guanosine triphosphate (GTP) is produced (step 5 of fig. 5.7), which donates a phosphate group to ADP to produce one ATP; (2) three molecules of NAD are reduced (steps 4, 5, and 8 of fig. 5.7); and (3) one molecule of FAD is reduced (step 6).

The production of reduced NAD and FAD (that is, NADH and $FADH_2$) by each "turn" of the Krebs cycle is far more significant, in terms of energy production, than the single GTP (converted to ATP) produced directly by the cycle. This is because NADH and $FADH_2$ eventually donate their

electrons to an energy-transferring process that results in the formation of a large number of molecules of ATP.

Electron Transport and Oxidative Phosphorylation

Built into the foldings, or cristae, of the inner mitochondrial membrane are a series of molecules that serve in **electron transport** during aerobic respiration. This electron-transport chain of molecules consists of a flavoprotein (derived from the vitamin riboflavin), coenzyme Q (derived from vitamin E), and a group of iron-containing pigments called *cytochromes*. The last of these cytochromes is cytochrome a_3, which donates

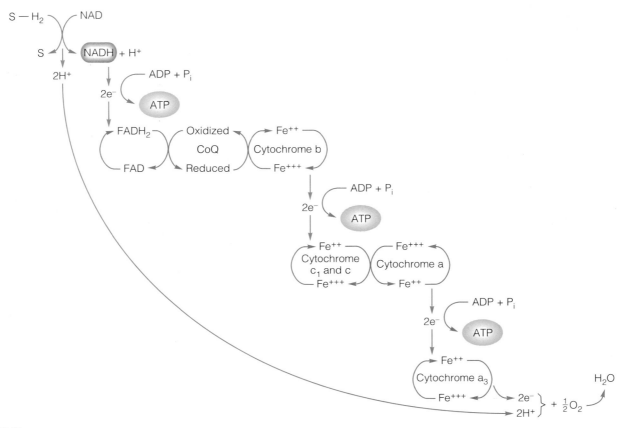

Figure 5.8

Electron transport and oxidative phosphorylation. Each element in the electron-transport chain alternately becomes reduced and oxidized as it transports electrons to the next member of the chain. This process provides energy for the formation of ATP. At the end of the electron-transport chain, the electrons are donated to oxygen, which becomes reduced (by the addition of two hydrogen atoms) to water.

electrons to oxygen in the final oxidation-reduction reaction (as will be described). These molecules of the electron-transport system are fixed in position within the inner mitochondrial membrane in such a way that they can pick up electrons from NADH and $FADH_2$ and transport them in a definite sequence and direction.

In aerobic respiration, NADH and $FADH_2$ become oxidized by transferring their pairs of electrons to the electron-transport system of the cristae. It should be noted that the protons (H^+) are not transported together with the electrons; their fate will be described a little later. The oxidized forms of NAD and FAD are thus regenerated and can continue to "shuttle" electrons from the Krebs cycle to the electron-transport chain. The first molecule of the electron-transport chain, in turn, becomes reduced when it accepts the electron pair from NADH. When the cytochromes receive a pair of electrons, two ferric ions (Fe^{+++}) become reduced to two ferrous ions (Fe^{++}). (Notice that the gain of an electron is indicated by the reduction of the number of positive charges.)

The electron-transport chain thus acts as an oxidizing agent for NAD and FAD. Each element in the chain, however, also functions as a reducing agent; one reduced

cytochrome transfers its electron pair to the next cytochrome in the chain (fig. 5.8). In this way, the iron ions in each cytochrome alternately become reduced (to ferrous ions) and oxidized (to ferric ions). This is an exergonic process, and the energy derived is used to phosphorylate ADP to ATP. The production of ATP in this manner is thus appropriately termed **oxidative phosphorylation.**

Coupling of Electron Transport to ATP Production

Under very high magnification with the electron microscope, lollipop-like structures can be seen protruding from the inner mitochondrial membrane into the matrix. Indeed, the cristae are covered with these structures (fig. 5.9). The "lollipops" are called *respiratory assemblies* and are critically important in the generation of ATP.

The **chemiosmotic theory** is the most widely accepted explanation of how the electron-transport system produces ATP. According to the chemiosmotic theory, three members of the electron-transport system become active **proton pumps** when they accept a pair of electrons. When these members become reduced, they move a pair of protons from the matrix to the space between the inner and outer mitochondrial membranes.

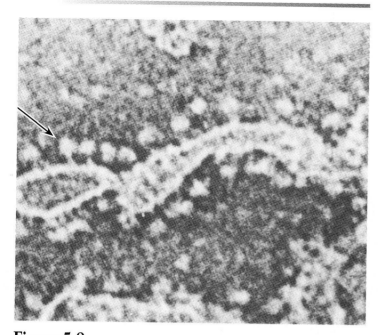

Figure 5.9

The ATP synthetase (*arrow*) at the end of the electron-transport system is attached to the cristae by a stalk. This unit produces ATP, using energy from electron transport (a process called *oxidative phosphorylation*).

Note that for every two transported electrons that originate from NADH, six protons will be pumped. Since $FADH_2$ donates electrons "further down the line" in the electron-transport system, electrons from $FADH_2$ activate only two of the three proton pumps. As a result of electron transport and activation of the proton pumps, the H^+ concentration in the space between the inner and outer membranes is much higher than in the matrix (fig. 5.10). A steep electrochemical gradient is thus established across the inner mitochondrial membrane.

The lollipop-like respiratory assemblies consist of a group of proteins that form a "stem" and a globular subunit. The stem contains a channel through the inner mitochondrial membrane that permits the passage of protons (H^+). The globular subunit, which protrudes into the matrix, contains an *ATP synthetase enzyme* that is capable of catalyzing the reaction $ADP + P_i \rightarrow ATP$.

The inner mitochondrial membrane is not permeable to H^+; however, the H^+ can diffuse through the channels provided by the stems of the respiratory assemblies. As each pair of protons moves into and through the bulbous portion of the respiratory assemblies, the ATP synthetase enzyme in this portion is activated, and a molecule of ATP is produced. Since each NADH results in the pumping of three pairs of protons by the electron-transport system, three ATP are produced. Since each $FADH_2$ results in the pumping of two pairs of protons, two ATP are produced as the protons diffuse back into the matrix through the respiratory assemblies.

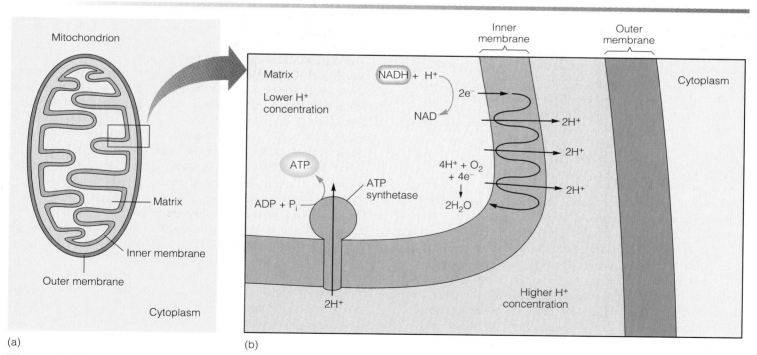

(a) (b)

Figure 5.10

A schematic representation of the chemiosmotic theory. (*a*) A mitochondrion. (*b*) The matrix and the compartment between the inner and outer mitochondrial membrane showing how the electron-transport system results in a steep H^+ gradient and the production of H_2O at the end of the electron-transport system. The enlarged view of ATP synthetase shows how the diffusion of H^+ results in the production of ATP.

Cell Respiration and Metabolism

Table 5.2 Maximum ATP Yield per Glucose in Aerobic Respiration*

Phase of Respiration	ATP Produced Directly	Reduced Coenzymes	ATP from Oxidative Phosphorylation
Glycolysis (glucose to pyruvic acid)	2 ATP (net gain)	2 NADH + H$^+$	6 ATP
Pyruvic acid to acetyl CoA	—	1 NADH + H$^+$	3 ATP
Pyruvic acid to acetyl CoA	—	1 NADH + H$^+$	3 ATP
Krebs cycle	1 ATP	3 NADH + H$^+$; 1 FADH$_2$	9 ATP; 2 ATP
Krebs cycle	1 ATP	3 NADH + H$^+$; 1 FADH$_2$	9 ATP; 2 ATP

*One glucose molecule yields two pyruvic acid molecules, and these result in two turns of the Krebs cycle.

Function of Oxygen

If the last cytochrome remained in a reduced state, it would be unable to accept more electrons. Electron transport would then progress only to the next-to-last cytochrome. This process would continue until all of the elements of the electron-transport chain remained in the reduced state. At this point, the electron-transport system would stop functioning and no ATP could be produced within the mitochondria. With the electron-transport system incapacitated, NADH and FADH$_2$ could not become oxidized by donating their electrons to the cytochrome chain, and through inhibition of Krebs cycle enzymes, no more NADH and FADH$_2$ could be produced in the mitochondria (the Krebs cycle would stop). Respiration would then become anaerobic.

Oxygen, from the air we breathe, allows electron transport to continue by functioning as the final electron acceptor of the electron-transport chain. This oxidizes cytochrome a$_3$ so that electron transport and oxidative phosphorylation can continue. At the very last step of aerobic respiration, therefore, oxygen becomes reduced by the two electrons that were passed to the chain from NADH and FADH$_2$. This reduced oxygen binds two protons, and a molecule of water is formed. Since the oxygen atom is part of a molecule of oxygen gas (O_2), this last reaction can be shown as follows:

$$O_2 + 4\,e^- + 4\,H^+ \rightarrow 2\,H_2O$$

ATP Balance Sheet

Each time the Krebs cycle turns, three molecules of NAD are reduced by electrons from three pairs of hydrogens removed from Krebs cycle intermediates. Each NADH donates a pair of electrons to the electron-transport chain. Transport of this pair of electrons to oxygen generates energy for the production of three molecules of ATP through oxidative phosphorylation. Electrons from FADH$_2$ enter the electron-transport chain "down the line" from where the first ATP is produced. Each pair of electrons from FADH$_2$, therefore, produces only two molecules of ATP from oxidative phosphorylation.

Since one NADH provides electrons for the production of three ATP, the three NADH produced per turn of the Krebs cycle result in the production of nine ATP molecules. The single FADH$_2$ per turn of the Krebs cycle results in the production of two ATP. Together with the single ATP made directly by the Krebs cycle, each turn of the Krebs cycle, therefore, yields a

Cyanide is a fast-acting lethal poison that produces symptoms of rapid heart rate, hypotension, coma, and ultimately death in the absence of quick treatment. The reason cyanide is so deadly is that it has one very specific action: it blocks the transfer of electrons from cytochrome a$_3$ to oxygen. The effects are thus the same as would occur if oxygen were completely removed—aerobic cell respiration and the production of ATP by oxidative phosphorylation comes to a halt.

total of twelve ATP molecules. Since one molecule of glucose produces two pyruvic acids, and thus two turns of the Krebs cycle, a total of twenty-four ATP molecules are produced by a single molecule of glucose, taking into account only the ATP made by the Krebs cycle and its NADH and FADH$_2$ products.

The conversion of pyruvic acid to acetyl CoA, however, also involves the reduction of one NAD (see fig. 5.5). Since two pyruvic acids are produced per glucose, two NADH are formed. And, since each NADH yields three ATP by oxidative phosphorylation, a total of twenty-four plus six, or thirty, ATP molecules are made in the mitochondrion from the steps that occur beyond pyruvic acid.

Now recall that two molecules of NADH are produced in the cytoplasm during glycolysis (conversion of glucose to pyruvic acid). These NADH cannot directly enter the mitochondria; instead, they donate their electrons to other molecules that "shuttle" these electrons into the mitochondria. Depending upon which shuttle is used, either two or three ATP can be produced from each pair of these cytoplasmic NADH electrons through oxidative phosphorylation. Thus, either four or six ATP molecules are produced. Added to the thirty ATP previously mentioned, this brings the total to thirty-four or thirty-six (depending upon which shuttle is used for the cytoplasmic electrons). Now, when we add the two molecules of ATP produced directly by glycolysis, we have a grand total of thirty-six to thirty-eight ATP produced by the aerobic respiration of glucose (table 5.2).

Glycogenesis and Glycogenolysis

Cells cannot accumulate very many separate glucose molecules, because these would exert an osmotic pressure (chapter 6) that would draw a dangerous amount of water into the cells. Instead,

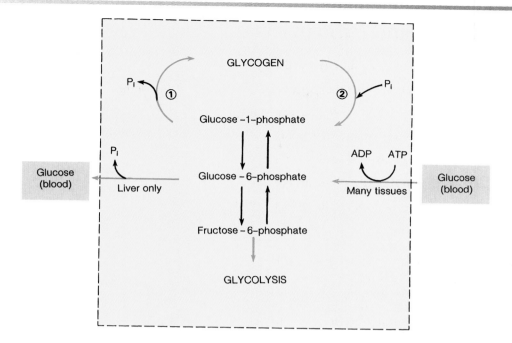

Figure 5.11

Blood glucose that enters tissue cells is rapidly converted to glucose–6–phosphate. This intermediate can be metabolized for energy in glycolysis, or it can be converted to glycogen (*1*) in a process called *glycogenesis*. Glycogen represents a storage form of carbohydrates, which can be used as a source for new glucose–6–phosphate (*2*) in a process called *glycogenolysis*. The liver contains an enzyme that can remove the phosphate from glucose–6–phosphate; liver glycogen thus serves as a source for new blood glucose.

many organs, particularly the liver, skeletal muscles, and heart, store carbohydrates in the form of glycogen.

The formation of glycogen from glucose is called **glycogenesis.** In this process, glucose is converted to glucose–6–phosphate by utilizing the terminal phosphate group of ATP. Glucose–6–phosphate is then converted into its isomer, glucose–1–phosphate. Finally, the enzyme *glycogen synthetase* removes these phosphate groups as it polymerizes glucose to form glycogen.

The reverse reactions are similar. The enzyme *glycogen phosphorylase* catalyzes the breakdown of glycogen to glucose–1–phosphate. (The phosphates are derived from inorganic phosphate, not from ATP, so glycogen breakdown does not require metabolic energy.) Glucose–1–phosphate is then converted to glucose–6–phosphate. The conversion of glycogen to glucose–6–phosphate is called **glycogenolysis.** In most tissues, glucose–6–phosphate can then be respired for energy (through glycolysis) or used to resynthesize glycogen. Only in the liver, for reasons that will now be explained, can the glucose–6–phosphate also be used to produce free glucose for secretion into the blood.

Organic molecules with phosphate groups cannot cross cell membranes. This has important consequences; the glucose derived from glycogen is in the form of glucose–1–phosphate and then glucose–6–phosphate, so it cannot leak out of the cell. Similarly, glucose that enters the cell from the blood is "trapped" within the cell by conversion to glucose–6–phosphate. Skeletal muscles, which have large amounts of glycogen, can generate glucose–6–phosphate for their own gly-

colytic needs but cannot secrete glucose into the blood because they lack the ability to remove the phosphate group.

Unlike skeletal muscles, the liver contains an enzyme—known as *glucose–6–phosphatase*—that can remove the phosphate groups and produce free glucose (fig. 5.11). This free glucose can then be transported through the cell membrane. The liver, then, can secrete glucose into the blood, whereas skeletal muscles cannot. Liver glycogen can thus supply blood glucose for use by other organs, including exercising skeletal muscles that may have depleted much of their own stored glycogen during exercise.

1. Compare the fate of pyruvic acid in aerobic respiration with its fate in anaerobic respiration.

2. Draw a simplified Krebs cycle using C_2 for acetic acid, C_4 for oxaloacetic acid, C_5 for alpha-ketoglutaric acid, and C_6 for citric acid. List the high-energy products that are produced at each turn of the Krebs cycle.

3. Using a diagram, show how electrons from NADH and $FADH_2$ are transferred by the cytochromes. Represent the oxidized and reduced forms of the cytochromes with Fe^{+++} and Fe^{++}, respectively.

4. Describe the pathways by which glucose and glycogen can be interconverted. Explain why only the liver can secrete glucose derived from its stored glycogen.

Metabolism of Lipids and Proteins

Triglycerides can be hydrolyzed into glycerol and fatty acids. The latter are of particular importance because they can be converted into numerous molecules of acetyl CoA that can enter Krebs cycles and generate a large amount of ATP. Amino acids derived from proteins may also be used for energy. This involves deamination (removal of the amine group) and the conversion of the remaining molecule into either pyruvic acid or one of the Krebs cycle molecules.

Energy can be derived by the cellular respiration of lipids and proteins, using the same aerobic pathway previously described for the metabolism of pyruvic acid. Indeed, some organs preferentially use molecules other than glucose as an energy source. Pyruvic acid and the Krebs cycle acids also serve as common intermediates in the interconversion of glucose, lipids, and amino acids.

When food intake by the body occurs at a faster rate than energy consumption, the cellular concentration of ATP rises. Cells, however, do not store extra energy in the form of extra ATP. When cellular ATP concentrations rise because more energy (from food) is available than can be immediately used, high ATP concentrations inhibit glycolysis. Under conditions of high cellular ATP concentrations, when glycolysis is inhibited, glucose is instead converted into glycogen and fat (fig. 5.12).

Lipid Metabolism

When glucose is going to be converted into fat, glycolysis occurs and pyruvic acid is converted into acetyl CoA. Some of the glycolytic intermediates—phosphoglyceraldehyde and dihydroxyacetone phosphate—do not complete their conversion to pyruvic acid, however, and acetyl CoA does not enter a Krebs cycle. The two-carbon acetic acid subunits of these acetyl CoA molecules can instead be used to produce a variety of lipids, including steroids such as cholesterol, ketone bodies and fatty acids (fig. 5.13). Acetyl CoA may thus be considered a branch point from which a number of different possible metabolic pathways may progress.

In the formation of fatty acids, a number of acetic acid (two-carbon) subunits are joined together to form the fatty acid chain. Six acetyl CoA molecules, for example, will produce a fatty acid that is twelve carbons long. When three of these fatty acids condense with one glycerol (derived from phosphoglyceraldehyde), a triglyceride molecule is formed. The formation of fat, or **lipogenesis,** occurs primarily in adipose tissue and in the liver when the concentration of blood glucose is elevated following a meal.

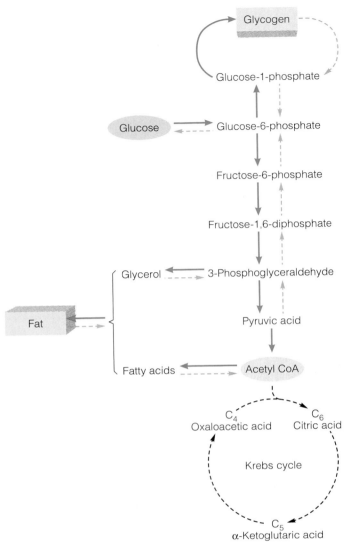

Figure 5.12

The conversion of glucose into glycogen and fat due to the allosteric inhibition of respiratory enzymes when the cell has adequate amounts of ATP. Favored pathways are indicated by blue arrows.

 It is common experience that the ingestion of excessive calories in the form of carbohydrates (cakes, ice cream, candy, and so on) increases fat production. The rise in blood glucose that follows carbohydrate-rich meals stimulates insulin secretion, and this hormone, in turn, promotes the entry of blood glucose into adipose cells. Increased availability of glucose within adipose cells, under conditions of high-insulin secretion, promotes the conversion of glucose to fat (fig. 5.12). The hormonal control of carbohydrate and fat metabolism is discussed in chapter 18.

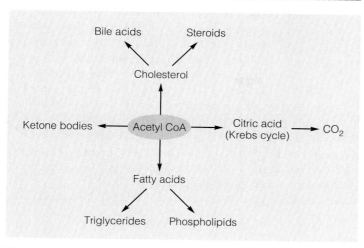

Figure 5.13

Divergent metabolic pathways for acetyl coenzyme A.

Fat represents the major form of energy storage in the body. One gram of fat contains 9 kilocalories of energy, compared to 4 kilocalories for a gram of carbohydrates or protein. In a nonobese 70-kilogram man, 80% to 85% of the body's energy is stored as fat, which amounts to about 140,000 kilocalories. Stored glycogen, by contrast, accounts for less than 2,000 kilocalories, most of which (about 350 g) is stored in skeletal muscles and is available only for use by the muscles. The liver contains from 80 to 90 grams of glycogen, which can be converted to glucose and used by other organs. Protein accounts for 15% to 20% of the stored calories in the body, but protein is usually not used extensively as an energy source because that would involve the loss of muscle mass.

Breakdown of Fat (Lipolysis)

When fat stored in adipose tissue is going to be used as an energy source, *lipase* enzymes hydrolyze triglycerides into glycerol and free fatty acids in a process called **lipolysis.** These molecules (primarily the free fatty acids) serve as *blood-borne energy carriers* that can be used by the liver, skeletal muscles, and other organs for aerobic respiration.

A few organs can utilize glycerol for energy by virtue of an enzyme that converts glycerol to phosphoglyceraldehyde. Free fatty acids, however, serve as the major energy source derived from triglycerides. Most fatty acids consist of a long hydrocarbon chain with a carboxylic acid group (COOH) at one end. In a process known as **β-oxidation** (β is the Greek letter *beta*), enzymes remove two-carbon acetic acid molecules from the acid end of a fatty acid chain. This results in the formation of acetyl CoA, as the third carbon from the end becomes oxidized to produce a new carboxylic acid group. The fatty acid chain is thus decreased in length by two carbons. The process of β-oxidation continues until the entire fatty acid molecule is converted to acetyl CoA (fig. 5.14).

A sixteen-carbon-long fatty acid, for example, yields eight acetyl CoA molecules. Each of these can enter a Krebs cycle

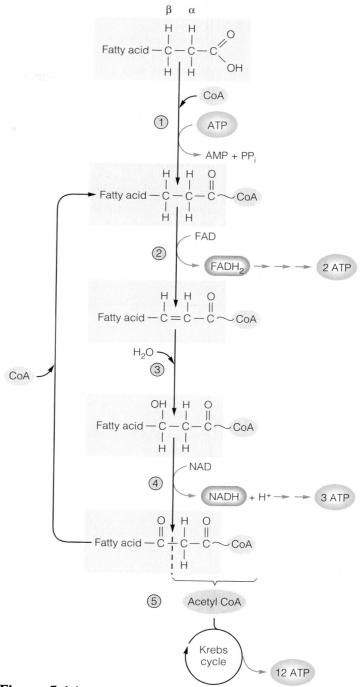

Figure 5.14

Beta-oxidation of a fatty acid. After the attachment of coenzyme A to the carboxylic acid group (*step 1*), a pair of hydrogens is removed from the fatty acid and used to reduce one molecule of FAD (*step 2*). When this electron pair is donated to the cytochrome chain, two ATP are produced. The addition of a hydroxyl group from water (*step 3*), followed by the oxidation of the β-carbon (*step 4*), results in the production of three ATP from the electron pair donated by NADH. The bond between the α and β carbons in the fatty acid is broken (*step 5*), releasing acetyl coenzyme A and a fatty acid chain that is two carbons shorter than the original. With the addition of a new coenzyme A to the shorter fatty acid, the process begins again (*step 2*), as acetyl CoA enters the Krebs cycle and generates twelve ATP.

The amount of **brown fat** in the body is greatest at the time of birth. Brown fat is the major site for thermiogenesis (heat production) in the newborn, and is especially prominent around the kidneys and adrenal glands. Smaller amounts are found around the blood vessels of the chest and neck. In response to regulation by thyroid hormone (chapter 11) and norepinephrine from sympathetic nerves (chapter 9), brown fat produces a unique *uncoupling protein*. This protein causes H^+ to leak out of the inner mitochondrial membrane, so that less H^+ is available to pass through the respiratory assemblies and drive ATP synthetase activity. Therefore, less ATP is made by the electron-transport system than would otherwise be the case. Lower ATP concentrations cause the electron-transport system to be more active and generate more heat from the respiration of fatty acids. Experimental evidence in other mammals suggests that this extra heat is needed to prevent hypothermia (low body temperature) in newborns.

Ketone bodies, which can be used for energy by many organs, are found in the blood under normal conditions. Under conditions of fasting or of diabetes mellitus, however, the increased liberation of free fatty acids from adipose tissue results in an elevated production of ketone bodies by the liver. The secretion of abnormally high amounts of ketone bodies into the blood produces **ketosis**, which is one of the signs of fasting or an uncontrolled diabetic state. A person in this condition may also have a sweet-smelling breath due to the presence of acetone, which is volatile and leaves the blood in the exhaled air.

and produce twelve ATP per turn of the cycle, so that eight times twelve, or ninety-six, ATP are produced. In addition, each time an acetyl CoA is formed and the end-carbon of the fatty acid chain is oxidized, one NADH and one $FADH_2$ are produced. Oxidative phosphorylation produces three ATP per NADH and two ATP per $FADH_2$. For a sixteen-carbon-long fatty acid, these five ATP molecules would be formed seven times (producing five times seven, or thirty-five, ATP). Not counting the single ATP used to start β-oxidation (fig. 5.14), this fatty acid could yield a grand total of 35 + 96, or 131, ATP molecules! Since one triglyceride molecule consists of three fatty acids and one glycerol, the aerobic respiration of one triglyceride molecule yields more than 400 ATP.

Ketone Bodies

Even when a person is not losing weight, there is a continuous turnover of triglycerides in adipose tissue. New triglycerides are produced, while others are hydrolyzed into glycerol and fatty acids. This turnover ensures that the blood will normally contain a sufficient level of fatty acids for aerobic respiration by skeletal muscles, the liver, and other organs. When the rate of lipolysis exceeds the rate of fatty acid utilization—as it may in starvation, dieting, and in diabetes mellitus—the blood concentrations of fatty acids increase.

If the liver cells contain sufficient amounts of ATP so that further production of ATP is not needed, some of the acetyl CoA derived from fatty acids is channeled into an alternate pathway. This pathway involves the conversion of two molecules of acetyl CoA into four-carbon-long acidic derivatives, *acetoacetic acid* and *β-hydroxybutyric acid*. Together with *acetone*, which is a three-carbon-long derivative of acetoacetic acid, these products are known as **ketone bodies.**

Amino Acid Metabolism

Nitrogen is ingested primarily as proteins, enters the body as amino acids, and is excreted mainly as urea in the urine. In childhood, the amount of nitrogen excreted may be less than the amount ingested because amino acids are incorporated into proteins during growth. Growing children are thus said to be in a state of *positive nitrogen balance*. People who are starving or suffering from prolonged wasting diseases, by contrast, are in a state of *negative nitrogen balance*; they excrete more nitrogen than they ingest because they are breaking down their tissue proteins.

Healthy adults maintain a state of nitrogen balance, in which the amount of nitrogen excreted is equal to the amount ingested. This does not imply that the amino acids ingested are unnecessary; on the contrary, they are needed to replace the protein that is "turned over" each day. When more amino acids are ingested than are needed to replace proteins, the excess amino acids are not stored as additional protein (one cannot build muscles simply by eating large amounts of protein). Rather, the amine groups can be removed, and the "carbon skeletons" of the organic acids that are left can be used for energy or converted to carbohydrate and fat.

Transamination

An adequate amount of all twenty amino acids is required to build proteins for growth and to replace the proteins that are turned over. Fortunately, only eight amino acids (in adults) or nine (in children) cannot be produced by the body and so must be obtained in the diet; these are the **essential amino acids** (table 5.3). The remaining amino acids are "nonessential" only in the sense that the body can produce them if provided with a sufficient amount of carbohydrates and the essential amino acids.

Pyruvic acid and the Krebs cycle acids are collectively termed *keto acids* because they have a ketone group; these should not be confused with the ketone bodies (derived from acetyl CoA) discussed in the previous section. Keto acids can be converted to amino acids by the addition of an amine (NH_2) group. This amine group is usually obtained by "cannibalizing" another amino acid; in this process, a new amino acid is formed as the one

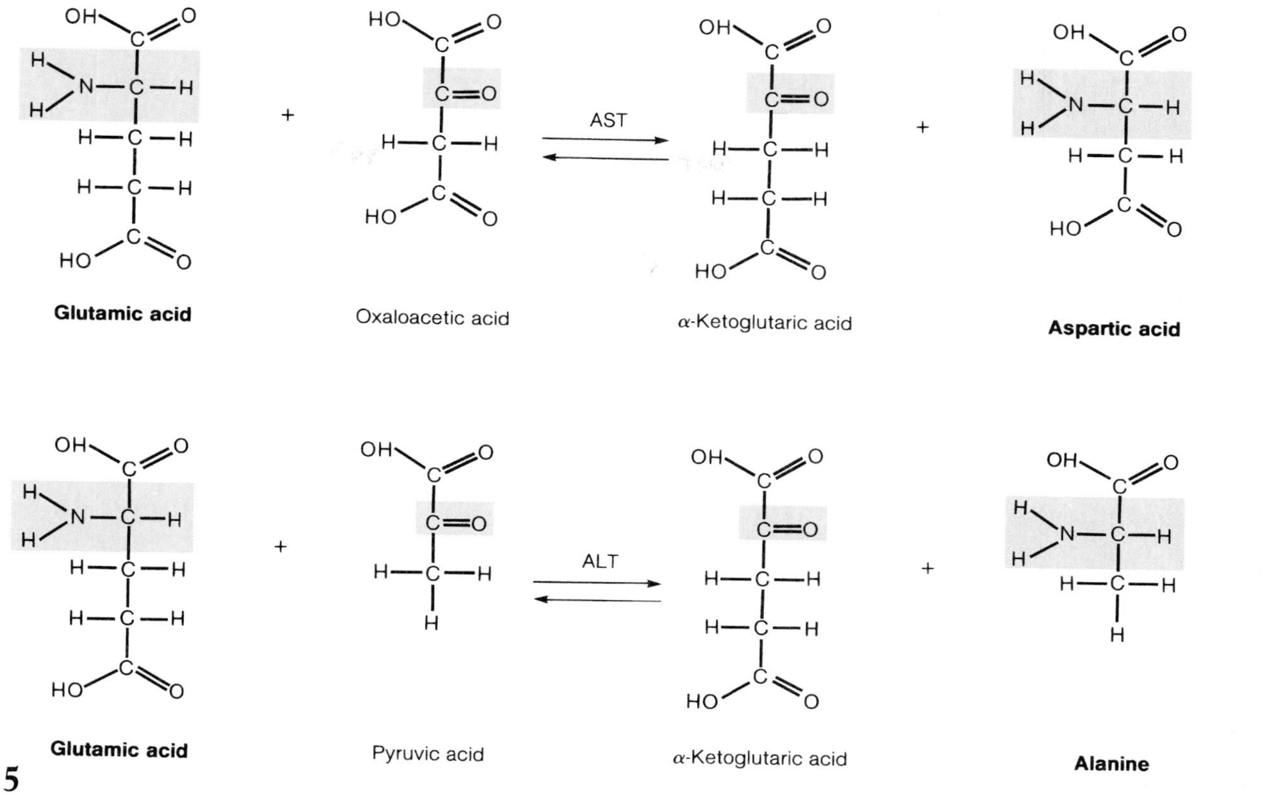

Figure 5.15

Two important transamination reactions. The areas shaded in blue indicate the parts of the molecules that are changed. (AST = aspartate transaminase; ALT = alanine transaminase. The amino acids are identified in boldface.)

Table 5.3 The Essential and Nonessential Amino Acids

Essential Amino Acids	Nonessential Amino Acids
Lysine	Aspartic acid
Tryptophan	Glutamic acid
Phenylalanine	Proline
Threonine	Glycine
Valine	Serine
Methionine	Alanine
Leucine	Cysteine
Isoleucine	Arginine
Histidine (children)	Asparagine
	Glutamine
	Tyrosine

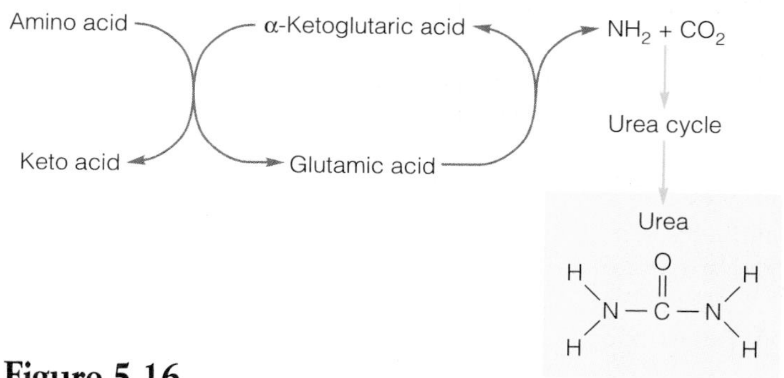

Figure 5.16

Oxidative deamination. Glutamic acid is converted to α-ketoglutaric acid as it donates its amine group to the metabolic pathway that results in the formation of urea.

that was cannibalized is converted to a new keto acid. This type of reaction, in which the amine group is transferred from one amino acid to form another, is called **transamination** (fig. 5.15).

Each transamination reaction is catalyzed by a specific enzyme (a transaminase) that requires vitamin B_6 (pyridoxine) as a coenzyme. The amine group from glutamic acid, for example, may be transferred to either pyruvic acid or oxaloacetic acid. The former reaction is catalyzed by the enzyme alanine transaminase (ALT); the latter reaction is catalyzed by aspartate transaminase (AST). The addition of an amine group to pyruvic acid produces the amino acid alanine; the addition of an amine group to oxaloacetic acid produces the amino acid known as aspartic acid (fig. 5.15).

Oxidative Deamination

As shown in figure 5.16, glutamic acid can be formed through transamination by the combination of an amine group with α-ketoglutaric acid. Glutamic acid is also produced in the liver

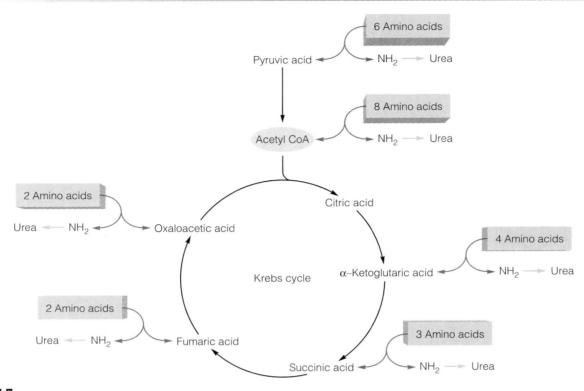

Figure 5.17

Pathways by which amino acids can be catabolized for energy. These pathways are indirect for some amino acids, which first must be transaminated into other amino acids before being converted into keto acids by deamination.

from the ammonia that is generated by intestinal bacteria and carried to the liver in the hepatic portal vein. Since free ammonia is very toxic, its removal from the blood and incorporation into glutamic acid is an important function of the healthy liver.

If there are more amino acids than are needed for protein synthesis, the amine group from glutamic acid may be removed and excreted as *urea* in the urine (fig. 5.16). The metabolic pathway that removes amine groups from amino acids—leaving a keto acid and ammonia (which is converted to urea)—is known as **oxidative deamination.**

A number of amino acids can be converted into glutamic acid by transamination. Since glutamic acid can donate amine groups to urea (through deamination), it serves as a channel through which other amino acids can be used to produce keto acids (pyruvic acid and Krebs cycle acids). These keto acids may then be used in the Krebs cycle as a source of energy (fig. 5.17).

Depending upon which amino acid is deaminated, the keto acid left over may be either pyruvic acid or one of the Krebs cycle acids. These can be respired for energy, converted to fat, or converted to glucose. In the last case, the amino acids are eventually changed to pyruvic acid, which is used to form glucose. This process—the formation of glucose from

amino acids or other noncarbohydrate molecules—is called gluconeogenesis, as mentioned previously in connection with the Cori cycle.

The main substrates for gluconeogenesis are the three-carbon long molecules of alanine (an amino acid), lactic acid, and glycerol. This illustrates the interrelationship between amino acids, carbohydrates, and fat, as shown in figure 5.18. Recent experiments in humans have suggested that, even in the early stages of fasting, most of the glucose secreted by the liver is derived through gluconeogenesis. Findings indicate that hydrolysis of liver glycogen (glycogenolysis) contributes only 36% of the glucose secreted during the early stages of a fast. By 42 hours of fasting, all of the glucose secreted by the liver is being produced by gluconeogenesis.

Uses of Different Energy Sources

The blood serves as a common trough from which all the cells in the body are fed. If all cells used the same energy source, such as glucose, this source would quickly be depleted and cellular starvation would occur. However, normally there are a variety of energy sources in the blood from which to draw: glucose and ketone bodies from the liver, fatty acids from adipose tissue, and lactic acid and amino acids from muscles. Some organs preferentially

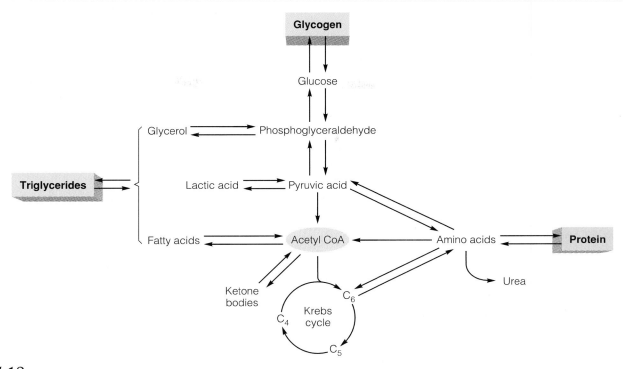

Figure 5.18

Simplified metabolic pathways showing how glycogen, fat, and protein can be interconverted.

Table 5.4 Relative Importance of Different Molecules in the Blood with Respect to the Energy Requirements of Different Organs

Organ	Glucose	Fatty Acids	Ketone Bodies	Lactic Acid
Brain	+++	+	−	−
Skeletal muscles (resting)	+	+++	+	−
Liver	+	+++	++	+
Heart	+	+	+	+++

use one energy source more than the others, so that each energy source is "spared" for organs with strict energy needs.

The brain uses blood glucose as its major energy source. Under fasting conditions, blood glucose is supplied primarily by the liver through glycogenolysis and gluconeogenesis. In addition, the blood glucose concentration is maintained because many organs spare glucose by using fatty acids, ketone bodies, and lactic acid as energy sources (table 5.4). During severe starvation, the brain also gains some ability to metabolize ketone bodies for energy.

Lactic acid produced anaerobically during exercise can be used for energy following exercise. The lactic acid, under aerobic conditions, is reconverted to pyruvic acid, which then enters the aerobic respiratory pathway. The extra oxygen required to metabolize lactic acid contributes to the *oxygen debt* following exercise (chapter 12).

1. Construct a flowchart to show the metabolic pathway by which glucose can be converted to fat. Indicate only the major intermediates involved (not all of the steps of glycolysis).

2. Define the terms *lipolysis* and β-*oxidation* and explain, in general terms, how fat can be used for energy.

3. Describe transamination and deamination and explain their functional significance.

4. List five blood-borne energy carriers and explain, in general terms, how these are used as sources of energy.

Summary

Glycolysis and the Lactic Acid Pathway   p. 102

I. Glycolysis refers to the conversion of glucose to two molecules of pyruvic acid.
   A. In the process, two molecules of ATP are consumed and four molecules of ATP are formed; there is thus a net gain of two ATP.
   B. In the steps of glycolysis, two pairs of hydrogens are released; electrons from these reduce two molecules of NAD.

II. When respiration is anaerobic, reduced NAD is oxidized by pyruvic acid, which accepts two hydrogen atoms and is thereby reduced to lactic acid.
   A. Skeletal muscles use anaerobic respiration and thus produce lactic acid during exercise; heart muscle respires anaerobically only for a short time, under conditions of ischemia.
   B. Lactic acid can be converted to glucose in the liver by a process called gluconeogenesis.

Aerobic Respiration   p. 106

I. The Krebs cycle begins when coenzyme A donates acetic acid to an enzyme that adds it to oxaloacetic acid to form citric acid.
   A. Acetyl CoA is formed from pyruvic acid by the removal of carbon dioxide and two hydrogens.
   B. The formation of citric acid begins a cyclic pathway that ultimately forms a new molecule of oxaloacetic acid.
   C. As the Krebs cycle progresses, one ATP is formed, and three molecules of NAD and one of FAD are reduced by hydrogens from the Krebs cycle.

II. Reduced NAD and FAD donate their electrons to an electron-transport chain of molecules located in the cristae.
   A. The electrons from NAD and FAD are passed from one cytochrome of the electron-transport chain to the next in a series of coupled oxidation-reduction reactions.
   B. As each cytochrome ion gains an electron, it becomes reduced; as it passes the electron to the next cytochrome, it becomes oxidized.
   C. The last cytochrome becomes oxidized by donating its electron to oxygen, which functions as the final electron acceptor.
   D. When one oxygen atom accepts two electrons and two protons, it becomes reduced to form water.
   E. The energy provided by electron transport is used to form ATP from ADP and P_i, which is a process called oxidative phosphorylation.

III. Thirty-six to thirty-eight ATP are produced by the aerobic respiration of one glucose molecule; of these, two are produced in the cytoplasm by glycolysis and the remainder are produced in the mitochondria.

IV. The formation of glycogen from glucose is called glycogenesis; the breakdown of glycogen is called glycogenolysis.
   A. Glycogenolysis yields glucose–6–phosphate, which can enter the pathway of glycolysis.
   B. The liver contains an enzyme (which skeletal muscles do not) that can produce free glucose from glucose–6–phosphate; the liver can thus secrete glucose derived from glycogen.

V. Carbohydrate metabolism is influenced by the availability of oxygen and by a negative feedback effect of ATP on glycolysis and the Krebs cycle.

Metabolism of Lipids and Proteins   p. 112

I. In lipolysis, triglycerides yield glycerol and fatty acids.
   A. Glycerol can be converted to phosphoglyceraldehyde and used for energy.
   B. In the process of β-oxidation of fatty acids, a number of acetyl CoA molecules are produced.
   C. Processes that operate in the reverse direction can convert glucose to triglycerides.

II. Amino acids derived from the hydrolysis of proteins can serve as sources of energy.
   A. Through transamination, a particular amino acid and a particular keto acid (pyruvic acid or one of the Krebs cycle acids) can serve as substrates to form a new amino acid and a new keto acid.
   B. In oxidative deamination, amino acids are converted into keto acids as their amino group is incorporated into urea.

III. Each organ uses certain blood-borne energy carriers as its preferred energy source.
   A. The brain has an almost absolute requirement for blood glucose as its energy source.
   B. During exercise, the needs of skeletal muscles for blood glucose can be met by glycogenolysis and by gluconeogenesis in the liver.

Clinical Investigation

A student is in training so that she can compete on the swim team. In the early stages of her training, she experienced great fatigue following a workout, and she found herself gasping and panting for air more than her teammates did. Her coach suggested that she eat less protein and fat and more carbohydrates than was her habit, and that she train more gradually. She also complained of a chronic pain in her arms and shoulders that began with her training. Following a particularly intense workout, she experienced severe pain in her left pectoral region and sought medical aid. What might be responsible for this student's symptoms?

Clue:

Study the sections "Lactic Acid Pathway," "Glycogenesis and Glycogenolysis," and "Uses of Different Energy Sources." Read the boxed information on ischemia.

Review Activities

Objective Questions

1. The net gain of ATP per glucose molecule in anaerobic respiration (lactic acid fermentation) is _____; the net gain in aerobic respiration is _____.
   a. 2;4
   b. 2;38
   c. 38;2
   d. 24;30

2. In anaerobic respiration in humans, the oxidizing agent for NADH (that is, the molecule that removes electrons from NADH) is
   a. pyruvic acid.
   b. lactic acid.
   c. citric acid.
   d. oxygen.

3. When skeletal muscles lack sufficient oxygen, there is an increased blood concentration of
   a. pyruvic acid.
   b. glucose.
   c. lactic acid.
   d. ATP.

4. The conversion of lactic acid to pyruvic acid occurs
   a. in anaerobic respiration.
   b. in the heart, where lactic acid is aerobically respired.
   c. in the liver, where lactic acid can be converted to glucose.
   d. in both a and b.
   e. in both b and c.

5. Which of the following statements about the oxygen in the air we breathe is *true*?
   a. It functions as the final electron acceptor of the electron-transport chain.
   b. It combines with hydrogen to form water.
   c. It combines with carbon to form CO_2.
   d. Both a and b are true.
   e. Both a and c are true.

6. In terms of the number of ATP molecules directly produced, the major energy-yielding process in the cell is
   a. glycolysis.
   b. the Krebs cycle.
   c. oxidative phosphorylation.
   d. gluconeogenesis.

7. Ketone bodies are derived from
   a. fatty acids.
   b. glycerol.
   c. glucose.
   d. amino acids.

8. The conversion of glycogen to glucose–6–phosphate occurs in
   a. the liver.
   b. skeletal muscles.
   c. both a and b.

9. The conversion of glucose–6–phosphate to free glucose, which can be secreted into the blood, occurs in
   a. the liver.
   b. skeletal muscles.
   c. both a and b.

10. The formation of glucose from pyruvic acid derived from lactic acid, amino acids, or glycerol is called
    a. glycogenesis.
    b. glycogenolysis.
    c. glycolysis.
    d. gluconeogenesis.

11. Which of the following organs has an almost absolute requirement for blood glucose as its energy source?
    a. liver
    b. brain
    c. skeletal muscles
    d. heart

12. When amino acids are used as an energy source,
    a. oxidative deamination occurs.
    b. pyruvic acid or one of the Krebs cycle acids (keto acids) is formed.
    c. urea is produced.
    d. all of the above occur.

13. Intermediates formed during fatty acid metabolism can enter the Krebs cycle as
    a. keto acids.
    b. acetyl CoA.
    c. Krebs cycle molecules.
    d. pyruvic acid.

Essay Questions

1. State the advantages and disadvantages of anaerobic respiration.[1]

2. What purpose is served by the formation of lactic acid during anaerobic respiration? How is this accomplished during aerobic respiration?

3. The poison cyanide blocks the transfer of electrons from the last cytochrome to oxygen. Describe the effect of this poison on oxidative phosphorylation and on the Krebs cycle. Why is cyanide deadly?

4. Describe the metabolic pathway by which glucose can be converted into fat and explain how end-product inhibition by ATP can favor this pathway.

5. Describe the metabolic pathway by which fat can be used as a source of energy and explain why the metabolism of fatty acids can yield more ATP than the metabolism of glucose.

6. Explain how energy is obtained from the metabolism of amino acids. Why does a starving person have high concentrations of urea in the blood?

7. Explain why the liver is the only organ able to secrete glucose into the blood. What are the possible sources of hepatic glucose?

8. Explain the two possible meanings of the term *anaerobic respiration*. Why is the production of lactic acid sometimes termed a "fermentation" pathway?

9. Explain the function of brown fat. What does its mechanism imply about the effect of ATP concentrations on the rate of cell respiration?

10. State the three major molecules used as substrates for gluconeogenesis and describe the situations in which each one would be involved in this process. Why can't fatty acids be used as a substrate for gluconeogenesis? (*Hint*: count the carbons in acetyl CoA and pyruvic acid.)

Selected Readings

Alberts, B. et al. 1989. *Molecular Biology of the Cell*. 2d ed. New York: Garland Publishing.

Baker, J. J. W., and G. E. Allen. 1982. *Matter, Energy, and Life*. 4th ed. Reading, MA: Addison-Wesley.

Berry, H. K. March 1984. The spectrum of metabolic disorders. *Diagnostic Medicine*, p. 39.

Devlin, T. M. 1992. *Textbook of Biochemistry with Clinical Correlations*. New York: Wiley.

Dyson, R. D. 1978. *Essentials of Cell Biology*. 2d ed. Boston: Allyn & Bacon.

Hinkle, P., and R. E. McCarty. March 1979. How cells make ATP. *Scientific American*.

Lehninger, A. L. September 1961. How cells transform energy. *Scientific American*.

Lehninger, A. L. 1982. *Principles of Biochemistry*. New York: Worth.

Nadel, E. R. 1985. Physiological adaptations to aerobic training. *American Scientist* 73:334.

Sheeler, P., and D. E. Bianchi. 1980. *Cell Biology: Structure, Biochemistry, and Function*. New York: Wiley.

Stryker, L., 1981. *Biochemistry*. 2d ed. New York: Freeman.

Explorations CD-ROM

The modules accompanying chapter 5 are #1 Cystic Fibrosis and #2 Active Transport.

Life Science Animations

The animations that relate to chapter 5 are #5 Glycolysis, #6 Oxidative Respiration, and #7 The Electron-Transport Chain and the Production of ATP.

---

[1]*Note:* This question is answered on page 52 of the Student Study Guide.

6

Membrane Transport and the Membrane Potential

OBJECTIVES   *After studying this chapter, you should be able to . . .*

1. describe diffusion and explain its physical basis.

2. explain how nonpolar molecules, inorganic ions, and water can diffuse through a cell membrane.

3. state the factors that influence the rate of diffusion through cell membranes.

4. define the term *osmosis* and describe the conditions required for osmosis to occur.

5. define the terms *osmolality* and *osmotic pressure* and explain how these factors relate to osmosis.

6. define the terms *isotonic*, *hypertonic*, and *hypotonic* and explain their physiological significance.

7. describe the characteristics of carrier-mediated transport.

8. describe the facilitated diffusion of glucose through cell membranes and give examples of its occurrence in the body.

9. define the term *active transport* and explain how active transport may be accomplished.

10. describe the active transport of Na^+ and K^+ by the Na^+/K^+ pumps and explain the physiological significance of this activity.

11. explain how an equilibrium potential is produced when only one ion is able to diffuse through a cell membrane.

12. explain why the resting membrane potential is slightly different than the potassium equilibrium potential and describe the effect of the extracellular potassium concentration on the resting membrane potential.

13. discuss the role of the Na^+/K^+ pumps in the maintenance of the resting membrane potential.

OUTLINE

Diffusion and Osmosis

Net diffusion of a molecule or ion through a cell membrane always occurs in the direction of its lower concentration. Nonpolar molecules can penetrate the phospholipid barrier, and small inorganic ions can pass through channels in the membrane. The net diffusion of water through a membrane is known as osmosis.

The cell (plasma) membrane separates the intracellular environment from the extracellular environment. Proteins, nucleotides, and other molecules needed for the structure and function of the cell cannot penetrate, or "permeate," the membrane. The cell membrane is, however, selectively permeable to certain molecules and many ions; this allows two-way traffic in nutrients and wastes needed to sustain metabolism and provides electrical currents created by the movements of ions through the membrane.

The mechanisms involved in the transport of molecules and ions through the cell membrane may be divided into two categories: (1) transport that requires the action of specific *carrier proteins* in the membrane (*carrier-mediated transport*) and (2) transport through the membrane that is not carrier-mediated. Carrier-mediated transport may further be subdivided into *facilitated diffusion* and *active transport*, both of which will be described later. Membrane transport that does not use carrier proteins involves the *simple diffusion* of ions, lipid-soluble molecules, and water through the membrane. The net diffusion of solvent (water) through a membrane is called *osmosis*.

Membrane transport processes may also be categorized on the basis of their energy requirements. Transport in which the net movement is from higher to lower concentration (down a concentration gradient) does not require metabolic energy; this is **passive transport.** Passive transport includes simple diffusion, osmosis, and facilitated diffusion. Transport that occurs against a concentration gradient (through a membrane to the region of higher concentration) is **active transport.** Active transport requires the expenditure of metabolic energy (ATP) and involves specific carrier proteins.

Diffusion

Molecules in a gas and molecules and ions dissolved in a solution are in a constant state of random motion as a result of their thermal (heat) energy. This random motion, called **diffusion,** tends to scatter the molecules evenly, or diffusely, within a given volume. Whenever a *concentration difference,* or *concentration gradient,* exists between two parts of a solution, therefore, random molecular motion tends to eliminate the gradient and to distribute the molecules uniformly (fig. 6.1). In terms of the second law of thermodynamics, the concentration difference represents an unstable state of high organization (low entropy), which changes to produce a uniformly distributed solution with maximum disorganization (high entropy).

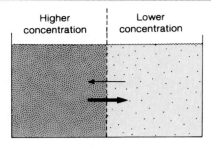

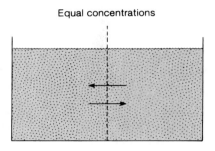

(a)

(b)

Figure 6.1

(*a*) Net diffusion occurs when there is a concentration difference (or concentration gradient) between two regions of a solution, provided that the membrane separating these regions is permeable to the diffusing substance. (*b*) Diffusion tends to equalize the concentrations of these solutions, and thus to eliminate the concentration differences.

Normally functioning kidneys remove waste products from the blood. After the blood is filtered through pores in capillary walls—which are large enough to permit the passage of wastes and other molecules—the molecules needed by the body are reabsorbed back into the blood. The wastes generally remain in the filtrate and are excreted in the urine. When the kidneys fail to perform this function, waste molecules must be removed from the blood artificially by the process of **dialysis.** In this process, particular molecules are removed from a solution by having them pass, by means of diffusion, through an artificial porous membrane. Since the pores in this dialysis membrane are large enough to permit the passage of some molecules but too small to permit the passage of others (the plasma proteins), small waste molecules can be removed from the blood by this technique.

As a result of random molecular motion, molecules in the part of the solution with a higher concentration will enter the area of lower concentration. Molecules will also move in the opposite direction, but not as frequently. As a result, there will be a *net movement* from the region of higher to the region of lower

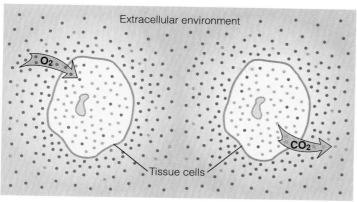

Figure 6.2

Gas exchange between the intracellular and extracellular compartments occurs by diffusion. The colored dots, which represent oxygen and carbon dioxide molecules, indicate relative concentrations inside the cell and in the extracellular environment.

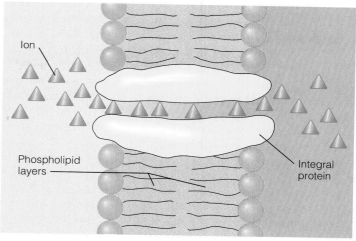

Figure 6.3

Inorganic ions (such as Na^+ and K^+) may be able to penetrate the membrane through pores within integral proteins that span the thickness of the double phospholipid layers.

concentration until the concentration difference no longer exists. This net movement is called **net diffusion.** Net diffusion is a physical process that occurs whenever there is a concentration difference; when the concentration difference exists across a membrane, diffusion becomes a type of membrane transport.

Diffusion Through the Cell Membrane

Since the cell membrane consists primarily of a double layer of phospholipids, molecules that are nonpolar, and thus lipid-soluble can easily pass from one side of the membrane to the other. The cell membrane, in other words, does not present a barrier to the diffusion of nonpolar molecules such as oxygen gas (O_2) or steroid hormones. Small molecules that have polar covalent bonds but which are uncharged, such as CO_2 (as well as ethanol and urea), are also able to penetrate the phospholipid bilayer. Net diffusion of these molecules can thus easily occur between the intracellular and extracellular compartments when concentration gradients exist.

The oxygen concentration is relatively high, for example, in the extracellular fluid because oxygen is carried from the lungs to the body tissues by the blood. Since oxygen is converted to water in aerobic cell respiration, the oxygen concentration within the cells is lower than in the extracellular fluid. The concentration gradient for carbon dioxide is in the opposite direction because cells produce CO_2. *Gas exchange* thus occurs by diffusion between the tissue cells and their extracellular environments (fig. 6.2).

Although water is not lipid-soluble, water molecules can diffuse through the cell membrane because of their small size and lack of net charge. In certain membranes, however, the passage of water is restricted to specific channels that can open or close in response to physiological regulation. The net diffusion of water molecules across the membrane is known as osmosis. Since osmosis is the simple diffusion of solvent instead of solute, a unique terminology (discussed in a later section) is used to describe it.

Cystic fibrosis occurs about once in every 2,500 births in the Caucasian population. As a result of a genetic defect, there is abnormal NaCl and water movement across wet epithelial membranes. Where such membranes line the pancreatic ductules and small airways, their secretions become more viscous and cannot be properly cleared, leading to pancreatic and pulmonary disorders. The genetic defect involves a particular glycoprotein that forms chloride (Cl⁻) channels in the apical membrane of the epithelial cells. This protein, known as *CFTR* (for cystic fibrosis transmembrane conductance regulator), is formed as normal in the endoplasmic reticulum but doesn't move into the Golgi apparatus for processing. It therefore doesn't get correctly processed and inserted into vesicles that would introduce it into the cell membrane (see chapter 3). The gene for CFTR has been identified and cloned. Recently, in a preliminary test of gene therapy for cystic fibrosis, the gene for CFTR was inserted into an adenovirus (cold virus) and introduced into the nasal passages of human volunteers. More research is required, however, before gene therapy for this disease becomes a real possibility.

Larger polar molecules, such as glucose, cannot pass through the double layer of phospholipid molecules and thus require special *carrier proteins* in the membrane for transport (described later). The phospholipid portion of the membrane is similarly impermeable to charged inorganic ions, such as Na^+ and K^+. Passage of these ions through the cell membrane may be permitted by tiny **ion channels** through the membrane that are too small to be seen even with an electron microscope. These channels are provided by some of the *integral proteins* that span the thickness of the membrane (fig. 6.3).

Rate of Diffusion

The rate of diffusion, measured by the number of diffusing molecules passing through the membrane per unit time, depends on (1) the magnitude of the concentration difference across the membrane (the "steepness" of the concentration gradient), (2) the permeability of the membrane to the diffusing substances, and (3) the surface area of the membrane through which the substances are diffusing.

The magnitude of the concentration difference across the membrane serves as the driving force for diffusion. Regardless of this concentration difference, however, the diffusion of a substance across a membrane will not occur if the membrane is not permeable to that substance. With a given concentration difference, the rate of diffusion through a membrane will vary directly with the degree of permeability. In a resting neuron, for example, the membrane is about twenty times more permeable to potassium (K^+) than to sodium (Na^+); consequently, K^+ diffuses much more rapidly than does Na^+. Changes in the protein structure of the membrane channels, however, can change the permeability of the membrane. This occurs during the production of a nerve impulse (chapter 7), when specific stimulation opens Na^+ channels temporarily and allows a faster diffusion rate for Na^+ than for K^+.

In areas of the body that are specialized for rapid diffusion, the surface area of the cell membranes may be increased by numerous folds. The rapid passage of the products of digestion across the epithelial membranes in the small intestine, for example, is aided by such structural adaptations. The surface area of the apical membranes (the part facing the lumen) in the small intestine is increased by many tiny folds that form finger-like projections called *microvilli* (discussed in chapter 3). Similar microvilli are also found in the kidney tubule epithelium, which must reabsorb various molecules that are filtered out of the blood.

Osmosis

Osmosis is the net diffusion of water (the solvent) across the membrane. In order for osmosis to occur, the membrane must be *semipermeable*; that is, it must be more permeable to water molecules than to solutes. Like the diffusion of solute molecules, the diffusion of water occurs when the water is more concentrated on one side of the membrane than on the other side; that is, when one solution is more dilute than the other (fig. 6.4). The more dilute solution has a higher concentration of water molecules and a lower concentration of solute. Although the terminology connected with osmosis can be awkward (because we are describing water instead of solute), the principles of osmosis are the same as those governing the diffusion of solute molecules through a membrane.

Imagine a cylinder is divided into two equal compartments by an artificial membrane partition that can freely move. One compartment initially contains 180 g/L (grams per liter) of glucose and the other compartment contains 360 g/L of glucose. If the membrane is permeable to glucose, glucose will diffuse from the 360-g/L compartment to the 180-g/L compartment

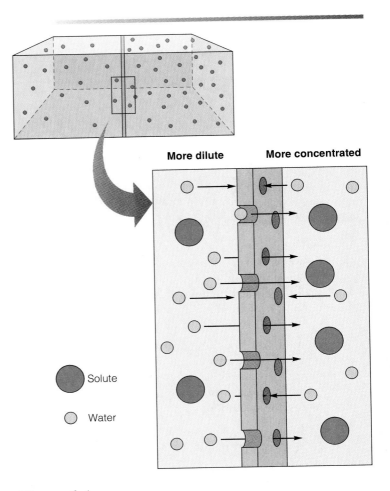

More dilute   **More concentrated**

Solute

Water

Figure 6.4

A model of osmosis—the net movement of water from the solution of lesser solute concentration to the solution of greater solute concentration.

until both compartments contain 270 g/L of glucose. If the membrane is not permeable to glucose but is permeable to water, the same result (270-g/L solutions on both sides of the membrane) will be achieved by the diffusion of water. As water diffuses from the 180-g/L compartment to the 360-g/L compartment, the former solution becomes more concentrated while the latter becomes more dilute. This is accompanied by volume changes, as illustrated in figure 6.5. Osmosis ceases when the concentrations become equal on both sides of the membrane.

Cell membranes behave in a similar manner because water is able to move to some degree through the lipid component of most cell membranes. The membranes of some cells, however, have special water channels that may be subject to regulation. This is particularly important in the functioning of the kidneys, as will be described in chapter 16.

Osmotic Pressure

Osmosis and the movement of the membrane partition could be prevented by an opposing force. If one compartment contained 180 g/L of glucose and the other compartment contained pure water, the osmosis of water into the glucose solution could be

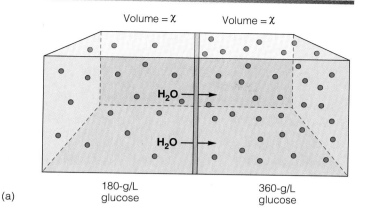

(a)

180-g/L
glucose

360-g/L
glucose

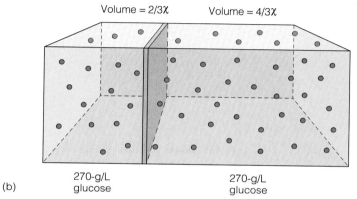

(b)

270-g/L
glucose

270-g/L
glucose

Figure 6.5

(a) A movable semipermeable membrane (permeable to water but not to glucose) separates two solutions of different glucose concentration. As a result, water moves by osmosis into the solution of greater concentration until (b) the volume changes equalize the concentrations on both sides of the membrane.

 In order for osmosis to occur between two solutions, the two solutions must have different concentrations, and the membrane must be relatively impermeable to the solutes that produce the differences in concentration. Such impermeable solutes are said to be *osmotically active*. Water, for example, returns from tissue fluid to blood capillaries as a result of the fact that the protein concentration of blood plasma is higher than the protein concentration of tissue fluid. This is because the plasma proteins, in contrast to other plasma solutes, cannot pass from the capillaries into the tissue fluid. The plasma proteins, in this case, are osmotically active. If a person has an abnormally low concentration of plasma proteins, as may occur, for example, in liver disease (such as cirrhosis), excessive accumulation of fluid in the tissues (edema) will result.

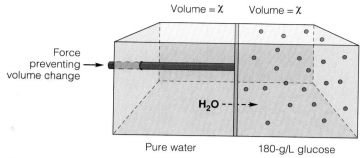

Figure 6.6

If a semipermeable membrane separates pure water from a 180-g/L glucose solution, water will tend to move by osmosis into the glucose solution, thus creating a hydrostatic pressure that will push the membrane to the left and expand the volume of the glucose solution. The amount of pressure that must be applied to just counteract this volume change is equal to the osmotic pressure of the glucose solution.

prevented by pushing against the membrane with a certain force (in this case, equal to 22.4 atmospheres pressure). This concept is illustrated in figure 6.6.

The force that would have to be exerted to prevent osmosis in the situation just described is the **osmotic pressure** of the solution. This backward measurement indicates how strongly the solution "draws" water into it by osmosis. The greater the solute concentration of a solution, the greater its osmotic pressure. Pure water, thus, has an osmotic pressure of zero, and a 360-g/L glucose solution has twice the osmotic pressure of a 180-g/L glucose solution.

Molarity and Molality

Glucose is a monosaccharide with a molecular weight of 180 (the sum of its atomic weights). Sucrose is a disaccharide of glucose and fructose, which have molecular weights of 180 each. When glucose and fructose join together by dehydration synthesis to form sucrose, a molecule of water (molecular weight = 18) is split off. Therefore, sucrose has a molecular weight of 342 (the sum of 180 + 180 − 18). Since the molecular weights of sucrose and glucose are in a ratio of 342/180, it follows that 342 grams of sucrose must contain the same number of molecules as 180 grams of glucose.

Notice that an amount of any compound equal to its molecular weight in grams must contain the same number of molecules as an amount of any other compound equal to its molecular weight in grams. This unit of weight is called a *mole*, and it always contains 6.02×10^{23} molecules (**Avogadro's number**). One mole of solute dissolved in water to make one liter of solution is described as a **one-molar solution** (abbreviated 1.0 M). Although this unit of measurement is commonly used in chemistry, it is not completely desirable in discussions of osmosis because the exact ratio of solute to water is not specified. For example, more water is needed to make a 1.0 M NaCl solution

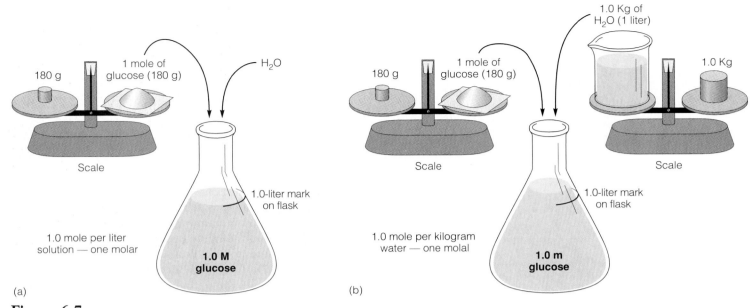

Figure 6.7

Diagrams illustrating the difference between (*a*) a one-molar (1.0 M) and (*b*) a one-molal (1.0 *m*) glucose solution.

(where a mole of NaCl weighs 58.5 grams) than is needed to make a 1.0 M glucose solution, since 180 grams of glucose takes up more volume than 58.5 grams of salt.

Since the ratio of solute to water molecules is of critical importance in osmosis, a more desirable measurement of concentration is **molality.** In a one-molal solution (abbreviated 1.0 *m*), one mole of solute (180 grams of glucose, for example) is dissolved in one kilogram of water (equal to one liter at 4°C). A 1.0 *m* NaCl solution and a 1.0 *m* glucose solution both contain a mole of solute dissolved in exactly the same amount of water (fig. 6.7).

Osmolality

If 180 grams of glucose and 180 grams of fructose were dissolved in the same kilogram of water, the osmotic pressure of the solution would be the same as that of a 360-g/L glucose solution. Osmotic pressure depends on the ratio of solute to solvent, *not* on the chemical nature of the solute molecules. The expression for the total molality of a solution is **osmolality (Osm).** Thus, the solution of 1.0 *m* glucose plus 1.0 *m* fructose has a total molality, or *osmolality*, of 2.0 osmol/L (abbreviated 2.0 Osm). This is the same as the 360-g/L glucose solution, which is 2.0 *m* and 2.0 Osm (fig. 6.8).

Unlike glucose, fructose, and sucrose, electrolytes such as NaCl ionize when they dissolve in water. One molecule of NaCl dissolved in water yields two ions (Na^+ and Cl^-); one mole of NaCl ionizes to form one mole of Na^+ and one mole of Cl^-. Thus, a 1.0 *m* NaCl solution has a total concentration of 2.0 Osm. The effect of this ionization on osmosis is illustrated in figure 6.9.

Measurement of Osmolality

Plasma and other biological fluids contain many organic molecules and electrolytes. The osmolality of such complex solutions can only be estimated by calculations. Fortunately,

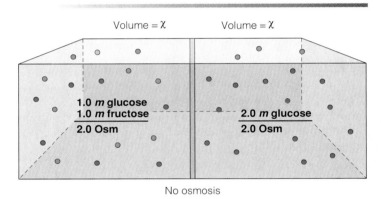

Figure 6.8

The osmolality (Osm) of a solution is equal to the sum of the molalities of each solute in the solution. If a semipermeable membrane separates two solutions with equal osmolalities, no osmosis will occur.

however, there is a relatively simple method for measuring osmolality. This method is based on the fact that the freezing point of a solution, like its osmotic pressure, is affected by the total concentration of the solution and not by the chemical nature of the solute.

One mole of solute per liter depresses the freezing point of water by –1.86°C. Accordingly, a 1.0 *m* glucose solution freezes at a temperature of –1.86°C, and a 1.0 *m* NaCl solution freezes at a temperature of 2 × –1.86 = –3.72°C, because of ionization. The *freezing-point depression* is, thus, a measure of the osmolality. Since plasma freezes at about –0.56°C, its osmolality is equal to 0.56/1.86, or 0.3 Osm, which is more commonly indicated as 300 milliosmolal (or 300 mOsm).

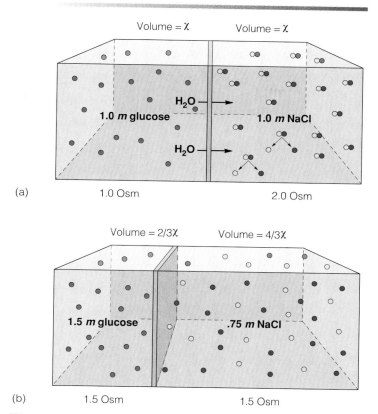

(a)

(b)

Figure 6.9

(*a*) If a semipermeable membrane (permeable to water but not to glucose, Na⁺, or Cl⁻) separates a 1.0 *m* glucose solution from a 1.0 *m* NaCl solution, water will move by osmosis into the NaCl solution. This is because NaCl can ionize to yield one-molal Na⁺ plus one-molal Cl⁻. (*b*) After osmosis, the total concentration, or osmolality, of the two solutions is equal.

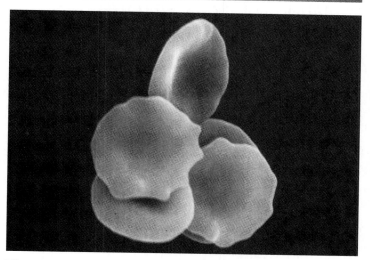

Figure 6.10

A scanning electron micrograph of normal and crenated red blood cells.

Fluids delivered intravenously must be isotonic to blood in order to maintain the correct osmotic pressure and prevent cells from either expanding or shrinking due to the gain or loss of water. Common fluids used for this purpose are normal saline and 5% dextrose, which, as previously described, have about the same osmolality as normal plasma (approximately 300 mOsm). Another isotonic solution frequently used in hospitals is *Ringer's lactate*. This solution contains glucose and lactic acid in addition to a number of different salts. Isotonic solutions are also used in heart-lung machines, which take the place of the heart and lungs during open-heart surgery.

Tonicity

A 0.3 *m* glucose solution, which is 0.3 Osm, or 300 milliosmolal (300 mOsm), has the same osmolality and osmotic pressure as plasma. The same is true of a 0.15 *m* NaCl solution, which ionizes to produce a total concentration of 300 mOsm. Both of these solutions are used clinically as intravenous infusions, labeled 5% *dextrose* (5 g of glucose per 100 ml, which is 0.3 *m*) and *normal saline* (0.9 g of NaCl per 100 ml, which is 0.15 *m*). Since 5% dextrose and normal saline have the same osmolality as plasma, they are said to be **isosmotic** to plasma.

The term *tonicity* is used to describe the effect of a solution on the osmotic movement of water. For example, if an isosmotic glucose or saline solution is separated from plasma by a membrane that is permeable to water but not to glucose or NaCl, osmosis will not occur. In this case, the solution is said to be **isotonic** (from the Greek *isos* = equal; *tonos* = tension) to plasma.

Red blood cells placed in an isotonic solution will neither gain nor lose water. It should be noted that a solution may be isosmotic but not isotonic; such is the case whenever the solute

in the isosmotic solution can freely penetrate the membrane. A 0.3 *m* urea solution, for example, is isosmotic but not isotonic because the cell membrane is permeable to urea. When red blood cells are placed in a 0.3 *m* urea solution, the urea diffuses into the cells until its concentration on both sides of the cell membranes becomes equal. Meanwhile, the solutes within the cells that cannot exit—and which are, therefore, osmotically active—cause osmosis of water into the cells. Red blood cells placed in 0.3 *m* urea will thus eventually burst.

Solutions that have a lower total concentration of osmotically active solutes and a lower osmotic pressure than plasma are said to be **hypotonic** to plasma. Red blood cells placed in hypotonic solutions gain water and may burst (*hemolysis*). When red blood cells are placed in a **hypertonic** solution (such as sea water), which has a higher osmolality and osmotic pressure than plasma, they shrink due to the osmosis of water out of the cells. In this process, called *crenation* (*crena* = notch), the cell surface takes on a scalloped appearance (fig. 6.10).

Regulation of Blood Osmolality

The osmolality of the blood plasma is normally maintained within very narrow limits by a variety of regulatory mechanisms. When a person becomes dehydrated, for example, the blood becomes more concentrated as the total blood volume is reduced. The increased blood osmolality and osmotic pressure stimulate *osmoreceptors*, which are neurons located in a part of the brain called the hypothalamus.

As a result of increased osmoreceptor stimulation, the person becomes thirsty and, if water is available, drinks. Along with increased water intake, a person who is dehydrated excretes a lower volume of urine. This occurs as a result of the following sequence of events: (1) increased plasma osmolality stimulates osmoreceptors in the hypothalamus of the brain; (2) the osmoreceptors stimulate the posterior pituitary, by means of a tract of nerve fibers, to secrete **antidiuretic hormone (ADH)**; and (3) ADH acts on the kidneys to promote water retention so that a lower volume of more concentrated urine is excreted.

A person who is dehydrated, therefore, drinks more and urinates less. This represents a negative feedback loop (fig. 6.11), which acts to maintain homeostasis of the plasma concentration (osmolality) and, in the process, helps to maintain a proper blood volume.

A person with a normal blood volume who eats salty food will also get thirsty and have an increased secretion of ADH. By drinking more and excreting less water in the urine, the salt from the food will become diluted to restore the normal blood concentration, but at a higher blood volume. The opposite occurs in salt deprivation. With a lower plasma osmolality, the osmoreceptors are less stimulated, and the posterior pituitary secretes less ADH. Consequently, more water is excreted in the urine to again restore the proper range of plasma concentration, but at a lower blood volume. Low blood volume and pressure as a result of continued salt deprivation can be fatal (refer to the discussion of blood volume and pressure in chapter 14).

1. Define the term *simple diffusion* and list the factors that influence the diffusion rate.
2. Define the terms *osmosis, osmolality,* and *osmotic pressure* and describe the conditions required for osmosis to occur.
3. Define the terms *isotonic, hypotonic,* and *hypertonic* and explain why hospitals use 5% dextrose and normal saline as intravenous infusions.
4. Explain how the body detects changes in the osmolality of plasma and describe the regulatory mechanisms by which a proper range of plasma osmolality is maintained.

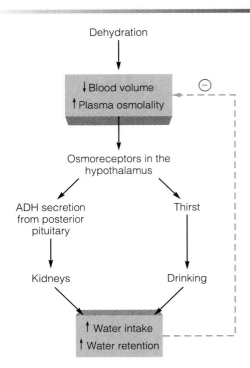

Figure 6.11

An increase in plasma osmolality (increased concentration and osmotic pressure) due to dehydration stimulates thirst and increased ADH secretion. These effects cause the person to drink more and urinate less. The blood volume, as a result, is increased while the plasma osmolality is decreased. These effects help to bring the blood volume back to the normal range and complete the negative feedback loop (indicated by a negative sign).

Carrier-Mediated Transport

Molecules such as glucose are transported across the cell membranes by special protein carriers. Carrier-mediated transport in which the net movement is down a concentration gradient, and which is therefore passive, is called facilitated diffusion. Carrier-mediated transport that occurs against a concentration gradient, and which therefore requires metabolic energy, is called active transport.

In order to sustain metabolism, cells must be able to take up glucose, amino acids, and other organic molecules from the extracellular environment. Molecules such as these, however, are too large and polar to pass through the lipid barrier of the

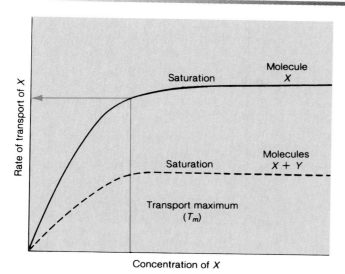

Figure 6.12

Carrier-mediated transport displays the characteristics of saturation (illustrated by the *transport maximum*) and competition. Molecules X and Y compete for the same carrier, so that when they are present together the rate of transport of each is lower than when either is present separately.

The kidneys transport a number of molecules from the blood filtrate (which will become urine) back into the blood. Glucose, for example, is normally completely reabsorbed so that urine normally is free of glucose. If the glucose concentration of the blood and filtrate is too high (a condition called *hyperglycemia*), however, the transport maximum will be exceeded. In this case, glucose will be found in the urine (a condition called *glycosuria*). This may result from the consumption of too much sugar or from inadequate action of the hormone *insulin* in the disease **diabetes mellitus**.

cell membrane by a process of simple diffusion. The transport of such molecules is mediated by **protein carriers** within the membrane. Although such carriers cannot be directly observed, their presence has been inferred by the observation that this transport has characteristics in common with enzyme activity. These characteristics include (1) *specificity,* (2) *competition,* and (3) *saturation.*

Like enzyme proteins, carrier proteins interact only with specific molecules. Glucose carriers, for example, can interact only with glucose and not with closely related monosaccharides. As a further example of specificity, particular carriers for amino acids transport some types of amino acids but not others. Two amino acids that are transported by the same carrier compete with each other, so that the rate of transport of each is lower when they are present together than it would be if each were present alone (fig. 6.12).

As the concentration of a transported molecule is increased, its rate of transport will also be increased—but only up to a maximum. Beyond this rate, called the *transport maximum* (T_m), further increases in concentration do not further increase the transport rate. This indicates that the carriers have become saturated (fig. 6.12).

As an example of saturation, imagine a bus stop that is serviced once per hour by a bus that can hold a maximum of forty people (its "transport maximum"). If there are ten people waiting at the bus stop, ten will be transported per hour. If twenty people are waiting, twenty will be transported per hour. This linear relationship will hold up to a maximum of forty people; if there are eighty people at the bus stop, the transport rate will still be forty per hour.

Facilitated Diffusion

The transport of glucose from the blood across the cell membranes of tissue cells occurs by **facilitated diffusion.** Facilitated diffusion, like simple diffusion, is powered by the thermal energy of the diffusing molecules and involves the net transport of substances through a cell membrane from the side of higher to the side of lower concentration. Active cellular metabolism is not required for either facilitated or simple diffusion.

Unlike simple diffusion of nonpolar molecules, water, and inorganic ions through a membrane, the diffusion of glucose through the cell membrane displays the properties of carrier-mediated transport: specificity, competition, and saturation. The diffusion of glucose through a cell membrane must therefore be mediated by carrier proteins. One conceptual model of the transport carriers is that each may be composed of two protein subunits that interact with glucose in such a way as to create a channel through the membrane (fig. 6.13), thus enabling the movement of glucose from the side of higher to the side of lower concentration.

The rate of the facilitated diffusion of glucose into tissue cells depends directly on the plasma glucose concentration. When the plasma glucose concentration is abnormally low—a condition called *hypoglycemia*—the rate of transport of glucose into brain cells may be inadequate for the metabolic needs of the brain. Severe hypoglycemia, as may be produced in a diabetic person by an overdose of insulin, can thus result in loss of consciousness or even death.

Active Transport

There are some aspects of cell transport that cannot be explained by simple or facilitated diffusion. The epithelial lining of the small intestine and of the kidney tubules, for example, moves glucose from the side of lower to the side of

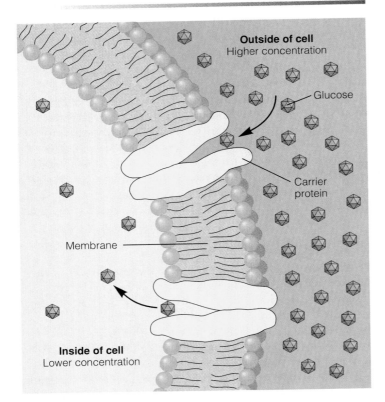

Figure 6.13

A model of facilitated diffusion in which a molecule is transported across the cell membrane by a carrier protein.

higher concentration (from the lumen to the blood). Similarly, all cells extrude Ca^{++} into the extracellular environment and, by this means, maintain an intracellular Ca^{++} concentration that is 1,000 to 10,000 times lower than the extracellular Ca^{++} concentration.

The movement of molecules and ions against their concentration gradients, from lower to higher concentrations, requires the expenditure of cellular energy that is obtained from ATP. This type of transport is termed **active transport.** If a cell is poisoned with cyanide (which inhibits oxidative phosphorylation), active transport will be inhibited. This is in contrast to passive transport, which can continue even if metabolic poisons kill the cell by preventing the formation of ATP.

Primary Active Transport

Primary active transport occurs when the hydrolysis of ATP is directly required for the function of the protein carrier. These carriers appear to be integral proteins that span the thickness of the membrane. The following sequence of events is believed to occur: (1) the molecule or ion to be transported binds to a specific "recognition site" on one side of the carrier protein; (2) this bonding stimulates the breakdown of ATP, which in turn results in phosphorylation of the carrier protein; (3) as a result of phosphorylation, the carrier protein undergoes a conformational (shape) change; and (4) a hingelike motion of the

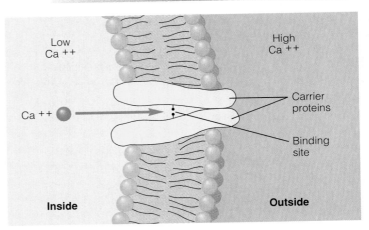

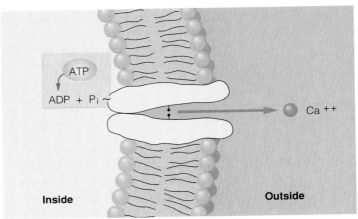

Figure 6.14

A model of active transport showing the hingelike motion of the integral protein subunits.

carrier protein releases the transported molecule or ion on the other side of the membrane. This model of active transport is illustrated in figure 6.14.

The Sodium-Potassium Pump

Primary active transport carriers are often referred to as "pumps." Although some of these carriers transport only one molecule or ion at a time, other carriers exchange one molecule or ion for another. The most important of the latter type of carriers is the **Na$^+$/ K$^+$ pump.** This carrier protein, which is also an ATPase enzyme that converts ATP to ADP and P$_i$, actively extrudes three Na$^+$ ions from the cell as it transports two K$^+$ into the cell. This transport is energy dependent because Na$^+$ is more highly concentrated outside the cell and K$^+$ is more concentrated within the cell. Both ions, in other words, are moved against their concentration gradients (fig. 6.15).

All cells have numerous Na$^+$/K$^+$ pumps that are constantly active. For example, there are about 200 Na$^+$/K$^+$ pumps per red blood cell, about 35,000 per white blood cell, and several million per cell in a part of the tubules within the kidney. This represents an enormous expenditure of energy used to maintain a steep gradient of Na⁺ and K⁺ across the cell membrane. This steep gradient serves three functions: (1) the steep Na⁺ gradient is used to provide energy for the "cotransport" of other molecules; (2) the activity of the Na⁺/K⁺ pumps can be adjusted (primarily by thyroid hormones) to regulate the resting calorie expenditure and basal metabolic rate of the body; and (3) the Na⁺ and K⁺ gradients across the cell membranes of nerve and muscle cells are used to produce electrical impulses. In addition, the active extrusion of Na⁺ is important for osmotic reasons. If the pumps stopped functioning, the increased Na⁺ within the cells would draw increased water and damage the cells.

Secondary Active Transport (Cotransport)

In **secondary active transport,** or **cotransport,** the energy needed for the "uphill" movement of a molecule or ion is obtained from the "downhill" transport of Na⁺ into the cell. Hydrolysis of ATP by the action of the Na⁺/K⁺ pumps is required indirectly, in order to maintain low intracellular Na⁺ concentrations. The diffusion of Na⁺ into the cell may power the uphill movement of a different ion or molecule into the cell, or it may power the movement of an ion or molecule out of the cell.

In the epithelial cells of the small intestine and kidney tubules, for example, glucose is transported against its concentration gradient by a carrier that requires the simultaneous bonding of Na⁺ (fig. 6.16). Glucose and Na⁺ move in the same direction (into the cell) as a result of the Na⁺ gradient created by the Na⁺/K⁺ pumps. Because of the distribution of the Na⁺/K⁺ pumps and glucose carriers in the epithelial cell membrane, the Na⁺ and glucose are moved from the lumina of the intestine and kidney tubules into the blood (fig. 6.17).

Cotransport can also occur in opposite directions. The uphill extrusion of Ca⁺⁺ from some cells, for example, is coupled to the passive diffusion of Na⁺ into the cell. Cellular energy, obtained from ATP, is not used to move Ca⁺⁺ directly out of the cell in this case, but energy is constantly required to maintain the steep Na⁺ gradient. The net movement of Ca⁺⁺ out of these cells is thus an example of secondary active transport.

Figure 6.15

The Na⁺/K⁺ pump actively exchanges intracellular Na⁺ for K⁺. The carrier itself is an ATPase that breaks down ATP for energy. Dashed arrows indicate the direction of passive transport (diffusion); solid arrows indicate the direction of active transport.

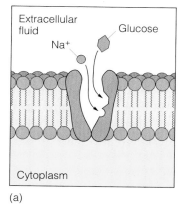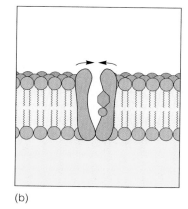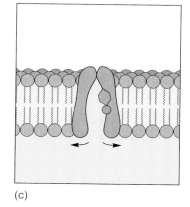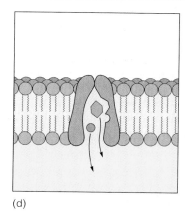

Figure 6.16

A model (*a–d*) for the cotransport of Na⁺ and glucose into a cell. This is secondary active transport because it is dependent upon the diffusion gradient for Na⁺ created by the Na⁺/K⁺ pumps.

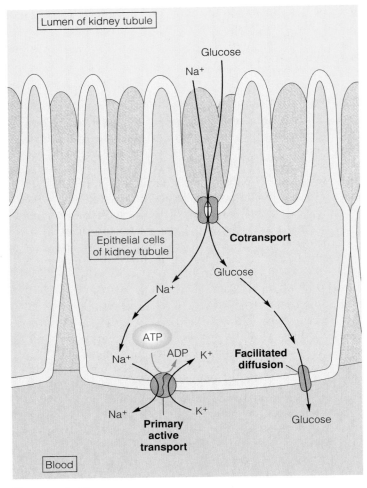

Figure 6.17

The transport of glucose from the fluid in the kidney tubules through the epithelial cells of the tubule and into the blood. All three types of carrier-mediated transport are used in this process. (A similar transport process occurs in the absorption of glucose from the intestine.)

 Severe diarrhea is responsible for almost half of all deaths worldwide of children under the age of 4 (amounting to about 4 million deaths per year). Because rehydration through intravenous therapy is often not practical, the World Health Organization (WHO) developed **oral rehydration therapy.** The therapy is based on the fact that absorption of water by osmosis across the intestine is proportional to the absorption of Na⁺ and on the fact that the intestinal epithelium cotransports Na⁺ and glucose. The WHO provides those in need with a mixture (which can be diluted with tap water in the home) containing both glucose and Na⁺, as well as other ions. The glucose in the mixture promotes the cotransport of Na⁺, and the Na⁺ transport promotes the osmotic movement of water from the intestine into the blood. It has been estimated that oral rehydration therapy saves the lives of more than a million small children each year.

1. List the three characteristics of facilitated diffusion that distinguish it from simple diffusion.
2. Draw a figure that illustrates two of the characteristics of carrier-mediated transport and explain how this type of movement differs from simple diffusion.
3. Describe active transport, including cotransport in your description. Explain how active transport differs from facilitated diffusion.
4. Explain the functional significance of the Na⁺/K⁺ pumps.

The Membrane Potential

The permeability properties of the cell membrane, the presence of nondiffusible negatively charged molecules in the cell, and the action of the Na⁺/K⁺ pumps produce an unequal distribution of charges across the membrane. As a result, the inside of the cell is negatively charged compared to the outside. This difference in charge, or potential difference, is known as the membrane potential.

In the preceding section, the action of the Na⁺/K⁺ pumps was discussed in conjunction with the topic of active transport, and it was noted that these pumps move Na⁺ and K⁺ against their concentration gradients. This action would, by itself, create and amplify the difference in concentration of these ions across the cell membrane. There is, however, another reason why the concentration of these ions would be unequal across the membrane.

Cellular proteins and the phosphate groups of ATP and other organic molecules are negatively charged at the pH of the cell cytoplasm. These negative ions (anions) are "fixed" within the cell because they cannot penetrate the cell membrane. Since these negatively charged organic molecules cannot leave the cell, they attract positively charged inorganic ions (cations) from the extracellular fluid that are small enough to diffuse through the membrane pores. The distribution of small inorganic cations (mainly K⁺, Na⁺, and Ca⁺⁺) between the intracellular and extracellular compartments, in other words, is influenced by the negatively charged fixed ions within the cell.

Since the cell membrane is much more permeable to K⁺ than to any other cation, K⁺ accumulates within the cell more than the others as a result of its electrical attraction for the fixed anions (fig. 6.18). So instead of being evenly distributed between the intracellular and extracellular compartments, K⁺ becomes more highly concentrated within the cell. In the human body, the intracellular K⁺ concentration is 150 mEq/L compared to an extracellular concentration of 5 mEq/L (mEq = milliequivalents, which is the millimolar concentration multiplied by the valence of the ion—in this case, by one).

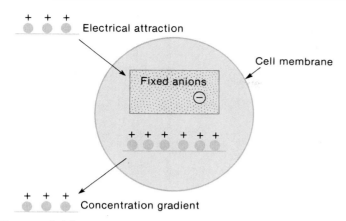

Figure 6.18

Proteins, organic phosphates, and other organic anions that cannot leave the cell create a fixed negative charge on the inside of the membrane. This negative charge attracts positively charged inorganic ions (cations), which therefore accumulate within the cell at a higher concentration than is found in the extracellular fluid. The amount of cations that accumulate within the cell is limited by the fact that a concentration gradient builds up, which favors the diffusion of the cations out of the cell.

As a result of the unequal distribution of charges between the inside and outside of cells, each cell is a tiny battery with the positive pole outside the cell membrane and the negative pole inside. The magnitude of this charge difference is measured in *voltage*. Although the voltage of this battery is very small (less than a tenth of a volt), it is of critical importance in muscle contraction, the regulation of the heartbeat, the generation of nerve impulses, and other physiological events. In order to understand these processes, then, we must first examine the electrical properties of cells.

Equilibrium Potentials

An equilibrium potential is a theoretical voltage that would be produced across a cell membrane if only one ion were able to diffuse through the membrane. Since the membrane is most permeable to K^+, we can construct a theoretical approximation of what would happen if K^+ were the *only* ion able to cross the membrane. If this were the case, K^+ would diffuse until its concentration inside and outside of a cell became stable, thus establishing an *equilibrium*. In this condition, if a certain amount of K^+ were to move inside the cell (by electrical attraction for the fixed anions), an identical amount of K^+ would diffuse out of the cell (down its concentration gradient). At equilibrium, in other words, the forces of electrical attraction and of the diffusion gradient are equal and opposite.

At this equilibrium, the concentration of K^+ would be higher inside the cell than outside the cell; a concentration difference would exist across the cell membrane that was stabilized by the attraction of K^+ to the fixed anions. At this point we could ask, Are the fixed anions neutralized . . . are the charges balanced? The answer depends on how much K^+ gets into the

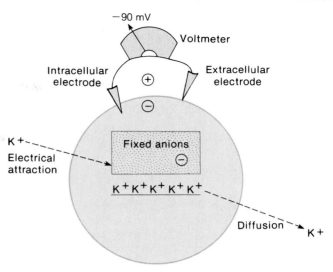

Figure 6.19

If K^+ were the only ion able to diffuse through the cell membrane, it would distribute itself between the intracellular and extracellular compartments until an equilibrium was established. At equilibrium, the K^+ concentration within the cell would be higher than outside the cell due to the attraction of K^+ for the fixed anions. Not enough K^+ would accumulate within the cell to neutralize these anions, however, so the inside of the cell would be 90 millivolts negative compared to the outside of the cell. This membrane voltage is the equilibrium potential (E_K) for potassium.

cell, which in turn depends on the K^+ concentration in the extracellular fluid. At the K^+ concentrations that are, in fact, found in the body, the answer to our question is no. Not enough K^+ is present in the cell to neutralize the fixed anions (fig. 6.19).

At equilibrium, therefore, the inside of the cell membrane would have a higher concentration of negative charges than the outside of the membrane. There is a difference in charge, as well as a difference in concentration, across the membrane. The magnitude of the difference in charge, or **potential difference,** on the two sides of the membrane under these conditions is 90 millivolts (mV). A sign (+ or –) placed in front of this number indicates the polarity within the cell. This is shown with a negative sign (as –90 mV) to indicate that the inside of the cell is the negative pole. The potential difference of –90 mV, which would be developed if K^+ were the only diffusible ion, is called the **K^+ equilibrium potential** (abbreviated E_K).

Nernst Equation

There is another way to look at the equilibrium potential: it is the membrane potential that would *exactly balance* the diffusion gradient and prevent the net movement of a particular ion. Since the diffusion gradient depends on the difference in concentration of the ion, the value of the equilibrium potential must depend on the ratio of the concentrations of the ion on the two sides of the membrane. The **Nernst equation** allows this theoretical equilibrium potential to be calculated for a particular ion when its concentrations are known. The following simplified form of the equation is valid at a temperature of 37°C:

$$E_x = \frac{61}{z} \log \frac{[X_o]}{[X_i]}$$

where:

E_x = equilibrium potential in millivolts (mV) for ion x

X_o = concentration of the ion outside the cell

X_i = concentration of the ion inside the cell

z = valence of the ion (+1 for Na^+ or K^+)

Note that, using the Nernst equation, the equilibrium potential for a cation will have a negative value when X_i is greater than X_o. If we substitute K^+ for X, this is indeed the case. As a hypothetical example, if the concentration of K^+ were ten times higher inside compared to outside the cell, the equilibrium potential would be 61 mV (log 1/10) = 61 × (–1) = –61 mV. In reality, the concentration of K^+ inside the cell is actually thirty times greater than outside (150 mEq/L inside compared to 5 mEq/L outside). Since the log of 1/30 is –1.477, the equilibrium potential for K^+ given realistic concentration values is –90 mV.

If we wish to calculate the equilibrium potential for Na^+, different values must be used. The concentration of Na^+ in the extracellular fluid is 145 mEq/L, whereas its concentration inside cells is only 12 mEq/L. The diffusion gradient thus promotes the movement of Na^+ into the cell, and, in order to oppose this diffusion, the membrane potential would have to have a positive polarity on the inside of the cell. This is indeed what the Nernst equation would provide, since $[X_o]/[X_i]$ is greater than one. Using this equation, the equilibrium potential for Na^+ can be calculated to be +60 mV.

A membrane potential of +60 mV, therefore, would prevent the diffusion of Na^+ into the cell, whereas a membrane potential of –90 mV would prevent the diffusion of K^+ out of the cell. It is clear that the membrane potential cannot be both values at the same time; indeed, it is seldom either value but instead is somewhere between these two extremes. We will call this the **resting membrane potential** to distinguish it from the theoretical equilibrium potentials. The actual value of the resting membrane potential depends upon the permeability of the membrane to each ion and on the equilibrium potential of each diffusible ion.

Resting Membrane Potential

The resting membrane potential of most cells in the body ranges from –65 mV to –85 mV (in neurons it averages –70 mV). This value is very close to the one that we had predicted if K^+ were the only ion able to move through the cell membrane and establish an equilibrium. The resting membrane potential, however, is quite different from the Na^+ equilibrium potential. The reason that the resting membrane potential is closer to the K^+ than to the Na^+ equilibrium potential is because, as previously discussed, the membrane is much more permeable to K^+ than to Na^+.

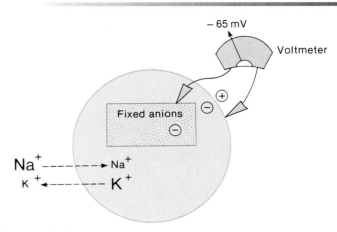

Figure 6.20

Because some Na^+ leaks into the cell by diffusion, the actual resting membrane potential is lower than the K^+ equilibrium potential. As a result, some K^+ diffuses out of the cell, as indicated by the dashed lines.

The resting membrane potential is particularly sensitive to changes in plasma potassium concentration. Since the maintenance of a particular membrane potential is critical for the generation of electrical events in the heart, mechanisms that act primarily through the kidneys maintain plasma K^+ concentrations within very narrow limits. An abnormal increase in the blood concentration of K^+ is called **hyperkalemia**. When hyperkalemia occurs, more K^+ can enter the cell. In terms of the Nernst equation, the ratio $[K^+_o] / [K^+_i]$ is decreased. This reduces the membrane potential (brings it closer to zero) and thus interferes with the proper function of the heart. For these reasons, the blood electrolyte concentrations are monitored very carefully in patients with heart or kidney disease.

The actual value of the membrane potential thus depends on the differential permeability of the membrane. When neurons produce nerve impulses, the permeability to Na^+ increases dramatically, sending the membrane potential toward the Na^+ equilibrium potential (nerve impulses are discussed in chapter 7). This is the reason that the term "resting" is used to describe the membrane potential when it is not producing impulses.

Role of the Na^+/K^+ Pumps

Since the membrane potential is less negative than E_K as a result of some Na^+ entry, some K^+ leaks out of the cell (fig. 6.20). The cell is *not* at equilibrium with respect to K^+ and Na^+ concentrations. Nonetheless, the concentrations of K^+ and Na^+ are maintained constant because of the constant expenditure

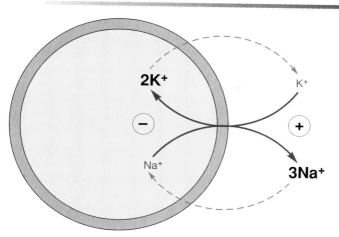

Figure 6.21

The concentrations of Na^+ and K^+ both inside and outside the cell do not change as a result of diffusion (*dashed arrows*) because of active transport (*solid arrows*) by the Na^+/K^+ pump. Since the pump transports three Na^+ for every two K^+, the pump itself helps to create a charge separation (a potential difference, or voltage) across the membrane.

the cell for every *two* K^+ ions that it moves in, its action helps generate a potential difference across the membrane (fig. 6.21). This electrogenic effect of the pumps adds a few more millivolts to the membrane potential. As a result of all of these activities, a real cell has (1) a relatively constant intracellular concentration of Na^+ and K^+ and (2) a constant membrane potential (in the absence of stimulation) in nerves and muscles of −65 mV to −85 mV.

of energy in active transport by the Na^+/K^+ pumps. The Na^+/K^+ pumps act to counter the leaks and thus maintain the membrane potential.

Actually, the Na^+/K^+ pump does more than simply work against the ion leaks; since it transports *three* Na^+ ions out of

---

1. Define the term *membrane potential* and explain how it is measured.
2. Explain how an equilibrium potential is produced when potassium is the only diffusible cation. State how the value of the equilibrium potential is affected by the potassium concentrations outside and inside the cell.
3. Explain why the resting membrane potential is close to but different from the potassium equilibrium potential.
4. Suppose a person has hyperkalemia such that the extracellular K^+ concentration increases from 5 mM to 10 mM (a potentially fatal condition). Use the Nernst equation to calculate the new E_K, and then verbally describe how the resting membrane potential would be changed.
5. Describe the role of the Na^+/K^+ pumps in the generation and maintenance of the resting membrane potential.

Summary

Diffusion and Osmosis   p. 124

I. Diffusion is the net movement of molecules or ions from regions of higher to regions of lower concentration.
   A. This is a type of passive transport—energy is provided by the thermal energy of the molecules, not by cellular metabolism.
   B. Net diffusion stops when the concentration is equal on both sides of the membrane.
II. The rate of diffusion is dependent on a variety of factors.
   A. The rate of diffusion depends on the concentration difference on the two sides of the membrane.
   B. The rate depends on the permeability of the cell membrane to the diffusing substance.

   C. The rate of diffusion through a membrane is also directly proportional to the surface area of the membrane, which can be increased by such adaptations as microvilli.
III. Simple diffusion is the type of passive transport in which small molecules and inorganic ions, such as Na^+ and K^+, move through the cell membrane.
   A. Inorganic ions pass through specific channels in the membrane.
   B. Lipids such as steroid hormones can pass directly through the phospholipid layers of the membrane by simple diffusion.
IV. Osmosis is the simple diffusion of solvent (water) through a membrane that is more permeable to the solvent than it is to the solute.

   A. Water moves from the solution that is more dilute to the solution that has a higher solute concentration.
   B. Osmosis depends on a difference in total solute concentration, not on the chemical nature of the solute.
      1. The concentration of total solute, in moles per kilogram (liter) of water, is measured in osmolality units.
      2. The solution with the higher osmolality has the higher osmotic pressure.
      3. Water moves by osmosis from the solution of lower osmolality and osmotic pressure to the solution of higher osmolality and osmotic pressure.

C. Solutions that have the same osmotic pressure as plasma (such as 0.9% NaCl and 5% glucose) are said to be isotonic.

   1. Solutions with a lower osmotic pressure are hypotonic; those with a higher osmotic pressure are hypertonic.

   2. Cells in a hypotonic solution gain water and swell; those in a hypertonic solution lose water and shrink (crenate).

D. The osmolality and osmotic pressure of the plasma is detected by osmoreceptors in the hypothalamus of the brain and maintained within a normal range by the action of antidiuretic hormone (ADH) secreted by the posterior pituitary.

   1. Increased osmolality of the blood stimulates the osmoreceptors.

   2. Stimulation of the osmoreceptors causes thirst and triggers the secretion of antidiuretic hormone (ADH) from the pituitary.

   3. ADH stimulates water retention by the kidneys, which serves to maintain a normal blood volume and osmolality.

Carrier-Mediated Transport   p. 130

I. The passage of glucose, amino acids, and other polar molecules through the cell membrane is mediated by carrier proteins in the cell membrane.

A. Carrier-mediated transport exhibits the properties of specificity, competition, and saturation.

B. The transport rate of molecules such as glucose reaches a maximum when the carriers are saturated. This maximum rate is called the transport maximum (T_m).

II. The transport of molecules such as glucose from the side of higher to the side of lower concentration by means of membrane carriers is called facilitated diffusion.

A. Like simple diffusion, this is passive transport—cellular energy is not required.

B. Unlike simple diffusion, facilitated diffusion displays the properties of specificity, competition, and saturation.

III. The active transport of molecules and ions across a membrane requires the expenditure of cellular energy (ATP).

A. In active transport, carriers move molecules or ions from the side of lower to the side of higher concentration.

B. One example of active transport is the action of the Na^+/K^+ pump.

   1. Sodium is more concentrated on the outside of the cell, whereas potassium is more concentrated on the inside of the cell.

   2. The Na^+/K^+ pump helps to maintain these concentration differences by transporting Na^+ out of the cell and K^+ into the cell.

The Membrane Potential   p. 134

I. The cytoplasm of the cell contains negatively charged organic ions (anions) that cannot leave the cell—they are "fixed anions."

A. These fixed anions attract K^+, which is the inorganic ion that can most easily pass through the cell membrane.

B. As a result of this electrical attraction, the concentration of K^+ within the cell is greater than the concentration of K^+ in the extracellular fluid.

C. If K^+ were the only diffusible ion, the concentrations of K^+ on the inside and outside of the cell would reach an equilibrium.

   1. At this point, the rate of K^+ entry (due to electrical attraction) would equal the rate of K^+ exit (due to diffusion).

   2. At this equilibrium, there would still be a higher concentration of negative charges within the cell (due to the fixed anions) than outside the cell.

   3. At this equilibrium, the inside of the cell would be 90 millivolts negative (–90 mV) compared to the outside of the cell. This potential difference is called the K^+ equilibrium potential (E_K).

D. The resting membrane potential is less than E_K (usually –65 mV to –85 mV) because some Na^+ can also enter the cell.

   1. Na^+ is more highly concentrated outside than inside the cell, and the inside of the cell is negative. These forces attract Na^+ into the cell.

   2. The rate of Na^+ entry is generally slow because the membrane is usually not very permeable to Na^+.

II. The slow rate of Na^+ entry is accompanied by a slow rate of K^+ leakage out of the cell.

A. The Na^+/K^+ pump counters this leakage, and thus maintains constant concentrations and a constant resting membrane potential.

B. There are numerous Na^+/K^+ pumps in all cells of the body that require a constant expenditure of energy.

C. The Na^+/K^+ pump itself contributes to the membrane potential because it pumps more Na^+ out than it pumps K^+ in (by a ratio of three to two).

Clinical Investigation

A student complains that he is constantly thirsty. During a laboratory exercise involving urinalysis, he discovers that his urine contains a significant amount of glucose. As a result of a medical examination, he later learns that he has hyperglycemia, hyperkalemia, and a high plasma osmolality. Further, he is told that his electrocardiogram shows some abnormalities. How might this student's symptoms and medical findings be interrelated?

Clues

Refer to the box on hyperglycemia in the section "Carrier-Mediated Transport" and review the characteristics of carrier-mediated transport. Think of glucose in urine as exerting an osmotic pressure and, using deductive reasoning, determine what effect this would have on the amount of water excreted in the urine and on the resulting blood volume and concentration. Read the boxed information on the effect of hyperkalemia on the heart in the section "Resting Membrane Potential."

Review Activities

Objective Questions

1. The movement of water across a cell membrane occurs by
   a. active transport.
   b. facilitated diffusion.
   c. simple diffusion (osmosis).
   d. all of the above.
2. Which of the following statements about the facilitated diffusion of glucose is *true*?
   a. There is a net movement from the region of lower to the region of higher concentration.
   b. Carrier proteins in the cell membrane are required for this transport.
   c. This transport requires energy obtained from ATP.
   d. It is an example of cotransport.
3. If a poison such as cyanide stopped the production of ATP, which of the following transport processes would cease?
   a. the movement of Na$^+$ out of a cell
   b. osmosis
   c. the movement of K$^+$ out of a cell
   d. all of the above
4. Red blood cells crenate in
   a. a hypotonic solution.
   b. an isotonic solution.
   c. a hypertonic solution.

5. Plasma has an osmolality of about 300 mOsm. Isotonic saline has an osmolality of
   a. 150 mOsm.
   b. 300 mOsm.
   c. 600 mOsm.
   d. none of the above.
6. Which of the following statements comparing a 0.5 *m* NaCl solution and a 1.0 *m* glucose solution is *true*?
   a. They have the same osmolality.
   b. They have the same osmotic pressure.
   c. They are isotonic to each other.
   d. All of the above are true.
7. The diffusible ion that is most important in the establishment of the membrane potential is
   a. K$^+$.
   b. Na$^+$.
   c. Ca^{++}.
   d. Cl$^-$.
8. Which of the following statements regarding an increase in blood osmolality is *true*?
   a. It can occur as a result of dehydration.
   b. It causes a decrease in blood osmotic pressure.
   c. It is accompanied by a decrease in ADH secretion.
   d. All of the above are true.
9. In hyperkalemia, the membrane potential
   a. increases.
   b. decreases.
   c. is not changed.

10. Which of the following statements about the Na$^+$/K$^+$ pump is *true*?
    a. Na$^+$ is actively transported into the cell.
    b. K$^+$ is actively transported out of the cell.
    c. An equal number of Na$^+$ and K$^+$ ions are transported with each cycle of the pump.
    d. The pumps are constantly active in all cells.
11. Which of the following statements about carrier-mediated facilitated diffusion is *true*?
    a. It uses cellular ATP.
    b. It is used for cellular uptake of blood glucose.
    c. It is a form of active transport.
    d. None of the above are true.
12. Which of the following is *not* an example of cotransport?
    a. movement of glucose and Na$^+$ through the apical epithelial membrane in the intestinal epithelium
    b. movement of Na$^+$ and K$^+$ through the action of the Na$^+$/K$^+$ pumps
    c. movement of Na$^+$ and glucose across the kidney tubules
    d. movement of Na+ into a cell while Ca^{++} moves out

Essay Questions

1. Describe the conditions required to produce osmosis and explain why osmosis occurs under these conditions.[1]
2. Explain how simple diffusion can be distinguished from facilitated diffusion and how active transport can be distinguished from passive transport.
3. Compare the theoretical membrane potential that occurs at K^+ equilibrium with the true resting membrane potential. Explain why these values differ.
4. Explain how the Na^+/K^+ pump contributes to the resting membrane potential.
5. Describe the cause-and-effect sequence whereby a genetic defect results in improper cellular transport and the symptoms of cystic fibrosis.
6. Using the principles of osmosis, explain why movement of Na^+ through a cell membrane is followed by movement of water. Use this concept to explain the rationale on which oral rehydration therapy is based.

Selected Readings

Benos, D. J. 1989. The biology of amiloride sensitive sodium channels. *Hospital Practice* 24:149.

Christensen, H. N. 1975. *Biological Transport.* 2d ed. Menlo Park, CA: W. A. Benjamen.

Collins, F. S. 1993. The molecular biology of cystic fibrosis. *Annual Review of Medicine* 44:133.

Dubinsky, W. P. Jr. 1989. The physiology of epithelial chloride channels. *Hospital Practice* 24:69.

Elsas, L. J., and N. Longo. 1992. Glucose transporters. *Annual Review of Medicine* 43:377.

Finean, J. B., R. Coleman, and R. H. Mitchell. 1978. *Membranes and Their Cellular Functions.* 2d ed. New York: Wiley.

Gadsby, G. N., Nagel, G., and Hwang, T. C. 1995. The CFTR chloride channel of mammalian heart. *Annual Review of Physiology* 57:387.

Kaplan, J. H. 1985. Ion movements through the sodium pump. *Annual Review of Physiology* 47:535.

Kotyk, A., and K. Janacek. 1975. *Cell Membrane Transport: Principles and Techniques.* 2d ed. New York: Plenum Publishing Corp.

Levitan, I. B. 1994. Modulation of ion channels by protein phosphorylation and dephosphorylation. *Annual Review of Physiology* 56:193.

Lienhard, G. E. et al. January 1992. How cells absorb glucose. *Scientific American.*

Silverman, M. 1989. Molecular biology of the Na^+–D-glucose cotransporter. *Hospital Practice* 24:180.

Skou, J. C. 1992. The Na^+/K^+ pump. *News in Physiological Sciences* 7:95.

Wheeler, T. J., and P. C. Hinkle. 1985. The glucose transporter of mammalian cells. *Annual Review of Physiology* 47:503.

---

[1]*Note:* This question is answered on page 62 of the Student Study Guide.

7

The Nervous System

Organization, Electrical Activity, and Synaptic Transmission

OBJECTIVES   *After studying this chapter, you should be able to . . .*

1. describe the structural and functional divisions of the nervous system.
2. describe the structure of a neuron and explain the functional significance of its principal regions.
3. describe the locations and functions of the different types of supporting cells.
4. explain what is meant by the blood-brain barrier and discuss its significance.
5. describe the sheath of Schwann and explain how this sheath functions in the regeneration of cut peripheral nerve fibers.
6. explain how a myelin sheath is formed.
7. define *depolarization, repolarization,* and *hyperpolarization.*
8. explain the actions of voltage-regulated Na^+ and K^+ channels and describe the events that occur during the production of an action potential.

9. describe the properties of action potentials and explain the significance of the all-or-none law and the refractory periods.
10. explain how action potentials are regenerated along a myelinated and a nonmyelinated axon.
11. describe the events that occur in the interval between the electrical excitation of an axon and the release of neurotransmitter.
12. describe the two general categories of chemically regulated ion channels and explain how these channels operate using nicotinic and muscarinic ACh receptors as examples.
13. explain how ACh produces EPSPs and IPSPs and discuss the significance of these processes.
14. compare the characteristics of EPSPs and action potentials.
15. compare the mechanisms that inactivate ACh with those that inactivate monoamine neurotransmitters.

16. explain the role of cyclic AMP in the action of monoamine neurotransmitters and describe some of the actions of monoamines in the nervous system.
17. explain the significance of the inhibitory effects of glycine and GABA in the central nervous system.
18. list some of the polypeptide neurotransmitters and explain the significance of the endogenous opioids in the nervous system.
19. explain the significance of nitric oxide as a neurotransmitter.
20. explain how EPSPs and IPSPs can interact and discuss the significance of spatial and temporal summation and of presynaptic and postsynaptic inhibition.
21. describe the nature of long-term potentiation and explain its significance.

OUTLINE

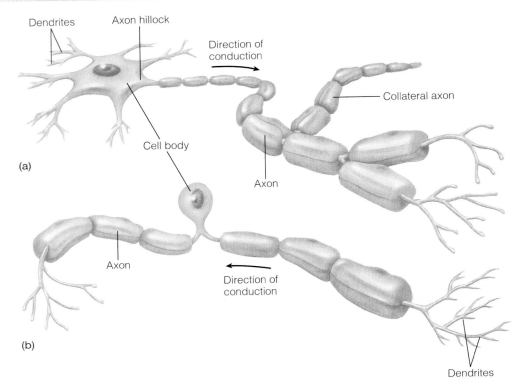

Figure 7.1

The structure of two kinds of neurons: (*a*) a motor neuron and (*b*) a sensory neuron.

Neurons and Supporting Cells

The nervous system is composed of neurons, which produce and conduct electrochemical impulses, and supporting cells, which assist the functions of neurons. Neurons are classified according to structure or function; the various types of supporting cells perform specialized functions.

The nervous system is divided into the **central nervous system (CNS),** which includes the brain and spinal cord, and the **peripheral nervous system (PNS),** which includes the *cranial nerves* arising from the brain and the *spinal nerves* arising from the spinal cord.

The nervous system is composed of only two principal types of cells—neurons and supporting cells. **Neurons** are the basic structural and functional units of the nervous system. They are specialized to respond to physical and chemical stimuli, conduct electrochemical impulses, and release specific chemical regulators. Through these activities, neurons perform such functions as the perception of sensory stimuli, learning, memory, and the control of muscles and glands. Neurons cannot divide by mitosis, although some neurons can regenerate a severed portion or sprout small new branches under some conditions.

Supporting cells aid the functions of neurons and are about five times more abundant than neurons. In the CNS, supporting cells are collectively called **neuroglia,** or simply **glial cells** (*glia* = glue). Unlike neurons, glial cells retain limited mitotic abilities (brain tumors that occur in adults are usually composed of glial cells rather than neurons).

Neurons

Although neurons vary considerably in size and shape, they generally have three principal regions: (1) a cell body, (2) dendrites, and (3) an axon (figs. 7.1 and 7.2). Dendrites and axons can be referred to generically as *processes,* or extensions from the cell body.

The **cell body,** or **perikaryon** (*peri* = around; *karyon* = nucleus), is the enlarged portion of the neuron that contains the nucleus. It serves as the "nutritional center" of the neuron, where macromolecules are produced. The perikaryon also contains densely staining areas of rough endoplasmic reticulum known as *Nissl bodies* that are not found in the dendrites or axon. The cell bodies within the CNS are frequently clustered into groups called *nuclei* (not to be confused with the nucleus of a cell). Cell bodies in the PNS usually occur in clusters called *ganglia* (table 7.1).

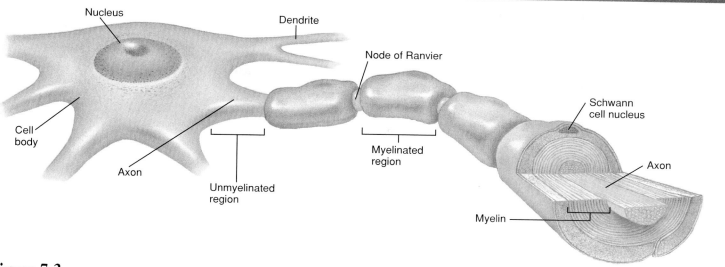

Figure 7.2

The parts of a neuron. The axon of this neuron is wrapped by Schwann cells, which form a myelin sheath.

Table 7.1   Terminology Pertaining to the Nervous System

Term	Definition
Central nervous system (CNS)	Brain and spinal cord
Peripheral nervous system (PNS)	Nerves and ganglia
Association neuron (interneuron)	Multipolar neuron located entirely within CNS
Sensory neuron	Neuron that transmits impulses from a sensory receptor into CNS (afferent fiber)
Motor neuron	Neuron that transmits impulses from CNS to an effector organ (e.g., muscle efferent fiber)
Nerve	Cablelike collection of nerve fibers; may be "mixed" (contain both sensory and motor fibers)
Somatic motor nerve	Nerve that stimulates contraction of skeletal muscles
Autonomic motor nerve	Nerve that stimulates contraction (or inhibits contraction) of smooth muscle and cardiac muscle and that stimulates glandular secretion
Ganglion	Grouping of neuron cell bodies located outside CNS
Nucleus	Grouping of neuron cell bodies within CNS
Tract	Grouping of nerve fibers that interconnect regions of CNS

Table 7.2   Comparison of Axoplasmic Flow and Axonal Transport

Axoplasmic Flow	Axonal Transport
Transport rate comparatively slow (1–2 mm/day)	Transport rate comparatively fast (200–400 mm/day)
Molecules transported only from cell body	Molecules transported from cell body to axon endings and in reverse direction
Bulk movement of proteins in axoplasm, including microfilaments and tubules	Transport of specific proteins, mainly of membrane proteins and acetylcholinesterase
Transport accompanied by peristaltic waves of axon membrane	Transport dependent on cagelike microtubule structure within axon and on actin and Ca^{++}

Dendrites (*dendron* = tree branch) are thin, branched processes that extend from the cytoplasm of the cell body. Dendrites serve as a receptive area that transmits electrical impulses to the cell body. The **axon** is a longer process that conducts impulses away from the cell body. Axons vary in length from only a millimeter long up to a meter or more (for those that extend from the CNS to the foot). The origin of the axon near the cell body is an expanded region called the *axon hillock*; it is here that nerve impulses originate. Side branches called *axon collaterals* may extend from the axon.

Proteins and other molecules are transported through the axon at faster rates than could be achieved by simple diffusion. This rapid movement is produced by two different mechanisms: axoplasmic flow and axonal transport (table 7.2). **Axoplasmic flow,** the slower of the two, results from rhythmic waves of contraction that push the cytoplasm from the axon hillock to the nerve endings. **Axonal transport,** which is more rapid and more selective, may occur in a reverse (retrograde) direction, as well as in a forward (orthograde) direction. Indeed, retrograde transport may be

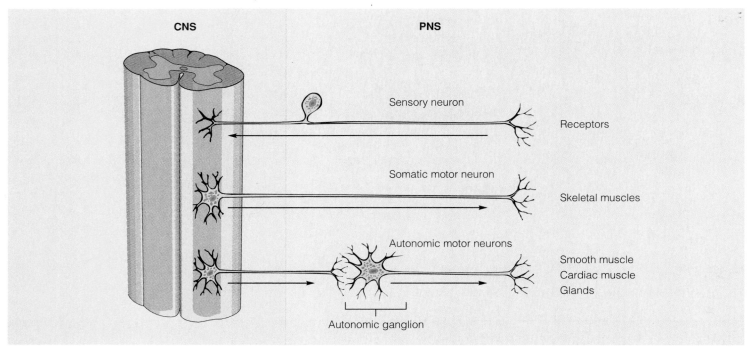

Figure 7.3

The relationship of sensory and motor neurons of the peripheral nervous system (PNS) to the central nervous system (CNS).

responsible for the movement of herpes virus, rabies virus, and tetanus toxin from the nerve terminals into cell bodies.

Classification of Neurons and Nerves

Neurons may be classified according to their structure or function. The functional classification is based on the direction in which they conduct impulses, as indicated in figure 7.3. **Sensory,** or **afferent, neurons** conduct impulses from sensory receptors into the CNS. **Motor,** or **efferent, neurons** conduct impulses out of the CNS to effector organs (muscles and glands). **Association neurons,** or **interneurons,** are located entirely within the CNS and serve the associative, or integrative, functions of the nervous system.

There are two types of motor neurons: somatic and autonomic. **Somatic motor neurons** are responsible for both reflex and voluntary control of skeletal muscles. **Autonomic motor neurons** innervate the involuntary effectors—smooth muscle, cardiac muscle, and glands. The cell bodies of the autonomic neurons that innervate these organs are located outside the CNS in autonomic ganglia (fig. 7.3). There are two subdivisions of autonomic neurons: *sympathetic* and *parasympathetic*. Autonomic motor neurons, together with their central control centers, constitute the *autonomic nervous system*, which will be discussed in chapter 9.

The structural classification of neurons is based on the number of processes that extend from the cell body of the neuron (fig. 7.4). **Bipolar neurons** have two processes, one at either end; this type is found in the retina of the eye. **Multipolar**

neurons, the most common type, have several dendrites and one axon extending from the cell body; motor neurons are good examples of this type. A **pseudounipolar neuron** has a single short process that divides like a T to form a longer process. Sensory neurons are pseudounipolar—one end of the longer process receives sensory stimuli and produces nerve impulses; the other end delivers these impulses to synapses within the brain or spinal cord. Anatomically, the part of the process that conducts impulses toward the cell body can be considered a dendrite and the part that conducts impulses away from the cell body can be considered an axon. Functionally, however, the entire process behaves as a single long axon and may be covered with a myelin sheath (discussed later), as are axons. Only the small branches at the receptive end of the process function as typical dendrites.

A **nerve** is a bundle of axons located outside the CNS. Most nerves are composed of both motor and sensory fibers and are thus called *mixed nerves*. Some of the cranial nerves, however, contain sensory fibers only. These are the nerves that serve the special senses of sight, hearing, taste, and smell.

Supporting Cells

Unlike other organs that are "packaged" in connective tissue derived from mesoderm (the middle layer of embryonic tissue), the supporting cells of the nervous system are derived from the same embryonic tissue layer (ectoderm) that produces neurons. There are six categories of supporting cells: (1) **Schwann cells,** which form myelin sheaths around peripheral axons; (2) **oligodendrocytes,** which form myelin sheaths around axons of the CNS;

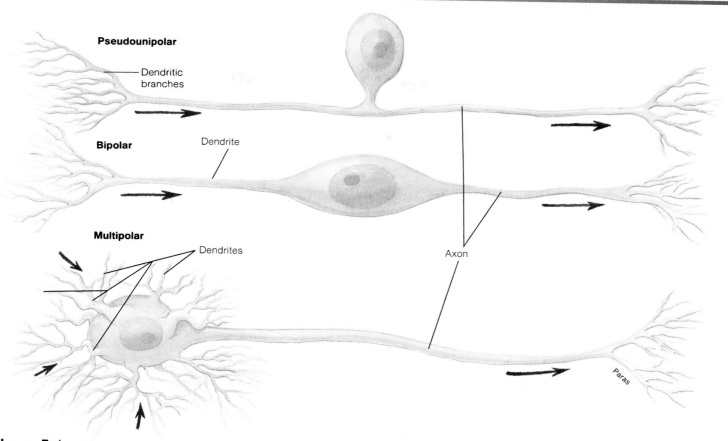

Figure 7.4

Three different types of neurons.

Table 7.3	Supporting Cells and Their Functions*	
Supporting Cells	**Location**	**Functions**
Schwann cells	PNS	Surround axons of all peripheral nerve fibers, forming neurilemmal sheath, or sheath of Schwann; wrap around many peripheral fibers to form myelin sheaths
Satellite cells	PNS	Support functions of neurons within sensory and autonomic ganglia
Oligodendrocytes	CNS	Form myelin sheaths around central axons, producing "white matter" of CNS
Astrocytes	CNS	Perivascular foot processes that cover capillaries within brain and contribute to the blood-brain barrier
Microglia	CNS	Amoeboid cells within CNS that are phagocytic
Ependymal cells	CNS	Form epithelial lining of brain cavities (ventricles) and central canal of spinal cord; cover tufts of capillaries to form choroid plexuses—structures that produce cerebrospinal fluid

*Supporting cells in the CNS are known as neuroglia.

(3) **microglia,** which are phagocytic cells that migrate through the CNS and remove foreign and degenerated material; (4) **astrocytes,** which help to regulate the external environment of neurons in the CNS; (5) **ependymal cells,** which line the ventricles of the brain and the central canal of the spinal cord; and (6) **satellite cells,** which support neuron cell bodies within the ganglia of the PNS (table 7.3).

Sheath of Schwann and Myelin Sheath

Some axons in the CNS and PNS are surrounded by a myelin sheath and are known as *myelinated axons*. Other axons do not have a myelin sheath and are *unmyelinated*. The myelin sheaths in the PNS are formed by Schwann cells, and the myelin sheaths in the CNS are formed by oligodendrocytes.

All axons in the PNS are surrounded by a living sheath of Schwann cells, known as the **sheath of Schwann.** The outer surface of this layer of Schwann cells is encased in a glycoprotein *basement membrane*, called the *neurilemma*, which is analogous to the basement membrane that underlies epithelial membranes. The axons of the CNS, by contrast, lack a sheath of Schwann (Schwann cells are found only in the PNS) and also lack a continuous basement membrane. This is significant in terms of nerve regeneration, as will be described in a later section.

The Nervous System

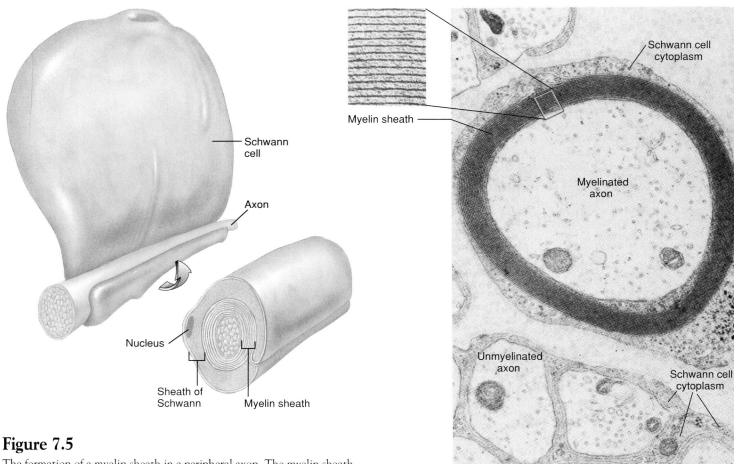

Figure 7.5

The formation of a myelin sheath in a peripheral axon. The myelin sheath is formed by successive wrappings of the Schwann cell membranes, leaving most of the Schwann cell cytoplasm outside the myelin. The sheath of Schwann is thus external to the myelin sheath.

Figure 7.6

An electron micrograph of unmyelinated and myelinated axons.

Axons that are smaller than 2 micrometers (2 μm) in diameter are usually unmyelinated. Larger axons are generally surrounded by a **myelin sheath,** which is composed of successive wrappings of the cell membrane of Schwann cells (in the PNS) or oligodendrocytes (in the CNS).

In the process of myelin formation in the PNS, Schwann cells roll around the axon, much like a roll of electrician's tape is wrapped around a wire. Unlike electrician's tape, however, the wrappings are made in the same spot, so that each wrapping overlaps the previous layers. The cytoplasm, meanwhile, becomes squeezed to the outer region of the Schwann cell, much as toothpaste is squeezed to the top of the tube as the bottom is rolled up (fig. 7.5). Each Schwann cell wraps only about 1 mm of axon, leaving gaps of exposed axon between the adjacent Schwann cells. These gaps in the myelin sheath are known as the **nodes of Ranvier.** The successive wrappings of Schwann cell membrane provide insulation around the axon, leaving only the nodes of Ranvier exposed to produce nerve impulses.

The Schwann cells remain alive as their cytoplasm is squeezed to the outside of the myelin sheath. As a result, myelinated axons of the PNS, like their unmyelinated counterparts, are surrounded by a living sheath of Schwann, to the outside of which is a continuous basement membrane (fig. 7.6).

The myelin sheaths of the CNS are formed by oligodendrocytes. Unlike a Schwann cell, which forms a myelin sheath around only one axon, each oligodendrocyte has extensions, like the tentacles of an octopus, that form myelin sheaths around several axons (fig. 7.7). Myelinated axons of the CNS, as a result, are not surrounded by a continuous basement membrane. The myelin sheaths around axons of the CNS give this tissue a white color; areas of the CNS that contain a high concentration of axons thus form the **white matter.** The **gray matter** of the CNS is composed of high concentrations of cell bodies and dendrites, which lack myelin sheaths. (The only dendrites that have myelin sheaths are those of the sensory neurons of the PNS.)

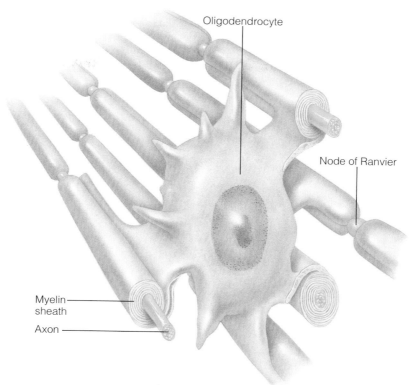

Oligodendrocyte

Node of Ranvier

Myelin sheath

Axon

Figure 7.7

The formation of myelin sheaths in the central nervous system by an oligodendrocyte. One oligodendrocyte forms myelin sheaths around several axons.

 **Multiple sclerosis (MS)** is a neurological disease usually diagnosed in people between the ages of 20 and 40. It is a chronic, degenerating, remitting, and relapsing disease that progressively destroys the myelin sheaths of neurons in multiple areas of the CNS. Initially, lesions form on the myelin sheaths and soon develop into hardened *scleroses,* or scars (*skleros* = hardened). Destruction of the myelin sheaths prohibits the normal conduction of impulses, resulting in a progressive loss of functions. Because myelin degeneration is widespread, MS has a wider variety of symptoms than any other neurological disease. This multiplicity of symptoms, coupled with remissions, frequently causes misdiagnosis of this disease.

Regeneration of a Cut Axon

When an axon in a peripheral nerve is cut, the distal portion of the axon that was severed from the cell body degenerates and is phagocytosed by Schwann cells. The Schwann cells, surrounded by the basement membrane, then form a *regeneration tube* (fig. 7.8), as the part of the axon that is connected to the cell body begins to grow and exhibit amoeboid movement. The Schwann cells of the regeneration tube are believed to secrete chemicals that attract the growing axon tip, and the regeneration tube helps to guide the regenerating axon to its proper des-

tination. Even a severed major nerve may be surgically reconnected and the function of the nerve largely reestablished if the surgery is performed before tissue death.

Injury in the CNS stimulates growth of axon collaterals, but central axons have a much more limited ability to regenerate than peripheral axons. This may be due in part to the absence of a continuous basement membrane (as is present in the PNS), which precludes the formation of a regeneration tube. Also, other aspects of the central nervous system environment make it inhospitable for axon regeneration.

 Experiments in vitro suggest that central axons can regenerate if they are provided with the appropriate environment. In a developing fetal brain, chemicals called **neurotrophins** promote neuron growth. *Nerve growth factor* was the first neurotrophin to be identified of the several currently known. Neurotrophins have also been shown to promote neuron regeneration in adult brains and spinal cords in some experimental animals. Additionally, some chemicals, including *myelin-associated inhibitory proteins,* have been shown to inhibit axon regeneration. In a recent experiment, axon regeneration in a rat spinal cord was improved by simultaneously blocking the inhibitory proteins with antibodies while providing a neurotrophin.

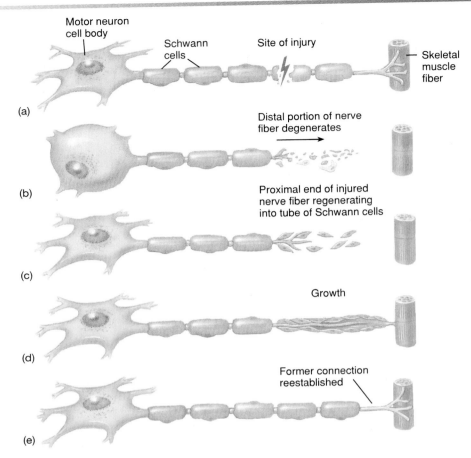

Figure 7.8

The process of peripheral neuron regeneration. (*a*) If a neuron is severed through a myelinated axon, the proximal portion may survive, but (*b*) the distal portion will degenerate through phagocytosis. The myelin sheath provides a pathway (*c* and *d*) for the regeneration of an axon, and (*e*) innervation is restored.

Astrocytes and the Blood-Brain Barrier

Astrocytes (*aster* = star) are large, stellate cells with numerous cytoplasmic processes that radiate outward. They are the most abundant of the neuroglial cells in the CNS, constituting up to 90% of the nervous tissue in some areas of the brain.

Astrocytes (fig. 7.9) are known to interact with neurons in two different ways. First, they have been shown to take up potassium ions from the extracellular fluid. Since K^+ is released from active neurons during the production of nerve impulses (discussed in a later section), this action of astrocytes may be very important in maintaining a proper ionic environment for the neurons. Second, astrocytes have been shown to take up specific neurotransmitter chemicals (which are released from the axon endings, as described in a later section). These neurotransmitters, known as glutamic acid and gamma-aminobutyric acid (GABA), are broken down within the astrocytes. The molecule produced from this breakdown—glutamine—is released from the astrocytes and made available to the neurons in order for them to resynthesize these particular neurotransmitters.

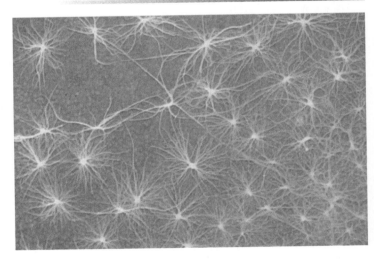

Figure 7.9

A photomicrograph showing the perivascular feet of the star-shaped neuroglial cells called astrocytes. Astrocyte processes cover most of the surface area of brain capillaries.

Chapter Seven

The blood-brain barrier presents difficulties in the chemotherapy of brain diseases because drugs that could enter other organs may not be able to enter the brain. In the treatment of *Parkinson's disease,* for example, patients who need a chemical called dopamine in the brain must be given a precursor molecule called levodopa (L-dopa) because L-dopa can cross the blood-brain barrier but dopamine cannot. Also, many antibiotics, including penicillin, cannot cross the blood-brain barrier; therefore, in treating infections such as *meningitis,* only those antibiotics that can cross the blood-brain barrier are used.

Astrocytes have also been shown to interact with blood capillaries within the brain. Indeed, the brain capillaries are almost entirely surrounded by extensions of the astrocytes commonly known as *perivascular feet.* This association between astrocytes and brain capillaries has important physiological consequences.

Capillaries in the brain, unlike those of most other organs, do not have pores between adjacent endothelial cells (the cells that compose the walls of capillaries). Instead, the endothelial cells of brain capillaries are joined together by tight junctions. Unlike other organs, therefore, the brain cannot obtain molecules from the blood plasma by a nonspecific filtering process. Instead, molecules within brain capillaries must be moved through the endothelial cells by diffusion, active transport, endocytosis, and exocytosis. This feature of brain capillaries imposes a very selective **blood-brain barrier.** There is evidence to suggest that the development of tight junctions between adjacent endothelial cells in brain capillaries, and thus the development of the blood-brain barrier, results from the effects of astrocytes on the brain capillaries.

1. Draw a neuron, label its parts, and describe the functions of these parts.

2. Distinguish between sensory neurons, motor neurons, and association neurons (interneurons) in terms of structure, location, and function.

3. Describe the structure of the sheath of Schwann and explain how it promotes nerve regeneration. Explain how a myelin sheath in the PNS is formed.

4. Explain how myelin sheaths in the CNS are formed. How does the presence or absence of myelin sheaths in the CNS determine the color of this tissue?

5. Explain what is meant by the blood-brain barrier. Describe its structure and discuss its clinical significance.

Electrical Activity in Axons

The permeability of the axon membrane to Na^+ and K^+ is regulated by gates, which open in response to stimulation. Net diffusion of these ions occurs in two stages: first Na^+ moves into the axon, then K^+ moves out. This flow of ions, and the changes in the membrane potential that result, constitute an event called an action potential or a nerve impulse.

All cells in the body maintain a potential difference (voltage) across the membrane, or *resting membrane potential,* in which the inside of the cell is negatively charged in comparison to the outside of the cell (for example, -65 mV). As explained in chapter 6, this potential difference is largely the result of the permeability properties of the cell membrane. The membrane traps large, negatively charged organic molecules within the cell and permits only limited diffusion of positively charged inorganic ions. These properties result in an unequal distribution of these ions across the membrane. The action of the Na^+/K^+ pumps also helps to maintain a potential difference because they pump out three Na^+ ions for every two K^+ ions that they transport into the cell. Na^+ is thus more highly concentrated in the extracellular fluid than in the cell, whereas K^+ is more highly concentrated within the cell.

Although all cells have a membrane potential, only a few types of cells have been shown to alter their membrane potential in response to stimulation. Such alterations in membrane potential are achieved by varying the membrane permeability to specific ions in response to stimulation. A central aspect of the physiology of neurons and muscle cells is their ability to produce and conduct these changes in membrane potential. Such an ability is termed *excitability* or *irritability.*

An increase in membrane permeability to a specific ion results in the diffusion of that ion down its concentration gradient, either into or out of the cell. These *ion currents* occur only across limited patches of membrane (located fractions of a millimeter apart), where specific ion channels are located. Changes in the potential difference across the membrane at these points can be measured by the voltage developed between two electrodes—one placed inside the cell, the other placed outside the cell membrane at the region being recorded. The voltage between these two *recording electrodes* can be visualized by connecting them to an oscilloscope (fig. 7.10).

In an oscilloscope, electrons from a cathode-ray "gun" are sprayed across a fluorescent screen, producing a line of light. Changes in the potential difference between the two recording electrodes cause this line to deflect. The oscilloscope can be calibrated in such a way that an upward deflection of the line indicates that the inside of the membrane has become less negative (or more positive) compared to the outside of the membrane. A downward deflection of the line, conversely, indicates

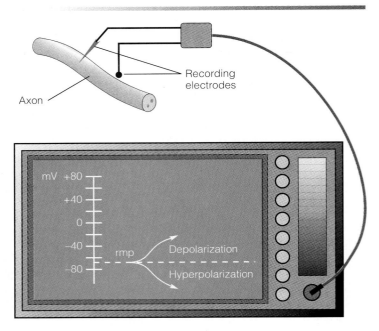

Figure 7.10

The difference in potential (in millivolts) between an intracellular and extracellular recording electrode is displayed on an oscilloscope screen. The resting membrane potential (rmp) of the axon may be reduced (depolarization) or increased (hyperpolarization).

that the inside of the cell has become more negative. The oscilloscope can thus function as a fast-responding voltmeter with an ability to display voltage changes as a function of time.

If both recording electrodes are placed outside of the cell, the potential difference between the two will be zero (because there is no charge separation). When one of the two electrodes penetrates the cell membrane, the oscilloscope will indicate that the intracellular electrode is electrically negative with respect to the extracellular electrode; a membrane potential is recorded. We will call this the **resting membrane potential** to distinguish it from events that are described in later sections. If appropriate stimulation causes positive charges to flow into the cell, the line will deflect upward. This change is called **depolarization,** or hypopolarization, because the potential difference between the two recording electrodes is reduced. A return to the resting membrane potential is known as **repolarization.** If stimulation causes negative charges to flow into the cell so that the inside of the cell becomes more negative than the resting membrane potential, the line on the oscilloscope will deflect downward. This change is called **hyperpolarization.**

Ion Gating in Axons

The changes in membrane potential just described—depolarization, repolarization, and hyperpolarization—are caused by changes in the net flow of ions through ion channels in the membrane. The permeability of the membrane to Na⁺, K⁺, and other ions is regulated by parts of the ion channels through the

membrane called **gates.** Gates are polypeptide chains that can open or close a membrane channel according to specific conditions. When the gates of specific ion channels are closed, the membrane is not very permeable to that ion, and when the gates are opened, the permeability to that ion can be greatly increased.

The ion channels for Na⁺ and K⁺ are fairly specific for each of these ions. It is believed that there are two types of channels for K⁺; one type lacks gates and is always open, whereas the other type has gates that are closed in the resting cell. Channels for Na⁺, by contrast, always have gates, and these gates are closed in the resting cell. The resting cell is thus more permeable to K⁺ than to Na⁺. (As described in chapter 6, some Na⁺ does leak into the cell; this leakage may occur in a nonspecific manner through open K⁺ channels.) The resting membrane potential is thus close to, but slightly less than, the equilibrium potential for K⁺ (described in chapter 6).

Depolarization of a small region of an axon can be experimentally induced by a pair of stimulating electrodes that act as if they were injecting positive charges into the axon. If two recording electrodes are placed in the same region (one electrode within the axon and one outside), an upward deflection of the oscilloscope line will be observed as a result of this depolarization. If a certain level of depolarization is achieved (from –65 mV to –55 mV, for example) by this artificial stimulation, a sudden and very rapid change in the membrane potential will be observed. This is because *depolarization to a threshold level causes the Na⁺ gates to open.* Now the permeability properties of the membrane are changed, and Na⁺ diffuses down its concentration gradient into the cell.

A fraction of a second after the Na⁺ gates open, they close again. Just before they do, *the depolarization stimulus causes the K⁺ gates to open.* This makes the membrane more permeable to K⁺ than it is at rest, and K⁺ diffuses down its concentration gradient out of the cell. The K⁺ gates will then close and the permeability properties of the membrane will return to what they were at rest.

The gated ion channels in the axon membrane are therefore said to be **voltage regulated.** These gates are closed at the resting membrane potential of –65 mV and open in response to depolarization of the membrane (fig. 7.11).

Action Potentials

In this section we will consider the events that occur at one point in an axon, when a small region of axon membrane is stimulated artificially and responds with changes in ion permeabilities. The resulting changes in membrane potential at this point are detected by recording electrodes placed in this region of the axon. The nature of the stimulus in vivo (in the body), and the manner by which electrical events are conducted to different points along the axon, will be described in later sections.

When the axon membrane has been depolarized to a threshold level—by stimulating electrodes, in the previous example—the Na⁺ gates open and the membrane becomes permeable to Na⁺. This permits Na⁺ to enter the axon by diffusion, which further depolarizes the membrane (makes the inside less negative, or more positive). Since the Na⁺ gates of the axon are voltage regulated, this further depolarization makes the membrane even more permeable to Na⁺, so that even more Na⁺ can enter the cell and open even more voltage-regulated Na⁺ gates. A *positive feedback loop* (fig. 7.12) is thus created, which causes the rate of Na⁺ entry and depolarization to accelerate in an explosive fashion.

After a slight delay following the opening of the Na⁺ gates, depolarization of the axon membrane also causes the opening of voltage-regulated K⁺ gates and the diffusion of K⁺ out of the cell. Since K⁺ is positively charged, the diffusion of K⁺ out of the cell makes the inside of the cell less positive, or more negative, and acts to restore the original resting membrane potential. This process is called **repolarization** and represents the completion of a *negative feedback loop* (fig. 7.12).

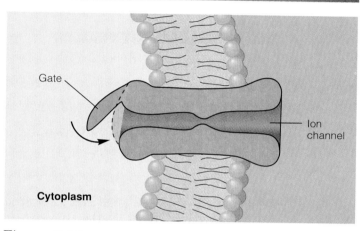

Figure 7.11

A diagram of a voltage-gated ion channel, which opens in response to depolarization.

Figure 7.13 (bottom) illustrates the movement of Na⁺ and K⁺ through the axon membrane in response to a depolarization stimulus. Notice that the explosive increase in Na⁺ diffusion causes rapid depolarization to 0 mV and then *overshoot* of the membrane potential so that the inside of the membrane actually becomes positively charged (almost +40 mV) compared to the outside (fig. 7.13 [top]). The greatly increased permeability to Na⁺ thus drives the membrane potential toward the equilibrium potential for Na⁺ (chapter 6). The Na⁺ permeability then rapidly decreases as the diffusion of K⁺ increases, resulting in repolarization to the resting membrane potential. These changes in Na⁺ and K⁺ diffusion and the resulting changes in the membrane potential that they produce constitute an event called the **action potential,** or **nerve impulse.**

Once an action potential has been completed, the Na⁺/K⁺ pumps will extrude the extra Na⁺ that has entered the axon and recover the K⁺ that has diffused out of the axon. This active transport of ions occurs very quickly because the events described occur across only a very small area of membrane, and so only a relatively small amount of Na⁺ and K⁺ actually diffuse through the membrane during the production of an action potential. The total concentrations of Na⁺ and K⁺ in the axon and in the extracellular fluid are not significantly changed during an action potential. Even during the overshoot phase, for example, the concentration of Na⁺ remains higher outside the axon; repolarization thus requires the outward diffusion of K⁺, the concentration gradient of which is opposite to that of the Na⁺.

Notice that active transport processes are not directly involved in the production of an action potential; both depolarization and repolarization are produced by the diffusion of ions down their concentration gradients. A neuron poisoned with cyanide, so that it cannot produce ATP, can still produce action potentials for a period of time. After awhile, however, the lack of ATP for active transport by the Na⁺/K⁺ pumps will result in a decline in the ability of the axon to produce action

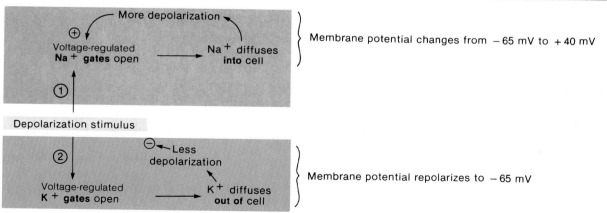

Figure 7.12

Depolarization of an axon has two effects: (1) Na⁺ gates open and Na⁺ diffuses into the cell and (2), after a brief period, K⁺ gates open and K⁺ diffuses out of the cell. An inward diffusion of Na⁺ causes further depolarization, which in turn causes further opening of Na⁺ gates in a positive feedback (+) fashion. The opening of K⁺ gates and outward diffusion of K⁺ makes the inside of the cell more negative, and thus has a negative feedback effect (−) on the initial depolarization.

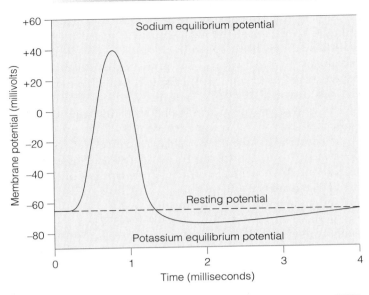

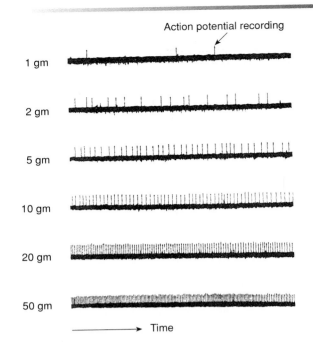

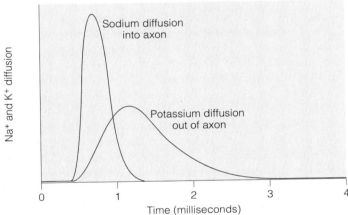

Figure 7.13

An action potential (*top*) is produced by an increase in sodium diffusion that is followed, with a short time delay, by an increase in potassium diffusion (*bottom*). This drives the membrane potential first toward the sodium equilibrium potential and then toward the potassium equilibrium potential.

Figure 7.14

Recordings from a single sensory fiber of a sciatic nerve of a frog stimulated by varying degrees of stretch of the gastrocnemius muscle. Notice that increasing degrees of stretch (indicated by increasing weights attached to the muscle) result in an increased frequency of action potentials.

potentials. This shows that the Na^+/K^+ pumps are not directly involved, but are instead required to maintain the concentration gradients needed for the diffusion of Na^+ and K^+ during action potentials.

All-or-None Law

Once a region of axon membrane has been depolarized to a threshold value, the positive feedback effect of depolarization on Na^+ permeability and of Na^+ permeability on depolarization causes the membrane potential to shoot toward about +40 mV. It does not normally become more positive because the Na^+ gates quickly close and the K^+ gates open. The length of time that the Na^+ and K^+ gates stay open is independent of the strength of the depolarization stimulus.

The amplitude of action potentials is therefore **all or none.** When depolarization is below a threshold value, the voltage-regulated gates are closed; when depolarization reaches threshold, a maximum potential change (the action potential) is produced. Since the change from −65 mV to +40 mV and back to −65 mV lasts only about 3 msec, the image of an action potential on an oscilloscope screen looks like a spike. Action potentials are therefore sometimes called *spike potentials*.

Since the gates are open for a fixed period of time, the duration of each action potential is about the same. Likewise, since the concentration gradient for Na^+ is relatively constant, the amplitude of each action potential is about the same in all axons at all times (from −65 mV to +40 mV, or about 100 mV in total amplitude).

Coding for Stimulus Intensity

If one depolarization stimulus is greater than another, the greater stimulus strength is not coded by a greater amplitude of action potentials (because action potentials are all-or-none events). The code for stimulus strength in the nervous system is not amplitude modulated (AM). When a greater stimulus strength is applied to a neuron, identical action potentials are produced more frequently (more are produced per minute). Therefore, the code for stimulus strength in the nervous system is frequency modulated (FM). This concept is illustrated in figure 7.14.

When an entire collection of axons (in a nerve) is stimulated, different axons will be stimulated at different stimulus intensities. A low-intensity stimulus will only activate those few

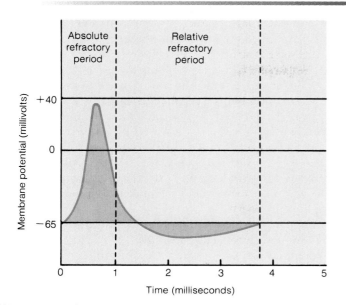

Figure 7.15

Absolute and relative refractory periods. While a segment of axon is producing an action potential, the membrane is absolutely or relatively resistant (refractory) to further stimulation.

fibers with low thresholds, whereas high-intensity stimuli can activate fibers with higher thresholds. As the intensity of stimulation increases, more and more fibers will become activated. This process, called **recruitment,** represents another mechanism by which the nervous system can code for stimulus strength.

Refractory Periods

If a stimulus of a given intensity is maintained at one point of an axon and depolarizes it to threshold, action potentials will be produced at that point at a given frequency (number per minute). As the stimulus strength is increased, the frequency of action potentials produced at that point will increase accordingly. As action potentials are produced with increasing frequency, the time between successive action potentials will decrease—but only up to a minimum time interval. The interval between successive action potentials will never become so short as to allow a new action potential to be produced before the preceding one has finished.

During the time that a patch of axon membrane is producing an action potential, it is incapable of responding—or *refractory*—to further stimulation. If a second stimulus is applied while the Na⁺ gates are open in response to a first stimulus, the second stimulus cannot have any effect (the gates are already open). During the time that the Na⁺ gates are open, therefore, the membrane is in an **absolute refractory period** and cannot respond to any subsequent stimulus. If a second stimulus is applied while the K⁺ gates are open (and the membrane is in the process of repolarizing), the membrane is in a **relative refractory period.** During this time, only a very strong stimulus can depolarize the membrane and produce a second action potential (fig. 7.15).

As a result of the fact that the cell membrane is refractory during the time it is producing an action potential, each action

potential remains a separate, all-or-none event. In this way, as a continuously applied stimulus increases in intensity, its strength can be coded strictly by the frequency of the action potentials it produces at each point of the axon membrane.

You might think that as an axon produces a large number of action potentials, the relative concentrations of Na⁺ and K⁺ would be changed in the extracellular and intracellular compartments. This is not the case. In a typical mammalian axon that is 1 μm in diameter, for example, only one intracellular K⁺ ion in 3,000 would be exchanged for a Na⁺. Since a typical neuron has about 1 million Na⁺/K⁺ pumps, which can transport nearly 200 million ions per second, these small changes can be quickly corrected.

Cable Properties of Neurons

If a pair of stimulating electrodes produces a depolarization that is too weak to cause the opening of voltage-regulated Na⁺ gates—that is, if the depolarization is below threshold (about –55 mV)—the change in membrane potential will be *localized* to within 1 to 2 mm of the point of stimulation. For example, if the stimulus causes depolarization from –65 mV to –60 mV at one point, and the recording electrodes are placed only 3 mm away from the stimulus, the membrane potential recorded will remain at –65 mV (the resting potential). The axon is thus a very poor conductor compared to a metal wire.

The term *cable properties* refers to the ability of a neuron to transmit charges through its cytoplasm. These cable properties are quite poor because there is a high internal resistance to the spread of charges and because many charges leak out of the axon through its membrane. If an axon had to conduct only through its cable properties, therefore, no axon could be more than a millimeter in length. The fact that some axons are a meter or more in length suggests that the conduction of nerve impulses does not rely on the cable properties of the axon.

Conduction of Nerve Impulses

When stimulating electrodes artificially depolarize one point of an axon membrane to a threshold level, voltage-regulated gates open and an action potential is produced at that small region of axon membrane containing those gates. For about the first millisecond of the action potential, when the membrane voltage changes from –65 mV to +40 mV, a current of Na⁺ enters the cell by diffusion due to the opening of the Na⁺ gates. Each action potential thus "injects" positive charges (sodium ions) into the axon.

These positively charged sodium ions are conducted, by the cable properties of the axon, to an adjacent region that still has a membrane potential of –65 mV. Within the limits of the cable properties of the axon (1 to 2 mm), this helps to depolarize the adjacent region of axon membrane. When this adjacent region of membrane reaches a threshold level of depolarization, it too produces an action potential as its voltage-regulated gates open.

Each action potential thus acts as a stimulus for the production of another action potential at the next region of membrane that contains voltage-regulated gates. In the description

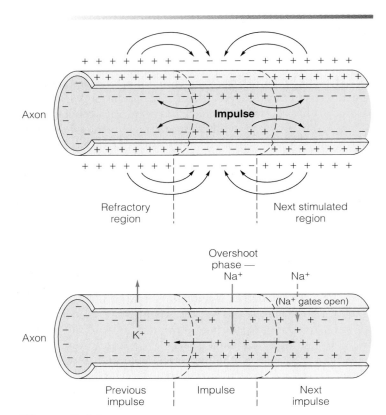

Figure 7.16

The conduction of a nerve impulse (action potential) in an unmyelinated nerve fiber (axon). Each action potential "injects" positive charges that spread to adjacent regions. The region that has just produced an action potential is refractory. The next region, not having been stimulated previously, is partially depolarized. As a result, its voltage-regulated Na+ gates open and the process is repeated.

of action potentials earlier in this chapter, the stimulus for their production was artificial—depolarization produced by a pair of stimulating electrodes. Now it can be seen that an action potential at a given point along an axon is produced by depolarization that results from the production of a preceding action potential. This explains how all action potentials along an axon are produced after the first action potential has been generated.

Conduction in an Unmyelinated Axon

In an unmyelinated axon, every patch of membrane that contains Na+ and K+ gates can produce an action potential. Action potentials are thus produced at locations only a fraction of a micrometer apart all along the length of the axon.

The cablelike spread of depolarization induced by the influx of Na+ during one action potential helps to depolarize the adjacent regions of membrane—a process that is also aided by movements of ions on the outer surface of the axon membrane (fig. 7.16). This process would depolarize the adjacent membranes on each side of the region to produce an action potential,

but the area that had previously produced one cannot produce another at this time because it is still in its refractory period. In this way, action potentials are passed in one direction only along the axon.

It is important to recognize that action potentials are not really "conducted," although it is convenient to use that word. Each action potential is a separate, complete event that is repeated, or *regenerated*, along the axon's length. The action potential produced at the end of the axon is thus a completely new event that was produced in response to depolarization from the previous action potential. The last action potential has the same amplitude as the first. Action potentials are thus said to be **conducted without decrement** (without decreasing in amplitude).

The spread of depolarization by the cable properties of an axon is fast compared to the time involved in producing an action potential. Thus, the more action potentials along a given stretch of axon that have to be produced, the slower the conduction. Since action potentials must be produced at every fraction of a micrometer in an unmyelinated axon, the conduction rate is relatively slow. This conduction rate is somewhat faster if the unmyelinated axon is thicker, since the ability of fibers to conduct charges by cable properties improves with increasing diameter. The conduction rate is substantially faster if the axon is myelinated because fewer action potentials are produced along a given length of myelinated axon.

Conduction in a Myelinated Axon

The myelin sheath provides insulation for the axon, preventing movements of Na+ and K+ through the membrane. If the myelin sheath were continuous, therefore, action potentials could not be produced. The myelin thus has interruptions known as the *nodes of Ranvier*.

Because the cable properties of axons can conduct depolarizations only over a very short distance (1–2 mm), the nodes of Ranvier must be very close together (usually they are about 1 mm apart). Studies have shown that Na+ channels are highly concentrated at the nodes (estimated at 10,000 per square micrometer) and almost absent in the regions of axon membrane between the nodes. Action potentials, therefore, occur only at the nodes of Ranvier (fig. 7.17) and seem to "leap" from node to node; this is called **saltatory conduction** (*saltario* = leap). The leaping is, of course, just a metaphor; the action potential at one node depolarizes the membrane at the next node to threshold, so that a new action potential is produced at the next node of Ranvier.

Since the cablelike spread of depolarization between the nodes is very fast and fewer action potentials need to be produced per given length of axon, saltatory conduction allows a *faster rate of conduction* than is possible in an unmyelinated fiber. Conduction rates in the human nervous system vary from 1.0 m/sec—in thin, unmyelinated fibers that mediate slow, visceral responses—to faster than 100 m/sec (225 miles per hour)—in thick, myelinated fibers involved in quick stretch reflexes in skeletal muscles (table 7.4).

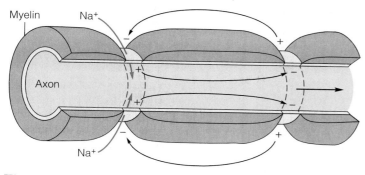

Figure 7.17

The conduction of the nerve impulse in a myelinated nerve fiber. Since the myelin sheath prevents inward Na⁺ current, action potentials can be produced only at gaps in the myelin sheath, called the nodes of Ranvier. This "leaping" of the action potential from node to node is known as *saltatory conduction*.

Table 7.4	Conduction Velocities and Functions of Mammalian Nerves of Different Diameters	
Diameter (μm)	**Conduction Velocity (m/sec)**	**Examples of Functions Served**
12–22	70–120	Sensory: muscle position
5–13	30–90	Somatic motor fibers
3–8	15–40	Sensory: touch, pressure
1–5	12–30	Sensory: pain, temperature
1–3	3–15	Autonomic fibers to ganglia
0.3–1.3	0.7–2.2	Autonomic fibers to smooth and cardiac muscles

1. Define the terms *depolarization* and *repolarization* and illustrate these processes graphically.

2. Describe how the permeability of the axon membrane to Na⁺ and K⁺ is regulated and how changes in permeability to these ions affect the membrane potential.

3. Describe how gating of Na⁺ and K⁺ in the axon membrane results in the production of an action potential.

4. Explain the all-or-none law of action potentials and describe the effect of increased stimulus strength on action potential production. How do the refractory periods affect the maximum frequency of action potential production?

5. Describe how action potentials are conducted by unmyelinated nerve fibers. Why is saltatory conduction in myelinated fibers more rapid?

The Synapse

Axons end close to, or in some cases at the point of contact with, another cell. Once action potentials reach the end of an axon, they directly or indirectly stimulate (or inhibit) the other cell. In specialized cases, action potentials can directly pass from one cell to another. In most cases, however, the action potentials stop at the axon ending where they stimulate the release of a chemical neurotransmitter that affects the next cell.

A **synapse** is the functional connection between a neuron and a second cell. In the CNS, this other cell is also a neuron. In the PNS the other cell may be a neuron or an *effector cell* within a muscle or a gland. Although the physiology of neuron-neuron synapses and neuron-muscle synapses is similar, the latter synapses are often called **myoneural,** or **neuromuscular, junctions.**

Neuron-neuron synapses usually involve a connection between the axon of one neuron and the dendrites, cell body, or axon of a second neuron. These are called, respectively, *axodendritic, axosomatic,* and *axoaxonic synapses* (fig. 7.18). In almost all synapses, transmission is in one direction only—from the axon of the first (or presynaptic) cell to the second (or postsynaptic) cell. *Dendrodendritic synapses* do not fit this classic pattern; in these synapses, two dendrites from different neurons make reciprocal innervations—some of these synapses conduct in one direction and others conduct in the opposite direction.

In the early part of the twentieth century, most physiologists believed that synaptic transmission was *electrical*—that is, that action potentials were conducted directly from one cell to the next. This was a logical assumption given that nerve endings appeared to touch the postsynaptic cells and that the delay in synaptic conduction was extremely short (about 0.5 msec). Improved histological techniques, however, revealed tiny gaps in the synapses, and experiments demonstrated that the actions of autonomic nerves could be duplicated by certain chemicals. This led to the hypothesis that synaptic transmission might be *chemical*—that the presynaptic nerve endings might release chemicals called **neurotransmitters** that stimulated action potentials in the postsynaptic cells.

In 1921, a physiologist named Otto Loewi published the results of an experiment suggesting that synaptic transmission was indeed chemical, at least at the junction between a branch of the vagus nerve (chapter 9) and the heart. He had isolated the heart of a frog and—while stimulating the branch of the vagus that innervates the heart—perfused the heart with an isotonic salt solution. Stimulation of this nerve slowed the heart rate, as expected. More importantly, application of this salt solution to the heart of a second frog caused the second heart to slow its rate of beat.

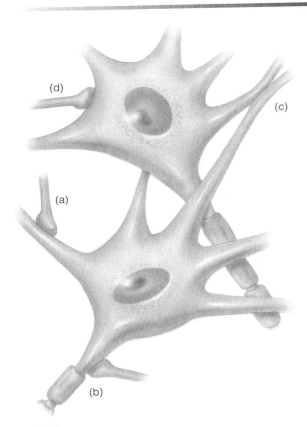

Figure 7.18
Different types of synapses: (*a*) axodendritic, (*b*) axoaxonic, (*c*) dendrodendritic, and (*d*) axosomatic.

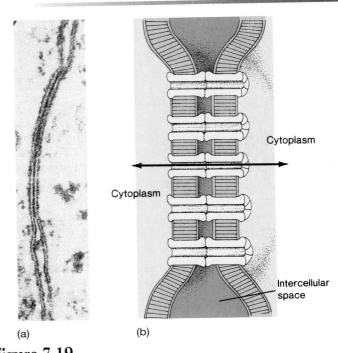

Figure 7.19
(*a*) An electron micrograph showing a gap junction. (*b*) An illustration of a gap junction, with the arrow indicating a channel through which ions and molecules may pass.

Loewi concluded that the nerve endings of the vagus must have released a chemical—which he called *vagusstoff*—that inhibited the heart rate. This chemical was subsequently identified as **acetylcholine,** or **ACh.** In the decades following Loewi's discovery, many other examples of chemical synapses were discovered, and the theory of electrical synaptic transmission fell into disrepute. More recent evidence, ironically, has shown that electrical synapses do exist in the nervous system (though they are the exception), within smooth muscles, and between cardiac cells in the heart.

Electrical Synapses: Gap Junctions

In order for two cells to be electrically coupled, they must be approximately equal in size and they must be joined by areas of contact with low electrical resistance. In this way, impulses can be regenerated from one cell to the next without interruption—and without Frankenstein-like sparks between cells.

Adjacent cells that are electrically coupled are joined together by **gap junctions.** In gap junctions, the membranes of the two cells are separated by only 2 nanometers (1 nanometer equals 10^{-9} meter). A surface view of gap junctions in the electron microscope reveals hexagonal arrays of particles that are believed to be channels through which ions and molecules may pass from one cell to the next (fig. 7.19).

Gap junctions are present in cardiac muscle and some smooth muscles, where they allow excitation and rhythmic contraction of large masses of muscle cells. Gap junctions have also been observed in various regions of the brain. Although their functional significance in the brain is unknown, it has been speculated that they may allow a two-way transmission of impulses (in contrast to chemical synapses, which are always one-way). Gap junctions have also been observed between neuroglial cells, which do not produce electrical impulses; these may act as channels for the passage of informational molecules between cells. It is interesting in this regard that many embryonic tissues have gap junctions, and that these gap junctions disappear as the tissue becomes more specialized.

Chemical Synapses

Transmission across the majority of synapses in the nervous system is one-way and occurs through the release of chemical neurotransmitters from presynaptic axon endings. These presynaptic endings, which are called **terminal boutons** (*bouton* = button) because of their swollen appearance, are separated from the postsynaptic cell by a **synaptic cleft** so narrow that it can be seen clearly only with an electron microscope (fig. 7.20).

Neurotransmitter molecules within the presynaptic neuron endings are contained within many small, membrane-enclosed **synaptic vesicles.** In order for the neurotransmitter within these vesicles to be released into the synaptic cleft, the vesicle membrane must fuse with the axon membrane in the process of *exocytosis.* The neurotransmitter is released *in*

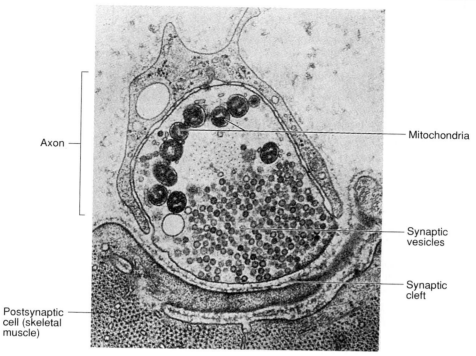

Axon

Mitochondria

Synaptic vesicles

Synaptic cleft

Postsynaptic cell (skeletal muscle)

Figure 7.20

An electron micrograph of a chemical synapse showing synaptic vesicles at the end of an axon.

quanta—that is, in multiples of the amount contained in one vesicle. The number of vesicles that undergo exocytosis is directly related to the frequency of action potentials produced at the presynaptic axon ending. Therefore, when stimulation to the presynaptic axon is increased, more vesicles will undergo exocytosis and release their neurotransmitters, so that the postsynaptic cell will be more greatly affected.

Action potentials in the terminal bouton of the presynaptic axon cause the opening of voltage-regulated Ca^{++} channels. There is thus a sudden, transient inflow of Ca^{++} into the presynaptic endings. Ca^{++} activates a regulatory protein within the cytoplasm called **calmodulin,** which in turn activates an enzyme called **protein kinase.** This enzyme phosphorylates (adds a phosphate group to) specific proteins known as **synapsins** in the membrane of the synaptic vesicle. This permits the vesicle to undergo exocytosis and release its content of neurotransmitter molecules (fig. 7.21). Since regulation by Ca^{++}, calmodulin, and protein kinase is also involved in the action of some hormones, it is discussed in more detail in chapter 11.

Once the neurotransmitter molecules have been released from the presynaptic axon terminals, they diffuse rapidly across the synaptic cleft and reach the membrane of the postsynaptic cell. The neurotransmitters then bind to specific **receptor proteins** that are part of the postsynaptic membrane. Like enzyme proteins, receptor proteins have high specificity for their neurotransmitter, which is the **ligand** of the receptor protein. The term *ligand* in this case refers to a smaller molecule that binds to and forms a complex with a larger protein molecule— the receptor. Binding of the neurotransmitter ligand

Botulinum toxins are neurotoxins released by the bacterium *Clostridium botulinum.* These toxins are *endopeptidases*—enzymes that cleave peptide bonds within a protein. Because of their specificity and location within axon terminals, these toxins cleave certain synapsins. This inhibits the exocytosis of synaptic vesicles and prevents the release of neurotransmitters. It is through this action that the symptoms of the disease **botulism** are produced.

to its receptor protein causes ion channels to open in the postsynaptic membrane. The gates that regulate these channels, therefore, can be called **chemically regulated gates** because they open in response to chemical changes in the postsynaptic cell membrane.

Note that two broad categories of gated ion channels have been described: voltage-regulated and chemically regulated. Voltage-regulated channels are found in the axons; chemically regulated channels are found in the postsynaptic membrane. Voltage-regulated channels open in response to depolarization; chemically regulated channels open in response to the binding of postsynaptic receptor proteins to their neurotransmitter ligands.

The chemically regulated channels are opened by a number of different mechanisms, and the effects of opening

The Nervous System

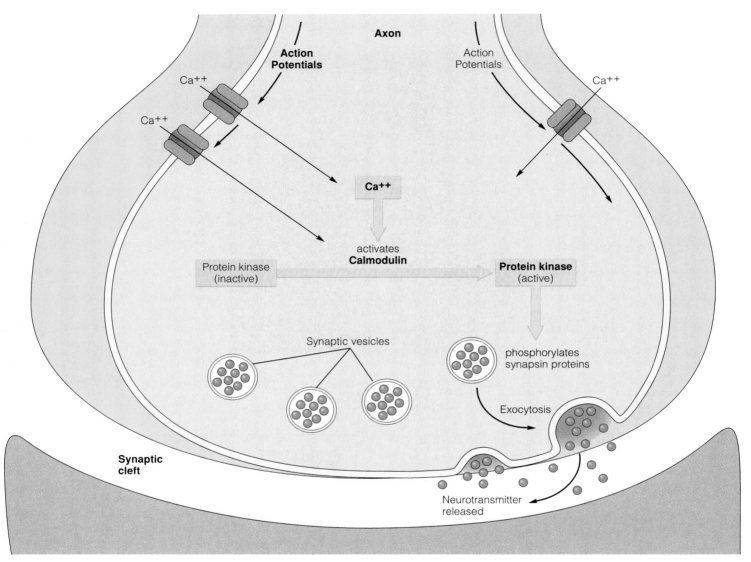

Figure 7.21

Action potentials at the axon terminals open Ca^{++} channels. The inward diffusion of Ca^{++} activates the protein calmodulin, which in turn activates the enzyme protein kinase. Activity of this enzyme results in the release of neurotransmitter molecules from the terminal bouton of the axon.

these channels vary. Opening of ion channels often produces a depolarization—the inside of the postsynaptic membrane becomes less negative. This depolarization is called an **excitatory postsynaptic potential (EPSP).** In other cases, a hyperpolarization occurs—the inside of the postsynaptic membrane becomes more negative. This hyperpolarization is called an **inhibitory postsynaptic potential (IPSP).** The mechanisms by which EPSPs and IPSPs are produced will be described in the following sections that deal with different types of neurotransmitters.

1. Describe the structure, locations, and functions of gap junctions.

2. Describe the location of neurotransmitters within an axon and explain the relationship between presynaptic axon activity and the amount of neurotransmitters released.

3. Describe the sequence of events by which action potentials stimulate the release of neurotransmitters from presynaptic axons.

4. Distinguish between voltage-regulated and chemically regulated ion channels.

Acetylcholine as a Neurotransmitter

When acetylcholine (ACh) binds to its receptor, it directly or indirectly causes the opening of chemically regulated gates. In most cases, this produces a depolarization called an excitatory postsynaptic potential, or EPSP. In some cases, however, ACh causes a hyperpolarization known as an inhibitory post-synaptic potential, or IPSP.

Acetylcholine (ACh) is used as an excitatory neurotransmitter by some neurons in the CNS and by somatic motor neurons at the neuromuscular junction. At autonomic nerve endings, ACh may be either excitatory or inhibitory, depending on the organ involved.

The varying responses of postsynaptic cells to the same chemical can be explained, in part, by the fact that different postsynaptic cells have different subtypes of ACh receptors. These receptor subtypes can be specifically stimulated by particular toxins, and are named according to these toxins. The stimulatory effect of ACh on skeletal muscle cells is produced by the binding of ACh to **nicotinic ACh receptors,** since these receptors can also be activated by nicotine. Effects of ACh on other cells occur when ACh binds to **muscarinic ACh receptors,** since these effects can also be produced by muscarine (from poisonous mushrooms).

Chemically Regulated Gated Channels

The binding of a neurotransmitter to its receptor protein can cause the opening of ion channels through two different mechanisms. These two mechanisms can be illustrated by the actions of ACh on the nicotinic and muscarinic subtypes of the ACh receptors.

Ligand-Operated Channels

This is the most direct mechanism by which chemically regulated gates can be opened. In this case, the ion channel runs through the receptor itself. The ion channel is opened by the binding of the receptor to the neurotransmitter ligand.

Such is the case when ACh binds to its nicotinic ACh receptor. This receptor consists of five polypeptide subunits that enclose the ion channel. Two of these subunits contain ACh binding sites, and the channel opens when both sites are bound to ACh (fig. 7.22). The opening of this channel permits the inward diffusion of Na^+, which produces a depolarization in the postsynaptic membrane. This acts as an excitatory postsynaptic potential (EPSP), as will be described in a separate section.

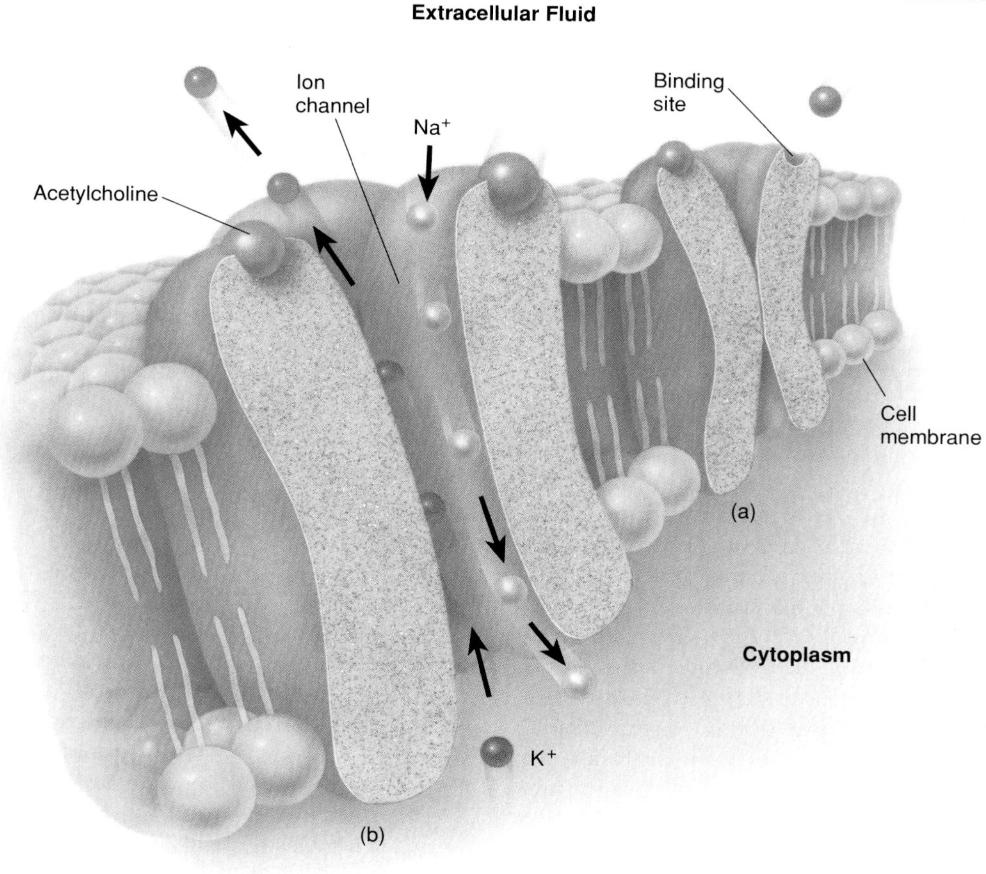

Figure 7.22

The nicotinic acetylcholine receptor contains a channel that is closed (*a*) until the receptor binds to ACh. (*b*) Na^+ and K^+ diffuse simultaneously, and in opposite directions, through the open ion channel.

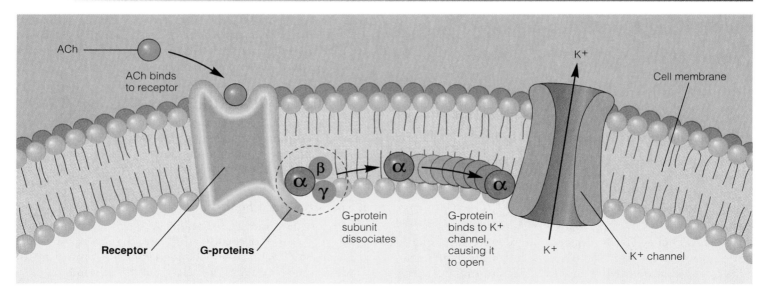

Figure 7.23

Binding of ACh to its muscarinic receptor causes the alpha G-protein subunit to dissociate from the other two. The alpha G-protein subunit then binds to a K⁺ channel, causing it to open.

G-Protein-Operated Channels

The muscarinic ACh receptors are formed from only a single subunit, which can bind to one ACh molecule. Unlike the nicotinic receptors, these receptors do not contain ion channels. The ion channels are separate proteins located at some distance from the muscarinic receptors. Binding of ACh (the ligand) to the muscarinic receptor causes it to activate a complex of proteins in the cell membrane known as the **G-proteins**—so named because their activity is influenced by guanosine nucleotides (GDP and GTP). There are three G-protein subunits, designated alpha, beta, and gamma. In response to the binding of ACh to its receptor, the alpha subunit of the G-proteins dissociates from the beta-gamma complex. It then diffuses through the membrane until it binds to an ion channel, causing the channel to open (fig. 7.23). A short time later, the G-protein alpha subunit dissociates from the channel and moves back to its previous position. This causes the ion channel to close. (It should be noted that, in some cases, it is the beta-gamma complex that serves as the activating agent.)

In the example shown in figure 7.23, opening of K⁺ channels leads to the diffusion of K⁺ out of the postsynaptic cell (because that is the direction of its concentration gradient). As a result, the cell becomes hyperpolarized, producing an inhibitory postsynaptic potential (IPSP). Such an effect is produced in the heart, for example, where autonomic nerve fibers (part of the vagus nerve) synapse with pacemaker cells and slow the rate of beat. It should be noted that inhibition also occurs in the CNS in response to other neurotransmitters, but those IPSPs are produced by a different mechanism that will be described in later sections.

There are other tissues with muscarinic receptors whose manner of response is different from that shown in figure 7.23. In the smooth muscle cells of the stomach, for example, the binding of ACh to its muscarinic receptors causes a different type of G-protein alpha subunit to dissociate and bind to the K⁺ channels. In this case, the binding of the G-protein subunit to the K⁺ channels causes the channels to close rather than to open. As a result, the outward diffusion of K⁺, which occurs at an ongoing rate in the resting cell, is reduced to below resting levels. Since the resting membrane potential is maintained by a balance between cations flowing into the cell and cations flowing out, a reduction in the outward flow of K⁺ produces a depolarization. The depolarization produced in these smooth muscle cells by this means results in contraction (chapter 12).

Acetylcholinesterase (AChE)

The bond between ACh and its receptor protein exists for only a brief instant. The ACh-receptor complex quickly dissociates but can be quickly reformed as long as free ACh is in the vicinity. In order for activity in the postsynaptic cell to be controlled, free ACh must be inactivated very soon after it is released. The inactivation of ACh is achieved by means of an enzyme called **acetylcholinesterase,** or **AChE,** which is present on the postsynaptic membrane or immediately outside the membrane, with its active site facing the synaptic cleft (fig. 7.24).

 Nerve gas exerts its odious effects by inhibiting AChE in skeletal muscles. Since ACh is not degraded, it can continue to combine with receptor proteins and can continue to stimulate the postsynaptic cell, leading to spastic paralysis. Clinically, cholinesterase inhibitors (such as *neostigmine*) are used to enhance the effects of ACh on muscle contraction when neuromuscular transmission is weak, as in the disease *myasthenia gravis*.

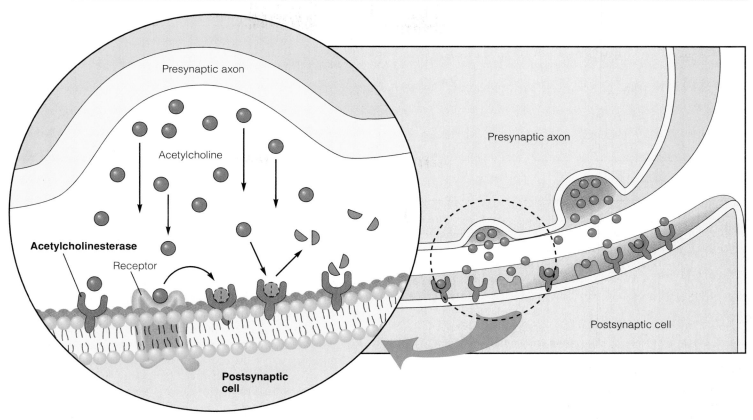

Figure 7.24

Acetylcholinesterase (AChE) in the postsynaptic cell membrane inactivates the ACh released into the synaptic cleft.

Table 7.5 Comparison of Action Potentials and Excitatory Postsynaptic Potentials (EPSPs)

Characteristic	Action Potential	Excitatory Postsynaptic Potential
Stimulus for opening of ionic gates	Depolarization	Acetylcholine (ACh)
Initial effect of stimulus	Na^+ gates open	Na^+ and K^+ gates open
Production of repolarization	Opening of K^+ gates	Loss of intracellular positive charges with time and distance
Conduction distance	Regenerated over length of the axon	1–2 mm; a localized potential
Positive feedback between depolarization and opening of Na^+ gates	Yes	No
Maximum depolarization	+40 mV	Close to zero
Summation	No summation—all-or-none event	Summation of EPSPs, producing graded depolarizations
Refractory period	Present	Absent
Effect of drugs	Inhibited by tetrodotoxin, not by curare	Inhibited by curare, not by tetrodotoxin

Excitatory Postsynaptic Potential (EPSP)

Binding of ACh to the nicotinic receptors causes opening of channels that allow Na^+ to diffuse into the postsynaptic cell and allow K^+ to diffuse out simultaneously and through the same membrane channel (see fig. 7.22). The two ion movements do not cancel each other out, however. The inward Na^+ diffusion predominates because of its greater electrochemical gradient, and a net depolarization is produced. The outflow of

K^+ does, however, prevent the depolarization from overshooting 0 mV. Therefore, the membrane polarity does not reverse in EPSPs as it does in action potentials. (Remember that action potentials are produced by voltage-regulated gates, where Na^+ and K^+ diffusion occur through different channels and at different steps.)

A comparison of EPSPs and action potentials is provided in table 7.5. Unlike action potentials, EPSPs have *no threshold*; a single quantum of ACh (released from a single synaptic vesicle) produces a tiny depolarization of the postsynaptic membrane. When more quanta of ACh are released, the

Table 7.6 Drugs That Affect the Neural Control of Skeletal Muscles

Drug	Origin	Effects
Botulinus toxin	Produced by *Clostridium botulinum* (bacteria)	Inhibits release of acetylcholine (ACh)
Curare	Resin from a South American tree	Prevents interaction of ACh with the postsynaptic receptor protein
α-Bungarotoxin	Venom of *Bungarus* snakes	Binds to ACh receptor proteins
Saxitoxin	Red tide (*Gonyaulax*) protozoa	Blocks Na^+ channels
Tetrodotoxin	Pufferfish	Blocks Na^+ channels
Nerve gas	Artificial	Inhibits acetylcholinesterase in postsynaptic cell
Neostigmine	Nigerian bean	Inhibits acetylcholinesterase in postsynaptic cell
Strychnine	Seeds of an Asian tree	Prevents IPSPs in spinal cord that inhibit contraction of antagonistic muscles

Muscle weakness in the disease **myasthenia gravis** is due to the fact that ACh receptors are blocked and destroyed by antibodies secreted by the immune system of the affected person. Paralysis in people who eat shellfish poisoned with saxitoxin, or pufferfish containing tetrodotoxin, results from the blockage of Na^+ gates. The effects of these and other poisons on neuromuscular transmission are summarized in table 7.6.

depolarization is correspondingly greater. EPSPs are therefore *graded* in magnitude, unlike all-or-none action potentials. Since EPSPs can be graded, and have *no refractory period*, they are capable of *summation*. That is, the depolarizations of several different EPSPs can be added together. Action potentials are prevented from summating by their all-or-none nature and by the presence of refractory periods.

Acetylcholine in the PNS

Somatic motor neurons form synapses with skeletal muscle cells (muscle fibers). At these synapses, or neuromuscular junctions, the postsynaptic membrane of the muscle fiber is known as a *motor end plate*. Therefore, the EPSPs produced by ACh acting at this postsynaptic membrane are often called **end-plate potentials.** This depolarization opens voltage-regulated gates, which are located adjacent to the end plate. Voltage-regulated channels produce action potentials in the muscle fiber, and these are reproduced by other voltage-regulated channels located all along the muscle cell membrane. This conduction is analogous to that of action potentials by axons; it is significant because action potentials in muscle fibers stimulate muscle contraction (as described in chapter 12).

If any stage in the process of neuromuscular transmission is blocked, muscle weakness—sometimes leading to paralysis and death—may result. The drug *curare*, for example, competes with ACh for attachment to the nicotinic ACh receptors and thus reduces the size of the end-plate potentials (table 7.6). This drug was first used on poison darts by South American Indians because it produced flaccid paralysis in their victims. Clinically, curare is used as a muscle relaxant during anesthesia and to prevent muscle damage during electroconvulsive shock therapy.

Autonomic motor neurons innervate cardiac muscle, smooth muscles in blood vessels and visceral organs, and glands. As previously mentioned, there are two classifications of autonomic nerves: sympathetic and parasympathetic. Most of the parasympathetic axons that innervate the effector organs use ACh as their neurotransmitter. In some cases, these axons have an inhibitory effect on the organs they innervate through the binding of ACh to muscarinic ACh receptors. The action of the vagus nerve in slowing the heart rate is an example of this inhibitory effect. In other cases, ACh released by autonomic neurons produces stimulatory effects as previously described. The structures and functions of the autonomic system are described in chapter 9.

Acetylcholine in the CNS

Within the central nervous system, the axon terminals of one neuron typically synapse with the dendrites or cell body of another. The dendrites and cell body thus serve as the receptive area of the neuron, and it is in these regions that receptor proteins for neurotransmitters and chemically regulated gates are located. The first voltage-regulated gates are located at the beginning of the axon, at the axon hillock. It is here that action potentials are first produced (fig. 7.25).

The cell body and dendrites lack voltage-regulated gates. The *initial segment* of the axon, which is the unmyelinated region of the axon around the axon hillock, on the other hand, has a high concentration of voltage-regulated gates. Depolarizations in the dendrites and cell body must therefore spread by cable properties to the initial segment of the axon in order to stimulate action potentials.

If the depolarization is at or above threshold by the time it reaches the initial segment of the axon, the EPSP will stimulate the production of action potentials, which can then regenerate themselves along the axon. If, however, the EPSP is below threshold at the initial segment, no action potentials will be produced in the postsynaptic cell (fig. 7.26). Gradations in the strength of the

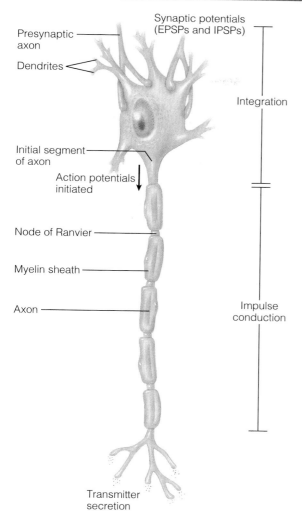

Figure 7.25

A diagram illustrating the functional specialization of different regions in a multipolar neuron.

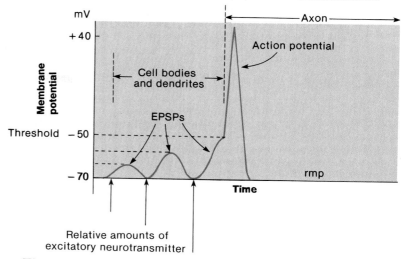

Figure 7.26

Excitatory postsynaptic potentials (EPSPs) are graded in nature; that is, stimuli of increasing strength produce increasing amounts of depolarization. When a threshold level of depolarization is produced, action potentials are generated in the axon.

EPSP above threshold determine the frequency with which action potentials will be produced at the axon hillock, and at each point in the axon where the impulse is conducted.

Earlier in this chapter, the action potential was introduced by describing the events that occurred when a depolarization stimulus was artificially produced by stimulating electrodes. Now you know that EPSPs, conducted from the dendrites and cell body, serve as the normal stimuli for the production of action potentials in the axon hillock, and that the action potentials at this point serve as the depolarization stimuli for the next region, and so on. This chain of events ends at the terminal boutons of the axon, where neurotransmitter is released.

Alzheimer's disease, the most common cause of senile dementia, often begins in middle age and produces progressive mental deterioration. The cause of Alzheimer's disease is not known, but there is evidence that it is associated with a loss of cholinergic neurons (those that use acetylcholine as a neurotransmitter) that terminate in the hippocampus and cerebral cortex of the brain (areas concerned with memory storage). Since acetylcholine is produced from acetyl coenzyme A and choline, attempts have been made to increase ACh in the brain by the ingestion of large amounts of lecithin (which contains choline). So far, such nutritional treatments and possible drug treatments to increase ACh by inhibiting acetylcholinesterase have met with only very limited success.

1. Distinguish between the two types of chemically regulated channels and explain how ACh opens each type.
2. State a location at which ACh has stimulatory effects. Where does it exert inhibitory effects? How are stimulation and inhibition accomplished?
3. Describe the function of acetylcholinesterase and discuss its physiological significance.
4. Compare the properties of EPSPs and action potentials and state where these events occur in a postsynaptic neuron.
5. Explain how EPSPs produce action potentials in the postsynaptic neuron.

Monoamines as Neurotransmitters

There are a variety of chemicals in the CNS besides ACh that function as neurotransmitters. Among these are dopamine, norepinephrine, and serotonin. Although these molecules are in the same chemical family and have similar mechanisms of action, they are used by different neurons for different functions.

The regulatory molecules epinephrine, norepinephrine, dopamine, and serotonin are in the chemical family known as **monoamines.** Serotonin is derived from the amino acid tryptophan. Epinephrine, norepinephrine, and dopamine are derived from the amino acid tyrosine and form a subfamily of monoamines called the **catecholamines.** Among the catecholamines, it should be noted that epinephrine is not a neurotransmitter; it is a widely known hormone, also called adrenalin, secreted by the adrenal glands. Norepinephrine is likewise secreted as a hormone, but also functions as a neurotransmitter. Dopamine is exclusively a neurotransmitter.

Like ACh, monoamine neurotransmitters are released by exocytosis from presynaptic vesicles and diffuse across the synaptic cleft to interact with specific receptor proteins in the membrane of the postsynaptic cell. The stimulatory effects of these monoamines, like those of ACh, are quickly inhibited. The inhibition of monoamine action is due to (1) reuptake of monoamines into the presynaptic neuron endings, (2) enzymatic degradation of monoamines in the presynaptic neuron endings by *monoamine oxidase* (MAO), and (3) the enzymatic degradation of catecholamines in the postsynaptic neuron by *catechol-O-methyltransferase* (COMT). This process is illustrated in figure 7.27. Drugs that inhibit MAO thus promote the effects of monoamine action.

The monoamine neurotransmitters do not directly cause opening of ionic channels in the postsynaptic membrane. Instead, these neurotransmitters act by means of an intermediate regulator, known as a **second messenger.** In the case of some synapses that use catecholamines for synaptic transmission, this second messenger is a compound known as **cyclic adenosine monophosphate (cAMP).** Although other synapses can use other second messengers, only the function of cAMP as a second messenger will be considered here. Other second-messenger systems are discussed in conjunction with hormone action in chapter 11.

Binding of norepinephrine, for example, with its receptor in the postsynaptic membrane stimulates the dissociation of the G-protein alpha subunit from the others in its complex (fig. 7.28). This subunit diffuses in the membrane until it binds to an enzyme known as *adenylate cyclase.* This enzyme converts ATP to cyclic AMP (cAMP) and pyrophosphate (two inorganic phosphates) within the postsynaptic cell cytoplasm. Cyclic AMP in turn activates another enzyme, called *protein kinase,* which phosphorylates (adds a phosphate group to) other proteins (fig. 7.28). Through this action, ion channels are opened in the postsynaptic membrane.

Monoamine oxidase (MAO) is an enzyme in the endings of presynaptic axons that breaks down catecholamines and serotonin after they have been taken up from the synaptic cleft. Drugs that act as **MAO inhibitors** thus increase transmission at these synapses and have been found to aid people suffering from clinical depression. This suggests that a deficiency in monoamine neural pathways may contribute to severe emotional depression. An MAO inhibitor (*Deprenyl*) has also been used to treat Parkinson's disease (discussed shortly) by promoting the activity of dopamine as a neurotransmitter.

Drugs that inhibit MAO promote the activity of all of the monoamines, and thus can produce undesired side effects. A newer drug, known as fluoxetine hydrochloride (*Prozac*), specifically blocks the reuptake of serotonin into presynaptic axons. This drug thus specifically promotes serotonin action, and has been found to be effective in the treatment of depression. The widespread use of this drug, however, has engendered considerable controversy.

Dopamine as a Neurotransmitter

Neurons that use **dopamine** as a neurotransmitter and postsynaptic neurons with dopamine receptor proteins in their membranes have been identified in postmortem brain tissue. More recently, the position of dopamine receptor proteins has been observed in the living brain using the technique of *positron emission tomography* (PET). These investigations have been spurred by the great clinical interest in the effects of **dopaminergic neurons** (those that use dopamine as a neurotransmitter).

Dopaminergic neurons are highly concentrated in the *substantia nigra* (literally, the "dark substance," so called because it contains melanin pigment) of the brain. Many neurons in the substantia nigra send fibers to the *basal nuclei,* which are large masses of cell bodies deep in the cerebrum that are involved in the coordination of skeletal movements. There is much evidence that **Parkinson's disease** is caused by degeneration of the dopaminergic neurons in the substantia nigra. Parkinson's disease is a major cause of neurological disability in people over

Cocaine—a stimulant related to the amphetamines in its action—is currently widely abused in the United States. Although early use of this drug produces feelings of euphoria and social adroitness, continued use leads to social withdrawal, depression, dependence upon ever-higher dosages, and serious cardiovascular and renal disease that can result in heart and kidney failure. The numerous effects of cocaine on the central nervous system appear to be mediated by one primary mechanism: cocaine blocks the reuptake of the monoamines into the presynaptic axon endings. This results in overstimulation of those neural pathways that use dopamine and other monoamines as neurotransmitters.

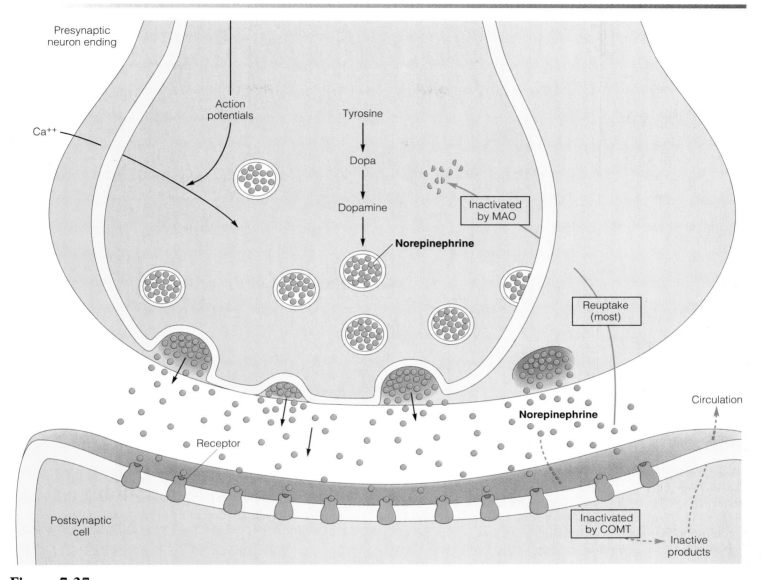

Figure 7.27

A diagram showing the production, release, and reuptake of catecholamine neurotransmitters from presynaptic nerve endings. The transmitters combine with receptor proteins in the postsynaptic membrane. (COMT = catecholamine O = methyltransferase; MAO = monoamine oxidase.)

the age of 60, and is associated with such symptoms as muscle tremors and rigidity, difficulty in initiating movements and speech, and other severe problems. These patients are treated with L-dopa to increase the production of dopamine in the brain, as described previously in this chapter.

Norepinephrine as a Neurotransmitter

Norepinephrine, like ACh, is used as a neurotransmitter in both the PNS and the CNS. Sympathetic neurons of the PNS use norepinephrine as a neurotransmitter at their synapse with smooth muscles, cardiac muscle, and glands. Some neurons in the CNS also use norepinephrine as a neurotransmitter; these neurons seem to be involved in general behavioral arousal. This would help to explain the mental arousal elicited by such drugs as *amphetamines*, which specifically stimulate pathways

A side effect of L-dopa treatment in some patients with Parkinson's disease is the appearance of symptoms characteristic of **schizophrenia.** This effect is not surprising in view of the fact that the drugs (called neuroleptics) used to treat schizophrenic patients act as specific antagonists of dopamine receptors. As might be predicted from these observations, schizophrenic patients treated with these drugs often develop symptoms of Parkinson's disease. Based on this evidence, it seems that schizophrenia may be caused, at least in part, by overactivity of the dopaminergic pathways. This hypothesis has recently received support from studies that link schizophrenia to increased amounts of a specific subtype of dopamine receptor that appears to be located in the forebrain rather than in the basal nuclei.

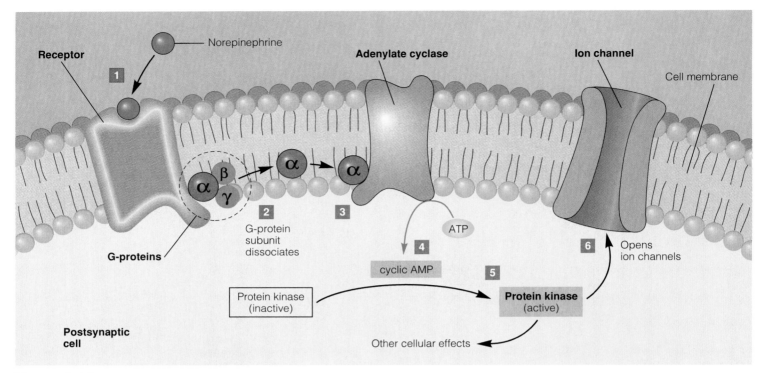

Figure 7.28

The binding of norepinephrine to its receptor (1) causes the dissociation of G-proteins (2). Binding of the alpha G-protein subunit to the enzyme adenylate cyclase (3) activates this enzyme, leading to the production of cyclic AMP (4). Cyclic AMP, in turn, activates protein kinase (5), which can open ion channels (6) and produce other effects.

that use norepinephrine as a neurotransmitter. Such drugs also stimulate the PNS pathways that use norepinephrine, however, and this duplicates the effects of sympathetic nerve activation. A rise in blood pressure, constriction of arteries, and other effects similar to the deleterious consequences of cocaine use can thereby be produced.

1. List the monoamines and indicate their chemical relationships.
2. Explain how monoamines are inactivated at the synapse and how this process can be clinically manipulated.
3. Describe the relationship between dopaminergic neurons, Parkinson's disease, and schizophrenia.
4. Explain how cocaine and amphetamines produce their effects in the brain. Discuss the dangers of these drugs.

Other Neurotransmitters

There are a surprisingly large number of different molecules that appear to function as neurotransmitters. These include some amino acids and their derivatives, many polypeptides, and even the gas nitric oxide.

Amino Acids as Neurotransmitters

The amino acids **glutamic acid** and **aspartic acid** function as excitatory neurotransmitters in the CNS. Glutamic acid (or *glutamate*), indeed, is the major excitatory neurotransmitter in the brain, producing excitatory postsynaptic potentials (EPSPs). Experiments using chemicals that differ slightly from glutamate but that mimic its actions have revealed different subcategories of glutamate receptors in brain neurons.

One subcategory of glutamate receptor is named after the glutamate analogue *NMDA* (*N-methyl-D-aspartate*). The NMDA receptors for glutamate have been implicated in the physiology of memory and will be described in more detail in a later section. Interestingly, the NMDA receptors are a major site of action for the street drug known as angel dust.

The amino acid **glycine** is inhibitory; instead of depolarizing the postsynaptic membrane and producing an EPSP, it hyperpolarizes the postsynaptic membrane and produces an inhibitory postsynaptic potential (IPSP). The binding of glycine to its receptor proteins causes the opening of chloride (Cl⁻) channels in the postsynaptic membrane. As a result, Cl⁻ diffuses into the postsynaptic neuron and produces the hyperpolarization. This inhibits the neuron by making the membrane potential even more negative than it is at rest, and therefore further from the threshold depolarization required to stimulate action potentials.

The inhibitory effects of glycine are very important in the spinal cord, where they help in the control of skeletal movements.

Benzodiazepines are drugs that act to increase the effectiveness of GABA in the brain and spinal cord. Since GABA is a neurotransmitter that inhibits the activity of spinal motor neurons that innervate skeletal muscles, the intravenous infusion of benzodiazepines acts to inhibit the muscular spasms in *status epilepticus* and seizures resulting from drug overdose and poisons. Probably as a result of its general inhibitory effects on the brain, GABA also appears to function as a neurotransmitter involved in mood and emotion. Benzodiazepines such as *Valium* are thus given orally as tranquilizers.

Table 7.7 Examples of Chemicals That Are Either Proven or Suspected Neurotransmitters

Category	Chemicals
Amines	Acetylcholine
	Histamine
	Serotonin
Catecholamines	Dopamine
	Epinephrine
	Norepinephrine
Amino acids	Aspartic acid
	GABA (gamma-aminobutyric acid)
	Glutamic acid
	Glycine
Polypeptides	Glucagon
	Insulin
	Somatostatin
	Substance P
	ACTH (adrenocorticotrophic hormone)
	Angiotensin II
	Endorphins
	LHRH (luteinizing hormone-releasing hormone)
	TRH (thyrotrophin-releasing hormone)
	Vasopressin (antidiuretic hormone)
	CCK (cholecystokinin)

Flexion of an arm, for example, involves stimulation of the flexor muscles by motor neurons in the spinal cord. The motor neurons that innervate the antagonistic extensor muscles are inhibited by IPSPs produced by glycine released from other neurons. The importance of the inhibitory actions of glycine is revealed by the deadly effects of the poison *strychnine,* which causes spastic paralysis by specifically blocking the glycine receptor proteins. Animals poisoned with strychnine die from asphyxiation due to their inability to relax the diaphragm muscle.

The neurotransmitter **GABA (gamma-aminobutyric acid)** is a derivative of another amino acid, glutamic acid. GABA is the most prevalent neurotransmitter in the brain; in fact, as many as one-third of all the neurons in the brain use GABA as a neurotransmitter. Like glycine, GABA is inhibitory—it hyperpolarizes the postsynaptic membrane. Also, the effects of GABA, like those of glycine, are involved in motor control. For example, the large *Purkinje cells* of the cerebellum mediate the motor functions of the cerebellum by producing IPSPs in their postsynaptic neurons. A deficiency in those neurons that release GABA as a neurotransmitter produces the uncontrolled movements seen in people with *Huntington's chorea.*

Polypeptides as Neurotransmitters

Many polypeptides of various sizes are found in the brain and are believed to function as neurotransmitters. Interestingly, some of the polypeptides that function as hormones secreted by the small intestine and other endocrine glands are also produced in the brain and may function there as neurotransmitters (table 7.7). For example, *cholecystokinin* (CCK) is secreted as a hormone from the small intestine and is also released from neurons and used as a neurotransmitter in the brain. Recent evidence suggests that CCK, acting as a neurotransmitter, may promote feelings of satiety in the brain following meals. Another polypeptide found in many organs, *substance P,* functions as a neurotransmitter in pathways in the brain that mediate sensations of pain.

Synaptic Plasticity

Although some of the polypeptides released from neurons may function as neurotransmitters in the traditional sense—by stimulating the opening of ionic gates and causing changes in the membrane potential—others may have more subtle and poorly understood effects. **Neuromodulators** has been proposed as a name for compounds with such alternative effects. An exciting recent discovery is that some neurons in both the PNS and CNS produce both a classical neurotransmitter (ACh or a catecholamine) and a polypeptide neurotransmitter. These are contained in different synaptic vesicles that can be distinguished using the electron microscope. The neuron can thus release either the classical neurotransmitter or the polypeptide neurotransmitter under different conditions.

Discoveries such as the one just described indicate that synapses have a greater capacity for alteration at the molecular level than was previously believed. This attribute has been termed **synaptic plasticity.** Synapses are also more plastic at the cellular level. There is evidence that sprouting of new axon branches can occur over short distances to produce a turnover of synapses, even in the mature CNS. This breakdown and reforming of synapses may occur within a time span of only a few hours. The physiological significance of these interesting discoveries is not yet fully understood.

Endogenous Opioids

The ability of opium and its analogues—that is, the *opioids*—to relieve pain (promote analgesia) has been known for centuries. Morphine, for example, has long been used for this purpose. The discovery in 1973 of opioid receptor proteins in the brain suggested that the effects of these drugs might be due to the stimulation of specific neuron pathways. This implied that opioids—along with LSD, mescaline, and other mind-altering drugs—might resemble neurotransmitters produced by the brain.

The analgesic effects of morphine are blocked in a specific manner by a drug called *naloxone*. In the same year that opioid receptor proteins were discovered, it was found that naloxone also blocked the analgesic effect of electrical brain stimulation. Subsequently, evidence suggested that the analgesic effects of hypnosis and acupuncture could also be blocked by naloxone. These experiments indicated that the brain might be producing its own endogenous morphinelike analgesic compounds.

These compounds have been identified as a family of polypeptides. One member is called β-endorphin (for "endogenously produced morphinelike compound") produced by the brain and pituitary gland. Another consists of a group of five-amino-acid peptides called **enkephalins,** which may function as neurotransmitters, and a third is a polypeptide neurotransmitter called **dynorphin.**

Endogenous opioids have been shown to block the transmission of pain. Current evidence for this effect includes results obtained both from neurophysiological studies—in which endogenous opioids blocked the release of substance P (the chemical transmitter believed to mediate pain pathways)—and from behavioral studies. The pain threshold of pregnant rats, for example, was found to decrease when they were treated with naloxone, which blocks the opioid receptors. Also, a "burst" in β-endorphin secretion recently was shown to occur in pregnant women during parturition (childbirth).

Opioids such as opium and morphine may also provide pleasant sensations, and thus mediate reward or positive reinforcement pathways. Overeating in genetically obese mice, for example, appears to be blocked by naloxone. It has also been found that blood levels of β-endorphin are increased in exercise. Some people have suggested that the "jogger's high" may thus be due to endogenous opioids. (Naloxone, however, does not appear to block the jogger's high.) Although evidence for this particular effect is poor, it does appear that endorphins may promote some type of psychic reward system as well as analgesia.

Nitric Oxide as a Neurotransmitter

Nitric oxide (NO) is the first gas known to act as a regulatory molecule in the body. Produced by nitric oxide synthetase in the cells of many organs from the amino acid L-arginine, nitric oxide's actions are very different from those of the more familiar nitrous oxide (N_2O), or laughing gas, sometimes used by dentists.

Nitric oxide has a number of different roles in the body. Within blood vessels, it acts as a local tissue regulator that causes the smooth muscles of those vessels to relax, so that the blood vessels dilate. This role will be described in conjunction with the cardiovascular system in chapter 14. Within macrophages and other cells, nitric oxide helps to kill bacteria. This activity is described in conjunction with the immune system in chapter 15. In addition, nitric oxide is a neurotransmitter of certain neurons in both the PNS and CNS. It diffuses out of the presynaptic axon and into neighboring cells by simply passing through the lipid portion of the cell membranes. Once in the target cells, NO exerts its effects by stimulating the production of cyclic guanosine monophosphate (cGMP), which acts as a second messenger.

In the PNS, nitric oxide is released by some neurons that innervate the gastrointestinal tract, penis, respiratory passages, and cerebral blood vessels. These are autonomic neurons that cause smooth muscle relaxation in their target organs. This can produce, for example, the engorgement of the spongy tissue of the penis with blood. In fact, scientists now believe that erection of the penis results from the action of nitric oxide released by specific parasympathetic nerves. Nitric oxide is also released as a neurotransmitter in the brain, and has been implicated in the processes of learning and memory. This will be discussed in more detail in a later section.

Although its importance in the body was recognized only recently, nitric oxide has already been exploited for medical use. The hypotension (low blood pressure) of septic shock, for example, is apparently mediated by nitric oxide and has been successfully treated with drugs that inhibit nitric oxide synthetase. Conversely, inhalation of nitric oxide has been used to treat pulmonary hypertension, as well as respiratory distress syndrome (discussed in chapter 16).

1. Explain the significance of glutamate in the brain and of NMDA receptors.
2. Describe the mechanism of action of glycine and GABA as neurotransmitters and discuss their significance.
3. Give examples of endogenous opioid polypeptides and discuss their significance.
4. Explain how nitric acid is produced in the body and describe its functions.

Synaptic Integration

The summation of a number of EPSPs may be needed to produce a depolarization of sufficient magnitude to stimulate the postsynaptic cell. The net effect of EPSPs on the postsynaptic neuron is reduced by hyperpolarization (IPSPs), which is produced by inhibitory neurotransmitters. The activity of neurons within the central nervous system is thus the net result of both excitatory and inhibitory effects.

Since voltage-regulated Na^+ and K^+ channels are absent in the dendrites and cell bodies, changes in membrane potential induced by neurotransmitters in these areas do not have the all-or-none characteristics of action potentials. Synaptic

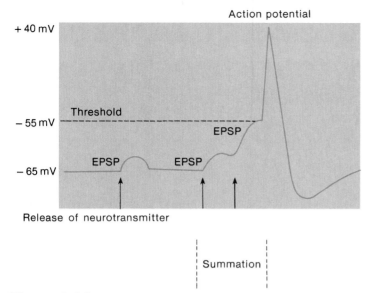

Figure 7.29

EPSPs can summate over distance (spatial summation) and time (temporal summation). When summation results in a threshold level of depolarization at the axon hillock, voltage-regulated Na+ gates are opened and an action potential is produced.

potentials are graded and can add together, or summate. **Spatial summation** occurs because numerous presynaptic nerve fibers (up to a thousand, in some cases) converge on a single postsynaptic neuron. In spatial summation, synaptic depolarizations (EPSPs) produced at different synapses may summate in the postsynaptic dendrites and cell body. In **temporal summation**, the successive activity of presynaptic axon terminals, causing successive waves of transmitter release, may result in the summation of EPSPs in the postsynaptic neuron. The summation of EPSPs helps to ensure that the depolarization that reaches the axon hillock will be of sufficient magnitude to generate new action potentials in the postsynaptic axon (fig. 7.29).

Long-Term Potentiation

When a presynaptic neuron is experimentally stimulated at a high frequency, even for as short a time as a few seconds, the excitability of the synapse is enhanced—or potentiated—when this neuron pathway is subsequently stimulated. The improved efficacy of synaptic transmission may last for hours or even weeks and is called **long-term potentiation (LTP).** Long-term potentiation may favor transmission along frequently used neural pathways and thus may represent a mechanism of neural "learning." It is interesting in this regard that LTP has been observed in the hippocampus of the brain, which is an area implicated in memory storage (chapter 8).

Most of the neural pathways in the hippocampus area of the brain use glutamate as their neurotransmitter. There are two subclasses of glutamate receptors: NMDA (as previously mentioned) and non-NMDA. Binding of glutamate to NMDA receptors is not required for normal synaptic transmis-

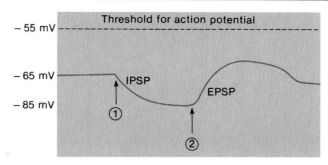

Figure 7.30

An inhibitory postsynaptic potential (IPSP) makes the inside of the postsynaptic membrane more negative than the resting potential—it hyperpolarizes the membrane. Subsequent or simultaneous excitatory postsynaptic potentials (EPSPs), which are depolarizations, must thus be stronger to reach the threshold required to generate action potentials at the axon hillock.

sion. It is the binding of glutamate to non-NMDA receptors that produces the depolarization of the postsynaptic neuron required for the production of action potentials in this neuron. However, when this depolarization occurs at a time when other glutamate molecules are binding to the NMDA receptors, channels for Ca++ open in the postsynaptic membrane. The diffusion of Ca++ into the postsynaptic neuron may induce LTP by making this neuron more sensitive to subsequent stimulation.

The inward diffusion of Ca++ also activates nitric oxide synthetase, causing the postsynaptic neuron to produce and release nitric oxide. The nitric oxide released from the postsynaptic neuron may then act as a "retrograde transmitter" and diffuse to the presynaptic neuron, where it could stimulate the release of more glutamate transmitter. In this way, synaptic transmission during LTP could be strengthened first postsynaptically and then presynaptically. While much evidence supports the existence of a retrograde transmitter in LTP, it should be noted that the identity of nitric oxide as that transmitter is currently controversial.

Synaptic Inhibition

Although many neurotransmitters depolarize the postsynaptic membrane (produce EPSPs), some transmitters do just the opposite. They hyperpolarize the postsynaptic membrane; that is, they make the inside of the membrane more negative than it is at rest. Since hyperpolarization (from –65 mV to, for example, –85 mV) drives the membrane potential farther from the threshold depolarization required to stimulate action potentials, this inhibits the activity of the postsynaptic neuron. Hyperpolarizations produced by neurotransmitters are therefore called *inhibitory postsynaptic potentials (IPSPs)*. The inhibition produced in this way is called **postsynaptic inhibition.**

Excitatory and inhibitory inputs (EPSPs and IPSPs) to a postsynaptic neuron can summate in an algebraic fashion (fig. 7.30). The effects of IPSPs in this way reduce, or may even

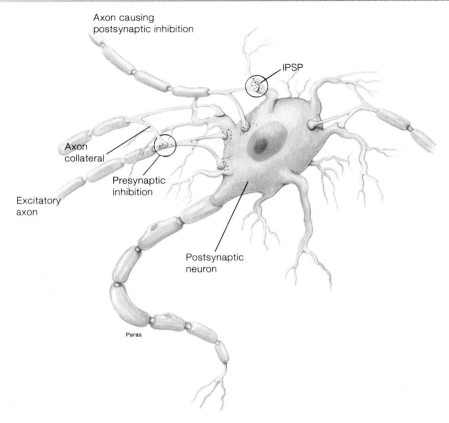

Axon causing postsynaptic inhibition

IPSP

Axon collateral

Presynaptic inhibition

Excitatory axon

Postsynaptic neuron

Paras

Figure 7.31

A diagram illustrating postsynaptic and presynaptic inhibition.

eliminate, the ability of EPSPs to generate action potentials in the postsynaptic cell. Considering that a given neuron may receive as many as 1,000 presynaptic inputs, the interactions of EPSPs and IPSPs can be tremendously variable.

In **presynaptic inhibition** (fig. 7.31), the amount of excitatory neurotransmitter released at the end of an axon is decreased by a second neuron, whose axon makes a synapse with the axon of the first neuron (axoaxonic synapse). The neurotransmitter released at this axoaxonic synapse partially depolarizes the axon of the first neuron, bringing it closer to threshold and making it easier to "fire" action potentials. The amplitude of these action potentials (subtracting the new lower potential from the +40-mV "top" of the action potential), however, is lower than normal. The smaller action

potentials that result cause the release of lesser amounts of excitatory neurotransmitter by the first neuron, resulting in inhibition of its effect on its postsynaptic cell.

---

1. Define *spatial summation* and *temporal summation* and explain their functional importance.
2. Describe long-term potentiation, explain how it is produced, and discuss its significance.
3. Explain how postsynaptic inhibition is produced and how IPSPs and EPSPs can interact.
4. Describe the mechanism of presynaptic inhibition.

Summary

Neurons and Supporting Cells   p. 144

I. The nervous system is divided into the central nervous system (CNS) and the peripheral nervous system (PNS).
   A. The central nervous system includes the brain and spinal cord, which contain nuclei and tracts.

B. The peripheral nervous system consists of nerves and ganglia.

II. A neuron consists of dendrites, a cell body, and an axon.
   A. The cell body contains the nucleus, Nissl bodies, neurofibrils, and other organelles.

B. Dendrites receive stimuli, and the axon conducts nerve impulses away from the cell body.

III. A nerve is a collection of axons in the PNS.
   A. A sensory, or afferent, neuron is pseudounipolar and conducts impulses from sensory receptors into the CNS.
   B. A motor, or efferent, neuron is multipolar and conducts impulses from the CNS to effector organs.
   C. Interneurons, or association neurons, are located entirely within the CNS.
   D. Somatic motor nerves innervate skeletal muscle; autonomic nerves innervate smooth muscle, cardiac muscle, and glands.
IV. There are six categories of supporting cells.
   A. Schwann cells form a sheath of Schwann around axons of the PNS.
   B. Some neurons are surrounded by successive wrappings of supporting cell membrane called a myelin sheath. This sheath is produced by Schwann cells in the PNS and by oligodendrocytes in the CNS.
   C. Astrocytes in the CNS may contribute to the blood-brain barrier.

Electrical Activity in Axons p. 151

I. The permeability of the axon membrane to Na^+ and K^+ is regulated by gates.
   A. At the resting membrane potential of –65 mV, the membrane is relatively impermeable to Na^+ and only slightly permeable to K^+.
   B. The voltage-regulated Na^+ and K^+ gates open in response to the stimulus of depolarization.
   C. When the membrane is depolarized to a threshold level, the Na^+ gates open first, followed quickly by opening of the K^+ gates.
II. The opening of voltage-regulated gates produces an action potential.
   A. The opening of Na^+ gates in response to depolarization allows Na^+ to diffuse into the axon, thus further depolarizing the membrane in a positive feedback fashion.
   B. The inward diffusion of Na^+ causes a reversal of the membrane potential from –65 mV to +40 mV.
   C. The opening of K^+ gates and outward diffusion of K^+ causes the reestablishment of the resting membrane potential. This is called repolarization.
   D. Action potentials are all-or-none events.
   E. The refractory periods of an axon membrane prevent action potentials from running together.
   F. Stronger stimuli produce action potentials with greater frequency.
III. One action potential serves as the depolarization stimulus for production of the next action potential in the axon.
   A. In unmyelinated axons, action potentials are produced fractions of a micrometer apart.
   B. In myelinated axons, action potentials are produced only at the nodes of Ranvier. This saltatory conduction is faster than conduction in an unmyelinated nerve fiber.

The Synapse p. 157

I. Gap junctions are electrical synapses found in cardiac muscle, smooth muscle, and some synapses in the CNS.
II. In chemical synapses, neurotransmitters are packaged in synaptic vesicles and released by exocytosis into the synaptic cleft.
   A. The neurotransmitter can be called the ligand of the receptor.
   B. Binding of the neurotransmitter to the receptor causes the opening of chemically regulated gates of ion channels.

Acetylcholine as a Neurotransmitter p. 161

I. There are two subtypes of ACh receptors: nicotinic and muscarinic.
   A. Nicotinic receptors enclose membrane channels and open when ACh binds to the receptor. This causes a depolarization called an excitatory postsynaptic potential (EPSP) in skeletal muscle cells.
   B. The binding of ACh to muscarinic receptors opens ion channels indirectly, through the action of G-proteins. This can cause a hyperpolarization called an inhibitory postsynaptic potential (IPSP).
   C. After ACh acts at the synapse, it is inactivated by the enzyme acetylcholinesterase (AChE).
II. EPSPs are graded and capable of summation, and they decrease in amplitude as they are conducted.
III. ACh is used in the PNS as the neurotransmitter of somatic motor neurons, which stimulate skeletal muscles to contract, and by some autonomic neurons.
IV. ACh in the CNS produces EPSPs at synapses in the dendrites or cell body. These EPSPs travel to the axon hillock, stimulate opening of voltage-regulated gates, and generate action potentials in the axon.

Monoamines as Neurotransmitters p. 166

I. Monoamines include serotonin, dopamine, norepinephrine, and epinephrine; the last three are included in the subcategory known as catecholamines.
   A. These neurotransmitters are inactivated after being released, primarily by reuptake into the presynaptic nerve endings.
   B. Catecholamines may activate adenylate cyclase in the postsynaptic cell, which catalyzes the formation of cyclic AMP.
II. Dopaminergic neurons (those that use dopamine as a neurotransmitter) are implicated in the development of Parkinson's disease and schizophrenia. Norepinephrine is used as a neurotransmitter by sympathetic neurons in the PNS and some neurons in the CNS.

Other Neurotransmitters *p. 168*

I. The amino acids glutamate and aspartate are excitatory in the CNS.
   A. The subclass of glutamate receptor designated as NMDA receptors are implicated in learning and memory.
   B. The amino acids glycine and GABA are inhibitory; they produce hyperpolarizations, causing IPSPs by opening Cl⁻ channels.

II. Numerous polypeptides function as neurotransmitters, including the endogenous opioids.
III. Nitric oxide functions as both a local tissue regulator and a neurotransmitter in the PNS and CNS. It promotes smooth muscle relaxation and is implicated in memory.

Synaptic Integration *p. 170*

I. Spatial and temporal summation of EPSPs allows a depolarization of sufficient magnitude to cause the stimulation of action potentials in the postsynaptic neuron.
   A. IPSPs and EPSPs from different synaptic inputs can summate.
   B. The production of IPSPs is called postsynaptic inhibition.

II. Long-term potentiation is a process that improves synaptic transmission as a result of the use of the synaptic pathway; thus this process may be a mechanism for learning.

Clinical Investigation

A student, whose grades were suffering due to a prolonged period of clinical depression, decided to treat herself to a dinner at a seafood restaurant. After eating a meal of mussels and clams, which were gathered from the local shore, she falls to the floor. Examination reveals that she has flaccid paralysis of her muscles and is experiencing difficulty breathing. Fortunately, quick emergency medical care saves her life. While this emergency care is being administered, a prescription bottle containing a monoamine oxidase inhibitor is found in her purse. Laboratory tests later reveal that her blood contains this drug at a concentration consistent with its prescribed therapeutic use. What might have happened to cause her medical emergency?

Clues

Read the section "Monoamine Neurotransmitters," including the box on MAO inhibitors, and examine table 7.6.

Review Activities

Objective Questions

1. The supporting cells that form myelin sheaths in the peripheral nervous system are
   a. oligodendrocytes.
   b. satellite cells.
   c. Schwann cells.
   d. astrocytes.
   e. microglia.
2. A collection of neuron cell bodies located outside the CNS is called
   a. a tract.
   b. a nerve.
   c. a nucleus.
   d. a ganglion.
3. Which of the following neurons are pseudounipolar?
   a. sensory neurons
   b. somatic motor neurons
   c. neurons in the retina
   d. autonomic motor neurons
4. Depolarization of an axon is produced by
   a. inward diffusion of Na^+.
   b. active extrusion of K^+.
   c. outward diffusion of K^+.
   d. inward active transport of Na^+.
5. Repolarization of an axon during an action potential is produced by
   a. inward diffusion of Na^+.
   b. active extrusion of K^+.
   c. outward diffusion of K^+.
   d. inward active transport of Na^+.
6. As the strength of a depolarizing stimulus to an axon is increased,
   a. the amplitude of action potentials increases.
   b. the duration of action potentials increases.
   c. the speed with which action potentials are conducted increases.
   d. the frequency with which action potentials are produced increases.
7. The conduction of action potentials in a myelinated nerve fiber is
   a. saltatory.
   b. without decrement.
   c. faster than in an unmyelinated fiber.
   d. all of the above.
8. Which of the following is *not* a characteristic of synaptic potentials?
   a. They are all or none in amplitude.
   b. They decrease in amplitude with distance.
   c. They are produced in dendrites and cell bodies.
   d. They are graded in amplitude.
   e. They are produced by chemically regulated gates.
9. Which of the following is *not* a characteristic of action potentials?
   a. They are produced by voltage-regulated gates.
   b. They are conducted without decrement.

c. Na$^+$ and K$^+$ gates open at the same time.

d. The membrane potential reverses polarity during depolarization.

10. A drug that inactivates acetylcholinesterase
a. inhibits the release of ACh from presynaptic endings.
b. inhibits the attachment of ACh to its receptor protein.
c. increases the ability of ACh to stimulate muscle contraction.
d. does all of the above.

11. Postsynaptic inhibition is produced by
a. depolarization of the postsynaptic membrane.
b. hyperpolarization of the postsynaptic membrane.
c. axoaxonic synapses.
d. post-tetanic potentiation.

12. Hyperpolarization of the postsynaptic membrane in response to glycine or GABA is produced by the opening of
a. Na$^+$ gates.
b. K$^+$ gates.
c. Ca^{++} gates.
d. Cl$^-$ gates.

13. The absolute refractory period of a neuron
a. is due to the high negative polarity of the inside of the neuron.
b. occurs only during the repolarization phase.
c. occurs only during the depolarization phase.
d. occurs during depolarization and the first part of the repolarization phase.

14. Which of the following statements about catecholamines is *false?*
a. They include norepinephrine, epinephrine, and dopamine.
b. Their effects are increased by action of the enzyme catechol-O-methyltransferase.
c. They are inactivated by monoamine oxidase.
d. They are inactivated by reuptake into the presynaptic axon.
e. They may stimulate the production of cyclic AMP in the postsynaptic axon.

15. The summation of EPSPs from numerous presynaptic nerve fibers converging onto one postsynaptic neuron is called
a. spatial summation.
b. post-tetanic potentiation.

c. temporal summation.
d. synaptic plasticity.

16. Which of the following statements about ACh receptors is *false?*
a. Skeletal muscles contain nicotinic ACh receptors.
b. The heart contains muscarinic ACh receptors.
c. G-proteins are needed to open ion channels for nicotinic receptors.
d. Stimulation of nicotinic receptors results in the production of EPSPs.

17. Hyperpolarization is caused by all of the following neurotransmitters *except:*
a. glutamic acid in the CNS.
b. ACh in the heart.
c. glycine in the spinal cord.
d. GABA in the brain.

18. Which of the following may be produced by the action of nitric oxide?
a. dilation of blood vessels
b. erection of the penis
c. relaxation of smooth muscles in the digestive tract
d. LTP among neighboring synapses in the brain
e. all of the above

Essay Questions

1. Compare the characteristics of action potentials with those of synaptic potentials.[1]
2. Explain how voltage-regulated gates produce an all-or-none action potential.
3. Explain how action potentials are regenerated along an axon.
4. Explain why conduction in a myelinated axon is faster than in an unmyelinated axon.
5. Describe the structure of nicotinic ACh receptors. Explain how ACh causes the production of an EPSP

and relate this process to the neural stimulation of skeletal muscle contraction.

6. Describe the nature of muscarinic ACh receptors and the function of G-proteins in the action of these receptors. Explain how stimulation of these receptors causes the production of hyperpolarization or a depolarization.

7. Trace the course of events in the interval between the production of an EPSP and the generation of action potentials at the axon

hillock. Describe the effect of spatial and temporal summation on this process.

8. Explain how an IPSP is produced and how IPSPs can inhibit activity of the postsynaptic neuron.

9. List the endogenous opioids in the brain and describe some of their proposed functions.

10. Define *long-term potentiation* and explain its significance. What may account for this process and what role might nitric oxide play?

---

[1]*Note:* This question is answered on page 76 of the Student Study Guide.

Selected Readings

Axelrod, J. 1974. Neurotransmitters. *Scientific American*.

Barchas, J. D. et al. 1978. Behavioral neurochemistry: Neuroregulators and behavioral states. *Science* 200:964.

Bartus, R. T. et al. 1982. The cholinergic hypothesis of geriatric memory dysfunction. *Science* 217:408.

Benfenati, F., and F. Valtorta. 1993. Synapsins and synaptic transmission. *News in Physiological Sciences* 8:18.

Block, I. B. et al. 1984. Neurotransmitter plasticity at the molecular level. *Science* 225:1266.

Blusztajn, J. K., and R. J. Wurtman. 1983. Choline and cholinergic neurons. *Science* 221:614.

Bredt, D. S., and S. H. Snyder. 1994. Nitric oxide: A physiologic messenger molecule. *Annual Review of Biochemistry* 63:175.

Brown, A. M. 1992. Ion channels in action potential generation. *Hospital Practice* 27:125.

Brown, A. M., and L. Birnbaumer, 1989. Ion channels and G-proteins. *Hospital Practice* 24:189.

Changeux, J.-P. November 1993. Chemical signaling in the brain. *Scientific American*.

Cottman, C. W., and M. Nieto-Sampedro. 1984. Cell biology of synaptic plasticity. *Science* 225:1287.

Coyle, J. T., D. L. Prince, and M. R. DeLong. 1983. Alzheimer's disease: A disorder of cortical cholinergic innervation. *Science* 219:1184.

Dunant, Y., and M. Israel. April 1985. The release of acetylcholine. *Scientific American*.

Edmonds, B., Gibb, A. J., and D. Colquhoun. 1995. Mechanisms of activation of glutamate receptors and the time course of excitatory synaptic currents. *Annual Review of Physiology* 57:495.

Freed, W. J., L. de Medinaceli, and R. J. Wyatt. 1985. Promoting functional plasticity in the damaged nervous system. *Science* 227:1544.

Gawin, F. H. 1991. Cocaine addiction: Psychology and neurophysiology. *Science* 251:1580.

Goldstein, G. W., and A. L. Betz. September 1986. The blood-brain barrier. *Scientific American*.

Gottlieb, D. I. February 1988. GABAergic neurons. *Scientific American*.

Greengard, P. et al. 1993. Synaptic vesicle phosphoproteins and regulation of synaptic function. *Science* 259:780.

Horn, J. P. 1992. The heroic age of neurophysiology. *Hospital Practice* 27:65.

Jacobs, B. L. 1994. Serotonin, motor activity, and depression-related disorders. *American Scientist* 82:456.

Jacobs, B. L., and M. E. Trulson. 1979. Mechanism of action of LSD. *American Scientist* 67:396.

Kaminski, H. J., and R. L. Ruff. 1992. Congenital disorders of neuromuscular transmission. *Hospital Practice* 27:73.

Kandel, E. R. 1979. Psychotherapy and the single synapse. *New England Journal of Medicine* 301:1028.

Katz, B. November 1952. The nerve impulse. *Scientific American*.

Kennedy, M. B. 1994. The biochemistry of synaptic regulation in the central nervous system. *Annual Review of Biochemistry* 63:571.

Keynes, R. D. March 1979. Ion channels in the nerve cell membrane. *Scientific American*.

Kimelberg, H. K., and M. D. Norenberg. April 1989. Astrocytes. *Scientific American*.

Krieger, D. T. 1983. Brain peptides: What, where, and why? *Science* 222:975.

Kuffler, S. W., and J. G. Nicholls. 1976. *From Neuron to Brain: A Cellular Approach to the Function of the Nervous System*. Sunderland, MA: Sinauer Associates.

Lester, H. A. February 1977. The response of acetylcholine. *Scientific American*.

Linder, M., and A. G. Gilman. July 1992. G proteins. *Scientific American*.

Lipton, S. A., and P. A. Rosenberg, 1994. Excitatory amino acids as a final common pathway for neurological disorders. *New England Journal of Medicine* 330:613.

Martin, A. R. 1992. Principles of neuromuscular transmission. *Hospital Practice* 27:147.

Moncada, S., and A. Higgs, 1993. The L-arginine-nitric oxide pathway. *New England Journal of Medicine* 329:2002.

Morrel, P., and W. Norton. May 1980. Myelin. *Scientific American*.

Nathanson, J. A., and P. Greengard. August 1977. "Second messengers" in the brain. *Scientific American*.

Neher, E., and B. Sakmann. March 1992. The patch clamp technique. *Scientific American*.

Orkland, R. K., and S. C. Opava. 1994. Glial function in homeostasis of the neuronal microenvironment. *News in Physiological Sciences* 9:265.

Ritz, M. C. et al. 1987. Cocaine receptors on dopamine transporters are related to self-administration of cocaine. *Science* 237:1219.

Schwartz, J. H. April 1980. The transport of substances in nerve cells. *Scientific American*.

Selkoe, D. J. November 1991. Amyloid protein and Alzheimer's disease. *Scientific American*.

Snyder, S. H. 1980. Brain peptides as neurotransmitters. *Science* 209:976.

Snyder, S. H. 1984. Drug and neurotransmitter receptors in the brain. *Science* 224:22.

Snyder, S. H., and D. S. Bredt. May 1992. Biological roles of nitric oxide. *Scientific American*.

Stevens, C. F. September 1979. The neuron. *Scientific American*.

Sweeney, P. J. 1991. New concepts in Parkinson's disease. *Hospital Practice* 26:84.

Toda, N., and T. Okamura, 1992. Regulation of nitroxidergic nerve of arterial tone. *News in Physiological Sciences* 7:148.

Wagner, H. N. 1984. Imaging CNS receptors: The dopaminergic system. *Hospital Practice* 19:187.

Waxman, S. G., and J. M. Ritchie. 1985. Organization of ion channels in the myelinated nerve fiber. *Science* 228:1502.

Wurtman, R. J. April 1982. Nutrients that modify brain function. *Scientific American*.

Wurtman, R. J. January 1985. Alzheimer's disease. *Scientific American*.

Explorations CD-ROM

The modules accompanying chapter 7 are #8 Nerve Conduction and #9 Synaptic Transmission.

Life Science Animations

The animations that relate to chapter 7 are #22 Formation of the Myelin Sheath, #23 Saltatory Nerve Conduction, #24 Signal Integration, and #25 Reflex Arcs.

The Central Nervous System

8

1. locate the major brain regions and describe the structures within each of these regions.

2. describe the organization of the cerebrum and the primary roles of its lobes.

3. describe the location and functions of the sensory cortex and motor cortex.

4. explain the lateralization of functions in the right and left cerebral hemispheres.

5. describe the structures involved in the control of speech and explain their interrelationships.

6. describe the different types of aphasias that result from damage to specific regions of the brain.

7. describe the structures involved in the limbic system and discuss the possible role of this system in emotion.

8. distinguish between different types of memory and describe the roles of different brain regions in memory.

9. describe the location of the thalamus and explain the significance of this organ.

10. describe the location of the hypothalamus and explain the significance of this organ.

11. describe the structures located in the midbrain and hindbrain and explain the role of the medulla oblongata in the control of visceral functions.

12. explain how the spinal cord is organized and how ascending and descending tracts are named.

13. describe the origin and pathways of the pyramidal motor tracts and explain the significance of these descending tracts.

14. explain the role of the basal nuclei and cerebellum in motor control by means of the extrapyramidal system and describe the pathways of this system.

15. describe the structures and pathways involved in a reflex arc.

OUTLINE

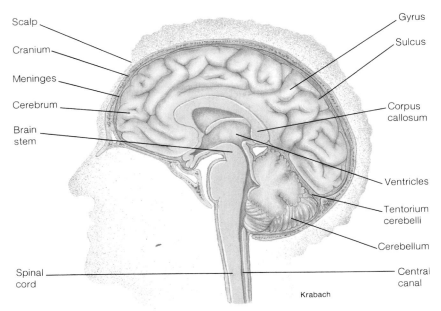

Figure 8.1

The CNS consists of the brain and the spinal cord, both of which are covered with meninges and bathed in cerebrospinal fluid.

Structural Organization of the Brain

The brain is composed of an enormous number of association neurons, with accompanying neuroglia, arranged in regions and subdivisions. These neurons receive sensory information, direct the activity of motor neurons, and perform such higher brain functions as learning and memory. Even self-awareness, emotions, and consciousness may derive from complex interactions of different brain regions.

The central nervous system (CNS), consisting of the brain and spinal cord (fig. 8.1), receives input from *sensory neurons* and directs the activity of *motor neurons*, which innervate muscles and glands. The *association neurons* within the brain and spinal cord are in a position, as their name implies, to associate appropriate motor responses with sensory stimuli, and thus to maintain homeostasis in the internal environment and the continued existence of the organism in a changing external environment. Further, the central nervous systems of all vertebrates (and most invertebrates) are capable of at least rudimentary forms of learning and retaining the memories of past experiences. This ability, which is most highly developed in the

human brain, permits behavior to be modified by experience and is thus of obvious benefit to survival. Perceptions, learning, memory, emotions, and perhaps even the self-awareness that forms the basis of consciousness are creations of the brain. Whimsical though it seems, the study of brain physiology is the process of the brain studying itself.

The study of the structure and function of the central nervous system requires a knowledge of its basic "plan," which is established during the course of embryonic development. The early embryo contains an embryonic tissue layer known as *ectoderm* on its surface; this will eventually form the epidermis of the skin, among other structures. As development progresses, a groove appears in this ectoderm along the dorsal midline of the embryo's body. This groove deepens, and by the twentieth day after conception, has fused to form a *neural tube*. The part of the ectoderm where the fusion occurs becomes a structure separate from the neural tube, called the *neural crest*, which is located between the neural tube and the surface ectoderm (fig. 8.2). Eventually, the neural tube will become the central nervous system and the neural crest will become the ganglia of the peripheral nervous system, among other structures.

By the middle of the fourth week after conception, three distinct swellings are evident on the anterior end of the neural tube, which is going to form the brain: the *prosencephalon (forebrain)*, *mesencephalon (midbrain)*, and *rhombencephalon (hindbrain)* (fig. 8.3). During the fifth week, these areas become modified to form five regions. The forebrain divides into the *telencephalon* and *diencephalon*; the mesencephalon remains unchanged; and the hindbrain divides into the *metencephalon* and

Chapter Eight

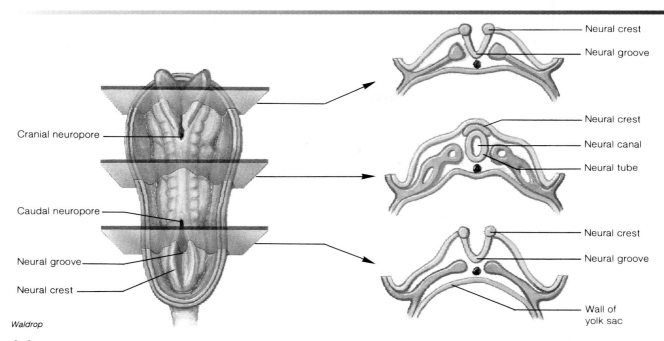

Figure 8.2

A dorsal view of a 22-day-old embryo showing transverse sections at three levels of the developing central nervous system.

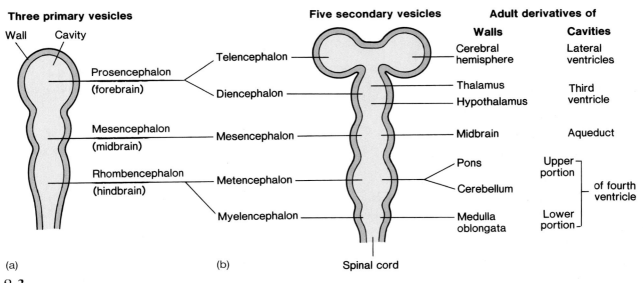

Figure 8.3

The developmental sequence of the brain. (*a*) During the fourth week, three principal regions of the brain are formed. (*b*) During the fifth week, a five-regioned brain develops and specific structures begin to form.

myelencephalon (fig. 8.3). These regions subsequently become greatly modified, but the terms described here are still used to indicate general regions of the adult brain.

The basic structural plan of the CNS can now be understood. The telencephalon (refer to fig. 8.3) grows dispropor-

tionately in humans, forming the two enormous hemispheres of the *cerebrum* that cover the diencephalon, the midbrain, and a portion of the hindbrain. Also, notice that the CNS begins as a hollow tube, and indeed remains hollow as the brain regions are formed. The cavities of the brain are known as

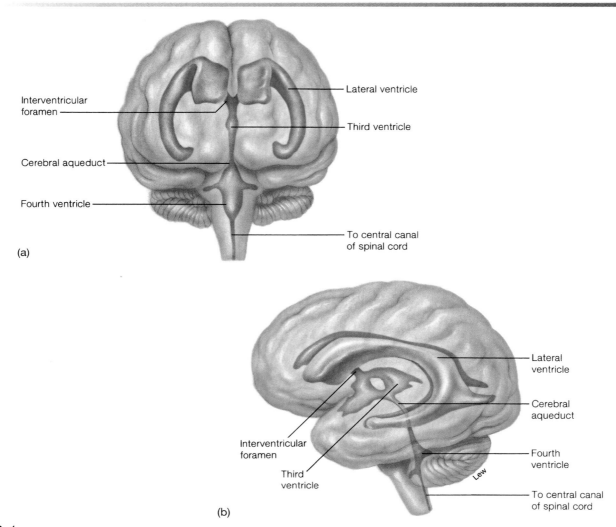

(a)

Interventricular foramen

Lateral ventricle

Third ventricle

Cerebral aqueduct

Fourth ventricle

To central canal of spinal cord

(b)

Interventricular foramen

Third ventricle

Lateral ventricle

Cerebral aqueduct

Fourth ventricle

To central canal of spinal cord

Figure 8.4

The ventricles of the brain. (*a*) An anterior view and (*b*) a lateral view.

ventricles and become filled with *cerebrospinal fluid* (CSF). The cavity of the spinal cord is called the *central canal*, and is also filled with CSF (fig. 8.4).

The CNS is composed of gray and white matter, as described in chapter 7. The gray matter, consisting of neuron cell bodies and dendrites, is found in the surface layer (the *cortex*) of the brain and deeper within the brain in aggregations known as *nuclei*. White matter consists of axon tracts (the myelin sheaths produce the white color) that underlie the cortex and surround the nuclei. The adult brain contains an estimated 100 billion (10^{11}) neurons, weighs approximately 1.5 kg (3–3.5 lb), and receives about 20% of the total blood flow to the body per minute. This high rate of blood flow is a conse-

quence of the high metabolic requirements of the brain; it is not, as Aristotle believed, because the brain's function is to cool the blood. (This fanciful notion—completely incorrect—is a striking example of prescientific thought, having no foundation in experimental evidence.)

1. Identify the three brain regions formed by the middle of the fourth week, and the five brain regions formed during the fifth week of gestation.

2. Describe the embryonic origin of the brain ventricles. Where are they located and what do they contain?

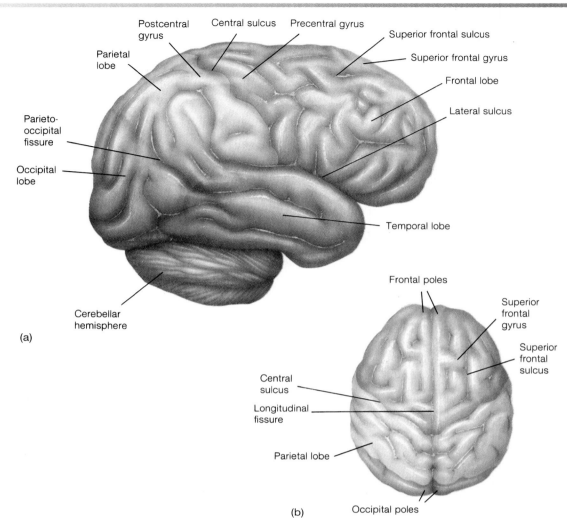

Figure 8.5

The cerebrum. (*a*) A lateral view and (*b*) a superior view.

Cerebrum

The cerebrum, consisting of five paired lobes within two convoluted hemispheres, contains gray matter in its cortex and in deeper cerebral nuclei. Most of what are considered to be the higher functions of the brain are performed by the cerebrum.

The **cerebrum** (fig. 8.5), which is the only structure of the telencephalon, is the largest portion of the brain (accounting for about 80% of its mass) and is believed to be the brain region primarily responsible for higher mental functions. The cerebrum consists of *right* and *left hemispheres,* which are connected internally by a large fiber tract called the *corpus callosum* (see fig. 8.1).

Cerebral Cortex

The cerebrum consists of an outer **cerebral cortex,** composed of 2 to 4 mm of gray matter and underlying white matter. The cerebral cortex is characterized by numerous folds and grooves called *convolutions*. The elevated folds of the convolutions are called *gyri,* and the depressed grooves are the *sulci*. Each cerebral hemisphere is subdivided by deep sulci, or *fissures,* into five lobes, four of which are visible from the surface (fig. 8.6). These lobes are the *frontal, parietal, temporal,* and *occipital,* which are visible from the surface, and the deep *insula* (table 8.1).

The **frontal lobe** is the anterior portion of each cerebral hemisphere. A deep fissure, called the *central sulcus,* separates the frontal lobe from the **parietal lobe.** The *precentral gyrus* (figs. 8.5 and 8.6), involved in motor control, is located in the frontal lobe, just in front of the central sulcus. The *postcentral gyrus,* which is located just behind the central sulcus in the parietal lobe, is the

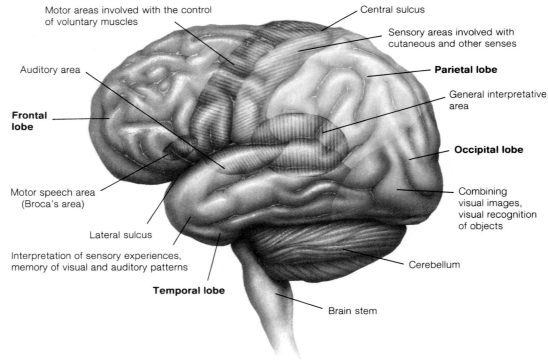

Motor areas involved with the control of voluntary muscles

Central sulcus

Sensory areas involved with cutaneous and other senses

Parietal lobe

Auditory area

General interpretative area

Frontal lobe

Occipital lobe

Motor speech area (Broca's area)

Combining visual images, visual recognition of objects

Lateral sulcus

Cerebellum

Interpretation of sensory experiences, memory of visual and auditory patterns

Temporal lobe

Brain stem

Figure 8.6

The lobes of the left cerebral hemisphere showing the principal motor and sensory areas of the cerebral cortex.

Table 8.1   Functions of the Cerebral Lobes

Lobe	Functions
Frontal	Voluntary motor control of skeletal muscles; personality; higher intellectual processes (e.g., concentration, planning, and decision making); verbal communication
Parietal	Somatesthetic interpretation (e.g., cutaneous and muscular sensations); understanding speech and formulating words to express thoughts and emotions; interpretation of textures and shapes
Temporal	Interpretation of auditory sensations; storage (memory) of auditory and visual experiences
Occipital	Integrates movements in focusing the eye; correlating visual images with previous visual experiences and other sensory stimuli; conscious perception of vision
Insula	Memory; integration of other cerebral activities

Source: From Kent M. Van De Graaff, *Human Anatomy,* 4th ed. Copyright © 1995 Wm. C. Brown Communications, Inc., Dubuque, Iowa. Reprinted by permission of Times Mirror Higher Education Group, Inc., Dubuque, Iowa. All Rights Reserved.

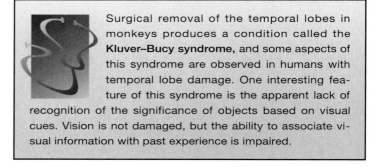

Surgical removal of the temporal lobes in monkeys produces a condition called the **Kluver–Bucy syndrome,** and some aspects of this syndrome are observed in humans with temporal lobe damage. One interesting feature of this syndrome is the apparent lack of recognition of the significance of objects based on visual cues. Vision is not damaged, but the ability to associate visual information with past experience is impaired.

primary area of the cortex responsible for the perception of *somatesthetic sensation*—sensation arising from cutaneous, muscle, tendon, and joint receptors.

The precentral (motor) and postcentral (sensory) gyri have been mapped in conscious patients undergoing brain surgery. Electrical stimulation of specific areas of the precentral gyrus causes specific movements, and stimulation of different areas of the postcentral gyrus evokes sensations in specific parts of the body. Typical maps of these regions (fig. 8.7) show an upside-down picture of the body, with the superior regions of cortex devoted to the toes and the inferior regions devoted to the head.

A striking feature of these maps is that the areas of cortex responsible for different parts of the body do not correspond to the size of the body parts being served. Instead, the body regions with the highest densities of receptors are represented with the largest areas of the sensory cortex, and the body regions with the greatest number of motor innervations are represented with the largest areas of motor cortex. The hands and face, therefore, which have a high density of sensory receptors and motor innervation, are served by larger areas of the precentral and postcentral gyri than is the rest of the body.

The **temporal lobe** contains auditory centers that receive sensory fibers from the cochlea of each ear. This lobe is also involved in the interpretation and association of auditory and visual information. The **occipital lobe** is the primary area responsible for vision and for the coordination of eye movements.

Chapter Eight

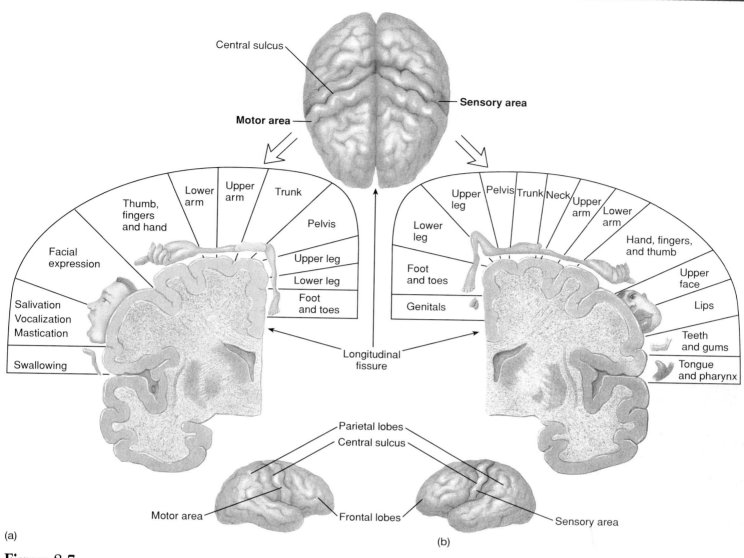

Figure 8.7

Motor and sensory areas of the cerebral cortex. (*a*) Motor areas that control skeletal muscles and (*b*) sensory areas that receive somatesthetic sensations.

Visualizing the Brain

Several relatively new techniques permit the brains of living people to be observed in detail for medical and research purposes. The first of these to be developed was *x-ray computed tomography* (*CT*). CT involves complex computer manipulation of data obtained from x-ray absorption by tissues of different densities. Using this technique, soft tissues such as the brain can be observed at different depths.

The next technique to be developed was *positron-emission tomography* (*PET*). In this technique, radioisotopes that emit positrons are injected into a person. Positrons are like electrons but carry a positive charge. The collision of a positron and an electron results in their mutual annihilation and the emission of gamma rays, which can be detected and used to pinpoint brain cells that are most active. Scientists have used PET to study brain metabolism, drug distribution in the brain, and changes in blood flow as a result of brain activity.

The newest technique for visualizing the living brain is *magnetic resonance imaging* (*MRI*). This technique is based on the concept that protons (H^+) respond to a magnetic field. The magnetic field is used to align the protons, which emit a detectable radio-wave signal when appropriately stimulated. This technique allows excellent images to be obtained (figs. 8.8 and 8.9) without subjecting the person to any known danger. Scientists are now using MRI together with other techniques to study the function of the brain (see fig. 8.8).

Electroencephalogram

The synaptic potentials (discussed in chapter 7) produced at the cell bodies and dendrites of the cerebral cortex create electrical currents that can be measured by electrodes placed on the scalp. A record of these electrical currents is called an *electroencephalogram*, or *EEG*. Deviations from normal EEG patterns can be used clinically to diagnose epilepsy and other abnormal states, and the absence of an EEG can be used to signify brain death.

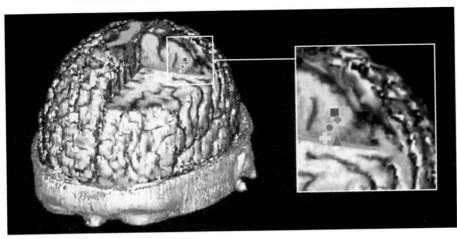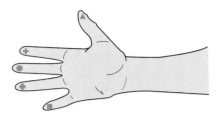

Figure 8.8

An MRI image of the brain reveals the sensory cortex. The integration of MRI and EEG information shows the location on the sensory cortex that corresponds to each of the digits of the hand.

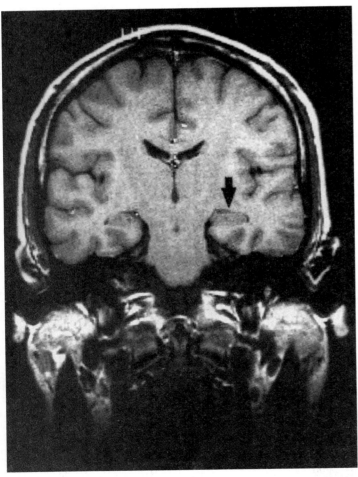

Figure 8.9

An MRI scan of a normal patient. In this coronal view of the brain, the lateral and third ventricles can be clearly seen. The arrow points to a part of the hippocampus.

There are normally four types of EEG patterns (fig. 8.10). **Alpha waves** are best recorded from the parietal and occipital regions while a person is awake and relaxed but with the eyes closed. These waves are rhythmic oscillations of 10 to 12 cycles/second. The alpha rhythm of a child under the age of 8 occurs at a slightly lower frequency of 4 to 7 cycles/second.

Beta waves are strongest from the frontal lobes, especially the area near the precentral gyrus. These waves are produced by visual stimuli and mental activity. Because they respond to stimuli from receptors and are superimposed on the continuous activity patterns, they constitute *evoked activity*. Beta waves occur at a frequency of 13 to 25 cycles per second.

Theta waves are emitted from the temporal and occipital lobes. They have a frequency of 5 to 8 cycles/second and are common in newborn infants. The recording of theta waves in adults generally indicates severe emotional stress and can be a forewarning of a nervous breakdown.

Delta waves are seemingly emitted in a general pattern from the cerebral cortex. These waves have a frequency of 1 to 5 cycles/second and are common during sleep and in an awake infant. The presence of delta waves in an awake adult indicates brain damage.

Basal Nuclei

The **basal nuclei,** also sometimes called basal ganglia, are masses of gray matter composed of neuron cell bodies located deep within the white matter of the cerebrum (fig. 8.11). The most prominent of the basal ganglia is the **corpus striatum,** which consists of several masses of nuclei (a *nucleus* is a collection of cell bodies in the CNS). The upper mass, called the *caudate nucleus,* is separated from two lower masses, collectively called the *lentiform nucleus.* The lentiform nucleus consists of a lateral portion, called the *putamen,* and a medial portion, called the *globus pallidus.* The basal ganglia function in the control of voluntary movements.

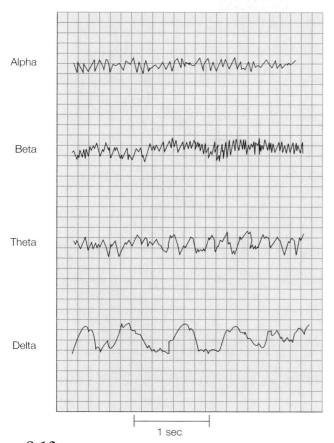

Figure 8.10

Different types of waves in an electroencephalogram (EEG).

 Degeneration of the caudate nucleus produces **chorea**—a hyperkinetic disorder characterized by rapid, uncontrolled, jerky movements. Degeneration of dopaminergic neurons to the caudate nucleus from the substantia nigra, a small nucleus considered a part of the basal ganglia, produces **Parkinson's disease.** As discussed in chapter 7, this disease is associated with rigidity, resting tremor, and difficulty in the initiation of voluntary movements.

Cerebral Lateralization

Each cerebral cortex controls, via motor fibers originating in the precentral gyrus, movements of the contralateral (opposite) side of the body. At the same time, somatesthetic sensation from each side of the body projects to the contralateral postcentral gyrus as a result of *decussation* (crossing over) of fibers. In a similar manner, images falling in the left half of each retina project to the right occipital lobe, and images in the right half of each retina project to the left occipital lobe. Each cerebral hemisphere, however, receives information from both sides of the body because the two hemispheres communicate with each other via the **corpus callosum,** a large tract composed of about 200 million fibers.

The corpus callosum has been surgically cut in some victims of severe epilepsy as a way of alleviating their symptoms. These *split-brain procedures* isolate each hemisphere from the other, but to a casual observer, surprisingly, split-brain patients do not show evidence of any disability as a result of the surgery. However, in specially designed experiments in which each hemisphere is separately presented with sensory images and the patient is asked to perform tasks (speech or writing or drawing with the contralateral hand), it has been learned that each hemisphere is good at certain categories of tasks and poor at others (fig. 8.12).

In a typical experiment, the image of an object may be presented to either the right or left hemisphere (by presenting it to either the left or right visual field only) and the person may be asked to name the object. Findings indicated that, in most people, the task can be performed successfully by the left hemisphere but not by the right. Similar experiments have shown that the left hemisphere is generally the one in which most of the language and analytical abilities reside. These findings have led to the concept of **cerebral dominance,** which is analogous to the concept of handedness—people generally have greater motor competence with one hand than with the other. Since most people are right-handed, and the right hand is also controlled by the left hemisphere, the left hemisphere was naturally considered to be the dominant hemisphere in most people. Further experiments have shown, however, that the right hemisphere is specialized along different, less obvious lines—rather than one hemisphere being dominant and the other subordinate, the two hemispheres appear to have complementary functions. The term **cerebral lateralization,** or specialization of function in one hemisphere or the other, is thus now preferred to the term *cerebral dominance,* although both terms are currently used.

Experiments have shown that the right hemisphere does have limited verbal ability; more noteworthy is the observation that the right hemisphere is most adept at *visuospatial tasks*. The right hemisphere, for example, can recognize faces better than the left, but it cannot describe facial appearances as well as the left. Acting through its control of the left hand, the right hemisphere is better than the left (controlling the right hand) at arranging blocks or drawing cubes. Patients with damage to the right hemisphere, as might be predicted from the results of split-brain research, have difficulty finding their way around a house and reading maps.

Perhaps as a result of the role of the right hemisphere in the comprehension of patterns and part-whole relationships, the ability to compose music, but not to critically understand it, appears to depend on the right hemisphere. Interestingly, damage to the left hemisphere may cause severe speech problems while leaving the ability to sing unaffected.

The lateralization of functions just described—with the left hemisphere specialized for language and analytical ability, and the right hemisphere specialized for visuospatial ability—is true for 97% of all people. It is true for all right-handers (who account for 90% of all people) and for 70% of all left-handers.

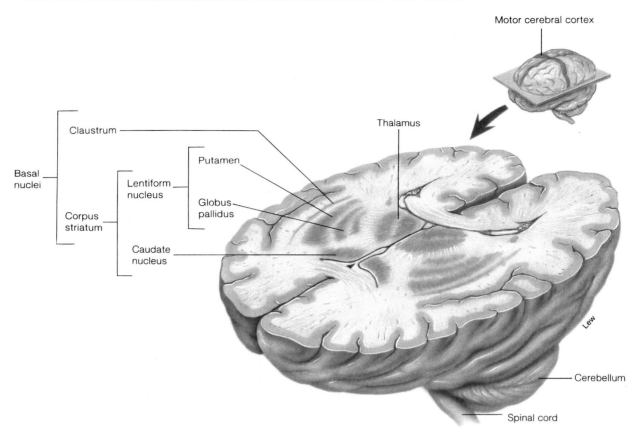

Figure 8.11

Structures of the cerebrum containing neurons involved in the control of skeletal muscles (higher motor neurons). The thalamus is a relay center between the motor cerebral cortex and other brain areas.

The remaining left-handers are split about equally into those who have language-analytical ability in the right hemisphere and those in whom this ability is present in both hemispheres.

It is interesting to speculate that the creative ability of a person may be related to the interaction of information between the right and left hemispheres. The finding of one study—that the number of left-handers among college art students is disproportionately higher than the number of left-handers in the general population—suggests that this interaction may be greater in left-handed people. The observation that Leonardo da Vinci and Michelangelo were both left-handed is interesting in this regard, but obviously does not constitute scientific proof of any hypothesis. Further research on the lateralization of function of the cerebral hemispheres may reveal much more about both brain function and the creative process.

Language

Knowledge of the brain regions involved in language has been gained primarily by the study of *aphasias*—speech and language disorders caused by damage to the brain through head injury and stroke. The language areas of the brain are primarily located in the left hemisphere of the cerebral cortex in most people, as previously described. Even in the nineteenth century,

two areas of the cortex—Broca's area and Wernicke's area (fig. 8.13)—were recognized as areas of particular importance in the production of aphasias.

Broca's aphasia is the result of damage to **Broca's area,** located in the left inferior frontal gyrus and surrounding areas. Common symptoms include weakness in the right arm and the right side of the face. People with Broca's aphasia are reluctant to speak, and when they try, their speech is slow and poorly articulated. Their comprehension of speech is unimpaired, however. People with this aphasia can understand a sentence but have difficulty repeating it. It should be noted that this is not simply due to a problem in motor control, since the neural control over the musculature of the tongue, lips, larynx, and so on is unaffected.

Wernicke's aphasia is caused by damage to **Wernicke's area,** located in the superior temporal gyrus. This results in speech that is rapid and fluid but without meaning. People with Wernicke's aphasia produce speech that has been described as a "word salad." The words used may be real words that are chaotically mixed together, or they may be made-up words. Language comprehension is destroyed; people with Wernicke's aphasia cannot understand either spoken or written language.

It appears that the concept of words to be spoken originates in Wernicke's area and is communicated to Broca's area in

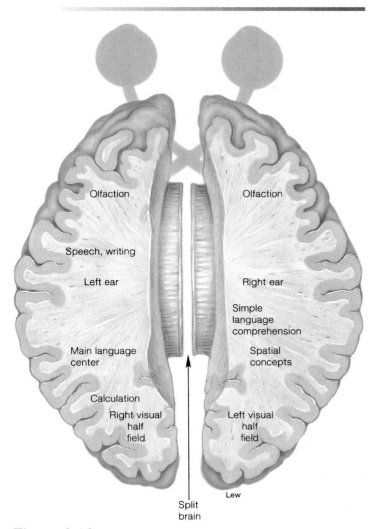

Olfaction

Speech, writing

Left ear

Main language center

Calculation

Right visual half field

Olfaction

Right ear

Simple language comprehension

Spatial concepts

Left visual half field

Lew

Split brain

Figure 8.12

Different functions of the right and left cerebral hemispheres as revealed by experiments with people who have had the tract connecting the two hemispheres (the corpus callosum) surgically split.

a fiber tract called the **arcuate fasciculus.** Broca's area, in turn, sends fibers to the motor cortex (precentral gyrus), which directly controls the musculature of speech. Damage to the arcuate fasciculus produces *conduction aphasia,* which is fluent but nonsensical speech as in Wernicke's aphasia, even though both Broca's and Wernicke's areas are intact.

The **angular gyrus,** located at the junction of the parietal, temporal, and occipital lobes, is believed to be a center for the integration of auditory, visual, and somatesthetic information. Damage to the angular gyrus produces aphasias, which suggests that this area projects to Wernicke's area. Some patients with damage to the left angular gyrus can speak and understand spoken language but cannot read or write. Other patients can write a sentence but cannot read it, presumably due to damage to the projections from the occipital lobe (involved in vision) to the angular gyrus.

Recovery of language ability, by transfer to the right hemisphere after damage to the left hemisphere, is very good in children but decreases after adolescence. Recovery is reported to be faster in left-handed people, possibly because language ability is more evenly divided between the two hemispheres in left-handed people. Some recovery usually occurs after damage to Broca's area, but damage to Wernicke's area produces more severe and permanent aphasias.

Emotion and Motivation

The parts of the brain that appear to be of paramount importance in the neural basis of emotional states are the hypothalamus (in the diencephalon) and the **limbic system.** The limbic system consists of a group of nuclei and fiber tracts that form a ring around the brain stem. The structures of the limbic system include the *cingulate gyrus* (part of the cerebral cortex), *amygdaloid nucleus* (or *amygdala*), *hippocampus,* and the *septal nuclei* (fig. 8.14).

The limbic system was once called the *rhinencephalon,* or "smell brain," because it is involved in the central processing of olfactory information. This may be its primary function in lower vertebrates, whose limbic system may constitute the entire forebrain. It is now known however, that the limbic system in humans is a center for basic emotional drives. The limbic system was derived early in the course of vertebrate evolution, and its tissue is phylogenetically older than the cerebral cortex. There are thus few synaptic connections between the cerebral cortex and the structures of the limbic system, which perhaps helps to explain why we have so little conscious control over our emotions.

There is a closed circuit of information flow between the limbic system and the thalamus and hypothalamus (fig. 8.14) called the *Papez circuit.* (The thalamus and hypothalamus are part of the diencephalon, and are described in a later section.) In the Papez circuit, a fiber tract, the *fornix,* connects the hippocampus to the mammillary bodies of the hypothalamus, which in turn project to the anterior nuclei of the thalamus. The nuclei of the thalamus, in turn, send fibers to the cingulate gyrus, which then completes the circuit by sending fibers to the hippocampus. Through these interconnections, the limbic system and the hypothalamus appear to cooperate in the neural basis of emotional states.

Studies of the functions of these regions include electrical stimulation of specific locations, destruction of tissue (producing *lesions*) in particular sites, and surgical removal, or *ablation,* of specific structures. These studies suggest that the hypothalamus and limbic system are involved in the following feelings and behaviors:

1. **Aggression.** Stimulation of certain areas of the amygdala produces rage and aggression, and lesions of the amygdala can produce docility in experimental animals. Stimulation of particular areas of the hypothalamus can produce similar effects.

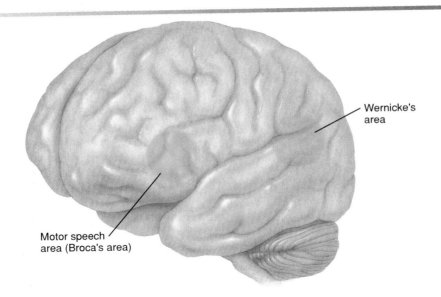

Figure 8.13

Brain areas involved in the control of speech.

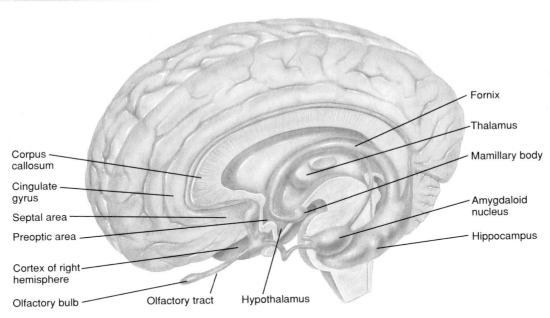

Figure 8.14

The limbic system and the pathways that interconnect the structures of the limbic system. (Note: The left temporal lobe of the cerebral cortex has been removed.)

2. **Fear.** Fear can be produced by electrical stimulation of the amygdala and hypothalamus, and surgical removal of the limbic system can result in an absence of fear. Monkeys are normally terrified of snakes, for example, but if they have had their limbic system removed they will handle snakes without fear.

3. **Feeding.** The hypothalamus contains both a *feeding center* and a *satiety center*. Electrical stimulation of the former causes overeating, and stimulation of the latter will stop feeding behavior in experimental animals.

4. **Sex.** The hypothalamus and limbic system are involved in the regulation of the sexual drive and sexual behavior, as

Chapter Eight

shown by stimulation and ablation studies in experimental animals. The cerebral cortex, however, is also critically important for the sex drive in lower animals, and the role of the cerebrum is even more important for the sex drive in humans.

5. **Goal-directed behavior (reward and punishment system).** Electrodes placed in particular sites between the frontal cortex and the hypothalamus can deliver shocks that function as a reward. In rats, this reward is more powerful than food or sex in motivating behavior. Similar studies have been done in humans, who report feelings of relaxation and relief from tension, but not of ecstasy. Electrodes placed in slightly different positions apparently stimulate a punishment system in experimental animals, who stop their behavior when stimulated in these regions.

One of the most dramatic examples of the role of higher brain areas in personality and emotion is the famous crowbar accident of 1848. A 25-year-old railroad foreman, Phineas P. Gage, was tamping gunpowder into a hole in a rock when it exploded. The rod—three feet, seven inches long and one and one-fourth inches thick—was driven through his left eye, passed through his brain, and emerged in the back of his skull.

After a few minutes of convulsions, Gage got up, rode a horse three-quarters of a mile into town, and walked up a long flight of stairs to see a doctor. He recovered well, with no noticeable sensory or motor deficits. His associates, however, noted striking personality changes. Before the accident Gage was a responsible, capable, and financially prudent man. Afterwards, he appeared to have lost his social inhibitions, engaging, for example, in gross profanity (which he did not do before the accident). He also seemed to be tossed about by chance whims. He was eventually fired from his job, and his old friends remarked that he was "no longer Gage."

Memory

Clinical studies of *amnesia* (loss of memory) suggest that several different brain regions are involved in memory storage and retrieval. Amnesia has been found to result from damage to the temporal lobe of the cerebral cortex, hippocampus, head of the caudate nucleus (in Huntington's disease), or the dorsomedial thalamus (in alcoholics suffering from Korsakoff's syndrome with thiamine deficiency). A number of researchers now believe that there are two different systems of information storage in the brain. One system relates to the simple learning of stimulus-response that even invertebrates can do to a lesser degree. This, together with skill learning and different kinds of conditioning and habits, are retained in people with amnesia.

People with amnesia have an impaired ability to remember facts and events, which some scientists have called "declarative memory." This system of memory can be divided into two major categories: **short-term memory** and **long-term memory.** People with head trauma, for example, and patients with suicidal depression who are treated by *electroconvulsive shock (ECS)*

therapy, may lose their memory of recent events but retain their older memories. The consolidation of short-term memory into long-term memory is the function of the **medial temporal lobe,** an area that includes the hippocampus, amygdaloid nucleus, and adjacent areas of the cerebral cortex (fig. 8.14). Once the memory is put into long-term storage, however, it is independent of the medial temporal lobe.

Surgical removal of the right and left medial temporal lobes was performed in one patient, designated "H. M.," in an effort to treat his epilepsy. After the surgery he was unable to consolidate any short-term memory. He could repeat a phone number and carry out a normal conversation; he could not remember the phone number if momentarily distracted, however, and if the person to whom he was talking left the room and came back a few minutes later, H. M. would have no recollection of seeing that person or of having had a conversation with that person before. Although his memory of events that occurred before the operation was intact, all subsequent events in his life seemed as if they were happening for the first time.

The effects of bilateral removal of the medial temporal lobes on H. M. are believed to be due to the fact that the hippocampus and amygdaloid nucleus (fig. 8.14) were also removed in the process. Surgical removal of the left medial temporal lobe impairs the consolidation of short-term verbal memories into long-term memory, and removal of the right medial temporal lobe impairs the consolidation of nonverbal memories.

On the basis of additional clinical experience, its appears that the hippocampus is a critical component of the memory system. Magnetic resonance imaging (MRI) reveals that the hippocampus is often shrunken in living amnesic patients. However, the degree of memory impairment is increased when other structures, as well as the hippocampus, are damaged. The hippocampus and associated structures of the medial temporal lobe are thus believed to be needed for the acquisition of new information about facts and events, and for the consolidation of short-term into long-term memory, which is stored in the cerebral cortex.

The cerebral cortex is thought to store factual information, with verbal memories lateralized to the left hemisphere and visuospatial information to the right hemisphere. The neurosurgeon Wilder Penfield has electrically stimulated various brain regions of awake patients, often evoking visual or auditory memories that were extremely vivid. Electrical stimulation of specific points in the temporal lobe evoked specific memories so detailed that the patients felt as if they were reliving the experience. The medial regions of the temporal lobes, however, cannot be the site where long-term memory is stored, since destruction of these areas in patients being treated for epilepsy did not destroy the memory of events prior to the surgery. The inferior region of the temporal lobe cortex, on the other hand, does appear to be a site of storage of visual memories.

The amount of memory destroyed by ablation (removal) of brain tissue appears to depend more on the amount of brain tissue removed than on the location of the surgery. On the basis of these observations, it was formerly believed that the memory was diffusely located in the brain; stimulation of the correct location of the cortex then retrieved the memory. According to current thinking, however, particular aspects of the memory—visual, auditory, olfactory, spatial, and so on—are stored in particular areas, and the cooperation of all of these areas is required to produce the complete memory.

Since long-term memory is not destroyed by electroconvulsive shock, it seems reasonable to conclude that the consolidation of memory depends on relatively permanent changes in the chemical structure of neurons and their synapses. Experiments suggest that protein synthesis is required for the consolidation of the "memory trace." The nature of the synaptic changes involved in memory storage has been studied using the phenomenon of **long-term potentiation (LTP)** in the hippocampus, as described in chapter 7. There also may be structural changes, including the growth of dendritic spines and the formation of new synapses.

The prefrontal lobes of the cerebral cortex are believed to be responsible for working memory, also called "scratchpad memory," because it is required for performing mental arithmetic or repeating a list of words. The function of the prefrontal lobes, however, is more complex than this description suggests; it involves the retrieval of parts of memories from different areas of the brain, where they are stored, into an interconnected and usable whole. Such a retrieval and indexing function is needed for language comprehension, planning, and reasoning ability.

---

1. Describe the locations of the sensory and motor areas of the cerebral cortex and explain how these areas are organized.

2. Describe the location and functions of the basal nuclei. Of what structures are the basal nuclei composed?

3. Describe the structures of the limbic system and explain the functional significance of this system.

4. Explain the differences in function of the right and left cerebral hemispheres.

5. List the areas of the brain believed to be involved in the production of speech and describe the different types of aphasias produced by damage to these areas.

6. Describe the different forms of memory, list the brain structures shown to be involved in memory, and discuss some of the experimental evidence on which this information is based.

Diencephalon

The diencephalon is the part of the forebrain that contains such important structures as the thalamus, hypothalamus, and pituitary gland. The hypothalamus performs numerous vital functions, most of which relate directly or indirectly to the regulation of visceral activities by way of other brain regions and the autonomic nervous system.

The diencephalon, together with the telencephalon (cerebrum) previously discussed, constitutes the forebrain. The diencephalon is almost completely surrounded by the cerebral hemispheres and contains a cavity known as the third ventricle.

Thalamus and Epithalamus

The **thalamus** is a mass of gray matter that composes about four-fifths of the diencephalon. It is located below the lateral ventricles and forms most of the walls of the third ventricle (fig. 8.15). The thalamus acts primarily as a relay center through which all sensory information (except smell) passes on the way to the cerebrum. The **epithalamus** is the dorsal portion of the diencephalon that contains a *choroid plexus* over the third ventricle, where cerebrospinal fluid is formed, and the *pineal gland* (*epiphysis*). The pineal gland secretes the hormone *melatonin*, which may play a role in the endocrine control of reproduction (discussed in chapter 20).

Hypothalamus and Pituitary Gland

The **hypothalamus** is a small portion of the diencephalon located below the thalamus, where it forms the floor and part of the lateral walls of the third ventricle (fig. 8.16). This extremely important brain region contains neural centers for hunger, thirst, and the regulation of body temperature and hormone secretion from the pituitary gland. In addition, centers in the hypothalamus contribute to the regulation of sleep, wakefulness, sexual arousal and performance, and such emotions as anger, fear, pain, and pleasure. Acting through its connections with the medulla oblongata of the brain stem, the hypothalamus helps to evoke the visceral responses to various emotional states. In its regulation of emotion, the hypothalamus works together with the limbic system, as was discussed in the previous section.

Experimental stimulation of different areas of the hypothalamus can evoke the autonomic responses characteristic of aggression, sexual behavior, eating, or satiety. Chronic stimulation of the lateral hypothalamus, for example, can make an animal eat and become obese, whereas stimulation of the medial hypothalamus inhibits eating. Other areas contain osmoreceptors that stimulate thirst and the secretion of antidiuretic hormone (ADH) from the posterior pituitary.

The hypothalamus is also where the body's "thermostat" is located. Experimental cooling of the preoptic-anterior hypothalamus causes shivering (a somatic response) and nonshivering thermogenesis (a sympathetic response). Experimental heating of this hypothalamic area results in hyperventilation (stimulated by somatic motor nerves), vasodilation, salivation, and sweat-gland secretion (stimulated by sympathetic nerves). These responses serve to correct the temperature deviations in a negative feedback fashion.

The coordination of sympathetic and parasympathetic reflexes is thus integrated with the control of somatic and endocrine responses by the hypothalamus. The activities of the hypothalamus are in turn influenced by higher brain centers.

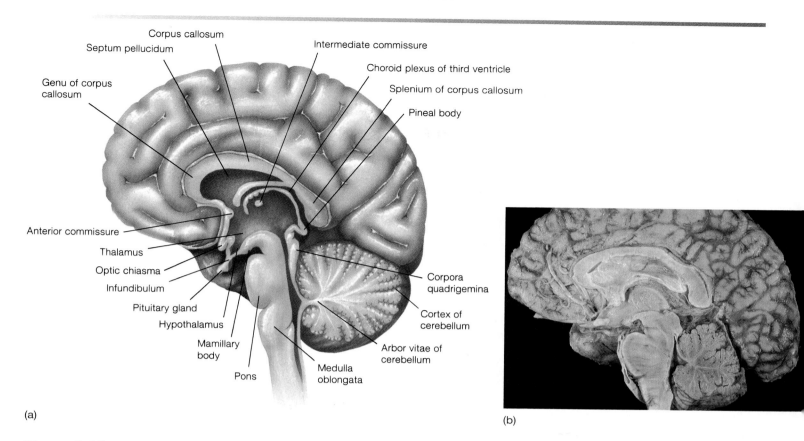

Corpus callosum

Septum pellucidum

Intermediate commissure

Choroid plexus of third ventricle

Genu of corpus callosum

Splenium of corpus callosum

Pineal body

Anterior commissure

Thalamus

Optic chiasma

Infundibulum

Pituitary gland

Hypothalamus

Mamillary body

Pons

Medulla oblongata

Corpora quadrigemina

Cortex of cerebellum

Arbor vitae of cerebellum

(a)

(b)

Figure 8.15

A midsagittal section through the brain. (*a*) A diagram and (*b*) a photograph.

The **pituitary gland** is located immediately inferior to the hypothalamus. Indeed, the posterior pituitary derives embryonically from a downgrowth of the diencephalon, and the entire pituitary remains connected to the diencephalon by means of a stalk (this relationship will be described in more detail in chapter 11). Neurons within the *supraoptic* and *paraventricular nuclei* of the hypothalamus (fig. 8.16) produce two hormones—**antidiuretic hormone (ADH),** which is also known as *vasopressin,* and **oxytocin.** These two hormones are transported in axons of the *hypothalamo-hypophyseal tract* to the **neurohypophysis** (posterior pituitary), where they are stored and secreted in response to hypothalamic stimulation. Oxytocin stimulates contractions of the uterus during labor, and ADH stimulates the kidneys to reabsorb water and thus to excrete a smaller volume of urine. Neurons in the hypothalamus also produce hormones known as **releasing hormones** and **inhibiting hormones,** which are transported by the blood to the **adenohypophysis** (anterior pituitary). These hypothalamic releasing and inhibiting hormones regulate the secretions of the anterior pituitary and, by this means, regulate the secretions of other endocrine glands (as described in chapter 11).

1. Describe the location of the diencephalon relative to the cerebrum and the brain ventricles.

2. List the functions of the hypothalamus and indicate the other brain regions that cooperate with the hypothalamus in the performance of these functions.

3. Explain the structural and functional relationships between the hypothalamus and the pituitary gland.

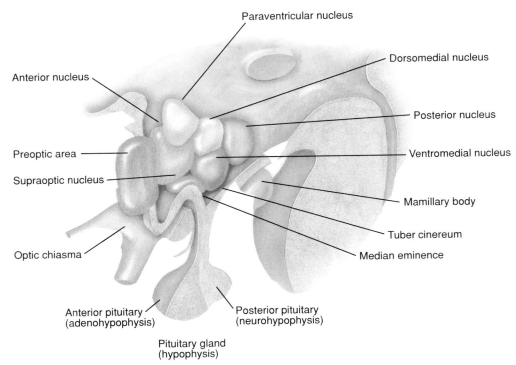

Figure 8.16

A diagram of the nuclei within the hypothalamus.

Midbrain and Hindbrain

The midbrain and hindbrain contain many important relay centers for sensory and motor pathways, and are particularly important in the control of skeletal movements by the brain. The medulla oblongata, a vital region of the hindbrain, contains centers for the control of breathing and cardiovascular function.

Midbrain

The *mesencephalon*, or **midbrain**, is located between the diencephalon and the pons. The **corpora quadrigemina** are four rounded elevations on the dorsal portion of the midbrain (see fig. 8.15). The upper two mounds, called the *superior colliculi*, are involved in visual reflexes. The posterior two, called the *inferior colliculi*, are relay centers for auditory information.

The mesencephalon also contains the cerebral peduncles, red nucleus, and the substantia nigra (not illustrated). The **cerebral peduncles** are a pair of structures composed of ascending and descending fiber tracts. The **red nucleus,** an area of gray matter deep in the midbrain, maintains connections with the cerebrum and the cerebellum and is involved in motor coordination. Another nucleus, the **substantia nigra,** is also involved in motor coordination through its synaptic connections with the basal ganglia, as previously discussed.

Hindbrain

The *rhombencephalon*, or **hindbrain,** is composed of two regions: the metencephalon and the myelencephalon. Each of these regions will be discussed separately.

Metencephalon

The metencephalon is composed of the pons and the cerebellum. The **pons** can be seen as a rounded bulge on the underside of the brain, between the midbrain and the medulla oblongata (fig. 8.17). Surface fibers in the pons connect to the cerebellum, and deeper fibers are part of motor and sensory tracts that pass from the medulla oblongata, through the pons, and on to the midbrain. Within the pons are several nuclei associated with specific cranial nerves—the trigeminal (V), abducens (VI), facial (VII), and vestibulocochlear (VIII). Other nuclei of the pons cooperate with nuclei in the medulla oblongata to regulate breathing. The two respiratory control centers in the pons are known as the *apneustic* and the *pneumotaxic centers*.

The **cerebellum,** which occupies the inferior and posterior aspect of the cranial cavity, is the second largest structure of the brain. Like the cerebrum, it contains outer gray and inner white matter. Fibers from the cerebellum pass through the red nucleus to the thalamus and then to the motor areas of the cerebral cortex. Other fiber tracts connect the cerebellum with the pons, medulla oblongata, and spinal cord. The cerebellum receives input from *proprioceptors* (joint, tendon, and muscle receptors) and, working together with the basal ganglia and motor areas of the cerebral cortex, participates in the coordination of movement.

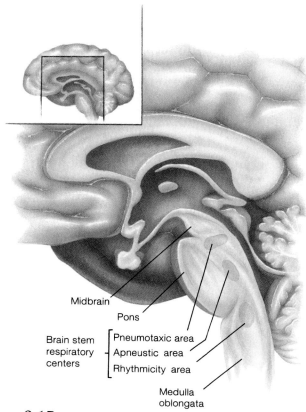

Figure 8.17

Nuclei within the pons and medulla oblongata that constitute the respiratory center.

Damage to the cerebellum produces **ataxia,** which is lack of coordination due to errors in the speed, force, and direction of movement. The movements and speech of a person afflicted with ataxia may resemble those of someone who is intoxicated. This condition is also characterized by *intention tremor,* which differs from the resting tremor of Parkinson's disease in that it occurs only when intentional movements are made. The person suffering cerebellar damage may reach for an object and miss it by placing the hand too far to the left or right, and then attempt to compensate by moving the hand in the opposite direction; this back-and-forth movement can result in oscillations of the limb.

Myelencephalon

The only structure within the myelencephalon is the **medulla oblongata,** often simply called the *medulla*. The medulla is about 3 cm long and is continuous with the pons superiorly and the spinal cord inferiorly. All of the descending and ascending fiber tracts that provide communication between the spinal cord and the brain must pass through the medulla. Many of

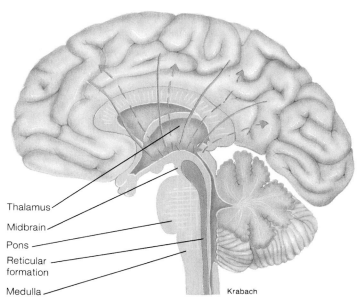

Figure 8.18

The reticular activating system. Arrows indicate the direction of impulses along pathways that connect with the RAS.

these fiber tracts cross to the contralateral side in elevated, triangular structures in the medulla called the **pyramids,** so that the left side of the brain receives sensory information from the right side of the body and vice versa. Similarly, because of the decussation of fibers, the right side of the brain controls motor activity in the left side of the body and vice versa.

There are many important nuclei within the medulla. Several nuclei are involved in motor control, giving rise to axons within cranial nerves VIII, IX, X, XI, and XII. The *vagus nuclei* (there is one on each lateral side of the medulla), for example, give rise to the highly important vagus (X) nerves. Other nuclei relay sensory information to the thalamus and then to the cerebral cortex.

The medulla contains groupings of neurons required for the regulation of breathing and of cardiovascular responses; hence, they are known as the *vital centers*. The **vasomotor center** controls the autonomic innervation of blood vessels; the **cardioinhibitory center** controls the parasympathetic innervation (via the vagus nerve) of the heart (there does not appear to be a separate cardioaccelerator center); and the **respiratory center** of the medulla acts together with centers in the pons to control breathing.

Reticular Formation

The reticular formation is a complex network of nuclei and nerve fibers within the medulla, pons, midbrain, thalamus, and hypothalamus that functions as the **reticular activating system,** or **RAS** (fig. 8.18). Because of its many interconnections, the RAS is activated in a nonspecific fashion by any modality of sensory information. Nerve fibers from the RAS, in turn, project diffusely to the cerebral cortex; this results in *nonspecific arousal* of the cerebral cortex to incoming sensory information.

Table 8.2   Principal Ascending Tracts of Spinal Cord

Tract	Funiculus	Origin	Termination	Function
Anterior spinothalamic	Anterior	Posterior horn on one side of cord but crosses to opposite side	Thalamus, then cerebral cortex	Conducts sensory impulses for crude touch and pressure
Lateral spinothalamic	Lateral	Posterior horn on one side of cord but crosses to opposite side	Thalamus, then cerebral cortex	Conducts pain and temperature impulses that are interpreted within cerebral cortex
Fasciculus gracilis and fasciculus cuneatus	Posterior	Peripheral afferent neurons; does not cross over	Nucleus gracilis and nucleus cuneatus of medulla; eventually thalamus, then cerebral cortex	Conducts sensory impulses from skin, muscles, tendons, and joints, which are interpreted as sensations of fine touch, precise pressures, and body movements
Posterior spinocerebellar	Lateral	Posterior horn; does not cross over	Cerebellum	Conducts sensory impulses from one side of body to same side of cerebellum; necessary for coordinated muscular contractions
Anterior spinocerebellar	Lateral	Posterior horn; some fibers cross, others do not	Cerebellum	Conducts sensory impulses from both sides of body to cerebellum; necessary for coordinated muscular contractions

Source: From Kent M. Van De Graaff, *Human Anatomy*, 4th ed. Copyright © 1995 Wm. C. Brown Communications, Inc., Dubuque, Iowa. Reprinted by permission of Times Mirror Higher Education Group, Inc., Dubuque, Iowa. All Rights Reserved.

 The RAS, through its nonspecific arousal of the cortex, helps to maintain a state of alert consciousness. Not surprisingly, there is evidence that general anesthetics may produce unconsciousness by depressing the RAS. Similarly, the ability to fall asleep may be due to the action of specific neurotransmitters that inhibit activity of the RAS.

1. List the structures of the midbrain and describe their functions.
2. Describe the functions of the medulla oblongata and pons.
3. Describe the composition of the reticular activating system. How does this system function?

Spinal Cord Tracts

Sensory information from receptors throughout most of the body is relayed to the brain by means of ascending tracts of fibers that conduct impulses up the spinal cord. When the brain directs motor activities, these directions are in the form of nerve impulses that travel down the spinal cord in descending tracts of fibers.

The spinal cord extends from the level of the foramen magnum of the skull to the first lumbar vertebra. Unlike the brain, in which the gray matter forms a cortex over white matter, the gray matter of the spinal cord is located centrally, surrounded by white matter. The central gray matter of the spinal cord is arranged in the form of an **H**, with two *dorsal horns* and two *ventral horns* (also called posterior and anterior horns, respectively). The white matter of the spinal cord is composed of ascending and descending fiber tracts. These are arranged into six columns of white matter called *funiculi*.

The fiber tracts within the white matter of the spinal cord are named to indicate whether they are ascending (sensory) or descending (motor) tracts. The names of the ascending tracts usually start with the prefix *spino-* and end with the name of the brain region where the spinal cord fibers first synapse. The anterior spinothalamic tract, for example, carries impulses conveying the sense of touch and pressure, and synapses in the thalamus. From there it is relayed to the cerebral cortex. The names of descending motor tracts, conversely, begin with a prefix denoting the brain region that gives rise to the fibers and end with the suffix *-spinal.* The lateral corticospinal tracts, for example, begin in the cerebral cortex and descend the spinal cord.

Ascending Tracts

The ascending fiber tracts convey sensory information from cutaneous receptors, proprioceptors (muscle and joint senses), and visceral receptors (table 8.2). Most of the sensory information that originates in the right side of the body crosses over to eventually reach the region on the left side of the brain, which analyzes this information. Similarly, the information arising in the left side of the body is ultimately analyzed by the right side of the brain. This decussation occurs in the medulla oblongata (fig. 8.19) for some sensory modalities, or in the spinal cord for other modalities of sensation. These neural pathways are discussed in more detail in chapter 10.

Table 8.3 Descending Motor Tracts to Spinal Interneurons and Motor Neurons

Tract	Category	Origin	Crossed/Uncrossed
Lateral corticospinal	Pyramidal	Cerebral cortex	Crossed
Anterior corticospinal	Pyramidal	Cerebral cortex	Uncrossed
Rubrospinal	Extrapyramidal	Red nucleus (midbrain)	Crossed
Tectospinal	Extrapyramidal	Superior colliculus (midbrain)	Crossed
Vestibulospinal	Extrapyramidal	Vestibular nuclei (medulla oblongata)	Uncrossed
Reticulospinal	Extrapyramidal	Brain stem reticular formation (medulla and pons)	Crossed

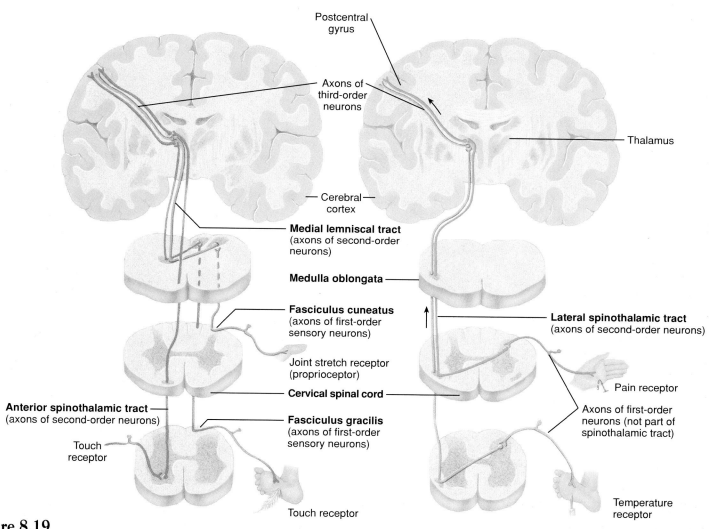

Figure 8.19

Ascending tracts carrying sensory information. This information is delivered by third-order neurons to the cerebral cortex.

Descending Tracts

There are two major groups of descending tracts from the brain: the **corticospinal,** or **pyramidal tracts,** and the **extrapyramidal tracts** (table 8.3). The pyramidal tracts descend directly, without synaptic interruption, from the cerebral cortex to the spinal cord. The cell bodies that contribute fibers to these pyramidal tracts are located primarily in the *precentral gyrus* (also called the *motor cortex*). Other areas of the cerebral cortex, however, also contribute to these tracts.

From 80% to 90% of the corticospinal fibers decussate in the pyramids of the medulla oblongata (hence the name "pyramidal tracts") and descend as the *lateral corticospinal tracts.* The remaining uncrossed fibers form the *anterior corticospinal tracts,*

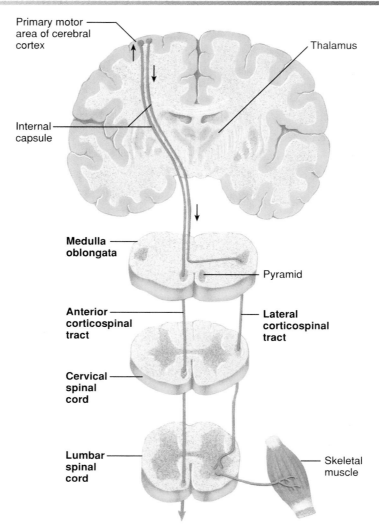

Labels for the figure:
Primary motor area of cerebral cortex
Thalamus
Internal capsule
Medulla oblongata
Pyramid
Anterior corticospinal tract
Lateral corticospinal tract
Cervical spinal cord
Lumbar spinal cord
Skeletal muscle

Figure 8.20

Descending corticospinal (pyramidal) motor tracts.

which decussate in the spinal cord. Because of the crossing of fibers, the right cerebral hemisphere controls the musculature on the left side of the body (fig. 8.20), whereas the left hemisphere controls the right musculature. The corticospinal tracts are primarily concerned with the control of fine movements that require dexterity.

The remaining descending tracts are extrapyramidal motor tracts, which originate in the midbrain and brain stem regions (table 8.3). If the pyramidal tracts of an experimental animal are cut, electrical stimulation of the cerebral cortex, cerebellum, and basal nuclei can still produce movements. The descending fibers that produce these movements must, by definition, be extrapyramidal motor tracts. The regions of the cerebral cortex, basal nuclei, and cerebellum that participate

The corticospinal tracts appear to be particularly important in voluntary movements that require complex interactions between sensory input and the motor cortex. Speech, for example, is impaired when the corticospinal tracts are damaged in the thoracic region of the spinal cord, whereas involuntary breathing continues. Damage to the pyramidal motor system can be detected clinically by a **Babinski sign,** in which stimulation of the sole of the foot causes extension of the big toe upward and fanning of the other toes. The Babinski sign is present in normal infants because neural control is not yet fully developed.

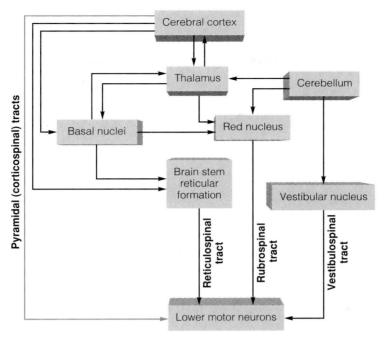

Figure 8.21

Pathways involved in the higher motor neuron control of skeletal muscles. The pyramidal (corticospinal) tracts are shown in red and the extrapyramidal tracts are shown in black.

in this motor control have numerous synaptic interconnections, and can influence movement only indirectly by means of stimulation or inhibition of the nuclei that give rise to the extrapyramidal tracts. Notice that this motor control differs from that by the neurons of the precentral gyrus, which send fibers directly down to the spinal cord in the pyramidal tracts.

The *reticulospinal tracts* are the major descending pathways of the extrapyramidal system. These tracts originate in the reticular formation of the brain stem, which receives either stimulatory or inhibitory input from the cerebrum and the cerebellum. There are no descending tracts from the cerebellum; the cerebellum can influence motor activity only indirectly by its effect on the vestibular nuclei, red nucleus, and basal nuclei (which send axons to the reticular formation). These nuclei, in turn, send axons down the spinal cord via the *vestibulospinal tracts, rubrospinal tracts,* and reticulospinal tracts, respectively (fig. 8.21). Neural control of skeletal muscle is explained in more detail in chapter 12.

1. Explain why each cerebral hemisphere receives sensory input from and directs motor output to the contralateral side of the body.
2. List the tracts of the pyramidal motor system and describe the function of the pyramidal system.
3. List the tracts of the extrapyramidal system and explain how this system differs from the pyramidal motor system.

Cranial and Spinal Nerves

The central nervous system communicates with the body by means of nerves that exit the CNS from the brain (cranial nerves) and spinal cord (spinal nerves). These nerves, together with aggregations of cell bodies located outside the CNS, constitute the peripheral nervous system.

As mentioned in chapter 7, the peripheral nervous system (PNS) consists of nerves (collections of axons) and their associated ganglia (collections of cell bodies). Although this chapter is devoted to the CNS, the CNS cannot function without the PNS. This section thus serves to complete the explanation of the CNS; it also introduces concepts concerning the PNS, which will be explored more thoroughly in later chapters (particularly chapters 9, 10, and 12).

Cranial Nerves

There are twelve pairs of cranial nerves. Two of these pairs arise from neuron cell bodies located in the forebrain and ten pairs arise from the midbrain and hindbrain. The cranial nerves are designated by Roman numerals and by names. The Roman numerals refer to the order in which the nerves are positioned

Table 8.4 Summary of Cranial Nerves

Number and Name	Composition	Function
I Olfactory	Sensory	Olfaction
II Optic	Sensory	Vision
III Oculomotor	Motor	Motor impulses to levator palpebrae superioris and extrinsic eye muscles except superior oblique and lateral rectus; innervation to muscles that regulate amount of light entering eye and that focus the lens
	Sensory: proprioception	Proprioception from muscles innervated with motor fibers
IV Trochlear	Motor	Motor impulses to superior oblique muscle of eyeball
	Sensory: proprioception	Proprioception from superior oblique muscle of eyeball
V Trigeminal		
Ophthalmic division	Sensory	Sensory impulses from cornea, skin of nose, forehead, and scalp
Maxillary division	Sensory	Sensory impulses from nasal mucosa, upper teeth and gums, palate, upper lip, and skin of cheek
Mandibular division	Sensory	Sensory impulses from temporal region, tongue, lower teeth and gums, and skin of chin and lower jaw
	Sensory: proprioception	Proprioception from muscles of mastication
	Motor	Motor impulses to muscles of mastication and muscle that tenses the tympanum
VI Abducens	Motor	Motor impulses to lateral rectus muscle of eyeball
	Sensory: proprioception	Proprioception from lateral rectus muscle of eyeball
VII Facial	Motor	Motor impulses to muscles of facial expression and muscle that tenses the stapes
	Motor: parasympathetic	Secretion of tears from lacrimal gland and salivation from sublingual and submandibular salivary glands
	Sensory	Sensory impulses from taste buds on anterior two-thirds of tongue; nasal and palatal sensation
	Sensory: proprioception	Proprioception from muscles of facial expression
VIII Vestibulocochlear	Sensory	Sensory impulses associated with equilibrium
		Sensory impulses associated with hearing
IX Glossopharyngeal	Motor	Motor impulses to muscles of pharynx used in swallowing
	Sensory: proprioception	Proprioception from muscles of pharynx
	Sensory	Sensory impulses from taste buds on posterior one-third of tongue; pharynx, middle-ear cavity, and carotid sinus
	Parasympathetic	Salivation from parotid salivary gland
X Vagus	Motor	Contraction of muscles of pharynx (swallowing) and larynx (phonation)
	Sensory: proprioception	Proprioception from visceral muscles
	Sensory	Sensory impulses from taste buds on rear of tongue; sensations from auricle of ear; general visceral sensations
	Motor:parasympathetic	Regulation of many visceral functions
XI Accessory	Motor	Laryngeal movement; soft palate
		Motor impulses to trapzius and sternocleidomastoid muscles for movement of head, neck, and shoulders
	Sensory: proprioception	Proprioception from muscles that move head, neck, and shoulders
XII Hypoglossal	Motor	Motor impulses to intrinsic and extrinsic muscles of tongue and infrahyoid muscles
	Sensory: proprioception	Proprioception from muscles of tongue

Source: From Kent M. Van De Graaff, *Human Anatomy,* 4th ed. Copyright © 1995 Wm. C. Brown Communications, Inc., Dubuque, Iowa. Reprinted by permission of Times Mirror Higher Education Group, Inc., Dubuque, Iowa. All Rights Reserved.

from the front of the brain to the back. The names indicate structures innervated by these nerves (e.g., facial) or the principal function of the nerves (e.g., occulomotor). A summary of the cranial nerves is presented in table 8.4.

Most cranial nerves are classified as *mixed nerves*. This term indicates that the nerve contains both sensory and motor fibers. Those cranial nerves associated with the special senses (e.g., olfactory, optic), however, consist of sensory fibers only. The cell bodies of these sensory neurons are not located in the brain, but instead are found in ganglia near the sensory organ.

Spinal Nerves

There are thirty-one pairs of spinal nerves. These nerves are grouped into eight cervical, twelve thoracic, five lumbar, five sacral, and one coccygeal according to the region of the vertebral column from which they arise (fig. 8.22).

Each spinal nerve is a mixed nerve composed of sensory and motor fibers. These fibers are packaged together in the nerve, but separate near the attachment of the nerve to the spinal cord. This produces two "roots" to each nerve. The **dorsal**

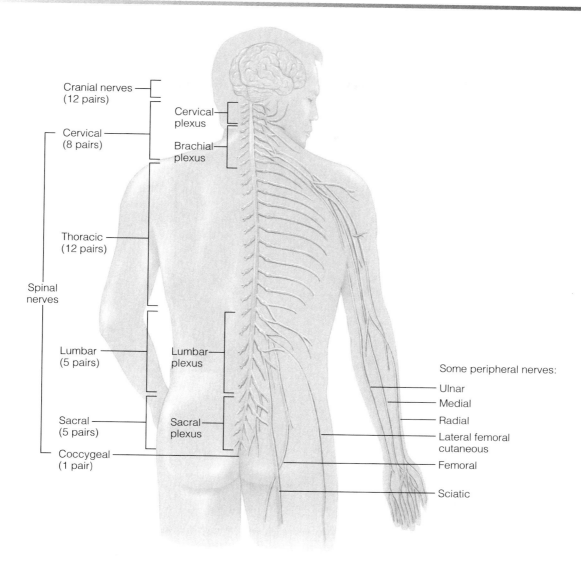

Figure 8.22

The distribution of the spinal nerves. These interconnect at plexuses (shown on the left) and form specific peripheral nerves.

root is composed of sensory fibers, and the **ventral root** is composed of motor fibers (fig. 8.23). The dorsal root contains an enlargement called the **dorsal root ganglion,** where the cell bodies of the sensory neurons are located. The motor neuron shown in figure 8.23 is a somatic motor neuron that innervates skeletal muscles; its cell body is not located in a ganglion, but instead is within the gray matter of the spinal cord. Some autonomic motor neurons (which innervate involuntary effectors), however, have their cell bodies in ganglia outside the spinal cord (the autonomic system is discussed separately in chapter 9).

Reflex Arc

The functions of the sensory and motor components of a spinal nerve can most easily be understood by examination of a simple reflex; that is, an unconscious motor response to a sensory stimulus. Figure 8.23 demonstrates the neural pathway involved in a **reflex arc.** Stimulation of sensory receptors evokes action potentials in sensory neurons, which are conducted into the spinal cord. In the example shown, a sensory neuron synapses with an association neuron (or interneuron), which in turn synapses with a somatic motor neuron. The somatic motor neuron then

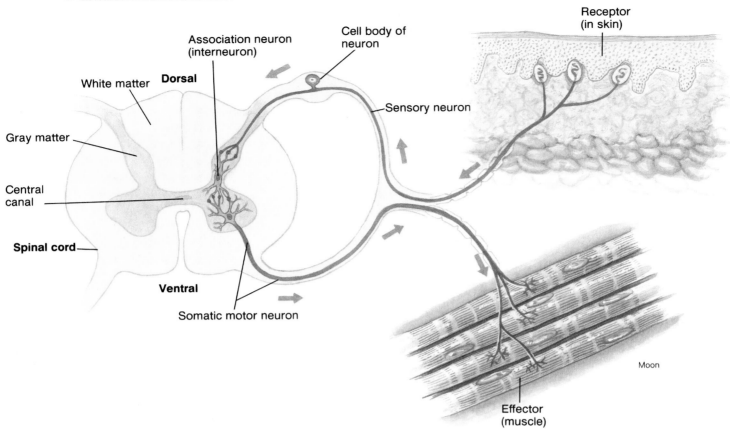

Labels on figure:
White matter — Dorsal
Gray matter
Central canal
Spinal cord
Ventral
Somatic motor neuron
Association neuron (interneuron)
Cell body of neuron
Sensory neuron
Receptor (in skin)
Effector (muscle)
Moon

Figure 8.23

A sensory neuron, an association neuron (interneuron), and a somatic motor neuron at the spinal cord level.

conducts impulses out of the spinal cord to the muscle and stimulates a reflex contraction. Notice that the brain is not directly involved in this reflex response to sensory stimulation. Some reflex arcs are even simpler than this; in a muscle stretch reflex (the knee-jerk reflex, for example) the sensory neuron synapses directly with a motor neuron. Other reflexes are more complex, involving a number of association neurons and resulting in motor responses on both sides of the spinal cord at different levels. These skeletal muscle reflexes are described together with muscle control in chapter 12.

1. Define the terms *dorsal root, dorsal root ganglion, ventral root,* and *mixed nerve.*
2. Describe the neural pathways and structures involved in a reflex arc.

Summary

Structural Organization of the Brain   p. 180

I. During embryonic development, five regions of the brain are formed: the telencephalon, diencephalon, mesencephalon, metencephalon, and myelencephalon.

A. The telencephalon and diencephalon constitute the forebrain; the mesencephalon is the midbrain, and the hindbrain is composed of the metencephalon and the myelencephalon.

B. The CNS begins as a hollow tube, and the brain and spinal cord are thus hollow; the cavities of the brain are known as ventricles.

Cerebrum   p. 183

I. The cerebrum consists of two hemispheres connected by a large fiber tract called the corpus callosum.
   A. The outer part of the cerebrum—the cerebral cortex—consists of gray matter.
   B. Under the gray matter is white matter, but there are nuclei of gray matter deep within the white matter of the cerebrum, known as the basal nuclei.
   C. Synaptic potentials within the cerebral cortex produce the electrical activity seen in an electroencephalogram (EEG).
II. The two cerebral hemispheres have a degree of specialization of function, which is termed cerebral lateralization.
   A. In most people, the left hemisphere is dominant in language and analytical ability, whereas the right hemisphere is important in pattern recognition, musical composition, singing, and the recognition of faces.
   B. The two hemispheres cooperate in their functions; this cooperation is aided by communication between the two via the corpus callosum.
III. Particular regions of the left cerebral cortex appear to be important in language ability; when these areas are damaged, characteristic types of aphasias result.
   A. Wernicke's area is involved in speech comprehension, whereas Broca's area is required for the mechanical performance of speech.
   B. Wernicke's area is believed to control Broca's area by means of the arcuate fasciculus.
   C. The angular gyrus is believed to integrate different sources of sensory information and project to Wernicke's area.
IV. The limbic system and hypothalamus are regions of the brain that have been implicated as centers for various emotions.
V. Memory can be divided into a short-term and a long-term form.
   A. The medial temporal lobes—in particular the hippocampus and perhaps the amygdaloid nucleus—appear to be required for the consolidation of short-term memory into long-term memory.
   B. Particular aspects of a memory may be stored in numerous brain regions.
   C. Long-term potentiation is a process that may explain in part how memory is produced.

Diencephalon   p. 192

I. The diencephalon is the region of the forebrain that includes the thalamus, epithalamus, hypothalamus, and pituitary gland.
   A. The thalamus serves as an important relay center for sensory information, among other functions.
   B. The epithalamus contains a choroid plexus for the formation of cerebrospinal fluid and the pineal gland.
   C. The hypothalamus forms the floor of the third ventricle, and the pituitary gland is located immediately inferior to the hypothalamus.
II. The hypothalamus has many important functions related to the control of visceral activities.
   A. The hypothalamus contains centers for the control of thirst, eating, body temperature, and (together with the limbic system) various emotions.
   B. The hypothalamus controls the pituitary gland. It controls the posterior pituitary by means of a fiber tract, and it controls the anterior pituitary by means of hormones.

Midbrain and Hindbrain   p. 194

I. The midbrain contains the superior and inferior colliculi, which are involved in visual and auditory reflexes, respectively.
II. The hindbrain consists of two regions: the metencephalon and the myelencephalon.
   A. The metencephalon contains the pons and cerebellum. The pons is the site of origination of some cranial nerves, and the cerebellum is an important organ involved in the control of skeletal movements.
   B. The myelencephalon consists of only one region, the medulla oblongata, which contains centers for the regulation of such vital functions as breathing and the control of the cardiovascular system.

Spinal Cord Tracts   p. 196

I. Ascending tracts carry sensory information from sensory organs up the spinal cord to the brain.
II. Descending tracts are motor tracts, and are divisible into the pyramidal and extrapyramidal systems.
   A. Pyramidal tracts are the corticospinal tracts. They begin in the precentral gyrus and descend, without making any synapses, into the spinal cord.
   B. Most of the corticospinal fibers decussate in the pyramids of the medulla oblongata.
   C. Regions of the cerebral cortex, the basal nuclei, and cerebellum control movements indirectly by making synapses with other regions that produce descending extrapyramidal fiber tracts.
   D. The major extrapyramidal motor tract is the reticulospinal tract, which originates in the reticular formation of the midbrain.

Cranial and Spinal Nerves   p. 199

I. There are twelve pairs of cranial nerves. Most of these are mixed, but some are exclusively sensory in function.
II. There are thirty-one pairs of spinal nerves. Each pair contains both sensory and motor fibers.
   A. The dorsal root of a spinal nerve contains sensory fibers, and the cell bodies of these neurons are contained in the dorsal root ganglion.
   B. The ventral root of a spinal nerve contains motor fibers.
III. A reflex arc is a neural pathway involving a sensory neuron and a motor neuron. One or more association neurons may also be involved in some reflexes.

Clinical Investigation

A 62-year-old man is brought to the hospital by his wife. As he leans on her for support, she explains to the doctor that her husband suddenly has become partially paralyzed. Upon examination, the patient displays a knee-jerk reflex and does not have a Babinski sign, but is paralyzed on the right side of his body. He does not voluntarily speak to the doctor, and when questioned, he answers slowly and with difficulty. The answers provided by the patient, however, are coherent. Magnetic resonance images (MRIs) of his brain reveal a blockage of blood flow in a branch of the middle cerebral artery. How might this patient's symptoms be explained?

Clues

Study the section "Cerebrum," particularly the material on language within this section. Also study the material on descending tracts in the section "Spinal Cord Tracts" and the material on the reflex arc in the section "Cranial and Spinal Nerves." Finally, examine figures 8.7 and 8.13.

Review Activities

Objective Questions

1. Which of the following statements about the precentral gyrus is *true*?
   a. It is involved in motor control.
   b. It is involved in sensory perception.
   c. It is located in the frontal lobe.
   d. Both *a* and *c* are true.
   e. Both *b* and *c* are true.

2. In most people, the right hemisphere controls movement of
   a. the right side of the body primarily.
   b. the left side of the body primarily.
   c. both the right and left sides of the body equally.
   d. the head and neck only.

3. Which of the following statements about the basal nuclei is *true*?
   a. They are located in the cerebrum.
   b. They contain the caudate nucleus.
   c. They are involved in motor control.
   d. They are part of the extrapyramidal system.
   e. All of the above are true.

4. Which of the following acts as a relay center for somatesthetic sensation?
   a. the thalamus
   b. the hypothalamus
   c. the red nucleus
   d. the cerebellum

5. Which of the following statements about the medulla oblongata is *false*?
   a. It contains nuclei for some cranial nerves.
   b. It contains the apneustic center.
   c. It contains the vasomotor center.
   d. It contains ascending and descending fiber tracts.

6. The reticular activating system
   a. is composed of neurons that are part of the reticular formation.
   b. is a loose arrangement of neurons with many interconnecting synapses.
   c. is located in the brain stem and midbrain.
   d. functions to arouse the cerebral cortex to incoming sensory information.
   e. is described correctly by all of the above.

7. In the control of emotion and motivation, the limbic system works together with
   a. the pons.
   b. the thalamus.
   c. the hypothalamus.
   d. the cerebellum.
   e. the basal nuclei.

8. Verbal ability predominates in
   a. the left hemisphere of right-handed people.
   b. the left hemisphere of most left-handed people.
   c. the right hemisphere of 97% of all people.
   d. both *a* and *b*.
   e. both *b* and *c*.

9. The consolidation of short-term memory into long-term memory appears to be a function of
   a. the substantia nigra.
   b. the hippocampus.
   c. the cerebral peduncles.
   d. the arcuate fasciculus.
   e. the precentral gyrus.

For questions 10–12, match the nature of the aphasia with its cause (choices are listed under question 12).

10. Comprehension good; can speak and write, but cannot read (although can see).

11. Comprehension good; speech is slow and difficult (but motor ability is not damaged).

12. Comprehension poor; speech is fluent but meaningless.
    a. damage to Broca's area
    b. damage to Wernicke's area
    c. damage to angular gyrus
    d. damage to precentral gyrus

13. Antidiuretic hormone (ADH) and oxytocin are synthesized by supraoptic and paraventricular nuclei, which are located in
    a. the thalamus.
    b. the pineal gland.
    c. the pituitary gland.
    d. the hypothalamus.
    e. the pons.

14. The superior colliculi are twin bodies within the corpora quadrigemina of the midbrain that are involved in
    a. visual reflexes.
    b. auditory reflexes.
    c. relaying of cutaneous information.
    d. release of pituitary hormones.

Essay Questions

1. Define the term *decussation* and explain its significance in terms of the pyramidal motor system.[1]

2. Electrical stimulation of the basal nuclei or cerebellum can produce skeletal movements. Describe the pathways by which these brain regions control motor activity.

3. Define the term *ablation*. Give two examples of how this experimental technique has been used to learn about the function of particular brain regions.

4. Explain how "split-brain" patients have contributed to research on the function of the cerebral hemispheres. Propose experiments that would reveal the lateralization of function in the two hemispheres.

5. What evidence do we have that Wernicke's area may control Broca's area? What evidence do we have that the angular gyrus has input to Wernicke's area?

6. Give two reasons why it is believed that there is a difference between short-term and long-term memory.

7. Describe evidence showing that the hippocampus is involved in the consolidation of short-term memory. Why may the hippocampus not be needed after long-term memory has been established?

8. Can we be aware of a reflex action involving our skeletal muscles? Is this awareness necessary for the response? Explain, identifying the neural pathways involved in the reflex response and the conscious awareness of a stimulus.

Selected Readings

Andreasen, N. C. 1988. Brain imaging: Applications in psychiatry. *Science* 239:1381.

Aoki, C., and P. Siekevitz. December 1988. Plasticity in brain development. *Scientific American*.

Benson, D. F., and N. Geschwind. 1972. Aphasia and related disturbances. In A. B. Baker, ed. *Clinical Neurology*. New York: Harper and Row.

Brown, T. H. et al. 1988. Long-term potentiation. *Science* 242:724.

Carpenter, W. T., and R. W. Buchanan. 1994. Schizophrenia. *New England Journal of Medicine* 330: 681.

Cote, L. 1981. Basal ganglia, the extrapyramidal motor system, and disease of transmitter metabolism. In E. R. Kandel and J. H. Schwartz, eds. *Principles of Neural Science*. New York: Elsevier North Holland.

Damasio, A. R. 1992. Aphasia. *New England Journal of Medicine* 326:531.

Damasio, A. R., and H. Damasio. September 1992. Brain and language. *Scientific American*.

Damasio, H. et al. 1994. The return of Phineas Gage: Clues about the brain from the skull of a famous patient. *Science* 264: 1102.

de Wied, D. 1989. Neuroendocrine aspects of learning and memory processes. *News in Physiological Sciences* 4:32.

Fine, A. August 1986. Transplantation in the central nervous system. *Scientific American*.

Ganong, W. F. 1985. *Review of Medical Physiology*. 12th ed. Los Altos, CA: Lange Medical Publishers.

Garthwaite, J., and C. L. Boulton. 1995. Nitric oxide signaling in the central nervous system. *Annual Review of Physiology* 57:683.

Gershon, E. S., and R. O. Rieder. September 1992. Major disorders of mind and brain. *Scientific American*.

Geschwind, N. April 1972. Language and the brain. *Scientific American*.

Ghez, C. 1981. Cortical control of voluntary movement. In E. R. Kandel and J. H. Schwartz, eds. *Principles of Neural Science*. New York: Elsevier North Holland.

Gilman, S. 1992. Advances in neurology. *New England Journal of Medicine* 326:1608.

Goldman-Rakic, P. S. September 1992. Working memory and the mind. *Scientific American*.

Graybiel, A. M. et al. 1994. The basal ganglia and adaptive motor control. *Science* 265:1826.

Heinze, H. J. et al. 1994. Combined spatial and temporal imaging of brain activity during visual selective attention in humans. *Nature* 372:543.

Hubel, D. H. September 1979. The brain. *Scientific American*.

Jacobs, B. L. 1994. Serotonin, motor activity, and depression-related disorders. *American Scientist* 82:456.

Kalin, N. H. May 1993. The neurobiology of fear. *Scientific American*.

Kandel, E. R., and R. D. Hawkins. September 1992. The biological basis of learning and individuality. *Scientific American*.

Kimura, D. September 1992. Sex differences in the brain. *Scientific American*.

Kupferman, I. 1981. Learning. In E. R. Kandel and J. H. Schwartz, eds. *Principles of Neural Science*. New York: Elsevier North Holland.

LeDoux, J. E. June 1994. Emotion, memory, and the brain. *Scientific American*.

Lemay, M., and N. Geschwind. 1978. Asymmetries of the human cerebral hemispheres. In A. Caramazza and E. Zurif, eds. *Language Acquisition and Language Breakdown*. Baltimore: Johns Hopkins University Press.

Lynch, G., and M. Baudry. 1984. The biochemistry of memory: A new and specific hypothesis. *Science* 224:1057.

Melzak, R. April 1992. Phantom limbs. *Scientific American*.

Mishkin, M., and T. Appenzeller. June 1987. The anatomy of memory. *Scientific American*.

Raichle, M. E. April 1994. Visualizing the mind. *Scientific American*.

Rand, M. J., and C. G. Li. 1995. Nitric oxide as a neurotransmitter in peripheral nerves: nature of transmitter and mechanisms of transmission. *Annual Review of Physiology* 57:659.

Routtenberg, A. November 1978. The reward system of the brain. *Scientific American*.

Selkoe, D. J. November 1991. Amyloid protein and Alzheimer's disease. *Scientific American*.

Shashoua, V. E. 1985. The role of extracellular proteins in learning and memory. *American Scientist* 73:364.

Shatz, C. J. September 1992. The developing brain. *Scientific American*.

Springer, S. P., and G. Deutch. 1985. *Left Brain, Right Brain*. Rev. ed. New York: Freeman.

Squire, L. R. 1986. Mechanisms of memory. *Science* 232:1612.

Squire, L. R., and S. Zola-Morgan. 1991. The medial temporal lobe memory system. *Science* 253:1380.

Thompson, R. F. 1985. *The Brain*. New York: Freeman.

Thompson, R. F. 1986. The neurobiology of learning and memory. *Science* 233:941.

Tuomanen, E. February 1993. Breaching the blood-brain barrier. *Scientific American*.

Witelson, S. F. 1976. Sex and the single hemisphere: Specialization of the right hemisphere for spatial processing. *Science* 193:425.

---

[1]*Note:* This question is answered on page 87 of the Student Study Guide.

9

The Autonomic Nervous System

OBJECTIVES

After studying this chapter, you should be able to . . .

1. compare the structures and pathways of the autonomic system with those involved in the control of skeletal muscle.

2. explain how autonomic innervation of involuntary effectors differs from the innervation of skeletal muscle.

3. describe the structure and general functions of the sympathetic division of the autonomic system.

4. describe the structure and general functions of the parasympathetic division of the autonomic system.

5. list the neurotransmitters of the preganglionic and postganglionic neurons of the sympathetic and parasympathetic systems.

6. describe the structural and functional relationships between the sympathetic system and the adrenal medulla.

7. distinguish between the different types of adrenergic receptors, give their anatomic locations, and explain the physiological and clinical significance of these receptors.

8. explain how the cholinergic receptors are divided into two categories and describe the effects produced by stimulation of these receptors.

9. explain the antagonistic, complementary, and cooperative effects of sympathetic and parasympathetic innervation in different organs.

10. describe the higher neural control of the autonomic system.

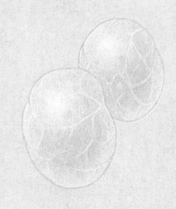

OUTLINE

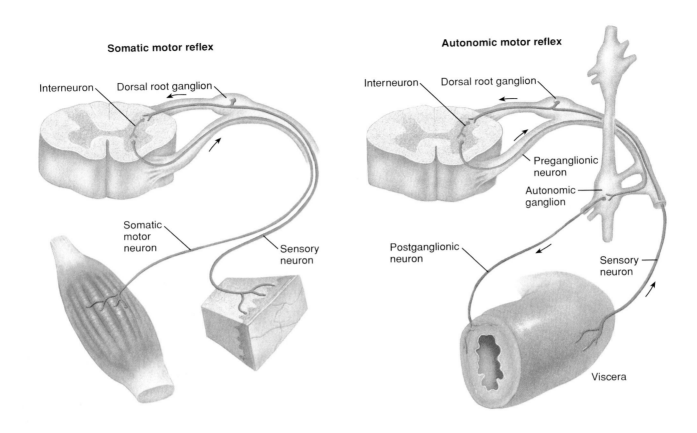

Somatic motor reflex

Interneuron Dorsal root ganglion

Somatic
motor
neuron

Sensory
neuron

Autonomic motor reflex

Interneuron Dorsal root ganglion

Preganglionic
neuron

Autonomic
ganglion

Postganglionic
neuron

Sensory
neuron

Viscera

Figure 9.1

Comparison of a somatic motor reflex, in which a skeletal muscle is stimulated, with an autonomic motor reflex, in which a smooth muscle may be the effector.

Neural Control of Involuntary Effectors

The autonomic nervous system helps to regulate the activities of cardiac muscle, smooth muscle, and glands. In this regulation, impulses are conducted from the CNS by an axon that synapses with a second autonomic neuron. It is the axon of this second neuron in the pathway that innervates the involuntary effectors.

Autonomic motor nerves innervate organs whose functions are not usually under voluntary control. The effectors that respond to autonomic regulation include **cardiac muscle** (the heart), **smooth** (visceral) **muscles,** and **glands.** These are part of the *visceral organs* (organs within the body cavities) and of blood vessels. The involuntary effects of autonomic innervation contrast with the voluntary control of skeletal muscles by way of somatic motor neurons.

Autonomic Neurons

As discussed in chapter 7, neurons of the peripheral nervous system (PNS) that conduct impulses away from the central nervous system (CNS) are known as *motor,* or *efferent, neurons.* There are two major categories of motor neurons: somatic and autonomic. Somatic motor neurons have their cell bodies within the CNS and send axons to skeletal muscles, which are usually under voluntary control. This was briefly described in chapter 8, in the section on the reflex arc, and is reviewed in the left half of figure 9.1. The control of skeletal muscles by somatic motor neurons is discussed in depth in chapter 12.

Table 9.1 Comparison of the Somatic Motor System and the Autonomic Motor System

Feature	Somatic Motor	Autonomic Motor
Effector organs	Skeletal muscles	Cardiac muscle, smooth muscle, and glands
Presence of ganglia	No ganglia	Cell bodies of postganglionic autonomic fibers located in paravertebral, prevertebral (collateral), and terminal ganglia
Number of neurons from CNS to effector	One	Two
Type of neuromuscular junction	Specialized motor-end plate	No specialization of postsynaptic membrane; all areas of smooth muscle cells contain receptor proteins for neurotransmitters
Effect of nerve impulse on muscle	Excitatory only	Either excitatory or inhibitory
Type of nerve fibers	Fast-conducting, thick (9–13 µm), and myelinated	Slow-conducting; preganglionic fibers lightly myelinated but thin (3 µm); postganglionic fibers unmyelinated and very thin (about 1.0 µm)
Effect of denervation	Flaccid paralysis and atrophy	Muscle tone and function persist; target cells show denervation hypersensitivity

Unlike somatic motor neurons, which conduct impulses along a single axon from the spinal cord to the neuromuscular junction, autonomic motor control involves two neurons in the efferent pathway (table 9.1). The first of these neurons has its cell body in the gray matter of the brain or spinal cord. The axon of this neuron does not directly innervate the effector organ but instead synapses with a second neuron within an *autonomic ganglion* (a ganglion is a collection of cell bodies outside the CNS). The first neuron is thus called a **preganglionic neuron.** The second neuron in this pathway, called a **postganglionic neuron,** has an axon that extends from the autonomic ganglion and synapses with the cells of an effector organ (fig. 9.1, right side).

Preganglionic autonomic fibers originate in the midbrain and hindbrain and in the upper thoracic to the fourth sacral levels of the spinal cord. Autonomic ganglia are located in the head, neck, and abdomen; chains of autonomic ganglia also parallel the right and left sides of the spinal cord. The origin of the preganglionic fibers and the location of the autonomic ganglia help to differentiate the **sympathetic** and **parasympathetic** divisions of the autonomic system, which will be discussed in later sections of this chapter.

Visceral Effector Organs

Since the autonomic nervous system helps to regulate the activities of glands, smooth muscles, and cardiac muscle, autonomic control is an integral aspect of the physiology of most of the body systems. Autonomic regulation, then, partly explains endocrine regulation (chapter 11), smooth muscle function (chapter 12), functions of the heart and circulation (chapters 13 and 14), and, in fact, all the remaining systems to be discussed. Although the functions of the target organs of autonomic innervation are described in subsequent chapters, at this point we will consider some of the common features of autonomic regulation.

Unlike skeletal muscles, which enter a state of flaccid paralysis and atrophy when their motor nerves are severed, the involuntary effectors are somewhat independent of their innervation. Smooth muscles maintain a resting tone (tension) in the absence of nerve stimulation, for example. In fact, damage to an autonomic nerve makes its target tissue more sensitive than normal to stimulating agents. This phenomenon is called **denervation hypersensitivity.** Such compensatory changes can explain why, for example, the ability of the mucosa of the stomach to secrete acid may be restored after its neural supply from the vagus nerve has been severed. (This procedure is called vagotomy, and is sometimes performed as a treatment for ulcers.)

In addition to their intrinsic ("built-in") muscle tone, cardiac muscle and many smooth muscles take their autonomy a step further. These muscles can contract rhythmically, even in the absence of nerve stimulation, in response to electrical waves of depolarization initiated by the muscles themselves. Autonomic innervation simply increases or decreases this intrinsic activity. Autonomic nerves also maintain a resting "tone," in the sense that they maintain a baseline firing rate that can be either increased or decreased. A decrease in the excitatory input to the heart, for example, will slow its rate of beat.

The release of the neurotransmitter (ACh) from somatic motor neurons always stimulates the effector organ (skeletal muscles). By contrast, some autonomic nerves release transmitters that inhibit the activity of their effectors. An increase in the activity of the vagus, a nerve that supplies inhibitory fibers to the heart, for example, will slow the heart rate, whereas a decrease in this inhibitory input will increase the heart rate.

---

1. Describe the preganglionic and postganglionic neurons in the autonomic system. Use a diagram to illustrate the difference in efferent outflow between somatic and autonomic nerves.

2. Describe how the regulation of the contraction of cardiac and smooth muscle cells differs from that of skeletal muscle cells. How are these muscles affected by the experimental removal of their innervation?

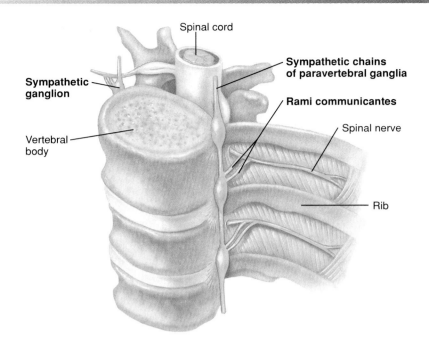

Figure 9.2

The relationship of the sympathetic chain of paravertebral ganglia to the vertebral column and the spinal cord.

Divisions of the Autonomic Nervous System

Preganglionic neurons of the sympathetic division of the autonomic system originate in the thoracic and lumbar levels of the spinal cord and send axons to sympathetic ganglia, which parallel the spinal cord. Preganglionic neurons of the parasympathetic division, by contrast, originate in the brain and in the sacral level of the spinal cord, and send axons to ganglia located in or near the effector organs.

The sympathetic and parasympathetic divisions of the autonomic system have some structural features in common. Both consist of preganglionic neurons that originate in the CNS and postganglionic neurons that originate outside of the CNS in ganglia. The specific origin of the preganglionic fibers and the location of the ganglia, however, are different in the two divisions of the autonomic system.

Sympathetic (Thoracolumbar) Division

The sympathetic system is also called the *thoracolumbar division* of the autonomic system because its preganglionic fibers exit the spinal cord from the first thoracic (T1) to the second lumbar (L2) levels. Most sympathetic nerve fibers, however, separate from the somatic motor fibers and synapse with postganglionic neurons within a double row of sympathetic ganglia, or **paravertebral ganglia,** located on either side of the spinal cord (fig. 9.2). Ganglia within each row are interconnected, forming a *sympathetic chain* of ganglia that parallels the spinal cord on each lateral side.

The myelinated preganglionic sympathetic axons exit the spinal cord in the ventral roots of spinal nerves, but soon diverge from the spinal nerves within *white rami communicantes.* The axons within each ramus enter the sympathetic chain of ganglia, where they can travel to ganglia at different levels and synapse with postganglionic sympathetic neurons. The axons of the postganglionic sympathetic neurons are unmyelinated and form the *grey rami communicantes* as they return to the spinal nerves and travel as part of the spinal nerves to their effector organs (fig. 9.3). Since sympathetic

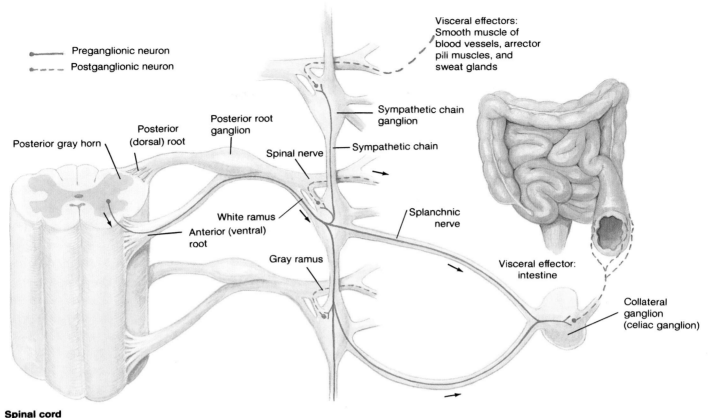

Spinal cord

Figure 9.3

The relationship between the rami communicantes, spinal nerves, sympathetic ganglia, and splanchnic nerves of the sympathetic division.

axons form a component of spinal nerves, they are widely distributed to the skeletal muscles and skin of the body, where they innervate blood vessels and other involuntary effectors.

Within the chain of paravertebral ganglia, *divergence* is apparent as preganglionic fibers branch to synapse with numerous postganglionic neurons located at different levels in the chain. *Convergence* is apparent also when a given postganglionic neuron receives synaptic input from a large number of preganglionic fibers. The divergence of impulses from the spinal cord to the ganglia and the convergence of impulses within the ganglia usually results in the **mass activation** of almost all of the postganglionic fibers. This explains why the sympathetic system is usually activated as a unit and affects all of its effector organs at the same time.

Many preganglionic fibers that exit the spinal cord in the upper thoracic level travel into the neck, where they synapse in cervical sympathetic ganglia (fig. 9.4). Postganglionic fibers from here innervate the smooth muscles and glands of the head and neck.

Collateral Ganglia

Many preganglionic fibers that exit the spinal cord below the level of the diaphragm pass through the sympathetic chain of ganglia without synapsing. Beyond the sympathetic chain, these preganglionic fibers form *splanchnic nerves*. Preganglionic fibers in the splanchnic nerves synapse in **collateral ganglia,** or *prevertebral ganglia.* These include the *celiac, superior mesenteric,* and *inferior mesenteric ganglia* (figs. 9.5 and 9.6). Postganglionic fibers that arise from the collateral ganglia innervate organs of the digestive, urinary, and reproductive systems.

Adrenal Glands

The paired adrenal glands are located above each kidney. Each adrenal is composed of two parts: an outer **cortex** and an inner **medulla.** These two parts are really two functionally different glands with different embryonic origins, different hormones, and different regulatory mechanisms. The adrenal

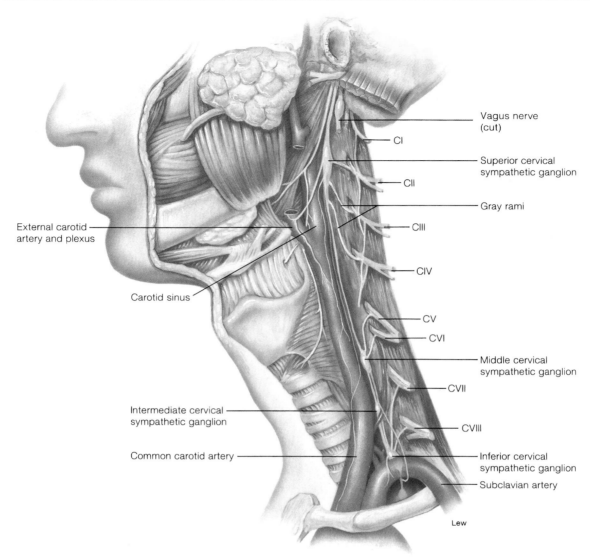

Labels on figure:
- Vagus nerve (cut)
- CI
- Superior cervical sympathetic ganglion
- CII
- Gray rami
- CIII
- External carotid artery and plexus
- CIV
- Carotid sinus
- CV
- CVI
- Middle cervical sympathetic ganglion
- CVII
- Intermediate cervical sympathetic ganglion
- CVIII
- Common carotid artery
- Inferior cervical sympathetic ganglion
- Subclavian artery
- Lew

Figure 9.4

The cervical sympathetic ganglia.

cortex secretes steroid hormones; the adrenal medulla secretes the hormone **epinephrine** (adrenaline) and, to a lesser degree, **norepinephrine,** when it is stimulated by the sympathetic system.

The adrenal medulla can be likened to a modified sympathetic ganglion; its cells are derived from the same embryonic tissue (the neural crest, chapter 8) that forms postganglionic sympathetic neurons. Like a sympathetic ganglion, the cells of the adrenal medulla are innervated by preganglionic sympathetic fibers. The adrenal medulla secretes epinephrine into the blood in response to this neural stimulation. The effects of epinephrine are complementary to those of the neurotransmitter norepinephrine, which is released from postganglionic sympathetic nerve endings. For this reason, and because the adrenal medulla is stimulated as part of the mass activation of the sympathetic system, the two are often grouped together as a single **sympathoadrenal system.**

Parasympathetic (Craniosacral) Division

The parasympathetic system is also known as the *craniosacral division* of the autonomic system. This is because its preganglionic fibers originate in the brain (specifically, in the midbrain, medulla oblongata, and pons) and in the second through fourth sacral levels of the spinal column. These preganglionic parasympathetic fibers synapse in ganglia that are located next to—or actually within—the organs innervated. These parasympathetic ganglia, which are called **terminal ganglia,** supply the postganglionic fibers that synapse with the effector cells. The comparative structures of the sympathetic and parasympathetic divisions are listed in tables 9.2 and 9.3. It should be noted that most parasympathetic fibers do not travel within spinal nerves, as do sympathetic fibers. As a result, cutaneous effectors (blood vessels, sweat glands, and arrector pili muscles) and blood vessels in skeletal muscles receive sympathetic but not parasympathetic innervation.

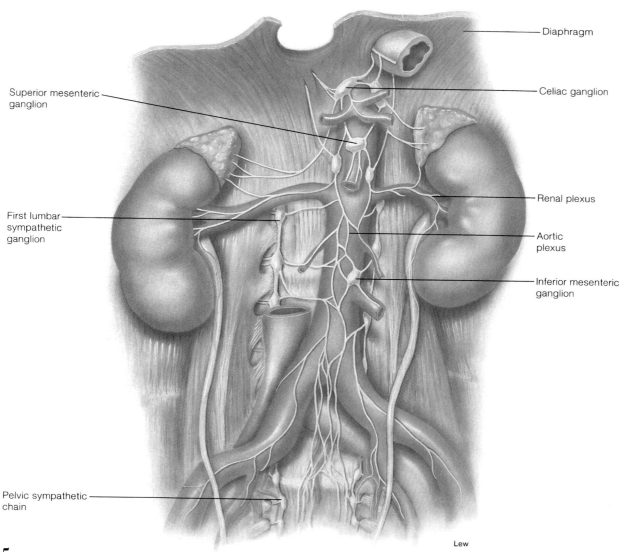

Superior mesenteric ganglion

Diaphragm

Celiac ganglion

First lumbar sympathetic ganglion

Renal plexus

Aortic plexus

Inferior mesenteric ganglion

Pelvic sympathetic chain

Lew

Figure 9.5

The collateral sympathetic ganglia: the celiac and the superior and inferior mesenteric ganglia.

Four of the twelve pairs of cranial nerves (described in chapter 8) contain preganglionic parasympathetic fibers. These are the oculomotor (III), facial (VII), glossopharyngeal (IX), and vagus (X) nerves. Parasympathetic fibers within the first three of these cranial nerves synapse in ganglia located in the head; fibers in the vagus nerve synapse in terminal ganglia located in many regions of the body.

The oculomotor nerve contains somatic motor and parasympathetic fibers that originate in the oculomotor nuclei of the midbrain. These parasympathetic fibers synapse in the *ciliary ganglion,* whose postganglionic fibers innervate the ciliary muscle and constrictor fibers in the iris of the eye. Preganglionic fibers that originate in the pons travel in the facial nerve to the *pterygopalatine ganglion,* which sends postganglionic fibers to the nasal mucosa, pharynx, palate, and lacrimal glands. Another group of fibers in the facial nerve terminates in the *submandibular ganglion,* which sends postganglionic fibers to the submandibular

and sublingual salivary glands. Preganglionic fibers of the glossopharyngeal nerve synapse in the *otic ganglion,* which sends postganglionic fibers to innervate the parotid salivary gland.

Nuclei in the medulla oblongata contribute preganglionic fibers to the very long *tenth cranial,* or *vagus nerves* (the "vagrant" or "wandering" nerves). These preganglionic fibers travel through the neck to the thoracic cavity, and through the esophageal opening in the diaphragm to the abdominal cavity (fig. 9.6). In each region, some of these preganglionic fibers branch from the main trunks of the vagus nerves and synapse with postganglionic neurons that are located *within* the innervated organs. The preganglionic vagus fibers are thus quite long, and provide parasympathetic innervation to the heart, lungs, esophagus, stomach, pancreas, liver, small intestine, and the upper half of the large intestine. Postganglionic parasympathetic fibers arise from terminal ganglia within these organs and synapse with effector cells (smooth muscles and glands).

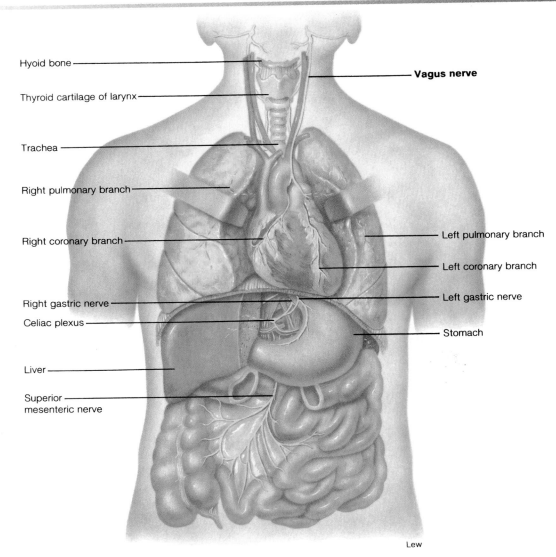

Hyoid bone

Thyroid cartilage of larynx

Trachea

Right pulmonary branch

Right coronary branch

Right gastric nerve

Celiac plexus

Liver

Superior
mesenteric nerve

Vagus nerve

Left pulmonary branch

Left coronary branch

Left gastric nerve

Stomach

Lew

Figure 9.6
The vagus nerves and their branches provide parasympathetic innervation to most organs within the thoracic and abdominal cavities.

Table 9.2	**The Sympathetic (Thoracolumbar) System**	
Parts of Body Innervated	**Spinal Origin of Preganglionic Fibers**	**Origin of Postganglionic Fibers**
Eye	C8 and T1	Cervical ganglia
Head and neck	T1 to T4	Cervical ganglia
Heart and lungs	T1 to T5	Upper thoracic (paravertebral) ganglia
Upper extremities	T2 to T9	Lower cervical and upper thoracic (paravertebral) ganglia
Upper abdominal viscera	T4 to T9	Celiac and superior mesenteric (collateral) ganglia
Adrenal	T10 and T11	—
Urinary and reproductive systems	T12 to L2	Celiac and interior mesenteric (collateral) ganglia
Lower extremities	T9 to L2	Lumbar and upper sacral (paravertebral) ganglia

Preganglionic fibers from the sacral levels of the spinal cord provide parasympathetic innervation to the lower half of the large intestine, rectum, and to the urinary and reproductive systems. These fibers, like those of the vagus, synapse with terminal ganglia located within the effector organs.

Parasympathetic nerves to the visceral organs thus consist of preganglionic fibers, whereas sympathetic nerves to these organs contain postganglionic fibers. A composite view of the sympathetic and parasympathetic systems is provided in figure 9.7, and these comparisons are summarized in table 9.4.

Chapter Nine

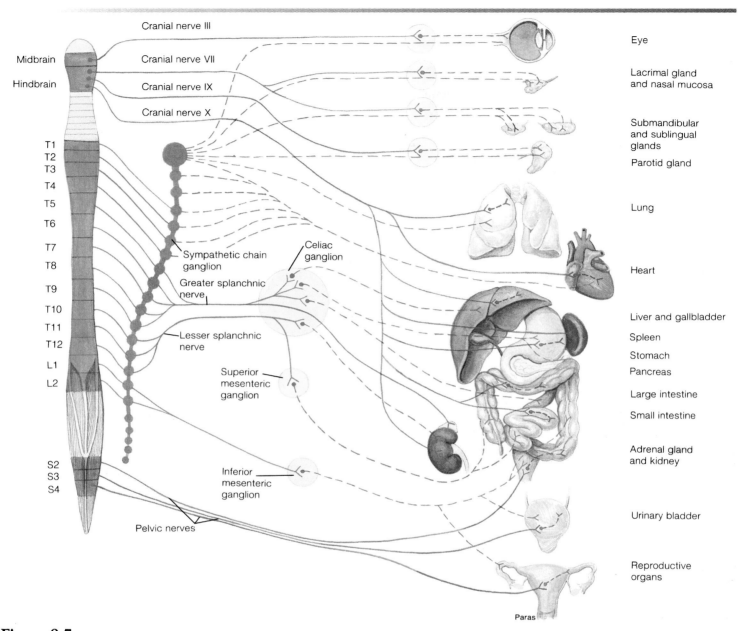

Figure 9.7

The autonomic nervous system. The sympathetic division is shown in red, the parasympathetic in blue. The solid lines indicate preganglionic fibers, and the dashed lines indicate postganglionic fibers.

Table 9.3	The Parasympathetic (Craniosacral) System		
Effector Organs	**Origin of Preganglionic Fibers**	**Nerve**	**Location of Terminal Ganglia**
Eye (ciliary and iris muscles)	Midbrain (cranial)	Oculomotor (third cranial) nerve	Ciliary ganglion
Lacrimal, mucous, and salivary glands in head	Pons (cranial)	Facial (seventh cranial) nerve	Pterygopalatine and submandibular ganglia
Parotid (salivary) gland	Medulla oblongata (cranial)	Glossopharyngeal (ninth cranial) nerve	Otic ganglion
Heart, lungs, gastrointestinal tract, liver, pancreas	Medulla oblongata (cranial)	Vagus (tenth cranial) nerve	Terminal ganglia in or near organ
Lower half of large intestine, rectum, urinary bladder, and reproductive organs	S2 to S4 (sacral)	Through pelvic spinal nerves	Terminal ganglia near organs

Table 9.4 Comparison of the Structural Features of the Sympathetic and Parasympathetic Systems

Feature	Sympathetic	Parasympathetic
Origin of preganglionic outflow	Thoracolumbar levels of spinal cord	Midbrain, hindbrain, and sacral levels of spinal cord
Location of ganglia	Chain of paravertebral ganglia and prevertebral (collateral) ganglia	Terminal ganglia in or near effector organs
Distribution of postganglionic fibers	Throughout the body	Mainly limited to the head and the viscera of the chest, abdomen, and pelvis
Divergence of impulses from pre- to postganglionic fibers	Great divergence (one preganglionic may activate twenty postganglionic fibers)	Little divergence (one preganglionic only activates a few postganglionic fibers)
Mass discharge of system as a whole	Yes	Not normally

1. Using a simple line diagram, illustrate the sympathetic pathway (a) from the spinal cord to the heart and (b) from the spinal cord to the adrenal gland. Label the preganglionic and postganglionic fibers and the ganglion.

2. Explain what is meant by the mass activation of the sympathetic system and discuss the significance of the term sympathoadrenal system.

3. Using a simple line diagram, illustrate the parasympathetic pathway from the brain to the heart. Compare the parasympathetic and sympathetic divisions in terms of the locations of the pre- and postganglionic fibers and their ganglia.

Functions of the Autonomic Nervous System

The sympathetic division activates the body to "fight or flight," largely through the release of norepinephrine from postganglionic fibers and the secretion of epinephrine from the adrenal medulla. The parasympathetic division often produces antagonistic effects through the release of acetylcholine from its postganglionic fibers. The actions of both divisions of the autonomic nervous system must be balanced in order to maintain homeostasis.

The sympathetic and parasympathetic divisions of the autonomic system affect the visceral organs in different ways. Mass activation of the sympathetic system prepares the body for intense physical activity in emergencies; the heart rate increases, blood glucose rises, and blood is diverted to the skeletal muscles (away from the visceral organs and skin). These and other effects are listed in table 9.5. The theme of the sympathetic system has been aptly summarized in a phrase: **"fight or flight."**

The effects of parasympathetic nerve stimulation are in many ways opposite to the effects of sympathetic stimulation. The parasympathetic system, however, is not normally activated as a whole. Stimulation of separate parasympathetic nerves can result in slowing of the heart, dilation of visceral blood vessels, and an increased activity of the digestive tract (table 9.5). Visceral organs respond differently to sympathetic and parasympathetic nerve activity because the postganglionic fibers of these two divisions release different neurotransmitters.

Adrenergic and Cholinergic Synaptic Transmission

Acetylcholine (ACh) is the neurotransmitter of all preganglionic fibers (both sympathetic and parasympathetic). Acetylcholine is also the transmitter released by most parasympathetic postganglionic fibers at their synapses with effector cells (fig. 9.8). Transmission at these synapses is thus said to be **cholinergic.**

The neurotransmitter released by most postganglionic sympathetic nerve fibers is **norepinephrine** (*noradrenaline*). Transmission at these synapses is thus said to be **adrenergic.** There are a few exceptions, however. Some sympathetic fibers that innervate blood vessels in skeletal muscles, as well as sympathetic fibers to sweat glands, release ACh (are cholinergic).

In view of the fact that the cells of the adrenal medulla are embryologically related to postganglionic sympathetic neurons, it is not surprising that the hormones they secrete should consist of epinephrine (about 85%) and norepinephrine (about 15%). Epinephrine differs from norepinephrine only in that the former has an additional methyl (CH_3) group, as shown in figure 9.9. Epinephrine, norepinephrine, and dopamine (a transmitter within the CNS) are all derived from the amino acid tyrosine, and are collectively termed **catecholamines.**

Table 9.5 Effects of Autonomic Nerve Stimulation on Various Effector Organs

Effector Organ	Sympathetic Effect	Parasympathetic Effect
Eye		
Iris (radial muscle)	Dilation of pupil	—
Iris (sphincter muscle)	—	Constriction of pupil
Ciliary muscle	Relaxation (for far vision)	Contraction (for near vision)
Glands		
Lacrimal (tear)	—	Stimulation of secretion
Sweat	Stimulation of secretion	—
Salivary	Decreased secretion; saliva becomes thick	Increased secretion; saliva becomes thin
Stomach	—	Stimulation of secretion
Intestine	—	Stimulation of secretion
Adrenal medulla	Stimulation of hormone secretion	
Heart		
Rate	Increased	Decreased
Conduction	Increased rate	Decreased rate
Strength	Increased	—
Blood Vessels	Mostly constriction; affects all organs	Dilation in a few organs (e.g., penis)
Lungs		
Bronchioles (tubes)	Dilation	Constriction
Mucous glands	Inhibition of secretion	Stimulation of secretion
Gastrointestinal Tract		
Motility	Inhibition of movement	Stimulation of movement
Sphincters	Closing stimulated	Closing inhibited
Liver	Stimulation of glycogen hydrolysis	—
Adipose (Fat) Cells	Stimulation of fat hydrolysis	—
Pancreas	Inhibition of exocrine secretions	Stimulation of exocrine secretions
Spleen	Contraction	—
Urinary Bladder	Muscle tone aided	Contraction
Arrector Pili Muscles	Erection of hair and goose bumps	—
Uterus	If pregnant: contraction; if not pregnant: relaxation	—
Penis	Ejaculation	Erection (due to vasodilation)

Responses to Adrenergic Stimulation

Adrenergic stimulation—by epinephrine in the blood and by norepinephrine released from sympathetic nerve endings—has both excitatory and inhibitory effects. The heart, dilatory muscles of the iris, and the smooth muscles of many blood vessels are stimulated to contract. The smooth muscles of the bronchioles and of some blood vessels, however, are inhibited from contracting; adrenergic chemicals, therefore, cause these structures to dilate.

Since excitatory and inhibitory effects can be produced in different tissues by the same chemical, the responses clearly depend on the biochemistry of the tissue cells rather than on the intrinsic properties of the chemical. Included in the biochemical differences among the target tissues for catecholamines are differences in the *membrane receptor proteins* for these chemical agents. The two major classes of these receptor proteins are designated **alpha- (α)** and **beta- (β) adrenergic receptors.** (The interaction of neurotransmitters and receptor proteins in the postsynaptic membrane was described in chapter 7.)

Experiments have revealed that there are two subtypes of each category of adrenergic receptor. These are designated by subscripts: α_1 and α_2; β_1 and β_2. Scientists have developed compounds that selectively bind to one or the other type of adrenergic receptor and, by this means, either promote or inhibit the normal action produced when epinephrine or norepinephrine binds to the receptor. As a result of its binding to an adrenergic receptor, a drug may either promote or inhibit the adrenergic effect. Also, by using these selective compounds, it has been possible to determine which subtype of adrenergic receptor is in each organ (table 9.6).

Both subtypes of beta receptors produce their effects by stimulating the production of cyclic AMP (discussed in chapter 7) within the target cells. The activation of the α_2 receptors has the opposite effect—cyclic AMP production is blocked and the cAMP concentration within the target cell is lowered,

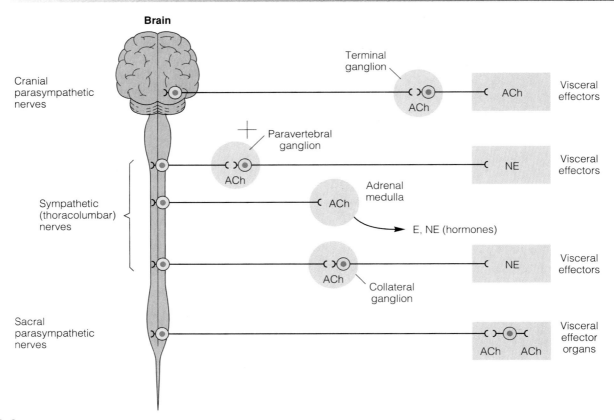

Figure 9.8

Neurotransmitters of the autonomic motor system (ACh = acetylcholine; NE = norepinephrine; E = epinephrine). Those nerves that release ACh are called cholinergic; those nerves that release NE are called adrenergic. The adrenal medulla secretes both epinephrine (85%) and norepinephrine (15%) as hormones into the blood.

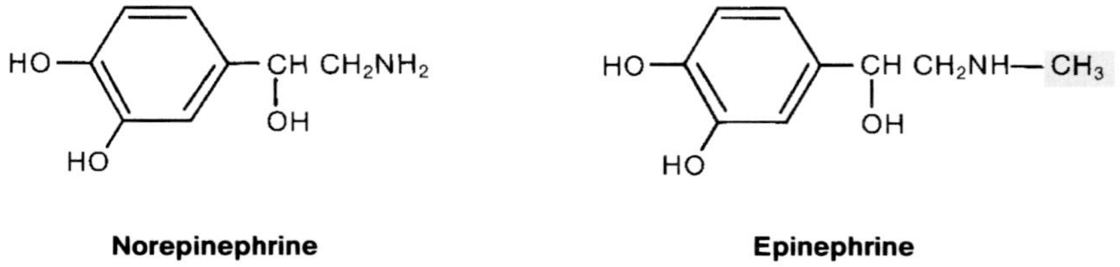

Norepinephrine   **Epinephrine**

Figure 9.9

The structure of the catecholamines norepinephrine and epinephrine.

inhibiting the effects of beta-adrenergic-receptor stimulation. The response of a target cell when norepinephrine binds to the α_1 receptors is mediated by a different second messenger system—a rise in the cytoplasmic concentration of Ca^{++}. This Ca^{++} second messenger system is similar, in many ways, to the cAMP second messenger system and is discussed together with endocrine regulation in chapter 11. It should be remembered that each of the intracellular changes following the binding of norepinephrine to its receptor ultimately results in the characteristic response of the tissue to the neurotransmitter.

Review of table 9.6 reveals certain generalities about the actions of adrenergic receptors. The stimulation of alpha-adrenergic receptors consistently causes contraction of smooth muscles. We can thus state that the vasoconstrictor effect of sympathetic nerves always results from the activation of alpha-adrenergic receptors. The effects of beta-adrenergic activation are more complex; these receptors promote the relaxation of smooth muscles (in the digestive tract, bronchioles, and uterus, for example), but stimulate contraction of cardiac muscle and promote an increase in cardiac rate.

A drug that binds to receptors for a neurotransmitter and that promotes the processes that are stimulated by that neurotransmitter is said to be an *agonist* of that neurotransmitter. A drug that blocks the action of a neurotransmitter, by contrast, is said to be an *antagonist*. The use of specific drugs that selectively stimulate or block α_1, α_2, β_1, and β_2 receptors has proven extremely useful in many medical applications.

The diverse effects of epinephrine and norepinephrine can be understood in terms of the "fight-or-flight" theme. Adrenergic stimulation wrought by activation of the sympathetic division produces an increase in cardiac pumping (a β_1 effect), vasoconstriction and thus reduced blood flow to the visceral organs (an α_1 effect), dilation of pulmonary bronchioles (a β_2 effect), and so on, preparing the body for physical exertion.

Many people with hypertension have been treated with a beta-blocking drug known as propranolol. This drug blocks β_1 receptors, which are located in the heart, and thereby has the desired effect of lowering the cardiac rate and blood pressure. Propranolol, however, also blocks β_2 receptors, which are located in the bronchioles of the lungs. This reduces the bronchodilation effect of epinephrine, producing bronchoconstriction and asthma in susceptible people. A more specific β_1 antagonist, atenolol, is now used instead to slow the cardiac rate and lower blood pressure. At one time, asthmatics inhaled an epinephrine spray, which stimulates β_1 receptors in the heart as well as β_2 receptors in the airways. Now, drugs such as terbutaline that selectively function as β_2 agonists are more commonly used.

Drugs that function as α_1 agonists, such as phenylephrine, are often part of nasal sprays because they promote vasoconstriction in the nasal mucosa. Clonidine is a drug that selectively stimulates α_2 receptors located on neurons in the brain. As a consequence of its action, clonidine reduces the activation of the sympathoadrenal system and thereby helps to lower the blood pressure. For reasons that are poorly understood, this drug is also helpful in treating patients with an addiction to opiates who are experiencing withdrawal symptoms.

Responses to Cholinergic Stimulation

Somatic motor neurons, all preganglionic autonomic neurons, and most postganglionic parasympathetic neurons are cholinergic—they release acetylcholine as a neurotransmitter. The cholinergic effects of somatic motor neurons and preganglionic autonomic neurons are always excitatory. The cholinergic effects of postganglionic parasympathetic fibers are usually excitatory, but there are notable exceptions. The parasympathetic fibers innervating the heart, for example, cause slowing of the heart rate. It is useful to remember that

the effects of parasympathetic stimulation are, in general, opposite to the effects of sympathetic stimulation.

Just as adrenergic receptors are divided into alpha and beta subtypes, cholinergic receptors are divided into nicotinic and muscarinic subtypes (described in chapter 7). The drug muscarine, derived from some poisonous mushrooms, stimulates the cholinergic receptors in the heart, digestive system, and other target organs of postganglionic parasympathetic nerve fibers. These axons must thus exert their effects on the target organs by stimulating the muscarinic subtype of cholinergic receptors (table 9.6). Muscarine, however, does not stimulate ACh receptor proteins in autonomic ganglia or at the neuromuscular junction of skeletal muscle fibers. The drug nicotine, derived from the tobacco plant, specifically stimulates these cholinergic receptors, which must therefore be the nicotinic subtype of ACh receptors. The drug *curare*, used clinically to cause skeletal muscle relaxation, specifically blocks nicotinic receptors but has little effect on muscarinic receptors.

The muscarinic effects of ACh are specifically inhibited by the drug **atropine**, derived from the deadly nightshade plant (*Atropa belladonna*). Indeed, extracts of this plant were used by women during the Middle Ages to dilate their pupils (atropine inhibits parasympathetic stimulation of the iris). This was thought to enhance their beauty (*belladonna* = beautiful lady). Atropine is used clinically today to dilate pupils during eye examinations, to dry mucous membranes of the respiratory tract prior to general anesthesia, to inhibit spasmodic contractions of the lower digestive tract and to inhibit stomach acid secretion in a person with gastritis.

Other Autonomic Neurotransmitters

Certain postganglionic autonomic axons produce their effects through mechanisms that do not involve either norepinephrine or acetylcholine. This can be demonstrated experimentally by the inability of drugs that block adrenergic and cholinergic effects from inhibiting the actions of those autonomic axons. These axons, consequently, have been termed "nonadrenergic noncholinergic fibers." Proposed neurotransmitters for these axons include ATP, a polypeptide called vasoactive intestinal peptide (VIP), and nitric oxide (NO).

The nonadrenergic noncholinergic parasympathetic axons that innervate the blood vessels of the penis cause the smooth muscles of these vessels to relax, thereby producing vasodilation and a consequent erection of the penis (chapter 20). These parasympathetic axons have been shown to use the gas nitric oxide (chapter 7) as their neurotransmitter. In a similar manner, nitric oxide appears to function as the autonomic neurotransmitter causing vasodilation of cerebral arteries. Studies suggest that nitric oxide is not stored in synaptic vesicles, as are other neurotransmitters, but instead is produced immediately when

Table 9.6 Adrenergic and Cholinergic Effects of Sympathetic and Parasympathetic Nerves

Organ	Sympathetic Action	Sympathetic Receptor*	Parasympathetic Action	Parasympathetic Receptor*
Eye				
Iris				
Radial Muscle	Contracts	α_1	. . .	. . .
Circular muscle	. . .	. . .	Contracts	M
Heart				
Sinoatrial node	Accelerates	β_1	Decelerates	M
Contractility	Increases	β_1	Decreases (atria)	M
Vascular Smooth Muscle				
Skin, splanchnic vessels	Contracts	α, β	. . .	. . .
Skeletal muscle vessels	Relaxes	β_2	. . .	. . .
	Relaxes	M**	. . .	. . .
Bronchiolar Smooth Muscle	Relaxes	β_2	Contracts	M
Gastrointestinal Tract				
Smooth Muscle				
Walls	Relaxes	β_2	Contracts	M
Sphincters	Contracts	α_1	Relaxes	M
Secretion	Decreases	α_1	Increases	M
Myenteric plexus	Inhibits	α_1	. . .	. . .
Genitourinary Smooth Muscle				
Bladder wall	Relaxes	β_2	Contracts	M
Sphincter	Contracts	α_1	Relaxes	M
Uterus, pregnant	Relaxes	β_2	. . .	. . .
	Contracts	α_1	. . .	. . .
Penis, seminal vesicles	Ejaculation	α_1		
Skin				
Pilomotor smooth muscle	Contracts	α_1	. . .	. . .
Sweat glands				
Thermoregulatory	Increases	M	. . .	. . .
Apocrine (stress)	Increases	α_1	. . .	. . .

Source: Reproduced, with permission, from Katzung, B. G.: *Basic and Clinical Pharmacology,* 4th edition, copyright Appleton & Lange, 1989.
*Adrenergic receptors are indicated as alpha (α) or beta (β); cholinergic receptors are indicated as muscarinic (M).
**Vascular smooth muscle in skeletal muscle has sympathetic cholinergic dilator fibers.

Ca^{++} enters the axon terminal in response to action potentials. This Ca^{++} indirectly activates nitric oxide synthetase, the enzyme that forms nitric oxide from the amino acid L-arginine. Nitric oxide then diffuses across the synaptic cleft and promotes relaxation of the postsynaptic smooth muscle cells.

Nitric oxide can produce relaxation of smooth muscles in many organs, including the stomach, small intestine, large intestine, and urinary bladder. There is some controversy, however, about whether the nitric oxide functions in each case as a neurotransmitter, or whether it is sometimes produced in the organ itself in response to autonomic stimulation. The latter alternative is a distinct possibility because different tissues, such as the endothelium of blood vessels, can produce nitric oxide (chapter 14). Indeed, nitric oxide is a member of a class of local tissue regulatory molecules called paracrine regulators (chapter 11). Regulation can therefore be a complex process involving the interacting effects of different neurotransmitters, hormones, and paracrine regulators.

Organs with Dual Innervation

Most visceral organs receive dual innervation—they are innervated by both sympathetic and parasympathetic fibers. In this condition, the effects of these two divisions may be antagonistic, complementary, or cooperative.

Antagonistic Effects

The effects of sympathetic and parasympathetic innervation of the pacemaker region of the heart is the best example of the antagonism of these two systems. In this case, sympathetic and parasympathetic fibers innervate the same cells. Adrenergic stimulation from sympathetic fibers increases the heart rate, whereas the release of acetylcholine from parasympathetic fibers decreases the heart rate. A reverse of this antagonism is seen in the digestive tract, where sympathetic nerves inhibit and parasympathetic nerves stimulate intestinal movements and secretions.

The effects of sympathetic and parasympathetic stimulation on the diameter of the pupil of the eye are analogous to the reciprocal innervation of flexor and extensor skeletal muscles by somatic motor neurons (chapter 12). This is because the iris contains antagonistic muscle layers. Contraction of the radial muscles, which are innervated by sympathetic nerves, causes dilation; contraction of the circular muscles, which are innervated by parasympathetic nerve endings, causes constriction of the pupils (chapter 10).

Complementary and Cooperative Effects

The effects of sympathetic and parasympathetic nerves are generally antagonistic, but, in a few cases, they can be complementary or cooperative. Their effects are complementary when sympathetic and parasympathetic stimulation produce similar effects. The effects are cooperative, or synergistic, when sympathetic and parasympathetic stimulation produce two different effects that cooperate to promote a single action.

The effects of sympathetic and parasympathetic stimulation on salivary-gland secretion are complementary. The secretion of watery saliva is stimulated through parasympathetic nerves, which also stimulate the secretion of other exocrine glands in the digestive tract. Sympathetic nerves stimulate the constriction of blood vessels throughout the digestive tract. The resultant decrease in blood flow to the salivary glands causes the production of a thicker, more viscous saliva.

The effects of sympathetic and parasympathetic stimulation on the urinary and reproductive systems are cooperative. Erection of the penis, for example, is due to vasodilation resulting from parasympathetic nerve stimulation; ejaculation is due to stimulation through sympathetic nerves. The two divisions of the autonomic system thus cooperate to promote reproduction. There is also cooperation between the two divisions in the micturition (urination) reflex. Although the contraction of the urinary bladder is largely independent of nerve stimulation, it is promoted in part by the action of parasympathetic nerves. This reflex is also enhanced by sympathetic nerve activity, which increases the tone of the bladder muscles. Emotional states that are accompanied by high sympathetic nerve activity (such as extreme fear) may thus result in reflex urination at bladder volumes that are normally too low to trigger this reflex.

Organs without Dual Innervation

Although most organs are innervated by both sympathetic and parasympathetic nerves, some—including the adrenal medulla, arrector pili muscles, sweat glands, and most blood vessels—receive only sympathetic innervation. In these cases, regulation is achieved by increases or decreases in the tone (firing rate) of the sympathetic fibers. Constriction of cutaneous blood vessels, for example, is produced by increased sympathetic activity that stimulates alpha-adrenergic receptors, and vasodilation results from decreased sympathetic nerve stimulation.

The sympathoadrenal system is required for *nonshivering thermogenesis*: animals deprived of their sympathetic system and adrenals cannot tolerate cold stress. The sympathetic system itself is required for proper thermoregulatory responses to heat. In a hot room, for example, decreased sympathetic stimulation produces dilation of the blood vessels in the surface of the skin, which increases cutaneous blood flow and provides better heat radiation. During exercise, by contrast, there is increased sympathetic activity, which causes constriction of the blood vessels in the skin of the limbs and stimulation of sweat glands in the trunk.

The sweat glands in the trunk secrete a watery fluid in response to cholinergic sympathetic stimulation. Evaporation of this dilute sweat helps to cool the body. The sweat glands also secrete a chemical called bradykinin in response to sympathetic stimulation. Bradykinin stimulates dilation of the surface blood vessels near the sweat glands, helping to radiate some heat despite the fact that other cutaneous blood vessels are constricted. At the conclusion of exercise, sympathetic stimulation is reduced, causing cutaneous blood vessels to dilate. This increases blood flow to the skin, which helps to eliminate metabolic heat. Notice that all of these thermoregulatory responses are achieved without the direct involvement of the parasympathetic system.

Autonomic dysreflexia, a serious condition producing rapid elevations in blood pressure that can lead to stroke (cerebrovascular accident), occurs in 85% of people with quadriplegia and others with spinal cord lesions above the sixth thoracic level. Lesions to the spinal cord first produce the symptoms of spinal shock, characterized by the loss of both skeletal muscle and autonomic reflexes. After a period of time, there is recovery of both types of reflexes, but in an exaggerated state. The skeletal muscles may become spastic due to absence of higher inhibitory influences, and the visceral organs experience denervation hypersensitivity. Patients in this condition have difficulty emptying their urinary bladders and must often be catheterized.

Noxious stimuli, such as overdistension of the urinary bladder, can result in reflex activation of the sympathetic nerves below the spinal cord lesion. This produces goose bumps, cold skin, and vasoconstriction in the regions served by the spinal cord below the level of the lesion. The rise in blood pressure resulting from this vasoconstriction activates pressure receptors that transmit impulses along sensory nerve fibers to the medulla. In response to this sensory input, the medulla directs a reflex slowing of the heart and vasodilation. Since descending impulses are blocked by the spinal lesion, however, the skin is warm and moist (due to vasodilation and sweat gland secretion) above the lesion, but cold below the level of spinal cord damage.

Table 9.7 Effects Resulting from Sensory Input from Afferent Fibers in the Vagus, Which Transmit This Input to Centers in the Medulla Oblongata

Organs	Type of Receptors	Reflex Effects
Lungs	Stretch receptors	Further inhalation inhibited; increase in cardiac rate and vasodilation stimulated
	Type J receptors	Stimulated by pulmonary congestion—produces feelings of breathlessness and causes a reflex fall in cardiac rate and blood pressure
Aorta	Chemoreceptors	Stimulated by rise in CO_2 and fall in O_2—produces increased rate of breathing, rise in heart rate, and vasoconstriction
	Baroreceptors	Stimulated by increased blood pressure—produces a reflex decrease in heart rate
Heart	Atrial stretch receptors	Antidiuretic hormone secretion inhibited, thus increasing the volume of urine excreted
	Stretch receptors in ventricles	Produces a reflex decrease in heart rate and vasodilation
Gastrointestinal tract	Stretch receptors	Feelings of satiety, discomfort, and pain

Control of the Autonomic Nervous System by Higher Brain Centers

Visceral functions are largely regulated by autonomic reflexes. In most autonomic reflexes, sensory input is transmitted to brain centers that integrate this information and respond by modifying the activity of preganglionic autonomic neurons. The neural centers that directly control the activity of autonomic nerves are influenced by higher brain areas, as well as by sensory input.

The **medulla oblongata** (chapter 8) of the brain stem is the area that most directly controls the activity of the autonomic system. Almost all autonomic responses can be elicited by experimental stimulation of the medulla, which contains centers for the control of the cardiovascular, pulmonary, urinary, reproductive, and digestive systems. Much of the sensory input to these centers travels in the afferent fibers of the vagus nerve, which is a mixed nerve containing both sensory and motor fibers. The reflexes that result are listed in table 9.7.

Although the medulla oblongata directly regulates the activity of autonomic motor fibers, the medulla is itself responsive to regulation by higher brain areas. One of these is the **hypothalamus** (chapter 8), which is the brain region that contains centers for the control of body temperature, hunger, thirst, regulation of the pituitary gland, and—together with the limbic system and cerebral cortex—various emotional states.

As described in chapter 8, the **limbic system** is a group of fiber tracts and nuclei that form a ring around the brain stem. It includes the cingulate gyrus of the cerebral cortex, the hypothalamus, the fornix (a fiber tract), the hippocampus, and the amygdaloid nucleus (see fig. 8.14). The limbic system is involved in basic emotional drives, such as anger, fear, sex, and hunger. The involvement of the limbic system with the control of autonomic function is responsible for the visceral responses that are characteristic of these emotional states. Blushing, pallor, fainting, breaking out in a cold sweat, a racing heartbeat, and "butterflies in the stomach" are only some of the many visceral reactions that accompany emotions as a result of autonomic activation.

1. Define the terms *adrenergic* and *cholinergic* and use these terms to describe the neurotransmitters of different autonomic nerve fibers.
2. List the effects of sympathoadrenal stimulation on different effector organs. In each case, indicate whether the effect is due to alpha- or beta-receptor stimulation.
3. Describe the effects of the drug atropine and explain these effects in terms of the actions of the parasympathetic system.
4. Explain how the sympathetic and parasympathetic systems can have antagonistic, cooperative, and complementary effects. Give examples of these different types of effects.
5. Explain the mechanisms involved when a person blushes. What structures are involved in this response?

Summary

Neural Control of Involuntary Effectors p. 208

I. Preganglionic autonomic neurons originate in the brain or spinal cord; postganglionic neurons originate in ganglia located outside the CNS.

II. Smooth muscle, cardiac muscle, and glands receive autonomic innervation.
   A. The involuntary effectors are somewhat independent of their innervation and become hypersensitive when their innervation is removed.
   B. Autonomic nerves can have either excitatory or inhibitory effects on their target organs.

Divisions of the Autonomic Nervous System *p. 210*

I. Preganglionic neurons of the sympathetic division originate in the spinal cord, between the thoracic and lumbar levels.

   A. Many of these fibers synapse with postganglionic neurons whose cell bodies are located in a double chain of sympathetic (paravertebral) ganglia outside the spinal cord.

   B. Some preganglionic fibers synapse in collateral ganglia; these are the celiac, superior mesenteric, and inferior mesenteric ganglia.

   C. Some preganglionic fibers innervate the adrenal medulla, which secretes epinephrine (and some norepinephrine) into the blood in response to this stimulation.

II. Preganglionic parasympathetic fibers originate in the brain and in the sacral levels of the spinal cord.

   A. Preganglionic parasympathetic fibers contribute to cranial nerves III, VII, IX, and X.

   B. Preganglionic fibers of the vagus (X) nerve are very long and synapse in terminal ganglia located next to or within the innervated organ. Short postganglionic fibers then innervate the effector cells.

   C. The vagus provides parasympathetic innervation to the heart, lungs, esophagus, stomach, liver, small intestine, and upper half of the large intestine.

   D. Parasympathetic outflow from the sacral levels of the spinal cord innervates terminal ganglia in the lower half of the large intestine, in the rectum, and in the urinary and reproductive systems.

Functions of the Autonomic Nervous System *p. 216*

I. The sympathetic division of the autonomic system activates the body to "fight or flight" through adrenergic effects. The parasympathetic division often produces antagonistic actions through cholinergic effects.

II. All preganglionic autonomic nerve fibers are cholinergic (use ACh as a neurotransmitter).

   A. All postganglionic parasympathetic fibers are cholinergic.

   B. Most postganglionic sympathetic fibers are adrenergic (use norepinephrine as a neurotransmitter).

   C. Sympathetic fibers that innervate sweat glands and those that innervate blood vessels in skeletal muscles are cholinergic.

III. Adrenergic effects include stimulation of the heart, vasoconstriction in the viscera and skin, bronchodilation, and glycogenolysis in the liver.

   A. There are two main groups of adrenergic receptor proteins: alpha and beta receptors.

   B. Some organs have only alpha or only beta receptors; other organs (such as the heart) have both receptors.

   C. There are two subtypes of alpha receptors (α_1 and α_2) and two subtypes of beta receptors (β_1 and β_2); these subtypes can be selectively stimulated or blocked by therapeutic drugs.

IV. Cholinergic effects of parasympathetic nerves are promoted by the drug muscarine and inhibited by atropine.

V. In organs with dual innervation, the actions of the sympathetic and parasympathetic divisions can be antagonistic, complementary, or cooperative.

   A. The effects are antagonistic in the heart and pupils of the eyes.

   B. The actions are complementary in the regulation of salivary gland secretion and are cooperative in the regulation of the reproductive and urinary systems.

VI. In organs without dual innervation (such as most blood vessels), regulation is achieved by variations in sympathetic nerve activity.

VII. The medulla oblongata of the brain stem is the area that most directly controls the activity of the autonomic system.

   A. The medulla oblongata is in turn influenced by sensory input and by input from the hypothalamus.

   B. The hypothalamus orchestrates somatic, autonomic, and endocrine responses during various behavioral states.

Clinical Investigation

A student laboratory assistant, studying for final examinations, finds that her pulse rate is faster than normal. Measurement of her blood pressure reveals that it is also elevated. Earlier that day, after mixing some drugs for a laboratory exercise dealing with autonomic control, she developed a severe headache and a very dry mouth. It was immediately apparent that her pupils were markedly dilated. What might be responsible for her symptoms?

Clues

Study the sections "Functions of the Autonomic Nervous System," "Responses to Adrenergic Stimulation," and "Responses to Cholinergic Stimulation." Examine table 9.6 and the boxed information on the effects of atropine.

Review Activities

Objective Questions

1. When a visceral organ is denervated,
   a. it ceases to function.
   b. it becomes less sensitive to subsequent stimulation by neurotransmitters.
   c. it becomes hypersensitive to subsequent stimulation.
2. Parasympathetic ganglia are located
   a. in a chain parallel to the spinal cord.
   b. in the dorsal roots of spinal nerves.
   c. next to or within the organs innervated.
   d. in the brain.
3. The neurotransmitter of preganglionic sympathetic fibers is
   a. norepinephrine.
   b. epinephrine.
   c. acetylcholine.
   d. dopamine.
4. Which of the following results from stimulation of alpha-adrenergic receptors?
   a. constriction of blood vessels
   b. dilation of bronchioles
   c. decreased heart rate
   d. sweat gland secretion

5. Which of the following fibers release norepinephrine?
   a. preganglionic parasympathetic fibers
   b. postganglionic parasympathetic fibers
   c. postganglionic sympathetic fibers in the heart
   d. postganglionic sympathetic fibers in sweat glands
   e. all of the above
6. The actions of sympathetic and parasympathetic fibers are cooperative in
   a. the heart.
   b. the reproductive system.
   c. the digestive system.
   d. the eyes.
7. Propranolol is a "beta-blocker." It would therefore cause
   a. vasodilation.
   b. slowing of the heart rate.
   c. increased blood pressure.
   d. secretion of saliva.
8. Atropine blocks parasympathetic nerve effects. It would therefore cause
   a. dilation of the pupils.
   b. decreased mucus secretion.
   c. decreased movements of the digestive tract.
   d. increased heart rate.
   e. all of the above.

9. Which area of the brain is most directly involved in the reflex control of the autonomic system?
   a. hypothalamus
   b. cerebral cortex
   c. medulla oblongata
   d. cerebellum
10. The two subtypes of cholinergic receptors are
    a. adrenergic and nicotinic.
    b. dopaminergic and muscarinic.
    c. nicotinic and muscarinic.
    d. nicotinic and dopaminergic.
11. A fall in cyclic AMP within the target cell occurs when norepinephrine binds to which of the following adrenergic receptors?
    a. α_1
    b. α_2
    c. β_1
    d. β_2
12. A drug that serves as an agonist for β_2 receptors can be used to
    a. increase the heart rate.
    b. decrease the heart rate.
    c. dilate the bronchioles.
    d. constrict the bronchioles.
    e. constrict the blood vessels.

Essay Questions

1. Compare the sympathetic and parasympathetic systems in terms of the location of their ganglia and the distribution of their nerves.[1]
2. Explain the anatomical and physiological relationship between the sympathetic nervous system and the adrenal glands.
3. Compare the effects of adrenergic and cholinergic stimulation on the cardiovascular and digestive systems.

4. Explain how effectors that receive only sympathetic innervation are regulated by the autonomic system.
5. Distinguish between the different types of adrenergic receptors and state where these receptors are located in the body.
6. Give examples of drugs that selectively stimulate or block different adrenergic receptors and explain how these drugs are used clinically.

7. Explain what is meant by nicotinic and muscarinic ACh receptors and describe their distribution in the body. List the effects of stimulation of these receptors.
8. Give examples of drugs that selectively stimulate and block the nicotinic and muscarinic receptors and explain how these drugs are used clinically.

---

[1]Note: This question is answered on page 98 of the Student Study Guide.

Selected Readings

Bredt, D. S., and S. H. Snyder. 1994. Nitric oxide: A physiologic messenger molecule. *Annual Review of Biochemistry* 63:175.

Decara, L. V. January 1970. Learning in the autonomic nervous system. *Scientific American*.

Goodman, L. S., and A. Gilman. 1980. *The Pharmacological Basis of Therapeutics*. 6th ed. New York: Macmillan.

Hoffman, B. B., and R. J. Lefkowitz. 1980. Alpha-adrenergic receptor subtypes. *New England Journal of Medicine* 302:1390.

Katzung, B. G., ed. 1984. *Basic and Clinical Pharmacology*. Los Altos, CA: Lange Medical Publications.

Lefkowitz, B. B. 1976. Beta-adrenergic receptors: Recognition and regulation. *New England Journal of Medicine* 295:323.

Makhlouf, G. M., and J. R. Grider. 1993. Nonadrenergic noncholinergic inhibitory transmitters of the gut. *News in Physiological Sciences* 8:195.

Moncada, S., and A. Higgs. 1993. The L-arginine–nitric oxide pathway. *New England Journal of Medicine* 329:2002.

Motulski, J. H., and P. A. Insel. 1982. Adrenergic receptors in man. *New England Journal of Medicine* 307:18.

Noback, C. E., and R. J. Demerest. 1975. *The Human Nervous System: Basic Principles of Neurobiology*. 2d ed. New York: McGraw-Hill.

Rand, M. J., and C. G. Li. 1995. Nitric oxide as a neurotransmitter in peripheral nerves: nature of transmitter and mechanism of transmission. *Annual Review of Physiology* 57:659.

Toda, N., and T. Okamura. 1992. Regulation by nitroxidergic nerve of arterial tone. *News in Physiological Sciences* 7:148.

Umans, J. G., and R. Levi. 1995. Nitric oxide in the regulation of blood flow and arterial pressure. *Annual Review of Physiology* 57:771.

Sensory Physiology

10

OBJECTIVES *After studying this chapter, you should be able to . . .*

1. explain how sensory receptors are categorized, give examples of functional categories, and explain how tonic and phasic receptors differ.

2. explain the law of specific nerve energies.

3. describe the characteristics of the generator potential.

4. give examples of different types of cutaneous receptors and describe the neural pathways for the cutaneous senses.

5. explain the concepts of receptive fields and lateral inhibition.

6. describe the distribution of taste receptors on the tongue and explain how the senses of salty, sour, sweet and bitter are produced.

7. describe the structure and function of the olfactory receptors and explain how odor discrimination might be accomplished.

8. describe the structure of the vestibular apparatus and explain how it provides information about acceleration of the body in different directions.

9. describe the functions of the outer and middle ear.

10. describe the structure of the cochlea and explain how movements of the stapes against the oval window result in vibrations of the basilar membrane.

11. explain how mechanical energy is converted into nerve impulses by the organ of Corti and how pitch perception is accomplished.

12. describe the structure of the eye and explain how images are brought to a focus on the retina.

13. explain how visual accommodation is achieved and describe the defects associated with myopia, hyperopia, and astigmatism.

14. describe the architecture of the retina and trace the pathways of light and nerve activity through the retina.

15. describe the function of rhodopsin in the rods and explain how dark adaptation is achieved.

16. explain how light affects the electrical activity of rods and their synaptic input to bipolar cells.

17. explain the trichromatic theory of color vision.

18. compare rods and cones with respect to their locations, synaptic connections, and functions.

19. describe the neural pathways from the retina, explaining the differences in pathways from different regions of the visual field.

20. describe the receptive fields of ganglion cells and explain the stimulus requirements of simple, complex, and hypercomplex cortical neurons.

OUTLINE

Characteristics of Sensory Receptors

Each type of sensory receptor responds to a particular modality of environmental stimulus by causing the production of action potentials in a sensory neuron. These impulses are conducted to parts of the brain that provide the proper interpretation of sensory perception when that particular neural pathway is activated.

Our perceptions of the world—its textures, colors, and sounds; its warmth, smells, and tastes—are created by the brain from electrochemical nerve impulses delivered to it from sensory receptors. These receptors **transduce** (change) different forms of energy in the "real world" into the energy of nerve impulses, which are conducted into the central nervous system by sensory neurons. Different *modalities* (forms) of sensation—sound, light, pressure, and so forth—result from differences in neural pathways and synaptic connections. The brain thus interprets impulses arriving from the auditory nerve as sound and from the optic nerve as sight, even though the impulses themselves are identical in the two nerves.

We know, through the use of scientific instruments, that our senses act as energy filters that allow us to perceive only a narrow range of energy. Vision, for example, is limited to light in the visible spectrum; ultraviolet and infrared light, X rays and radio waves, which are the same type of energy as visible light, cannot normally excite the photoreceptors in the eyes. The perception of cold is entirely a product of the nervous system—there is no such thing as cold in the physical world, only varying degrees of heat. The perception of cold, however, has obvious survival value. Although filtered and distorted by the limitations of sensory function, our perceptions of the world allow us to interact effectively with the environment.

Categories of Sensory Receptors

Sensory receptors can be categorized on the basis of their structure or various functional criteria. Structurally, the sensory receptors may be the dendritic endings of sensory neurons, which are either free (such as those in the skin that mediate pain and temperature) or are encapsulated within nonneural structures, such as pressure receptors in the skin (fig. 10.1). The photoreceptors in the retina of the eyes (rods and cones) are highly specialized neurons that synapse with other neurons in the retina. In the case of taste buds and of hair cells in the inner ears, modified epithelial cells respond to an environmental stimulus and activate sensory neurons.

Functional Categories

Sensory receptors can be grouped according to the type of stimulus energy they transduce. These categories include (1) *chemoreceptors*, such as the taste buds, olfactory epithelium, and the aortic and carotid bodies, which sense chemical stimuli in the environment or the blood; (2) *photoreceptors*—the rods and cones in the retina of the eye; (3) *thermoreceptors*, which respond to heat and cold; and (4) *mechanoreceptors*, which are stimulated by mechanical deformation of the receptor cell membrane—these include touch and pressure receptors in the skin and hair cells within the inner ear.

Nociceptors—or pain receptors—have a higher threshold for activation than do the other cutaneous receptors, so that a more intense stimulus is required for their activation. Their firing rate then increases with stimulus intensity. Receptors that subserve other sensations may also become involved in pain transmission when the stimulus is more prolonged, particularly when tissue damage occurs.

Receptors can also be grouped according to the type of sensory information they deliver to the brain. *Proprioceptors* include the muscle spindles, Golgi tendon organs, and joint receptors. These provide a sense of body position and allow fine control of skeletal movements (as discussed in chapter 12). *Cutaneous (skin) receptors* include (1) touch and pressure receptors, (2) hot and cold receptors, and (3) pain receptors. The receptors that mediate sight, hearing, and equilibrium are grouped together as the *special senses*.

Tonic and Phasic Receptors: Sensory Adaptation

Some receptors respond with a burst of activity when a stimulus is first applied, but then quickly decrease their firing rate—adapt to the stimulus—when the stimulus is maintained. Receptors with this response pattern are called *phasic receptors*. Receptors that produce a relatively constant rate of firing as long as the stimulus is maintained are known as *tonic receptors* (fig. 10.2).

Phasic receptors alert us to changes in sensory stimuli and are in part responsible for the fact that we can cease paying attention to constant stimuli. This ability is called **sensory adaptation.** Odor, touch, and temperature, for example, adapt rapidly; bathwater feels hotter when we first enter it. Sensations of pain, by contrast, adapt little if at all.

Law of Specific Nerve Energies

Stimulation of a sensory nerve fiber produces only one sensation—touch, cold, pain, and so on. According to the **law of specific nerve energies,** the sensation characteristic of each sensory neuron is that produced by its normal, or *adequate, stimulus* (table 10.1). The adequate stimulus for the photoreceptors of the eye, for example, is light. If these receptors are

Table 10.1 Classification of Receptors Based on Their Normal (or "Adequate") Stimulus

Receptor	Normal Stimulus	Mechanisms	Examples
Mechanoreceptors	Mechanical force	Deforms cell membranes of sensory dendrites; or deforms hair cells that activate sensory nerve endings	Cutaneous touch and pressure receptors; vestibular apparatus and cochlea
Pain receptors	Tissue damage	Damaged tissues release chemicals that excite sensory endings	Cutaneous pain receptors
Chemoreceptors	Dissolved chemicals	Chemical interaction affects ionic permeability of sensory cells	Smell and taste (exteroceptors) osmoreceptors and carotid body chemoreceptors (interoceptors)
Photoreceptors	Light	Photochemical reaction affects ionic permeability of receptor cell	Rods and cones in retina of eye

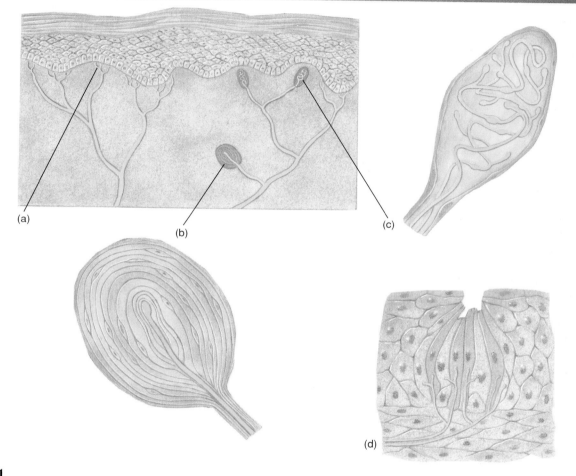

Figure 10.1

Different types of sensory receptors. Free nerve endings (*a*) mediate many cutaneous sensations, including heat. Some nerve endings are encapsulated within associated structures: e.g., the pacinian, or lamellated, corpuscle (*b*) for deep pressure and the tactile (Meissner's) corpuscle (*c*) for light touch. Some receptors, such as the taste bud (*d*), are modified epithelial cells that are innervated by sensory neurons.

stimulated by some other means—such as by pressure produced by a punch to the eye—a flash of light (the adequate stimulus) may be perceived.

The effect of *paradoxical cold* provides another example of the law of specific nerve energies. When the tip of a cold metal rod is touched to the skin, the perception of cold gradually dis-

appears as the rod warms to body temperature. Then, when the tip of a rod heated to 45°C is applied to the same spot, the sensation of cold is perceived once again. This paradoxical cold is produced because the heat slightly damages receptor endings, and by this means produces an "injury current" that stimulates the receptor.

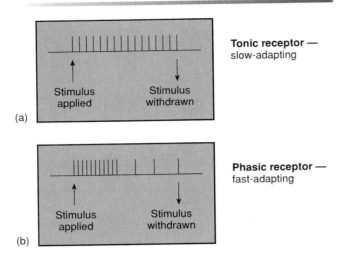

(a)

Tonic receptor — slow-adapting

Stimulus applied

Stimulus withdrawn

(b)

Phasic receptor — fast-adapting

Stimulus applied

Stimulus withdrawn

Figure 10.2

Tonic receptors (*a*) continue to fire at a relatively constant rate as long as the stimulus is maintained. These produce slow-adapting sensations. Phasic receptors (*b*) respond with a burst of action potentials when the stimulus is first applied, but then quickly reduce their rate of firing while the stimulus is maintained. This produces fast-adapting sensations.

Regardless of how a sensory neuron is stimulated, therefore, only one sensory modality will be perceived. This specificity is due to the synaptic pathways within the brain that are activated by the sensory neuron. The ability of receptors to function as sensory filters and be stimulated by only one type of stimulus (the adequate stimulus) allows the brain to perceive the stimulus accurately under normal conditions.

Generator (Receptor) Potential

The electrical behavior of sensory nerve endings is similar to that of the dendrites of other neurons. In response to an environmental stimulus, the sensory endings produce local, graded changes in the membrane potential. In most cases these potential changes are depolarizations analogous to excitatory postsynaptic potentials (EPSPs), as described in chapter 7. In the sensory endings, however, these potential changes in response to environmental stimulation are called **receptor,** or **generator, potentials** because they serve to generate action potentials in response to the sensory stimulation. Since sensory neurons are pseudounipolar (chapter 7), the action potentials produced in response to the generator potential are conducted continuously from the periphery into the CNS.

The *pacinian corpuscle,* a cutaneous receptor for pressure (see fig. 10.1), can serve as an example of sensory transduction. When a light touch is applied to the receptor, a small depolarization (the generator potential) is produced. Increasing the pressure on the pacinian corpuscle increases the magnitude of the generator potential until it reaches the threshold required to produce an action potential (fig. 10.3). The pacinian corpuscle, however, is a phasic receptor; if the pressure is maintained, the size of the generator potential produced quickly diminishes.

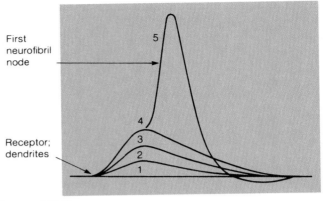

First neurofibril node

Receptor; dendrites

Figure 10.3

Sensory stimuli result in the production of local graded potential changes known as receptor, or generator, potentials (numbers 1–4). If the receptor potential reaches a threshold value of depolarization, it generates action potentials (number 5) in the sensory neuron.

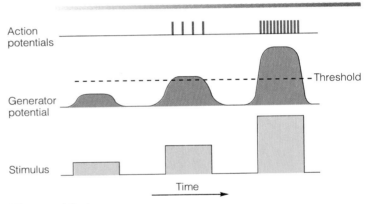

Action potentials

Threshold

Generator potential

Stimulus

Time

Figure 10.4

The response of tonic receptors to stimuli. Three successive stimuli of increasing strengths are delivered to a receptor. The increasing amplitude of the generator potential results in increases in the frequency of action potentials, which last as long as the stimulus is maintained.

It is interesting to note that this phasic response is a result of the onionlike covering around the dendritic nerve ending; if the layers are peeled off and the nerve ending is stimulated directly, it will respond in a tonic fashion.

When a tonic receptor is stimulated, the generator potential it produces is proportional to the intensity of the stimulus. After a threshold depolarization is produced, increases in the amplitude of the generator potential result in increases in the *frequency* with which action potentials are produced (fig. 10.4). In this way, the frequency of action potentials that are conducted into the central nervous system serves as the code for the strength of the stimulus. As described in chapter 7, this frequency code is needed because the amplitude of action potentials is constant (all or none). Acting through changes in action potential frequency, tonic receptors thus provide information about the relative intensity of a stimulus.

Table 10.2 Cutaneous Receptors

Anatomical Class (Structure)	Type	Sensation
Encapsulated (dendrites within associated structures)	Pacinian (lamellated) corpuscles; Meissner's corpuscles; Krause's end bulbs; Ruffini's end organs	All serve touch and pressure
Free nerve endings	—	Touch, pressure, heat, cold, pain

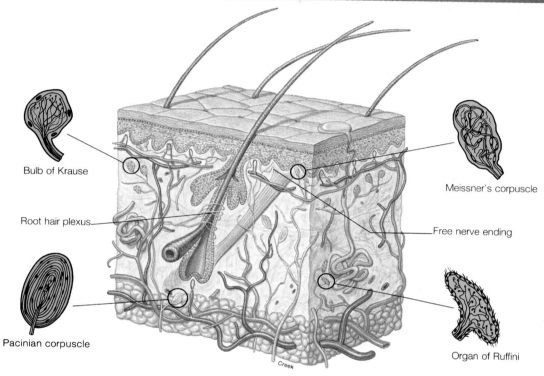

Figure 10.5

The cutaneous sensory receptors.

---

1. Our perceptions are products of our brains; they relate to physical reality only indirectly and incompletely. Explain this statement, using examples of vision and the perception of cold.

2. Explain what is meant by the law of specific nerve energies and the adequate stimulus and relate these concepts to your answer for question 1.

3. Describe sensory adaptation in olfactory and pain receptors. Using a line drawing, relate sensory adaptation to the responses of phasic and tonic receptors.

4. Describe how the magnitude of a sensory stimulus is transduced into a receptor potential and how the magnitude of the receptor potential is coded in the sensory nerve fiber.

Cutaneous Sensations

There are several different types of sensory receptors in the skin, each of which is specialized to be maximally sensitive to one modality of sensation. A receptor will be activated when a given area of the skin is stimulated; this is the receptive field of that receptor. A process known as lateral inhibition helps to sharpen the perceived location of the stimulus on the skin.

The cutaneous sensations of touch, pressure, hot and cold, and pain are mediated by the dendritic nerve endings of different sensory neurons. The receptors for hot, cold, and pain are the naked endings of sensory neurons. Sensations of touch and pressure are mediated by both naked dendritic endings and dendrites that are encapsulated within various structures (table 10.2). In pacinian corpuscles, for example, the dendritic endings are encased within thirty to fifty onionlike layers of connective tissue (fig. 10.5).

These layers absorb some of the pressure when a stimulus is maintained, and thus help to accentuate the phasic response of this receptor.

Neural Pathways for Somatesthetic Sensations

The conduction pathways for the **somatesthetic senses**—a term that includes sensations from cutaneous receptors and proprioceptors—are shown in chapter 8, figure 8.19. Sensory information from proprioceptors and pressure receptors are carried by large, myelinated nerve fibers that ascend in the *dorsal columns* of the spinal cord on the same (ipsilateral) side. These fibers do not synapse until they reach the *medulla oblongata* of the brain stem; hence, fibers that carry these sensations from the feet are incredibly long. After synapsing in the medulla with other, second-order sensory neurons, information in the latter neurons crosses over to the contralateral side as it ascends via a fiber tract, called the **medial lemniscus,** to the *thalamus.* Third-order sensory neurons in the thalamus that receive this input in turn project to the **postcentral gyrus** (the sensory cortex, as described in chapter 8).

Sensations of hot, cold, and pain are carried by thin, unmyelinated sensory neurons into the spinal cord. These synapse with second-order association neurons within the spinal cord, which cross over to the contralateral side and ascend to the brain in the **lateral spinothalamic tract.** Fibers that mediate *touch* and *pressure* ascend in the **anterior spinothalamic tract.** Fibers of both spinothalamic tracts synapse with third-order neurons in the thalamus, which in turn project to the postcentral gyrus. Notice that somatesthetic information is always carried to the postcentral gyrus in third-order neurons. Also, because of crossing over, somatesthetic information from each side of the body is projected to the postcentral gyrus of the contralateral cerebral hemisphere.

All somatesthetic information from the same area of the body projects to the same area of the postcentral gyrus, so that a "map" of the body can be drawn on the postcentral gyrus to represent sensory projection points (see chapter 8, fig. 8.7). This map is very distorted, however, because it shows larger areas of cortex devoted to sensation in the face and hands than in other areas in the body. This disproportionately larger area of the cortex devoted to the face and hands reflects the fact that there is a higher density of sensory receptors in these regions.

Receptive Fields and Sensory Acuity

The **receptive field** of a neuron serving cutaneous sensation is the area of skin whose stimulation results in changes in the firing rate of the neuron. Changes in the firing rate of primary sensory neurons affect the firing of second- and third-order neurons, which in turn affects the firing of those neurons in the postcentral gyrus that receive input from the third-order neurons. Indirectly, therefore, neurons in the postcentral gyrus can be said to have receptive fields in the skin.

The phenomenon of the **phantom limb** was first described by a neurologist during the Civil War. In this account, a veteran with amputated legs asked for someone to massage his cramped leg muscle. It is now known that this phenomenon is common in amputees, who may experience complete sensations from the missing limbs. They perceive the phantom as being very real, especially with their eyes closed, and they sense it moving in accordance with the way the limb would naturally move if it were real. These sensations are sometimes useful; for example, in fitting prostheses into which the phantom has seemingly entered. However, pain in the phantom is experienced by 70% of amputees, and the pain can be severe and persistent.

One explanation for phantom limbs is that the nerves remaining in the stump can grow into nodules called neuromas, which generate nerve impulses that are transmitted to the brain and interpreted as arising from the missing limb. However, phantom limbs may occur in cases where the limb is not amputated, but the nerves that normally enter from the limb are severed in an accident. Or it may occur in individuals with spinal cord injuries above the level of the limb, so that the sensations from the limb do not enter the brain. In these cases, the phantom limb phenomenon requires a different explanation. Current theories propose that the source of the phantom may arise in several brain regions whose activity is somehow changed by the absence of the sensations that would normally arise from the missing limb.

The area of each receptive field in the skin varies inversely with the density of receptors in the region. In the back and legs, where a large area of skin is served by relatively few sensory endings, the receptive field of each neuron is correspondingly large. In the fingertips, where a large number of cutaneous receptors serve a small area of skin, the receptive field of each sensory neuron is correspondingly small.

Two-Point Touch Threshold

The approximate size of the receptive fields serving light touch can be measured by the *two-point touch threshold test.* In this procedure, two points of a pair of calipers are lightly touched to the skin at the same time. If the distance between the points is sufficiently great, each point will stimulate a different receptive field and a different sensory neuron—two separate points of touch will thus be felt. If the distance is sufficiently small, both points will touch the receptive field of only one sensory neuron, and only one point of touch will be felt (fig. 10.6).

The two-point touch threshold, which is the minimum distance that can be distinguished between two points of touch, is a measure of the distance between receptive fields. If the distance between the two points of the calipers is less than this minimum distance, only one "blurred" point of touch can be felt. The two-point touch threshold is thus an indication of tactile *acuity* (*acus* = needle) or the sharpness of touch perception.

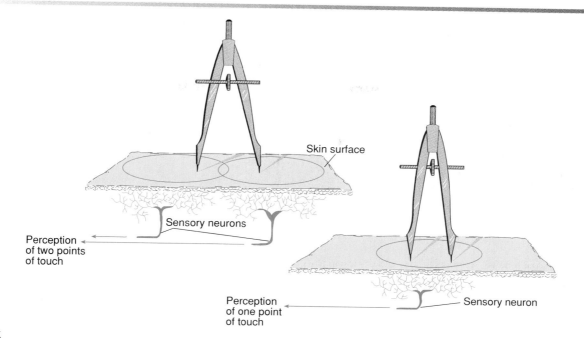

Skin surface

Sensory neurons

Perception
of two points
of touch

Perception
of one point
of touch

Sensory neuron

Figure 10.6

The two-point touch threshold test. If each point touches the receptive fields of different sensory neurons, two separate points of touch will be felt. If both caliper points touch the receptive field of one sensory neuron, only one point of touch will be felt.

The tactile acuity of the fingertips is exploited in the reading of *Braille*. Braille symbols are formed by raised dots on the page that are separated from each other by 2.5 mm, which is slightly greater than the two-point touch threshold in the fingertips (table 10.3). Experienced Braille readers can scan words at about the same speed that a sighted person can read aloud—a rate of about 100 words per minute.

Table 10.3 The Two-Point Touch Threshold for Different Regions of the Body

Body Region	Two-Point Touch Threshold (mm)
Big toe	10
Sole of foot	22
Calf	48
Thigh	46
Back	42
Abdomen	36
Upper arm	47
Forehead	18
Palm of hand	13
Thumb	3
First finger	2

Source: From S. Weinstein and D. R. Kenshalo, editors, *The Skin Senses,* © 1968. Courtesy of Charles C Thomas, Publisher, Springfield, Illinois.

Lateral Inhibition

When a blunt object touches the skin, a number of receptive fields are stimulated—some more than others. The receptive fields in the center areas where the touch is strongest will be stimulated more than those in the neighboring fields where the touch is lighter. Stimulation will gradually diminish from the point of greatest contact, without a clear, sharp boundary. What we perceive, however, is not the fuzzy sensation that might be predicted. Instead, only a single touch with well-defined borders is felt. This sharpening of sensation is due to a process called **lateral inhibition** (fig. 10.7).

Lateral inhibition and the resultant sharpening of sensation occur within the central nervous system. Those sensory neurons whose receptive fields are stimulated most strongly inhibit—via interneurons that pass "laterally" within the CNS—sensory neurons that serve neighboring receptive fields. Lateral inhibition similarly plays a prominent role in the ability of the ears and brain to discriminate sounds of different pitch, as described in a later section.

1. Using a flow diagram, describe the neural pathways leading from cutaneous pain and pressure receptors to the postcentral gyrus. Indicate where crossing over occurs.

2. Define the term *sensory acuity* and explain how acuity is related to the density of receptive fields in different parts of the body.

3. Explain the mechanism of lateral inhibition in cutaneous sensory perception and discuss its significance.

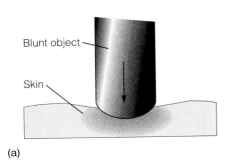
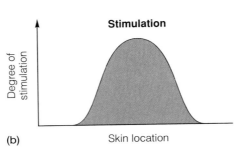
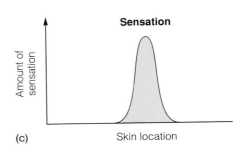

Lateral inhibition
within central nervous system

(a)

(b)

(c)

Figure 10.7

When an object touches the skin (*a*), receptors in the center of the touched skin are stimulated more than neighboring receptors (*b*). Lateral inhibition within the central nervous system reduces the input from these neighboring sensory neurons. Sensation, as a result, is sharpened within the areas of skin that was stimulated the most (*c*).

Taste and Olfaction

The receptors for taste and olfaction respond to molecules that are dissolved in fluid, and are thus classified as chemoreceptors. Although there are only four basic modalities of taste, they combine in various ways and are influenced by sensations of olfaction, thus permitting a wide variety of different sensory experiences.

Chemoreceptors that respond to chemical changes in the internal environment are called **interoceptors;** those that respond to chemical changes in the external environment are **exteroceptors.** Included in the latter category are *taste receptors,* which respond to chemicals dissolved in food or drink, and *olfactory receptors,* which respond to gaseous molecules in the air. This distinction is somewhat arbitrary, however, because odorant molecules in air must first dissolve in fluid within the olfactory mucosa before the sense of smell can be stimulated. Also, the sense of olfaction strongly influences the sense of taste, as can easily be verified by eating an onion (or almost anything else) with the nostrils pinched together.

Taste

Taste receptors consist of barrel-shaped arrangements of specialized epithelial cells called *taste buds* that are located primarily on the surface of the tongue (fig. 10.8). Long microvilli project from the tips of these epithelial cells and extend through a pore in the surface of the taste bud to the external environment, where they are bathed in saliva. Although the taste cells are not neurons, they can become depolarized under appropriate stimulation and release chemical transmitters that stimulate associated sensory neurons. Taste buds in the posterior third of the tongue are innervated by the *glossopharyngeal* (*ninth cranial*) *nerve;* those in the anterior two-thirds of the tongue are innervated by the *facial* (*seventh cranial*) *nerve*.

There are only four basic modalities of taste, which are sensed most acutely in particular regions of the tongue. These are *sweet* (tip of the tongue), *sour* (sides of the tongue), *bitter* (back of the tongue), and *salty* (over most of the tongue). This distribution is illustrated in figure 10.9. All the different tastes that we can perceive are combinations of these four, together with nuances provided by the sense of smell.

The salty taste of food is due to the presence of Na^+ ions. These pass into the sensitive receptor cells through channels in the apical membranes. This depolarizes the cells, causing them to release their transmitter. The anion associated with the Na^+, however, modifies the perceived saltiness to a surprising degree: NaCl tastes much more salty than most other sodium salts (such as sodium acetate) that have been tested. Recent evidence suggests that the anions can pass through the tight junctions between the receptor cells, and that the Cl^- anion passes through this barrier more readily than the other anions. This is presumably related to the ability of Cl^- to impart a saltier taste to the Na^+ than do the other anions.

Sour taste, like salty taste, is produced by ion movement through membrane channels. Sour taste, however, is due to the presence of hydrogen ions (H^+); all acids therefore taste sour. In contrast to the salty and sour tastes, the sweet and bitter tastes are produced by interaction of taste molecules with specific membrane receptor proteins.

Most organic molecules, particularly sugars, taste sweet to varying degrees. Bitter taste is evoked by quinine and seemingly unrelated molecules. Both sweet and bitter are mediated by receptors that are coupled to G-proteins (chapter 7). Dissociation of the G-protein subunit activates second-messenger systems, leading to depolarization of the receptor cell. In each case, the stimulated receptor cell, in turn, activates an associated sensory neuron that transmits impulses to the brain, where they are interpreted as the corresponding taste perception.

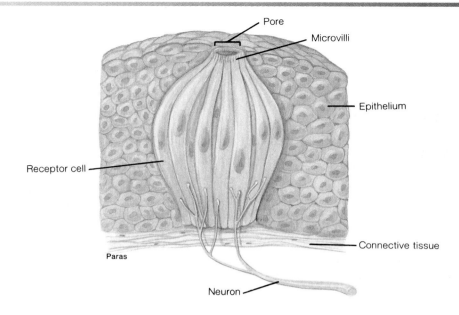

Figure 10.8

A taste bud.

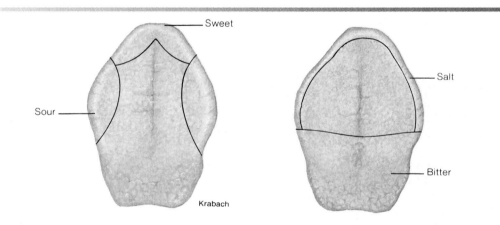

Figure 10.9

Patterns of taste receptor distribution on the dorsum of the tongue.

Olfaction

The olfactory receptors consist of several million bipolar sensory neurons located within a pseudostratified epithelium. Unique among the neurons of an adult, these sensory neurons divide mitotically and replace themselves every 1 to 2 months. Each bipolar sensory neuron has one unmyelinated axon that projects up into the olfactory bulb of the cerebrum, where it synapses with second-order neurons, and one dendrite that projects into the nasal cavity, where it terminates in a knob containing cilia (figs. 10.10 and 10.11). Therefore, unlike other sensory modalities that are relayed to the cerebrum from the thalamus, the sense of smell is transmitted directly to the cerebral cortex. The olfactory bulb is part of the limbic system, which was described in chapter 8 as having an important role in generating emotions and in memory. Perhaps this explains why the smell of a particular odor, more powerfully than other sensations, can evoke emotionally charged memories.

The molecular basis of olfaction is complex. At least in some cases, odorant molecules bind to receptors and act through G-proteins to increase the cyclic AMP within the cell. This, in turn, opens membrane channels and causes depolarization. There is evidence that up to fifty G-proteins may be associated with a single receptor protein. Dissociation of these G-proteins would release many G-protein subunits, thereby amplifying the effect many times. This amplification could account for the extreme sensitivity of the sense of smell: the human nose can detect a billionth of an ounce of perfume in air, and our sense of smell is not nearly as keen as that of many other mammals.

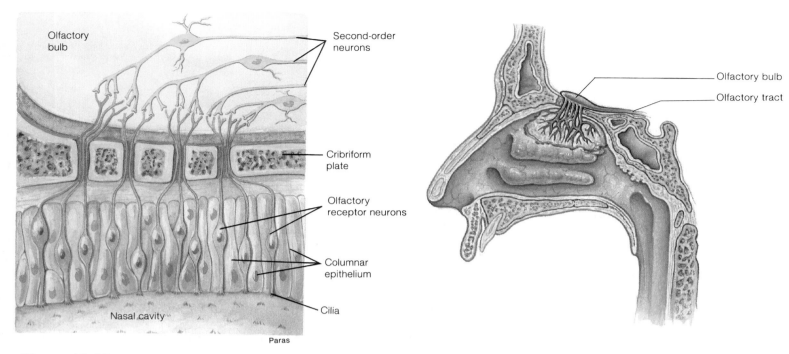

Figure 10.10

The olfactory epithelium contains receptor neurons that synapse with neurons in the olfactory bulb of the brain.

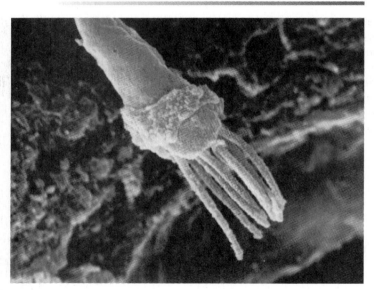

Figure 10.11

A scanning electron micrograph of an olfactory neuron, showing the tassel of cilia.

A family of genes that codes for the olfactory receptor proteins has been discovered. This is a large family, which may include as many as a thousand genes. The large number may reflect the importance of the sense of smell to mammals in general. Even a thousand different genes coding for a thousand different receptor proteins, however, cannot account for the fact that humans can distinguish up to 10,000 different odors. Clearly, the brain must integrate the signals from several different receptors and then interpret the pattern as a characteristic "fingerprint" for a particular odor.

1. Describe the distribution of taste receptors in the tongue, and explain how damage to the facial nerve might affect taste sensation.

2. Compare the mechanism by which salty and sour foods stimulate the taste receptors to those mechanisms by which the sweet and bitter tastes stimulate the taste receptors.

3. Explain how odorant molecules stimulate the olfactory receptors and why we can perceive very tiny concentrations of odors.

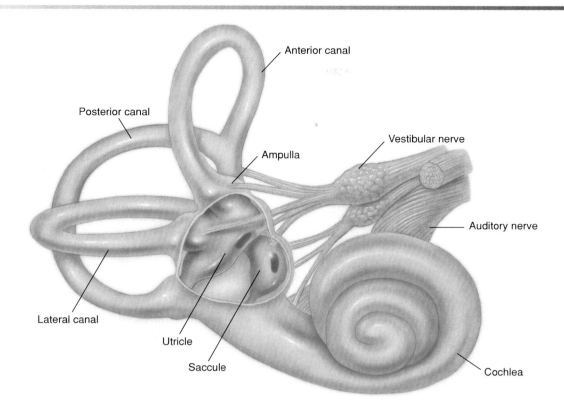

Figure 10.12

Structures within the inner ear include the cochlea and vestibular apparatus. The vestibular apparatus consists of the utricle and saccule (together called the otolith organs) and the three semicircular canals. The base of each semicircular canal is expanded into an ampulla that contains sensory hair cells.

Vestibular Apparatus and Equilibrium

The sense of equilibrium is provided by structures in the inner ear, collectively known as the vestibular apparatus. Movements of the head cause fluid within these structures to bend extensions of sensory hair cells, and this mechanical bending results in the production of action potentials.

The sense of equilibrium, which provides orientation with respect to gravity, is due to the function of an organ called the **vestibular apparatus.** The vestibular apparatus and a snaillike structure called the *cochlea*, which is involved in hearing, form the *inner ear* within the temporal bones of the skull. The vestibular apparatus consists of two parts: (1) the *otolith organs,* which include the *utricle* and *saccule,* and (2) the *semicircular canals* (fig. 10.12).

The sensory structures of the vestibular apparatus and cochlea are located within a tubular structure called the **membranous labyrinth** (fig. 10.13), which is filled with a fluid that is similar in composition to intracellular fluid. This fluid is called *endolymph.* The membranous labyrinth is located within a bony cavity in the skull. Within this cavity, between the membranous labyrinth and the bone, is a fluid called *perilymph.* Perilymph is similar in composition to cerebrospinal fluid.

Sensory Hair Cells of the Vestibular Apparatus

The utricle and saccule provide information about *linear acceleration*—changes in velocity when traveling horizontally or vertically. We therefore have a sense of acceleration and deceleration when riding in a car or when skipping rope. A sense of *rotational,* or *angular, acceleration* is provided by the semicircular canals, which are oriented in three planes like the faces of a cube. This helps us maintain balance when turning the head, spinning, or tumbling.

The receptors for equilibrium are modified epithelial cells. They are known as **hair cells** because they contain twenty to fifty hairlike extensions. All but one of these hairlike extensions are **stereocilia**—processes containing filaments of protein surrounded by part of the cell membrane. There is one larger extension that has the structure of a true cilium (chapter 3) that is known as a **kinocilium** (fig. 10.14). When the stereocilia

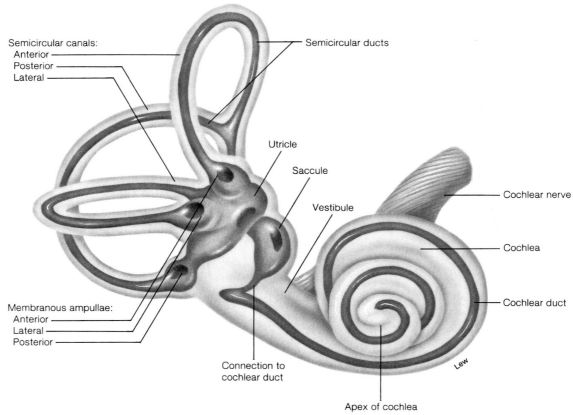

Figure 10.13

The labyrinths of the inner ear. The membranous labyrinth (darker color) is contained within the bony labyrinth.

are bent in the direction of the kinocilium, the cell membrane is depressed and becomes depolarized. This causes the hair cell to release a synaptic transmitter that stimulates the dendrites of sensory neurons that are part of the vestibulocochlear (eighth cranial) nerve (chapter 8). When the stereocilia are bent in the opposite direction, the membrane of the hair cell becomes hyperpolarized (fig. 10.14) and, as a result, releases less synaptic transmitter. In this way, the frequency of action potentials in the sensory neurons that innervate the hair cells carries information about movements that cause the hair cell processes to bend.

Utricle and Saccule

The hair cells of the utricle and saccule protrude into the endolymph-filled membranous labyrinth, with their hairs embedded within a gelatinous **otolith membrane.** The otolith membrane contains microscopic crystals of calcium carbonate

from which it derives its name (*oto* = ear; *lith* = stone). These stones increase the mass of the membrane, which results in a higher *inertia* (resistance to change in movement).

Because of the orientation of their hair cell processes into the otolith membrane, the utricle is more sensitive to horizontal acceleration and the saccule is more sensitive to vertical acceleration. During forward acceleration, the otolith membrane lags behind the hair cells, so the hairs of the utricle are pushed backward (fig. 10.15). This is similar to the backward thrust of the body when a car quickly accelerates forward. The inertia of the otolith membrane similarly causes the hairs of the saccule to be pushed upward when a person descends rapidly in an elevator. These effects, and the opposite ones that occur when a person accelerates backward or upward, produce a changed pattern of action potentials in sensory nerve fibers that allows us to maintain our equilibrium with respect to gravity during linear acceleration.

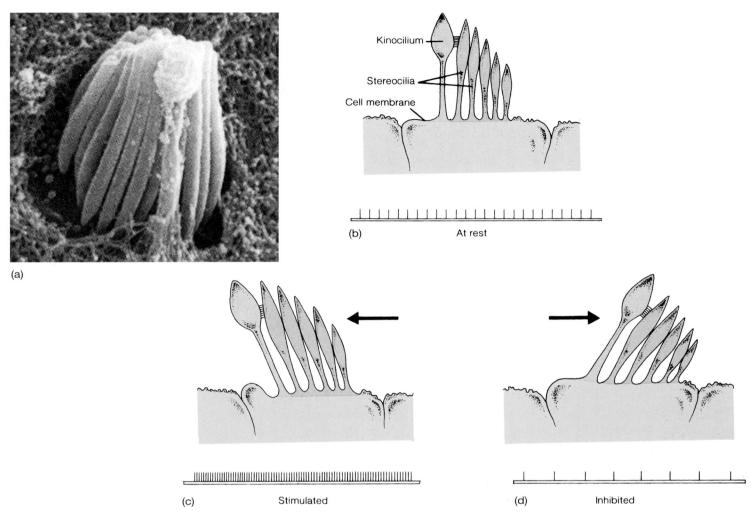

Figure 10.14

Sensory hair cells within the vestibular apparatus. (*a*) A scanning electron photograph of a kinocilium. (*b*) Each sensory hair cell contains a single kinocilium and several stereocilia. (*c*) When stereocilia are displaced toward the kinocilium (*arrows*), the cell membrane is depressed and the sensory neuron innervating the hair cell is stimulated. (*d*) When the stereocilia are bent in the opposite direction, away from the kinocilium, the sensory neuron is inhibited.

Semicircular Canals

The three semicircular canals are oriented at right angles to each other. Each canal contains a bulge, called the *ampulla*, where the sensory hair cells are located. The processes of these sensory cells are embedded within a gelatinous membrane called the **cupula** (fig. 10.16) that projects into the endolymph of the membranous canals. Like a sail in the wind, the cupula can be pushed in one direction or the other by movements of the endolymph.

The endolymph of the semicircular canals serves a function analogous to that of the otolith membrane—it provides inertia so that the sensory processes will be bent in a direction opposite to that of the angular acceleration. As the head rotates to the right, for example, the endolymph causes the cupula to be bent toward the left, thereby stimulating the hair cells.

Neural Pathways

Stimulation of hair cells in the vestibular apparatus activates sensory neurons of the *vestibulocochlear* (*eighth cranial*) *nerve*. These fibers transmit impulses to the cerebellum and to the vestibular nuclei of the medulla oblongata. The vestibular nuclei, in turn, send fibers to the oculomotor center of the brain stem and to the spinal cord (fig. 10.17). Neurons in the oculomotor center control eye movements, and neurons in the spinal

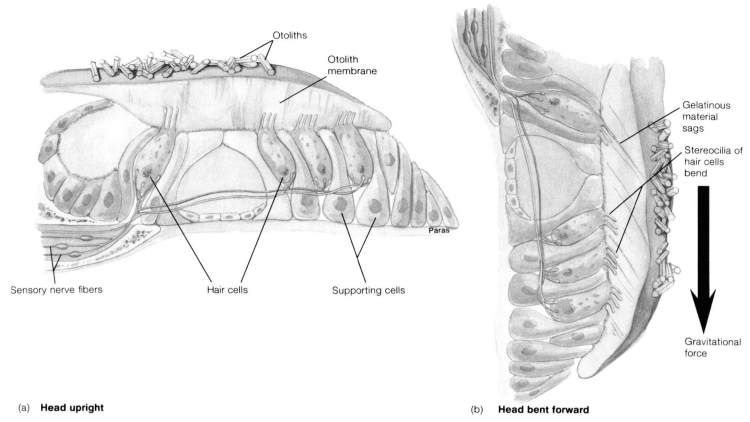

Otoliths

Otolith membrane

Gelatinous material sags

Stereocilia of hair cells bend

Gravitational force

Paras

Sensory nerve fibers

Hair cells

Supporting cells

(a) **Head upright**

(b) **Head bent forward**

Figure 10.15

The otolith organ. (*a*) When the head is in an upright position, the weight of the otoliths applies direct pressure to the sensitive cytoplasmic extensions of the hair cells. (*b*) As the head is tilted forward, the extensions of the hair cells bend in response to gravitational force and cause the sensory nerve fibers to be stimulated.

cord stimulate movements of the head, neck, and limbs. Movements of the eyes and body produced by these pathways serve to maintain balance and "track" the visual field during rotation.

Nystagmus and Vertigo

When a person first begins to spin, the inertia of endolymph within the semicircular canals causes the cupula to bend in the opposite direction. As the spin continues, however, the inertia of the endolymph is overcome and the cupula straightens. At this time, the endolymph and the cupula are moving in the same direction and at the same speed. If movement is suddenly stopped, the greater inertia of the endolymph causes it to continue moving in the previous direction of spin and to bend the cupula in that direction.

Bending of the cupula after movement has stopped affects muscular control of the eyes and body through the neural pathways previously discussed. The eyes slowly drift in the direction of the previous spin, and then are rapidly jerked back to the midline position, producing involuntary oscillations. These movements are called **vestibular nystagmus,** and people experiencing this effect may feel that they, or the room, are spinning. The loss of equilibrium that results is called *vertigo.* If

the vertigo is sufficiently severe, or the person particularly susceptible, the autonomic system may become involved. This can produce dizziness, pallor, sweating, and nausea.

Vestibular nystagmus is one of the symptoms of an inner-ear disease called **Ménière's disease.** The early symptom of this disease is often "ringing in the ears," or *tinnitus.* Since the endolymph of the cochlea and the endolymph of the vestibular apparatus are continuous through a tiny canal called the duct of Hensen, vestibular symptoms of vertigo and nystagmus often accompany hearing problems in this disease.

1. Describe the structure of the utricle and saccule and explain how linear acceleration results in stimulation of the hair cells within these organs.

2. Describe the structure of the semicircular canals and explain how they provide a sense of angular acceleration.

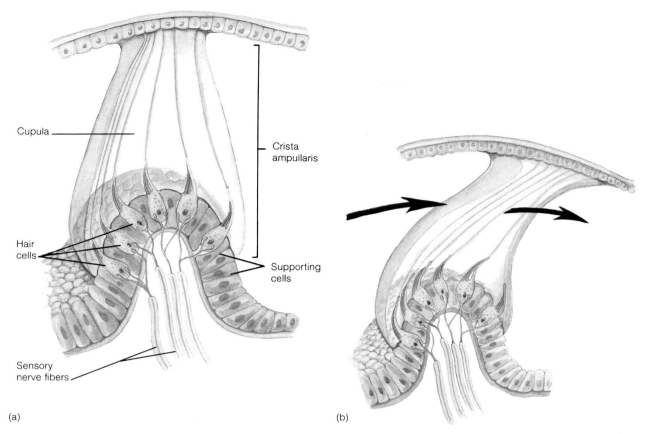

Figure 10.16

(*a*) The cupula and hair cells within the semicircular canals. (*b*) Movement of the endolymph during rotation causes the cupula to displace, thus stimulating the hair cells.

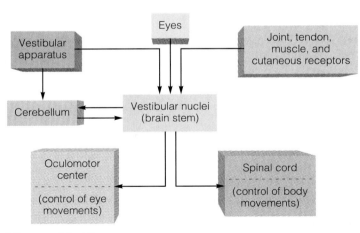

Figure 10.17

Neural pathways involved in the maintenance of equilibrium and balance.

The Ears and Hearing

Sound causes vibrations of the tympanic membrane. These vibrations, in turn, produce movements of the middle-ear ossicles, which press against a membrane called the oval window in the cochlea. The movements of the oval window produce pressure waves within the fluid of the cochlea, which in turn cause movements of a membrane called the basilar membrane. Sensory hair cells are located on the basilar membrane, and the movements of this membrane in response to sound result in the bending of the hair cell processes. This stimulates action potentials in sensory fibers, which are transmitted to the brain and interpreted as sound.

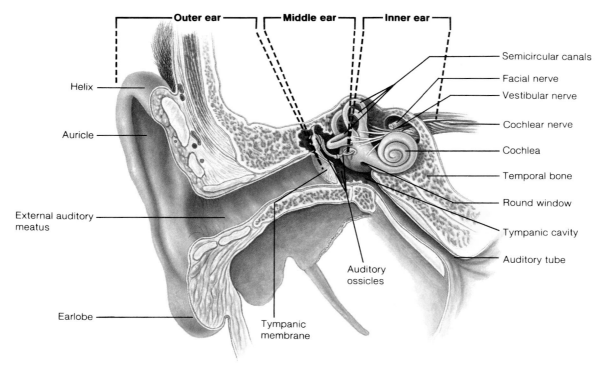

Figure 10.18

The ear. (Note the external, middle, and inner regions.)

Sound waves travel in all directions from their source, like ripples in a pond into which a stone has been dropped. These waves are characterized by their frequency and their intensity. The **frequency,** or distances between crests of the sound waves, is measured in *hertz* (*Hz*), which is the modern designation for *cycles per second* (*cps*). The *pitch* of a sound is directly related to its frequency—the greater the frequency of a sound, the higher its pitch.

The **intensity,** or loudness of a sound, is directly related to the amplitude of the sound waves. This is measured in units known as *decibels* (*dB*). A sound that is barely audible—at the threshold of hearing—has an intensity of zero decibels. Every 10 decibels indicates a tenfold increase in sound intensity; a sound is ten times louder than threshold at 10 dB, 100 times louder at 20 dB, a million times louder at 60 dB, and 10 billion times louder at 100 dB.

The ear of a trained, young individual can hear sound over a frequency range of 20,000 to 30,000 Hz, yet can distinguish between two pitches that have only a 0.3% difference in frequency. The human ear can detect differences in sound intensities of only 0.1 to 0.5 dB, while the range of audible intensities covers twelve orders of magnitude (10^{12}), from the barely audible to the limits of painful loudness.

The Outer Ear

Sound waves are funneled by the *pinna*, or *auricle* (flap), into the *external auditory meatus* (fig. 10.18). These two structures form the *outer ear*. The external auditory meatus channels the sound waves (while increasing their intensity) to the eardrum, or **tympanic membrane.** Sound waves in the external auditory meatus produce extremely small vibrations of the tympanic membrane; movements of the eardrum during speech (with an average sound intensity of 60 dB) are estimated to be about equal to the diameter of a molecule of hydrogen!

The Middle Ear

The middle ear is the cavity between the tympanic membrane on the outer side and the cochlea on the inner side (fig. 10.19). Within this cavity are three **middle-ear ossicles**—the *malleus* (hammer), *incus* (anvil), and *stapes* (stirrup). The malleus is attached to the tympanic membrane, so that vibrations of this membrane are transmitted, via the malleus and incus, to the stapes. The stapes, in turn, is attached to a membrane in the cochlea called the *oval window*, which thus vibrates in response to vibrations of the tympanic membrane.

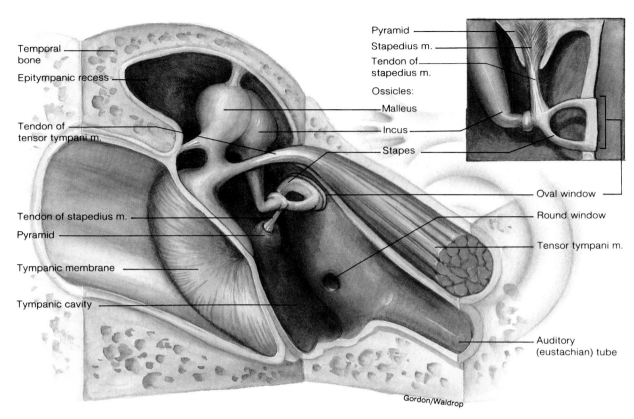

Figure 10.19

A medial view of the middle ear showing the position of the auditory muscles.

The *auditory (eustachian) tube* is a passageway leading from the middle ear to the nasopharynx (a cavity that is posterior to the palate of the oral cavity). The auditory tube is usually collapsed, so that debris and infectious agents are prevented from traveling from the oral cavity to the middle ear. In order to open the auditory tube, the *tensor tympani muscle,* attaching to the auditory tube and the malleus (fig. 10.19), must contract. This occurs when you swallow, yawn, or sneeze. People sense a "popping" sensation in the ears as they swallow when driving up a mountain because the opening of the auditory canal permits air to move from the region of higher pressure in the middle ear to the lower pressure in the nasopharynx.

The fact that vibrations of the tympanic membrane are transferred through three bones instead of just one affords protection. If the sound is too intense, the ossicles may buckle. This protection is increased by the action of the *stapedius muscle,* which attaches to the neck of the stapes (fig. 10.19). When sound becomes too loud, the stapedius muscle contracts and dampens the movements of the stapes against the oval window. This action helps to prevent nerve damage within the cochlea. If sounds reach loud amplitudes extremely rapidly—as in gunshots—the stapedius muscle may not respond rapidly enough to prevent nerve damage.

Damage to the tympanic membrane or middle-ear ossicles produces **conduction deafness.** This impairment can result from a variety of causes, including *otitis media* and *otosclerosis.* In otitis media, which sometimes follows allergic reactions or respiratory disease, inflammation produces excessive fluid accumulation within the middle ear. This, in turn, can result in the excessive growth of epithelial tissue and damage to the eardrum. In otosclerosis, bone is reabsorbed and replaced by "sclerotic bone" that grows over the oval window and immobilizes the footplate of the stapes. In conduction deafness, these pathological changes hinder the transmission of sound waves from the air to the cochlea of the inner ear.

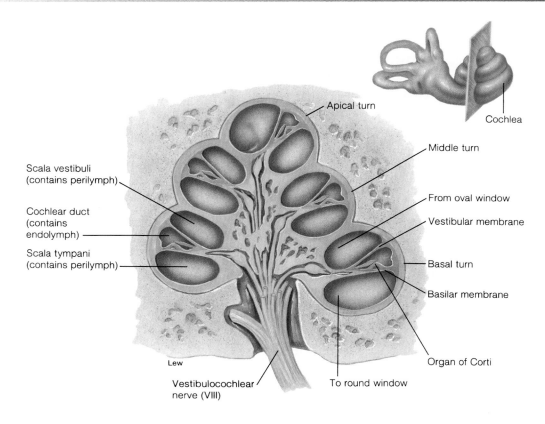

Figure 10.20

A cross section of the cochlea showing its three turns and its three compartments—the scala vestibuli, cochlear duct (scala media), and scala tympani.

The Cochlea

Encased within dense skull bone is a snail shell-shaped organ called the **cochlea,** about the size of a pea. Together with the vestibular apparatus (previously described), it composes the **inner ear**.

Vibrations of the stapes and oval window indirectly displace endolymph within a part of the membranous labyrinth known as the **cochlear duct,** also called the **scala media** because it is in the middle part of the cochlea. Like the cochlea as a whole, the cochlear duct coils to form three levels (fig. 10.20), similar to the basal, middle, and apical portions of a snail shell. The part of the cochlea above the cochlear duct is called the *scala vestibuli,* and the part below is called the *scala tympani* (fig. 10.20). Unlike the central cochlear duct, which contains endolymph, the scala vestibuli and scala tympani are filled with perilymph.

The perilymph of the scala vestibuli and scala tympani is continuous at the apex of the cochlea because the cochlear duct ends blindly, leaving a small space called the *helicotrema* between the end of the cochlear duct and the wall of the cochlea. Vibrations of the oval window produced by movements of the stapes cause pressure waves within the scala vestibuli, which pass to the scala tympani. Movements of perilymph within the scala tympani, in turn, travel to the base of the cochlea where they cause displacement of a membrane called the *round window* into the middle ear cavity (see fig. 10.19). This occurs because fluid, such as perilymph, cannot be compressed; an inward movement of the oval window is thus compensated for by an outward movement of the round window.

When the sound frequency (pitch) is sufficiently low, there is adequate time for the pressure waves of perilymph within the upper scala vestibuli to travel through the helicotrema to the scala tympani. As the sound frequency increases, however, pressure waves of perilymph within the scala vestibuli do not have time to travel all the way to the apex of the cochlea. Instead, they are transmitted through the *vestibular membrane,* which separates the scala vestibuli from the cochlear duct, and through the **basilar membrane,** which separates the cochlear duct from the scala tympani, to the perilymph of the scala tympani (fig. 10.20). The distance that these pressure waves travel, therefore, decreases as the sound frequency increases.

The Organ of Corti

Movements of perilymph from the scala vestibuli to the scala tympani thus produce displacement of the vestibular membrane and the basilar membrane. Although the movement of the vestibular membrane does not directly contribute to hearing, displacement of the basilar membrane is central to pitch discrimination. The

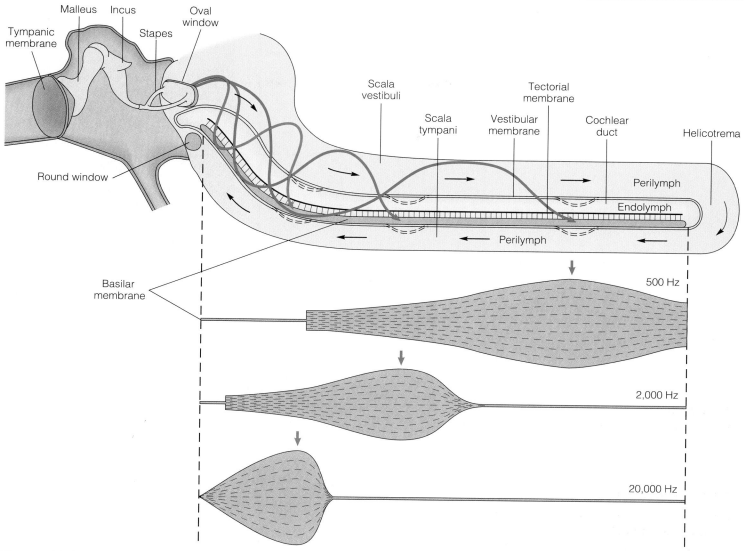

Malleus Incus Oval window
Tympanic membrane Stapes
Scala vestibuli Tectorial membrane
Scala tympani Vestibular membrane Cochlear duct Helicotrema
Round window Perilymph
Endolymph
Basilar membrane Perilymph
500 Hz
2,000 Hz
20,000 Hz

Figure 10.21

The cochlea is shown "unwound" in this diagram. Sounds of low frequency cause pressure waves of perilymph to pass through the helicotrema. Sounds of higher frequency cause pressure waves to "shortcut" through the cochlear duct. This causes displacement of the basilar membrane, which is central to the transduction of sound waves into nerve impulses. (The frequency of sound waves is measured in hertz (Hz), or cycles per second.)

basilar membrane is fixed on the inner side of the cochlear wall to a bony ridge, and is supported at its free end by a ligament.

Vibrations of the stapes against the oval window set up moving waves of perilymph in the scala vestibuli, which cause displacement of the basilar membrane into the scala tympani. This produces vibrations of the basilar membrane, each region of which vibrates with maximum amplitude to a different sound frequency. Sounds of higher frequency (pitch) cause maximum vibrations of the basilar membrane closer to the stapes, as illustrated in figure 10.21.

The sensory *hair cells* are located on the basilar membrane, with their "hairs" (actually stereocilia) projecting into the endolymph of the cochlear duct. These hair cells are arranged to form one row of inner cells, which extend the length of the basilar membrane, and multiple rows of outer hair cells: three rows in the basal turn, four in the middle turn, and five in the apical turn of the cochlea (fig. 10.22).

The stereocilia of the outer hair cells are embedded within a gelatinous **tectorial membrane,** which overhangs the hair cells within the cochlear duct (fig. 10.23). The association of the basilar membrane, hair cells with sensory fibers, and tectorial membrane forms a functional unit called the **organ of Corti** (fig. 10.23). When the cochlear duct is displaced by pressure waves of perilymph, a shearing force is created between the basilar membrane and the tectorial membrane. This causes the stereocilia to move and bend. Such movement causes ion channels in the membrane to open, which in turn depolarizes the hair cells. Each depolarized hair cell then releases a transmitter chemical, which stimulates an associated sensory neuron.

Figure 10.22

A scanning electron micrograph of the hair cells of the organ of Corti.

Reproduced from: R. G. Kessel and R. H. Kardon, *Tissues and Organs: A Text-Atlas of Scanning Electron Microscopy*, W. H. Freeman & Co., 1979.

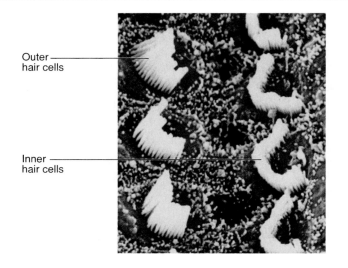

Outer hair cells

Inner hair cells

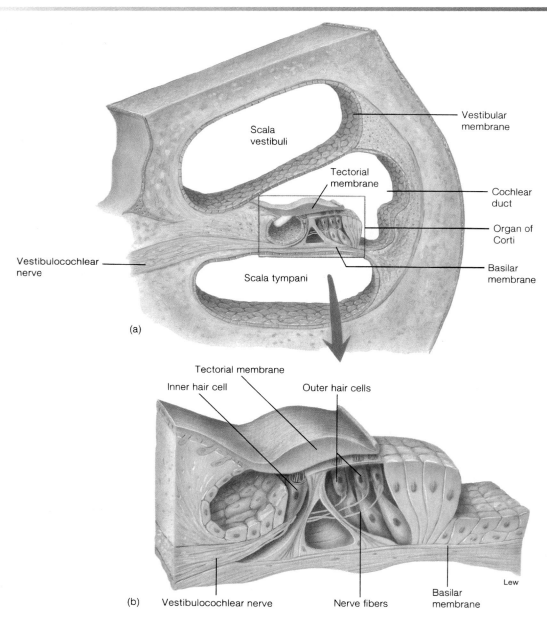

(a)

Scala vestibuli

Tectorial membrane

Vestibular membrane

Cochlear duct

Organ of Corti

Vestibulocochlear nerve

Scala tympani

Basilar membrane

Tectorial membrane

Inner hair cell

Outer hair cells

(b) Vestibulocochlear nerve

Nerve fibers

Basilar membrane

Lew

Figure 10.23

The organ of Corti (*a*) within the cochlear duct and (*b*) isolated to show greater detail.

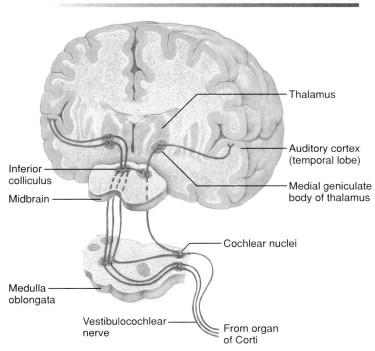

Figure 10.24
Neural pathways from the organ of Corti of the cochlea to the auditory cortex.

The greater the displacement of the basilar membrane and the bending of the stereocilia, the greater the amount of transmitter released by the hair cell, and therefore the greater the generator potential produced in the sensory neuron. In other words, a greater bending of the stereocilia will increase the frequency of action potentials produced by the fibers of the cochlear (eighth cranial) nerve that are stimulated by the hair cells. Experiments suggest that the stereocilia need only bend 0.3 nanometers to be detected at the threshold of hearing! A greater bending will result in a higher frequency of action potentials, which will be perceived as a louder sound.

Traveling waves in the basilar membrane reach a peak in different regions, depending on the pitch of the sound. High-pitch sounds produce a peak displacement closer to the base, while sounds of lower pitch cause peak displacement further toward the apex (see fig. 10.21). Those neurons that originate in hair cells located where the displacement is greatest will be stimulated more than neurons that originate in other regions. This mechanism provides a neural code for **pitch discrimination.**

Neural Pathways for Hearing

Sensory neurons in the vestibulocochlear (eighth cranial) nerve synapse with neurons in the medulla (fig. 10.24), which project to the inferior colliculus of the midbrain. Neurons in this area, in turn, project to the thalamus, which sends axons to the auditory cortex of the temporal lobe. By means of this pathway, neurons in different regions of the basilar membrane stimulate neurons in corresponding areas of the auditory cortex. Each area of this cortex thus represents a different part of the basilar membrane and a different pitch (fig. 10.25).

Hearing Impairments

There are two major causes of hearing loss: (1) **conductive deafness,** in which the transmission of sound waves from air through the middle ear to the oval window is impaired, and (2) **nerve** or **sensory deafness,** in which the transmission of nerve impulses anywhere from the cochlea to the auditory cortex is impaired. Conductive deafness can be caused by middle-ear damage from otitis media or otosclerosis, as previously discussed. Sensory deafness may result from a wide variety of pathological processes and by exposure to extremely loud sounds. Unfortunately, the hair cells in the inner ears of mammals cannot regenerate once they are destroyed. Experiments have shown, however, that the hair cells of reptiles and birds can regenerate, and scientists are trying to determine the reasons for this difference.

Conduction deafness impairs hearing at all sound frequencies. Sensory deafness, by contrast, often impairs the ability to hear some pitches more than others. This may be due to pathological processes or to changes that occur during aging. Age-related hearing deficits—called *presbycusis*—begin after age 20 when the ability to hear high frequencies (18,000 to 20,000 Hz) diminishes. Although the progression is variable, and occurs in men to a greater degree than in women, these deficits may gradually extend into the 4,000- to 8,000-Hz range. These impairments can be detected by a technique called *audiometry*, in which the threshold intensity of different pitches is determined. The ability to hear speech is particularly affected by hearing loss in the higher frequencies. People with these types of impairments can be helped by *hearing aids*, which amplify sounds and conduct the sound waves through bone to the inner ear.

1. Use a flowchart to describe how sound waves in air within the external auditory meatus are transduced into movements of the basilar membrane.

2. Explain how movements of the basilar membrane affect hair cells, and how hair cells can stimulate associated sensory neurons.

3. Explain how sounds of different degrees of loudness affect the function of the cochlea, and how different pitches of sounds are discriminated by the cochlea.

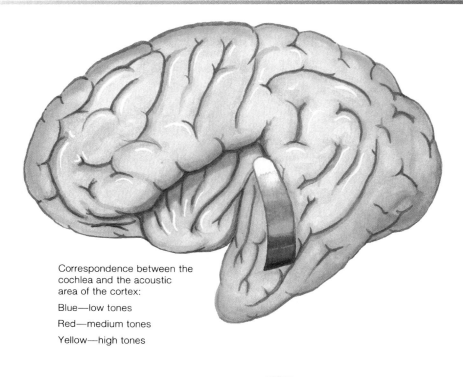

Correspondence between the
cochlea and the acoustic
area of the cortex:

Blue—low tones

Red—medium tones

Yellow—high tones

Figure 10.25

Sounds of different frequencies (pitches) excite different sensory neurons in the cochlea; these in turn send their input to different regions of the auditory cortex.

The Eyes and Vision

Light from an observed object is focused by the cornea and lens onto the photoreceptive retina at the back of the eye. The focus is maintained on the retina at different distances between the eyes and the object by variations in the curvature of the lens.

The eyes transduce energy in the electromagnetic spectrum (fig. 10.26) into nerve impulses. Only a limited part of this spectrum can excite the photoreceptors—electromagnetic energy with wavelengths between 400 and 700 nanometers (nm) constitutes *visible light*. Light of longer wavelengths in the infrared regions of the spectrum does not have sufficient energy to excite the receptors but is felt as heat. Ultraviolet light, which has shorter wavelengths and more energy than visible light, is filtered out by the yellow color of the eye's lens.

Honeybees—and people who have had their lenses removed—can see light in the ultraviolet range.

The parts of the eye are summarized in table 10.4. The outermost layer of the eye is a tough coat of connective tissue called the *sclera*, which can be seen externally as the white of the eyes. The tissue of the sclera is continuous with the transparent *cornea*. Light passes through the cornea to enter the *anterior chamber* of the eye. Light then passes through an opening, called the *pupil*, within a pigmented muscle known as the *iris*. After passing through the pupil, light enters the *lens* (fig. 10.27).

The iris is like the diaphragm of a camera; it can increase or decrease the diameter of its aperture (the pupil) to admit more or less light. Constriction of the pupils is produced by contraction of circular muscles within the iris; dilation is produced by contraction of radial muscles. Constriction of the pupils results from parasympathetic stimulation, whereas dilation results from sympathetic stimulation (fig. 10.28). Variations in the diameter of the pupil are similar in effect to variations in the f-stop of a camera.

Table 10.4 Location and Functions of Structures of the Eyeball

Tunic and Structure	Location	Composition	Function
Fibrous tunic	Outer layer of eyeball	Avascular connective tissue	Gives shape to eyeball
Sclera	Posterior, outer layer; white of the eye	Tightly bound elastic and collagen fibers	Supports and protects eyeball
Cornea	Anterior surface of eyeball	Tightly packed dense connective tissue—transparent and convex	Transmits and refracts light
Vascular tunic (uvea)	Middle layer of eyeball	Highly vascular pigmented tissue	Supplies blood; prevents reflection
Choroid	Middle layer in posterior portion of eyeball	Vascular layer	Supplies blood to eyeball
Ciliary body	Anterior portion of vascular tunic	Smooth muscle fibers and glandular epithelium	Supports the lens through suspensory ligament and determines its thickness; secretes aqueous humor
Iris	Anterior portion of vascular tunic continuous with ciliary body	Pigment cells and smooth muscle fibers	Regulates the diameter of the pupil, and hence the amount of light entering the vitreous chamber
Internal tunic	Inner layer of eyeball	Tightly packed photoreceptors, neurons, blood vessels, and connective tissue	Provides location and support for rods and cones
Retina	Principal portion of internal tunic	Photoreceptor neurons (rods and cones), bipolar neurons, and ganglion neurons	Photoreception; transmits impulses
Lens (not part of any tunic)	Between posterior and vitreous chambers; supported by suspensory ligament of ciliary body	Tightly arranged protein fibers; transparent	Refracts light and focuses onto fovea centralis

Source: From Kent M. Van De Graaff, *Human Anatomy*, 4th ed. Copyright © 1995 Wm. C. Brown Communications, Inc., Dubuque, Iowa. Reprinted by permission of Times Mirror Higher Education Group, Inc., Dubuque, Iowa. All Rights Reserved.

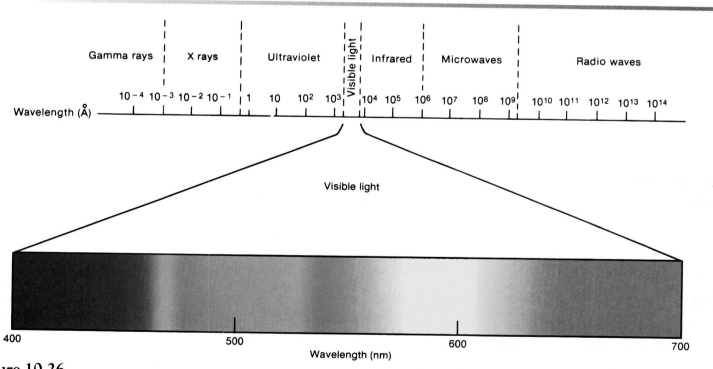

Figure 10.26

The electromagnetic spectrum (*top*) is shown in Angstrom units (1 Å = 10^{-10} meter). The visible spectrum comprises only a small range of this spectrum (*bottom*), shown in nanometer units (1 nm = 10^{-9} meter).

Figure 10.27

The internal anatomy of the eyeball.

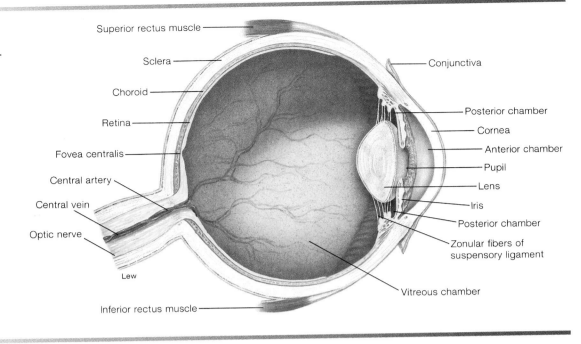

Superior rectus muscle

Sclera

Choroid

Retina

Fovea centralis

Central artery

Central vein

Optic nerve

Lew

Inferior rectus muscle

Conjunctiva

Posterior chamber

Cornea

Anterior chamber

Pupil

Lens

Iris

Posterior chamber

Zonular fibers of suspensory ligament

Vitreous chamber

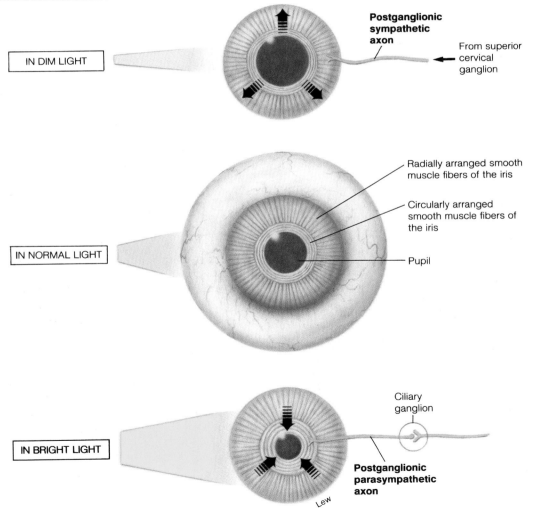

IN DIM LIGHT

Postganglionic sympathetic axon

From superior cervical ganglion

Radially arranged smooth muscle fibers of the iris

Circularly arranged smooth muscle fibers of the iris

IN NORMAL LIGHT

Pupil

Ciliary ganglion

IN BRIGHT LIGHT

Postganglionic parasympathetic axon

Lew

Figure 10.28

Dilation and constriction of the pupil. In dim light, the radially arranged smooth muscle fibers are stimulated to contract by sympathetic stimulation, dilating the pupil. In bright light, the circularly arranged smooth muscle fibers are stimulated to contract by parasympathetic stimulation, constricting the pupil.

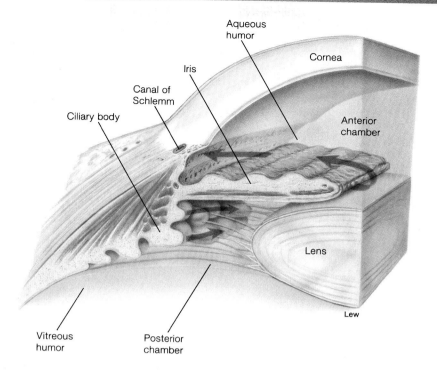

Figure 10.29

Aqueous humor maintains the intraocular pressure within the anterior and posterior chambers. It is secreted into the posterior chamber, flows through the pupil into the anterior chamber, and drains from the eyeball through the canal of Schlemm.

The posterior part of the iris contains a pigmented epithelium that gives the eye its color. The color of the eye is determined by the amount of pigment—blue eyes have the least pigment, brown eyes have more, and black eyes have the greatest amount of pigment. In the condition of *albinism*—a congenital defect in the ability to produce melanin pigment— the eyes appear pink because the absence of pigment allows blood vessels to be seen.

The lens is suspended from a muscular process called the **ciliary body,** which is connected to the sclera and encircles the lens. *Zonular fibers* (*zon* = girdle) suspend the lens from the ciliary body, forming a **suspensory ligament** that supports the lens. The space between the cornea and iris is the *anterior chamber,* and the space between the iris and the ciliary body and lens is the *posterior chamber* (fig. 10.29).

The anterior and posterior chambers are filled with a fluid called the **aqueous humor.** This fluid is secreted by the ciliary body into the posterior chamber and passes through the pupil into the anterior chamber, where it provides nourishment to the avascular lens and cornea. Aqueous humor is drained from the anterior chamber into the *canal of Schlemm,* which returns this fluid to the venous blood (fig. 10.29). Inadequate drainage of aqueous humor can lead to excessive accumulation of fluid, which in turn results in increased intraocular pressure.

This condition is called *glaucoma,* and may produce serious damage to the retina and loss of vision.

The portion of the eye located behind the lens is filled with a thick, viscous substance known as the **vitreous body,** or **vitreous humor.** Light from the lens that passes through the vitreous body enters the neural layer, which contains photoreceptors, at the back of the eye. This neural layer is called the **retina.** Light that passes through the retina is absorbed by a darkly pigmented *choroid layer* underneath. While passing through the retina, some of this light stimulates photoreceptors, which in turn activate other neurons. Neurons in the retina contribute fibers that are gathered together at a region called the *optic disc* (fig. 10.30) to exit the retina as the optic nerve. The optic disc is also the site of entry and exit of blood vessels.

Refraction

Light that passes from a medium of one density into a medium of a different density is *refracted,* or bent. The degree of refraction depends on the comparative densities of the two media, as indicated by their *refractive index.* The refractive index of air is set at 1.00; the refractive index of the cornea, by comparison, is 1.38; and the refractive indices of the aqueous humor and lens are 1.33 and 1.40, respectively. Since the

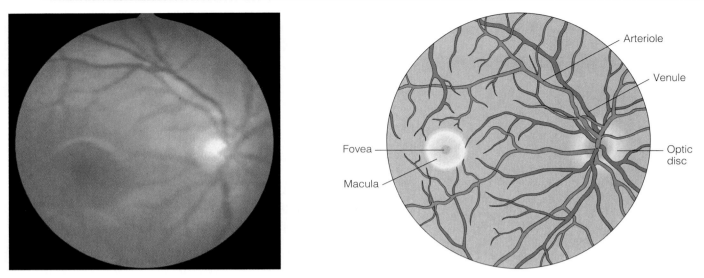

Figure 10.30

A view of the retina as seen with an ophthalmoscope. Optic nerve fibers leave the eyeball at the optic disc to form the optic nerve. (Note the blood vessels that can be seen entering the eyeball at the optic disc.)

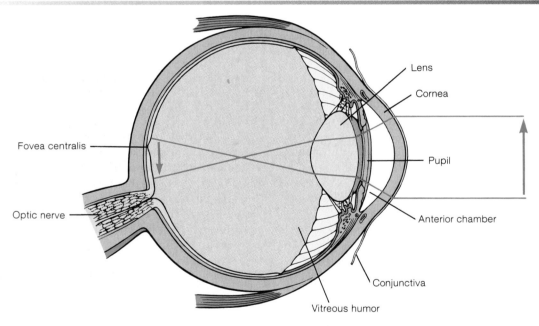

Figure 10.31

The refraction of light waves within the eyeball causes the image of an object to be inverted on the retina.

greatest difference in refractive index occurs at the air-cornea interface, the light is refracted most at the cornea.

The degree of refraction also depends on the curvature of the interface between two media. The curvature of the cornea is constant, however, while the curvature of the lens can be varied. The refractive properties of the lens can thus provide fine control for focusing light on the retina. As a result of light refraction, the image formed on the retina is upside down and right to left (fig. 10.31).

The *visual field*—which is the part of the external world projected onto the retina—is thus reversed in each eye. The cornea and lens focus the right part of the visual field on the left half of the retina of each eye, while the left half of the visual field is focused on the right half of each retina (fig. 10.32). The medial (or nasal) half-retina of the left eye therefore receives the same image as the lateral (or temporal) half-retina of the right eye. The nasal half-retina of the right eye receives the same image as the temporal half-retina of the left eye.

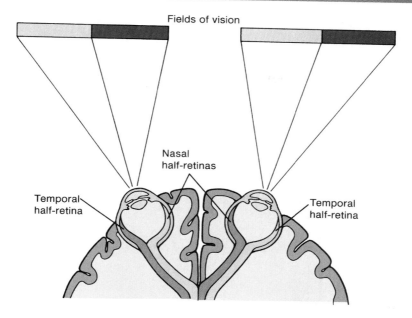

Figure 10.32

Refraction of light in the cornea and lens produces a right-to-left image on the retina. The left side of the visual field is projected to the right half of each retina, while the right side of each visual field is projected to the left half of each retina.

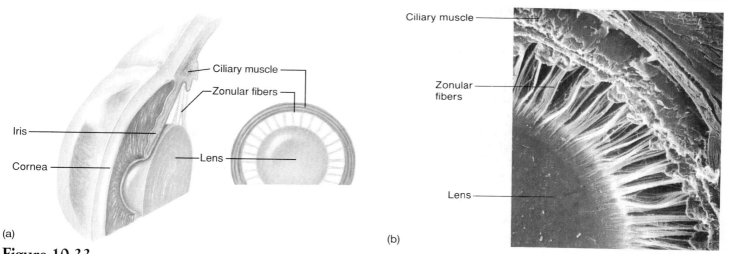

(a)

(b)

Figure 10.33

(a) A diagram, and (b) scanning electron micrograph (from the eye of a 17-year-old-boy) showing the relationship between the lens, zonular fibers, and ciliary muscle of the eye.

(Part [b] from "How the Eye Focuses" by James F. Koretz and George H. Handleman. Copyright © 1988 by Scientific American, Inc. All rights reserved.)

Accommodation

When a normal eye views an object, parallel rays of light are refracted to a point, or *focus*, on the retina (see fig. 10.35). If the degree of refraction remained constant, movement of the object closer to or farther from the eye would cause corresponding movement of the focal point, so that the focus would either be behind or in front of the retina.

The ability of the eyes to keep the image focused on the retina as the distance between the eyes and object is changed is called **accommodation.** Accommodation results from contraction of the ciliary muscle, which is like a sphincter muscle that can vary its aperture (fig. 10.33). When the ciliary muscle is relaxed, its aperture is wide. Relaxation of the ciliary muscle thus places tension on the zonular fibers of the suspensory ligament and pulls the lens taut. These are the conditions that prevail

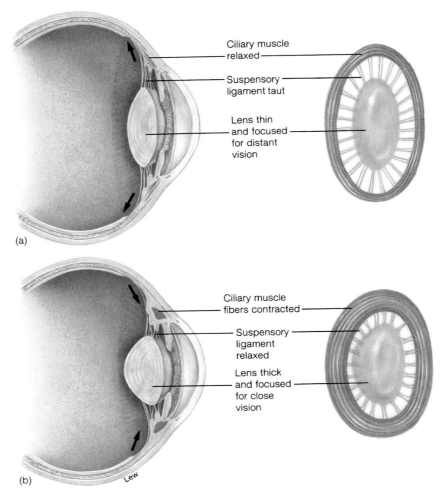

Figure 10.34

Changes in the shape of the lens during accommodation. (*a*) The lens is flattened for distant vision when the ciliary muscle fibers are relaxed and the suspensory ligament is taut. (*b*) The lens is more spherical for close-up vision when the ciliary muscle fibers are contracted and the suspensory ligament is relaxed.

The ability of a person's eyes to accommodate can be measured by the *near-point-of-vision test,* which is the minimum distance from the eyes that an object can be maintained in focus. This distance increases with age, and indeed accommodation in almost everyone over the age of 45 is significantly impaired. Loss of accommodating ability with age is known as **presbyopia** (*presby* = old). This loss appears to have a number of causes, including thickening of the lens and a forward movement of the attachments of the zonular fibers to the lens. As a result of these changes, the zonular fibers and lens are pulled taut even when the ciliary muscle contracts. The lens is thus not able to thicken and increase its refraction when, for example, a printed page is brought close to the eyes.

when viewing an object that is 20 feet or more from a normal eye; the image is focused on the retina and the lens is in its most flat, least convex form. As the object moves closer to the eyes, the muscles of the ciliary body contract. This muscular contraction narrows the aperture of the ciliary body and thus reduces the tension on the zonular fibers that suspend the lens. When the tension is reduced, the lens becomes more rounded and convex as a result of its inherent elasticity (fig. 10.34).

Visual Acuity

Visual acuity refers to the sharpness of vision. The sharpness of an image depends on the *resolving power* of the visual system—that is, on the ability of the visual system to distinguish (resolve) two closely spaced dots. The better the resolving power of the system, the closer together these dots can be and still be seen as separate. When the resolving power of the system is exceeded, the dots blur and are perceived as a single image.

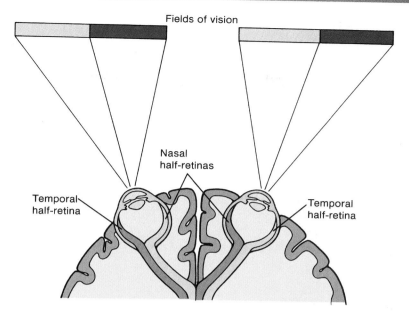

Fields of vision

Nasal half-retinas

Temporal half-retina

Temporal half-retina

Figure 10.32

Refraction of light in the cornea and lens produces a right-to-left image on the retina. The left side of the visual field is projected to the right half of each retina, while the right side of each visual field is projected to the left half of each retina.

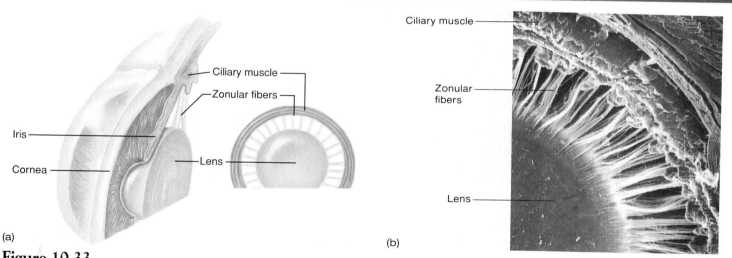

Ciliary muscle

Zonular fibers

Iris

Lens

Cornea

(a)

Ciliary muscle

Zonular fibers

Lens

(b)

Figure 10.33

(*a*) A diagram, and (*b*) scanning electron micrograph (from the eye of a 17-year-old-boy) showing the relationship between the lens, zonular fibers, and ciliary muscle of the eye.

(*Part [b] from "How the Eye Focuses" by James F. Koretz and George H. Handleman. Copyright © 1988 by Scientific American, Inc. All rights reserved.*)

Accommodation

When a normal eye views an object, parallel rays of light are refracted to a point, or *focus*, on the retina (see fig. 10.35). If the degree of refraction remained constant, movement of the object closer to or farther from the eye would cause corresponding movement of the focal point, so that the focus would either be behind or in front of the retina.

The ability of the eyes to keep the image focused on the retina as the distance between the eyes and object is changed is called **accommodation**. Accommodation results from contraction of the ciliary muscle, which is like a sphincter muscle that can vary its aperture (fig. 10.33). When the ciliary muscle is relaxed, its aperture is wide. Relaxation of the ciliary muscle thus places tension on the zonular fibers of the suspensory ligament and pulls the lens taut. These are the conditions that prevail

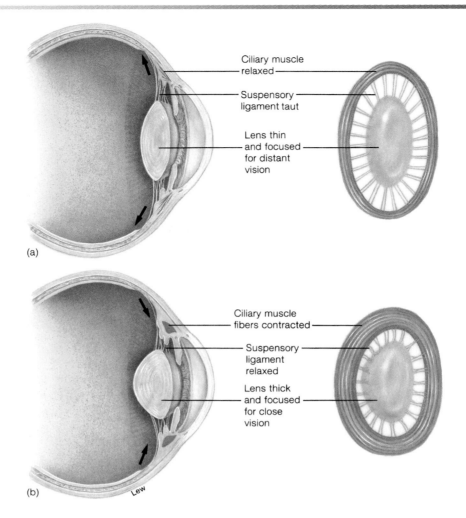

Figure 10.34

Changes in the shape of the lens during accommodation. (*a*) The lens is flattened for distant vision when the ciliary muscle fibers are relaxed and the suspensory ligament is taut. (*b*) The lens is more spherical for close-up vision when the ciliary muscle fibers are contracted and the suspensory ligament is relaxed.

 The ability of a person's eyes to accommodate can be measured by the *near-point-of-vision test,* which is the minimum distance from the eyes that an object can be maintained in focus. This distance increases with age, and indeed accommodation in almost everyone over the age of 45 is significantly impaired. Loss of accommodating ability with age is known as **presbyopia** (*presby* = old). This loss appears to have a number of causes, including thickening of the lens and a forward movement of the attachments of the zonular fibers to the lens. As a result of these changes, the zonular fibers and lens are pulled taut even when the ciliary muscle contracts. The lens is thus not able to thicken and increase its refraction when, for example, a printed page is brought close to the eyes.

when viewing an object that is 20 feet or more from a normal eye; the image is focused on the retina and the lens is in its most flat, least convex form. As the object moves closer to the eyes, the muscles of the ciliary body contract. This muscular contraction narrows the aperture of the ciliary body and thus reduces the tension on the zonular fibers that suspend the lens. When the tension is reduced, the lens becomes more rounded and convex as a result of its inherent elasticity (fig. 10.34).

Visual Acuity

Visual acuity refers to the sharpness of vision. The sharpness of an image depends on the *resolving power* of the visual system— that is, on the ability of the visual system to distinguish (resolve) two closely spaced dots. The better the resolving power of the system, the closer together these dots can be and still be seen as separate. When the resolving power of the system is exceeded, the dots blur and are perceived as a single image.

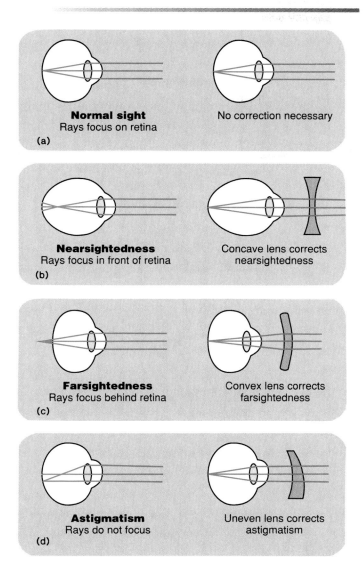

Figure 10.35

In a normal eye (*a*), parallel rays of light are brought to a focus on the retina by refraction in the cornea and lens. If the eye is too long, as in myopia (*b*), the focus is in front of the retina. This can be corrected by a concave lens. If the eye is too short, as in hyperopia (*c*), the focus is behind the retina. This is corrected by a convex lens. In astigmatism (*d*), light refraction is uneven due to an abnormal shape of the cornea or lens.

Myopia and Hyperopia

When a person with normal visual acuity stands 20 feet from a *Snellen eye chart* (so that accommodation is not a factor influencing acuity), the line of letters marked "20/20" can be read. If a person has **myopia** (nearsightedness), this line will appear blurred because the image will be brought to a focus in front of the retina. This is usually due to the fact that the eyeball is too long. Myopia is corrected by glasses with concave lenses that cause the light rays to diverge, so that the point of focus is farther from the lens and is thus pushed back to the retina (fig. 10.35).

If the eyeballs are too short, the line marked "20/20" will appear blurred because the focal length of the lens is longer than the distance to the retina. The image will thus be brought to a focus behind the retina, and the object will have to be placed farther from the eyes to be seen clearly. This condition is called **hyperopia** (farsightedness). Hyperopia is corrected by glasses with convex lenses that increase the convergence of light, so that the focus is brought closer to the lens and falls on the retina.

Astigmatism

Because the curvature of the cornea and lens is not perfectly symmetrical, light passing through some parts of these structures may be refracted to a different degree than light passing through other parts. When the asymmetry of the cornea and/or lens is significant, the person is said to have **astigmatism.** If a person with astigmatism views a circle of lines radiating from the center, like the spokes of a wheel, the image of these lines will not appear clear in all 360 degrees. The parts of the circle that appear blurred can thus be used to map the astigmatism. This condition is corrected by cylindrical lenses that compensate for the asymmetry in the cornea or lens of the eye.

---

1. Using a line diagram, explain why an inverse image is produced on the retina. Also explain how the image in one eye corresponds to the image in the other eye.

2. Using a line diagram, show how parallel rays of light are brought to a focus on the retina. Explain how this focus is maintained as the distance from the object to the eye is increased or decreased (that is, explain accommodation).

3. Explain why a blurred image is produced in each of the following conditions: presbyopia, myopia, hyperopia, and astigmatism.

The Retina

There are two types of photoreceptor neurons: rods and cones. Both receptor cell types contain pigment molecules that undergo dissociation in response to light, and it is this photochemical reaction that eventually results in the production of action potentials in the optic nerve. Rods provide black-and-white vision under conditions of low light intensities, whereas cones provide sharp color vision when light intensities are greater.

The retina consists of a pigment epithelium, photoreceptor neurons called *rods* and *cones,* and layers of other neurons.

The neural layers of the retina are actually a forward extension of the brain. In this sense the optic nerve can be considered a tract, and indeed the myelin sheaths of its fibers are derived from oligodendrocytes (like other CNS axons) rather than from Schwann cells.

Since the retina is an extension of the brain, the neural layers face outward, toward the incoming light. Light, therefore, must pass through several neural layers before striking the photoreceptors (fig. 10.36). The photoreceptors then synapse with other neurons, so that nerve impulses are conducted outward in the retina.

The outer layers of neurons that contribute axons to the optic nerve are called *ganglion cells*. These neurons receive synaptic input from *bipolar cells* underneath, which in turn receive input from rods and cones. In addition to the flow of information from photoreceptors to bipolar cells to ganglion cells, neurons called *horizontal cells* synapse with several photoreceptors (and possibly also with bipolar cells), and neurons called *amacrine cells* synapse with several ganglion cells.

Effect of Light on the Rods

The photoreceptors—rods and cones (fig. 10.37)—are activated when light produces a chemical change in molecules of pigment contained within the membranous lamellae of the outer segments of the receptor cells. Rods contain a purple pigment known as **rhodopsin.** The pigment appears purple (a combination of red and blue), because it transmits light in the red and blue regions of the spectrum, while absorbing light energy in the green region. The wavelength of light that is absorbed best—the *absorption maximum*—is about 500 nm (a green-colored light).

Cars and other objects that are green in color are seen more easily at night (when rods are used for vision) than are red objects. This is because red light is not well absorbed by rhodopsin, and only absorbed light can produce the photochemical reaction that results in vision. In response to absorbed light, rhodopsin dissociates into its two components: a pigment called **retinene,** derived from vitamin A, and a protein called **opsin.** This reaction is known as the *bleaching reaction*.

Retinene can exist in two possible configurations (shapes)—one known as the all-*trans* form and one called the 11-*cis* form (fig. 10.38). The all-*trans* form is more stable, but only the 11-*cis* form is found attached to opsin. In response to absorbed light energy, the 11-*cis* retinene is converted to the all-*trans* form, causing it to dissociate from the opsin. This dissociation reaction in response to light initiates changes in the ionic permeability of the rod cell membrane and ultimately results in the production of nerve impulses in the ganglion cells. As a result of these effects, rods provide black-and-white vision under conditions of low light intensity (as described in a later section).

Dark Adaptation

The bleaching reaction that occurs in the light results in a lowered amount of rhodopsin in the rods and lowered amounts of visual pigments in the cones. When a light-adapted person first enters a darkened room, therefore, sensitivity to light is low and vision is poor. A gradual increase in photoreceptor sensitivity,

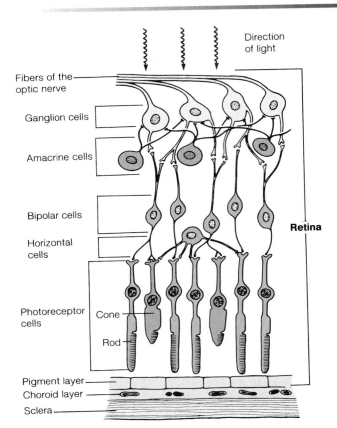

Figure 10.36

The layers of the retina. Since the retina is inverted, light must pass through various layers of nerve cells before reaching the photoreceptors (rods and cones).

 Scientists have recently discovered the genetic basis for blindness in the disease **dominant retinitis pigmentosa.** People with this disease inherit a gene for the opsin protein in which a single base change in the gene (substitution of adenine for cytosine) causes the amino acid histidine to be substituted for proline at a specific point in the polypeptide chain. This abnormal opsin leads to degeneration of the photoreceptors.

known as *dark adaptation*, then occurs, reaching maximal sensitivity in about 20 minutes. The increased sensitivity to low light intensity is due partly to increased amounts of visual pigments produced in the dark. Increased pigments in the cones produce a slight dark adaptation in the first 5 minutes. Increased rhodopsin in the rods produces a much greater increase in sensitivity to low light levels and is partly responsible for the adaptation that occurs after about 5 minutes in the dark. In addition to the increased concentration of rhodopsin, other more subtle (and less well understood) changes occur in the rods that ultimately result in a 100,000-fold increase in light sensitivity in dark-adapted as compared to light-adapted eyes.

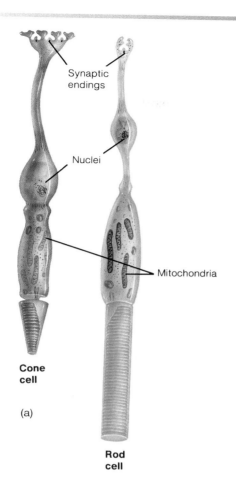

Synaptic endings

Nuclei

Mitochondria

Cone cell

(a)

(b)

Rod cell

Figure 10.37

(*a*) A diagram showing the structure of a rod and a cone. (*b*) A scanning electron micrograph of rods and cones.

Electrical Activity of Retinal Cells

The only neurons in the retina that produce all-or-none action potentials are ganglion cells and amacrine cells. The photoreceptors, bipolar cells, and horizontal cells instead produce only graded depolarizations or hyperpolarizations, analogous to EPSPs and IPSPs.

In the dark, the photoreceptors have a resting membrane potential that is less negative (closer to zero) than that of most other neurons. This is caused by a constant current of Na⁺ into the cell, called a **dark current,** through special Na⁺ channels. These Na⁺ channels are kept open by cyclic GMP (cGMP). When light is absorbed by rhodopsin, there is a conformational change in the opsin. The opsin is associated with G-proteins (chapter 7) known as *transducins*, and the conformational change in the opsin induced by light causes the alpha subunits of the G-proteins to dissociate. These G-protein subunits bind to and activate the enzyme phosphodiesterase, which converts cGMP to GMP. As a result of the decline in cGMP, the dark current is blocked and the photoreceptors become less depolarized than they are in the dark. Put another way, light causes the photoreceptors to become *hyperpolarized* in comparison to their membrane potential in the dark. Since light must ulti-

mately have a stimulatory effect on the optic nerve, this hyperpolarization (which is associated with inhibition—chapter 7) is certainly surprising.

The translation of the effect of light on photoreceptors to the production of nerve impulses may be explained in the following manner. Photoreceptors in the dark release a neurotransmitter chemical at a constant rate at their synapses with bipolar cells. Hyperpolarization of the photoreceptors in response to light decreases the release of neurotransmitter by the photoreceptors. If the neurotransmitter is inhibitory, the secretion of lower amounts of this chemical will stimulate the bipolar cells. These neurons in turn activate ganglion cells, and action potentials will thus be produced in fibers of the optic nerve in response to light.

Light can also inhibit the production of action potentials in ganglion cells. The purpose of this inhibition is to sharpen the image produced by those ganglion cells that are stimulated in response to the light (this is similar to the process of lateral inhibition discussed earlier in this chapter). The inhibition is achieved when light causes a decrease in the amount of excitatory neurotransmitter released in the dark by photoreceptors. The light-induced decrease in excitatory neurotransmitter inhibits the activity of the bipolar cells

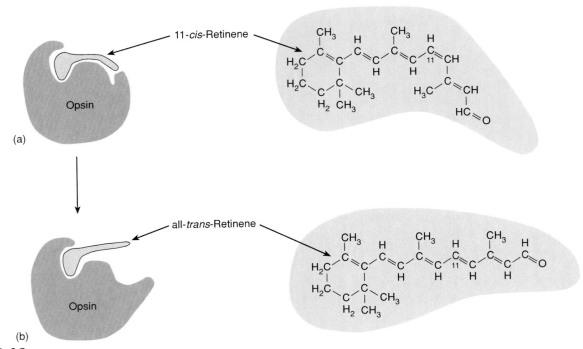

Figure 10.38

(a) The photopigment rhodopsin consists of the protein opsin combined with 11-*cis* retinene. (b) Upon exposure to light, the retinene is converted to a different form, called all-*trans*, and dissociates from the opsin. This photochemical reaction induces changes in ionic permeability that ultimately result in stimulation of ganglion cells in the retina.

and thus of the ganglion cells. These effects produce the characteristics of ganglion cell receptive fields—a topic discussed later in this chapter.

Cones and Color Vision

Cones are less sensitive than rods to light, but provide color vision and greater visual acuity, as described in the next section. During the day, therefore, the high light intensity bleaches out the rods, and color vision with high acuity is provided by the cones. According to the **trichromatic theory of color vision,** our perception of a multitude of colors is due to stimulation of only three types of cones. Each type of cone contains retinene, as in rhodopsin, but the retinene in the cones is associated with proteins other than opsin. The protein is different for each of the three cone pigments, and as a result, each of the pigments absorbs light of a given wavelength (color) to a different degree. The three types of cones are designated *blue*, *green*, and *red*, according to the region of the visible spectrum in which each cone pigment absorbs light maximally (fig. 10.39). Our perception of any given color is produced by the relative degree to which each cone is stimulated by any given wavelength of visible light.

Suppose a person has become dark adapted in a photographic darkroom over a period of 20 minutes or longer, but needs light to examine some prints. Since rods do not absorb

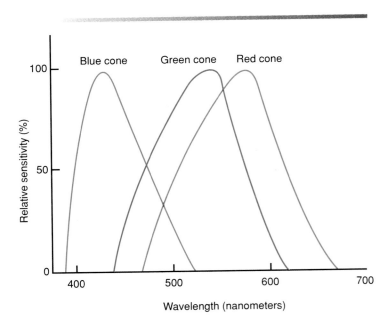

Figure 10.39

There are three types of cones. Each type contains retinene, but the protein with which the retinene is combined is different in each case. Thus, each different pigment absorbs light maximally at a different wavelength. Color vision is produced by the activity of these blue cones, green cones, and red cones.

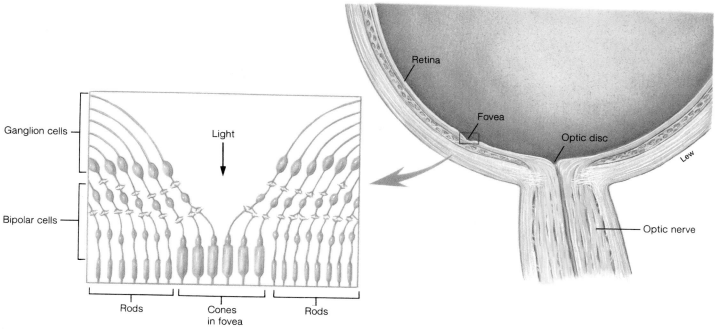

Figure 10.40

When the eyes "track" an object, the image is cast upon the fovea centralis of the retina. The fovea is literally a "pit" formed by parting of the neural layers, so that light falls directly on the photoreceptors (cones) in this region.

red light but red cones do, a red light in a photographic dark-room allows vision (because of the red cones), but does not cause bleaching of the rods. When the light is turned off, there-fore, the rods will still be dark adapted and the person will still be able to see.

Visual Acuity and Sensitivity

While reading or similarly focusing visual attention on objects in daylight, each eye is oriented so that the image falls within a tiny area of the retina called the **fovea centralis.** The fovea is a pinhead-sized pit (*fovea* = pit) within a yellow area of the retina called the *macula lutea*. The pit is formed as a result of the dis-placement of neural layers around the periphery; therefore light falls directly on photoreceptors in the center (fig. 10.40). Light falling on other areas by contrast, must pass through several lay-ers of neurons, as previously described.

There are approximately 120 million rods and 6 million cones in each retina, but only about 1.2 million nerve fibers enter the optic nerve of each eye. This gives an overall conver-gence ratio of photoreceptors on ganglion cells of about 105:1. This number is misleading, however, because the degree of con-vergence is much lower for cones than for rods, and it is 1:1 in the fovea.

The photoreceptors are distributed in such a way that the fovea contains only cones, whereas more peripheral regions of the retina contain a mixture of rods and cones. Approximately 4,000 cones in the fovea provide input to approximately 4,000 ganglion cells; each ganglion cell in this region, therefore, has

Color blindness is due to a congenital lack of one or more types of cones. People with nor-mal color vision are *trichromats;* people with only two types of cones are *dichromats*. They may be missing red cones (have *protanopia*), or green cones (have *deuteranopia*), or blue cones (have *tritanopia*). They may have difficulty, for exam-ple, distinguishing red from green. People who are *mono-chromats* have only one cone system and can only see black, white, and shades of gray. Color blindness is a trait carried on the X chromosome. Since men have only one X chromosome per cell, whereas women have two X chromo-somes (chapter 20), men are far more likely to be color blind than women (who can carry this trait in a recessive state).

a private line to the visual field. Each ganglion cell thus re-ceives input from an area of retina corresponding to the diame-ter of one cone (about 2 μm). Peripheral to the fovea, however, many rods synapse with a single bipolar cell, and many bipolar cells synapse with a single ganglion cell. A single ganglion cell outside the fovea thus may receive input from large numbers of rods, corresponding to an area of about 1 mm^2 on the retina (fig. 10.41).

Since each cone in the fovea has a private line to a gan-glion cell, and since each ganglion cell receives input from only a tiny region of the retina, visual acuity is greatest and sensitivity

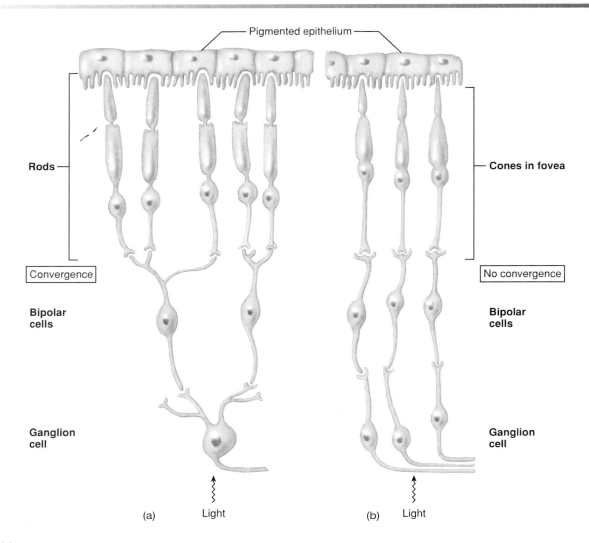

Pigmented epithelium

Rods

Convergence

Bipolar cells

Ganglion cell

(a)   Light

Cones in fovea

No convergence

Bipolar cells

Ganglion cell

(b)   Light

Figure 10.41

Since bipolar cells receive input from the convergence of many rods (*a*), and since a number of such bipolar cells converge on a single ganglion cell, rods maximize sensitivity to low levels of light at the expense of visual acuity. By contrast, the 1:1:1 ratio of cones to bipolar cells to ganglion cells in the fovea (*b*) provides high visual acuity, but sensitivity to light is decreased.

to low light is poorest when light falls on the fovea. In dim light only the rods are activated, and vision is best out of the corners of the eye when the image falls away from the fovea. Under these conditions, the convergence of large numbers of rods on a single bipolar cell and the convergence of large numbers of bipolar cells on a single ganglion cell increase sensitivity to dim light at the expense of visual acuity. Night vision is therefore less distinct than day vision.

The difference in visual sensitivity between cones in the fovea centralis and rods in the periphery of the retina can easily be demonstrated. If you go out on a clear night and stare hard at a very dim star, it will disappear. This is because the light falls on the fovea and is not sufficiently bright to activate the cones. If you then look slightly off to the side, the star will reappear because the light falls on the rods. Astronomers commonly practice this technique, which is known as averted vision. The

technique is also used by mountaineers who are forced by circumstances to hike under conditions of failing light.

Neural Pathways from the Retina

As a result of light refraction by the cornea and lens, the right half of the visual field is projected to the left half of the retina of both eyes (the temporal half of the left retina and the nasal half of the right retina). The left half of the visual field is projected to the right half of the retina of both eyes. The temporal half of the left retina and the nasal half of the right retina therefore see the same image. Axons from ganglion cells in the left (temporal) half of the left retina pass to the left **lateral geniculate body** of the thalamus. Axons from ganglion cells in the nasal half of the right retina cross (decussate) in the *optic chiasma*, also to synapse in the left lateral geniculate body. The

Table 10.5 Muscles of the Eye

Extrinsic Muscles (Striated)	Innervation	Action	Intrinsic Muscles (Smooth)	Innervation	Action
Superior rectus	Oculomotor nerve (III)	Rotates eye upward and toward midline	Ciliary muscles	Oculomotor nerve (III) parasympathetic fibers	Cause suspensory ligaments to relax
Inferior rectus	Oculomotor nerve (III)	Rotates eye downward and toward midline	Iris, circular muscles	Oculomotor nerve (III) parasympathetic fibers	Cause pupil to constrict
Medial rectus	Oculomotor nerve (III)	Rotates eye toward midline	Iris, radial muscles	Sympathetic fibers	Cause pupil to dilate
Lateral rectus	Abducens nerve (VI)	Rotates eye away from midline			
Superior oblique	Trochlear nerve (IV)	Rotates eye downward and away from midline			
Inferior oblique	Oculomotor nerve (III)	Rotates eye upward and away from midline			

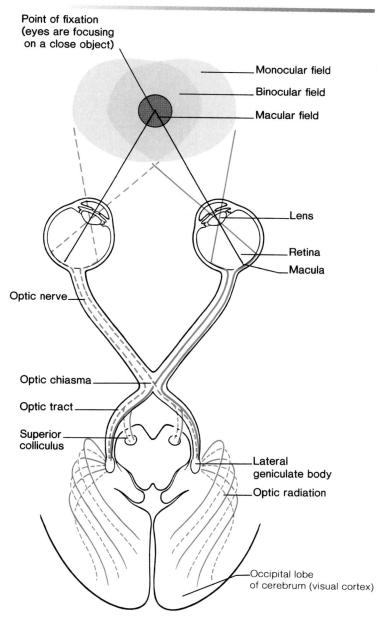

Point of fixation (eyes are focusing on a close object)

Monocular field

Binocular field

Macular field

Lens

Retina

Macula

Optic nerve

Optic chiasma

Optic tract

Superior colliculus

Lateral geniculate body

Optic radiation

Occipital lobe of cerebrum (visual cortex)

Figure 10.42

The neural pathway leading from the retina to the lateral geniculate body to the visual cortex. As a result of the crossing of optic fibers, the visual cortex of each cerebral hemisphere receives input from the opposite (contralateral) visual field.

left lateral geniculate, therefore, receives input from both eyes that relates to the right half of the visual field (fig. 10.42).

The right lateral geniculate body, similarly, receives input from both eyes relating to the left half of the visual field. Neurons in both lateral geniculate bodies of the thalamus in turn project to the **striate cortex** of the occipital lobe in the cerebral cortex (fig. 10.42). This area is also called area 17, in reference to a numbering system developed by K. Brodmann in 1906. Neurons in area 17 synapse with neurons in areas 18 and 19 of the occipital lobe (fig. 10.43).

Approximately 70% to 80% of the axons from the retina pass to the lateral geniculate bodies and to the striate cortex. This **geniculostriate system** is involved in perception of the visual field. Put another way, the geniculostriate system is needed to answer the question, What is it? Approximately 20% to 30% of the fibers from the retina, however, follow a different path to the *superior colliculus* of the midbrain (also called the *optic tectum*). Axons from the superior colliculus activate motor pathways leading to eye and body movements. The **tectal system,** in other words, is needed to answer the question, Where is it?

Superior Colliculus and Eye Movements

Neural pathways from the superior colliculus to motor neurons in the spinal cord help mediate the startle response to the sight of an unexpected intruder. Other nerve fibers from the superior colliculus stimulate the **extrinsic eye muscles** (table 10.5), which are the striated muscles that move the eyes.

There are two types of eye movements coordinated by the superior colliculus. *Smooth pursuit movements* track moving objects and keep the image focused on the fovea centralis. *Saccadic eye movements* are quick (lasting 20 to 50 msec), jerky movements that occur while the eyes appear to be still. These saccadic movements continuously move the image to different photoreceptors; if they were to stop, the image would disappear as the photoreceptors became bleached.

The tectal system is also involved in the control of the intrinsic eye muscles—the iris and the muscles of the ciliary body. Shining a light into one eye stimulates the *pupillary reflex* in which both pupils constrict. This is caused by activation of parasympathetic neurons by fibers from the superior colliculus. Postganglionic neurons in the ciliary ganglia behind the eyes, in turn, stimulate constrictor fibers in the iris. Contraction of the ciliary body during accommodation also involves stimulation of the superior colliculus.

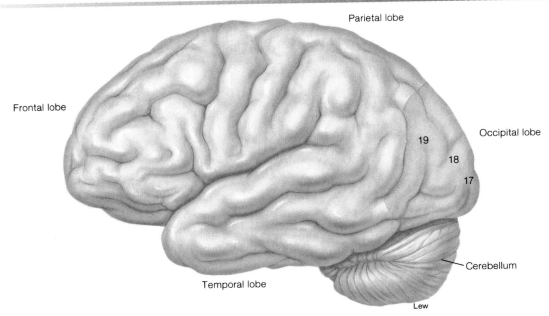

Figure 10.43

The striate cortex (area 17) and the visual association areas (18 and 19).

1. List the different layers of the retina and describe the path of light and of nerve activity through these layers.

2. Describe the photochemical reaction in the rods and explain how dark adaptation occurs.

3. Describe the electrical state of photoreceptors in the dark. Explain how light affects the electrical activity of retinal cells.

4. Explain what is meant by the trichromatic theory of color vision.

5. Compare the architecture of the fovea centralis with more peripheral regions of the retina. How does this architecture relate to visual acuity and sensitivity?

6. Describe how different parts of the visual field are projected on the retinas of both eyes. Trace the neural pathways of this information in the geniculostriate system.

7. Describe the neural pathways involved in the tectal system. What are the functions of these pathways?

Neural Processing of Visual Information

Electrical activity in ganglion cells of the retina, and in neurons of the lateral geniculate nucleus and cerebral cortex, is evoked in response to light on the retina. The way in which each neuron type responds to light at a particular point on the retina provides information about the way the brain interprets visual information.

Light that is cast on the retina directly affects the activity of photoreceptors and indirectly affects the neural activity in bipolar and ganglion cells. The part of the visual field that affects the activity of a particular ganglion cell can be considered to be its **receptive field.** Since each cone in the fovea has a private line to a ganglion cell, the receptive fields of these ganglion cells are equal to the width of one cone (about 2 μm). Ganglion cells in more peripheral parts of the retina receive input from hundreds of photoreceptors, and thus are influenced by a larger area of the retina (about 1 mm in diameter).

Ganglion Cell Receptive Fields

Studies of the electrical activity of ganglion cells have yielded some surprising results. In the dark, each ganglion cell discharges spontaneously at a slow rate. When the room lights are turned on, the firing rate of many (but not all) ganglion cells increases slightly. A small spot of light that is directed at the center of some ganglion cells' receptive fields, however, stimulates a large increase in firing rate. A small spot of light can thus be a more effective stimulus than larger areas of light.

When the spot of light is moved only a short distance away from the center of the receptive field, the ganglion cell responds in the opposite manner. The ganglion cell that was stimulated with light at the center of its receptive field is inhibited by light in the periphery of its field. The response produced by light in the center and by light in the "surround" of the visual field is *antagonistic*. Those ganglion cells that are stimulated by light at the center of their visual fields are said to have **on-center fields.** Ganglion cells that are inhibited by light in the center and stimulated by light in the surround have **off-center fields.**

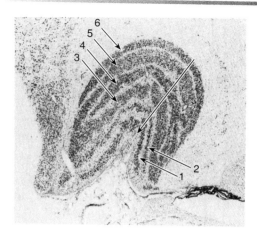

Figure 10.44

A scanning electron micrograph of the lateral geniculate nucleus. Each lateral geniculate consists of six layers (numbered 1 through 6 in this figure). Each of these layers receives input from only one eye, with right and left eyes alternating. For example, the long arrow through these six layers of the left lateral geniculate encounters corresponding projections from a part of the right visual field in right and left eyes, alternately, as it passes from the outer to the inner layers.

The reason wide illumination of the retina has less effect than pinpoint illumination is now clear; diffuse illumination gives the ganglion cell conflicting orders—on and off. Because of the antagonism between the center and surround of ganglion cell receptive fields, the activity of each ganglion is a result of the *difference in light intensity* between the center and surround of its visual field. This is a form of *lateral inhibition* that helps to accentuate the contours of images and improve visual acuity.

Lateral Geniculate Bodies

Each of the two lateral geniculate bodies receives input from ganglion cells in both eyes. The right lateral geniculate receives input from the right half of each retina (corresponding to the left half of the visual field); the left lateral geniculate receives input from the left half of each retina (corresponding to the right half of the visual field). Each neuron in the lateral geniculate, however, is activated only by input from one eye. Neurons that are activated by ganglion cells from the left eye are in separate layers within the lateral geniculate from those that are activated by the right eye (fig. 10.44).

The receptive field of each ganglion cell, as previously described, is the part of the retina it "sees" through its photoreceptor input. The receptive field of lateral geniculate neurons, similarly, is the part of the retina it "sees" through its ganglion cell input. Experiments in which the lateral geniculate receptive fields are mapped with a spot of light reveal that they are circular, with an antagonistic center and surround, much like the ganglion cell receptive fields.

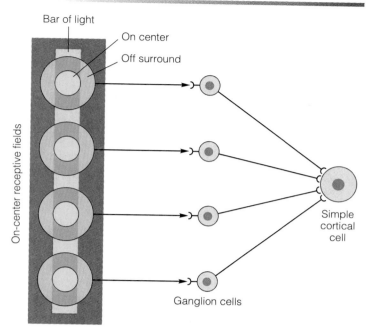

Figure 10.45

Cortical neurons described as simple cells have rectangular receptive fields that are best stimulated by slits of light of particular orientations. This may be due to the fact that these simple cells receive input from ganglion cells that have circular receptive fields along a particular line.

The Cerebral Cortex

Projections of nerve fibers from the lateral geniculate bodies to area 17 of the occipital lobe form the *optic radiation* (see fig. 10.42). Because these fiber projections give area 17 a striped or striated appearance, this area is also known as the striate cortex. Neurons in area 17, in turn, project to areas 18 and 19 of the occipital lobe. Cortical neurons in areas 17, 18, and 19 are thus stimulated indirectly by light on the retina. On the basis of their stimulus requirements, these cortical neurons are classified as simple, complex, and hypercomplex.

Simple Neurons

The receptive fields of simple cortical neurons are rectangular rather than circular. This results from the fact that they receive input from lateral geniculate neurons whose receptive fields are aligned in a particular way (as illustrated in fig. 10.45). Simple cortical neurons are best stimulated by a slit or bar of light that is located in a precise part of the visual field (of either eye) at a precise orientation (fig. 10.46).

The striate cortex (area 17) contains simple, complex, and hypercomplex neurons. The other visual association areas, designated areas 18 and 19, contain only complex and hypercomplex cells. Complex neurons receive input from simple cells, and hypercomplex neurons receive input from complex cells.

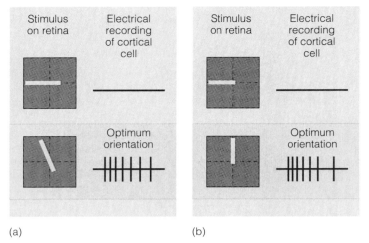

(a)   (b)

Figure 10.46

Simple cells are best stimulated by a slit or bar of light along a particular orientation within a particular region of the receptive field. The behavior of two different cortical cells is illustrated in (a) and (b).

Source: Data from Gunther Stent, "Cellular Communication" in *Scientific American*, 1972.

Complex and Hypercomplex Neurons

Complex neurons respond best to straight lines with a specific orientation that move in a particular direction through the receptive field. Unlike simple neurons, complex neurons do not require that the stimulus have a particular position within the receptive field. Hypercomplex neurons require that the stimulus be of a particular length or have a particular bend or corner.

The dotlike information from ganglion and lateral geniculate cells is thus transformed in the occipital lobe into information about edges—their position, length, orientation, and movement. Although this information is highly abstract, the visual association areas of the occipital lobe probably represent only an early stage in the integration of visual information. Other areas of the brain receive input from the visual association areas and provide meaning to visual perception.

---

1. Describe the response characteristics of ganglion cells to light on the retina. Why is a small spot of light a more effective stimulus than general illumination of the retina?

2. Explain how the arrangement of ganglion cells' receptive fields can enhance visual acuity.

3. Describe the stimulus requirements of simple, complex, and hypercomplex cortical neurons.

Summary

Characteristics of Sensory Receptors   p. 228

I. Sensory receptors may be categorized on the basis of their structure, the stimulus energy they transduce, or the nature of their response.
   A. Receptors may be dendritic nerve endings, specialized neurons, or specialized epithelial cells associated with sensory nerve endings.
   B. Receptors may be chemoreceptors, photoreceptors, thermoreceptors, mechanoreceptors, or nociceptors.
      1. Proprioceptors include receptors in the muscles, tendons, and joints.
      2. The senses of sight, hearing, taste, olfaction, and equilibrium are grouped as special senses.

   C. Tonic receptors fire continuously as long as the stimulus is applied; they monitor the presence and intensity of a stimulus.
      1. Phasic receptors respond to stimulus changes; they do not respond to a sustained stimulus.
      2. Phasic receptors therefore help to provide sensory adaptation to sustained stimuli.

II. According to the law of specific nerve energies, each sensory receptor responds with lowest threshold to only one modality of sensation.
   A. That stimulus modality is called the adequate stimulus.
   B. Stimulation of the sensory nerve from a receptor by any means is interpreted in the brain as the adequate stimulus modality of that receptor.

III. Generator potentials are graded changes (usually depolarizations) in the membrane potential of the dendritic endings of sensory neurons.
   A. The magnitude of the potential change of the generator potential is directly proportional to the strength of the stimulus applied to the receptor.
   B. After the generator potential reaches a threshold value, increases in the magnitude of the depolarization result in increased frequency of action potential production in the sensory neuron.

Cutaneous Sensations   p. 231

I. Somatesthetic information—from cutaneous receptors and proprioceptors—project by third-order neurons to the postcentral gyrus of the cerebrum.
   A. Proprioception and pressure sensations ascend on the

ipsilateral side of the spinal cord, synapse in the medulla and cross to the contralateral side, and ascend in the medial lemniscus to the thalamus; neurons from the thalamus in turn project to the postcentral gyrus.

B. Sensory neurons from other cutaneous receptors synapse and cross to the contralateral side in the spinal cord and ascend in the lateral and ventral spinothalamic tracts to the thalamus; neurons in the thalamus then project to the postcentral gyrus.

II. The receptive field of a cutaneous sensory neuron is the area of skin that, when stimulated, produces responses in the neuron.

A. The receptive fields are smaller where the skin has a greater density of cutaneous receptors.

B. The two-point touch threshold test reveals that the fingertips and tip of the tongue have a greater density of touch receptors and thus greater sensory acuity than other areas of the body.

III. Lateral inhibition acts to sharpen the sensation by inhibiting the activity of sensory neurons coming from areas of the skin around the area that is most greatly stimulated.

Taste and Olfaction p. 234

I. The sense of taste is mediated by taste buds.

A. A particular taste bud is most sensitive to one of four taste modalities: sweet, sour, bitter, and salty.

B. Taste buds are located in characteristic regions of the tongue according to the modality to which they are most sensitive.

C. Salty and sour taste are produced by the movement of Na^+ and H^+ ions, respectively, through membrane channels; sweet and bitter tastes are produced by binding of molecules to protein receptors that are coupled to G-proteins.

II. The olfactory receptors are neurons that synapse with the olfactory bulb of the brain.

A. Odorant molecules bind to membrane protein receptors. There may be as many as 1,000 different receptor proteins responsible for the ability to detect as many as 10,000 different odors.

B. Binding of an odorant molecule to its receptor causes the dissociation of large numbers of G-protein subunits, which amplify the signal and contribute to the extreme sensitivity of the sense of smell.

Vestibular Apparatus and Equilibrium p. 237

I. The structures for equilibrium and hearing are located in the inner ear in the membranous labyrinth.

A. The structure involved in equilibrium, known as the vestibular apparatus, consists of the otolith organs (utricle and saccule) and the semicircular canals.

B. The utricle and saccule provide information about linear acceleration, whereas the semicircular canals provide information about angular acceleration.

C. The sensory receptors for equilibrium are hair cells that support numerous stereocilia and one kinocilium.

1. When the stereocilia are bent in the direction of the kinocilium, the cell membrane becomes depolarized.

2. When the stereocilia are bent in the opposite direction, the membrane becomes hyperpolarized.

II. The stereocilia of the hair cells in the utricle and saccule protrude into the endolymph of the membranous labyrinth and are embedded within a gelatinous otolith membrane.

A. When the person is upright, the stereocilia of the utricle are oriented vertically and those of the saccule are oriented horizontally.

B. Linear acceleration causes a shearing force between the hairs of the otolith membrane, thus bending the stereocilia and electrically stimulating the sensory endings.

III. The three semicircular canals are oriented at right angles to each other, like the faces of a cube.

A. The hair cells are embedded within a gelatinous membrane called the cupula, which projects into the endolymph.

B. Movement along one of the planes of a semicircular canal causes the endolymph to bend the cupula and stimulate the hair cells.

C. Stimulation of the hair cells in the vestibular apparatus activates the sensory neurons of the vestibulocochlear (eighth cranial) nerve, which projects to the cerebellum and the vestibular nuclei of the medulla oblongata.

1. The vestibular nuclei in turn send fibers to the oculomotor center, which controls eye movements.

2. Spinning and then stopping abruptly can thus cause oscillatory movements of the eyes (nystagmus).

The Ears and Hearing p. 241

I. The outer ear funnels sound waves of a given frequency (measured in hertz) and intensity (measured in decibels) to the tympanic membrane, causing it to vibrate.

II. Vibrations of the tympanic membrane cause movement of the middle-ear ossicles—malleus, incus, and stapes—which in turn produces vibrations of the oval window of the cochlea.

III. Vibrations of the oval window set up a traveling wave of perilymph in the scala vestibuli.

A. This wave can pass around the helicotrema to the scala tympani, or it can reach the scala tympani by passing through the scala media (cochlear duct).

B. The scala media is filled with endolymph.
  1. The membrane of the cochlear duct that faces the scala vestibuli is called the vestibular membrane.
  2. The membrane that faces the scala tympani is called the basilar membrane.

IV. The sensory structure of the cochlea is called the organ of Corti.
  A. The organ of Corti consists of sensory hair cells on the basilar membrane.
    1. The stereocilia of the hair cells project upward into an overhanging tectorial membrane.
    2. The hair cells are innervated by the vestibulocochlear (eighth cranial) nerve.
  B. Sounds of high frequency cause maximum displacement of the basilar membrane closer to its base, near the stapes; sounds of lower frequency produce maximum displacement of the basilar membrane closer to its apex, near the helicotrema.
    1. Displacement of the basilar membrane causes the hairs to bend against the tectorial membrane and stimulate the production of nerve impulses.
    2. Pitch discrimination is thus dependent on the region of the basilar membrane that vibrates maximally to sounds of different frequencies.
    3. Pitch discrimination is enhanced by lateral inhibition.

The Eyes and Vision p. 248

I. Light enters the cornea of the eye, passes through the pupil (the opening of the iris), and then through the lens, from which it is projected to the retina in the back of the eye.

A. Light rays are bent, or refracted, by the cornea and lens.
B. Because of refraction, the image on the retina is upside down and right to left.
C. The right half of the visual field is projected to the left half of the retina in each eye, and vice versa.

II. Accommodation is the ability to maintain a focus on the retina as the distance between the object and the eyes is changed.
  A. Accommodation is achieved by changes in the shape and refractive power of the lens.
  B. When the muscles of the ciliary body are relaxed, the suspensory ligament is tight and the lens is pulled to its least convex form.
    1. This gives the lens a low refractive power for distance vision.
    2. As an object is brought closer than 20 feet from the eyes, the ciliary body contracts, the suspensory ligament becomes less tight, and the lens becomes more convex and more powerful.

III. Visual acuity refers to the sharpness of the image and depends in part on the ability of the lens to bring the image to a focus on the retina.
  A. People with myopia have an eyeball that is too long, so that the image is brought to a focus in front of the retina; this is corrected by a concave lens.
  B. People with hyperopia have an eyeball that is too short, so that the image is brought to focus behind the retina; this is corrected by a convex lens.
  C. Astigmatism is the uneven refraction of light around 360 degrees of a circle, caused by uneven curvature of the cornea and/or lens.

The Retina p. 255

I. The retina contains photoreceptor neurons called rods and cones, which synapse with bipolar cells.

A. When light strikes the rods, it causes the photodissociation of rhodopsin into retinene and opsin.
    1. This bleaching reaction occurs maximally with a light wavelength of 500 nm.
    2. Photodissociation is caused by the conversion of the 11-*cis* to the all-*trans* form of retinene, which cannot bind to opsin.
  B. In the dark, more rhodopsin can be produced, and increased rhodopsin in the rods makes the eyes more sensitive to light. The increased concentration of rhodopsin in the rods is partly responsible for dark adaptation.
  C. The rods provide black-and-white vision under conditions of low light intensity; at higher light intensity the rods are bleached out and the cones provide color vision.

II. In the dark, there is a constant movement of Na^+ into the rods that produces what is known as a "dark current."
  A. When light causes the dissociation of rhodopsin, the Na^+ channels become blocked and the rods become hyperpolarized in comparison to their membrane potential in the dark.
  B. When the rods are hyperpolarized, they release less neurotransmitter at their synapses with bipolar cells.
  C. Neurotransmitters from rods cause depolarization of bipolar cells in some cases, and hyperpolarization of bipolar cells in other cases; so when the rods are in light and release less neurotransmitter, these effects are inverted.

III. According to the trichromatic theory of color vision, there are three systems of cones, each of which responds to one of three colors: red, blue, or green.
  A. Each type of cone contains retinene attached to a different type of protein.

B. The names for the cones signify the region of the spectrum in which the cones absorb light maximally.

IV. The fovea centralis contains only cones; more peripheral parts of the retina contain both cones and rods.

A. Each cone in the fovea synapses with one bipolar cell, which in turn synapses with one ganglion cell.

1. The ganglion cell that receives input from the fovea thus has a visual field limited to that part of the retina that activated its cone.

2. As a result of this 1:1 ratio of cones to bipolar cells, visual acuity is high in the fovea but sensitivity to low light levels is lower than in other regions of the retina.

B. In other regions of the retina, where rods predominate, large numbers of rods provide input to each ganglion cell (there is a great convergence). As a result, visual acuity is impaired, but sensitivity to low light levels is improved.

V. The right half of the visual field is projected to the left half of the retina of each eye.

A. The left half of the left retina sends fibers to the left lateral geniculate body of the thalamus.

B. The left half of the right retina also sends fibers to the left lateral geniculate body; this is because these fibers decussate in the optic chiasma.

C. The left lateral geniculate body thus receives input from the left half of the retina of both eyes, corresponding to the right half of the visual field; the right lateral geniculate receives information about the left half of the visual field.

1. Neurons in the lateral geniculate bodies send fibers to the striate cortex of the occipital lobes.

2. The geniculostriate system is involved in providing meaning to the images that form on the retina.

D. Instead of synapsing in the geniculate bodies, some fibers from the ganglion cells of the retina synapse in the superior colliculus of the midbrain, which controls eye movement.

1. Since this brain region is also called the optic tectum, this pathway is called the tectal system.

2. The tectal system enables the eyes to move and track an object; it is also responsible for the pupillary reflex and the changes in lens shape that are needed for accommodation.

Neural Processing of Visual Information p. 262

I. The area of the retina that provides input to a ganglion cell is called the receptive field of the ganglion cell.

A. The receptive field of a ganglion cell is roughly circular, with an "on" or "off" center and an antagonistic surround.

1. A spot of light in the center of an "on" receptive field stimulates the ganglion cell; a spot of light in its surround inhibits the ganglion cell.

2. The opposite is true for ganglion cells with "off" receptive cells.

3. Wide illumination that stimulates both the center and the surround of a receptive field affects a ganglion cell to a lesser degree than a pinpoint of light that only illuminates either the center or the surround.

B. The antagonistic center and surround of the receptive field of ganglion cells provide lateral inhibition, which enhances contours and provides better visual acuity.

II. Each lateral geniculate body receives input from both eyes relating to the same part of the visual field.

A. The neurons receiving input from each eye are arranged in layers within the lateral geniculate.

B. The receptive fields of neurons in the lateral geniculate are circular with an antagonistic center and surround, much like the receptive field of ganglion cells.

III. Cortical neurons involved in vision are either simple, complex, or hypercomplex.

A. Simple neurons receive input from neurons in the lateral geniculate; complex neurons receive input from simple cells; and hypercomplex neurons receive input from complex cells.

B. Simple neurons are best stimulated by a slit or bar of light that is located in a precise part of the visual field and that has a precise orientation.

C. Complex cells respond best to a straight line that has a particular orientation and that moves in a particular direction; the position of the line in the visual field is not important.

D. Hypercomplex cells respond best to lines that have a particular length or a particular bend or corner.

Clinical Investigation

During a routine eye exam, a 45-year-old woman mentions to her optometrist that her glasses no longer allow her to see small print, although they do help her see street signs and other distant objects while driving. This problem, she adds, began about a year ago, but in the last few months has gotten worse. The optometrist recommends bifocals and also tests her eyes with a device that blows a puff of air on the cornea to measure the intraocular pressure. The results of this test are normal. What can you conclude about this woman's vision?

Clues

Examine the section "The Eyes and Vision." In particular, study the sections on refraction and accommodation.

Review Activities

Objective Questions

Match the vestibular organ on the left with its correct component on the right.

1. utricle and saccule
2. semicircular canals
3. cochlea

   a. cupula
   b. ciliary body
   c. basilar membrane
   d. otolith membrane

4. The dissociation of rhodopsin in the rods in response to light causes
   a. the Na^+ channels to become blocked.
   b. the rods to secrete less neurotransmitter.
   c. the bipolar cells to become either stimulated or inhibited.
   d. all of the above.
5. Tonic receptors
   a. are fast-adapting.
   b. do not fire continuously to a sustained stimulus.
   c. produce action potentials at a greater frequency as the generator potential is increased.
   d. are described by all of the above.
6. Cutaneous receptive fields are smallest in
   a. the fingertips.
   b. the back.
   c. the thighs.
   d. the arms.
7. The process of lateral inhibition
   a. increases the sensitivity of receptors.
   b. promotes sensory adaptation.
   c. increases sensory acuity.
   d. prevents adjacent receptors from being stimulated.

8. The receptors for taste are
   a. naked sensory nerve endings.
   b. encapsulated sensory nerve endings.
   c. modified epithelial cells.
9. Which of the following statements about the utricle and saccule are *true?*
   a. They are otolith organs.
   b. They are located in the middle ear.
   c. They provide a sense of linear acceleration.
   d. Both *a* and *c* are true.
   e. Both *b* and *c* are true.
10. Since fibers of the optic nerve that originate in the nasal halves of each retina cross at the optic chiasma, each lateral geniculate receives input from
    a. both the right and left sides of the visual field of both eyes.
    b. the ipsilateral visual field of both eyes.
    c. the contralateral visual field of both eyes.
    d. the ipsilateral field of one eye and the contralateral field of the other eye.
11. When a person with normal vision views an object from a distance of at least 20 feet,
    a. the ciliary muscles are relaxed.
    b. the suspensory ligament is tight.
    c. the lens is in its most flat, least convex shape.
    d. all of the above apply.
12. Glasses with concave lenses help correct
    a. presbyopia.
    b. myopia.
    c. hyperopia.
    d. astigmatism.

13. Parasympathetic nerves that stimulate constriction of the iris (in the pupillary reflex) are activated by neurons in
    a. the lateral geniculate.
    b. the superior colliculus.
    c. the inferior colliculus.
    d. the striate cortex.
14. A bar of light in a specific part of the retina, with a particular length and orientation, is the most effective stimulus for
    a. ganglion cells.
    b. lateral geniculate cells.
    c. simple cortical cells.
    d. complex cortical cells.
15. The ability of the lens to increase its curvature and maintain a focus at close distances is called
    a. convergence.
    b. accommodation.
    c. astigmatism.
    d. ambylopia.
16. Which of the following sensory modalities is transmitted directly to the cerebral cortex without being relayed through the thalamus?
    a. taste
    b. sight
    c. smell
    d. hearing
    e. touch
17. Stimulation of membrane protein receptors by binding to specific molecules is *not* responsible for
    a. the sense of smell.
    b. the sense of sweet taste.
    c. the sense of sour taste.
    d. the sense of bitter taste.
18. Epithelial cells release transmitter chemicals that excite sensory neurons in all of the following senses *except*
    a. taste.
    b. smell.
    c. equilibrium.
    d. hearing.

Essay Questions

1. Define the term *lateral inhibition* and give examples of its effects in three sensory systems.[1]

2. Describe the nature of the generator potential and explain its relationship to stimulus intensity and to frequency of action potential production.

3. Describe the phantom limb phenomenon and give a possible explanation for its occurrence.

4. Explain the relationship between smell and taste. How are these senses similar? How do they differ?

5. Explain how the vestibular apparatus provides information about changes in the position of our body in space.

6. Define *accommodation* and explain how it is accomplished. Why is it more of a strain on the eyes to look at a small object close to the eyes than large objects far away?

7. Describe the effects of light on the photoreceptors and explain how these effects influence the bipolar cells.

8. Explain why images that fall on the fovea centralis are seen more clearly than images that fall on the periphery of the retina. Why are the "corners of the eyes" more sensitive to light than the fovea?

9. Explain why rods provide only black-and-white vision. Include a discussion of different types of color blindness in your answer.

10. Explain why green-colored objects can be seen better at night than objects of other colors. What effect does red light in a darkroom have on a dark-adapted eye?

11. Describe the receptive fields of ganglion cells and explain how the nature of these fields helps to improve visual acuity.

12. How many genes code for the sense of color vision? How many for taste? How many for smell? What does this information say about the level of integration required by the brain for the perception of these senses?

Selected Readings

Barinaga, M. 1991. How the nose knows: Olfactory receptors cloned. *Science* 252:209.

Borg, E., and S. A. Counter. August 1989. The middle-ear muscles. *Scientific American.*

Boynton, R. M. 1979. *Human Color Vision.* New York: Holt, Rinehart & Winston.

Cervero, F., and J. M. A. Laird. 1991. One pain or many pains? A new look at pain mechanisms. *News in Physiological Sciences* 6:268.

Fireman, P. 1987. Newer concepts in otitis media. *Hospital Practice* 23:85.

Freedman, D. H. June 1993. In the realm of the chemical. *Discover* p. 69.

Freeman, W. J. February 1991. The physiology of perception. *Scientific American.*

Glickstein, M. September 1988. The discovery of the visual cortex. *Scientific American.*

Goldberg, J. M., and C. Fernandez. 1975. Vestibular mechanisms. *Annual Review of Physiology* 37:129.

Hamill, O. P., and D. W. McBride Jr. 1994. Mechanoreceptive membrane channels. *American Scientist* 83:30.

Hubel, D. H. 1979. The visual cortex of normal and deprived monkeys. *American Scientist* 67:532.

Hubel, D. H., and T. Wiesel. September 1979. Brain mechanisms of vision. *Scientific American.*

Hudspeth, A. J. February 1983. The hair cells of the inner ear. *Scientific American.*

Hudspeth, A. J. 1989. How the ear's works work. *Nature* 341:397.

Koretz, J. F., and G. H. Handelman. July 1988. How the human eye focuses. *Scientific American.*

Loeb, G. E. February 1985. The functional replacement of the ear. *Scientific American.*

MacNichol, E. F., Jr. December 1964. Three pigment color vision. *Scientific American.*

Melzack, R. April 1992. Phantom limbs. *Scientific American.*

Nathans, J. February 1989. The genes for color vision. *Scientific American.*

O'Brian, D. F. 1982. The chemistry of vision. *Science* 218:961.

Parker, D. E. November 1980. The vestibular apparatus. *Scientific American.*

Pfaffmann, C., M. Frank, and R. Norgren. 1979. Neural mechanisms and behavioral aspects of taste. *Annual Review of Physiology* 30:283.

Rhode, W. S. 1984. Cochlear mechanics. *Annual Review of Physiology* 46:231.

Rushton, W. A. H. March 1975. Visual pigments and color blindness. *Scientific American.*

Sackin, H. 1995. Review of mechanosensitive channels. *Annual Review of Physiology* 57:333.

Shreeve, J. June 1993. Touching the phantom. *Discover,* p. 35.

Van Essen, D. C. 1979. Visual areas of the mammalian cerebral cortex. *Annual Review of Neurosciences* 2:277.

Ye, Q., G. L. Heck, and J. A. DeSimone. 1991. The anion paradox in sodium taste reception: Resolution by voltage-clamp studies. *Science* 254:724.

Life Science Animations

The animations that relate to chapter 10 are #26 Organ of Static Equilibrium and #27 The Organ of Corti.

---

[1]Note: This question is answered on pages 112 of the Student Study Guide.

11

Endocrine Glands

Secretion and Action of Hormones

OBJECTIVES   *After studying this chapter, you should be able to . . .*

1. define the terms *hormone* and *endocrine gland* and describe how chemical transformations in the endocrine gland or target cells can activate certain hormones.

2. list the general chemical categories of hormones and give examples of hormones within each category.

3. explain how different hormones can exert synergistic, permissive, or antagonistic effects.

4. explain how the concentrations of a hormone in the blood are regulated and how the effects of a hormone are influenced by its concentration.

5. describe the mechanisms of hormone action for steroid and thyroid hormones.

6. describe the mechanism of hormone action when cAMP is used as a second messenger.

7. describe the mechanism of hormone action when Ca^{++} is used as a second messenger.

8. describe the structure of the pituitary gland and the functional relationship between the pituitary and the hypothalamus.

9. list the hormones secreted by the posterior pituitary, state the origin of these hormones, and explain how the hypothalamus regulates their secretion.

10. list the hormones of the anterior pituitary and explain how their secretion is regulated by the hypothalamus.

11. describe the production and actions of the thyroid hormones and explain how thyroid secretion is regulated.

12. describe the location of the parathyroid glands and explain the actions of PTH and the regulation of its secretion.

13. describe the types and actions of corticosteroids and explain how the secretions of the adrenal cortex are regulated.

14. describe the actions of epinephrine and norepinephrine and explain how the secretions of the adrenal medulla are regulated.

15. explain why the pancreas is both an exocrine and an endocrine gland and describe the structure and functions of the islets of Langerhans.

16. describe the actions of insulin and glucagon and explain the regulation of their secretion.

17. list the hormones secreted by the pineal gland and thymus and explain the significance of these hormones in general terms.

18. list the hormones secreted by the gonads and placenta.

19. describe autocrine and paracrine regulation with reference to the blood vessels and the immune system.

20. describe the chemical nature and physiological roles of the prostaglandins and explain how the nonsteroidal anti-inflammatory drugs work.

OUTLINE

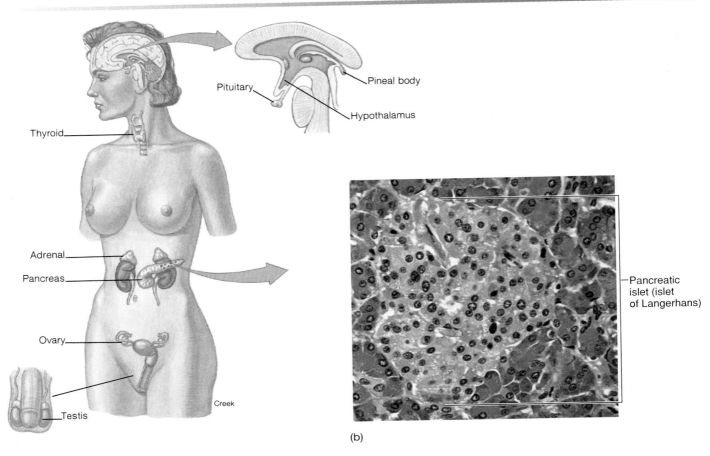

(a)

(b)

Figure 11.1

(*a*) The anatomy of some of the endocrine glands. (*b*) A pancreatic islet (islet of Langerhans) within the pancreas.

Endocrine Glands and Hormones

Hormones are regulatory molecules secreted into the blood by endocrine glands. Chemical categories of hormones include steroids, catecholamines, and polypeptides. Interactions between the various hormones produce effects described as synergistic, permissive, or antagonistic.

Endocrine glands lack the ducts present in exocrine glands (chapter 1). The endocrine glands secrete their products, which are biologically active chemicals called **hormones,** directly into the blood. The blood carries the hormones to target organs that respond in a specific fashion to them. Many endocrine glands are discrete organs (fig. 11.1*a*) whose primary functions are the production and secretion of hormones. The pancreas functions as both an exocrine and an endocrine gland; the endocrine portion of the pancreas is composed of microscopic structures called the islets of Langerhans (fig. 11.1*b*). The concept of the **endocrine system,** however, must be extended beyond these organs. In recent years, it has been discovered that

many other organs in the body secrete hormones. When these hormones can be demonstrated to have significant physiological functions, the organs that produce them may be categorized as endocrine glands, although they serve other functions as well. It is appropriate, then, that a partial list of the endocrine glands (table 11.1) should include the heart, liver, hypothalamus, and kidneys.

Hormones affect the metabolism of their target organs and, by this means, help to regulate total body metabolism, growth, and reproduction. The effects of hormones on body metabolism and growth are discussed in chapter 19; the regulation of reproductive functions by hormones is included in chapter 20.

Chemical Classification of Hormones

Hormones secreted by different endocrine glands are diverse in chemical structure. All hormones, however, can be grouped into three general chemical categories: (1) **catecholamines** (epinephrine and norepinephrine—see chapter 9, fig. 9.9); (2) **polypeptides** and **glycoproteins,** which include shorter chain polypeptides such as antidiuretic hormone and insulin and large glycoproteins such as thyroid-stimulating hormone (table 11.2); and (3) **steroids,** such as cortisol and testosterone.

Table 11.1 A Partial Listing of the Endocrine Glands

Endocrine Gland	Major Hormones	Primary Target Organs	Primary Effects
Adrenal cortex	Glucocorticoids Aldosterone	Liver, muscles Kidneys	Glucocorticoids influence glucose metabolism; aldosterone promotes Na^+ retention, K^+ excretion
Adrenal medulla	Epinephrine	Heart, bronchioles, blood vessels	Causes adrenergic stimulation
Heart	Atrial natriuretic hormone	Kidneys	Promotes excretion of Na^+ in the urine
Hypothalamus	Releasing and inhibiting hormones	Anterior pituitary	Regulates secretion of anterior pituitary hormones
Small intestine	Secretin and cholecystokinin	Stomach, liver, and pancreas	Inhibits gastric motility and stimulates bile and pancreatic juice secretion
Islets of Langerhans (pancreas)	Insulin Glucagon	Many organs Liver and adipose tissue	Insulin promotes cellular uptake of glucose and formation of glycogen and fat; glucagon stimulates hydrolysis of glycogen and fat
Kidneys	Erythropoietin	Bone marrow	Stimulates red blood cell production
Liver	Somatomedins	Cartilage	Stimulates cell division and growth
Ovaries	Estradiol-17β and progesterone	Female reproductive tract and mammary glands	Maintains structure of reproductive tract and promotes secondary sex characteristics
Parathyroid glands	Parathyroid hormone	Bone, small intestine, and kidneys	Increases Ca^{++} concentration in blood
Pineal gland	Melatonin	Hypothalamus and anterior pituitary	Affects secretion of gonadotrophic hormones
Pituitary, anterior	Trophic hormones	Endocrine glands and other organs	Stimulates growth and development of target organs; stimulates secretion of other hormones
Pituitary, posterior	Antidiuretic hormone Oxytocin	Kidneys, blood vessels Uterus, mammary glands	Antidiuretic hormone promotes water retention and vasoconstriction; oxytocin stimulates contraction of uterus and mammary secretory units
Skin	1,25-Dihydroxyvitamin D_3	Small intestine	Stimulates absorption of Ca^{++}
Stomach	Gastrin	Stomach	Stimulates acid secretion
Testes	Testosterone	Prostate, seminal vesicles, other organs	Stimulates secondary sexual development
Thymus	Thymosin	Lymph nodes	Stimulates white blood cell production
Thyroid gland	Thyroxine (T_4) and triiodothyronine (T_3)	Most organs	Promotes growth and development and stimulates basal rate of cell respiration (basal metabolic rate or BMR)

Table 11.2 Examples of Polypeptide and Glycoprotein Hormones

Hormone	Structure	Gland	Primary Effects
Antidiuretic hormone	8 amino acids	Posterior pituitary	Water retention and vasoconstriction
Oxytocin	8 amino acids	Posterior pituitary	Uterine and mammary contraction
Insulin	21 and 30 amino acids (double chain)	β cells in islets of Langerhans	Cellular glucose uptake, lipogenesis, and glycogenesis
Glucagon	29 amino acids	α cells in islets of Langerhans	Hydrolysis of stored glycogen and fat
ACTH	39 amino acids	Anterior pituitary	Stimulation of adrenal cortex
Parathyroid hormone	84 amino acids	Parathyroid	Increase in blood Ca^{++} concentration
FSH, LH, TSH	Glycoproteins	Anterior pituitary	Stimulation of growth, development, and secretion of target glands

Steroid hormones, which are derived from cholesterol (fig. 11.2), are lipids and thus are not water-soluble. The gonads—testes and ovaries—secrete *sex steroids*; the adrenal cortex secretes *corticosteroids*, including cortisol and aldosterone among others.

The major thyroid hormones are composed of two derivatives of the amino acid tyrosine bonded together. These hormones are unique because they contain iodine (fig. 11.3). When the hormone contains four iodine atoms, it is called *tetraiodothyronine* (T_4), or *thyroxine*. When it contains three atoms of iodine, it is called *triiodothyronine* (T_3). Although these hormones are not steroids, they are like steroids in that they are relatively small, nonpolar molecules. Steroid and thyroid hormones are active when taken orally (as a pill). Sex steroids are a component of contraceptive pills, and thyroid hormone pills are taken by people whose thyroid is deficient

(who are hypothyroid). Other types of hormones cannot be taken orally because they would be digested into inactive fragments before being absorbed into the blood.

Prohormones and Prehormones

Hormone molecules that affect the metabolism of target cells are often derived from less active "parent," or *precursor*, molecules. In the case of polypeptide hormones, the precursor may be a longer chained **prohormone** that is cut and spliced together to make the hormone. Insulin, for example, is produced from *proinsulin* within the endocrine beta cells of the islets of Langerhans of the pancreas (see chapter 3, fig. 3.26). In some cases, the prohormone itself is derived from an even larger precursor molecule; in the case of insulin, this molecule is

Endocrine Gland	Prehormone	Active Products	Comments
Skin	Vitamin D_3	1,25-Dihydroxyvitamin D_3	Hydroxylation reactions occur in the liver and kidneys.
Testes	Testosterone	Dihydrotestosterone (DHT)	DHT and other 5α-reduced androgens are formed in most androgen-dependent tissue.
		Estradiol-17β (E_2)	E_2 is formed in the brain from testosterone, where it is believed to affect both endocrine function and behavior; small amounts of E_2 are also produced in the testes.
Thyroid	Thyroxine (T_4)	Triiodothyronine (T_3)	Conversion of T_4 to T_3 occurs in almost all tissues.

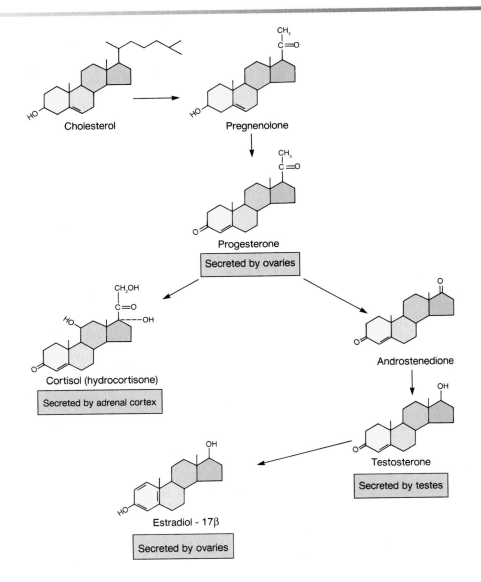

Figure 11.2

Simplified biosynthetic pathways for steroid hormones. Notice that progesterone (a hormone secreted by the ovaries) is a common precursor of all other steroid hormones and that testosterone (the major androgen secreted by the testes) is a precursor of estradiol-17β, the major estrogen secreted by the ovaries.

called *pre-proinsulin*. The term *prehormone* is sometimes used to indicate such precursors of prohormones.

In some cases, the molecule secreted by the endocrine gland (and considered to be the hormone of that gland) is actually inactive in the target cells. In order to become active, the target cells must modify the chemical structure of the secreted hormone. Thyroxine (T_4), for example, must be changed into T_3 within the target cells to affect the metabolism of the target cells. Similarly, testosterone (secreted by the testes) and vitamin D_3 (secreted by the skin) are converted into more active molecules within their target cells (table 11.3). In this text, the term **prehormone** will be used to designate those molecules secreted by endocrine glands that are inactive until changed by their target cells.

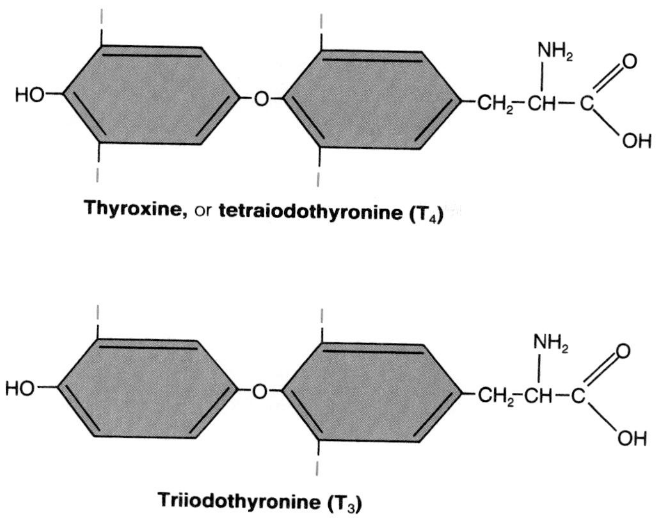

Thyroxine, or tetraiodothyronine (T₄)

Triiodothyronine (T₃)

Figure 11.3

The thyroid hormones, thyroxine (T_4) and triiodothyronine (T_3), are secreted in a ratio of 9 to 1.

Common Aspects of Neural and Endocrine Regulation

One might believe that, since endocrine regulation is chemical in nature, it is therefore fundamentally different from neural control systems. This assumption is incorrect. Electrical nerve impulses are in fact chemical events produced by the diffusion of ions through the neuron cell membrane (chapter 7). Interestingly, the action of some hormones (such as insulin) is accompanied by ion diffusion and electrical changes in the target cells, so changes in membrane potential are not unique to the nervous system. Also, most nerve fibers stimulate the cells they innervate through the release of a chemical neurotransmitter. Neurotransmitters do not travel in the blood as do hormones; instead, they diffuse only a very short distance across a synapse. In other respects, however, the actions of neurotransmitters are very similar to the actions of hormones.

Indeed, many polypeptide hormones, including those secreted by the pituitary gland and by the digestive tract, have been discovered in the brain. In certain locations in the brain, some of these compounds are produced and secreted as hormones. In other brain locations, some of these compounds apparently serve as neurotransmitters. The discovery of some of these polypeptides in unicellular organisms, which of course lack a nervous and endocrine system, suggests that these regulatory molecules appeared early in evolution and were incorporated into the function of nervous and endocrine tissue as these systems evolved. This fascinating theory would help to explain, for example, why insulin, a polypeptide hormone produced in the pancreas of vertebrates, is found in neurons of invertebrates (which lack a distinct endocrine system).

Regardless of whether a particular chemical is acting as a neurotransmitter or as a hormone, in order for it to function in physiological regulation: (1) target cells must have specific **receptor proteins** that combine with the regulatory molecule; (2) the combination of the regulatory molecule with its receptor proteins must cause a specific sequence of changes in the target cells; and (3) there must be a mechanism to quickly turn off the action of the regulator. This mechanism, which involves rapid removal and/or chemical inactivation of the regulator molecules, is essential because without an "off-switch" physiological control would be impossible.

Hormone Interactions

A given target tissue is usually responsive to a number of different hormones. These hormones may antagonize each other or work together to produce effects that are additive or complementary. The responsiveness of a target tissue to a particular hormone is thus affected not only by the concentration of that hormone, but also by the effects of other hormones on that tissue. Terms used to describe hormone interactions include *synergistic, permissive,* and *antagonistic.*

Synergistic and Permissive Effects

When two or more hormones work together to produce a particular result, their effects are said to be **synergistic.** These effects may be additive or complementary. The action of epinephrine and norepinephrine on the heart is a good example of an additive effect. Each of these hormones separately produces an increase in cardiac rate; acting together in the same concentrations, they stimulate an even greater increase in cardiac rate. The synergistic action of FSH and testosterone is an example of a complementary effect; each hormone separately stimulates a different stage of spermatogenesis during puberty, so that both hormones together are needed at that time to complete sperm development. Likewise, the ability of mammary glands to produce and secrete milk requires the synergistic action of many hormones—estrogen, cortisol, prolactin, oxytocin, and others.

A hormone is said to have a **permissive effect** on the action of a second hormone when it enhances the responsiveness of a target organ to the second hormone or when it increases the activity of the second hormone. Prior exposure of the uterus to estrogen, for example, induces the formation of receptor proteins for progesterone, which improves the response of the uterus when it is subsequently exposed to progesterone. Estrogen thus has a permissive effect on the responsiveness of the uterus to progesterone. Glucocorticoids (a class of corticosteroids including cortisol) exert permissive effects on the actions of catecholamines (epinephrine and norepinephrine). When there is an absence of these permissive effects due to abnormally low glucocorticoids, the catecholamines will not be as effective as they are normally. One symptom of this condition may be an abnormally low blood pressure.

Vitamin D_3 is a prehormone that must first be converted by enzymes in the kidneys and liver, where two hydroxyl

(OH⁻) groups are added to form the active hormone 1,25-dihydroxyvitamin D_3. This hormone helps to raise blood calcium levels. Parathyroid hormone (PTH) has a permissive effect on the actions of vitamin D_3 because it stimulates the production of the hydroxylating enzymes in the kidneys and liver. By this means, an increased secretion of PTH has a permissive effect on the ability of vitamin D_3 to stimulate the intestinal absorption of calcium.

Antagonistic Effects

In some situations, the actions of one hormone antagonize the effects of another. Lactation during pregnancy, for example, is inhibited because the high concentration of estrogen in the blood inhibits the secretion and action of prolactin. Another example of antagonism is the action of insulin and glucagon (two hormones from the islets of Langerhans) on adipose tissue; the formation of fat is promoted by insulin, whereas glucagon promotes fat breakdown.

Effects of Hormone Concentrations on Tissue Response

The concentration of hormones in the blood primarily reflects the rate of secretion by the endocrine glands. Hormones do not generally accumulate in the blood because they are rapidly removed by target organs and by the liver. The **half-life** of a hormone, which is the time required for the plasma concentration of a given amount of a hormone to be reduced to half its reference level, ranges from minutes to hours for most hormones (thyroid hormone, however, has a half-life of several days). Hormones removed from the blood by the liver are converted by enzymatic reactions into less active products. Steroids, for example, are converted to more polar derivatives. These less active, more water-soluble polar derivatives are released into the blood and are excreted in the urine and bile.

The effects of hormones are very dependent on concentration. Normal tissue responses are produced only when the hormones are present within their normal, or *physiological*, range of concentrations. When some hormones are taken in abnormally high, or *pharmacological*, concentrations (as when they are taken as drugs), their effects may be different from those produced physiologically. This may be partly due to the fact that abnormally high concentrations of a hormone may cause the hormone to bind to tissue receptor proteins of different but related hormones. Also, since some steroid hormones can be converted by their target cells into products that have different biological effects (such as the conversion of androgens into estrogens), the administration of large quantities of one steroid can result in the production of a significant quantity of other steroids with different effects.

Pharmacological doses of hormones, particularly of steroids, can thus have widespread and often damaging side effects. People with inflammatory diseases who are treated with high doses of cortisone over long periods of time, for example, may develop characteristic changes in bone and soft tissue structure. Contraceptive pills, which contain sex steroids, have

Anabolic steroids are synthetic androgens (male hormones) that promote protein synthesis in muscles and other organs. Use of these drugs by body builders and other athletes became widespread in the 1960s and, although prohibited by most athletic organizations, the practice is still common today. Although administration of exogenous androgens does promote muscle growth, it can also cause a number of undesirable side effects. Since the liver and some other organs can change androgens into estrogens, male athletes who take exogenous androgens often develop *gynecomastia*—an abnormal growth of femalelike mammary tissue. For some, this tissue is so excessive that it must be surgically removed. Damaging effects attributed to anabolic steroids include liver cancer, shrinkage of the testes and temporary sterility, stunted growth in teenage users, masculinization in female users, and antisocial behavior. Anabolic steroids may also raise the blood levels of cholesterol and LDL, thus predisposing users to atherosclerosis of the coronary arteries and heart disease (chapter 13).

a number of potential side effects that could not have been predicted in 1960, when "the pill" was first introduced. At that time, the concentrations of sex steroids were much higher than they are in the pills presently being marketed.

Priming Effects

Variations in hormone concentration within the normal, physiological range can affect the responsiveness of target cells. This is due in part to the effects of polypeptide and glycoprotein hormones on the number of their receptor proteins in target cells. More receptors may be formed in the target cells in response to particular hormones. Small amounts of gonadotropin-releasing hormone (GnRH), secreted by the hypothalamus, for example, increase the sensitivity of anterior pituitary cells to further GnRH stimulation. This is a priming effect, sometimes also called *upregulation*. Subsequent stimulation by GnRH thus causes a greater response from the anterior pituitary.

Desensitization and Downregulation

Prolonged exposure to high concentrations of polypeptide hormones has been found to *desensitize* the target cells. Subsequent exposure to the same concentration of the same hormone thus produces less of a target tissue response. This desensitization may be partially due to the fact that high concentrations of these hormones cause a decrease in the number of receptor proteins in their target cells—a phenomenon called *downregulation*. Such desensitization and downregulation of receptors has been shown to occur, for example, in adipose cells exposed to high concentrations of insulin and in testicular cells exposed to high concentrations of luteinizing hormone (LH).

In order to prevent desensitization from occurring under normal conditions, many polypeptide and glycoprotein hormones are secreted in pulses rather than continuously. This *pulsatile secretion* is an important aspect, for example, in the hormonal control of the reproductive system. The pulsatile secretion of GnRH and LH is needed to prevent desensitization; when these hormones are artificially presented in a continuous fashion, they produce a decrease (rather than the normal increase) in gonadal function. This effect has important clinical implications, as will be described in chapter 20.

---

1. List the chemical categories of hormones and give examples of hormones in each category.
2. Define the terms *prohormone* and *prehormone,* and give examples of each.
3. Describe the common characteristics of hormones and neurotransmitters.
4. List the terms used to describe hormone interactions and give examples of these effects.
5. Explain how the response of the body to a given hormone can be affected by the concentration of that hormone in the blood.

Mechanisms of Hormone Action

Each hormone exerts its characteristic effects on target organs through the actions it has on the cells of these organs. The mechanisms of action are similar for hormones that have similar chemical natures. Lipid-soluble hormones pass through the target cell membrane, bind to intracellular receptor proteins, and act directly within the target cell. Polar hormones do not enter the target cells, but instead bind to receptors on the cell membrane. This results in the activation of intracellular second-messenger systems that mediate the actions of the hormone.

Although each hormone exerts its own characteristic effects on specific target cells, hormones that are in the same chemical category have similar mechanisms of action. These similarities involve the location of cellular receptor proteins and the events that occur in the target cells after the hormone has combined with its receptor protein.

Hormones are delivered by the blood to every cell in the body, but only the **target cells** are able to respond to these hormones. In order to respond to any given hormone, a target cell must have specific receptor proteins for that hormone. Receptor protein–hormone interaction is highly specific. In addition to this property of *specificity*, hormones bind to receptors with a *high affinity* (high bond strength) and with a *low capacity*. The latter characteristic refers to the possibility of saturating receptors with hormones because of the limited number of receptors per target cell (usually a few thousand). Notice that the characteristics of specificity and saturation of receptor proteins are similar to the characteristics of enzyme and carrier proteins discussed in previous chapters.

The location of a hormone's receptor proteins in its target cells depends on the chemical nature of the hormone. Based on the location of the receptor proteins, hormones can be grouped into three categories: (1) receptor proteins within the nucleus of target cells—*thyroid hormones and some steroid hormones*; (2) receptor proteins within the cytoplasm of target cells—*steroid hormones*; and (3) receptor proteins on the outer surface of the target cell membrane—*catecholamine* and *polypeptide hormones*. This information is summarized in table 11.4.

Mechanisms of Steroid and Thyroid Hormone Action

Steroid and thyroid hormones are similar in size. Both groups are also nonpolar, and thus they are not very water-soluble. Unlike other hormones, therefore, steroids and thyroid hormones (primarily thyroxine) do not travel dissolved in the aqueous portion of the plasma but instead are transported to their target cells attached to plasma carrier proteins. These hormones then dissociate from the carrier proteins in the blood and easily pass through the lipid component of the target cell's membrane.

Steroid and thyroid hormones bind to receptor proteins within the target cells. These receptor proteins may be originally

---

Table 11.4 Functional Categories of Hormones Based on the Location of Their Receptor Proteins and Mechanisms of Action

Types of Hormones	Secreted by	Location of Receptors	Effects of Hormone-Receptor Interaction
Catecholamines, polypeptides, glycoproteins	All glands except adrenal cortex, gonads, and thyroid	Outer surface of cell membrane	Stimulates production of intracellular "second messenger," which activates previously inactive enzymes
Steroids	Adrenal cortex, testes, ovaries	Cytoplasm or nucleus of target cells	Stimulates translocation of hormone-receptor complex to nucleus and activation of specific genes
Thyroxine (T_4)	Thyroid	Nucleus of target cells	After conversion to triiodothyronine (T_3), activates specific genes

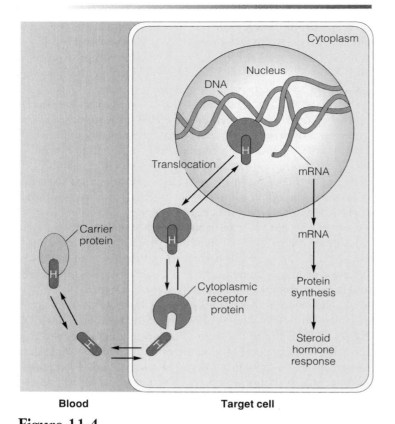

Figure 11.4

The mechanism of action of a steroid hormone (H) on the target cells.

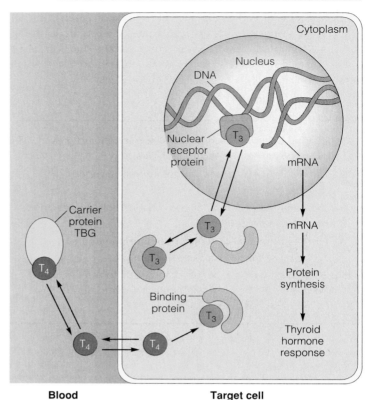

Figure 11.5

The mechanism of action of thyroid hormones on the target cells.

located in either the cytoplasm or nucleus, depending on the specific receptor. After binding to its hormone, the receptor undergoes a change that allows it to also bind to DNA. This promotes genetic transcription, or RNA synthesis. The synthesis of specific proteins coded by this newly formed messenger RNA produces the effects of the hormone within the target cell.

Steroid Hormones

Once through the cell membrane, most steroid hormones attach to *cytoplasmic receptor proteins* in the target cells. The steroid hormone–receptor protein complex then *translocates* to the nucleus and attaches by means of the receptor proteins to the chromatin. The sites of attachment in the chromatin, termed *acceptor sites,* are specific for the target tissue. This specificity is believed to be determined by acidic (nonhistone) proteins in the chromatin (chapter 3). According to one theory, part of the receptor binds to an acidic protein while a different part of the receptor bonds to DNA.

The attachment of the steroid hormone–receptor protein complex to the acceptor site "turns on" genes. Specific genes become activated by this process and produce nuclear RNA, which is then processed into messenger RNA (mRNA). This new mRNA enters ribosomes and codes for the production of new proteins. Since some of these newly synthesized proteins may be enzymes, the metabolism of the target cell is changed in a specific manner (fig. 11.4).

Thyroxine

As previously discussed, the major hormone secreted by the thyroid gland is thyroxine, or tetraiodothyronine (T_4). Like steroid hormones, thyroxine travels in the blood attached to carrier proteins (primarily attached to *thyroxine-binding globulin,* or *TBG*). The thyroid also secretes a small amount of triiodothyronine, or T_3. The carrier proteins have a higher affinity for T_4 than for T_3, however, and as a result, the amount of unbound (or "free") T_3 in the plasma is about ten times greater than the amount of free T_4.

Approximately 99.96% of the thyroxine in the blood is attached to carrier proteins in the plasma; the rest is free. Only the thyroxine and T_3 that is free can enter target cells; the protein-bound thyroxine serves as a reservoir of this hormone in the blood (this is why it takes a couple of weeks after surgical removal of the thyroid for the symptoms of hypothyroidism to develop). Once the free thyroxine passes into the target cell cytoplasm, it is enzymatically converted into T_3. As previously discussed, it is the T_3 rather than T_4 that is active within the target cells.

Inactive T_3 receptor proteins are already in the nucleus attached to chromatin. These receptors are inactive until T_3 enters the nucleus from the cytoplasm. The attachment of T_3 to the chromatin-bound receptor proteins activates genes and results in the production of new mRNA and new proteins. This sequence of events is summarized in figure 11.5.

Chapter Eleven

Second-Messenger Mechanisms in Hormone Action

Catecholamine hormones (epinephrine and norepinephrine) and polypeptide hormones cannot pass through the lipid barrier of the target cell membrane. Although some of these hormones may enter the cell by pinocytosis, most of their effects result from their binding to receptor proteins on the outer surface of the target cell membrane. Since they exert their effects without entering the target cells, the actions of these hormones must be mediated by other molecules within the target cells. If you think of hormones as "messengers" from the endocrine glands, the intracellular mediators of the hormone's action can be called **second messengers.** (The concept of second messengers was introduced in connection with synaptic transmission in chapter 7.)

Hormones That Use Cyclic AMP as a Second Messenger

Cyclic adenosine monophosphate (abbreviated **cAMP**) was the first "second messenger" to be discovered and is the best understood. The β-adrenergic effects (chapter 9) of epinephrine and norepinephrine are due to cAMP production within the target cells. It was later discovered that the effects of many (but not all) polypeptide and glycoprotein hormones are also mediated by cAMP (see table 11.5).

When one of these hormones binds to its receptor protein, it causes the dissociation of a subunit from the complex of G-proteins (discussed in chapter 7). This G-protein subunit moves through the membrane until it reaches the enzyme **adenylate cyclase.** The G-protein then binds to and activates this enzyme, which catalyzes the following reaction within the cytoplasm of the cell:

$$ATP \rightarrow cAMP + PP_i$$

Adenosine triphosphate (ATP) is thus converted into cyclic AMP (cAMP) and two inorganic phosphates (*pyrophosphate*, abbreviated PP_i). As a result of the interaction of the hormone with its receptor and the activation of adenylate cyclase, therefore, the intracellular concentration of cAMP is increased. Cyclic AMP activates a previously inactive enzyme in the cytoplasm called **protein kinase.** The inactive form of this enzyme consists of two subunits: a catalytic subunit and an inhibitory subunit. The enzyme is produced in an inactive form and becomes active only when cAMP attaches to the inhibitory subunit. Binding of cAMP to the inhibitory subunit causes it to dissociate from the catalytic subunit, which then becomes active (fig. 11.6). In summary, the hormone—acting through an increase in cAMP production—causes an increase in protein kinase enzyme activity within its target cells.

Active protein kinase catalyzes the phosphorylation of (attachment of phosphate groups to) different proteins in the target cells. This causes some enzymes to become activated and others to become inactivated. Cyclic AMP, acting through protein kinase, thus modulates the activity of enzymes that are already present in the target cell. This alters the metabolism of the target tissue in a manner characteristic of the actions of that specific hormone (table 11.6).

Table 11.5	Hormones That Use cAMP as a Second Messenger and Hormones That Use Other Second Messengers
Hormones That Use cAMP as a Second Messenger	**Hormones That Use Other Second Messengers**
Adrenocorticotropic hormone (ACTH)	Catecholamines (α-adrenergic)
Calcitonin	Growth hormone (GH)
Catecholamines (β-adrenergic)	Insulin
Follicle-stimulating hormone (FSH)	Oxytocin
Glucagon	Prolactin
Luteinizing hormone (LH)	Somatomedin
Parathyroid hormone	Somatostatin
Thyrotropin-releasing hormone (TRH)	
Thyroid-stimulating hormone (TSH)	
Antidiuretic hormone (ADH)	

Table 11.6	Sequence of Events Involving Cyclic AMP as a Second Messenger

1. The hormone binds to its receptor on the outer surface of the target cell membrane.
2. Hormone-receptor interaction stimulates the activity of adenylate cyclase on the cytoplasmic side of the membrane.
3. Activated adenylate cyclase catalyzes the conversion of ATP to cyclic AMP (cAMP) within the cytoplasm.
4. Cyclic AMP activates protein kinase enzymes that were already present in the cytoplasm in an inactive state.
5. Activated cAMP-dependent protein kinase transfers phosphate groups (phosphorylates) to other enzymes in the cytoplasm.
6. The activity of specific enzymes is either increased or inhibited by phosphorylation.
7. Altered enzyme activity mediates the target cell's response to the hormone.

Like all biologically active molecules, cAMP must be rapidly inactivated for it to function effectively as a second messenger in hormone action. This inactivation is accomplished by an enzyme within the target cells called **phosphodiesterase,** which hydrolyzes cAMP into inactive fragments. Through the action of phosphodiesterase, the stimulatory effect of a hormone that uses cAMP as a second messenger depends upon the continuous generation of new cAMP molecules, and thus depends upon the level of secretion of the hormone.

In addition to cyclic AMP, **cyclic guanosine monophosphate (cGMP)** functions as a second messenger in certain cases. For example, the regulatory molecule nitric oxide (discussed in chapter 7 and later in this chapter) exerts its effects on smooth muscle relaxation by stimulating the production of cGMP in its target cells. In different regulatory systems cGMP and cAMP may interact, producing effects that are either antagonistic or

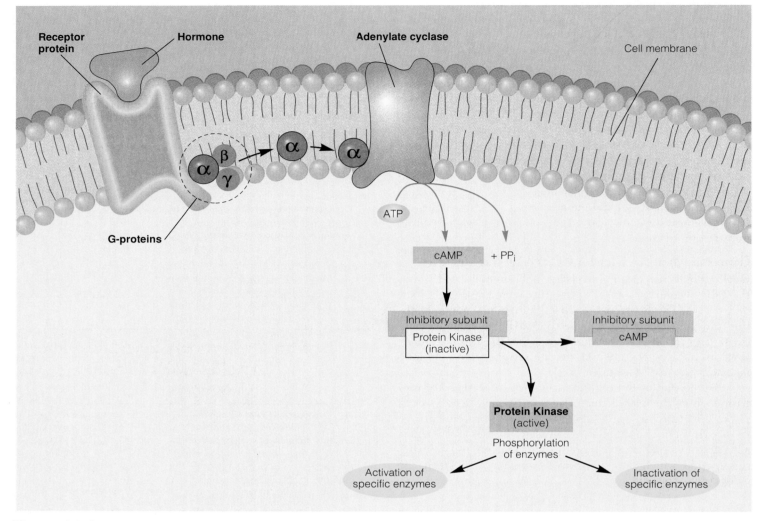

Figure 11.6
The function of cyclic AMP (cAMP) as a second messenger in the action of many hormones.

 Drugs that inhibit the activity of phosphodiesterase prevent the breakdown of cAMP and thus result in increased concentrations of cAMP within the target cells. The drug **theophylline** and its derivatives, for example, are used clinically to raise cAMP levels within bronchiolar smooth muscle. This duplicates and enhances the effect of epinephrine on the bronchioles (producing dilation) in people who suffer from asthma. **Caffeine**, a compound related to theophylline, is also a phosphodiesterase inhibitor, and thus exerts its effects by raising the cAMP concentrations within tissue cells.

complementary. For example, the control of cell division and the cell cycle (chapter 3) is related to the ratio of cAMP to cGMP in the cell.

Hormones That Use the Ca++ Second-Messenger System
The concentration of Ca^{++} in the cytoplasm is kept very low by the action of active transport carriers—calcium pumps—in the cell membrane. Through the action of these pumps, the concentration of calcium in the cytoplasm is 5,000 to 10,000 times lower in the cytoplasm than in the extracellular fluid. In addition, the endoplasmic reticulum (chapter 3) of many cells contains calcium pumps that actively transport Ca^{++} from the cytoplasm into the cisternae of the endoplasmic reticulum. The steep concentration gradient for Ca^{++} that results allows various stimuli to evoke

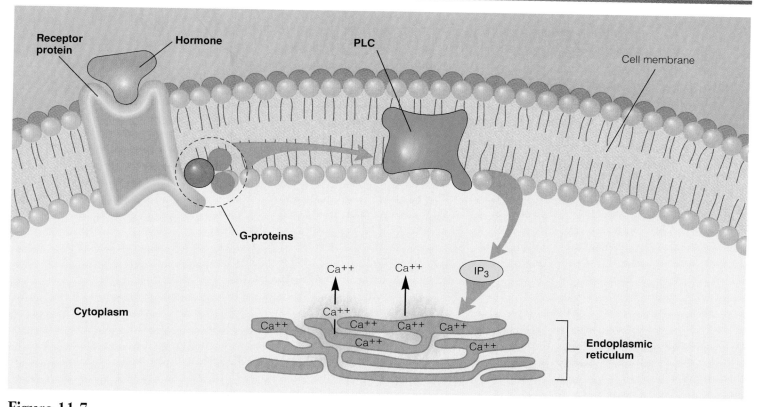

Figure 11.7

Some hormones, when they bind to their membrane receptors, activate phospholipase C (PLC). This enzyme catalyzes the formation of inositol triphosphate (IP_3), which causes Ca^{++} channels to open in the endoplasmic reticulum. Ca^{++} is thus released and acts as a second messenger in the action of the hormone.

a rapid, though brief, diffusion of Ca^{++} into the cytoplasm, which can serve as a signal in different control systems.

At the terminal boutons of axons, for example, the entry of Ca^{++} through voltage-regulated Ca^{++} channels in the cell membrane serves as a signal for the release of neurotransmitters (chapter 7). Similarly, when muscles are stimulated to contract, Ca^{++} couples electrical excitation of the muscle cell to the mechanical processes of contraction (chapter 12). Additionally, it is now known that Ca^{++} serves as a part of a second-messenger system in the action of a number of hormones.

When epinephrine stimulates its target organs, it must first bind to adrenergic receptor proteins in the membrane of its target cells. As discussed in chapter 9, there are two types of adrenergic receptors—alpha and beta. Stimulation of the beta-adrenergic receptors by epinephrine results in activation of adenylate cyclase and the production of cAMP. Stimulation of alpha-adrenergic receptors by epinephrine, in contrast, activates the target cell via the Ca^{++} second-messenger system.

The binding of epinephrine to its alpha-adrenergic receptor activates, via a G-protein intermediate, an enzyme known as **phospholipase C** in the cell membrane. The substrate of this enzyme is a class of membrane phospholipid, which is split by the active enzyme into **inositol triphosphate (IP_3)** and another derivative (diacylglycerol). IP_3 leaves the cell membrane

and diffuses through the cytoplasm to the endoplasmic reticulum. The membrane of the endoplasmic reticulum contains receptor proteins for IP_3, so that the IP_3 is a second messenger in its own right, carrying the hormone's message from the cell membrane to the endoplasmic reticulum. Binding of IP_3 to its receptors causes specific Ca^{++} channels to open, so that Ca^{++} diffuses out of the endoplasmic reticulum and into the cytoplasm (fig. 11.7).

As a result of these events, there is a rapid and transient rise in the cytoplasmic Ca^{++} concentration. This signal is augmented, through mechanisms that are incompletely understood, by the opening of Ca^{++} channels in the cell membrane. This may be due to the action of yet a different (and currently unknown) messenger sent from the endoplasmic reticulum to the cell membrane. The Ca^{++} that enters the cytoplasm from the endoplasmic reticulum and extracellular fluid binds to a cytoplasmic protein called **calmodulin.** Once Ca^{++} binds to calmodulin, the now-active calmodulin in turn activates specific protein kinase enzymes (those that add phosphate groups to proteins) that modify the actions of other enzymes in the cell (fig. 11.8). Activation of specific calmodulin-dependent enzymes is analogous to the activation of enzymes by cAMP-dependent protein kinase. The steps of the Ca^{++} second-messenger system are summarized in table 11.7.

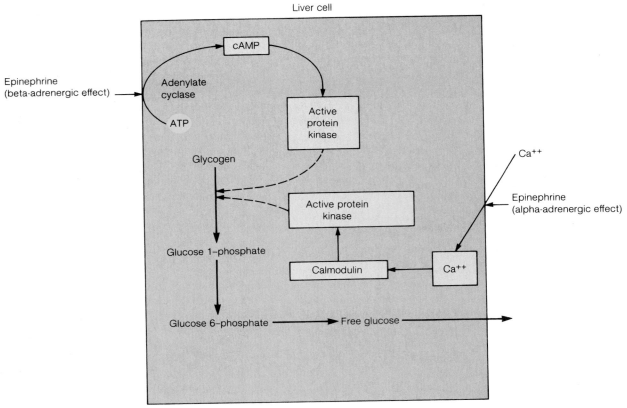

Figure 11.8

Epinephrine stimulates glycogenolysis by way of two second messengers. The stimulation of β-adrenergic receptors invokes the cAMP second messenger system, and the stimulation of the α-adrenergic receptors invokes the Ca^{++} second-messenger system.

Table 11.7. Sequence of Events Involving the Ca^{++} Second-Messenger System

1. The hormone binds to its receptor on the outer surface of the target cell membrane.
2. Hormone-receptor interaction stimulates the activity of a membrane enzyme, phospholipase C.
3. Activated phospholipase C catalyzes the conversion of particular phospholipids in the membrane to inositol triphosphate (IP_3) and another derivative, diacylglycerol.
4. Inositol triphosphate enters the cytoplasm and diffuses to the endoplasmic reticulum, where it binds to its receptor proteins and causes the opening of Ca^{++} channels.
5. Since the endoplasmic reticulum accumulates Ca^{++} by active transport, there exists a steep Ca^{++} concentration gradient favoring the diffusion of Ca^{++} into the cytoplasm.
6. Ca^{++} that enters the cytoplasm binds to and activates a protein called calmodulin.
7. Activated calmodulin, in turn, activates protein kinase, which phosphorylates other enzyme proteins.
8. Altered enzyme activity mediates the target cell's response to the hormone.

Different hormones can act on the same target cell and produce different, and even antagonistic, effects. For example, insulin stimulates the synthesis of fat while the hormone glucagon stimulates hydrolysis of fat in adipose cells. Clearly, these two hormones cannot both use the same second-messenger

system. The Ca^{++} system, which is stimulated by insulin, promotes lipogenesis, while the cAMP system, which is stimulated by glucagon, promotes lipolysis.

Also, the way in which a target cell responds to a given second messenger can be different for different types of cells. In adipose cells, for example, cAMP produces lipolysis; in myocardial cells, by contrast, cAMP produces an increase in contraction strength; and, in the Leydig cells of the testis, cAMP stimulates testosterone secretion. These responses are different because the cAMP-protein kinase system affects different proteins, due to differences in genetic expression, in these different types of cells.

1. Using diagrams, describe how steroid hormones and thyroxine exert their effects on their target cells.
2. Using a diagram, describe how cyclic AMP is produced within a target cell in response to hormone stimulation and how cAMP functions as a second messenger.
3. Describe the sequence of events by which a hormone can cause a rise in the cytoplasmic Ca^{++} concentration, and how Ca^{++} can function as a second messenger.
4. Give an example of how antagonistic effects may be produced through the use of different second-messenger systems.

Chapter Eleven

Pituitary Gland

The pituitary gland includes the anterior pituitary and the posterior pituitary. The posterior pituitary secretes hormones that are actually produced by the hypothalamus, whereas the anterior pituitary produces and secretes its own hormones. The anterior pituitary, however, is regulated by hormones secreted by the hypothalamus, as well as by feedback influences exerted by hormones from other endocrine glands.

The pituitary gland, or *hypophysis*, is located on the inferior aspect of the brain in the region of the diencephalon (chapter 8). The pituitary is a rounded, pea-shaped gland measuring about 1.3 cm (0.5 in.) in diameter, and is attached to the hypothalamus by a stalklike structure called the *infundibulum* (fig. 11.9).

The pituitary gland is structurally and functionally divided into an anterior lobe, or **adenohypophysis,** and a posterior lobe called the **neurohypophysis.** These two parts have different embryonic origins. The adenohypophysis is derived from a pouch of epithelial tissue (*Rathke's pouch*) that migrates upward from the embryonic mouth, whereas the neurohypophysis is formed as a downgrowth of the brain. The adenohypophysis consists of three parts: (1) the *pars distalis*, also known as the **anterior pituitary,** is the rounded portion and the major endocrine part of the gland; (2) the *pars tuberalis* is the thin extension in contact with the infundibulum; and (3) the *pars intermedia* is located between the anterior and posterior parts of the pituitary. These parts are illustrated in figure 11.9.

The neurohypophysis is the neural part of the pituitary gland. It consists of the *pars nervosa*, also called the **posterior pituitary,** which is in contact with the pars intermedia of the adenohypophysis, and the *infundibulum*, which is the connecting stalk to the hypothalamus. Nerve fibers extend through the infundibulum along with small neuroglia-like cells, called *pituicytes*.

Pituitary Hormones

The hormones secreted by the anterior pituitary (the pars distalis of the adenohypophysis) are called **trophic hormones.** The term *trophic* means "food." Although the anterior pituitary hormones are not food for their target organs, this term is used because high concentrations of the anterior pituitary hormones cause their target organs to hypertrophy, while low levels cause their target organs to atrophy. When names are applied to the hormones of the anterior pituitary, therefore, "trophic" (conventionally shortened to *tropic*, meaning "attracted to") is incorporated into these names. Also, the shortened forms of the names for the anterior pituitary hormones end in the suffix *-tropin*. The hormones of the pars distalis (table 11.8) are:

1. **Growth hormone (GH, or somatotropin).** This hormone promotes the movement of amino acids into tissue cells and the incorporation of these amino acids into tissue proteins, thus stimulating growth of organs.

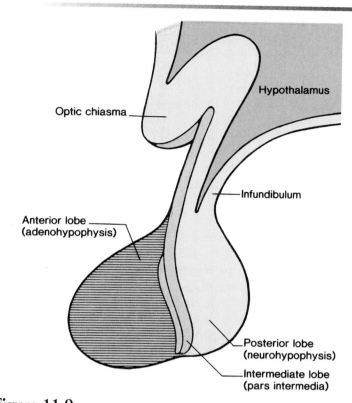

Figure 11.9

The structure of the pituitary gland as seen in sagittal view.

2. **Thyroid-stimulating hormone (TSH, or thyrotropin).** This hormone stimulates the thyroid gland to produce and secrete thyroxine (tetraiodothyronine, or T_4).

3. **Adrenocorticotropic hormone (ACTH, or corticotropin).** This hormone stimulates the adrenal cortex to secrete the glucocorticoids, such as hydrocortisone (cortisol).

4. **Follicle-stimulating hormone (FSH, or folliculotropin).** This hormone stimulates the growth of ovarian follicles in females and the production of sperm in the testes of males.

5. **Luteinizing hormone (LH, or luteotropin).** This hormone and FSH are collectively called **gonadotropic hormones.** In females, LH stimulates ovulation and the conversion of the ovulated ovarian follicle into an endocrine structure called a corpus luteum. In males, LH (which is sometimes also called *interstitial cell-stimulating hormone*, or *ICSH*) stimulates the secretion of male sex hormones (mainly testosterone) from the interstitial cells of Leydig in the testes.

6. **Prolactin.** This hormone is secreted in both males and females. Its best known function is the stimulation of milk production by the mammary glands of women after the birth of their babies. Prolactin plays a supporting role in the regulation of the male reproductive system by the gonadotropins (FSH and LH) and acts on the kidneys to help regulate water and electrolyte balance.

Table 11.8 Anterior Pituitary Hormones

Hormone	Target Tissue	Stimulated by Hormone	Regulation of Secretion
ACTH (adrenocorticotropic hormone)	Adrenal cortex	Secretion of glucocorticoids	Stimulated by CRH (corticotropin-releasing hormone); inhibited by glucocorticoids
TSH (thyroid-stimulating hormone)	Thyroid gland	Secretion of thyroid hormones	Stimulated by TRH (thyrotropin-releasing hormone); inhibited by thyroid hormones
GH (growth hormone)	Most tissue	Protein synthesis and growth; lipolysis and increased blood glucose	Inhibited by somatostatin; stimulated by growth hormone–releasing hormone
FSH (follicle-stimulating hormone) and LH (luteinizing hormone)	Gonads	Gamete production and sex steroid hormone secretion	Stimulated by GnRH (gonadotropin-releasing hormone); inhibited by sex steroids
Prolactin	Mammary glands and other sex accessory organs	Milk production; controversial actions in other organs	Inhibited by PIH (prolactin-inhibiting hormone)
LH (luteinizing hormone)	Gonads	Sex hormone secretion; ovulation and corpus luteum formation	Stimulated by GnRH

 Inadequate growth hormone secretion during childhood causes *pituitary dwarfism.* Hyposecretion of growth hormone in an adult produces a rare condition called *pituitary cachexia (Simmonds' disease).* One of the symptoms of this disease is premature aging caused by tissue atrophy. Oversecretion of growth hormone during childhood, by contrast, causes *gigantism.* Excessive growth hormone secretion in an adult does not cause further growth in length because the cartilaginous epiphyseal discs have ossified. Hypersecretion of growth hormone in an adult instead causes *acromegaly,* in which the person's appearance gradually changes as a result of thickening of bones and the growth of soft tissues, particularly in the face, hands, and feet.

The pars intermedia of the adenohypophysis in an adult human is not well defined, and its function, if any, is poorly understood. It produces different forms of **melanocyte-stimulating hormone (MSH),** which in lower vertebrates (fish, amphibians, and reptiles) cause a darkening of the skin to provide camouflage against a dark background. Although the exogenous administration of MSH to humans stimulates melanin (pigment) production in the melanocytes of skin, MSH does not appear to have a significant effect in normal human physiology. It is interesting that ACTH, secreted from the pars distalis, contains the amino acid sequence of MSH as part of its structure, and abnormally high amounts of ACTH secretion (as in Addison's disease) cause a darkening of the skin. The pars intermedia also produces large amounts of β-endorphin (chapter 7), as do other structures, but the physiological significance of this is not understood.

The posterior pituitary, or pars nervosa, secretes only two hormones, both of which are produced in the hypothalamus and merely stored in the posterior lobe of the pituitary:

1. **Antidiuretic hormone (ADH),** also known as *arginine vasopressin* (AVP). Antidiuretic hormone stimulates the kidneys to retain water so that less water is excreted in the urine and more water is retained in the blood. At high doses this hormone also has a "pressor" effect; that is, it causes vasoconstriction in experimental animals. The physiological significance of this pressor effect in humans, however, is controversial.

2. **Oxytocin.** In females, oxytocin stimulates contractions of the uterus during labor and for this reason is needed for parturition (childbirth). Oxytocin also stimulates contractions of the mammary gland alveoli and ducts, which result in the milk-ejection reflex in a lactating woman. In men, a rise in oxytocin secretion at the time of ejaculation has been measured, but the physiological significance of this hormone in males remains to be demonstrated.

 Injections of oxytocin may be given to a woman to induce labor if the pregnancy is past term, or if the fetal membranes have ruptured and there is a danger of infection. Labor may also be induced by injections of oxytocin if there is pregnancy-induced hypertension, or **preeclampsia.** Also, oxytocin is commonly administered after delivery, when, by stimulating contractions of the uterine muscle, it promotes the regression in the size of the uterus and squeezes the blood vessels, thus minimizing the danger of hemorrhage.

Hypothalamic Control of the Posterior Pituitary

Although the posterior pituitary secretes two hormones—antidiuretic hormone (ADH) and oxytocin—these hormones are actually produced in neuron cell bodies of the *supraoptic nuclei* and *paraventricular nuclei* of the hypothalamus. These nuclei within the hypothalamus are thus endocrine glands. The hormones they produce are transported along axons of the

hypothalamo-hypophyseal tract (fig. 11.10) to the posterior pituitary, which stores and later secretes these hormones. The posterior pituitary is thus more a storage organ than a true gland.

The secretion of ADH and oxytocin from the posterior pituitary is controlled by **neuroendocrine reflexes.** In nursing mothers, for example, the stimulus of suckling acts via sensory nerve impulses to the hypothalamus to stimulate the reflex secretion of oxytocin. The secretion of ADH is stimulated by osmoreceptor neurons in the hypothalamus in response to a rise in blood osmotic pressure (chapter 6); its secretion is inhibited by sensory impulses from stretch receptors in the left atrium of the heart in response to a rise in blood volume. These reflexes are discussed in more detail in later chapters.

Hypothalamic Control of the Anterior Pituitary

At one time the anterior pituitary was called the "master gland" because it secretes hormones that regulate some other endocrine glands (fig. 11.11 and table 11.8). Adrenocorticotropic hormone (ACTH), thyroid-stimulating hormone (TSH), and the gonadotropic hormones (FSH and LH) stimulate the adrenal cortex, thyroid, and gonads, respectively, to secrete their hormones. The anterior pituitary hormones also have a "trophic" effect on their target glands in that the health of the target glands depends on adequate stimulation by anterior pituitary hormones. The anterior pituitary, however, is not really the master gland, since secretion of its hormones is in turn controlled by hormones secreted by the hypothalamus.

Releasing and Inhibiting Hormones

Since axons do not enter the anterior pituitary, hypothalamic control of the anterior pituitary is achieved through hormonal rather than neural regulation. Neurons in the hypothalamus produce releasing and inhibiting hormones, which are transported to axon endings in the basal portion of the hypothalamus. This region, known as the *median eminence,* contains blood capillaries that are drained by venules in the stalk of the pituitary.

The venules that drain the median eminence deliver blood to a second capillary bed in the anterior pituitary. Since this second capillary bed is downstream from the capillary bed in the median eminence and receives venous blood from it, the vascular link between the median eminence and the anterior pituitary forms a *portal system.* (This is analogous to the hepatic portal system that delivers venous blood from the intestine to the liver—chapter 18.) The vascular link between the hypothalamus and the anterior pituitary is thus called the **hypothalamo-hypophyseal portal system.**

Polypeptide hormones are secreted into the hypothalamo-hypophyseal portal system by neurons of the hypothalamus. These hormones regulate the secretions of the anterior pituitary (fig. 11.12 and table 11.9). Thyrotropin-releasing hormone **(TRH)** stimulates the secretion of TSH, and corticotropin-releasing hormone **(CRH)** stimulates the secretion of ACTH from the anterior pituitary. A single releasing hormone, gonadotropin-releasing hormone, or **GnRH,** appears to stimulate

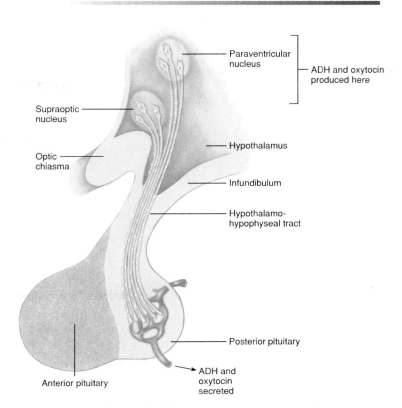

Figure 11.10

The posterior pituitary, or neurohypophysis, stores and secretes hormones (vasopressin and oxytocin) produced in neuron cell bodies within the supraoptic and paraventricular nuclei of the hypothalamus. These hormones are transported to the posterior pituitary by nerve fibers of the hypothalamo-hypophyseal tract.

the secretion of both gonadotropic hormones (FSH and LH) from the anterior pituitary. The secretion of prolactin and of growth hormone from the anterior pituitary is regulated by hypothalamic inhibitory hormones, known as **PIH** (prolactin-inhibiting hormone) and **somatostatin,** respectively.

A specific **growth hormone–releasing hormone (GHRH)** that stimulates growth hormone secretion has been identified as a polypeptide consisting of forty-four amino acids. Experiments suggest that a releasing hormone for prolactin may also exist, but no such specific releasing hormone has yet been discovered.

Feedback Control of the Anterior Pituitary

In view of its secretion of releasing and inhibiting hormones, the hypothalamus might be considered the "master gland." The chain of command, however, is not linear; the hypothalamus and anterior pituitary are controlled by the effects of their own actions. In the endocrine system, to use an analogy, the general takes orders from the private. The hypothalamus and anterior pituitary are not master glands because their secretions are controlled by the target glands they regulate.

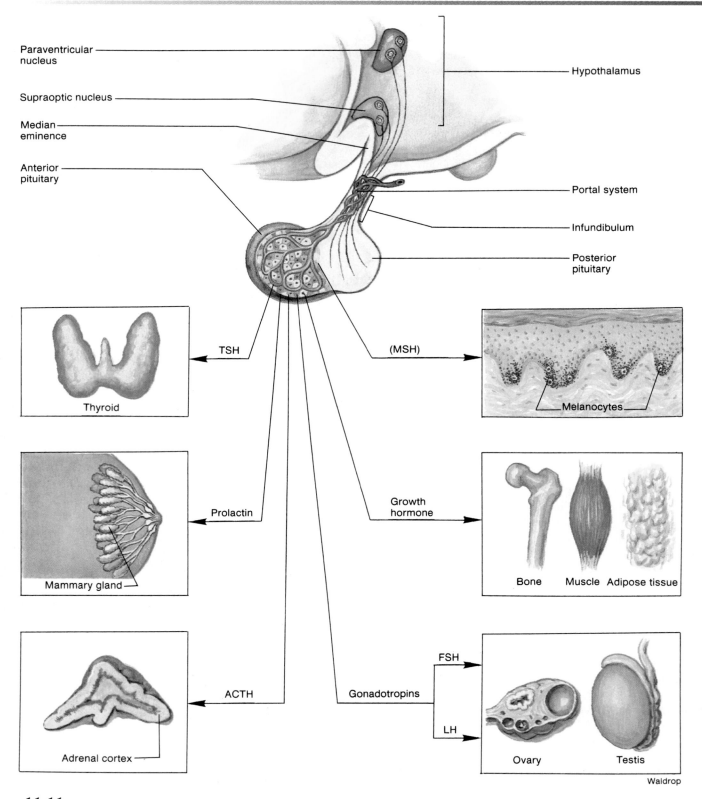

Paraventricular nucleus

Supraoptic nucleus

Median eminence

Anterior pituitary

Hypothalamus

Portal system

Infundibulum

Posterior pituitary

TSH

Thyroid

(MSH)

Melanocytes

Prolactin

Mammary gland

Growth hormone

Bone Muscle Adipose tissue

ACTH

Adrenal cortex

Gonadotropins

FSH

LH

Ovary Testis

Waldrop

Figure 11.11

Hormones secreted by the anterior pituitary and the target organs for these hormones.

Table 11.9 Hypothalamic Hormones Involved in the Control of the Anterior Pituitary

Hypothalamic Hormone	Structure	Effect on Anterior Pituitary	Action of Anterior Pituitary Hormone
Corticotropin-releasing hormone (CRH)	41 amino acids	Stimulates secretion of adrenocorticotropic hormone (ACTH)	Stimulates secretions of adrenal cortex
Gonadotropin-releasing hormone (GnRH)	10 amino acids	Stimulates secretion of follicle-stimulating hormone (FSH) and luteinizing hormone (LH)	Stimulates gonads to produce gametes (sperm and ova) and secrete sex steroids
Prolactin-inhibiting hormone (PIH)	Dopamine	Inhibits prolactin secretion	Stimulates production of milk in mammary glands
Somatostatin	14 amino acids	Inhibits secretion of growth hormone	Stimulates anabolism and growth in many organs
Thyrotropin-releasing hormone (TRH)	3 amino acids	Stimulates secretion of thyroid-stimulating hormone (TSH)	Stimulates secretion of thyroid gland
Growth hormone–releasing hormone (GHRH)	44 amino acids	Stimulates growth hormone secretion	Stimulates anabolism and growth in many organs

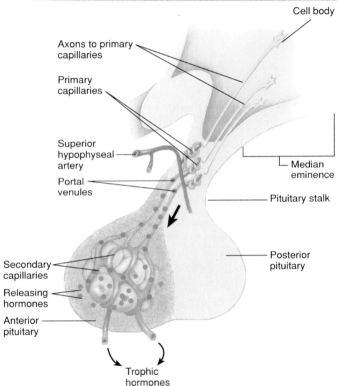

Figure 11.12

Neurons in the hypothalamus secrete releasing hormones (shown as dots) into the blood vessels of the hypothalamo-hypophyseal portal system. These releasing hormones stimulate the anterior pituitary to secrete its hormones into the general circulation.

Anterior pituitary secretion of ACTH, TSH, and the gonadotropins (FSH and LH) is controlled by **negative feedback inhibition** from the target gland hormones. Secretion of ACTH is inhibited by a rise in corticosteroid secretion, for example, and TSH is inhibited by a rise in the secretion of thyroxine from the thyroid. These negative feedback relationships are easily demonstrated by removal of the target glands. Castration (surgical removal of the gonads), for example, produces a rise in the secretion of FSH and LH. In a similar manner, removal of the adrenals or the thyroid results in an abnormal increase in ACTH or TSH secretion from the anterior pituitary.

The effects of removal of the target glands demonstrate that under normal conditions these glands exert an inhibitory effect on the anterior pituitary. This inhibitory effect can occur at two levels: (1) the target gland hormones can act on the hypothalamus and inhibit the secretion of releasing hormones, and (2) the target gland hormones can act on the anterior pituitary and inhibit its response to the releasing hormones. Thyroxine, for example, appears to inhibit the response of the anterior pituitary to TRH and thus acts to reduce TSH secretion (fig. 11.13). Sex steroids, in contrast, reduce the secretion of gonadotropins by inhibiting both GnRH secretion and the ability of the anterior pituitary to respond to stimulation by GnRH (fig. 11.14).

Evidence suggests that there may be retrograde transport of blood from the anterior pituitary to the hypothalamus. This may permit a *short feedback loop* in which a particular trophic hormone inhibits the secretion of its releasing hormone from the hypothalamus. A high secretion of TSH, for example, may inhibit further secretion of TRH by this means.

In addition to negative feedback control of the anterior pituitary, there is an example of a hormone from a target organ whose action actually stimulates the secretion of an anterior pituitary hormone. Toward the middle of the menstrual cycle, the rising secretion of estradiol from the ovaries stimulates the anterior pituitary to secrete a "surge" of LH, which results in ovulation. This case is commonly referred to as a *positive feedback* effect to distinguish it from the more usual negative feedback inhibition of target gland hormones on anterior pituitary secretion. Interestingly, higher levels of estradiol at a later stage of the menstrual cycle exert the opposite effect—negative feedback inhibition—on LH secretion. The control of gonadotropin secretion is discussed in more detail in chapter 20.

Higher Brain Function and Pituitary Secretion

The relationship between the anterior pituitary and a particular target gland is described as an *axis*; the pituitary-gonad axis, for example, refers to the action of gonadotropic hormones on the testes and ovaries. This axis is stimulated by GnRH from the hypothalamus, as previously described. Since the hypothalamus

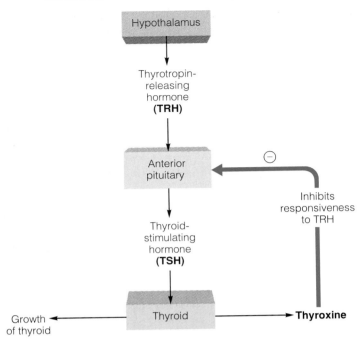

Figure 11.13

The secretion of thyroxine from the thyroid is stimulated by thyroid-stimulating hormone (TSH) from the anterior pituitary. The secretion of TSH is stimulated by thyrotropin-releasing hormone (TRH) secreted from the hypothalamus. This stimulation is balanced by negative feedback inhibition from thyroxine, which decreases the responsiveness of the anterior pituitary to stimulation by TRH.

receives neural input from "higher brain centers," however, it is not surprising that the pituitary-gonad axis can be affected by emotions. Indeed, the ability of intense emotions to alter the timing of ovulation or menstruation is well known. Psychological stress, as another example, also stimulates another axis—the pituitary-adrenal axis (described in the next section).

 The influence of higher brain centers on the pituitary-gonad axis helps to explain the "dormitory effect"—that is, a tendency for the menstrual cycles of female roommates to synchronize. This synchronization of menstrual cycles will not occur in a new roommate if her nasal cavity is plugged with cotton, suggesting that the dormitory effect is due to the action of **pheromones**. Pheromones are chemicals excreted by an individual that act through the olfactory sense and modify the physiology or behavior of another member of the same species. Pheromones are important regulatory molecules in the urine, vaginal fluid, and other secretions of most mammals, and help to regulate their reproductive cycles and behavior. Although pheromones have a more limited function in human physiology and behavior, their significance is difficult to assess because of our frequent cleansing, use of deodorants, and applications of artificial scents.

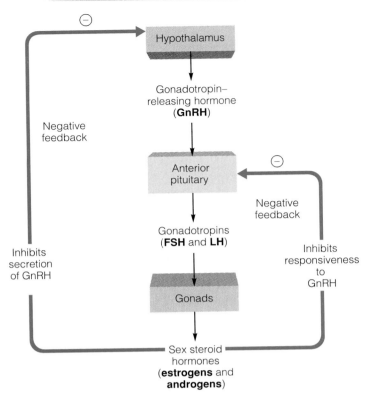

Figure 11.14

Negative feedback control of gonadotropin secretion.

Stressors, as described later in this chapter, produce an increase in CRH secretion from the hypothalamus, which in turn results in elevated ACTH and corticosteroid secretion. In addition, the influence of higher brain centers produces *circadian* ("about a day") *rhythms* in the secretion of many anterior pituitary hormones. The secretion of growth hormone, for example, is highest during sleep and decreases during wakefulness, although its secretion is also stimulated by the absorption of particular amino acids following a meal.

1. Describe the embryonic origins of the adenohypophysis and neurohypophysis, and list the parts of each. Which of these parts is also called the anterior pituitary? Which is called the posterior pituitary?

2. List the hormones secreted by the posterior pituitary. State the site of origin of these hormones and describe the mechanisms by which their secretions are regulated.

3. List the hormones secreted by the anterior pituitary and describe how the hypothalamus controls the secretion of each hormone.

4. Draw a negative feedback loop showing the control of ACTH secretion. Explain how this system would be affected by (a) an injection of ACTH, (b) surgical removal of the pituitary, (c) an injection of corticosteroids, and (d) surgical removal of the adrenal glands.

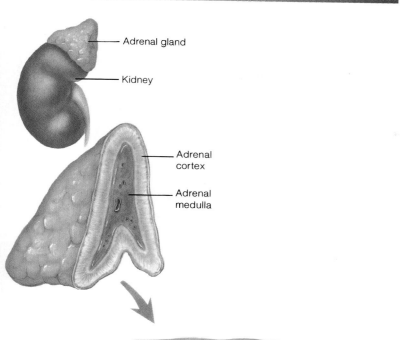

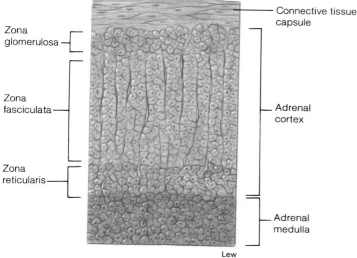

Figure 11.15

The structure of the adrenal gland, showing the three zones of the adrenal cortex.

Adrenal Glands

The adrenal cortex and adrenal medulla are structurally and functionally different. The adrenal medulla secretes catecholamine hormones, which complement the sympathetic nervous system in the "fight-or-flight" reaction. The adrenal cortex secretes steroid hormones that participate in the regulation of mineral and energy balance.

The adrenal glands are paired organs that cap the superior borders of the kidneys (fig. 11.15). Each adrenal consists of an outer cortex and inner medulla, which function as separate glands. The differences in function of the adrenal cortex and medulla are related to the differences in their embryonic derivation. The adrenal medulla is derived from embryonic neural crest ectoderm (the same tissue that produces the sympathetic ganglia), whereas the adrenal cortex is derived from a different embryonic tissue (mesoderm).

As a consequence of its embryonic derivation, the adrenal medulla secretes catecholamine hormones (mainly epinephrine, with lesser amounts of norepinephrine) into the blood in response to stimulation by preganglionic sympathetic nerve fibers (chapter 9). The adrenal cortex does not receive neural innervation, and so must be stimulated hormonally (by ACTH secreted from the anterior pituitary). The cortex consists of three zones: an outer *zona glomerulosa*, a middle *zona fasciculata*, and an inner *zona reticularis* (fig. 11.15). These zones are believed to have different functions.

Functions of the Adrenal Cortex

The adrenal cortex secretes steroid hormones called **corticosteroids,** or **corticoids,** for short. There are three functional categories of corticosteroids: (1) **mineralocorticoids,** which regulate Na^+ and K^+ balance; (2) **glucocorticoids,** which regulate the metabolism of glucose and other organic molecules; and (3) **sex steroids,** which are weak androgens (and lesser amounts of estrogens) that supplement the sex steroids secreted by the gonads. These hormones are secreted by the different zones of the adrenal cortex.

Adrenogenital syndrome is caused by the hypersecretion of adrenal sex hormones, particularly the androgens. Adrenogenital syndrome in young children causes premature puberty and enlarged genitals, especially the penis in males and the clitoris in females. An increase in body hair and a deeper than normal voice are other characteristics. This condition in a mature woman can cause the growth of a beard. A related condition, called **congenital adrenal hyperplasia,** can cause masculinization of a female fetus prior to birth.

Aldosterone is the most potent mineralocorticoid. The mineralocorticoids are produced in the zona glomerulosa (fig. 11.16). The predominant glucocorticoid in humans is *cortisol* (*hydrocortisone*), which is secreted by the zona fasciculata and perhaps also by the zona reticularis. The secretion of cortisol by the zona fasciculata is stimulated by ACTH (fig. 11.17). The secretion of aldosterone is controlled by other mechanisms related to blood volume and electrolyte balance (chapter 14).

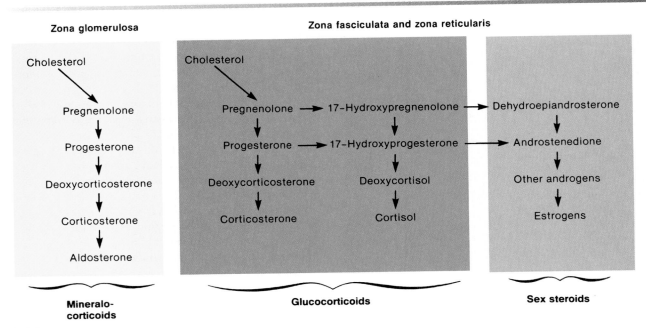

Figure 11.16
Simplified pathways for the synthesis of steroid hormones in the adrenal cortex. The adrenal cortex produces steroids that regulate Na^+ and K^+ balance (mineralocorticoids), steroids that regulate glucose balance (glucocorticoids), and small amounts of sex steroid hormones.

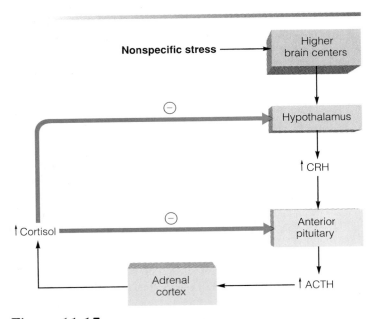

Figure 11.17
The activation of the pituitary-adrenal axis by nonspecific stress. Negative feedback control of the adrenal cortex is also shown.

Hypersecretion of corticosteroids results in **Cushing's syndrome.** This is generally caused by a tumor of the adrenal cortex or by over-secretion of ACTH from the anterior pituitary. Cushing's syndrome is characterized by changes in carbohydrate and protein metabolism, hyperglycemia, hypertension, and muscular weakness. Metabolic problems give the body a puffy appearance and can cause structural changes characterized as "buffalo hump" and "moon face." Similar effects are seen when people with chronic inflammatory diseases receive prolonged treatment with corticosteroids, which are given to reduce inflammation and inhibit the immune response.

Addison's disease is caused by inadequate secretion of both glucocorticoids and mineralocorticoids, which results in hypoglycemia, sodium and potassium imbalance, dehydration, hypotension, rapid weight loss, and generalized weakness. A person with this condition who is not treated with corticosteroids will die within a few days because of the severe electrolyte imbalance and dehydration. President John F. Kennedy had Addison's disease, but few knew of it because it was well controlled by corticosteroids.

Functions of the Adrenal Medulla

The cells of the adrenal medulla secrete **epinephrine** and **norepinephrine** in an approximate ratio of 4 to 1, respectively. The effects of these hormones are similar to those caused by stimulation of the sympathetic nervous system, except that the hormonal effect lasts about ten times longer. The hormones from the adrenal medulla increase cardiac output and heart rate, dilate coronary blood vessels, increase mental alertness, increase the respiratory rate, and elevate metabolic rate. The effects of epinephrine and norepinephrine are compared in table 11.10. The metabolic effects of these hormones are discussed in chapter 19.

The adrenal medulla is innervated by sympathetic nerve fibers. Many stressors, therefore, activate the adrenal medulla as

Table 11.10	Comparison of Adrenal Medullary Hormones	
Epinephrine	**Norepinephrine**
Elevates blood pressure because of increased cardiac output and peripheral vasoconstriction	Elevates blood pressure because of generalized vasoconstriction
Accelerates respiratory rate and dilates respiratory passageways	Similar effect but to a lesser degree
Increases efficiency of muscular contraction	Similar effect but to a lesser degree
Increases rate of glycogen breakdown into glucose, so level of blood glucose rises	Similar effect but to a lesser degree
Increases rate of fatty acid released from fat, so level of blood fatty acids rises	Similar effect but to a lesser degree
Increases release of ACTH and TSH from the adenohypophysis of the pituitary gland	No effect

Source: From Kent M. Van De Graaff, *Human Anatomy,* 4th ed. Copyright © 1995 Wm. C. Brown Communications, Inc., Dubuque, Iowa. Reprinted by permission of Times Mirror Higher Education Group, Inc., Dubuque, Iowa. All Rights Reserved.

A tumor of the adrenal medulla is referred to as a **pheochromocytoma.** This tumor causes hypersecretion of epinephrine and norepinephrine, which produces an effect similar to continuous sympathetic nerve stimulation. The symptoms of this condition are hypertension, elevated metabolism, hyperglycemia and sugar in the urine, nervousness, digestive problems, and sweating. It does not take long for the body to become totally fatigued under these conditions, making the patient susceptible to other diseases.

well as the adrenal cortex. Activation of the adrenal medulla together with the sympathetic nervous system prepares the body for greater physical performance—the *fight-or-flight* response.

Stress and the Adrenal Gland

In 1936, a Canadian physiologist, Hans Selye, discovered that injections of a cattle ovary extract into rats (1) stimulated growth of the adrenal cortex; (2) caused atrophy of the lymphoid tissue of the spleen, lymph nodes, and thymus; and (3) produced bleeding peptic ulcers. At first he thought that these ovarian extracts contained a specific hormone that caused these effects. He later discovered that injections of a variety of substances, including foreign chemicals such as formaldehyde, could produce the same effects. Indeed, the same pattern occurred when he subjected rats to cold environments or when he dropped them into water and made them swim until they were exhausted.

The specific pattern of effects produced by these procedures suggested that these effects were the result of something that the procedures shared in common. Selye reasoned that all of the procedures were stressful. Stress, according to Selye, is the reaction of an organism to stimuli called *stressors,* which may produce damaging effects. The pattern of changes he observed represented a specific response to any stressful agent. He later discovered that stressors produce these effects because they stimulate the pituitary-adrenal axis. Under stressful conditions, there is increased secretion of ACTH from the anterior pituitary, and thus there is increased secretion of glucocorticoids from the adrenal cortex.

On this basis, Selye stated that there is "a nonspecific response of the body to readjust itself following any demand made upon it." A rise in the plasma glucocorticoid levels results from the demands of the stressors. Selye termed this nonspecific response the **general adaptation syndrome (GAS).** Stress, in other words, produces GAS. There are three stages in the response to stress: (1) the *alarm reaction,* when the adrenal glands are activated; (2) the *stage of resistance,* in which readjustment occurs; and (3) if the readjustment is not complete, the *stage of exhaustion,* which may lead to sickness and possibly death.

Selye's concept of stress has been refined by subsequent research. These investigations demonstrate that the sympathoadrenal system becomes activated, with increased secretion of epinephrine and norepinephrine, in response to stressors that challenge the organism to respond physically. This is the "fight-or-flight" reaction described in chapter 9. Different emotions, however, are accompanied by different endocrine responses. The pituitary-adrenal axis, with rising levels of glucocorticoids, becomes more active when the stress is more of a chronic nature and when the person feels less in control and is more passive.

Glucocorticoids, such as hydrocortisone, can inhibit the immune system. Indeed, these steroids are often given medically for this reason to treat various inflammatory diseases and to suppress the immune rejection of a transplanted organ. It seems reasonable, therefore, that the elevated glucocorticoid secretion that can accompany stress may inhibit the ability of the immune system to protect against disease. Indeed, studies suggest that prolonged stress results in an increased incidence of cancer and other diseases.

1. List the categories of corticosteroids and the zones of the adrenal cortex that secrete these hormones.
2. Identify the hormones of the adrenal medulla and describe their effects.
3. Explain how the secretions of the adrenal cortex and adrenal medulla are regulated.
4. Describe how stress affects the secretions of the adrenal cortex and medulla. Why does hypersecretion of the adrenal medullary hormones make a person more susceptible to disease?

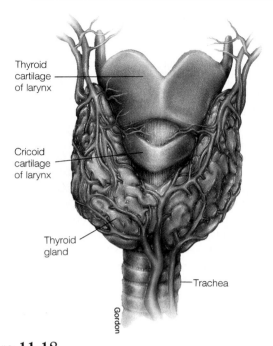

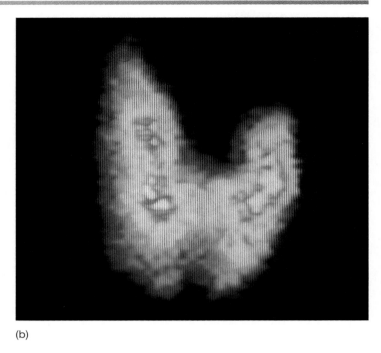

(a)

(b)

Figure 11.18

The thyroid gland. (*a*) Its relationship to the larynx and trachea. (*b*) A scan of the thyroid gland 24 hours after the intake of radioactive iodine.

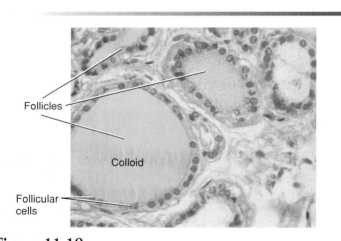

Figure 11.19

A photomicrograph (250×) of a thyroid gland showing numerous thyroid follicles. Each follicle consists of follicular cells surrounding the fluid known as colloid, which contains thyroglobulin.

Thyroid and Parathyroid Glands

The thyroid secretes thyroxine (T₄) and triiodothyronine (T₃), which are needed for proper growth and development and which are primarily responsible for determining the basal metabolic rate (BMR). The parathyroid glands secrete parathyroid hormone, which helps to raise the blood Ca⁺⁺ concentration.

The thyroid gland is located just below the larynx (fig. 11.18). Its two lobes are positioned on either side of the trachea and are connected anteriorly by a medial mass of thyroid tissue called the *isthmus*. The thyroid is the largest of the endocrine glands, weighing between 20 and 25 g.

On a microscopic level, the thyroid gland consists of many spherical hollow sacs called **thyroid follicles** (fig. 11.19). These follicles are lined with a simple cuboidal epithelium composed of *follicular cells* that synthesize the principal thyroid hormone, thyroxine. The interior of the follicles contains *colloid*, a protein-rich fluid. Between the follicles are epithelial cells called *parafollicular cells*; these cells produce a hormone called calcitonin (or thyrocalcitonin).

Production and Action of Thyroid Hormones

The thyroid follicles actively accumulate iodide (I⁻) from the blood and secrete it into the colloid. Once the iodide is in the colloid, it is oxidized to iodine and attached to specific amino acids (tyrosines) within the polypeptide chain of a protein called **thyroglobulin.** The attachment of one iodine to tyrosine produces *monoiodotyrosine (MIT)*; the attachment of two iodines produces *diiodotyrosine (DIT)*.

Within the colloid, enzymes modify the structure of MIT and DIT and couple them together (fig. 11.20). When two DIT molecules that are appropriately modified are coupled together, a molecule of **tetraiodothyronine (T₄),** or **thyroxine,** is produced. The combination of one MIT with one DIT forms **triiodothyronine (T₃).** Note that within the colloid T₄ and T₃

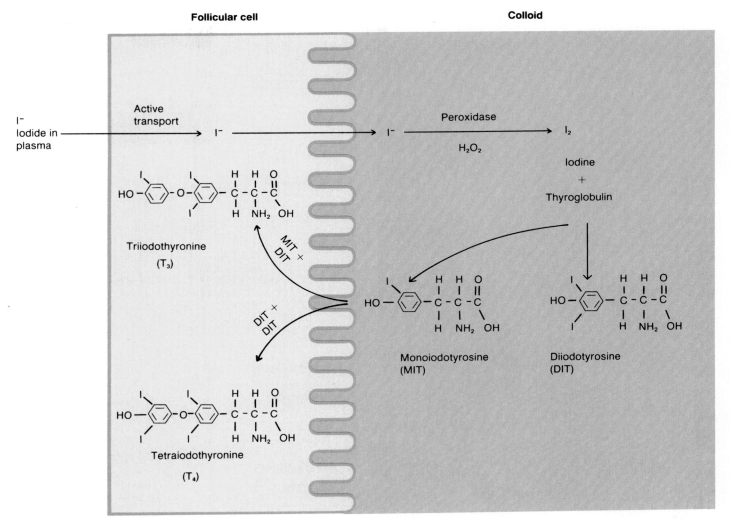

Follicular cell   **Colloid**

Figure 11.20

Stages in the formation and secretion of thyroid hormones. Iodide is actively accumulated by the follicular cells. In the colloid it is converted into iodine and attached to tyrosine amino acids within the thyroglobulin protein. Pinocytosis of iodinated thyroglobulin, coupling of MIT and DIT, and the release of thyroid hormones are stimulated by TSH from the anterior pituitary.

are still attached to thyroglobulin. Upon stimulation by TSH, the cells of the follicle take up a small volume of colloid by pinocytosis, hydrolyze the T_3 and T_4 from the thyroglobulin, and secrete the free hormones into the blood.

The transport of thyroid hormones in the blood and their mechanism of action at the cellular level has been previously described. Through the activation of genes, thyroid hormones stimulate protein synthesis, promote maturation of the nervous system, and increase the rate of energy utilization by the body.

The development of the central nervous system is particularly dependent on thyroid hormones, and a deficiency of these hormones during development can cause serious mental retardation. The basal metabolic rate (BMR)—which is the minimum rate of caloric expenditure by the body—is determined to a large degree by the level of thyroid hormones in the blood. The physiological functions of thyroid hormones are described in more detail in chapter 19.

Diseases of the Thyroid

Thyroid-stimulating hormone (TSH) from the anterior pituitary stimulates the thyroid to secrete thyroxine and exerts a trophic effect on the thyroid gland. This trophic effect is immediately apparent in people who develop an **iodine-deficiency (endemic) goiter** (fig. 11.21). In the absence of sufficient dietary iodine the thyroid cannot produce adequate amounts of T_4 and T_3. The resulting lack of negative feedback inhibition causes abnormally high levels of TSH secretion, which in turn stimulate the abnormal growth of the thyroid (a goiter). These events are summarized in figure 11.22.

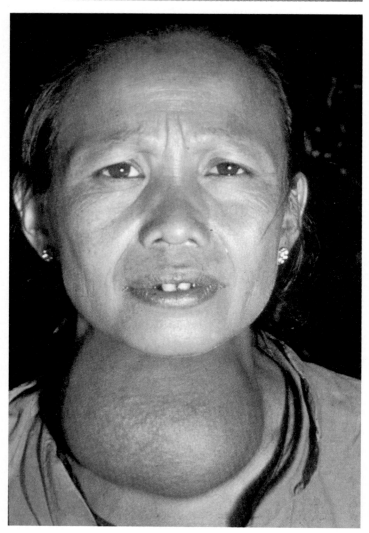

Figure 11.21

A simple or endemic goiter is caused by insufficient iodine in the diet.

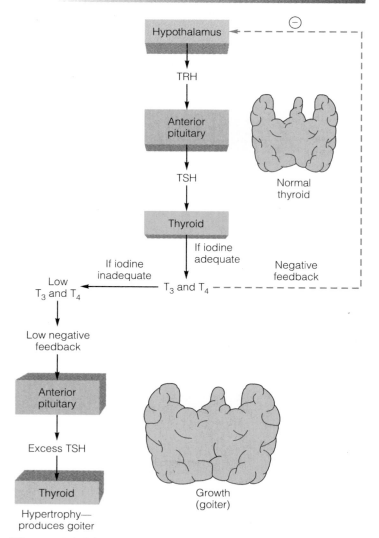

Figure 11.22

Lack of adequate iodine in the diet interferes with the negative feedback control of TSH secretion, resulting in the formation of an endemic goiter.

The infantile form of **hypothyroidism** is known as *cretinism*. An affected child usually appears normal at birth because thyroxine is received from the mother through the placenta. The clinical symptoms of cretinism are stunted growth, thickened facial features, abnormal bone development, mental retardation, low body temperature, and general lethargy. If cretinism is diagnosed early, it can be successfully treated by administering thyroxine.

Hypothyroidism in an adult causes *myxedema*. This disorder affects body fluids, causing an edema characterized by the accumulation of mucopolysaccharides. Symptoms of myxedema include a low metabolic rate, lethargy, and a tendency to gain weight. This condition is treated with thyroxine or with tri-iodothyronine, which are taken orally (as pills).

Graves' disease, also called **toxic goiter,** involves growth of the thyroid associated with hypersecretion of thyroxine. This hyperthyroidism is produced by antibodies that act like TSH and stimulate the thyroid; it is an autoimmune disease. As a consequence of high levels of thyroxine secretion, the metabolic rate and heart rate increase, there is loss of weight, and the autonomic nervous system induces excessive sweating. In about half of the cases, *exophthalmos* (bulging of the eyes) also develops (fig. 11.23) because of edema in the tissues of the eye sockets and swelling of the extrinsic eye muscles. Graves' disease is more common in women than in men, and in smokers than in nonsmokers.

Radiotherapy of Thyroid Cancer

The treatment of thyroid cancer illustrates how the principles previously presented find practical application. Thyroid cancer, particularly when it occurs in younger adults, has a more optimistic prognosis than most other forms of cancer. It is extremely slow growing and usually spreads only to the lymph nodes of the neck (although other sites can be invaded). Surgical treatment involves removing most of the thyroid and the

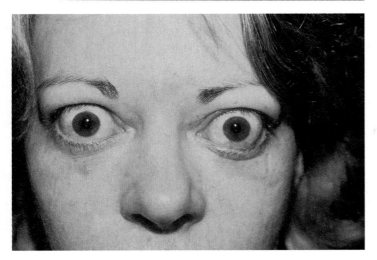

Figure 11.23

Hyperthyroidism is characterized by an increased metabolic rate, weight loss, muscular weakness, and nervousness. The eyes may also protrude.

affected cervical lymph nodes. This surgical treatment is followed by having the patient swallow solutions containing radioactive iodine (^{131}I), which is selectively transported only into cells of the thyroid gland and into cancerous thyroid cells that have metastasized (traveled) to other regions of the body. These cells are then killed by the radioactive iodine.

Thyroid-stimulating hormone (TSH) stimulates the active accumulation of iodine into thyroid cells so that these cells can produce and secrete thyroxine. Therefore, in order for the ingested radioactive iodine to be maximally effective in the treatment of thyroid cancer, the blood levels of TSH must be raised to high levels.

When most of the patient's thyroid gland is surgically removed, the blood levels of thyroxine gradually decline. Owing to the fact that most of the thyroxine is bound to protein in the plasma, it has an extremely long half-life. As the thyroxine levels in the blood decline, the negative feedback inhibition of TSH secretion lessens and TSH levels consequently rise. A period of about 4 to 5 weeks is required after removal of the thyroid for the TSH levels to rise sufficiently. At this point, there is enough TSH to stimulate accumulation of radioactive iodine into thyroid cells and promote the destruction of both normal and metastasized cells.

Parathyroid Glands

The small, flattened parathyroid glands are embedded in the posterior surfaces of the lateral lobes of the thyroid gland, as shown in figure 11.24. There are usually four parathyroid glands: a *superior* and an *inferior pair*, although the precise number can vary. Each parathyroid gland is a small yellowish-brown body 3 to 8 mm (0.1 to 0.3 in.) long, 2 to 5 mm (0.07 to 0.2 in.) wide, and about 1.5 mm (0.05 in.) deep.

Parathyroid hormone (PTH) is the only hormone secreted by the parathyroid glands. PTH, however, is the single most important hormone in the control of the calcium levels of

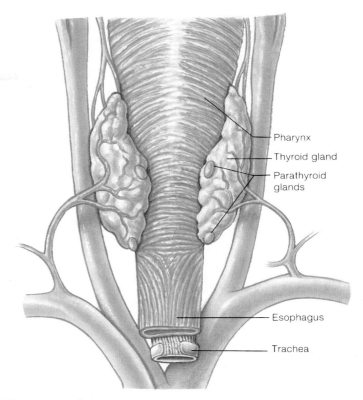

Figure 11.24

A posterior view of the parathyroid glands.

the blood. It promotes a rise in blood calcium levels by acting on the bones, kidneys, and intestine (fig. 11.25). Regulation of calcium balance is described in more detail in chapter 19.

1. Describe the structure of the thyroid gland and list the effects of thyroid hormones.
2. Describe how thyroid hormones are produced and how their secretion is regulated.
3. Explain the consequences of an inadequate dietary intake of iodine.

Pancreas and Other Endocrine Glands

The islets of Langerhans in the pancreas secrete two hormones, insulin and glucagon. Insulin promotes the lowering of blood glucose and the storage of energy in the form of glycogen and fat. Glucagon has antagonistic effects that act to raise the blood glucose concentration. Additionally, many other organs secrete hormones that help to regulate digestion, metabolism, growth, immune function, and reproduction.

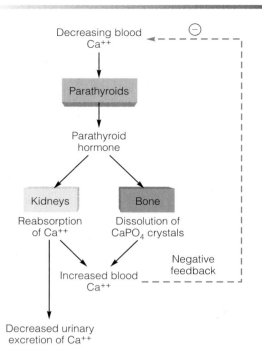

Figure 11.25

The actions of parathyroid hormone and the negative feedback control of its secretion. An increased level of parathyroid hormone causes the bones to release calcium and the kidneys to conserve calcium that would otherwise be lost through the urine.

The pancreas is both an endocrine and an exocrine gland. The gross structure of this gland and its exocrine functions in digestion are described in chapter 17. The endocrine portion of the pancreas consists of scattered clusters of cells called the **islets of Langerhans,** or **pancreatic islets.** These endocrine structures are most common in the body and tail of the pancreas (fig. 11.26).

Islets of Langerhans

On a microscopic level, the most conspicuous cells in the islets are the *alpha* and *beta* cells (fig. 11.27). The alpha cells secrete the hormone **glucagon,** and the beta cells secrete **insulin.**

Alpha cells secrete glucagon in response to a fall in the blood glucose concentrations. Glucagon stimulates the liver to hydrolyze glycogen to glucose (*glycogenolysis*), which causes the blood glucose level to rise. This effect represents the completion of a negative feedback loop. Glucagon also stimulates the hydrolysis of stored fat (*lipolysis*) and the consequent release of free fatty acids into the blood. This effect helps to provide energy substrates for the body during fasting, when blood glucose levels decrease. Glucagon, together with other hormones, also stimulates the conversion of fatty acids to ketone bodies, which can be secreted by the liver into the blood and used by other organs as an energy source. Glucagon is thus a hormone that helps to maintain homeostasis during times of fasting, when the body's energy reserves must be utilized (chapter 19).

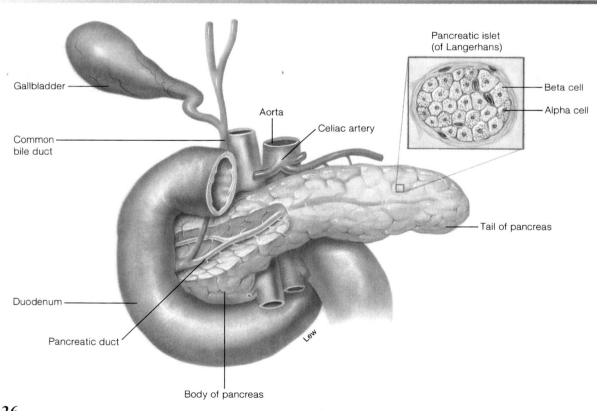

Figure 11.26

The pancreas and the associated pancreatic islets (islets of Langerhans).

Chapter Eleven

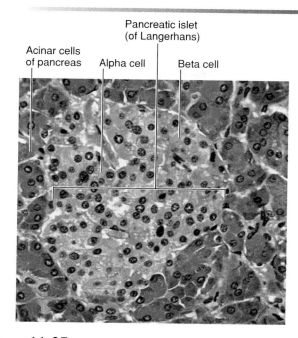

Acinar cells of pancreas Alpha cell Beta cell Pancreatic islet (of Langerhans)

Figure 11.27

The histology of the pancreatic islets (of Langerhans).

Beta cells secrete insulin in response to a rise in the blood glucose concentrations. Insulin promotes the entry of glucose into tissue cells, and the conversion of this glucose into energy storage molecules of glycogen and fat. Insulin also aids the entry of amino acids into cells and the production of cellular protein. The actions of insulin and glucagon are thus antagonistic. After a meal, insulin secretion is increased and glucagon secretion is decreased; fasting, by contrast, causes a rise in glucagon and a fall in insulin secretion.

Pineal Gland

The small, cone-shaped pineal gland is located in the roof of the third ventricle near the corpora quadrigemina, where it is encapsulated by the meninges covering the brain. The pineal gland of a child weighs about 0.2 g and is 5 to 8 mm (0.2 to 0.3 in.) long and 9 mm wide. The gland begins to regress in size at about age 7 and in the adult appears as a thickened strand of fibrous tissue. Although the pineal gland lacks direct nervous connections to the rest of the brain, it is highly innervated by the sympathetic nervous system from the superior cervical ganglion.

The principal hormone of the pineal is **melatonin.** Production and secretion of this hormone is stimulated by activity of the *suprachiasmatic nucleus (SCN)* in the hypothalamus of the brain via activation of sympathetic neurons to the pineal. Activity of the SCN, and thus secretion of melatonin, is highest at night. During the day, neural pathways from the retina of the eyes depress the activity of the SCN, reducing sympathetic stimulation of the pineal and thus decreasing melatonin secretion.

The pineal gland has been implicated in a variety of physiological processes. One of the most widely studied is the ability of melatonin to inhibit the pituitary-gonad axis in some species. Indeed, a decrease in melatonin secretion in many species is responsible for maturation of the gonads during their reproductive season. Excessive melatonin secretion in humans

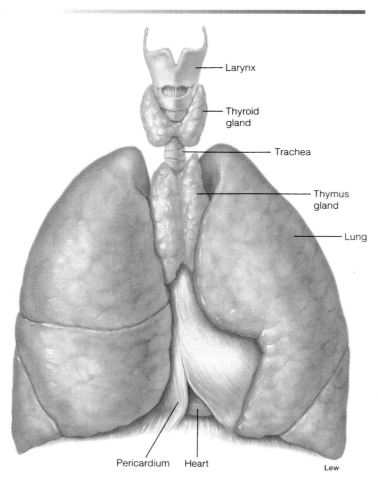

Figure 11.28

The thymus is a bilobed organ within the mediastinum of the thorax.

is associated with a delay in the onset of puberty. Research findings indicate that melatonin secretion is highest in children between the ages of 1 and 5 and decreases thereafter, reaching its lowest levels at the end of puberty, when concentrations are 75% lower than during early childhood. Because of much conflicting data, however, the importance of melatonin in influencing the time of puberty is highly controversial.

Thymus

The thymus is a bilobed organ positioned in front of the aorta and behind the manubrium of the sternum (fig. 11.28). Although the size of the thymus varies considerably from person to person, it is relatively large in newborns and children and sharply regresses in size after puberty. Besides decreasing in size, the thymus of adults becomes infiltrated with strands of fibrous and fatty connective tissue.

The thymus serves as the site of production of **T cells** (*thymus-dependent cells*), which are the lymphocytes involved in cell-mediated immunity (chapter 15). In addition to providing T cells, the thymus secretes a number of hormones that are believed to stimulate T cells after they leave the thymus.

Gastrointestinal Tract

The stomach and small intestine secrete a number of hormones that act on the gastrointestinal tract itself and on the pancreas and gallbladder (chapter 18). The effects of these hormones act together with regulation by the autonomic nervous system to coordinate the activities of different regions of the digestive tract and the secretions of pancreatic juice and bile.

Gonads and Placenta

The gonads (**testes and ovaries**) secrete **sex steroids.** These include male sex hormones, or **androgens,** and female sex hormones—**estrogens** and **progestogens.** The principal hormones in each of these categories are *testosterone, estradiol-17β,* and *progesterone,* respectively.

The testes consist of two compartments: *seminiferous tubules,* which produce sperm, and *interstitial tissue* between the convolutions of the tubules. Within the interstitial tissue are *Leydig cells,* which secrete testosterone. Testosterone is needed for the development and maintenance of the male genitalia (penis and scrotum) and male sex accessory organs (prostate, seminal vesicles, epididymides, and vas deferens), as well as for the development of male secondary sexual characteristics.

During the first half of the menstrual cycle, estrogen is secreted by many small structures within the ovary called *ovarian follicles.* These follicles contain the egg cell, or *ovum,* and *granulosa cells* that secrete estrogen. By about midcycle, one of these follicles grows very large and, in the process of ovulation, extrudes its ovum from the ovary. The empty follicle, under the influence of luteinizing hormone (LH) from the anterior pituitary, then becomes a new endocrine structure called a *corpus luteum.* The corpus luteum secretes progesterone as well as estradiol-17β.

The **placenta**—the organ responsible for nutrient and waste exchange between the fetus and mother—is also an endocrine gland in that it secretes large amounts of estrogens and progesterone. In addition, it secretes a number of polypeptide and protein hormones that are similar to some hormones secreted by the anterior pituitary. These hormones include **human chorionic gonadotropin (hCG),** which is similar to LH, and **somatomammotropin,** which is similar in action to both growth hormone and prolactin. The physiology of the placenta and other aspects of reproductive endocrinology are covered in chapter 20.

1. Describe the structure of the endocrine pancreas and state the sites of origin of insulin and glucagon.

2. Describe how insulin and glucagon secretion are affected by eating and by fasting and explain the actions of these two hormones.

3. Describe the location of the pineal gland and the possible functions of melatonin.

4. Describe the location and function of the thymus.

5. Explain how the hormones secreted by the gonads and placenta are categorized and list the hormones secreted by each gland.

Table 11.11 Examples of Autocrine and Paracrine Regulators

Autocrine or Paracrine Regulator	Major Sites of Production	Major Actions
Insulin-like growth factors (somatomedins)	Many organs, particularly the liver and cartilages	Growth and cell division
Nitric oxide	Endothelium of blood vessels; neurons; macrophages	Dilation of blood vessels; neural messenger; antibacterial agent
Endothelins	Endothelium of blood vessels; other organs	Constriction of blood vessels; other effects
Platelet-derived growth factor	Platelets; macrophages; vascular smooth muscle cells	Cell division within blood vessels
Epidermal growth factors	Epidermal tissues	Cell division in wound healing
Neurotrophins	Schwann cells; neurons	Regeneration of peripheral nerves
Bradykinin	Endothelium of blood vessels	Dilation of blood vessels
Interleukins (cytokines)	Macrophages; lymphocytes	Regulation of immune system
Prostaglandins	Many tissues	Wide variety (see text)

Autocrine and Paracrine Regulation

Many regulatory molecules produced throughout the body act within the organs that produce them. These molecules may regulate different cells within one tissue, or they may be produced within one tissue and regulate a different tissue within the same organ.

Thus far in this text, two types of regulatory molecules have been considered—neurotransmitters in chapter 7 and hormones in the present chapter. These two classes of regulatory molecules cannot be defined simply by differences in chemical structure, since on this basis the same molecule (such as norepinephrine) may be included in both categories; rather, they must be defined by function. Neurotransmitters are released by axons, travel across a narrow synaptic cleft, and affect a postsynaptic cell. Hormones are secreted into the blood by an endocrine gland and, through transport in the blood, influence the activities of one or more target organs.

There are yet other classes of regulatory molecules. These molecules are distinguished by the fact that they are produced in many different organs and are active within the organ in which they are produced. Molecules of this type are called **autocrine regulators** if they are produced and act within the same tissue of an organ. They are called **paracrine regulators** if they are produced within one tissue and regulate a different tissue of the same organ (table 11.11). In the following discussion, for the sake of simplicity and because the same chemical can function as an autocrine or a paracrine regulator in different situations, the term *autocrine* will be used in a generic sense to refer to both types of local regulation.

Examples of Autocrine Regulation

Many autocrine regulatory molecules are also known as **cytokines,** particularly if they regulate different cells of the immune system, and as **growth factors,** if they promote growth and cell division in any organ. This distinction is somewhat blurred, however, because some cytokines may also function as growth factors. Cytokines produced by lymphocytes (the type of white blood cell involved in specific immunity—see chapter 15) are also known as **lymphokines,** and the specific molecules involved are called *interleukins*. The terminology can be confusing because new regulatory molecules, and new functions for previously named regulatory molecules, are being discovered at a rapid pace.

Cytokines secreted by macrophages (a type of phagocytic cell found in connective tissues) and lymphocytes stimulate proliferation of specific cells involved in the immune response (chapter 15). **Neurotrophins,** such as *nerve growth factor*, guide regenerating peripheral neurons that have been injured (chapter 7). Nitric oxide, which can function as a neurotransmitter involved in memory (chapters 7 and 8) and other functions, is also produced by the endothelium of blood vessels. In this context, it is a paracrine regulator because it diffuses to the smooth muscle layer of the blood vessel and promotes smooth muscle relaxation, leading to dilation of the blood vessel. In this action, nitric oxide functions as the regulator previously known as *endothelium-derived relaxation factor*. Neural and paracrine regulation interact in this case, since autonomic axons that release acetylcholine in blood vessels cause dilation by stimulating the synthesis of nitric oxide in those vessels.

The endothelium of blood vessels also produces other paracrine regulators. These include the *endothelins* (specifically *endothelin-1* in humans), which directly promote vasoconstriction, and *bradykinin*, which promotes vasodilation. These regulatory molecules are thus very important in the control of blood flow and blood pressure (chapter 14). They are also involved in the development of atherosclerosis, the leading cause of heart disease and stroke (chapter 13). In addition, endothelin-1 is

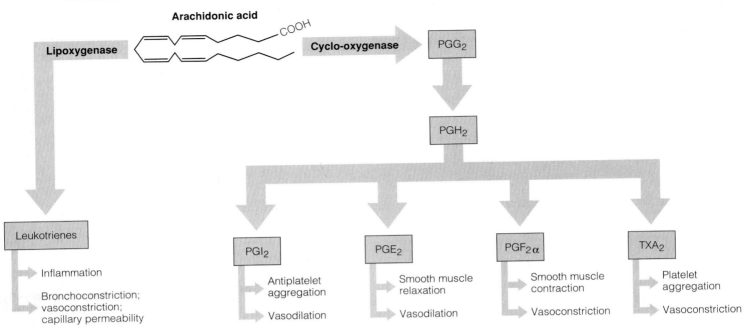

Figure 11.29
The formation of leukotrienes and prostaglandins and the actions of these autocrine regulators (PG = prostaglandin; TX = thromboxane).

produced by the epithelium of the airways and may be important in the embryological development and function of the respiratory system.

All autocrine regulators control gene expression in their target cells to some degree. This is very clearly the case with the various growth factors, such as *platelet-derived growth factor, epidermal growth factor,* and the *insulin-like growth factors* that stimulate cell division and proliferation of their target cells. Regulators in the last group interact with the endocrine system in a number of ways, as will be described in chapter 19.

Prostaglandins

The most diverse group of autocrine regulators are the **prostaglandins.** A prostaglandin is a twenty-carbon-long fatty acid that contains a five-membered carbon ring. This molecule is derived from the precursor molecule *arachidonic acid,* which can be released from phospholipids in the cell membrane under hormonal or other stimulation. Arachidonic acid can then enter one of two possible metabolic pathways. In one case, the arachidonic acid is converted by the enzyme *cyclo-oxygenase* into a prostaglandin, which can then be changed by other enzymes into other prostaglandins. In the other case, arachidonic acid is converted by the enzyme *lipoxygenase* into **leukotrienes,** which are compounds that are closely related to the prostaglandins (fig. 11.29).

Prostaglandins are produced in almost every organ and have been implicated in a wide variety of regulatory functions. The study of prostaglandin function can be confusing, in part because a given prostaglandin may have opposite effects in different organs. Prostaglandins of the E series (PGE), for example, cause smooth muscle to relax in the bladder, bronchioles, intestine, and uterus, but the same molecules cause vascular smooth muscle to contract. A different prostaglandin, designated $PGF_{2\alpha}$, has exactly the opposite effects.

The antagonistic effects of prostaglandins on blood clotting make good physiological sense. Blood platelets, which are required for blood clotting, produce *thromboxane A2.* This prostaglandin promotes clotting by stimulating platelet aggregation and vasoconstriction. The endothelial cells of blood vessels, by contrast, produce a different prostaglandin, known as PGI_2, in *prostacyclin,* which has the opposite effects—it inhibits platelet aggregation and causes vasodilation. These antagonistic effects help to promote clotting, while at the same time ensuring that clots will not normally form on the walls of intact blood vessels.

Examples of Prostaglandin Actions

The following are some of the regulatory functions proposed for prostaglandins in different systems of the body:

1. **Immune system.** Prostaglandins promote many aspects of the inflammatory process, including the development of pain and fever. Drugs that inhibit prostaglandin synthesis help to alleviate these symptoms.

2. **Reproductive system.** Prostaglandins may play a role in ovulation and corpus luteum function in the ovaries and in contraction of the uterus. Excessive prostaglandin production may be involved in premature labor, endometriosis, dysmenorrhea (painful menstrual cramps), and other gynecological disorders.

Chapter Eleven

3. **Digestive system.** The stomach and intestines produce prostaglandins, which are believed to inhibit gastric secretions and influence intestinal motility and fluid absorption. Since prostaglandins inhibit gastric secretion, drugs that suppress prostaglandin production may make a patient more susceptible to peptic ulcers.

4. **Respiratory system.** Some prostaglandins cause constriction whereas others cause dilation of blood vessels in the lungs and of bronchiolar smooth muscle. The leukotrienes are potent bronchoconstrictors, and these compounds, together with some prostaglandins, may cause respiratory distress and contribute to bronchoconstriction in asthma.

5. **Circulatory system.** Some prostaglandins are vasoconstrictors and others are vasodilators. In a fetus, PGE_2 is believed to promote dilation of the ductus arteriosus—a short vessel that connects the pulmonary artery with the aorta. After birth, the ductus arteriosus normally closes as a result of a rise in blood oxygen when the baby breathes. If the ductus remains patent (open), however, it can be closed by the administration of drugs that inhibit prostaglandin synthesis. Thromboxane A_2, a vasoconstrictor, and prostacyclin, a vasodilator, play a role in blood clotting, as previously described.

6. **Urinary system.** Prostaglandins are produced in the renal medulla and cause vasodilation, resulting in increased renal blood flow and increased excretion of water and electrolytes in the urine.

Inhibitors of Prostaglandin Synthesis

Aspirin is the most widely used member of a class of drugs known as **nonsteroidal anti-inflammatory drugs (NSAID).** Other members of this class are indomethacin and ibuprofen. These drugs produce their effects because they specifically inhibit the cyclo-oxygenase enzyme that is needed for prostaglandin synthesis. Through this action, the drugs inhibit inflammation but produce some unwanted side effects, including gastric bleeding, possible kidney problems, and prolonged clotting time.

It is now known that there are two isoenzyme forms (chapter 4) of cyclo-oxygenase. The type I form is produced constitutively (that is, in a constant fashion) by cells of the stomach and kidneys and by blood platelets, which are cellular structures involved in blood clotting (chapter 13). The type II form of the enzyme is induced in a number of cells in response to cytokines involved in inflammation, and the prostaglandins produced by this isoenzyme promote the inflammatory condition.

When aspirin and the other drugs of its class inhibit the type I isoenzyme of cyclo-oxygenase, the synthesis of prostacyclin is inhibited. Since prostacyclin protects the stomach lining, inhibition of its synthesis with aspirin may be responsible for the stomach irritation caused by aspirin and indomethacin, the two most potent inhibitors of the type I isoenzyme. Inhibition of the type I isoenzyme is thus responsible for the negative side effects of these drugs. Inhibition of the type II isoenzyme is responsible for the intended anti-inflammatory benefits of the drugs, and research is currently underway to develop new drugs that more selectively inhibit the type II isoenzyme of cyclo-oxygenase.

There is, however, one important benefit derived from the inhibition of the type I isoenzyme by aspirin. The type I isoenzyme is the form of cyclo-oxygenase present in blood platelets, where it is needed for the production of thromboxane A_2. Since this prostaglandin is needed for platelet aggregation, inhibition of its synthesis by aspirin reduces the ability of the blood to clot. While this can have negative consequences in some circumstances, low doses of aspirin have been shown to significantly reduce the risk of heart attacks and strokes (chapter 13) by reducing platelet function. It should be noted that this beneficial effect is produced by lower doses of aspirin than are commonly taken to reduce inflammation.

---

1. Explain the nature of autocrine regulation. How does it differ from regulation by hormones and neurotransmitters?

2. List some of the autocrine regulators produced by blood vessels and describe their actions. Also, identify specific growth factors and describe their actions.

3. Describe the chemical nature of prostaglandins. List some of the different forms of prostaglandins and describe their actions.

4. Explain the significance of the isoenzymatic forms of cyclo-oxygenase in the action of nonsteroidal anti-inflammatory drugs.

Summary

Endocrine Glands and Hormones   p. 272

I. Hormones are chemicals, including steroids, catecholamines, and polypeptides, that are secreted into the blood by endocrine glands.

II. Precursors of active hormones may be called either prohormones or prehormones.
   A. Prohormones are relatively inactive precursor molecules made in the endocrine cells.

B. Prehormones are the normal secretions of an endocrine gland that, in order to be active, must be converted to other derivatives by target cells.

III. Hormones can interact in permissive, synergistic, or antagonistic ways.

IV. The effects of a hormone in the body depend on its concentration.
   A. Abnormally high amounts of a hormone can result in atypical effects.

B. Target tissues can become desensitized by high hormone concentrations.

Mechanisms of Hormone Action   p. 277

I. Steroid and thyroid hormones enter their target cells and bind to receptor proteins.
   A. Thyroid hormones attach to chromatin-bound receptors located in the nucleus.

B. Steroid hormones bind to cytoplasmic receptor proteins and translocate to the nucleus.

C. Attachment of the hormone-receptor protein complex to the chromatin activates genes and thereby stimulates RNA and protein synthesis.

II. Polar hormones bind to receptor proteins on the outer surface of the target cell membrane, indirectly activating a specific enzyme in the cell membrane.

A. In the case of some hormones, activation of adenylate cyclase occurs, resulting in the intracellular production of cyclic AMP.

B. Cyclic AMP activates a cytoplasmic enzyme called protein kinase. Protein kinase phosphorylates specific enzyme proteins and thereby changes the metabolism of the target cell for those hormones that use cAMP as a second messenger.

C. In the case of other hormone actions, the enzyme phospholipase C is activated in the cell membrane, causing the release of inositol triphosphate into the cell.

D. Inositol triphosphate stimulates the release of Ca^{++} from the endoplasmic reticulum of the target cell.

E. Ca^{++} binds to regulatory proteins, such as calmodulin, that can influence the metabolism of the target cell.

Pituitary Gland   p. 283

I. The pituitary gland secretes eight hormones.

A. The anterior pituitary secretes growth hormone, thyroid-stimulating hormone, adrenocorticotropic hormone, follicle-stimulating hormone, luteinizing hormone, and prolactin.

B. The posterior pituitary secretes antidiuretic hormone (also called vasopressin) and oxytocin.

II. The hormones of the posterior pituitary are produced in the hypothalamus and transported to the posterior pituitary by the hypothalamo-hypophyseal nerve tract.

III. Secretions of the anterior pituitary are controlled by hypothalamic hormones that stimulate or inhibit secretions of the anterior pituitary.

A. Hypothalamic hormones include TRH, CRH, GnRH, PIH, somatostatin, and a growth hormone–releasing hormone.

B. These hormones are carried to the anterior pituitary by the hypothalamo-hypophyseal portal system.

IV. Secretions of the anterior pituitary are also regulated by the feedback (usually negative feedback) of hormones from the target glands.

V. Higher brain centers, acting through the hypothalamus, can influence pituitary secretion.

Adrenal Glands   p. 289

I. The adrenal cortex secretes mineralocorticoids (mainly aldosterone), glucocorticoids (mainly cortisol), and sex steroids (primarily weak androgens).

A. The glucocorticoids help to regulate energy balance; they also can inhibit inflammation and suppress immune function.

B. The pituitary-adrenal axis is stimulated by stress as part of the general adaptation syndrome.

II. The adrenal medulla secretes epinephrine and lesser amounts of norepinephrine. These hormones complement the action of the sympathetic nervous system.

Thyroid and Parathyroid Glands   p. 292

I. The thyroid follicles secrete tetraiodothyronine (T_4, or thyroxine) and lesser amounts of triiodothyronine (T_3).

A. These hormones are formed within the colloid of the thyroid follicles.

B. The parafollicular cells of the thyroid secrete the hormone calcitonin, which may act to lower blood calcium levels.

II. The parathyroids are small structures embedded within the thyroid gland. They secrete a hormone that promotes a rise in blood calcium levels.

Pancreas and Other Endocrine Glands   p. 295

I. Beta cells in the islets secrete insulin; alpha cells secrete glucagon.

A. Insulin lowers blood glucose and stimulates the production of glycogen, fat, and protein.

B. Glucagon raises blood glucose by stimulating the breakdown of liver glycogen. It also promotes lipolysis and the formation of ketone bodies.

C. The secretion of insulin is stimulated by a rise in blood glucose following meals; the secretion of glucagon is stimulated by a fall in blood glucose during periods of fasting.

II. The pineal gland, located on the roof of the third ventricle of the brain, secretes melatonin. This hormone may play a role in regulating reproductive function.

III. The thymus is the site of T cell lymphocyte production and secretes a number of hormones that may help to regulate the immune system.

IV. The gastrointestinal tract secretes a number of hormones that help to regulate functions of the digestive system.

V. The gonads secrete sex steroid hormones.

A. Leydig cells in the interstitial tissue of the testes secrete testosterone and other androgens.

B. Granulosa cells of the ovarian follicles secrete estrogen.

C. The corpus luteum of the ovaries secretes progesterone, as well as estrogen.

VI. The placenta secretes estrogen, progesterone, and a variety of polypeptide hormones that have actions similar to some anterior pituitary hormones.

Autocrine and Paracrine Regulation p. 299

I. Autocrine regulators are produced and act within the same tissue of an organ, whereas paracrine regulators are produced within one tissue and regulate a different tissue of the same organ. Both types are local regulators—they do not travel in the blood.

II. Prostaglandins are special twenty-carbon-long fatty acids produced by many different organs. They usually have regulatory functions within the organ in which they are produced.

Clinical Investigation

A male patient is found to have hypertension and hyperglycemia. The results of an oral glucose tolerance test for insulin action are within the normal range. Blood tests reveal that the patient has normal catecholamine levels and normal levels of T_4 and T_3, but his blood cortisol levels are abnormally high. He has a generalized "puffiness," but not myxedema. He does not have a history of chronic inflammation and has not been taking immunosuppressive drugs. Assay of the blood ACTH concentrations reveals levels that are only about one-fiftieth of normal. What might account for this patient's symptoms?

Clues

Read the sections "Feedback Control of the Anterior Pituitary," "Functions of the Adrenal Cortex," "Functions of the Adrenal Medulla," "Production and Action of Thyroid Hormones," and "Islets of Langerhans."

Review Activities

Objective Questions

1. Which of the following statements about hypothalamic-releasing hormones is *true?*
   a. They are secreted into capillaries in the median eminence.
   b. They are transported by portal veins to the anterior pituitary.
   c. They stimulate the secretion of specific hormones from the anterior pituitary.
   d. All of the above are true.

2. The hormone primarily responsible for setting the basal metabolic rate and for promoting the maturation of the brain is
   a. cortisol.
   b. ACTH.
   c. TSH.
   d. thyroxine.

3. Which of the following statements about the adrenal cortex is *true?*
   a. It is not innervated by nerve fibers.
   b. It secretes some androgens.
   c. The zona glomerulosa secretes aldosterone.
   d. The zona fasciculata is stimulated by ACTH.
   e. All of the above are true.

4. Which of the following statements about the hormone insulin is *true?*
   a. It is secreted by alpha cells in the islets of Langerhans.
   b. It is secreted in response to a rise in blood glucose.
   c. It stimulates the production of glycogen and fat.
   d. Both *a* and *b* are true.
   e. Both *b* and *c* are true.

Match the hormone with the primary agent that stimulates its secretion.

5. epinephrine
6. thyroxine
7. corticosteroids
8. ACTH

   a. TSH
   b. ACTH
   c. growth hormone
   d. sympathetic nerves
   e. CRH

9. Steroid hormones are secreted by
   a. the adrenal cortex.
   b. the gonads.
   c. the thyroid.
   d. both *a* and *b*.
   e. both *b* and *c*.

10. The secretion of which of the following hormones would be *increased* in a person with endemic goiter?
    a. TSH
    b. thyroxine
    c. triiodothyronine
    d. all of the above

11. Which of the following hormones uses cAMP as a second messenger?
    a. testosterone
    b. cortisol
    c. insulin
    d. epinephrine

12. Which of the following terms best describes the interactions of insulin and glucagon?
    a. synergistic
    b. permissive
    c. antagonistic
    d. cooperative

13. Which of the following correctly describes the role of inositol triphosphate in hormone action?
    a. It activates adenylate cyclase.
    b. It stimulates the release of Ca^{++} from the endoplasmic reticulum.
    c. It activates protein kinase.
    d. It opens Ca^{++} channels in the cell membrane.

14. Which of the following hormones has been implicated in the establishment of circadian rhythms?
    a. estradiol
    b. insulin
    c. adrenocorticotropic hormone
    d. melatonin

15. Human chorionic gonadotropin (hCG) is secreted by
    a. the anterior pituitary.
    b. the posterior pituitary.
    c. the placenta.
    d. the thymus.
    e. the pineal gland.

16. What do insulin-like growth factors, neurotrophins, nitric oxide, and lymphokines have in common?
    a. They are hormones.
    b. They are autocrine regulators.
    c. They are neurotransmitters.
    d. They all use cAMP as a second messenger.
    e. They all use Ca^{++} as a second messenger.

Essay Questions

1. Explain how the regulation of the neurohypophysis and adrenal medulla are related to their embryonic origins.[1]
2. Explain the mechanism of action of steroid hormones and thyroxine.
3. Explain why polar hormones cannot regulate their target cells without using second messengers. Also explain how cyclic AMP is used as a second messenger in hormone action.
4. Describe the sequence of events by which a hormone can cause an increase in the Ca^{++} concentration within a target cell. How can this increased Ca^{++} affect the metabolism of the target cell?
5. Explain the significance of the term *trophic* with respect to the actions of anterior pituitary hormones.
6. Suppose a drug blocks the conversion of T_4 to T_3. Explain what the effects of this drug would be on (a) TSH secretion, (b) thyroxine secretion, and (c) the size of the thyroid gland.
7. Explain why the anterior pituitary is sometimes referred to as the "master gland" and why this reference is misleading.
8. Suppose a person's immune system made antibodies against insulin receptor proteins. Describe the possible effect of this condition on carbohydrate and fat metabolism.
9. Explain how light affects the function of the pineal gland. What is the relationship between pineal gland function and circadian rhythms?
10. Distinguish between endocrine and paracrine regulation. List some autocrine regulators and describe their functions.

Selected Readings and Multimedia

Anderson, S. et al. 1991. Deletion of steroid 5α-reductase 2 gene in male pseudohermaphroditism. *Nature* 354:159.

Austin, L. A., and H. Heath III. 1981. Calcitonin: Physiology and pathophysiology. *New England Journal of Medicine* 304:269.

Axelrod, J., and T. D. Reisine. 1984. Stress hormones: Their interaction and regulation. *Science* 224:452.

Baxter, J. D., and W. J. Funder. 1979. Hormone receptors. *New England Journal of Medicine* 300:117.

Berridge, M. J., and R. F. Irvine. 1989. Inositol phosphates and cell signalling. *Nature* 341:197.

Biglieri, E. G. 1989. ACTH effects on aldosterone, cortisol, and other steroids. *Hospital Practice* 24:145.

Bredt, D. S., and S. H. Snyder. 1994. Nitric oxide: A physiologic messenger molecule. *Annual Review of Biochemistry* 63:175.

Brent, G. A. 1994. The molecular bases of thyroid hormone action. *New England Journal of Medicine* 331:847.

Brent, G. A., D. D. Moore, and P. R. Larsen. 1991. Thyroid hormone regulation of gene expression. *Annual Review of Physiology* 53:17.

Brownstein, M. J. et al. 1980. Synthesis, transport and release of posterior pituitary hormones. *Science* 207:373.

Carmichael, S. W., and H. Winkler. August 1985. The adrenal chromaffin cell. *Scientific American*.

Cohick, W. S., and D. R. Clemmons. 1993. The insulin-like growth factors. *Annual Review of Physiology* 55:131.

Demers, L. M. September 1984. The effects of prostaglandins. *Diagnostic Medicine*, p. 37.

Ebadi, M. et al. 1993. Pineal gland in synchronizing and refining physiological events. *News in Physiological Sciences* 8:30.

Exton, J. H. 1994. Phosphoinositide phospholipases and G proteins in hormone action. *Annual Review of Physiology* 56:349.

Ganong, W. F. 1988. The stress concept—A dynamic overview. *Hospital Practice* 23:155.

Ganong, W. F., L. C. Alpert, and T. C. Lee. 1974. ACTH and the regulation of adrenocortical secretion. *New England Journal of Medicine* 290:1006.

Gershengorn, M. C. 1986. Mechanism of thyrotrophin releasing hormone stimulation of pituitary hormone secretion. *Annual Review of Physiology* 48:515.

Gillie, R. B. June 1971. Endemic goiter. *Scientific American*.

Guillemin, R., and R. Burgus. November 1972. The hormones of the hypothalamus. *Scientific American*.

Harrison, R. W. III, and S. S. Lippman. 1989. How steroid hormones work. *Hospital Practice* 24:63.

Henry, J. P. 1993. Biological basis of the stress response. *News in Physiological Sciences* 8:69.

Hoberman, J. M., and C. E. Yesalis. February 1995. The history of synthetic testosterone. *Scientific American*.

Houben, H., and C. Denef. November/December 1990. Regulatory peptides produced in the anterior pituitary. *Trends in Endocrinology and Metabolism*, p. 398.

[1]Note: This question is answered on page 126 of the Student Study Guide.

Ingami, T., Naruse, M., and R. Hoover. 1995. Endothelium as an endocrine organ. *Annual Review of Physiology* 57:171.

Katzenellenbogen, B. S. 1980. Dynamics of steroid hormone receptor action. *Annual Review of Physiology* 42:17.

Krieger, D. T. 1984. Brain peptides: What, where, and why? *Science* 222:975.

Levitzki, A. 1988. From epinephrine to cyclic AMP. *Science* 241:800.

Linder, M. E., and A. G. Gilman. July 1992. G proteins. *Scientific American.*

McEwen, B. S. July 1976. Interactions between hormones and nerve tissue. *Scientific American.*

Martin, C. R. *Endocrine Physiology.* 1985. New York: Oxford University Press.

Monado, S., and A. Higgs. 1993. The L-arginine-nitric oxide pathway. *New England Journal of Medicine* 329: 2002.

O'Malley, B., and W. T. Shrader. February 1976. The receptors of steroid hormones. *Scientific American.*

Petersen, O. H., and H. Kasai. 1994. Calcium and hormone action. *Annual Review of Physiology* 56: 297.

Putney, J. W. 1993. Excitement about calcium signaling in inexcitable cells. *Science* 262: 676.

Rasmussen, H. 1986. The calcium messenger system. *New England Journal of Medicine* 314: First part, p. 1094; second part, p. 1164.

Rasmussen, H. October 1989. The cycling of calcium as an intracellular messenger. *Scientific American.*

Reichlin, S. et al. 1976. Hypothalamic hormones. *Annual Review of Physiology* 39:389.

Reisine, T. 1988. Neurohumoral aspects of ACTH release. *Hospital Practice* 23:77.

Reiter, R. J. 1991. Melatonin: That ubiquitously acting pineal hormone. *News in Physiological Sciences* 6:223.

Reiter, R. J. January/February 1991. Pineal gland: Interface between the photoperiodic environment and the endocrine system. *Trends in Endocrinology and Metabolism*, p. 13.

Roth, J. 1980. Insulin receptors in diabetes. *Hospital Practice* 15:98.

Roth, J., and S. I. Taylor. 1982. Receptors for peptide hormones: Alterations in diseases of humans. *Annual Review of Physiology* 44:639.

Schally, A. V. 1978. Aspects of the hypothalamic control of the pituitary gland. *Science* 202:18.

Selye, H. 1973. The evolution of the stress concept. *American Scientist* 61:692.

Spiegel, A. M. 1988. G proteins in clinical medicine. *Hospital Practice* 23:93.

Strauss, R. H., and C. E. Yesalis. 1991. Anabolic steroids in the athlete. *Annual Review of Medicine* 42:449.

Taylor, A. L., and L. M. Fishman. 1988. Corticotropin-releasing hormone. *New England Journal of Medicine* 319:213.

Toft, A. D. 1994. Thyroxine therapy. *New England Journal of Medicine* 331:174.

Tsai, M-J., and B. W. O'Malley. 1994. Molecular mechanisms of action of steroid/thyroid receptor superfamily members. *Annual Review of Biochemistry* 61:451.

Utiger, R. D. 1992. Melatonin—The hormone of darkness. *New England Journal of Medicine* 327:1377.

Vane, J. 1994. Towards a better aspirin. *Nature* 367:215.

Vanhoutte, P. M. 1994. Endothelin-1: A matter of life and breath. *Nature* 368:693.

Villanua, M. A., C. Agrasai, and A. I. Esquifino. 1990. New perspectives in the research of pineal gland. *News in Physiological Sciences* 5:11.

White, P. C. 1994. Disorders of aldosterone biosynthesis and action. *New England Journal of Medicine* 331:250.

Explorations CD-ROM

 The module accompanying chapter 11 is #11 Hormone Action.

Life Sciences Animation

The animation that relates to chapter 11 is #28 Peptide Hormone Action.

Muscle 12

Mechanisms of Contraction and Neural Control

OBJECTIVES

After studying this chapter, you should be able to . . .

1. describe the gross and microscopic structure of skeletal muscles.

2. describe the all-or-none contraction of skeletal muscle fibers and explain how muscles can produce graded and sustained contractions.

3. distinguish between isometric and isotonic contraction.

4. explain how the series-elastic component affects muscle contraction.

5. define the term *motor unit* and explain how motor units are used to control muscle contraction.

6. describe the structure of myofibrils and explain how this structure accounts for the striated appearance of skeletal muscle fibers.

7. explain what is meant by the sliding filament theory of contraction.

8. list the events that occur during cross-bridge cycles and describe the role of ATP in muscle contraction.

9. describe the physiological roles of tropomyosin and troponin and the role of Ca^{++} in excitation-contraction coupling.

10. explain how action potentials in a nerve and muscle fiber affect the contractile mechanism and describe the role of the sarcoplasmic reticulum in muscle contraction and relaxation.

11. describe the structure and functions of muscle spindles and explain the mechanisms involved in a stretch reflex.

12. describe the function of Golgi tendon organs and explain why a slow, gradual muscle stretch might avoid the spasm that could result from a rapid stretch.

13. explain what is meant by reciprocal innervation and describe the neural pathways involved in a crossed-extensor reflex.

14. explain the significance of gamma motoneurons in the neural control of muscle contraction and in the maintenance of muscle tone.

15. describe the neural pathways involved in the pyramidal and extrapyramidal systems.

16. explain the significance of the maximal oxygen uptake and describe the function of phosphocreatine in muscles.

17. explain how slow-twitch, fast-twitch, and intermediate fibers differ in structure and function.

18. explain how muscles fatigue and how muscle fibers change as a result of physical training.

19. compare cardiac muscle and skeletal muscle in terms of strucutre and physiology.

20. describe the structure of smooth muscle and explain how its contraction is regulated.

OUTLINE

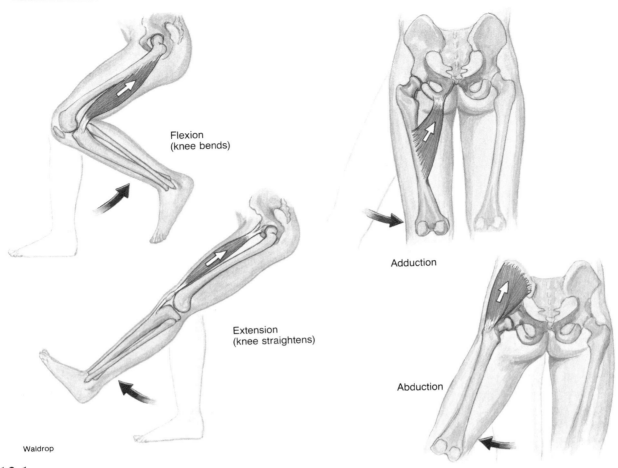

Figure 12.1

Actions of antagonistic muscles in the leg.

Structure and Actions of Skeletal Muscles

Skeletal muscles are composed of individual muscle fibers that contract when they are stimulated by a motor neuron. Each motor neuron branches to innervate a number of muscle fibers, and all of these fibers contract when their motor neuron is activated. Activation of varying numbers of motor neurons, and thus varying numbers of muscle fibers, results in gradations in the strength of contraction of the whole muscle.

Skeletal muscles are usually attached to bone on each end by tough connective tissue tendons. When a muscle contracts, it shortens, and this places tension on its tendons and attached bones. The muscle tension causes movement of the bones at a joint, where one of the attached bones generally moves more than the other. The more movable bony attachment of the muscle, known as its *insertion*, is pulled toward its less movable attachment known as its *origin*. A variety of skeletal movements

Table 12.1	Skeletal Muscle Actions
Category	**Action**
Extensor	Increases the angle at a joint
Flexor	Decreases the angle at a joint
Abductor	Moves limb away from the midline of the body
Adductor	Moves limb toward the midline of the body
Levator	Moves insertion upward
Depressor	Moves insertion downward
Rotator	Rotates a bone along its axis
Sphincter	Constricts an opening

are possible, depending on the type of joint involved and the attachments of the muscles (table 12.1 and fig. 12.1). When *flexor muscles* contract, for example, they decrease the angle of a joint. Contraction of *extensor muscles* increases the angle of their attached bones at the joint. Flexors and extensors that act on the same joint are thus *antagonistic muscles*.

The position of the limbs, for example, is determined by the actions of a variety of antagonistic muscles. In addition to the movements of flexion and extension, a limb can be moved away from the midline of the body by *abductor muscles* and brought inward toward the midline by contraction of *adductor*

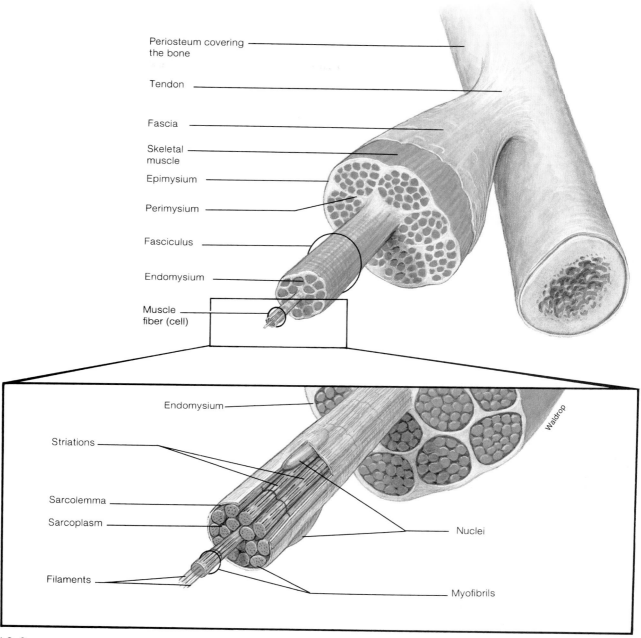

Labels in figure:
- Periosteum covering the bone
- Tendon
- Fascia
- Skeletal muscle
- Epimysium
- Perimysium
- Fasciculus
- Endomysium
- Muscle fiber (cell)
- Endomysium
- Striations
- Sarcolemma
- Sarcoplasm
- Filaments
- Nuclei
- Myofibrils
- Waldrop

Figure 12.2

The relationship between muscle fibers and the connective tissues of the tendon, epimysium, perimysium, and endomysium. Below is a close-up of a single muscle fiber.

muscles. In all cases, these skeletal movements are produced by the shortening of the appropriate muscle groups—the *agonists*—while the antagonist muscles remain relaxed.

Structure of Skeletal Muscles

The fibrous connective tissue proteins within the tendons continue in an irregular arrangement around the muscle to form a sheath known as the *epimysium* (*epi* = above; *my* = muscle). Connective tissue from this outer sheath extends into the body of the muscle, subdividing it into columns, or *fascicles* (these are

the "strings" in stringy meat). Each of these fascicles is thus surrounded by its own connective tissue sheath, known as the *perimysium* (*peri* = around).

Dissection of a muscle fascicle under a microscope reveals that it, in turn, is composed of many **muscle fibers**, or *myofibers*. Each is surrounded by a cell membrane, or sarcolemma, enveloped by a thin connective tissue layer called an *endomysium* (fig. 12.2). The muscle fibers are actually the muscle cells. Since the connective tissue of the tendons, epimysium, perimysium, and endomysium is continuous, muscle fibers do not normally pull out of the tendons when they contract.

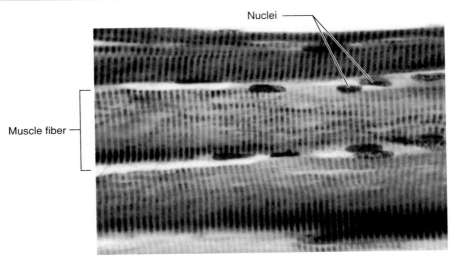

Nuclei

Muscle fiber

Figure 12.3

The appearance of skeletal muscle fibers through the light microscope. The striations are produced by alternating dark A bands and light I bands.

Duschenne muscular dystrophy is the most severe of the muscular dystrophies, afflicting one out of 3,500 boys each year. This disease, inherited as an X-linked recessive trait, involves progressive muscular wasting and usually results in death by the age of 20. The product of the defective gene is a protein named *dystrophin* that is associated with the sarcolemma. Using this information, scientists have recently developed laboratory tests that can detect this disease in fetal cells obtained by amniocentesis. In the future, genetic therapy may be possible. This research has been aided by the development of a strain of mice that exhibit an equivalent form of this disease. When the "good genes" for dystrophin are inserted into mouse embryos of this strain, the mice do not develop the disease. Insertion of the gene into large numbers of mature muscle cells, however, is more difficult, and so far has met with only limited success.

Despite their unusual fiber shape, muscle cells have the same organelles that are present in other cells: mitochondria, intracellular membranes, glycogen granules, and others. Unlike most other cells in the body, skeletal muscle fibers are multinucleate—that is, they contain multiple nuclei. The most distinctive feature of the microscopic appearance of skeletal muscle fibers, however, is their **striated** appearance (fig. 12.3). The striations (stripes) are produced by alternating dark and light bands that appear to cross the width of the fiber.

The dark bands are called **A bands** and the light bands are called **I bands.** At high magnification in an electron microscope, thin, dark lines can be seen in the middle of the I bands. These are called **Z lines.** The labels A, I, and Z are useful for

describing the functional architecture of muscle fibers, and were derived during the history of muscle research. The letters A and I stand for *anisotropic* and *isotropic*, respectively, which indicate the behavior of polarized light as it passes through these regions; the letter Z comes from the German word *Zwischenscheibe*, which translates to "between disc." These derivations are of historical interest only.

Types of Muscle Contractions

The contractile behavior of skeletal muscles is more easily studied in vitro (outside the body) than in vivo (within the body). When a muscle—for example, the gastrocnemius (calf muscle) of a frog—is studied in vitro, it is usually mounted so that one end is fixed and the other is movable. The mechanical force of the muscle contraction is transduced into an electric current, which can be amplified and displayed as pen deflections in a multichannel recorder (fig. 12.4). In this way, the contractile behavior of the whole muscle in response to experimentally administered electric shocks can be studied.

Twitch, Summation, and Tetanus

When the muscle is stimulated with a single electric shock of a sufficient voltage, it quickly contracts and relaxes. This response is called a **twitch.** Increasing the stimulus voltage increases the strength of the twitch up to a maximum. The strength of a muscle contraction can thus be *graded*, or varied—an obvious requirement for the proper control of skeletal movements. If a second electric shock is delivered immediately after the first, it will produce a second twitch that may partially "ride piggyback" on the first. This response is called **summation.**

Stimulation of fibers within a muscle in vitro with an electric stimulator, or in vivo by motor axons, usually results in the full contraction of the individual fibers. Stronger muscle

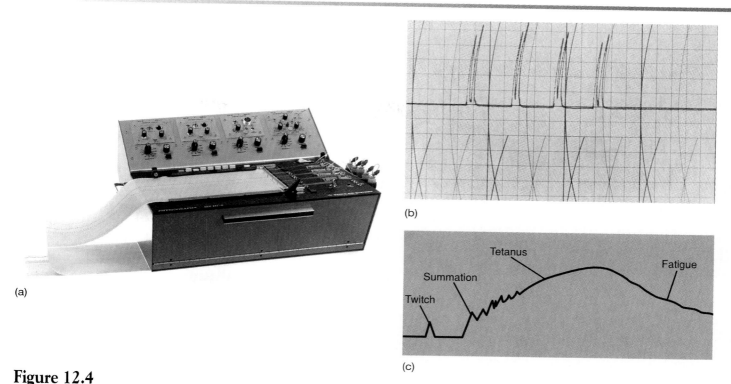

Figure 12.4

(*a*) A physiograph recorder. (*b*) A photograph and (*c*) an illustration of the behavior of an isolated gastrocnemius muscle of a frog in response to electrical shocks.

contractions are produced by the stimulation of greater numbers of muscle fibers. Skeletal muscles can thus produce **graded contractions,** the strength of which depends on the number of fibers stimulated rather than on the strength of the contractions of individual muscle fibers.

If the stimulator is set to deliver an increasing frequency of electric shocks automatically, the relaxation time between successive twitches will get shorter and shorter, as the strength of contraction increases in amplitude. This is **incomplete tetanus.** Finally, at a particular "fusion frequency" of stimulation, there is no visible relaxation between successive twitches (fig. 12.4). Contraction is smooth and sustained, as it is during normal muscle contraction in vivo. This smooth, sustained contraction is called **complete tetanus.** (The term *tetanus* should not be confused with the disease of the same name, which is accompanied by a painful state of muscle contracture, or *tetany.*)

Treppe

If the voltage of the electrical shocks delivered to an isolated muscle in vitro is gradually increased from zero, the strength of the muscle twitches will increase accordingly, up to a maximal value at which all of the muscle fibers are stimulated. This demonstrates the graded nature of the muscle contraction. If a series of electrical shocks at this maximal voltage is given to a fresh muscle so that each shock produces a separate twitch, each of the twitches evoked will be sucessively stronger, up to a higher maximum. This demonstrates **treppe,** or the *staircase effect.* Treppe may represent a warm-up effect, and is believed to

be due to an increase in intracellular Ca^{++}, which is needed for muscle contraction (as discussed in a later section).

Isotonic and Isometric Contractions

In order for muscle fibers to shorten when they contract, they must generate a force that is greater than the opposing forces that act to prevent movement of the muscle's insertion. When a weight is lifted by flexing the elbow joint, for example, the force produced by contraction of the biceps brachii muscle is greater than the force of gravity on the object being lifted (fig. 12.5). The tension produced by the contraction of each muscle fiber separately is insufficient to overcome the opposing force, but the combined contractions of many muscle fibers may be sufficient to overcome the opposing force and flex the forearm. In this case, the muscle and all of its fibers shorten in length.

Contraction that results in muscle shortening is called **isotonic contraction,** so called because the force of contraction remains relatively constant throughout the shortening process (*iso* = same; *tonic* = strength). If the opposing forces are too great or the number of muscle fibers activated is too few to shorten the muscle, however, the contraction is called an **isometric** (literally, "same length") **contraction.**

Isometric contraction can be voluntarily produced, for example, by lifting a weight and maintaining the forearm in a partially flexed position. We can then increase the amount of muscle tension produced by recruiting more muscle fibers until the muscle begins to shorten; at this point, isometric contraction is converted to isotonic contraction.

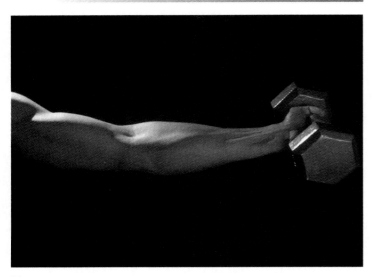

(a)

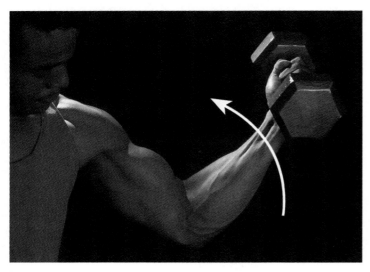

(b)

Figure 12.5

(*a*) Isometric and (*b*) isotonic contraction.

Series-Elastic Component

In order for a muscle to shorten when it contracts, and thus to move its insertion toward its origin, the noncontractile parts of the muscle and the connective tissue of its tendons must first be pulled tight. These structures, particularly the tendons, have elasticity—they resist distension, and when the distending force is released, they tend to spring back to their resting lengths. Since the tendons are in series with the force of muscle contraction, they provide what is called the **series-elastic component** of muscle contraction. The series-elastic component absorbs some of the tension as a muscle contracts, and must be pulled tight before muscle contraction can result in muscle shortening.

When the gastrocnemius muscle was stimulated with a single electric shock as described earlier, the amplitude of the twitch was reduced because some of the force of contraction was used to stretch the series-elastic component. Quick delivery of a second shock thus produced a greater degree of muscle shortening than the first shock, culminating at the fusion frequency of stimulation with complete tetanus, in which the strength of contraction was much greater than that of individual twitches.

Some of the energy used to stretch the series-elastic component during muscle contraction is released by elastic recoil when the muscle relaxes. This elastic recoil helps the muscles to return to their resting length, and is of particular importance for the muscles involved in breathing. As we will see in chapter 15, inspiration is produced by muscle contraction and expiration is produced by the elastic recoil of the thoracic structures that were stretched during inspiration.

Motor Units

In vivo, each muscle fiber receives a single axon terminal from a somatic motor neuron (fig. 12.6) that stimulates it to contract by liberating acetylcholine at the neuromuscular junction (described in chapter 7). The cell body of a somatic motor neuron is located in the ventral horn of the gray matter of the spinal cord and gives rise to a single axon that emerges in the ventral root of a spinal nerve (chapter 8). Each axon, however, can produce a number of collateral branches to innervate an equal number of muscle fibers. Each somatic motor neuron together with all of the muscle fibers that it innervates is known as a **motor unit** (fig. 12.7).

Whenever a somatic motor neuron is activated, all of the muscle fibers that it innervates are stimulated to contract with all-or-none twitches. In vivo, graded contractions of whole muscles are produced by variations in the number of motor units that are activated. In order for these graded contractions to be smooth and sustained, as in complete tetanus, different motor units must be activated by rapid, asynchronous stimulation.

Fine neural control over the strength of muscle contraction is optimal when there are many small motor units involved. In the extraocular muscles that position the eyes, for example, the *innervation ratio* (motor neuron:muscle fibers) of an average motor unit is one neuron per twenty-three muscle fibers. This affords a fine degree of control. The innervation ratio of the gastrocnemius, in contrast, averages one neuron per thousand muscle fibers. Stimulation of these motor units results in more powerful contractions at the expense of finer gradations in contraction strength.

All of the motor units controlling the gastrocnemius, however, are not the same size. Innervation ratios vary from 1:100 to 1:2,000. A neuron that innervates fewer muscle fibers has a smaller cell body and is stimulated by lower levels of excitatory input than a larger neuron that innervates a greater number of muscle fibers. The smaller motor units, as a result, are the ones that are used most often. When contractions of greater strength are required, larger and larger motor units are activated in a process known as **recruitment** of motor units.

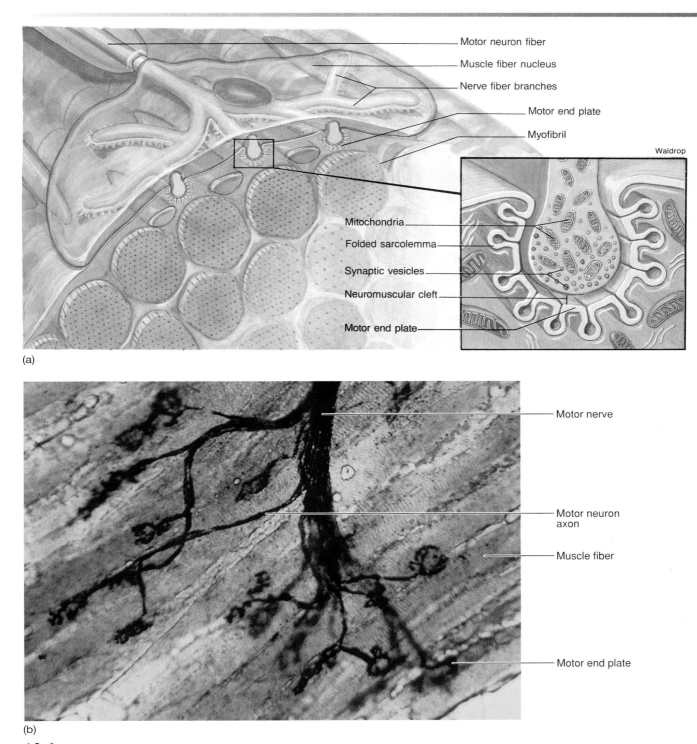

(a) Motor neuron fiber

Muscle fiber nucleus

Nerve fiber branches

Motor end plate

Myofibril

Waldrop

Mitochondria

Folded sarcolemma

Synaptic vesicles

Neuromuscular cleft

Motor end plate

(b) Motor nerve

Motor neuron axon

Muscle fiber

Motor end plate

Figure 12.6

(*a*) Motor end plates at the neuromuscular junction. It is at the neuromuscular junction that the nerve fiber and muscle fiber meet. The motor end plate is the specialized portion of the sarcolemma of a muscle fiber surrounding the terminal end of the axon. Notice the slight gap between the membrane of the axon and that of the muscle fiber. (*b*) A photomicrograph of muscle fibers and motor end plates. A motor neuron and the muscle fibers it innervates constitute a motor unit.

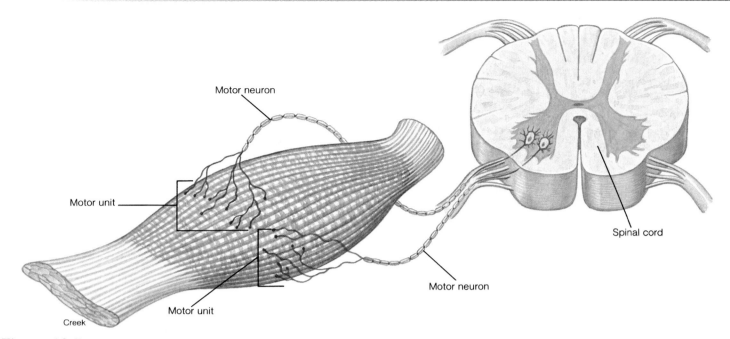

Figure 12.7

A diagram illustrating the innervation of muscle fibers by different motor units. (Actually, many more muscle fibers would be included in a single motor unit than are shown in this drawing.)

---

1. Describe muscle twitch, summation, and tetanus and use a graph to illustrate these effects.

2. Describe how graded muscle contractions result from fiber twitches that are all-or-none and how a smooth muscle contraction (tetanus) is produced in vivo.

3. Graphically depict a motor unit with an innervation ratio of 1:5. Describe the functional significance of the innervation ratio.

4. While trying to "make a muscle" with your arm, feel your biceps brachii and triceps brachii. Are these muscles contracting isotonically or isometrically? Explain.

Mechanisms of Contraction

The A bands within each muscle fiber are composed of thick filaments and the I bands contain thin filaments. Movement of cross bridges that extend from the thick to the thin filaments causes sliding of the filaments, and thus muscle tension and shortening. The activity of the cross bridges is regulated by the availability of Ca++, which is increased by electrical stimulation of the muscle fiber. Electrical stimulation produces contraction of the muscle through the binding of Ca++ to regulatory proteins within the thin filaments.

When muscle cells are viewed in the electron microscope, which can produce images at several thousand times the magnification possible in an ordinary light microscope, each cell is seen to be composed of many subunits known as **myofibrils** (*fibrils* = little fibers—fig. 12.8). These myofibrils are approximately 1 micrometer (1 μm) in diameter and extend in parallel rows from one end of the muscle fiber to the other. The myofibrils are so densely packed that other organelles, such as mitochondria and intracellular membranes, are restricted to the narrow cytoplasmic spaces that remain between adjacent myofibrils.

With the electron microscope, it can be seen that the muscle fiber does not have striations that extend from one side of the fiber to the other. It is the myofibrils that are striated with dark A *bands* and light I *bands* (fig. 12.9). The striated appearance of the entire muscle fiber when seen with a light microscope is an illusion created by the alignment of the dark and light bands of the myofibrils from one side of the fiber to the other. Since the separate myofibrils are not clearly seen at low magnification, the dark and light bands appear to be continuous across the width of the fiber.

When a myofibril is observed at high magnification in longitudinal section (side view), the A bands are seen to contain **thick filaments** (about 110 angstroms [110 Å] thick; 1 Å = 10^{-10} m) that are stacked in register. It is these thick filaments that give the A band its dark appearance. The lighter I band, by contrast, contains **thin filaments** (from 50 to 60 Å thick). The thick filaments are composed of the protein **myosin,** and the thin filaments are composed primarily of the protein **actin.**

The I bands within a myofibril are the lighter areas that extend from the edge of one stack of thick myosin filaments to

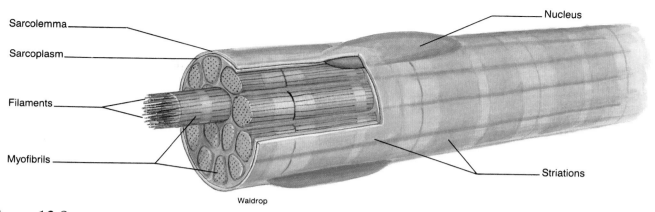

Sarcolemma

Sarcoplasm

Filaments

Myofibrils

Nucleus

Striations

Waldrop

Figure 12.8

A skeletal muscle fiber is composed of numerous myofibrils that contain filaments of actin and myosin. A skeletal muscle fiber is striated and multinucleated.

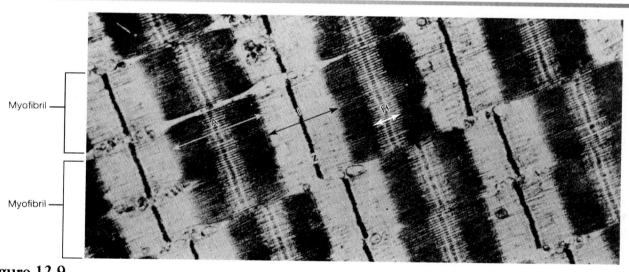

Myofibril

Myofibril

Figure 12.9

An electron micrograph of a longitudinal section of myofibrils showing A, H, and I bands. Notice how the dark and light bands of each myofibril are stacked in register.

the edge of the next stack of thick filaments. They are light in appearance because they contain only thin filaments. The thin filaments, however, do not end at the edges of the I bands. Instead, each thin filament continues partway into the A bands on each side (between the stack of thick filaments on each side of an I band). Since thick and thin filaments overlap at the edges of each A band, the edges of the A band are darker in appearance than the central region. These central lighter regions of the A bands are called the *H bands* (for *helle*, a German word meaning "bright"). The central H bands thus contain only thick filaments that are not overlapped with thin filaments.

In the center of each I band is a thin dark Z line. The arrangement of thick and thin filaments between a pair of Z lines forms a repeating pattern that serves as the basic subunit

of striated muscle contraction. These subunits, from Z to Z, are known as **sarcomeres.** A longitudinal section of a myofibril thus presents a side view of successive sarcomeres.

This side view is, in a sense, misleading; there are numerous sarcomeres within each myofibril that are out of the plane of the section (and out of the picture). A better appreciation of the three-dimensional structure of a myofibril can be obtained by viewing the myofibril in cross section. In this view, it can be seen that the Z lines are actually disc shaped, and that the thin filaments that penetrate these Z discs surround the thick filaments in a hexagonal arrangement (fig. 12.10c). If we concentrate on a single row of dark thick filaments in this cross section, the alternating pattern of thick and thin filaments seen in longitudinal section becomes apparent.

Muscle

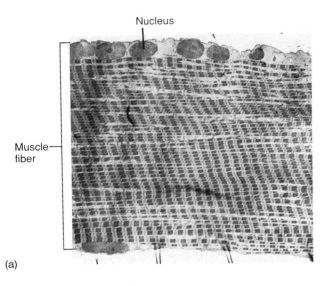

(a)

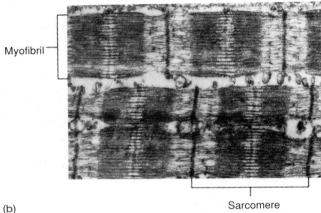

(b)

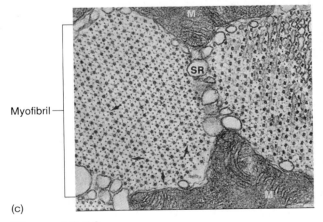

(c)

Figure 12.10

Electron micrographs of myofibrils of a muscle fiber: (*a*) at low power (1,600×), a single muscle fiber containing numerous myofibrils and (*b*), at high power (53,000×), myofibrils in longitudinal section. Notice the sarcomeres and overlapping thick and thin filaments. (*c*) The hexagonal arrangement of thick and thin filaments as seen in cross section (arrows point to cross bridges; SR = sarcoplasmic reticulum; M = mitochondria).

Part [c] reproduced from: R. G. Kessel and R. H. Kardon, *Tissues and Organs: A Text-Atlas of Scanning Electron Microscopy*, W. H. Freeman and Co., 1979.

Sliding Filament Theory of Contraction

When a muscle contracts isotonically, it decreases in length as a result of the shortening of its individual fibers. Shortening of the muscle fibers, in turn, is produced by shortening of their myofibrils, which occurs as a result of the shortening of the distance from Z line to Z line. As the sarcomeres shorten in length, however, the A bands do *not* shorten but instead move closer together. The I bands—which represent the distance between A bands of successive sarcomeres—decrease in length (table 12.2).

The thin actin filaments composing the I band, however, do not shorten. Close examination reveals that the thick and thin filaments remain the same length during muscle contraction. Shortening of the sarcomeres is produced not by shortening of the filaments, but rather by the *sliding* of thin filaments over and between the thick filaments. In the process of contraction, the thin filaments on either side of each A band slide deeper and deeper toward the center, producing increasing amounts of overlap with the thick filaments. The I bands (containing only thin filaments) and H bands (containing only thick filaments) thus get shorter during contraction (fig. 12.11).

Cross Bridges

Sliding of the filaments is produced by the action of numerous **cross bridges** that extend out from the myosin toward the actin. These cross bridges are part of the myosin proteins that extend from the axis of the thick filaments to form "arms" that terminate in globular "heads" (fig. 12.12). The orientation of cross bridges on one side of a sarcomere is opposite to that of the cross bridges on the other side, so that when the myosin cross bridges attach to actin on each side of the sarcomere they can pull the actin from each side toward the center.

Isolated muscles in vitro are easily stretched (although this is opposed in vivo by the stretch reflex, described in a later section), demonstrating that the myosin cross bridges are not attached to actin when the muscle is at rest. Each globular head of a cross bridge contains an ATP-binding site closely associated with an actin-binding site (fig. 12.12). The globular heads function as **myosin ATPase** enzymes, splitting ATP into ADP and P_i. This reaction occurs before the cross bridges combine with actin, and indeed is required for activating the cross bridges so that they can attach to actin. The ADP and P_i remain bonded to the myosin heads until the cross bridges attach to the actin.

The myosin heads are able to bind to specific attachment sites in the actin subunits. When the cross bridges bind to actin, they undergo a conformational change. This has two effects: (1) ADP and P_i are released and (2) the cross bridges change their orientation, resulting in a *power stroke*, that pulls the thin filaments toward the center of the A bands. At the end of the power stroke, each cross bridge binds to a fresh ATP molecule. This bonding of the cross bridge to a new ATP causes the cross bridge to break its bond with actin. The myosin ATPase will then split ATP and become activated as in the previous cycle. Note that the splitting of ATP is required *before* a cross

Table 12.2   Summary of the Sliding Filament Theory of Contraction

1. A myofiber, together with all its myofibrils, shortens by movement of the insertion toward the origin of the muscle.
2. Shortening of the myofibrils is caused by shortening of the sarcomeres—the distance between Z lines (or discs) is reduced.
3. Shortening of the sarcomeres is accomplished by sliding of the myofilaments—the length of each filament remains the same during contraction.
4. Sliding of the filaments is produced by asynchronous power strokes of myosin cross bridges, which pull the thin filaments (actin) over the thick filaments (myosin).
5. The A bands remain the same length during contraction, but are pulled toward the origin of the muscle.
6. Adjacent A bands are pulled closer together as the I bands between them shorten.
7. The H bands shorten during contraction as the thin filaments from each end of the sarcomeres are pulled toward the middle.

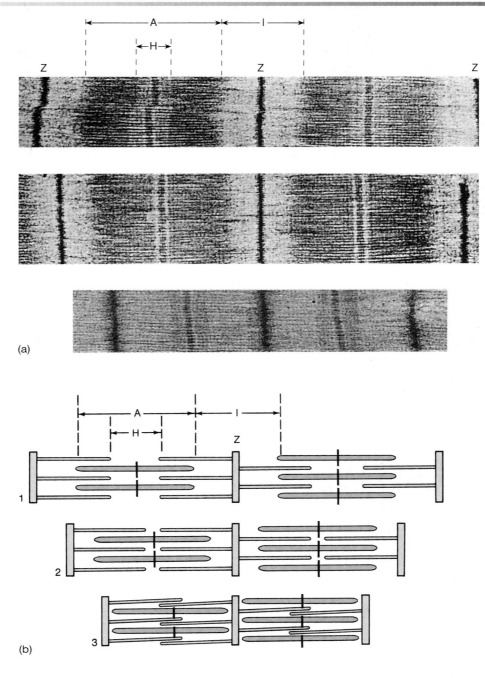

Figure 12.11

(a) An electron micrograph and (b) a diagram of the sliding filament model of contraction. As the filaments slide, the Z lines are brought closer together.
(1) Relaxed muscle; (2) partially contracted muscle; (3) fully contracted muscle.

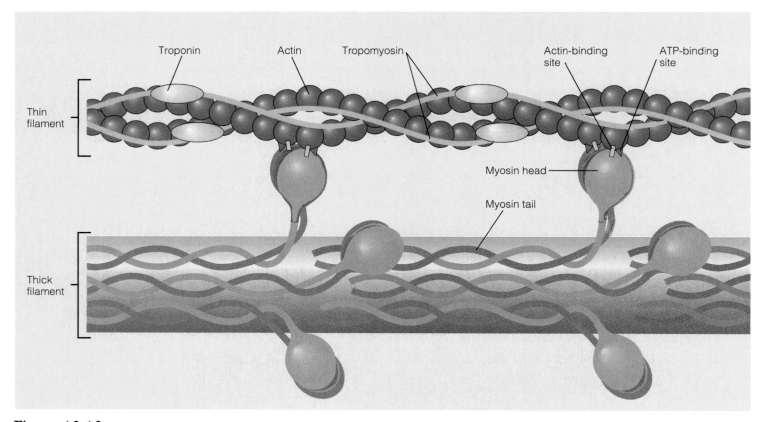

Figure 12.12

The structure of myosin, showing its binding sites for ATP and for actin.

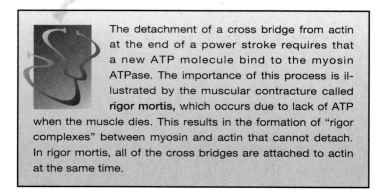

The detachment of a cross bridge from actin at the end of a power stroke requires that a new ATP molecule bind to the myosin ATPase. The importance of this process is illustrated by the muscular contracture called **rigor mortis,** which occurs due to lack of ATP when the muscle dies. This results in the formation of "rigor complexes" between myosin and actin that cannot detach. In rigor mortis, all of the cross bridges are attached to actin at the same time.

bridge can attach to actin and undergo a power stroke, and that the attachment of a *new ATP* is needed for the cross bridge to release from actin at the end of a power stroke (fig. 12.13).

Because the cross bridges are quite short, a single contraction cycle and power stroke of all the cross bridges in a muscle would shorten the muscle by only about 1% of its resting length. Since muscles can shorten up to 60% of their resting lengths, it is obvious that the contraction cycles must be repeated many times. In order for this to occur, the cross bridges must detach from the actin at the end of a power stroke, reassume their resting orientation, and then reattach to the actin and repeat the cycle.

During normal contraction, however, only about 50% of the cross bridges are attached at any given time. The power strokes are thus not in synchrony, as the strokes of a competitive rowing team would be. Rather, they are like the actions of a team engaged in tug-of-war, where the pulling action of the members is asynchronous. Some cross bridges are engaged in power strokes at all times during the contraction.

Regulation of Contraction

When the cross bridges attach to actin, they undergo power strokes and cause muscle contraction. In order for a muscle to relax, therefore, the attachment of myosin cross bridges to actin must be prevented. The regulation of cross bridge attachment to actin is a function of two proteins that are associated with actin in the thin filaments.

The actin filament—or F-actin—is a polymer formed of 300 to 400 globular subunits (G-actin), arranged in a double row and twisted to form a helix (fig. 12.14). A different type of protein, known as **tropomyosin,** lies within the groove between the double row of G-actin. There are forty to sixty tropomyosin molecules per thin filament, with each tropomyosin spanning a distance of approximately seven actin subunits.

Attached to the tropomyosin, rather than directly to the actin, is a third type of protein called **troponin.** Troponin and

Chapter Twelve

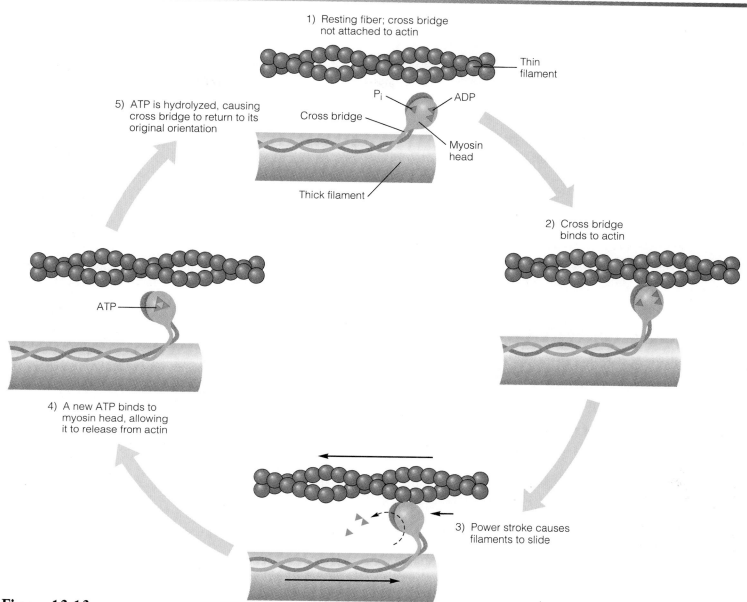

1) Resting fiber; cross bridge not attached to actin

Thin filament

P_i ADP

Cross bridge

Myosin head

Thick filament

5) ATP is hydrolyzed, causing cross bridge to return to its original orientation

2) Cross bridge binds to actin

ATP

4) A new ATP binds to myosin head, allowing it to release from actin

3) Power stroke causes filaments to slide

Figure 12.13

The cross-bridge cycle that causes sliding of the filaments and muscle contraction.

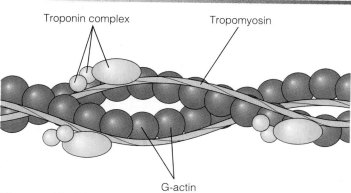

Troponin complex

Tropomyosin

G-actin

Figure 12.14

The relationship of troponin and tropomyosin to actin in the thin filaments. The tropomyosin is attached to actin, whereas the troponin complex of three subunits is attached to tropomyosin (not directly to actin).

tropomyosin work together to regulate the attachment of cross bridges to actin, and thus serve as a switch for muscle contraction and relaxation. In a relaxed muscle, the position of the tropomyosin in the thin filaments is such that it physically blocks the cross bridges from binding to specific attachment sites in the actin. Thus, in order for the myosin cross bridges to attach to actin, the tropomyosin must be moved. This requires the interaction of troponin with Ca^{++}, as described in the next section.

Role of Ca^{++} in Muscle Contraction

In a relaxed muscle, when tropomyosin blocks the attachment of cross bridges to actin, the concentration of Ca^{++} in the sarcoplasm (cytoplasm of muscle cells) is very low. When the muscle cell is stimulated to contract, mechanisms that will be

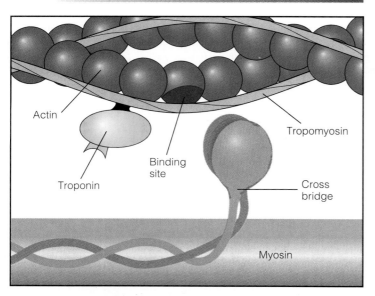

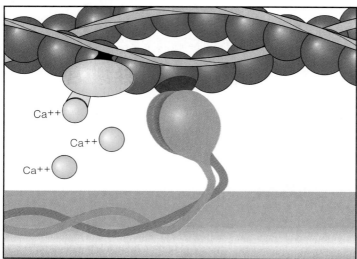

Figure 12.15

The attachment of Ca^{++} to troponin causes movement of the troponin-tropomyosin complex, which exposes binding sites on the actin. The myosin cross bridges can then attach to actin and undergo a power stroke.

discussed shortly cause the concentration of Ca^{++} in the sarcoplasm to quickly rise. Some of this Ca^{++} attaches to a subunit of troponin, causing a conformational change that moves the troponin *and* its attached tropomyosin out of the way so that the cross bridges can attach to actin (fig. 12.15). Once the attachment sites on the actin are exposed, the cross bridges can bind to actin, undergo power strokes, and produce muscle contraction.

The position of the troponin-tropomyosin complexes in the thin filaments is thus adjustable. When Ca^{++} is not attached to troponin, the tropomyosin is in a position that inhibits attachment of cross bridges to actin, preventing muscle contraction. When Ca^{++} attaches to troponin, the troponin-tropomyosin complexes shift position. The cross bridges can

then attach to actin, produce a power stroke, and detach from actin. Moreover, these contraction cycles can continue as long as Ca^{++} is attached to troponin.

Excitation-Contraction Coupling

Muscle contraction is turned on when sufficient amounts of Ca^{++} bind to troponin. This occurs when the Ca^{++} concentration of the sarcoplasm rises above 10^{-6} molar. In order for muscle relaxation to occur, therefore, the Ca^{++} concentration of the sarcoplasm must be lowered below this level. Muscle relaxation is produced by the active transport of Ca^{++} out of the sarcoplasm into the **sarcoplasmic reticulum** (fig. 12.16). The sarcoplasmic reticulum is a modified endoplasmic reticulum, consisting of interconnected sacs and tubes that surround each myofibril within the muscle cell.

Most of the Ca^{++} in a relaxed muscle fiber is stored within expanded portions of the sarcoplasmic reticulum known as *terminal cisternae*. When a muscle fiber is stimulated to contract by either a motor neuron in vivo or electric shocks in vitro, the stored Ca^{++} is released from the sarcoplasmic reticulum so that it can attach to troponin. When a muscle fiber is no longer stimulated, the Ca^{++} from the sarcoplasm is actively transported back into the sarcoplasmic reticulum. Now, in order to understand how the release and uptake of Ca^{++} is regulated, one more organelle within the muscle fiber must be described.

The terminal cisternae of the sarcoplasmic reticulum are separated by only a very narrow gap from **transverse tubules** (or **T tubules**), which are narrow membranous "tunnels" formed from and continuous with the *sarcolemma* (muscle cell membrane). The transverse tubules thus open to the extracellular environment through pores in the cell surface, and are capable of conducting action potentials. The stage is now set to explain exactly how a motor neuron stimulates a muscle fiber to contract.

The release of acetylcholine from axon terminals at the neuromuscular junctions (motor end plates), as previously described, causes electrical activation of skeletal muscle fibers. End-plate potentials (analogous to EPSPs—chapter 7) are produced that generate action potentials. Muscle action potentials, like the action potentials of axons, are all-or-none events that are regenerated along the cell membrane. It must be remembered that action potentials involve the flow of ions between the extracellular and intracellular environments across a cell membrane that separates these two compartments. In muscle cells, therefore, action potentials can be conducted into the interior of the fiber across the membrane of the transverse tubules.

Action potentials in the transverse tubules cause the release of Ca^{++} from the sarcoplasmic reticulum. This process is known as **excitation-contraction coupling.** Since the transverse tubules are not physically continuous with the sarcoplasmic reticulum, there must be some mechanism to permit communication between these two organelles. It has been proposed, for example, that action potentials stimulate the formation of inositol triphosphate (IP₃—see chapter 11), which may serve as a second messenger from the transverse tubules to the sarcoplasmic reticulum. Although this explanation is consistent with the way in which hormones stimulate Ca^{++} release from the endoplasmic reticulum

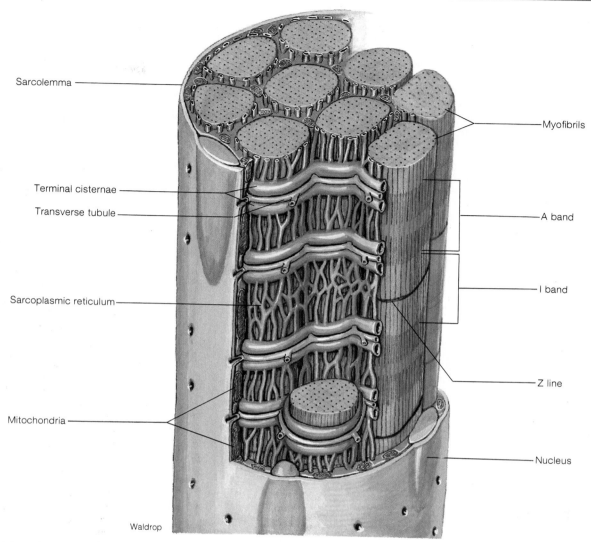

Sarcolemma

Terminal cisternae

Transverse tubule

Sarcoplasmic reticulum

Mitochondria

Myofibrils

A band

I band

Z line

Nucleus

Waldrop

Figure 12.16

The relationship between myofibrils, the transverse tubules, and the sarcoplasmic reticulum.

of other cells (chapter 11), evidence for a similar mechanism operating in skeletal muscle fibers is inconclusive. Other substances have also been proposed to act as second messengers in excitation-contraction coupling.

Evidence is accumulating that there may be a direct coupling, on a molecular level, between the Ca^{++} channel in the sarcoplasmic reticulum and the transverse tubules. The Ca^{++} channel proteins have a part that protrudes into the cytoplasm. This part, which looks like a "foot" in the electron microscope, may be able to interact directly with the specific proteins of the transverse tubules that are responsible for stimulating the sarcoplasmic reticulum. If this is indeed the mechanism responsible for excitation-contraction coupling in skeletal muscles, no second messenger is needed.

By whatever mechanism it is accomplished, action potentials in the transverse tubules cause release of Ca^{++} from the sarcoplasmic reticulum. The released Ca^{++} then diffuses into the sarcomeres and binds to troponin, causing the displacement of tropomyosin and allowing the actin to bind to the myosin cross bridges. Muscle contraction is thus stimulated.

As long as action potentials continue to be produced—which is as long as the neural stimulation of the muscle is maintained—Ca^{++} will remain attached to troponin and cross bridges will be able to undergo contraction cycles. When neural activity and action potentials in the muscle fiber cease, the sarcoplasmic reticulum actively accumulates Ca^{++} and muscle relaxation occurs. Note that the return of Ca^{++} to the sarcoplasmic reticulum involves active transport and thus requires the hydrolysis of ATP. ATP is therefore needed for muscle relaxation as well as for muscle contraction. The events that occur in excitation-contraction coupling are summarized in table 12.3.

Table 12.3 Summary of Events in Excitation-Contraction Coupling

1. Action potentials in a somatic motor nerve cause the release of acetylcholine neurotransmitter at the myoneural junction (one myoneural junction per myofiber).
2. Acetylcholine, through its interaction with receptors in the muscle cell membrane (sarcolemma), produces action potentials that are regenerated across the sarcolemma.
3. The membranes of the transverse tubules (T tubules) are continuous with the sarcolemma and conduct action potentials deep into the muscle fiber.
4. Action potentials in the T tubules, by a mechanism that is poorly understood, stimulate the release of Ca^{++} from the terminal cisternae of the sarcoplasmic reticulum.
5. Ca^{++} released into the sarcoplasm attaches to troponin, causing a change in its structure.
6. The shape change in troponin causes its attached tropomyosin to shift position in the actin filament, thus exposing bonding sites for the myosin cross bridges.
7. Myosin cross bridges, previously activated by the hydrolysis of ATP, attach to actin.
8. Once the previously activated cross bridges attach to actin, they undergo a power stroke and pull the thin filaments over the thick filaments.
9. Attachment of fresh ATP allows the cross bridges to detach from actin and repeat the contraction cycle as long as Ca^{++} remains attached to troponin.
10. When action potentials stop being produced, the sarcoplasmic reticulum actively accumulates Ca^{++} and tropomyosin moves again to its inhibitory position.

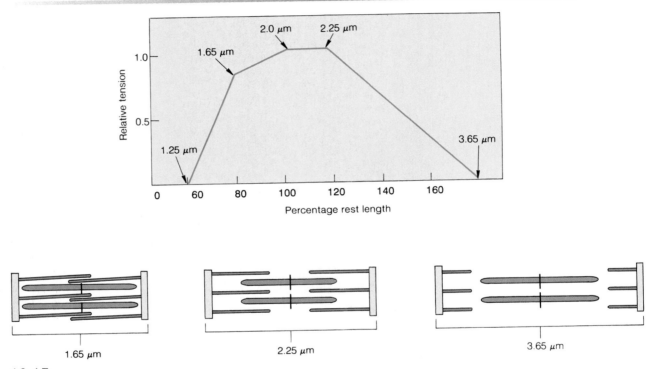

Figure 12.17

The length-tension relationship in skeletal muscles. Maximum relative tension (1.0) is achieved when the muscle is 100% to 120% of its resting length (sarcomere lengths from 2.0 to 2.25 µm). Increases or decreases in muscle (and sarcomere) lengths result in rapid decreases in tension.

Length–Tension Relationship

A number of factors influence the strength of a muscle's contraction. These include the number of fibers within the muscle that are stimulated to contract, the thickness of each muscle fiber (thicker fibers have more myofibrils and thus can exert more power), and the initial length of the muscle fibers when they are at rest.

There is an "ideal" resting length for striated muscle fibers. This is the length at which they can generate maximum force. When the resting length exceeds this ideal, the overlap between actin and myosin is so small that few cross bridges can attach. When the muscle is stretched to the point that there is no overlap of actin with myosin, no cross bridges can attach to

the thin filaments and the muscle cannot contract. When the muscle is shortened to about 60% of its resting length, the Z lines abut the thick filaments so that further contraction cannot occur.

The strength of a muscle's contraction can be measured by the force required to prevent it from shortening. Under these isometric conditions, the strength of contraction, or *tension*, can be measured when the muscle length at rest is varied. Maximum tension of skeletal muscle is produced when the muscle is at its normal resting length in vivo (fig. 12.17). If the muscle were any shorter or longer than its normal length, in other words, its strength of contraction would be reduced. This resting length is maintained by reflex contraction in response to passive stretching, as described in a later section of this chapter.

1. Describe how the lengths of the A, I, and H bands change during contraction and use the sliding filament theory to explain these changes.
2. Describe a cycle of cross-bridge activity during contraction and discuss the role of ATP in this cycle.
3. Draw a sarcomere in a relaxed muscle and a sarcomere in a contracted muscle and label the bands in each. What is the significance of the differences in your drawings?
4. Describe the molecular structure of myosin and actin and explain the positions and functions of tropomyosin and troponin.
5. Draw a flowchart (using arrows) of the events that occur from the time that ACh is released from a nerve ending to the time that Ca^{++} is released from the sarcoplasmic reticulum.
6. Explain the requirements for Ca^{++} and ATP in muscle contraction and relaxation.

The disease known as **amyotrophic lateral sclerosis (ALS)** involves degeneration of the lower motor neurons, leading to muscle paralysis. This disease is sometimes called *Lou Gehrig's disease*, after the baseball player who suffered from it, and also includes the famous physicist Steven Hawking among its victims. Scientists have recently learned that the inherited form of this disease is caused by a defect in the gene for a specific enzyme—superoxide dismutase. This enzyme is responsible for eliminating superoxide free radicals, which are highly toxic products that can damage the motor neurons. Although most cases of ALS are not inherited, the hereditary and nonhereditary forms of the disease are clinically the same and may therefore be treatable through the same means, possibly involving mechanisms to detoxify the free radicals.

Neural Control of Skeletal Muscles

Skeletal muscles contain stretch receptors called muscle spindles that stimulate the production of impulses in sensory neurons when the muscle is stretched. These sensory neurons can synapse with alpha motoneurons, which stimulate the muscle to contract in response to the stretch. Other motor neurons, called gamma motoneurons, stimulate the tightening of the spindles and thus increase their sensitivity.

Motor neurons in the spinal cord, or **lower motor neurons** (often shortened to *motoneurons*), are those previously described that have cell bodies in the spinal cord and axons within nerves that stimulate muscle contraction (table 12.4). The activity of these neurons is influenced by (1) sensory feedback from the muscles and tendons and (2) facilitory and inhibitory effects from **upper motor neurons** in the brain that contribute axons to descending motor tracts. Lower motor neurons are thus said to be the *final common pathway* by which sensory stimuli and higher brain centers exert control over skeletal movements.

The cell bodies of lower motor neurons are located in the ventral horn of the gray matter of the spinal cord (chapter 8). Axons from these cell bodies leave the ventral side of the spinal cord to form the *ventral roots* of spinal nerves (see fig. 8.23). The *dorsal roots* of spinal nerves contain sensory fibers whose cell bodies are located in the *dorsal root ganglia*. Both sensory (*afferent*) and motor (*efferent*) fibers join in a common connective tissue sheath to form the spinal nerves at each segment of the spinal cord. In the lumbar region there are about 12,000 sensory and 6,000 motor fibers per spinal nerve.

About 375,000 cell bodies have been counted in a lumbar segment—a number far larger than can be accounted for by the number of motor neurons. Most of these neurons do not contribute fibers to the spinal nerve, but rather serve as *interneurons* whose fibers conduct impulses up, down, and across the central nervous system. Those fibers that conduct impulses to higher spinal cord segments and the brain form *ascending tracts* and those that conduct to lower spinal segments contribute to *descending tracts*. Those fibers that cross the midline of the CNS to synapse on the opposite side are part of *commissural tracts*. Interneurons can thus conduct impulses up and down on the same, or *ipsilateral*, side, and can affect neurons on the opposite, or *contralateral*, side of the central nervous system.

Muscle Spindle Apparatus

In order for the nervous system to control skeletal movements properly, it must receive continuous sensory feedback information concerning the effects of its actions. This sensory information includes (1) the tension that the muscle exerts on its tendons, provided by the **Golgi tendon organs** and (2) muscle length, provided by the **muscle spindle apparatus.** The spindle apparatus, so called because it is wider in the center and tapers toward the ends, functions as a length detector. Muscles that require the finest degree of control, such as the muscles of the hand, have the highest density of spindles (table 12.5).

Each spindle apparatus contains several thin muscle cells, called *intrafusal fibers* (*fusus* = spindle), packaged within a connective tissue sheath. Like the stronger and more numerous "ordinary" muscle fibers outside the spindles—the *extrafusal fibers*—the spindles insert into tendons on each end of the muscle. Spindles are therefore said to be in parallel with the extrafusal fibers.

Table 12.4 A Partial Listing of Terms Used to Describe the Neural Control of Skeletal Muscles

Term	Description
1. Lower motoneurons	Neurons whose axons innervate skeletal muscles—also called the "final common pathway" in the control of skeletal muscles
2. Higher motoneurons	Neurons in the brain that are involved in the control of skeletal movements and that act by facilitating or inhibiting (usually by way of interneurons) the activity of the lower motoneurons
3. Alpha motoneurons	Lower motoneurons whose fibers innervate ordinary (extrafusal) muscle fibers
4. Gamma motoneurons	Lower motoneurons whose fibers innervate the muscle spindle fibers (intrafusal fibers)
5. Agonist/antagonist	Pair of muscles or muscle groups that insert on the same bone, the agonist being the muscle of reference
6. Synergist	A muscle whose action facilitates the action of the agonist
7. Ipsilateral/contralateral	Ipsilateral—to the same side, or the side of reference; contralateral—the opposite side
8. Afferent/efferent	Afferent neurons—sensory; efferent neurons—motor

Table 12.5 Spindle Apparatus Content of Selected Skeletal Muscles

Muscle		Muscle Weight (g)	Average Number of Spindles	Number of Spindles per Gram Muscle
Gastrocnemius		7.6	35	5
Rectus femoris		8.36	104	12
Tibialis anterior	Leg	4.57	71	15
Semitendinosis		6.41	114	18
Soleus		2.49	56	23
Fifth interossei—foot		0.33	29	88
Fifth interossei—hand		0.21	25	119

Unlike the extrafusal fibers, which contain myofibrils along their entire length, the contractile apparatus is absent from the central regions of the intrafusal fibers. The central, noncontracting part of an intrafusal fiber contains nuclei. There are two types of intrafusal fibers. One type, the *nuclear bag fibers*, have their nuclei arranged in a loose aggregate in the central regions of the fibers. The other type of intrafusal fibers have their nuclei arranged in rows and are called *nuclear chain fibers*. There are likewise two types of sensory neurons that serve these intrafusal fibers. **Primary,** or **annulospiral, sensory endings** wrap around the central regions of the nuclear bag and chain fibers (fig. 12.18), and **secondary,** or **flower-spray, endings** are located over the contracting poles of the nuclear chain fibers.

Since the spindles are arranged in parallel with the extrafusal muscle fibers, stretching a muscle causes its spindles to stretch. This stimulates both the primary and secondary sensory endings. The spindle apparatus thus serves as a length detector because the frequency of impulses produced in the primary and secondary endings is proportional to the length of the muscle. The primary endings, however, are stimulated most at the onset of stretch, whereas the secondary endings respond in a more tonic (sustained) fashion as stretch is maintained. Sudden, rapid stretching of a muscle activates both types of sensory endings and is thus a more powerful stimulus for the muscle spindles than slower, more gradual stretching, which has less of an effect on the primary sensory endings. Since the activation of the sensory endings in muscle spindles produces a reflex contraction, the force of this reflex contraction is greater in response to rapid stretch than to gradual stretch.

Alpha and Gamma Motoneurons

There are two types of lower motor neurons in the spinal cord that innervate skeletal muscles. The motor neurons that innervate the extrafusal muscle fibers are called **alpha motoneurons;** those that innervate the intrafusal fibers are called **gamma motoneurons** (fig. 12.18). The alpha motoneurons are larger and faster conducting (60–90 meters per second) than the thinner, slower conducting (10–40 meters per second) gamma motoneurons. Since only the extrafusal muscle fibers are sufficiently strong and numerous to cause a muscle to shorten, only stimulation by the alpha motoneurons can cause muscle contraction that results in skeletal movements.

The intrafusal fibers of the muscle spindle are stimulated to contract by gamma motoneurons, which represent one-third of all efferent fibers in spinal nerves. The intrafusal fibers are too few in number and their contraction is too weak, however, to cause a muscle to shorten. Stimulation by gamma motoneurons thus results in only isometric contraction of the spindles. Since myofibrils are present in the poles but absent in the central regions of intrafusal fibers, the more distensible central region of the intrafusal fiber is pulled toward the ends in response to stimulation by gamma motoneurons. As a result, the spindle is tightened. This effect of gamma motoneurons, which is sometimes termed *active stretch* of the spindles, functions to increase the sensitivity of the spindles when the entire muscle is passively stretched by external forces. The activation of gamma motoneurons thus enhances the stretch reflex (described in the next section), and is an important feature in the voluntary control of skeletal movements.

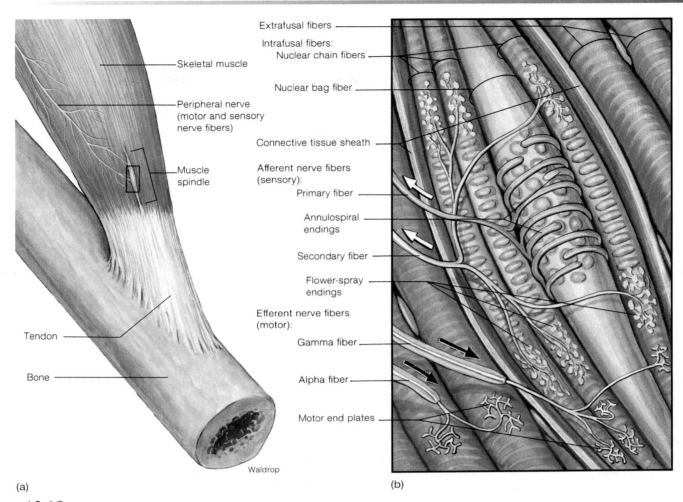

Extrafusal fibers
Intrafusal fibers:
Nuclear chain fibers
Nuclear bag fiber
Skeletal muscle
Connective tissue sheath
Peripheral nerve
(motor and sensory
nerve fibers)
Afferent nerve fibers
(sensory):
Muscle
spindle
Primary fiber
Annulospiral
endings
Secondary fiber
Flower-spray
endings
Efferent nerve fibers
(motor):
Tendon
Gamma fiber
Bone
Alpha fiber
Motor end plates

Waldrop

(a)   (b)

Figure 12.18

(*a*) A muscle spindle within a skeletal muscle. (*b*) The structure and innervation of a muscle spindle.

Coactivation of Alpha and Gamma Motoneurons

Most of the fibers in the descending motor tracts synapse with interneurons in the spinal cord; only about 10% of the descending fibers synapse directly with the lower motor neurons. It is likely that very rapid movements are produced by direct synapses with the lower motor neurons, whereas most other movements are produced indirectly via synapses with spinal interneurons, which in turn stimulate the motor neurons.

Upper motor neurons—neurons in the brain that contribute fibers to descending motor tracts—usually stimulate both alpha and gamma motoneurons simultaneously. This is known as **coactivation** of the alpha and gamma motoneurons. Stimulation of alpha motoneurons results in muscle contraction and shortening; stimulation of gamma motoneurons stimulates contraction of the intrafusal fibers and thus "takes out the slack" that would otherwise be present in the spindles as the muscles shorten. In this way, the spindles remain under

tension and provide information about the length of the muscle even while the muscle is shortening.

When upper motor neurons stimulate the contraction of one muscle, they inhibit, via synapses with spinal interneurons, the alpha and gamma motoneurons of the antagonist muscles. Inhibition of the gamma motoneurons makes the antagonist less sensitive to stretch, so that its stretch reflex will not oppose the intended action. Flexion of the arm by contraction of the biceps, for example, stretches the triceps muscle. The triceps is not normally stimulated to contract by this action, however, because its stretch reflex is dampened by inhibition of its gamma motoneurons.

Under normal conditions, the activity of gamma motoneurons is maintained at the level needed to keep the muscle spindles under proper tension while the muscles are relaxed. Undue relaxation of the muscles is prevented by stretch and activation of the spindles, which in turn elicits a reflex contraction (described in the next section). This mechanism produces a normal resting muscle length and state of tension, or **muscle tone.**

Table 12.6 Summary of Events in a Monosynaptic Stretch Reflex

1. Passive stretch of a muscle (produced by tapping its tendon) stretches the spindle (intrafusal) fibers.
2. Stretching of a spindle distorts its central (bag or chain) region, which stimulates dendritic endings of sensory nerves.
3. Action potentials are conducted by afferent (sensory) nerve fibers into the spinal cord on the dorsal roots of spinal nerves.
4. Axons of sensory neurons synapse with dendrites and cell bodies of somatic motor neurons located in the ventral horn gray matter of the spinal cord.
5. Efferent nerve impulses in the axons of somatic motor neurons (which form the ventral roots of spinal nerves) are conducted to the ordinary (extrafusal) muscle fibers. These neurons are alpha motoneurons.
6. Release of acetylcholine from the endings of alpha motoneurons stimulates contraction of the extrafusal fibers, and thus of the whole muscle.
7. Contraction of the muscle relieves the stretch of its spindles, thus decreasing electrical activity in the spindle afferent nerve fibers.

Table 12.7 Effects of Lesions (Damage) at Different Levels on the Neural Control of Skeletal Muscles

Dorsal roots of spinal nerves (sensory)	Uncoordinated movements; difficulty in walking (ataxia)
Transsection of spinal cord	First—*spinal shock,* characterized by lack of stretch reflexes and low muscle tone. After a few weeks—*decerebrate rigidity,* characterized by hyperactive stretch reflexes, flexion of the arms, and extension of the legs (spasticity); also, forced flexion of hand or foot, producing oscillating extension and flexion (clonus)
Cutting of pyramids	Voluntary movements requiring great effort; movements lacking precision; low muscle tone on side opposite lesion
Transsection between midbrain and cerebral cortex	Animal placed on side can right itself, stand, and walk; clumsy movements and some rigidity of extensor muscles
Removal of motor cortex	No paralysis, but uncoordinated movements

Skeletal Muscle Reflexes

Although skeletal muscles are often called voluntary muscles, because they are controlled by descending motor pathways that are under conscious control, they often contract in an unconscious, reflex fashion in response to particular stimuli. In the simplest type of reflex, a skeletal muscle contracts in response to the stimulus of muscle stretch. More complex reflexes involve inhibition of antagonistic muscles and regulation of a number of muscles on both sides of the body.

The Monosynaptic Stretch Reflex

Reflex contraction of skeletal muscles occurs in response to sensory input and is not dependent upon the activation of upper motor neurons. The **reflex arc,** which describes the nerve impulse pathway from sensory to motor endings in such reflexes, involves only a few synapses within the CNS. The simplest of all reflexes—the *muscle stretch reflex*—consists of only one synapse within the CNS. The sensory neuron directly synapses with the motor neuron, without involving spinal cord interneurons. The stretch reflex is thus *monosynaptic* in terms of the individual reflex arcs (many sensory neurons, of course, are activated at the same time, leading to the activation of many motor neurons). Resting skeletal muscles are maintained at an optimal length, as previously described under the heading "Length–Tension Relationship," by stretch reflexes.

The stretch reflex is present in all muscles, but it is most dramatic in the extensor muscles of the limbs. The **knee-jerk reflex**—the most commonly evoked stretch reflex—is initiated by striking the patellar ligament with a rubber mallet. This stretches the entire body of the muscle, and thus passively stretches the spindles within the muscle so that sensory nerves with primary (annulospiral) endings in the spindles are

activated. Axons of these sensory neurons synapse within the ventral gray matter of the spinal cord with *alpha motoneurons.* These large, fast-conducting motor nerve fibers stimulate the extrafusal fibers of the extensor muscle, resulting in isotonic contraction and the knee jerk. This is an example of negative feedback—stretching of the muscles (and spindles) stimulates shortening of the muscles (and spindles). These events are summarized in table 12.6 and in figure 12.19.

Damage to spinal nerves, or to the cell bodies of lower motor neurons (by poliovirus, for example), produces a *flaccid paralysis,* characterized by reduced muscle tone, depressed stretch reflexes, and atrophy. Damage to upper motor neurons or descending motor tracts (table 12.7) at first produces *spinal shock* in which there is a flaccid paralysis. This is followed in a few weeks by *spastic paralysis,* characterized by increased muscle tone, exaggerated stretch reflexes, and other signs of hyperactive lower motor neurons.

The appearance of spastic paralysis suggests that upper motor neurons normally exert an inhibitory effect on lower alpha and gamma motor neurons. When this inhibition is removed, the gamma motoneurons become hyperactive and the spindles thus become overly sensitive to stretch. This can be demonstrated dramatically by forcefully dorsiflecting the patient's foot (pushing it up) and then releasing it. Forced extension stretches the antagonistic flexor muscles, which contract and produce the opposite movement (plantar flexion). Alternating activation of antagonistic stretch reflexes produces a flapping motion known as **clonus.**

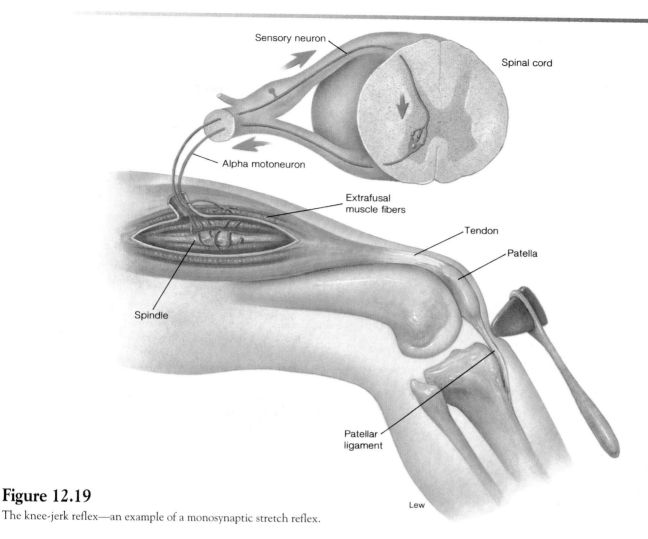

Figure 12.19

The knee-jerk reflex—an example of a monosynaptic stretch reflex.

Labels on figure: Sensory neuron, Spinal cord, Alpha motoneuron, Extrafusal muscle fibers, Tendon, Patella, Spindle, Patellar ligament, Lew

Golgi Tendon Organs

The Golgi tendon organs continuously monitor tension in the tendons produced by muscle contraction or passive stretching of a muscle. Sensory neurons from these receptors synapse with interneurons in the spinal cord; these interneurons, in turn, have *inhibitory synapses* (via IPSPs and postsynaptic inhibition—chapter 7) with motor neurons that innervate the muscle (fig. 12.20). This inhibitory Golgi tendon organ relex is called a **disynaptic reflex** (because two synapses are crossed in the CNS), and it helps to prevent excessive muscle contractions or excessive passive muscle stretching. Indeed, if a muscle is stretched extensively, it will actually relax as a result of the inhibitory effects produced by the Golgi tendon organs.

Reciprocal Innervation and the Crossed-Extensor Reflex

In the knee-jerk and other stretch reflexes, the sensory neuron that stimulates the motor neuron of a muscle also stimulates interneurons within the spinal cord via collateral branches. These interneurons inhibit the motor neurons of antagonist muscles via inhibitory postsynaptic potentials (IPSPs). This dual stimulatory and inhibitory activity is called **reciprocal innervation** (fig. 12.21).

Rapid stretching of skeletal muscles produces very forceful muscle contractions as a result of the activation of primary and secondary endings in the muscle spindles and the monosynaptic stretch reflex. This can result in painful muscle spasms, as may occur, for example, when muscles are forcefully pulled in the process of setting broken bones. Painful muscle spasms may be avoided in physical exercise by stretching slowly and thereby stimulating mainly the secondary endings in the muscle spindles. A slower rate of stretch also provides time for the inhibitory Golgi tendon organ reflex to occur and promote muscle relaxation.

When a limb is flexed, for example, the antagonistic extensor muscles are passively stretched. Extension of a limb similarly stretches the antagonistic flexor muscles. If the monosynaptic stretch reflexes were not inhibited, reflex contraction of the antagonistic muscles would always interfere with

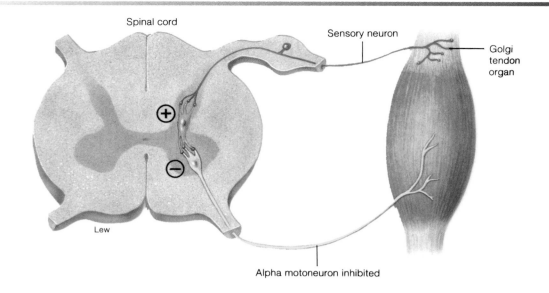

Figure 12.20

An increase in muscle tension stimulates the activity of sensory nerve endings in the Golgi tendon organ. This sensory input stimulates ⊕ an interneuron, which in turn inhibits ⊖ the activity of a motor neuron innervating that muscle. This is therefore a disynaptic reflex.

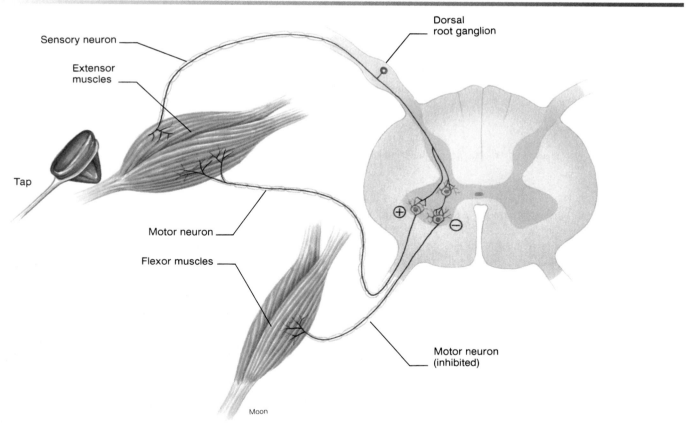

Figure 12.21

A diagram of reciprocal innervation. Afferent impulses from muscle spindles stimulate alpha motoneurons to the agonist muscle (the extensor) directly, but (via an inhibitory interneuron) they inhibit activity in the alpha motoneuron to the antagonist muscle.

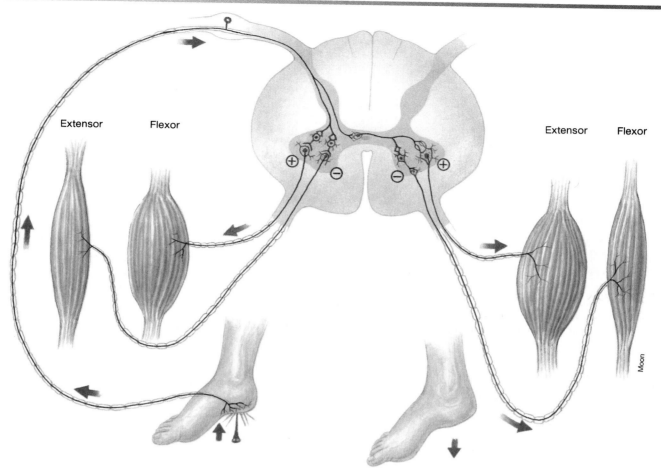

Extensor **Flexor**

Extensor **Flexor**

Figure 12.22

The crossed-extensor reflex, demonstrating double reciprocal innervation.

the intended movement. Fortunately, whenever the "intended," or agonist muscles, are stimulated to contract, the alpha and gamma motoneurons that stimulate the antagonist muscles are inhibited.

The stretch reflex, with its reciprocal innervations, involves the muscles of one limb only and is controlled by only one segment of the spinal cord. More complex reflexes involve muscles controlled by numerous spinal cord segments and affect muscles on the contralateral side of the cord. Such reflexes involve **double reciprocal innervation** of muscles.

Double reciprocal innervation is illustrated by the **crossed-extensor reflex.** If you step on a tack with the right foot, for example, this foot is withdrawn by contraction of its flexors and relaxation of its extensors. The contralateral left leg, by contrast, extends to help support the body during this withdrawal reflex. The extensors of the left leg contract while its flexors relax. These events are shown in figure 12.22.

Upper Motor Neuron Control of Skeletal Muscles

As previously described, upper motor neurons are neurons in the brain that influence the control of skeletal muscle by lower motor neurons (alpha and gamma motoneurons). Neurons in the precentral gyrus of the cerebral cortex contribute axons that cross to the contralateral sides in the pyramids of the medulla oblongata; these tracts are thus called **pyramidal tracts** (chapter 8). The pyramidal tracts include the *lateral* and *ventral corticospinal tracts.* Neurons in other areas of the brain produce the **extrapyramidal tracts.** The major extrapyramidal tract is the *reticulospinal tract,* which originates in the reticular formation of the medulla oblongata and pons. Brain areas that influence the activity of extrapyramidal tracts are believed to produce the inhibition of lower motor neurons described in the preceding section.

Table 12.8 Symptoms of Upper Motor Neuron Damage

Babinski reflex—Extension of the big toe when the sole of the foot is rubbed along the lateral border
Spastic paralysis—High muscle tone and hyperactive stretch reflexes; flexion of arms and extension of legs
Hemiplegia—Paralysis of upper and lower limbs on one side—commonly produced by damage to motor tracts as they pass through internal capsule (such as by cerebrovascular accident—stroke)
Paraplegia—Paralysis of lower limbs on both sides as a result of lower spinal cord damage
Quadriplegia—Paralysis of both upper and lower limbs on both sides as a result of damage to the upper region of the spinal cord or brain
Chorea—Random uncontrolled contractions of different muscle groups (such as Saint Vitus' dance) as a result of damage to basal nuclei
Resting tremor—Shaking of limbs at rest; disappears during voluntary movements; produced by damage to basal nuclei
Intention tremor—Oscillations of arm following voluntary reaching movements; produced by damage to cerebellum

Cerebellum

The cerebellum, like the cerebrum, receives sensory input from muscle spindles and Golgi tendon organs. It also receives fibers from areas of the cerebral cortex devoted to vision, hearing, and equilibrium.

There are no descending tracts from the cerebellum. The cerebellum can influence motor activity only indirectly, through its output to the vestibular nuclei, red nucleus, and basal nuclei. These structures, in turn, affect lower motor neurons via the vestibulospinal tract, rubrospinal tract, and reticulospinal tract. It is interesting that all output from the cerebellum is inhibitory; these inhibitory effects aid motor coordination by eliminating inappropriate neural activity. Damage to the cerebellum interferes with the ability to coordinate movements with spatial judgment. Under- or overreaching for an object may occur, followed by *intention tremor,* in which the limb moves back and forth in a pendulum-like motion.

Basal Nuclei

The basal nuclei, sometimes called the basal ganglia, include the *caudate nucleus, putamen,* and *globus pallidus* (chapter 8). Often included in this group are other nuclei of the *thalamus, subthalamus, substantia nigra,* and *red nucleus.* Acting directly via the rubrospinal tract and indirectly via synapses in the reticular formation and thalamus, the basal nuclei have profound effects on the activity of lower motor neurons.

In particular, through their synapses in the reticular formation, the basal nuclei exert an inhibitory influence on the activity of lower motor neurons. Damage to the basal nuclei thus results in increased muscle tone, as previously described. People with such damage display *akinesia* (lack of desire to use the affected limb) and *chorea*—sudden and uncontrolled random movements (table 12.8).

Parkinson's disease (or *paralysis agitans*) is a disorder of the basal nuclei involving degeneration of fibers from the substantia nigra. These fibers, which use dopamine as a neurotransmitter, are required to antagonize the effects of other fibers that use acetylcholine (ACh) as a transmitter. The relative deficiency of dopamine compared to ACh is believed to produce the symptoms of Parkinson's disease, including *resting tremor*. This "shaking" of the limbs tends to disappear during voluntary movements and then reappear when the limb is again at rest.

1. Draw a muscle spindle surrounded by a few extrafusal fibers, indicating the location of primary and secondary sensory endings. Describe how these endings respond to muscle stretch.

2. Describe all of the events that occur from the time the patellar tendon is struck with a mallet to the time the leg kicks.

3. Explain how a Golgi tendon organ is stimulated and describe the disynaptic reflex that occurs.

4. Explain the significance of reciprocal innervation and double reciprocal innervation in muscle reflexes.

5. Describe the functions of gamma motoneurons and explain why they are stimulated at the same time as alpha motoneurons during voluntary muscle contractions.

6. Explain how a person with spinal cord damage might develop clonus.

Energy Requirements of Skeletal Muscles

Skeletal muscles generate ATP through aerobic and anaerobic respiration and through the use of phosphate groups donated by creatine phosphate. The aerobic and anaerobic abilities of skeletal muscle fibers differ according to muscle fiber type. Slow-twitch (type I) fibers are adapted for aerobic respiration; fast-twitch (type II) fibers are adapted for anaerobic respiration.

Skeletal muscles at rest obtain most of their energy from the aerobic respiration of fatty acids. During exercise, muscle glycogen and blood glucose are also used as energy sources. Energy obtained by cell respiration is used to make ATP, which serves as the immediate source of energy for (1) the movement of the cross bridges for muscle contraction and (2) the pumping of Ca^{++} into the sarcoplasmic reticulum for muscle relaxation.

Metabolism of Skeletal Muscles

Skeletal muscles respire anaerobically for the first 45 to 90 seconds of moderate-to-heavy exercise, because the cardiopulmonary system requires this amount of time to sufficiently increase the oxygen supply to the exercising muscles. If exercise is moderate and the person is in good physical condition, aerobic respiration contributes the major portion of the skeletal muscle energy requirements following the first 2 minutes of exercise (fig. 12.23).

Maximal Oxygen Uptake and Oxygen Debt

The maximum rate of oxygen consumption (by aerobic respiration) in the body is called the **maximal oxygen uptake,** and it is determined primarily by a person's age, size, and sex. It is from 15% to 20% higher for males than for females and highest at age 20 for both sexes. Some world-class athletes have maximal oxygen uptakes that are twice the average for their age and sex—this appears to be due largely to genetic factors, but training can increase the maximum oxygen uptake by about 20%.

When a person stops exercising, the rate of oxygen uptake does not immediately go back to pre-exercise levels; it returns slowly (the person continues to breathe heavily for some time afterwards). This extra oxygen is used to repay the **oxygen debt** incurred during exercise. The oxygen debt includes oxygen that was withdrawn from savings deposits—hemoglobin in blood and myoglobin in muscle (chapter 16); the extra oxygen required for metabolism by tissues warmed during exercise; and the oxygen needed for the metabolism of the lactic acid produced during anaerobic respiration.

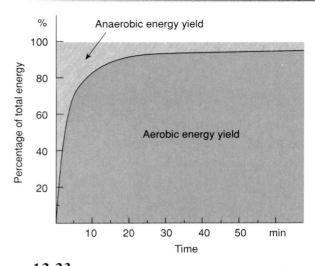

Figure 12.23

The relative contributions of anaerobic and aerobic respiration to the total energy in a well-trained person performing at maximal effort.

Phosphocreatine

During sustained muscle activity, ATP may be used faster than it can be produced through cell respiration. At these times the rapid renewal of ATP is extremely important. This is accomplished by combining ADP with phosphate derived from another high-energy phosphate compound called **phosphocreatine,** or **creatine phosphate.**

The phosphocreatine concentration within muscle cells is more than three times the concentration of ATP and represents a ready reserve of high-energy phosphate that can be donated directly to ADP (fig. 12.24). During times of rest, the depleted reserve of phosphocreatine can be restored by the reverse reaction—phosphorylation of creatine with phosphate derived from ATP.

The enzyme that transfers phosphate between creatine and ATP is called **creatine kinase,** or **creatine phosphokinase.** Skeletal muscle and heart muscle have two different forms of this enzyme (they have different isoenzymes, as described in chapter 4). The skeletal muscle isoenzyme is found to be elevated in the blood of people with *muscular dystrophy* (degenerative disease of skeletal muscles). The plasma concentration of the isoenzyme characteristic of heart muscle is elevated as a result of *myocardial infarction* (damage to heart muscle), and measurements of this enzyme are thus used as a means of diagnosing this condition.

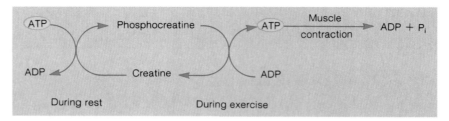

Figure 12.24

The production and utilization of phosphocreatine in muscles.

Slow- and Fast-Twitch Fibers

Skeletal muscle fibers can be divided on the basis of their contraction speed (time required to reach maximum tension) into **slow-twitch, or type I, fibers,** and **fast-twitch, or type II, fibers.** These differences are associated with different myosin ATPase isoenzymes, which can also be designated as "slow" and "fast." The two fiber types can be distinguished by their ATPase isoenzyme when they are appropriately stained (fig. 12.25). The extraocular muscles that position the eyes, for example, have a high proportion of fast-twitch fibers and reach maximum tension in about 7.3 msec (milliseconds—thousandths of a second); the soleus muscle in the leg, by contrast, has a high proportion of slow-twitch fibers and requires about 100 msec to reach maximum tension (fig. 12.26).

Muscles like the soleus are *postural muscles* that must be able to sustain a contraction for a long period of time without fatigue. The resistance to fatigue demonstrated by these muscles is aided by other characteristics of slow-twitch (type I) fibers that endow them with a high oxidative capacity for aerobic respiration. Slow-twitch fibers have a rich capillary supply, numerous mitochondria and aerobic respiratory enzymes, and a high concentration of *myoglobin* pigment. Myoglobin is a red pigment, similar to the hemoglobin in red blood cells, that improves the delivery of oxygen to the slow-twitch fibers. Because of their high myoglobin content, slow-twitch fibers are also called *red fibers*.

The thicker, fast-twitch (type II) fibers have fewer capillaries and mitochondria than slow-twitch fibers and not as much myoglobin; hence, these fibers are also called *white fibers*. Fast-twitch fibers are adapted to respire anaerobically by a large store of glycogen and a high concentration of glycolytic enzymes. In addition to the type I (slow-twitch) and type II (fast-twitch) fibers, human muscles also have an intermediate form of fibers. These intermediate fibers are fast-twitch but also have a high oxidative capacity and so are relatively resistant to fatigue. These are called **type IIA fibers,** to distinguish them from the anaerobically adapted fast-twitch **type IIB** fibers, which have a low oxidative capacity and fatigue rapidly. The three fiber types are compared in table 12.9.

Interestingly, the conduction rate of motor neurons that innervate fast-twitch fibers is faster (80–90 meters per second) than the conduction rate to slow-twitch fibers (60–70 meters per second). The fiber type indeed seems to be determined by

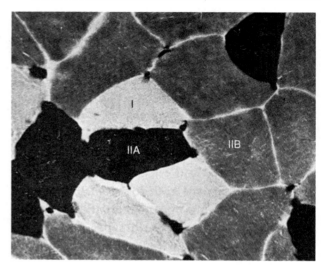

Figure 12.25

Skeletal muscle (of a cat) stained to indicate the activity of myosin ATPase. ATPase activity is greater in the type II fibers than in the type I fibers.

the motor neuron. When the motor neurons to different fiber types are switched in experimental animals, the previously fast-twitch fibers become slow and the slow-twitch fibers become fast. As expected from these observations, all of the muscle fibers innervated by the same motor neuron (that are part of the same motor unit) are of the same type.

A muscle such as the gastrocnemius contains both fast- and slow-twitch fibers, although fast-twitch fibers predominate. A given somatic motor axon, however, innervates muscle fibers of only one type. The size of these motor units differ; the motor units composed of slow-twitch fibers tend to be smaller (have fewer fibers) than the motor units of fast-twitch fibers. Since motor units are recruited from smaller to larger when increasing effort is required, as previously described, the smaller motor units with slow-twitch fibers would be used most often in routine activities. Larger motor units with fast-twitch fibers, which can exert a great deal of force but which respire anaerobically and thus fatigue quickly, would be used relatively infrequently and for only short periods of time.

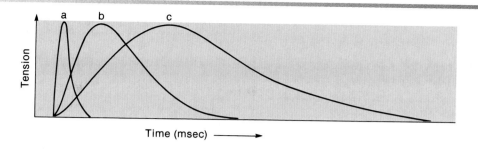

Figure 12.26

A comparison of the rates with which maximum tension is developed in three muscles: (*a*) the relatively fast-twitch extraocular and (*b*) gastrocnemius muscles and (*c*) the slow-twitch soleus muscle.

Table 12.9 Characteristics of Red, Intermediate, and White Muscle Fibers

Feature	Red (Type I)	Intermediate (Type IIA)	White (Type IIB)
Diameter	Small	Intermediate	Large
Z-line thickness	Wide	Intermediate	Narrow
Glycogen content	Low	Intermediate	High
Resistance to fatigue	High	Intermediate	Low
Capillaries	Many	Many	Few
Myoglobin content	High	High	Low
Respiration type	Aerobic	Aerobic	Anaerobic
Oxidative capacity	High	High	Low
Glycolytic ability	Low	High	High
Twitch rate	Slow	Fast	Fast
Myosin ATPase content	Low	High	High

Muscle Fatigue

Muscle fatigue may be defined as the inability to maintain a particular muscle tension when the contraction is sustained or to reproduce a particular tension during rhythmic contraction over time. Fatigue during a sustained maximal contraction, when all the motor units are used and the rate of neural firing is maximal—as when lifting an extremely heavy weight—appears to be due to an accumulation of extracellular K^+. (Remember that K^+ leaves axons and muscle fibers during the repolarization phase of action potentials.) This reduces the membrane potential of muscle fibers and interferes with their ability to produce action potentials. Fatigue under these circumstances lasts only a short time, and maximal tension can again be produced after less than a minute's rest.

Fatigue during moderate exercise occurs as the slow-twitch fibers deplete their reserve glycogen and fast-twitch fibers are increasingly recruited. Fast-twitch fibers obtain their energy through anaerobic respiration, converting glucose to lactic acid, and this results in a rise in intracellular H^+ and a fall in pH. The decrease in muscle pH, in turn, promotes muscle fatigue, but the exact physiological mechanisms by which this occurs are not well understood. One possibility is that there may be a reduced ability of the sarcoplasmic reticulum to accumulate Ca^{++} by active transport, or there may be a reduced ability of the sarcoplasmic reticulum to release Ca^{++} in response to stimulation. By either mechanism, the decrease cellular pH would produce muscle fatigue by interfering with excitation-contraction coupling.

Adaptations to Exercise

The maximal oxygen uptake, obtained during very strenuous exercise, averages 50 ml O_2 per minute per kilogram body weight in males between the ages of 20 and 25 (females average 25% lower). For trained endurance athletes (swimmers, long-distance runners), maximal oxygen uptakes can be as high as 86 ml O_2 per minute per kilogram. When exercise is performed at low levels of effort, such that the oxygen consumption rate is below 50% of its maximum, the energy for muscle contraction is obtained almost entirely from aerobic cell respiration. Anaerobic cell respiration, with its consequent production of lactic acid, contributes to the energy requirements as the exercise level rises and more than 60% of the maximal oxygen uptake is required. Highly trained endurance athletes, however, can continue to respire aerobically, with little lactic acid production, at up to 80% of their maximal oxygen uptake. These athletes thus produce less lactic acid at a given level of exercise than the average person, and therefore they are less subject to fatigue than the average person.

All fiber types adapt to endurance training by an increase in mitochondria, and thus in aerobic respiratory enzymes. In fact, the maximal oxygen uptake can be increased by as much as 20% through endurance training. There is a decrease in type IIB fibers, which have a low oxidative capacity, accompanied by an increase in type IIA fibers, which have a high oxidative capacity. Although these fibers are still classified as fast-twitch (type IIA), they show an increase in the slow myosin ATPase

Table 12.10 Effects of Endurance Training (Long-Distance Running, Swimming, Bicycling, etc.) on Skeletal Muscles

1. Improved ability to obtain ATP from oxidative phosphorylation
2. Increased size and number of mitochondria
3. Less lactic acid produced per given amount of exercise
4. Increased myoglobin content
5. Increased intramuscular triglyceride content
6. Increased lipoprotein lipase (enzyme needed to utilize lipids from blood)
7. Increased proportion of energy derived from fat; less from carbohydrates
8. Lower rate of glycogen depletion during exercise
9. Improved efficiency in extracting oxygen from blood
10. Decreased number of type IIB fibers; increased number of type IIA fibers

isoenzyme form, indicating that they are in a transitional state between the type II and type I fibers.

In addition to changes in aerobic capacity, muscle fibers show an increase in their content of triglycerides, which serve as an alternate energy source and help to spare their stores of glycogen. A summary of the changes that occur as a result of endurance training is presented in table 12.10.

Endurance training does not increase the size of muscles. Muscle enlargement is produced only by frequent periods of high-intensity exercise in which muscles work against a high resistance, as in weight lifting. As a result of resistance training, type II muscle fibers become thicker, and the muscle therefore grows by hypertrophy (an increase in cell size, rather than number of cells). This happens first because the myofibrils within a muscle fiber thicken, due to the synthesis of actin and myosin proteins and the addition of new sarcomeres. After a myofibril attains a certain thickness it may split into two myofibrils, each of which may then become thicker due to the addition of sarcomeres. Muscle hypertrophy, in short, is associated with an increase in the size and then the number of myofibrils within the muscle fibers.

1. Draw a figure illustrating the relationship between ATP and creatine phosphate and explain the physiological significance of this relationship.
2. Describe the characteristics of slow- and fast-twitch fibers (including intermediate fibers). Explain how the fiber types are determined and list the functions of different fiber types.
3. Explain the different causes of muscle fatigue with reference to the various fiber types.
4. Describe the effects of endurance training and resistance training on the fiber characteristics of the muscles.

Cardiac and Smooth Muscle

Cardiac muscle, like skeletal muscle, is striated and contains sarcomeres that shorten by sliding of thin and thick filaments. But while skeletal muscle requires nervous stimulation to contract, cardiac muscle can produce impulses and contract spontaneously. Smooth muscles lack sarcomeres, but contain actin and myosin, which produce contractions in response to a unique regulatory mechanism.

Unlike skeletal muscles, which are voluntary effectors regulated by somatic motor neurons, cardiac and smooth muscles are involuntary effectors regulated by autonomic motor neurons. Although there are important differences between skeletal muscle and cardiac and smooth muscle, there are also significant similarities. All types of muscle are believed to contract by means of sliding of thin actin filaments over thick myosin filaments. The sliding of the filaments is produced by the action of myosin cross bridges in all types of muscles, and excitation-contraction coupling in all types of muscles involves Ca^{++}.

Cardiac Muscle

Like skeletal muscle cells, cardiac (heart) muscle cells, or *myocardial cells*, are striated; they contain actin and myosin filaments arranged in the form of sarcomeres, and they contract by means of the sliding filament mechanism. The long, fibrous skeletal muscle cells, however, are structurally and functionally separated from each other, whereas the myocardial cells are short, branched, and interconnected. Adjacent myocardial cells are joined by electrical synapses, or **gap junctions** (described in chapter 7). Gap junctions in cardiac muscle have an affinity for stain that makes them appear as dark lines between adjacent cells when viewed in the light microscope. These dark-staining lines are known as *intercalated discs* (fig. 12.27).

Electrical impulses that originate at any point in a mass of myocardial cells, called a *myocardium*, can spread to all cells in the mass that are joined by gap junctions. Because all cells in a myocardium are electrically joined, a myocardium behaves as a single functional unit. Thus, unlike skeletal muscles that produce contractions that are graded depending on the number of cells stimulated, a myocardium contracts to its full extent each time because all of its cells contribute to the contraction. The ability of the myocardial cells to contract, however, can be increased by epinephrine and by stretching of the heart chambers. The heart contains two distinct myocardia (atria and ventricles), as will be described in chapter 13.

Unlike skeletal muscles, which require external stimulation by somatic motor nerves before they can produce action potentials and contract, cardiac muscle is able to produce action

Figure 12.27

Cardiac muscle. Notice that the cells are short, branched, and striated and that they are interconnected by intercalated discs.

potentials automatically. Cardiac action potentials normally originate in a specialized group of cells called the *pacemaker.* However, the rate of this spontaneous depolarization and, thus, the rate of the heartbeat, are regulated by autonomic innervation. Regulation of the cardiac rate is described more fully in chapter 14.

Smooth Muscle

Smooth (visceral) muscles are arranged in circular layers around the walls of blood vessels and bronchioles (small air passages in the lungs). Both circular and longitudinal smooth muscle layers occur in the tubular digestive tract, the ureters (which transport urine), the vasa deferentia (which transport sperm), and the uterine tubes (which transport ova). The alternate contraction of circular and longitudinal smooth muscle layers in the intestine produces **peristaltic waves,** which propel the contents of these tubes in one direction.

Although smooth muscle cells do not contain sarcomeres (which produce striations in skeletal and cardiac muscle), they do contain a great deal of actin and some myosin, which produces a ratio of thin-to-thick filaments of about 16:1 (in striated muscles the ratio is 2:1). Unlike striated muscles, in which the thick filaments are short and stacked between Z discs in sarcomeres, the myosin filaments in smooth muscle cells are quite long (fig. 12.28).

The long length of myosin filaments and the fact that they are not organized into sarcomeres may be of advantage in smooth muscle function. Smooth muscles must be able to contract even when greatly stretched—in the urinary bladder, for example, the smooth muscle cells may be stretched up to two and a half times their resting length. The smooth muscle cells of the uterus may be stretched up to eight times their original length by the end of pregnancy. Striated muscles, because of their structure, lose their ability to contract when the sarcomeres are stretched to the point where actin and myosin no longer overlap.

As in striated muscles, the contraction of smooth muscles is triggered by a sharp rise in the Ca^{++} concentration within the cytoplasm of the muscle cells. However, the sarcoplasmic reticulum of smooth muscles is less developed than that of

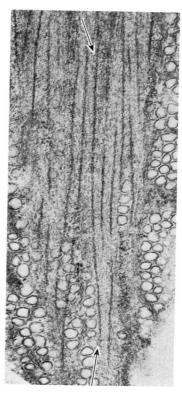

Figure 12.28

An electron micrograph of the thick and thin filaments of smooth muscle. A longitudinal section of a complete long myosin filament is shown between the arrows (32,000×).

skeletal muscles, and Ca^{++} released from this organelle may account for only the initial phase of smooth muscle contraction. Extracellular Ca^{++} diffusing into the smooth muscle cell through its cell membrane is responsible for sustained contractions. This Ca^{++} enters primarily through voltage-sensitive Ca^{++} gates in the cell membrane. The opening of these gates is graded by the amount of depolarization; the greater the depolarization, the more Ca^{++} will enter the cell and the stronger will be the smooth muscle contraction.

 Drugs such as *verapamil* and related compounds are **calcium-channel blockers.** These drugs block Ca^{++} channels in the membrane of smooth muscle cells within the walls of blood vessels, causing the muscles to relax and the vessels to dilate. This effect, called vasodilation, may be helpful in treating hypertension (high blood pressure). Calcium-channel-blocking drugs are also used when spasm of the coronary arteries (vasospasm) produces angina pectoris, which is pain caused by insufficient blood flow to the heart.

Table 12.11 Comparison of Skeletal, Cardiac, and Smooth Muscle

Skeletal Muscle	Cardiac Muscle	Smooth Muscle
Striated; actin and myosin arranged in sarcomeres	Striated; actin and myosin arranged in sarcomeres	Not striated; more actin than myosin; actin inserts into dense bodies and cell membrane
Well-developed sarcoplasmic reticulum and transverse tubules	Moderately developed sarcoplasmic reticulum and transverse tubules	Poorly developed sarcoplasmic reticulum; no transverse tubules
Contains troponin in the thin filaments	Contains troponin in the thin filaments	Contains a Ca^{++} binding protein; may be located in thick filaments
Ca^{++} released into cytoplasm from sarcoplasmic reticulum	Ca^{++} enters cytoplasm from sarcoplasmic reticulum and extracellular fluid	Ca^{++} enters cytoplasm from extracellular fluid, sarcoplasmic reticulum, and perhaps mitochondria
Cannot contract without nerve stimulation; denervation results in muscle atrophy	Can contract without nerve stimulation; action potentials originate in pacemaker cells of heart	Maintains tone in absence of nerve stimulation; visceral smooth muscle produces pacemaker potentials; denervation results in hypersensitivity to stimulation
Muscle fibers stimulated independently; no gap junctions	Gap junctions present as intercalated discs	Gap junctions generally present

The events that follow the entry of Ca^{++} into the cytoplasm are somewhat different in smooth muscles than in striated muscles. In striated muscles, Ca^{++} combines with troponin. Troponin, however, is not present in smooth muscle cells. In smooth muscles, Ca^{++} combines with a protein in the cytoplasm called **calmodulin,** which is structurally similar to troponin. Calmodulin was previously discussed in relation to the function of Ca^{++} as a second messenger in hormone action (chapter 11). The calmodulin-Ca^{++} complex thus formed combines with and activates an enzyme called *myosin light chain kinase,* which catalyzes the phosphorylation of (addition of phosphate groups to) the myosin cross bridges. In smooth muscle cells, the cross bridges must be phosphorylated before they can bond to actin, which is not the case in striated muscles.

In the preceding sequence of events, the concentration of Ca^{++} in the smooth muscle cell cytoplasm determines how many cross bridges will combine with actin, and thus determines the strength of contraction. The concentration of Ca^{++} is in turn regulated by the degree of depolarization. Unlike the situation in striated muscle cells, which produce all-or-none action potentials, smooth muscle cells can produce graded depolarizations and contractions without producing action potentials. (This is discussed in more detail in conjunction with intestinal contractions in chapter 18.) Indeed, only these graded depolarizations are conducted from cell to cell in many smooth muscles.

In addition to being graded, the contractions of smooth muscle cells are slow and sustained. The slowness of contraction is related to the fact that myosin ATPase in smooth muscle is slower in its action (splitting ATP for the cross-bridge cycle) than it is in striated muscle. The sustained nature of smooth muscle contraction is explained by the theory that cross bridges in smooth muscles can enter a *latch state*.

The latch state allows smooth muscle to maintain its contraction in a very energy-efficient manner, hydrolyzing less ATP than would otherwise be required. This ability is obviously important for smooth muscles, given that they encircle the walls of hollow organs and must sustain contractions for very long time periods. The mechanisms by which the latch state is produced, however, are complex and poorly understood.

Despite their differences, it is currently believed that both smooth muscles and striated muscles contract by means of a sliding filament mechanism. The three muscle types—skeletal, cardiac, and smooth—are compared in table 12.11.

Single-Unit and Multiunit Smooth Muscles

Smooth muscles are often grouped into two functional categories: **single-unit** and **multiunit.** Single-unit smooth muscles have numerous gap junctions (electrical synapses) between adjacent cells that weld them together electrically; they thus behave as a single unit, much like cardiac muscle. Most smooth muscles—including those in the gastrointestinal tract and uterus—are single-unit.

Only some cells of single-unit smooth muscles receive autonomic innervation; the ACh released by the axon can diffuse to other smooth muscle cells. Binding of ACh to its muscarinic receptors causes depolarization by closing K^+ channels, as described in chapter 7. Such stimulation, however, only modifies the automatic behavior of single-unit smooth muscles. Single-unit smooth muscles display *pacemaker* activity in which certain cells stimulate others in the mass. This is similar to the situation in cardiac muscle. Single-unit smooth muscles also display intrinsic, or *myogenic*, electrical activity and contraction in response to stretch. For example, the stretch induced by an increase in the volume of a ureter or a section of the gastrointestinal tract can stimulate myogenic contraction. Such contraction does not require stimulation by autonomic nerves.

Contraction of multiunit smooth muscles, in contrast, requires nerve stimulation. Multiunit smooth muscles have few, if any, gap junctions. The individual cells must thus be stimulated separately by nerve fibers. Examples of multiunit smooth muscles are the arrector pili muscles in the skin and the ciliary muscles attached to the lens of the eye. Single-unit and multiunit smooth muscles are compared in table 12.12.

Autonomic Innervation of Smooth Muscles

There are significant differences between the neural control of skeletal muscles and that of smooth muscles. A skeletal muscle fiber has only one junction with a somatic nerve fiber, and the receptors for the neurotransmitter are located only at the neuromuscular junction. By contrast, the entire surface of smooth

Table 12.12   Comparison of Single-Unit and Multiunit Smooth Muscles

Feature	Single-Unit Muscle	Multiunit Muscle
Location	Gastrointestinal tract; uterus; ureter; small arteries (arterioles)	Arrector pili muscles of hair follicles; ciliary muscle (attached to lens); iris; vas deferens; large arteries
Origin of electrical activity	Spontaneous activity by pacemakers (myogenic)	Not spontaneously active; potentials are neurogenic
Type of potentials	Action potentials	Graded depolarizations
Response to stretch	By contraction; not dependent on nerve stimulation	No inherent response
Presence of gap junctions	Numerous; join all cells together electrically	Few (if any)
Type of contraction	Slow and sustained	Slow and sustained

muscle cells contains neurotransmitter receptor proteins. Neurotransmitter molecules are released along a stretch of an autonomic nerve fiber that is located some distance from the smooth muscle cells. The regions of the autonomic fiber that release transmitters appear as bulges, or *varicosities,* and the neurotransmitters released from these varicosities stimulate a number of smooth muscle cells. Since there are a number of varicosities along a stretch of an autonomic nerve ending, they form synapses "in passing"—or *synapses en passant*—with the smooth muscle cells.

1. Explain how cardiac muscle differs from skeletal muscle in its structure and regulation of contraction.
2. Contrast the structure of a smooth muscle cell with that of a skeletal muscle fiber and discuss the advantages of each type of structure.
3. Describe the events by which depolarization of a smooth muscle cell results in contraction and explain why smooth muscle contractions are slow and sustained.
4. Distinguish between single-unit and multiunit smooth muscles.

Summary

Structure and Actions of Skeletal Muscles p. 308

I. Skeletal muscles are attached by tendons to bones.
   A. Skeletal muscles are composed of separate cells, or fibers, that are attached in parallel to the tendons.
   B. Skeletal muscle fibers are striated.
      1. The dark striations are called A bands and the light regions are called I bands.
      2. Z lines are located in the middle of each I band.
II. Muscles in vitro can exhibit twitch, summation, and tetanus.
   A. The rapid contraction and relaxation of muscle fibers is called a twitch.
   B. A whole muscle also produces a twitch in response to a single electrical pulse in vitro.
      1. The stronger the electric shock, the stronger the muscle twitch—whole muscle can produce a graded contraction.

   2. The graded contraction of whole muscles is due to different numbers of fibers participating in the contraction.
   C. The summation of fiber twitches can occur so rapidly that the muscle produces a smooth, sustained contraction known as tetanus.
   D. When a muscle exerts tension without shortening, the contraction is termed isometric; when shortening does occur, the contraction is isotonic.
III. The contraction of muscle fibers in vivo is stimulated by somatic motor neurons.
   A. Each somatic motor axon branches to innervate numerous muscle fibers.
   B. The motor neuron and the muscle fibers it innervates are called a motor unit.
      1. When a muscle is composed of a relatively large number of motor units (such as in the hand),

   there is fine control of muscle contraction.
   2. The large muscles of the leg have relatively few motor units, which are correspondingly large in size.
   3. Sustained contractions are produced by the asynchronous stimulation of different motor units.

Mechanisms of Contraction p. 314

I. Skeletal muscle cells, or fibers, contain structures called myofibrils.
   A. Each myofibril is striated with dark (A) and light (I) bands. In the middle of each I band are Z lines.
   B. The A bands contain thick filaments composed of myosin.
      1. The edges of each A band also contain thin filaments overlapped with the thick filaments.
      2. The central regions of the A bands contain only thick filaments—these regions are the H bands.

C. The I bands contain only thin filaments, which are composed primarily of the protein called actin.

D. Thin filaments are composed of globular actin subunits known as G-actin. A protein known as tropomyosin is also located at intervals in the thin filaments. Another protein—troponin—is attached to the tropomyosin.

II. Myosin cross bridges extend out from the thick filaments to the thin filaments.

A. At rest, the cross bridges are not attached to actin.
1. The cross-bridge heads function as ATPase enzymes.
2. ATP is split into ADP and P_i, activating the cross bridge.

B. When the activated cross bridges attach to actin, they undergo a power stroke and in the process release ADP and P_i.

C. At the end of a power stroke, the cross bridge binds to a new ATP.
1. This allows the cross bridge to detach from actin and repeat the cycle.
2. Rigor mortis is caused by the inability of cross bridges to detach from actin due to a lack of ATP.

III. The activity of the cross bridges causes the thin filaments to slide toward the centers of the sarcomeres.

A. The filaments slide—they do not shorten—during muscle contraction.

B. The lengths of the H and I bands decrease, whereas the A bands stay the same length during contraction.

IV. When a muscle is at rest, the Ca^{++} concentration of the sarcoplasm is very low and cross bridges are prevented from attaching to actin.

A. The Ca^{++} is actively transported into the sarcoplasmic reticulum.

B. The sarcoplasmic reticulum is a modified endoplasmic reticulum that surrounds the myofibrils.

V. Action potentials are conducted by transverse tubules into the muscle fiber.

A. Transverse tubules are invaginations of the cell membrane that almost touch the sarcoplasmic reticulum.

B. Action potentials in the transverse tubules stimulate the release of Ca^{++} from the sarcoplasmic reticulum.

VI. When action potentials cease, Ca^{++} is removed from the sarcoplasm and stored in the sarcoplasmic reticulum.

Neural Control of Skeletal Muscles p. 323

I. The somatic motor neurons that innervate the muscles are called the lower motor neurons.

A. Alpha motoneurons innervate the ordinary, or extrafusal, muscle fibers. These are the fibers that produce muscle shortening during contraction.

B. Gamma motoneurons innervate the intrafusal fibers of the muscle spindles.

II. Muscle spindles are length detectors in the muscle.

A. Spindles consist of several intrafusal fibers wrapped together. The spindles are in parallel with the extrafusal fibers.

B. Stretching of the muscle stretches the spindles, which excites sensory endings in the spindle apparatus.
1. Impulses in the sensory neurons travel into the spinal cord in the dorsal roots of spinal nerves.
2. The sensory neuron synapses directly with an alpha motoneuron within the spinal cord—this produces a monosynaptic reflex.
3. The alpha motoneuron stimulates the extrafusal muscle fibers to contract, thus relieving the stretch—this is called the stretch reflex.

C. The activity of gamma motoneurons tightens the spindles, thus making them more sensitive to stretch and better able to monitor the length of the muscle, even during muscle shortening.

III. The Golgi tendon organs monitor the tension that the muscle exerts on its tendons.

A. As the tension increases, sensory neurons from Golgi tendon organs inhibit the activity of alpha motoneurons.

B. This is a disynaptic reflex, since the sensory neurons synapse with interneurons, which in turn make inhibitory synapses with motoneurons.

IV. A crossed-extensor reflex occurs when a foot steps on a tack.

A. Sensory input from the injured foot causes stimulation of flexor muscles and inhibition of the antagonistic extensor muscles.

B. The sensory input also crosses the spinal cord to cause stimulation of extensor and inhibition of flexor muscles in the contralateral leg.

V. Most of the fibers of descending tracts synapse with spinal interneurons, which in turn synapse with the lower motor neurons.

A. Alpha and gamma motoneurons are usually stimulated at the same time, or coactivated.

B. The stimulation of gamma motoneurons keeps the muscle spindles under tension and sensitive to stretch.

C. Upper motor neurons, primarily in the basal nuclei, also exert inhibitory effects on gamma motoneurons.

VI. Neurons in the brain that affect the lower motor neurons are called upper motor neurons.

A. The fibers of neurons in the precentral gyrus, or motor cortex, descend to the lower motor neurons as the lateral and ventral corticospinal tracts.
1. Most of these fibers cross to the contralateral side in the brain stem, forming structures called the pyramids; therefore, this system is called the pyramidal system.
2. The left side of the brain thus controls the musculature on the right side, and vice versa.

B. Other descending motor tracts are part of the extrapyramidal system.
   1. The neurons of the extrapyramidal system make numerous synapses in different areas of the brain, including the midbrain, brain stem, basal nuclei, and cerebellum.
   2. Damage to the cerebellum produces intention tremor; degeneration of dopaminergic neurons in the basal nuclei produces Parkinson's disease.

Energy Requirements of Skeletal Muscles p. 331

I. Aerobic cell respiration is ultimately required for the production of ATP needed for cross-bridge activity.
   A. New ATP can be quickly produced, however, from the combination of ADP with phosphate derived from phosphocreatine.
      1. The phosphocreatine represents a ready reserve of high-energy phosphate during sustained muscle contractions.
      2. Phosphocreatine is produced at rest from creatine, and phosphate that is derived from ATP.
   B. Muscle fibers are of three types.
      1. Slow-twitch red fibers are adapted for aerobic respiration and are resistant to fatigue.
      2. Fast-twitch white fibers are adapted for anaerobic respiration.
      3. Intermediate fibers are fast-twitch but adapted for aerobic respiration.

II. Muscle fatigue may be caused by a number of mechanisms.
   A. Fatigue during sustained maximal contraction may be produced by the accumulation of extracellular K^+ as a result of high levels of nerve activity.
   B. Fatigue during rhythmic, moderate exercise is primarily a result of anaerobic respiration by fast-twitch fibers.
      1. The production of lactic acid lowers the intracellular pH, which inhibits glycolysis and decreases ATP concentrations.
      2. Decreased ATP inhibits excitation-contraction coupling, possibly due to a cellular loss of Ca^{++}.

III. Physical training affects the characteristics of the muscle fibers.
   A. Endurance training increases the aerobic capacity of all muscle fiber types, so that their reliance on anaerobic respiration—and thus their susceptibility to fatigue—is reduced.
   B. Resistance training causes hypertrophy of the muscle fibers due to an increase in the size and number of myofibrils.

Cardiac and Smooth Muscle p. 334

I. Cardiac muscle is striated and contains sarcomeres.
   A. In contrast to skeletal muscles, which require neural stimulation to contract, action potentials in the heart originate in myocardial cells; stimulation by neurons is not required.
   B. Also unlike the situation in skeletal muscles, action potentials can cross from one myocardial cell to another.

II. Smooth muscle cells lack sarcomeres and are not striated.
   A. Smooth muscle cells contain myosin and actin, but not arranged in sarcomeres.
   B. Because myosin filaments are very long, smooth muscle cells can contract even when they are greatly stretched.
   C. When stimulated by graded depolarizations, Ca^{++} enters smooth muscle cells and combines with calmodulin. This activates an enzyme that phosphorylates myosin cross bridges.
   D. Unlike the situation in striated muscles, phosphorylation of cross bridges is required for their binding to actin.
   E. Depending upon their neural regulation, smooth muscles can be classified as single-unit or multiunit.

Clinical Investigation

A woman who has been athletically active for most of her life complains that she is experiencing fatigue and muscle pain. Upon exercise testing, it is found that she has a high maximal oxygen uptake. Her muscles are not large, but appear to be well toned—perhaps excessively so. Laboratory tests reveal that this woman has a normal blood concentration of creatine phosphokinase but an elevated blood Ca^{++} concentration. She has a history of hypertension, and is currently taking a calcium-channel-blocking drug for this condition. What might be responsible for her fatigue and muscle pain?

Clues

Study the sections "Maximal Oxygen Uptake and Oxygen Debt" and "Adaptations to Exercise" and look at the boxed information on calcium-channel blockers in the "Smooth Muscle" section. Also read the boxed calcium-channel information in the section "Phosphocreatine" and study the section "Role of Ca^{++} in Muscle Contraction."

Review Activities

Objective Questions

1. A graded whole muscle contraction is produced in vivo primarily by variations in
   a. the strength of the fiber's contraction.
   b. the number of fibers that are contracting.
   c. both of the above.
   d. neither of the above.

2. The series elastic component of muscle contraction is responsible for
   a. increased muscle shortening to successive twitches.
   b. a time delay between contraction and shortening.
   c. the lengthening of muscle after contraction has ceased.
   d. all of the above.

3. Which of the following muscles have motor units with the highest innervation ratio?
   a. leg muscles
   b. arm muscles
   c. muscles that move the fingers
   d. muscles of the trunk

4. The stimulation of gamma motoneurons produces
   a. isotonic contraction of intrafusal fibers.
   b. isometric contraction of intrafusal fibers.
   c. either isotonic or isometric contraction of intrafusal fibers.
   d. contraction of extrafusal fibers.

5. In a single reflex arc involved in the knee-jerk reflex, how many synapses are activated within the spinal cord?
   a. thousands
   b. hundreds
   c. dozens
   d. two
   e. one

6. Spastic paralysis may occur when there is damage to
   a. the lower motor neurons.
   b. the upper motor neurons.
   c. either the lower or the upper motor neurons.

7. When a skeletal muscle shortens during contraction, which of the following statements is *false*?
   a. The A bands shorten.
   b. The H bands shorten.
   c. The I bands shorten.
   d. The sarcomeres shorten.

8. Electrical excitation of a muscle fiber *most directly* causes
   a. movement of tropomyosin.
   b. attachment of the cross bridges to actin.
   c. release of Ca^{++} from the sarcoplasmic reticulum.
   d. splitting of ATP.

9. The energy for muscle contraction is *most directly* obtained from
   a. phosphocreatine.
   b. ATP.
   c. anaerobic respiration.
   d. aerobic respiration.

10. Which of the following statements about cross bridges is *false*?
    a. They are composed of myosin.
    b. They bind to ATP after they detach from actin.
    c. They contain an ATPase.
    d. They split ATP before they attach to actin.

11. When a muscle is stimulated to contract, Ca^{++} binds to
    a. myosin.
    b. tropomyosin.
    c. actin.
    d. troponin.

12. Which of the following statements about muscle fatigue is *false*?
    a. It may result when ATP is no longer available for the cross-bridge cycle.
    b. It may be caused by a loss of muscle cell Ca^{++}.
    c. It may be caused by the accumulation of extracellular K^+.
    d. It may be a result of lactic acid production.

13. Which of the following types of muscle cells are *not* capable of spontaneous depolarization?
    a. single-unit smooth muscle
    b. multiunit smooth muscle
    c. cardiac muscle
    d. skeletal muscle
    e. both *b* and *d*
    f. both *a* and *c*

14. Which of the following muscle types is striated and contains gap junctions?
    a. single-unit smooth muscle
    b. multiunit smooth muscle
    c. cardiac muscle
    d. skeletal muscle

15. In an isotonic muscle contraction,
    a. the length of the muscle remains constant.
    b. the muscle tension remains constant.
    c. both muscle length and tension are changed.
    d. movement of bones does not occur.

Essay Questions

1. Using the concept of motor units, explain how skeletal muscles in vivo produce graded and sustained contractions.[1]

2. Describe how an isometric contraction can be converted into an isotonic contraction using the concepts of motor unit recruitment and the series-elastic component of muscles.

3. Trace the sequence of events in which the cross bridges attach to the thin filaments when a muscle is stimulated by a nerve. Why don't the cross bridges attach to the thin filaments when a muscle is relaxed?

4. Using the sliding filament theory of contraction, explain why the contraction strength of a muscle is maximal at a particular muscle length.

5. Explain why muscle tone is first decreased and then increased when descending motor tracts are damaged. How is muscle tone maintained?

[1]*Note:* This question is answered on page 138 of the Student Study Guide.

6. Explain the role of ATP in muscle contraction and muscle relaxation.
7. Why are all the muscle fibers of a given motor unit of the same type? Why are smaller motor units and slow-twitch muscle fibers used more frequently than larger motor units and fast-twitch fibers?
8. Describe the changes that occur as a result of endurance training and explain how these changes raise the level of exercise that can be performed before the onset of muscle fatigue.
9. Compare the mechanism of excitation-contraction coupling in striated muscle with that in smooth muscle.
10. Compare cardiac muscle, single-unit smooth muscle, and multiunit smooth muscle in terms of the regulation of their contraction.

Selected Readings

Astrand, P. O., and K. Rodahl. 1977. *Textbook of Work Physiology: Physiological Basis of Exercise.* New York: McGraw-Hill.

Baumann, H. et al. 1987. Exercise training induces transitions of myosin isoform subunits within histochemically typed human muscle fibers. *Pflugers Archives* 409:349.

Booth, F. W., and B. S. Tseng. 1993. Olympic goal: Molecular and cellular approaches to understanding muscle adaptation. *News in Physiological Sciences* 8:165.

Bourne, G. H., ed. 1973. *The Structure and Function of Muscle.* 2d ed. 4 vols. New York: Academic Press.

Cohen, C. November 1975. The protein switch of muscle contraction. *Scientific American.*

Cox, G. A. et al. 1993. Overexpression of dystrophin in transgenic mdx mice eliminates dystrophic symptoms without toxicity. *Nature* 364:725.

Drachman, D. B. 1978. Myasthenia gravis. *New England Journal of Medicine* 298:136.

Edington, D. W., and V. R. Edgerton. 1976. *The Biology of Physical Activity.* Boston: Houghton Mifflin.

Eisenberg, E., and L. E. Greene. 1980. The relation of muscle biochemistry to muscle physiology. *Annual Review of Physiology* 42:293.

Emont-Denand, F., C. C. Hunt, and Y. Laporte. 1988. How muscle spindles signal changes in muscle length. *News in Physiological Sciences* 3:105.

Felig, P., and J. Wahren. 1975. Fuel homeostasis in exercise. *New England Journal of Medicine* 293:1078.

Finer, J. R. et al. 1984. Single myosin molecule mechanics: Piconewton forces and nanometre steps. *Nature* 368:113.

Franzini-Armstrong, C., and A. O. Jorgensen. 1994. Structure and development of e-c coupling units in skeletal muscle. *Annual Review of Physiology* 56:509.

Geng, H. -X. et al. 1993. Amyotrophic lateral sclerosis and structural defects in Cu, zn superoxide dismutase. *Science* 261:1047.

Grinnel, A. D., and M. A. B. Brazier, eds. 1981. *Regulation of Muscle Contraction: Excitation-Contraction Coupling.* New York: Academic Press.

Hartshorne, D. J., and T. Kawamura. 1992. Regulation of contraction-relaxation in smooth muscle. *News in Physiological Sciences* 7:59.

Hermann, H. 1989. The unification of muscle structure and function: A semicentennial anniversary. *Perspectives in Biology and Medicine* 33:1.

Hoh, J. F. Y. 1991. Myogenic regulation of mammalian skeletal muscle fibers. *News in Physiological Sciences* 6:1.

Hoyle, G. April 1970. How is muscle turned on and off? *Scientific American.*

Huxley, H. E. 1969. The mechanism of muscular contraction. *Science* 164:1356.

Margaria, R. March 1972. The sources of muscular energy. *Scientific American.*

Meissner, G. 1994. Ryanodine receptor/Ca^{2+} release channels and their regulation by endogenous effectors. *Annual Review of Physiology* 56:485.

Merton, P. A. May 1972. How we control the contraction of our muscles. *Scientific American.*

Murphy, R. A. 1988. Muscle cells of hollow organs. *News in Physiological Sciences* 3:124.

Murray, J. H., and A. Weber. February 1974. The cooperative action of muscle proteins. *Scientific American.*

Nadel, E. R. 1985. Physiological adaptations to aerobic training. *American Scientist* 73:334.

Porter, K. R., and C. Franzini-Armstrong. March 1965. The sarcoplasmic reticulum. *Scientific American.*

Prosser, C. L. 1992. Smooth muscle: diversity and rhythmicity. *News in Physiological Sciences* 7:100.

Rayment, I. et al. 1993. Three-dimensional structure of myosin subfragment-1: A molecular motor. *Science* 261:50.

Schneider, M. F. 1994. Control of calcium release in functioning skeletal muscle fibers. *Annual Review of Physiology* 56:463.

Shepard, R. J. 1982. *Physiology and Biochemistry of Exercise.* New York: Praeger.

Sims, S. M., and L. J. Janssen. 1993. Cholinergic excitation of smooth muscle. *News in Physiological Sciences* 8:207.

Somlyo, A. P. and A. V. Somlyo. 1994. Signal transduction and regulation in smooth muscle. *Nature* 372:231.

Stein, R. B., ed. 1980. *Nerve and Muscle: Membranes, Cells, and Systems.* New York: Plenum Publishing Corp.

Westerblad, H. et al. 1991. Cellular mechanisms of fatigue in skeletal muscle. *American Journal of Physiology* 261:C195.

Explorations CD-ROM

 The module accompanying chapter 12 is #4 Muscle Contraction.

Life Science Animations

The animations that relate to chapter 12 are #29 Levels of Muscle Structure, #30 Sliding Filament Model of Muscle Contraction, and #31 Regulation of Muscle Contraction.

13

Heart and Circulation

OBJECTIVES

After studying this chapter, you should be able to . . .

1. describe the composition of blood plasma and the classification of the formed elements of the blood.
2. describe the ABO system of red blood cell antigens and explain the significance of the blood types.
3. explain how a blood clot is formed and how it is ultimately destroyed.
4. explain how the acid-base balance of the blood is affected by carbon dioxide and bicarbonate, and describe the roles of the lungs and kidneys in maintaining acid-base balance.
5. describe the path of the blood through the heart and the function of the atrioventricular and semilunar valves.
6. describe the structures and pathways of the pulmonary and systemic circulations.

7. describe the structures and pathways of electrical impulse conduction in the heart.
8. describe the electrical activity in the sinoatrial node and explain why this tissue functions as the normal pacemaker of the heart.
9. relate the time involved in the production of an action potential to the time involved in the contraction of myocardial cells and explain the significance of this relationship.
10. describe the pressure changes that occur in the ventricles during the cardiac cycle and relate these changes to the action of the valves and the flow of blood.
11. explain the origin of the heart sounds and state when these sounds are produced in the cardiac cycle.

12. explain how electrocardiogram waves are produced and relate these waves to other events in the cardiac cycle.
13. compare the structure of an artery and vein, and explain how the structure of each type of vessel relates to its function.
14. describe the structure of capillaries and explain the physiological significance of this structure.
15. explain the nature and significance of atherosclerosis and describe how this condition may be produced.
16. define *ischemia* and explain how myocardial ischemia may be produced.
17. describe some common arrhythmias that can be detected with an ECG.
18. describe the components and functions of the lymphatic system.

OUTLINE

Functions and Components of the Circulatory System

Blood serves numerous functions, including the transport of respiratory gases, nutritive molecules, metabolic wastes, and hormones. Blood is transported through the body in a system of vessels leading from and returning to the heart.

A unicellular organism can provide for its own maintenance and continuity by performing the wide variety of functions needed for life. By contrast, the complex human body is composed of trillions of specialized cells that demonstrate a division of labor. Cells of a multicellular organism depend on one another for the very basics of their existence. The majority of the cells of the body are firmly implanted in tissues and are incapable of procuring food and oxygen or even moving away from their own wastes. Therefore, a highly specialized and effective means of transporting materials within the body is needed.

The blood serves this transportation function. An estimated 60,000 miles of vessels throughout the body of an adult ensures that continued sustenance reaches each of the trillions of living cells. But then, too, the blood can serve to transport disease-causing viruses, bacteria, and their toxins. To guard against this, the circulatory system has protective mechanisms— the white blood cells and lymphatic system. In order to perform its various functions, the circulatory system works together with the respiratory, urinary, digestive, endocrine, and integumentary systems in maintaining homeostasis.

Functions of the Circulatory System

The functions of the circulatory system can be divided into three broad areas: transportation, regulation, and protection.

1. **Transportation.** All of the substances essential for cellular metabolism are transported by the circulatory system. These substances can be categorized as follows:
   a. *Respiratory*. Red blood cells, or *erythrocytes*, transport oxygen to the tissue cells. In the lungs, oxygen from the inhaled air attaches to hemoglobin molecules within the erythrocytes and is transported to the cells for aerobic respiration. Carbon dioxide produced by cell respiration is carried by the blood to the lungs for elimination in the exhaled air.
   b. *Nutritive*. The digestive system is responsible for the mechanical and chemical breakdown of food so that it can be absorbed through the intestinal wall and into the blood vessels of the circulatory system. The blood then carries these absorbed products of digestion through the liver and to the cells of the body.
   c. *Excretory*. Metabolic wastes, excessive water and ions, as well as other molecules in plasma (the fluid portion of blood), are filtered through the capillaries of the kidneys and excreted in urine.

2. **Regulation.** The blood carries hormones from their site of origin to distant target tissues. Diversion of blood from deeper vessels to superficial vessels in the skin helps to regulate body temperature.
3. **Protection.** The circulatory system protects against injury and foreign microbes or toxins introduced into the body. The clotting mechanism protects against blood loss when vessels are damaged, and *leukocytes* (white blood cells) provide immunity against many disease-causing agents.

Major Components of the Circulatory System

The circulatory system is frequently divided into the **cardiovascular system,** which consists of the heart and blood vessels, and the **lymphatic system,** which consists of lymph vessels, lymph nodes, and lymphoid organs (spleen, thymus, and tonsils).

The **heart** is a four-chambered double pump. Its pumping action creates the pressure head needed to push blood in the vessels to the lungs and body cells. At rest, the heart of an adult pumps about 5 liters of blood per minute. It takes about 1 minute for blood to be circulated to the most distal extremity and back to the heart.

Blood vessels form a tubular network that permits blood to flow from the heart to all the living cells of the body and then back to the heart. *Arteries* carry blood away from the heart, whereas *veins* return blood to the heart. Arteries and veins are continuous with each other through smaller blood vessels.

Arteries branch extensively to form a "tree" of progressively smaller vessels. Those that are microscopic in diameter are called *arterioles*. Blood passes from the arterial to the venous system in *capillaries*, which are the thinnest and most numerous of the blood vessels. All exchanges of fluid, nutrients, and wastes between the blood and tissues occur across the walls of capillaries. Blood flows through capillaries into microscopic-sized veins, called *venules*, which deliver blood into progressively larger veins that eventually return the blood to the heart.

As blood plasma passes through capillaries, the hydrostatic pressure of the blood forces some of this fluid out of the capillary walls. Fluid derived from plasma that passes out of capillary walls into the surrounding tissues is called *tissue fluid*, or *interstitial fluid*. Some of this fluid returns directly to capillaries, and some enters into **lymph vessels** located in the connective tissues around the blood vessels. Fluid in lymph vessels is called *lymph*. This fluid is returned to the venous blood at specific sites. **Lymph nodes,** positioned along the way, cleanse the lymph prior to its return to the venous blood. The lymphatic system is thus considered a part of the circulatory system and is discussed at the end of this chapter.

---

1. State the components of the circulatory system that function in oxygen transport, in the transport of nutrients from the digestive system, and in protection.
2. Describe the structures and functions of arteries, veins, and capillaries.
3. Define the terms *interstitial fluid* and *lymph*. How do these fluids relate to blood plasma?

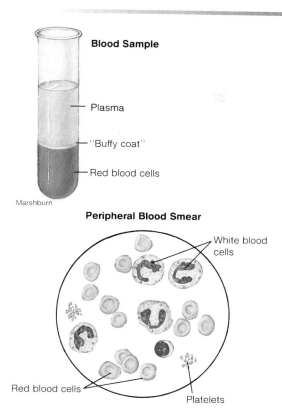

Blood Sample

- Plasma
- "Buffy coat"
- Red blood cells

Marshburn

Peripheral Blood Smear

White blood cells

Red blood cells

Platelets

Figure 13.1

Blood cells become packed at the bottom of the test tube when whole blood is centrifuged, leaving the fluid plasma at the top of the tube. Red blood cells are the most abundant of the blood cells—white blood cells and platelets form only a thin, light-colored "buffy coat" at the interface between the packed red blood cells and the plasma.

Table 13.1	Representative Normal Plasma Values
Measurement	**Normal Range**
Blood volume	80–85 ml/kg body weight
Blood osmolality	280–296 mOsm
Blood pH	7.35–7.45
Enzymes	
Creatine phosphokinase (CPK)	Female: 10–79 U/L
	Male: 17–148 U/L
Lactic dehydrogenase (LDH)	45–90 U/L
Phosphatase (acid)	Female: 0.01–0.56 Sigma U/ml
	Male: 0.13–0.63 Sigma U/ml
Hematology Values	
Hematocrit	Female: 37%–48%
	Male: 45%–52%
Hemoglobin	Female: 12–16 g/100 ml
	Male:13–18 g/100 ml
Red blood cell count	4.2–5.9 million/mm³
White blood cell count	4,300–10,880/mm³
Hormones	
Testosterone	Male: 300–1,100 ng/100 ml
	Female: 25–90 ng/100 ml
Adrenocorticotrophic hormone (ACTH)	15–70 pg/ml
Growth hormone	Children: over 10 ng/ml
	Adult male: below 5 ng/ml
Insulin	6–26 µU/ml (fasting)
Ions	
Bicarbonate	24–30 mmol/l
Calcium	2.1–2.6 mmol/l
Chloride	100–106 mmol/l
Potassium	3.5–5.0 mmol/l
Sodium	135–145 mmol/l
Organic Molecules (Other)	
Cholesterol	120–220 mg/100 ml
Glucose	70–110 mg/100 ml (fasting)
Lactic acid	0.6–1.8 mmol/l
Protein (total)	6.0–8.4 g/100 ml
Triglyceride	40–150 mg/100 ml
Urea nitrogen	8–25 mg/100 ml
Uric acid	3–7 mg/100 ml

Source: Excerpted from material appearing in *The New England Journal of Medicine,* "Case Records of the Massachusetts General Hospital," 302:37–48 and 314:39–49. Copyright © 1980, 1986.

Composition of the Blood

Blood consists of formed elements that are suspended and carried in the plasma. These formed elements and their major functions include erythrocytes (oxygen transport), leukocytes (immune defense), and platelets (blood clotting). Plasma contains different types of proteins and many water-soluble molecules.

The average-sized adult has about 5 liters of blood, constituting about 8% of the total body weight. Blood leaving the heart is referred to as *arterial blood*. Arterial blood, with the exception of that going to the lungs, is bright red in color due to a high concentration of oxyhemoglobin (the combination of oxygen and hemoglobin) in the red blood cells. *Venous blood* is blood returning to the heart and, except for the venous blood from the lungs, has a darker color (due to hemoglobin that is no longer combined with oxygen).

Blood is composed of a cellular portion, called **formed elements,** and a fluid portion, called **plasma.** When a blood sample is centrifuged, the heavier formed elements are packed into the bottom of the tube, leaving plasma at the top (fig. 13.1). The formed elements constitute approximately 45% of the total blood volume (a measurement called the *hematocrit*), and the plasma accounts for the remaining 55%.

Plasma

Plasma is a straw-colored liquid consisting of water and dissolved solutes. The major solute of the plasma in terms of its concentration is Na⁺. In addition to Na⁺, plasma contains many other ions, as well as organic molecules such as metabolites, hormones, enzymes, antibodies, and other proteins. The values of some of these constituents of plasma are shown in table 13.1.

Plasma Proteins

Plasma proteins constitute 7% to 9% of the plasma. The three types of proteins are albumins, globulins, and fibrinogen. **Albumins** account for most (60% to 80%) of the plasma proteins and are the smallest in size. They are produced by the liver and serve to provide the osmotic pressure needed to draw water from the surrounding tissue fluid into the capillaries. This action is needed to maintain blood volume and pressure. **Globulins** are grouped into three subtypes: **alpha globulins, beta globulins, and gamma globulins.** The alpha and beta globulins are produced by the liver and function to transport lipids and fat-soluble vitamins in the blood. Gamma globulins are antibodies produced by lymphocytes (one of the formed elements found in blood and lymphoid tissues) and function in immunity. **Fibrinogen,** which accounts for only about 4% of the total plasma proteins, is an important clotting factor produced by the liver. During the process of clot formation (described later in this chapter), fibrinogen is converted into insoluble threads of *fibrin.* The fluid from clotted blood, which is called **serum,** thus does not contain fibrinogen but is otherwise identical to plasma.

Plasma Volume

A number of regulatory mechanisms in the body maintain homeostasis of the plasma volume. If the body should lose water, the remaining plasma becomes excessively concentrated (its osmolality increases—chapter 6). This is detected by osmoreceptors in the hypothalamus, resulting in a sensation of thirst and the secretion of antidiuretic hormone (ADH) from the posterior pituitary (chapter 11). This hormone promotes water retention by the kidneys, which—together with increased drinking of fluids—helps to compensate for the dehydration and lowered blood volume. This regulatory mechanism is very important in maintaining blood pressure (chapter 14).

The Formed Elements of Blood

The formed elements of blood include two types of blood cells: **erythrocytes,** or **red blood cells,** and **leukocytes,** or **white blood cells.** Erythrocytes are by far the most numerous of the two. A cubic millimeter of blood contains 5.1 million to 5.8 million erythrocytes in males and 4.3 million to 5.2 million erythrocytes in females. The same volume of blood, by contrast, contains only 5,000 to 9,000 leukocytes.

Erythrocytes

Red blood cells are flattened, biconcave discs, about 7 μm in diameter and 2.2 μm thick. Their unique shape relates to their function of transporting oxygen and provides an increased surface area through which gas can diffuse (fig. 13.2). Erythrocytes lack nuclei and mitochondria (they obtain energy through anaerobic respiration). Because of these deficiencies, erythrocytes have a circulating life span of only about 120 days before they are destroyed by phagocytic cells in the liver, spleen, and bone marrow.

Each erythrocyte contains approximately 280 million *hemoglobin* molecules, which give blood its red color. Each hemoglobin molecule consists of a protein, called globin, and an

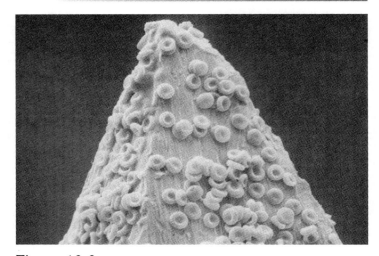

Figure 13.2

A scanning electron micrograph of red blood cells clinging to a hypodermic needle.

 Anemia refers to any condition in which there is an abnormally low hemoglobin concentration and/or red blood cell count. The most common type is **iron-deficiency anemia,** caused by a deficiency in iron, which is an essential component of the hemoglobin molecule. In **pernicious anemia,** there is inadequate availability of vitamin B_{12}, which is needed for red blood cell production. In most cases, this results from atrophy of the glandular mucosa of the stomach, which normally secretes a substance, called *intrinsic factor.* In the absence of intrinsic factor, the vitamin B_{12} obtained in the diet cannot be absorbed by intestinal cells. **Aplastic anemia** is anemia due to destruction of the bone marrow, which may be caused by chemicals (including benzene and arsenic), X rays, or by chemotherapy for cancer.

iron-containing pigment, called heme. The iron group of heme is able to combine with oxygen in the lungs and release oxygen in the tissues.

Leukocytes

Leukocytes differ from erythrocytes in several ways. Leukocytes contain nuclei and mitochondria and can move in an amoeboid fashion (erythrocytes are not able to move independently). Because of their amoeboid ability, leukocytes can squeeze through pores in capillary walls and move to a site of infection, whereas erythrocytes usually remain confined within blood vessels. The movement of leukocytes through capillary walls is called *diapedesis.*

White blood cells are almost invisible under the microscope unless they are stained, and are classified according to

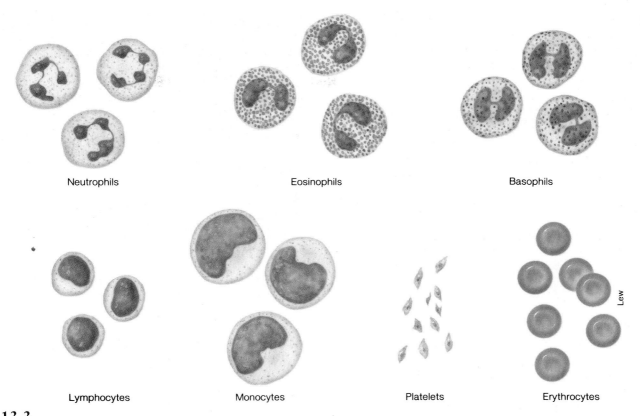

Neutrophils

Eosinophils

Basophils

Lymphocytes

Monocytes

Platelets

Erythrocytes

Lew

Figure 13.3

The blood cells and platelets.

their staining properties. Those leukocytes that have granules in their cytoplasm are called **granular leukocytes;** those that do not are called **agranular** (or nongranular) **leukocytes.**

The stain used to identify white blood cells is usually a mixture of a pink-to-red stain called eosin and a blue-to-purple stain called a "basic stain." Granular leukocytes with pink-staining granules are therefore called *eosinophils,* and those with blue-staining granules are called *basophils.* Those with granules that have little affinity for either stain are *neutrophils.* Neutrophils are the most abundant type of leukocyte, accounting for 50% to 70% of the leukocytes in the blood. Because of their oddly shaped nuclei, with lobes and strands, they are sometimes called **polymorphonuclear neutrophils (PMNs).**

There are two types of agranular leukocytes: lymphocytes and monocytes. *Lymphocytes* are usually the second most numerous type of leukocyte; they are small cells with round nuclei and little cytoplasm. *Monocytes,* in contrast, are the largest of the leukocytes and generally have kidney- or horseshoe-shaped nuclei. In addition to these two cell types, there are smaller numbers of cells, derived from lymphocytes, called *plasma cells.* Plasma cells produce and secrete large amounts of antibodies. The immune functions of the different white blood cells are described in more detail in chapter 15.

Platelets

Platelets, or **thrombocytes,** are the smallest of the formed elements and are actually fragments of large cells called *megakaryocytes,* found in bone marrow. (This is why the term *formed elements* is used instead of *blood cells* to describe erythrocytes, leukocytes, and platelets.) The fragments that enter the circulation as platelets lack nuclei but, like leukocytes, are capable of amoeboid movement. The platelet count per cubic millimeter of blood is 250,000 to 450,000. Platelets survive about 5 to 9 days before being destroyed by the spleen and liver.

Platelets play an important role in blood clotting. They constitute the major portion of the mass of the clot, and phospholipids in their cell membranes serve to activate the clotting factors in plasma that result in threads of fibrin, which reinforce the platelet plug. Platelets that attach together in a blood clot release a chemical called *serotonin,* that stimulates constriction of the blood vessels and thus reduces the flow of blood to the injured area. Platelets also secrete growth factors (autocrine regulators—see chapter 11) that are important in maintaining the integrity of blood vessels. These regulators also may be involved in the development of atherosclerosis, as described in a later section.

The formed elements of the blood are illustrated in figure 13.3, and their characteristics are summarized in table 13.2.

Table 13.2 Formed Elements of the Blood

Component	Description	Number Present	Function
Erythrocyte (red blood cell)	Biconcave disc without nucleus; contains hemoglobin; survives 100 to 120 days	4,000,000 to 6,000,000 / mm^3	Transports oxygen and carbon dioxide
Leukocytes (white blood cells)		5,000 to 10,000 / mm^3	Aid in defense against infections by microorganisms
Granulocytes	About twice the size of red blood cells; cytoplasmic granules present; survive 12 hours to 3 days		
1. Neutrophil	Nucleus with 2 to 5 lobes; cytoplasmic granules stain slightly pink	54% to 62% of white cells present	Phagocytic
2. Eosinophill	Nucleus bilobed; cytoplasmic granules stain red in eosin stain	1% to 3% of white cells present	Helps to detoxify foreign substances; secretes enzymes that dissolve clots; fights parasitic infections
3. Basophil	Nucleus lobed; cytoplasmic granules stain blue in hematoxylin stain	Less than 1% of white cells present	Releases anticoagulant heparin
Agranulocytes	Cytoplasmic granules absent; survive 100 to 300 days		
1. Monocyte	2 to 3 times larger than red blood cell; nuclear shape varies from round to lobed	3% to 9% of white cells present	Phagocytic
2. Lymphocyte	Only slightly larger than red blood cell; nucleus nearly fills cell	25% to 33% of white cells present	Provides specific immune response (including antibodies)
Platelet (thrombocyte)	Cytoplasmic fragment; survives 5 to 9 days	250,000 to 450,000 / mm^3	Enables clotting; provides vascular protection

Source: Modified from Kent M. Van De Graaff, *Human Anatomy,* 4th ed. Copyright © 1995 Wm. C. Brown Communications, Inc., Dubuque, Iowa. Reprinted by permission of Times Mirror Higher Education Group, Inc., Dubuque, Iowa. All Rights Reserved.

Blood cell counts are an important source of information in determining the health of a person. An abnormal increase in erythrocytes, for example, is termed **polycythemia** and an abnormally low red blood cell count is termed **anemia.** (Polycythemia and anemia are described in detail in chapter 16.) An elevated leukocyte count, called **leukocytosis,** is often associated with infection (see chapter 15). A large number of immature leukocytes in a blood sample is diagnostic of the disease **leukemia.** A low white blood cell count is called **leukopenia.** This may be due to a variety of factors; low numbers of lymphocytes, for example, may occur as a result of poor nutrition.

The major purpose of **bone marrow transplantation** is to provide competent hematopoietic stem cells to the recipient. If the donor and recipient are one and the same, the procedure is called an *autotransplantation*. If the donor and recipient are different people it is an *allogeneic transplantation*. The benefit of autotransplantation is that it avoids the danger of immunological rejection that can occur in allogeneic transplantation.

Hemopoiesis

Blood cells are constantly formed through a process called **hemopoiesis.** The hematopoietic stem cells—those that give rise to blood cells—originate in the yolk sac of the human embryo and then migrate to the liver. Hemopoiesis thus occurs in the liver of the fetus. The stem cells then migrate to the bone marrow, and shortly after birth the liver ceases to be a source of blood cell production.

The term *erythropoiesis* refers to the formation of erythrocytes, and *leukopoiesis* to the formation of leukocytes. These processes occur in two classes of tissues after birth, myeloid and lymphoid. **Myeloid tissue** is the red bone marrow of the long bones, ribs, sternum, bodies of vertebrae, and portions of the skull. **Lymphoid tissue** includes the lymph nodes, tonsils, spleen, and thymus. The bone marrow produces all of the different types of blood cells, and one type—lymphocytes—is also produced in the lymphoid tissue.

Erythropoiesis is an extremely active process. It is estimated that about 2.5 million erythrocytes are produced every second in order to replace those that are continuously destroyed by the spleen and liver. During the destruction of erythrocytes, iron is salvaged and returned to the red bone marrow, where it is used again in the formation of erythrocytes. The life span of an erythrocyte is approximately 120 days. Agranular leukocytes remain functional for 100 to 300 days under normal conditions. Granular leukocytes, in contrast, have an extremely short life span of 12 hours to 3 days.

Hemopoiesis begins the same way in both myeloid and lymphoid tissue. A population of undifferentiated (unspecialized) cells gradually differentiate (specialize) to become "stem cells," which give rise to the blood cells. At each step along the way, stem cells can duplicate themselves by mitosis, thus ensuring that the parent population never becomes depleted. As the cells become differentiated, they develop membrane receptors for chemical signals that cause further development along particular lines. The earliest stem cells that can be distinguished under a microscope are the *erythroblasts* (which become erythrocytes), *myeloblasts* (which become granular leukocytes), *lymphoblasts* (which form lymphocytes), and *monoblasts* (which form monocytes).

The production of different subtypes of lymphocytes is stimulated by chemicals called **lymphokines,** which are discussed as part of autocrine regulation in chapter 11 and in conjunction with the immune system in chapter 15. The production of red blood cells is stimulated by a hormone called **erythropoietin,** which is secreted by the kidneys. Erythropoietin, a glycoprotein hormone consisting of 166 amino acids, stimulates cell division and differentiation of erythrocyte stem cells in bone marrow. The secretion of erythropoietin by the kidneys is stimulated whenever the delivery of oxygen to the kidneys and other organs is lower than normal. Under these conditions—which can occur, for example, when a person lives at high altitude—the increased production of red blood cells allows the blood to carry a higher concentration of oxygen to the tissues.

Scientists have recently identified a specific cytokine that stimulates proliferation of megakaryocytes and their maturation into platelets (thrombocytes). By analogy with erythropoietin, they named this newly discovered regulatory molecule **thrombopoietin.** The gene that codes for thrombopoietin has been cloned, so that recombinant thrombopoietin may soon be commercially produced. This cytokine may eventually be used to treat the thrombocytopenia (low platelet count) that occurs, for example, in patients undergoing chemotherapy or bone marrow transplantation for cancer.

Red Blood Cell Antigens and Blood Typing

On the surfaces of all cells in the body, there are certain molecules that can be recognized as foreign by the immune system of another individual. These molecules are known as *antigens.* As part of the immune response, particular lymphocytes secrete a class of proteins called *antibodies* that bond in a specific fashion with antigens. The specificity of antibodies for antigens is analogous to the specificity of enzymes for their substrates, and of receptor proteins for neurotransmitters and hormones. A complete description of antibodies and antigens is provided in chapter 15.

ABO System

The distinguishing antigens on other cells are far more varied than the antigens on red blood cells. Red blood cell antigens, however, are of extreme clinical importance because their types must be matched between donors and recipients for blood transfusions. There are several groups of red blood cell antigens, but the major group is known as the ABO system. In terms of the antigens present on the red blood cell surface, a person may be *type A* (with only A antigens), *type B* (with only B antigens), *type AB* (with both A and B antigens), or *type O* (with neither

Considering that the kidneys produce erythropoietin, and that erythropoietin is needed to stimulate red blood cell production, it is not surprising that patients with kidney disease may also suffer from anemia. Many people who require renal dialysis treatment are now given human erythropoietin. This exciting therapeutic breakthrough was made possible by the commercial production of recombinant erythropoietin, using cultured mammalian cells that had incorporated the human gene for erythropoietin through genetic engineering techniques.

A nor B antigens). It should again be noted that the blood type denotes the class of antigens found on the red blood cell surface.

Each person inherits two genes (one from each parent) that control the production of the ABO antigens. The genes for A or B antigens are dominant to the gene for O, since O simply means the absence of A or B. The genes for A and B are usually shown as I^A and I^B, and the recessive gene for O is shown as the lowercase i. A person who is type A, therefore, may have inherited the A gene from each parent (may have the genotype $I^A I^A$), or the A gene from one parent and the O gene from the other parent (and thus have the genotype $I^A i$). Likewise, a person who is type B may have the genotype $I^B I^B$ or $I^B i$. It follows that a type O person inherited the O gene from each parent (has the genotype ii), whereas a type AB person inherited the A gene from one parent and the B gene from the other (there is no dominant-recessive relationship between A and B).

The immune system is tolerant to its own red blood cell antigens. A person who is type A, for example, does not produce anti-A antibodies. Surprisingly, however, people with type A blood do make antibodies against the B antigen, and conversely, people with blood type B make antibodies against the A antigen. This is believed to result from the fact that antibodies made in response to some common bacteria cross-react with the A or B antigens. A person who is type A, therefore, acquires antibodies that can react with B antigens by exposure to these bacteria but does not develop antibodies that can react with A antigens, because this is prevented by tolerance mechanisms.

People who are type AB develop tolerance to both these antigens and thus do not produce either anti-A or anti-B antibodies. Those who are type O, in contrast, do not develop tolerance to either antigen and, therefore, have both anti-A and anti-B antibodies in their plasma (table 13.3).

Table 13.3	The ABO System of Red Blood Cell Antigens	
Genotype	**Antigen on RBCs**	**Antibody in Plasma**
$I^A I^A$; $I^A i$	A	Anti-B
$I^B I^B$; $I^B i$	B	Anti-A
ii	O	Anti-A and anti-B
$I^A I^B$	AB	Neither anti-A nor anti-B

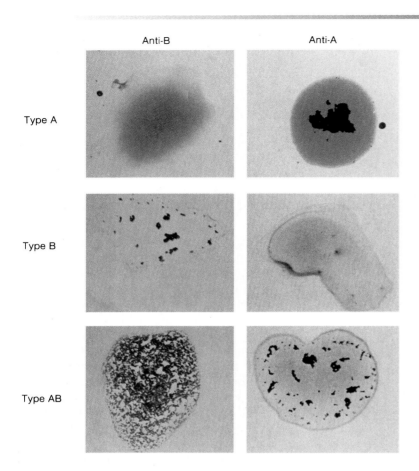

Anti-B Anti-A

Type A

Type B

Type AB

Figure 13.4

Agglutination (clumping) of red blood cells occurs when cells with A-type antigens are mixed with anti-A antibodies and when cells with B-type antigens are mixed with anti-B antibodies. No agglutination would occur with type O blood (not shown).

Transfusion Reactions

Before transfusions are performed, a *major crossmatch* is made by mixing serum from the recipient with blood cells from the donor. If the types do not match—if the donor is type A, for example, and the recipient is type B—the recipient's antibodies attach to the donor's red blood cells and form bridges that cause the cells to clump together, or **agglutinate** (fig. 13.4). Because of this agglutination reaction, the A and B antigens are sometimes called *agglutinogens*, and the antibodies against them are called *agglutinins*. Transfusion errors that result in such agglutination in the blood can produce a blockage of small blood vessels and cause hemolysis, which may damage the kidneys and other organs.

In emergencies, type O blood has been given to people who are type A, B, AB, or O. Since type O red blood cells lack A and B antigens, the recipient's antibodies cannot cause agglutination of the donor red blood cells. Type O is, therefore, a *universal donor*, but only as long as the volume of plasma donated is small, because plasma from a type O person would agglutinate type A, type B, and type AB red blood cells. Likewise, type AB people are *universal recipients* because they lack anti-A and anti-B antibodies and thus cannot agglutinate donor red blood cells. (Donor plasma could agglutinate recipient red blood cells if the transfusion volume were too large.) Because of the dangers involved, the universal donor and recipient concept in blood transfusions is strongly discouraged.

Rh Factor

Another group of antigens found in most red blood cells is the *Rh factor* (Rh stands for rhesus monkey, in which these antigens were first discovered). People who have these antigens are said to be **Rh positive,** whereas those who do not are **Rh negative.** There are fewer Rh negative people because this condition is recessive to Rh positive. The Rh factor is of particular significance when Rh negative mothers give birth to Rh positive babies.

Since the fetal and maternal blood are normally kept separate across the placenta (chapter 20), the Rh negative mother is not usually exposed to the Rh antigen of the fetus during the pregnancy. At the time of birth, however, a variable degree of exposure may occur, and the mother's immune system may become sensitized and produce antibodies against the Rh antigen. This does not always occur, however, because the exposure may be minimal and because Rh negative women vary in their sensitivity to the Rh factor. If the woman does produce antibodies against the Rh factor, these antibodies can cross the placenta in subsequent pregnancies and cause hemolysis of the Rh positive red blood cells of the fetus. The baby is therefore born anemic, with a condition called *erythroblastosis fetalis*, or *hemolytic disease of the newborn*.

Erythroblastosis fetalis can be prevented by injecting the Rh negative mother with an antibody preparation against the Rh factor (a trade name for this preparation is RhoGAM—the GAM is short for gamma globulin, the class of plasma proteins in which antibodies are found) within 72 hours after the birth of each Rh positive baby. This is a type of passive immunization in which the injected antibodies inactivate the Rh antigens and thus prevent the mother from becoming actively immunized to them.

Blood Clotting

When a blood vessel is injured, a number of physiological mechanisms are activated that promote **hemostasis,** or the cessation of bleeding (*hemo* = blood; *stasis* = standing). Breakage of the endothelial lining of a vessel exposes collagen proteins from the subendothelial connective tissue to the blood. This initiates three separate, but overlapping, hemostatic mechanisms: (1) vasoconstriction, (2) the formation of a platelet plug, and (3) the production of a web of fibrin proteins around the platelet plug.

Functions of Platelets

In the absence of vessel damage, platelets are repelled from each other and from the endothelial lining of vessels. The repulsion of platelets from an intact endothelium is believed to be due to *prostacyclin,* a derivative of prostaglandins, produced within the endothelium. Mechanisms that prevent platelets from sticking to the blood vessels and to each other are obviously needed to prevent inappropriate blood clotting.

Damage to the endothelium of vessels exposes subendothelial tissue to the blood. Platelets are able to stick to exposed collagen proteins that have become coated with a protein (*von Willebrand factor*) secreted by endothelial cells. Platelets contain secretory granules; when platelets stick to collagen, they *degranulate* as the secretory granules release their products. These products include *ADP* (adenosine diphosphate), *serotonin,* and a prostaglandin called *thromboxane* A_2. This event is known as the **platelet release reaction.**

Serotonin and thromboxane A_2 stimulate vasoconstriction, which helps to decrease blood flow to the injured vessel. Phospholipids that are exposed on the platelet membrane participate in the activation of clotting factors.

The release of ADP and thromboxane A_2 from platelets that are stuck to exposed collagen makes other platelets in the vicinity "sticky," so that they adhere to those stuck to the collagen. The second layer of platelets, in turn, undergoes a platelet release reaction, and the ADP and thromboxane A_2 that are secreted cause additional platelets to aggregate at the site of injury. This produces a **platelet plug** in the damaged vessel, which is strengthened by the activation of plasma clotting factors.

Clotting Factors: Formation of Fibrin

The platelet plug is strengthened by a meshwork of insoluble protein fibers known as **fibrin** (fig. 13.5). Blood clots therefore contain platelets and fibrin, and they usually contain trapped red blood cells that give the clot a red color (clots formed in arteries generally lack red blood cells and are gray in color). Finally, contraction of the platelet mass in the process of *clot retraction* forms a more compact and effective plug. Fluid squeezed from the clot as it retracts is called *serum,* which is

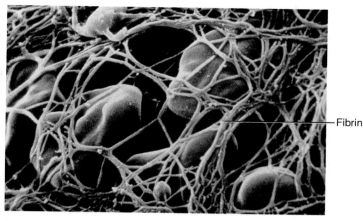

Figure 13.5

A scanning electron micrograph showing threads of fibrin.

 In order to undergo a release reaction, the production of prostaglandins by the platelets is required. **Aspirin** inhibits the cyclo-oxygenase enzyme that catalyzes the conversion of arachidonic acid (a cyclic fatty acid) into prostaglandins (see chapter 11), thereby inhibiting the release reaction and consequent formation of a platelet plug. Since platelets are not complete cells, they cannot regenerate new enzymes, and so the enzymes remain inhibited for the life of the platelets. The ingestion of excessive amounts of aspirin can thus significantly prolong bleeding time for several days, which is why blood donors and women in the last trimester of pregnancy are advised to avoid aspirin. Slight inhibition of platelet aggregation by low doses of aspirin, however, can significantly reduce the risk of atherosclerotic heart disease (described in a later section), and is often recommended for patients diagnosed with this condition.

plasma without fibrinogen, the soluble precursor of fibrin. (Serum is obtained in laboratories by allowing blood to clot in a test tube and then centrifuging the tube so that the clot and blood cells become packed at the bottom of the tube.)

The conversion of fibrinogen into fibrin may occur via either of two pathways. Blood left in a test tube will clot without the addition of any external chemicals; the pathway that produces this clot is thus called the **intrinsic pathway.** The intrinsic pathway also produces clots in damaged blood vessels when collagen is exposed to plasma. Damaged tissues, however, release a chemical that initiates a "shortcut" to the formation of fibrin. Since this chemical is not part of blood, the shorter pathway is called the **extrinsic pathway.**

The intrinsic pathway is initiated by the exposure of plasma to a negatively charged surface, such as that provided by collagen at the site of a wound or by the glass of a test tube. This activates a plasma protein called factor XII (table 13.4),

Table 13.4 The Plasma Clotting Factors

Factor	Name	Function	Pathway
I	Fibrinogen	Converted to fibrin	Common
II	Prothrombin	Enzyme	Common
III	Tissue thromboplastin	Cofactor	Extrinsic
IV	Calcium ions (Ca^{++})	Cofactor	Intrinsic, extrinsic, and common
V	Proaccelerin	Cofactor	Common
VII*	Proconvertin	Enzyme	Extrinsic
VIII	Antihemophilic factor	Cofactor	Intrinsic
IX	Plasma thromboplastin component; Christmas factor	Enzyme	Intrinsic
X	Stuart–Prower Factor	Enzyme	Common
XI	Plasma thromboplastin antecedent	Enzyme	Intrinsic
XII	Hageman factor	Enzyme	Intrinsic
XIII	Fibrin stabilizing factor	Enzyme	Common

*The number VI is no longer used; the substance is now believed to be the same as activated factor V.

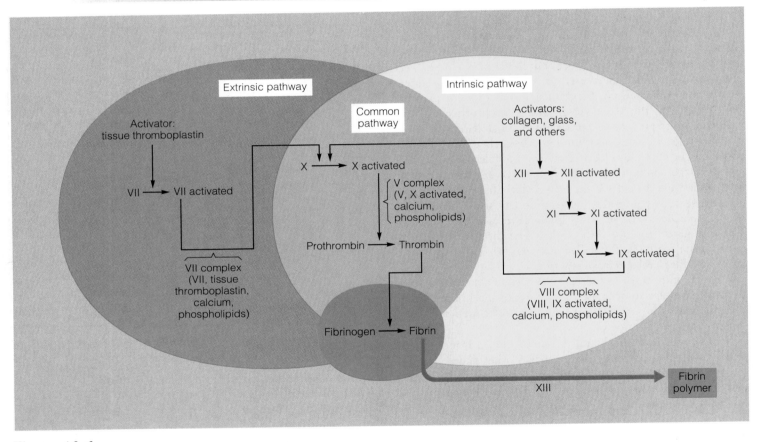

Figure 13.6

The extrinsic and intrinsic clotting pathways that lead to the formation of insoluble threads of fibrin polymers.

which is a protein-digesting enzyme (protease). Active factor XII in turn activates another clotting factor, which activates yet another. The plasma clotting factors are numbered in order of their discovery, which does not reflect the actual sequence of reactions.

The next steps in the sequence require the presence of phospholipids, which are provided by platelets, and Ca^{++}. These steps result in the conversion of an inactive enzyme, called **prothrombin,** into the active enzyme **thrombin.** Throm-

bin converts the soluble protein **fibrinogen** into **fibrin** monomers. These monomers are joined together to produce the insoluble fibrin polymers that form a meshwork supporting the platelet plug. The intrinsic clotting sequence is shown on the right side of figure 13.6.

The formation of fibrin can occur more rapidly as a result of the release of **tissue thromboplastin** from damaged tissue cells. This extrinsic clotting pathway is shown on the left side of figure 13.6. Notice that the intrinsic and extrinsic

Chapter Thirteen

Table 13.5 Some Acquired and Inherited Clotting Disorders and a Listing of Anticoagulant Drugs

Category	Cause of Disorder	Comments
Acquired clotting disorders	Vitamin K deficiency	Inadequate formation of prothrombin and other clotting factors in the liver
Inherited clotting disorders	Hemophilia A (defective factor VIII$_{AHF}$)	Recessive trait carried on X chromosome; results in delayed formation of fibrin
	von Willebrand's disease (defective factor VIII$_{VWF}$)	Dominant trait carried on autosomal chromosome; impaired ability of platelets to adhere to collagen in subendothelial connective tissue
	Hemophilia B (defective factor IX), also called Christmas disease	Recessive trait carried on X chromosome; results in delayed formation of fibrin

Anticoagulants

Aspirin	Inhibits prostaglandin production, resulting in defective platelet release reaction
Coumarin	Competes with the action of vitamin K
Heparin	Inhibits activity of thrombin
Citrate	Combines with Ca^{++}, and thus inhibits activity of many clotting factors

clotting pathways overlap, as both result in the activation of thrombin, which converts fibrinogen to fibrin.

Dissolution of Clots

As the damaged blood vessel wall is repaired, activated factor XII promotes the conversion of an inactive molecule in plasma into the active form called *kallikrein*. Kallikrein, in turn, catalyzes the conversion of inactive *plasminogen* into the active molecule called **plasmin**. Plasmin is an enzyme that digests fibrin into "split products," thus promoting dissolution of the clot.

In addition to kallikrein, a number of other plasminogen activators are used clinically to promote dissolution of clots. An exciting recent development in genetic engineering technology is the commercial availability of an endogenous compound, **tissue plasminogen activator (TPA)**, which is the product of human genes introduced into bacteria. **Streptokinase**, a natural bacterial product, is a potent and more widely used activator of plasminogen. Streptokinase and TPA may be injected into the general circulation or injected specifically into a coronary vessel that has become occluded by a thrombus (blood clot).

Anticoagulants

Clotting of blood in test tubes can be prevented by the addition of *sodium citrate* or *ethylenediaminetetraacetic acid* (EDTA), both of which chelate (bind to) calcium. By this means, Ca^{++} levels in the blood that can participate in the clotting sequence are lowered, and clotting is inhibited. A mucoprotein called *heparin* can also be added to the tube to prevent clotting. Heparin activates a plasma protein called *antithrombin III*, which combines with and inactivates thrombin. Heparin is also given intravenously during certain medical procedures to prevent clotting. Patients may also be given *coumarins* as anticoagulants. The coumarins prevent blood clotting by competing with vitamin K.

A number of hereditary diseases involve the clotting system. Examples of hereditary clotting disorders include two different genetic defects in factor VIII. A defect in one subunit of factor VIII prevents this factor from participating in the intrinsic clotting pathway. This genetic disease, called **hemophilia A**, is an X-linked recessive trait that is prevalent in the royal families of Europe. A defect in another subunit of factor VIII results in **von Willebrand's disease**. In this disease, rapidly circulating platelets are unable to stick to collagen, and a platelet plug cannot be formed. Some acquired and inherited defects in the clotting system are summarized in table 13.5.

Vitamin K is needed for the conversion of glutamic acid, an amino acid found within many of the clotting factor proteins, into a derivative called *gamma-carboxyglutamic acid*. This derivative is more effective than glutamic acid at binding to Ca^{++}, and such binding is needed for proper function of clotting factors II, VII, IX, and X. Because of the indirect action of vitamin K on blood clotting, coumarin must be given to a patient for several days before it is effective as an anticoagulant.

1. Distinguish between the different types of formed elements of the blood in terms of their origin, appearance, and function.
2. Describe how the rate of erythropoiesis is regulated.
3. Explain what is meant by "type A positive" and describe what can happen in a blood transfusion if donor and recipient are not properly matched.
4. List the steps common to both the intrinsic and extrinsic clotting pathways.
5. Explain the meaning of the terms *intrinsic* and *extrinsic* in terms of the clotting pathways and describe how these pathways differ from each other.

Table 13.6   Terms Used to Describe Acid-Base Balance

Term	Definition
Acidosis, respiratory	Increased carbon dioxide retention (due to hypoventilation), which can result in the accumulation of carbonic acid and thus a fall in blood pH below normal
Acidosis, metabolic	Increased production of "nonvolatile" acids such as lactic acid, fatty acids, and ketone bodies, or loss of blood bicarbonate (such as by diarrhea) resulting in a fall in blood pH below normal
Alkalosis, respiratory	A rise in blood pH due to loss of CO_2 and carbonic acid (through hyperventilation)
Alkalosis, metabolic	A rise in blood pH produced by loss of nonvolatile acids (as in excessive vomiting) or by excessive accumulation of bicarbonate base
Compensated acidosis or alkalosis	Metabolic acidosis or alkalosis are partially compensated by opposite changes in blood carbonic acid levels (through changes in ventilation). Respiratory acidosis or alkalosis are partially compensated by increased retention or excretion of bicarbonate in the urine.

Acid-Base Balance of the Blood

The pH of blood plasma is maintained very constant through the functions of the lungs and kidneys. The lungs regulate the carbon dioxide concentration of the blood, and the kidneys regulate the bicarbonate concentration.

The blood plasma within arteries normally has a pH between 7.35 and 7.45, with an average of 7.40. Using the definition of pH described in chapter 2, this means that arterial blood has a H^+ concentration of about $10^{-7.4}$ molar. Some of these hydrogen ions are derived from carbonic acid, which is formed in the blood plasma from carbon dioxide and which can ionize, as indicated in the following equations:

$$CO_2 + H_2O \rightleftharpoons H_2CO_3$$
$$H_2CO_3 \rightleftharpoons H^+ + HCO_3^-$$

Carbon dioxide is produced by tissue cells through aerobic cell respiration and is transported by the blood to the lungs, where it can be exhaled. As will be described in more detail in chapter 16, carbonic acid can be reconverted to carbon dioxide, which is a gas. Because it can be converted to a gas, carbonic acid is referred to as a *volatile acid*, and its concentration in the blood is controlled by the lungs through proper ventilation (breathing). All other acids in the blood—including lactic acid, fatty acids, ketone bodies, and so on—are *nonvolatile acids*.

Under normal conditions, the H^+ released by nonvolatile acids do not affect the blood pH because these hydrogen ions are bound to molecules that function as *buffers*. The major buffer in the plasma is *bicarbonate* (HCO_3^-), and it buffers H^+ as described in the following equation:

$$HCO_3^- + H^+ \rightarrow H_2CO_3$$

This buffering reaction could not go on forever because the free HCO_3^- would eventually disappear. If this were to occur, the H^+ concentration would increase and the pH of the blood would decrease. Under normal conditions, however, excessive H^+ is eliminated in the urine by the kidneys. Through this action, and through their ability to produce bicarbonate, the kidneys are responsible for maintaining a normal concentration of free bicarbonate in the plasma. The role of the kidneys in acid-base balance is described in chapter 17.

A fall in blood pH below 7.35 is called **acidosis** because the pH is to the acid side of normal. Acidosis does not mean acidic (pH less than 7); a blood pH of 7.2, for example, represents serious acidosis. Similarly, a rise in blood pH above 7.45 is called **alkalosis.** Each of these categories is subdivided into respiratory and metabolic components of acid-base balance (table 13.6).

Respiratory acidosis is caused by inadequate ventilation (hypoventilation), which results in a rise in the plasma concentration of carbon dioxide, and thus carbonic acid. **Respiratory alkalosis,** in contrast, is caused by excessive ventilation (hyperventilation). **Metabolic acidosis** can result from excessive production of nonvolatile acids; for example, it can result from excessive production of ketone bodies in uncontrolled diabetes mellitus (chapter 19). It can also result from the loss of bicarbonate, in which case there would not be sufficient free bicarbonate to buffer the nonvolatile acids. (This occurs in diarrhea because of the loss of bicarbonate derived from pancreatic juice—see chapter 18.) **Metabolic alkalosis,** in contrast, can be caused by either too much bicarbonate (perhaps from an intravenous infusion) or inadequate nonvolatile acids (perhaps as a result of excessive vomiting and a consequent loss of gastric acid).

Since the *respiratory component* of acid-base balance is represented by the plasma carbon dioxide concentration and the *metabolic component* is represented by the free-bicarbonate concentration, the study of acid-base balance can be simplified. A normal arterial blood pH is obtained when there is a proper ratio of bicarbonate to carbon dioxide. Indeed, the pH can be calculated given these values, and a normal pH is obtained when the ratio of these concentrations is 20 to 1. This is given by the **Henderson–Hasselbalch** equation:

$$pH = 6.1 + \log \frac{[HCO_3^-]}{[CO_2]}$$

Respiratory acidosis or alkalosis occurs when the carbon dioxide concentrations are abnormal. Metabolic acidosis and alkalosis occur when the bicarbonate concentrations are abnormal (table 13.7). Often, however, a primary disturbance in one area, for example, metabolic acidosis, will be accompanied by

Table 13.7   Classification of Metabolic and Respiratory Components of Acidosis and Alkalosis

Plasma CO_2	Plasma HCO_3^-	Condition	Causes
Normal	Low	Metabolic acidosis	Increased production of "nonvolatile" acids (lactic acid, ketone bodies, and others), or loss of HCO_3^- in diarrhea
Normal	High	Metabolic alkalosis	Vomiting of gastric acid; hypokalemia; excessive steroid administration
Low	Low	Respiratory alkalosis	Hyperventilation
High	High	Respiratory acidosis	Hypoventilation

secondary changes in another area, for example, respiratory alkalosis. It is important for hospital personnel to identify and treat the area of primary disturbance, but such analysis lies outside the scope of this discussion.

A more complete description of the respiratory and metabolic components of acid-base balance requires the study of pulmonary and renal function, and so will be presented with these topics in chapters 16 and 17.

---

1. State the normal pH range of arterial blood plasma and explain how it is affected by the concentration of carbon dioxide in the blood.
2. Explain how bicarbonate helps to maintain the acid-base balance and describe the conditions that may result in metabolic acidosis or alkalosis.

Structure of the Heart

The heart contains four chambers: two upper atria, which receive venous blood, and two lower ventricles, which eject blood into arteries. The right ventricle pumps blood to the lungs, where the blood becomes oxygenated; the left ventricle pumps oxygenated blood to the entire body. The proper flow of blood within the heart is aided by two pairs of one-way valves.

The heart is divided into four chambers. The right and left **atria** (singular, *atrium*) receive blood from the venous system; the right and left **ventricles** pump blood into the arterial system. The right atrium and ventricle (sometimes called the *right pump*) are separated from the left atrium and ventricle (the *left pump*) by a muscular wall, or *septum*. This septum normally prevents mixture of the blood from the two sides of the heart.

Between the atria and ventricles there is a layer of dense connective tissue known as the **fibrous skeleton** of the heart. Bundles of myocardial cells (described in chapter 12) in the atria attach to the upper margin of this fibrous skeleton and form a single functioning unit, or *myocardium*. The myocardial cell bundles of the ventricles attach to the lower margin and form a different myocardium. As a result, the myocardia of the

atria and ventricles are structurally and functionally separated from each other, and special conducting tissue is needed to carry action potentials from the atria to the ventricles (this will be described in a later section). The connective tissue of the fibrous skeleton also forms rings, called *annuli fibrosi*, around the four heart valves, providing a foundation for the support of the valve flaps.

Pulmonary and Systemic Circulations

Blood in which the oxygen content has become partially depleted and the carbon dioxide content has increased as a result of tissue metabolism returns to the right atrium. This blood then enters the right ventricle, which pumps it into the *pulmonary trunk* and *pulmonary arteries*. The pulmonary arteries branch to transport blood to the lungs, where gas exchange occurs between the lung capillaries and the air sacs (alveoli) of the lungs. Oxygen diffuses from the air to the capillary blood, while carbon dioxide diffuses in the opposite direction.

The blood that returns to the left atrium by way of the *pulmonary veins* is therefore enriched in oxygen and partially depleted in carbon dioxide. The path of blood from the heart (right ventricle), through the lungs, and back to the heart (left atrium) completes one circuit: the **pulmonary circulation.**

Oxygen-rich blood in the left atrium enters the left ventricle and is pumped into a very large, elastic artery—the *aorta*. The aorta ascends for a short distance, makes a U-turn, and then descends through the thoracic (chest) and abdominal cavities. Arterial branches from the aorta supply oxygen-rich blood to all of the organ systems and are thus part of the **systemic circulation.**

As a result of cellular respiration, the oxygen concentration is lower and the carbon dioxide concentration is higher in the tissues than in the capillary blood. Blood that drains into the systemic veins is thus partially depleted of oxygen and increased in carbon dioxide content. These veins ultimately empty into two large veins—the *superior* and *inferior venae cavae*—that return the oxygen-poor blood to the right atrium. This completes the systemic circulation: from the heart (left ventricle), through the organ systems, and back to the heart (right atrium). The characteristics of the systemic and pulmonary circulations are summarized in table 13.8 and illustrated in figure 13.7.

The numerous small muscular arteries and arterioles of the systemic circulation present greater resistance to blood flow than that in the pulmonary circulation. Despite the differences

Table 13.8 Summary of the Pulmonary and Systemic Circulations

	Source	Arteries	O₂ Content of Arteries	Veins	O₂ Content of Veins	Termination
Pulmonary Circulation	Right ventricle	Pulmonary arteries	Low	Pulmonary veins	High	Left atrium
Systemic Circulation	Left ventricle	Aorta and its branches	High	Superior and inferior venae cavae and their branches*	Low	Right atrium

*Blood from the coronary circulation does not enter the venae cavae, but instead returns directly to the right atrium via the coronary sinus.

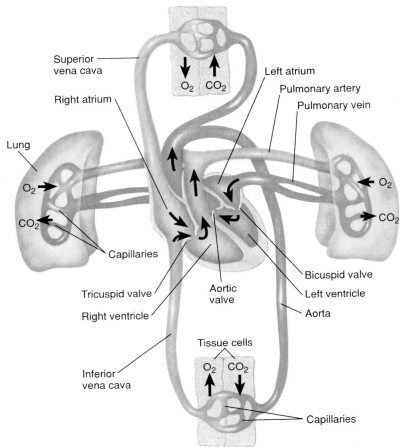

Figure 13.7
A diagram of the circulatory system.

in resistance, the rate of blood flow through the systemic circulation must be matched to the flow rate of the pulmonary circulation. Since the amount of work performed by the left ventricle is greater (by a factor of 5 to 7) than that performed by the right ventricle, it is not surprising that the muscular wall of the left ventricle is thicker (8–10 mm) than that of the right ventricle (2–3 mm).

Atrioventricular and Semilunar Valves

Although adjacent myocardial cells are joined together mechanically and electrically by intercalated discs, the atria and ventricles are separated into two functional units by a sheet of connective tissue—the fibrous skeleton previously mentioned.

Embedded within this sheet of tissue are one-way **atrioventricular (AV) valves.** The AV valve located between the right atrium and right ventricle has three flaps, and is therefore called the *tricuspid valve.* The AV valve between the left atrium and left ventricle has two flaps and is thus called the *bicuspid valve,* or, alternatively, the *mitral valve* (fig. 13.8).

The AV valves allow blood to flow from the atria to the ventricles, but normally prevent the backflow of blood into the atria. Opening and closing of these valves occur as a result of pressure differences between the atria and ventricles. When the ventricles are relaxed, the venous return of blood to the atria causes the pressure in the atria to exceed that in the ventricles. The AV valves therefore open, allowing blood

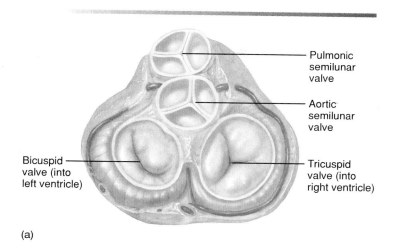

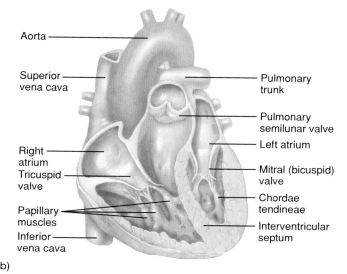

(a)

(b)

Figure 13.8

Diagrams (a) of the structure of the heart and (b) of the atrioventricular (AV) and semilunar valves.

to enter the ventricles. As the ventricles contract, the intraventricular pressure rises above the pressure in the atria and pushes the AV valves closed.

There is a danger, however, that the high pressure produced by contraction of the ventricles could push the valve flaps too much and evert them. This is normally prevented by contraction of the *papillary muscles* within the ventricles, which are connected to the AV valve flaps by the *chordae tendineae*. Contraction of the papillary muscles occurs at the same time as contraction of the muscular walls of the ventricles and serves to keep the valve flaps tightly closed.

One-way **semilunar valves** (fig. 13.9) are located at the origin of the pulmonary artery and aorta. These valves open during ventricular contraction, allowing blood to enter the pulmonary and systemic circulations. During ventricular relaxation, when the pressure in the arteries is greater than the pressure in the ventricles, the semilunar valves snap shut and thus prevent the backflow of blood into the ventricles.

---

1. Using a flow diagram (arrows), describe the pathway of the pulmonary circulation. Indicate the relative amounts of oxygen and carbon dioxide in the vessels involved.

2. Use a flow diagram to describe the systemic circulation and indicate the relative amounts of oxygen and carbon dioxide in the blood vessels.

3. List the AV valves and the valves of the pulmonary artery and aorta. Describe how these valves ensure a one-way flow of blood.

4. Explain the structure of the fibrous skeleton of the heart and explain its significance.

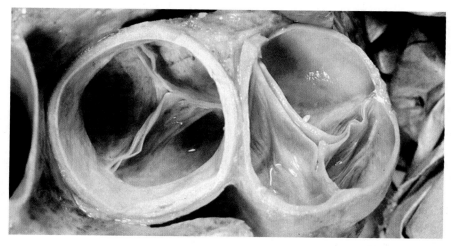

Figure 13.9

A photograph of the aortic and pulmonary semilunar valves.

Cardiac Cycle and Heart Sounds

The two atria fill with blood and then contract simultaneously. This is followed by simultaneous contraction of both ventricles, which sends blood through the pulmonary and systemic circulations. Contraction of the ventricles closes the AV valves and opens the semilunar valves; relaxation of the ventricles allows the semilunar valves to close. The closing of first the AV valves and then the semilunar valves produces the "lub-dub" sounds heard with a stethoscope.

The *cardiac cycle* refers to the repeating pattern of contraction and relaxation of the heart. The phase of contraction is called **systole,** and the phase of relaxation is called **diastole.** When these terms are used without reference to specific chambers, they refer to contraction and relaxation of the ventricles. It should be noted, however, that the atria also contract and relax. There is an atrial systole and diastole. Atrial contraction occurs toward the end of diastole, when the ventricles are relaxed; when the ventricles contract during systole, the atria are relaxed.

The heart thus has a two-step pumping action. The right and left atria contract almost simultaneously, followed by contraction of the right and left ventricles 0.1 to 0.2 second later. During the time when both the atria and ventricles are relaxed, the venous return of blood fills the atria. The buildup of pressure that results causes the AV valves to open and blood to flow from atria to ventricles. It has been estimated that the ventricles are about 80% filled with blood even before the atria contract. Contraction of the atria adds the final 20% to the *end-diastolic volume,* which is the total volume of blood in the ventricles at the end of diastole.

Contraction of the ventricles in systole ejects about two-thirds of the blood that they contain—an amount called the *stroke volume*—leaving one-third of the initial amount left in the ventricles as the *end-systolic volume.* The ventricles then fill with blood during the next cycle. At an average *cardiac rate* of 75 beats per minute, each cycle lasts 0.8 second; 0.5 second is spent in diastole, and systole takes 0.3 second (fig. 13.10).

 Interestingly, the blood contributed by contraction of the atria does not appear to be essential for life. Elderly people who have atrial fibrillation (a condition in which the atria fail to contract) do not appear to have a higher mortality than those who have normally functioning atria. People with atrial fibrillation, however, become fatigued more easily during exercise because the reduced filling of the ventricles compromises the ability of the heart to sufficiently increase its output during exercise. Cardiac output and blood flow during rest and exercise are discussed in chapter 14.

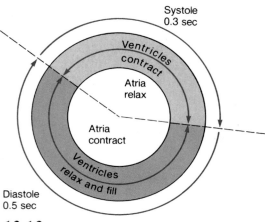

Figure 13.10

The cardiac cycle of ventricular systole and diastole. Contraction of the atria occurs in the last 0.1 second of ventricular diastole. Relaxation of the atria occurs during ventricular systole. The durations of systole and diastole given relate to a cardiac rate of 75 beats per minute.

Pressure Changes during the Cardiac Cycle

When the heart is in diastole, pressure in the systemic arteries averages about 80 mmHg (millimeters of mercury). The following events in the cardiac cycle then occur:

1. As the ventricles begin their contraction, the intraventricular pressure rises, causing the AV valves to snap shut. At this time, the ventricles are neither being filled with blood (because the AV valves are closed) nor ejecting blood (because the intraventricular pressure has not risen sufficiently to open the semilunar valves). This is the phase of *isovolumetric contraction.*

2. When the pressure in the left ventricle becomes greater than the pressure in the aorta, the phase of *ejection* begins as the semilunar valves open. The pressure in the left ventricle and aorta rises to about 120 mmHg (fig. 13.11) when ejection begins and the ventricular volume decreases.

3. As the pressure in the left ventricle falls below the pressure in the aorta, the back pressure causes the semilunar valves to snap shut. The pressure in the aorta falls to 80 mmHg while pressure in the left ventricle falls to 0 mmHg.

4. During *isovolumetric relaxation,* the AV and semilunar valves are closed. This phase lasts until the pressure in the ventricles falls below the pressure in the atria.

5. When the pressure in the ventricles falls below the pressure in the atria, a phase of *rapid filling* of the ventricles occurs.

6. *Atrial contraction* (*atrial systole*) empties the final amount of blood into the ventricles immediately before the next phase of isovolumetric contraction of the ventricles.

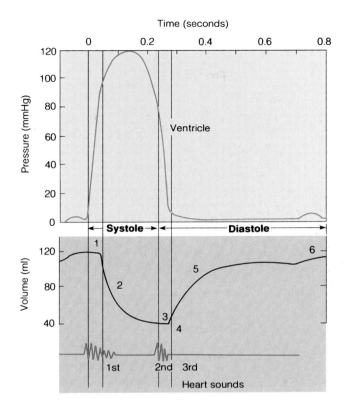

Time (seconds)

Figure 13.11

The relationship between the heart sounds and the intraventricular pressure and volume. The numbers refer to the events described in the text.

Similar events occur in the right ventricle and pulmonary circulation, but the pressures are lower. The maximum pressure produced at systole in the right ventricle is 25 mmHg, which falls to a low of 8 mmHg at diastole.

Heart Sounds

Closing of the AV and semilunar valves produces sounds that can be heard at the surface of the chest with a stethoscope. The words used to imitate these sounds are often *lub-dub*. The "lub," or **first sound,** is produced by closing of the AV valves during isovolumetric contraction of the ventricles. The "dub," or **second sound,** is produced by closing of the semilunar valves when the pressure in the ventricles falls below the pressure in the arteries. The first sound is thus heard when the ventricles contract at systole, and the second sound is heard when the ventricles relax at the beginning of diastole.

The first sound may be heard to split into tricuspid and mitral components, particularly during inhalation. Closing of the tricuspid is best heard at the fifth intercostal space (between the ribs), just to the right of the sternum; closing of the mitral valve is best heard at the fifth left intercostal space at the apex of the heart (fig. 13.12). The second sound may also be split under certain conditions. Closing of the pulmonary and aortic semilunar valves is best heard at the second left and right intercostal spaces, respectively.

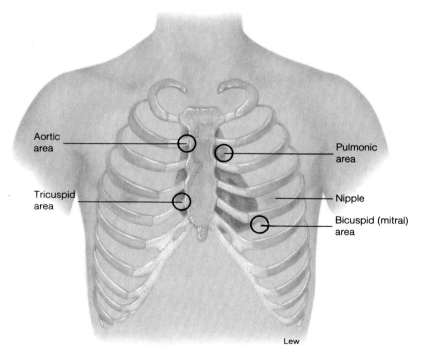

Figure 13.12

Routine stethoscope positions for listening to the heart sounds.

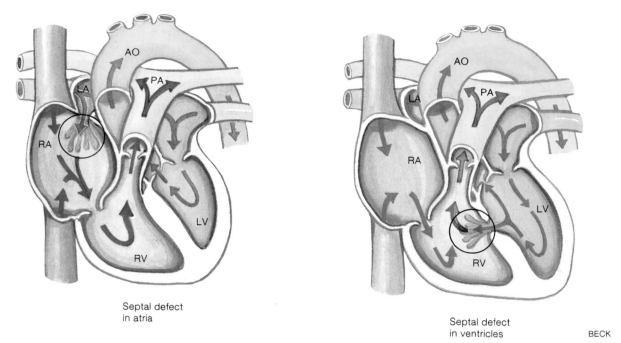

Figure 13.13

Abnormal patterns of blood flow due to septal defects. Left-to-right shunting of blood is shown (*circled areas*) because the left pump is at a higher pressure than the right pump. Under certain conditions, however, the pressure in the right atrium may exceed that of the left, causing right-to-left shunting of blood through a septal defect in the atria (patent foramen ovale).

Heart Murmurs

Murmurs are abnormal heart sounds produced by abnormal patterns of blood flow in the heart. Many murmurs are caused by defective heart valves. Defective heart valves may be congenital, or they can occur as a result of *rheumatic endocarditis*, associated with rheumatic fever. In this disease, the valves become damaged by antibodies made in response to an infection caused by streptococcus bacteria (the same bacteria that produce strep throat). Many people have small defects that produce detectable murmurs but do not seriously compromise the pumping ability of the heart. Larger defects, however, may have dangerous consequences and thus may require surgical correction.

In *mitral stenosis*, for example, the mitral valve becomes thickened and calcified. This can impair the blood flow from the left atrium to the left ventricle. An accumulation of blood in the left atrium may cause a rise in left atrial and pulmonary vein pressure, resulting in pulmonary hypertension.

As a compensation for the increased pulmonary pressure, the right ventricle grows thicker and stronger.

Valves are said to be incompetent when they do not close properly, and murmurs may be produced as blood regurgitates through the valve flaps. One important cause of incompetent AV valves is damage to the papillary muscles (see fig. 13.8). When this occurs, the tension in the chordae tendineae may not be sufficient to prevent the valve from everting as pressure in the ventricle rises during systole.

Murmurs can also be produced by the flow of blood through *septal defects*—holes in the septum between the right and left sides of the heart. These are usually congenital and may occur either in the interatrial or interventricular septum (fig. 13.13). When a septal defect is not accompanied by other abnormalities, blood will usually pass through the defect from the left to the right side, due to the higher pressure on the left side. The buildup of blood and pressure on the right side of the heart that results may lead to pulmonary hypertension and edema (fluid in the lungs).

The lungs of a fetus are collapsed, and blood is routed away from the pulmonary circulation by an opening in the interatrial septum called the **foramen ovale** (fig. 13.13) and by a connection between the pulmonary trunk and aorta called the **ductus arteriosus** (fig. 13.14). These shunts normally close after birth, but when they remain open (are *patent*), murmurs can result. Since blood usually goes from left to right through these shunts, the left ventricle still pumps blood that is high in oxygen. When other defects are present that increase the pressure in the right pump (as in the *tetralogy of Fallot*), however, a significant amount of oxygen-depleted blood from the right side of the heart may enter the left side. The mixture of oxygen-poor blood from the right side with oxygen-rich blood in the left side of the heart lowers the oxygen concentration of the blood ejected into the systemic circulation. Since blood low in oxygen imparts a bluish tinge to the skin, the baby may be born *cyanotic* (blue).

1. Using a figure or a summary statement, describe the sequence of events that occurs during the cardiac cycle. Indicate when atrial and ventricular filling occur and when atrial and ventricular contraction occur.

2. Describe, in words, how the pressure in the left ventricle and in the systemic arteries varies during the cardiac cycle.

3. Draw a figure to illustrate the pressure variations described in question 2, and indicate in your figure when the AV and semilunar valves close. Discuss the origin of the heart sounds.

4. Explain why blood usually flows from left to right through a septal defect. What must occur through a septal defect to produce cyanosis?

Electrical Activity of the Heart and the Electrocardiogram

The pacemaker region of the heart (SA node) exhibits a spontaneous depolarization that causes action potentials and results in the automatic beating of the heart. Electrical impulses are conducted by myocardial cells in the atria and are transmitted to the ventricles by special conducting tissue. Electrocardiogram waves correspond to the electrical events in the heart as follows: P wave (depolarization of the atria); QRS wave (depolarization of the ventricles); and T wave (repolarization of the ventricles).

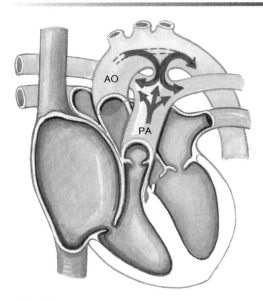

Figure 13.14

The flow of blood through a patent (open) ductus arteriosus.

As described in chapter 12, myocardial cells are short, branched, and interconnected by gap junctions that function as electrical synapses. The entire mass of cells interconnected by gap junctions is known as a *myocardium*. A myocardium is a single functioning unit, or *functional syncitium*, since action potentials that originate in any cell in the mass can be transmitted to all the other cells. The atria and ventricles are different myocardia that are separated by the fibrous skeleton of the heart, as previously described. Since the impulse normally originates in the atria, the atrial myocardium is excited before that of the ventricles.

Electrical Activity of the Heart

If the heart of a frog is removed from the body and all neural innervations are severed, it will still continue to beat as long as the myocardial cells remain alive. The automatic nature of the heartbeat is referred to as *automaticity*. As a result of experiments with isolated myocardial cells and clinical experience with patients who have specific heart disorders, many regions within the heart have been shown to be capable of originating action potentials and functioning as pacemakers. In a normal heart, however, only one region demonstrates spontaneous electrical activity and by this means functions as a pacemaker. This pacemaker region is called the **sinoatrial node,** or **SA node.** The SA node is located in the right atrium, near the opening of the superior vena cava.

The cells of the SA node do not keep a resting membrane potential in the manner of resting neurons or skeletal muscle cells. Instead, during the period of diastole, the SA node exhibits a slow, spontaneous depolarization called the **pacemaker potential.** The membrane potential begins at about –60 mV

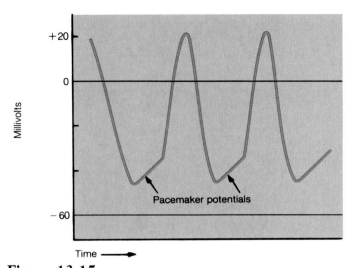

Figure 13.15

Pacemaker potentials and action potentials in the SA node.

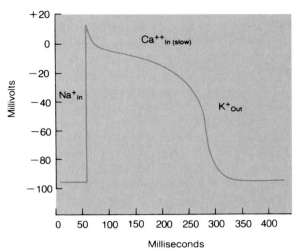

Figure 13.16

An action potential in a myocardial cell from the ventricles. The plateau phase of the action potential is maintained by a slow inward diffusion of Ca^{++}. The cardiac action potential, as a result, has a duration that is about 100 times longer than the "spike potential" of an axon.

and gradually depolarizes to –40 mV, which is the threshold for producing an action potential in these cells. This spontaneous depolarization is produced by the diffusion of Ca^{++} through openings in the membrane called *slow calcium channels*. At the threshold level of depolarization, other channels, called *fast calcium channels*, open, and Ca^{++} rapidly diffuses into the cells. The opening of voltage-regulated Na^+ gates, and the inward diffusion of Na^+ that results, may also contribute to the upshoot phase of the action potential in pacemaker cells (fig. 13.15). Repolarization is produced by the opening of K^+ gates and outward diffusion of K^+, as in the other excitable tissues previously discussed. Once repolarization to –60 mV has been achieved, a new pacemaker potential begins, again culminating with a new action potential at the end of diastole.

Some other regions of the heart, including the area around the SA node and the atrioventricular bundle, can potentially produce pacemaker potentials. The rate of spontaneous depolarization of these cells, however, is slower than that of the SA node. The potential pacemaker cells are, therefore, stimulated by action potentials from the SA node before they can stimulate themselves through their own pacemaker potentials. If action potentials from the SA node are prevented from reaching these areas (through blockage of conduction), they will generate pacemaker potentials at their own rate and serve as sites for the origin of action potentials; they will function as pacemakers. A pacemaker other than the SA node is called an *ectopic pacemaker* (also called an *ectopic focus*). From this discussion, it is clear that the rhythm set by such an ectopic pacemaker is usually slower than that normally set by the SA node.

Once another myocardial cell has been stimulated by action potentials originating in the SA node, it produces its own action potentials. The majority of myocardial cells have resting membrane potentials of about –90 mV. When stimulated by action potentials from a pacemaker region, these cells become depolarized to threshold, at which point their voltage-regulated Na^+ gates open. The upshoot phase of the action potential of

nonpacemaker cells is due to the inward diffusion of Na^+. Following the rapid reversal of the membrane polarity, the membrane potential quickly declines to about –15 mV. Unlike the action potential of other cells, however, this level of depolarization is maintained for 200 to 300 msec before repolarization (fig. 13.16). This *plateau phase* results from a slow inward diffusion of Ca^{++}, which balances a slow outward diffusion of cations. Rapid repolarization at the end of the plateau phase is achieved, as in other cells, by the opening of K^+ gates and the rapid outward diffusion of K^+ that results.

Conducting Tissues of the Heart

Action potentials that originate in the SA node spread to adjacent myocardial cells of the right and left atria through the gap junctions between these cells. Since the myocardium of the atria is separated from the myocardium of the ventricles by the fibrous skeleton of the heart, however, the impulse cannot be conducted directly from the atria to the ventricles. Specialized conducting tissue, composed of modified myocardial cells, is thus required. These specialized myocardial cells form the AV node, bundle of His, and Purkinje fibers.

Once the impulse spreads through the atria, it passes to the **atrioventricular node (AV node),** which is located on the inferior portion of the interatrial septum (fig. 13.17). From here, the impulse continues through the **atrioventricular bundle,** or **bundle of His** (pronounced "hiss"), beginning at the top of the interventricular septum. This conducting tissue pierces the fibrous skeleton of the heart and continues to descend along the interventricular septum. The atrioventricular bundle divides into right and left bundle branches, which are continuous with the **Purkinje fibers** within the ventricular walls. Stimulation of these conduction myofibers causes both ventricles to contract simultaneously and eject blood into the pulmonary and systemic circulation.

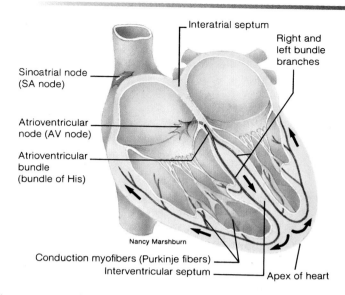

Figure 13.17

The conduction system of the heart.

Nancy Marshburn

Interatrial septum

Right and left bundle branches

Sinoatrial node (SA node)

Atrioventricular node (AV node)

Atrioventricular bundle (bundle of His)

Conduction myofibers (Purkinje fibers)

Interventricular septum

Apex of heart

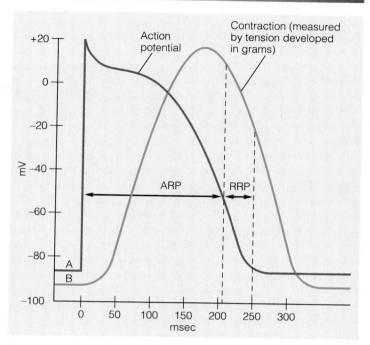

Figure 13.18

The time course for the myocardial action potential (A) is compared with the duration of contraction (B). Notice that the long action potential results in a correspondingly long absolute refractory period (ARP) and relative refractory period (RRP). These refractory periods last almost as long as the contraction, so that the myocardial cells cannot be stimulated a second time until they have completed their contraction from the first stimulus.

Conduction of the Impulse

Action potentials from the SA node spread very quickly—at a rate of 0.8 to 1.0 meter per second (m/sec)—across the myocardial cells of both atria. The conduction rate then slows considerably as the impulse passes into the AV node. Slow conduction of impulses (0.03 to 0.05 m/sec) through the AV node accounts for over half of the time delay between excitation of the atria and ventricles. After the impulses spread through the AV node, the conduction rate increases greatly in the atrioventricular bundle and reaches very high velocities (5 m/sec) in the Purkinje fibers. As a result of this rapid conduction of impulses, ventricular contraction begins 0.1 to 0.2 second after the contraction of the atria.

Unlike skeletal muscles, the heart cannot sustain a contraction. This is because the atria and ventricles behave as if each were composed of only one muscle cell; the entire myocardium of each is electrically stimulated as a single unit and contracts as a unit. This contraction, which corresponds in time to the long action potential of myocardial cells and lasts almost 300 msec, is analogous to the twitch produced by a single skeletal muscle fiber (which lasts only 20 to 100 msec in comparison). The heart normally cannot be stimulated again until after it has relaxed from its previous contraction because myocardial cells have *long refractory periods* (fig. 13.18) that correspond to the long duration of their action potentials. Summation of contractions is thus prevented, and the myocardium must relax after each contraction. The rhythmic pumping action of the heart is thus ensured.

The Electrocardiogram

A pair of surface electrodes placed directly on the heart will record a repeating pattern of potential changes. As action potentials spread from the atria to the ventricles, the voltage measured between these two electrodes will vary in a way that provides a "picture" of the electrical activity of the heart.

Abnormal patterns of electrical conduction in the heart can produce abnormalities of the cardiac cycle and seriously compromise the function of the heart. These *arrhythmias* may be treated with a variety of drugs that inhibit specific aspects of the cardiac action potentials and, in this way, inhibit the production or conduction of impulses along abnormal pathways. Drugs used to treat arrhythmias may (1) block the fast Na$^+$ channel (quinidine, procainamide, lidocaine); (2) block the slow Ca^{++} channel (verapamil); or (3) block β-adrenergic receptors (propranolol, atenolol) because the rates of impulse production and conduction are stimulated by catecholamines.

The body is a good conductor of electricity because tissue fluids contain a high concentration of ions that move (creating a current) in response to potential differences. Potential differences generated by the heart are thus conducted to the body surface, where they can be recorded by surface electrodes placed on the skin. The recording thus obtained is called an **electrocardiogram**

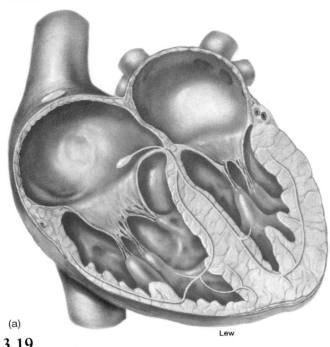

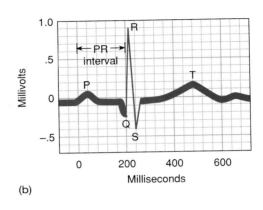

(a)

(b)

Lew

Figure 13.19

The electrocardiogram indicates the conduction of electrical impulses through the heart (a) and measures and records both the intensity of this electrical activity (in millivolts) and the time intervals involved (b).

(ECG or EKG) (fig. 13.19); the recording device is called an *electrocardiograph*. By changing the position of the recording electrodes, an observer can gain a more complete picture of the electrical events.

There are two types of ECG recording electrodes, or "leads." The *bipolar limb leads* record the voltage between electrodes placed on the wrists and legs. These bipolar leads include lead I (right arm to left arm), lead II (right arm to left leg), and lead III (left arm to left leg). In the *unipolar leads*, voltage is recorded between a single "exploratory electrode" placed on the body and an electrode that is built into the electrocardiograph and maintained at zero potential (ground).

The unipolar limb leads are placed on the right arm, left arm, and left leg, and are abbreviated AVR, AVL, and AVF, respectively. The unipolar chest leads are labeled 1 through 6, starting from the midline position (fig. 13.20). There are thus a total of twelve standard ECG leads that "view" the changing pattern of the heart's electrical activity from different perspectives (table 13.9). This is important because certain abnormalities are best seen with particular leads and may not be visible at all with other leads.

Each cardiac cycle produces three distinct ECG waves, designated P, QRS, and T. It should be noted that these waves are not action potentials; they represent changes in potential between two regions on the surface of the heart that are produced by the composite effects of action potentials in numerous myocardial cells. For example, the spread of depolarization through the atria causes a potential difference that is indicated by an upward deflection of the ECG line. When about half the mass of the atria is depolarized, this upward deflection reaches a maximum value, since the potential difference between the depolarized and unstimulated portions of the atria is at a maximum. When the entire mass of the atria is depolarized, the ECG returns to baseline because all regions of the atria have the same polarity. The spread of atrial depolarization thus creates the **P wave.**

Conduction of the impulse into the ventricles similarly creates a potential difference that results in a sharp upward deflection of the ECG line, which then returns to the baseline as the entire mass of the ventricles becomes depolarized. The spread of the depolarization into the ventricles is thus represented by the **QRS wave.** The atria repolarize during this time, but this event is hidden by the greater depolarization occurring in the ventricles. Finally, repolarization of the ventricles produces the **T wave** (fig. 13.21).

Correlation of the ECG with Heart Sounds

Depolarization of the ventricles, as indicated by the QRS wave, stimulates contraction by promoting the uptake of Ca^{++} into the regions of the sarcomeres. The QRS wave is thus seen to occur at the beginning of systole. The rise in intraventricular pressure that results causes the AV valves to close, so that the first heart sound (S_1, or lub) is produced immediately after the QRS wave (fig. 13.22).

Repolarization of the ventricles, as indicated by the T wave, occurs at the same time that the ventricles relax at the beginning of diastole. The resulting fall in intraventricular pressure causes the aortic and pulmonary semilunar valves to close, so that the second heart sound (S_2, or dub) is produced shortly after the T wave begins in an electrocardiogram.

Chapter Thirteen

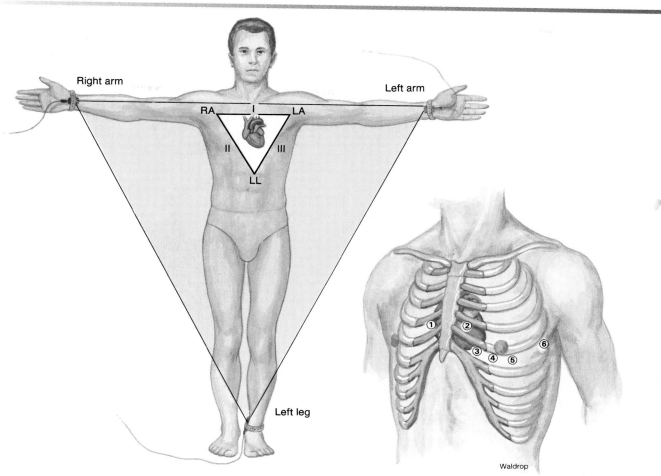

Figure 13.20

The placement of the bipolar limb leads and the exploratory electrode for the unipolar chest leads in an electrocardiogram (ECG). (RA = right arm; LA = left arm; LL = left leg.)

Table 13.9 Electrocardiograph (ECG) Leads

Name of Lead	Placement of Electrodes
Bipolar limb leads	
I	Right arm and left arm
II	Right arm and left leg
III	Left arm and left leg
Unipolar limb leads	
AVR	Right arm
AVL	Left arm
AVF	Left leg
Unipolar chest leads	
V_1	4th intercostal space right of sternum
V_2	4th intercostal space left of sternum
V_3	5th intercostal space to the left of the sternum
V_4	5th intercostal space in line with the middle of the clavicle (collarbone)
V_5	5th intercostal space to the left of V_4
V_6	5th intercostal space in line with the middle of the axilla (underarm)

1. Describe the electrical activity of the cells of the SA node and explain how the SA node functions as the normal pacemaker.

2. Using a line diagram, illustrate a myocardial action potential and the time course for myocardial contraction. Describe how the relationship between these two events prevents the heart from sustaining a contraction and how it normally prevents abnormal rhythms of electrical activity.

3. Draw an ECG and label the waves. Indicate the electrical events in the heart that produce these waves.

4. Draw a figure illustrating the relationship between ECG waves and the heart sounds. Explain this relationship.

5. Using a flowchart (arrows), describe the pathway of electrical conduction of the heart, starting with the SA node. Explain how damage to the AV node affects this conduction pathway and the ECG.

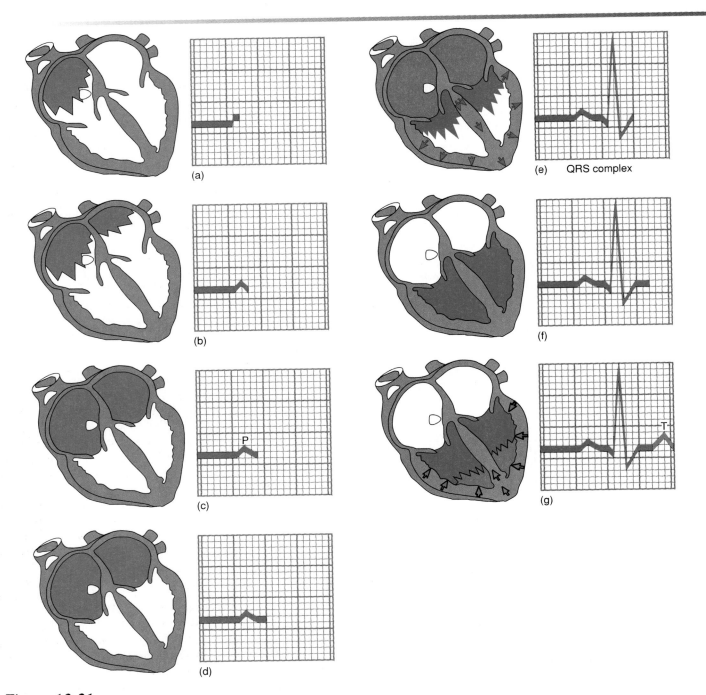

(a)

(b)

(c) P

(d)

(e) QRS complex

(f)

(g) T

Figure 13.21

The conduction of electrical impulses in the heart as indicated by the electrocardiogram (ECG). The direction of the arrows in (*e*) indicates that depolarization of the ventricles occurs from the inside (endocardium) out (to the epicardium), whereas the arrows in (*g*) indicate that repolarization of the ventricles occurs in the opposite direction.

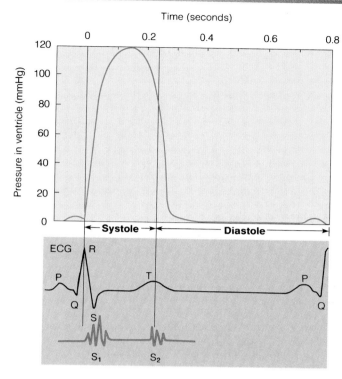

Figure 13.22

The relationship between changes in intraventricular pressure and the electrocardiogram during the cardiac cycle. The QRS wave (representing depolarization of the ventricles) occurs at the beginning of systole, whereas the T wave (representing repolarization of the ventricles) occurs at the beginning of diastole.

Blood Vessels

The thick muscle layer of arteries allows them to transmit blood ejected from the heart under high pressure, and the elastic recoil of the large arteries further contributes to blood flow. The thinner muscle layer of veins allows them to distend when an increased amount of blood enters them, and their one-way valves ensure that blood flows back to the heart. Capillaries are composed of only a single layer of endothelium, which allows water and other molecules to move across the capillary walls and thus permits exchanges between the blood and tissue fluid.

Blood vessels form a tubular network throughout the body that permits blood to flow from the heart to all the living cells of the body and then back to the heart. Blood leaving the heart passes through vessels of progressively smaller diameters, referred to as arteries, arterioles, and capillaries. Capillaries are microscopic vessels that join the arterial flow to the venous flow. Blood returning to the heart from the capillaries passes through vessels of progressively larger diameters, called venules and veins.

The walls of arteries and veins are composed of three coats, or "tunics." The outermost layer is the **tunica externa,** the middle layer is the **tunica media,** and the inner layer is the **tunica intima.** The tunica externa is composed of connective tissue. The tunica media is composed primarily of smooth muscle. The tunica intima consists of three parts: (1) an innermost simple squamous epithelium, the *endothelium,* which lines the lumina of all blood vessels; (2) the basement membrane (a layer of glycoproteins) overlying some connective tissue fibers; and (3) a layer of elastic fibers, or *elastin,* forming an *internal elastic lamina*.

Although both arteries and veins have the same basic structure (fig. 13.23), there are some important differences between these vessels. Arteries have more muscle for their diameters than do comparably sized veins. As a result, arteries appear more rounded in cross section, whereas veins are usually partially collapsed. In addition, many veins have valves, which are absent in arteries.

Arteries

The aorta and other large arteries contain numerous layers of elastin fibers between smooth muscle cells in the tunica media. These large **elastic arteries** expand when the pressure of the blood rises as a result of the heart's contraction; they recoil, like a stretched rubber band, when the blood pressure falls during relaxation of the ventricles. This elastic recoil helps to produce a smoother, less pulsatile flow of blood through the smaller arteries and arterioles.

The small arteries and arterioles are less elastic than the larger arteries and have a thicker layer of smooth muscle for their diameters. Unlike the larger elastic arteries, therefore, the diameter of the smaller **muscular arteries** changes only slightly as the pressure of the blood rises and falls during the heart's pumping activity. Since arterioles and small muscular arteries have narrow lumina, they provide the greatest resistance to blood flow through the arterial system.

Small muscular arteries that are 100 μm or less in diameter branch to form smaller arterioles (20–30 μm in diameter). In some tissues, blood from the arterioles can enter the venules directly through *arteriovenous anastomoses*. In most cases, however, blood from arterioles passes into capillaries (fig. 13.24). Capillaries are the narrowest of blood vessels (7–10 μm in diameter), and serve as the "business end" of the circulatory system in which exchanges of gases and nutrients between the blood and the tissues occur.

Capillaries

The arterial system branches extensively (table 13.10) to deliver blood to over 40 billion capillaries in the body. As evidence of the extensiveness of these branchings, consider the fact that all tissue cells are located within a distance of only 60 to 80 μm of a capillary and that capillaries provide a total surface area of 1,000 square miles for exchanges between blood and tissue fluid.

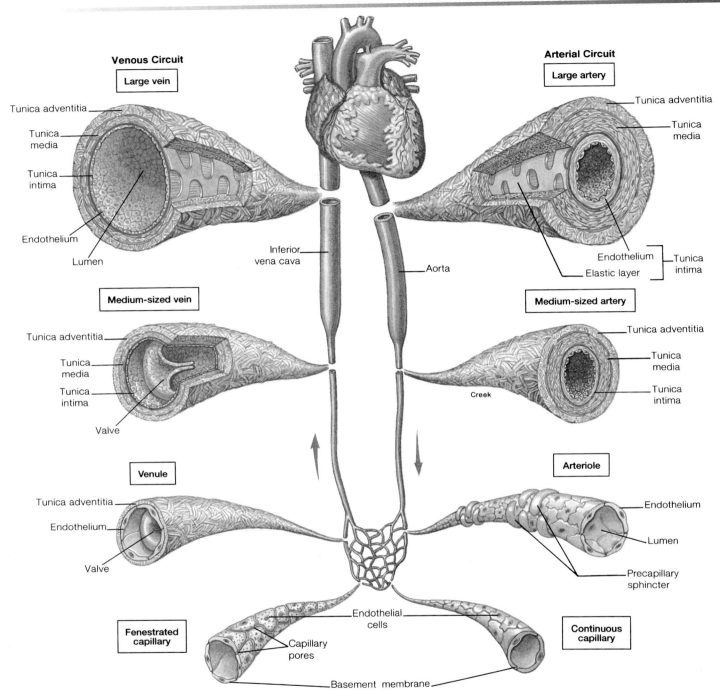

Figure 13.23

Relative thickness and composition of the tunicas in comparable arteries and veins.

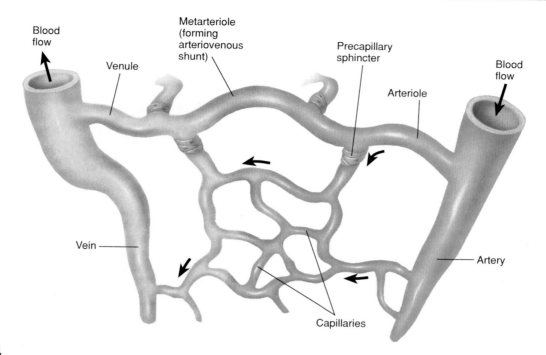

Figure 13.24

The microcirculation. Metarterioles (arteriovenous anastomoses) provide a path of least resistance between arterioles and venules. Precapillary sphincter muscles regulate the flow of blood through the capillaries.

Table 13.10 Characteristics of the Vascular Supply to the Mesenteries in a Dog

Kind of Vessels	Diameter (mm)	Number	Total Cross-Sectional Area (cm²)	Length(cm)	Total Volume (cm³)
Aorta	10	1	0.8	40	30
Large arteries	3	40	3.0	20	60
Main artery branches	1	600	5.0	10	50
Terminal branches	0.6	1,800	5.0	1	25
Arterioles	0.02	40,000,000	125	0.2	25
Capillaries	0.008	1,200,000,000	600	0.1	60
Venules	0.03	80,000,000	570	0.2	110
Terminal veins	1.5	1,800	30	1	30
Main venous branches	2.4	600	27	10	270
Large veins	6.0	40	11	20	220
Vena cava	12.5	1	1.2	40	50
					930

Note: The pattern of vascular supply is similar in dogs and humans.

Source: Reprinted with the permission of Simon & Schuster, Inc. from the Macmillan College text Animal Physiology 4/E by Malcolm S. Gordon. Copyright © 1968, 1972,1977, 1982 by Malcolm S. Gordon.

Despite their large number, capillaries contain only about 250 ml of blood at any time, out of a total blood volume of about 5,000 ml (most is contained in the venous system). The amount of blood flowing through a particular capillary bed is determined in part by the action of the **precapillary sphincter muscles** (fig. 13.24). These muscles allow only 5% to 10% of the capillary beds in skeletal muscles, for example, to be open at rest. Blood flow to an organ is regulated by the action of these precapillary sphincters and by the degree of resistance to blood flow (due to constriction or dilation) provided by the small arteries and arterioles in the organ.

Unlike the vessels of the arterial and venous systems, the walls of capillaries are composed of only one cell layer—a simple squamous epithelium, or endothelium (fig. 13.25). The absence of smooth muscle and connective tissue layers permits a more rapid rate of transport of materials between the blood and the tissues.

Types of Capillaries

Different organs have different types of capillaries, which are distinguished by significant differences in structure. In terms of their endothelial lining, these capillary types include those that are continuous, those that are discontinuous, and those that are fenestrated.

Continuous capillaries are those in which adjacent endothelial cells are closely joined together. These are found in muscles, lungs, adipose tissue, and in the central nervous system. The lack of intercellular channels in continuous capillaries in the CNS contributes to the blood-brain barrier (chapter 7). Continuous capillaries in other organs have narrow intercellular channels (from 40 to 45 Å in width), which allow the passage of molecules other than protein between the capillary blood and tissue fluid (fig. 13.25).

Examination of endothelial cells with an electron microscope has revealed the presence of pinocytotic vesicles (fig. 13.25), which suggests that the intracellular transport of material may occur across the capillary walls. This type of transport appears to be the only mechanism of capillary exchange available within the central nervous system and may account, in part, for the selective nature of the blood-brain barrier.

The kidneys, endocrine glands, and intestines have *fenestrated capillaries*, characterized by wide intercellular pores (800 to 1,000 Å) that are covered by a layer of mucoprotein which may serve as a diaphragm. In the bone marrow, liver, and spleen, the distance between endothelial cells is so great that these *discontinuous capillaries* appear as little cavities (*sinusoids*) in the organ.

Veins

Most of the total blood volume is contained in the venous system. Unlike arteries, which provide resistance to the flow of blood from the heart, veins are able to expand as they accumulate additional amounts of blood. The average pressure in the veins is only 2 mmHg, compared to a much higher average arterial

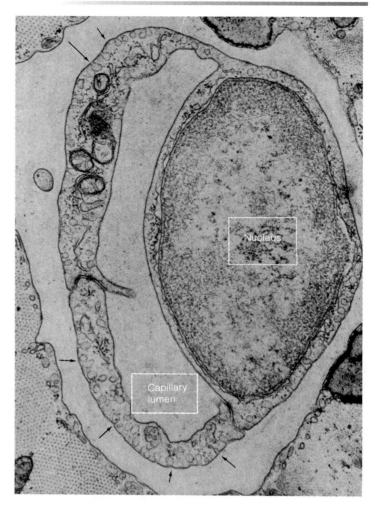

Figure 13.25

An electron micrograph of a capillary in the heart. Notice the thin intercellular channel (*middle left*) and the capillary wall, composed of only one cell layer. Arrows show some of the many pinocytotic vesicles.

pressure of about 100 mmHg. These values, expressed in millimeters of mercury, represent the hydrostatic pressure that the blood exerts on the walls of the vessels.

The low venous pressure is insufficient to return blood to the heart, particularly from the lower limbs. Veins, however, pass between skeletal muscle groups that produce a massaging action as they contract (fig. 13.26). As the veins are squeezed by contracting skeletal muscles, a one-way flow of blood to the heart is ensured by the presence of **venous valves.** The ability of these valves to prevent the flow of blood away from the heart was demonstrated in the seventeenth century by William Harvey (fig. 13.27). After applying a tourniquet to a subject's arm, Harvey found that he could push the blood in a bulging vein toward the heart but not in the reverse direction.

The effect of the massaging action of skeletal muscles on venous blood flow is often described as the **skeletal muscle**

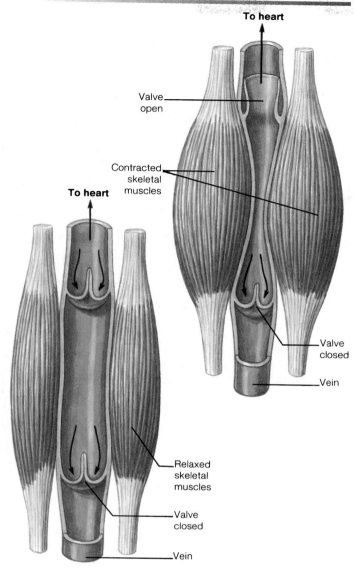

Figure 13.26

The action of the one-way venous valves. Contraction of skeletal muscles helps to pump blood toward the heart but is prevented from pushing blood away from the heart by closure of the venous valves.

The accumulation of blood in the veins of the legs over a long period of time, as may occur in people with occupations that require standing still all day, can cause the veins to stretch to the point where the venous valves are no longer efficient. This can produce **varicose veins**. During walking, the movements of the foot activate the soleus muscle pump. This effect can be produced in bedridden people by upward and downward manipulations of the feet.

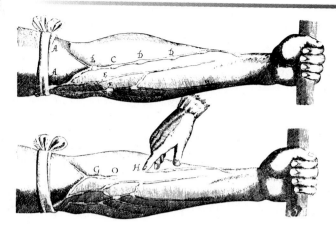

Figure 13.27

A classical demonstration by William Harvey of the existence of venous valves that prevent the flow of blood away from the heart.

Source: After William Harvey, *On the Motion of the Heart and Blood in Animals,* 1628.

pump. The rate of venous return to the heart is dependent, in large part, on the action of skeletal muscle pumps. When these pumps are less active, as when a person stands still or is bedridden, blood accumulates in the veins and causes them to bulge. When a person is more active, blood returns to the heart at a faster rate and less is left in the venous system.

Action of the skeletal muscle pumps aid the return of venous blood from the lower limbs to the large abdominal veins. Movement of venous blood from abdominal to thoracic veins, however, is aided by an additional mechanism—breathing. When a person inhales, the diaphragm—a muscular sheet separating the thoracic and abdominal cavities—contracts. As the diaphragm contracts, it changes from a dome-shaped to a more flattened form and thus protrudes more into the abdomen. This has the dual effect of increasing the pressure in the abdomen, thus squeezing the abdominal veins, and decreasing the pressure in the thoracic cavity. The pressure difference in the veins created by this inspiratory movement of the diaphragm forces blood into the thoracic veins that return the venous blood to the heart.

1. Describe the basic structural pattern of arteries and veins. Describe how arteries and veins differ in structure and how these differences contribute to their differences in function.

2. Describe the functional significance of the skeletal muscle pump and illustrate the action of venous valves.

3. Explain the functions of capillaries and describe the structural differences between capillaries in different organs.

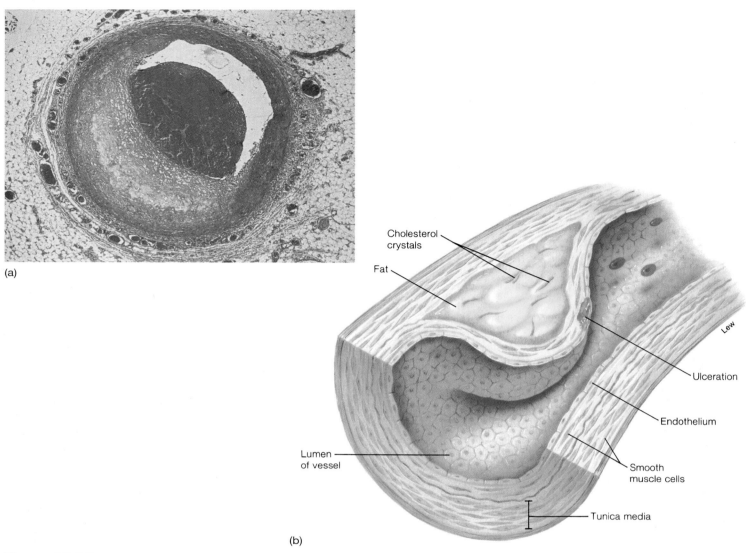

(a)

Cholesterol crystals

Fat

Lumen of vessel

Ulceration

Endothelium

Smooth muscle cells

Tunica media

(b)

Figure 13.28

(*a*) A photograph of the lumen (cavity) of a human coronary artery that is partially occluded by an atherosclerotic plaque and a thrombus. (*b*) A diagram of the structure of an atherosclerotic plaque.

Atherosclerosis and Cardiac Arrhythmias

Atherosclerosis is a disease process that can lead to obstruction of coronary blood flow. As a result, the electrical properties of the heart and its ability to function as a pump may be seriously compromised. Abnormal cardiac rhythms, or arrhythmias, can be detected by the abnormal electrocardiogram patterns they produce.

Atherosclerosis

Atherosclerosis is the most common form of arteriosclerosis (hardening of the arteries) and, through its contribution to heart disease and stroke, is responsible for about 50% of the deaths in the United States, Europe, and Japan. In atherosclerosis, localized **plaques,** or *atheromas,* protrude into the lumen of the artery and thus reduce blood flow. The atheromas additionally serve as sites for *thrombus* (blood clot) formation, which can further occlude the blood supply to an organ (fig. 13.28).

It is currently believed that the process of atherosclerosis begins as a result of damage, or "insult," to the endothelium. Such insults are produced by smoking, hypertension (high blood pressure), high blood cholesterol, and diabetes. The first anatomically recognized change is the appearance of "fatty streaks," which are gray-white areas that protrude into the lumen of arteries, particularly at arterial branch points. These are aggregations of lipid-filled macrophages and lymphocytes within the tunica intima. They are present to a small degree in the aorta and coronary arteries of children aged 10 to 14, but progress to more advanced stages at different rates in different people. In the intermediate stage, the area contains layers of macrophages and smooth muscle cells. The more advanced lesions are called fibrous plaques, and consist of a cap of connective tissue with smooth muscle cells over accumulated lipid and debris, macrophages that have been derived from monocytes (chapter 15), and lymphocytes.

The process may be instigated by damage to the endothelium, but its progression appears to be a result of a wide variety of cytokines and other autocrine regulators (chapter 11) secreted by the endothelium and by the other participating cells, including platelets, macrophages, and lymphocytes. Some of these regulators attract monocytes and lymphocytes to the damaged endothelium and cause them to penetrate into the tunica intima. The monocytes then become macrophages, engulf lipids, and take on the appearance of "foamy cells." Smooth muscle cells change from a contractile state to a "synthetic" state, in which they produce and secrete connective tissue matrix proteins. (This is unique; in other tissues, connective tissue matrix is secreted by cells called fibroblasts.) The changed smooth muscle cells respond to chemical attractants and migrate from the tunica media to the tunica intima, where they can proliferate.

Endothelial cells normally prevent this progression by presenting a physical barrier to the penetration of monocytes and lymphocytes and by secreting autocrine regulators. Hypertension, smoking, and high blood cholesterol, among other risk factors, interfere with this protective function. The role of cholesterol in this process is described in the next section.

Cholesterol and Plasma Lipoproteins

There is much evidence that high blood cholesterol is associated with an increased risk of atherosclerosis. This high blood cholesterol can be produced by a diet rich in cholesterol and saturated fat, or it may be the result of an inherited condition known as *familial hypercholesteremia*. This condition is inherited as a single dominant gene; individuals who inherit two of these genes have extremely high cholesterol concentrations (regardless of diet) and usually suffer heart attacks during childhood.

Lipids, including cholesterol, are carried in the blood attached to protein carriers (this topic is covered in detail in chapter 18). Cholesterol is carried to the arteries by plasma proteins called **low-density lipoproteins (LDL).** These particles, produced by the liver, consist of a core of cholesterol surrounded by a layer of phospholipids and a protein (to make the particle water-soluble). Cells in various organs contain receptors for the proteins in LDL; when LDL proteins attach to their receptors, the cell engulfs the LDL by receptor-mediated endocytosis (described in chapter 3) and utilizes the cholesterol for different purposes. Most of the LDL particles in the blood are removed in this way by the liver.

People who eat a diet high in cholesterol and saturated fat, and people with familial hypercholesteremia, have a high blood LDL concentration because their livers have a low number of LDL receptors. With fewer LDL receptors, the liver is less able to remove the LDL from the blood and thus more LDL is available to enter the endothelial cells of arteries.

When endothelial cells engulf LDL, they oxidize it to a product called *oxidized LDL*. Recent evidence suggests that oxidized LDL contributes to endothelial cell injury, migration of monocytes and lymphocytes into the tunica intima, conversion of monocytes into macrophages, and other events that occur in the progression of atherosclerosis.

Since oxidized LDL appears to be so important in the progression of atherosclerosis, it seems logical that antioxidant compounds may aid in the prevention or treatment of this condition. The antioxidant drug *probucol,* as well as vitamin C, vitamin E, and beta-carotene, which are antioxidants (chapter 19), have been shown to be effective in this regard in experimental animals. This research may lead to clinical trials of antioxidant therapy for atherosclerosis.

Excessive cholesterol may be released from cells and travel in the blood as **high-density lipoproteins (HDL),** which are removed by the liver. The cholesterol in HDL is not taken into the artery wall because these cells lack the membrane receptor required for endocytosis of the HDL particles, and therefore this cholesterol does not contribute to atherosclerosis. Indeed, a high proportion of cholesterol in HDL as compared to LDL is beneficial, since it indicates that cholesterol may be traveling away from the blood vessels to the liver. The concentration of HDL-cholesterol appears to be higher and the risk of atherosclerosis lower in people who exercise regularly. The HDL-cholesterol concentration, for example, is higher in marathon runners than in joggers and is higher in joggers than in sedentary individuals. Women in general have higher HDL-cholesterol concentrations and a lower risk of atherosclerosis than men.

Many people can significantly lower their blood cholesterol concentration through a regimen of exercise and diet. Since saturated fat in the diet raises blood cholesterol, such foods as fatty meat, egg yolks, and internal animal organs (liver, brain, etc.) should be eaten only sparingly. The American Heart Association recommends that fat contribute less than 30% to the total calories of a diet, and many experts argue for an even lower percentage. By way of comparison, the typical fast-food meal contains from 40% to 58% of its calories from fat. The single most effective action that smokers can take to lower their risk of atherosclerosis, however, is to stop smoking.

	Table 13.11 Changes in the Enzyme Activity in Plasma Following a Myocardial Infarction			
Serum Enzyme	Earliest Increase (hr)	Maximum Concentration (hr)	Return to Normal (days)	Amplitude of Increase (× normal)
Creatine kinase	3 to 6	24 to 36	3	7
Malate dehydrogenase	4 to 6	24 to 48	5	4
AST	6 to 8	24 to 48	4 to 6	5
Lactate dehydrogenase	10 to 12	48 to 72	11	3
α-Hydroxbutyrate dehydrogenase	10 to 12	48 to 72	13	3 to 4
Aldolase	6 to 8	24 to 48	4	4

Source: From Rex Montgomery, Robert L. Dryer, Thomas W. Conway, Arthur A. Spector, *Biochemistry: A Case-Oriented Approach*, 4th ed. Copyright 1983 Mosby/Yearbook. Used by permission.

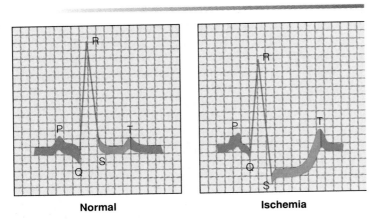

Figure 13.29

Depression of the S-T segment of the electrocardiogram as a result of myocardial ischemia.

Ischemic Heart Disease

A tissue is said to be **ischemic** when it receives an inadequate supply of oxygen because of an inadequate blood flow. The most common cause of myocardial ischemia is atherosclerosis of the coronary arteries. The adequacy of blood flow is relative—it depends on the metabolic requirements of the tissue for oxygen. An obstruction in a coronary artery, for example, may allow sufficient coronary blood flow at rest but not when the heart is stressed by exercise or emotional conditions. In these cases, the increased activity of the sympathoadrenal system causes increased heart rate and blood pressure, increasing the work of the heart and raising its oxygen requirements. Recent evidence also suggests that mental stress can cause constriction of atherosclerotic coronary arteries, leading to ischemia of the heart muscle. The vasoconstriction is believed to result from abnormal function of a damaged endothelium, which normally prevents constriction (through secretion of paracrine regulators) in response to mental stress. The control of vasoconstriction and vasodilation is covered in more detail in chapter 14.

Myocardial ischemia is associated with increased concentrations of blood lactic acid produced by anaerobic respiration of the ischemic tissue. This condition often causes substernal pain, which may also be referred to the left shoulder and arm,

as well as to other areas. This referred pain is called **angina pectoris.** People with angina frequently take nitroglycerin or related drugs that help to relieve the ischemia and pain. These drugs are effective because they produce vasodilation, which improves circulation to the heart and decreases the work that the ventricles must perform to eject blood into the arteries.

Myocardial cells are adapted to respire aerobically and cannot respire anaerobically for more than a few minutes. If ischemia and anaerobic respiration continue for more than a few minutes, *necrosis* (cellular death) may occur in the areas most deprived of oxygen. A sudden, irreversible injury of this kind is called a **myocardial infarction,** or **MI.** The lay term "heart attack," though imprecise, usually refers to a myocardial infarction.

Myocardial ischemia may be detected by changes in the S-T segment of the electrocardiogram (fig. 13.29). The diagnosis of myocardial infarction is aided by measurement of the concentration of enzymes in the blood that are released by the infarcted tissue. Plasma concentrations of *creatine phosphokinase* (*CPK*), for example, increase within 3 to 6 hours after the onset of symptoms and return to normal after 3 days. Plasma levels of *lactate dehydrogenase* (*LDH*) reach a peak within 48 to 72 hours after the onset of symptoms and remain elevated for about 11 days (table 13.11).

Arrythmias Detected by the Electrocardiograph

Arrhythmias, or abnormal heart rhythms, can be detected and described by the abnormal ECG tracings they produce. Although proper clinical interpretation of electrocardiograms requires knowledge of technical information not covered in this chapter, some knowledge of abnormal rhythms is interesting in itself and is useful in gaining an understanding of normal physiology.

Since a heartbeat occurs whenever a normal QRS complex is seen, and since the ECG chart paper moves at a known speed so that its x-axis indicates time, the cardiac rate (beats per minute) can be easily obtained from examination of the ECG recording. A cardiac rate slower than 60 beats per minute indicates **bradycardia;** a rate faster than 100 beats per minute is described as **tachycardia** (fig. 13.30).

Both bradycardia and tachycardia can occur normally. Endurance-trained athletes, for example, often have heart rates ranging from 40 to 60 beats per minute. This *athlete's bradycardia*

Chapter Thirteen

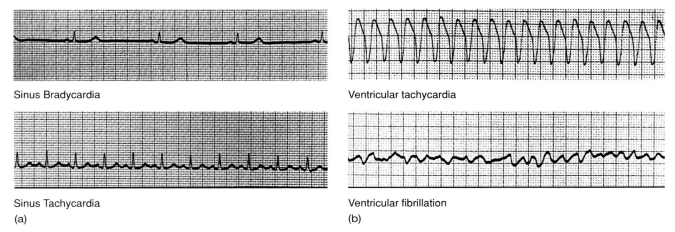

Sinus Bradycardia

Ventricular tachycardia

Sinus Tachycardia

Ventricular fibrillation

(a)

(b)

Figure 13.30

In (a) the heartbeat is paced by the normal pacemaker—the SA node (hence the name *sinus rhythm*). This can be abnormally slow (bradycardia—42 beats per minute in this example) or fast (tachycardia—125 beats per minute in this example). Compare the pattern of tachycardia in (a) with the tachycardia in (b). Ventricular tachycardia is produced by an ectopic pacemaker in the ventricles. This dangerous condition can quickly lead to ventricular fibrillation, also shown in (b).

occurs as a result of higher levels of parasympathetic inhibition of the SA node and is a beneficial adaptation. Activation of the sympathetic system, during exercise or emergencies ("fight or flight"), causes a normal tachycardia to occur.

Abnormal tachycardia occurs when a person is at rest. This may result from abnormally fast pacing by the atria due to drugs, or to the development of abnormally fast *ectopic pacemakers*—cells located outside the SA node that assume a pacemaker function. This abnormal atrial tachycardia thus differs from normal "sinus" (SA node) tachycardia. *Ventricular tachycardia* results when abnormally fast ectopic pacemakers in the ventricles cause them to beat rapidly and independently of the atria. This is very dangerous because it can quickly degenerate into a lethal condition known as *ventricular fibrillation*.

Flutter and Fibrillation

Extremely rapid rates of electrical excitation and contraction of either the atria or the ventricles may produce flutter or fibrillation. In **flutter,** the contractions are very rapid (200–300 per minute) but are coordinated. In **fibrillation,** contractions of different groups of myocardial fibers occur at different times so that a coordinated pumping action of the chambers is impossible.

Atrial flutter usually degenerates quickly into *atrial fibrillation*. This causes the pumping action of the atria to stop. Since the ventricles fill to about 80% of their end-diastolic volume before atrial contraction normally occurs, however, the heart is still able to eject a sufficient quantity of blood into the circulation. People who have atrial fibrillation can thus live for many years. People who have *ventricular fibrillation* (fig. 13.30), by contrast, can live for only a few minutes before the brain and heart—which are very dependent upon oxygen for their metabolism—cease to function.

Fibrillation is caused by a continuous recycling of electrical waves, known as **circus rhythms,** through the myocardium.

This recycling is normally prevented by the fact that the entire myocardium enters a refractory period (due to the long duration of action potentials, as previously discussed). If some cells emerge from their refractory period before others, however, electrical waves can be continuously regenerated and conducted. Recycling of electrical waves along continuously changing pathways produces uncoordinated contraction and an impotent pumping action.

Circus rhythms are thus produced whenever impulses can be conducted without interruption by nonrefractory tissue. This may occur when the conduction pathway is longer than normal, as in a dilated heart. It can also be produced by an electric shock delivered at the middle of the T wave, when different myocardial cells are in different stages of recovery from their refractory period. Finally, circus rhythms and fibrillation may be produced by damage to the myocardium, which slows the normal rate of impulse conduction.

Fibrillation can sometimes be stopped by a strong electric shock delivered to the chest. This procedure is called **electrical defibrillation.** The electric shock depolarizes all of the myocardial cells at the same time, causing them all to enter a refractory state. Conduction of circus rhythms thus stops, and the SA node can begin to stimulate contraction in a normal fashion. This does not correct the initial problem that caused circus rhythms and fibrillation, but it does keep the person alive long enough to take other corrective measures.

AV Node Block

The time interval between the beginning of atrial depolarization—indicated by the P wave—and the beginning of ventricular depolarization (as shown by the Q part of the QRS complex) is called the *P-R interval* (see fig. 13.19). In the normal heart, this time interval is 0.12 to 0.20 second in duration. Damage to the AV node causes slowing of impulse conduction and is reflected by changes in the P-R interval. This condition is known as *AV node block* (fig. 13.31).

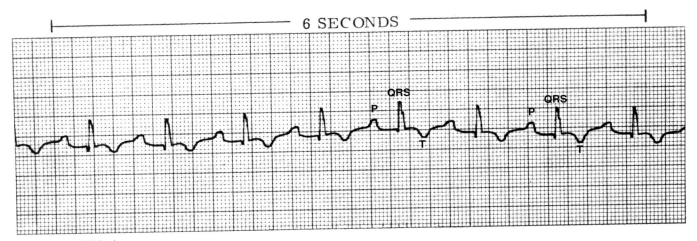

First-degree AV block

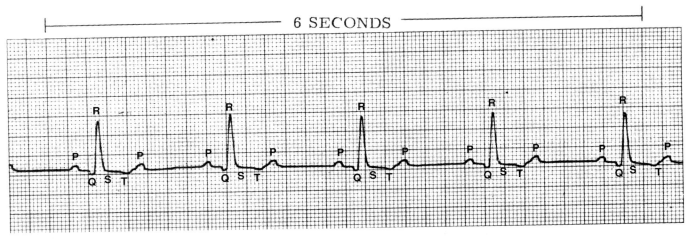

Second-degree AV block

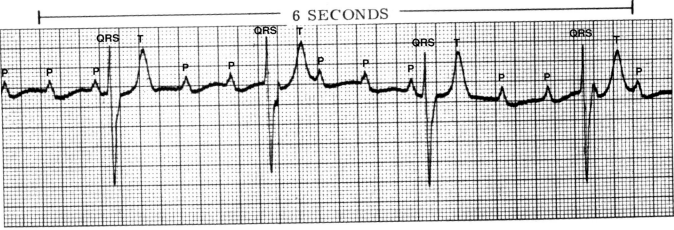

Third-degree AV block

Figure 13.31

Atrioventricular (AV) node block. In first-degree block, the P-R interval is greater than 0.20 second (in the example here, the P-R interval is 0.26–0.28 second). In second-degree block, P waves are seen that are not accompanied by QRS waves (in this example, the atria are beating 90 times per minute [as represented by the P waves], while the ventricles are beating 50 times per minute [as represented by the QRS waves]). In third-degree block, the ventricles are paced independently of the atria by an ectopic pacemaker. Ventricular depolarization (QRS) and repolarization (T) therefore have a variable position in the electrocardiogram relative to the P waves (atrial depolarization).

A number of abnormal conditions, including a blockage in conduction of the impulse along the bundle of His, require the insertion of an **artificial pacemaker.** This is a battery-powered device, about the size of a locket, which may be placed in permanent position under the skin. The electrodes from the pacemaker are guided by means of a fluoroscope through a vein to the right atrium, through the tricuspid valve, and into the right ventricle. The electrodes are fixed to the trabeculae carnae and are in contact with the wall of the ventricle. When these electrodes deliver shocks—either at a continuous pace or on demand (when the heart's own impulse doesn't arrive on time)—both ventricles are depolarized and contract and then repolarize and relax, just as they do in response to endogenous stimulation.

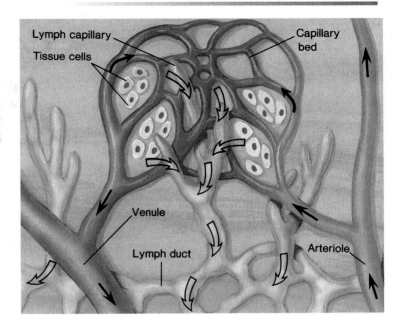

Figure 13.32

A diagram showing the structural relationship of a capillary bed and a lymph capillary.

First-degree AV node block occurs when the rate of impulse conduction through the AV node (as reflected by the P-R interval) exceeds 0.20 second. **Second-degree AV node block** occurs when the AV node is damaged so severely that only one out of every two, three, or four atrial electrical waves can pass through to the ventricles. This is indicated in an ECG by the presence of P waves without associated QRS waves.

In **third-degree,** or **complete, AV node block,** none of the atrial waves can pass through the AV node to the ventricles. The atria are paced by the SA node (follow a normal "sinus rhythm"), but the ventricles are paced by a different, ectopic pacemaker (usually located in the bundle of His or Purkinje fibers). Since the SA node is the normal pacemaker by virtue of the fact that it has the fastest cycle of electrical activity, the ectopic pacemaker in the ventricles causes the ventricles to beat at an abnormally slow rate. The bradycardia that results is usually corrected by insertion of an artificial pacemaker.

1. Explain how cholesterol is carried in the plasma and how the concentrations of cholesterol carriers are related to the risk of developing atherosclerosis.

2. Explain how angina pectoris is produced and discuss the significance of this symptom.

3. Define bradycardia and tachycardia, and give normal and pathological examples of each. Also, describe how flutter and fibrillation are produced.

4. Explain the effects of first-, second-, and third-degree AV node block on the electrocardiogram.

Lymphatic System

Lymphatic vessels absorb excess tissue fluid and transport this fluid—now called lymph—to ducts that drain into veins. Lymph nodes, and lymphoid tissue in the thymus, spleen, and tonsils produce lymphocytes, which are white blood cells involved in immunity.

The lymphatic system has three basic functions: (1) it transports interstitial (tissue) fluid, initially formed as a blood filtrate, back to the blood; (2) it transports absorbed fat from the small intestine to the blood; and (3) its cells—called *lymphocytes*—help to provide immunological defenses against disease-causing agents.

The smallest vessels of the lymphatic system are the **lymph capillaries** (fig. 13.32). Lymph capillaries are microscopic, closed-ended tubes that form vast networks in the intercellular spaces within most organs. Because the walls of lymph capillaries are composed of endothelial cells with porous junctions, interstitial fluid, proteins, microorganisms, and absorbed fat (in the intestine—chapter 17) can easily enter. Once fluid enters the lymphatic capillaries, it is referred to as **lymph.**

Figure 13.33

The schematic relationship of the circulatory and lymphatic systems. Lymphatic vessels transport lymph fluid from interstitial spaces to the venous bloodstream.

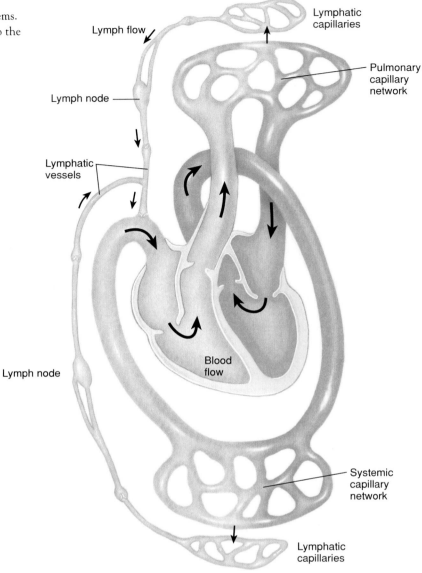

From merging lymph capillaries, the lymph is carried into larger **lymphatic vessels.** The walls of lymphatic vessels are similar to those of veins in that they have the same three layers and contain valves to prevent backflow. Fluid movement within these vessels occurs as a result of peristaltic waves of contraction (chapter 12). The lymphatic vessels eventually empty into one of two principal vessels: the *thoracic duct* or the *right lymphatic duct*. These ducts drain the lymph into the left and right subclavian veins, respectively. Thus interstitial fluid, which is formed by filtration of plasma out of blood capillaries (a process described in chapter 14), is ultimately returned back to the cardiovascular system (fig. 13.33).

Before the lymph is returned to the cardiovascular system, it is filtered through **lymph nodes** (fig. 13.34). Lymph nodes contain phagocytic cells, which help to remove pathogens, and *germinal centers,* which are the sites of lymphocyte production. The tonsils, thymus, and spleen—together called **lymphoid organs**—likewise contain germinal centers and are sites of lymphocyte production. Lymphocytes are the cells of the immune system that respond in a specific fashion to antigens, and their functions are described as part of the immune system in chapter 15.

1. Compare the composition of lymph and blood and describe the relationship between blood capillaries and lymphatic capillaries.

2. Explain how the lymphatic system and the cardiovascular system are related. How do these systems differ?

3. Describe the functions of lymph nodes and lymphoid organs.

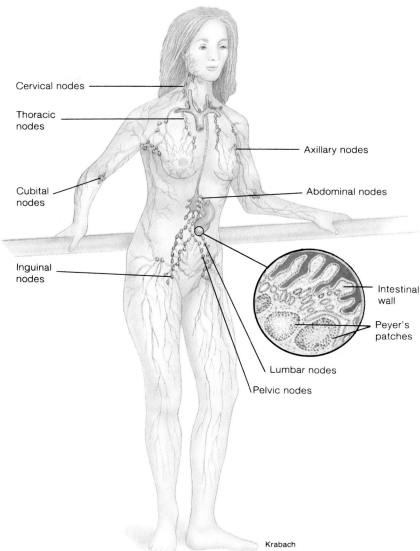

Figure 13.34
Primary locations of lymph nodes.

Summary

Functions and Components of the Circulatory System   p. 344

I.   The blood transports oxygen and nutrients to the tissue cells and removes waste products from the tissues. It also serves a regulatory function through its transport of hormones.

   A.   Oxygen is carried by red blood cells, or erythrocytes.

   B.   The white blood cells, or leukocytes, serve to protect the body from disease.

II. The circulatory system consists of the cardiovascular system (heart and blood vessels) and the lymphatic system.

Composition of the Blood   p. 345

I.   Plasma is the fluid part of the blood, containing dissolved ions and various organic molecules.

   A.   Hormones are found in the plasma portion of the blood.

   B.   Plasma proteins are divided into albumins and alpha, beta, and gamma globulins.

II. The formed elements of the blood include erythrocytes, leukocytes, and platelets.

   A.   Erythrocytes, or red blood cells, contain hemoglobin and transport oxygen.

   B.   Leukocytes may be granular (also called polymorphonuclear) or agranular. They function in immunity.

   C.   Thrombocytes, or platelets, are required for blood clotting.

III. Production of red blood cells is stimulated by the hormone erythropoietin, and development of different kinds of white blood cells is controlled by chemicals called lymphokines.

IV. The major blood typing groups are the ABO system and the Rh system.
   A. Blood type refers to the kind of antigens found on the red blood cells' surface.
   B. When different types of blood are mixed, antibodies against the red blood cell antigens cause the red blood cells to agglutinate.

V. When a blood vessel is damaged, platelets adhere to the exposed subendothelial collagen proteins.
   A. Platelets that stick to collagen undergo a release reaction in which they secrete ADP, serotonin, and thromboxane A_2.
   B. Serotonin and thromboxane A_2 cause vasoconstriction. ADP and thromboxane A_2 attract other platelets and cause them to stick to the growing mass of platelets that are stuck to the collagen in the broken vessel.

VI. In the formation of a blood clot, a soluble protein called fibrinogen is converted into insoluble threads of fibrin.
   A. This reaction is catalyzed by the enzyme thrombin.
   B. Thrombin is derived from its inactive precursor, called prothrombin, by either an intrinsic or an extrinsic pathway.
      1. The intrinsic pathway is the longer of the two and requires the activation of more clotting factors.
      2. The shorter extrinsic pathway is initiated by the secretion of tissue thromboplastin.
   C. The clotting sequence requires Ca^{++} and phospholipids present in the platelet cell membranes.

VII. Dissolution of the clot eventually occurs by the action of plasmin, which cleaves fibrin into split products.

Acid-Base Balance of the Blood p. 354

I. The normal pH of arterial blood is 7.40, with a range of 7.35 to 7.45.
   A. Carbonic acid is formed from carbon dioxide and contributes to the blood pH. It is referred to as a volatile acid because it can be eliminated in the exhaled breath.
   B. Nonvolatile acids, such as lactic acid and the ketone bodies, are buffered by bicarbonate.

II. The blood pH is maintained by a proper ratio of carbon dioxide to bicarbonate.
   A. The lungs maintain the correct carbon dioxide concentration. An increase in carbon dioxide, due to inadequate ventilation, produces respiratory acidosis, for example.
   B. The kidneys maintain the free-bicarbonate concentration. An abnormally low plasma bicarbonate concentration produces metabolic acidosis, for example.

Structure of the Heart p. 355

I. The right and left sides of the heart pump blood through the pulmonary and systemic circulations.
   A. The right ventricle pumps blood to the lungs; this blood then returns to the left atrium.
   B. The left ventricle pumps blood into the aorta and systemic arteries; this blood then returns to the right atrium.

II. The heart contains two pairs of one-way valves.
   A. The atrioventricular valves allow blood to go from the atria to the ventricles, but not in the reverse direction.
   B. The semilunar valves allow blood to leave the ventricles and enter the pulmonary and systemic circulations, but they prevent blood from returning from the arteries to the ventricles.

III. The electrical impulse begins in the sinoatrial node and spreads through both atria by electrical conduction from one myocardial cell to another.
   A. The impulse then excites the atrioventricular node, from which it is conducted by the bundle of His into the ventricles.
   B. The Purkinje fibers transmit the impulse into the ventricular muscle and cause it to contract.

Cardiac Cycle and Heart Sounds p. 358

I. The heart is a two-step pump; first the atria contract and then the ventricles contract.
   A. During diastole, first the atria and then the ventricles fill with blood.
   B. The ventricles are about 80% filled before the atria contract and add the final 20% to the end-diastolic volume.
   C. Contraction of the ventricles ejects about two-thirds of their blood, leaving about one-third as the end-systolic volume.

II. When the ventricles contract at systole, the pressure within them first rises sufficiently to close the AV valves and then rises sufficiently to open the semilunar valves.
   A. Blood is ejected from the ventricles until the pressure within them falls below the pressure in the arteries. At this point the semilunar valves close and the ventricles begin relaxation.
   B. When the pressure in the ventricles falls below the pressure in the atria, a phase of rapid filling of the ventricles occurs, followed by the final filling caused by contraction of the atria.

III. Closing of the AV valves produces the first heart sound, or "lub," at systole; closing of the semilunar valves produces the second heart sound, or "dub," at diastole. Abnormal valves can cause abnormal sounds called murmurs.

Electrical Activity of the Heart and the Electrocardiogram *p. 361*

I. In the normal heart, action potentials originate in the SA node as a result of spontaneous depolarization called the pacemaker potential.
   A. When this spontaneous depolarization reaches a threshold value, an action potential is produced due to opening of the voltage-regulated Na^+ gates and fast Ca^{++} channels.
   B. Repolarization is produced by the outward diffusion of K^+, but a stable resting membrane potential is not attained because spontaneous depolarization once again occurs.
   C. Other myocardial cells are capable of spontaneous activity, but the SA node is the normal pacemaker because its rate of spontaneous depolarization is the fastest.
   D. When the action potential produced by the SA node reaches other myocardial cells, they produce action potentials with a long plateau phase due to a slow inward diffusion of Ca^{++}.
   E. The long action potential and long refractory period of myocardial cells allows the entire mass of cells to be in a refractory period while it contracts. This prevents the myocardium from being stimulated again until after it relaxes.
II. The regular pattern of conduction in the heart produces a changing pattern of potential differences between two points on the body surface.
   A. The recording of this changing pattern caused by the heart's electrical activity is called an electrocardiogram (ECG).
   B. The P wave is caused by depolarization of the atria; the QRS wave is caused by depolarization of the ventricles; the T wave is produced by repolarization of the ventricles.

Blood Vessels *p. 367*

I. Arteries contain three layers, or tunics: the intima, media, and externa.
   A. The tunica intima consists of a layer of endothelium, which is separated from the tunica media by a band of elastin fibers.
   B. The tunica media consists of smooth muscle.
   C. The tuncia externa is the outermost layer.
   D. Large arteries have many layers of elastin and are elastic. Medium and small arteries and arterioles are less distensible and thus provide greater resistance to blood flow.
II. Capillaries are the narrowest but the most numerous of the blood vessels.
   A. Capillary walls consist of only a single layer of endothelial cells; they provide for the exchange of molecules between the blood and the surrounding tissues.
   B. The flow of blood from arterioles to capillaries is regulated by precapillary sphincter muscles.
   C. The capillary wall may be continuous, fenestrated, or discontinuous.
III. Veins have the same three tunics as arteries, but generally have a thinner muscular layer than comparably sized arteries.
   A. Veins are more distensible than arteries and can expand to hold a larger quantity of blood.
   B. Many veins have venous valves that permit a one-way flow of blood to the heart.
   C. The flow of blood back to the heart is aided by contraction of the skeletal muscles that surround veins. This action is called the skeletal muscle pump.

Atherosclerosis and Cardiac Arrhythmias *p. 372*

I. Atherosclerosis of arteries can occlude blood flow to the heart and brain and is a causative factor in up to 50% of all deaths in the United States, Europe, and Japan.
   A. Atherosclerosis begins with injury to the endothelium, the movement of monocytes and lymphocytes into the tunica intima, and the conversion of monocytes into macrophages that engulf lipids. Smooth muscle cells then proliferate and secrete extracellular matrix.
   B. Atherosclerosis is promoted by such risk factors as smoking, hypertension, and high plasma cholesterol concentration. Low-density lipoproteins (LDLs), which carry cholesterol into the artery wall, are oxidized by the endothelium and are a major contributor to atherosclerosis.
II. Occlusion of blood flow in the coronary arteries by atherosclerosis may produce ischemia of the heart muscle and angina pectoris, leading to myocardial infarction.
III. The ECG can be used to detect abnormal cardiac rates, abnormal conduction between the atria and ventricles, and other abnormal patterns of electrical conduction in the heart.

Lymphatic System *p. 377*

I. Lymphatic capillaries are blind-ended but highly permeable, and can drain excess tissue fluid into lymphatic vessels.
II. Lymph passes through lymph nodes and is returned by way of lymph ducts to the venous blood.

Clinical Investigation

A heart murmur is detected in a teenage girl who has sought medical care because of feelings of chronic fatigue. Her radial pulse is rapid and weak. An echocardiogram and coronary angiography reveal that she has a ventricular septal defect and mitral stenosis. Her ECG indicates sinus tachycardia, and laboratory findings indicate a very high plasma cholesterol concentration with a high LDL/HDL ratio. What can be concluded from these findings?

Clues:

Study the sections "Cardiac Cycle and Heart Sounds," "Arrhythmias Detected by the Electrocardiograph," and "Atherosclerosis."

Review Activities

Objective Questions

1. Which of the following statements is *false?*
   a. Most of the total blood volume is contained in veins.
   b. Capillaries have a greater total surface area than any other type of vessel.
   c. Exchanges between blood and tissue fluid occur across the walls of venules.
   d. Small arteries and arterioles present great resistance to blood flow.

2. All arteries in the body contain oxygen-rich blood with the exception of
   a. the aorta.
   b. the pulmonary artery.
   c. the renal artery.
   d. the coronary arteries.

3. The "lub," or first heart sound, is produced by closing of
   a. the aortic semilunar valve.
   b. the pulmonary semilunar valve.
   c. the tricuspid valve.
   d. the bicuspid valve.
   e. both AV valves.

4. The first heart sound is produced at
   a. the beginning of systole.
   b. the end of systole.
   c. the beginning of diastole.
   d. the end of diastole.

5. Changes in the cardiac rate primarily reflect changes in the duration of
   a. systole.
   b. diastole.

6. The QRS wave of an ECG is produced by
   a. depolarization of the atria.
   b. repolarization of the atria.
   c. depolarization of the ventricles.
   d. repolarization of the ventricles.

7. The second heart sound immediately follows the occurrence of
   a. the P wave.
   b. the QRS wave.
   c. the T wave.

8. The cells that normally have the fastest rate of spontaneous diastolic depolarization are located in
   a. the SA node.
   b. the AV node.
   c. the bundle of His.
   d. the Purkinje fibers.

9. Which of the following statements is *true?*
   a. The heart can produce a graded contraction.
   b. The heart can produce a sustained contraction.
   c. The action potentials produced at each cardiac cycle normally travel around the heart in circus rhythms.
   d. All of the myocardial cells in the ventricles are normally in a refractory period at the same time.

10. An ischemic injury to the heart that destroys myocardial cells is
    a. angina pectoris.
    b. a myocardial infarction.
    c. fibrillation.
    d. heart block.

11. The activation of factor X is
    a. part of the intrinsic pathway only.
    b. part of the extrinsic pathway only.
    c. part of both the intrinsic and extrinsic pathways.
    d. not part of either the intrinsic or extrinsic pathways.

12. Platelets
    a. form a plug by sticking to each other.
    b. release chemicals that stimulate vasoconstriction.
    c. provide phospholipids needed for the intrinsic pathway.
    d. serve all of the above functions.

13. Antibodies against both type A and type B antigens are found in the plasma of a person who is
    a. type A.
    b. type B.
    c. type AB.
    d. type O.
    e. any of the above types.

14. Production of which of the following blood cells is stimulated by a hormone secreted by the kidneys?
    a. lymphocytes
    b. monocytes
    c. erythrocytes
    d. neutrophils
    e. thrombocytes

15. Which of the following statements about plasmin is *true?*
    a. It is involved in the intrinsic clotting system.
    b. It is involved in the extrinsic clotting system.
    c. It functions in fibrinolysis.
    d. It promotes the formation of emboli.

16. During the phase of isovolumetric relaxation of the ventricles, the pressure in the ventricles is
    a. rising.
    b. falling.
    c. first rising, then falling.
    d. constant.

17. Peristaltic waves of contraction move fluid within which of the following vessels?
    a. arteries
    b. veins
    c. capillaries
    d. lymphatic vessels
    e. all of the above

18. Excessive diarrhea may cause
    a. respiratory acidosis.
    b. respiratory alkalosis.
    c. metabolic acidosis.
    d. metabolic alkalosis.

Essay Questions

1. Explain why the beat of the heart is automatic and why the SA node functions as the normal pacemaker.[1]

2. Compare the duration of the heart's contraction with the myocardial action potential and refractory period. Explain the significance of these relationships.

3. Describe the pressure changes that occur during the cardiac cycle and relate these changes to the occurrence of the heart sounds.

4. Describe the causes of the P, QRS, and T waves of an ECG and indicate when each of these waves occurs in the cardiac cycle. Explain why the first heart sound occurs immediately after the QRS wave and why the second sound occurs at the time of the T wave.

5. Can a defective valve be detected by an ECG? Can a partially damaged AV node be detected by auscultation (listening) with a stethoscope? Explain.

6. Explain how a cut in the skin initiates both the intrinsic and extrinsic clotting pathways. Which pathway finishes first? Why?

7. Distinguish between the respiratory and metabolic components of acid-base balance. What are some of the causes of acid-base disturbances?

8. Explain how aspirin, coumarin, EDTA, and heparin function as anticoagulants. Which of these are effective when added to a test tube? Which are not? Why?

9. How does blood move through arteries, capillaries, and veins? Explain how exercise would affect this movement.

10. Explain the processes involved in the development of atherosclerosis. How might antioxidants help reduce the progression of this disease? How might exercise help? What other changes in lifestyle might help prevent or reduce atherosclerotic plaques?

Selected Readings

Armitage, J. O. 1994. Bone marrow transplantation. *New England Journal of Medicine* 330:827.

Benditt, D. G. et al. 1988. Supraventricular tachycardias: Mechanisms and therapies. *Hospital Practice* 23:161.

Benditt, E. P. February 1977. The origin of atherosclerosis. *Scientific American*.

Bennet, J. S. 1992. Mechanisms of platelet adhesion and aggregation: An update. *Hospital Practice* 27:124.

Berne, R. M., and M. N. Levy. 1981. *Cardiovascular Physiology*. 4th ed. St. Louis: Mosby.

Boltz, M. A. April 1994. Identifying cardiac arrythmias. *Nursing*, p. 54.

Brown, M. S., and J. L. Goldstein. 1984. How LDL receptors influence cholesterol and atherosclerosis. *Scientific American*.

Broze, G. J., Jr. 1992. Why do hemophiliacs bleed? *Hospital Practice* 27:71.

Conover, M. B. 1980. *Understanding Electrocardiography*. 3d ed. St. Louis: Mosby.

Deitschy, J. 1990. LDL cholesterol: Its regulation and manipulation. *Hospital Practice* 25:67.

Del Zoppo, G. J., and L. A. Harker. 1984. Blood/vessel interaction in coronary disease. *Hospital Practice* 19:163.

Dubin, D. 1981. *Rapid Interpretation of EKG's*. 3d ed. Tampa: Cover Publishing Co.

Ellsworth, M. L. et al. 1994. Role of microvessels in oxygen supply to tissue. *News in Physiological Sciences* 9:119.

Fitzgerald, D., and R. Lazzara. 1988. Functional anatomy of the conducting system. *Hospital Practice* 23:81.

Fulkow, B., and E. Neill. 1971. *Circulation*. London: Oxford University Press.

Fuster, V. et al. 1992. The pathogenesis of coronary artery disease and the acute coronary symptoms. *New England Journal of Medicine* 326:242.

Garrard, J. M. 1988. Platelet aggregation: Cellular regionation and physiologic role. *Hospital Practice* 23:89.

Glasser, S. P., and R. G. Zoble. 1985. Management of cardiac arrhythmias. *Hospital Practice* 20:127.

Golde, D. W. December 1991. The stem cell. *Scientific American*.

Golde, D. W., and J. C. Gasson. July 1988. Hormones that stimulate the growth of blood cells. *Scientific American*.

Groopman, J. E., J. M. Molina, and D. T. Scadden. 1989. Hematopoietic growth factors: Biology and clinical applications. *New England Journal of Medicine* 321:1449.

Hainsworth, R. 1990. The importance of vascular capacitance in cardiovascular control. *News in Physiological Sciences* 5:250.

Harken, A. H. July 1993. Surgical treatment of cardiac arrhythmias. *Scientific American*.

Hills, D., and E. Braunwald. 1977. Myocardial ischemia. *New England Journal of Medicine* 296: first part, p. 971; second part, p. 1033; third part, p. 1093.

Hoyer, L. W. 1994. Hemophilia A. *New England Journal of Medicine* 330:38.

Jagannath, S. et al. 1993. Hematopoietic stem cell transplantation. *Hospital Practice* 28:79.

Jennings, R. B., and K. A. Reimer. 1991. The cell biology of acute myocardial ischemia. *Annual Review of Medicine* 42:225.

Jongsma, H. J., and D. Gros. 1991. The cardiac connection. *News in Physiological Sciences* 6:43.

[1] *Note:* This question is answered on page 154 of the Student Study Guide.

Klatsky, A. L. 1995. Cardiovascular effects of alcohol. *Science and Medicine* 2:28.

Knopp, R. H. 1988. New approaches to cholesterol lowering: Efficacy and safety. *Hospital Practice, Supplement 1* 23:22.

Koury, M. J., and M. C. Bondurant. 1993. Prevention of programmed death in hematopoietic progenitor cells by hematopoietic growth factors. *News in Physiological Sciences* 8:170.

Kurachi, K. et al. 1992. Deficiencies in factors IX and VIII: What is now known. *Hospital Practice* 27:41.

Lawn, R. M. June 1992. Lipoprotein (a) in heart disease. *Scientific American*.

Lawn, R. M., and G. A. Vehar. March 1986. The molecular genetics of hemophilia. *Scientific American*.

Leon, A. S. 1983. Exercise and coronary heart disease. *Hospital Practice* 18:38.

Libby, P. 1992. Do vascular wall cytokines promote atherosclerosis? *Hospital Practice* 27:51.

Little, R. C. 1981. *Physiology of the Heart and Circulation*. 2d ed. Chicago: Year Book Medical Publishers.

Metcalf, D. 1989. The molecular control of cell division, differentiation commitment and maturation of haemopoietic cells. *Nature* 339:27.

Metcalf, D. 1994. Thrombopoietin—at last. *Nature* 369: 519.

Porzig, H. 1991. Signaling mechanisms in erythropoiesis: New insights. *News in Physiological Sciences* 6:247.

Roberts, H. R., and J. N. Lozier. 1992. New perspectives on the coagulation cascade. *Hospital Practice* 27:97.

Robinson, T. F., S. M. Factor, and E. H. Sonnenblick. June 1986. The heart as a suction pump. *Scientific American*.

Roddie, I. C. 1990. Lymph transport mechanisms in peripheral lymphatics. *News in Physiological Sciences* 5:85.

Rodgers, G. P., C. T. Noguchi, and A. N. Schechter. 1994. Sickle cell anemia. *Science and Medicine* 1:48.

Rosenberg, R. D., and K. A. Bauer. 1986. New insights into hypercoaguable states. *Hospital Practice* 21:131.

Rosenshtraukh, L. V., and A. V. Zaitsev. 1990. Atrial tachycardias: A new look. *News in Physiological Sciences* 5:187.

Ross, R. 1993. The pathogenesis of atherosclerosis: A perspective for the 1990's. *Nature* 362:801.

Ross, R. 1995. Cell biology of atherosclerosis. *Annual Review of Physiology* 57:791.

Schmid-Schonbein, G. W., and B. W. Zweifach. 1994. Fluid pump mechanisms in initial lymphatics. *News in Physiological Sciences* 9:67.

Shen, W. K., and H. C. Strauss. 1988. Mechanisms of bradyarrythmias and blocks. *Hospital Practice* 23:93.

Smith, J. J., and J. P. Kampine. 1980. *Circulatory Physiology: The Essentials*. Baltimore: Williams & Wilkins.

Spear, J. F., and E. N. Moore. 1982. Mechanisms of cardiac arrhythmias. *Annual Review of Physiology* 44:485.

Stein, B. et al. 1988. Pathogenesis of coronary occlusion. *Hospital Practice* 23:87.

Weber, K. T., C. G. Brilla, and J. S. Janicki. 1991. Myocardial remodeling and pathological hypertrophy. *Hospital Practice* 26:73.

Wood, J. E. January 1968. The venous system. *Scientific American*.

Woolf, N., and M. J. Davies. 1994. Arterial plaque and thrombus formation. *Science and Medicine* 1:38.

Yamamoto, F. et al. 1990. Molecular genetic basis of the histo-blood group ABO system. *Nature* 345:229.

Young, A. C. et al. 1991. The effect of atherosclerosis on the vasomotor response of coronary arteries to mental stress. *New England Journal of Medicine* 325:1551.

Zucker, M. B. June 1980. The function of blood platelets. *Scientific American*.

Explorations CD-ROM

The module accompanying chapter 13 is #5 Evolution of the Human Heart.

Life Science Animations

The animations that relate to chapter 13 are #32 The Cardiac Cycle and Productions of Sounds, #37 Blood Circulation, #38 Production of Electrocardiogram, #39 Common Congenital Defects of the Heart, and #40 A, B, O Blood Types.

14
Cardiac Output, Blood Flow, and Blood Pressure

OBJECTIVES
After studying this chapter, you should be able to . . .

1. define *cardiac output* and explain how cardiac rate and stroke volume affect the cardiac output.

2. explain how autonomic nerves regulate the cardiac rate and the strength of ventricular contraction.

3. explain the intrinsic regulation of stroke volume (the Frank–Starling law of the heart).

4. list the factors that affect the venous return of blood to the heart.

5. explain how tissue fluid is formed and how it is returned to the capillary blood.

6. explain how edema may be produced.

7. explain how antidiuretic hormone helps to regulate the blood volume, plasma osmolality, and the blood pressure.

8. explain the role of aldosterone in the regulation of blood volume and pressure.

9. describe the renin-angiotensin system and discuss its significance in cardiovascular regulation.

10. use Poiseuille's law to explain how blood flow is regulated.

11. define total *peripheral resistance* and explain how vascular resistance is regulated by extrinsic control mechanisms.

12. describe the functions of nitric oxide and endothelin-1 in the paracrine regulation of blood flow.

13. describe the intrinsic mechanisms involved in the autoregulation of blood flow.

14. explain the mechanisms by which blood flow to the heart and skeletal muscles is regulated.

15. describe the changes that occur in the cardiac output and in the distribution of blood flow during exercise.

16. describe the cutaneous circulation and explain how circulation in the skin is regulated.

17. list the factors that regulate the arterial blood pressure.

18. describe the baroreceptor reflex and explain its significance in blood pressure regulation.

19. explain how the sounds of Korotkoff are produced and how these sounds are used to measure blood pressure.

20. describe how the pulse pressure and mean arterial pressure are calculated and explain the significance of these measurements.

21. explain the mechanisms that contribute to and that help compensate for the conditions of hypertension, circulatory shock, and congestive heart failure.

OUTLINE

Cardiac Output

The pumping ability of the heart is a function of the beats per minute (cardiac rate) and the volume of blood ejected per beat (stroke volume). The cardiac rate and stroke volume are regulated by autonomic nerves and by mechanisms that are intrinsic to the cardiovascular system.

The **cardiac output** is equal to the volume of blood pumped per minute by each ventricle. The average resting **cardiac rate** in an adult is 70 beats per minute; the average **stroke volume** (volume of blood pumped per beat by each ventricle) is 70 to 80 ml per beat. The product of these two variables gives an average cardiac output of 5,500 ml (5.5 L) per minute:

Cardiac output = stroke volume × cardiac rate
(ml/min)   (ml/beat)   (beats/min)

The **total blood volume** also averages about 5.5 L. This means that each ventricle pumps the equivalent of the total blood volume each minute under resting conditions. Put another way, it takes about a minute for a drop of blood to complete the systemic and pulmonary circuits. An increase in cardiac output, as occurs during exercise, must thus be accompanied by an increased rate of blood flow through the circulation. This is accomplished by factors that regulate the cardiac rate and stroke volume.

Regulation of Cardiac Rate

In the complete absence of neural influences, the heart will continue to beat according to the rhythm set by the SA node. This automatic rhythm is produced by the spontaneous decay of the resting membrane potential to a threshold depolarization, at which point voltage-regulated membrane gates are opened and action potentials are produced. As described in chapter 13, Ca^{++} enters the myocardial cytoplasm during the action potential, attaches to troponin, and causes contraction.

Normally, however, sympathetic and vagus (parasympathetic) nerve fibers to the heart are continuously active and modify the rate of spontaneous depolarization of the SA node. Norepinephrine, released primarily by sympathetic nerve endings, and epinephrine, secreted by the adrenal medulla, stimulate an increase in the spontaneous rate of firing of the SA node. Acetylcholine, released from parasympathetic endings, hyperpolarizes the SA node and thus decreases the rate of its spontaneous firing (fig. 14.1). The actual pace set by the SA node at any time depends on the net effect of these antagonistic influences. Mechanisms that affect the cardiac rate are said to have a **chronotropic effect** (*chrono* = time). Those that increase cardiac rate have a positive chronotropic effect; those that decrease the rate have a negative chronotropic effect.

Autonomic innervation of the SA node represents the major means by which cardiac rate is regulated. However, other

Table 14.1	Effects of Autonomic Nerve Activity on the Heart	
Region Affected	**Sympathetic Nerve Effects**	**Parasympathetic Nerve Effects**
SA node	Increased rate of diastolic depolarization; increased cardiac rate	Decreased rate of diastolic depolarization; decreased cardiac rate
AV node	Increased conduction rate	Decreased conduction rate
Atrial muscle	Increased strength of contraction	Decreased strength of contraction
Ventricular muscle	Increased strength of contraction	No significant effect

autonomic control mechanisms also affect cardiac rate to a lesser degree. Sympathetic endings in the musculature of the atria and ventricles increase the strength of contraction and cause a slight decrease in the time spent in systole when the cardiac rate is high (table 14.1).

During exercise, the cardiac rate increases as a result of decreased vagus nerve inhibition of the SA node. Further increases in cardiac rate are achieved by increased sympathetic nerve stimulation. The resting bradycardia (slow heart rate) of endurance-trained athletes is due largely to high vagus nerve activity. The activity of the autonomic innervation of the heart is coordinated by **cardiac control centers** in the medulla oblongata of the brain stem. The question of whether there are separate cardioaccelerator and cardioinhibitory centers in the medulla is currently controversial. These cardiac control centers, in turn, are affected by higher brain areas and by sensory feedback from pressure receptors, or *baroreceptors*, in the aorta and carotid arteries. In this way, a rise in blood pressure can produce a reflex slowing of the heart. The *baroreceptor reflex* is discussed in more detail in relation to blood pressure regulation later in this chapter.

Regulation of Stroke Volume

The stroke volume is regulated by three variables: (1) the **end-diastolic volume (EDV)**, which is the volume of blood in the ventricles at the end of diastole; (2) the **total peripheral resistance**, which is the frictional resistance, or impedance, to blood flow in the arteries; and (3) the **contractility**, or strength, of ventricular contraction.

The end-diastolic volume is the amount of blood in the ventricles just prior to contraction. This is a workload imposed on the ventricles prior to contraction, and thus is sometimes called a *preload*. The stroke volume is directly proportional to the preload; an increase in EDV results in an increase in stroke volume. The stroke volume is also directly proportional to contractility; when the ventricles contract more forcefully, they pump more blood.

In order to eject blood, the pressure generated in a ventricle when it contracts must be greater than the pressure in the arteries (since blood only flows from a location of higher pressure

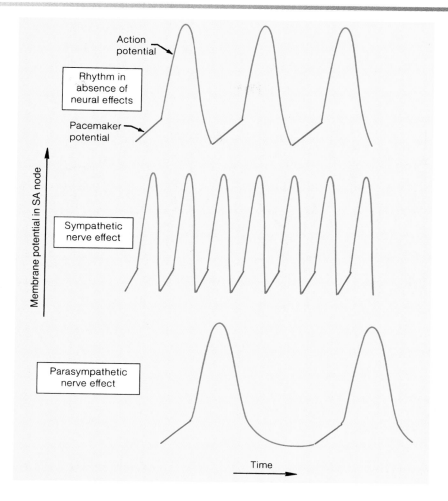

Figure 14.1

The rhythm set by the pacemaker potentials in the SA node. Sympathetic nerve effects increase the rate of spontaneous depolarization, thus influencing the rate at which action potentials are produced.

to one of lower pressure). The pressure in the arterial system before the ventricle contracts is, in turn, a function of the total peripheral resistance—the higher the peripheral resistance, the higher the pressure. As blood begins to be ejected from the ventricle, the added volume of blood in the arteries causes a rise in mean arterial pressure against the "bottleneck" presented by the peripheral resistance; ejection of blood stops shortly after the aortic pressure becomes equal to the intraventricular pressure. The total peripheral resistance thus presents an impedance to the ejection of blood from the ventricle, or an *afterload* imposed on the ventricle after contraction has begun.

In summary, the stroke volume is inversely proportional to the total peripheral resistance; the greater the peripheral resistance, the lower the stroke volume. It should be noted that this lowering of stroke volume in response to a raised peripheral resistance occurs for only a few beats. Thereafter, a healthy heart is able to compensate for the increased peripheral resistance by beating more strongly. This compensation occurs by means of a mechanism called the Frank–Starling law, to be described shortly.

The proportion of the end-diastolic volume that is ejected against a given afterload depends upon the strength of ventricular contraction. Normally, contraction strength is sufficient to eject 70 to 80 ml of blood out of a total end-diastolic volume of 110 to 130 ml. The *ejection fraction* is thus about 60%. More blood is pumped per beat as the EDV increases, and thus the ejection fraction remains relatively constant over a range of end-diastolic volumes. In order for this to be true, the strength of ventricular contraction must increase as the end-diastolic volume increases.

Frank–Starling Law of the Heart

Two physiologists, Otto Frank and Ernest Starling, demonstrated that the strength of ventricular contraction varies directly with the end-diastolic volume. Even in experiments where the heart is removed from the body (and is thus not subject to neural or hormonal regulation) and where the heart is filled with blood flowing from a reservoir, an increase in EDV within the physiological range results in increased contraction strength and, therefore, in increased stroke volume. This relationship between

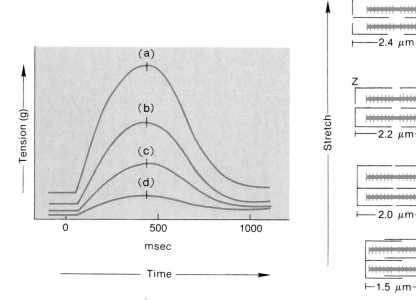

Resting sarcomere lengths

Figure 14.2

The Frank–Starling mechanism (law of the heart). When the heart muscle is subjected to an increasing degree of stretch, it contracts more forcefully. As a result of the increased contraction strength (shown as tension), the time required to reach maximum contraction remains constant, regardless of the degree of stretch.

EDV, contraction strength, and stroke volume is thus a built-in, or *intrinsic*, property of heart muscle, and is known as the **Frank–Starling law of the heart.**

Intrinsic Control of Contraction Strength

The intrinsic control of contraction strength and stroke volume is due to variations in the degree to which the myocardium is stretched by the end-diastolic volume. As the EDV rises within the physiological range, the myocardium is increasingly stretched and, as a result, contracts more forcefully.

Stretch can also increase the contraction strength of skeletal muscles. The resting length of skeletal muscles, however, is close to ideal, so that significant stretching decreases contraction strength. This is not true of the heart. Prior to filling with blood during diastole, the sarcomere lengths of myocardial cells are only about 1.5 μm. At this length, the actin filaments from each side overlap in the middle of the sarcomeres, and the cells can contract only weakly (fig. 14.2).

As the ventricles fill with blood, the myocardium stretches so that the actin filaments overlap with myosin only at the edges of the A bands (fig. 14.2). This allows more force to be developed during contraction. Since this more advantageous overlapping of actin and myosin is produced by stretching of the ventricles, and since the degree of stretching is controlled by the degree of filling (the end-diastolic volume), the strength of contraction is intrinsically adjusted by the end-diastolic volume.

The Frank–Starling law explains how the heart can adjust to a rise in total peripheral resistance: (1) a rise in peripheral resistance causes a decrease in the stroke volume of the ventricle, so that (2) more blood remains in the ventricle and the end-diastolic volume is greater for the next cycle; as a result, (3) the ventricle is stretched to a greater degree in the next cycle and contracts more strongly to eject more blood. This allows a healthy ventricle to sustain a normal cardiac output. A very important consequence of these events is that the cardiac output of the left ventricle, which pumps blood into the ever-changing resistances of the systemic circulation, can be adjusted to match the output of the right ventricle (which pumps blood into the pulmonary circulation). Clearly, the rate of blood flow through the pulmonary and systemic circulations must be equal in order to prevent fluid accumulation in the lungs and to deliver fully oxygenated blood to the body.

Extrinsic Control of Contractility

The *contractility* is the contraction strength at any given fiber length. At any given degree of stretch, the strength of ventricular contraction depends on the activity of the sympathoadrenal system. Norepinephrine from sympathetic endings and epinephrine from the adrenal medulla produce an increase in contraction strength. This **positive inotropic effect** is believed to result from an increase in the amount of Ca^{++} available to the sarcomeres.

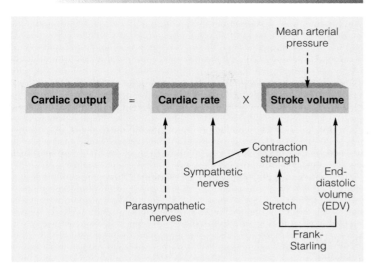

Figure 14.3

The regulation of cardiac output. Factors that stimulate cardiac output are shown as solid arrows; factors that inhibit cardiac output are shown as dashed arrows.

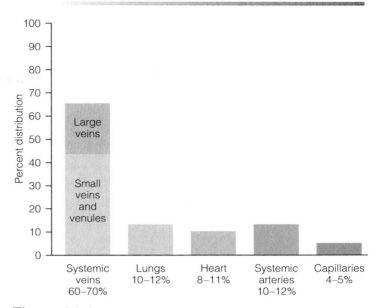

Figure 14.4

The distribution of blood within the circulatory system at rest.

The cardiac output is thus affected in two ways by the activity of the sympathoadrenal system: (1) through a positive inotropic effect on contractility and (2) through a positive chronotropic effect on cardiac rate (fig. 14.3). Stimulation through the parasympathetic nerve to the heart has a negative chronotropic effect but does not directly affect the contraction strength of the ventricles.

Venous Return

The end-diastolic volume—and thus the stroke volume and cardiac output—is controlled by factors that affect the *venous return*, which is the return of blood via veins to the heart. The rate at which the atria and ventricles are filled with venous blood depends on the total blood volume and the venous pressure (pressure in the veins), which serves as the driving force for the return of blood to the heart.

Veins have thinner, less muscular walls than do arteries and thus have a higher *compliance*. This means that a given amount of pressure will cause more distension (expansion) in veins than in arteries, so that the veins can hold more blood. Approximately two-thirds of the total blood volume is located in the veins (fig. 14.4). Veins are therefore called *capacitance vessels*, after capacitors in electronics, which can accumulate electrical charges. Muscular arteries and arterioles expand less under pressure (are less compliant), and thus are called *resistance vessels*.

Although veins contain almost 70% of the total blood volume, the mean venous pressure is only 2 mmHg, compared to a mean arterial pressure of 90 to 100 mmHg. The lower venous pressure is due in part to a pressure drop between arteries and capillaries and in part to the high venous compliance.

The venous pressure is highest in the venules (10 mmHg) and lowest at the junction of the venae cavae with the right atrium (0 mmHg). In addition to this pressure difference, the venous return to the heart is aided by (1) sympathetic nerve activity, which stimulates smooth muscle contraction in the venous walls and thus reduces compliance; (2) the skeletal muscle pump, which squeezes veins during muscle contraction; and (3) the pressure difference between the thoracic and abdominal cavities, which promotes the flow of venous blood back to the heart.

Contraction of the skeletal muscles functions as a "pump" by virtue of its squeezing action on veins (described in chapter 13). Contraction of the diaphragm during inhalation also improves venous return. As the diaphragm contracts, it lowers to increase the thoracic volume and decrease the abdominal volume. This creates a partial vacuum in the thoracic cavity and a higher pressure in the abdominal cavity. The pressure difference thus produced favors blood flow from abdominal to thoracic veins (fig. 14.5).

---

1. Describe how the stroke volume is intrinsically regulated by the end-diastolic volume. Why is this regulation significant?

2. Describe the effects of autonomic nerve stimulation on the cardiac rate and stroke volume.

3. Define the terms *preload* and *afterload* and explain how these factors affect the cardiac output.

4. List the factors that affect venous return. Using a flowchart, show how an increased venous return can result in an increased cardiac output.

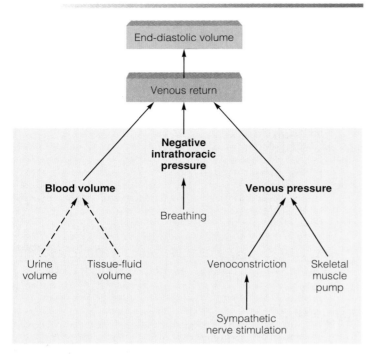

Figure 14.5

Variables that affect venous return and thus end-diastolic volume. Direct relationships are indicated by solid arrows; inverse relationships are shown with dashed arrows.

Blood Volume

Fluid in the extracellular environment of the body is distributed between the blood and the tissue fluid compartments by filtration and osmotic forces acting across the walls of capillaries. The function of the kidneys influences blood volume because urine is derived from blood plasma. Through their actions on the kidneys, ADH and aldosterone help to regulate the blood volume.

Blood volume represents one part, or compartment, of the total body water. Approximately two-thirds of the total body water is contained within cells—in the intracellular compartment. The remaining one-third is in the **extracellular compartment.** This extracellular fluid is normally distributed so that about 80% is contained in the tissues—as *tissue* or *interstitial fluid*—with the blood plasma accounting for the remaining 20% (fig. 14.6).

The distribution of water between the intracellular and extracellular fluid compartments is determined by a balance between opposing forces acting at the capillaries. Blood pressure, for example, promotes the formation of tissue fluid from plasma, whereas osmotic forces draw water from the tissues into the vascular system. The total volume of intracellular and extracellular

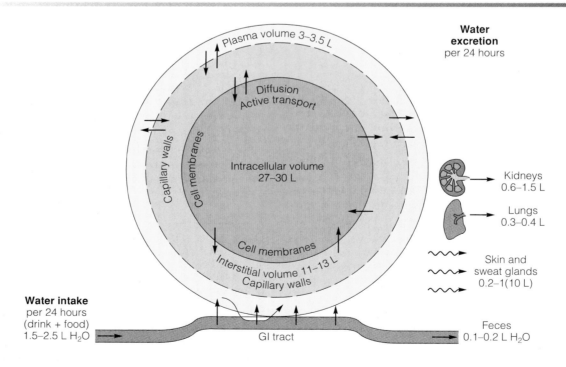

Figure 14.6

The distribution of body water between the intracellular and extracellular compartments.

fluid is normally maintained constant by a balance between water loss and water gain. Mechanisms that affect drinking, urine volume, and the distribution of water between plasma and tissue fluid thus help to regulate blood volume and, by this means, help to regulate cardiac output and blood flow.

Exchange of Fluid Between Capillaries and Tissues

The distribution of extracellular fluid between the plasma and interstitial compartments is in a state of dynamic equilibrium. Tissue fluid is not normally a "stagnant pond" but rather a continuously circulating medium, formed from and returning to the vascular system. In this way, the tissue cells receive a continuously fresh supply of glucose and other plasma solutes that are filtered through tiny endothelial channels in the capillary walls.

Filtration results from the blood pressure within the capillaries. This hydrostatic pressure, which is exerted against the inner capillary wall, is equal to about 37 mmHg at the arteriolar end of systemic capillaries and drops to about 17 mmHg at the venular end of the capillaries. The **net filtration pressure** is equal to the hydrostatic pressure of the blood in the capillaries minus the hydrostatic pressure of tissue fluid outside the capillaries, which opposes filtration. If, as an extreme example, these two values were equal, there would be no filtration. The magnitude of the tissue hydrostatic pressure varies from organ to organ. With a hydrostatic pressure in the tissue fluid of 1 mmHg, as it is outside the capillaries of skeletal muscles, the net filtration pressure would be 37 − 1 = 36 mmHg at the arteriolar end of the capillary and 17 − 1 = 16 mmHg at the venular end.

Glucose, comparably sized organic molecules, inorganic salts, and ions are filtered along with water through the capillary channels. The concentrations of these substances in tissue fluid are thus the same as in plasma. The protein concentration of tissue fluid (2 g/100 ml), however, is less than the protein concentration of plasma (6 to 8 g/100 ml). This difference is due to the restricted filtration of proteins by the capillary pores. The osmotic pressure exerted by plasma proteins—called the **colloid osmotic pressure** of the plasma (because proteins are present as a colloidal suspension)—is therefore much greater than the colloid osmotic pressure of tissue fluid. The difference between these two pressures is called the **oncotic pressure.** Since the colloid osmotic pressure of the tissue fluid is sufficiently low to be neglected, the oncotic pressure is essentially equal to the colloid osmotic pressure of the plasma. This value has been estimated to be 25 mmHg. Since water will move by osmosis from the solution of lower to the solution of higher osmotic pressure (chapter 6), this oncotic pressure favors the movement of water into the capillaries.

Whether fluid will move out of or into the capillary depends on the magnitude of the net filtration pressure, which varies from the arteriolar to the venular end of the capillary, and on the oncotic pressure. These opposing forces that affect

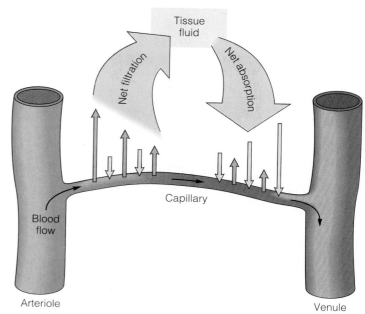

Figure 14.7

Tissue, or interstitial, fluid is formed by filtration (*orange arrows*) as a result of blood pressures at the arteriolar ends of capillaries and is returned to venular ends of capillaries by the colloid osmotic pressure of plasma proteins (*yellow arrows*).

the distribution of fluid across the capillary are known as **Starling forces,** and their effects can be calculated according to the following equation:

Fluid movement is proportional to:

$$(P_c + \pi_i) - (P_i + \pi_p)$$

where

P_c = hydrostatic pressure in the capillary
π_i = colloid osmotic pressure of the interstitial (tissue) fluid
P_i = hydrostatic pressure of interstitial fluid
π_p = colloid osmotic pressure of the blood plasma

The expression to the left of the minus sign represents the sum of forces acting to move fluid out of the capillary. The expression to the right represents the sum of forces acting to move fluid into the capillary. In the capillaries of skeletal muscles, the values are as follows: At the arteriolar end of the capillary, (37 + 0) − (1 + 25) = 11 mmHg; at the venular end of the capillary, (17 + 0) − (1 + 25) = −9 mmHg. The positive value at the arteriolar end indicates that the force favoring the extrusion of fluid from the capillary predominates. The negative value at the venular end indicates that the net Starling forces favor the return of fluid to the capillary. Fluid thus leaves the capillaries at the arteriolar end and returns to the capillaries at the venular end (fig. 14.7).

Figure 14.8

Parasitic larvae that block lymphatic drainage produce tissue edema and the tremendous enlargement of the limbs and external genitalia in elephantiasis.

In the tropical disease *filariasis,* mosquitoes transmit a nematode worm parasite to humans. The larvae of these worms invade lymphatic vessels and block lymphatic drainage. The edema that results can be so severe that the tissues swell to produce an elephant-like appearance, with thickening and cracking of the skin. This condition is thus aptly named **elephantiasis** (fig. 14.8).

This "classic" view of capillary dynamics has been challenged by some investigators who believe that, when capillaries are open, the net filtration force exceeds the force for the osmotic return of water throughout the length of the capillary. They believe that the opposite is true in closed capillaries (capillaries can be opened or closed by the action of precapillary sphincter muscles). Net filtration, in summary, would occur in open capillaries, whereas net absorption of water would occur in closed capillaries.

By either proposed mechanism, plasma and tissue fluid are continuously interchanged. The return of fluid to the vascular system at the venular ends of the capillaries, however, does not exactly equal the amount filtered at the arteriolar ends. According to some estimates, approximately 85% of the capillary filtrate is returned directly to the capillaries; the remaining 15% (amounting to at least 2 L per day) is returned to the vascular system by way of the lymphatic system. Lymphatic capillaries, it may be recalled from chapter 13, drain excess tissue fluid and proteins and, by way of lymphatic vessels, ultimately return this fluid to the venous system.

Causes of Edema

Excessive accumulation of tissue fluid is known as **edema.** This condition is normally prevented by a proper balance between capillary filtration and osmotic uptake of water and by proper

| Table 14.2 | Causes of Edema |

Cause	Comments
Increased blood pressure or venous obstruction	Increases capillary filtration pressure so that more tissue fluid is formed at the arteriolar ends of capillaries.
Increased tissue protein concentration	Decreases osmosis of water into the venular ends of capillaries. Usually a localized tissue edema due to leakage of plasma proteins through capillaries during inflammation and allergic reactions. Myxedema due to hypothyroidism is also in this category.
Decreased plasma protein concentration	Decreases osmosis of water into the venular ends of capillaries. May be caused by liver disease (which can be associated with insufficient plasma protein production), kidney disease (due to leakage of plasma protein into urine), or protein malnutrition.
Obstruction of lymphatic vessels	Infections by filaria roundworms (nematodes) transmitted by a certain species of mosquito block lymphatic drainage, causing edema and tremendous swelling of the affected areas.

lymphatic drainage. Edema may thus result from (1) *high arterial blood pressure,* which increases capillary pressure and causes excessive filtration; (2) *venous obstruction*—as in phlebitis (where a thrombus forms in a vein) or mechanical compression of veins (during pregnancy, for example)—which produces a congestive increase in capillary pressure; (3) *leakage of plasma proteins into tissue fluid,* which causes reduced osmotic flow of water into the capillaries (this occurs during inflammation and allergic reactions as a result of increased capillary permeability); (4) *myxedema*—the excessive production of particular glycoproteins (mucin) in the interstitial spaces caused by hypothyroidism; (5) *decreased plasma protein concentration* as a result of liver disease (the liver makes most of the plasma proteins) or kidney disease, in which proteins are excreted in the urine; and (6) *obstruction of the lymphatic drainage* (table 14.2).

Regulation of Blood Volume by the Kidneys

The formation of urine begins in the same manner as the formation of tissue fluid—by filtration of plasma through capillary pores. These capillaries are known as *glomeruli,* and the filtrate they produce enters a system of tubules that transports and modifies the filtrate (by mechanisms discussed in chapter 16). The kidneys produce about 180 L per day of blood filtrate, but since there is only 5.5 L of blood in the body, it is clear that most of this filtrate must be returned to the vascular system and recycled. Only about 1.5 L of urine is excreted daily; 98% to 99% of the amount filtered is **reabsorbed** back into the vascular system.

The volume of urine excreted can be varied by changes in the reabsorption of filtrate. If 99% of the filtrate is reabsorbed, for example, 1% must be excreted. Decreasing the reabsorption by only 1%—from 99% to 98%—would double the volume of urine excreted (an increase to 2% of the amount filtered). Carrying the logic further, a doubling of urine volume from, for example, 1 to 2

Chapter Fourteen

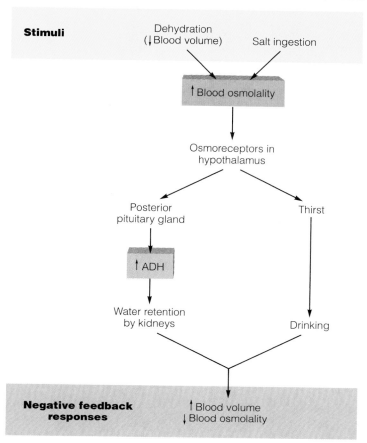

Dehydration
(↓Blood volume)   Salt ingestion

↑Blood osmolality

Osmoreceptors in
hypothalamus

Posterior
pituitary gland   Thirst

↑ADH

Water retention
by kidneys   Drinking

**Negative feedback
responses**
↑Blood volume
↓Blood osmolality

Figure 14.9

The negative feedback control of blood volume and blood osmolality.

During prolonged exercise, particularly on a warm day, a substantial amount of water may be lost from the body through sweating (up to 900 ml per hour or more). The lowering of blood volume that results decreases the ability of the body to dissipate heat, and the consequent overheating of the body can cause ill effects and put an end to the exercise. The need for athletes to remain well hydrated is commonly recognized, but drinking pure water may not be the answer. This is because blood sodium is lost in sweat, so that a lesser amount of water is required to dilute the blood osmolality back to normal. When the blood osmolality is normal, the urge to drink is extinguished. For these reasons, athletes performing prolonged endurance exercise should drink solutions containing sodium (as well as carbohydrates for energy), and they should drink at a predetermined rate rather than at a rate determined only by thirst.

tuitary. Through mechanisms that will be discussed in conjunction with kidney physiology in chapter 16, ADH stimulates water reabsorption from the filtrate. A smaller volume of urine is thus excreted as a result of the action of ADH (fig. 14.9).

A person who is dehydrated or who consumes excessive amounts of salt thus drinks more and urinates less. This raises the blood volume and, in the process, dilutes the plasma to lower its previously elevated osmolality. The rise in blood volume that results from these mechanisms is extremely important in stabilizing the condition of a dehydrated person with low blood volume and pressure.

Drinking excessive amounts of water without excessive amounts of salt does not result in a prolonged increase in blood volume and pressure. The water does enter the blood from the intestine and momentarily raises the blood volume; at the same time, however, it dilutes the blood. Dilution of the blood decreases the plasma osmolality and thus inhibits ADH secretion. With less ADH there is less reabsorption of filtrate in the kidneys—a larger volume of urine is excreted. Water is therefore a *diuretic*—a substance that promotes urine formation—because it inhibits the secretion of antidiuretic hormone.

In addition to the activity of osmoreceptors, another mechanism operates to inhibit ADH secretion when fluid intake is excessive. An excessively high blood volume stimulates stretch receptors located in the left atrium of the heart. Stimulation of these stretch receptors, in turn, activates a reflex inhibition of ADH secretion, which promotes a lowering of blood volume through increased urine production.

Regulation by Aldosterone

From the preceding discussion, it is clear that a certain amount of dietary salt is required to maintain blood volume and pressure. Since Na^+ and Cl^- are easily filtered in the kidneys, a mechanism must exist to promote the reabsorption and retention of salt when the dietary salt intake is too low. **Aldosterone,** a steroid hormone secreted by the adrenal cortex, stimulates the

liters, would result in the loss of one additional liter of blood volume. The percentage of the glomerular filtrate reabsorbed—and thus the urine volume and blood volume—is adjusted according to the needs of the body by the action of specific hormones on the kidneys. Through their effects on the kidneys, and the resulting changes in blood volume, these hormones serve important functions in the regulation of the cardiovascular system.

Regulation by Antidiuretic Hormone (ADH)

One of the major hormones involved in the regulation of blood volume is **antidiuretic hormone (ADH),** also known as *vasopressin.* As described in chapter 11, this hormone is produced by neurons in the hypothalamus, transported by axons into the posterior pituitary, and released from this storage gland in response to hypothalamic stimulation. The secretion of ADH from the posterior pituitary occurs when neurons called **osmoreceptors** in the hypothalamus detect an increase in plasma osmolality (osmotic pressure).

An increase in plasma osmolality occurs when the plasma becomes more concentrated (chapter 6). This can be produced either by *dehydration* or by *excessive salt intake.* Stimulation of osmoreceptors produces sensations of thirst, leading to increased water intake and increased ADH secretion from the posterior pi-

reabsorption of salt by the kidneys. Aldosterone is thus a "salt-retaining hormone." Retention of salt indirectly promotes retention of water (in part, by the action of ADH, as previously discussed). The action of aldosterone thus causes an increase in blood volume, but, unlike ADH, it does not cause a change in plasma osmolality. This is because aldosterone promotes the reabsorption of salt and water in proportionate amounts, whereas ADH promotes only the reabsorption of water. Thus, unlike ADH, aldosterone does not act to dilute the blood.

The secretion of aldosterone is stimulated during salt deprivation, when the blood volume and pressure are reduced. The adrenal cortex, however, is not directly stimulated to secrete aldosterone by these conditions. Instead, a decrease in blood volume and pressure activates an intermediate mechanism, described in the next section.

Renin-Angiotensin System

When the blood flow and pressure are reduced in the renal artery (as they would be in the low blood volume state of salt deprivation), a group of cells in the kidneys called the *juxtaglomerular apparatus* secretes the enzyme **renin** into the blood. This enzyme cleaves a ten-amino-acid polypeptide called *angiotensin I* from a plasma protein called *angiotensinogen*. As angiotensin I passes through the capillaries of the lungs and other organs, an *angiotensin-converting enzyme* removes two amino acids. This leaves an eight-amino-acid polypeptide called **angiotensin II** (fig. 14.10). Conditions of salt deprivation, low blood volume, and low blood pressure, in summary, cause increased production of angiotensin II in the blood.

Angiotensin II exerts numerous effects that produce a rise in blood pressure. This rise in pressure is partly due to vasoconstriction and partly to increases in blood volume. Vasoconstriction of arterioles and small muscular arteries is produced directly by the effects of angiotensin II on the smooth muscle layers of these vessels. The increased blood volume is an indirect effect of angiotensin II.

Angiotensin II promotes a rise in blood volume by means of two mechanisms: (1) thirst centers in the hypothalamus are stimulated by angiotensin II, and thus more water is ingested, and (2) secretion of aldosterone from the adrenal cortex is stimulated by angiotensin II, and higher aldosterone secretion causes more salt and water to be retained by the kidneys. The relationship between angiotensin II and aldosterone is sometimes described as the *renin-angiotensin-aldosterone system*.

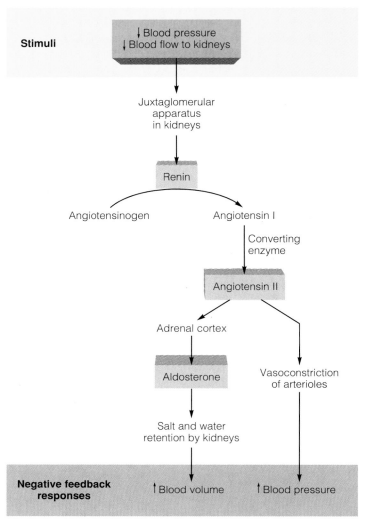

Figure 14.10

The negative feedback control of blood volume and pressure by the renin-angiotensin-aldosterone system.

One of the newer classes of drugs that can be used to treat hypertension (high blood pressure) are the **angiotensin converting enzyme, or ACE, inhibitors.** These drugs (such as *captopril*) block the formation of angiotension II, thus reducing its vasoconstrictor effect. The ACE inhibitors also increase the activity of bradykinin, a polypeptide that promotes vasodilation. The reduced formation of angiotensin II and increased action of bradykinin result in vasodilation, which decreases the total peripheral resistance. Because this reduces the afterload of the heart, the ACE inhibitors are also used to treat left ventricular hypertrophy and congestive heart failure.

The renin-angiotensin-aldosterone system can also work in the opposite direction: high salt intake, leading to high blood volume and pressure, normally inhibits renin secretion. With less angiotensin II formation and less aldosterone secretion, less salt is retained by the kidneys and more is excreted in the urine. Unfortunately, many people with chronically high blood pressure may have normal or even elevated levels of renin secretion. In these cases, the intake of salt must be lowered to match the impaired ability to excrete salt in the urine.

Atrial Natriuretic Factor

As described in the previous section, a fall in blood volume is compensated for by renal retention of fluid through activation of the renin-angiotensin-aldosterone system. An increase in blood volume, conversely, is compensated for by renal excretion of a larger volume of urine. Experiments suggest that the increase in water excretion under conditions of high blood volume is at least partly due to an increase in the excretion of Na$^+$ in the urine, or *natriuresis* (*natrium* = sodium; *uresis* = making water).

Increased Na$^+$ excretion (natriuresis) may be produced by a decline in aldosterone secretion, but there is evidence that there is a separate hormone that stimulates natriuresis. This **natriuretic hormone** would thus be antagonistic to aldosterone and would promote Na$^+$ and water excretion in the urine in response to a rise in blood volume. The atria of the heart produce a polypeptide hormone with these properties, which has been identified as *atrial natriuretic factor* (*ANF*). By promoting salt and water excretion in the urine, ANF can act to lower the blood volume and pressure. This is analogous to the action of diuretic drugs taken by people with hypertension, as described later in this chapter.

In addition to its stimulation of salt and water excretion by the kidneys, ANF also antagonizes various actions of angiotensin II. As a result of this action, ANF decreases the secretion of aldosterone and promotes vasodilation.

1. Describe the composition of tissue fluid. Using a flow diagram, explain how tissue fluid is formed and how it is returned to the vascular system.
2. Define the term *edema* and describe four different mechanisms that can produce this condition.
3. Describe the effects of dehydration on blood and urine volumes. What cause-and-effect mechanism is involved?
4. Explain why salt deprivation causes increased salt and water retention by the kidneys.
5. Describe the actions of atrial natriuretic factor and explain their significance.

Vascular Resistance to Blood Flow

The rate of blood flow to an organ is related to the resistance to flow in the small arteries and arterioles that serve the organ. Vasodilation decreases resistance and increases flow, whereas vasoconstriction increases resistance and decreases flow. Vasodilation and vasoconstriction occur in response to intrinsic and extrinsic regulatory mechanisms.

The amount of blood that the heart pumps per minute is equal to the rate of venous return and thus is equal to the rate of blood flow through the entire circulation. The cardiac output of 5 to 6 L per minute is distributed unequally to the different organs. At rest, blood flow is about 2,500 ml/min through the liver, kidneys, and gastrointestinal tract; 1,200 ml/min through the skeletal muscles; 750 ml/min through the brain; and 250 ml/min through the coronary arteries of the heart. The balance of the cardiac output (500 to 1,100 ml/min) is distributed to the other organs (table 14.3).

Physical Laws Describing Blood Flow

The flow of blood through the vascular system, like the flow of any fluid through a tube, depends in part on the difference in pressure at the two ends of the tube. If the pressure at both ends of the tube is the same there will be no flow. If the pressure at one end is greater than at the other, blood will flow from the region of higher to the region of lower pressure. The rate of blood flow is proportional to the pressure difference ($P_1 - P_2$) between the two ends of the tube. The term **pressure difference** is abbreviated ΔP, in which the Greek letter Δ (*delta*) means "change in" (fig. 14.11).

If the systemic circulation is pictured as a single tube leading from and back to the heart (fig. 14.11), blood flow through this system would occur as a result of the pressure difference between the beginning of the tube (the aorta) and the end of the tube (the junction of the venae cavae with the right atrium). The average, or mean, arterial pressure is about 100 mmHg; the pressure at the right atrium is 0 mmHg. The "pressure head," or driving force (ΔP), is therefore about $100 - 0 = 100$ mmHg.

Blood flow is directly proportional to the pressure difference between the two ends of the tube (ΔP) but is *inversely proportional* to the frictional resistance to blood flow through the vessels. Inverse proportionality is expressed by showing one of the factors in the denominator of a fraction, since a fraction decreases when the denominator increases:

$$\text{Blood flow} \propto \frac{\Delta P}{\text{resistance}}$$

The **resistance** to blood flow through a vessel is directly proportional to the length of the vessel and to the viscosity of the blood (the "thickness," or ability of molecules to "slip over"

Table 14.3 Estimated Distribution of the Cardiac Output at Rest

Organs	Blood Flow	
	Milliliters per Minute	Percent Total
Gastrointestinal tract and liver	1,400	24
Kidneys	1,100	19
Brain	750	13
Heart	250	4
Skeletal muscles	1,200	21
Skin	500	9
Other organs	600	10
Total organs	5,800	100

Source: From O. L. Wade and J. M. Bishop, *Cardiac Output and Regional Blood Flow.* Copyright ©1962 Blackwell Scientific Publications, Ltd. Used with permission.

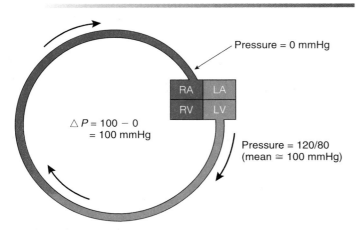

Pressure = 0 mmHg

RA | LA
RV | LV

$\triangle P = 100 - 0$
$= 100$ mmHg

Pressure = 120/80
(mean ≅ 100 mmHg)

Figure 14.11

The flow of blood in the systemic circulation is ultimately dependent on the pressure difference (ΔP) between the mean pressure of about 100 mmHg at the origin of flow in the aorta and the pressure at the end of the circuit—zero mmHg in the vena cava, where it joins the right atrium (RA). (LA = left atrium; RV = right ventricle; LV = left ventricle.)

each other). Vascular resistance is inversely proportional to the fourth power of the radius of the vessel:

$$\text{Resistance} \propto \frac{L\eta}{r^4}$$

where

$$L = \text{length of vessel}$$
$$\eta = \text{viscosity of blood}$$
$$r = \text{radius of vessel}$$

For example, if one vessel has half the radius of another and if all other factors are the same, the smaller vessel will have sixteen times (2^4) the resistance of the larger vessel. Blood flow through the larger vessel, as a result, will be sixteen times greater than in the smaller vessel (fig. 14.12).

When physical constants are added to this relationship, the rate of blood flow can be calculated according to **Poiseuille's** (*pwǎ-zuh´yez*) **law:**

$$\text{Blood flow} = \frac{\Delta P r^4 (\pi)}{\eta L(8)}$$

Vessel length and blood viscosity do not vary significantly in normal physiology, although blood viscosity is increased in severe dehydration and in the polycythemia (high red blood cell count) that occurs as an adaptation to life at high altitudes. The major physiological regulators of blood flow through an organ are the mean arterial pressure (driving the flow) and the vascular resistance to flow. At a given mean arterial pressure, blood can be diverted from one organ to another by variations in the degree of vasoconstriction and vasodilation. Vasoconstriction in one organ and vasodilation in another result in a diversion of blood to the second organ. Since arterioles are the smallest arteries and can become narrower by vasoconstriction, they provide the greatest resistance to blood flow (fig. 14.13). Blood flow to an organ is thus largely determined by the degree of vasoconstriction or vasodilation of its arterioles. The rate of blood flow to an organ can be increased by dilation of its arterioles and can be decreased by constriction of its arterioles.

Figure 14.12

Relationships between blood flow, vessel radius, and resistance. (*a*) The resistance and blood flow are equally divided between two branches of a vessel. (*b*) A doubling of the radius of one branch and halving of the radius of the other produces a sixteenfold increase in blood flow in the former and a sixteenfold decrease of blood flow in the latter.

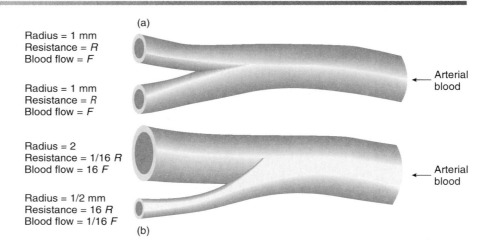

(a)

Radius = 1 mm
Resistance = *R*
Blood flow = *F*

Radius = 1 mm
Resistance = *R*
Blood flow = *F*

Arterial blood

Radius = 2
Resistance = 1/16 *R*
Blood flow = 16 *F*

Radius = 1/2 mm
Resistance = 16 *R*
Blood flow = 1/16 *F*

(b)

Arterial blood

Total Peripheral Resistance

The sum of all the vascular resistances within the systemic circulation is called the *total peripheral resistance*. The arteries that supply blood to the organs are generally in parallel rather than in series with each other. That is, arterial blood passes through only one set of resistance vessels (arterioles) before returning to the heart (fig. 14.14). Since one organ is not "downstream" from another in terms of its arterial supply, changes in resistance within one organ directly affect blood flow in only that organ.

Vasodilation in a large organ might, however, significantly decrease the total peripheral resistance and, by this means, might decrease the mean arterial pressure. In the absence of compensatory mechanisms, the driving force for blood flow through all organs might be reduced. This situation is normally prevented by an increase in the cardiac output and by vasoconstriction in other areas. During exercise of the large muscles, for example, the arterioles in the exercising muscles are dilated. This would cause a great fall in mean arterial pressure if there were no compensations. The blood pressure actually rises during exercise, however, because the cardiac output is increased and because there is constriction of arterioles in the viscera and skin.

Extrinsic Regulation of Blood Flow

The term *extrinsic regulation* refers to control by the autonomic nervous system and endocrine system. **Angiotensin II,** for example, directly stimulates vascular smooth muscle to produce generalized vasoconstriction. Antidiuretic hormone (ADH) also has a vasoconstrictor effect at high concentrations; this is why it is also called **vasopressin.** This vasopressor effect of ADH is not believed to be significant under physiological conditions in humans.

Regulation by Sympathetic Nerves

Stimulation of the sympathoadrenal system produces an increase in the cardiac output (as previously discussed) and an increase in total peripheral resistance. The latter effect is due to alpha-adrenergic stimulation (chapter 9) of vascular smooth

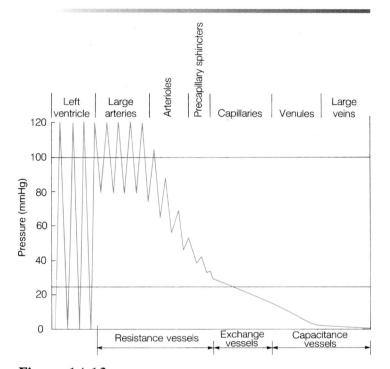

Figure 14.13

Blood pressure in different vessels of the systemic circulation.

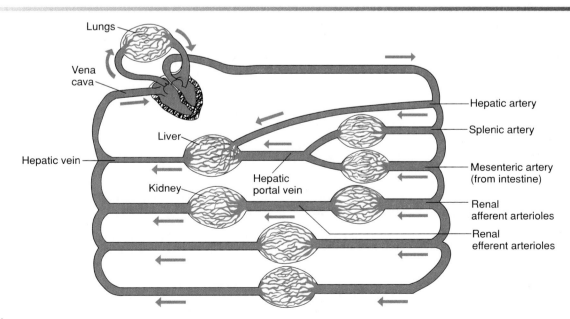

Figure 14.14

A diagram of the systemic and pulmonary circulations. Notice that with few exceptions (such as blood flow in the renal circulation) the flow of arterial blood is in parallel rather than in series (arterial blood does not usually flow from one organ to another).

Cardiac Output, Blood Flow, and Blood Pressure

Table 14.4 Extrinsic Control of Vascular Resistance and Blood Flow

Extrinsic Agent	Effect	Comments
Sympathetic nerves		
Alpha-adrenergic	Vasoconstriction	Vasoconstriction is the dominant effect of sympathetic nerve stimulation on the vascular system, and it occurs throughout the body.
Beta-adrenergic	Vasodilation	There is some activity in arterioles in skeletal muscles and in coronary vessels, but effects are masked by dominant alpha-receptor-mediated constriction.
Cholinergic	Vasodilation	Effects are localized to arterioles in skeletal muscles and are produced only during defense (fight-or-flight) reaction.
Parasympathetic nerves	Vasodilation	Effects are restricted primarily to the gastrointestinal tract, external genitalia, and salivary glands and have little effect on total peripheral resistance.
Angiotensin II	Vasoconstriction	A powerful vasoconstrictor produced as a result of secretion of renin from the kidneys, it may function to help maintain adequate filtration pressure in the kidneys when systemic blood flow and pressure are reduced.
ADH (vasopressin)	Vasoconstriction	Although the effects of this hormone on vascular resistance and blood pressure in anesthetized animals are well documented, the importance of these effects in conscious humans is controversial.
Histamine	Vasodilation	Histamine promotes localized vasodilation during inflammation and allergic reactions.
Bradykinins	Vasodilation	Bradykinins are polypeptides secreted by endothelium and sweat glands that promote local vasodilation.
Prostaglandins	Vasodilation or vasoconstriction	Prostaglandins are cyclic fatty acids that can be produced by most tissues, including blood vessel walls. Prostaglandin I_2 is a vasodilator, whereas thromboxane A_2 is a vasoconstrictor. The physiological significance of these effects is presently controversial.

muscle by norepinephrine and, to a lesser degree, by epinephrine. This produces vasoconstriction of the arterioles in the viscera and skin.

Even when a person is calm, the sympathoadrenal system is active to a certain degree and helps set the "tone" of vascular smooth muscles. In this case, **adrenergic sympathetic fibers** (those that release norepinephrine) cause a basal level of vasoconstriction throughout the body. During the fight-or-flight reaction, an increase in the activity of adrenergic fibers produces vasoconstriction in the digestive tract, kidneys, and skin.

Arterioles in skeletal muscles receive **cholinergic sympathetic fibers,** which release acetylcholine as a neurotransmitter. During the fight-or-flight reaction, the activity of these cholinergic fibers increases. This causes vasodilation. Vasodilation in skeletal muscles is also produced by epinephrine secreted by the adrenal medulla, which stimulates beta-adrenergic receptors. During the fight-or-flight reaction, therefore, blood flow is decreased to the viscera and skin due to the alpha-adrenergic effects of vasoconstriction in these organs, whereas blood flow to the skeletal muscles is increased. This diversion of blood flow to the skeletal muscles during emergency conditions may produce an "extra edge" for the skeletal muscles' response to the emergency.

Parasympathetic Control of Blood Flow

Parasympathetic endings in arterioles are always cholinergic and always promote vasodilation. Parasympathetic innervation of blood vessels, however, is limited to the digestive tract, external genitalia, and salivary glands. Because of this limited distribution, the parasympathetic system is less important than the sympathetic system in the control of total peripheral resistance. The extrinsic control of blood flow is summarized in table 14.4.

Paracrine Regulation of Blood Flow

Paracrine regulators, as described in chapter 11, are molecules produced by one tissue that help regulate another tissue of the same organ. A blood vessel is an organ that is particularly subject to paracrine regulation. Specifically, the endothelium of the tunica intima produces a number of paracrine regulators that cause the smooth muscle of the tunica media to either relax or contract.

The endothelium produces several molecules that promote smooth muscle relaxation, including nitric oxide, bradykinin, and prostacyclin (chapter 11). Nitric oxide appears to be the endothelium-derived relaxation factor that earlier research had shown to be required for the vasodilation response to nerve stimulation. The steps in this process are as follows: (1) the parasympathetic axons release ACh, which stimulates the opening of Ca^{++} channels in the endothelial cell membrane; (2) the Ca^{++} then binds to and activates calmodulin (chapter 11); (3) calmodulin, in turn, activates the enzyme nitric oxide synthetase, which converts L-arginine into nitric oxide; and (4) the nitric oxide then diffuses into the smooth muscle cells of the vessel to produce the vasodilation response to the nerve stimulation. It is interesting in this regard that vasodilator drugs often given to treat angina pectoris—including nitroglycerin and sodium nitroprusside—promote vasodilation indirectly through their conversion into nitric oxide.

The endothelium also produces paracrine regulators that stimulate vasoconstriction. Notable among these is the polypeptide *endothelin-1*. Although the precise physiological role of this regulator is incompletely understood, it is currently believed that it works together with the vasodilator regulators to help maintain normal vessel diameter and blood pressure.

Table 14.5 Intrinsic Control of Vascular Resistance and Blood Flow

Category	Agent ($\uparrow$ = increase; $\downarrow$ = decrease)	Mechanisms	Comments
Myogenic	$\uparrow$Blood pressure	Stretching of the arterial wall as the blood pressure rises directly stimulates increased smooth muscle tone (vasoconstriction).	It helps to maintain relatively constant rates of blood flow and pressure within an organ despite changes in systemic arterial pressure (autoregulation).
Metabolic	$\downarrow$Oxygen $\uparrow$Carbon dioxide $\downarrow$pH $\uparrow$Adenosine $\uparrow$K	Local changes in gas and metabolite concentrations act directly on vascular smooth muscle walls to produce vasodilation in the systemic circulation. The importance of different agents varies in different organs.	It aids in autoregulation of blood flow and also helps to shunt increased amounts of blood to organs with higher metabolic rates (active hyperemia).

Experiments with isolated atherosclerotic arteries suggest that they may have a reduced response to paracrine regulators that cause vasodilation. This could predispose them to vasoconstriction and helps to explain the observation that emotional stress sometimes causes vasoconstriction in atherosclerotic coronary arteries. Also, a reduced ability of the endothelium to produce nitric oxide may contribute to hypertension. Drugs that increase nitric oxide production, therefore, may help to lower blood pressure.

Intrinsic Regulation of Blood Flow

Intrinsic, or "built-in," mechanisms within individual organs provide a localized regulation of vascular resistance and blood flow. Intrinsic mechanisms are classified as *myogenic* or *metabolic*. Some organs, the brain and kidneys in particular, utilize these intrinsic mechanisms to maintain relatively constant flow rates despite wide fluctuations in blood pressure. This ability is termed **autoregulation.**

Myogenic Control Mechanisms

If the arterial blood pressure and flow through an organ are inadequate—if the organ is inadequately *perfused* with blood—the metabolism of the organ cannot be maintained beyond a limited period of time. Excessively high blood pressure can also be dangerous, particularly in the brain, because this may cause fine blood vessels to burst (causing cerebrovascular accident— CVA, or stroke).

Changes in systemic arterial pressure are compensated for in the brain and some other organs by the appropriate responses of vascular smooth muscle. A decrease in arterial pressure causes cerebral vessels to dilate, so that adequate rates of blood flow can be maintained despite the decreased pressure. High blood pressure, by contrast, causes cerebral vessels to constrict, so that finer vessels downstream are protected from the elevated

pressure. These responses are myogenic; they are direct responses by the vascular smooth muscle to changes in pressure.

Metabolic Control Mechanisms

Local vasodilation within an organ can occur as a result of the chemical environment created by the organ's metabolism. The localized chemical conditions that promote vasodilation include (1) *decreased oxygen concentrations* that result from increased metabolic rate; (2) *increased carbon dioxide concentrations*; (3) *decreased tissue pH* (due to CO_2, lactic acid, and other metabolic products); and (4) the *release of adenosine or K^+* from the tissue cells.

The vasodilation that occurs in response to tissue metabolism can be demonstrated by constricting the blood supply to an area for a short time and then removing the constriction. The constriction allows metabolic products to accumulate by preventing venous drainage of the area. When the constriction is removed and blood flow resumes, the metabolic products that have accumulated cause vasodilation. The tissue thus appears red. This response is called **reactive hyperemia.** A similar increase in blood flow occurs in skeletal muscles and other organs as a result of increased metabolism. This is called **active hyperemia.** Intrinsic control mechanisms are summarized in table 14.5.

1. Describe the relationship between blood flow, arterial blood pressure, and vascular resistance.
2. Describe the relationship between vascular resistance and the radius of a vessel. Explain how blood flow can be diverted from one organ to another.
3. Explain how vascular resistance and blood flow are regulated by (a) sympathetic adrenergic fibers, (b) sympathetic cholinergic fibers, and (c) parasympathetic fibers.
4. Describe the formation and action of nitric oxide and explain why it is considered a paracrine regulator.
5. Define *autoregulation* and explain how this process occurs through myogenic and metabolic mechanisms.

Blood Flow to the Heart and Skeletal Muscles

Blood flow to the heart and skeletal muscles is regulated by both extrinsic and intrinsic mechanisms. These mechanisms provide increased rates of blood flow when the metabolic requirements of these tissues are raised during exercise.

Survival requires that the heart and brain receive adequate blood flow at all times. The ability of skeletal muscles to respond quickly in emergencies and to maintain continued high levels of activity may also be critically important for survival. During such times, high rates of blood flow to the skeletal muscles must be maintained without compromising blood flow to the heart and brain. This is accomplished by mechanisms that increase the cardiac output and that divert a higher proportion of the cardiac output to the heart, skeletal muscles, and brain and away from the viscera and skin.

Aerobic Requirements of the Heart

The coronary arteries supply an enormous number of capillaries, which are packed within the myocardium at a density ranging from 2,500 to 4,000 per cubic millimeter of tissue. Fast-twitch skeletal muscles, by contrast, have a capillary density of 300 to 400 per cubic millimeter of tissue. Each myocardial cell, as a consequence, is within 10 μm of a capillary (compared to an average distance in other organs of 70 μm). The exchange of gases by diffusion between myocardial cells and capillary blood thus occurs very quickly.

Contraction of the myocardium squeezes the coronary arteries. Unlike blood flow in all other organs, flow in the coronary vessels thus decreases in systole and increases during diastole. The myocardium, however, contains large amounts of *myoglobin*, a pigment related to hemoglobin (the molecules in red blood cells that carry oxygen). Myoglobin in the myocardium stores oxygen during diastole and releases its oxygen during systole. In this way, the myocardial cells can receive a continuous supply of oxygen even though coronary blood flow is temporarily reduced during systole.

In addition to containing large amounts of myoglobin, heart muscle contains numerous mitochondria and aerobic respiratory enzymes. This indicates that—even more than slow-twitch skeletal muscles—the heart is extremely specialized for aerobic respiration (table 14.6). The normal heart always respires aerobically, even during heavy exercise when the metabolic demand for oxygen can rise to five times resting levels. This increased oxygen requirement is met by a corresponding increase in coronary blood flow, from about 80 ml at rest to about 400 ml per minute per 100 g tissue during heavy exercise.

Regulation of Coronary Blood Flow

Sympathetic nerve fibers, through stimulation of alpha-adrenergic receptors in the coronary arterioles, produce a relatively high vascular resistance in the coronary circulation at rest. Vasodilation of coronary vessels may be produced in part by sympathoadrenal activation of beta-adrenergic receptors. Most of the vasodilation that occurs during exercise, however, is due to intrinsic metabolic control mechanisms. As the metabolism of the myocardium increases, local accumulation of carbon dioxide, K^+, and adenosine, together with depletion of oxygen, acts directly on the vascular smooth muscle to cause relaxation and vasodilation.

Table 14.6	Vascular and Metabolic Comparisons of Heart and Skeletal Muscle	
Feature	**Cardiac Muscle**	**Skeletal Muscle**
Number of capillaries	4×	1×
Mean blood flow	10–20×	1×
Myolemma (sarcolemma)	Thin, low resistance	Thicker, higher resistance
Sarcomere length	1×	1.7×
Glycogen concentration	Maintained	Depleted by fasting, diabetes
Glycolytic enzyme systems	1×	2×, strongly developed
Creatine phosphate concentration	1×	6×
Anaerobic energy production	Beating: 2 min Arrested: 30–90 min	Up to 40% total energy
Lactic acid production	Terminal mechanism	Frequent when incurring O_2 debt
Ability to incur O_2 debt	1×	4×
Increased O_2 requirement met primarily by	Increased flow	Increased extraction
Oxygen consumption	3×	1×
Oxygen extraction at rest	Near maximal	Significant reserve
Increased O_2 consumption with increased work	2×	30×
Myoglobin	Present	Present in red skeletal muscle
Mitochondria	Abundant, giant	Fewer, smaller
Krebs cycle enzymes	2–3×	1×
Cytochrome-C	6×	1×
Myosin ATPase activity	1×	3×

Source: From N. Brachfeld, "The Physiology of Muscular Exercise," *Primary Cardiology,* June 1979, p.112. Reproduced with permission.

Under abnormal conditions the blood flow to the myocardium may be inadequate, resulting in myocardial ischemia (described in chapter 13). This can result from blockage by atheromas and/or blood clots or from muscular spasm of a coronary artery (fig. 14.15). Occlusion of a coronary artery can be visualized by a technique called *selective coronary arteriography*. In this procedure, a catheter (plastic tube) is inserted into a brachial or femoral artery all the way to the opening of the coronary arteries in the aorta, whereupon radiographic contrast material is injected. The picture thus obtained is called an **angiogram.**

If the occlusion is sufficiently great, a *coronary bypass* may be performed. In this procedure, a length of blood vessel, usually taken from the saphenous vein in the leg, is sutured to the aorta and to the coronary artery at a location beyond the site of the occlusion (fig. 14.16).

Regulation of Blood Flow through Skeletal Muscles

The arterioles in skeletal muscles, like those of the coronary circulation, have a high vascular resistance at rest as a result of alpha-adrenergic sympathetic stimulation. This produces a relatively low rate of blood flow, but because muscles have such a large mass, this still accounts for 20% to 25% of the total blood flow in the body at rest. Also, as in the heart, blood flow in a skeletal muscle decreases when the muscle contracts and squeezes its arterioles, and in fact blood flow stops entirely when the muscle contracts beyond about 70% of its maximum. Pain and fatigue thus occur much more quickly when an isometric contraction is sustained than when rhythmic isotonic contractions are performed.

In addition to adrenergic fibers, which promote vasoconstriction by stimulation of alpha-adrenergic receptors, there are also sympathetic cholinergic fibers in skeletal muscles. These cholinergic fibers, together with the stimulation of beta-adrenergic receptors by the hormone epinephrine, stimulate vasodilation as part of the "fight-or-flight" response to any stressful state, including that existing just prior to exercise (table 14.7). These extrinsic controls have been previously discussed and function to regulate blood flow through muscles at rest and upon anticipation of exercise.

As exercise progresses, the vasodilation and increased skeletal muscle blood flow that occur are almost entirely due to intrinsic metabolic control. The high metabolic rate of skeletal muscles during exercise causes local changes, such as increased carbon dioxide concentrations, decreased pH (due to carbonic acid and lactic acid), decreased oxygen, increased extracellular K^+, and the secretion of adenosine. As in the intrinsic control

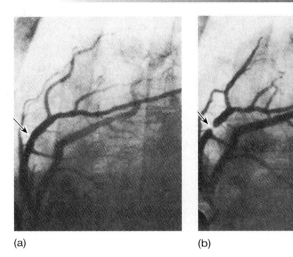

Figure 14.15

An angiogram of the left coronary artery in a patient (*a*) when the ECG was normal and (*b*) when the ECG showed evidence of myocardial ischemia. Notice that a coronary artery spasm (see arrow in [*b*]) appears to accompany the ischemia.

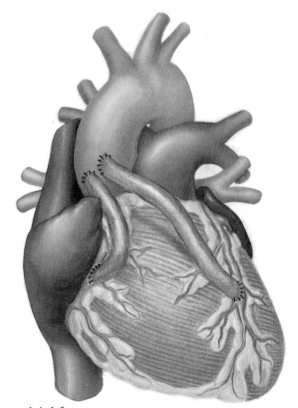

Figure 14.16

A diagram of coronary artery bypass surgery.

Table 14.7 Changes in Skeletal Muscle Blood Flow Under Conditions of Rest and Exercise

Condition	Blood Flow (ml/min)	Mechanism
Rest	1,000	High adrenergic sympathetic stimulation of vascular alpha-receptors, causing vasoconstriction
Beginning exercise	Increased	Dilation of arterioles in skeletal muscles due to cholinergic sympathetic nerve activity and stimulation of β-adrenergic receptors by the hormone epinephrine
Heavy exercise	20,000	Fall in alpha-adrenergic activity Increased sympathetic cholinergic activity Increased metabolic rate of exercising muscles, producing intrinsic vasodilation

Table 14.8 Relationship Between Age and Average Maximum Cardiac Rate

Age	Maximum Cardiac Rate
20–29	190 beats/min
30–39	160 beats/min
40–49	150 beats/min
50–59	140 beats/min
60+	130 beats/min

of the coronary circulation, these changes cause vasodilation of arterioles in skeletal muscles. This decreases the vascular resistance and increases the rate of blood flow. This effect is combined with the recruitment of capillaries by the opening of precapillary sphincter muscles (only 5% to 10% of the skeletal muscle capillaries are open at rest). As a result of these changes, skeletal muscles can receive as much as 85% of the total blood flow in the body during maximal exercise.

Circulatory Changes during Exercise

While the vascular resistance in skeletal muscles decreases during exercise, the resistance to flow through visceral organs and skin increases. This increased resistance occurs because of vasoconstriction stimulated by adrenergic sympathetic fibers and results in decreased rates of blood flow through these organs. During exercise, therefore, the blood flow to skeletal muscles increases because of three simultaneous changes: (1) increased total blood flow (cardiac output); (2) metabolic vasodilation in the exercising muscles; and (3) the diversion of blood away from the viscera and skin. Blood flow to the heart also increases during exercise, whereas blood flow to the brain does not appear to change significantly (fig. 14.17).

During exercise, the cardiac output can increase fivefold—from about 5 L per minute to about 25 L per minute. This is primarily due to an increase in cardiac rate. The cardiac rate, however, can increase only up to a maximum value (table 14.8), which is determined mainly by a person's age. In well-trained athletes, the stroke volume can also increase significantly, allowing these individuals to achieve cardiac outputs during strenuous exercise up to six or seven times greater than their resting values.

In most people, the increase in stroke volume that occurs during exercise will not exceed 35%. The fact that the stroke volume can increase at all during exercise may at first be surprising, in view of the fact that the heart has less time to fill with blood between beats when it is pumping faster. Despite the faster beat, however, the end-diastolic volume during exercise is not decreased. This is because the venous return is aided by the improved action of the skeletal muscle pumps and by increased respiratory movements during exercise (fig. 14.18). Since the end-diastolic volume is not significantly changed during exercise, any increase in stroke volume that occurs must be due to an increase in the proportion of blood ejected per stroke.

The proportion of the end-diastolic volume ejected per stroke can increase from 60% at rest to as much as 90% during heavy exercise. This increased *ejection fraction* is produced by the increased contractility that results from sympathoadrenal stimulation. There also may be a decrease in total peripheral resistance as a result of vasodilation in the exercising skeletal muscles, which decreases the afterload and thus further augments the increase in stroke volume. The cardiovascular changes that occur during exercise are summarized in table 14.9.

Endurance training often results in a lowering of the resting cardiac rate and an increase in the resting stroke volume. The lowering of the resting cardiac rate results from a greater degree of inhibition of the SA node by the vagus nerve. The increased resting stroke volume is believed to be due to an increase in blood volume; indeed, studies have shown that the blood volume can increase by about 500 ml after only 8 days of training. These adaptations enable the trained athlete to produce a larger proportionate increase in cardiac output and achieve a higher absolute cardiac output during exercise. This large cardiac output is the major factor in the improved oxygen delivery to skeletal muscles that occurs as a result of endurance training.

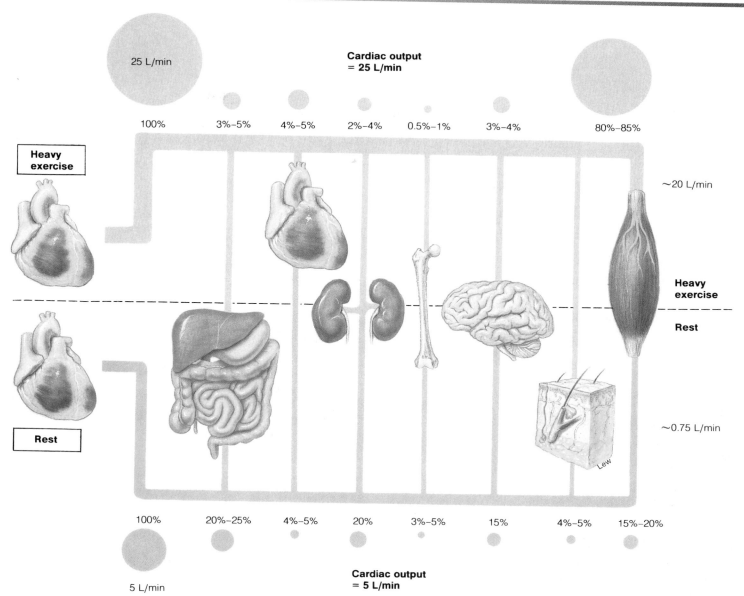

Cardiac output = 25 L/min

25 L/min

100%   3%–5%   4%–5%   2%–4%   0.5%–1%   3%–4%   80%–85%

Heavy exercise

~20 L/min

Heavy exercise

Rest

~0.75 L/min

Rest

100%   20%–25%   4%–5%   20%   3%–5%   15%   4%–5%   15%–20%

Cardiac output = 5 L/min

5 L/min

Lew

Figure 14.17

The distribution of blood flow (cardiac output) during periods of rest and heavy exercise. At rest, the cardiac output is 5 L per minute (*bottom of figure*); during heavy exercise the cardiac output increases to 25 L per minute (*top of figure*). At rest, for example, the brain receives 15% of 5 L per minute (= 750 ml/min), whereas during exercise it receives 3% to 4% of 25 L per minute (0.03 × 25 = 750 ml/min). Flow to the skeletal muscles increases more than twentyfold because the total cardiac output increases (from 5 L/min to 25 L/min) and because the percentage of the total received by the muscles increases from 15% to 80%.

Table 14.9 Cardiovascular Changes During Moderate Exercise

Variable	Change	Mechanisms
Cardiac output	Increased	Cardiac rate and stroke volume increased
Cardiac rate	Increased	Increased sympathetic nerve activity; decreased activity of the vagus nerve
Stroke volume	Increased	Increased myocardial contractility due to stimulation by sympathoadrenal system; decreased total peripheral resistance
Total peripheral resistance	Decreased	Vasodilation of arterioles in skeletal muscles (and in skin when thermoregulatory adjustments are needed)
Arterial blood pressure	Increased	Increased systolic and pulse pressure due primarily to increased cardiac output; diastolic pressure arises less due to decreased total peripheral resistance
End-diastolic volume	Unchanged	Decreased filling time at high cardiac rates is compensated by increased venous pressure, increased activity of the skeletal muscle pump, and decreased intrathoracic pressure aiding the venous return
Blood flow to heart and muscles	Increased	Increased muscle metabolism produces intrinsic vasodilation; aided by increased cardiac output and increased vascular resistance in visceral organs
Blood flow to visceral organs	Decreased	Vasoconstriction in digestive tract, liver, and kidneys due to sympathetic nerve stimulation
Blood flow to skin	Increased	Metabolic heat produced by exercising muscles produces reflex (involving hypothalamus) that reduces sympathetic constriction of arteriovenous shunts and arterioles
Blood flow to brain	Unchanged	Autoregulation of cerebral vessels, which maintains constant cerebral blood flow despite increased arterial blood pressure

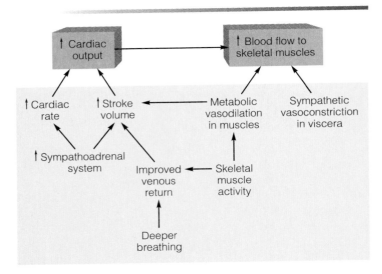

Figure 14.18

Cardiovascular adaptations to exercise.

1. Describe blood flow and oxygen delivery to the myocardium during systole and diastole.
2. Describe how blood flow to the heart is affected by exercise. Explain how blood flow to the heart is regulated at rest and during exercise.
3. Describe the mechanisms that produce vasodilation of the arterioles in skeletal muscles during exercise. Give two other reasons for the increased blood flow to muscles during exercise.
4. Explain how the stroke volume can increase during exercise despite the fact that the filling times are reduced at high cardiac rates.

Blood Flow to the Brain and Skin

Intrinsic control mechanisms help to maintain a relatively constant blood flow to the brain. Blood flow to the skin, in contrast, can vary tremendously in response to regulation by sympathetic nerve stimulation.

The examination of cerebral and cutaneous blood flow is a study in contrasts. Cerebral blood flow is regulated primarily by intrinsic mechanisms; cutaneous blood flow is regulated by extrinsic mechanisms. Cerebral blood flow is relatively constant; cutaneous blood flow exhibits more variation than blood flow in any other organ. The brain is the organ that can least tolerate low rates of blood flow; the skin is the organ that can tolerate low rates the most.

Cerebral Circulation

When the brain is deprived of oxygen for a few seconds, a person loses consciousness; irreversible brain injury may occur after a few minutes. For these reasons, the cerebral blood flow is remarkably constant at about 750 ml per minute. This amounts to about 15% of the total cardiac output at rest.

Unlike the coronary and skeletal muscle blood flow, cerebral blood flow is not normally influenced by sympathetic nerve activity. Only when the mean arterial pressure rises to about 200 mmHg do sympathetic nerves cause a significant degree of vasoconstriction in the cerebral circulation. This vasoconstriction helps to protect small, thin-walled arterioles from bursting under the pressure and thus helps to prevent cerebrovascular accident (stroke).

In the normal range of arterial pressures, cerebral blood flow is regulated almost exclusively by intrinsic mechanisms.

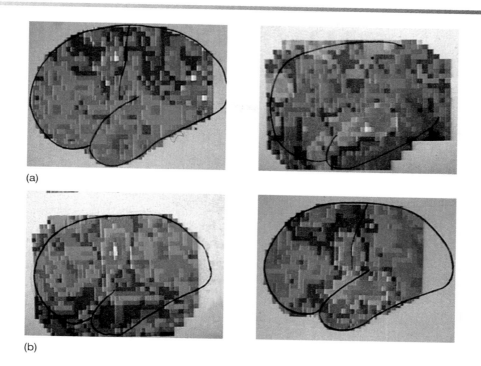

Figure 14.19

A computerized picture of blood-flow distribution in the brain after injecting the carotid artery with a radioactive isotope. In (a), on the left, the subject followed a moving object with his eyes. High activity is seen over the occipital lobe of the brain. In (a), on the right, the subject listened to spoken words. Notice that the high activity is seen over the temporal lobe (the auditory cortex). In (b), on the left, the subject moved his fingers on the side of the body opposite to the cerebral hemisphere being studied. In (b), on the right, the subject counted to 20. High activity is shown over the mouth area of the motor cortex, the supplementary motor area, and the auditory cortex.

These mechanisms help ensure a constant rate of blood flow despite changes in systemic arterial pressure—a process called *autoregulation*. The autoregulation of cerebral blood flow is achieved by both myogenic and metabolic mechanisms.

Myogenic Regulation

Myogenic regulation occurs when there is variation in systemic arterial pressure. The cerebral arteries automatically dilate when the blood pressure falls and constrict when the pressure rises. This helps to maintain a constant flow rate during the normal pressure variations that occur during rest, exercise, and emotional states.

The cerebral vessels are also sensitive to the carbon dioxide concentration of arterial blood. When the carbon dioxide concentration rises, as a result of inadequate ventilation (hypoventilation), the cerebral arterioles dilate. This is believed to be due to decreases in the pH of cerebrospinal fluid rather than to a direct effect of CO_2 on the cerebral vessels. Conversely, when the arterial CO_2 falls below normal during hyperventilation, the cerebral vessels constrict. The resulting decrease in cerebral blood flow is responsible for the dizziness that occurs during hyperventilation.

Metabolic Regulation

The cerebral arterioles are exquisitely sensitive to local changes in metabolic activity, so that those brain regions with the highest metabolic activity get the most blood. Indeed, areas of the brain that control specific processes have been mapped by the changing patterns of blood flow that result when these areas are activated. Visual and auditory stimuli, for example, increase blood flow to the appropriate sensory areas of the cerebral cortex, whereas motor activities, such as movements of the eyes, arms, and organs of speech, result in different patterns of blood flow (fig. 14.19).

The exact mechanisms by which increases in neural activity in a particular area of the brain elicit local vasodilation are not completely understood. There is evidence, however, that local cerebral vasodilation may be caused by K^+, which is released from active neurons during repolarization. It has been proposed that astrocytes may take up this extruded K^+ near the active neurons and then release the K^+ through their perivascular feet (chapter 7) surrounding arterioles, thereby causing the arterioles to dilate.

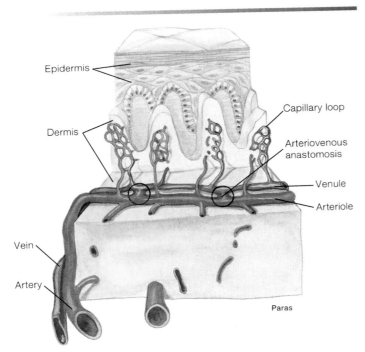

Epidermis

Dermis

Vein

Artery

Capillary loop

Arteriovenous anastomosis

Venule

Arteriole

Paras

Figure 14.20

Circulation in the skin showing arteriovenous anastomoses. These vessels function as shunts, allowing blood to be diverted directly from the arteriole to the venule and thus bypass superficial capillary loops.

Cutaneous Blood Flow

The skin is the outer covering of the body and as such serves as the first line of defense against invasion by disease-causing organisms. The skin, as the interface between the internal and external environments, also helps to maintain a constant deep-body temperature despite changes in the ambient (external) temperature—a process called *thermoregulation*. The thinness and extensiveness of the skin (1.0–1.5 mm thick; 1.7–1.8 square meters in surface area) make it an effective radiator of heat when the body temperature is greater than the ambient temperature. The transfer of heat from the body to the external environment is aided by the flow of warm blood through capillary loops near the surface of the skin.

Blood flow through the skin is adjusted to maintain deep-body temperature at about 37°C (98.6°F). These adjustments are made by variations in the degree of constriction or dilation of ordinary arterioles and of unique **arteriovenous anastomoses** (fig. 14.20). These latter vessels, found predominantly in the fingertips, palms of the hands, toes, soles of the feet, ears, nose, and lips, shunt (divert) blood directly from arterioles to deep venules, thus bypassing superficial capillary loops. Both the ordinary arterioles and the arteriovenous anastomoses are innervated by sympathetic nerve fibers. When the ambient temperature is low, sympathetic nerves stimulate cutaneous vasoconstriction; cutaneous blood flow is thus decreased, so that less heat will be lost from the body. Since the arteriovenous anastomoses also constrict, the skin may appear rosy as a

result of the fact that blood is diverted to the superficial capillary loops. In spite of this rosy appearance, however, the total cutaneous blood flow and rate of heat loss is lower than under usual conditions.

Skin can tolerate an extremely low blood flow in cold weather because its metabolic rate decreases when the ambient temperature decreases. In cold weather, therefore, the skin requires less blood. As a result of exposure to extreme cold, however, blood flow to the skin can be so severely restricted that the tissue dies—a condition known as *frostbite*. Blood flow to the skin can vary from less than 20 ml per minute at maximal vasoconstriction to as much as 3 to 4 L per minute at maximal vasodilation.

As the temperature warms, cutaneous arterioles in the hands and feet dilate as a result of decreased sympathetic nerve activity. Continued warming causes dilation of arterioles in other areas of the skin. If the resulting increase in cutaneous blood flow is not sufficient to cool the body, secretion of the sweat glands may be stimulated. Sweat helps to cool the body as it evaporates from the surface of the skin. Also, the sweat glands secrete **bradykinin,** a polypeptide that stimulates vasodilation. This increases blood flow to the skin and to the sweat glands, so that larger volumes of more dilute sweat are produced.

Under the usual conditions of ambient temperature, the cutaneous vascular resistance is high and the blood flow is low when a person is not exercising. In the pre-exercise state of fight or flight, sympathetic nerve activity reduces cutaneous blood flow still further. During exercise, however, the need to maintain a deep-body temperature takes precedence over the need to maintain an adequate systemic blood pressure. As the body temperature rises during exercise, vasodilation in cutaneous vessels occurs together with vasodilation in the exercising muscles. This can produce an even greater lowering of total peripheral resistance. If exercise is performed in hot and humid weather and if restrictive clothing is worn that increases skin temperature and cutaneous vasodilation, a dangerously low blood pressure may be produced after exercise has ceased and the cardiac output has declined. People have lost consciousness and have even died as a result.

Changes in cutaneous blood flow occur as a result of changes in sympathetic nerve activity. Since the activity of the sympathetic nervous system is controlled by the brain, emotional states, acting through control centers in the medulla oblongata, can affect sympathetic activity and cutaneous blood flow. During fear reactions, for example, vasoconstriction in the skin, along with activation of the sweat glands, can produce a pallor and a "cold sweat." Other emotions may cause vasodilation and blushing.

---

1. Define the term *autoregulation* and describe how this process is accomplished in the cerebral circulation.

2. Explain how hyperventilation can cause dizziness.

3. Explain how cutaneous blood flow is adjusted to maintain a constant deep-body temperature.

Blood Pressure

The pressure of the arterial blood is regulated by the blood volume, total peripheral resistance, and the cardiac rate. Regulatory mechanisms adjust these factors in a negative feedback manner to compensate for deviations. Arterial pressure rises and falls as the heart goes through systole and diastole.

Resistance to flow in the arterial system is greatest in the arterioles because these vessels have the smallest diameters. Although the total blood flow through a system of arterioles must be equal to the flow in the larger vessel that gave rise to those arterioles, the narrow diameter of each arteriole reduces the flow rate in each according to Poiseuille's law. Blood flow rate and pressure are thus reduced in the capillaries, which are located downstream of the high resistance imposed by the arterioles. The blood pressure upstream of the arterioles—in the medium and large arteries—is correspondingly increased (fig. 14.21).

The blood pressure and flow rate within the capillaries are further reduced by the fact that their total cross-sectional area is much greater, due to their large number, than the cross-sectional areas of the arteries and arterioles (fig. 14.22). Thus, although each capillary is much narrower than each arteriole, the capillary beds served by arterioles do not provide as great a resistance to blood flow as do the arterioles.

Variations in the diameter of arterioles due to vasoconstriction and vasodilation thus simultaneously affect both blood flow through capillaries and the *arterial blood pressure* "upstream" from the capillaries. An increase in total peripheral resistance due to vasoconstriction of arterioles can raise arterial blood pressure. Blood pressure can also be raised by an increase in the cardiac output. This may be due to elevations in cardiac rate or stroke volume, which in turn are affected by other factors. The three most important variables affecting blood pressure are the **cardiac rate, stroke volume** (determined primarily by the **blood volume**), and **total peripheral resistance.** An increase in any of these, if not compensated for by a decrease in another variable, will result in an increased blood pressure.

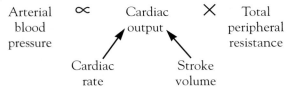

Blood pressure can thus be regulated by the kidneys, which control blood volume and thus stroke volume, and by the sympathoadrenal system. Increased activity of the sympathoadrenal system can raise blood pressure by stimulating vasoconstriction of arterioles (thus raising total peripheral resistance) and by promoting an increased cardiac output. Sympathetic stimulation can also affect blood volume indirectly, by stimulating constriction of renal blood vessels and thus reducing urine output.

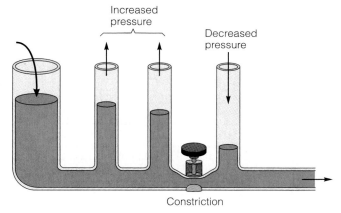

Figure 14.21

A constriction increases blood pressure upstream (analogous to the arterial pressure) and decreases pressure downstream (analogous to capillary and venous pressure).

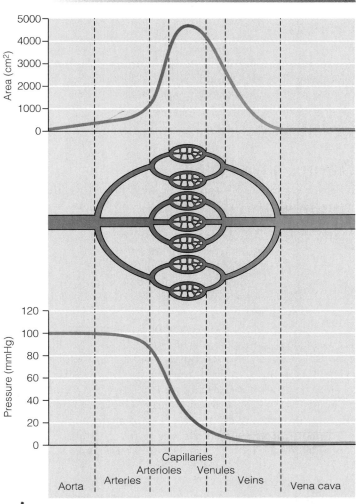

Figure 14.22

As blood passes from the aorta to the smaller arteries, arterioles, and capillaries, the cross-sectional area increases as the pressure decreases.

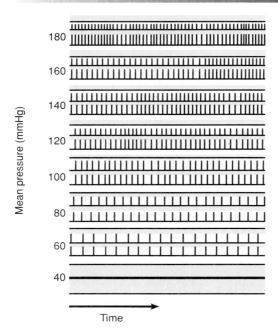

Figure 14.23

Action potential frequency in sensory nerve fibers from baroreceptors in the carotid sinus and aortic arch. As the blood pressure increases, the baroreceptors become increasingly stretched. This results in an increase in the frequency of action potentials that are transmitted to the cardiac and vasomotor control centers in the medulla oblongata.

Baroreceptor Reflex

In order for blood pressure to be maintained within limits, specialized receptors for pressure are needed. These **baroreceptors** are stretch receptors located in the *aortic arch* and in the *carotid sinuses*. An increase in pressure causes the walls of these arterial regions to stretch, increasing the frequency of action potentials along sensory nerve fibers (fig. 14.23). A fall in pressure below the normal range, by contrast, causes a decrease in the frequency of action potentials produced by these sensory nerve fibers.

Sensory nerve activity from the baroreceptors ascends via the vagus and glossopharyngeal nerves to the medulla oblongata, which directs the autonomic system to respond appropriately. **Vasomotor control centers** in the medulla control vasoconstriction/vasodilation, and hence help regulate total peripheral resistance. **Cardiac control centers** in the medulla regulate the cardiac rate (fig. 14.24). Acting through the activity of motor fibers within the vagus and sympathetic nerves controlled by these brain centers, the baroreceptors function to buffer blood pressure changes so that fluctuations in pressure are minimized.

The baroreceptor reflex is activated whenever blood pressure either increases or decreases. The reflex is somewhat more sensitive to decreases in pressure than to increases, and is more sensitive to sudden changes in pressure than to more gradual changes. A good example of the importance of the baroreceptor

Since the baroreceptor reflex may require a few seconds to be fully effective, many people feel dizzy and disoriented if they stand up too rapidly. If the baroreceptor sensitivity is abnormally reduced, perhaps by atherosclerosis, an uncompensated fall in pressure may occur upon standing. This condition—called **postural,** or **orthostatic, hypotension** (hypotension = low blood pressure)—can make a person feel extremely dizzy or even faint because of inadequate perfusion of the brain.

reflex in normal physiology is its activation whenever a person goes from a lying to a standing position.

When a person goes from a lying to a standing position, there is a shift of 500 to 700 ml of blood from the veins of the thoracic cavity to veins in the lower extremities, which expand to contain the extra volume of blood. This pooling of blood reduces the venous return and cardiac output. The resulting fall in blood pressure is almost immediately compensated for by the baroreceptor reflex. A decrease in baroreceptor sensory information, traveling in the glossopharyngeal (ninth cranial) and the vagus (tenth cranial) nerves to the medulla oblongata, inhibits parasympathetic activity and promotes sympathetic nerve activity. This produces an increase in cardiac rate and vasoconstriction, which help to maintain an adequate blood pressure upon standing (fig. 14.25).

The baroreceptor reflex can also mediate the opposite response. When the blood pressure rises above an individual's normal range, the baroreceptor reflex causes a slowing of the cardiac rate and vasodilation. Manual massage of the carotid sinus, a procedure sometimes employed by physicians to reduce tachycardia and lower blood pressure, also evokes this reflex.

Valsalva's maneuver is the term used to describe an expiratory effort against a closed glottis (which prevents the air from escaping—see chapter 15). This maneuver, commonly performed during forceful defecation or when lifting heavy weights, increases the intrathoracic pressure. Compression of the thoracic veins causes a fall in venous return and cardiac output, thus lowering arterial blood pressure. The lowering of arterial pressure stimulates the baroreceptor reflex, resulting in tachycardia and increased total peripheral resistance. When the glottis is finally opened and the air is exhaled, the cardiac output returns to normal but the total peripheral resistance is still elevated, causing a rise in blood pressure. The blood pressure is then brought back to normal by the baroreceptor reflex, which causes a slowing of the heart rate. These fluctuations in cardiac output and blood pressure can be dangerous in people with cardiovascular disease. Even healthy people are advised to exhale normally when lifting weights.

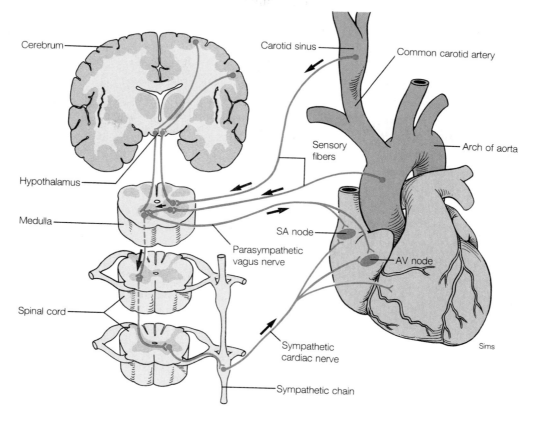

Figure 14.24

The baroreceptor reflex. Sensory stimuli from baroreceptors in the carotid sinus and the aortic arch, acting via control centers in the medulla oblongata, affect the activity of sympathetic and parasympathetic nerve fibers in the heart.

Such carotid massage should be used cautiously, however, because the intense vagus-nerve-induced slowing of the cardiac rate could cause loss of consciousness (as occurs in emotional fainting). Manual massage of both carotid sinuses simultaneously can even cause cardiac arrest in susceptible people.

Atrial Stretch Reflexes

There are several other reflexes, in addition to the baroreceptor reflex, that help to regulate blood pressure. The reflex control of ADH secretion by osmoreceptors in the hypothalamus and the control of angiotensin II production and aldosterone secretion by the juxtaglomerular apparatus of the kidneys have been previously discussed. Antidiuretic hormone and aldosterone increase blood pressure by increasing blood volume, and angiotensin II stimulates vasoconstriction to cause an increase in blood pressure.

Other reflexes that are important to blood pressure regulation are initiated by **atrial stretch receptors** located in the atria of the heart. These receptors are activated by increased venous return to the heart and, in response, stimulate (1) reflex tachycardia, as a result of increased sympathetic nerve activity; (2) inhibition of ADH secretion, resulting in larger volumes of urine excretion and a lowering of blood volume; and (3) increased secretion of atrial natriuretic factor (ANF). The ANF, as previously

discussed, lowers blood volume by increasing urinary salt and water excretion and by antagonizing the actions of angiotensin II.

Measurement of Blood Pressure

The first documented measurement of blood pressure was accomplished by Stephen Hales (1677–1761), an English physiologist. Hales inserted a cannula into the artery of a horse and measured the height to which blood would rise in the vertical tube. The height of this blood column bounced between the **systolic pressure** at its highest and the **diastolic pressure** at its lowest, as the heart went through its cycle of systole and diastole. Modern clinical blood pressure measurements, fortunately, are less direct. The indirect, or **auscultatory,** method of blood pressure measurement is based on the correlation of blood pressure and arterial sounds.

In the auscultatory method, an inflatable rubber bladder within a cloth cuff is wrapped around the upper arm and a stethoscope is applied over the brachial artery (fig. 14.26). The artery is normally silent before inflation of the cuff because blood normally travels in a smooth *laminar flow* through the arteries. The term *laminar* means layered—blood in the central axial stream moves the fastest, and blood flowing closer to the artery wall moves more slowly. There is little transverse movement between these layers that would produce mixing.

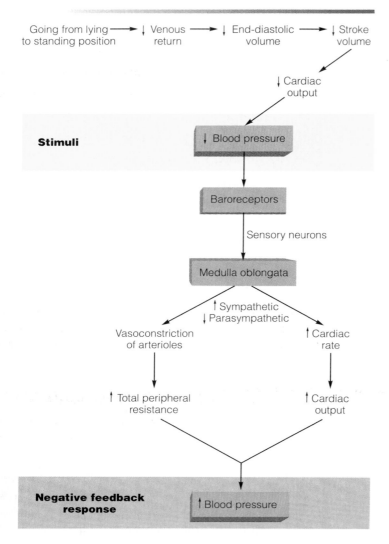

Figure 14.25

The negative feedback control of blood pressure by the baroreceptor reflex. This reflex helps to maintain an adequate blood pressure upon standing.

The laminar flow that normally occurs in arteries produces little vibration and is thus silent. When the artery is pinched, however, blood flow through the constriction becomes turbulent. This causes the artery to vibrate and produce sounds, much like the sounds produced by water through a kink in a garden hose. The tendency of the cuff pressure to constrict the artery is opposed by the blood pressure. Thus, in order to constrict the artery, the cuff pressure must be greater than the diastolic blood pressure. If the cuff pressure is also greater than the systolic blood pressure, the artery will be pinched off and silent. *Turbulent flow* and sounds produced by vibrations of the artery as a result of this flow, therefore, occur only when the cuff pressure is greater than the diastolic blood pressure and less than the systolic pressure.

Let's say that a person has a systolic pressure of 120 mmHg and a diastolic pressure of 80 mmHg (the average normal values). When the cuff pressure is between 80 and 120 mmHg, the artery will be closed during diastole and open during systole.

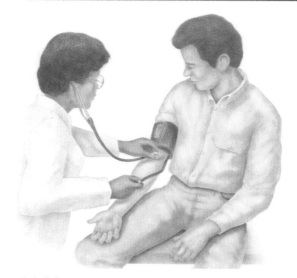

Figure 14.26

The use of a pressure cuff and sphygmomanometer to measure blood pressure.

As the artery begins to open with every systole, turbulent flow of blood through the constriction will create vibrations that are known as the **sounds of Korotkoff,** as shown in figure 14.27. These are usually "tapping" sounds because the artery becomes constricted, blood flow stops, and silence resumes with every diastole. It should be understood that the sounds of Korotkoff are *not* "lub-dub" sounds produced by closing of the heart valves (those sounds can only be heard on the chest, not on the brachial artery).

Initially, the cuff is usually inflated to produce a pressure greater than the systolic pressure so that the artery is pinched off and silent. The pressure in the cuff is read from an attached meter called a *sphygmomanometer*. A valve is then turned to allow the release of air from the cuff, causing a gradual decrease in cuff pressure. When the cuff pressure is equal to the systolic pressure, the **first Korotkoff sound** is heard as blood passes in a turbulent flow through the constricted opening of the artery.

Korotkoff sounds will continue to be heard at every systole as long as the cuff pressure remains greater than the diastolic pressure. When the cuff pressure becomes equal to or less than the diastolic pressure, the sounds disappear because the artery remains open, laminar flow occurs, and the vibrations of the artery stop (fig. 14.28). The **last Korotkoff sound** thus occurs when the cuff pressure is equal to the diastolic pressure.

Different phases in the measurement of blood pressure are identified on the basis of the quality of the Korotkoff sounds (fig. 14.29). In some people, the Korotkoff sounds do not disappear even when the cuff pressure is reduced to zero (zero pressure means that it is equal to atmospheric pressure). In these cases— and often routinely—the onset of muffling of the sounds (phase 4 in fig. 14.29) is used as an indication of diastolic pressure rather than the onset of silence (phase 5). Normal blood pressure values are shown in table 14.10.

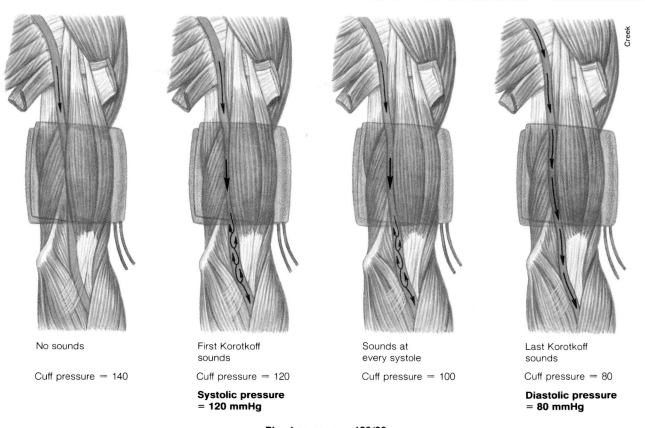

No sounds

Cuff pressure = 140

First Korotkoff
sounds

Cuff pressure = 120

**Systolic pressure
= 120 mmHg**

Sounds at
every systole

Cuff pressure = 100

Last Korotkoff
sounds

Cuff pressure = 80

**Diastolic pressure
= 80 mmHg**

Blood pressure = 120/80

Figure 14.27

Korotkoff sounds are produced by the turbulent flow of blood through the partially constricted brachial artery. This occurs when the cuff pressure is greater than the diastolic pressure but less than the systolic pressure.

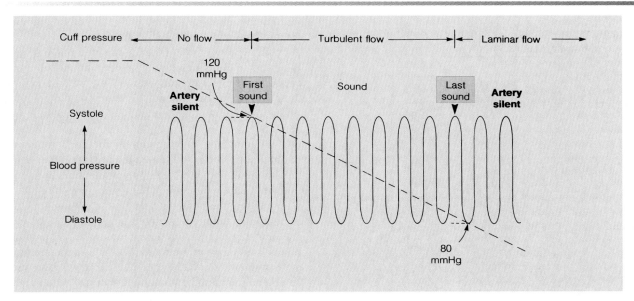

Figure 14.28

The indirect, or auscultatory, method of blood pressure measurement. Korotkoff sounds, produced by turbulent blood flow through a constricted artery, occur whenever the cuff pressure is lower than the systolic blood pressure and greater than the diastolic blood pressure. As a result, the first Korotkoff sound is heard when the cuff pressure is equal to the systolic blood pressure, and the last sound is heard when the cuff pressure and diastolic blood pressure are equal.

Table 14.10 Normal Arterial Blood Pressure at Different Ages

Age	Systolic Men	Systolic Women	Diastolic Men	Diastolic Women	Age	Systolic Men	Systolic Women	Diastolic Men	Diastolic Women
1 day	70				16 years	118	116	73	72
3 days	72				17 years	121	116	74	72
9 days	73				18 years	120	116	74	72
3 weeks	77				19 years	122	115	75	71
3 months	86				20–24 years	123	116	76	72
6–12 months	89	93	60	62	25–29 years	125	117	78	74
1 year	96	95	66	65	30–34 years	126	120	79	75
2 years	99	92	64	60	35–39 years	127	124	80	78
3 years	100	100	67	64	40–44 years	129	127	81	80
4 years	99	99	65	66	45–49 years	130	131	82	82
5 years	92	92	62	62	50–54 years	135	137	83	84
6 years	94	94	64	64	55–59 years	138	139	84	84
7 years	97	97	65	66	60–64 years	142	144	85	85
8 years	100	100	67	68	65–69 years	143	154	83	85
9 years	101	101	68	69	70–74 years	145	159	82	85
10 years	103	103	69	70	75–79 years	146	158	81	84
11 years	104	104	70	71	80–84 years	145	157	82	83
12 years	106	106	71	72	85–89 years	145	154	79	82
13 years	108	108	72	73	90–94 years	145	150	78	79
14 years	110	110	73	74	95–106 years	145	149	78	81
15 years	112	112	75	76					

Source: From K. Diem and C. Lentner, editors, *Documenta Geigy Scientific Tables,* 7th ed. Copyright ©1970 J. R. Geigy S. A., Basel, Switzerland. Used by permission.

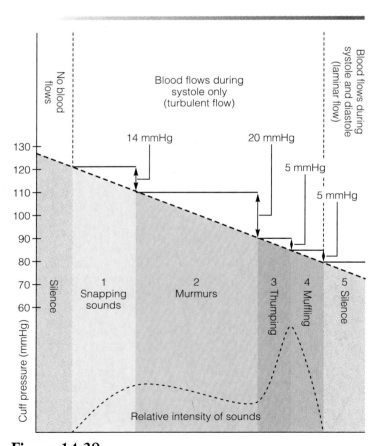

Figure 14.29
The five phases of blood-pressure measurement.

Pulse Pressure and Mean Arterial Pressure

When someone "takes a pulse," he or she palpates an artery (for example, the radial artery) and feels the expansion of the artery occur in response to the beating of the heart; the pulse rate is thus a measure of the cardiac rate. The expansion of the artery with each pulse occurs as a result of the rise in blood pressure within the artery as the artery receives the volume of blood ejected by a stroke of the left ventricle.

The pulse is thus produced by the **pulse pressure,** which is equal to the difference between the systolic and diastolic pressures. If a person has a blood pressure of 120/80 (systolic/diastolic), therefore, the pulse pressure would be 40 mmHg.

Pulse pressure = systolic pressure – diastolic pressure

At diastole in this example the aortic pressure equals 80 mmHg. When the left ventricle contracts, the intraventricular pressure rises above 80 mmHg and ejection begins. As a result, the amount of blood in the aorta increases by the amount ejected from the left ventricle (the stroke volume). Due to the increase in volume, there is an increase in blood pressure. The pressure in the brachial artery, where blood pressure measurements are commonly taken, therefore increases to 120 mmHg in this example. The rise in pressure from diastolic to systolic levels (pulse pressure) is thus a reflection of the stroke volume.

The **mean arterial pressure** represents the average arterial pressure during the cardiac cycle. This value is significant because it is the difference between this pressure and the venous pressure that drives blood through the capillary beds of organs. The mean arterial pressure is not a simple arithmetic average because the period of diastole is longer than the period of systole. Mean arterial pressure can most correctly be approximated by adding one-third of the pulse pressure to the diastolic pressure. If a person has a blood pressure of 120/80, for example, the mean arterial pressure would be 80 + 1/3 (40) = 93 mmHg.

Mean arterial pressure = diastolic pressure + 1/3 pulse pressure

A rise in total peripheral resistance and cardiac rate increases the diastolic pressure more than it increases the systolic pressure. When the baroreceptor reflex is activated by going from a lying to a standing position, for example, the diastolic pressure usually increases by 5 to 10 mmHg, whereas the systolic pressure is either unchanged or is slightly reduced (as a result of decreased venous return). People with hypertension (high blood pressure), who usually have elevated total peripheral resistance and cardiac rates, likewise have a greater increase in diastolic than in systolic pressure. Dehydration or blood loss results in decreased cardiac output, and thus also produces a decrease in pulse pressure.

An increase in cardiac output, in contrast, raises the systolic pressure more than it raises the diastolic pressure (although both pressures do rise). This occurs during exercise, for example, when the blood pressure may rise to values as high as 200/100 (yielding a pulse pressure of 100 mmHg).

---

1. Describe the relationship between blood pressure and the total cross-sectional area of arteries, arterioles, and capillaries. Describe how arterioles influence blood flow through capillaries and arterial blood pressure.

2. Describe how the baroreceptor reflex helps to compensate for a fall in blood pressure. Explain why a person who is severely dehydrated will have a rapid pulse.

3. Describe how the sounds of Korotkoff are produced and how these sounds are used to measure blood pressure.

4. Define *pulse pressure* and explain the physiological significance of this measurement.

Hypertension, Shock, and Congestive Heart Failure

An understanding of the normal physiology of the cardiovascular system is prerequisite to the study of its pathophysiology, or mechanisms of abnormal function. Since the mechanisms that regulate cardiac output, blood flow, and blood pressure are highlighted in particular disease states, a study of pathophysiology at this time can strengthen your understanding of the mechanisms involved in normal function.

Table 14.11	Classification of Blood Pressure for Adults Age 18 Years and Older	
Category	**Systolic mmHg**	**Diastolic mmHg**
Normal	<130	<85
High normal	130–139	85–89
Hypertension*		
STAGE 1 (Mild)	140–159	90–99
STAGE 2 (Moderate)	160–179	110–109
STAGE 3 (Severe)	180–209	110–119
STAGE 4 (Very Severe)	≥210	≥120

Source: *Fifth Report of the Joint National Committee on Detection, Evaluation, and Treatment of High Blood Pressure,* National Institutes of Health, Washington, D.C., 1993.

*Based on the average of two or more readings taken at each of two or more visits following an initial screening.

Hypertension

Approximately 20% of all adults in the United States have *hypertension*—blood pressure in excess of the normal range for a person's age and sex. Hypertension that is a result of (secondary to) known disease processes is logically called **secondary hypertension.** Of the hypertensive population, secondary hypertension accounts for only about 10%. Hypertension that is the result of complex and poorly understood processes is not so logically called **primary,** or **essential, hypertension.** Hypertension in adults is defined by a systolic pressure greater than 140 mmHg and/or a diastolic pressure greater than 90 mmHg (table 14.11).

Diseases of the kidneys and arteriosclerosis of the renal arteries can cause secondary hypertension because of high blood volume. More commonly, the reduction of renal blood flow can raise blood pressure by stimulating the secretion of vasoactive chemicals from the kidneys. Experiments in which the renal artery is pinched, for example, produce hypertension that is associated (at least initially) with elevated renin secretion. These and other causes of secondary hypertension are summarized in table 14.12.

Essential Hypertension

The vast majority of people with hypertension have essential hypertension. An increased total peripheral resistance is a universal characteristic of this condition. Cardiac rate and the cardiac output are elevated in many, but not all, of these cases.

The secretion of renin, which is correlated with angiotensin II production and aldosterone secretion, is likewise variable. Although some people with essential hypertension have low renin secretion, most have either normal or elevated levels of renin secretion. Renin secretion in the normal range is inappropriate for people with hypertension, since high blood pressure should inhibit renin secretion and, through a lowering of aldosterone, result in greater excretion of salt and water. Inappropriately high levels of renin secretion could thus contribute to hypertension by promoting (via stimulation of aldosterone secretion) salt and water retention and high blood volume.

Table 14.12   Possible Causes of Secondary Hypertension

System Involved	Examples	Mechanisms
Kidneys	Kidney disease	Decreased urine formation
	Renal artery disease	Secretion of vasoactive chemicals
Endocrine	Excess catecholamines (tumor of adrenal medulla)	Increased cardiac output and total peripheral resistance
	Excess aldosterone (Conn's syndrome)	Excess salt and water retention by the kidneys
Nervous	Increased intracranial pressure	Activation of sympathoadrenal system
	Damage to vasomotor center	Activation of sympathoadrenal system
Cardiovascular	Complete heart block; patent ductus arteriosus	Increases stroke volume
	Arteriosclerosis of aorta; coarctation of aorta	Decreased distensibility of aorta

Table 14.13   Mechanisms of Action of Selected Antihypertensive Drugs

Category of Drugs	Examples	Mechanisms
Extracellular fluid volume depletors	Thiazide diuretics	Increase volume of urine excreted, thus lowering blood volume
Sympathoadrenal system inhibitors	Clonidine; alpha-methyldopa	Act to decrease sympathoadrenal stimulation by binding to α_2-adrenergic receptors in the brain
	Guanethidine; reserpine	Deplete norepinephrine from sympathetic nerve endings
	Propranolol; atenolol	Block beta-adrenergic receptors, decreasing cardiac output and/or renin secretion
	Phentolamine	Blocks alpha-adrenergic receptors, decreasing sympathetic vasoconstriction
Direct vasodilators	Hydralazine; minoxidil sodium nitroprusside	Cause vasodilation by acting directly on vascular smooth muscle
Calcium channel blockers	Verapamil; diltiazem	Inhibit diffusion of Ca^{++} into vascular smooth muscle cells, causing vasodilation and reduced peripheral resistance
Angiotensin converting enzyme (ACE) inhibitors	Captopril; benazepril	Inhibit the conversion of angiotension I into angiotension II

Sustained high stress (acting via the sympathetic nervous system) and high salt intake appear to act synergistically in the development of hypertension. There is some evidence that Na^+ enhances the vascular response to sympathetic stimulation. Further, sympathetic nerve stimulation can cause constriction of the renal blood vessels and thus decrease the excretion of salt and water.

As an adaptive response to prolonged high blood pressure, the arterial wall becomes thickened. This response can lead to arteriosclerosis and results in an even greater increase in total peripheral resistance, thus raising blood pressure still more in a positive feedback fashion.

The interactions between salt intake, sympathetic nerve activity, cardiovascular responses to sympathetic nerve activity, kidney function, and genetics make it difficult to sort out the cause-and-effect sequence that leads to essential hypertension. Many researchers have suggested that there is no single cause-and-effect sequence but rather a web of causes and effects. This view is currently controversial.

Dangers of Hypertension

If other factors remain constant, blood flow increases as arterial blood pressure increases. People with hypertension thus have adequate perfusion of their organs with blood until the hypertension causes vascular damage. Hypertension, as a result, is usually asymptomatic (without symptoms) until a dangerous amount of vascular damage is produced.

Hypertension is dangerous for a number of reasons. First, high arterial pressure increases the afterload, making it more difficult for the ventricles to eject blood. The heart, then, must work harder, which can result in pathological changes in heart structure and function, leading to congestive heart failure. Additionally, high pressure may damage cerebral blood vessels, leading to cerebrovascular accident (stroke). Finally, hypertension contributes to the development of atherosclerosis, which can itself lead to heart disease and stroke as previously described.

Preeclampsia is a condition of pregnancy characterized by high blood pressure together with proteinuria (the presence of proteins in the urine). For reasons discussed in chapter 17, only negligible amounts of proteins are normally found in urine, and the loss of plasma proteins in the urine can cause edema. The danger of preeclampsia is that it could quickly degenerate into a state called eclampsia, during which seizures occur. This can be life-threatening, and so the woman with preeclampsia is immediately treated for her symptoms and the fetus is delivered as quickly as possible.

Table 14.14 Signs of Shock

	Early Sign	Late Sign
Blood Pressure	Decreased pulse pressure Increased diastolic pressure	Decreased systolic pressure
Urine	Decreased Na^+ concentration Increased osmolality	Decreased volume
Blood pH	Increased pH (alkalosis) due to hyperventilation	Decreased pH (acidosis) due to "metabolic" acids
Effects of Poor Tissue Perfusion	Slight restlessness; occasionally warm, dry skin	Cold, clammy skin; "cloudy" senses

Source: From *Principles and Techniques of Critical Care*, Vol. 1, edited by R. F. Wilson. Copyright © 1977 F. A. Davis Company, Philadelphia. Used by permission.

Table 14.15 Cardiovascular Reflexes That Help to Compensate for Circulatory Shock

Organ(s)	Compensatory Mechanisms
Heart	Sympathoadrenal stimulation produces increased cardiac rate and increased stroke volume due to "positive inotropic effect" on myocardial contractility
Digestive tract and skin	Decreased blood flow due to vasoconstriction as a result of sympathetic nerve stimulation (alpha-adrenergic effect)
Kidneys	Decreased urine production as a result of sympathetic-nerve-induced constriction of renal arterioles; increased salt and water retention due to increased aldosterone and antidiuretic hormone (ADH) secretion

Treatment of Hypertension

The first form of treatment that is usually attempted is modification of lifestyle. This modification includes cessation of smoking, moderation of alcohol intake, and weight reduction, if applicable. It can also include the addition of regular physical exercise and a reduction in sodium intake. People with essential hypertension may have a potassium deficiency, and there is evidence that eating food that is rich in potassium may help to lower blood pressure. There is also evidence that supplementation of the diet with Ca^{++} may be of benefit, but this is more controversial.

If lifestyle modifications alone are insufficient, various drugs may be prescribed. Most commonly, these are *diuretics* that increase urine volume, thus decreasing blood volume and pressure. Drugs that block β_1-adrenergic receptors (such as atenolol) lower blood pressure by decreasing the cardiac rate and are also frequently prescribed. ACE inhibitors, calcium antagonists, and various vasodilators (table 14.13) may also be used in particular situations. Methyldopa, for example, may be given to treat hypertension of a pregnant woman.

Circulatory Shock

Circulatory shock occurs when there is inadequate blood flow and/or oxygen utilization by the tissues. Some of the signs of shock (table 14.14) are a result of inadequate tissue perfusion; other signs of shock are produced by cardiovascular responses that help to compensate for the poor tissue perfusion (table 14.15). When these compensations are effective, they (together with emergency medical care) are able to reestablish adequate tissue perfusion. In some cases, however, and for reasons that are not clearly understood, the shock may progress to an irreversible stage and death may result.

Hypovolemic Shock

The term **hypovolemic shock** refers to circulatory shock due to low blood volume, as might be caused by hemorrhage (bleeding), dehydration, or burns. This is accompanied by decreased blood pressure and decreased cardiac output. In response to these changes, the sympathoadrenal system is activated by means of the baroreceptor reflex. As a result, tachycardia is produced and vasoconstriction occurs in the skin, digestive tract, kidneys, and muscles. Decreased blood flow through the kidneys stimulates renin secretion and activation of the renin-angiotensin-aldosterone system. A person in hypovolemic shock thus has low blood pressure; a rapid pulse, cold, clammy skin; and a reduced urine output.

Since the resistance in the coronary and cerebral circulations is not increased, blood is diverted to the heart and brain at the expense of other organs. Interestingly, a similar response occurs in diving mammals and, to a lesser degree, in Japanese pearl divers during prolonged submersion. These responses help to deliver blood to the two organs that have the highest requirements for aerobic metabolism.

Vasoconstriction in organs other than the brain and heart raises total peripheral resistance, which helps (along with the reflex increase in cardiac rate) to compensate for the drop in blood pressure due to low blood volume. Constriction of arterioles also decreases capillary blood flow and capillary filtration pressure. As a result, less filtrate is formed. At the same time, the osmotic return of fluid to the capillaries is either unchanged or increased (during dehydration). The blood volume is thus raised at the expense of tissue fluid volume. Blood volume is also conserved by decreased urine production, which occurs as a result of vasoconstriction in the kidneys and the water-conserving effects of ADH and aldosterone, which are secreted in increased amounts during shock.

Septic Shock

Septic shock refers to a dangerously low blood pressure (hypotension) that may result from sepsis, or infection. This can occur through the action of a bacterial lipopolysaccharide called *endotoxin*. The mortality associated with septic shock is presently very high, estimated at 50% to 70%. According to recent information, endotoxin activates the enzyme nitric oxide synthetase within macrophages, which are cells that play an important role in the immune response (chapter 15). As previously discussed, nitric oxide synthetase produces nitric oxide, which promotes vasodilation and, as a result, a fall in blood pressure. Septic shock has recently been successfully treated with drugs that inhibit the production of nitric oxide.

Other Causes of Circulatory Shock

A rapid fall in blood pressure occurs in **anaphylactic shock** as a result of a severe allergic reaction (usually to bee stings or penicillin). This results from the widespread release of histamine, which causes vasodilation and thus decreases total peripheral resistance. A rapid fall in blood pressure also occurs in **neurogenic shock,** in which sympathetic tone is decreased, usually because of upper spinal cord damage or spinal anesthesia. **Cardiogenic shock** results from cardiac failure, as defined by a cardiac output that is inadequate to maintain tissue perfusion. This commonly results from infarction that causes the loss of a significant proportion of the myocardium.

Congestive Heart Failure

Cardiac failure occurs when the cardiac output is insufficient to maintain the blood flow required by the body. This may be due to heart disease—resulting from myocardial infarction or congenital defects—or to hypertension, which increases the afterload of the heart. The most common causes of left ventricular heart failure are myocardial infarction, aortic valve stenosis, and incompetence of the aortic and bicuspid (mitral) valves. Failure of the right ventricle is usually caused by prior failure of the left ventricle.

Heart failure can also result from disturbance in the electrolyte concentrations of the blood. Excessive plasma K^+ concentration decreases the resting membrane potential of myocardial cells; low blood Ca^{++} reduces excitation-contraction coupling. High blood K^+ and low blood Ca^{++} can thus cause the heart to stop in diastole. Conversely, low blood K^+ and high blood Ca^{++} can arrest the heart in systole.

The term *congestive* is often used in describing heart failure because of the increased venous volume and pressure that results. Failure of the left ventricle, for example, raises the left atrial pressure and produces pulmonary congestion and edema. This causes shortness of breath and fatigue; if severe, pulmonary

People with congestive heart failure are often treated with the drug *digitalis*. Digitalis appears to bind to and inhibit the action of Na^+/K^+ pumps in the cell membranes, causing a rise in the intracellular concentrations of Na^+. The increased availability of Na^+, in turn, stimulates the activity of another membrane transport carrier, which exchanges Na^+ for extracellular Ca^{++}. As a result, the intracellular concentrations of Ca^{++} are increased, which strengthens the contractions of the heart.

edema can be fatal. Failure of the right ventricle results in increased right atrial pressure, which produces congestion and edema in the systemic circulation.

The compensatory responses that occur during congestive heart failure are similar to those that occur during hypovolemic shock. Activation of the sympathoadrenal system stimulates cardiac rate, contractility of the ventricles, and constriction of arterioles. As in hypovolemic shock, renin secretion is increased and urine output is reduced.

As a result of these compensations, chronically low cardiac output is associated with elevated blood volume and dilation and hypertrophy of the ventricles. These changes can themselves be dangerous. Elevated blood volume places a work overload on the heart, and the enlarged ventricles have a higher metabolic requirement for oxygen. These problems are often treated with drugs that increase myocardial contractility (such as digitalis), drugs that are vasodilators (such as nitroglycerin), and diuretic drugs that lower blood volume by increasing the volume of urine excreted.

1. Explain how stress and a high-salt diet can contribute to hypertension. Also, explain how different drugs may act to lower blood pressure.
2. Using a flowchart to show cause and effect, explain why a person in hypovolemic shock may have a fast pulse and cold, clammy skin.
3. Describe the compensatory mechanisms that act to raise blood volume during cardiovascular shock.
4. Explain how septic shock may be produced.
5. Describe congestive heart failure and explain the compensatory responses that occur during this condition.

Summary

Cardiac Output p. 388

I. Cardiac rate is increased by sympathoadrenal stimulation and decreased by the effects of parasympathetic fibers that innervate the SA node.

II. Stroke volume is regulated both extrinsically and intrinsically.

   A. The Frank–Starling law of the heart describes the way the end-diastolic volume, through various degrees of myocardial stretching, influences the contraction strength of the myocardium and thus the stroke volume.

   B. The end-diastolic volume is called the preload. The total peripheral resistance, through its effect on arterial blood pressure, provides an afterload that acts to reduce the stroke volume.

   C. At a given end-diastolic volume, the amount of blood ejected depends on contractility. Strength of contraction is increased by sympathoadrenal stimulation.

III. The venous return of blood to the heart is dependent largely on the total blood volume and mechanisms that improve the flow of blood in veins.

   A. The total blood volume is regulated by the kidneys.

   B. The venous flow of blood to the heart is aided by the action of skeletal muscle pumps and the effects of breathing.

Blood Volume p. 392

I. Tissue fluid is formed from and returns to the blood.

   A. The hydrostatic pressure of the blood forces fluid from the arteriolar ends of capillaries into the interstitial spaces of the tissues.

   B. Since the colloid osmotic pressure of plasma is greater than that of tissue fluid, water returns by osmosis to the venular ends of capillaries.

   C. Excess tissue fluid is returned to the venous system by lymphatic vessels.

   D. Edema occurs when there is an accumulation of tissue fluid.

II. The kidneys control the blood volume by regulating the amount of filtered fluid that will be reabsorbed.

   A. Antidiuretic hormone stimulates reabsorption of water from the kidney filtrate and thus acts to maintain the blood volume.

   B. A decrease in blood flow through the kidneys activates the renin-angiotensin system.

   C. Angiotensin II stimulates vasoconstriction and the secretion of aldosterone by the adrenal cortex.

   D. Aldosterone acts on the kidneys to promote the retention of salt and water.

Vascular Resistance to Blood Flow p. 397

I. According to Poiseuille's law, blood flow is directly related to the pressure difference between the two ends of a vessel and is inversely related to the resistance to blood flow through the vessel.

II. Extrinsic regulation of vascular resistance is provided mainly by the sympathetic nervous system, which stimulates vasoconstriction of arterioles in the viscera and skin.

III. Intrinsic control of vascular resistance allows organs to autoregulate their own blood flow rates.

   A. Myogenic regulation occurs when vessels constrict or dilate as a direct response to a rise or fall in blood pressure.

   B. Metabolic regulation occurs when vessels dilate in response to the local chemical environment within the organ.

Blood Flow to the Heart and Skeletal Muscles p. 402

I. The heart normally respires aerobically because of its extensive capillary supply and high myoglobin and enzyme content.

II. During exercise, when the heart's metabolism increases, intrinsic metabolic mechanisms stimulate vasodilation of the coronary vessels, and thus increase coronary blood flow.

III. Just prior to exercise and at the start of exercise, blood flow through skeletal muscles increases due to vasodilation caused by the activity of cholinergic sympathetic nerve fibers. During exercise, intrinsic metabolic vasodilation occurs.

IV. Since cardiac output can increase by a factor of five or more during exercise, the heart and skeletal muscles receive an increased proportion of a higher total blood flow.

   A. The cardiac rate increases due to lower activity of the vagus nerve and higher activity of the sympathetic nerve.

   B. The venous return is greater because of higher activity of the skeletal muscle pumps and increased breathing.

   C. Increased contractility of the heart, combined with a decrease in total peripheral resistance, can result in a higher stroke volume.

Blood Flow to the Brain and Skin p. 406

I. Cerebral blood flow is regulated both myogenically and metabolically.

   A. Cerebral vessels automatically constrict if the systemic blood pressure rises too high.

   B. Metabolic products cause local vessels to dilate and supply more active areas with more blood.

II. The skin has unique arteriovenous anastomoses, which can shunt the blood away from surface capillary loops.
   A. The activity of sympathetic nerve fibers causes constriction of cutaneous arterioles.
   B. As a thermoregulatory response, there is increased cutaneous blood flow and increased flow through surface capillary loops when the body temperature rises.

Blood Pressure   p. 409

I. Baroreceptors in the aortic arch and carotid sinuses affect the cardiac rate and the total peripheral resistance via the sympathetic nervous system.
   A. The baroreceptor reflex causes pressure to be maintained when an upright posture is assumed. This reflex can cause a lowered pressure when the carotid sinuses are massaged.
   B. Other mechanisms that affect blood volume help to regulate blood pressure.

II. Blood pressure is commonly measured indirectly by auscultation of the brachial artery when a pressure cuff is inflated and deflated.
   A. The first sound of Korotkoff, caused by turbulent flow of blood through a constriction in the artery, occurs when the cuff pressure equals the systolic pressure.
   B. The last sound of Korotkoff is heard when the cuff pressure equals the diastolic blood pressure.

III. The mean arterial pressure represents the driving force for blood flow through the arterial system.

Hypertension, Shock, and Congestive Heart Failure   p. 415

I. Hypertension, or high blood pressure, is classified as either primary or secondary.
   A. Primary hypertension, also called essential hypertension, may be the result of the interaction of many

mechanisms that raise the blood volume, cardiac output, and/or peripheral resistance.
   B. Secondary hypertension is the direct result of known specific diseases.

II. Circulatory shock occurs when there is inadequate delivery of oxygen to the organs of the body.
   A. In hypovolemic shock, low blood volume causes low blood pressure that may progress to an irreversible state.
   B. The fall in blood volume and pressure stimulates various reflexes that produce a rise in cardiac rate, a shift of fluid from the tissues into the vascular system, a decrease in urine volume, and vasoconstriction.

III. Congestive heart failure occurs when the cardiac output is insufficient to supply the blood flow required by the body. The term *congestive* is used to describe the increased venous volume and pressure that result.

Clinical Investigation

A young man who was participating in a class project in the Mojave Desert wandered off on his own and became lost. Thirty-six hours later, he was found crawling along a seldom-used one lane road. He was very weak, his skin was cold, and he was found to have low blood pressure and a rapid pulse. Intravenous albumin was administered in the hospital, where it was further observed that he had a low urine output. Analysis of his urine revealed a high total solute concentration (osmolality), but a virtual absence of sodium. What could account for his symptoms and laboratory findings?

Clues

Study the sections "Regulation of Blood Pressure," "Exchange of Fluid between Capillaries and Tissues," and "Regulation of Blood Volume by the Kidneys." Also note the description of hypovolemic shock in the last section of the chapter.

Review Activities

Objective Questions

1. According to the Frank–Starling law, the strength of ventricular contraction is
   a. directly proportional to the end-diastolic volume.
   b. inversely proportional to the end-diastolic volume.
   c. independent of the end-diastolic volume.

2. In the absence of compensations, the stroke volume will decrease when
   a. blood volume increases.
   b. venous return increases.
   c. contractility increases.
   d. arterial blood pressure increases.

3. Which of the following statements about tissue fluid is *false?*
   a. It contains the same glucose and salt concentration as plasma.
   b. It contains a lower protein concentration than plasma.
   c. Its colloid osmotic pressure is greater than that of plasma.
   d. Its hydrostatic pressure is less than that of plasma.

4. Edema may be caused by
   a. high blood pressure.
   b. decreased plasma protein concentration.
   c. leakage of plasma protein into tissue fluid.
   d. blockage of lymphatic vessels.
   e. all of the above.
5. Both ADH and aldosterone act to
   a. increase urine volume.
   b. increase blood volume.
   c. increase total peripheral resistance.
   d. produce all of the above effects.
6. The greatest resistance to blood flow occurs in
   a. large arteries.
   b. medium-sized arteries.
   c. arterioles.
   d. capillaries.
7. If a vessel were to dilate to twice its previous radius, and if pressure remained constant, blood flow through this vessel would
   a. increase by a factor of 16.
   b. increase by a factor of 4.
   c. increase by a factor of 2.
   d. decrease by a factor of 2.
8. The sounds of Korotkoff are produced by
   a. closing of the semilunar valves.
   b. closing of the AV valves.
   c. the turbulent flow of blood through an artery.
   d. elastic recoil of the aorta.
9. Vasodilation in the heart and skeletal muscles during exercise is primarily due to the effects of
   a. alpha-adrenergic stimulation.
   b. beta-adrenergic stimulation.
   c. cholinergic stimulation.
   d. products released by the exercising muscle cells.
10. Blood flow in the coronary circulation
    a. increases during systole.
    b. increases during diastole.
    c. remains constant throughout the cardiac cycle.

11. Blood flow in the cerebral circulation
    a. varies with systemic arterial pressure.
    b. is regulated primarily by the sympathetic system.
    c. is maintained constant within physiological limits.
    d. increases during exercise.
12. Which of the following organs is able to tolerate the greatest restriction in blood flow?
    a. brain
    b. heart
    c. skeletal muscles
    d. skin
13. Which of the following statements about arteriovenous shunts in the skin is *true*?
    a. They divert blood to superficial capillary loops.
    b. They are closed when the ambient temperature is very cold.
    c. They are closed when the deep-body temperature rises much above 37°C.
    d. All of the above are true.
14. An increase in blood volume will cause
    a. a decrease in ADH secretion.
    b. an increase in Na^+ excretion in the urine.
    c. a decrease in renin secretion.
    d. all of the above.
15. The volume of blood pumped per minute by the left ventricle is
    a. greater than the volume pumped by the right ventricle.
    b. less than the volume pumped by the right ventricle.
    c. the same as the volume pumped by the right ventricle.
    d. either less or greater than the volume pumped by the right ventricle, depending on the strength of contraction.

16. Blood pressure is lowest in
    a. arteries.
    b. arterioles.
    c. capillaries.
    d. venules.
    e. veins.
17. Stretch receptors in the aortic arch and carotid sinus
    a. stimulate secretion of atrial natriuretic factor.
    b. serve as baroreceptors that affect activity of the vagus and sympathetic nerves.
    c. serve as osmoreceptors to stimulate secretion of ADH.
    d. stimulate secretion of renin, thus increasing angiotensin II formation.
18. Angiotensin II
    a. stimulates vasoconstriction.
    b. stimulates the adrenal cortex to secrete aldosterone.
    c. inhibits the action of bradykinin.
    d. does all of the above.
19. Which of the following is a paracrine regulator that stimulates vasoconstriction?
    a. nitric oxide
    b. prostacyclin
    c. bradykinin
    d. endothelin-1
20. The pulse pressure is a measure of
    a. the number of heartbeats per minute.
    b. the sum of the diastolic and systolic pressures.
    c. the difference between the systolic and diastolic pressures.
    d. the difference between the arterial and venous pressures.

Essay Questions

1. Define the terms *contractility*, *preload*, and *afterload* and explain how these factors affect the cardiac output.[1]

2. Using the Frank–Starling law, explain how the stroke volume is affected by (a) bradycardia and (b) a "missed beat."

3. Which part of the cardiovascular system contains the most blood? Which part provides the greatest resistance to blood flow? Which part provides the greatest cross-sectional area? Explain.

---

[1]*Note:* This question is answered on page 170 of the Student Study Guide.

4. Explain how the kidneys regulate blood volume.
5. A person who is dehydrated drinks more and urinates less. Explain the mechanisms involved.
6. Using Poiseuille's law, explain how arterial blood flow can be diverted from one organ system to another.
7. Describe the mechanisms that increase the cardiac output during exercise and that increase the rate of blood flow to the heart and skeletal muscles.
8. Explain why an anxious person may have a cold, clammy skin and why the skin becomes hot and flushed on a hot, humid day.
9. Explain the different ways in which a drug that acts as an inhibitor of angiotensin converting enzyme (ACE) can lower the blood pressure. Also, explain how diuretics and β_1-adrenergic-blocking drugs work to lower the blood pressure.
10. Explain how hypotension may be produced in (a) hypovolemic shock and (b) septic shock. Also, explain the mechanisms whereby people in shock have a rapid but weak pulse, cold, clammy skin, and low urine output.

Selected Readings

Atlas, S. A. 1986. Atrial natriuretic factor: Renal and systemic effects. *Hospital Practice* 21:67.

Belardinelli, L. and J. C. Shryock. 1992. Does adenosine function as a retaliatory metabolite in the heart? *News in Physiological Sciences* 7:52.

Berne, R. M., and M. N. Levy. 1981. *Cardiovascular Physiology.* 4th ed. St. Louis: Mosby.

Bevan, J. A., and R. D. Bevan. 1993. Is innervation a prime regulator of cerebral blood flow? *News in Physiological Sciences* 8:149.

Braunwald, E. 1974. Regulation of the circulation. *New England Journal of Medicine* 290: first part, p. 1124; second part, p. 1420.

Bredt, D. S., and S. H. Snyder. 1994. Nitric oxide: A physiologic messenger molecule. *Annual Review of Biochemistry* 63:175.

Brody, M. J., J. R. Haywood, and K. B. Toun. 1980. Neural mechanisms in hypertension. *Annual Review of Physiology* 42:441.

Brunner, H. R. 1990. The renin-angiotensin system in hypertension: An update. *Hospital Practice* 25:71.

Califf, R. M., and J. R. Bengtson. 1994. Cardiogenic shock. *New England Journal of Medicine* 330:1724.

Cantin, M., and J. Genest. February 1986. The heart as an endocrine gland. *Scientific American.*

Carafol, E., and J. T. Penniston. November 1985. The calcium signal. *Scientific American.*

Conger, J. D. 1994. Endothelial regulation of vascular tone. *Hospital Practice* 29:117.

Del Zoppo, G. J., and L. A. Harker. 1984. Blood vessel interaction in coronary disease. *Hospital Practice* 19:163.

Donald, D. E., and J. T. Shepard. 1980. Autonomic regulation of the peripheral circulation. *Annual Review of Physiology* 42:429.

Eckberg, D. L., and J. M. Fritsch. 1993. How should human baroreflexes be tested? *News in Physiological Sciences* 8:7.

Folkow, B. 1990. Salt and hypertension. *News in Physiological Sciences* 5:220.

Folkow, B., and E. Neil. 1971. *Circulation.* London: Oxford University Press.

Franciosa, J. A. 1981. Hypertensive left heart failure: Pathogenesis and therapy. *Hospital Practice* 16:165.

Garcia, R. 1993. Atrial natriuretic factor in experimental and human hypertension. *News in Physiological Sciences* 8:161.

Gewirtz, H. 1991. The coronary circulation: Limitations of current concepts of metabolic control. *News in Physiological Sciences* 6:265.

Granger, D. E., and P. R. Kvietys. 1981. The splanchnic circulation: Intrinsic regulation. *Annual Review of Physiology* 43:409.

Haber, E., and M.-E. Lee. 1994. Endothelin to the rescue? *Nature* 370:252.

Hainsworth, R. 1990. The importance of vascular capacitance in cardiovascular function. *News in Physiological Sciences* 5:250.

Herd, J. A. 1984. Cardiovascular response to stress in man. *Annual Review of Physiology* 46:177.

Hilton, P. J. 1986. Cellular sodium transport in essential hypertension. *New England Journal of Medicine* 314:222.

Hilton, S. M., and K. M. Spyer. 1980. Central nervous regulation of vascular resistance. *Annual Review of Physiology* 42:399.

Johnson, P. C. 1991. The myogenic response. *News in Physiological Sciences* 6:41.

Kaplan, N. M. 1980. The control of hypertension: A therapeutic breakthrough. *American Scientist* 68:537.

Katz, A. M. 1975. Congestive heart failure. *New England Journal of Medicine* 293:1184.

Kayz, M. H. 1987. A physiologic approach to the treatment of heart failure. *Hospital Practice* 22:117.

Kontos, H. A. 1981. Regulation of the cerebral circulation. *Annual Review of Physiology* 43:397.

Laragh, J. H. 1985. Atrial natriuretic hormone, the renin-aldosterone axis, and blood pressure-electrolyte homeostasis. *New England Journal of Medicine* 313:1330.

Light, K. C. et al. 1983. Psychological stress induces sodium and fluid retention in men at high risk for hypertension. *Science* 220:249.

Luscher, T. F. 1994. The endothelium and cardiovascular disease: A complex relationship. *New England Journal of Medicine* 330:1081.

McCarron, D. A. et al. 1984. Blood pressure and nutrient intake in the United States. *Science* 224:1392.

Nadel, E. R. 1985. Physiological adaptations to aerobic training. *American Scientist* 73:334.

National Institutes of Health—Heart, Lung, and Blood Institute. 1993. The Fifth report of the joint national committee on detection, evaluation, and treatment of high blood pressure. NIH Publication No. 93–1088. Washington, DC.

Needleman, P., and J. E. Greenwald. 1986. Atriopeptin: A cardiac hormone intimately involved in fluid, electrolyte, and blood pressure homeostasis. *New England Journal of Medicine* 314:828.

Olsson, R. A. 1981. Local factors regulating cardiac and skeletal muscle blood flow. *Annual Review of Physiology* 43:385.

Oparil, S., and J. M. Wyss. 1993. Atrial natriuretic factor in central cardiovascular control. *News in Physiological Sciences* 8:223.

Page, I. H. 1991. Salt in hypertension— important to control or not? *Perspectives in Biology and Medicine* 34:159.

Pelleg, A. 1993. Adenosine in the heart: Its emerging roles. *Hospital Practice* 28:71.

Petros, A. et. al. 1991. Effect of nitric oxide synthetase inhibitors on hypotension in patients with septic shock. *The Lancet* 338:1557.

Robertson, D. et. al. 1993. The diagnosis and treatment of baroreflex failure. *New England Journal of Medicine* 329:1449.

Robinson, T. F., S. M. Factor, and E. H. Sonnenblick. June 1986. The heart as a suction pump. *Scientific American*.

Ross, J., Jr. 1983. The failing heart and circulation. *Hospital Practice* 18:151.

Ruskoaho, H., and O. Vuolteenaho. 1993. Regulation of atrial natriuretic peptide secretion. *News in Physiological Sciences* 8:261.

Schatz, I. J. 1983. Orthostatic hypotension: Diagnosis and treatment. *Hospital Practice* 18:59.

Stella, A., R. Golin, and A. Zanchetti. 1990. Sympathoadrenal interactions in the control of cardiovascular functions. *News in Physiological Sciences* 5:237.

Stephensen, R. B. 1984. Modification of reflex regulation of blood pressure by behavior. *Annual Review of Physiology* 46:133.

Umans, J. G., and R. Levi. 1995. Nitric oxide in the regulation of blood flow and arterial pressure. *Annual Review of Physiology* 57:771.

Vanhoutte, P. M. 1994. Endothelin-1: A matter of life and breath. *Nature* 368:693.

Vatner, S. F., and E. Braunwald. 1975. Cardiovascular control mechanisms in the conscious state. *New England Journal of Medicine* 293:970.

Weber, K. T., J. S. Janicki, and W. Laskey. 1983. The mechanics of ventricular function. *Hospital Practice* 18:113.

Weinberger, M. H. 1986. Dietary sodium and blood pressure. *Hospital Practice* 21:55.

Zelis, R. et al. 1981. Cardiovascular dynamics in the normal and failing heart. *Annual Review of Physiology* 43:455.

15

The Immune System

1. describe some of the mechanisms of nonspecific immunity and distinguish between nonspecific and specific immune defenses.

2. describe how B lymphocytes respond to antigens and define the terms *memory cell* and *plasma cell*.

3. describe the structure and classification of antibodies and the nature of antigens.

4. describe the complement system and explain how antigen-antibody reactions lead to the destruction of an invading pathogen.

5. describe the events that occur during a local inflammation.

6. describe the process of active immunity and explain how the clonal selection theory may account for this process.

7. describe the mechanisms of passive immunity and give natural and clinical examples of this form of immunization.

8. explain how monoclonal antibodies are produced and describe some of their clinical uses.

9. explain how T lymphocytes are classified and describe the function of the thymus.

10. define the term *lymphokines* and list some of these molecules, together with their functions.

11. describe the histocompatibility antigens and explain the importance of these antigens in the function of the T cell receptor proteins.

12. describe the interaction between macrophages and helper T lymphocytes and explain how the helper T cells affect immunological defense by killer T cells and B cells.

13. describe the possible role of suppressor T lymphocytes in the negative feedback control of the immune response.

14. describe the possible mechanisms responsible for tolerance of self-antigens.

15. describe some of the characteristics of cancer and explain how natural killer cells and killer T lymphocytes provide immunological surveillance against cancer.

16. define the term *autoimmune disease*, give examples of different kinds of autoimmune diseases, and explain some of the mechanisms by which these diseases are produced.

17. explain how immune complex diseases may be produced and give examples of these diseases.

18. distinguish between immediate hypersensitivity and delayed hypersensitivity and describe the mechanisms responsible for each form of allergy.

OUTLINE

Table 15.1 Structures and Defense Mechanisms of Nonspecific Immunity

	Structure	Mechanisms
External	Skin	Anatomic barrier to penetration by pathogens; secretions have lysozyme (enzyme that destroys bacteria)
	Digestive tract	High acidity of stomach; protection by normal bacterial population of colon
	Respiratory tract	Secretion of mucus; movement of mucus by cilia; alveolar macrophages
	Genitourinary tract	Acidity of urine; vaginal lactic acid
Internal	Phagocytic cells	Ingest and destroy bacteria, cellular debris, denatured proteins, and toxins
	Interferons	Inhibit replication of viruses
	Complement proteins	Promote destruction of bacteria and other effects of inflammation
	Endogenous pyrogen	Secreted by leukocytes and other cells; produces fever

Defense Mechanisms

Nonspecific immune protection is provided by such mechanisms as phagocytosis, fever, and the release of interferons. Specific immunity involves the functions of lymphocytes and is directed at specific molecules, or parts of molecules, known as antigens.

The immune system includes all of the structures and processes that provide a defense against potential pathogens. These defenses can be grouped into *nonspecific* and *specific* categories. Nonspecific, or *innate*, defense mechanisms are inherited as part of the structure of each organism. Epithelial membranes that cover the body surfaces, for example, restrict infection by most pathogens. The strong acidity of gastric juice (pH 1–2) also helps to kill many microorganisms before they can invade the body. These external defenses are backed by internal defenses, such as phagocytosis, which function in both a specific and nonspecific manner (table 15.1).

Each individual can acquire the ability to defend against specific pathogens (disease-causing agents) by a prior exposure to those pathogens. This specific, or *acquired*, immune response is a function of lymphocytes. Internal specific and nonspecific defense mechanisms function together to combat infection, with lymphocytes interacting in a coordinated effort with phagocytic cells.

Nonspecific Immunity

Invading pathogens, such as bacteria, that have crossed epithelial barriers enter connective tissues. These invaders—or chemicals, called *toxins*, secreted from them—may enter blood or lymphatic capillaries and be carried to other areas of the body. To counter the invasion and spread of infection, nonspecific immunological defenses are first employed. If these defenses are not sufficient to destroy the pathogens, lymphocytes may be recruited, and their specific actions used to reinforce the nonspecific immune defenses.

Table 15.2 Phagocytic Cells and Their Locations

Phagocyte	Location
Neutrophils	Blood and all tissues
Monocytes	Blood and all tissues
Tissue macrophages (histiocytes)	All tissues (including spleen, lymph nodes, bone marrow)
Kupffer cells	Liver
Alveolar macrophages	Lungs
Microglia	Central nervous system

Phagocytosis

There are three major groups of phagocytic cells: (1) **neutrophils;** (2) the cells of the **mononuclear phagocyte system,** including *monocytes* in the blood and *macrophages* (derived from monocytes) in the connective tissues; and (3) **organ-specific phagocytes** in the liver, spleen, lymph nodes, lungs, and brain (table 15.2). Organ-specific phagocytes, such as the microglia of the brain, are embryologically and functionally related to macrophages and may be considered part of the mononuclear phagocyte system.

The *Kupffer cells* in the liver, as well as phagocytic cells in the spleen and lymph nodes, are **fixed phagocytes.** This term refers to the fact that these cells are immobile ("fixed") in the walls of the sinusoids (chapter 13) within these organs. As blood flows through these wide capillaries of the liver and spleen, foreign chemicals and debris are removed by phagocytosis and chemically inactivated within the phagocytic cells. Invading pathogens are very effectively removed in this manner, so that blood is usually sterile after a few passes through the liver and spleen. Fixed phagocytes in lymph nodes similarly help to remove foreign particles from the lymph.

Connective tissues have a resident population of all leukocyte types. Neutrophils and monocytes in particular can be highly mobile within connective tissues as they scavenge for invaders and cellular debris. These leukocytes are recruited to the site of an infection by a process known as **chemotaxis**—movement toward chemical attractants. Neutrophils are the first to arrive at

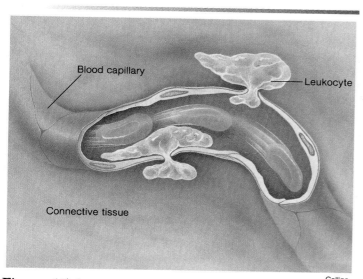

Figure 15.1

Diapedesis, White blood cells squeeze through openings between capillary endothelial cells to enter underlying connective tissues.

the site of an infection; monocytes arrive later and can be transformed into macrophages as the battle progresses.

If the infection is sufficiently large, new phagocytic cells from the blood may join those already in the connective tissue. These new neutrophils and monocytes are able to squeeze through the tiny gaps between adjacent endothelial cells in the capillary wall and enter the connective tissues. This process, called **diapedesis,** is illustrated in figure 15.1.

Phagocytic cells engulf particles in a manner similar to the way an amoeba eats. The particle becomes surrounded by cytoplasmic extensions called pseudopods, which ultimately fuse. The particle thus becomes surrounded by a membrane derived from the plasma membrane (fig. 15.2) and contained within an organelle analogous to a food vacuole in an amoeba. This vacuole then fuses with lysosomes (organelles that contain digestive enzymes), so that the ingested particle and the digestive enzymes still are separated from the cytoplasm by a continuous membrane. Often, however, lysosomal enzymes are released before the food vacuole has completely formed. When this occurs, free lysosomal enzymes may be released into the infected area and contribute to inflammation.

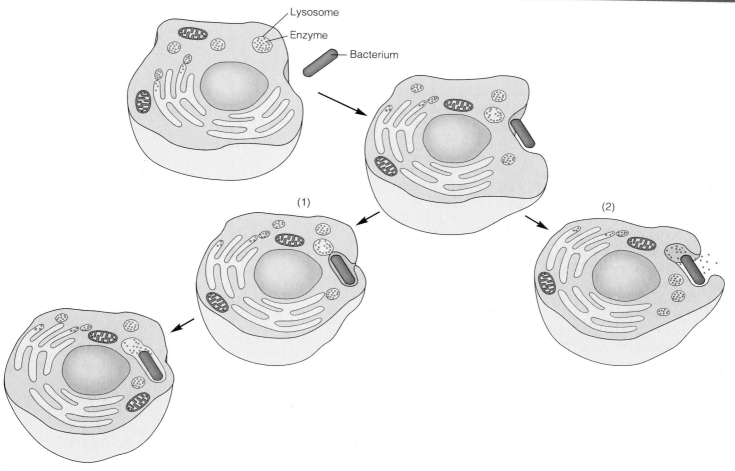

Figure 15.2

Phagocytosis by a neutrophil or macrophage. A phagocytic cell extends its pseudopods around the object to be engulfed (such as a bacterium). (Blue dots represent lysosomal enzymes.) (1) If the pseudopods fuse to form a complete food vacuole, lysosomal enzymes are restricted to the organelle formed by the lysosome and food vacuole. (2) If the lysosome fuses with the vacuole before fusion of the pseudopods is complete, lysosomal enzymes are released into the infected area of tissue.

Figure 15.3

The life cycle of the human immunodeficiency virus (HIV). This virus, like others of its family, contains RNA instead of DNA. Once inside the host cell, the viral RNA is transcribed by reverse transcriptase into complementary DNA (cDNA). The genes in the cDNA then direct the synthesis of new virus particles.

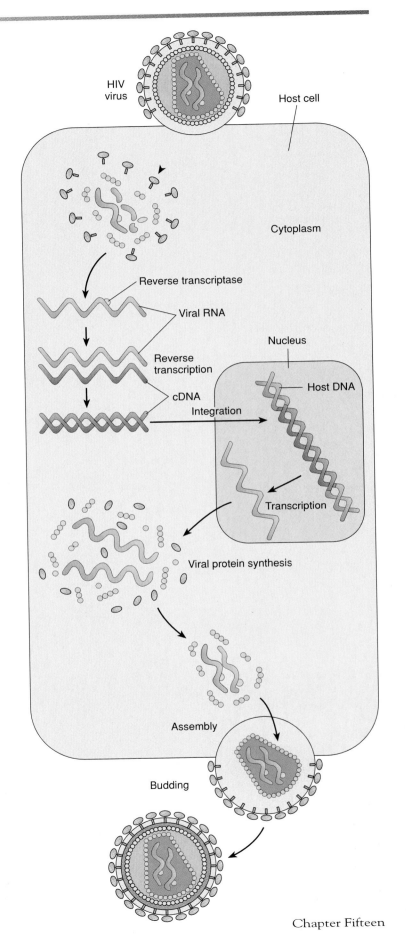

Fever

Fever may be a component of the nonspecific defense system. Body temperature is regulated by the hypothalamus, which contains a thermoregulatory control center (a "thermostat") that coordinates skeletal muscle shivering and the activity of the sympathoadrenal system to maintain body temperature at about 37°C. This thermostat is reset upward in response to a chemical called **endogenous pyrogen,** secreted by leukocytes. Endogenous pyrogen secretion is stimulated by a chemical called *endotoxin,* which is released by certain bacteria.

Although high fevers are definitely dangerous, many believe that a mild to moderate fever may be a beneficial response that aids recovery from bacterial infections. There is some evidence to support this view, but the mechanisms involved are not clearly understood. One theory is that elevated body temperature may interfere with the uptake of iron by some bacteria.

Interferons

In 1957, researchers demonstrated that cells infected with a virus produced polypeptides that interfered with the ability of a second, unrelated strain of virus to infect other cells in the same culture. These **interferons,** as they were called, thus produced a nonspecific, short-acting resistance to viral infection. This discovery generated a great deal of excitement, but further research in this area was hindered by the fact that human interferons could be obtained only in very small quantities and animal interferons were shown to have little effect in humans. In 1980, however, technological breakthroughs allowed researchers to introduce human interferon genes into bacteria—through a technique called *genetic recombination* (chapter 3)—enabling the bacteria to act as interferon factories.

There are three major categories of interferons: *alpha, beta,* and *gamma interferons.* Almost all cells in the body make alpha interferon and beta interferon. These polypeptides act as messengers that protect other cells in the vicinity from viral infection. The viruses are still able to penetrate these other cells, but the ability of the viruses to replicate and assemble new virus particles is inhibited. Viral infection, replication, and dispersal are shown in figure 15.3, using the virus that causes AIDS (discussed later) as an example. Gamma interferon is produced only by particular lymphocytes and a related type of cell called natural killer cells. The secretion of gamma interferon by these cells is part of the immunological defense against infection and cancer, as will be described later. Some of the proposed effects of interferons are summarized in table 15.3.

The Food and Drug Administration (FDA) has currently approved the use of interferons to treat a number of diseases.

Table 15.3	Proposed Effects of Interferons	
Stimulation	**Inhibition**
Macrophage phagocytosis	Cell division
Activity of cytotoxic ("killer") T cells	Tumor growth
Activity of natural killer cells	Maturation of adipose cells
Production of antibodies	Maturation of erythrocytes

Among these are the use of alpha interferon to treat chronic hepatitis A and B, hairy-cell leukemia, virally induced genital warts, and Kaposi's sarcoma. The FDA has also approved the use of beta interferon to treat relapsing-remitting multiple sclerosis and the use of gamma interferon to treat chronic granulomatous disease. Interferon treatment of numerous forms of cancer is currently in various stages of clinical trials.

Specific Immunity

In 1890, a German bacteriologist, Emil Adolf von Behring, demonstrated that a guinea pig that had been previously injected with a sublethal dose of diphtheria toxin could survive subsequent injections of otherwise lethal doses of that toxin. Further, von Behring showed that this immunity could be transferred to a second, nonexposed animal by injections of serum from the immunized guinea pig. He concluded that the immunized animal had chemicals in its serum—which he called **antibodies**—that were responsible for the immunity. He also showed that these antibodies conferred immunity only to subsequent diphtheria infections; the antibodies were *specific* in their actions. It was later learned that antibodies are proteins produced by a particular type of lymphocyte.

Antigens

Antigens are molecules that stimulate antibody production and combine with these specific antibodies. Most antigens are large molecules (such as proteins) with a molecular weight greater than about 10,000, although there are important exceptions, as will be described shortly. Also, most antigens are foreign to the blood and other body fluids. This is because the immune system can distinguish self from nonself, and normally mounts an immune response only against nonself antigens. This topic is discussed in a later section on immune tolerance. The ability of a molecule to function as an antigen depends not only on its size but also on the complexity of its structure. Proteins are therefore more antigenic than polysaccharides, which have a simpler structure. Plastics used in artificial implants are composed of large molecules but are not very antigenic because of their simple, repeating structures.

A large, complex, foreign molecule can have a number of different **antigenic determinant sites,** which are areas of the molecule that stimulate production of and combine with different antibodies. Most naturally occurring antigens have many antigenic determinant sites and stimulate the production of different antibodies with specificities for these sites.

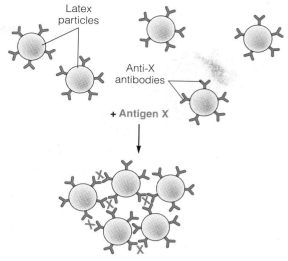

Antibodies attached to latex particles

Latex particles

Anti-X antibodies

+ Antigen X

Agglutination (clumping) of latex particles

Figure 15.4

Immunoassay using the agglutination technique. Antibodies against a particular antigen are adsorbed to latex particles. When these are mixed with a solution that contains the appropriate antigen, the formation of the antigen-antibody complexes produces clumping (agglutination) that can be seen with the unaided eye.

Haptens

Many small organic molecules are not antigenic in and of themselves but can become antigens if they bind to proteins (and thus become antigenic determinant sites on the proteins). This discovery was made by Karl Landsteiner, who is also credited with the discovery of the ABO blood groups (chapter 13). By binding these small molecules—which Landsteiner called **haptens**—to proteins in the laboratory, new antigens could be created for research or diagnostic purposes. The binding of foreign haptens to a person's own proteins can also occur in the body. By this means, derivatives of penicillin, for example, that would otherwise be harmless can produce fatal allergic reactions in susceptible people.

Immunoassays

When the antigen or antibody is attached to the surface of a cell or to particles of latex rubber (in commercial diagnostic tests), the antigen-antibody reaction becomes visible because the particles *agglutinate* (clump) as a result of antigen-antibody bonding (fig. 15.4). These agglutinated particles can be used to assay a variety of antigens, and tests that utilize this procedure are called *immunoassays*. Blood typing and modern pregnancy tests are examples of such immunoassays. A latex agglutination test for detecting AIDS using fingertip blood may also soon be available.

Table 15.4　Comparison of B and T Lymphocytes

Characteristic	B Lymphocytes	T Lymphocytes
Site where processed	Bone marrow	Thymus
Type of immunity	Humoral (secretes antibodies)	Cell-mediated
Subpopulations	Memory cells and plasma cells	Cytotoxic (killer) T cells, helper cells, suppressor cells
Presence of surface antibodies	Yes—IgM or IgD	Not detectable
Receptors for antigens	Present—are surface antibodies	Present— are related to immunoglobulins
Life span	Short	Long
Tissue distribution	High in spleen, low in blood	High in blood and lymph
Percentage of blood lymphocytes	10%–15%	75%–80%
Transformed by antigens to	Plasma cells	Small lymphocytes
Secretory product	Antibodies	Lymphokines
Immunity to viral infections	Enteroviruses, poliomyelitis	Most others
Immunity to bacterial infections	*Streptococcus, Staphylococcus,* many others	Tuberculosis, leprosy
Immunity to fungal infections	None known	Many
Immunity to parasitic infections	Trypanosomiasis, maybe to malaria	Most others

Lymphocytes

Leukocytes, erythrocytes, and blood platelets are all ultimately derived from ("stem from") unspecialized cells in the **bone marrow.** These *stem cells* produce the specialized blood cells, and they replace themselves by cell division so that the stem cell population is not exhausted. Lymphocytes produced in this manner seed the thymus, spleen, and lymph nodes, producing self-replacing lymphocyte colonies in these organs.

The lymphocytes that become seeded in the **thymus** become **T lymphocytes.** These cells have surface characteristics and an immunological function that differ from those of other lymphocytes. The thymus, in turn, seeds other organs; about 65% to 85% of the lymphocytes in blood and most of the lymphocytes in the germinal centers of the lymph nodes and spleen are T lymphocytes. T lymphocytes, therefore, either come from or had an ancestor that came from the thymus gland.

Most of the lymphocytes that are not T lymphocytes are called **B lymphocytes.** The letter *B* is derived from immunological research performed in chickens. Chickens have an organ called the *bursa of Fabricius* that processes B lymphocytes. Since mammals do not have a bursa, the *B* is often translated as the "bursa equivalent" for humans and other mammals. It is currently believed that the B lymphocytes in mammals are processed in the bone marrow, which conveniently also begins with the letter *B.*

Both B and T lymphocytes function in specific immunity. The B lymphocytes combat bacterial and some viral infections by secreting antibodies into the blood and lymph. Because blood and lymph are body fluids (humors), the B lymphocytes are said to provide **humoral immunity,** although the term *antibody-mediated immunity* is also used. T lymphocytes attack host cells that have become infected with viruses or fungi, transplanted human cells, and cancerous cells. The T lymphocytes do not secrete antibodies; they must come in close proximity to or have actual physical contact with the victim cell in order to destroy it. T lymphocytes are therefore said to provide **cell-mediated immunity** (table 15.4).

1. List the phagocytic cells found in blood and lymph and indicate which organs contain fixed phagocytes.
2. Describe the actions of interferons.
3. List four properties characteristic of most antigens. Explain why proteins are more antigenic than polysaccharides.
4. Define the term *hapten,* and describe how haptens may be used clinically.
5. Distinguish between B and T lymphocytes in terms of their origins and immune functions.

Functions of B Lymphocytes

B lymphocytes secrete antibodies that can bond in a specific fashion with antigens. Binding of these secreted antibodies to antigens stimulates a cascade of reactions whereby a system of proteins in the plasma called complement is activated. Some of these activated complement proteins kill the cells containing the antigen; others promote phagocytosis and other activity, resulting in a more effective defense against pathogens.

Exposure of a B lymphocyte to the appropriate antigen results in cell growth followed by many cell divisions. Some of the progeny become memory cells, which are indistinguishable from the original cell; others are transformed into **plasma cells** (fig. 15.5). Plasma cells are protein factories that produce about 2,000 antibody proteins per second in their brief life span of about 5 to 7 days.

The antibodies that are produced by plasma cells when B lymphocytes are exposed to a particular antigen react specifically

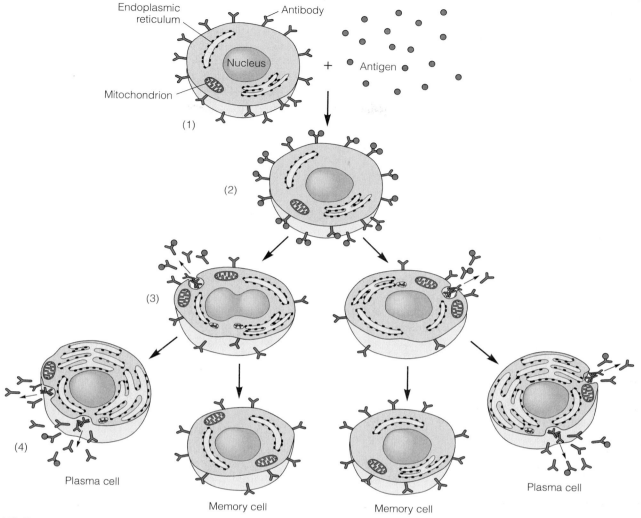

Figure 15.5

B lymphocytes have antibodies on their surface that function as receptors for specific antigens. The interaction of antigens and antibodies on the surface stimulates cell division and the maturation of the B cell progeny into memory cells and plasma cells. Plasma cells produce and secrete large amounts of the antibody. (Note the extensive rough endoplasmic reticulum in these cells.)

with that antigen. Such antigens may be isolated molecules, or they may be molecules at the surface of an invading foreign cell. The specific binding of antibodies to antigens serves to identify the enemy and to activate defense mechanisms that lead to the invader's destruction.

Antibodies

Antibody proteins are also known as **immunoglobulins.** These are found in the gamma globulin class of plasma proteins, as identified by a technique called *electrophoresis* in which classes of plasma proteins are separated by their movement in an electric field (fig. 15.6). The five distinct bands of proteins that appear are albumin, alpha-1 globulin, alpha-2 globulin, beta globulin, and gamma globulin.

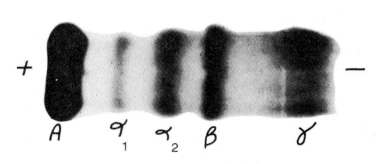

Figure 15.6

The separation of serum protein by electrophoresis. (A = albumin; α_1 = alpha-1 globulin; α_2 = alpha-2 globulin; β = beta globulin; γ = gamma globulin.)

Table 15.5 The Immunoglobulins

Immunoglobulin	Functions
IgG	Main form of antibodies in circulation: production increased after immunization
IgA	Main antibody type in external secretions, such as saliva and mother's milk
IgE	Responsible for allergic symptoms in immediate hypersensitivity reactions
IgM	Function as antigen receptors on lymphocyte surface prior to immunization; secreted during primary response
IgD	Function as antigen receptors on lymphocyte surface prior to immunization; other functions unknown

The gamma globulin band is wide and diffuse because it represents a heterogeneous class of molecules. Since antibodies are specific in their actions, it follows that different types of antibodies should have different structures. An antibody against smallpox, for example, does not confer immunity to poliomyelitis and, therefore, must have a slightly different structure than an antibody against polio. Despite these differences, antibodies are structurally related and form only a few subclasses.

There are five immunoglobulin (abbreviated Ig) subclasses: *IgG, IgA, IgM, IgD,* and *IgE*. Most of the antibodies in serum are in the IgG subclass, whereas most of the antibodies in external secretions (saliva and milk) are IgA (table 15.5). Antibodies in the IgE subclass are involved in allergic reactions.

Antibody Structure

All antibody molecules consist of four interconnected polypeptide chains. Two long, heavy chains (the *H chains*) are joined to two shorter, lighter *L chains*. Research has shown that these four chains are arranged in the form of a Y. The stalk of the Y has been called the "crystallizable fragment" (abbreviated F_c), whereas the top of the Y is the "antigen-binding fragment" (F_{ab}). This structure is shown in figure 15.7.

The amino acid sequences of some antibodies have been determined through the analysis of antibodies sampled from people with multiple myelomas. These lymphocyte tumors arise from the division of a single B lymphocyte, forming a population of genetically identical cells (a clone) that secretes identical antibodies. Clones and the antibodies they secrete are different, however, from one patient to another. Analyses of these antibodies have shown that the F_c regions of different antibodies are the same (are constant), whereas the F_{ab} regions are variable. Variability of the antigen-binding regions is required for the specificity of antibodies for antigens. Thus, it is the F_{ab} region of an antibody that provides a specific site for bonding with a particular antigen (fig. 15.8).

B lymphocytes have antibodies on their cell membrane that serve as **receptors** for antigens. Combination of antigens with these antibody receptors stimulates the B cell to divide and produce more of these antibodies, which are secreted. Exposure to a given antigen thus results in increased amounts of the specific type of antibody that can attack that antigen. This provides active immunity, as described in the next major section.

Diversity of Antibodies

It is estimated that there are about 100 million trillion (10^{20}) antibody molecules in each individual, representing a few million different specificities for different antigens. Considering that antibodies to particular antigens can cross-react to some degree with closely related antigens, this tremendous antibody diversity usually ensures that there are some antibodies that can combine with almost any antigen a person might encounter. These observations evoke a question that has long fascinated scientists: How can a few million different antibodies be produced? A person cannot possibly inherit a correspondingly large number of genes devoted to antibody production.

Two mechanisms have been proposed to explain antibody diversity. First, since different combinations of heavy and light chains can produce different antibody specificities, a person does not have to inherit a million different genes to code for a million different antibodies. If a few hundred genes code for different H chains and a few hundred code for different L chains, different combinations of these polypeptide chains could produce millions of different antibodies. Second, the diversity of antibodies could increase during development if, when some lymphocytes divided, the progeny received antibody genes that had been slightly altered by mutations. Such mutations are called *somatic mutations* because they occur in body cells rather than in sperm or ova. Antibody diversity would thus increase as the lymphocyte population increased.

The Complement System

The combination of antibodies with antigens does not itself cause destruction of the antigens or the pathogenic organisms that contain these antigens. Antibodies, rather, serve to identify the targets for immunological attack and to activate nonspecific immune processes that destroy the invader. Bacteria that are buttered with antibodies, for example, are better targets for phagocytosis by neutrophils and macrophages. The ability of antibodies to stimulate phagocytosis is termed **opsonization.** Immune destruction of bacteria is also promoted by antibody-induced activation of a system of serum proteins known as *complement*.

In the early part of the twentieth century, it was learned that rabbit antibodies to sheep red blood cell antigens could not lyse (destroy) these cells unless certain protein components of serum were present. These proteins, called **complement,** are a nonspecific defense system that is activated by the binding of antibodies to antigens and by this means is directed against specific invaders that have been identified by antibodies.

There are eleven complement proteins, designated C1 (which has three protein components) through C9. These proteins are present in an inactive state within plasma and other body fluids and become activated by the attachment of antibodies to antigens. In terms of their functions, the complement proteins can be subdivided into three components: (1) recognition (C1); (2) activation (C4, C2, and C3, in that order); and (3) attack (C5–C9). The attack phase consists of **complement fixation,** in which complement proteins attach to the cell membrane and destroy the victim cell.

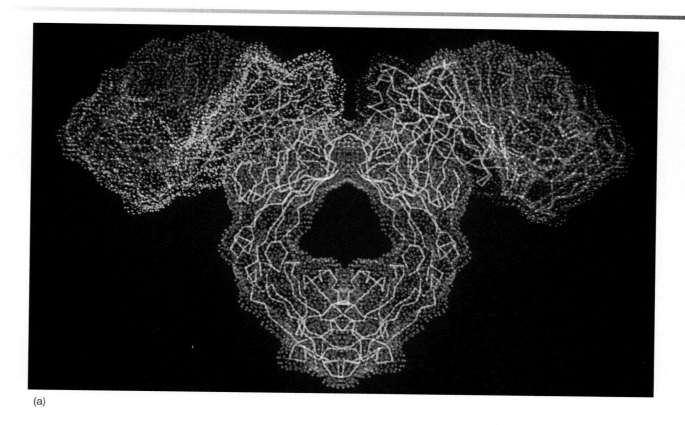

(a)

Antigen molecule

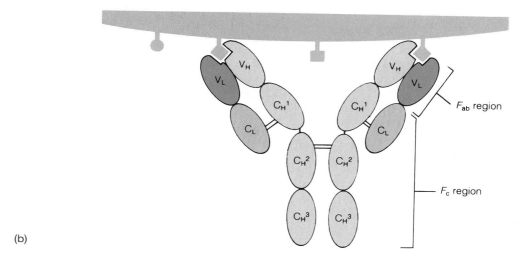

F_{ab} region

F_c region

(b)

Figure 15.7

Antibodies are composed of four polypeptide chains—two are heavy (H) and two are light (L). (a) A computer-generated model of antibody structure. (b) A simplified diagram showing the constant and variable regions. (The variable regions are abbreviated V, and the constant regions are abbreviated C.) Antigens combine with the variable regions. Each antibody molecule is divided into an F_{ab} (antigen-binding) fragment and an F_c (crystallizable) fragment.

Antibodies of the IgG and IgM subclasses attach to antigens on the invading cell's membrane, bind to C1, and by this means activate its enzyme activity. Activated C1 catalyzes the hydrolysis of C4 into two fragments (fig. 15.9), designated $C4_a$ and $C4_b$. The $C4_b$ fragment binds to the cell membrane (is "fixed") and becomes an active enzyme that splits C2 into two fragments, $C2_a$ and $C2_b$. The $C2_a$ becomes attached to $C4_b$ and cleaves C3 into $C3_a$ and $C3_b$. Fragment $C3_b$ becomes attached to the growing complex of complement proteins on the cell membrane. The $C3_b$ converts C5 to $C5_a$ and $C5_b$. The $C5_b$ and, eventually, C6 through C9 become fixed to the cell membrane.

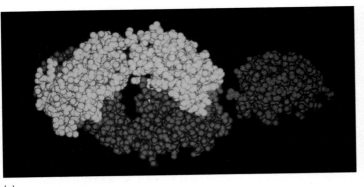

(a)

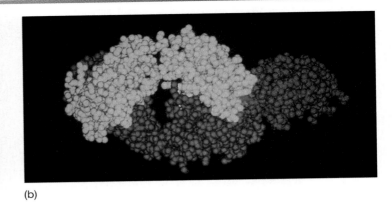

(b)

Figure 15.8

The structure of the F_{ab} portion of an antibody molecule and the antigen with which it combines as determined by X-ray diffraction. The heavy and light chains of the antibody are shown in blue and yellow, respectively, and the antigen is shown in green. (Note the complementary shape at the region where the two join together in [b].)

Photos from *"Three-Dimensional Structure of an Antigen-Antibody Complex at 2.8A Resolution,"* Science, vol. 233, p. 747. © 1986 AAAS.

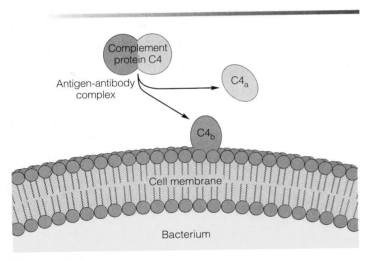

Figure 15.9

The fixation of complement proteins. The formation of an antibody-antigen complex causes complement protein C4 to be split into two subunits—$C4_a$ and $C4_b$. The $C4_b$ subunit attaches (is fixed) to the membrane of the cell to be destroyed (such as a bacterium). This event triggers the activation of other complement proteins, some of which attach to the $C4_b$ on the membrane surface.

Complement proteins C5 through C9 create large pores in the membrane (fig. 15.10). These pores permit the osmotic influx of water, so that the victim cell swells and bursts. Note that the complement proteins, not the antibodies directly, kill the cell; antibodies serve only as activators of this process. Other molecules can also activate the complement system in an alternate nonspecific pathway that bypasses the early phases of the specific pathway described here.

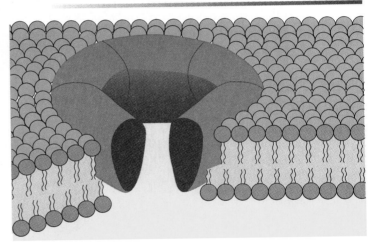

Figure 15.10

Complement proteins C5 through C9 (illustrated as a doughnut-shaped ring) puncture the membrane of the cell to which they are attached (fixed). This aids destruction of the cell.

Complement fragments that are liberated into the surrounding fluid rather than becoming fixed have a number of effects. These effects include (1) *chemotaxis*—the liberated complement fragments attract phagocytic cells to the site of complement activation; (2) *opsonization*—phagocytic cells have receptors for $C3_b$, so that this fragment may form bridges between the phagocyte and the victim cell, thus facilitating phagocytosis; and (3) *stimulation of the release of histamine* from mast cells (a connective tissue cell type) and basophils by fragments $C3_a$ and $C5_a$. As a result of histamine release, there is increased blood flow to the infected area due to vasodilation and increased capillary permeability. The latter effect can result in the leakage of plasma proteins into the surrounding tissue fluid, producing local edema.

Table 15.6 Summary of Events in a Local Inflammation

Category	Events
Nonspecific Immunity	Bacteria enter through break in anatomic barrier of skin.
	Resident phagocytic cells—neutrophils and macrophages—engulf bacteria.
	Nonspecific activation of complement proteins occurs.
Specific Immunity	B cells are stimulated to produce specific antibodies.
	Phagocytosis is enhanced by antibodies attached to bacterial surface antigens.
	Specific activation of complement proteins occurs, which stimulates phagocytosis, chemotaxis of new phagocytes to the infected area, and secretion of histamine from tissue mast cells.
	Diapedesis allows new phagocytic leukocytes (neutrophils and monocytes) to invade the infected area.
	Vasodilation and increased capillary permeability (as a result of histamine secretion) produce redness and edema.

Local Inflammation

Aspects of the nonspecific and specific immune responses and their interactions are well illustrated by the events that occur when bacteria enter a break in the skin and produce a local inflammation (table 15.6). The inflammatory reaction is initiated by the nonspecific mechanisms of phagocytosis and complement activation. Activated complement further increases this nonspecific response by attracting new phagocytes to the area and by stimulating their activity.

After some time, B lymphocytes are stimulated to produce antibodies against specific antigens that are part of the invading bacteria. Attachment of these antibodies to antigens in the bacteria greatly amplifies the previously nonspecific response. This occurs because of greater activation of complement, which directly destroys the bacteria and which—together with the antibodies themselves—promotes the phagocytic activity of neutrophils, macrophages, and monocytes (fig. 15.11).

As inflammation progresses, the release of lysosomal enzymes from macrophages causes the destruction of leukocytes and other tissue cells. These effects, together with those produced by histamine and other chemicals released from mast cells, produce the characteristic symptoms of a local inflammation: *redness* and *warmth* (due to vasodilation); *swelling* (edema); and *pus* (the accumulation of dead leukocytes). If the infection continues, the release of endogenous pyrogen from leukocytes and macrophages may produce a fever.

1. Illustrate the structure of an antibody molecule. Label the constant and variable regions, the F_c and F_{ab} parts, and the heavy and light chains.
2. Define *opsonization* and identify two types of molecules that promote this process.
3. Describe complement fixation and explain the roles of complement fragments that do not become fixed.
4. Explain how nonspecific and specific immune mechanisms cooperate during a local inflammation.

Active and Passive Immunity

When a person is first exposed to a pathogen, the immune response may be insufficient to combat the disease. In the process, however, the lymphocytes that have specificity for those antigens are stimulated to divide many times and produce a clone. This is active immunity, and it can protect the person from getting the disease upon subsequent exposures.

It first became known in Western Europe in the mid-eighteenth century that the fatal effects of smallpox could be prevented by inducing mild cases of the disease. This was accomplished at that time by rubbing needles into the pustules of people who had mild forms of smallpox and injecting these needles into healthy people. Understandably, this method of immunization did not gain wide acceptance.

Acting on the observation that milkmaids who contracted cowpox—a disease similar to smallpox but less *virulent* (less pathogenic)—were immune to smallpox, an English physician named Edward Jenner inoculated a healthy boy with cowpox. When the boy recovered, Jenner inoculated him with what was considered a deadly amount of smallpox, from which he proved to be immune. (This was fortunate for both the boy—who was an orphan—and Jenner; Jenner's fame spread, and as the boy grew into manhood he proudly gave testimonials on Jenner's behalf.) This experiment, performed in 1796, began the first widespread immunization program.

A similar, but more sophisticated, demonstration of the effectiveness of immunizations was performed by Louis Pasteur almost a century later. Pasteur isolated the bacteria that cause anthrax and heated them until their ability to cause disease was greatly reduced (their virulence was *attenuated*), although the nature of their antigens was not significantly changed (fig. 15.12). He then injected these attenuated bacteria into twenty-five cows, leaving twenty-five unimmunized. Several weeks later, before a gathering of scientists, he injected all fifty cows

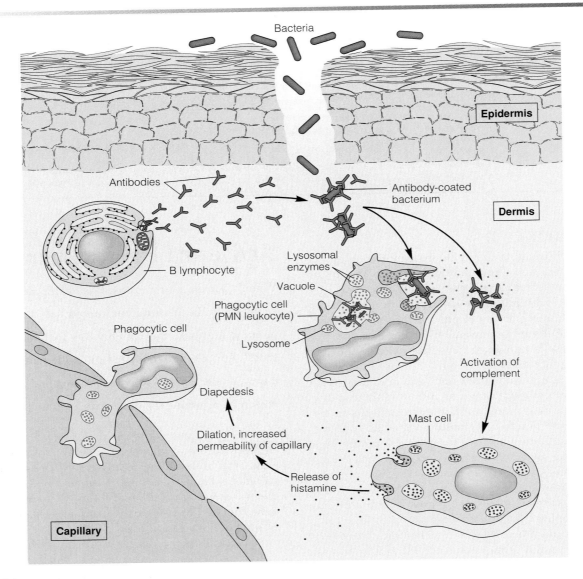

Figure 15.11

The entry of bacteria through a cut in the skin produces a local inflammatory reaction. In this reaction, antigens on the bacterial surface are coated with antibodies and ingested by phagocytic cells. Symptoms of inflammation are produced by the release of lysosomal enzymes and by the secretion of histamine and other chemicals from tissue mast cells.

with the completely active anthrax bacteria. All twenty-five of the unimmunized cows died—all twenty-five of the immunized animals survived.

Active Immunity and the Clonal Selection Theory

When a person is exposed to a particular pathogen for the first time, there is a latent period of 5 to 10 days before measurable amounts of specific antibodies appear in the blood. This sluggish **primary response** may not be sufficient to protect the individual against the disease caused by the pathogen. Antibody concentrations in the blood during this primary response reach a plateau in a few days and decline after a few weeks.

A subsequent exposure of the same individual to the same antigen results in a **secondary response** (fig. 15.13). Compared to the primary response, antibody production during the secondary response is much more rapid. Maximum antibody concentrations in the blood are reached in less than 2 hours and are maintained for a longer time than in the primary response. This rapid rise in antibody production is usually sufficient to prevent the disease.

Clonal Selection Theory

The immunization procedures of Jenner and Pasteur were effective because the people who were inoculated produced a secondary rather than a primary response when exposed to the virulent pathogens. The type of protection they were afforded does not depend on accumulations of antibodies in the blood,

Chapter Fifteen

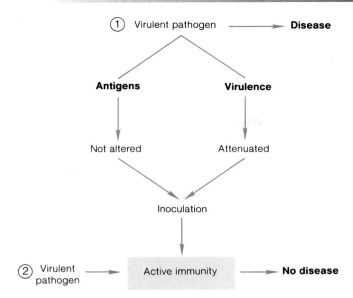

Figure 15.12

Active immunity to a pathogen can be gained by exposure to the fully virulent form or by inoculation with a pathogen whose virulence (ability to cause disease) has been attenuated (reduced) without alteration to the antigens.

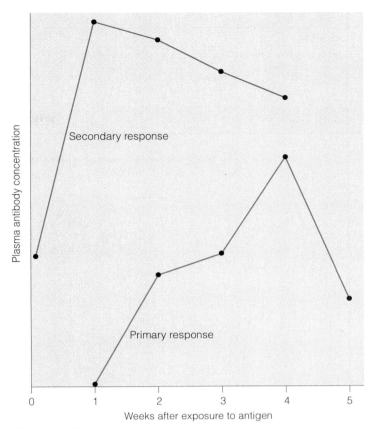

Figure 15.13

A comparison of antibody production in the primary response (upon first exposure to an antigen) to antibody production in the secondary response (upon subsequent exposure to the antigen). The greater secondary response is believed to be due to the development of lymphocyte clones produced during the primary response.

since secondary responses occur even after antibodies produced by the primary response have disappeared. Immunizations, therefore, seem to produce a type of "learning" in which the ability of the immune system to combat a particular pathogen is improved by prior exposure.

The mechanisms by which secondary responses are produced are not completely understood; the **clonal selection theory,** however, appears to account for most of the evidence. According to this theory, B lymphocytes *inherit* the ability to produce particular antibodies (and T lymphocytes inherit the ability to respond to particular antigens). A given B lymphocyte can produce only one type of antibody, with specificity for one antigen. Since this ability is genetically inherited rather than acquired, some lymphocytes, for example, can respond to smallpox and produce antibodies against it even if the person has never been previously exposed to this disease.

The inherited specificity of each lymphocyte is reflected in the antigen receptor proteins on the surface of the lymphocyte's plasma membrane. Exposure to smallpox antigens thus stimulates these specific lymphocytes to divide many times until a large population of genetically identical cells—a clone—is produced. Some of these cells become plasma cells that secrete antibodies for the primary response; others become memory cells that can be stimulated to secrete antibodies during the secondary response (fig. 15.14).

Notice that, according to the clonal selection theory (table 15.7), antigens do not induce lymphocytes to make the appropriate antibodies. Rather, antigens select lymphocytes (through interaction with surface receptors) that are already able to make antibodies against that antigen. This is analogous to evolution by natural selection. An environmental agent (in this case, antigens) acts on the genetic diversity already present in a population of organisms (lymphocytes) to cause an increase in number of the individuals that are selected.

Active Immunity

The development of a secondary response provides **active immunity** against the specific pathogens. The development of active immunity requires prior exposure to the specific antigens, at which time the primary response may cause the person to develop symptoms of the disease. Some parents, for example, deliberately expose their children to others who have measles, chickenpox, and mumps so that their children will be immune to these diseases in later life, when the diseases are potentially more serious.

Clinical immunization programs induce primary responses by inoculating people with pathogens whose virulence has been attenuated or destroyed (such as Pasteur's heat-inactivated anthrax bacteria) or by using closely related strains of microorganisms that are antigenically similar but less pathogenic (such as Jenner's cowpox inoculations). The name for these procedures—**vaccinations** (after the Latin word *vacca*, meaning "cow")—reflects the history of this

Table 15.7 Summary of the Clonal Selection Theory (as Applied to B Cells)

Process	Results
Lymphocytes inherit the ability to produce specific antibodies.	Prior to antigen exposure, lymphocytes that can make the appropriate antibodies are already present in the body.
Antigens interact with antibody receptors on the lymphocyte surface.	Antigen-antibody interaction stimulates cell division and the development of lymphocyte clones that contain memory cells and plasma cells that secrete antibodies.
Subsequent exposure to the specific antigens produces a more efficient response.	Exposure of lymphocyte clones to specific antigens results in greater and more rapid production of specific antibodies.

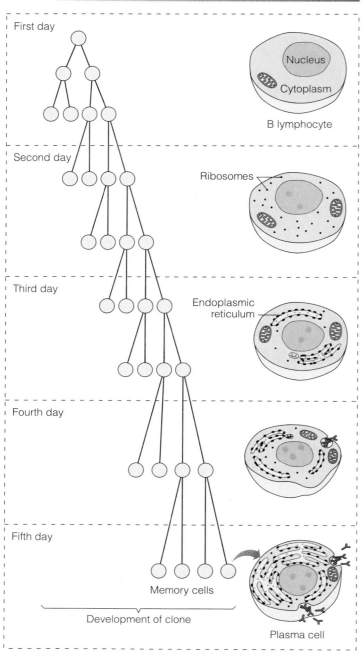

Figure 15.14

The clonal selection theory as applied to B lymphocytes. Most members of the B lymphocyte clone become memory cells, but some become antibody-secreting plasma cells.

There is always a danger when vaccines are prepared using attenuated viruses or toxins that some virulence may remain and cause disease in vaccinated people. A case in point is the commonly used vaccine for **pertussis** (whooping cough)—a disease caused by a toxin released from a species of bacteria. Pertussis is responsible for about 1 million infant deaths annually. Although the vaccine against pertussis prepared from the bacterial cells or their toxin is relatively effective, it occasionally produces severe side effects. Using genetic engineering techniques, scientists have recently produced a protein subunit of pertussis toxin that appears to confer immunity without the virulent action of the toxin. In the future, production of specific proteins from cloned DNA may provide other vaccines that are safer and more effective than those prepared by traditional methods.

technique. All of these procedures cause the development of lymphocyte clones that can combat the virulent pathogens by producing secondary responses.

The first successful polio vaccine (the Salk vaccine) was composed of viruses that had been inactivated by treatment with formaldehyde. These "killed" viruses were injected into the body, in contrast to the currently used oral (Sabin) vaccine. The oral vaccine contains "living" viruses that have attenuated virulence. These viruses invade the epithelial lining of the intestine and multiply but do not invade nerve tissue. The immune system can, therefore, become sensitized to polio antigens and produce a secondary response if polio viruses that attack the nervous system are later encountered.

Passive Immunity

The term **passive immunity** refers to the immune protection that can be produced by the transfer of antibodies to a recipient from another person or from an animal. The donor person or animal has been actively immunized, as explained by the clonal selection theory. The person who receives these ready-made antibodies is thus passively immunized to the same antigens. Passive immunity also occurs naturally in the transfer of immunity from mother to fetus during pregnancy.

Table 15.8 Comparison of Active and Passive Immunity

Characteristic	Active Immunity	Passive Immunity
Injection of person with	Antigens	Antibodies
Source of antibodies	The person inoculated	Natural—the mother; artificial—injection with antibodies
Method	Injection with killed or attenuated pathogens or their toxins	Natural—transfer of antibodies across the placenta; artificial—injection with antibodies
Time to develop resistance	5 to 14 days	Immediately after injection
Duration of resistance	Long (perhaps years)	Short (days to weeks)
When used	Before exposure to pathogen	Before or after exposure to pathogen

The ability to mount a specific immune response—called **immunological competence**—does not develop until about a month after birth. The fetus, therefore, cannot immunologically reject its mother. The immune system of the mother is fully competent but does not usually respond to fetal antigens for reasons that are not completely understood. Some IgG antibodies from the mother do cross the placenta and enter the fetal circulation, however, and these serve to confer passive immunity to the fetus.

The fetus and the newborn baby are, therefore, immune to the same antigens as the mother. However, since the baby did not itself produce the lymphocyte clones needed to form these antibodies, such passive immunity disappears when the infant is about 1 month old. If the baby is breast-fed it can receive additional antibodies of the IgA subclass in its mother's first milk (the *colostrum*).

Passive immunizations are used clinically to protect people who have been exposed to extremely virulent infections or toxins, such as snake venom, tetanus, and others. In these cases the affected person is injected with *antiserum* (serum containing antibodies), also called *antitoxin,* from an animal that has been previously exposed to the pathogen. The animal develops the lymphocyte clones and active immunity and thus has a high concentration of antibodies in its blood. Since the person who is injected with these antibodies does not develop active immunity, he or she must again be injected with antitoxin upon subsequent exposures. Active and passive immunity are compared in table 15.8.

Monoclonal Antibodies

In addition to their use in passive immunity, antibodies are also commercially prepared for use in research and clinical laboratory tests. In the past, antibodies were obtained by chemically purifying a specific antigen and then injecting this antigen into animals. Since an antigen typically has many different antigenic determinant sites, however, the antibodies obtained by this method were polyclonal; they had different specificities. This decreased their sensitivity to a particular antigenic site and resulted in some degree of cross-reaction with closely related antigen molecules.

Monoclonal antibodies, by contrast, exhibit specificity for one antigenic determinant only. In the preparation of monoclonal antibodies, an animal (frequently, a mouse) is injected with an antigen and subsequently killed. B lymphocytes are then obtained from the animal's spleen and placed in thousands of different in vitro incubation vessels. These cells soon die, however, unless they are hybridized with cancerous multiple myeloma cells. Cell fusion is promoted by a chemical—polyethylene glycol. The fusion of a B lymphocyte with a cancerous cell produces a hybrid that undergoes cell division and produces a clone, called a *hybridoma.* Each hybridoma secretes large amounts of identical, monoclonal antibodies. From among the thousands of hybridomas produced in this way, the one that produces the desired antibody is cultured for large-scale production, and the rest are discarded (fig. 15.15).

The availability of large quantities of pure monoclonal antibodies has resulted in the development of much more sensitive clinical laboratory tests (for pregnancy, for example). These pure antibodies have also been used to pick one molecule (the specific antigen interferon, for example) out of a solution of many molecules and thus isolate and concentrate it. In the future, monoclonal antibodies against specific tumor antigens may aid the diagnosis of cancer. Even more exciting, drugs that can kill normal as well as cancerous cells might be aimed directly at a tumor by combining these drugs with monoclonal antibodies against specific tumor antigens.

1. Describe three methods used to induce active immunity.
2. Using graphs to illustrate your discussion, explain the characteristics of the primary and secondary immune responses.
3. Explain the clonal selection theory and indicate how this theory accounts for the secondary response.
4. Describe passive immunity and give examples of how it may occur naturally and how it may be conferred by artificial means.

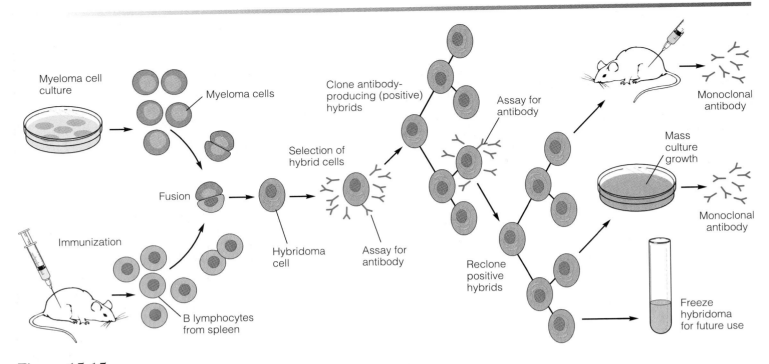

Figure 15.15
The production of monoclonal antibodies that are directed against a specific antigen.

Functions of T Lymphocytes

T cells assist all aspects of the immune system, including cell-mediated destruction by killer T cells and supporting roles by helper and suppressor T cells. T cells are activated only by antigens presented to them by macrophages; the activated T cells in turn produce lymphokines, which activate other cells of the immune system.

The thymus processes lymphocytes in such a way that their functions become quite distinct from those of B cells. Lymphocytes residing in the thymus or orginating from the thymus, or those derived from cells that came from the thymus, are all T lymphocytes. These cells can be distinguished from B cells by specialized techniques. Unlike B cells, the T lymphocytes provide specific immune protection without secreting antibodies. This is accomplished in different ways by the three subpopulations of T lymphocytes, which will be described shortly.

Thymus

The thymus extends from below the thyroid in the neck into the thoracic cavity. As mentioned in chapter 11, this organ grows during childhood but gradually regresses after puberty. Lymphocytes from the fetal liver and spleen and from the bone marrow postnatally seed the thymus and become transformed into T cells. These lymphocytes, in turn, enter the blood and seed lymph nodes and other organs, where they divide to produce new T cells when stimulated by antigens.

Small T lymphocytes that have not yet been stimulated by antigens have very long life spans—months or perhaps years. Still, new T cells must be continuously produced to provide efficient cell-mediated immunity. Since the thymus atrophies after puberty, this organ may not be able to provide new T cells in later life. Colonies of T cells in the lymph nodes and other organs are apparently able to produce new T cells under the stimulation of various **thymus hormones.**

Two hormones that are believed to be secreted by the thymus—*thymopoietin I* and *thymopoietin II*—may promote the transformation of lymphocytes into T cells. Another thymus hormone, called *thymosin,* may promote the maturation of T lymphocytes.

Killer, Helper, and Suppressor T Lymphocytes

The **killer,** or **cytotoxic, T lymphocytes** destroy specific victim cells that are identified by specific antigens on their surface. In order to effect this *cell-mediated destruction,* the T lymphocytes must be in actual contact with their victim cells (in contrast to B cells, which kill at a distance). Although the mechanisms by which the cytotoxic lymphocytes kill their victims are not completely understood, there is evidence that they accomplish this task by secreting certain molecules at the region of contact. Among these molecules, specific polypeptides called *perforins* have been identified. Perforins polymerize in the cell membrane of the victim cell and form cylindrical channels through the membrane. This process is similar to the formation of channels by complement proteins previously discussed, and can result in osmotic destruction of the victim cell.

The killer T lymphocytes defend against viral and fungal infections and are also responsible for transplant rejection reactions and for immunological surveillance against cancer. Although most bacterial infections are fought by B lymphocytes, some are the targets of cell-mediated attack by killer T lymphocytes. This is the case with the tubercle bacilli that cause tuberculosis. Injections of some of these bacteria under the skin produce inflammation after a latent period of 48 to 72 hours. This *delayed hypersensitivity reaction* is cell mediated rather than humoral, as shown by the fact that it can be induced in an unexposed guinea pig by an infusion of lymphocytes, but not of serum, from an exposed animal.

The **helper T lymphocytes** and **suppressor T lymphocytes** indirectly participate in the specific immune response by regulating the responses of the B cells (fig. 15.16) and the killer T cells. The activity of B cells and killer T cells is increased by helper T lymphocytes and decreased by suppressor T lymphocytes. The amount of antibodies secreted in response to antigens is thus affected by the relative numbers of helper to suppressor T cells that develop in response to a given antigen.

Lymphokines

The T lymphocytes, as well as some other cells such as macrophages, secrete a number of polypeptides that serve in an autocrine fashion (chapter 11) to regulate many aspects of the immune system. These products are generally called **cytokines;** the term **lymphokine** is often used to refer to the cytokines of lymphocytes. When a cytokine is first discovered, it is named according to its biological activity (e.g., *B cell–stimulating factor*). Since each cytokine has many different actions (table 15.9), however, such names can be misleading. Scientists have thus agreed to use the name *interleukin,* followed by a number, to indicate a cytokine once its amino acid sequence has been determined.

Interleukin-1, for example, is secreted by macrophages and other cells and can activate the T cell system. B cell-stimulating factor, now called *interleukin-4,* is secreted by T lymphocytes and is required for the proliferation and clone development of B cells. *Interleukin-2* is released by helper T lymphocytes and is required for activation of killer T lymphocytes, among other functions.

Macrophage colony–stimulating factor is secreted by helper T lymphocytes and promotes the activity of macrophages.

Current research has demonstrated that there are two subtypes of helper T lymphocytes, designated T_H1 and T_H2. Helper T lymphocytes of the T_H1 subtype produce interleukin-2 and gamma interferon. Because they secrete these cytokines, T_H1 lymphocytes activate killer T lymphocytes and promote cell-mediated immunity. The T_H2 lymphocytes secrete interleukin-4, interleukin-5, and interleukin-10, which stimulates B lymphocytes to promote humoral immunity.

Scientists have recently discovered that "uncommitted" helper T lymphocytes are changed into the T_H1 subtype in response to a cytokine called interleukin-12, which is secreted by macrophages under appropriate conditions. This process could thus provide a switch for determining how much of the immune response to an antigen will be cell mediated and how much will be humoral.

T Cell Receptor Proteins

Unlike B cells, T cells do not make antibodies and thus do not have antibodies on their surfaces to serve as receptors for antigens. The T cells do, however, have specific receptors for antigens on their membrane surfaces, and these T cell receptors have recently been identified as molecules closely related to immunoglobulins. The T cell receptors differ from the antibody receptors on B cells

Table 15.9 Some Cytokines That Regulate the Immune System

Cytokine	Biological Functions	Secreted By
Interleuken-1	Activates resting T cells	Macrophages and others
Interleuken-2	Serves as growth factor for activated T cells; activates cytotoxic T lymphocytes	Helper T cells
Interleuken-3	Promotes the growth of bone marrow stem cells; serves as growth factor for mast cells	Helper T cells
Interleuken-4 (B cell-stimulating factor)	Promotes growth of activated B cells; promotes growth of resting T cells; enhances activity of cytotoxic T cells	Helper T cells
B cell–differentiating factor	Induces the conversion of activated B cells into antibody-secreting plasma cells	T cells and others
Colony-stimulating factors	Different colony-stimulating factors stimulate the proliferation of granulocytic leukocytes and of macrophages	T cells and others
Interferons	Activate macrophages; augment natural killer cell body activity; exhibit antiviral activity	T cells and others
Tumor necrosis factors	Exert direct cytotoxic effect on some tumor cells; stimulate production of other lymphokines	Macrophages and others

Source: Adapted from C. A. Dinarello, et al., *N Engl J Med,* 1987, Volume 317, page 941. Copyright 1987. Massachusetts Medical Society. All rights reserved.

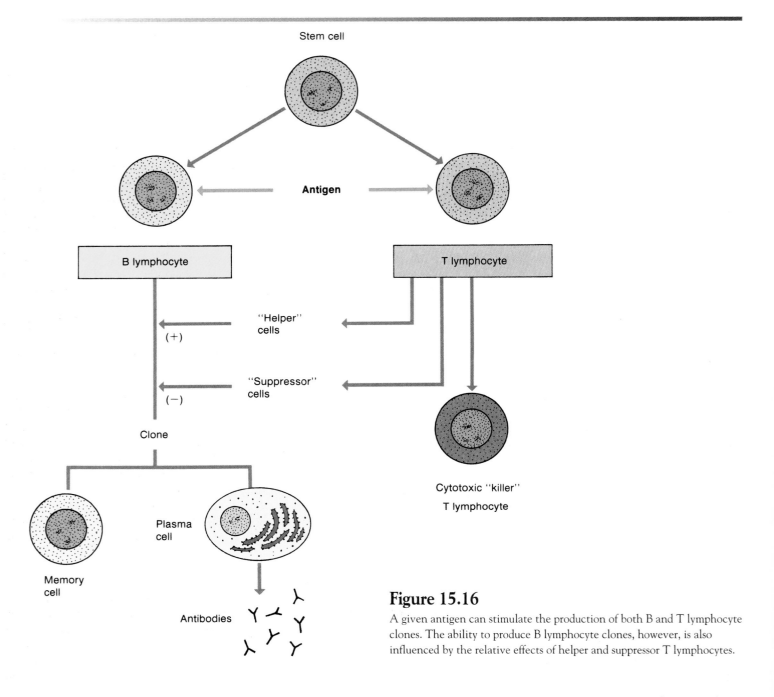

Figure 15.16

A given antigen can stimulate the production of both B and T lymphocyte clones. The ability to produce B lymphocyte clones, however, is also influenced by the relative effects of helper and suppressor T lymphocytes.

in another, and very important, respect: the T cell receptors *cannot bind to free antigens*. In order for a T lymphocyte to respond to a foreign antigen, the antigen must be presented to the T lymphocyte on the membrane of an *antigen-presenting cell*. The chief antigen-presenting cells are macrophages, which present the foreign antigen together with other surface antigens, called *histocompatibility antigens*, to the T lymphocytes. Some knowledge of the histocompatibility antigens is thus required before T cell–macrophage interactions and T cell functions can be understood.

Histocompatibility Antigens

Tissue that is transplanted from one person to another contains antigens that are foreign to the host. This is because all tissue cells, with the exception of mature red blood cells, are genetically marked with a characteristic combination of **histocompatibility antigens** on the membrane surface. The greater the variance in these antigens between the donor and the recipient in a transplant, the greater will be the chance of transplant rejection. Prior to organ transplantation, therefore, the "tissue type" of the recipient is matched to that of potential donors. Since the person's white blood cells are used for this purpose, histocompatibility antigens in humans are also called *human leukocyte antigens*, abbreviated *HLAs*. They are also called **MHC molecules,** after the name of the genes that code for them.

The histocompatibility antigens are proteins that are coded by a group of genes, called the **major histocompatibility complex (MHC),** located on chromosome number 6. These four genes are labeled A, B, C, and D. Each of them can code for only one protein in a given individual, but this protein can be different in different people. Two people, for example, could both have antigen A3, but one might have antigen B17 and the other antigen B21. The closer two people are related, the more similar their histocompatibility antigens will be.

Interactions Between Macrophages and T Lymphocytes

The major histocompatibility complex of genes produces two classes of MHC molecules, designated *class 1* and *class 2*. The class-1 molecules are made by all cells in the body except red blood cells. Class-2 MHC molecules are produced only by macrophages and B lymphocytes and promote the interactions between T cells and these other cells of the immune system.

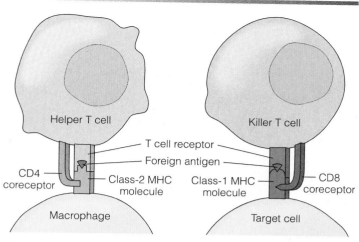

Figure 15.17

A foreign antigen is presented to T lymphocytes in association with MHC molecules. The CD4 and CD8 coreceptors permit the T lymphocyte receptors to interact with only a specific class of MHC molecule.

Killer T lymphocytes can interact only with antigens presented with class-1 MHC molecules, whereas helper T lymphocytes can interact only with antigens presented with class-2 MHC molecules. These restrictions result from the presence of *coreceptors*, which are proteins associated with the T cell receptors. The coreceptor known as *CD8* is associated with the killer T lymphocyte receptor and interacts only with the class-1 MHC molecules; the coreceptor known as *CD4* is associated with the helper T lymphocyte receptor and interacts only with the class-2 MHC molecules. These structures are illustrated in figure 15.17.

When a foreign particle, such as a virus, infects the body, it is taken into macrophages by phagocytosis and partially digested. Within the macrophage, the partially digested virus particles provide foreign antigens that are moved to the surface of the cell membrane. At the membrane, these foreign antigens form a complex with the class-2 MHC molecules. This combination of MHC molecules and foreign antigens is required for interaction with the receptors on the surface of helper T cells.

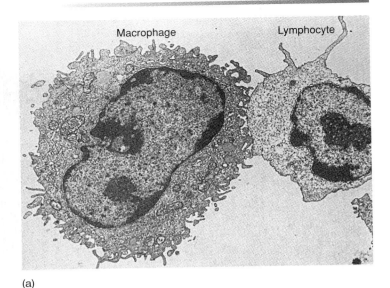

(a)

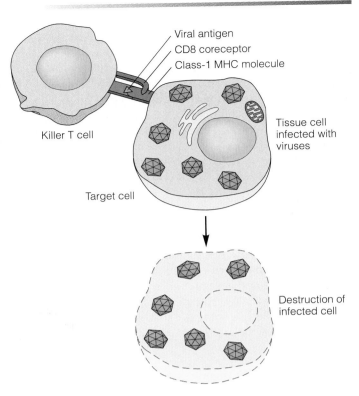

Viral antigen
CD8 coreceptor
Class-1 MHC molecule

Killer T cell

Tissue cell infected with viruses

Target cell

Destruction of infected cell

Figure 15.19

In order for a killer T cell to destroy a tissue cell infected with viruses, the T cell must interact with both the foreign antigen and the class-1 MHC molecule on the surface of the infected cell.

The macrophages thus "present" the antigens to the helper T cells and, in this way, stimulate activation of the T cells (fig. 15.18). It should be remembered that T cells are "blind" to free antigens; they can respond only to antigens presented to them by macrophages (and some other cells) in combination with class-2 MHC molecules.

The first phase of macrophage–T cell interaction then occurs: the macrophage is stimulated to secrete the cytokine known as interleukin-1. As previously discussed, interleukin-1 stimulates cell division and proliferation of T lymphocytes. The activated helper T cells, in turn, secrete macrophage colony–stimulating factor and gamma interferon, which promote the activity of macrophages. In addition, interleukin-2 is secreted by the T lymphocytes and stimulates the macrophages to secrete *tumor necrosis factor*, which is particularly effective in killing cancer cells.

Killer T cells can destroy infected cells only if those cells display the foreign antigen together with their class-1 MHC molecules (fig. 15.19). Such interaction of killer T cells with the foreign antigen–MHC class-1 complex also stimulates proliferation of those killer T cells. In addition, proliferation of the killer T lymphocytes is stimulated by interleukin-2 secreted by the helper T lymphocytes that were activated by macrophages, as previously described (fig. 15.20).

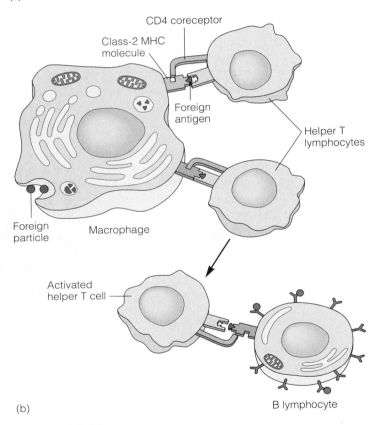

CD4 coreceptor

Class-2 MHC molecule

Foreign antigen

Helper T lymphocytes

Foreign particle

Macrophage

Activated helper T cell

B lymphocyte

(b)

Figure 15.18

(a) An electron micrograph showing contact between a macrophage (*left*) and a lymphocyte (*right*). As illustrated in (*b*), such contact between a macrophage and a T cell requires that the helper T cell interact with both the foreign antigen and the class-2 MHC molecule on the surface of the macrophage.

Chapter Fifteen

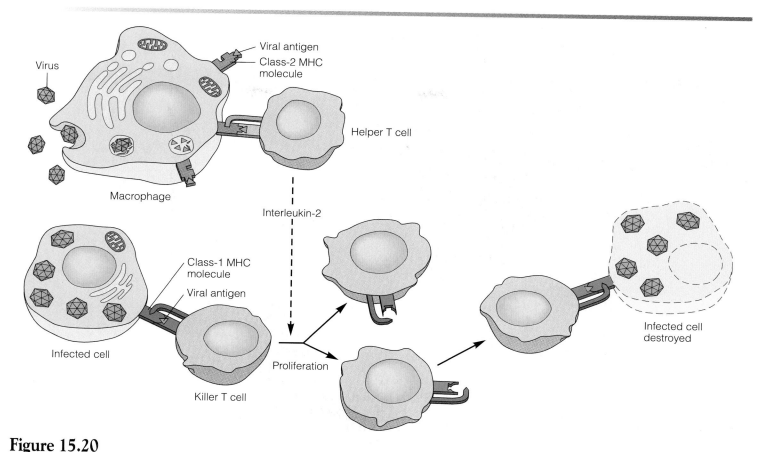

Virus

Viral antigen

Class-2 MHC molecule

Macrophage

Helper T cell

Interleukin-2

Class-1 MHC molecule

Viral antigen

Infected cell

Killer T cell

Proliferation

Infected cell destroyed

Figure 15.20

Interaction between macrophages, helper T lymphocytes, cytotoxic T lymphocytes, and infected cells in the immunological defense against viral infections.

The network of interaction among different cell types of the immune system now spreads outward. Helper T cells, activated to an antigen by macrophages, can also promote the humoral immune response of B cells. In order to do this, the membrane receptor proteins on the surface of the helper T lymphocytes must interact with molecules on the surface of the B cells. This occurs when the foreign antigen attaches to the immunoglobulin receptors on the B cells, so that the B cells can present this antigen together with its class-2 MHC molecules to the receptors on the helper T cells (fig. 15.21). This interaction stimulates proliferation of the B cells, their conversion to plasma cells, and their secretion of antibodies against the foreign antigens.

Tolerance

The ability to produce antibodies against foreign, **nonself antigens,** while tolerating (not producing antibodies against) **self-antigens** occurs during the first month or so of postnatal life, when immunological competence is established. If a fetal mouse of one strain receives transplanted antigens from a different strain, therefore, it will not recognize tissue transplanted later in life from the other strain as foreign and, as a result, will not immunologically reject the transplant.

Glucocorticoids (such as hydrocortisone), secreted by the adrenal cortex, can act to suppress the activity of the immune system and inflammation. This is why **cortisone** and its analogues are used clinically to treat inflammatory disorders and to inhibit the immune rejection of transplanted organs. The immunosuppressive effect of these hormones may result from the fact that they inhibit the secretion of the cytokines. It is interesting in this regard that interleukin-1, which can be produced by microglial cells in the brain, has been shown to stimulate the pituitary-adrenal axis by promoting CRH, ACTH, and glucocorticoid secretion (chapter 11). In a negative feedback fashion, the glucocorticoids then inhibit the immune system and suppress interleukin-1 production. It should be noted that growth hormone, prolactin, and other hormones have also been demonstrated to influence the immune system. Such observations have opened up a new scientific field in which the interactions of the nervous, endocrine, and immune systems are studied.

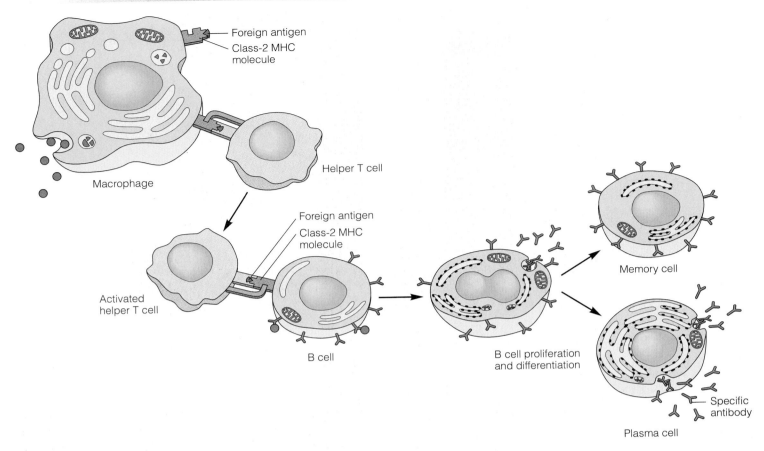

Figure 15.21

A schematic representation of the events that are believed to occur in the interactions of macrophages, helper T lymphocytes, and B lymphocytes.

The ability of an individual's immune system to recognize and tolerate self-antigens requires continuous exposure of the immune system to those antigens. If this exposure begins when the immune system is weak—such as in fetal and early postnatal life—tolerance is more complete and long lasting than that produced by exposure beginning later in life. Some self-antigens, however, are normally hidden from the blood, such as thyroglobulin within the thyroid gland and lens protein in the eye. An exposure to these self-antigens results in antibody production just as if these proteins were foreign. Antibodies made against self-antigens are called **autoantibodies.** Killer T cells that attack self-antigens are called **autoreactive T cells.**

The mechanisms of tolerance are not well understood. Two general types of theories have been proposed to account for immunological tolerance: **clonal deletion** and **clonal anergy.** According to the *clonal deletion theory,* tolerance to self-antigens is achieved by destruction of the lymphocytes that recognize self-antigens. This occurs primarily during fetal life, when those lymphocytes that have receptors on their surface for self-antigens are recognized and destroyed. There is much evidence for clonal deletion in the thymus, and this mechanism is believed to largely account for T cell tolerance. *Anergy* (which means "without working") occurs when lymphocytes directed

 A spontaneous mutation in mice leads to the development of a strain that suffers from *severe combined immunodeficiency (SCID)*. This condition is similar to a rare congenital condition in humans and is characterized by the absence of both B and T lymphocytes. Grafts in SCID mice are therefore not rejected. This inability to reject transplants has recently been exploited by reconstituting a human immune system in the SCID mice using lymphocytes from peripheral human blood or human fetal liver, thymus, and lymph node grafts. This technique may provide a means for studying the function of the immune system and diseases of the human immune system. Since the HIV virus, for example, infects only human (and chimpanzee) lymphocytes, the mouse-human chimera may provide an animal model for experimental investigation of AIDS.

against self-antigens are present throughout life but, for complex and poorly understood reasons, do not attack those antigens. Clonal anergy is believed to be largely responsible for tolerance in B cells, and there is some evidence that it may also contribute to tolerance in T cells.

Chapter Fifteen

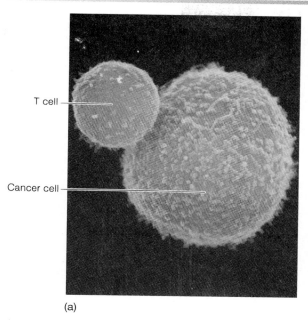

(a)

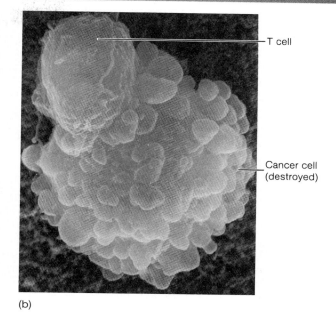

T cell

Cancer cell
(destroyed)

(b)

Figure 15.22

A killer T cell (*a*) contacts a cancer cell (the larger cell), in a manner that requires specific interaction with antigens on the cancer cell. The killer T cell releases lymphokines, including toxins that cause the death of the cancer cell, as shown in (*b*).

Scanning electron micrographs © Andrejs Liepens.

1. Describe the role of the thymus in cell-mediated immunity.

2. Define the term *cytokines*. State the origin of these molecules and describe their different functions.

3. Define the term *histocompatibility antigens* and explain the importance of class-1 and class-2 MHC molecules in the function of T cells.

4. Describe the requirements for activation of helper T cells by macrophages. Explain how helper T cells promote the immunological defenses provided by killer T cells and by B cells.

5. Describe the mechanisms that have been proposed to explain tolerance of self-antigens.

Tumor Immunology

Tumor cells can reveal antigens that activate an immune reaction that destroys the tumor. When cancers develop, this immunological surveillance system—primarily the function of T cells and natural killer cells—has failed to prevent the growth and metastasis of the tumor.

Oncology (the study of tumors) has revealed that tumor biology is similar to and interrelated with the functions of the immune system. Most tumors appear to be clones of single cells that have become transformed in a process similar to the development of lymphocyte clones in response to specific antigens. Lymphocyte clones, however, are under complex inhibitory control systems—such as those exerted by suppressor T lymphocytes and negative feedback by antibodies. The division of tumor cells, by contrast, is not effectively controlled by normal inhibitory mechanisms. Tumor cells are also relatively unspecialized—they *dedifferentiate*, which means that they become similar to the less specialized cells of an embryo.

Tumors are described as *benign* when they are relatively slow growing and limited to a specific location (warts, for example). *Malignant* tumors grow more rapidly and undergo **metastasis,** a term that refers to the dispersion of tumor cells and the resultant seeding of new tumors in different locations. The term **cancer,** as it is generally applied, refers to malignant tumors.

As tumors dedifferentiate, they reveal surface antigens that can stimulate the immune destruction of the tumor cells. Consistent with the concept of dedifferentiation, some of these antigens are proteins produced in embryonic or fetal life that are not normally produced postnatally. Since they are absent at the time immunological competence is established, they are treated as foreign and fit subjects for immunological attack when they are produced by cancerous cells. The release of two such antigens into the blood has provided the basis for laboratory diagnosis of some cancers. *Carcinoembryonic antigen tests* are useful in the diagnosis of colon cancer, for example, and tests for *alpha-fetoprotein* (normally produced only by the fetal liver) help in the diagnosis of liver cancer.

Tumor antigens activate the immune system, initiating an attack primarily by killer T lymphocytes (fig. 15.22) and natural killer cells (described in the next section). The concept of **immunological surveillance** against cancer was introduced in the early 1970s to describe the proposed role of the immune system in fighting cancer. According to this concept, tumor cells frequently appear in the body but are normally recognized and destroyed by the immune system before they can cause cancer. There is evidence that immunological surveillance does prevent some types of cancer; this explains why, for example, people with AIDS (who have a depressed immune system) have a high incidence of Kaposi's sarcoma. It is not clear, however, why all types of cancers do not appear with high frequency in AIDS patients and others whose immune systems are suppressed. For these reasons, the generality of the immunological surveillance system concept is currently controversial.

Immune Therapy of Cancer

The production of human interferons by genetically engineered bacteria has made large amounts of these substances available for the experimental treatment of cancer. Thus far, interferons have proven to be a useful addition to the treatment of particular forms of cancer, including some types of lymphomas, renal carcinoma, melanoma, Kaposi's sarcoma, and breast cancer. They have not, however, proved to be the "magic bullet" against cancer (a term coined by Paul Ehrlich) as had previously been hoped.

A team of scientists led by Dr. S. A. Rosenberg at the National Cancer Institute has pioneered the use of another lymphokine that is now available through genetic engineering techniques. This is **interleukin-2 (IL-2),** which activates both killer T lymphocytes and B lymphocytes. These investigators removed some of the blood from cancer patients who could not be successfully treated by conventional means and isolated a population of their lymphocytes. They treated these lymphocytes with IL-2 to produce *lymphokine-activated killer (LAK) cells* and then reinfused these cells, together with IL-2 and interferons, into the patients. Depending on the combinations and dosages, they obtained remarkable success (but not a complete cure for all cancers) in many of these patients.

The research group next identified a subpopulation of lymphocytes that had invaded solid tumors in mice. These *tumor-infiltrating lymphocyte (TIL) cells* were allowed to replicate in tissue culture, whereupon they were reintroduced into the mice with excellent effects. Recently, the same techniques were used to treat an experimental group of people with metastatic melanoma, a cancer that claims the lives of 6,000 Americans annually. The patients were first given conventional chemotherapy and radiation therapy. They were then treated with their own TIL cells and interleukin-2. Some of the preliminary results of this treatment seem promising, but, like gamma interferon, IL-2 does not seem to be a magic bullet against cancer.

Besides interleukin-2 and gamma interferon, other cytokines may be useful in the treatment of cancer and are currently undergoing experimental investigations. Interleukin-12, for example, seems promising because it is needed for the changing of uncommitted helper T lymphocytes into the T_H1 subtype that bolsters cell-mediated immunity. Scientists are also attempting to identify specific antigens and their genes that may become uniquely expressed in cancer cells, in an effort to help the immune system to better target cancer cells for destruction.

Natural Killer Cells

In a particular strain of hairless mice, a thymus and T lymphocytes are genetically lacking, yet these mice do not appear to have an especially high incidence of tumor production. This surprising observation led to the discovery of **natural killer (NK) cells,** which are lymphocytes that are related to, but distinct from, T lymphocytes. Unlike killer T cells, NK cells destroy tumors in a nonspecific fashion and do not require prior exposure for sensitization to the tumor antigens. The NK cells thus provide a first line of cell-mediated defense, which is subsequently backed up by a specific response mediated by killer T cells. These two cell types interact, however; the activity of NK cells is stimulated by interferon, released as one of the lymphokines from T lymphocytes.

Effects of Aging and Stress

Susceptibility to cancer varies greatly. The Epstein–Barr virus that causes Burkitt's lymphoma in some individuals (mainly in Africa), for example, can also be found in healthy people throughout the world. Most often the virus is harmless; in some cases, it causes mononucleosis (involving a limited proliferation of white blood cells). Only rarely does this virus cause the uncontrolled proliferation of leukocytes characteristic of Burkitt's lymphoma. The reasons for these differences in response to the Epstein–Barr virus, and indeed for the differing susceptibilities of people to other forms of cancer, are not well understood.

It is known that cancer risk increases with age. According to one theory, this is due to the fact that aging lymphocytes gradually accumulate genetic errors that decrease their effectiveness. The secretion of thymus hormones also decreases with age in parallel with a decrease in cell-mediated immune competence. Both of these changes, and perhaps others not yet discovered, could increase susceptibility to cancer.

Numerous experiments have demonstrated that tumors grow faster in experimental animals subjected to stress than in unstressed control animals. This is generally attributed to the fact that stressed animals, including humans, have increased secretion of corticosteroid hormones that act to suppress the immune system (which is why cortisone is given to people who receive organ transplants and to people with chronic inflammatory diseases). Some recent experiments, however, suggest that the stress-induced suppression of the immune system may also be due to other factors that do not involve the adrenal cortex. Future advances in cancer therapy may incorporate methods of strengthening the immune system together with methods that directly destroy tumors.

Table 15.10 Examples of Autoimmune Diseases

Disease	Antigen	Ig and/or T Cell Response
Postvaccinal and postinfectious encephalomyelitis	Myelin, cross-reactive	T cell
Aspermatogenesis	Sperm	T cell
Sympathetic ophthalmia	Uvea	T cell
Hashimoto's disease	Thyroglobulin	IgG and T cell
Graves' disease	Receptor proteins for TSH	Thyroid-stimulating antibody (TSAb)
Autoimmune hemolytic disease	I, Rh, and others on surface of RBCs	IgM and IgG
Thrombocytopenic purpura	Hapten-platelet or hapten-absorbed antigen complex	IgG
Myasthenia gravis	Acetylcholine receptors	IgG
Rheumatic fever	Streptococcal, cross-reactive with heart	IgG and IgM
Glomerulonephritis	Streptococcal, cross-reactive with kidney	IgG and IgM
Rheumatoid arthritis	IgG	IgM to Fc(γ)
Systemic lupus erythematosus	DNA, nucleoprotein, RNA, etc.	IgG

Source: From James T. Barrett, *Textbook of Immunology,* 5th ed. Copyright © 1988 Mosby/Yearbook. Reprinted by permission.

1. Explain why cancer cells are believed to be dedifferentiated and describe some of the clinical applications of this concept.
2. Define the term *immunological surveillance* against cancer and identify the cells involved in this function.
3. Explain the possible relationships between stress and susceptibility to cancer.

Diseases Caused by the Immune System

Immune mechanisms that normally protect the body are very complex and subject to errors that can result in diseases. Autoimmune diseases and allergies are two categories of disease that are not caused by an invading pathogen, but rather by a derangement in the normal functions of the immune system.

The ability of the normal immune system to tolerate self-antigens while it identifies and attacks foreign antigens provides a specific defense against invading pathogens. In every individual, however, this system of defense against invaders at times commits domestic offenses. This can result in diseases that range in severity from the sniffles to sudden death.

Diseases caused by the immune system can be grouped into three interrelated categories: (1) *autoimmune diseases,* (2) *immune complex diseases,* and (3) *allergy,* or *hypersensitivity.* It is important to remember that these diseases are not caused by foreign pathogens but by abnormal responses of the immune system.

Autoimmunity

Autoimmune diseases are those produced by failure of the immune system to recognize and tolerate self-antigens. This failure results in the activation of autoreactive T cells and the production of autoantibodies by B cells, causing inflammation and organ damage (table 15.10). There are over forty known or suspected autoimmune diseases that affect 5% to 7% of the population. Two-thirds of the people with autoimmune diseases are women. Autoimmune diseases can result from a variety of mechanisms.

1. **An antigen that does not normally circulate in the blood may become exposed to the immune system.** Thyroglobulin protein that is normally trapped within the thyroid follicles, for example, can stimulate the production of autoantibodies that cause the destruction of the thyroid (fig. 15.23); this occurs in *Hashimoto's thyroiditis.* Similarly, autoantibodies developed against lens protein in a damaged eye may cause the destruction of a healthy eye (in *sympathetic ophthalmia*).

2. **A self-antigen that is otherwise tolerated may be altered by combining with a foreign hapten.** The disease *thrombocytopenia* (low platelet count), for example, can be caused by the autoimmune destruction of thrombocytes (platelets). This occurs when drugs such as aspirin, sulfonamide, antihistamines, digoxin, and others combine with platelet proteins to produce new antigens. The symptoms of this disease usually disappear when the person stops taking these drugs.

3. **Antibodies may be produced that are directed against other antibodies.** Such interactions may be necessary for the prevention of autoimmunity, but imbalances may actually cause autoimmune diseases. *Rheumatoid arthritis,* for example, is an autoimmune disease associated with the abnormal production of one group of antibodies (of the IgM type) that attack other antibodies (of the IgG type). This contributes to an inflammation reaction of the joints characteristic of the disease.

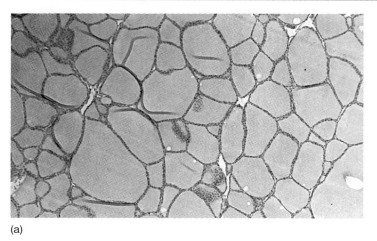
(a)

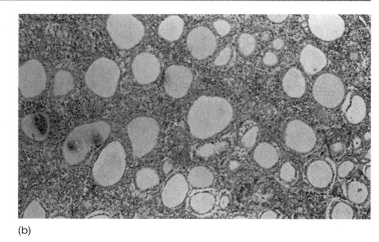
(b)

Figure 15.23

Autoimmune thyroiditis in a rabbit, induced experimentally by injection with thyroglobulin. (Compare the picture of a normal thyroid (*a*) with that of the diseased thyroid (*b*). The grainy appearance of the diseased thyroid is due to the infiltration of large numbers of lymphocytes and macrophages.)

4. **Antibodies produced against foreign antigens may cross-react with self-antigens.** Autoimmune diseases of this sort can occur, for example, as a result of *Streptococcus* bacterial infections. Antibodies produced in response to antigens in this bacterium may cross-react with self-antigens in the heart and kidneys. The inflammation induced by such autoantibodies can produce heart damage (including the valve defects characteristic of *rheumatic fever*) and damage to the glomerular capillaries in the kidneys (*glomerulonephritis*).

5. **Self-antigens, such as receptor proteins, may be presented to the helper T lymphocytes together with class-2 MHC molecules.** Normally, only macrophages and B lymphocytes produce class-2 MHC molecules, which are associated with foreign antigens and are recognized by helper T cells. Perhaps as a result of viral infection, however, cells that do not normally produce class-2 MHC molecules may start to do so and, in this way, present a self-antigen to the helper T cells. In *Graves' disease*, for example, the thyroid cells produce class-2 MHC molecules, and the immune system produces autoantibodies against the TSH receptor proteins in the thyroid cells. These autoantibodies, called *TSAb* for "thyroid-stimulating antibody," interact with the TSH receptors and overstimulate the thyroid gland. Similarly, in *type I diabetes mellitus*, the beta cells of the pancreatic islets abnormally produce class-2 MHC molecules, resulting in autoimmune destruction of the insulin-producing cells.

Immune Complex Diseases

The term *immune complexes* refers to combinations of antibodies with antigens that are free rather than attached to bacterial or other cells. The formation of such complexes activates complement proteins and promotes inflammation. This inflammation normally is self-limiting because the immune complexes are removed by phagocytic cells. When large numbers of immune complexes are continuously formed, however, the inflammation may be prolonged. Also, the dispersion of immune complexes to other sites can lead to widespread inflammations and organ damage. The damage produced by the inflammatory response to antigens is called **immune complex disease.**

Immune complex diseases can result from infections by bacteria, parasites, and viruses. In viral hepatitis B, for example, an immune complex that consists of viral antigens and antibodies can cause widespread inflammation of arteries (*periarteritis*). Arterial damage is not caused by the hepatitis virus itself but by the inflammatory process.

Immune complex diseases can also result from the formation of complexes between self-antigens and autoantibodies. This is the case in rheumatoid arthritis, where the inflammation is produced by complexes of altered IgG antibodies (the antigens in this case) and IgM antibodies. Another immune complex disease that has an autoimmune basis is *systemic lupus erythematosus (SLE)*. People with SLE produce antibodies against their own DNA and nuclear proteins. This can result in the formation of immune complexes throughout the body, including the glomerular capillaries, where glomerulonephritis may be produced.

Allergy

The term *allergy*, often used interchangeably with *hypersensitivity*, refers to particular types of abnormal immune responses to antigens, which are called *allergens* in these cases. There are two major forms of allergy: (1) **immediate hypersensitivity,** which is due to an abnormal B lymphocyte response to an allergen that produces symptoms within seconds or minutes, and (2) **delayed hypersensitivity,** which is an abnormal T cell response that produces symptoms within about 48 hours after exposure to an allergen. These two types of hypersensitivity are compared in table 15.11.

Table 15.11	Allergy: Comparison of Immediate and Delayed Hypersensitivity Reactions	
Characteristic	**Immediate Reaction**	**Delayed Reaction**
Time for onset of symptoms	Within several minutes	Within 1 to 3 days
Lymphocytes involved	B cells	T cells
Immune effector	IgE antibodies	Cell-mediated immunity
Allergies most commonly produced	Hay fever, asthma, and most other allergic conditions	Contact dermatitis (such as to poison ivy and poison oak)
Therapy	Antihistamines and adrenergic drugs	Corticosteroids (such as cortisone)

Table 15.12	Chemicals That Produce the Symptoms of Allergic Reactions	
Chemical	**Derivation**	**Action**
Histamine	From histidine	Contracts smooth muscles in bronchioles; dilates blood vessels; increases capillary permeability
Serotonin	From tryptophane	Contracts smooth muscles
Prostaglandins and leukotrienes	Arachidonic acid	Prolonged contraction of smooth muscle; increased capillary permeability
Eosinophil chemotactic factor	Polypeptides	Attracts eosinophils
Bradykinin and related compounds	Polypeptides	Slow smooth muscle contraction

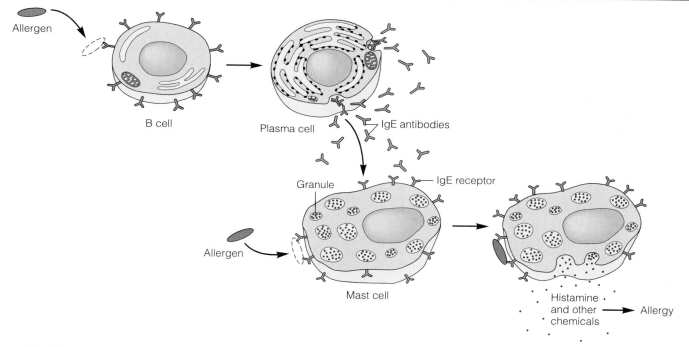

Figure 15.24

Allergy (immediate hypersensitivity) is produced when antibodies of the IgE subclass attach to tissue mast cells. The combination of these antibodies with allergens (antigens that provoke an allergic reaction) causes the mast cell to secrete histamine and other chemicals that produce the symptoms of allergy.

Immediate Hypersensitivity

Immediate hypersensitivity can produce allergic rhinitis (chronic runny or stuffy nose); conjunctivitis (red eyes); allergic asthma; atopic dermatitis (urticaria, or hives); and other symptoms. These symptoms result from the production of antibodies of the IgE subclass instead of the normal IgG antibodies.

Unlike IgG antibodies, IgE antibodies do not circulate in the blood. Instead they attach to tissue mast cells and basophils, which have membrane receptors for these antibodies. When the person is again exposed to the same allergen, the allergen binds to the antibodies attached to the mast cells and basophils. This stimulates these cells to secrete various chemicals, including **histamine** (fig. 15.24). During this process, leukocytes may also secrete **prostaglandin D** and related molecules called **leukotrienes.** These chemicals (table 15.12) produce the symptoms of the allergic reactions. Examination of table 15.12 reveals that histamine stimulates smooth muscle contraction in the respiratory tract but stimulates smooth muscle relaxation in

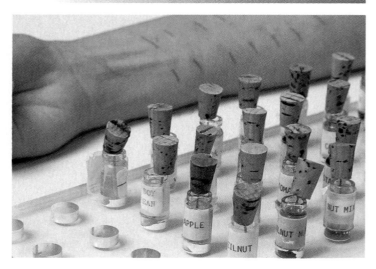

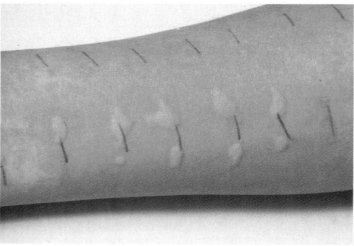

Figure 15.25

A skin test for allergy. If an allergen is injected into the skin of a sensitive individual, a typical flare-and-wheal response occurs within several minutes.

The symptoms of immediate hypersensitivity are produced, to a large degree, by the binding of histamine to a subtype of histamine receptor called the H_1 receptor. The "antihistamine" drugs used to treat allergies compete with histamine for binding to these H_1 receptors, but do not themselves stimulate the receptors. It should be noted that these drugs do not affect the other subtype of histamine receptors, called the H_2 receptors. These receptors are located in the stomach and are involved in stimulation of gastric acid secretion. A different category of drugs is used to block the H_2 receptors as a treatment for ulcers (described in chapter 18).

the bronchioles in the lungs as a result of leukotrienes and other molecules released in an allergic reaction. Asthma is treated with epinephrine and more specific β-adrenergic stimulating drugs (chapter 9), which cause bronchodilation, and with corticosteroids, which inhibit inflammation and leukotriene synthesis. Asthma and its treatment are discussed more completely in chapter 16.

Immediate hypersensitivity to a particular antigen is commonly tested for by injecting various antigens under the skin (fig. 15.25). Within a short time a *flare-and-wheal reaction* is produced if the person is allergic to that antigen. This reaction is due to the release of histamine and other chemical mediators: the flare (spreading flush) is due to vasodilation, and the wheal (elevated area) results from local edema.

Allergens that provoke immediate hypersensitivity include various foods, bee stings, and pollen grains. The most common allergy of this type is seasonal hay fever, which may be provoked by ragweed (*Ambrosia*) pollen grains (fig. 15.26*a*). People with chronic allergic rhinitis and asthma due to an allergy to dust or feathers are usually allergic to a tiny mite (fig. 15.26*b*) that lives in dust and eats the scales of skin that are constantly shed from the body. Actually, most of the antigens from the dust mite are not in its body but rather in its feces, which are tiny particles that can enter the nasal mucosa, much like pollen grains.

Delayed Hypersensitivity

In delayed hypersensitivity, as the name implies, symptoms take a longer time (hours to days) to develop than in immediate hypersensitivity. This may be due to the fact that immediate hypersensitivity is mediated by antibodies, whereas delayed hypersensitivity is a cell-mediated T lymphocyte response. Since the symptoms are caused by the secretion of lymphokines rather than by the secretion of histamine, treatment with antihistamines provides little benefit. At present, corticosteroids are the only drugs that can effectively treat delayed hypersensitivity.

the walls of blood vessels. This seemingly paradoxical effect is because the smooth muscle relaxation in the wall of blood vessels is actually produced by nitric oxide (chapter 14), which is synthesized in response to histamine stimulation.

The symptoms of hay fever (itching, sneezing, tearing, runny nose) are produced largely by histamine and can be treated effectively by antihistamine drugs. Food allergies, causing diarrhea and colic, are mediated primarily by prostaglandins and can be treated with aspirin, which inhibits prostaglandin synthesis (these are the only allergies that respond positively to aspirin). In a certain type of asthma, the difficulty in breathing is caused by inflammation and smooth muscle constriction in

(a)

(b)

Figure 15.26

(*a*) A scanning electron micrograph of ragweed (*Ambrosia* pollen), which is responsible for hay fever. (*b*) A scanning electron micrograph of the house dust mite (*Dermatophagoides farinae*). Waste-product particles produced by the dust mite are often responsible for chronic allergic rhinitis and asthma.

Part [a]: Reproduced by permission from R. G. Kessel and C. Y. Shih,"Scanning Electron Microscopy" Springer-Verlag, 1976.

One of the best-known examples of delayed hypersensitivity is **contact dermatitis,** caused by poison ivy, poison oak, and poison sumac. The skin tests for tuberculosis—the tine test and the Mantoux test—also rely on delayed hypersensitivity reactions. If a person has been exposed to the tubercle bacillus and consequently has developed T cell clones, skin reactions appear within a few days after the tubercle antigens are rubbed into the skin with small needles (tine test) or are injected under the skin (Mantoux test).

1. Explain the mechanisms that may be responsible for autoimmune diseases.
2. Distinguish between immediate and delayed hypersensitivity.
3. Describe the sequence of events by which allergens can produce symptoms of runny nose, skin rash, and asthma.

Summary

Defense Mechanisms p. 426

I. Nonspecific defense mechanisms include barriers to penetration of the body and internal defenses.
   A. Phagocytic cells engulf invading pathogens.
   B. Interferons are polypeptides secreted by cells infected with viruses that help to protect other cells from viral infections.
II. Specific immune responses are directed against antigens.
   A. Antigens are molecules or parts of molecules that are usually large, complex, and foreign.
   B. A given molecule can have a number of antigenic determinant sites that stimulate the production of different antibodies.
III. Specific immunity is a function of lymphocytes.
   A. B lymphocytes secrete antibodies and provide humoral immunity.
   B. T lymphocytes provide cell-mediated immunity.

Functions of B Lymphocytes p. 430

I. There are five subclasses of antibodies, or immunoglobulins.
   A. These subclasses differ with respect to the polypeptides in the constant region of the heavy chains.
   B. The five subclasses are IgG, IgA, IgM, IgD, and IgE.
   C. Each type of antibody has two variable regions that combine with specific antigens.
   D. The combination of antibodies with antigens promotes phagocytosis.
II. Antigen-antibody complexes activate a system of proteins called the complement system.
   A. This results in complement fixation, where complement proteins attach to a cell membrane and promote the destruction of the cell.

B. Free complement proteins promote opsonization and chemotaxis and stimulate the release of histamine from tissue mast cells.

III. Specific and nonspecific immune mechanisms cooperate in the development of a local inflammation.

Active and Passive Immunity   p. 435

I. A primary response is produced when a person is first exposed to a pathogen; a subsequent exposure results in a secondary response.
A. During the primary response, IgM antibodies are produced slowly and the person is likely to get sick.
B. During the secondary response, IgG antibodies are produced quickly and the person is able to resist the pathogen.
C. In active immunizations, the person is exposed to pathogens of attenuated virulence that have the same antigenicity as the virulent pathogen.
D. The secondary response is believed to be due to the development of lymphocyte clones as a result of the antigen-stimulated proliferation of appropriate lymphocytes.

II. Passive immunity is provided by the transfer of antibodies from an immune to a nonimmune organism.
A. Passive immunity occurs naturally in the transfer of antibodies from mother to fetus.
B. Injections of antiserum provide passive immunity to some pathogenic organisms and toxins.

III. Monoclonal antibodies are made by hybridomas, which are formed artificially by the fusion of B lymphocytes and multiple myeloma cells.

Functions of T Lymphocytes   p. 440

I. The thymus processes T lymphocytes and secretes hormones that are believed to be required for the proper function of the immune response of T lymphocytes throughout the body.

II. There are three subcategories of T lymphocytes.
A. Killer T lymphocytes kill victim cells by a mechanism that does not involve antibodies but that does require close contact between the killer T cell and the victim cell.
B. Killer T lymphocytes are responsible for transplant rejection and for the immunological defense against fungal and viral infections, as well as for the defense against some bacterial infections.
C. Helper T lymphocytes stimulate, and suppressor T lymphocytes suppress, the function of B lymphocytes and killer T lymphocytes.
D. The T lymphocytes secrete a family of compounds called lymphokines that promote the action of lymphocytes and macrophages.
E. Receptor proteins on the cell membrane of T lymphocytes must bind to a foreign antigen in combination with a histocompatibility antigen in order for the T cell to become activated.
F. Histocompatibility antigens, or MHC molecules, are a family of molecules on the membranes of cells that are present in different combinations in different individuals.

III. Macrophages partially digest a foreign body, such as a virus, and present the antigens to the lymphocytes on the surface of the macrophage in combination with class-2 MHC antigens.
A. Helper T lymphocytes require such interaction with macrophages in order to be activated by a foreign antigen; when activated in this way, the helper T cells secrete interleukin-2.
B. Interleukin-2 stimulates proliferation of killer T lymphocytes that are specific for the foreign antigen.

C. In order for the killer T lymphocytes to attack a victim cell, the victim cell must present the foreign antigen in combination with a class-1 MHC molecule.
D. Interleukin-2 also stimulates proliferation of B lymphocytes, and thus promotes the secretion of antibodies in response to the foreign antigen.

IV. Tolerance to self-antigens may be due to the destruction of lymphocytes that can recognize the self-antigens, or it may be due to suppression of the immune response by the action of specific suppressor T lymphocytes.

Tumor Immunology   p. 447

I. Immunological surveillance against cancer is provided mainly by killer T lymphocytes and natural killer cells.
A. Cancerous cells dedifferentiate and may produce fetal antigens; these or other antigens may be presented to lymphocytes in association with abnormally produced class-2 MHC antigens.
B. Natural killer cells are nonspecific; T lymphocytes are directed against specific antigens on the cancer cell surface.
C. Immunological surveillance against cancer is weakened by stress.

Diseases Caused by the Immune System   p. 449

I. Autoimmune diseases may be caused by the production of autoantibodies against self-antigens, or they may result from the development of autoreactive T lymphocytes.

II. Immune complex diseases are those caused by the inflammation that results when free antigens are bonded to antibodies.

III. There are two types of allergic responses: immediate hypersensitivity and delayed hypersensitivity.
A. Immediate hypersensitivity results when an allergen provokes the production of antibodies in the IgE class.

These antibodies attach to tissue mast cells and stimulate the release of chemicals from the mast cells.

B. Mast cells secrete histamine, leukotrienes, and prostaglandins, which are believed to produce the symptoms of allergy.

C. Delayed hypersensitivity, as in contact dermatitis, is a cell-mediated response of T lymphocytes.

Clinical Investigation

A 6-year-old girl was playing by crawling through the underbrush in the surrounding hills while her parents were picnicking. When she returned to their campsite, she tearfully showed them a bee sting, the first she had ever received. The next day she developed an itching rash on her chest and abdomen, which was not relieved by antihistamines. The family physician prescribed oral corticosteroids, which alleviated her symptoms. Three weeks later, she was again stung by a bee, and developed severe swelling that responded to antihistamine treatment. What happened to the little girl?

Clues

Read the section "Active Immunity and the Clonal Selection Theory" and note the descriptions of immediate and delayed hypersensitivity under the heading "Allergy." Also refer to table 15.11.

Review Activities

Objective Questions

1. Which of the following offers a nonspecific defense against viral infection?
   a. antibodies
   b. leukotrienes
   c. interferon
   d. histamine

Match the cell type with its secretion.

2. killer T cells   a. antibodies
3. mast cells   b. perforins
4. plasma cells   c. lysosomal
5. macrophages   enzymes
               d. histamine

6. Which of the following statements about the F_{ab} portion of antibodies is *true*?
   a. It binds to antigens.
   b. Its amino acid sequences are variable.
   c. It consists of both H and L chains.
   d. All of the above are true.

7. Which of the following statements about complement proteins $C3_a$ and $C5_a$ is *false*?
   a. They are released during the complement fixation process.
   b. They stimulate chemotaxis of phagocytic cells.
   c. They promote the activity of phagocytic cells.
   d. They produce pores in the victim cell membrane.

8. Mast cell secretion during an immediate hypersensitivity reaction is stimulated when antigens combine with
   a. IgG antibodies.
   b. IgE antibodies.
   c. IgM antibodies.
   d. IgA antibodies.

9. During a secondary immune response,
   a. antibodies are made quickly and in great amounts.
   b. antibody production lasts longer than in a primary response.
   c. antibodies of the IgG class are produced.
   d. lymphocyte clones are believed to develop.
   e. all of the above apply.

10. Which of the following cell types aids the activation of T lymphocytes by antigens?
    a. macrophages
    b. neutrophils
    c. mast cells
    d. natural killer cells

11. Which of the following statements about T lymphocytes is *false*?
    a. Some T cells promote the activity of B cells.
    b. Some T cells suppress the activity of B cells.

    c. Some T cells secrete interferon.
    d. Some T cells produce antibodies.

12. Delayed hypersensitivity is mediated by
    a. T cells.
    b. B cells.
    c. plasma cells.
    d. natural killer cells.

13. Active immunity may be produced by
    a. contracting a disease.
    b. receiving a vaccine.
    c. receiving gamma globulin injections.
    d. both *a* and *b*.
    e. both *b* and *c*.

14. Which of the following statements about class-2 MHC molecules is *false*?
    a. They are found on the surface of B lymphocytes.
    b. They are found on the surface of macrophages.
    c. They are required for B cell activation by a foreign antigen.
    d. They are needed for interaction of helper and killer T cells.
    e. They are presented together with foreign antigens by macrophages.

Match the cytokine with its description:

15. interleukin-1
16. interleukin-2
17. interleukin-12

a. stimulates formation of T$_H$1 helper T lymphocytes
b. stimulates ACTH secretion
c. stimulates proliferation of killer T lymphocytes
d. stimulates proliferation of B lymphocytes

18. Which of the following statements about gamma interferon is *false?*
   a. It is a polypeptide autocrine regulator.
   b. It can be produced in response to viral infections.
   c. It stimulates the immune system to attack infected cells and tumors.
   d. It is produced by almost all cells in the body.

Essay Questions

1. Explain how antibodies help destroy invading bacterial cells.[1]
2. Identify the different types of interferons and describe their origin and actions.
3. Distinguish between the class-1 and class-2 MHC molecules in terms of their locations and functions.
4. Describe the role of macrophages in activating the specific immune response to antigens.
5. Distinguish between the two subtypes of helper T lymphocytes and explain how they may be produced.
6. Describe how plasma cells attack antigens and how they can destroy an invading foreign cell. Compare this mechanism with that by which killer T lymphocytes destroy a target cell.
7. Explain how tolerance to self-antigens may be produced. Also, give two examples of autoimmune diseases and explain their possible causes.
8. Use the clonal selection theory to explain how active immunity is produced by vaccinations.
9. Describe the nature of passive immunity and explain how antitoxins are produced and used.
10. Distinguish between immediate and delayed hypersensitivity. What drugs are used to treat immediate hypersensitivity and how do these drugs work? Why don't these compounds work in treating delayed hypersensitivity?

Selected Readings

Acuto, O., and E. Reinherz. 1985. The human T cell receptor: Structure and function. *New England Journal of Medicine* 312:1100.

Ada, G. L., and G. Nossal. August 1987. The clonal selection theory. *Scientific American.*

Akbar, A. N., and M. Salmon. 1995. Selection and survival of activated lymphocytes. *Science and Medicine* 2:48.

Amit, A. G. et al. 1986. Three-dimensional structure of an antigen-antibody complex at 2.8 Å resolution. *Science* 233:747.

Baglioni, C., and T. W. Nilsen. 1981. The action of interferon at the molecular level. *American Scientist* 69:392.

Biusseret, P. D. August 1982. Allergy. *Scientific American.*

Boon, T. March 1993. Teaching the immune system to fight cancer. *Scientific American.*

Border, W. A., and N. A. Noble. 1995. TGF-β. *Science and Medicine* 2:68.

Capra, J. D., and A. B. Edmunson. January 1977. The antibody combining site. *Scientific American.*

Clark, E. A., and J. A. Ledbetter. 1994. How B and T cells talk to each other. *Nature* 367:425.

Cohen, I. R. April 1988. The self, the world, and autoimmunity. *Scientific American.*

Croce, C. M. 1985. Chromosomal translocations, oncogenes, and B-cell tumors. *Hospital Practice* 20:41.

Cunningham, B. A. October 1977. The structure and function of histocompatibility antigens. *Scientific American.*

Dausset, J. 1981. The major histocompatibility complex in man: Past, present, and future concepts. *Science* 213:1469.

DiNome, M. A., and D. E. Young. August 1987. The clonal selection theory. *Scientific American.*

Drachman, D. B. 1994. Myasthenia gravis. *New England Journal of Medicine* 330:1797.

Engelhard, V. E. August 1994. How cells process antigens. *Scientific American.*

Greene, W. C. September 1993. AIDS and the immune system. *Scientific American.*

Grey, H. M., A. Sette, and S. Buus. November 1989. How T cells see antigens. *Scientific American.*

Hamburger, R. N. 1976. Allergy and the immune system. *American Scientist* 64:157.

Henderson, B. E. et al. 1991. Toward the primary prevention of cancer. *Science* 254:1131.

Herberman, R. B., and J. R. Ortaldo. 1981. Natural killer cells: Their role in defense against disease. *Science* 214:24.

Janeway, C. A., Jr. September 1993. How the immune system recognizes invaders. *Scientific American.*

Johnson, H. M. et al. May 1994. How interferons fight disease. *Scientific American.*

Kapp, J. A., C. W. Pierce, and C. M. Sorensen. August 1984. Antigen-specific suppressor T cell factors. *Hospital Practice* 19:85.

Koffler, D. July 1980. Systemic lupus erythematosus. *Scientific American.*

Krensky, A. M. et al. 1990. T-lymphocyte–antigen interactions in transplant rejection. *New England Journal of Medicine* 322:510.

Laurence, J. December 1985. The immune system in AIDS. *Scientific American.*

Leder, P. November 1982. The genetics of antibody diversity. *Scientific American.*

[1]*Note:* This question is answered on page 184 of the Student Study Guide.

Levine, A. J. 1995. Tumor supressor genes. *Science and Medicine* 2:28.

Lichtenstein, L. M. September 1993. Allergy and the immune system. *Scientific American*.

Liotta, L. A. February 1992. Cancer cell invasion and metastasis. *Scientific American*.

Marrack, P., and J. W. Kappler. February 1986. The T cell and its receptor. *Scientific American*.

Marrack, P., and J. W. Kappler. September 1993. How the immune system recognizes the body. *Scientific American*.

Matthews, T. J., and D. P. Bolognesi. October 1988. AIDS vaccines. *Scientific American*.

McDevitt, H. O. 1985. The MHC system and its relation to disease. *Hospital Practice* 20:57.

Metcalf, D. 1991. Control of granulocytes and macrophages: Molecular, cellular, and clinical aspects. *Science* 254:529.

Milstein, C. October 1980. Monoclonal antibodies. *Scientific American*.

Milstein, C. 1986. From antibody structure to immunological diversification of immune response. *Science* 231:1261.

Nossal, G. J. V. 1987. The basic components of the immune system. *New England Journal of Medicine* 316:1320.

Nossal, G. J. V. September 1993. Life, death, and the immune system. *Scientific American*.

Oettgen, H. F. 1981. Immunological aspects of cancer. *Hospital Practice* 16:93.

Old, L. J. May 1977. Cancer immunology. *Scientific American*.

Old, L. J. May 1988. Tumor necrosis factor. *Scientific American*.

Oldham, R. K. 1985. Biologicals for cancer treatment: Interferons. *Hospital Practice* 20:71.

Ovary, Z. 1989. The history of immediate hypersensitivity. *Hospital Practice* 24:169.

Ramsdell, F., and B. J. Fowlkes. 1990. Clonal deletion versus clonal anergy: The role of the thymus in inducing self tolerance. *Science* 248:1342.

Reichlin, S. 1993. Neuroendocrine-immune interactions. *New England Journal of Medicine* 329: 1246.

Rennie, J. December 1990. The body against itself. *Scientific American*.

Rose, N. R. February 1981. Autoimmune diseases. *Scientific American*.

Sachs, L. January 1986. Growth, differentiation, and the reversal of malignancy. *Scientific American*.

Samuelsson, B. 1983. Leukotrienes: Mediators of immediate hypersensitivity and inflammation. *Science* 220:568.

Schwartz, R. H. August 1993. T cell anergy. *Scientific American*.

Scott, P. 1993. IL-12: Initiation cytokine for cell-mediated immunity. *Science* 260:496.

Sinha, A. A., M. T. Lopez, and H. O. McDevitt. 1990. Autoimmune diseases: The failure of self-tolerance. *Science* 248:1380.

Smith, K. A. March 1990. Interleukin-2. *Scientific American*.

Sprent, J., and D. F. Taugh. 1994. Lymphocyte life-span and memory. *Science* 265:1395.

Steinman, L. September 1993. Autoimmune disease. *Scientific American*.

Steinmetz, M., and L. Hood. 1983. Genes of the major histocompatibility complex in mice and men. *Science* 222:727.

Tannock, I. F. 1983. Biology of tumor growth. *Hospital Practice* 18:81.

Tonegawa, S. October 1985. The molecules of the immune system. *Scientific American*.

Unanue, E. R., and P. M. Allen. 1987. The immunoregulatory role of the macrophage. *Hospital Practice* 22:87.

Vaughan, J. A. 1984. Rheumatoid arthritis: Evidence of a defect in T-cell function. *Hospital Practice* 19:101.

von Boehmer, H., and P. Kisielow. 1990. Non-self discrimination by T cells. *Science* 248:1369.

Weber, J. N., and R. A. Weiss. October 1988. HIV infection: The cellular picture. *Scientific American*.

Weigle, W. O. 1995. Immunologic tolerance: development and disruption. *Hospital Practice* 30:81.

Weissman, G. January 1991. Aspirin. *Scientific American*.

Wigzell, H. September 1993. The immune system as a therapeutic agent. *Scientific American*.

Winter, G., and C. Milstein. 1991. Man-made antibodies. *Nature* 349:293.

Wolde-Mariam, W., and J. B. Peter. March 1984. Recent diagnostic advances in cellular immunology. *Diagnostic Medicine*, p. 25.

Yelton, D. E., and M. D. Scharff. 1980. Monoclonal antibodies. *American Scientist* 63:510.

Young, J. D., and Z. A. Cohen. January 1988. How killer cells kill. *Scientific American*.

Life Science Animations

The animations that relate to chapter 15 are #41 B-cell Immune Response, #42 Structure and Function of Antibodies, #43 Types of T-cells, and #44 Relationship of Helper T-cells and Killer T-cells.

16

Respiratory Physiology

OBJECTIVES   *After studying this chapter, you should be able to . . .*

1. describe the functions of the respiratory system and the structures of the lungs, and explain the significance of the thoracic membranes.

2. explain how the intrapulmonary and intrapleural pressures vary during ventilation and relate these pressure changes to Boyle's law.

3. define the terms *compliance* and *elasticity* and explain how these lung properties affect ventilation.

4. discuss the significance of surface tension in lung mechanics, explain how the law of LaPlace applies to lung function, and describe the role of pulmonary surfactant.

5. explain how inspiration and expiration are accomplished in unforced breathing and describe the accessory respiratory muscles that are used in forced breathing.

6. define the various lung volumes and capacities that can be measured by spirometry and explain how obstructive diseases can be detected by the FEV test.

7. describe the nature of some pulmonary disorders, including asthma, bronchitis, emphysema, and fibrosis.

8. explain Dalton's law and illustrate how the partial pressure of a gas in a mixture of gases is calculated.

9. explain Henry's law, describe how the partial pressure of oxygen and carbon dioxide in a fluid (such as blood) is measured, and explain the clinical significance of these measurements.

10. describe the roles of the medulla oblongata, pons, and cerebral cortex in the regulation of breathing.

11. explain why changes in the P_{CO_2} and pH of blood rather than in its oxygen content serve as the primary stimuli in the control of breathing.

12. explain how the chemoreceptors in the medulla oblongata and the peripheral chemoreceptors in the aortic and carotid bodies respond to changes in P_{CO_2}, pH, and P_{O_2}.

13. describe the Hering–Breuer reflex and explain its significance.

14. describe the different forms of hemoglobin and explain the significance of these different forms.

15. describe the loading and unloading reactions and explain how the extent of these reactions is influenced by the P_{O_2} and affinity of hemoglobin for oxygen.

16. describe the oxyhemoglobin dissociation curve, explain the significance of its shape, and demonstrate how the curve is used to derive the unloading percentage for oxygen.

17. explain how oxygen transport is influenced by changes in blood pH and temperature and explain the effect and physiological significance of 2,3-DPG on oxygen transport.

18. list the different forms of carbon dioxide transport in the blood and explain the chloride shift in the tissues and the reverse chloride shift in the lungs.

19. explain how carbon dioxide affects blood pH and describe how hypoventilation and hyperventilation affect acid-base balance.

20. describe the hyperpnea of exercise and explain how the anaerobic threshold is affected by endurance training.

21. explain the adjustments of the respiratory system to life at a high altitude.

OUTLINE

The Respiratory System

The respiratory system is divided into a **respiratory zone**, which is the site of gas exchange between air and blood, and a **conducting zone**, which conducts the air to the respiratory zone. The exchange of gases between air and blood occurs across the walls of tiny air sacs called **alveoli**, which are only a single cell layer thick to permit very rapid rates of gas diffusion.

The term *respiration* includes three separate but related functions: (1) **ventilation** (breathing); (2) **gas exchange,** which occurs between the air and blood in the lungs and between the blood and other tissues of the body; and (3) **oxygen utilization** by the tissues in the energy-liberating reactions of cell respiration. Ventilation and the exchange of gases (oxygen and carbon dioxide) between the air and blood are collectively called *external respiration*. Gas exchange between the blood and other tissues and oxygen utilization by the tissues are collectively known as *internal respiration*.

Ventilation is the mechanical process that moves air into and out of the lungs. Since the oxygen concentration of air is higher in the lungs than in the blood, oxygen diffuses from air to blood. Carbon dioxide, conversely, moves from the blood to the air within the lungs by diffusing down its concentration gradient. As a result of this gas exchange, the inspired air contains more oxygen and less carbon dioxide than the expired air. More importantly, blood leaving the lungs (in the pulmonary veins) has a higher oxygen and a lower carbon dioxide concentration than the blood delivered to the lungs in the pulmonary arteries. This results from the fact that the lungs function to bring the blood into gaseous equilibrium with the air.

Gas exchange between the air and blood occurs entirely by diffusion through lung tissue. This diffusion occurs very rapidly because there is a large surface area within the lungs and a very short diffusion distance between blood and air.

Structure of the Respiratory System

Gas exchange in the lungs occurs across about 300 million tiny (0.25–0.50 mm in diameter) air sacs, known as **alveoli.** The enormous number of these structures provides a large surface area (60–80 square meters, or about 760 square feet) for diffusion of gases. The diffusion rate is further increased by the fact that each alveolus is only one cell-layer thick, so that the total "air-blood barrier" is only two cells across (an alveolar cell and a

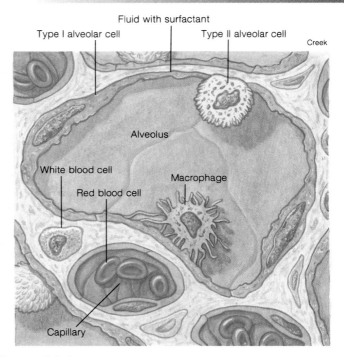

Figure 16.1

A diagram showing the relationship between lung alveoli and pulmonary capillaries.

capillary endothelial cell), or about 2 μm. This is an average distance because the type II alveolar cells are thicker than the type I cells (fig. 16.1). Where the basement membranes of capillary endothelial cells fuse with those of type I alveolar cells, the diffusion distance can be as small as 0.3 μm (fig. 16.2), which is about 1/100th the width of a human hair.

Alveoli are polyhedral in shape and are usually clustered together, like the units of a honeycomb. Air within one member of a cluster can enter other members through tiny pores. These clusters of alveoli usually occur at the ends of *respiratory bronchioles,* which are the very thin air tubes that end blindly in alveolar sacs. Individual alveoli also occur as separate outpouchings along the length of respiratory bronchioles. Although the distance between each respiratory bronchiole and its terminal alveoli is only about 0.5 mm, these units together constitute most of the mass of the lungs.

The air passages of the respiratory system are divided into two functional zones. The **respiratory zone** is the region where gas exchange occurs, and it therefore includes the respiratory bronchioles (because they contain separate outpouchings of

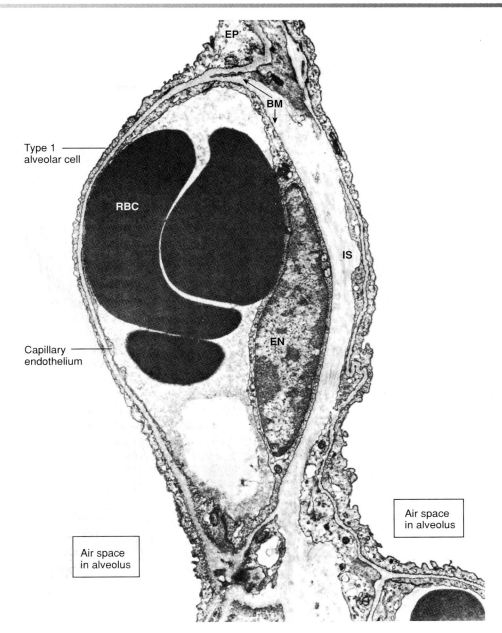

Labels on figure: EP, BM, Type 1 alveolar cell, RBC, IS, EN, Capillary endothelium, Air space in alveolus, Air space in alveolus

Figure 16.2

An electron micrograph of a capillary within the thin interalveolar septum that separates two adjacent alveoli. Notice the short distance separating the alveolar space on one side (*left*, in this figure) from the capillary. (EP = epithelial cell of alveolus; RBC = red blood cell; BM = basement membrane; IS = interstitial connective tissue.)

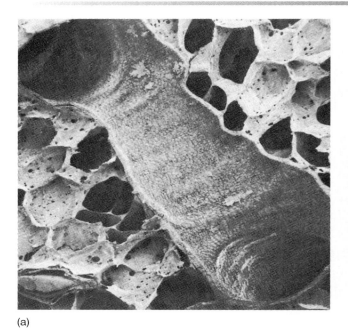

(a)

(b)

Figure 16.3

(*a*) A scanning electron micrograph showing lung alveoli and a small bronchiole. (*b*) The alveoli under higher power, with an arrow indicating an alveolar pore through which air can pass from one alveolus to another.

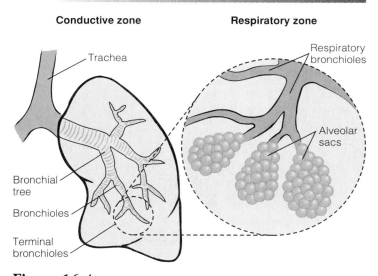

Conductive zone   **Respiratory zone**

Figure 16.4

The conducting and respiratory zones of the respiratory system.

alveoli) and the terminal clusters of alveolar sacs (fig. 16.3). The **conducting zone** includes all of the anatomical structures through which air passes before reaching the respiratory zone (fig. 16.4; see also fig. 16.22).

Air enters the respiratory bronchioles from *terminal bronchioles*, which are narrow airways formed from many successive divisions of the right and left *primary bronchi*. These two large

If the trachea becomes occluded through inflammation, excessive secretion, trauma, or aspiration of a foreign object, it may be necessary to create an emergency opening into this tube so that ventilation can still occur. A **tracheotomy** is the process of surgically opening the trachea, and a **tracheostomy** is the procedure of inserting a tube into the trachea to permit breathing and to keep the passageway open. A tracheotomy should be performed only by a competent physician because of the great risk of cutting a recurrent laryngeal nerve or the common carotid artery.

air passages, in turn, are continuous with the *trachea*, or windpipe, which is located in the neck in front of the esophagus (a muscular tube that carries food to the stomach). The trachea is a sturdy tube supported by rings of cartilage (fig. 16.5).

Air enters the trachea from the *pharynx*, which is the cavity located behind the palate that receives the contents of both the oral and nasal passages. In order for air to enter or leave the trachea and lungs, however, it must pass through a valvelike opening called the *glottis* between the vocal cords. The vocal cords are part of the *larynx*, or voice box, which guards the entrance to the trachea (fig. 16.6). The projection at the front of the throat, commonly called the "Adam's apple," is formed by the largest cartilage of the larynx.

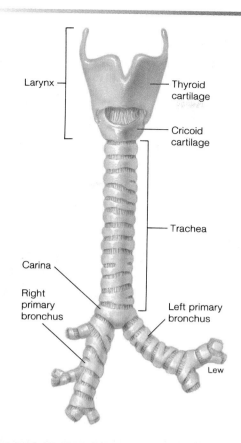

Larynx

Thyroid cartilage

Cricoid cartilage

Trachea

Carina

Right primary bronchus

Left primary bronchus

Lew

(a)

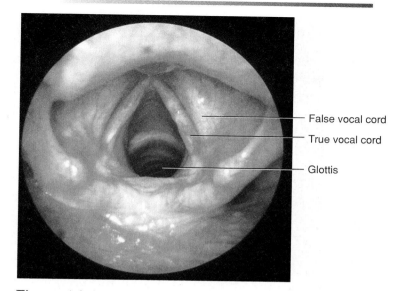

False vocal cord

True vocal cord

Glottis

Figure 16.6

A photograph of the larynx showing the true and false vocal cords and the glottis.

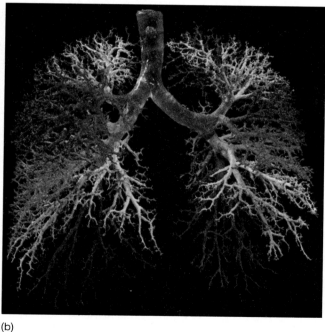

(b)

Figure 16.5

The conducting zone of the respiratory system. (*a*) An anterior view extending from the larynx to the terminal bronchi and (*b*) the airway from the trachea to the terminal bronchioles, as represented by a plastic cast.

The conducting zone of the respiratory system, in summary, consists of the mouth, nose, pharynx, larynx, trachea, primary bronchi, and all successive branchings of the bronchioles up to and including the terminal bronchioles. In addition to conducting air into the respiratory zone, these structures serve additional functions: *warming* and *humidification* of the inspired air and *filtration* and *cleaning*.

Regardless of the temperature and humidity of the atmosphere, when the inspired air reaches the respiratory zone it is at a temperature of 37°C (body temperature), and it is saturated with water vapor. This ensures that a constant internal body temperature will be maintained and that delicate lung tissue will be protected from desiccation.

Mucus secreted by cells of the conducting zone serves to trap small particles in the inspired air and thereby performs a filtration function. This mucus is moved along at a rate of 1 to 2 centimeters per minute by cilia projecting from the tops of epithelial cells that line the conducting zone (fig. 16.7). There are about 300 cilia per cell that beat in a coordinated fashion to move mucus toward the pharynx, where it can either be swallowed or expectorated.

As a result of this filtration function, particles larger than about 6 μm do not normally enter the respiratory zone of the lungs. The importance of this function is evidenced by the disease called *black lung,* which occurs in miners who inhale too much carbon dust and therefore develop pulmonary fibrosis (as described in a later section). The alveoli themselves are normally kept clean by the action of resident macrophages (chapter 15). The cleansing action of cilia and macrophages in the lungs is diminished by cigarette smoke.

Figure 16.7
A scanning electron micrograph of a bronchial wall. The cilia projecting from the tops of the epithelial cells help to cleanse the lung by moving trapped particles.

Thoracic Cavity

The *diaphragm*, a dome-shaped sheet of striated muscle, divides the body cavity into two parts. The area below the diaphragm, or the *abdominal cavity*, contains the liver, pancreas, gastrointestinal tract, spleen, genitourinary tract, and other organs. Above the diaphragm, the chest, or *thoracic cavity*, contains the heart, large blood vessels, trachea, esophagus, and thymus in the central region, and is filled elsewhere by the right and left lungs.

The structures in the central region—or *mediastinum*—are enveloped by a double layer of wet epithelial membranes called the *pleural membranes*. One membrane of this double layer is continuous with the *parietal pleural membrane*, a wet epithelial membrane that lines the inside of the thoracic wall. The other layer is continuous with the *visceral pleural membranes* that cover the surface of the lungs (fig. 16.8).

The lungs normally fill the thoracic cavity so that the visceral pleural membranes covering the lungs are pushed against the parietal pleural membrane lining the thoracic wall. There is thus, under normal conditions, little or no air between the visceral and parietal pleural membranes. There is, however, a "potential space"—called the *intrapleural space*—that can become a real space if the visceral and parietal pleural membranes separate when a lung collapses. The normal position of the lungs in the thoracic cavity is shown in the radiograph in figure 16.9.

---

1. Describe the structures involved in gas exchange in the lungs and explain how gas exchange occurs.
2. Describe the structures and functions of the conducting zone of the respiratory system.
3. Describe how each lung is packaged separately in pleural membranes. What is the relationship between the visceral and parietal pleural membranes?

Physical Aspects of Ventilation

The movement of air into and out of the lungs occurs as a result of pressure differences induced by changes in lung volumes. Ventilation is thus influenced by the physical properties of the lungs, including their compliance, elasticity, and surface tension.

Movement of air from the conducting zone to the terminal bronchioles occurs as a result of the pressure difference between the two ends of the airways. Air flow through bronchioles, like blood flow through blood vessels, is directly proportional to the pressure difference and inversely proportional to the frictional resistance to flow. The pressure differences in the pulmonary system are induced by changes in lung volumes. The compliance, elasticity, and surface tension of the lungs are physical properties that affect their functioning.

Intrapulmonary and Intrapleural Pressures

The wet, serous membranes of the visceral and parietal pleurae are normally flush against each other, so that the lungs are stuck to the chest wall in the same manner as two wet pieces of glass sticking to each other. The *intrapleural space* between the two wet membranes contains only a thin layer of fluid secreted by the pleural membranes. The pleural cavity in a healthy, living

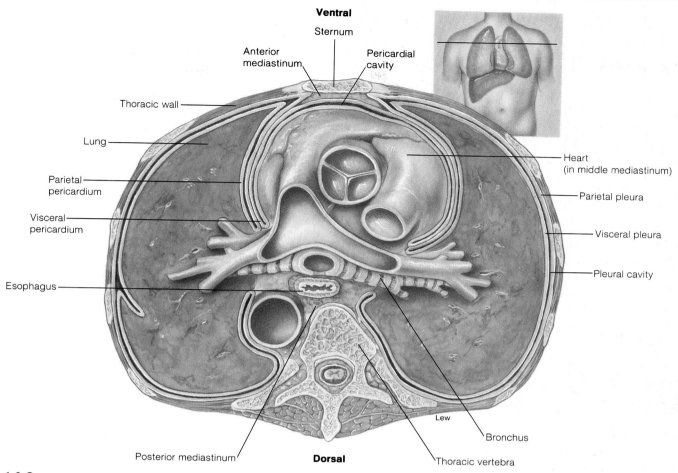

Ventral
Sternum
Anterior mediastinum
Pericardial cavity
Thoracic wall
Lung
Parietal pericardium
Visceral pericardium
Esophagus
Heart (in middle mediastinum)
Parietal pleura
Visceral pleura
Pleural cavity
Lew
Bronchus
Posterior mediastinum
Dorsal
Thoracic vertebra

Figure 16.8

A cross section of the thoracic cavity showing the mediastinum and pleural membranes.

organism is thus potential rather than real; it can become real only in abnormal situations when air enters the intrapleural space. Since the lungs normally remain in contact with the chest wall, they get larger and smaller together with the thoracic cavity during respiratory movements.

Air enters the lungs during inspiration because the atmospheric pressure is greater than the **intrapulmonary, or alveolar, pressure.** Since the atmospheric pressure does not usually change, the intrapulmonary pressure must fall below atmospheric pressure to cause inspiration. A pressure below that of the atmosphere is called a *subatmospheric pressure*, or *negative pressure*. During quiet inspiration, for example, the intrapulmonary pressure may decrease to 3 mmHg below the pressure of the atmosphere. This subatmospheric pressure is shown as ⁻3 mmHg.

Expiration, conversely, occurs when the intrapulmonary pressure is greater than the atmospheric pressure. During quiet expiration, for example, the intrapulmonary pressure may rise to at least +3 mmHg over the atmospheric pressure.

The lack of air in the intrapleural space produces a subatmospheric **intrapleural pressure** that is lower than the intrapulmonary pressure (table 16.1). There is thus a pressure difference across the wall of the lung—called the **transpulmonary pressure**—which is the difference between the intrapulmonary pressure and the intrapleural pressure. Since the pressure within the lungs (intrapulmonary pressure) is greater than outside the lungs (intrapleural pressure), the difference in pressure (transpulmonary pressure) acts to expand the lungs as the thoracic volume expands during inspiration.

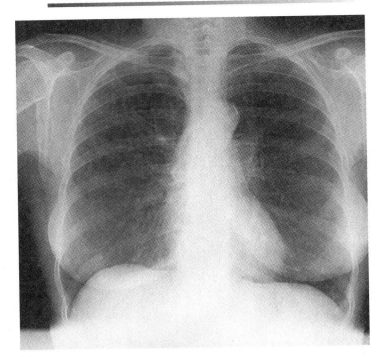

(a)

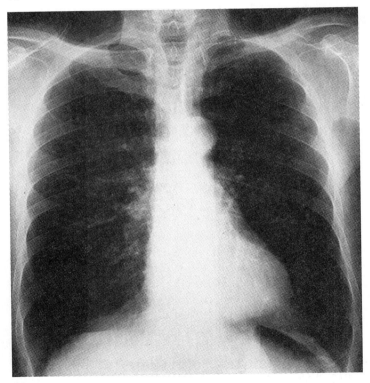

(b)

Figure 16.9

Radiographic (X-ray) views of the chest (a) of a normal female and (b) of a normal male.

Table 16.1	Intrapulmonary and Intrapleural Pressures in Normal, Quiet Breathing, and the Transpulmonary Pressure	
	Inspiration	**Expiration**
Intrapulmonary pressure (mmHg)	–3	+3
Intrapleural pressure (mmHg)	–6	–3
Transpulmonary pressure (mmHg)	+3	+6

Note: Pressures indicate mmHg below or above atmospheric pressure.

Boyle's Law

Changes in intrapulmonary pressure occur as a result of changes in lung volume. This follows from **Boyle's law,** which states that the pressure of a given quantity of gas is inversely proportional to its volume. An increase in lung volume during inspiration decreases intrapulmonary pressure to subatmospheric levels; air therefore goes in. A decrease in lung volume raises the intrapulmonary pressure above that of the atmosphere, pushing air out. These changes in lung volume occur as a consequence of changes in thoracic volume, as will be described in a later section on the mechanics of breathing.

Physical Properties of the Lungs

In order for inspiration to occur, the lungs must be able to expand when stretched; they must have high *compliance*. In order for expiration to occur, the lungs must get smaller when this stretching force is released; they must have *elasticity*. The tendency to get smaller is also aided by *surface tension* forces within the alveoli.

Compliance

The lungs are very distensible (stretchable)—they are, in fact, about one hundred times more distensible than a toy balloon. Another term for distensibility is **compliance,** which here refers to the ease with which the lungs can expand under pressure. Lung compliance can be defined as the change in lung volume per change in transpulmonary pressure, which is expressed symbolically as $\Delta V/\Delta P$. A given transpulmonary pressure, in other words, will cause greater or lesser expansion, depending on the compliance of the lungs.

The compliance of the lungs is reduced by factors that produce a resistance to distension. If the lungs were filled with concrete (as an extreme example), a given transpulmonary pressure would produce no increase in lung volume and no air would enter; the compliance would be zero. The infiltration of lung tissue with connective tissue proteins, a condition called *pulmonary fibrosis*, similarly decreases lung compliance.

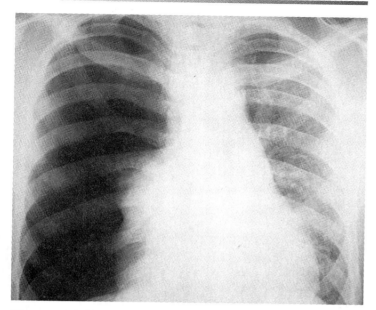

Figure 16.10

A pneumothorax of the right lung. The right side of the thorax appears uniformly dark because it is filled with air; the spaces between the ribs are also greater on the left due to release from the elastic tension of the lungs. The left lung appears denser (less dark) because of shunting of blood from the right to the left lung.

 The elastic nature of lung tissue is revealed when air enters the intrapleural space (as a result of an open chest wound, for example). This condition is called a **pneumothorax**, which is shown in figure 16.10. As air enters the intrapleural space, the intrapleural pressure rises until it is equal to the atmospheric pressure. When the intrapleural pressure is the same as the intrapulmonary pressure, the lung can no longer expand. Not only does the lung not expand during inspiration, it actually collapses away from the chest wall as a result of elastic recoil. Fortunately, a pneumothorax usually causes only one lung to collapse, since each lung is contained in a separate pleural compartment.

Elasticity

The term **elasticity** refers to the tendency of a structure to return to its initial size after being distended. The lungs are very elastic, due to a high content of elastin proteins, and resist distension. Since the lungs are normally stuck to the chest wall, they are always in a state of elastic tension. This tension increases during inspiration when the lungs are stretched and is reduced by elastic recoil during expiration. The elasticity of the lungs and of other thoracic structures thus aids in pushing the air out during expiration.

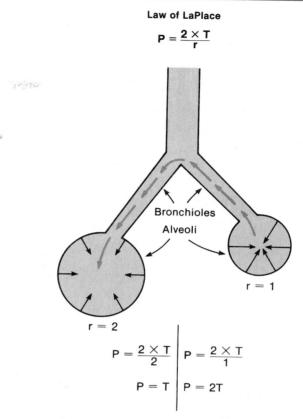

Figure 16.11

According to the law of LaPlace, the pressure created by surface tension should be greater in the smaller alveolus (*right*) than in the larger alveolus (*left*). This implies that (without surfactant) smaller alveoli would collapse and empty their air into larger alveoli.

Surface Tension

The forces that act to resist distension include elastic resistance and the **surface tension** that is exerted by fluid in the alveoli. The lungs both secrete and absorb fluid in two antagonistic processes that normally leave only a very thin film of fluid on the alveolar surface. Fluid absorption is driven (through osmosis) by the active transport of Na^+, while fluid secretion is driven by the active transport of Cl^- out of the alveolar epithelial cells. Research has demonstrated that people with cystic fibrosis have a genetic defect in one of the Cl^- carriers (called the cystic fibrosis transmembrane regulator, or CFTR, as described in chapter 6). This results in an imbalance of fluid absorption and secretion, so that the airway fluid becomes excessively viscous and difficult to clear.

The thin film of fluid normally present in the alveolus has a surface tension, which is due to the fact that water molecules at the surface are attracted more to other water molecules than to air. As a result, the surface water molecules are pulled tightly together by attractive forces from underneath. This surface tension produces a force that is directed inward, raising the pressure within the alveolus. As described by the **law of LaPlace,** the pressure thus created is directly proportional to the surface tension and inversely proportional to the radius of the alveolus (fig. 16.11).

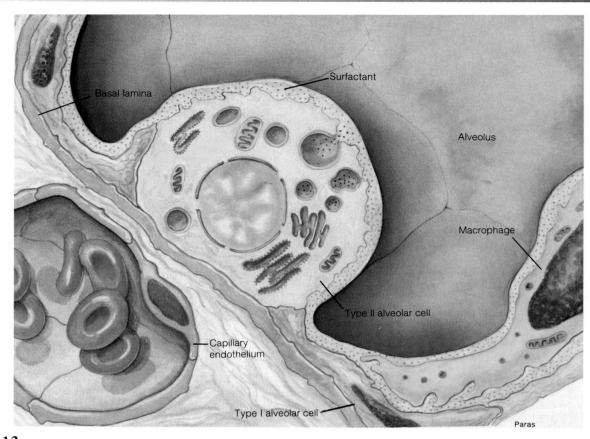

Figure 16.12

The production of pulmonary surfactant by type II alveolar cells. Surfactant appears to be composed of a derivative of lecithin combined with protein.

According to this law, the pressure in a smaller alveolus would be greater than in a larger alveolus if the surface tension were the same in both. The greater pressure of the smaller alveolus would then cause it to empty its air into the larger one (fig. 16.11). This does not normally occur because, as an alveolus decreases in size, its surface tension (the numerator in the equation) is decreased at the same time that its radius (the denominator) is reduced. The reason for the decreased surface tension, which prevents the alveoli from collapsing, is described in the next section.

Surfactant and the Respiratory Distress Syndrome

Alveolar fluid contains a phospholipid known as dipalmitoyl lecithin, probably attached to a protein, which functions to lower surface tension. This compound is called **surfactant**—a contraction of the term *surface active agent*. Because of the presence of surfactant, the surface tension in the alveoli is lower than would be predicted if surfactant were absent. Further, the

ability of surfactant to lower surface tension improves as the alveoli get smaller during expiration. This may be because the surfactant molecules become more concentrated as the alveoli get smaller. Surfactant thus prevents the alveoli from collapsing during expiration, as would be predicted from the law of LaPlace. Even after a forceful expiration, the alveoli remain open and a *residual volume* of air remains in the lungs. Since the alveoli do not collapse, less surface tension has to be overcome to inflate them at the next inspiration.

Surfactant is produced by type II alveolar cells (fig. 16.12) in late fetal life. Because no surfactant is produced until about the eighth month, premature babies are sometimes born with lungs that lack sufficient surfactant, and their alveoli are collapsed as a result. This condition is called **respiratory distress syndrome.** It is also called **hyaline membrane disease** because the high surface tension causes plasma fluid to leak into the alveoli, producing a glistening "membrane" appearance (and pulmonary edema). This condition does not occur in all premature babies; the rate of lung development depends on hormonal conditions (thyroxine and hydrocortisone primarily) and on genetic factors.

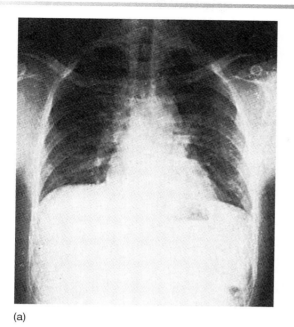
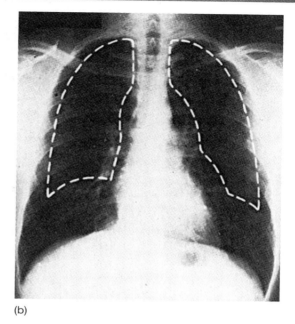

(a) (b)

Figure 16.13

A change in lung volume, as shown by radiographs (*a*) during expiration and (*b*) during inspiration. The increase in lung volume during full inspiration is shown by comparison with the lung volume in full expiration (*dashed lines*).

Even under normal conditions, the first breath of life is a difficult one because the newborn must overcome great surface tension forces in order to inflate its partially collapsed alveoli. The transpulmonary pressure required for the first breath is fifteen to twenty times that required for subsequent breaths, and an infant with respiratory distress syndrome must duplicate this effort with every breath. Fortunately, many babies with this condition can be saved by mechanical ventilators, and by exogenous surfactant delivered to the baby's lungs by means of an endotracheal tube. The exogenous surfactant may be a synthetic mixture of phospholipids, or it may be surfactant obtained from bovine lungs. The mechanical ventilator and exogenous surfactant help to keep the baby alive long enough for its lungs to mature so that it can manufacture sufficient surfactant on its own.

1. Describe how the intrapulmonary and intrapleural pressures change during inspiration and use Boyle's law to explain the reasons for these changes.

2. Define the terms *compliance* and *elasticity* and explain how these lung properties affect inspiration and expiration.

3. Describe lung surfactant and explain why the alveoli would collapse in the absence of surfactant.

Mechanics of Breathing

Normal, quiet inspiration results from muscle contraction, and normal expiration from muscle relaxation and elastic recoil. These actions can be forced by contractions of the accessory respiratory muscles. The amount of air inspired and expired can be measured in a number of ways to test pulmonary function.

The thorax must be sufficiently rigid so that it can protect vital organs and provide attachments for many short, powerful muscles. Breathing, or **pulmonary ventilation**, requires that the thorax also be flexible to function as a bellows during the ventilation cycle. The rigidity and the surfaces for muscle attachment are provided by the bony composition of the rib cage. The rib cage is pliable, however, because the ribs are separate from one another and because most ribs (the upper ten of the twelve pairs) are attached to the sternum by resilient costal cartilages. The vertebral attachments likewise provide considerable mobility. The structure of the rib cage and associated cartilages provides continuous elastic tension, so that when stretched by muscle contraction during inspiration, the rib cage can return passively to its resting dimensions when the muscles relax. This elastic recoil is greatly aided by the elasticity of the lungs.

Pulmonary ventilation consists of two phases: *inspiration* and *expiration.* Inspiration (inhalation) and expiration (exhalation) are accomplished by alternately increasing and decreasing the volumes of the thorax and lungs (fig. 16.13).

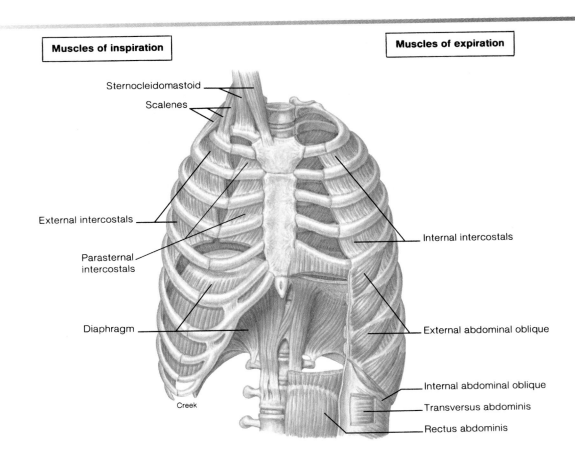

Muscles of inspiration

Muscles of expiration

Sternocleidomastoid

Scalenes

External intercostals

Parasternal intercostals

Diaphragm

Creek

Internal intercostals

External abdominal oblique

Internal abdominal oblique

Transversus abdominis

Rectus abdominis

Figure 16.14

The principal muscles of inspiration (*left*) and expiration (*right*).

Inspiration and Expiration

Between the bony portions of the rib cage are two layers of intercostal muscles: the **external intercostal muscles** and the **internal intercostal muscles** (fig. 16.14). Between the costal cartilages, however, there is only one muscle layer, and it has fibers oriented in a manner similar to those of the internal intercostals. These muscles may thus be called the *interchondral part* of the internal intercostals. Another name for them is the **parasternal intercostals.**

An unforced, or quiet, inspiration results primarily from contraction of the dome-shaped diaphragm, which lowers and flattens when it contracts. This increases thoracic volume in a vertical direction. Inspiration is aided by contraction of the parasternal and external intercostals, which raise the ribs when they contract and increase thoracic volume laterally. Other thoracic muscles become involved in forced (deep) inspiration. The most important of these is the *scalenus*, followed by the *pectoralis minor*, and in extreme cases the *sternocleidomastoid*. Contraction of these muscles elevates the ribs in an anteroposterior direction; at the same time, the upper rib cage is stabilized so that the intercostals become more effective.

Quiet expiration is a passive process. After becoming stretched by contractions of the diaphragm and thoracic muscles, the thorax and lungs recoil as a result of their elastic tension when the respiratory muscles relax. The decrease in lung volume raises the pressure within the alveoli above the atmospheric pressure and pushes the air out. During forced expiration, the internal intercostal muscles (excluding the interchondral part) contract and depress the rib cage. The abdominal muscles also aid expiration because, when they contract, they force abdominal organs up against the diaphragm and further decrease the volume of the thorax. By this means, the intrapulmonary pressure can rise 20 or 30 mmHg above the atmospheric pressure. The events that occur during inspiration and expiration are summarized in table 16.2 and shown in figure 16.15.

Pulmonary Function Tests

Pulmonary function may be assessed clinically by means of a technique known as *spirometry*. In this procedure, a subject breathes in a closed system in which air is trapped within a light plastic bell floating in water. The bell moves up when the subject exhales and down when the subject inhales. The movements of

Table 16.2	Summary of the Mechanisms Involved in Normal, Quiet Ventilation and Forced Ventilation	
	Inspiration	**Expiration**
Normal, quiet breathing	Contraction of the diaphragm and external intercostal muscles increases the thoracic and lung volume, decreasing intrapulmonary pressure to about –3 mmHg.	Relaxation of the diaphragm and external intercostals, plus elastic recoil of lungs, decreases lung volume and increases intrapulmonary pressure to about +3 mmHg.
Forced ventilation	Inspiration, aided by contraction of accessory muscles such as the scalenes and sternocleidomastoid, decreases intrapulmonary pressure to –20 mmHg or lower.	Expiration, aided by contraction of abdominal muscles and internal intercostal muscles, increases intrapulmonary pressure to +30 mmHg or higher.

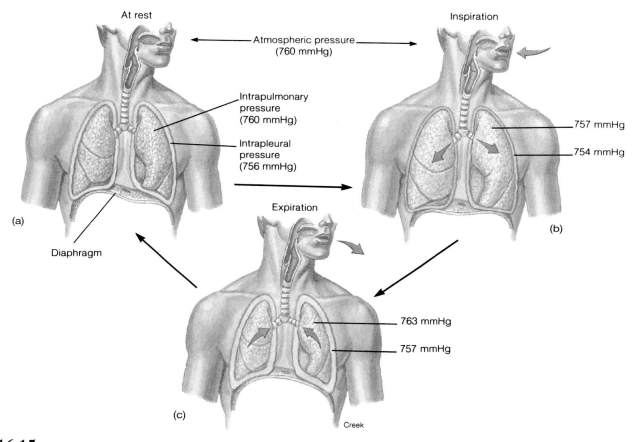

Figure 16.15

The mechanics of pulmonary ventilation. Pressures are shown (*a*) before inspiration, (*b*) during inspiration, and (*c*) during expiration.

the bell cause corresponding movements of a pen, which traces a record of the breathing on a rotating drum recorder (fig. 16.16). More sophisticated computerized devices are also commonly employed to assess lung function.

Lung Volumes and Capacities

An example of a spirogram is shown in figure 16.17, and the various lung volumes and capacities are defined in table 16.3. A lung capacity is equal to the sum of two or more lung volumes. During quiet breathing, for example, the amount of air expired in each breath is the **tidal volume.** The maximum amount of air

that can be forcefully exhaled after a maximum inhalation is called the **vital capacity,** and is equal to the sum of the **inspiratory reserve volume, tidal volume,** and **expiratory reserve volume** (fig. 16.17). Multiplying the tidal volume at rest times the number of breaths per minute yields a **total minute volume** of about 6 L per minute. During exercise, the tidal volume and the number of breaths per minute increase to produce a total minute volume as high as 100 to 200 L per minute.

It should be noted that not all of the inspired volume reaches the alveoli with each breath. As the fresh air is inhaled, it is mixed with air in the **anatomical dead space** (table 16.4). This

Table 16.3 Terms Used to Describe Lung Volumes and Capacities

Term	Definition
Lung Volumes	The four nonoverlapping components of the total lung capacity
Tidal volume	The volume of gas inspired or expired in an unforced respiratory cycle
Inspiratory reserve volume	The maximum volume of gas that can be inspired during forced breathing in addition to tidal volume
Expiratory reserve volume	The maximum volume of gas that can be expired during forced breathing in addition to tidal volume
Residual volume	The volume of gas remaining in the lungs after a maximum expiration
Lung Capacities	Measurements that are the sum of two or more lung volumes
Total lung capacity	The total amount of gas in the lungs at the end of a maximum inspiration
Vital capacity	The maximum amount of gas that can be expired after a maximum inspiration
Inspiratory capacity	The maximum amount of gas that can be inspired at the end of a tidal expiration
Functional residual capacity	The amount of gas remaining in the lungs at the end of a tidal expiration

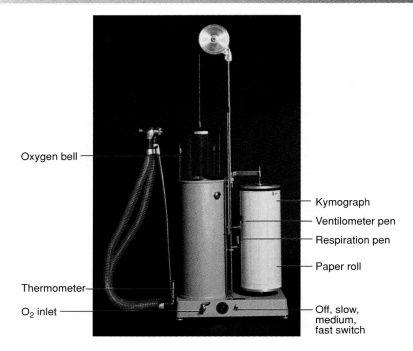

Oxygen bell

Kymograph
Ventilometer pen
Respiration pen

Paper roll

Thermometer

O_2 inlet

Off, slow, medium, fast switch

Figure 16.16

A spirometer (Collins 9L respirometer) used to measure lung volumes and capacities.

dead space comprises the conducting zone of the respiratory system—nose, mouth, larynx, trachea, bronchi, and bronchioles—where no gas exchange occurs. Air within the anatomical dead space has a lower oxygen concentration and a higher carbon dioxide concentration than the external air. Since the air in the dead space enters the alveoli first, the amount of fresh air reaching the alveoli with each breath is less than the tidal volume. But, since the volume of air in the dead space is an anatomical constant, the percentage of fresh air entering the alveoli is increased with increasing tidal volumes. For example, if the anatomical dead space is 150 ml and the tidal volume is 500 ml, the percentage of fresh air reaching the alveoli is $350/500 \times 100\% = 70\%$. If the tidal volume is increased to 2,000 ml, the percentage of fresh air reaching the alveoli is $1,850/2,000 \times 100\% = 93\%$. An increase in tidal volume can thus be a factor in the respiratory adaptations to exercise and high altitude, as will be described in later sections.

Restrictive and Obstructive Disorders

Spirometry is useful in the diagnosis of lung diseases. On the basis of pulmonary function tests, lung disorders can be classified as *restrictive* or *obstructive*. In restrictive disorders, such as pulmonary fibrosis, the vital capacity is reduced below normal. The rate at which the vital capacity can be forcibly exhaled, however, is normal. In disorders that are exclusively obstructive, such as asthma, the vital capacity is normal because lung tissue is not damaged. In asthma the bronchioles constrict, and this bronchoconstriction increases the resistance to air flow. Although the vital capacity is normal, the increased airway resistance makes expiration more difficult and take a longer time.

Table 16.4 Ventilation Terminology

Term	Definition
Air spaces	Alveolar ducts, alveolar sacs, and alveoli
Airways	Structures that conduct air from the mouth and nose to the respiratory bronchioles
Alveolar ventilation	Removal and replacement of gas in alveoli; equal to the tidal volume minus the volume of dead space times the breathing rate
Anatomical dead space	Volume of the conducting airways to the zone where gas exchange occurs
Apnea	Cessation of breathing
Dyspnea	Unpleasant, subjective feeling of difficult or labored breathing
Eupnea	Normal, comfortable breathing at rest
Hyperventilation	Alveolar ventilation that is excessive in relation to metabolic rate; results in abnormally low alveolar CO_2
Hypoventilation	Alveolar ventilation that is low in relation to metabolic rate; results in abnormally high alveolar CO_2
Physiological dead space	Combination of anatomical dead space and underventilated or underperfused alveoli that do not contribute normally to blood gas exchange
Pneumothorax	Presence of gas in the pleural space (the space between the visceral and parietal pleural membranes) causing lung collapse
Torr	Unit of pressure very nearly equal to the millimeter of mercury (760 mmHg = 760 torr)

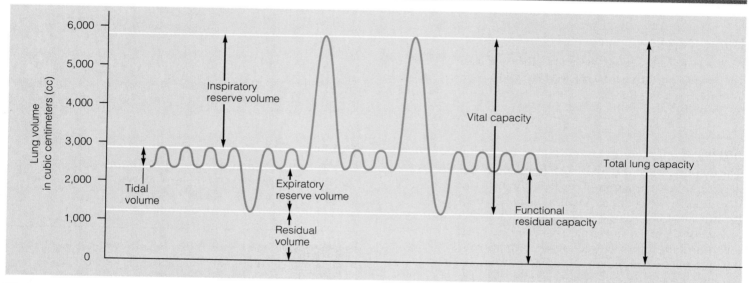

Figure 16.17

A spirogram showing lung volumes and capacities.

Obstructive disorders are thus diagnosed by tests that measure the rate of expiration. One such test is the **forced expiratory volume (FEV),** in which the percentage of the vital capacity that can be exhaled in the first second ($FEV_{1.0}$) is measured (fig. 16.18). An $FEV_{1.0}$ that is significantly less than 75% suggests the presence of obstructive pulmonary disease.

Pulmonary Disorders

People with pulmonary disorders frequently complain of **dyspnea,** which is a subjective feeling of "shortness of breath." Dyspnea may occur even when ventilation is normal, however, and may not occur even when total minute volume is very high, as in exercise. Some of the terms used to describe ventilation are defined in table 16.4.

 Bronchoconstriction often occurs in response to the inhalation of noxious agents in the air, such as from smoke or smog. The $FEV_{1.0}$ has therefore been used by researchers to determine the effects of various components of smog and of passive cigarette smoke inhalation on pulmonary function. These studies have shown that it is unhealthy to exercise on very smoggy days and that inhalation of smoke from other people's cigarettes in a closed environment can measurably affect pulmonary function.

There is normally a decline in the $FEV_{1.0}$ with age, but research suggests that this decline may be accelerated in cigarette smokers. Smokers under the age of 35 who quit have improved lung function; those who quit after the age of 35 slow their age-related decline in $FEV_{1.0}$ to normal rates.

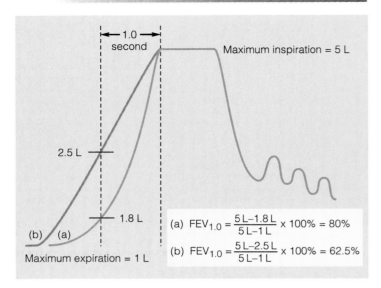

Figure 16.18

An illustration of the one-second forced expiratory volume ($FEV_{1.0}$) test. The percentage in (a) is normal, whereas that in (b) may indicate an obstructive pulmonary disorder.

In the figure:
- 1.0 second
- Maximum inspiration = 5 L
- 2.5 L
- 1.8 L
- (b) (a)
- Maximum expiration = 1 L

(a) $FEV_{1.0} = \dfrac{5\,L - 1.8\,L}{5\,L - 1\,L} \times 100\% = 80\%$

(b) $FEV_{1.0} = \dfrac{5\,L - 2.5\,L}{5\,L - 1\,L} \times 100\% = 62.5\%$

Asthma

The dyspnea, wheezing, and other symptoms of asthma are produced by the obstruction of air flow through the bronchioles that occurs in episodes or "attacks." This obstruction is caused by inflammation, mucous secretion, and bronchoconstriction. Inflammation of the airways is characteristic of asthma and itself contributes to increased airway responsiveness to agents that promote bronchiolar constriction. Bronchoconstriction further increases airway resistance and makes breathing difficult. Constriction of bronchiolar smooth muscles is stimulated by leukotrienes and histamine released by mast cells and leukocytes (chapter 15), which can be provoked by an allergic reaction or by the release of acetylcholine from parasympathetic nerve endings.

Emphysema

Alveolar tissue is destroyed in emphysema, resulting in fewer but larger alveoli (fig. 16.19). This reduces the surface area for gas exchange and decreases the ability of the bronchioles to remain open during expiration. Collapse of the bronchioles as a result of the compression of the lungs during expiration produces *air trapping*, which further decreases the efficiency of gas exchange in the alveoli.

Among the different types of emphysema, the most common occurs almost exclusively in people who have smoked cigarettes heavily over a period of years. A component of cigarette smoke apparently stimulates the macrophages and leukocytes to secrete proteolytic (protein-digesting) enzymes that destroy lung tissues. A less common type of emphysema results from the genetic inability to produce a plasma protein called α_1-antitrypsin.

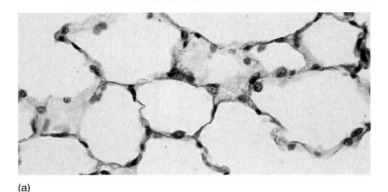

(a)

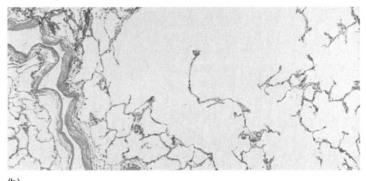

(b)

Figure 16.19

Photomicrographs of tissue (a) from a normal lung and (b) from the lung of a person with emphysema. In emphysema, lung tissue is destroyed, resulting in fewer and larger alveoli.

Asthma is often treated with glucocorticoid drugs, which inhibit inflammation. Epinephrine and related compounds stimulate beta-adrenergic receptors in the bronchioles and by this means promote bronchodilation. Therefore, epinephrine was used in the past as an inhaled spray to relieve the symptoms of an asthma attack. It has since been learned that there are two subtypes of beta receptors for epinephrine, and that the subtype in the heart (called β_1) is different from the one in the bronchioles (β_2). Capitalizing on these differences, compounds such as terbutaline have been developed. These compounds can more selectively stimulate the β_2-adrenergic receptors and cause bronchodilation without affecting the heart to the extent that epinephrine does.

This protein normally inhibits proteolytic enzymes such as trypsin, and thus normally protects the lungs against the effects of enzymes that are released from alveolar macrophages.

Chronic bronchitis and emphysema, the two most common causes of respiratory failure, are together called **chronic obstructive pulmonary disease (COPD).** In addition to the more direct obstructive and restrictive aspects of these conditions, other pathological changes may occur. These include edema, inflammation, hyperplasia (increased cell number), zones of pulmonary fibrosis, pneumonia, pulmonary emboli (traveling blood clots), and heart failure. Patients with severe chronic bronchitis or emphysema may develop *cor pulmonale*— pulmonary hypertension with hypertrophy and the eventual failure of the right ventricle. COPD is the fifth leading cause of death in the United States.

Pulmonary Fibrosis

Under certain conditions, for reasons that are poorly understood, lung damage leads to pulmonary fibrosis instead of emphysema. In this condition, the normal structure of the lungs is disrupted by the accumulation of fibrous connective tissue proteins. Fibrosis can result, for example, from the inhalation of particles less than 6 μm in size that can accumulate in the respiratory zone of the lungs. Included in this category is *anthracosis*, or black lung, which is produced by the inhalation of carbon particles from coal dust.

---

1. Describe the actions of the diaphragm and external intercostal muscles during inspiration. Explain how quiet expiration is produced.
2. Describe how forced inspiration and forced expiration are produced.
3. Define the terms *tidal volume* and *vital capacity*. Describe how the total minute volume is calculated and explain how this value is affected by exercise.
4. Describe how the vital capacity and the forced expiratory volume measurements are affected by asthma and pulmonary fibrosis. Give the reasons for these effects.

Gas Exchange in the Lungs

Gas exchange between the alveolar air and the blood in pulmonary capillaries results in an increased oxygen concentration and a decreased carbon dioxide concentration in the blood leaving the lungs. This blood enters the systemic arteries, where blood gas measurements are taken to assess the effectiveness of lung function.

The atmosphere is an ocean of gas that exerts pressure on all objects within it. The amount of this pressure can be measured with a glass U-tube filled with fluid. One end of the U-tube is exposed to the atmosphere, while the other side is continuous with a sealed vacuum tube. Since the atmosphere presses on the open-ended side, but not on the side connected to the vacuum tube, atmospheric pressure pushes fluid in the U-

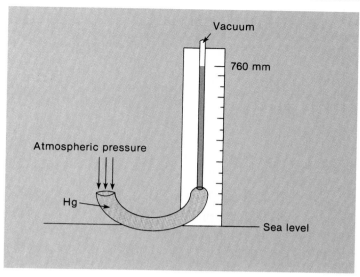

Figure 16.20

Atmospheric pressure at sea level can push a column of mercury to a height of 760 millimeters. This is also described as 760 torr, or one atmospheric pressure.

tube up on the vacuum side to a height determined by the atmospheric pressure and the density of the fluid. Water, for example, will be pushed up to a height of 33.9 feet (10,332 mm) at sea level, whereas mercury (Hg)—which is more dense—will be raised to a height of 760 mm. As a matter of convenience, therefore, devices used to measure atmospheric pressure (barometers) use mercury rather than water. The atmospheric pressure at sea level is thus said to be equal to 760 mmHg (or 760 *torr*), which is also described as a pressure of *one atmosphere* (fig. 16.20).

According to **Dalton's law**, the total pressure of a gas mixture (such as air) is equal to the sum of the pressures that each gas in the mixture would exert independently. The pressure that a particular gas in a mixture exerts independently is the **partial pressure** of that gas, which is equal to the product of the total pressure and the fraction of that gas in the mixture. The total pressure of the gas mixture is thus equal to the sum of the partial pressures of the constituent gases. Since oxygen constitutes about 21% of the atmosphere, for example, its partial pressure (abbreviated P_{O_2}) is 21% of 760, or about 159 mmHg. Since nitrogen constitutes about 78% of the atmosphere, its partial pressure is equal to $0.78 \times 760 = 593$ mmHg. These two gases thus contribute about 99% of the total pressure of 760 mmHg:

$$P_{dry\ atmosphere} = P_{N_2} + P_{O_2} + P_{CO_2} = 760\ mmHg$$

Calculation of P_{O_2}

With increasing altitude, the total atmospheric pressure and the partial pressure of the constituent gases decrease (table 16.5). At Denver, for example (5,000 feet above sea level), the atmospheric pressure is decreased to 619 mmHg, and the P_{O_2} is therefore

Table 16.5 Effect of Altitude on Partial Oxygen Pressure (P_{O_2})

Altitude (Feet above Sea Level)	Atmospheric Pressure (mmHg)	P_{O_2} in Air (mmHg)	P_{O_2} in Alveoli (mmHg)	P_{O_2} in Arterial Blood (mmHg)
0	760	159	105	100
2,000	707	148	97	92
4,000	656	137	90	85
6,000	609	127	84	79
8,000	564	118	79	74
10,000	523	109	74	69
20,000	349	73	40	35
30,000	226	47	21	19

reduced to $619 \times 0.21 = 130$ mmHg. At the peak of Mount Everest (at 29,000 feet), the P_{O_2} is only 42 mmHg. As one descends below sea level, as in ocean diving, the total pressure increases by one atmosphere for every 33 feet. At 33 feet therefore, the pressure equals $2 \times 760 = 1,520$ mmHg. At 66 feet, the pressure equals three atmospheres.

Inspired air contains variable amounts of moisture. By the time the air has passed into the respiratory zone of the lungs, however, it is normally saturated with water vapor (has a relative humidity of 100%). The capacity of air to contain water vapor depends on its temperature; since the temperature of the respiratory zone is constant at 37°C, its water vapor pressure is also constant (at 47 mmHg).

Water vapor, like the other constituent gases, contributes a partial pressure to the total atmospheric pressure. Since the total atmospheric pressure is constant (depending only on the height of the air mass), the water vapor "dilutes" the contribution of other gases to the total pressure:

$$P_{\text{wet atmosphere}} = P_{N_2} + P_{O_2} + P_{CO_2} + P_{H_2O}$$

When the effect of water vapor pressure is considered, the partial pressure of oxygen in the inspired air is decreased at sea level to

$$P_{O_2} \text{ (sea level)} = 0.21 (760 - 47) = 150 \text{ mmHg}$$

As a result of gas exchange in the alveoli, the P_{O_2} of alveolar air is further diminished to about 105 mmHg. The partial pressures of the inspired air and the partial pressures of alveolar air are compared in figure 16.21.

Partial Pressures of Gases in Blood

The enormous surface area of alveoli and the short diffusion distance between alveolar air and the capillary blood quickly help to bring the blood into gaseous equilibrium with the alveolar air. This function is further aided by the tremendous number of capillaries that surround each alveolus, forming an almost continuous sheet of blood around the alveoli (fig. 16.22).

When a liquid and a gas, such as blood and alveolar air, are at equilibrium, the amount of gas dissolved in the fluid reaches a maximum value. According to **Henry's law** (table 16.6), this value depends on (1) the solubility of the gas in the fluid, which is a physical constant; (2) the temperature of the

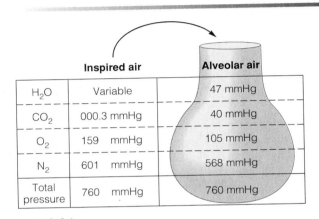

	Inspired air	Alveolar air
H_2O	Variable	47 mmHg
CO_2	000.3 mmHg	40 mmHg
O_2	159 mmHg	105 mmHg
N_2	601 mmHg	568 mmHg
Total pressure	760 mmHg	760 mmHg

Figure 16.21

Partial pressures of gases in the inspired air and the alveolar air.

fluid—more gas can be dissolved in cold water than warm water; and (3) the partial pressure of the gas. Since the temperature of the blood does not vary significantly, *the concentration of a gas dissolved in a fluid (such as plasma) depends directly on its partial pressure in the gas mixture*. When water—or plasma—is brought into equilibrium with air at a P_{O_2} of 100 mmHg, for example, the fluid will contain 0.3 ml of O_2 per 100 ml fluid at 37°C. If the P_{O_2} of the gas were reduced by half, the amount of dissolved oxygen would also be reduced by half.

Blood Gas Measurements

Measurement of the oxygen content of blood (in ml of O_2 per 100 ml blood) is a laborious procedure. Fortunately, an **oxygen electrode** that produces an electric current in proportion to the concentration of *dissolved oxygen* has been developed. If this electrode is placed in a fluid while oxygen is artificially bubbled into it, the current produced by the oxygen electrode will increase up to a maximum value. At this maximum value the fluid is *saturated* with oxygen—that is, all of the oxygen that can be dissolved at that temperature and P_{O_2} is dissolved. At a constant temperature, the amount dissolved, and thus the electric current, depend only on the P_{O_2} of the gas.

As a matter of convenience, it can now be said that *the fluid has the same P_{O_2} as the gas*. If it is known that the gas has a P_{O_2} of 152 mmHg, for example, the deflection of a needle by

Table 16.6	Some Physical Laws of Importance in Ventilation and Gas Exchange
Physical Law	**Description**
Boyle's law	The volume of a gas is inversely proportional to its pressure when the temperature and mass are constant (PV = constant).
Dalton's law	The total pressure of a gas mixture is equal to the sum of the partial pressures of its constituent gases. The partial pressure of each gas is the pressure it would exert if it alone occupied the total volume of the mixture.
Graham's law	The rate of diffusion of a gas through a liquid is directly proportional to its solubility and inversely proportional to its density (or gram molecular weight).
Henry's law	The volume of a gas that will dissolve in a liquid at a given temperature is proportional to the partial pressure of the gas.
LaPlace's law	The inward pressure tending to collapse a bubble is equal to two times its surface tension divided by the radius of the bubble.

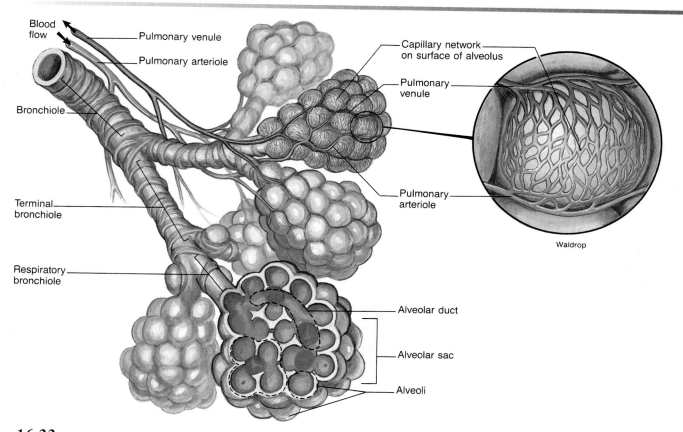

Figure 16.22

The extensive surface area of contact between the pulmonary capillaries and the alveoli allows for rapid exchange of gases between the air and blood.

the oxygen electrode can be calibrated on a scale at 152 mmHg (fig. 16.23). The actual amount of dissolved oxygen under these circumstances is not particularly important (it can be looked up in solubility tables, if desired); it is simply a linear function of the P_{O_2}. A lower P_{O_2} indicates that less oxygen is dissolved; a higher P_{O_2} indicates that more oxygen is dissolved.

If the oxygen electrode is next inserted into an unknown sample of blood, the P_{O_2} of that sample can be read directly from the previously calibrated scale. Suppose, as illustrated in figure 16.23, the blood sample has a P_{O_2} of 100 mmHg. Since alveolar air has a P_{O_2} of about 105 mmHg, this reading indicates that the blood is almost in complete equilibrium with the alveolar air.

The oxygen electrode responds only to oxygen dissolved in water or plasma; it cannot respond to oxygen that is bound to hemoglobin in red blood cells. Most of the oxygen in blood, however, is located in the red blood cells attached to hemoglobin. The oxygen content of whole blood thus depends on both its P_{O_2} and its red blood cell and hemoglobin content. At a P_{O_2} of about 100 mmHg, whole blood normally contains almost 20 ml O_2 per 100 ml blood; of this amount, only 0.3 ml O_2 is dissolved in the plasma and 19.7 ml O_2 is found within the red blood cells. Since only the 0.3 ml O_2 per 100 ml blood affects the P_{O_2} measurement, this measurement would be unchanged if the red blood cells were removed from the sample.

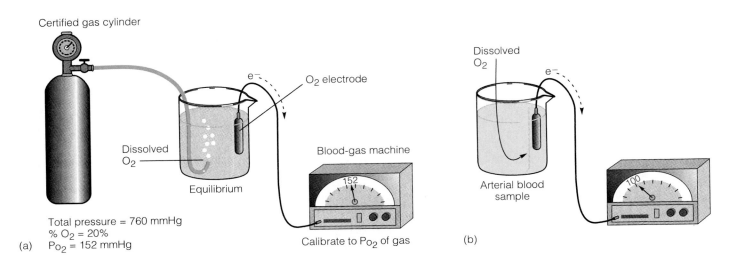

Figure 16.23

Blood gas measurements using the P_{O_2} electrode. (*a*) The electrical current generated by the oxygen electrode is calibrated so that the needle of the blood gas machine points to the P_{O_2} of the gas with which the fluid is in equilibrium. (*b*) Once standardized in this way, the electrode can be inserted in a fluid, such as blood, and the P_{O_2} of this solution can be measured.

Significance of Blood P_{O_2} and P_{CO_2} Measurements

Since blood P_{O_2} measurements are not directly affected by the oxygen in red blood cells, the P_{O_2} does not provide a measurement of the total oxygen content of whole blood. It does, however, provide a good index of *lung function*. If the inspired air had a normal P_{O_2} but the arterial P_{O_2} was below normal, for example, gas exchange in the lungs would have to be impaired. Measurements of arterial P_{O_2} thus provide valuable information in treating people with pulmonary diseases, in performing surgery (when breathing may be depressed by anesthesia), and in caring for premature babies with respiratory distress syndrome.

When the lungs are functioning properly, the P_{O_2} of systemic arterial blood is only 5 mmHg less than the P_{O_2} of alveolar air. At a normal P_{O_2} of about 100 mmHg, hemoglobin is almost completely loaded with oxygen. An increase in blood P_{O_2}—produced, for example, by breathing 100% oxygen from a gas tank—thus cannot significantly increase the amount of oxygen contained in the red blood cells. It can, however, significantly increase the amount of oxygen dissolved in the plasma (because the amount dissolved is directly determined by the P_{O_2}). If the P_{O_2} doubles, the amount of oxygen dissolved in the plasma also doubles, but the total oxygen content of whole blood increases only slightly, since most of the oxygen by far is not in plasma but in the red blood cells.

Since the oxygen carried by red blood cells must first dissolve in plasma before it can diffuse to the tissue cells, however, a doubling of the blood P_{O_2} means that the *rate of oxygen diffusion* to the tissues would double under these conditions. For this

reason, breathing from a tank of 100% oxygen (with a P_{O_2} of 760 mmHg) would significantly increase oxygen delivery to the tissues, although it would have little effect on the total oxygen content of blood.

An electrode that produces a current in response to dissolved carbon dioxide is also used, so that the P_{CO_2} of blood can be measured together with its P_{O_2}. Blood in the systemic veins, which is delivered to the lungs by the pulmonary arteries, usually has a P_{O_2} of 40 mmHg and a P_{CO_2} of 46 mmHg. After gas exchange in the alveoli of the lungs, blood in the pulmonary veins and systemic arteries has a P_{O_2} of about 100 mmHg and a P_{CO_2} of 40 mmHg (fig. 16.24). The values in arterial blood are relatively constant and clinically significant because they reflect lung function. Blood gas measurements of venous blood are not as clinically useful because these values are far more variable. Venous P_{O_2} is much lower and P_{CO_2} much higher after exercise, for example, than at rest, whereas arterial values are not significantly affected by physical activity.

Pulmonary Circulation and Ventilation/Perfusion Ratios

In a fetus, the pulmonary circulation has a high vascular resistance because the lungs are partially collapsed. This high vascular resistance helps to shunt blood from the right to the left atrium through the foramen ovale, and from the pulmonary artery to the aorta through the ductus arteriosus (described in chapter 13). After birth, the foramen ovale and ductus arteriosus close, and the vascular resistance of the pulmonary circulation falls sharply. This fall in vascular resistance at birth is due

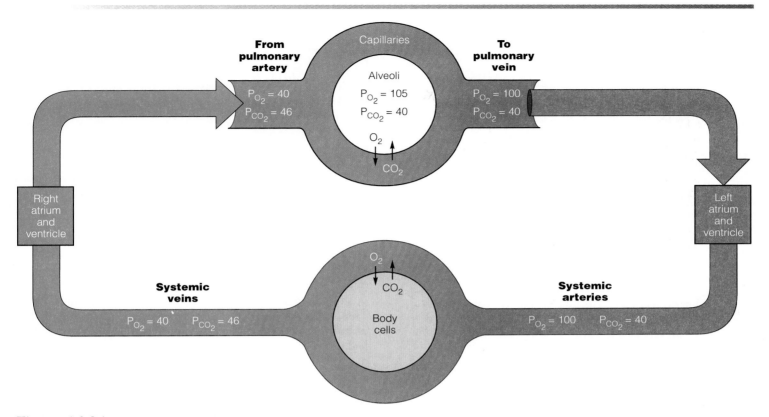

Figure 16.24

The P_{O_2} and P_{CO_2} of blood as a result of gas exchange in the lung alveoli and gas exchange between systemic capillaries and body cells.

to (1) opening of the vessels as a result of the subatmospheric intrapulmonary pressure and physical stretching of the lungs during inspiration and (2) dilation of the pulmonary arterioles in response to increased alveolar P_{O_2}.

In the adult, the right ventricle (like the left) has an output of about 5.5 L per minute. The rate of blood flow through the pulmonary circulation is thus equal to the flow rate through the systemic circulation. Blood flow, as described in chapter 14, is directly proportional to the pressure difference between the two ends of a vessel and inversely proportional to the vascular resistance. In the systemic circulation, the mean arterial pressure is 90 to 100 mmHg, and the pressure of the right atrium is 0 mmHg; the pressure difference is thus about 100 mmHg. The mean pressure of the pulmonary artery, by contrast, is only 15 mmHg and the pressure of the left atrium is 5 mmHg. The driving pressure in the pulmonary circulation is thus 15 − 5, or 10 mmHg.

Since the driving pressure in the pulmonary circulation is only one-tenth that of the systemic circulation, and since the flow rates are equal, it follows that the pulmonary vascular resistance must be one-tenth that of the systemic vascular resistance. The pulmonary circulation, in other words, is a low-resistance, low-pressure pathway. The low pulmonary blood pressure produces less filtration pressure than in the systemic capillaries, and thus helps to protect against *pulmonary edema*,

which is a dangerous condition that can impede ventilation and gas exchange. Pulmonary edema can occur when there is pulmonary hypertension, which may be produced, for example, by left heart failure.

Pulmonary arterioles constrict when the alveolar P_{O_2} is low and dilate as the alveolar P_{O_2} is raised. This response is opposite to that of systemic arterioles, which dilate in response to low tissue P_{O_2} (as described in chapter 14). Dilation of the systemic arterioles when the P_{O_2} is low helps to supply more blood and oxygen to the tissues; constriction of the pulmonary arterioles when the alveolar P_{O_2} is low helps to decrease blood flow to alveoli that are inadequately ventilated.

Constriction of the pulmonary arterioles where the alveolar P_{O_2} is low and their dilation where the alveolar P_{O_2} is high helps to *match ventilation to perfusion* (the term *perfusion* refers to blood flow). If this autoregulation of blood flow did not occur, blood from poorly ventilated alveoli would mix with blood from well-ventilated alveoli and the blood leaving the lungs would have a lowered P_{O_2} as a result of this dilution effect.

Dilution of the P_{O_2} of pulmonary vein blood actually does occur to some degree despite these regulatory mechanisms. When a person stands upright, the force of gravity causes a greater blood flow to the base of the lungs than to the apex (top). Ventilation likewise increases from apex to base, but this increase is not proportionate to the increase in blood flow. The

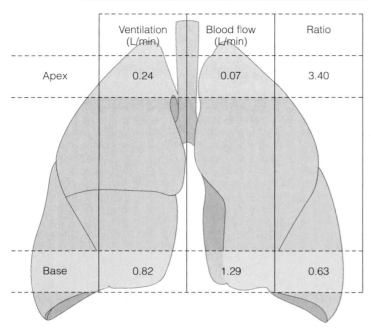

	Ventilation (L/min)	Blood flow (L/min)	Ratio
Apex	0.24	0.07	3.40
Base	0.82	1.29	0.63

Figure 16.25

Ventilation, blood flow, and ventilation/perfusion ratios at the apex and base of the lungs. The ratios indicate that the apex is relatively overventilated and the base underventilated in relation to their blood flows. As a result of such uneven matching of ventilation to perfusion, the blood leaving the lungs has a P_{O_2} that is slightly less (by about 4.5 mmHg) than the P_{O_2} of alveolar air.

ventilation/perfusion ratio at the apex is thus high (0.24 L air divided by 0.07 L blood per minute is equivalent to a ratio of 3.4/1.0), while at the base of the lungs it is low (0.82 L air divided by 1.29 L blood per minute is equivalent to a ratio of 0.6/1.0). This is illustrated in figure 16.25.

Functionally, the alveoli at the apex of the lungs are thus overventilated (or underperfused) and are actually larger than alveoli at the base. This mismatch of ventilation/perfusion ratios is normal, but is largely responsible for the 5 mmHg difference in P_{O_2} between alveolar air and arterial blood. Abnormally large mismatches of ventilation/perfusion ratios can occur in cases of pneumonia, pulmonary emboli, edema, or other pulmonary disorders.

Disorders Caused by High Partial Pressures of Gases

The total atmospheric pressure increases by one atmosphere (760 mmHg) for every 10 m (33 ft) below sea level. If a diver descends 10 m below sea level, therefore, the partial pressures and amounts of dissolved gases in the plasma will be twice those values at sea level. At 20 m they are three times, and at 30 m they are four times the values at sea level. The increased amounts of nitrogen and oxygen dissolved in the blood plasma under these conditions can have serious effects on the body.

Hyperbaric oxygen—oxygen at greater than one atmosphere pressure—is often used to treat conditions such as carbon monoxide poisoning, circulatory shock, and gas gangrene. Before the dangers of oxygen toxicity were realized, these hyperbaric oxygen treatments sometimes resulted in tragedy. Particularly tragic were the cases of **retrolental fibroplasia**, in which damage to the retina and blindness resulted from hyperbaric oxygen treatment of premature babies with hyaline membrane disease.

Oxygen Toxicity

Although breathing 100% oxygen at one or two atmospheres pressure can be safely tolerated for a few hours, higher partial oxygen pressures can be very dangerous. *Oxygen toxicity* develops rapidly when the P_{O_2} rises above about 2.5 atmospheres. This is apparently caused by the oxidation of enzymes and other destructive changes that can damage the nervous system and lead to coma and death. For these reasons, deep-sea divers commonly use gas mixtures in which oxygen is diluted with inert gases such as nitrogen (as in ordinary air) or helium.

Nitrogen Narcosis

Although at sea level nitrogen is physiologically inert, larger amounts of dissolved nitrogen under hyperbaric conditions have deleterious effects. Since it takes time for the nitrogen to dissolve, these effects usually do not appear until the person has remained submerged over an hour. *Nitrogen narcosis* resembles alcohol intoxication; depending on the depth of the dive, the diver may experience "rapture of the deep" or may become so drowsy as to be totally incapacitated.

Decompression Sickness

The amount of nitrogen dissolved in the plasma decreases as the diver ascends to sea level as a result of the progressive decrease in the P_{N_2}. If the diver surfaces slowly, a large amount of nitrogen can diffuse through the alveoli and be eliminated in the expired breath. If decompression occurs too rapidly, however, bubbles of nitrogen gas (N_2) can form in the blood. This process is analogous to the formation of carbon dioxide bubbles in a champagne bottle when the cork is removed. The bubbles of N_2 gas in the blood can block small blood channels, producing muscle and joint pain, as well as more serious damage. These effects are known as *decompression sickness*, or the bends.

The cabins of airplanes that fly long distances at high altitudes (30,000 to 40,000 ft) are pressurized so that the passengers and crew do not experience the very low atmospheric pressures of these altitudes. If a cabin were to become rapidly depressurized at high altitude, much less nitrogen could remain dissolved at the greatly lowered pressure. People in this situation, like the divers that ascend too rapidly, would thus experience decompression sickness.

1. Describe how the P_{O_2} of air is calculated and how this value is affected by altitude, diving, and water vapor pressure.

2. Describe how blood P_{O_2} measurements are taken, and explain the physiological and clinical significance of these measurements.

3. Explain how the arterial P_{O_2} and the oxygen content of whole blood are affected by (a) hyperventilation, (b) breathing from a tank containing 100% oxygen, (c) anemia (low red blood cell count and hemoglobin concentration), and (d) high altitude.

4. Describe the ventilation/perfusion ratios of the lungs and explain why systemic arterial blood has a slightly lower P_{O_2} than alveolar air.

5. Explain how decompression sickness is produced in divers who ascend too rapidly.

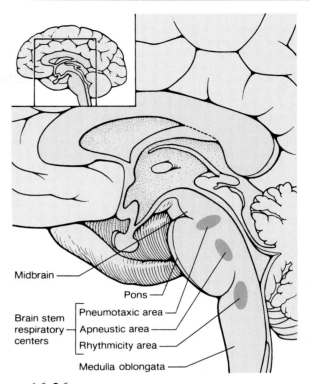

Figure 16.26

Approximate locations of the brain stem respiratory centers.

Regulation of Breathing

The motor neurons that stimulate the respiratory muscles are controlled by two major descending pathways: one that controls voluntary breathing, and one that controls involuntary breathing. The unconscious, rhythmic control of breathing is influenced by sensory feedback from receptors sensitive to the P_{CO_2}, pH, and P_{O_2} of arterial blood.

Inspiration and expiration are produced by the contraction and relaxation of skeletal muscles in response to activity in somatic motor neurons in the spinal cord. The activity of these motor neurons, in turn, is controlled by descending tracts from neurons in the respiratory control centers in the medulla oblongata and from neurons in the cerebral cortex.

Brain Stem Respiratory Centers

A loose aggregation of neurons in the reticular formation of the *medulla oblongata* forms the **rhythmicity center** that controls automatic breathing. The rhythmicity center consists of interacting pools of neurons that fire either during inspiration (*I neurons*) or expiration (*E neurons*). The I neurons project to and stimulate spinal motoneurons that innervate the respiratory muscles. Expiration is a passive process that occurs when the I neurons are inhibited by the activity of the E neurons. The activity of I and E neurons varies in a reciprocal way, so that a rhythmic pattern of breathing is produced. The cycle of inspiration and expiration is thus intrinsic to the neural activity of the medulla. The rhythmicity center in the medulla is divided into a dorsal group of neurons, which regulates the activity of the phrenic nerves to the diaphragm, and a ventral group, which controls the motor neurons to the intercostal muscles.

 The automatic control of breathing is regulated by nerve fibers that descend in the lateral and ventral white matter of the spinal cord from the medulla oblongata. The voluntary control of breathing is a function of the cerebral cortex and involves nerve fibers that descend in the corticospinal tracts (chapter 8). The separation of the voluntary and involuntary pathways is dramatically illustrated in the condition called **Ondine's curse.** In this condition, neurological damage abolishes the automatic but not the voluntary control of breathing. People with Ondine's curse must consciously force themselves to breathe and be put on artificial respirators when they sleep.

The activity of the medullary rhythmicity center is influenced by centers in the *pons*. As a result of research in which the brain stem is destroyed at different levels, two respiratory control centers have been identified in the pons. One area—the **apneustic center**—appears to promote inspiration by stimulating the I neurons in the medulla. The other pons area—called the **pneumotaxic center**—seems to antagonize the apneustic center and inhibit inspiration (fig. 16.26). The apneustic center is believed to provide a tonic, or constant, stimulus for inspiration, which is cyclically inhibited by the activity of the pneumotaxic center.

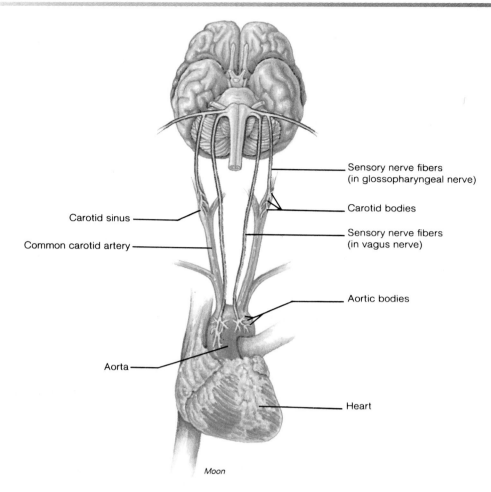

Moon

Figure 16.27

The peripheral chemoreceptors (aortic and carotid bodies) regulate the brain stem respiratory centers by means of sensory nerve stimulation.

The automatic control of breathing is also influenced by input from receptors sensitive to the chemical composition of the blood. There are two groups of *chemoreceptors* that respond to changes in blood P_{CO_2}, pH, and P_{O_2}. These are the **central chemoreceptors** in the medulla oblongata and the **peripheral chemoreceptors.** The peripheral chemoreceptors are contained within small nodules associated with the aorta and the carotid arteries, and receive blood from these critical arteries via small arterial branches. The peripheral chemoreceptors include the **aortic bodies,** located around the aortic arch, and the **carotid bodies,** located in each common carotid artery at the point at which it branches into the internal and external carotid arteries (fig. 16.27). The aortic and carotid bodies should not be confused with the aortic and carotid sinuses (chapter 14) that are located within these arteries. The aortic and carotid sinuses contain receptors that monitor the blood pressure.

The peripheral chemoreceptors control breathing indirectly via sensory nerve fibers to the medulla. The aortic bodies send sensory information to the medulla in the vagus (tenth cranial) nerve; the carotid bodies stimulate sensory fibers in the glossopharyngeal (ninth cranial) nerve. The neural and sensory control of ventilation is summarized in figure 16.28.

Effects of Blood P_{CO_2} and pH on Ventilation

Chemoreceptor input to the brain stem modifies the rate and depth of breathing so that, under normal conditions, arterial P_{CO_2}, pH, and P_{O_2} remain relatively constant. If hypoventilation (inadequate ventilation) occurs, P_{CO_2} quickly rises and pH falls. The fall in pH is due to the fact that carbon dioxide can combine with water to form carbonic acid, which in turn can release H^+ into the solution. This is shown in the following equations:

$$CO_2 + H_2O \rightarrow H_2CO_3$$
$$H_2CO_3 \rightarrow HCO_3^- + H^+$$

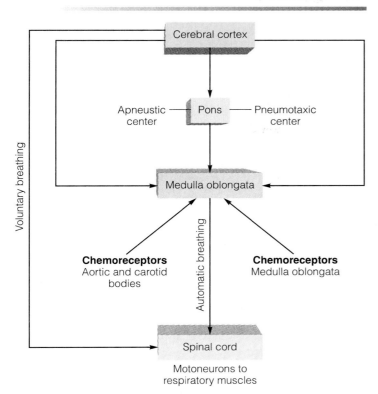

Figure 16.28

A diagram showing the control of ventilation by the central nervous system. (The feedback effects of pulmonary stretch receptors and "irritant" receptors are not shown.)

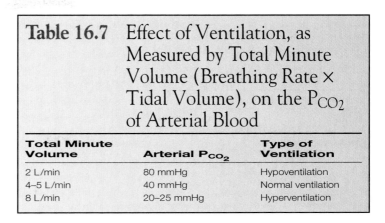

Table 16.7 Effect of Ventilation, as Measured by Total Minute Volume (Breathing Rate × Tidal Volume), on the P_{CO_2} of Arterial Blood

Total Minute Volume	Arterial P_{CO_2}	Type of Ventilation
2 L/min	80 mmHg	Hypoventilation
4–5 L/min	40 mmHg	Normal ventilation
8 L/min	20–25 mmHg	Hyperventilation

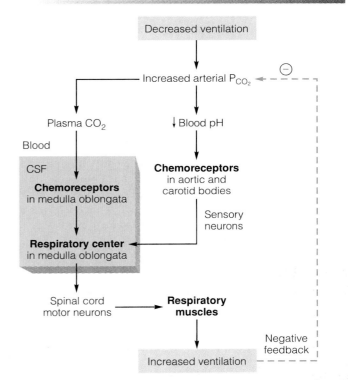

Figure 16.29

Negative feedback control of ventilation through changes in blood P_{CO_2} and pH. The orange box represents the blood-brain barrier, which allows CO_2 to pass into the cerebrospinal fluid but prevents the passage of H^+.

The oxygen content of the blood decreases much more slowly, because there is a large "reservoir" of oxygen attached to hemoglobin. During hyperventilation, conversely, blood P_{CO_2} quickly falls and pH rises due to the excessive elimination of carbonic acid. The oxygen content of blood, on the other hand, is not significantly increased by hyperventilation (hemoglobin in arterial blood is 97% saturated with oxygen during normal ventilation).

The blood P_{CO_2} and pH are, therefore, more immediately affected by changes in ventilation than is the oxygen content. Indeed, changes in P_{CO_2} provide a sensitive index of ventilation, as shown in table 16.7. In view of these facts, it is not surprising that changes in P_{CO_2} provide the most potent stimulus for the reflex control of ventilation. Ventilation, in other words, is adjusted to maintain a constant P_{CO_2}; proper oxygenation of the blood occurs naturally as a side product of this reflex control.

The rate and depth of ventilation are normally adjusted to maintain an arterial P_{CO_2} of 40 mmHg. Hypoventilation causes a rise in P_{CO_2}—a condition called *hypercapnia*. Hyperventilation, conversely, results in *hypocapnia*. Chemoreceptor regulation of breathing in response to changes in P_{CO_2} is illustrated in figure 16.29.

Chemoreceptors in the Medulla

The chemoreceptors most sensitive to changes in the arterial P_{CO_2} are located in the ventral area of the medulla oblongata, near the exit of the ninth and tenth cranial nerves. These chemoreceptor neurons are anatomically separate from, but synaptically communicate with, the neurons of the respiratory control center in the medulla.

An increase in arterial P_{CO_2} causes a rise in the H^+ concentration of the blood as a result of increased carbonic acid

Table 16.8	Terms Used to Describe Blood Oxygen and Carbon Dioxide Levels	
Term	**Definition**
Hypoxemia	A lower than normal oxygen content or P_{O_2} in arterial blood.
Hypoxia	A lower than normal oxygen content or P_{O_2} in the lungs, blood, or tissues. This is a more general term than hypoxemia. Tissues can be hypoxic, for example, even though there is no hypoxemia (as when the blood flow is occluded).
Hypercapnia, or hypercarbia	An increase in the P_{CO_2} of systemic arteries to above 40 mmHg. Usually this occurs when the ventilation is inadequate for a given metabolic rate (hypoventilation). Antonyms are *hypocapnia* and *hypocarbia* (usually produced by hyperventilation).

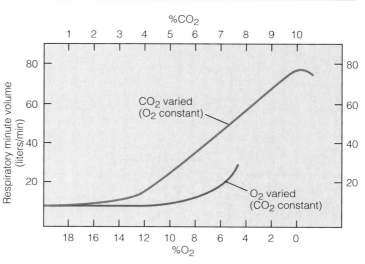

Figure 16.30

A comparison of the effects on respiration of increasing concentrations of CO_2 with decreasing concentrations of O_2 in air. Notice that respiration increases linearly with increasing CO_2 concentration, whereas O_2 concentrations must decrease to half the normal value before respiration is stimulated.

 People who hyperventilate during psychological stress are sometimes told to breathe into a paper bag so that they rebreathe their expired air that is enriched in CO_2. This procedure helps to raise their blood P_{CO_2} back up to the normal range. This is needed because hypocapnia causes cerebral vasoconstriction. In addition to producing dizziness, the cerebral ischemia that results can lead to acidotic conditions in the brain that, through stimulation of the medullary chemoreceptors, can cause further hyperventilation. Breathing into a paper bag can thus relieve the hypocapnia and stop the hyperventilation.

concentrations. The H^+ in the blood, however, cannot cross the blood-brain barrier and, therefore, cannot influence the medullary chemoreceptors. Carbon dioxide in the arterial blood *can* cross the blood-brain barrier and, through the formation of carbonic acid, can lower the pH of cerebrospinal fluid. This fall in cerebrospinal fluid pH directly stimulates the chemoreceptors in the medulla when there is a rise in arterial P_{CO_2}.

The chemoreceptors in the medulla are ultimately responsible for 70% to 80% of the increased ventilation that occurs in response to a sustained rise in arterial P_{CO_2}. This response, however, takes several minutes. The immediate increase in ventilation that occurs when P_{CO_2} rises is produced by stimulation of the peripheral chemoreceptors.

Peripheral Chemoreceptors

The aortic and carotid bodies are not stimulated directly by blood CO_2. Instead, they are stimulated by a rise in the H^+ concentration (fall in pH) of arterial blood, which occurs when the blood CO_2, and thus carbonic acid, is raised. The retention of CO_2 during hypoventilation thus stimulates the medullary chemoreceptors through a lowering of cerebrospinal fluid pH and stimulates peripheral chemoreceptors through a lowering of blood pH.

Effects of Blood P_{O_2} on Ventilation

Under normal conditions, blood P_{O_2} affects breathing only indirectly, by influencing the chemoreceptor sensitivity to changes in P_{CO_2}. Chemoreceptor sensitivity to P_{CO_2} is augmented by a low P_{O_2} (so ventilation is increased at a high altitude, for example) and is decreased by a high P_{O_2}. If the blood P_{O_2} is raised by breathing 100% oxygen, therefore, the breath can be held longer because the response to increased P_{CO_2} is blunted.

When the blood P_{CO_2} is held constant by experimental techniques, the P_{O_2} of arterial blood must fall from 100 mmHg to below 50 mmHg before ventilation is significantly stimulated (fig. 16.30). This stimulation is apparently due to a direct effect of P_{O_2} on the carotid bodies. Since this degree of *hypoxemia*, or

low blood oxygen (table 16.8), does not normally occur even in breath holding, P_{O_2} does not normally exert this direct effect on breathing.

In emphysema, when there is a chronic retention of carbon dioxide, the chemoreceptor response to the carbon dioxide becomes blunted. This is because the choroid plexus in the brain (chapter 8) secretes more bicarbonate into the cerebrospinal fluid, buffering the fall in cerebrospinal fluid pH. The abnormally high P_{CO_2}, however, enhances the sensitivity of the carotid bodies to a fall in P_{O_2}. For people with emphysema, therefore, breathing may thus be stimulated by a *hypoxic drive* rather than by increases in blood P_{CO_2}.

The effects of changes in the blood P_{CO_2}, pH, and P_{O_2} on chemoreceptors and the regulation of ventilation are summarized in table 16.9.

Table 16.9 Sensitivity of Chemoreceptors to Changes in Blood Gases and pH

Stimulus	Chemoreceptor	Comments
$\uparrow P_{CO_2}$	Medulla oblongata; aortic and carotid bodies	Medullary chemoreceptors are sensitive to the pH of cerebrospinal fluid (CSF). Diffusion of CO_2 from the blood into the CSF lowers the pH of CSF by forming carbonic acid. Similarly, the aortic and carotid bodies are stimulated by a fall in blood pH induced by increases in blood CO_2.
$\downarrow pH$	Aortic and carotid bodies	Peripheral chemoreceptors are stimulated by decreased blood pH independent of the effect of blood CO_2. Chemoreceptors in the medulla are not affected by changes in blood pH because H^+ cannot cross the blood-brain barrier.
$\downarrow P_{O_2}$	Carotid bodies	Low blood P_{O_2} (hypoxemia) augments the chemoreceptor response to blood P_{CO_2} and can stimulate ventilation directly when the P_{O_2} falls below 50 mmHg.

A variety of disease processes can produce cessation of breathing during sleep, or *sleep apnea.* **Sudden infant death syndrome (SIDS)** is an especially tragic form of sleep apnea that claims the lives of about ten thousand babies annually in the United States. Victims of this condition are apparently healthy 2-to-5-month-old babies who die in their sleep without apparent reason—hence, the layperson's term "crib death." These deaths seem to be caused by failure of the respiratory control mechanisms in the brain stem and/or by failure of the carotid bodies to be stimulated by reduced arterial oxygen.

1. Describe the effects of voluntary hyperventilation and breath holding on arterial P_{CO_2}, pH, and oxygen content. Indicate the relative degree of changes in these values.
2. Using a flowchart to show a negative feedback loop, explain the relationship between ventilation and arterial P_{CO_2}.
3. Explain the effect of increased arterial P_{CO_2} on
   (a) chemoreceptors in the medulla oblongata and
   (b) chemoreceptors in the aortic and carotid bodies.
4. Explain the role of arterial P_{O_2} in the regulation of breathing. Explain why ventilation increases when a person goes to a high altitude.

Pulmonary Stretch and Irritant Reflexes

The lungs contain various types of receptors that influence the brain stem respiratory control centers via sensory fibers in the vagus. These receptors are involved in both regulatory and defensive reflexes. Irritant receptors in the lungs, for example, stimulate reflex constriction of the bronchioles in response to smoke, ozone, and smog. A chemical called *capsaicin*, which is the ingredient of hot peppers that creates the burning sensation, has been shown to be a potent stimulator of sensory fibers in the vagus that promote reflex bronchosecretion and bronchoconstriction. These reflexes are presumed to be beneficial in situations where they improve the propulsive force, and thus the effectiveness, of coughing.

The **Hering–Breuer reflex** is stimulated by pulmonary stretch receptors. The activation of these receptors during inspiration inhibits the respiratory control centers, making further inspiration increasingly difficult. This helps to prevent undue distension of the lungs and may contribute to the smoothness of the ventilation cycles. A similar inhibitory reflex may occur during expiration. The Hering–Breuer reflex appears to be important in maintaining normal ventilation in the newborn. Pulmonary stretch receptors in adults, however, are probably not active at normal resting tidal volumes (500 ml per breath) but may contribute to respiratory control at high tidal volumes, as during exercise.

Hemoglobin and Oxygen Transport

Hemoglobin without oxygen, or deoxyhemoglobin, can bond with oxygen to form oxyhemoglobin. This "loading" reaction occurs in the capillaries of the lungs. The dissociation of oxyhemoglobin, or "unloading" reaction, occurs in the tissue capillaries. The bond strength between hemoglobin and oxygen, and thus the extent of the unloading reaction, is adjusted by various factors to ensure an adequate delivery of oxygen to the tissues.

If the lungs are functioning properly, blood leaving in the pulmonary veins and traveling in the systemic arteries has a P_{O_2} of about 100 mmHg, indicating a plasma oxygen concentration of about 0.3 ml O_2 per 100 ml blood. The total oxygen content of the blood, however, cannot be derived if only the P_{O_2} of plasma is known. The total oxygen content depends not only on the P_{O_2} but also on the hemoglobin concentration. If the P_{O_2} and hemoglobin concentration are normal, arterial blood contains about 20 ml of O_2 per 100 ml of blood (fig. 16.31).

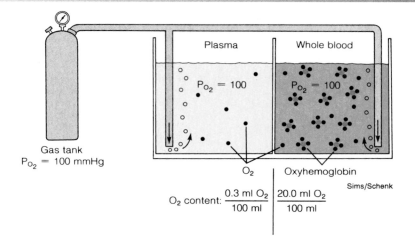

Figure 16.31

Plasma and whole blood that are brought into equilibrium with the same gas mixture have the same P_{O_2} and thus the same number of dissolved oxygen molecules (shown as black dots). The oxygen content of whole blood, however, is much higher than that of plasma because of the binding of oxygen to hemoglobin.

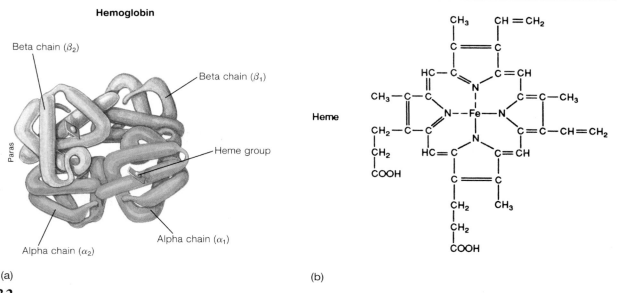

(a)   (b)

Figure 16.32

(a) An illustration of the three-dimensional structure of hemoglobin in which the two alpha and two beta polypeptide chains are shown. The four heme groups are represented as flat structures with iron (*spheres*) in the centers. (b) The chemical structure of heme.

Hemoglobin

Most of the oxygen in the blood is contained within the red blood cells, where it is chemically bonded to **hemoglobin.** Each hemoglobin molecule consists of a protein globin part, composed of four polypeptide chains, and four nitrogen-containing, disc-shaped organic pigment molecules called *hemes* (fig. 16.32).

The protein part of hemoglobin is composed of two identical *alpha chains*, each 141 amino acids long, and two identical *beta chains*, each 146 amino acids long. Each of the four polypeptide chains is combined with one heme group. In the center of each heme group is one atom of iron, which can combine with one molecule of oxygen (O_2). One hemoglobin molecule can thus combine with four molecules of oxygen; since there are about 280 million hemoglobin molecules per red blood cell, each red blood cell can carry over a billion molecules of oxygen.

Normal heme contains iron in the reduced form (Fe^{++}, or ferrous iron). In this form, the iron can share electrons and bond with oxygen to form **oxyhemoglobin.** When oxyhemoglobin dissociates to release oxygen to the tissues, the heme iron is still in the reduced (Fe^{++}) form and the hemoglobin is called **deoxyhemoglobin,** or **reduced hemoglobin.** The term *oxyhemoglobin* is thus not equivalent to *oxidized* hemoglobin; hemoglobin does not lose an electron (and become oxidized) when it combines with oxygen. Oxidized hemoglobin, or **methemoglobin,** has iron in the oxidized (Fe^{+++}, or ferric) state. Methemoglobin thus lacks the electron it needs to form a bond with oxygen and cannot participate in oxygen transport. Blood normally contains only a small amount of methemoglobin, but certain drugs can increase this amount.

In **carboxyhemoglobin,** another abnormal form of hemoglobin, the reduced heme is combined with *carbon monoxide* instead of oxygen. Since the bond with carbon monoxide is about 210 times stronger than the bond with oxygen, carbon monoxide tends to displace oxygen in hemoglobin and remains attached to hemoglobin as the blood passes through systemic capillaries. The transport of oxygen to the tissues is thus reduced in carbon monoxide poisoning.

Hemoglobin Concentration

The *oxygen-carrying capacity* of whole blood is determined by its concentration of normal hemoglobin. If the hemoglobin concentration is below normal—a condition called **anemia**—the oxygen concentration of the blood falls below normal. Conversely, when the hemoglobin concentration rises above the normal range—as occurs in **polycythemia** (high red blood cell count)—the oxygen-carrying capacity of blood is increased accordingly. This can occur as an adaptation to life at a high altitude.

The production of hemoglobin and red blood cells in bone marrow is controlled by a hormone called **erythropoietin,** produced by the kidneys. The production of erythropoietin—and thus the production of red blood cells—is stimulated when the amount of oxygen delivered to the kidneys and other organs is lower than normal. Red blood cell production is also promoted by androgens, which explains why the hemoglobin concentration in men is from 1 to 2 g per 100 ml higher than in women.

The Loading and Unloading Reactions

Deoxyhemoglobin and oxygen combine to form oxyhemoglobin; this is called the **loading reaction.** Oxyhemoglobin, in turn, dissociates to yield deoxyhemoglobin and free oxygen molecules; this is the **unloading reaction.** The loading reaction occurs in the lungs and the unloading reaction occurs in the systemic capillaries.

Loading and unloading can thus be shown as a reversible reaction:

$$\text{Deoxyhemoglobin} + O_2 \overset{\text{(lungs)}}{\underset{\text{(tissues)}}{\rightleftharpoons}} \text{Oxyhemoglobin}$$

The extent to which the reaction will go in each direction depends on two factors: (1) the P_{O_2} of the environment and (2) the *affinity,* or bond strength, between hemoglobin and oxygen. High P_{O_2} drives the equation to the right (favors the loading reaction); at the high P_{O_2} of the pulmonary capillaries,

almost all the deoxyhemoglobin molecules combine with oxygen. Low P_{O_2} in the systemic capillaries drives the reaction in the opposite direction to promote unloading. The extent of this unloading depends on how low the P_{O_2} values are.

The affinity (bond strength) between hemoglobin and oxygen also influences the loading and unloading reactions. A very strong bond would favor loading but inhibit unloading; a weak bond would hinder loading but improve unloading. The bond strength between hemoglobin and oxygen is normally strong enough so that 97% of the hemoglobin leaving the lungs is in the form of oxyhemoglobin, yet the bond is sufficiently weak so that adequate amounts of oxygen are unloaded to sustain aerobic respiration in the tissues.

The Oxyhemoglobin Dissociation Curve

Blood in the systemic arteries, at a P_{O_2} of 100 mmHg, has a *percent oxyhemoglobin saturation* of 97% (which means that 97% of the hemoglobin is in the form of oxyhemoglobin). This blood is delivered to the systemic capillaries, where oxygen diffuses into the tissue cells and is consumed in aerobic respiration. Blood leaving in the systemic veins is thus reduced in oxygen; it has a P_{O_2} of about 40 mmHg and a percent oxyhemoglobin saturation of about 75% when a person is at rest (table 16.10). Expressed another way, blood entering the tissues contains 20 ml O_2 per 100 ml blood, and blood leaving the tissues contains 15.5 ml O_2 per 100 ml blood (fig. 16.33). Thus, 22%, or 4.5 ml of O_2 out of 20 ml O_2 per 100 ml blood, is unloaded to the tissues.

A graphic illustration of the percent oxyhemoglobin saturation at different values of P_{O_2} is called an **oxyhemoglobin dissociation curve** (fig. 16.33). The values in this graph are obtained by subjecting samples of blood in vitro to different partial oxygen pressures. The percent oxyhemoglobin saturations obtained by this procedure, however, can be used to predict what the unloading percentages would be in vivo with a given difference in arterial and venous P_{O_2} values.

Figure 16.33 shows the difference between the arterial and venous P_{O_2} and the percent oxyhemoglobin saturation at rest. The relatively large amount of oxyhemoglobin remaining

Table 16.10	Relationship Between Percent Oxyhemoglobin Saturation and P_{O_2} (at pH = 7.40 and Temperature = 37° C)										
P_{O_2} (mmHg)	100	80	61	45	40	36	30	26	23	21	19
Percent Oxyhemoglobin	97	95	90	80	75	70	60	50	40	35	30
	Arterial Blood				Venous Blood						

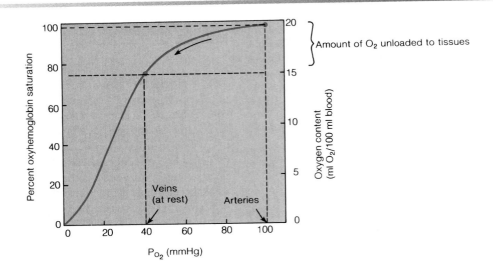

Figure 16.33

The percentage of oxyhemoglobin saturation and the blood oxygen content are shown at different values of P_{O_2}. Notice that the percent oxyhemoglobin decreases by about 25% as the blood passes through the tissue from arteries to veins, resulting in the unloading of approximately 5 ml O_2 per 100 ml to the tissues.

in the venous blood at rest functions as an oxygen reserve. If a person stops breathing, there will be a sufficient reserve of oxygen in the blood to keep the brain and heart alive for about 4.5 minutes without using cardiopulmonary resuscitation (CPR) techniques. This reserve supply of oxygen can also be tapped when the tissue's requirements for oxygen are raised.

The oxyhemoglobin dissociation curve is S-shaped, or *sigmoidal*. The fact that it is relatively flat at high P_{O_2} values indicates that changes in P_{O_2} within this range have little effect on the loading reaction. One would have to ascend as high as 10,000 feet, for example, before the oxyhemoglobin saturation of arterial blood would decrease from 97% to 93%. At more common elevations, the percent oxyhemoglobin saturation would not be significantly different from the 97% value at sea level.

At the steep part of the sigmoidal curve, however, small changes in P_{O_2} values produce large differences in percent saturation. A decrease in *venous* P_{O_2} from 40 mmHg to 30 mmHg, as might occur during mild exercise, corresponds to a change in percent saturation from 75% to 58%. Since the *arterial* percent saturation is usually still 97% during exercise, this change in venous percent saturation indicates that more oxygen has been unloaded to the tissues. The difference between the arterial and venous percent saturations indicates the percent unloading. In

the preceding example, 97% − 75% = 22% unloading at rest, and 97% − 58% = 39% unloading during mild exercise. During heavier exercise, the venous P_{O_2} can drop to 20 mmHg or less, indicating a percent unloading in excess of 70%.

Effect of pH and Temperature on Oxygen Transport

In addition to changes in P_{O_2}, the loading and unloading reactions are influenced by changes in the bond strength, or *affinity*, of hemoglobin for oxygen. Such changes ensure that active skeletal muscles, as a result of their higher metabolism, receive more oxygen from the blood than they do at rest.

The affinity is decreased when the pH is lowered and increased when the pH is raised; this is called the **Bohr effect.** When the affinity of hemoglobin for oxygen is reduced, there is slightly less loading of the blood with oxygen in the lungs but greater unloading of oxygen in the tissues. The net effect is that the tissues receive more oxygen when the blood pH is lowered (table 16.11). Since the pH can be decreased by carbon dioxide (through the formation of carbonic acid), the Bohr effect helps to provide more oxygen to the tissues when their carbon dioxide output (and metabolism) is increased.

Chapter Sixteen

Table 16.11	Effect of pH on Hemoglobin Affinity for Oxygen and Unloading of Oxygen to the Tissues			
pH	Affinity	Arterial O_2 Content per 100 ml	Venous O_2 Content per 100 ml	O_2 Unloaded to Tissues per 100 ml
7.40	Normal	19.8 ml O_2	14.8 ml O_2	5.0 ml O_2
7.60	Increased	20.0 ml O_2	17.0 ml O_2	3.0 ml O_2
7.20	Decreased	19.2 ml O_2	12.6 ml O_2	6.6 ml O_2

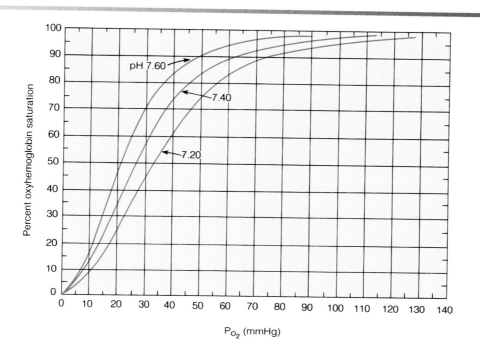

Figure 16.34

A decrease in blood pH (an increase in H^+ concentration) decreases the affinity of hemoglobin for oxygen at each P_{O_2} value, resulting in a "shift to the right" of the oxyhemoglobin dissociation curve. A curve that is shifted to the right has a lower percent oxyhemoglobin saturation at each P_{O_2}, but the effect is more marked at lower P_{O_2} values. This is called the *Bohr effect*.

When the percent oxyhemoglobin saturation at different pH values is graphed as a function of P_{O_2}, the dissociation curve is shown to be shifted to the right by a lowering of pH and shifted to the left by a rise in pH (fig. 16.34). If the percent unloading is calculated by subtracting the percent oxyhemoglobin saturation at given P_{O_2} values for arterial and venous blood, it will be clear that a *shift to the right* of the curve indicates a greater oxygen unloading, whereas a *shift to the left* indicates less unloading but slightly more oxygen loading in the lungs.

When oxyhemoglobin dissociation curves are constructed at constant pH values but at different temperatures, it can be seen that the affinity of hemoglobin for oxygen is decreased by a rise in temperature. An increase in temperature weakens the bond between hemoglobin and oxygen and thus has the same effect as a fall in pH—the oxyhemoglobin dissociation curve is shifted to the right. At higher temperatures, therefore, more oxygen is unloaded to the tissues than would be the case if the bond strength were constant. This effect can significantly increase the delivery of oxygen to muscles that are warmed during exercise.

Effect of 2,3-DPG on Oxygen Transport

Mature red blood cells lack both nuclei and mitochondria. Without mitochondria they cannot respire aerobically; the very cells that carry oxygen are the only cells in the body that cannot use it! Red blood cells, therefore, must obtain energy through the anaerobic respiration of glucose. At a certain point in the glycolytic pathway there is a "side reaction" in red blood cells that results in a unique product—**2,3-diphosphoglyceric acid (2,3-DPG)**.

The enzyme that produces 2,3-DPG is inhibited by oxyhemoglobin. When the oxyhemoglobin concentration is decreased, therefore, the production of 2,3-DPG is increased. This increase in 2,3-DPG production can occur when the total hemoglobin concentration is low (in anemia) or when the P_{O_2} is low (at a high altitude, for example). The bonding of 2,3-DPG with deoxyhemoglobin makes the deoxyhemoglobin more stable. At

Table 16.12 Factors That Affect the Affinity of Hemoglobin for Oxygen and the Position of the Oxyhemoglobin Dissociation Curve

Factor	Affinity	Position of Curve	Comments
↓pH	Decreased	Shift to the right	Called the Bohr effect; increases oxygen delivery during hypercapnia
↑Temperature	Decreased	Shift to the right	Increases oxygen unloading during exercise and fever
↑2,3-DPG	Decreased	Shift to the right	Increases oxygen unloading when there is a decrease in total hemoglobin or total oxygen content; an adaptation to anemia and high-altitude living

 The importance of 2,3-DPG within red blood cells is now recognized in blood banking. Old, stored red blood cells can lose their ability to produce 2,3-DPG as they lose their ability to metabolize glucose. Modern techniques for blood storage, therefore, include the addition of energy substrates for respiration and phosphate sources needed for the production of 2,3-DPG.

the P_{O_2} values in the tissue capillaries, therefore, a higher proportion of the oxyhemoglobin will be converted to deoxyhemoglobin by the unloading of its oxygen. An increased concentration of 2,3-DPG in red blood cells thus increases oxygen unloading (table 16.12) and shifts the oxyhemoglobin dissociation curve to the right.

Anemia

When the total blood hemoglobin concentration falls below normal in anemia, each red blood cell produces increased amounts of 2,3-DPG. A normal hemoglobin concentration of 15 g per 100 ml unloads about 4.5 ml O_2 per 100 ml at rest, as previously described. If the hemoglobin concentration were reduced by half, you might expect that the tissues would receive only half the normal amount of oxygen (2.25 ml O_2 per 100 ml). It has been shown, however, that an amount as great as 3.3 ml O_2 per 100 ml is unloaded to the tissues under these conditions. This occurs as a result of a rise in 2,3-DPG production that produces a lowering of hemoglobin affinity for oxygen.

Hemoglobin F

The effects of 2,3-DPG are also important in the transfer of oxygen from maternal to fetal blood. The mother has hemoglobin molecules composed of two alpha and two beta chains, as previously described, whereas the fetal hemoglobin contains two alpha and two *gamma* chains in place of beta chains (gamma chains differ from beta chains in thirty-seven of their amino acids). Normal adult hemoglobin in the mother (*hemoglobin A*) is able to bind to 2,3-DPG. Fetal hemoglobin (*hemoglobin F*), by contrast, cannot bind to 2,3-DPG, and thus has a higher affinity for oxygen at a given P_{O_2} than does hemoglobin A. Since hemoglobin F can have a higher percent oxyhemoglobin saturation than hemoglobin A at a given P_{O_2}, oxygen is transferred from the maternal to the fetal blood as these two come into close proximity in the placenta.

Inherited Defects in Hemoglobin Structure and Function

A number of hemoglobin diseases are produced by inherited (congenital) defects in the protein part of hemoglobin. **Sickle-cell anemia**—a disease occurring almost exclusively in blacks, and carried in a recessive state by 8% to 11% of the black population of the United States—for example, is caused by an abnormal form of hemoglobin called *hemoglobin S*. Hemoglobin S differs from normal hemoglobin A in only one amino acid: valine is substituted for glutamic acid in position 6 on the beta chains. This amino acid substitution is caused by a single base change in the region of DNA that codes for the beta chains.

Under conditions of low blood P_{O_2}, hemoglobin S comes out of solution and cross-links to form a "paracrystalline gel" within the red blood cells. This causes the characteristic sickle shape of red blood cells (fig. 16.35) and makes them less flexible. Since red blood cells must be able to bend in the middle to pass through many narrow capillaries, a decrease in their flexibility may cause them to block small blood channels and produce organ ischemia. The decreased solubility of hemoglobin S in solutions of low P_{O_2} is used in the diagnosis of sickle-cell anemia and sickle-cell trait (the carrier state, in which a person has the genes for both hemoglobin A and hemoglobin S).

Thalassemia is any of a family of hemoglobin diseases found predominantly among people of Mediterranean ancestry. In *alpha thalassemia,* there is decreased synthesis of the alpha chains of hemoglobin, whereas in *beta thalassemia* the synthesis of the beta chains is impaired. One of the compensations for thalassemia is increased synthesis of gamma chains, resulting in the retention of large amounts of hemoglobin F (fetal hemoglobin) into adulthood. Although this may partially compensate for the anemia, it has a major drawback; hemoglobin F has a higher affinity for oxygen than hemoglobin A, and thus cannot unload as much oxygen to the tissues.

Some types of abnormal hemoglobins have been shown to be advantageous in the environments in which they evolved. A person who is a carrier for sickle-cell anemia, for example (and who therefore has both hemoglobin A and hemoglobin S), has a high resistance to malaria. This is because the parasite that causes malaria cannot live in red blood cells that contain hemoglobin S.

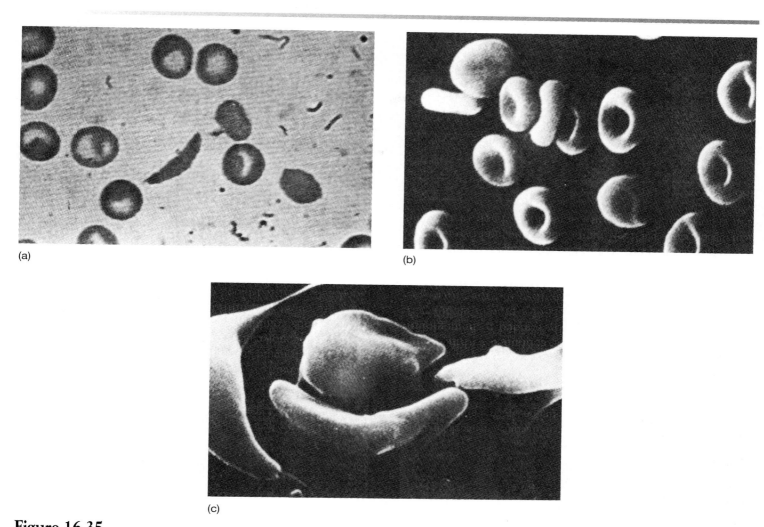

Figure 16.35

(a) A sickled red blood cell as seen in the light microscope. (b) Normal cells. (c) Sickled red blood cells as seen in the scanning electron microscope.

Muscle Myoglobin

As described in chapter 12, **myoglobin** is a red pigment found exclusively in striated muscle cells. In particular, slow-twitch, aerobically respiring skeletal fibers and cardiac muscle cells are rich in myoglobin. Myoglobin is similar to hemoglobin, but it has one rather than four hemes and, therefore, can combine with only one molecule of oxygen.

Myoglobin has a higher affinity for oxygen than does hemoglobin, and its dissociation curve is therefore to the left of the oxyhemoglobin dissociation curve (fig. 16.36). The shape of the myoglobin curve is also different from the oxyhemoglobin dissociation curve. The myoglobin curve is rectangular, indicating that oxygen will be released only when the P_{O_2} gets very low.

Since the P_{O_2} in mitochondria is very low (because oxygen is incorporated into water here), myoglobin may act as a "middleman" in the transfer of oxygen from blood to the mitochondria within muscle cells. Myoglobin may also have an oxygen-storage function, which is of particular importance in the heart. During diastole, when the coronary blood flow is greatest, myoglobin can load up with oxygen. This stored oxygen can then be released during systole, when the coronary arteries are squeezed closed by the contracting myocardium.

---

1. Use a graph to illustrate the effects of P_{O_2} on the loading and unloading reactions.

2. Draw an oxyhemoglobin dissociation curve and label the P_{O_2} values for arterial and venous blood under resting conditions. Use this graph to show the changes in unloading that occur during exercise.

3. Explain how changes in pH and temperature affect oxygen transport and state when such changes occur.

4. Explain how a person who is anemic or a person at high altitude could have an increase in the percent unloading of oxygen by hemoglobin.

Respiratory Physiology

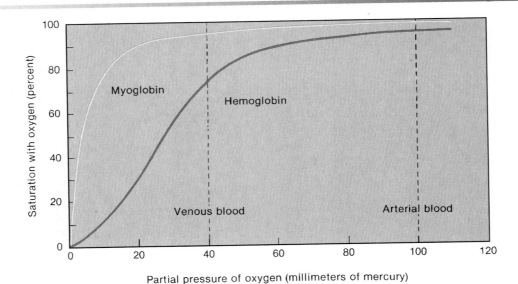

Figure 16.36

A comparison of the dissociation curves for hemoglobin and for myoglobin. At the P_{O_2} of venous blood, the myoglobin retains almost all of its oxygen, indicating a higher affinity than hemoglobin for oxygen. The myoglobin does, however, release its oxygen at the very low P_{O_2} values found inside the mitochondria.

Carbon Dioxide Transport and Acid-Base Balance

Carbon dioxide is transported in the blood primarily in the form of bicarbonate (HCO_3^-), which is released when carbonic acid dissociates. Bicarbonate can buffer H^+, and thus helps to maintain a normal arterial pH. Hypoventilation raises, and hyperventilation lowers, the carbonic acid concentration of the blood.

Carbon dioxide is carried by the blood in three forms: (1) as *dissolved* CO_2—carbon dioxide is about twenty-one times more soluble than oxygen in water, and about one-tenth of the total blood CO_2 is dissolved in plasma; (2) as *carbaminohemoglobin*—about one-fifth of the total blood CO_2 is carried attached to an amino acid in hemoglobin (carbaminohemoglobin should not be confused with carboxyhemoglobin, which is a combination of hemoglobin and carbon monoxide); and (3) as *bicarbonate*, which accounts for most of the CO_2 carried by the blood.

Carbon dioxide is able to combine with water to form carbonic acid. This reaction occurs spontaneously in the plasma at a slow rate but occurs much more rapidly within the red blood cells due to the catalytic action of the enzyme **carbonic anhydrase.** Since this enzyme is confined to the red blood cells, most of the carbonic acid is produced there rather than in the plasma. The formation of carbonic acid from CO_2 and water is favored by the high P_{CO_2} found in tissue capillaries (this is an example of the *law of mass action*, described in chapter 4).

$$CO_2 + H_2O \xrightarrow[\text{high } P_{CO_2}]{\text{carbonic anhydrase}} H_2CO_3$$

The Chloride Shift

As a result of catalysis by carbonic anhydrase within the red blood cells, large amounts of carbonic acid are produced as blood passes through the systemic capillaries. The buildup of carbonic acid concentrations within the red blood cells favors the dissociation of these molecules into *hydrogen ions* (protons, which contribute to the acidity of a solution) and HCO_3^- (bicarbonate). The equation describing this reaction has been previously introduced (see chapter 2, under the heading "Blood pH").

The hydrogen ions (H^+) released by the dissociation of carbonic acid are largely buffered by their combination with deoxyhemoglobin within the red blood cells. Although the unbuffered hydrogen ions are free to diffuse out of the red blood cells, more bicarbonate diffuses outward into the plasma than does H^+. As a result of the "trapping" of hydrogen ions within the red blood cells by their attachment to hemoglobin and the outward diffusion of bicarbonate, the inside of the red blood cell gains a net positive charge. This attracts chloride ions (Cl^-), which move into the red blood cells as HCO_3^- moves out. This exchange of anions as blood travels through the tissue capillaries is called the **chloride shift** (fig. 16.37).

The ability of the blood to transport carbon dioxide is affected by the transport of oxygen. The unloading of oxygen is increased by the bonding of H^+, released from carbonic acid, to oxyhemoglobin. This is the Bohr effect, and results in increased

Chapter Sixteen

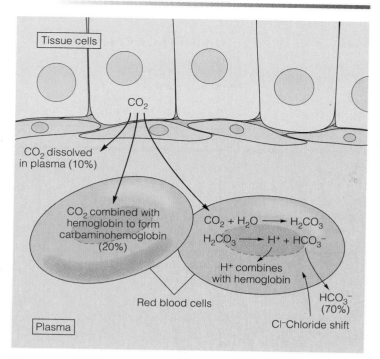

Figure 16.37

An illustration of carbon dioxide transport by the blood and the "chloride shift." Carbon dioxide is transported in three forms: as dissolved CO_2 gas, attached to hemoglobin as carbaminohemoglobin, and as carbonic acid and bicarbonate. Percentages indicate the proportion of CO_2 in each of the forms.

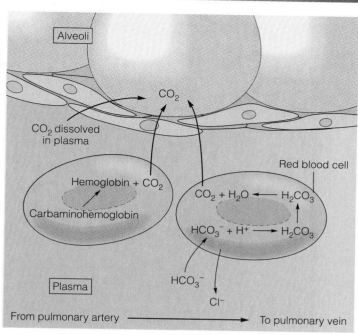

Figure 16.38

Carbon dioxide is released from the blood as it travels through the pulmonary capillaries. During this time a "reverse chloride shift" occurs, and carbonic acid is transformed into CO_2 and H_2O.

conversion of oxyhemoglobin to deoxyhemoglobin. Now, since deoxyhemoglobin bonds H^+ more strongly than does oxyhemoglobin, the act of unloading its oxygen improves the ability of hemoglobin to buffer the H^+ released by carbonic acid. Removal of H^+ from solution by combining with hemoglobin (through the law of mass action), in turn, favors the continued production of carbonic acid and thus improves the ability of the blood to transport carbon dioxide.

When blood reaches the pulmonary capillaries, deoxyhemoglobin is converted to oxyhemoglobin. Since oxyhemoglobin has a lower affinity for H^+ than does deoxyhemoglobin, hydrogen ions are released within the red blood cells. This attracts HCO_3^- from the plasma, which combines with H^+ to form carbonic acid:

$$H^+ + HCO_3^- \rightarrow H_2CO_3$$

Under conditions of lower P_{CO_2}, as occurs in the pulmonary capillaries, carbonic anhydrase catalyzes the conversion of carbonic acid to carbon dioxide and water:

$$H_2CO_3 \xrightarrow[\text{low } P_{CO_2}]{\text{carbonic anhydrase}} CO_2 + H_2O$$

In summary, the carbon dioxide produced by the tissue cells is converted within the systemic capillaries, mostly through the action of carbonic anhydrase in the red blood cells, to carbonic acid. With the buildup of carbonic acid concentra-

tions in the red blood cells, the carbonic acid dissociates into bicarbonate and H^+, which results in the chloride shift. A *reverse chloride shift* operates in the pulmonary capillaries to convert carbonic acid to H_2O and CO_2 gas, which is eliminated in the expired breath (fig. 16.38). The P_{CO_2}, carbonic acid, H^+, and bicarbonate concentrations in the systemic arteries are thus maintained relatively constant by normal ventilation. This is required to maintain the acid-base balance of the blood (fig. 16.39), as discussed in chapter 13 and in the next section.

Ventilation and Acid-Base Balance

The basic concepts and terminology relating to the acid-base balance of the blood were introduced in chapter 13. In brief review, acidosis refers to an arterial pH below 7.35, and alkalosis refers to an arterial pH above 7.45. There are two components of each: respiratory and metabolic. The respiratory component refers to the carbon dioxide concentration of the blood, as measured by the P_{CO_2}. As implied by its name, the respiratory component is regulated by the respiratory system. The metabolic component is controlled by the kidneys, and is discussed in chapter 17.

Ventilation is normally adjusted to keep pace with the metabolic rate, so that the arterial P_{CO_2} remains in the normal range. In **hypoventilation,** the ventilation is insufficient to "blow off" carbon dioxide and maintain a normal P_{CO_2}. Indeed, hypoventilation can be operationally defined as an abnormally high arterial P_{CO_2}. Under these conditions, carbonic acid production is excessively high and **respiratory acidosis** occurs.

Table 16.13 Effect of Lung Function on Blood Acid-Base Balance

Condition	pH	P_{CO_2}	Ventilation	Cause or Compensation
Normal	7.35–7.45	39–41 mmHg	Normal	Not applicable
Respiratory acidosis	Low	High	Hypoventilation	Cause of the acidosis
Respiratory alkalosis	High	Low	Hyperventilation	Cause of the alkalosis
Metabolic acidosis	Low	Low	Hyperventilation	Compensation for acidosis
Metabolic alkalosis	High	High	Hypoventilation	Compensation for alkalosis

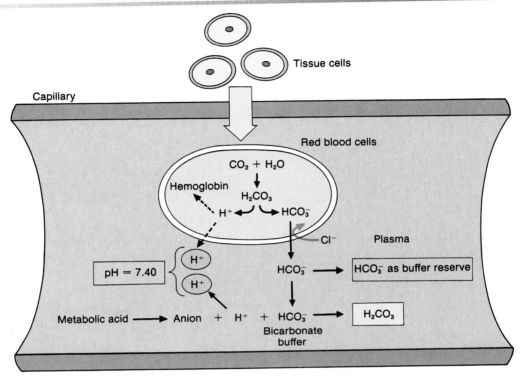

Figure 16.39

Bicarbonate released into the plasma from red blood cells functions to buffer H^+ produced by the ionization of metabolic acids (lactic acid, fatty acids, ketone bodies, and others). Binding of H^+ to hemoglobin also promotes the unloading of O_2.

In **hyperventilation,** conversely, the rate of ventilation is greater than the rate of CO_2 production. Arterial P_{CO_2} therefore decreases, so that less carbonic acid is formed than under normal conditions. The depletion of carbonic acid raises the pH, and **respiratory alkalosis** occurs.

A change in blood pH, produced by alterations in either the respiratory or metabolic component of acid-base balance, can be partially compensated for by a change in the other component. For example, a person with metabolic acidosis will hyperventilate. This is because the aortic and carotid bodies are stimulated by an increased blood H^+ concentration (fall in pH). As a result of the hyperventilation, a secondary respiratory alkalosis is produced. The person is still acidotic, but not as much so as would be the case without the compensation. People with partially compensated metabolic acidosis would thus have a low pH, which would be accompanied by a low blood P_{CO_2} as a result of the hyperventilation. Metabolic alkalosis, similarly, is partially compensated for by the retention of carbonic acid due to hypoventilation (table 16.13).

1. List the ways in which carbon dioxide is carried by the blood. Using equations, show how carbonic acid and bicarbonate are formed.

2. Describe the events that occur in the chloride shift in the systemic capillaries; also describe the reverse chloride shift that occurs in the pulmonary capillaries.

3. Describe the functions of bicarbonate and carbonic acid in blood.

4. Describe the effects of hyperventilation and hypoventilation on the blood pH and explain the mechanisms involved.

5. Explain why a person with ketoacidosis hyperventilates. What are the potential benefits of hyperventilation under these conditions?

Chapter Sixteen

Effect of Exercise and High Altitude on Respiratory Function

The arterial blood gases and pH do not significantly change during moderate exercise because ventilation increases to keep pace with increased metabolism. This increased ventilation requires neural feedback from the exercising muscles and chemoreceptor stimulation. Adjustments are made at high altitude in both the control of ventilation and the oxygen transport ability of the blood to permit adequate delivery of oxygen to the tissues.

Changes in ventilation and oxygen delivery occur during exercise and during acclimatization to a high altitude. These changes help to compensate for the increased metabolic rate during exercise and for the decreased arterial P_{O_2} at a high altitude.

Ventilation during Exercise

As soon as a person begins to exercise, the rate and depth of breathing increase to produce a total minute volume that is many times the resting value. This increased ventilation, particularly in well-trained athletes, is exquisitely matched to the simultaneous increase in oxygen consumption and carbon dioxide production by the exercising muscles. The arterial blood P_{O_2}, P_{CO_2}, and pH thus remain surprisingly constant during exercise (fig. 16.40).

It is tempting to suppose that ventilation increases during exercise as a result of the increased CO_2 production by the exercising muscles. Ventilation increases together with increased CO_2 production, however, so that blood measurements of P_{CO_2} during exercise are not significantly higher than at rest. The mechanisms responsible for the increased ventilation during exercise must therefore be more complex.

Two kinds of mechanisms—*neurogenic* and *humoral*—have been proposed to explain the increased ventilation that occurs during exercise. Possible neurogenic mechanisms include the following: (1) sensory nerve activity from the exercising limbs may stimulate the respiratory muscles, either through spinal reflexes or via the brain stem respiratory centers, and/or (2) input from the cerebral cortex may stimulate the brain stem centers to modify ventilation. These neurogenic theories help to explain the immediate increase in ventilation that occurs as exercise begins.

Rapid and deep ventilation continues after exercise has stopped, suggesting that humoral (chemical) factors in the blood may also stimulate ventilation during exercise. Since the P_{O_2}, P_{CO_2}, and pH of the blood samples from exercising subjects

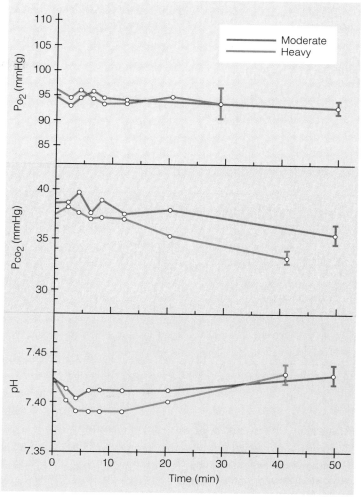

Figure 16.40

The effect of moderate and heavy exercise on arterial blood gases and pH. Notice that there are no consistent and significant changes in these measurements during the first several minutes of moderate and heavy exercise and that only the P_{CO_2} changes (actually decreases) during more prolonged exercise.

are within the resting range, these humoral theories propose that (1) the P_{CO_2} and pH in the region of the chemoreceptors may be different from these values "downstream," where blood samples are taken, and/or (2) cyclic variations in these values that cannot be detected by blood samples may stimulate the chemoreceptors. The evidence suggests that both neurogenic and humoral mechanisms are involved in the **hyperpnea,** or increased ventilation, of exercise. (Note that hyperpnea differs from hyperventilation in that the blood P_{CO_2} remains within the normal range during hyperpnea but is decreased in hyperventilation.)

Table 16.14 Changes in Respiratory Function During Exercise

Variable	Change	Comments
Ventilation	Increased	In moderate exercise, ventilation is matched to increased metabolic rate. Mechanisms responsible for increased ventilation are not well understood.
Blood gases	No change	Blood gas measurements during light and moderate exercise show little change because ventilation is increased to match increased muscle oxygen consumption and carbon dioxide production.
Oxygen delivery to muscles	Increased	Although the total oxygen content and P_{O_2} do not increase during exercise, there is an increased rate of blood flow to the exercising muscles.
Oxygen extraction by muscles	Increased	Increased oxygen consumption lowers the tissue P_{O_2} and lowers the affinity of hemoglobin for oxygen (due to the effect of increased temperature). More oxygen, as a result, is unloaded so that venous blood contains a lower oxyhemoglobin saturation than at rest. This effect is enhanced by endurance training.

Table 16.15 Blood Gas Measurements at Different Altitudes

Altitude	Arterial P_{O_2} (mmHg)	Percent Oxyhemoglobin Saturation	Arterial P_{CO_2} (mmHg)
Sea level	90–95	96%	40
1,524 m (5,000 ft)	75–81	95%	32–33
2,286 m (7,500 ft)	69–74	92%–93%	31–33
4,572 m (15,000 ft)	48–53	86%	25
6,096 m (20,000 ft)	37–45	76%	20
7,620 m (25,000 ft)	32–39	68%	13
8,848 m (29,029 ft)	26–33	58%	9.5–13.8

Source: From P. H. Hackett, et al., "High Altitude Medicine" in *Management of Wilderness and Environmental Emergencies*, 2d ed., edited by Paul S. Auerbach and Edward C. Geehr. Copyright © 1989 Mosby/Yearbook. Reprinted by permission.

Anaerobic Threshold and Endurance Training

The ability of the cardiopulmonary system to deliver adequate amounts of oxygen to the exercising muscles at the beginning of exercise may be insufficient because of the time lag required to make proper cardiovascular adjustments. During this time, therefore, the muscles respire anaerobically and a "stitch in the side"—possibly due to hypoxia of the diaphragm—may develop. After numerous cardiovascular and pulmonary adjustments have been made, a person may experience a "second wind" when the muscles are receiving sufficient oxygen for their needs.

Continued heavy exercise can cause a person to reach the **anaerobic threshold,** which is the maximum rate of oxygen consumption that can be attained before blood lactic acid levels rise as a result of anaerobic respiration. This occurs when 50% to 60% of the person's maximal oxygen uptake has been reached. The rise in lactic acid levels is due to the aerobic limitations of the muscles; it is not due to a malfunction of the cardiopulmonary system. Indeed, the arterial oxygen hemoglobin saturation remains at 97% and venous blood draining the muscles contains unused oxygen.

The rise in blood lactic acid that occurs when the anaerobic threshold is exceeded is due to the inability of the exercising muscles to increase their oxygen consumption rate sufficiently to prevent anaerobic respiration. The anaerobic threshold, however, is higher in endurance-trained athletes than it is in other people. Endurance training increases the skeletal muscle content of mitochondria and Krebs cycle enzymes. These muscles, therefore, are able to utilize more of the oxygen delivered to them by the arterial blood. At a given level of exercise, consequently, the venous blood that drains from muscles in endurance-trained people contains a lower percentage of oxyhemoglobin than in other people. The effects of exercise and endurance training on respiratory function are summarized in table 16.14.

Acclimatization to High Altitude

When a person from a region near sea level moves to a significantly higher elevation, several adjustments in respiratory function must be made to compensate for the decreased P_{O_2} at the higher altitude. These adjustments include changes in ventilation, in the hemoglobin affinity for oxygen, and in the total hemoglobin concentration.

Reference to table 16.15 indicates that at an altitude of 7,500 feet, for example, the P_{O_2} of arterial blood is in the range of 69 to 74 mmHg (compared to 100 mmHg at sea level). This table also indicates that the percent oxyhemoglobin saturation at this altitude is from 92% to 93%, compared to about 96% at sea level. The amount of oxygen attached to hemoglobin, and thus the total oxygen content of blood, is therefore decreased. In addition, the rate at which oxygen can be delivered to the tissue cells (by the plasma-derived tissue fluid) after it dissociates from oxyhemoglobin is reduced at the higher altitude. This is because the maximum concentration of oxygen that can be dissolved in the plasma decreases in a linear fashion with the fall in P_{O_2}. People may thus experience rapid fatigue even at more

Table 16.16	Changes in Respiratory Function During Acclimatization to High Altitude	
Variable	**Change**	**Comments**
Partial pressure of oxygen	Decreased	Due to decreased total atmospheric pressure
Partial pressure of carbon dioxide	Decreased	Due to hyperventilation in response to low arterial P_{O_2}
Percent oxyhemoglobin saturation	Decreased	Due to lower P_{O_2} in pulmonary capillaries
Ventilation	Increased	Due to lower P_{O_2}
Total hemoglobin	Increased	Due to stimulation by erythropoietin; raises oxygen capacity of blood to partially or completely compensate for the reduced partial pressure
Oxyhemoglobin affinity	Decreased	Due to increased DPG within the red blood cells; results in a higher percent unloading of oxygen to the tissues

moderate elevations (for example, 5,000 to 6,000 feet), at which the oxyhemoglobin saturation is only slightly decreased. Compensations made by the respiratory system gradually reduce the amount of fatigue caused by a given amount of exertion at high altitudes.

Changes in Ventilation

Starting at altitudes as low as 1,500 m (5,000 ft), the decreased arterial P_{O_2} stimulates an increase in ventilation. This *hypoxic ventilatory response* produces hyperventilation, which lowers the arterial P_{CO_2} (see table 16.5) and thus produces a respiratory alkalosis. The rise in arterial pH helps to blunt the hyperventilation, which after a few days becomes stabilized at about 2.5 L more per minute than the total minute volume at sea level.

Hyperventilation at high altitude increases tidal volume, thus reducing the proportionate contribution of air from the anatomical dead space and increasing the proportion of fresh air brought to the alveoli. This improves the oxygenation of the blood over what it would be in the absence of the hyperventilation. Hyperventilation, however, cannot increase blood P_{O_2} above that of the inspired air. The P_{O_2} of arterial blood decreases with increasing altitude (table 16.15), regardless of the ventilation. In the Peruvian Andes, for example, the normal arterial P_{O_2} is reduced from 100 mmHg (at sea level) to 45 mmHg. The loading of hemoglobin with oxygen is therefore incomplete, producing an oxyhemoglobin saturation that is decreased from 97% (at sea level) to 81%.

Hemoglobin Affinity for Oxygen

Normal arterial blood at sea level unloads only about 22% of its oxygen to the tissues at rest; the percent saturation is reduced from 97% in arterial blood to 75% in venous blood. As a partial compensation for the decrease in oxygen content at high altitude, the affinity of hemoglobin for oxygen is reduced, so that a higher proportion of oxygen is unloaded. This occurs because the low oxyhemoglobin content of red blood cells stimulates the production of 2,3-DPG, which in turn decreases the hemoglobin affinity for oxygen.

The ability of 2,3-DPG to decrease the affinity of hemoglobin for oxygen thus predominates over the ability of respiratory alkalosis (caused by the hyperventilation) to increase the affinity. At very high altitudes, however, the story becomes more complex. In a 1984 study by J. B. West, the very low arterial P_{O_2} (28 mmHg) of subjects at the summit of Mount Everest stimulated intense hyperventilation, so that the arterial P_{CO_2} was decreased to 7.5 mmHg. The resultant respiratory al-

kalosis (arterial pH greater than 7.7) caused a shift to the left of the oxyhemoglobin dissociation curve (indicating greater affinity of hemoglobin for oxygen) despite the antagonistic effects of increased 2,3-DPG concentrations. It was suggested that the increased affinity of hemoglobin for oxygen caused by the respiratory alkalosis may have been beneficial at such a high altitude, since it increased the loading of hemoglobin with oxygen in the lungs.

Increased Hemoglobin and Red Blood Cell Production

In response to tissue hypoxia, the kidneys secrete the hormone erythropoietin (chapter 13). Erythropoietin stimulates the bone

Acute Mountain Sickness (AMS) is common in people who arrive at altitudes in excess of 5,000 feet. Cardinal symptoms of AMS are headache, malaise, anorexia, nausea and fragmented sleep. Headache is the most common symptom, and may result from changes in blood flow to the brain. Low arterial P_{O_2} stimulates vasodilation of vessels in the pia mater, increasing blood flow and pressure within the skull. The hypocapnia produced by hyperventilation, however, causes cerebral vasoconstriction. Whether there is a net cerebral vasoconstriction or vasodilation depends on the balance between these two antagonistic effects. Pulmonary edema is common at altitudes above 9,000 feet, and can produce shortness of breath, coughing, and a mild fever. Cerebral edema, which generally occurs above an altitude of 10,000 feet, can produce mental confusion and even hallucinations. Pulmonary and cerebral edema are potentially dangerous and should be alleviated by descending to a lower altitude.

marrow to increase its production of hemoglobin and red blood cells. In the Peruvian Andes, for example, people have a total hemoglobin concentration that is increased from 15 g per 100 ml (at sea level) to 19.8 g per 100 ml. Although the percent oxyhemoglobin saturation is still lower than at sea level, the total oxygen content of the blood is actually greater—22.4 ml O_2 per 100 ml compared to a sea-level value of about 20 ml O_2 per 100 ml. These adjustments of the respiratory system to high altitude are summarized in table 16.16.

It should be noted that these changes are not unalloyed benefits. Polycythemia (high red blood cell count) increases the viscosity of blood; hematocrits of 55% to 60% have been measured in people who live in the Himalayas, and higher values are reached if dehydration accompanies the polycythemia. The increased blood viscosity contributes to pulmonary hypertension, which can cause accompanying edema and ventricular hypertrophy that can lead to heart failure.

1. Describe the effect of exercise on the P_{O_2}, P_{CO_2}, and pH blood values, and explain how ventilation might be increased during exercise.

2. Explain why endurance-trained athletes have a higher anaerobic threshold than other people.

3. Describe the changes that occur in the respiratory system during acclimatization to life at a high altitude.

Summary

The Respiratory System p. 460

I. Alveoli are small, numerous, thin-walled air sacs that provide an enormous surface area for gas diffusion.
   A. The region of the lungs where gas exchange with the blood occurs is known as the respiratory zone.
   B. The trachea, bronchi, and bronchioles that deliver air to the respiratory zone constitute the conducting zone.

II. The thoracic cavity is delimited by the chest wall and diaphragm.
   A. The structures of the thoracic cavity are covered by thin, wet pleural membranes.
   B. The lungs are covered by a visceral pleura that is normally flush against the parietal pleura that lines the chest wall.
   C. The potential space between the two pleural membranes is called the intrapleural space.

Physical Aspects of Ventilation p. 464

I. The intrapleural and intrapulmonary pressures vary during ventilation.
   A. The intrapleural pressure is always less than the intrapulmonary pressure.
   B. The intrapulmonary pressure is subatmospheric during inspiration and greater than the atmospheric pressure during expiration.
   C. Pressure changes in the lungs are produced by variations in

lung volume in accordance with the inverse relationship between the volume and pressure of a gas described by Boyle's law.

II. The mechanics of ventilation are influenced by the physical properties of the lungs.
   A. The compliance of the lungs is the change in lung volume per change in transpulmonary pressure (the difference between intrapulmonary and intrapleural pressure).
   B. The elasticity of the lungs refers to their tendency to recoil after distension.
   C. The surface tension of the fluid in the alveoli exerts a force directed inward, which acts to resist distension.

III. On first consideration, it would seem that the surface tension in the alveoli would create a pressure that would cause smaller alveoli to collapse and empty their air into large alveoli.
   A. This would occur because the pressure caused by a given amount of surface tension would be greater in smaller alveoli than in larger alveoli, as described by the law of LaPlace.
   B. Collapse of alveoli due to surface tension does not normally occur, however, because pulmonary surfactant (a combination of phospholipid and protein) functions to lower surface tension.

   C. In hyaline membrane disease, the lungs of premature infants collapse due to the absence of surfactant.

Mechanics of Breathing p. 469

I. Inspiration and expiration are accomplished by contraction and relaxation of striated muscles.
   A. During quiet inspiration, the diaphragm and external intercostal muscles contract, and thus increase the volume of the thorax.
   B. During quiet expiration, these muscles relax, and the elastic recoil of the lungs and thorax causes a decrease in thoracic volume.
   C. Forced inspiration and expiration are aided by contraction of the accessory respiratory muscles.

II. Spirometry aids the diagnosis of a number of pulmonary disorders.
   A. In restrictive disease, such as pulmonary fibrosis, the vital capacity measurement is decreased below normal.
   B. In obstructive disease, such as asthma and bronchitis, the forced expiratory volume is reduced below normal because of increased airway resistance to air flow.

III. Asthma results from bronchoconstriction; emphysema and chronic bronchitis are frequently referred to as chronic obstructive pulmonary disease.

Gas Exchange in the Lungs *p. 475*

I. According to Dalton's law, the total pressure of a gas mixture is equal to the sum of the pressures that each gas in the mixture would exert independently.
  A. The partial pressure of a gas in a dry gas mixture is thus equal to the total pressure times the percent composition of that gas in the mixture.
  B. Since the total pressure of a gas mixture decreases with altitude above sea level, the partial pressures of the constituent gases likewise decrease with altitude.
  C. When the partial pressure of a gas in a wet gas mixture is calculated, the water vapor pressure must be taken into account.

II. According to Henry's law, the amount of gas that can be dissolved in a fluid is directly proportional to the partial pressure of that gas in contact with the fluid.
  A. The concentrations of oxygen and carbon dioxide that are dissolved in plasma are proportional to an electric current generated by special electrodes that react with these gases.
  B. Normal arterial blood has a P_{O_2} of 100 mmHg, indicating a concentration of dissolved oxygen of 0.3 ml per 100 ml of blood; the oxygen contained in red blood cells (about 19.7 ml per 100 ml of blood) does not affect the P_{O_2} measurement.

III. The P_{O_2} and P_{CO_2} measurements of arterial blood provide information about lung function.

IV. In addition to proper ventilation of the lungs, blood flow (perfusion) in the lungs must be adequate and matched to air flow (ventilation) in order for adequate gas exchange to occur.

V. Abnormally high partial pressures of gases in blood can cause a variety of disorders, including oxygen toxicity, nitrogen narcosis, and decompression sickness.

Regulation of Breathing *p. 481*

I. The rhythmicity center in the medulla oblongata directly controls the muscles of respiration.
  A. Activity of the inspiratory and expiratory neurons varies in a reciprocal way to produce an automatic breathing cycle.
  B. Activity in the medulla is influenced by the apneustic and pneumotaxic centers in the pons, as well as by sensory feedback information.
  C. Conscious breathing involves direct control by the cerebral cortex via corticospinal tracts.

II. Breathing is affected by chemoreceptors sensitive to the P_{CO_2}, pH, and P_{O_2} of the blood.
  A. The P_{CO_2} of the blood and consequent changes in pH are usually of greater importance than the blood P_{O_2} in the regulation of breathing.
  B. Central chemoreceptors in the medulla oblongata are sensitive to changes in blood P_{CO_2} because of the resultant changes in the pH of cerebrospinal fluid.
  C. The peripheral chemoreceptors in the aortic and carotid bodies are sensitive to changes in blood P_{CO_2} indirectly, because of consequent changes in blood pH.

III. Decreases in blood P_{O_2} directly stimulate breathing only when the blood P_{O_2} is less than 50 mmHg. A drop in P_{O_2} also stimulates breathing indirectly, by making the chemoreceptors more sensitive to changes in P_{CO_2} and pH.

IV. At tidal volumes of 1 L or more, inspiration is inhibited by stretch receptors in the lungs (the Hering–Breuer reflex); there is a similar deflation reflex.

Hemoglobin and Oxygen Transport *p. 485*

I. Hemoglobin is composed of two alpha and two beta polypeptide chains and four heme groups that contain a central atom of iron.
  A. When the iron is in the reduced form and not attached to oxygen, the hemoglobin is called deoxyhemoglobin, or reduced hemoglobin; when it is attached to oxygen, it is called oxyhemoglobin.
  B. If the iron is attached to carbon monoxide, the hemoglobin is called carboxyhemoglobin; when the iron is in an oxidized state and unable to transport any gas, the hemoglobin is called methemoglobin.
  C. Deoxyhemoglobin combines with oxygen in the lungs (the loading reaction) and breaks its bonds with oxygen in the tissue capillaries (the unloading reaction). The extent of each reaction is determined by the P_{O_2} and the affinity of hemoglobin for oxygen.

II. A graph of percent oxyhemoglobin saturation at different values of P_{O_2} is called an oxyhemoglobin dissociation curve.
  A. At rest, the difference between arterial and venous oxyhemoglobin saturations indicates that about 22% of the oxyhemoglobin unloads its oxygen to the tissues.
  B. During exercise, the venous P_{O_2} and percent oxyhemoglobin saturation are decreased, indicating that a higher percentage of the oxyhemoglobin has unloaded its oxygen to the tissues.

III. The pH and temperature of the blood influence the affinity of hemoglobin for oxygen, and thus the extent of loading and unloading.
  A. A fall in pH decreases the affinity, and a rise in pH increases the affinity of hemoglobin for oxygen. This is called the Bohr effect.
  B. A rise in temperature decreases the affinity of hemoglobin for oxygen.
  C. When the affinity is decreased, the oxyhemoglobin dissociation curve is shifted to the right. This indicates a greater unloading percentage of oxygen to the tissues.

IV. The affinity of hemoglobin for oxygen is also decreased by an organic molecule in the red blood cells called 2,3-diphosphoglyceric acid (2,3-DPG).

    A. Since oxyhemoglobin inhibits 2,3-DPG production, there will be higher 2,3-DPG concentrations when the oxyhemoglobin is decreased due to anemia or low P_{O_2} (as in high altitude).

    B. If a person is anemic, the lowered hemoglobin concentration is partially compensated for by the fact that a higher percentage of the oxyhemoglobin will unload its oxygen due to the effect of 2,3-DPG.

    C. Fetal hemoglobin cannot bind to 2,3-DPG, and thus, it has a higher affinity for oxygen than the mother's hemoglobin. This facilitates the transfer of oxygen to the fetus.

V. Inherited defects in the amino acid composition of hemoglobin are responsible for such diseases as sickle-cell anemia and thalassemia.

VI. Striated muscles contain myoglobin, a pigment related to hemoglobin that can combine with oxygen and deliver it to the muscle cell mitochondria at low P_{O_2} values.

Carbon Dioxide Transport and Acid-Base Balance   p. 492

I. Red blood cells contain an enzyme called carbonic anhydrase that catalyzes the reversible reaction whereby carbon dioxide and water are used to form carbonic acid.

    A. This reaction is favored by the high P_{CO_2} in the tissue capillaries, and as a result, carbon dioxide produced by the tissues is converted into carbonic acid in the red blood cells.

    B. Carbonic acid then ionizes to form H^+ and HCO_3^- (bicarbonate).

    C. Since much of the H^+ is buffered by hemoglobin, but more bicarbonate is free to diffuse outward, an electrical gradient is established that draws Cl^- into the red blood cells. This is called the chloride shift.

    D. A reverse chloride shift occurs in the lungs. In this process, the low P_{CO_2} favors the conversion of carbonic acid to carbon dioxide, which can be exhaled.

II. By adjusting the blood concentration of carbon dioxide, and thus of carbonic acid, the process of ventilation helps to maintain proper acid-base balance of the blood.

    A. Normal arterial blood pH is 7.40. A pH below 7.35 is termed acidosis; a pH above 7.45 is termed alkalosis.

    B. Hyperventilation causes respiratory alkalosis, and hypoventilation causes respiratory acidosis.

    C. Metabolic acidosis stimulates hyperventilation, which can cause a respiratory alkalosis as a partial compensation.

Effect of Exercise and High Altitude on Respiratory Function   p. 495

I. During exercise there is increased ventilation, or hyperpnea, which can be matched to the increased metabolic rate so that the arterial blood P_{CO_2} remains normal.

    A. This hyperpnea may be caused by proprioceptor information, cerebral input, and/or changes in arterial P_{CO_2} and pH.

    B. During heavy exercise, the anaerobic threshold may be reached at about 55% of the maximal oxygen uptake; at this point, lactic acid is released into the blood by the muscles.

    C. Endurance training enables the muscles to utilize oxygen more effectively, so that greater levels of exercise can be performed before the anaerobic threshold is reached.

II. Acclimatization to a high altitude involves changes that help to deliver oxygen more effectively to the tissues despite reduced arterial P_{O_2}.

    A. Hyperventilation occurs in response to the low P_{O_2}.

    B. The red blood cells produce more 2,3-DPG, which lowers the affinity of hemoglobin for oxygen and improves the unloading reaction.

    C. The kidneys produce the hormone erythropoietin, which stimulates the bone marrow to increase its production of red blood cells, so that more oxygen can be carried by the blood at given values of P_{O_2}.

Clinical Investigation

A taxi driver in New York City is found parked at the curb of a street, complaining of great pain in his right thoracic region where he had been stabbed with the tip of an umbrella by an irate passenger. The wound has punctured the thorax, and radiographs reveal that his right lung is collapsed, though his left lung is still functional. His arterial blood has a high P_{CO_2} and a pH of 7.15. He is treated surgically, and upon recovery his blood gases are again analyzed. Although the arterial P_{CO_2} and pH have been restored to normal, he has a carboxyhemoglobin concentration of 20%. Pulmonary function tests reveal that his vital capacity is slightly low, and that his $FEV_{1.0}$ is significantly decreased below normal. What do the radiographs and laboratory tests disclose about the health of this man?

Clues

Examine the sections "Pulmonary Function Tests," "Partial Pressures of Gases in Blood," and "Ventilation and Acid-Base Balance." Read the boxed information pertaining to a pneumothorax and to carboxyhemoglobin in the major sections "Physical Aspects of Ventilation" and "Hemoglobin and Oxygen Transports," respectively.

Review Activities

Objective Questions

1. Which of the following statements about intrapulmonary and intrapleural pressure is *true*?
   a. The intrapulmonary pressure is always subatmospheric.
   b. The intrapleural pressure is always greater than the intrapulmonary pressure.
   c. The intrapulmonary pressure is greater than the intrapleural pressure.
   d. The intrapleural pressure equals the atmospheric pressure.

2. If the transpulmonary pressure equals zero,
   a. a pneumothorax has probably occurred.
   b. the lungs cannot inflate.
   c. elastic recoil causes the lungs to collapse.
   d. all of the above apply.

3. The maximum amount of air that can be expired after a maximum inspiration is the
   a. tidal volume.
   b. forced expiratory volume.
   c. vital capacity.
   d. maximum expiratory flow rate.

4. If the blood lacked red blood cells but the lungs were functioning normally,
   a. the arterial P_{O_2} would be normal.
   b. the oxygen content of arterial blood would be normal.
   c. both *a* and *b* would apply.
   d. neither *a* nor *b* would apply.

5. If a person were to dive with scuba equipment to a depth of 66 feet, which of the following statements would be *false*?
   a. The arterial P_{O_2} would be three times normal.
   b. The oxygen content of plasma would be three times normal.
   c. The oxygen content of whole blood would be three times normal.

6. Which of the following would be most affected by a decrease in the affinity of hemoglobin for oxygen?
   a. arterial P_{O_2}
   b. arterial percent oxyhemoglobin saturation
   c. venous oxyhemoglobin saturation
   d. arterial P_{CO_2}

7. If a person with normal lung function were to hyperventilate for several seconds, there would be a significant
   a. increase in the arterial P_{O_2}.
   b. decrease in the arterial P_{CO_2}.
   c. increase in the arterial percent oxyhemoglobin saturation.
   d. decrease in the arterial pH.

8. Erythropoietin is produced by
   a. the kidneys.
   b. the liver.
   c. the lungs.
   d. the bone marrow.

9. The affinity of hemoglobin for oxygen is decreased under conditions of
   a. acidosis.
   b. fever.
   c. anemia.
   d. acclimatization to a high altitude.
   e. all of the above.

10. Most of the carbon dioxide in the blood is carried in the form of
    a. dissolved CO_2.
    b. carbaminohemoglobin.
    c. bicarbonate.
    d. carboxyhemoglobin.

11. The bicarbonate concentration of the blood would be decreased during
    a. metabolic acidosis.
    b. respiratory acidosis.
    c. metabolic alkalosis.
    d. respiratory alkalosis.

12. The chemoreceptors in the medulla are directly stimulated by
    a. CO_2 from the blood.
    b. H^+ from the blood.

c. H^+ in cerebrospinal fluid that is derived from blood CO_2.
   d. decreased arterial P_{O_2}.

13. The rhythmic control of breathing is produced by the activity of inspiratory and expiratory neurons in
    a. the medulla oblongata.
    b. the apneustic center of the pons.
    c. the pneumotaxic center of the pons.
    d. the cerebral cortex.

14. Which of the following occur(s) during hypoxemia?
    a. increased ventilation
    b. increased production of 2,3-DPG
    c. increased production of erythropoietin
    d. all of the above

15. During exercise, which of the following statements is *true*?
    a. The arterial percent oxyhemoglobin saturation is decreased.
    b. The venous percent oxyhemoglobin saturation is decreased.
    c. The arterial P_{CO_2} is measurably increased.
    d. The arterial pH is measurably decreased.

16. All of the following can bond with hemoglobin *except*
    a. HCO_3^-.
    b. O_2.
    c. H^+.
    d. CO_2.

17. Which of the following statements about the partial pressure of carbon dioxide is true?
    a. It is higher in the alveoli than in the pulmonary arteries.
    b. It is higher in the systemic arteries than in the tissues.
    c. It is higher in the systemic veins than in the systemic arteries.
    d. It is higher in the pulmonary veins than in the pulmonary arteries.

Essay Questions

1. Using a flow diagram to show cause and effect, explain how contraction of the diaphragm produces inspiration.[1]

2. Radiographic (X-ray) pictures show that the rib cage of a person with a pneumothorax is expanded and the ribs are farther apart. Explain why this should be so.

3. Explain, using a flowchart, how a rise in blood P_{CO_2} stimulates breathing. Include both the central and peripheral chemoreceptors in your answer.

4. Explain why a person with ketoacidosis may hyperventilate. What benefit might it provide? Also explain why this hyperventilation can be stopped by an intravenous fluid containing bicarbonate.

5. What blood measurements can be performed to detect (a) anemia, (b) carbon monoxide poisoning, and (c) poor lung function?

6. Explain how measurements of blood P_{CO_2}, bicarbonate, and pH are affected by hypoventilation and hyperventilation.

7. Describe the changes in ventilation that occur during exercise. Explain how these changes are produced and how they affect arterial blood gases and pH.

8. How would an increase in the red blood cell content of 2,3-DPG affect the P_{O_2} of venous blood? Explain your answer.

9. What are the mechanisms that affect ventilation at a high altitude? Explain why the change in ventilation is beneficial and how it can also be detrimental. What other factors operate at a high altitude to improve the oxygen delivery to the tissues?

10. Compare asthma and emphysema in terms of their characteristics and the effects they have on pulmonary function tests.

11. Explain the mechanisms involved in quiet inspiration and in forced inspiration, and in quiet expiration and forced expiration. What muscles are involved in each case?

12. Describe the formation, composition, and function of pulmonary surfactant. What happens when surfactant is absent? How is this condition treated?

Selected Readings

Avery, M. E., N. S. Wang, and H. W. Taeusch Jr. March 1975. The lung of the newborn infant. *Scientific American*.

Berger, A. J., R. A. Mitchel, and J. W. Severinghaus. 1977. Regulation of respiration. *New England Journal of Medicine* 297: first part, p. 92; second part, p. 138; third part, p. 194.

Brample, D. M., and D. R. Carrier. 1983. Running and breathing in mammals. *Science* 219:251.

Browning, R. J. 1982. Pulmonary disease: Back to basics (part 1); Putting blood gases to work (part 2). *Diagnostic Medicine*: first part, Jan/Feb, p. 39; second part, March/April, p. 59.

Cherniak, N. S. 1986. Breathing disorders during sleep. *Hospital Practice* 21:81.

Cockcroft, D. W. 1990. Airway hyperresponsiveness in asthma. *Hospital Practice* 25:111.

Coleridge, H. M., and J. C. G. Coleridge. 1994. Pulmonary reflexes: Neural mechanisms of pulmonary defense. *Annual Review of Physiology* 56:69.

Crapo, R. O. 1994. Pulmonary function testing. *New England Journal of Medicine* 331:25.

Dantzker, D. R. 1986. Physiology and pathophysiology of pulmonary gas exchange. *Hospital Practice* 21:135.

Davis, J. M. et al. 1988. Changes in pulmonary mechanics after the administration of surfactant to infants with respiratory distress syndrome. *New England Journal of Medicine* 319:476.

Decramer, M. 1993. Respiratory muscle interaction. *News in Physiological Sciences* 8:121.

Drazen, J. M., Gaston, B., and S. A. Shore. 1995. Chemical regulation of pulmonary airway tone. *Annual Review of Physiology* 57:151.

Finch, C. A., and C. Lenfant. 1972. Oxygen transport in man. *New England Journal of Medicine* 286:407.

Flenley, D. C., and P. M. Warren. 1983. Ventilatory response to O_2 and CO_2 during exercise. *Annual Review of Physiology* 45:415.

Fraser, R. G., and J. A. P. Pare. 1977. *Structure and Function of the Lung*. 2d ed. Philadelphia: W. B. Saunders Co.

Guz, A. 1975. Regulation of respiration in man. *Annual Review of Physiology* 37:303.

Haagsman, H. P., and L. M. G. van Golde. 1991. Synthesis and assembly of lung surfactant. *Annual Review of Physiology* 53:441.

Haddad, G. G., and R. B. Mellins. 1984. Hypoxia and respiratory control in early life. *Annual Review of Physiology* 46:629.

Houston, C. S. October 1992. Mountain sickness. *Scientific American*.

Irsigler, G. B., and J. W. Severinghaus. 1980. Clinical problems of ventilatory control. *Annual Review of Medicine* 31:109.

Kassirer, J. P., and N. E. Madias. 1980. Respiratory acid-base disorders. *Hospital Practice* 15:57.

Knowles, M. R. 1994. Modulation of the ionic milieu of the airways in health and disease. *Annual Review of Medicine* 45:421.

Kumar, A., and W. Busse. 1995. Airway inflammation in asthma. *Science and Medicine* 2:38.

Macklem, P. T. 1986. Respiratory muscle dysfunction. *Hospital Practice* 21:83.

Massaro, D. 1986. Oxygen: Toxicity and tolerance. *Hospital Practice* 21:95.

Massaro, D. 1990. Regulation of alveolar formation. *Hospital Practice* 25:81.

---

[1]*Note:* This question is answered on page 202 of the Student Study Guide.

Mendelson. C. R., and V. Boggaram. 1991. Hormonal control of the surfactant system in fetal lung. *Annual Review of Physiology* 53:415.

Murray, J. F. 1985. The lungs and heart failure. *Hospital Practice* 20:55.

Nadel, E. R. 1985. Physiological adaptations to aerobic training. *American Scientist* 73:334.

Naeye, R. L. April 1980. Sudden infant death. *Scientific American*.

Naquet, R., and J. C. Rostain. 1988. Deep-diving humans. *News in Physiological Sciences* 3:72.

Perutz, M. F. December 1978. Hemoglobin structure and respiratory transport. *Scientific American*.

Rigatto, H. 1984. Control of ventilation in the newborn. *Annual Review of Physiology* 46:661.

Roussos, C., and P. T. Macklem. 1982. The respiratory muscles. *New England Journal of Medicine* 307:786.

Shannon, D. C., and D. H. Kelly. 1982. SIDS and near-SIDS. *New England Journal of Medicine* 306: first part, p. 959; second part, p. 1022.

Snapper, J. R., and K. L. Bingham. 1986. Pulmonary edema. *Hospital Practice* 21:87.

Tobin, M. J. 1986. Update on strategies in mechanical ventilation. *Hospital Practice* 21:69.

Undem, B. J. 1994. Neural-immunologic interactions in asthma. *Hospital Practice* 29:59.

Walker, D. W. 1984. Peripheral and central chemoreceptors in the fetus and newborn. *Annual Review of Physiology* 46:687.

West, J. B. 1984. Human physiology at extreme altitudes on Mount Everest. *Science* 223:784.

Whipp, B. J. 1983. Ventilatory control during exercise in humans. *Annual Review of Physiology* 45:393.

Widdicombe, J. G., and S. E. Webber. 1990. Airway mucus secretion. *News in Physiological Sciences* 5:2.

Wright, J. R., and L. G. Dobbs. 1991. Regulation of pulmonary surfactant secretion and clearance. *Annual Review of Physiology* 53:395.

Physiology of the Kidneys

17

OBJECTIVES

After studying this chapter, you should be able to . . .

1. describe the different regions of the nephron tubules and explain the anatomic relationship between the tubules and the gross structure of the kidney.

2. describe the structural and functional relationships between the nephron tubules and their associated blood vessels.

3. describe the composition of glomerular ultrafiltrate and explain how it is produced.

4. explain how the proximal convoluted tubule reabsorbs salt and water.

5. describe active transport and osmosis in the loop of Henle and explain how these processes produce a countercurrent multiplier system.

6. explain how the vasa recta function in countercurrent exchange.

7. explain how antidiuretic hormone (ADH) functions to regulate the final urine volume.

8. describe the mechanisms of glucose reabsorption and define the terms *transport maximum* and *renal plasma threshold*.

9. define the term *renal plasma clearance* and explain why the clearance of inulin is equal to the glomerular filtration rate.

10. explain how the clearance of different molecules is determined and how the processes of reabsorption and secretion affect the clearance measurement.

11. describe the mechanism of Na^+ reabsorption in the distal tubule and explain why this reabsorption occurs together with the secretion of K^+.

12. describe the effects of aldosterone on the distal convoluted tubule and explain how aldosterone secretion is regulated.

13. explain how activation of the renin-angiotensin system results in the stimulation of aldosterone secretion.

14. explain how the interaction between plasma K^+ and H^+ concentrations affects the tubular secretion of these ions.

15. describe the role of the kidneys in the regulation of acid-base balance.

16. describe the different mechanisms by which substances can act as diuretics and explain why some diuretics cause excessive loss of K^+.

OUTLINE

Structure and Function of the Kidneys

Each kidney contains many tiny tubules that empty into a cavity drained by the ureter. Each of the tubules receives a blood filtrate from a capillary bed called the glomerulus. The filtrate is similar to tissue fluid, but is modified as it passes through different regions of the tubule and is thereby changed into urine. The tubules and associated blood vessels thus form the functioning units of the kidneys, which are known as nephrons.

The primary function of the kidneys is regulation of the extracellular fluid (plasma and tissue fluid) environment in the body. This function is accomplished through the formation of urine, which is a modified filtrate of plasma. In the process of urine formation, the kidneys regulate (1) the volume of blood plasma (and thus contribute significantly to the regulation of blood pressure); (2) the concentration of waste products in the blood; (3) the concentration of electrolytes (Na^+, K^+, HCO_3^-, and other ions) in the plasma; and (4) the pH of plasma. In order to understand how these functions are performed by the kidneys, a knowledge of kidney structure is required.

Gross Structure of the Urinary System

The paired kidneys lie on either side of the vertebral column, below the diaphragm and liver, just behind the peritoneum of the abdominal cavity (they are retroperitoneal). Each adult kidney weighs about 160 g and is about 12 cm long and 6 cm wide—about the size of a fist. Urine produced in the kidneys is drained into a cavity known as the *renal pelvis* (= basin), and from there it is channeled via two long ducts—the *ureters*—to the single *urinary bladder* (fig. 17.1).

A coronal section of the kidney shows two distinct regions (fig. 17.2). The outer **cortex,** in contact with the capsule, is reddish brown and granular in appearance because of its many capillaries. The deeper region, or **medulla,** is lighter in color and appears to be striped because of the presence of microscopic tubules and blood vessels. The medulla is composed of eight to fifteen conical *renal pyramids,* separated by *renal columns.*

The cavity of the kidney collects and transports urine from the kidney to the ureter. It is divided into several portions. Each pyramid projects into a small depression called a *minor calyx* (the plural form is *calyces*). Several minor calyces unite to form a *major calyx.* In turn, the major calyces join to form the funnel-shaped *renal pelvis.* The renal pelvis is actually an expanded portion of the ureter in the kidney and serves to collect urine from the calyces and transport it to the ureter and urinary bladder (fig. 17.3).

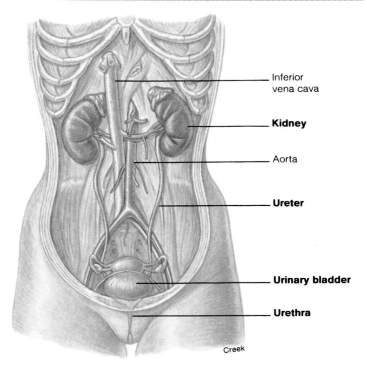

Inferior vena cava

Kidney

Aorta

Ureter

Urinary bladder

Urethra

Creek

Figure 17.1

Organs of the urinary system (labeled in boldface).

Kidney stones are composed of crystals and proteins that grow in the renal medulla until they break loose and pass into the urine collection system. Small stones that are anchored in place are not usually noticed, but large stones in the calyces or pelvis may obstruct the flow of urine. When a stone breaks loose and passes into a ureter it produces a steadily increasing sensation of pain, which often becomes so intense that the patient requires narcotic drugs. Most kidney stones contain crystals of calcium oxalate, but stones may also be composed of crystals of calcium phosphate, uric acid, or cystine. These substances are normally present in urine in a supersaturated state, from which they can crystalize for a variety of reasons.

The **urinary bladder** is a storage sac for urine, and its shape is determined by the amount of urine it contains. As the urinary bladder fills, it changes from a triangular to an ovoid shape as it bulges upward into the abdominal cavity. The urinary bladder is drained inferiorly by the tubular **urethra.** In females, the urethra is 4 cm (1.5 in.) long and opens into the space between the labia minora (chapter 20). In males, the urethra is about 20 cm (8 in.) long and opens at the tip of the penis, from which it can discharge either urine or semen.

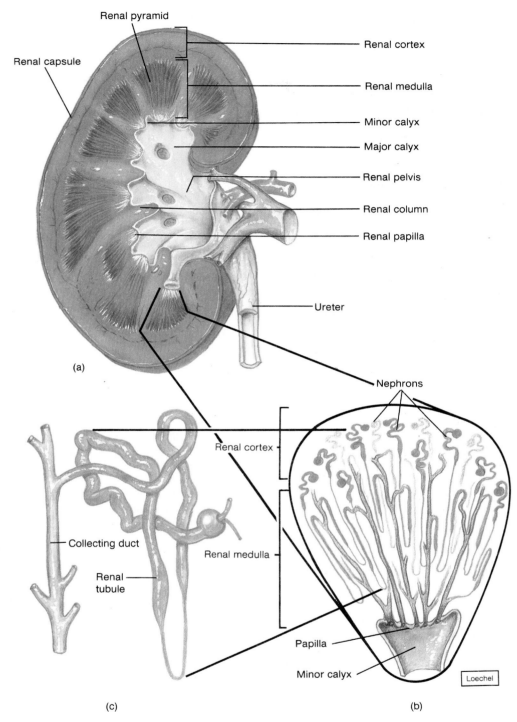

Figure 17.2

(*a*) A coronal section of the kidney and (*b*) a magnified view of the contents of a renal pyramid. (*c*) A single nephron tubule.

Micturition Reflex

Two muscular sphincters surround the urethra. The upper sphincter is composed of smooth muscle and is called the *internal urethral sphincter;* the lower sphincter is composed of voluntary skeletal muscle and is called the *external urethral sphincter.* The actions of these sphincters are regulated in the process of urination, which is also known as **micturition.**

Micturition is controlled by a reflex center located in the second, third, and fourth sacral levels of the spinal cord. Filling of the urinary bladder activates stretch receptors that send impulses to this micturition center. As a result, parasympathetic neurons are activated, causing rhythmic contractions of the detrusor muscle of the urinary bladder and relaxation of the internal urethral sphincter. At this point, a sense of urgency is perceived by the

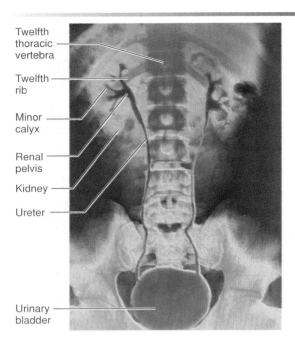

Twelfth thoracic vertebra

Twelfth rib

Minor calyx

Renal pelvis

Kidney

Ureter

Urinary bladder

Figure 17.3

A pseudocolor radiograph in which shades of gray are assigned colors. The calyces of the kidneys, the renal pelvises, the ureters, and the urinary bladder are shown.

brain, but there is still voluntary control over the external urethral sphincter. When urination is consciously allowed to occur, descending motor tracts to the micturition center inhibit somatic motor fibers to the external urethral sphincter. This muscle then relaxes, and urine is expelled. The ability to voluntarily inhibit micturition generally develops between the ages of 2 and 3.

Microscopic Structure of the Kidney

The **nephron** is the functional unit of the kidney that is responsible for the formation of urine. Each kidney contains more than a million nephrons. A nephron consists of **tubules** and associated small blood vessels. Fluid formed by capillary filtration enters the tubules and is subsequently modified by transport processes; the resulting fluid that leaves the tubules is urine.

Renal Blood Vessels

Arterial blood enters the kidney through the *renal artery,* which divides into *interlobar arteries* (fig. 17.4) that pass between the pyramids through the renal columns. *Arcuate arteries* branch from the interlobar arteries at the boundary of the cortex and medulla. A number of *interlobular arteries* radiate from the arcuate arteries into the cortex and subdivide into numerous **afferent arterioles** (fig. 17.5), which are microscopic in size. The afferent arterioles deliver blood into capillary networks, called **glomeruli,** which produce a blood filtrate that enters the urinary tubules.

The blood remaining in the glomerulus leaves through an **efferent arteriole,** which delivers the blood into another capillary network, the **peritubular capillaries,** that surrounds the tubules.

This arrangement of blood vessels is unique. It is the only one in the body in which a capillary bed (the glomerulus) is drained by an arteriole rather than by a venule and delivered to a second capillary bed located downstream (the peritubular capillaries). Blood from the peritubular capillaries is drained into veins that parallel the course of the arteries in the kidney. These veins are called the *interlobular veins, arcuate veins,* and *interlobar veins.* The interlobar veins descend between the pyramids, converge, and leave the kidney as a single *renal vein* that empties into the inferior vena cava.

Nephron Tubules

The tubular portion of a nephron consists of a glomerular capsule, a proximal convoluted tubule, a descending limb of the loop of Henle, an ascending limb of the loop of Henle, and a distal convoluted tubule (fig. 17.5).

The **glomerular (Bowman's) capsule** surrounds the glomerulus. The glomerular capsule and its associated glomerulus are located in the cortex of the kidney and together constitute the *renal corpuscle.* The glomerular capsule contains an inner visceral layer of epithelium around the glomerular capillaries and an outer parietal layer. The space between these two layers is continuous with the lumen of the tubule and receives the glomerular filtrate, as will be described in the next section.

Filtrate in the glomerular capsule passes into the lumen of the **proximal convoluted tubule.** The wall of the proximal convoluted tubule consists of a single layer of cuboidal cells, containing millions of microvilli; these microvilli serve to increase the surface area for reabsorption. In the process of reabsorption, salt, water, and other molecules needed by the body are transported from the lumen, through the tubular cells, and into the surrounding peritubular capillaries.

The glomerulus, glomerular capsule, and proximal convoluted tubule are located in the renal cortex. Fluid passes from the proximal convoluted tubule to the **loop of Henle.** This fluid is carried into the medulla in the **descending limb** of the loop and returns to the cortex in the **ascending limb** of the loop. Back in the cortex, the tubule becomes coiled again and is called the **distal convoluted tubule.** The distal convoluted tubule is shorter than the proximal tubule and has relatively few microvilli. The distal convoluted tubule is the last segment of the nephron and terminates as it empties into a collecting duct.

The two principal types of nephrons are classified according to their position in the kidney and the lengths of their loops of Henle. Nephrons that originate in the inner one-third of the cortex—called *juxtamedullary nephrons*—have longer loops of Henle than the more numerous *cortical nephrons,* which originate in the outer two-thirds of the cortex (fig. 17.6). The juxtamedullary nephrons are involved in the process of concentrating the urine by means of a mechanism described in a later section.

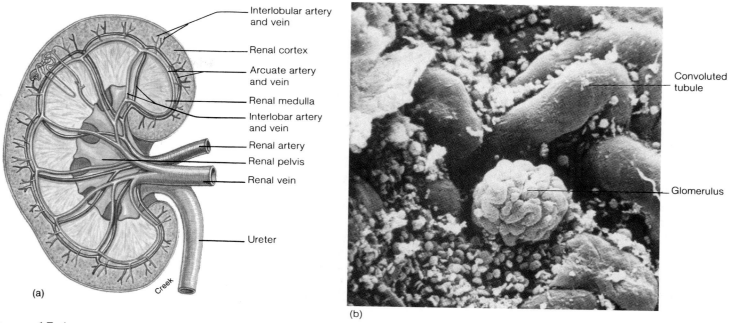

Figure 17.4

The vascular structure of the kidneys. (*a*) An illustration of the major arterial supply and (*b*) a scanning electron micrograph of the glomeruli (300 ×).

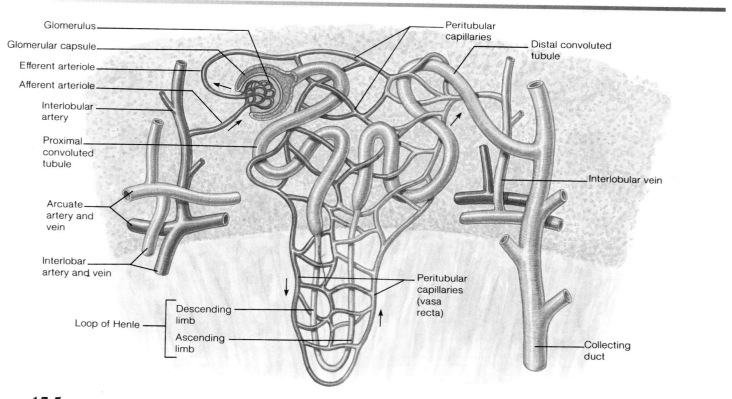

Figure 17.5

A simplified illustration of blood flow from a glomerulus to an efferent arteriole, to the peritubular capillaries, and to the venous drainage of the kidneys.

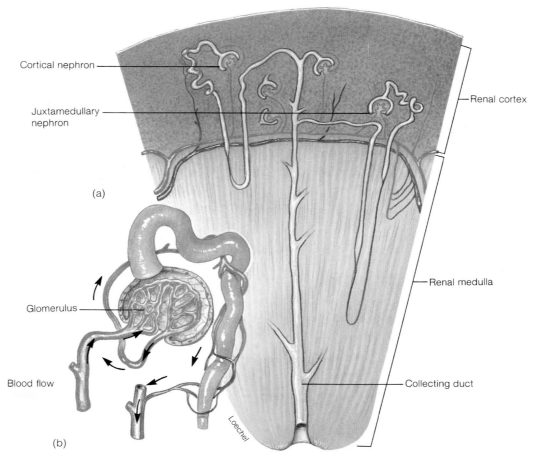

Figure 17.6

(*a*) The position of cortical and juxtamedullary nephrons within the kidney. (*b*) The direction of blood flow in the vessels of the nephron.

Polycystic kidney disease is a condition inherited as an autosomal dominant trait (chapter 20) that affects one in every 600 to 1,000 people. This disease is thus more common than sickle-cell anemia, cystic fibrosis, or muscular dystrophy, which are also genetic diseases. In 50% of the people who inherit the defective gene (located on the short arm of chromosome 16), progressive renal failure develops during middle age to the point that dialysis or kidney transplants are required. The cysts that develop are expanded portions of the renal tubule. Cysts that originate in the proximal tubule contain fluid that resembles glomerular filtrate and plasma. Cysts that originate in the distal tubule contain a lower NaCl concentration and a higher potassium and urea concentration than plasma as a result of the transport processes that occur during the passage of fluid through the tubules.

A **collecting duct** receives fluid from the distal convoluted tubules of several nephrons. Fluid is then drained by the collecting duct from the cortex to the medulla as the collecting duct passes through a renal pyramid. This fluid, now called urine, passes into a minor calyx. Urine is then funneled through the renal pelvis and out of the kidney in the ureter.

1. Describe the "theme" of kidney function in a single sentence. List the components of this functional theme.
2. Trace the course of blood flow through the kidney from the renal artery to the renal vein.
3. Trace the course of tubular fluid from the glomerular capsules to the ureter.
4. Draw a diagram of the tubular component of a nephron. Label the segments, and indicate which parts are in the cortex and which are in the medulla.

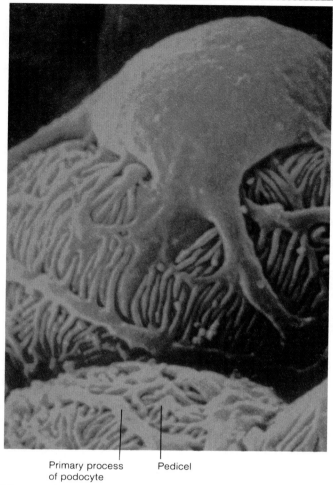

Figure 17.7

The inner (visceral) layer of Bowman's capsule is composed of podocytes, as shown in this scanning electron micrograph. Very fine extensions of these podocytes form foot processes, or pedicels, that interdigitate around the glomerular capillaries. Spaces between adjacent pedicels form the "filtration slits."

Glomerular Filtration

The glomerular capillaries have large pores, and the layer of Bowman's capsule in contact with the glomerulus has filtration slits. Water, together with dissolved solutes (but not proteins), can thus pass from the blood plasma to the inside of the capsule and the lumina of the nephron tubules. The volume of this filtrate produced per minute by both kidneys is called the glomerular filtration rate (GFR).

The glomerular capillaries have extremely large pores (200–500 Å in diameter) called fenestrae, and are thus said to be *fenestrated*. As a result of these large pores, glomerular

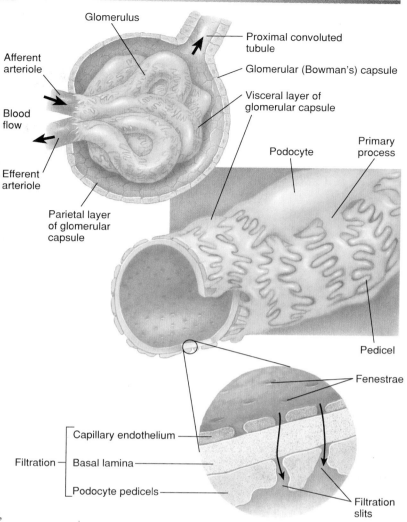

Figure 17.8

An illustration of the relationship between glomerular capillaries and the inner layer of Bowman's capsule.

capillaries are one hundred to four hundred times more permeable to plasma water and dissolved solutes than are the capillaries of skeletal muscles. Although the pores of glomerular capillaries are large, they are still small enough to prevent the passage of red blood cells, white blood cells, and platelets into the filtrate.

Before the filtrate can enter the interior of the glomerular capsule, it must pass through the capillary pores, the basement membrane (a thin layer of glycoproteins immediately outside the endothelial cells), and the inner, visceral layer of the glomerular capsule. The inner layer of the glomerular capsule is composed of unique cells, called *podocytes*, with numerous cytoplasmic extensions known as *pedicels*, or "foot processes" (fig. 17.7). These pedicels interdigitate, like fingers of clasped hands, as they wrap around the glomerular capillaries. The narrow slits between adjacent pedicels provide the passageways through which filtered molecules must pass to enter the interior of the glomerular capsule (fig. 17.8).

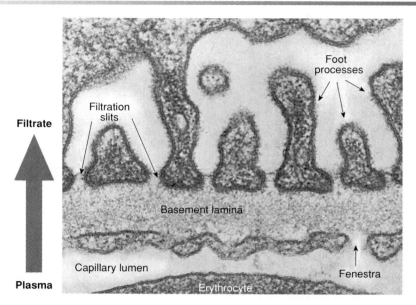

Figure 17.9

An electron micrograph of the "filtration barrier" between the capillary lumen and the cavity of the glomerular (Bowman's) capsule.

Although the glomerular capillary pores are apparently large enough to permit the passage of proteins, the fluid that enters the capsular space contains only a small amount of plasma proteins. This relative exclusion of plasma proteins from the filtrate is partially a result of their negative charges, which hinder their passage through the negatively charged glycoproteins in the basement membrane of the capillaries (fig. 17.9). The large size and negative charges of plasma proteins may also restrict their movement through the filtration slits between pedicels.

Glomerular Ultrafiltrate

The fluid that enters the glomerular capsule is called *ultrafiltrate* (fig. 17.10) because it is formed under pressure (the hydrostatic pressure of the blood). This process is similar to the formation of tissue fluid by other capillary beds in the body in response to Starling forces (chapter 14). The force favoring filtration is opposed by a counterforce developed by the hydrostatic pressure of fluid in the glomerular capsule. Also, since the protein concentration of the tubular fluid is low (less than 2 to 5 mg per 100 ml) compared to that of plasma (6 to 8 g per 100 ml), the greater colloid osmotic pressure of plasma promotes the osmotic return of filtered water. When these opposing forces are subtracted from the hydrostatic pressure of the glomerular capillaries, a *net filtration pressure* of approximately 10 mmHg is obtained.

Because glomerular capillaries are extremely permeable and have an extensive surface area, this modest net filtration pressure produces an extraordinarily large volume of filtrate. The **glomerular filtration rate (GFR)** is the volume of filtrate produced per minute by both kidneys. The GFR averages 115 ml per minute in women and 125 ml per minute in men. This is

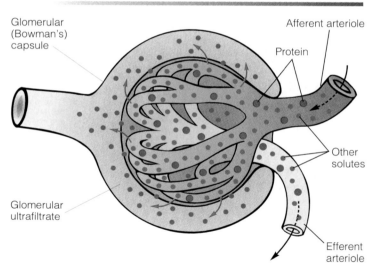

Figure 17.10

The formation of glomerular ultrafiltrate. Only a very small proportion of plasma proteins (*green circles*) are filtered, but smaller plasma solutes (*purple dots*) easily enter the glomerular ultrafiltrate. Arrows indicate the direction of filtration.

equivalent to 7.5 L per hour or 180 L per day (about 45 gallons)! Since the total blood volume averages about 5.5 L, this means that the total blood volume is filtered into the urinary tubules every 40 minutes. Most of the filtered water must obviously be returned immediately to the vascular system, or a person would literally urinate to death within several minutes.

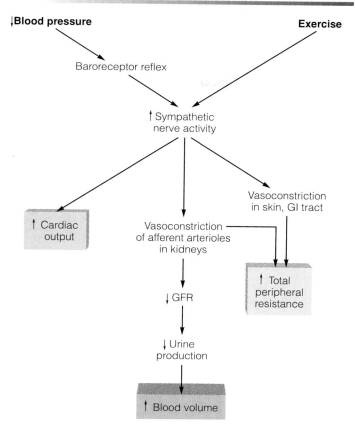

↓Blood pressure   **Exercise**

Baroreceptor reflex

↑ Sympathetic nerve activity

↑ Cardiac output

Vasoconstriction of afferent arterioles in kidneys

Vasoconstriction in skin, GI tract

↑ Total peripheral resistance

↓ GFR

↓ Urine production

↑ Blood volume

Figure 17.11

The effect of increased sympathetic nerve activity on kidney function and other physiological processes.

Regulation of Glomerular Filtration Rate

Vasoconstriction or dilation of afferent arterioles affects the rate of blood flow to the glomerulus and thus affects the glomerular filtration rate. Changes in the diameter of the afferent arterioles result from both extrinsic (sympathetic innervation) and intrinsic regulatory mechanisms. These mechanisms are needed to ensure that the GFR is sufficiently high so that the kidneys can eliminate wastes and regulate blood pressure, but not so high as to cause excessive water loss and accompanying problems.

Sympathetic Nerve Effects

An increase in sympathetic nerve activity, as occurs during the fight-or-flight reaction and exercise, stimulates constriction of afferent arterioles. This is an alpha-adrenergic effect, which helps to preserve blood volume and to divert blood to the muscles and heart. A similar effect occurs during cardiovascular shock, in which sympathetic nerve activity stimulates vasoconstriction. The decreased GFR and the resulting decreased rate of urine formation help to compensate for the rapid drop of blood pressure under these circumstances (fig. 17.11).

Table 17.1	Regulation of the Glomerular Filtration Rate (GFR)		
Regulation	**Stimulus**	**Afferent Arteriole**	**GFR**
Sympathetic nerves	Activation by baroreceptor reflex or by higher brain centers	Constricts	Decreases
Autoregulation	Decreased blood pressure	Dilates	No change
Autoregulation	Increased blood pressure	Constricts	No change

Renal Autoregulation

When the direct effect of sympathetic stimulation is experimentally removed, the effect of systemic blood pressure on GFR can be observed. Under these conditions, surprisingly, the GFR remains relatively constant despite changes in mean arterial pressure within a range of 70 to 80 mmHg (normal mean arterial pressure is 100 mmHg). The ability of the kidneys to maintain a relatively constant GFR in the face of fluctuating blood pressures is called **renal autoregulation.**

Renal autoregulation is achieved through the effects of locally produced chemicals on the afferent arterioles (effects on the efferent arterioles are believed to be of secondary importance). When systemic arterial pressure falls toward a mean of 70 mmHg, the afferent arterioles dilate, and when the pressure rises, the afferent arterioles constrict. Blood flow to the glomeruli and GFR can thus remain relatively constant within the autoregulatory range of blood pressure values. The effects of different regulatory mechanisms on the GFR are summarized in table 17.1.

Autoregulation is also achieved through a negative feedback relationship between the afferent arterioles and the volume of fluid in the filtrate. An increased flow of filtrate is sensed by a special group of cells called the *macula densa* in the thick portion of the ascending limb (see fig. 17.24). The macula densa is part of the juxtaglomerular apparatus, as discussed in chapter 14 and in a later section of this chapter. When the macula densa senses an increased flow of filtrate, it signals the afferent arteriole to constrict. This lowers the GFR, thereby decreasing the formation of filtrate in a process called **tubuloglomerular feedback.**

1. Describe the structures that plasma fluid must pass through to enter the glomerular capsule. Explain how proteins are excluded from the filtrate.
2. Describe the forces that affect the formation of glomerular ultrafiltrate.
3. Describe the effect of sympathetic innervation on the glomerular filtration rate and explain what is meant by renal autoregulation.

Reabsorption of Salt and Water

Most of the salt and water filtered from the blood is reabsorbed across the wall of the proximal tubule. The reabsorption of water occurs by osmosis, in which water follows the active extrusion of NaCl from the tubule and into the surrounding peritubular capillaries. Most of the remaining water in the filtrate is reabsorbed across the wall of the collecting duct in the renal medulla. This occurs as a result of the high osmotic pressure of the surrounding tissue fluid, which is produced by transport processes in the loop of Henle.

Although about 180 L of glomerular ultrafiltrate are produced per day, the kidneys normally excrete only 1 to 2 L of urine per day. Approximately 99% of the filtrate must thus be returned to the vascular system, while 1% is excreted in the urine. The urine volume, however, varies according to the needs of the body. When a well-hydrated person drinks a liter or more of water, urine volume increases to 16 ml per minute (the equivalent of 23 L per day if this were to continue for 24 hours). In severe dehydration, when the body needs to conserve water, only 0.3 ml of urine per minute, or 400 ml per day, are produced. A volume of 400 ml of urine per day is needed to excrete the amount of metabolic wastes produced by the body; this is called the *obligatory water loss*. When water in excess of this amount is excreted, the urine volume is increased and its concentration is decreased.

Regardless of the body's state of hydration, it is clear that most of the filtered water must be returned to the vascular system to maintain blood volume and pressure. The return of filtered molecules from the tubules to the blood is called **reabsorption** (fig. 17.12). It is important to realize that the transport of water always occurs passively by *osmosis*; there is no such thing as active transport of water. A concentration gradient must thus be created between tubular fluid and blood that favors the osmotic return of water to the vascular system.

Reabsorption in the Proximal Tubule

Since all plasma solutes, with the exception of proteins, are able to enter the glomerular ultrafiltrate freely, the total solute concentration (osmolality—see chapter 6) of the filtrate is essentially the same as that of plasma. This total solute concentration is equal to 300 milliosmoles per liter (300 mOsm). The filtrate is thus said to be *isosmotic* with the plasma. Osmosis cannot occur unless the concentration of plasma in the peritubular capillaries and the concentration of filtrate are altered by active transport processes. This is achieved by the active transport of Na^+ from the filtrate to the peritubular blood.

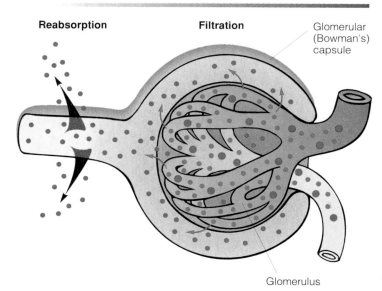

Figure 17.12

Plasma water and its dissolved solutes (except proteins) enter the glomerular ultrafiltrate, but most of these filtered molecules are reabsorbed. The term *reabsorption* refers to the transport of molecules out of the tubular filtrate back into the blood.

Active and Passive Transport

The epithelial cells that compose the wall of the proximal tubule are joined together by tight junctions only on their apical sides—that is, the sides of each cell that are closest to the lumen of the tubule (fig. 17.13). Each cell therefore has four exposed surfaces: the apical side facing the lumen, which contains microvilli; the opposite, basal side facing the peritubular capillaries; and the lateral sides, facing the narrow clefts between adjacent epithelial cells.

The concentration of Na^+ in the glomerular ultrafiltrate—and thus in the fluid entering the proximal tubule—is the same as that in plasma. The epithelial cells of the tubule, however, have a much lower Na^+ concentration. This lower Na^+ concentration is partially due to the low permeability of the cell membrane to Na^+ and partially due to the active transport of Na^+ out of the cell by Na^+/K^+ pumps, as described in chapter 6. In the cells of the proximal tubule, the Na^+/K^+ pumps are located in the basal and lateral sides of the cell membrane but not in the apical membrane. As a result of the action of these active transport pumps, a concentration gradient is created that favors the diffusion of Na^+ from the tubular fluid across the apical cell membranes and into the epithelial cells of the proximal tubule. The Na^+ is then extruded into the surrounding tissue fluid by the Na^+/K^+ pumps.

The transport of Na^+ from the tubular fluid to the interstitial (tissue) fluid surrounding the epithelial cells of the proximal tubule creates a potential difference across the wall of the tubule. This electrical gradient favors the passive transport of

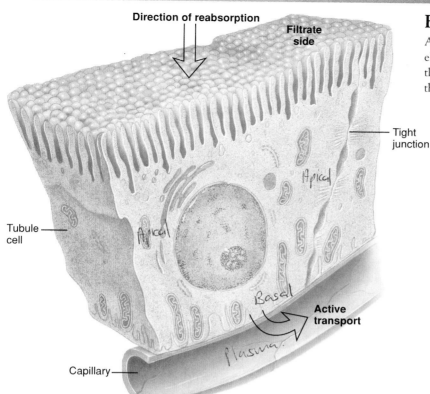

Direction of reabsorption

Filtrate side

Tight junction

Apical

Apical

Tubule cell

Basal

Active transport

Plasma.

Capillary

Figure 17.13

An illustration of the appearance of tubule cells in the electron microscope. Molecules that are reabsorbed pass through the tubule cells from the apical membrane (facing the filtrate) to the basolateral membrane (facing the blood).

Cl⁻ toward the higher Na⁺ concentration in the tissue fluid. Chloride ions, therefore, passively follow sodium ions out of the filtrate to the interstitial fluid. As a result of the accumulation of NaCl, the osmolality and osmotic pressure values of the tissue fluid surrounding the epithelial cells are increased above those of the tubular fluid. This is particularly true of the tissue fluid between the lateral membranes of adjacent epithelial cells, where the narrow spaces permit the accumulated NaCl to achieve a higher concentration.

An osmotic gradient is thus created between the tubular fluid and the tissue fluid surrounding the proximal tubule. Since the cells of the proximal tubule are permeable to water, water moves by osmosis from the tubular fluid into the epithelial cells and then across the basal and lateral sides of the epithelial cells into the tissue fluid. The salt and water that were reabsorbed from the tubular fluid can then move passively into the surrounding peritubular capillaries, and in this way be returned to the blood (fig. 17.14).

Significance of Proximal Tubule Reabsorption

Approximately 65% of the salt and water in the original glomerular ultrafiltrate is reabsorbed across the proximal tubule and returned to the vascular system. The volume of tubular fluid remaining is reduced accordingly, but this fluid is still isosmotic with the blood (has a concentration of 300 mOsm). This

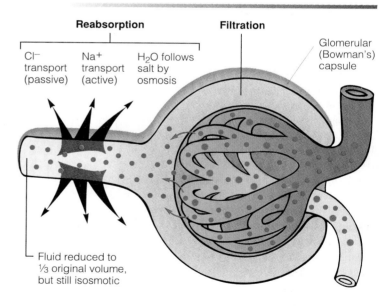

Reabsorption — **Filtration**

Cl⁻ transport (passive)   Na⁺ transport (active)   H₂O follows salt by osmosis

Glomerular (Bowman's) capsule

Fluid reduced to ⅓ original volume, but still isosmotic

Figure 17.14

Mechanisms of salt and water reabsorption in the proximal tubule. Sodium is actively transported out of the filtrate and chloride follows passively by electrical attraction. Water follows the salt out of the tubular filtrate into the peritubular capillaries by osmosis.

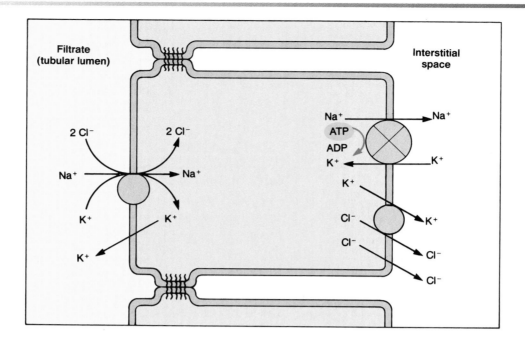

Figure 17.15

In the thick segment of the ascending limb of the loop, Na^+ and K^+, together with two Cl^-, enter the tubule cells. Na^+ is then actively transported out into the interstitial space and Cl^- follows passively. The K^+ diffuses back into the filtrate, and some also enters the interstitial space.

is because the cell membranes in the proximal tubule are freely permeable to water, so that water and salt are removed in proportionate amounts.

An additional smaller amount of salt and water is returned to the vascular system by reabsorption in the loop of Henle. This reabsorption, like that in the proximal tubule, occurs constantly, regardless of the person's state of hydration. Unlike reabsorption in later regions of the nephron (distal tubule and collecting duct), it is not subject to hormonal regulation. Approximately 85% of the filtered salt and water is, therefore, reabsorbed in a constant fashion in the early regions of the nephron (proximal tubule and loop of Henle). This reabsorption is very costly in terms of energy expenditures, accounting for as much as 6% of the calories consumed by the body at rest.

Since 85% of the original glomerular ultrafiltrate is immediately reabsorbed in the early region of the nephron, only 15% of the initial filtrate remains to enter the distal convoluted tubule and collecting duct. This is still a large volume of fluid—15% × GFR (180 L per day) = 27 L per day—that must be reabsorbed to varying degrees in accordance with the body's state of hydration. This "fine tuning" of the percentage of reabsorption and urine volume is accomplished by the action of hormones on the later regions of the nephron.

The Countercurrent Multiplier System

Water cannot be actively transported across the tubule wall, and osmosis of water cannot occur if the tubular fluid and surrounding tissue fluid are isotonic to each other. In order for water to be reabsorbed by osmosis, the surrounding tissue fluid must be hypertonic. The osmotic pressure of the tissue fluid in

the renal medulla is, in fact, raised to over four times that of plasma. This results partly from the fact that the tubule bends; the geometry of the loop of Henle permits interaction between the descending and ascending limbs. Since the ascending limb is the active partner in this interaction, its properties will be described before those of the descending limb.

Ascending Limb of the Loop of Henle

Salt (NaCl) is actively extruded from the ascending limb into the surrounding tissue fluid. This is not accomplished, however, by the same process that occurs in the proximal tubule. Instead, Na^+, K^+, and Cl^- passively diffuse from the filtrate into the ascending limb cells, in a ratio of 1 Na^+ to 1 K^+ to 2 Cl^-. The Na^+ is then actively transported across the basolateral membrane to the tissue fluid by the Na^+/K^+ pump. Cl^- follows the Na^+ passively because of electrical attraction, and K^+ diffuses back into the filtrate (fig. 17.15).

The ascending limb is structurally divisible into two regions: a *thin segment*, nearest to the tip of the loop, and a *thick segment* of varying lengths, which carries the filtrate outward into the cortex and into the distal convoluted tubule. It is currently believed that only the cells of the thick segments of the ascending limb are capable of actively transporting NaCl from the filtrate into the surrounding tissue fluid.

Although the mechanism of NaCl transport is different in the ascending limb than in the proximal tubule, the net effect is the same: salt (NaCl) is extruded into the surrounding tissue fluid. Unlike the epithelial walls of the proximal tubule, however, the walls of the ascending limb of the loop of Henle are *not permeable to water*. The tubular fluid thus becomes increasingly dilute as it ascends toward the cortex, whereas the tissue

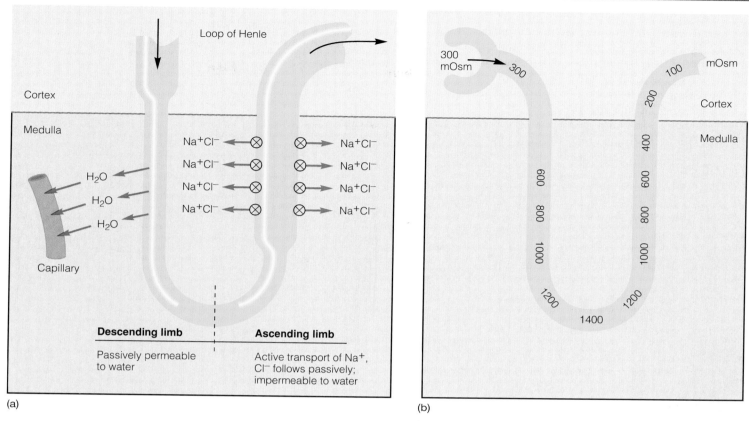

Figure 17.16

The countercurrent multiplier system. The extrusion of sodium chloride from the ascending limb makes the surrounding tissue fluid more concentrated. This concentration is multiplied by the fact that the descending limb is passively permeable so that its fluid increases in concentration as the surrounding tissue fluid becomes more concentrated. The transport properties of the loop and their effect on tubular fluid concentration are shown in (*a*). The values of these changes in osmolality, together with the effect on surrounding tissue fluid concentration, are shown in (*b*).

fluid around the loops of Henle in the medulla becomes increasingly more concentrated. By means of these processes, the tubular fluid that enters the distal tubule in the cortex is made hypotonic (with a concentration of about 100 mOsm), whereas the tissue fluid in the medulla is made hypertonic.

Descending Limb of the Loop of Henle

The deeper regions of the medulla, around the tips of the loops of juxtamedullary nephrons, reach a concentration of 1,200 to 1,400 mOsm. In order to reach this high a concentration, the salt pumped out of the ascending limb must accumulate in the tissue fluid of the medulla. This occurs as a result of the properties of the descending limb, to be discussed next, and because blood vessels around the loop do not carry back all of the extruded salt to the general circulation. The capillaries in the medulla are uniquely arranged to trap NaCl in the tissue fluid, as will be discussed in a later section.

The descending limb does not actively transport salt, and indeed is believed to be impermeable to the passive diffusion of salt. It is, however, permeable to water. Since the surrounding interstitial fluid is hypertonic to the filtrate in the descending limb, water is drawn out of the descending limb by osmosis and

enters blood capillaries. The concentration of tubular fluid is thus increased, and its volume is decreased, as it descends toward the tips of the loops.

As a result of these passive transport processes in the descending limb, the fluid that "rounds the bend" at the tip of the loop has the same osmolality as that of the surrounding tissue fluid (1,200 to 1,400 mOsm). There is, therefore, a higher salt concentration arriving in the ascending limb than there would be if the descending limb simply delivered isotonic fluid. Salt transport by the ascending limb is increased accordingly, so that the "saltiness" of the tissue fluid is multiplied (fig. 17.16).

Countercurrent Multiplication

Countercurrent flow (flow in opposite directions) in the ascending and descending limbs and the close proximity of the two limbs allow interaction to take place. Since the concentration of the tubular fluid in the descending limb reflects the concentration of surrounding tissue fluid, and since the concentration of this tissue fluid is raised by the active extrusion of salt from the ascending limb, a *positive feedback mechanism* is created. The more salt the ascending limb extrudes, the more concentrated will be the fluid that is delivered to it from the descending limb. This positive feedback mechanism multiplies the concentration of tissue fluid and descending limb fluid, and is thus called the **countercurrent multiplier system.**

The countercurrent multiplier system recirculates salt and thus traps some of the salt that enters the loop of Henle in the tissue fluid of the renal medulla. This system results in a gradually increasing concentration of renal tissue fluid from the cortex to the inner medulla; the osmolality of tissue fluid increases from 300 mOsm (isotonic) in the cortex to between 1,200 and 1,400 mOsm in the deepest part of the medulla.

Vasa Recta

In order for the countercurrent multiplier system to be effective, most of the salt that is extruded from the ascending limbs must remain in the tissue fluid of the medulla, while most of the water that leaves the descending limbs must be removed by the blood. This is accomplished by vessels known as the **vasa recta.** The vasa recta are peritubular capillaries that form long capillary loops that parallel the long loops of Henle of the juxtamedullary nephrons (see fig. 17.19).

The vasa recta maintain the hypertonicity of the renal medulla by means of a mechanism known as **countercurrent exchange.** Salt and other solutes (such as urea, described in the next section) that are present at high concentrations in the medullary tissue fluid diffuse into the blood as the blood descends into the capillary loops of the vasa recta, but then passively diffuse out of the ascending vessels and back into the descending vessels (where the concentration is lower). Solutes are thus recirculated and trapped within the medulla. Since the walls of the vasa recta are freely permeable to dissolved solutes, the concentration of these solutes becomes the same inside the vasa recta as in the surrounding interstitial fluid at each level within the medulla. The colloid osmotic pressure within the vasa recta, however, is higher than in the interstitial fluid because plasma proteins do not easily pass through the capillary walls. This is similar to the situation in other capillary beds (chapter 14), and results in the osmotic movement of water into both the descending and ascending limbs of the vasa recta. The vasa recta thus trap salt and urea within the interstitial fluid but transport water out of the renal medulla (fig. 17.17).

Effects of Urea

Countercurrent multiplication of the NaCl concentration is the mechanism traditionally cited to explain the hypertonicity of the interstitial fluid in the medulla. It is currently believed, however, that urea also contributes significantly to the total osmolality of the interstitial fluid.

The role of urea was inferred from experimental evidence showing that active transport of Na⁺ occurs only in the thick segments of the ascending limbs. The thin segments of the ascending limbs, which are located in the deeper regions of the medulla, are not able to extrude salt actively. But since salt does indeed leave the thin segments, a diffusion gradient for salt must exist, despite the fact that the surrounding tissue fluid has the same osmolality as the tubular fluid. Investigators therefore concluded that molecules other than salt—specifically urea—contribute to the hypertonicity of the tissue fluid.

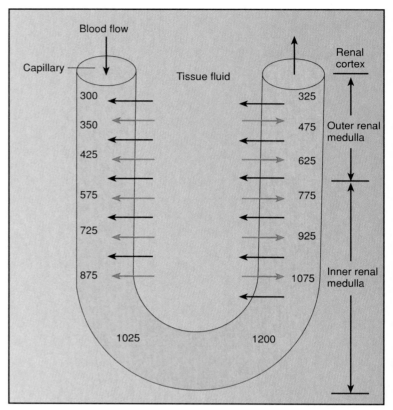

Black arrows = diffusion of NaCl and urea
Blue arrows = movement of water by osmosis

Figure 17.17

Countercurrent exchange in the vasa recta. The diffusion of salt and water first into and then out of these blood vessels helps to maintain the "saltiness" (hypertonicity) of the interstitial fluid in the renal medulla. (Numbers indicate osmolality.)

It was later shown that the ascending limb of the loop of Henle and the collecting duct are permeable to urea. Urea can thus diffuse out of the collecting duct and into the ascending limb (fig. 17.18). In this way, a certain amount of urea is recycled through these two segments of the nephron and is thus trapped in the interstitial fluid.

The transport properties of different tubule segments with respect to the concentrating-diluting mechanisms of the kidney are summarized in table 17.2.

Collecting Duct: Effect of Antidiuretic Hormone (ADH)

As a result of the recycling of salt between the ascending and descending limbs and the recycling of urea between the collecting duct and the loop of Henle, the medullary tissue fluid is made very hypertonic. The collecting ducts must transport their fluid through this hypertonic environment in order to empty their contents of urine into the calyces. Whereas the

Table 17.2 Transport Properties of Different Segments of the Renal Tubules and the Collecting Ducts

Nephron Segment	Active Transport	Passive Transport		
		Salt	Water	Urea
Proximal tubule	Na$^+$	Cl$^-$	Yes	Yes
Descending limb of Henle's loop	None	Maybe	Yes	No
Thin segment of ascending limb	None	NaCl	No	Yes
Thick segment of ascending limb	Na$^+$	Cl$^-$	No	No
Distal tubule	Na$^+$	No	No	No
Collecting duct*	Slight Na$^+$	No	Yes (ADH) or slight (no ADH)	Yes

*The permeability of the collecting duct to water depends on the presence of ADH.

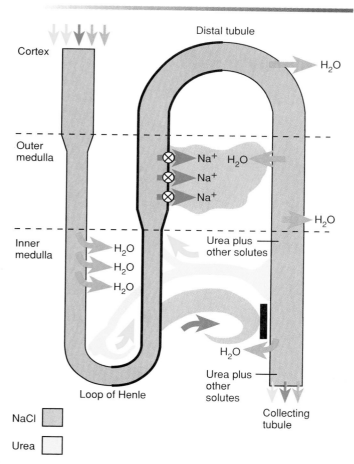

Figure 17.18

According to some authorities, urea diffuses out of the collecting duct and contributes significantly to the concentration of the interstitial fluid in the renal medulla. The active transport of Na$^+$ out of the thick segments of the ascending limbs also contributes to the hypertonicity of the medulla, so that water is reabsorbed by osmosis from the collecting ducts.

fluid surrounding the collecting ducts in the medulla is hypertonic, the fluid that passes into the collecting ducts in the cortex is hypotonic as a result of the active extrusion of salt by the ascending limbs of the loops.

There is some reabsorption of Na$^+$ and secretion of K$^+$ in the cortical region of the collecting duct, but the medullary region is relatively impermeable to the high concentration of NaCl that surrounds it. The wall of the collecting duct, however, is permeable to water. Since the surrounding tissue fluid in the renal medulla is very hypertonic, as a result of the countercurrent multiplier system, water is drawn out of the collecting ducts by osmosis. This water does not dilute the surrounding tissue fluid because it is transported by capillaries to the general circulation. In this way, most of the water remaining in the filtrate is returned to the vascular system (fig. 17.19).

Note that the osmotic gradient created by the countercurrent multiplier system provides the force for water reabsorption through the collecting ducts. The rate at which this osmotic movement occurs, however, is determined by the permeability of the collecting duct cell membranes to water. This depends on the number of *water channels* in the cell membranes of the collecting duct epithelial cells.

The water channels are produced as proteins within the membranes of vesicles that bud from the Golgi apparatus (chapter 3). In the absence of stimulation, these vesicles are present in the cytoplasm of the collecting duct cells. When **antidiuretic hormone (ADH)** binds to its membrane receptors on the collecting duct, it acts (via cAMP as a second messenger) to stimulate the fusion of these vesicles with the cell membrane. This is identical to exocytosis (chapter 3), except that, in this case, there is no secretion of product. The importance of this process in the collecting duct is that, by the fusion of the vesicles with the cell membrane, the water channels become incorporated into the cell membrane. In response to ADH, therefore, the collecting duct becomes more permeable to water. When ADH is no longer secreted, so that it no longer binds to its membrane receptors, the water channels are removed from the cell membrane by a process of endocytosis (chapter 3). Endocytosis is the opposite of exocytosis; the cell membrane invaginates to reform vesicles that again contain the water channels. Alternating exocytosis and endocytosis in response to the presence and absence of ADH, respectively, is believed to result in the recycling of water channels within the cell.

When the concentration of ADH is increased, the collecting ducts become more permeable to water, and more water is reabsorbed. A decrease in ADH, conversely, results in less

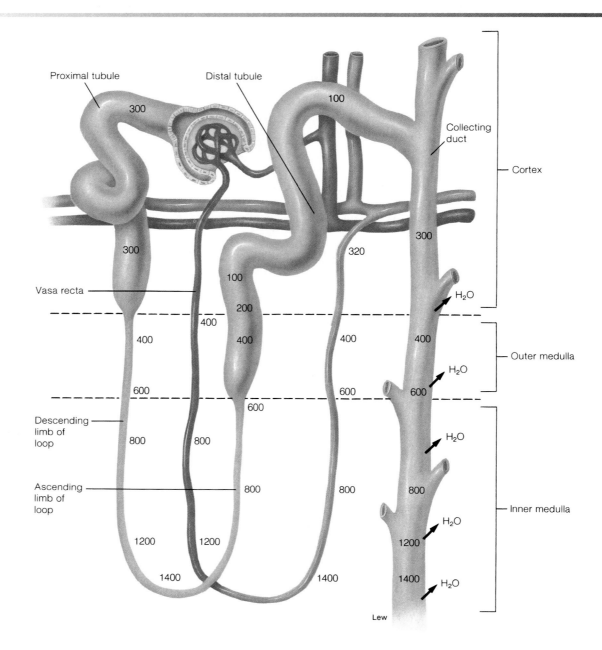

Figure 17.19

The countercurrent multiplier system in the loop of the nephron and countercurrent exchange in the vasa recta help to create a hypertonic renal medulla. Under the influence of antidiuretic hormone (ADH) the collecting duct becomes more permeable to water, and thus more water is drawn by osmosis out into the hypertonic renal medulla and into the peritubular capillaries.

reabsorption of water and thus in the excretion of a larger volume of more dilute urine. ADH is produced by neurons in the hypothalamus and is secreted from the posterior pituitary (chapter 11). The secretion of ADH is stimulated when osmoreceptors in the hypothalamus respond to an increase in blood osmotic pressure. During dehydration, therefore, when the plasma becomes more concentrated, increased secretion of ADH promotes increased permeability of the collecting ducts to water. In severe dehydration, only the minimal amount of water needed to eliminate the body's wastes is excreted. This minimum, about 400 ml per day, is limited by the fact that urine

cannot become more concentrated than the medullary tissue fluid surrounding the collecting ducts. Under these conditions, about 99.8% of the initial glomerular ultrafiltrate is reabsorbed.

A person in a state of normal hydration excretes about 1.5 L of urine per day, indicating that 99.2% of the glomerular ultrafiltrate volume is reabsorbed. Notice that small changes in percent reabsorption translate into large changes in urine volume. Increasing water consumption—and thus decreasing ADH secretion (table 17.3)—results in correspondingly larger volumes of urine excretion. It should be noted, however, that even in the complete absence of ADH some water is still reabsorbed through the collecting ducts.

Table 17.3   Antidiuretic Hormone Secretion and Action

Stimulus	Receptors	Secretion of ADH	Effects on Urine Volume	Effects on Blood
↑Osmolality (dehydration)	Osmoreceptors in hypothalamus	Increased	Decreased	Increased water retention; decreased blood osmolality
↓Osmolality	Osmoreceptors in hypothalamus	Decreased	Increased	Water loss increases blood osmolality
↑Blood volume	Stretch receptors in left atrium	Decreased	Increased	Decreased blood volume
↓Blood volume	Stretch receptors in left atrium	Increased	Decreased	Increased blood volume

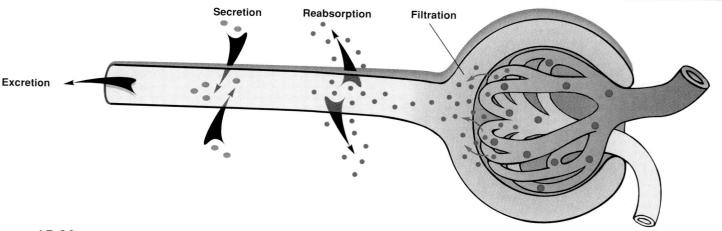

Figure 17.20

Secretion refers to the active transport of substances from the peritubular capillaries into the tubular fluid. The direction of this transport is opposite to that of reabsorption.

Diabetes insipidus is a disease associated with the inadequate secretion or action of ADH. The collecting ducts are thus not very permeable to water and, therefore, a large volume (5 to 10 L per day) of dilute urine is produced. The dehydration that results causes intense thirst, but a person with this condition has difficulty drinking enough to compensate for the large volumes of water lost in the urine.

1. Describe the mechanisms for salt and water reabsorption in the proximal tubule.
2. Compare the transport of Na⁺, Cl⁻, and water across the walls of the proximal tubule, ascending and descending limbs of the loop of Henle, and collecting duct.
3. Describe the interaction between the ascending and descending limbs of the loop and explain how this interaction results in a hypertonic renal medulla.
4. Explain how ADH helps the body to conserve water. How do variations in ADH secretion affect the volume and concentration of urine?

Renal Plasma Clearance

As blood passes through the kidneys, some of the constituents of the plasma are removed and excreted in the urine. The blood is thus "cleared," to one degree or another, of particular solutes in the process of urine formation. These solutes may be removed from the blood by filtration through the glomerular capillaries or by secretion by the tubular cells into the filtrate. At the same time, certain molecules in the tubular fluid can be reabsorbed back into the blood.

One of the major functions of the kidneys is the elimination of waste products such as urea, creatinine, and other molecules from the blood by excreting them in the urine. These molecules are filtered through the glomerulus into the glomerular capsule along with water, salt, and other plasma solutes. In addition, some waste products can gain access to the urine by a process called **secretion** (fig. 17.20). Secretion is the opposite of reabsorption. Molecules that are secreted move out of the peritubular capillaries and into the tubular cells, from which they are actively transported into the tubular lumen. In this way, molecules that were not filtered out of the blood in the

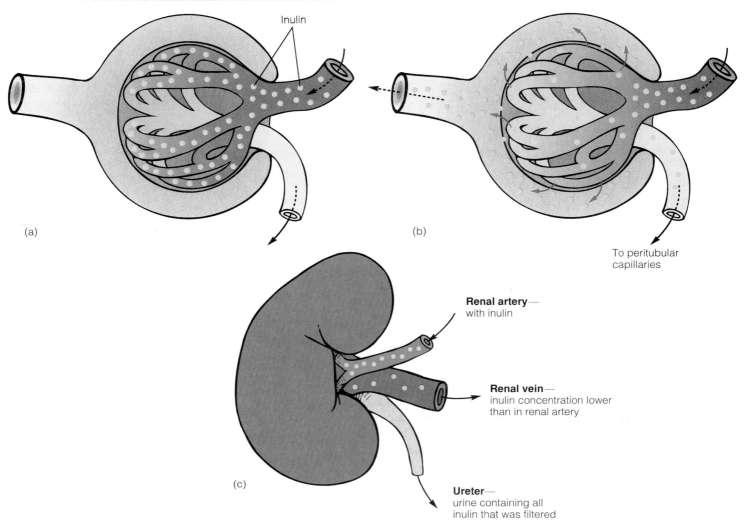

Figure 17.21

The renal clearance of inulin. (*a*) Inulin is present in the blood entering the glomeruli, and (*b*) some of this blood, together with its dissolved inulin, is filtered. All of this filtered inulin enters the urine, whereas most of the filtered water is returned to the vascular system (is reabsorbed). (*c*) The blood leaving the kidneys in the renal vein, therefore, contains less inulin than the blood that entered the kidneys in the renal artery. Since inulin is filtered but neither reabsorbed nor secreted, the inulin clearance rate equals the glomerular filtration rate (GFR).

glomerulus, but instead passed through the efferent arterioles to the peritubular capillaries, can still be excreted in the urine. This is useful because it allows the kidneys to rapidly eliminate certain potential toxins. Also, K^+ and H^+ ions, which are first filtered and reabsorbed, are then secreted into the filtrate to varying degrees. This allows the kidneys to more finely control the amount of K^+ and H^+ excreted in the urine.

Although most (about 99%) of the filtered water is returned to the vascular system by reabsorption, most of the unneeded molecules that are filtered or secreted are eliminated in the urine. The concentration of these substances in the renal vein leaving the kidney is therefore lower than their concentrations in the blood entering the kidney in the renal artery. Some of the blood that passes through the kidneys, in other words, is "cleared" of these waste products.

Renal Clearance of Inulin: Measurement of GFR

If a substance is neither reabsorbed nor secreted by the tubules, the amount excreted per minute in the urine will be equal to the amount that is filtered out of the glomeruli. There does not seem to be a single substance produced by the body, however, that is not reabsorbed or secreted to some degree. Plants such as artichokes, dahlias, onions, and garlic, fortunately, do produce such a compound. This compound, a polymer of the monosaccharide fructose, is *inulin*. Once injected into the blood, inulin is filtered by the glomeruli, and the amount of inulin excreted per minute is exactly equal to the amount that was filtered per minute (fig. 17.21).

Table 17.4 Effects of Filtration, Reabsorption, and Secretion on Renal Clearance Rates

Term	Definition	Effect on Renal Clearance
Filtration	A substance enters the glomerular ultrafiltrate	Some or all of a filtered substance may enter the urine and be "cleared'" from the blood.
Reabsorption	A substance is transported from the filtrate, through tubular cells, and into the blood	Reabsorption decreases the rate at which a substance is cleared; clearance rate is less than the glomerular filtration rate (GFR).
Secretion	A substance is transported from peritubular blood, through tubular cells, and into the filtrate	When a substance is secreted by the nephrons, its clearance rate is greater than the GFR.

If the concentration of inulin in urine is measured and the rate of urine formation is determined, the rate of inulin excretion can easily be calculated:

$$\text{Quantity excreted per minute} = \underset{\left(\frac{ml}{min}\right)}{V} \times \underset{\left(\frac{mg}{ml}\right)}{U}$$
(mg/min)

where

V = rate of urine formation

U = inulin concentration in urine

The rate at which a substance is filtered by the glomeruli (in mg per minute) can be calculated by multiplying the ml per minute of plasma that is filtered (the **glomerular filtration rate, or GFR**) by the concentration of that substance in the plasma. This is shown in the following equation:

$$\text{Quantity filtered per minute} = \underset{\left(\frac{ml}{min}\right)}{GFR} \times \underset{\left(\frac{mg}{ml}\right)}{P}$$
(mg/min)

where

P = inulin concentration in plasma

Since inulin is neither reabsorbed nor secreted, the amount filtered equals the amount excreted:

$$\underset{\text{(amount filtered)}}{GFR \times P} = \underset{\text{(amount excreted)}}{V \times U}$$

If the preceding equation is now solved for the glomerular filtration rate,

$$GFR_{(ml/min)} = \frac{V_{(ml/min)} \times U_{(mg/ml)}}{P_{(mg/ml)}}$$

Suppose, for example, that inulin is infused into a vein and its concentrations in the urine and plasma are found to be 30 mg per ml and 0.5 mg per ml, respectively. If the rate of urine formation is 2 ml per minute, the GFR can be calculated as follows:

$$GFR = \frac{2\ ml/min \times 30\ mg/ml}{0.5\ mg/ml} = 120\ ml/min$$

This equation states that 120 ml of plasma per minute must have been filtered in order to excrete the measured

Measurements of the plasma concentration of **creatinine** are often used clinically as an index of kidney function. Creatinine, produced as a waste product of muscle creatine, is secreted to a slight degree by the renal tubules so that its excretion rate is a little above that of inulin. Since it is released into the blood at a constant rate and since its excretion is closely matched to the GFR, an abnormal decrease in GFR causes the plasma creatinine concentration to rise. Thus, a simple measurement of blood creatinine concentration can indicate whether the GFR is normal and provide information about the health of the kidneys.

amount of inulin that appeared in the urine. The glomerular filtration rate is thus 120 ml per minute in this example.

Clearance Calculations

The **renal plasma clearance** is the volume of plasma from which a substance is completely removed in one minute by excretion in the urine. Notice that the units for renal plasma clearance are ml/min. In the case of inulin, which is filtered but neither reabsorbed nor secreted, the amount of inulin that enters the urine is that which is contained in the volume of plasma filtered. The clearance of inulin is thus equal to the GFR (120 ml/min in the previous example). This volume of filtered plasma, however, also contains other solutes that may be reabsorbed to varying degrees. If a portion of a filtered solute is reabsorbed, the amount excreted in the urine is less than that which was contained in the 120 ml of plasma filtered. Thus, the renal plasma clearance of a substance that is reabsorbed must be less than the GFR (table 17.4).

If a substance is not reabsorbed, all of the filtered amount will be cleared. If this substance is, in addition, secreted by active transport into the renal tubules from the peritubular blood, an additional amount of plasma can be cleared of that substance. The renal plasma clearance of a substance that is filtered and secreted is therefore greater than the GFR (table 17.5). Thus, in order to compare the renal "handling" of various substances in terms of their reabsorption or secretion, the

Table 17.5 Renal "Handling" of Different Plasma Molecules

If Substance Is:	Example	Concentration in Renal Vein	Renal Clearance Rate
Not filtered	Proteins	Same as in renal artery	Zero
Filtered, not reabsorbed or secreted	Inulin	Less than in renal artery	Equal to GFR (115–125 ml/min)
Filtered, partially reabsorbed	Urea	Less than in renal artery	Less than GFR
Filtered, completely reabsorbed	Glucose	Same as in renal artery	Zero
Filtered and secreted	PAH	Less than in renal artery; approaches zero	Greater than GFR; up to total plasma flow rate (~625 ml/min)
Filtered, reabsorbed, and secreted	K⁺	Variable	Variable

renal plasma clearance is calculated using the same formula used for the determination of the GFR:

$$\text{Renal plasma clearance} = \frac{V \times U}{P}$$

where

V = urine volume per minute
U = concentration of substance in urine
P = concentration of substance in plasma

Clearance of Urea

Urea may be used as an example of how the clearance calculations can reveal the way the kidneys handle a molecule. Urea is a waste product of amino acid metabolism that is secreted by the liver into the blood and filtered into the glomerular capsules. Using the formula for renal clearance previously described and the following sample values, the following clearance may be calculated:

V = 2 ml/min
U = 7.5 mg/ml of urea
P = 0.2 mg/ml of urea

$$\text{Urea clearance} = \frac{2 \text{ ml/min} \times 7.5 \text{ mg/ml}}{0.2 \text{ mg/ml}} = 75 \text{ ml/min}$$

The clearance of urea in this example (75 ml/min) is less than the clearance of inulin (120 ml/min). Thus, even though 120 ml of plasma filtrate entered the nephrons, only the amount of urea contained in 75 ml of filtrate is excreted per minute. The kidneys therefore must reabsorb some of the urea that is filtered. Despite the fact that it is a waste product, a significant portion of the filtered urea (ranging from 40% to 60%) is always reabsorbed. This is a passive reabsorption that cannot be avoided because of the high permeability of cell membranes to urea.

Clearance of PAH: Measurement of Renal Blood Flow

Not all of the blood delivered to the glomeruli is filtered into the glomerular capsules; most of the glomerular blood passes through to the efferent arterioles and peritubular capillaries. The inulin and urea in this unfiltered blood are not excreted but instead return to the general circulation. Blood must therefore make many passes through the kidneys before it can be completely cleared of a given amount of inulin or urea.

Many antibiotics are secreted by the renal tubules and thus have clearance rates greater than the glomerular filtration rate. Penicillin, for example, is rapidly removed from the blood by renal clearance; hence, large amounts must be administered to be effective. The kidneys can be better visualized in radiographs by the injection of Diodrast. This substance is secreted into the tubules and improves contrast by absorbing X rays. Many drugs and some hormones are inactivated in the liver by chemical transformations and are rapidly cleared from the blood by active secretion in the nephrons.

In order for compounds in the unfiltered renal blood to be cleared, they must be secreted into the tubules by active transport from the peritubular capillaries. In this way, all of the blood going to the kidneys can potentially be cleared of a secreted compound in a single pass. This is the case for a molecule called **para-aminohippuric acid,** or **PAH** (fig. 17.22). The clearance (in ml/min) of PAH can be used to measure the *total renal blood flow*. The normal PAH clearance has been found to average 625 ml/min. Since the glomerular filtration rate averages about 120 ml/min, this indicates that only about 120/625, or roughly 20%, of the renal plasma flow is filtered. The remaining 80% passes on to the efferent arterioles.

Since filtration and secretion clear only the molecules dissolved in plasma, the PAH clearance measures the renal plasma flow. In order to convert this to the total renal blood flow, the volume of blood occupied by erythrocytes must be taken into account. If the hematocrit (chapter 13) is 45, for example, erythrocytes occupy 45% of the blood volume and plasma accounts for the remaining 55% of the blood volume. The **total renal blood flow** is calculated by dividing the PAH clearance by the fractional blood volume occupied by plasma (0.55, in this example). The total renal blood flow in this example is thus 625 ml/min divided by 0.55, or 1.1 L/min.

Reabsorption of Glucose

Glucose and amino acids in the blood are easily filtered by the glomeruli into the renal tubules. These molecules, however, are usually not present in the urine. It can be concluded, therefore, that filtered glucose and amino acids are normally completely reabsorbed by the nephrons. This occurs

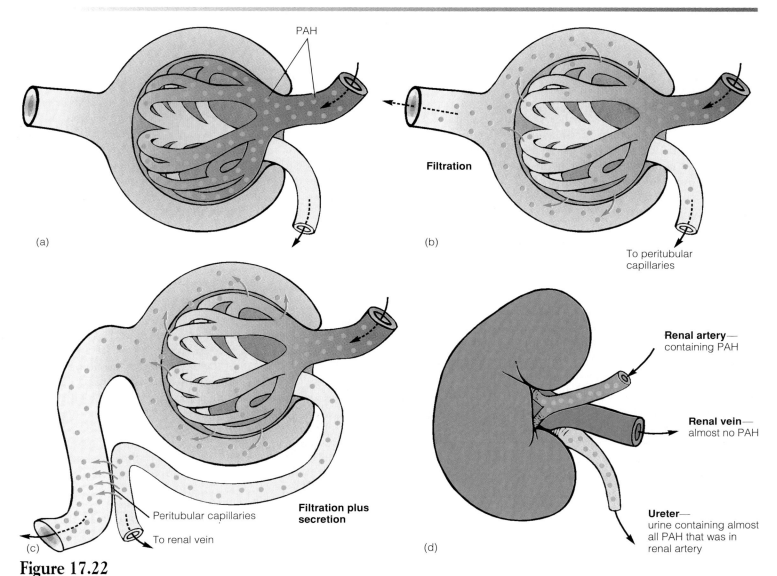

PAH

(a)

(b) **Filtration**

To peritubular capillaries

(c) Peritubular capillaries

To renal vein

Filtration plus secretion

(d)

Renal artery— containing PAH

Renal vein— almost no PAH

Ureter— urine containing almost all PAH that was in renal artery

Figure 17.22

Some of the para-aminohippuric acid (PAH) in glomerular blood (*a*) is filtered into the glomerular (Bowman's) capsules (*b*). The PAH present in the unfiltered blood is secreted from the peritubular capillaries into the nephron (*c*), so that all of the blood leaving the kidneys is free of PAH (*d*). The clearance rate of PAH therefore equals the total plasma flow to the glomeruli.

by secondary active transport, which is mediated by membrane carriers that cotransport glucose and Na^+ (chapter 6).

Carrier-mediated transport displays the property of *saturation*. This means that when the transported molecule (such as glucose) is present in sufficiently high concentrations, all of the carriers become "busy," and the transport rate reaches a maximal value. The concentration of transported molecules needed to just saturate the carriers and to just achieve the maximal transport rate is called the **transport maximum** (abbreviated T_m).

The carriers for glucose and amino acids in the renal tubules are not normally saturated and so are able to remove the filtered molecules completely. The T_m for glucose, for example, averages 375 mg per minute, which is well above the rate at which glucose is delivered to the tubules. The rate of glucose delivery can be calculated by multiplying the plasma glucose concentration (about 1 mg per ml) by the GFR (about 125 ml per minute). Approximately 125 mg per minute are thus delivered to the tubules, whereas a rate of 375 mg per minute is required to reach saturation.

Glycosuria

Glucose appears in the urine—a condition called *glycosuria*—when more glucose passes through the tubules than can be reabsorbed. This occurs when the plasma glucose concentration reaches 180 to 200 mg per 100 ml. Since the rate of glucose delivery under these conditions is still below the average T_m for glucose, we must conclude that some nephrons have considerably lower T_m values than the average.

The **renal plasma threshold** is the minimum plasma concentration of a substance that results in the excretion of that substance in the urine. The renal plasma threshold for glucose, for example, is 180 to 200 mg per 100 ml. Glucose is normally absent from urine because plasma glucose concentrations normally remain below this threshold value. The appearance of glucose in the urine (glycosuria) thus occurs only when the plasma glucose concentration is abnormally high (hyperglycemia) and exceeds the renal plasma threshold.

Fasting hyperglycemia is caused by the inadequate secretion or action of insulin. When this hyperglycemia results in glycosuria, the disease is called **diabetes mellitus.** A person with uncontrolled diabetes mellitus also excretes a large volume of urine because the excreted glucose carries water with it as a result of the osmotic pressure it generates in the tubules. This condition should not be confused with diabetes insipidus, in which a large volume of dilute urine is excreted as a result of inadequate ADH secretion.

1. Define *renal plasma clearance* and describe how this volume is measured. Explain why the glomerular filtration rate is equal to the clearance rate of inulin.
2. Define the terms *reabsorption* and *secretion.* Using examples, describe how the renal plasma clearance is affected by the processes of reabsorption and secretion.
3. Explain why the total renal blood flow can be measured by the renal plasma clearance of PAH.
4. Define *transport maximum* and *renal plasma threshold.* Explain why people with diabetes mellitus have glycosuria.

Renal Control of Electrolyte and Acid-Base Balance

The kidneys regulate the blood concentration of Na⁺, K⁺, HCO₃⁻, and H⁺. Aldosterone stimulates the reabsorption of Na⁺ in exchange for K⁺ in the distal tubule. Aldosterone thus promotes the renal retention of Na⁺ and the excretion of K⁺. Secretion of aldosterone from the adrenal cortex is stimulated directly by a high blood K⁺ concentration, and indirectly by a low Na⁺ concentration via the renin-angiotensin system.

The kidneys help regulate the concentrations of plasma electrolytes—sodium, potassium, chloride, bicarbonate, and phosphate—by matching the urinary excretion of these compounds to the amounts ingested. The control of plasma Na⁺ is important in the regulation of blood volume and pressure; the control of plasma K⁺ is required to maintain proper function of cardiac and skeletal muscles.

Role of Aldosterone in Na⁺/K⁺ Balance

Approximately 90% of the filtered Na⁺ and K⁺ is reabsorbed in the early part of the nephron before the filtrate reaches the distal tubule. This reabsorption occurs at a constant rate and is not subject to hormonal regulation. The final concentration of Na⁺ and K⁺ in the urine is varied according to the needs of the body by processes that occur in the late distal tubule and in the cortical region of the collecting duct (the portion of the collecting duct within the medulla does not participate in this regulation). Renal reabsorption of Na⁺ and secretion of K⁺ are regulated by **aldosterone,** the principal mineralocorticoid secreted by the adrenal cortex (chapter 11).

Sodium Reabsorption

Although 90% of the filtered sodium is reabsorbed in the early region of the nephron, the amount left in the filtrate delivered to the distal convoluted tubule is still quite large. In the absence of aldosterone, 80% of this amount is automatically reabsorbed through the wall of the distal tubule into the peritubular blood; this is 8% of the amount filtered. The amount of sodium excreted without aldosterone is thus 2% of the amount filtered. Although this percentage seems small, the actual amount of sodium this represents is an impressive 30 g per day excreted in the urine. When aldosterone is secreted in maximal amounts, in contrast, all of the sodium delivered to the distal tubule is reabsorbed. Under these conditions, urine contains no Na⁺ at all.

Potassium Secretion

About 90% of the filtered K⁺ is reabsorbed in the early regions of the nephron (mainly in the proximal tubule). When aldosterone is absent, all of the filtered K⁺ that remains is reabsorbed in the distal tubule. In the absence of aldosterone, therefore, no K⁺ is excreted in the urine. The presence of aldosterone stimulates the secretion of K⁺ from the peritubular blood into the late distal tubule and cortical collecting duct (fig. 17.23). This aldosterone-induced secretion is thus the only means by which K⁺ can be eliminated in the urine. When aldosterone secretion is maximal, as much as fifty times more K⁺ is excreted in the urine, because of secretion into the distal tubule, than was originally filtered through the glomeruli.

In summary, aldosterone promotes sodium retention and potassium loss from the blood by stimulating the reabsorption of Na⁺ and the secretion of K⁺ across the wall of the late distal convoluted tubules and cortical portions of the collecting ducts. Since aldosterone promotes the retention of Na⁺, it contributes to an increased blood volume and pressure.

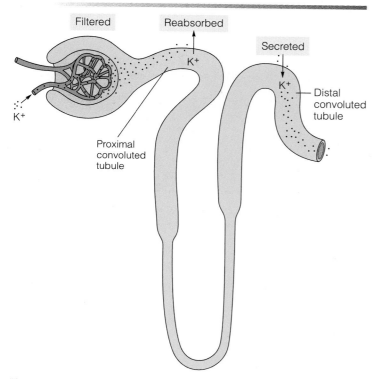

Figure 17.23

Potassium is almost completely reabsorbed in the proximal tubule, but under aldosterone stimulation it is secreted into the distal tubule. All of the K⁺ in urine is derived from secretion rather than from filtration.

The body cannot get rid of excess K⁺ in the absence of aldosterone-stimulated secretion of K⁺ into the distal tubules. Indeed, when both adrenal glands are removed from an experimental animal, the **hyperkalemia** (high blood K⁺) that results can produce fatal cardiac arrhythmias. Abnormally low plasma K⁺ concentrations, as might result from excessive aldosterone secretion, can also produce arrhythmias, as well as muscle weakness.

Control of Aldosterone Secretion

Since aldosterone promotes Na⁺ retention and K⁺ loss, one might predict (on the basis of negative feedback) that aldosterone secretion would be increased when there was a low Na⁺ or a high K⁺ concentration in the blood. This indeed is the case. A rise in blood K⁺ *directly* stimulates the secretion of aldosterone from the adrenal cortex. Decreases in plasma Na⁺ concentrations also promote aldosterone secretion, but they do so indirectly.

Juxtaglomerular Apparatus

The juxtaglomerular apparatus is the region in each nephron where the afferent arteriole and distal tubule come into contact (fig. 17.24). The microscopic appearance of the afferent arteriole and distal tubule in this small region differs from the appearance in other regions. *Granular cells* within the afferent arteriole secrete the enzyme **renin** into the blood; this enzyme catalyzes the conversion of angiotensinogen (a protein) into angiotensin I (a ten-amino-acid polypeptide).

Secretion of renin into the blood thus results in the formation of angiotensin I, which is then converted to **angiotensin II** by angiotensin converting enzyme (ACE). This conversion occurs primarily as blood passes through the capillaries of the lungs, where most of the converting enzyme is present. Angiotensin II, in addition to other effects (described in chapter 14), stimulates the adrenal cortex to secrete aldosterone. Secretion of renin from the granular cells of the juxtaglomerular apparatus is thus said to initiate the **renin-angiotensin-aldosterone system.** Conditions that result in renin secretion thus cause increased aldosterone secretion and, by this means, promote the reabsorption of Na⁺ in the distal convoluted tubules.

Regulation of Renin Secretion

A fall in plasma Na⁺ concentration is always accompanied by a fall in blood volume. This is because ADH secretion is inhibited by the decreased plasma concentration (osmolality); with less ADH, less water is reabsorbed through the collecting ducts and more is excreted in the urine. The fall in blood volume and the fall in renal blood flow that result cause increased renin secretion. Increased renin secretion is believed to be due in part to the direct effect of blood flow on the granular cells, which may function as baroreceptors in the afferent arterioles. Renin secretion is also stimulated by sympathetic nerve activity, which is increased when the blood volume and pressure fall.

An increased secretion of renin acts, via the increased production of angiotensin II, to stimulate aldosterone secretion. Consequently, less sodium is excreted in the urine and more is retained in the blood. This negative feedback system is illustrated in figure 17.25.

Role of the Macula Densa

The region of the distal tubule in contact with the granular cells of the afferent arteriole is called the **macula densa** (fig. 17.24). There is evidence that this region helps to inhibit renin secretion when the blood Na⁺ concentration is raised.

According to the proposed mechanism, the cells of the macula densa respond to Na⁺ within the filtrate delivered to the distal tubule. When the plasma Na⁺ concentration is raised, the rate of Na⁺ delivered to the distal tubule is also increased. Through an effect on the macula densa, this increase in filtered Na⁺ may inhibit the granular cells from secreting renin. Aldosterone secretion thus decreases, and since less Na⁺ is reabsorbed in the distal tubule, more Na⁺ is excreted in the urine. The regulation of renin and aldosterone secretion is summarized in table 17.6.

Table 17.6 Regulation of Renin and Aldosterone Secretion

Stimulus	Effect on Renin Secretion	Angiotensin II Production	Aldosterone Secretion	Mechanisms
↓Na$^+$	Increased	Increased	Increased	Low blood volume stimulates renal baroreceptors; granular cells release renin.
↑Na$^+$	Decreased	Decreased	Decreased	Increased blood volume inhibits baroreceptors; increased Na$^+$ in distal tubule acts via macula densa to inhibit release of renin from granular cells.
↑K$^+$	None	Not changed	Increased	Direct stimulation of adrenal cortex
↑Sympathetic nerve activity	Increased	Increased	Increased	α-adrenergic effect stimulates constriction of afferent arterioles; β-adrenergic effect stimulates renin secretion directly.

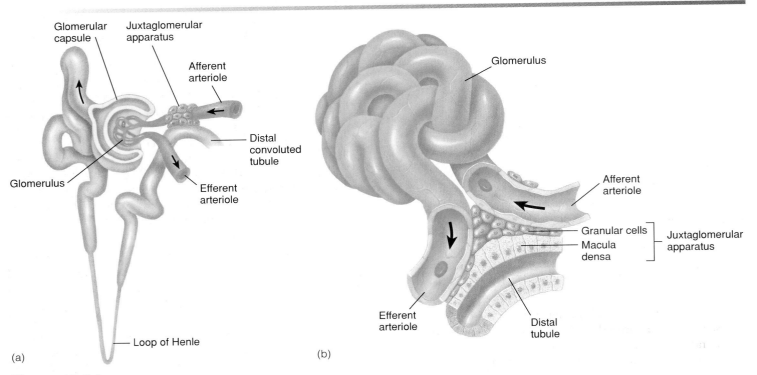

(a)

(b)

Figure 17.24

The juxtaglomerular apparatus (*a*) includes the region of contact of the afferent arteriole with the distal tubule. The afferent arterioles in this region contain granular cells that secrete renin, and the distal tubule cells in contact with the granular cells form an area called the macula densa (*b*).

Natriuretic Hormone

Expansion of the blood volume causes increased salt and water excretion in the urine. This is due in part to an inhibition of aldosterone secretion, as previously described. There is much experimental evidence, however, that the increased salt excretion that occurs under these conditions is due not only to the inhibition of aldosterone secretion, but also to the increased secretion of another substance with hormone properties. This other substance is called **natriuretic hormone,** and is so named because it stimulates salt excretion (the opposite of aldosterone's action). The source and chemical nature of natriuretic hormone remained elusive for many years, but recent evidence has shown that the atria of the heart produce a polypeptide that appears to fit the description of the natriuretic hormone proposed by renal physiologists. This polypeptide is called *atrial natriuretic factor.*

Relationship between Na$^+$, K$^+$, and H$^+$

The reabsorption of Na$^+$ in the distal convoluted tubules is accompanied by K$^+$ secretion. This occurs because the aldosterone-stimulated reabsorption of Na$^+$ creates a large potential difference between the two sides of the tubular wall, with the lumen side

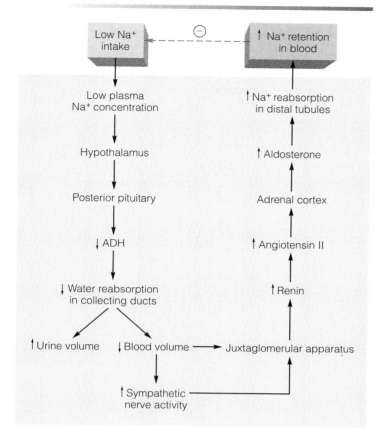

Figure 17.25

The sequence of events by which a low sodium (salt) intake leads to increased sodium reabsorption by the kidneys. The dashed arrow and negative sign show the completion of the negative feedback loop.

being very negative (−50 mV) compared to the basolateral side. The secretion of K⁺ into the tubular fluid is driven by this electrical gradient. Because of the Na⁺/K⁺ exchange in the distal tubule, an increase in Na⁺ reabsorption in the distal tubule results in an increase in K⁺ secretion.

Some diuretic drugs inhibit Na⁺ reabsorption in the loop of Henle and therefore increase the delivery of Na⁺ to the distal tubule. As a result, there is an increased reabsorption of Na⁺ and secretion of K⁺ in the distal convoluted tubule when a person takes these types of diuretics. People who take these diuretics, therefore, tend to have excessive K⁺ loss in the urine. The actions of different diuretics and their side effects on blood K⁺ are discussed in the last section of this chapter.

The plasma K⁺ concentration indirectly affects the plasma H⁺ concentration (pH). Changes in plasma pH likewise affect the K⁺ concentration of the blood. These effects serve to stabilize the ratio of K⁺ to H⁺. When the extracellular H⁺ concentration increases, for example, some of the H⁺ moves into the tissue cells and causes cellular K⁺ to diffuse outward into the extracellular fluid. The plasma concentration of H⁺ is thus decreased while the K⁺ increases, helping to reestablish the proper ratio of these ions in the extracellular fluid. A similar effect occurs in the cells of the distal region of the nephron.

Complications may arise from the use of diuretics as a result of the K⁺ loss that occurs. If K⁺ secretion into the distal convoluted tubules is significantly increased, a condition of **hypokalemia** (low blood K⁺) may be produced, which must be compensated for by the increased ingestion of potassium. People who take diuretics for the treatment of high blood pressure are usually on a low-sodium diet and often must supplement their meals with potassium chloride (KCl).

In the cells of the late distal tubule and cortical collecting duct, positively charged ions (K⁺ and H⁺) are secreted in response to the negative polarity produced by reabsorption of Na⁺ (fig. 17.26). When a person has severe acidosis, there is an increased amount of H⁺ secretion at the expense of a decrease in the amount of K⁺ secreted. Acidosis may thus be accompanied by a rise in blood K⁺. If, on the other hand, hyperkalemia is the primary problem, there is an increased secretion of K⁺ and thus a decreased secretion of H⁺. Hyperkalemia can thus cause an increase in the blood concentration of H⁺ and acidosis.

If a person is suffering from potassium deprivation, according to recent evidence, the collecting duct may be able to partially compensate by reabsorbing some K⁺. This occurs in the outer medulla, and results in the reabsorption of some of the K⁺ that was secreted into the cortical collecting duct.

Aldosterone stimulates the secretion of H⁺ as well as K⁺ into the distal tubules. Abnormally high aldosterone secretion, in **primary aldosteronism**, or **Conn's syndrome**, therefore, results in both hypokalemia and metabolic alkalosis. Conversely, abnormally low aldosterone secretion, as occurs in **Addison's disease**, can produce hyperkalemia, which is accompanied by metabolic acidosis.

Renal Acid-Base Regulation

The kidneys help regulate the blood pH by excreting H⁺ in the urine and by reabsorbing bicarbonate. The H⁺ enter the filtrate in two ways: by filtration through the glomeruli and by secretion into the tubules. Most of the H⁺ secretion occurs across the wall of the proximal tubule in exchange for the reabsorption of Na⁺; a similar exchange also occurs in the late distal tubule and cortical collecting duct, as previously described. Since the kidneys normally reabsorb almost all of the filtered bicarbonate and excrete H⁺, normal urine contains little bicarbonate and is slightly acidic (with a pH range of between 5 and 7). The mechanisms involved in acidification of the urine and reabsorption of bicarbonate are summarized in figure 17.27.

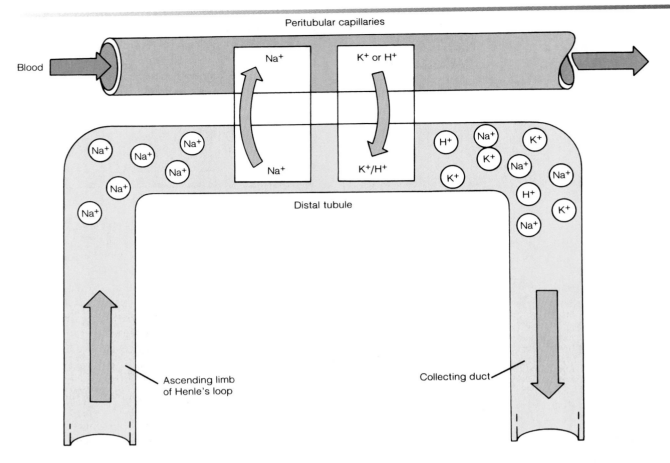

Peritubular capillaries

Blood

Na⁺

K⁺ or H⁺

Na⁺

K⁺/H⁺

Distal tubule

H⁺ Na⁺ K⁺ K⁺ Na⁺ Na⁺ K⁺ H⁺ Na⁺ K⁺

Ascending limb
of Henle's loop

Collecting duct

Figure 17.26

In the distal tubule, K^+ and H^+ are secreted in exchange for Na^+. High concentrations of H^+ may therefore decrease K^+ secretion, and vice versa.

Reabsorption of Bicarbonate in the Proximal Tubule

The apical membranes of the tubule cells (facing the lumen) are impermeable to bicarbonate. The reabsorption of bicarbonate must therefore occur indirectly. When the urine is acidic, HCO_3^- combines with H^+ to form carbonic acid. Carbonic acid in the filtrate is then converted to CO_2 and H_2O by the action of **carbonic anhydrase.** This enzyme is located in the apical cell membrane of the proximal tubule in contact with the filtrate. Notice that the reaction that occurs in the filtrate is the same one that occurs within the red blood cells in pulmonary capillaries (discussed in chapter 16).

The tubule cell cytoplasm also contains carbonic anhydrase. Under the conditions of high CO_2 that prevail within the cytoplasm, carbonic anhydrase catalyzes the reverse reaction (similar to that which occurs within red blood cells in tissue capillaries). The CO_2 that enters the tubule cells is converted to carbonic acid, which in turn dissociates to HCO_3^- and H^+ within the tubule cells. The bicarbonate within the tubule cell can then diffuse through the basolateral membrane and enter the blood (fig. 17.28). When conditions are normal, the same amount of HCO_3^- passes into the blood as was removed from the filtrate. The H^+, which was produced at the same time as HCO_3^- in the cytoplasm of the tubule cell, can either pass back into the filtrate or pass into the blood. Under acidotic conditions, almost all of the H^+ goes back into the filtrate and is used to help reabsorb all of the filtered bicarbonate.

During alkalosis, less H^+ is secreted into the filtrate. Since the reabsorption of filtered bicarbonate requires the combination of HCO_3^- with H^+ to form carbonic acid, less bicarbonate is reabsorbed. This results in urinary excretion of bicarbonate, which helps to partially compensate for the alkalosis.

When people go to the high elevations of the mountains they hyperventilate, as discussed in chapter 16. This lowers the arterial P_{CO_2} and produces a respiratory alkalosis. The kidneys participate in this acclimatization by excreting a larger amount of bicarbonate, which helps to partially compensate for the alkalosis and bring the pH back down toward normal. It is interesting in this regard that the drug *acetazolamide*, which inhibits renal carbonic anhydrase, is often used to treat acute mountain sickness (AMS; see chapter 16). The inhibition of renal carbonic anhydrase causes the loss of bicarbonate and water in the urine, producing a metabolic acidosis and diuresis that apparently help to alleviate the symptoms of AMS.

Chapter Seventeen

Table 17.7 Categories of Disturbances in Acid-Base Balance, Including Those That Involve Both Respiratory and Metabolic Components

Pco₂ (mmHg)	Bicarbonate (mEq/L)*		
	Less than 21	*21–26*	*More than 26*
More than 45	Combined metabolic and respiratory acidosis	Respiratory acidosis	Metabolic alkalosis and respiratory acidosis
35–45	Metabolic acidosis	Normal	Metabolic alkalosis
Less than 35	Metabolic acidosis and respiratory alkalosis	Respiratory alkalosis	Combined metabolic and respiratory alkalosis

*mEq/L = milliequivalents per liter. This is the millimolar concentration of HCO₃⁻ multiplied by its valence (×1).

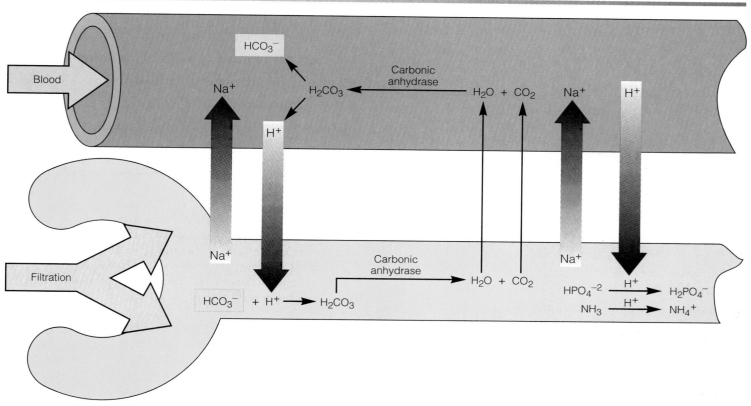

Figure 17.27

A diagram summarizing how the urine becomes acidified and how bicarbonate is reabsorbed from the filtrate.

In this way, disturbances in acid-base balance caused by respiratory problems can be partially compensated for by changes in plasma bicarbonate concentrations. Metabolic acidosis or alkalosis—in which changes in bicarbonate concentrations occur as the primary disturbance—similarly can be partially compensated for by changes in ventilation. These interactions of the respiratory and metabolic components of acid-base balance are shown in table 17.7.

Urinary Buffers

When a person has a blood pH of less than 7.35 (acidosis), the urine pH almost always falls below 5.5. The nephron, however, cannot produce a urine pH that is significantly less than 4.5. In order for more H^+ to be excreted, the acid must be buffered. Actually, most of the H^+ excreted even in normal urine is in a buffered form. Bicarbonate cannot serve this function because it is normally completely reabsorbed. Instead, the buffering action of phosphates (mainly HPO_4^{-2}) and ammonia (NH_3) provide the means for excreting most of the H^+ in the urine. Phosphate enters the urine by filtration. Ammonia (whose presence is strongly evident in a diaper pail or kitty litter box) is produced in the tubule cells by deamination of amino acids. These molecules buffer H^+ as described in the following equations:

$$NH_3 + H^+ \rightarrow NH_4^+ \text{ (ammonium ion)}$$
$$HPO_4^{-2} + H^+ \rightarrow H_2PO_4^-$$

Table 17.8 Actions of Different Classes of Diuretics

Category of Diuretic	Example	Mechanism of Action	Major Site of Action
Carbonic anhydrase inhibitors	Acetazolamide	Inhibits reabsorption of bicarbonate	Proximal tubule
Loop diuretics	Furosemide	Inhibits sodium transport	Thick segments of ascending limbs
Thiazides	Hydrochlorothiazide	Inhibits sodium transport	Last part of ascending limb and first part of distal tubule
Potassium-sparing diuretics	Spironolactone	Inhibits action of aldosterone	Last part of distal tubule and cortical collecting duct
	Triamterene	Inhibits Na^+/K^+ exchange	Last part of distal tubule and cortical collecting duct

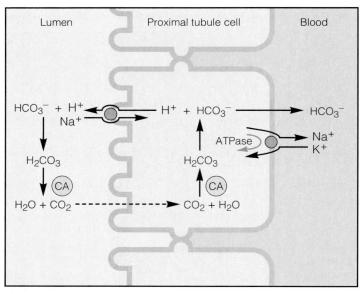

Figure 17.28

The mechanism of bicarbonate reabsorption. Through this mechanism, the cells of the proximal convoluted tubule can reabsorb bicarbonate while secreting H^+. (CA = carbonic anhydrase.)

1. Describe the effects of aldosterone on the renal nephrons and explain how aldosterone secretion is regulated.

2. Explain how changes in plasma Na^+ concentrations regulate renin secretion and how the secretion of renin helps to regulate the plasma Na^+ concentration.

3. Explain the mechanisms by which the distal tubule secretes K^+ and H^+. How might hyperkalemia affect the blood pH?

4. Explain how the kidneys reabsorb filtered bicarbonate and how this process is affected by acidosis and alkalosis.

5. Suppose a person with diabetes mellitus had an arterial pH of 7.30, an abnormally low arterial P_{CO_2}, and an abnormally low bicarbonate concentration. What would be the type of acid-base disturbance and how might these values have been produced?

Clinical Applications

Different types of diuretic drugs act on specific segments of the nephron tubule to indirectly inhibit the reabsorption of water and thus promote the lowering of blood volume. A knowledge of the mechanism of action of diuretics thus promotes better understanding of the physiology of the nephron. Clinical analysis of the urine, similarly, is meaningful only when the mechanisms that produce normal urine composition are understood.

The importance of renal function in maintaining homeostasis and the ease with which urine can be collected and used as a mirror of the plasma's chemical composition make the clinical study of renal function and urine composition particularly significant. Further, the ability of the kidneys to regulate blood volume is exploited clinically in the management of high blood pressure.

Use of Diuretics

People who need to lower their blood volume because of hypertension, congestive heart failure, or edema take medications that increase the volume of urine excreted. Such medications are called **diuretics.** The various diuretic drugs act on the renal nephron in different ways (table 17.8). Based on their chemical structure or aspects of their actions, commonly used diuretics are categorized as carbonic acid inhibitors, loop diuretics, thiazides, osmotic diuretics, or potassium-sparing diuretics.

The most powerful diuretics, which inhibit salt and water reabsorption by as much as 25%, are the drugs that act to inhibit active salt transport out of the ascending limb of the loop of Henle. Examples of these loop diuretics include *furosemide* and *ethacrynic acid*. The thiazide diuretics, such as *hydrochlorothiazide*, inhibit salt and water reabsorption by as much as 8% through inhibition of salt transport by the first segment of the distal convoluted tubule. The carbonic anhydrase inhibitors (*acetazolamide*) are much weaker diuretics and act primarily in the proximal tubule to prevent the water reabsorption that occurs when bicarbonate is reabsorbed.

When extra solutes are present in the filtrate, they increase the osmotic pressure of the filtrate and in this way decrease the osmotic reabsorption of water throughout the nephron. *Mannitol* is sometimes used clinically for this purpose. Osmotic diuresis can occur in diabetes mellitus due to the presence of glucose in the filtrate and urine; this extra solute causes the excretion of excessive amounts of water in the urine and can result in severe dehydration of a person with uncontrolled diabetes.

The previously mentioned diuretics can, as discussed in an earlier section, result in the excessive secretion of K^+ into the filtrate and its excessive elimination in the urine. For this reason, potassium-sparing diuretics are sometimes used. *Spironolactones* are aldosterone antagonists that compete with aldosterone for cytoplasmic receptor proteins in the cells of the late distal tubule. These drugs, therefore, block the aldosterone stimulation of Na^+ reabsorption and K^+ secretion. *Triamterene* is a different type of potassium-sparing diuretic that appears to act more directly on the Na^+/K^+ pumps in the distal tubule.

Renal Function Tests and Kidney Disease

Renal function can be tested by techniques that include the PAH clearance rate, which measures total blood flow to the kidneys, and the measurement of GFR by the inulin clearance rate. The plasma creatinine concentration, as previously described, also provides an index of renal function. These tests aid the diagnosis of kidney diseases such as glomerulonephritis and renal insufficiency.

Glomerulonephritis

Inflammation of the glomeruli, or glomerulonephritis, is currently believed to be an *autoimmune disease* that involves the person's own antibodies (as described in chapter 15). These antibodies may have been raised against the basement membrane of the glomerular capillaries, but more commonly, they appear to have been produced in response to streptococcus infections (such as strep throat). A variable number of glomeruli are destroyed in this condition, and the remaining glomeruli become more permeable to plasma proteins. Leakage of proteins into the urine results in decreased plasma colloid osmotic pressure and can therefore lead to edema.

Renal Insufficiency

When nephrons are destroyed—as in chronic glomerulonephritis, infection of the renal pelvis and nephrons (*pyelonephritis*), or loss of a kidney—or when kidney function is reduced by damage caused by diabetes mellitus, arteriosclerosis, or blockage by kidney stones, a condition of *renal insufficiency*

may develop. This can cause hypertension, due primarily to the retention of salt and water, and *uremia* (high plasma urea concentrations). The inability to excrete urea is accompanied by elevated plasma H^+ concentrations (acidosis) and elevated K^+ concentrations, which are more immediately dangerous than the high levels of urea. Uremic coma appears to result from these associated changes.

Patients with uremia or the potential for developing uremia are often placed on *dialysis* machines. The term *dialysis* refers to the separation of molecules on the basis of size by their ability to diffuse through an artificial semipermeable membrane. This principle is used in the "artificial kidney machine" for hemodialysis. Urea and other wastes in the patient's blood can easily pass through the membrane pores, whereas plasma proteins are left behind (just as occurs across glomerular capillaries). The plasma is thus cleansed of these wastes as they pass from the blood into the dialyzing fluid. Unlike the tubules, however, the dialysis membrane cannot reabsorb Na^+, K^+, glucose, and other needed molecules. These substances are prevented from diffusing through the membrane by including them in the dialysis fluid. Hemodialysis is commonly performed three times a week for several hours each session.

More recent techniques include the use of the patient's own peritoneal membranes (which line the abdominal cavity—see chapter 17) for dialysis. Dialysis fluid is introduced into the peritoneal cavity, and then, after a period of time, discarded after wastes have accumulated. This procedure, called *continuous ambulatory peritoneal dialysis* (*CAPD*), can be performed several times a day by the patients themselves on an outpatient basis.

The many dangers presented by renal insufficiency and the difficulties encountered in attempting to compensate for this condition are stark reminders of the importance of renal function in maintaining homeostasis. The ability of the kidneys to regulate blood volume and chemical composition in accordance with the body's changing needs requires great complexity of function. Homeostasis is maintained in large part by coordination of renal functions with those of the cardiovascular and pulmonary systems, as described in the preceding chapters.

---

1. List the different categories of clinical diuretics and explain how each exerts its diuretic effect.

2. Explain why most diuretics can cause excessive loss of K^+. How is this prevented by the potassium-sparing diuretics?

3. Define *uremia* and discuss the dangers associated with this condition. Explain how uremia can be corrected through the use of renal dialysis.

Summary

Structure and Function of the Kidneys   p. 506

I. The kidney is divided into an outer cortex and inner medulla.
   A. The medulla is composed of renal pyramids, separated by renal columns.
   B. The renal pyramids empty urine into the calyces that drain into the renal pelvis, and from there to the ureter.

II. Each kidney contains more than a million microscopic functional units called nephrons; nephrons consist of vascular and tubular components.
   A. Filtration occurs in the glomerulus, which receives blood from an afferent arteriole.
   B. Glomerular blood is drained by an efferent arteriole, which delivers blood to peritubular capillaries that surround the nephron tubules.
   C. The glomerular (Bowman's) capsule and the proximal and distal convoluted tubules are located in the cortex.
   D. The loop of Henle is located in the medulla.
   E. Filtrate from the distal convoluted tubule is drained into collecting ducts, which plunge through the medulla to empty urine into the calyces.

Glomerular Filtration   p. 511

I. A filtrate derived from plasma in the glomerulus must pass through a basement membrane of the glomerular capillaries and through slits in the processes of the podocytes—the cells that compose the inner layer of the glomerular (Bowman's) capsule.
   A. The glomerular ultrafiltrate is formed under the force of blood pressure and has a low protein concentration.
   B. The glomerular filtration rate (GFR) is the volume of filtrate produced per minute by both kidneys; it ranges from 115 to 125 ml/min.

II. The GFR can be regulated by constriction or dilation of the afferent arterioles.
   A. Sympathetic innervation causes constriction of the afferent arterioles.
   B. Intrinsic mechanisms help to autoregulate the rate of renal blood flow and the GFR.

Reabsorption of Salt and Water   p. 514

I. Approximately 65% of the filtered salt and water is reabsorbed across the proximal convoluted tubules.
   A. Sodium is actively transported, chloride follows passively by electrical attraction, and water follows the salt out of the proximal tubule.
   B. Salt transport in the proximal tubules is not under hormonal regulation.

II. The reabsorption of most of the remaining water occurs as a result of the action of the countercurrent multiplier system.
   A. Sodium is actively extruded from the ascending limb, followed passively by chloride.
   B. Since the ascending limb is impermeable to water, the remaining filtrate becomes hypotonic.
   C. Because of this salt transport and because of countercurrent exchange in the vasa recta, the tissue fluid of the medulla becomes hypertonic.
   D. The hypertonicity of the medulla is multiplied by a positive feedback mechanism involving the descending limb, which is passively permeable to water and perhaps to salt.

III. The collecting duct is permeable to water but not to salt.
   A. As the collecting ducts pass through the hypertonic renal medulla, water leaves by osmosis and is carried away in surrounding capillaries.

B. The permeability of the collecting ducts to water is stimulated by antidiuretic hormone (ADH).

Renal Plasma Clearance   p. 521

I. Inulin is filtered but neither reabsorbed nor secreted; its clearance is thus equal to the glomerular filtration rate.

II. Some of the filtered urea is reabsorbed; its clearance is therefore less than the glomerular filtration rate.

III. Since almost all the PAH in blood going through the kidneys is cleared by filtration and secretion, the PAH clearance is a measure of the total renal blood flow.

IV. Normally all of the filtered glucose is reabsorbed; glycosuria occurs when the transport carriers for glucose become saturated due to hyperglycemia.

Renal Control of Electrolyte and Acid-Base Balance   p. 526

I. Aldosterone stimulates sodium reabsorption and potassium secretion in the distal convoluted tubule.

II. Aldosterone secretion is stimulated directly by a rise in blood potassium and indirectly by a fall in blood sodium.
   A. Decreased blood flow through the kidneys stimulates the secretion of the enzyme renin from the juxtaglomerular apparatus.
   B. Renin catalyzes the formation of angiotensin I, which is then converted to angiotensin II.
   C. Angiotensin II stimulates the adrenal cortex to secrete aldosterone.

III. Aldosterone stimulates the secretion of H^+ as well as potassium into the filtrate in exchange for sodium.

IV. The nephrons filter bicarbonate and reabsorb the amount required to maintain acid-base balance; reabsorption of bicarbonate, however, is indirect.

   A. Filtered bicarbonate combines with H$^+$ to form carbonic acid in the filtrate.

   B. Carbonic anhydrase in the membranes of microvilli in the tubules catalyzes the conversion of carbonic acid to carbon dioxide and water.

   C. Carbon dioxide is reabsorbed and converted in either the tubule cells or the red blood cells to carbonic acid, which dissociates to bicarbonate and H$^+$.

   D. In addition to reabsorbing bicarbonate, the nephrons filter and secrete H$^+$, which is excreted in the urine buffered by ammonium and phosphate buffers.

Clinical Applications *p. 532*

I. Diuretic drugs are used clinically to increase the urine volume and thus to lower the blood volume and pressure.

   A. Loop diuretics and the thiazides inhibit active Na$^+$ transport in the ascending limb and early portion of the distal tubule, respectively.

   B. Osmotic diuretics are extra solutes in the filtrate that increase the osmotic pressure of the filtrate and inhibit the osmotic reabsorption of water.

   C. The potassium-sparing diuretics act on the distal tubule to inhibit the reabsorption of Na$^+$ and secretion of K$^+$.

II. In glomerulonephritis, the glomeruli can permit the leakage of plasma proteins into the urine.

III. The technique of renal dialysis is used to treat people with renal insufficiency.

Clinical Investigation

A teenage boy visits his family physician complaining of pain in his back between the twelfth rib and the lumbar vertebrae. The boy's urine is noticeably discolored, and he is referred to a urologist. Urinalysis reveals that he has hematuria; his urine sediment contains casts with associated red blood cells. Only trace amounts of protein are detected in the urine. Further analysis shows mild oliguria and an elevated plasma creatinine concentration.

Edema is present. The boy stated that he had had a sore throat for about a month, but that he had continued to compete on his cross-country running team. A throat culture demonstrates the presence of a streptococcus infection. The boy is placed on antibiotics and given hydrochloro-thiazide. Within a few weeks, his symptoms are gone. What was responsible for the boy's symptoms, and why did they disappear with this treatment?

Clues

Examine the section "Renal Clearance of Inulin: Measurement of GFR" and the boxed information on creatinine. Look up hydrochlorothiazide and glomerulonephritis in the "Clinical Applications" section.

Review Activities

Objective Questions

1. Which of the following statements about the renal pyramids is *false?*
   a. They are located in the medulla.
   b. They contain glomeruli.
   c. They contain collecting ducts.
   d. They empty urine into the calyces.

Match the following:

2. active transport of sodium; water follows passively
3. active transport of sodium; impermeable to water
4. passively permeable to water and possibly salt
5. passively permeable to water only

   a. proximal tubule
   b. descending limb
   c. ascending limb
   d. distal tubule
   e. collecting duct

6. Antidiuretic hormone promotes the retention of water by stimulating
   a. the active transport of water.
   b. the active transport of chloride.
   c. the active transport of sodium.
   d. the permeability of the collecting duct to water.

7. Aldosterone stimulates sodium reabsorption and potassium secretion in
   a. the proximal convoluted tubule.
   b. the descending limb of the loop.
   c. the ascending limb of the loop.
   d. the distal convoluted tubule.
   e. the collecting duct.

8. Substance X has a clearance greater than zero but less than that of inulin. What can you conclude about substance X?
   a. It is not filtered.
   b. It is filtered, but neither reabsorbed nor secreted.
   c. It is filtered and partially reabsorbed.
   d. It is filtered and secreted.
9. Substance Y has a clearance greater than that of inulin. What can you conclude about substance Y?
   a. It is not filtered.
   b. It is filtered, but neither reabsorbed nor secreted.
   c. It is filtered and partially reabsorbed.
   d. It is filtered and secreted.
10. About 65% of the glomerular ultrafiltrate is reabsorbed in
    a. the proximal tubule.
    b. the distal tubule.
    c. the loop of Henle.
    d. the collecting duct.

11. Diuretic drugs that act in the loop of Henle
    a. inhibit active sodium transport.
    b. cause an increased flow of filtrate to the distal convoluted tubule.
    c. cause an increased secretion of potassium into the tubule.
    d. promote the excretion of salt and water.
    e. do all of the above.
12. The appearance of glucose in the urine
    a. occurs normally.
    b. indicates the presence of kidney disease.
    c. occurs only when the transport carriers for glucose become saturated.
    d. is a result of hypoglycemia.
13. Reabsorption of water through the tubules occurs by
    a. osmosis.
    b. active transport.
    c. facilitated diffusion.
    d. all of the above.
14. Which of the following factors oppose(s) filtration from the glomerulus?
    a. plasma oncotic pressure
    b. hydrostatic pressure in glomerular (Bowman's) capsule
    c. plasma hydrostatic pressure
    d. both *a* and *b*
    e. both *b* and *c*

15. The countercurrent exchange in the vasa recta
    a. removes Na^+ from the extracellular fluid.
    b. maintains high concentrations of NaCl in the extracellular fluid.
    c. raises the concentration of Na^+ in the blood leaving the kidneys.
    d. causes large quantities of Na^+ to enter the filtrate.
    e. does all of the above.
16. The kidneys help maintain acid-base balance by
    a. K^+/H^+ exchange in the distal regions of the nephron.
    b. the action of carbonic anhydrase within the apical cell membranes.
    c. the action of carbonic anhydrase within the cytoplasm of the tubule cells.
    d. the buffering action of phosphates and ammonia in the urine.
    e. all of the above means.

Essay Questions

1. Explain how glomerular ultrafiltrate is produced and why it has a low protein concentration.[1]
2. Explain how the countercurrent multiplier system works and discuss its functional significance.
3. Explain how countercurrent exchange occurs in the vasa recta and discuss its functional significance.
4. Explain how an increase in ADH secretion promotes increased water reabsorption and how water reabsorption decreases when ADH secretion falls.

5. Explain how the structure of the epithelial wall of the proximal tubule and the distribution of Na^+/K^+ pumps in the epithelial cell membranes contribute to the ability of the proximal tubule to reabsorb salt and water.
6. Describe how the thiazide diuretics, loop diuretics, and osmotic diuretics work. How do these substances cause hypokalemia?
7. Which diuretic drugs do not produce hypokalemia? How do these drugs work?

8. What happens to urinary bicarbonate excretion when a person hyperventilates? How might this response be helpful?
9. Describe the location of the macula densa and explain its role in the regulation of renin secretion and in tubuloglomerular feedback.
10. Describe how the nephron handles K^+, how the urinary excretion of K^+ changes under different conditions, and how this process is regulated by aldosterone.

---

[1]*Note:* This question is answered on page 217 of the Student Study Guide.

Chapter Seventeen

Selected Readings

Alexander, E. 1986. Metabolic acidosis: Recognition and etiologic diagnosis. *Hospital Practice* 21:100E.

Anderson, B. 1977. Regulation of body fluids. *Annual Review of Physiology* 39:185.

Bauman, J. W., and F. P. Chinard. 1975. *Renal Function: Physiological and Medical Aspects.* St. Louis: Mosby.

Beeuwkes, R., III. 1980. The vascular organization of the kidney. *Annual Review of Physiology* 42:531.

Brenner, B. M., and R. Beeuwkes III. 1978. The renal circulation. *Hospital Practice* 13:35.

Brenner, B. M., T. H. Hostetter, and H. D. Humes. 1978. Molecular basis of proteinuria of glomerular origin. *New England Journal of Medicine* 298:826.

Breyer, M. D., and Y. Ando. 1994. Hormonal signaling and regulation of salt and water transport in the collecting duct. *Annual Review of Physiology* 56:711.

Buckalew, V. M., Jr., and K. A. Gruber. 1984. Natriuretic hormone. *Annual Review of Physiology* 46:343.

Coe, F. L. et al. 1992. The pathogenesis and treatment of kidney stones. *New England Journal of Medicine* 327:1141.

Culpepper, R. M. 1989. Na^+-K^+-$2Cl^-$ cotransport in the thick ascending limb of the loop of Henle. *Hospital Practice* 24:217.

deBold, A. 1985. Atrial natriuretic factor: A hormone produced by the heart. *Science* 230:767.

Dempster, J. A. et al. 1992. The quest for water channels. *News in Physiological Sciences* 7:172.

Epstein, F. H., and R. S. Brown. 1988. Acute renal failure: A collection of paradoxes. *Hospital Practice* 23:171.

Galla, J. H., and R. G. Luke. 1987. Pathophysiology of metabolic alkalosis. *Hospital Practice* 22:123.

Giebisch, G. H., and B. Stanton. 1979. Potassium transport in the nephron. *Annual Review of Physiology* 41:241.

Glassock, R. J. 1987. Pathophysiology of acute glomerulonephritis. *Hospital Practice* 22:163.

Grabow, P. A. 1993. Autosomal dominant polycystic kidney disease. *New England Journal of Medicine* 329:332.

Grantham, J. J. 1992. Polycystic kidney disease: I. Etiology and pathogenesis. *Hospital Practice* 27:51.

Gluck, S. L. 1989. Cellular and molecular aspects of renal H^+ transport. *Hospital Practice* 24:149.

Hays, R. M. 1978. Principles of ion and water transport in the kidneys. *Hospital Practice* 13:79.

Hollenberg, N. K. 1986. The kidney in heart failure. *Hospital Practice* 21:81.

Holtzman, /e., and D. A. Ausiello. 1994. Nephrogenic diabetes insipidus: Causes revealed. *Hospital Practice* 29:89.

Hostetter, T. H. 1995. Progression of renal disease and renal hypertrophy. *Annual Review of Physiology* 57:263.

Ito, S. 1994. Role of nitric oxide in glomerular arterioles and macula densa. *News in Physiological Sciences* 9:115.

Jacobson, H. R. 1987. Diuretics: Mechanisms of action and uses. *Hospital Practice* 22:129.

Kokko, J. S. 1979. Renal concentrating and diluting mechanisms. *Hospital Practice* 14:110.

Kurtzman, N. A. 1987. Renal tubular acidosis: A constellation of syndromes. *Hospital Practice* 22:173.

Morsing, P. 1993. A role for tubuloglomerular feedback in chronic partial uereteral obstruction. *News in Physiological Sciences* 8:74.

Peart, W. S. 1975. Renin-angiotensin system. *New England Journal of Medicine* 292:302.

Rector, F. C., Jr., and M. G. Cogan. 1980. The renal acidoses. *Hospital Practice* 15:99.

Reid, I. A., B. J. Morris, and W. F. Ganong. 1978. The renin-angiotensin system. *Annual Review of Physiology* 40:377.

Renkin, E. M., and R. R. Robinson. 1974. Glomerular filtration. *New England Journal of Medicine* 290:79.

Sabolic, I., and D. Brown. 1995. Water channels in renal and nonrenal tissues. *News in Physiological Sciences* 10:12.

Steinmetz, P. R., and B. M. Koeppen. 1984. Cellular mechanisms of diuretic action along the nephron. *Hospital Practice* 19:125.

Tanner, R. L. 1980. Control of acid excretion by the kidney. *Annual Review of Medicine* 31:35.

Their, S. O. 1987. Diuretic mechanisms as a guide to therapy. *Hospital Practice* 22:81.

Thompson, C. 1988. The spectrum of renal cystic diseases. *Hospital Practice* 23:165.

Vander, A. J. 1980. *Renal Physiology.* 2d ed. New York: McGraw-Hill.

Verkman, A. 1992. Water channels in cell membranes. *Annual Review of Physiology* 54:97.

Walker, L. A., and H. Valtin. 1982. Biological importance of nephron heterogeneity. *Annual Review of Physiology* 44:203.

Warnock, D. G., and F. C. Rector Jr. 1979. Proton secretion by the kidney. *Annual Review of Physiology* 41:197.

Weinman, E. J., and S. Shenolikar. 1993. Regulation of the renal brush border membrane Na^+–H^+ exchanger. *Annual Review of Physiology* 55:289.

Wingo, C. S., and B. D. Cain. 1993. The renal H-K-ATPase: Physiological significance and role in potassium homeostasis. *Annual Review of Physiology* 55:323.

18 The Digestive System

OBJECTIVES

After studying this chapter, you should be able to . . .

1. describe the functions of the digestive system and list its structures and regions.

2. describe the layers of the gastrointestinal tract and state the function(s) of each.

3. describe the mechanism of peristalsis and explain the significance of the lower esophageal sphincter.

4. describe the structure of the gastric mucosa, list the secretions of the mucosa and their functions, and identify the cells that produce each of these secretions.

5. describe the roles of HCl and pepsin in digestion and explain why the stomach does not normally digest itself.

6. describe the structure and function of the villi, microvilli, and crypts of Lieberkühn in the small intestine.

7. describe the location and functions of the brush border enzymes of the intestine.

8. explain the electrical activity that occurs in the intestine and describe the nature of peristalsis and segmentation.

9. explain how the large intestine absorbs fluid and electrolytes.

10. describe the flow of blood in the liver and explain how the liver functions to modify the chemical composition of the blood.

11. describe the composition and functions of bile and explain how bile is kept separate from blood in the liver.

12. trace the pathway of the formation, conjugation, and excretion of bilirubin and explain how jaundice may be produced.

13. explain the significance of the enterohepatic circulation of various compounds and describe the enterohepatic circulation of bile pigment.

14. identify the endocrine and exocrine structures of the pancreas and describe the composition and functions of pancreatic juice.

15. explain how gastric secretion is regulated during the cephalic, gastric, and intestinal phases.

16. identify the neurons in the wall of the intestine and explain their functions.

17. explain how pancreatic juice and bile secretion is regulated by nerves and hormones.

18. discuss the trophic effects of gastrointestinal hormones on the digestive tract.

19. describe the enzymes involved in the digestion of carbohydrates, lipids, and proteins and explain how monosaccharides and amino acids are absorbed.

20. describe the roles of bile and pancreatic lipase in fat digestion and trace the pathways and structures involved in the absorption of lipids.

OUTLINE

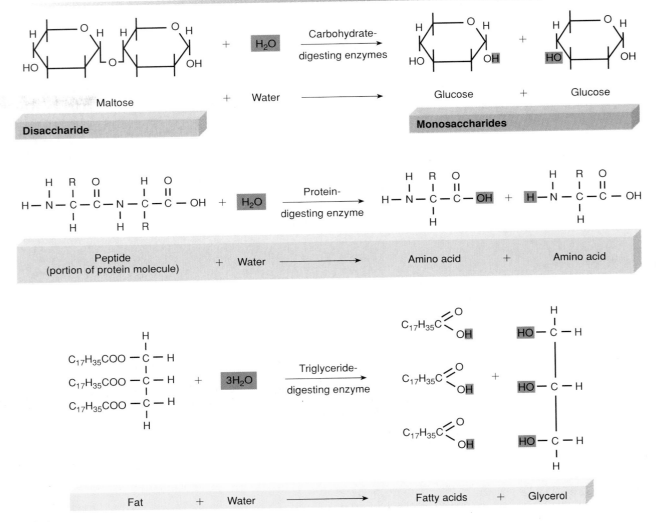

Figure 18.1

The digestion of food molecules occurs by means of hydrolysis reactions.

Introduction to the Digestive System

Within the lumen of the gastrointestinal tract, large food molecules are hydrolyzed into their monomers (subunits). These monomers pass through the inner layer, or mucosa, of the small intestine to enter the blood or lymph in a process called absorption. Digestion and absorption are aided by specializations of the mucosa and by characteristic movements caused by contractions of the muscle layers of the gastrointestinal tract.

Unlike plants, which can form organic molecules using inorganic compounds such as carbon dioxide, water, and ammonia, humans and other animals must obtain their basic organic molecules from food. Some of the ingested food molecules are needed for their energy (caloric) value—obtained by the reactions of cell respiration and used in the production of ATP—and the balance is used to make additional tissue.

Most of the organic molecules that are ingested are similar to the molecules that form human tissues. These are generally large molecules (*polymers*), which are composed of subunits (*monomers*). Within the gastrointestinal tract, the **digestion** of these large molecules into their monomers occurs by means of *hydrolysis reactions* (reviewed in fig. 18.1). The monomers thus formed are transported across the wall of the small intestine into the blood and lymph in the process of **absorption.** Digestion and absorption are the primary functions of the digestive system.

Since the composition of food is similar to the composition of body tissues, enzymes that digest food are also capable of digesting a person's own tissues. This does not normally occur, however, because a variety of protective devices inactivate digestive enzymes in the body and keep them away from the cytoplasm of tissue cells. The fully active digestive enzymes are normally confined to the lumen (cavity) of the gastrointestinal tract.

The lumen of the gastrointestinal tract is continuous with the environment because it is open at both ends (mouth and anus). Indigestible materials, such as cellulose from plant walls, pass from one end to the other without crossing the epithelial lining of the digestive tract (that is, without being absorbed). In this sense, these indigestible materials never enter the body, and the harsh conditions required for digestion thus occur *outside* the body.

In *planaria* (a type of flatworm), the gastrointestinal tract has only one opening—the mouth is also the anus. Each cell that lines the gastrointestinal tract is thus exposed to food, absorbable digestion products, and waste products. The two open ends of the digestive tract of higher organisms, in contrast, allow one-way transport, which is ensured by wavelike muscle contractions and by the action of sphincter muscles. This one-way transport allows different regions of the gastrointestinal tract to be specialized for different functions, as a "dis-assembly line." These functions of the digestive system include the following:

1. **Motility.** This refers to the movement of food through the digestive tract through the processes of:
   a. *Ingestion:* Taking food into the mouth.
   b. *Mastication:* Chewing the food and mixing it with saliva.
   c. *Deglutition:* Swallowing food.
   d. *Peristalsis:* Rhythmic, wavelike contractions that move food through the gastrointestinal tract.
2. **Secretion.** This includes both exocrine and endocrine secretions.
   a. *Exocrine secretions:* Water, hydrochloric acid, bicarbonate, and many enzymes are secreted into the lumen of the gastrointestinal tract. The stomach alone, for example, secretes 2 to 3 liters of gastric juice a day.
   b. *Endocrine secretions:* The stomach and small intestine secrete a number of hormones that help to regulate the digestive system.
3. **Digestion.** This refers to the breakdown of food molecules into their smaller subunits, which can be absorbed.
4. **Absorption.** This refers to the passage of food molecules after their digestion into the blood or lymph.

Anatomically and functionally, the digestive system can be divided into the tubular *gastrointestinal (GI) tract,* or *alimentary canal,* and *accessory organs.* The GI tract is approximately 9 m (30 ft) long and extends from the mouth to the anus. It traverses the thoracic cavity and enters the abdominal cavity at the level of the diaphragm. The anus is located at the inferior portion of the pelvic cavity. The organs of the GI tract include

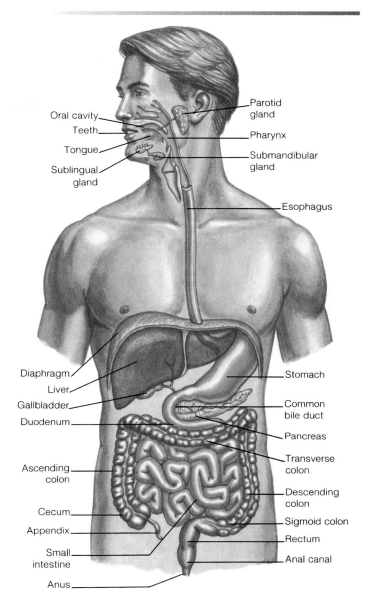

Figure 18.2

The digestive system, including the gastrointestinal tract and the accessory digestive organs.

the *oral (buccal) cavity, pharynx, esophagus, stomach, small intestine,* and *large intestine* (fig. 18.2). The accessory digestive organs include the *teeth, tongue, salivary glands, liver, gallbladder,* and *pancreas.* The term *viscera* is frequently used to refer to the abdominal organs of digestion, but it also can be used in reference to any of the organs in the thoracic and abdominal cavities.

Layers of the Gastrointestinal Tract

The GI tract from the esophagus to the anal canal is composed of four layers, or *tunics.* Each tunic contains a dominant tissue type that performs specific functions in the digestive process. The four tunics of the GI tract, from the inside out, are the **mucosa, submucosa, muscularis,** and **serosa** (fig. 18.3a).

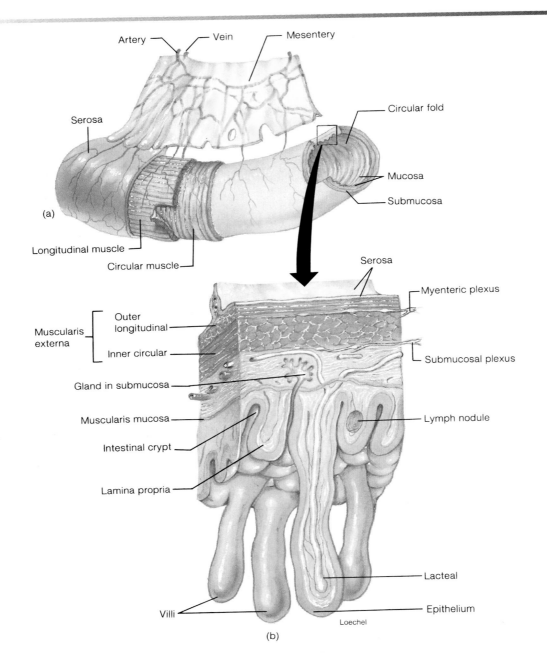

Figure 18.3

(a) An illustration of the major tunics, or layers, of the intestine. The insert shows how folds of mucosa form projections called villi in the small intestine.
(b) An illustration of a cross section of the small intestine showing layers and glands.

Mucosa

The mucosa surrounds the lumen of the GI tract and is the absorptive and major secretory layer. It consists of a simple columnar epithelium supported by the *lamina propria*, which is a thin layer of connective tissue. The lamina propria contains numerous lymph nodules that are important in protecting against disease (fig. 18.3b). External to the lamina propria is a thin layer of smooth muscle called the *muscularis mucosa*. This is the muscle layer responsible for the numerous small folds in certain portions of the GI tract. These folds greatly increase the absorptive surface area. Specialized goblet cells in the mucosa secrete mucus throughout most of the GI tract.

Submucosa

The relatively thick submucosa is a highly vascular layer of connective tissue serving the mucosa. Absorbed molecules that pass through the columnar epithelial cells of the mucosa enter into blood vessels of the submucosa. In addition to blood vessels, the

Chapter Eighteen

submucosa contains glands and nerve plexuses. The *submucosal plexus (Meissner's plexus)* (fig. 18.3*b*) provides an autonomic nerve supply to the muscularis mucosa.

Muscularis

The muscularis (also called the muscularis externa) is responsible for segmental contractions and peristaltic movement through the GI tract. The muscularis has an inner circular and an outer longitudinal layer of smooth muscle. Contractions of these layers move the food through the tract and physically pulverize and churn the food with digestive enzymes. The *myenteric plexus (Auerbach's plexus)* located between the two muscle layers provides the major nerve supply to the GI tract and includes fibers and ganglia from both the sympathetic and parasympathetic divisions of the autonomic nervous system.

Serosa

The outer serosa layer completes the wall of the GI tract. It is a binding and protective layer consisting of areolar connective tissue covered with a layer of simple squamous epithelium.

Innervation of the Gastrointestinal Tract

The GI tract is innervated by the sympathetic and parasympathetic divisions of the autonomic nervous system. As discussed in chapter 9, the parasympathetic division in general stimulates motility and secretions of the gastrointestinal tract. The vagus nerve is the source of parasympathetic activity in the esophagus, stomach, pancreas, gallbladder, small intestine, and upper portion of the large intestine. The lower portion of the large intestine receives parasympathetic innervation from spinal nerves in the sacral region. The submucosal plexus and myenteric plexus are the sites where parasympathetic preganglionic fibers synapse with postganglionic neurons that innervate the smooth muscle of the GI tract.

Postganglionic sympathetic fibers pass through the submucosal and myenteric plexuses and innervate the GI tract. The effects of sympathetic nerves reduce peristalsis and secretions and stimulate the contraction of sphincter muscles along the GI tract; therefore, they are antagonistic to the effects of parasympathetic nerve stimulation.

---

1. Define the terms *digestion* and *absorption,* describe how molecules are digested, and indicate which molecules are absorbed.
2. Describe the structure and function of the mucosa, submucosa, and muscularis layers of the gastrointestinal tract.
3. Describe the location and composition of the submucosal and myenteric plexuses and explain the actions of autonomic nerves on the gastrointestinal tract.

Esophagus and Stomach

Swallowed food is passed through the esophagus to the stomach by wavelike contractions known as peristalsis. The mucosa of the stomach secretes hydrochloric acid and pepsinogen. Upon entering the lumen of the stomach, pepsinogen is activated to become the protein-digesting enzyme known as pepsin. The stomach partially digests proteins and functions to store its contents, called chyme, for later processing by the small intestine.

Mastication of food mixes it with saliva, which contains *salivary amylase,* or *ptyalin,* an enzyme that can catalyze the partial digestion of starch (as discussed in a later section). Saliva also contains mucus and various antimicrobial agents. Deglutition (swallowing) begins as a voluntary activity in which the larynx is raised so that the epiglottis covers the entrance to the respiratory tract (chapter 16), preventing ingested material from entering. Involuntary muscular contractions and relaxations in the esophagus follow, as food is passed from the esophagus to the stomach. Once in the stomach, the ingested material is churned and mixed with hydrochloric acid and pepsin, a protein-digesting enzyme. The mixture thus produced is pushed by muscular contractions of the stomach past the pyloric sphincter (*pylorus* = gatekeeper), which guards the junction of the stomach and the duodenum of the small intestine.

Esophagus

The **esophagus** is that portion of the GI tract which connects the pharynx to the stomach. It is a collapsible muscular tube approximately 25 cm (10 in.) long, located posterior to the trachea within the mediastinum of the thorax. The esophagus passes through the diaphragm by means of an opening called the **esophageal hiatus** before terminating at the stomach. The esophagus is lined with a nonkeratinized stratified squamous epithelium; its walls contain either skeletal or smooth muscle, depending on the location. The upper third of the esophagus contains skeletal muscle, the middle third contains a mixture of skeletal and smooth muscle, and the terminal portion contains only smooth muscle.

Swallowed food is pushed from one end of the esophagus to the other by a wavelike muscular contraction called **peristalsis** (fig. 18.4). Peristaltic waves are produced by constriction of the lumen as a result of circular muscle contraction, followed by shortening of the tube by longitudinal muscle contraction. These contractions progress from the superior end of the esophagus to the *gastroesophageal junction* at a rate of 2 to 4 cm per second as they empty the contents of the esophagus into the cardiac region of the stomach.

The lumen of the terminal portion of the esophagus is slightly narrowed because of a thickening of the circular muscle

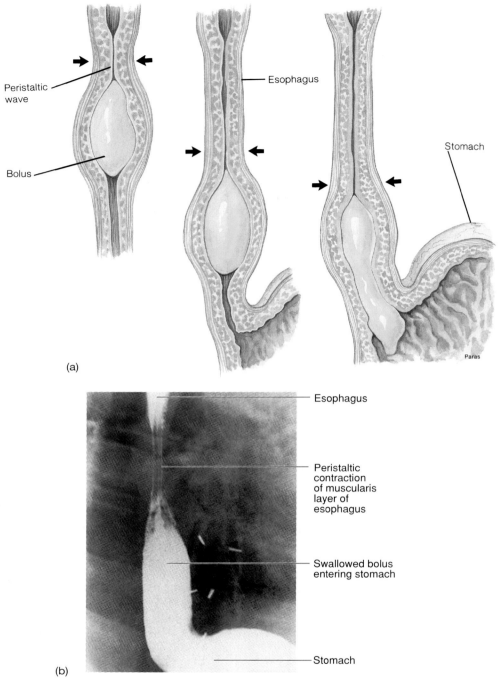

Peristaltic wave

Esophagus

Bolus

Stomach

Paras

(a)

Esophagus

Peristaltic contraction of muscularis layer of esophagus

Swallowed bolus entering stomach

Stomach

(b)

Figure 18.4

(a) A diagram and (b) a radiograph of peristalsis in the esophagus.

fibers in its wall. This portion is referred to as the **lower esophageal (gastroesophageal) sphincter.** The muscle fibers of this region constrict after food passes into the stomach to help prevent the stomach contents from regurgitating into the esophagus. Regurgitation would occur because the pressure in the abdominal cavity is greater than the pressure in the thoracic cavity as a result of respiratory movements. The lower esophageal sphincter must remain closed, therefore, until food is pushed through it by peristalsis into the stomach.

Stomach

The J-shaped stomach is the most distensible part of the GI tract. It is continuous with the esophagus superiorly and empties into the duodenum of the small intestine inferiorly. The functions of the stomach are to store food, to initiate the digestion of proteins, to kill bacteria with the strong acidity of gastric juice, and to move the food into the small intestine as a pasty material called **chyme.**

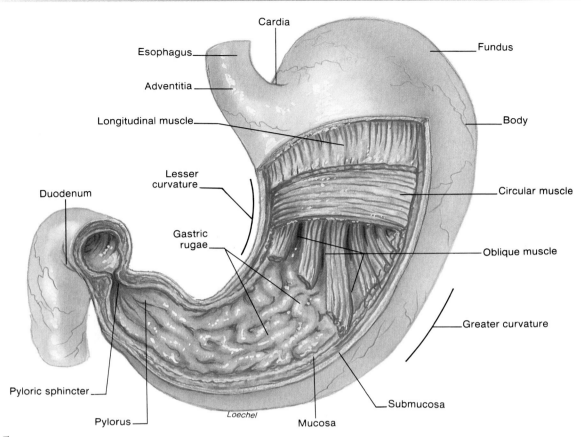

Cardia

Esophagus

Adventitia

Longitudinal muscle

Fundus

Body

Lesser
curvature

Duodenum

Gastric
rugae

Circular muscle

Oblique muscle

Greater curvature

Pyloric sphincter

Pylorus

Loechel

Submucosa

Mucosa

Figure 18.5

The primary regions and structures of the stomach.

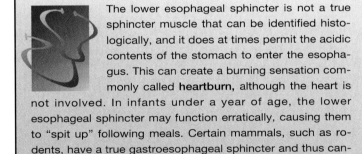

The lower esophageal sphincter is not a true sphincter muscle that can be identified histologically, and it does at times permit the acidic contents of the stomach to enter the esophagus. This can create a burning sensation commonly called **heartburn,** although the heart is not involved. In infants under a year of age, the lower esophageal sphincter may function erratically, causing them to "spit up" following meals. Certain mammals, such as rodents, have a true gastroesophageal sphincter and thus cannot regurgitate. This is why poison grains are effective in killing mice and rats.

The inner surface of the stomach is thrown into long folds called *rugae,* which can be seen with the unaided eye. Microscopic examination of the gastric mucosa shows that it is likewise folded. The openings of these folds into the stomach lumen are called **gastric pits.** The cells that line the folds deeper in the mucosa secrete various products into the stomach; these cells form the exocrine **gastric glands** (figs. 18.7 and 18.8).

Gastric glands have several types of cells that secrete different products: (1) **goblet cells,** which secrete mucus; (2) **parietal cells,** which secrete hydrochloric acid (HCl); (3) **chief** (or **zymogenic**) **cells,** which secrete pepsinogen, an inactive form of the protein-digesting enzyme pepsin; (4) **argentaffin cells,** which secrete serotonin and histamine; and (5) **G cells,** which secrete the hormone

Swallowed food is delivered from the esophagus to the *cardiac region* of the stomach (figs. 18.5 and 18.6). An imaginary horizontal line drawn through the cardiac region divides the stomach into an upper *fundus* and a lower *body,* which together compose about two-thirds of the stomach. The distal portion of the stomach is called the *pyloric region,* or *antrum.* Contractions of the stomach churn the chyme, mixing it more thoroughly with the gastric secretions. These contractions also push partially digested food from the antrum through the pyloric sphincter and into the first part of the small intestine.

The only stomach function that appears to be essential for life is the secretion of intrinsic factor. This polypeptide is needed for the absorption of vitamin B$_{12}$ in the terminal portion of the ileum in the small intestine, and vitamin B$_{12}$ is required for maturation of red blood cells in the bone marrow. A patient with a gastrectomy has to receive B$_{12}$ orally (together with intrinsic factor) or through injections to prevent the development of **pernicious anemia.**

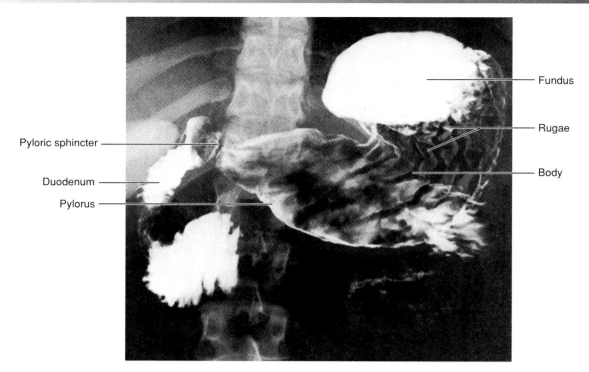

Figure 18.6

A radiograph of a stomach.

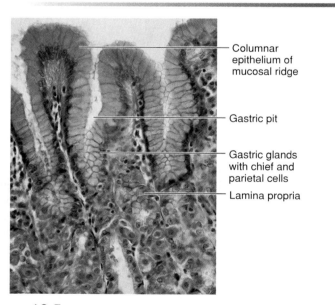

Figure 18.7

Microscopic structures of the mucosa of the stomach.

gastrin. In addition to these products (table 18.1), the gastric mucosa (probably the parietal cells) secretes a polypeptide called *intrinsic factor,* which is required for absorption of vitamin B_{12} in the intestine. These secretions, together with a large amount of water (2 to 3 L/day), form what is known as **gastric juice.**

Table 18.1	Secretions of the Fundus and Pyloric Regions of the Stomach	
Stomach Region	**Cell Type**	**Secretions**
Fundus	Parietal cells	Hydrochloric acid; intrinsic factor
	Chief cells	Pepsinogen
	Goblet cells	Mucus
	Argentaffin cells	Histamine, serotonin
Pyloric	G cells	Gastrin
	Chief cells	Pepsinogen
	Goblet cells	Mucus

Pepsin and Hydrochloric Acid Secretion

The secretion of hydrochloric acid by parietal cells makes gastric juice very acidic, with a pH less than 2. This strong acidity serves three functions: (1) ingested proteins are denatured at low pH—that is, their tertiary structure (chapter 2) is altered so that they become more digestible; (2) under acidic conditions, weak pepsinogen enzymes partially digest each other—this frees the active pepsin enzyme as small peptide fragments are removed (fig. 18.9); and (3) pepsin is more active under acidic conditions—it has a pH optimum (chapter 4) of about 2.0. The peptide bonds of ingested protein are broken (through hydrolysis reactions) by pepsin under acidic conditions; the HCl itself does not directly digest proteins.

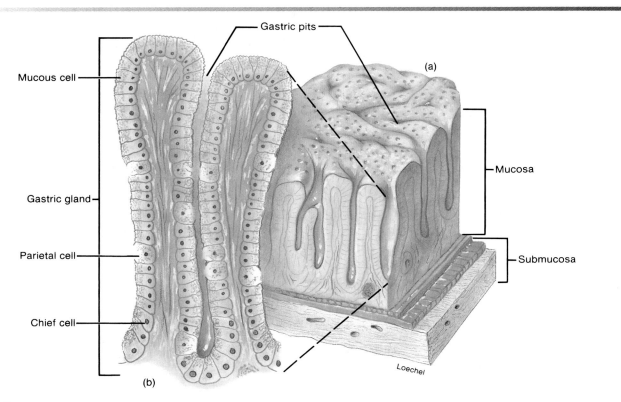

Mucous cell

Gastric pits

(a)

Mucosa

Gastric gland

Parietal cell

Submucosa

Chief cell

Loechel

(b)

Figure 18.8

Gastric pits and gastric glands of the mucosa. (*a*) Gastric pits are the openings of the gastric glands. (*b*) Gastric glands consist of three types of cells—mucous cells, chief cells, and parietal cells—each of which produces a specific secretion.

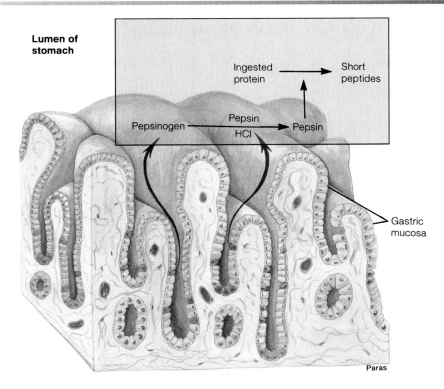

Lumen of
stomach

Ingested
protein → Short
peptides

Pepsinogen → Pepsin → Pepsin
HCl

Gastric
mucosa

Paras

Figure 18.9

The gastric mucosa secretes the inactive enzyme pepsinogen and hydrochloric acid (HCl). In the presence of HCl, the active enzyme pepsin is produced. Pepsin digests proteins into shorter polypeptides.

Digestion and Absorption in the Stomach

Proteins are only partially digested in the stomach by the action of pepsin. Carbohydrates and fats are not digested at all in the stomach. (Digestion of starch begins in the mouth with the action of salivary amylase, but this enzyme is inactivated by the strong acidity of gastric juice.) The complete digestion of food molecules occurs later, when chyme enters the small intestine. People who have had partial gastric resections, therefore, and even those who have had complete gastrectomies (removal of the stomach), can still adequately digest and absorb their food.

Almost all of the products of digestion are absorbed through the wall of the intestine; the only commonly ingested substances that can be absorbed across the stomach wall are alcohol and aspirin. Absorption occurs as a result of the lipid solubility of these molecules. The passage of aspirin through the gastric mucosa has been shown to cause bleeding, which may be significant if aspirin is taken in large doses.

Gastritis and Peptic Ulcers

Peptic ulcers are erosions of the mucous membranes of the stomach or duodenum produced by the action of HCl. In *Zollinger–Ellison syndrome*, ulcers of the duodenum are produced by excessive gastric acid secretion in response to very high levels of the hormone gastrin. Gastrin is normally produced by the stomach but, in this case, it may be secreted by a pancreatic tumor. This is a rare condition, but it does demonstrate that excessive gastric acid can cause ulcers of the duodenum. Ulcers of the stomach, however, are not believed to be due to excessive acid secretion, but rather to mechanisms that reduce the barriers of the gastric mucosa to self-digestion.

Recent experiments demonstrate that the parietal and chief cells of the gastric mucosa are extremely impermeable to the acid in the gastric lumen. In fact, they are even impermeable to CO_2 and NH_3, which are molecules that can freely pass through most other cell membranes. This impermeability of the cell membrane of the gastric epithelium is but one of the mechanisms that work to protect the gastric mucosa. Other protective mechanisms include a layer of alkaline mucus, containing bicarbonate, covering the gastric mucosa; tight junctions between adjacent epithelial cells, preventing acid from leaking into the submucosa; a rapid rate of cell division, allowing damaged cells to be replaced (the entire epithelium is replaced every 3 days); and several protective effects provided by prostaglandins that are produced by the gastric mucosa. Indeed, a common cause of gastric ulcers is believed to be the use of nonsteroidal anti-inflammatory drugs (NSAID). This class of drugs, including aspirin and ibuprofen, acts to inhibit the production of prostaglandins (see chapter 11).

When the gastric barriers to self-digestion are broken down, acid can leak through the mucosa to the submucosa, causing direct damage and stimulating inflammation. The histamine released from mast cells (chapter 15) during inflammation may stimulate further acid secretion and result in further damage to the mucosa. The inflammation that occurs during these events is called **acute gastritis.**

The duodenum is normally protected from gastric acid by the buffering action of bicarbonate in alkaline pancreatic juice, as well as by secretion of bicarbonate by Brunner's glands in the submucosa of the duodenum. However, people who develop duodenal ulcers produce excessive amounts of gastric acid that are not neutralized by the bicarbonate. People with gastritis and peptic ulcers must avoid substances that stimulate acid secretion, including coffee and wine, and often must take antacids.

It has been known for some time that most people who have peptic ulcers are infected with a type of bacterium known as *Helicobacter pylori*, which resides in the gastrointestinal tract. Also, clinical trials have demonstrated that antibiotics that eliminate this infection help in the treatment of the peptic ulcers. Many people, however, have *H. pylori* infections without having ulcers. Therefore, the infection may not itself cause the ulcer, but perhaps only contribute to the weakening of the mucosal barriers to gastric acid damage. Further research is needed to gain a more complete understanding of the causes of peptic ulcers.

The secretion of gastric acid is stimulated by *acetylcholine* (from parasympathetic nerve endings), *gastrin* (a hormone secreted by the stomach), and *histamine* (secreted by mast cells in the connective tissue). In response to these stimulants, the parietal cells secrete H^+ into the gastric lumen by active transport against a million-to-one concentration gradient, using a carrier that functions as a H^+/K^+ ATPase pump. People with ulcers, consequently, can be treated with drugs that block ACh, gastrin, or histamine action. Drugs in the last category (e.g., *Tagamet*) are more commonly used, and specifically block the H_2 histamine receptors in the gastric mucosa. This is a different receptor subtype than that blocked by antihistamines commonly used to treat cold and allergy symptoms.

1. Describe the structure and function of the lower esophageal sphincter.
2. List the secretory cells of the gastric mucosa and the products they secrete.
3. Describe the functions of hydrochloric acid in the stomach.
4. Explain how peptic ulcers are produced and why they are more likely to occur in the duodenum than in the stomach.
5. Identify the stimulants that cause the parietal cells to secrete H^+ and explain how various drugs inhibit acid secretion.

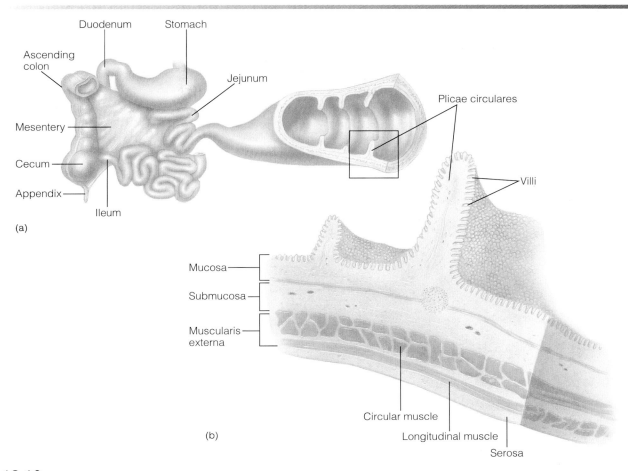

Figure 18.10

(*a*) The small intestine. (*b*) A section of the intestinal wall showing the tissue layers, plicae circulares, and villi.

Small Intestine

The mucosa of the small intestine is folded into villi that project into the lumen. In addition, the cells that line these villi have foldings of their plasma membrane called microvilli. This arrangement greatly increases the surface area for absorption. It also improves digestion, since the digestive enzymes of the small intestine are embedded within the cell membrane of the microvilli.

The *small intestine* (fig. 18.10) is that portion of the GI tract between the pyloric sphincter of the stomach and the ileocecal valve opening into the large intestine. It is called "small" because of its relatively small diameter compared to that of the large intestine. The small intestine is the longest part of the GI tract, however. It is approximately 3 m (12 ft) long in a living person, but it will measure nearly twice this length in a cadaver when the muscle wall is relaxed. The first 20 to 30 cm (10 in.) extending from the pyloric sphincter is the **duodenum** (fig.

18.11). The next two-fifths of the small intestine is the **jejunum,** and the last three-fifths is the **ileum.** The ileum empties into the large intestine through the ileocecal valve.

The products of digestion are absorbed across the epithelial lining of intestinal mucosa. Absorption of carbohydrates, lipids, amino acids, calcium, and iron occurs primarily in the duodenum and jejunum. Bile salts, vitamin B_{12}, water, and electrolytes are absorbed primarily in the ileum. Absorption occurs at a rapid rate as a result of the extensive mucosal surface area in the small intestine, provided by folds. The mucosa and submucosa form large folds called the *plicae circulares*, which can be observed with the unaided eye. The surface area is further increased by the microscopic folds of mucosa, called *villi*, and by the foldings of the apical cell membrane of epithelial cells (which can only be seen with an electron microscope), called *microvilli*.

Villi and Microvilli

Each **villus** is a fingerlike fold of mucosa that projects into the intestinal lumen (fig. 18.12). The villi are covered with columnar epithelial cells, among which are interspersed mucus-secreting *goblet cells*. The lamina propria forms a connective core of each villus and contains numerous lymphocytes, blood capillaries, and

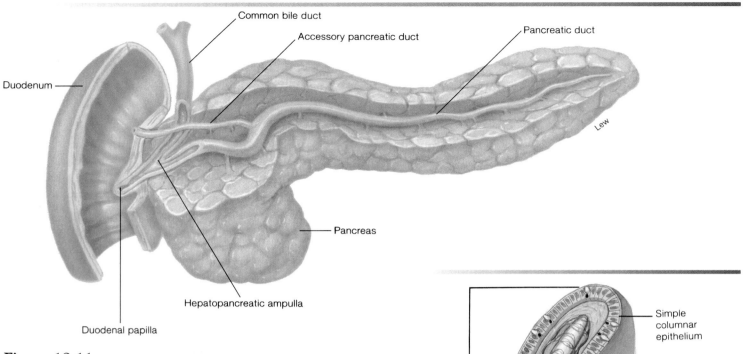

Figure 18.11

The duodenum and associated structures.

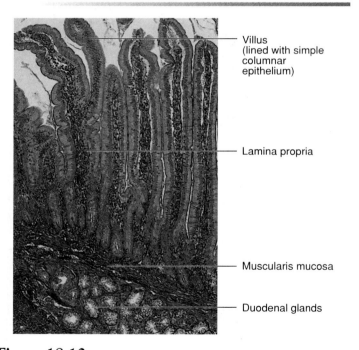

Figure 18.12

The histology of the duodenum.

Figure 18.13

A diagram of the structure of an intestinal villus.

a lymphatic vessel called the *central lacteal* (fig. 18.13). Absorbed monosaccharides and amino acids are secreted into the blood capillaries; absorbed fat enters the central lacteals.

Epithelial cells at the tips of the villi are continuously exfoliated (shed) and are replaced by cells that are pushed up from the bases of the villi. The epithelium at the base of the

villi invaginates downward at various points to form narrow pouches that open through pores to the intestinal lumen. These structures are called **intestinal crypts,** or *crypts of Lieberkühn* (fig. 18.13).

Microvilli are fingerlike projections formed by foldings of the cell membrane that can be clearly seen only in an electron

Chapter Eighteen

Table 18.2 Brush Border Enzymes Attached to the Cell Membrane of Microvilli in the Small Intestine

Category	Enzyme	Comments
Disaccharidase	Sucrase	Digests sucrose to glucose and fructose; deficiency produces gastrointestinal disturbances
	Maltase	Digests maltose to glucose
	Lactase	Digests lactose to glucose and galactose; deficiency produces gastrointestinal disturbances (lactose intolerance)
Peptidase	Aminopeptidase	Produces free amino acids, dipeptides, and tripeptides
	Enterokinase	Activates trypsin (and indirectly other pancreatic juice enzymes); deficiency results in protein malnutrition
Phosphatase	Ca^{++}, Mg^{++}—ATPase	Needed for absorption of dietary calcium; enzyme activity regulated by vitamin D
	Alkaline phosphatase	Removes phosphate groups from organic molecules; enzyme activity may be regulated by vitamin D

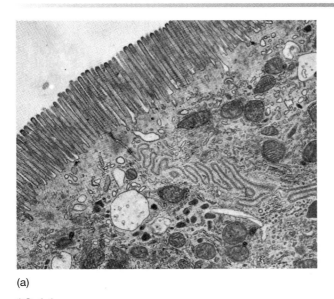

(a)

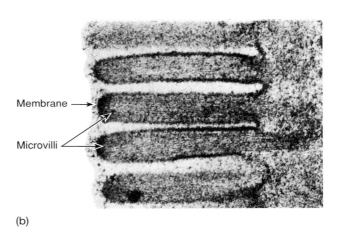

(b)

Figure 18.14

Electron micrographs of microvilli at the apical surface of the columnar epithelial cells in the small intestine: (*a*) lower magnification and (*b*) higher magnification.

microscope. In a light microscope, the microvilli produce a somewhat vague **brush border** on the edges of the columnar epithelial cells. The terms *brush border* and *microvilli* are thus often used interchangeably in describing the small intestine (fig. 18.14).

Intestinal Enzymes

In addition to providing a large surface area for absorption, the cell membranes of the microvilli contain digestive enzymes. These enzymes are not secreted into the lumen, but instead remain attached to the cell membrane with their active sites exposed to the chyme. These **brush border enzymes** hydrolyze disaccharides, polypeptides, and other substrates (table 18.2). One brush border enzyme, *enterokinase,* is required for activation of the protein-digesting enzyme *trypsin,* which enters the small intestine in pancreatic juice.

 The ability to digest milk sugar, or lactose, depends on the presence of a brush border enzyme called *lactase.* This enzyme is present in all children under the age of 4 but becomes inactive to some degree in most adults (people of Asiatic or African heritage are more often lactase deficient than Caucasians). In some cases, lactase deficiency can result in **lactose intolerance.** The presence of large amounts of undigested lactose in the intestine causes diarrhea, gas, cramps, and other unpleasant symptoms. Yogurt is better tolerated than milk because it contains lactase produced by the yogurt bacteria, which becomes activated in the duodenum and digests lactose.

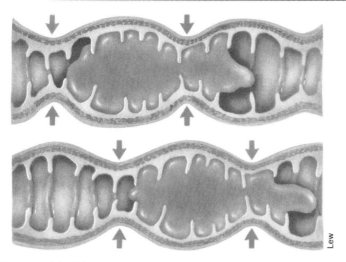

Figure 18.15

Segmentation of the small intestine. Simultaneous contractions of numerous segments of the intestine help to mix the chyme with digestive enzymes and mucus.

Intestinal Contractions and Motility

Two major types of contractions occur in the small intestine: peristalsis and segmentation. Peristalsis is much weaker in the small intestine than in the esophagus and stomach. Intestinal motility—the movement of chyme through the intestine—is relatively slow and is due primarily to the fact that the pressure at the pyloric end of the small intestine is greater than at the distal end.

The major contractile activity of the small intestine is **segmentation.** This term refers to muscular constrictions of the lumen, which occur simultaneously at different intestinal segments (fig. 18.15). This action serves to mix the chyme more thoroughly.

Like cardiac muscle, intestinal smooth muscle is capable of spontaneous electrical activity and automatic rhythmic contractions. Spontaneous depolarizations begin in pacemaker smooth muscle cells at the boundary of the circular muscle layer and spread through both the circular and longitudinal muscle layers across *nexuses*. The term *nexus* is used here to indicate an electrical synapse between smooth muscle cells. The spontaneous depolarizations, called **pacesetter potentials,** or *slow waves*, decrease in amplitude as they are conducted from one muscle cell to another, much like excitatory postsynaptic potentials (EPSPs). The pacesetter potentials can stimulate the production of action potentials when they reach a plateau level of depolarization in the smooth muscle cells through which they are conducted (fig. 18.16).

The nexuses conduct the pacesetter potentials, not the action potentials. Action potentials are therefore limited to those smooth muscle cells that are depolarized to threshold by the spreading pacesetter potentials. These action potentials stimulate smooth muscle contraction in only limited regions of the small intestine, producing the localized contractions of segmentation. Although this activity is automatic in nature, the

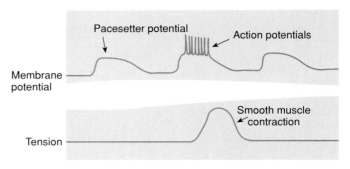

Figure 18.16

The smooth muscle of the gastrointestinal tract produces and conducts spontaneous pacesetter potentials. As these potential changes reach a threshold level of depolarization, they stimulate the production of action potentials, which in turn stimulate smooth muscle contraction.

excitability of the intestinal smooth muscle cells is increased by acetylcholine released by parasympathetic stimulation and is decreased by sympathetic nerves. Relaxation of intestinal smooth muscle is promoted by nitric oxide produced within the smooth muscle cells in response to neurotransmitters released by autonomic neurons. The neurotransmitters that cause intestinal relaxation include nitric oxide itself, norepinephrine, and vasoactive intestinal peptide, or VIP (chapter 9).

---

1. Describe the structures that increase the surface area of the small intestine and explain the function of the crypts of Lieberkühn.

2. Explain what is meant by the term *brush border* and give some examples of brush border enzymes. Why is it that many adults cannot tolerate milk?

3. Describe how smooth muscle contraction in the small intestine is regulated and explain the function of segmentation.

Large Intestine

The large intestine absorbs water and electrolytes from the chyme it receives from the small intestine and, in a process regulated by the action of sphincter muscles, passes waste products out of the body through the rectum and anal canal.

Chyme from the ileum passes into the cecum, which is a blind pouch (open only at one end) at the beginning of the large intestine, or *colon*. Waste material then passes in sequence through the ascending colon, transverse colon, descending colon, sigmoid colon, rectum, and anal canal (fig. 18.17). Waste material (feces) is excreted through the anus.

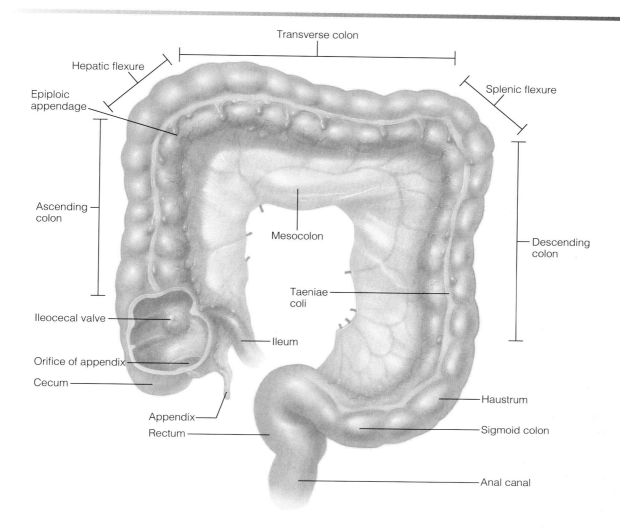

Figure 18.17

The large intestine.

As in the small intestine, the mucosa of the large intestine contains many scattered lymphocytes and lymphatic nodules and is covered by columnar epithelial cells and mucus-secreting goblet cells. Although this epithelium does form crypts of Lieberkühn, there are no villi in the large intestine—the intestinal mucosa therefore appears flat. The outer surface of the colon bulges outward to form pouches, or **haustra** (fig. 18.18). Occasionally, the muscularis externa of the haustra may become so weakened that the wall forms a more elongated outpouching, or diverticulum (*divert* = turned aside). Inflammation of one or more of these structures is called *diverticulitis*.

Fluid and Electrolyte Absorption in the Intestine

Most of the fluid and electrolytes in the lumen of the GI tract are absorbed by the small intestine. Although a person may drink only about 1.5 L of water per day, the small intestine re-

 The *appendix* is a short, thin outpouching of the cecum. It does not function in digestion, but like the tonsils, it contains numerous lymphatic nodules (fig. 18.19) and is subject to inflammation—a condition called **appendicitis.** This is commonly detected in its later stages by pain in the lower right quadrant of the abdomen. Rupture of the appendix can cause inflammation of the surrounding body cavity—*peritonitis.* This dangerous event may be prevented by surgical removal of the inflamed appendix (appendectomy).

ceives 7 to 9 L per day as a result of the fluid secreted into the GI tract by the salivary glands, stomach, pancreas, liver, and gallbladder. The small intestine absorbs most of this fluid and passes 1.5 to 2.0 L of fluid per day to the large intestine. The large intestine absorbs about 90% of this remaining volume, leaving less than 200 ml of fluid to be excreted in the feces.

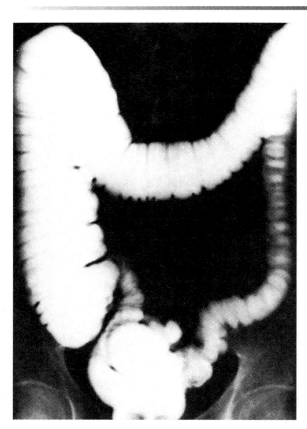

Figure 18.18

A radiograph after a barium enema, showing the haustra of the large intestine.

Absorption of water in the intestine occurs passively as a result of the osmotic gradient created by the active transport of ions. The epithelial cells of the intestinal mucosa are joined together much like those of the kidney tubules and, like the kidney tubules, contain Na^+/K^+ pumps in the basolateral membrane. The analogy with kidney tubules is emphasized by the observation that aldosterone, which stimulates salt and water reabsorption in the renal tubules, also appears to stimulate salt and water absorption in the ileum.

The handling of salt and water transport in the large intestine is made more complex by the fact that the large intestine can secrete, as well as absorb, water. The secretion of water by the mucosa of the large intestine occurs by osmosis as a result of the active transport of Na^+, followed passively by Cl^-, out of the epithelial cells into the intestinal lumen. Secretion in this way is normally minor compared to the far greater amount of salt and water absorption, but this balance may be altered in some disease states.

Defecation

After electrolytes and water have been absorbed, the waste material that is left passes to the rectum, leading to an increase in rectal pressure and the urge to defecate. If the urge to defecate is denied, feces are prevented from entering the anal canal by the internal anal sphincter. In this case the feces remain in the rectum, and may even back up into the sigmoid colon. The **defecation reflex** normally occurs when the rectal pressure rises to a particular level that is determined, to a large degree, by habit. At this point, the internal anal sphincter relaxes to admit feces into the anal canal.

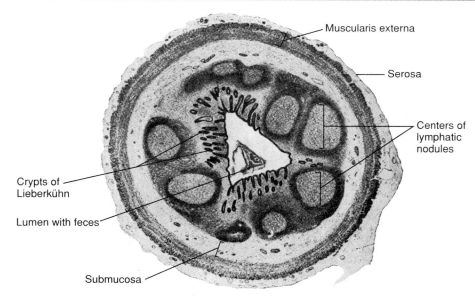

Figure 18.19

The microscopic appearance of a cross section of the human appendix.

Diarrhea is characterized by excessive fluid excretion in the feces. There are three different mechanisms, illustrated by three different diseases, which can cause diarrhea. In *cholera*, severe diarrhea results from a chemical called *enterotoxin*, released from the infecting bacteria. Enterotoxin stimulates active NaCl transport followed by the osmotic movement of water into the lumen of the intestine. Diarrhea caused by damage to the intestinal mucosa occurs in *celiac sprue*, a disease produced in susceptible people by eating foods that contain gluten (proteins from grains such as wheat). In *lactose intolerance*, diarrhea is produced by the increased osmolarity of the contents of the intestinal lumen as a result of the presence of undigested lactose.

During the act of defecation, the longitudinal rectal muscles contract to increase rectal pressure, and the internal and external anal sphincter muscles relax. Excretion is aided by contractions of abdominal and pelvic skeletal muscles, which raise the intra-abdominal pressure and help push the feces from the rectum through the anal canal and out the anus.

1. Describe how electrolytes and water are absorbed in the large intestine and explain how diarrhea may be produced.
2. Describe the structures and mechanisms involved in defecation.

Liver, Gallbladder, and Pancreas

The liver regulates the chemical composition of the blood in numerous ways. In addition, the liver produces and secretes bile, which is stored and concentrated in the gallbladder prior to its discharge into the duodenum. The pancreas produces pancreatic juice, an exocrine secretion containing bicarbonate and important digestive enzymes that is passed into the duodenum via the pancreatic duct.

The liver, which is the largest internal organ, lies immediately beneath the diaphragm in the abdominal cavity. Attached to the inferior surface of the liver, between its right and quadrate lobes, is the pear-shaped gallbladder. This organ is approximately 10 cm long by 3.5 cm wide. The pancreas is located behind the stomach along the posterior abdominal wall.

Structure of the Liver

Although the liver is the largest internal organ, it is, in a sense, only one to two cells thick. This is because the liver cells, or **hepatocytes,** form **hepatic plates** that are one to two cells thick. The plates are separated from each other by large capillary spaces, called **sinusoids** (fig. 18.20). The sinusoids are lined with phagocytic **Kupffer cells,** which are part of the reticuloendothelial system (chapter 15). There are large intercellular gaps

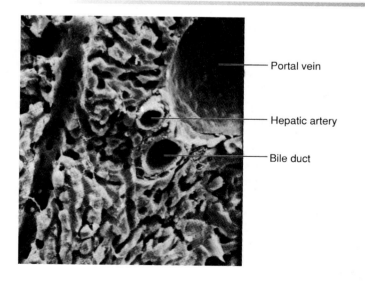

(a)

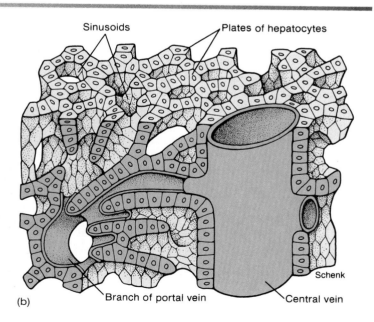

(b)

Figure 18.20

The structure of the liver. (*a*) A scanning electron micrograph of the liver. Hepatocytes are arranged in plates so that blood that passes through sinusoids (*b*) will be in contact with each liver cell.

Part [a] reproduced from R. G. Kessel and R. H. Kardon, *Tissues and Organs: A Text Atlas of Scanning Electron Microscopy,* W. H. Freeman & Co., 1979.

The Digestive System

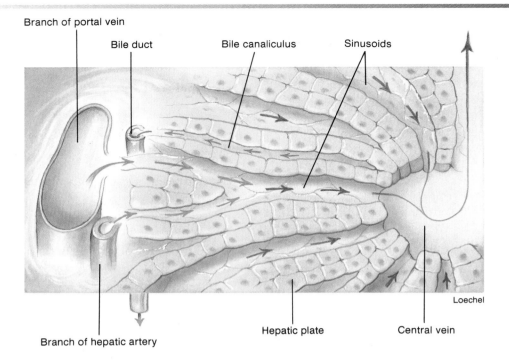

Branch of portal vein
Bile duct
Bile canaliculus
Sinusoids
Loechel
Branch of hepatic artery
Hepatic plate
Central vein

Figure 18.21

The flow of blood and bile in a liver lobule. Blood flows within sinusoids from a portal vein to the central vein (from the periphery to the center of a lobule). Bile flows within hepatic plates from the center to bile ducts at the periphery of a lobule.

between adjacent Kupffer cells that make the hepatic sinusoids more highly permeable than most other capillaries. The plate structure of the liver and the high permeability of the sinusoids allow each hepatocyte to have direct contact with the blood.

Hepatic Portal System

The products of digestion that are absorbed into blood capillaries in the intestine do not directly enter the general circulation. Instead, this blood is delivered first to the liver. Capillaries in the digestive tract drain into the *hepatic portal vein,* which carries this blood to capillaries in the liver. It is not until the blood has passed through this second capillary bed that it enters the general circulation through the *hepatic vein* that drains the liver. The term **portal system** is used to describe this unique pattern of circulation: capillaries → vein → capillaries → vein. In addition to receiving venous blood from the intestine, the liver also receives arterial blood via the *hepatic artery.*

Liver Lobules

The hepatic plates are arranged into functional units called **liver lobules** (fig. 18.21). In the middle of each lobule is a *central vein,* and at the periphery of each lobule are branches of the hepatic portal vein and of the hepatic artery, which open into the sinusoids *between* hepatic plates. Arterial blood and portal venous blood, containing molecules absorbed in the GI tract, thus mix as the blood flows within the sinusoids from the periphery of the lobule to the central vein. The central veins of different liver lobules converge to form the hepatic vein, which carries blood from the liver to the inferior vena cava.

Bile is produced by the hepatocytes and secreted into thin channels called **bile canaliculi,** located *within* each hepatic plate (fig. 18.21). These bile canaliculi are drained at the periphery of each lobule by *bile ducts,* which in turn drain into *hepatic ducts* that carry bile away from the liver. Since blood travels in the sinusoids and bile travels in the opposite direction within the hepatic plates, blood and bile do not mix in the liver lobules.

 In **cirrhosis,** large numbers of liver lobules are destroyed and replaced with permanent connective tissue and "regenerative nodules" of hepatocytes. These regenerative nodules don't have the platelike structure of normal liver tissue and are therefore less functional. One indication of this decreased function is the entry of ammonia from the hepatic portal blood into the general circulation. Cirrhosis may be caused by chronic alcohol abuse, viral hepatitis, or by other agents that attack liver cells.

Enterohepatic Circulation

In addition to the normal constituents of bile, a wide variety of exogenous compounds (drugs) are secreted by the liver into the bile ducts (table 18.3). The liver can thus "clear" the blood of particular compounds by removing them from the blood and excreting them into the intestine with the bile. (The liver can also clear the blood by other mechanisms that will be described in a later section.) Molecules that are cleared from the blood by secretion

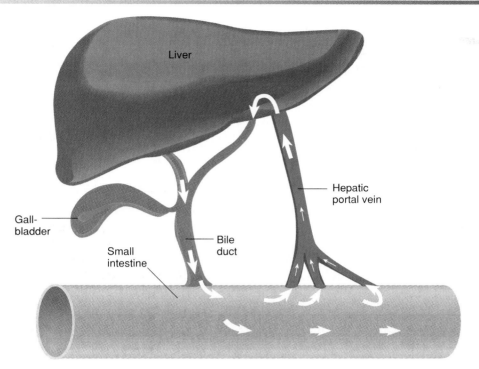

Figure 18.22

Enterohepatic circulation. Substances secreted in the bile may be absorbed by the intestinal epithelium and recycled to the liver via the hepatic portal vein.

Table 18.3	Compounds Excreted by the Liver into the Bile Ducts	
Category	**Compound**	**Comments**
Endogenous (Naturally Occurring)	Bile salts, urobilinogen, cholesterol	Higher percentage is absorbed and has an enterohepatic circulation*
	Lecithin	Small percentage is absorbed and has an enterohepatic circulation
	Bilirubin	No enterohepatic circulation
Exogenous (Drugs)	Ampicillin, streptomycin, tetracycline	High percentage is absorbed and has an enterohepatic circulation
	Sulfonamides, penicillin	Small percentage is absorbed and has an enterohepatic circulation

*Compounds with an enterohepatic circulation are absorbed to some degree by the intestine and are returned to the liver in the hepatic portal vein.

into the bile are excreted in the feces; this is analogous to renal clearance of blood through excretion in the urine (chapter 16).

Many compounds that are released with the bile into the intestine are not excreted with the feces, however. Some of these can be absorbed through the small intestine and enter the hepatic portal blood. These absorbed molecules are thus carried back to the liver, where they can be again secreted by hepatocytes into the bile ducts. Compounds that recirculate between the liver and intestine in this way are said to have an **enterohepatic circulation** (fig. 18.22).

Functions of the Liver

As a result of its very large and diverse enzymatic content and its unique structure, and because it receives venous blood from the intestine, the liver has a wider variety of functions than any other organ in the body. The major categories of liver function are summarized in table 18.4.

Bile Production and Secretion

The liver produces and secretes 250 to 1,500 ml of bile per day. The major constituents of bile include bile salts, bile pigment (bilirubin), phospholipids (mainly lecithin), cholesterol, and inorganic ions (table 18.5).

Bile pigment, or bilirubin, is produced in the spleen, liver, and bone marrow as a derivative of the heme groups (minus the iron) from hemoglobin. The **free bilirubin** is not very water-soluble, and thus most is carried in the blood attached to albumin proteins. This protein-bound bilirubin can neither be filtered by the kidneys into the urine nor directly excreted by the liver into the bile.

The liver can take some of the free bilirubin out of the blood and conjugate (combine) it with glucuronic acid. This **conjugated bilirubin** is water-soluble and can be secreted into the bile. Once in the bile, the conjugated bilirubin can enter the intestine where it is converted by bacteria into another pigment—**urobilinogen**—which is partially responsible for the

Table 18.4 Major Categories of Liver Function

Functional Category	Actions
Detoxication of Blood	Phagocytosis by Kupffer cells
	Chemical alteration of biologically active molecules (hormones and drugs)
	Production of urea, uric acid, and other molecules that are less toxic than parent compounds
	Excretion of molecules in bile
Carbohydrate Metabolism	Conversion of blood glucose to glycogen and fat
	Production of glucose from liver glycogen and from other molecules (amino acids, lactic acid) by gluconeogenesis
	Secretion of glucose into the blood
Lipid Metabolism	Synthesis of triglyceride and cholesterol
	Excretion of cholesterol in bile
	Production of ketone bodies from fatty acids
Protein Synthesis	Production of albumin
	Production of plasma transport proteins
	Production of clotting factors (fibrinogen, prothrombin, and others)
Secretion of Bile	Synthesis of bile salts
	Conjugation and excretion of bile pigment (bilirubin)

Table 18.5 Composition of Bile

Component	Concentration
pH	5.7–8.6
Bile salts	140–2,230 mg/100 ml
Lecithin	140–810 mg/100 ml
Cholesterol	97–320 mg/100 ml
Bilirubin	12–70 mg/100 ml
Urobilinogen	5–45 mg/100 ml
Sodium	145–165 mEq/L
Potassium	2.7–4.9 mEq/L
Chloride	88–115 mEq/L
Bicarbonate	27–55 mEq/L

Jaundice is a yellow staining of the tissues produced by high blood concentrations of either free or conjugated bilirubin. Jaundice due to high blood levels of conjugated bilirubin in adults may result when bile excretion is blocked by gallstones. Since free bilirubin is derived from heme, jaundice due to high blood levels of free bilirubin is usually caused by an excessively high rate of red blood cell destruction. This is the cause of jaundice in newborn babies who suffer from erythroblastosis fetalis (described in chapter 13). *Physiological jaundice of the newborn* is due to high levels of free bilirubin in otherwise healthy neonates. This type of jaundice may be caused by the rapid fall in blood hemoglobin concentrations that normally occurs at birth. In premature infants, it may be caused by inadequate amounts of hepatic enzymes that are needed to conjugate bilirubin so that it can be excreted in the bile.

Newborn infants with jaundice are usually treated by *phototherapy,* in which they are placed under blue light in the wavelength range of 400 to 500 nm. This light is absorbed by bilirubin in cutaneous vessels and results in the conversion of bilirubin to a more polar isomer, which is soluble in plasma without having to be conjugated with glucuronic acid. The more water-soluble photoisomer of bilirubin can then be excreted in the bile and urine.

color of the feces. About 30% to 50% of the urobilinogen, however, is absorbed by the intestine and enters the hepatic portal vein. Of the urobilinogen that enters the liver sinusoids, some is secreted into the bile and is thus returned to the intestine in an enterohepatic circulation; the rest enters the general circulation (fig. 18.23). The urobilinogen in plasma, unlike free bilirubin, is not attached to albumin and, therefore, is easily filtered by the kidneys into the urine, giving urine its characteristic yellow color.

Bile salts are derivatives of cholesterol that have two to four polar groups on each molecule. The principal bile salts in humans are *cholic acid* and *chenodeoxycholic acid* (fig. 18.24). In aqueous solutions these molecules "huddle" together to form aggregates known as **micelles.** The nonpolar parts are located in the central region of the micelle (away from water), whereas the polar groups face water around the periphery of the micelle. Lecithin, cholesterol, and other lipids in the small intestine enter these micelles in a process that aids the digestion and absorption of fats (described in a later section).

Detoxication of the Blood

The liver can remove hormones, drugs, and other biologically active molecules from the blood by (1) excretion of these compounds in the bile, (2) phagocytosis by the Kupffer cells that line the sinusoids, and (3) chemical alteration of these molecules within the hepatocytes.

Ammonia, for example, is a very toxic molecule produced by deamination of amino acids (chapter 19) in the liver and by

the action of bacteria in the intestine. Since the ammonia concentration of portal vein blood is four to fifty times greater than that of blood in the hepatic vein, it is clear that the ammonia is removed by the liver. The liver has the enzymes needed to convert ammonia into less toxic **urea** molecules, which are secreted by the liver into the blood and excreted by the kidneys in the urine. Similarly, the liver converts toxic porphyrins into **bilirubin** and toxic purines into **uric acid.**

Steroid hormones and many drugs are inactivated in their passage through the liver by modifications of their chemical structures. The liver has enzymes that convert these nonpolar molecules into more polar (more water-soluble) forms by *hydroxylation* (the addition of OH^- groups) and by *conjugation* with highly polar groups, such as sulfate and glucuronic acid. Polar derivatives of steroid hormones and drugs are less biologically

Chapter Eighteen

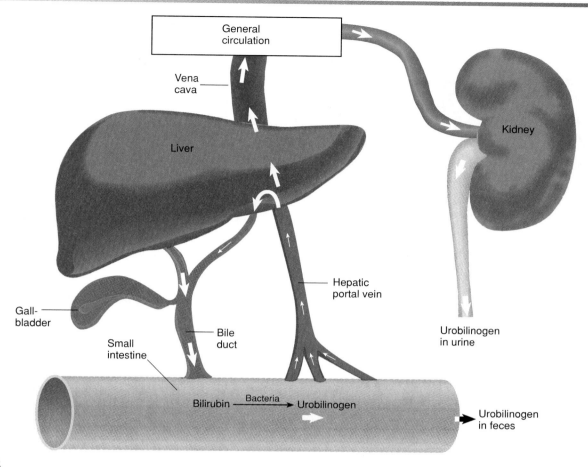

Figure 18.23

The enterohepatic circulation of urobilinogen. Bacteria in the intestine convert bile pigment (bilirubin) into urobilinogen. Some of this pigment leaves the body in the feces; some is absorbed by the intestine and is recycled through the liver. A portion of the urobilinogen that is absorbed enters the general circulation and is filtered by the kidneys into the urine.

The liver cells contain enzymes for the metabolism of steroid hormones and other endogenous molecules, as well as for the detoxication of such exogenous toxic compounds as benzopyrene (a carcinogen from tobacco smoke and charbroiled meat), polychlorinated biphenyls (PCBs), and dioxin. These enzymes are members of a class called the **cytochrome P450 enzymes** (not related to the cytochromes of cell respiration), that comprises a few dozen enzymes with varying specificities. Together, these enzymes can metabolize thousands of toxic compounds. Since people vary in their hepatic content of the different cytochrome P450 enzymes, one person's sensitivity to a drug may be greater than another's because of a relative deficiency in the appropriate cytochrome P450 enzyme needed to metabolize that drug.

active and, because of their increased water solubility, are more easily excreted by the kidneys into the urine.

Secretion of Glucose, Triglycerides, and Ketone Bodies

The liver helps to regulate the blood glucose concentration by either removing glucose from the blood or adding glucose to it, according to the needs of the body. After a carbohydrate-rich meal, the liver can remove some glucose from the hepatic portal blood and convert it into glycogen and triglycerides through the processes of **glycogenesis** and **lipogenesis,** respectively. During fasting, the liver secretes glucose into the blood. This glucose can be derived from the breakdown of stored glycogen in a process called **glycogenolysis,** or it can be produced by the conversion of noncarbohydrate molecules (such as amino acids) into glucose in a process called **gluconeogenesis.** The liver also contains the enzymes required to convert free fatty acids into ketone bodies **(ketogenesis),** which are secreted into the blood in large amounts during fasting. These processes are controlled by hormones and are explained in more detail in chapter 19.

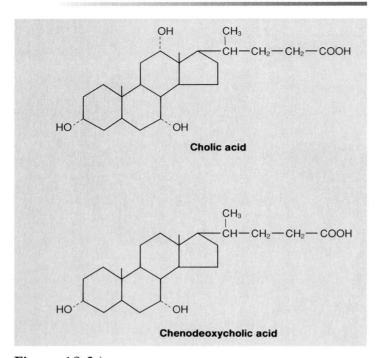

Figure 18.24

The two major bile acids (which form bile salts) in humans.

Production of Plasma Proteins

Plasma albumin and most of the plasma globulins (with the exception of immunoglobulins, or antibodies) are produced by the liver. Albumin constitutes about 70% of the total plasma protein and contributes most to the colloid osmotic pressure of the blood (chapter 14). The globulins produced by the liver have a wide variety of functions, including transport of cholesterol and triglycerides, transport of steroid and thyroid hormones, inhibition of trypsin activity, and blood clotting. Clotting factors I (fibrinogen), II (prothrombin), III, V, VII, IX, and XI, as well as angiotensinogen, are all produced by the liver.

Gallbladder

The gallbladder is a saclike organ attached to the inferior surface of the liver. This organ stores and concentrates bile, which drains to it from the liver by way of the bile ducts, hepatic ducts, and *cystic duct*, respectively. A sphincter valve at the neck of the gallbladder allows a storage capacity of from 35 to 100 ml. The inner mucosal layer of the gallbladder is arranged in rugae similar to those of the stomach. When the gallbladder fills with bile, it expands to the size and shape of a small pear. Bile is a yellowish-green fluid containing bile salts, bilirubin, cholesterol, and other compounds as previously discussed. Contraction of the muscularis layer of the gallbladder ejects bile through the cystic duct into the *common bile duct*, which conveys bile into the duodenum (fig. 18.25).

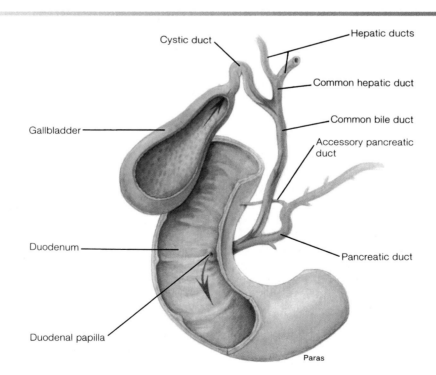

Figure 18.25

The pancreatic duct joins the common bile duct to empty its secretions through the duodenal papilla into the duodenum. The release of bile and pancreatic juice into the duodenum is controlled by the sphincter of ampulla (Oddi).

Approximately 20 million Americans have **gallstones**—small, hard mineral deposits (calculi) that can produce painful symptoms by obstructing the cystic or common bile ducts. Gallstones commonly contain cholesterol as their major component. Cholesterol normally has an extremely low water solubility (20 μg/L), but it can be present in bile at 2 million times its water solubility (40 g/L) because cholesterol molecules cluster together with bile salts and lecithin in the hydrophobic centers of micelles. In order for gallstones to be produced, the liver must secrete enough cholesterol to create a supersaturated solution, and some substance within the gallbladder must serve as a nucleus for the formation of cholesterol crystals. The gallstone is formed from cholesterol crystals that become hardened by the precipitation of inorganic salts (fig. 18.26). Gallstones may be removed surgically; cholesterol gallstones, however, may be dissolved by oral ingestion of bile acids. This may be combined with a newer treatment that involves fragmentation of the gallstones by high-energy shock waves delivered to a patient immersed in a water bath. This procedure is called *extracorporeal shock-wave lithotripsy.*

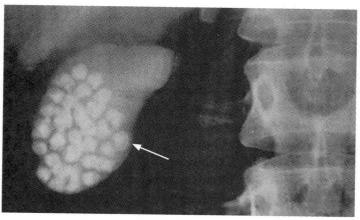

(a)

(b)

Figure 18.26

(*a*) A radiograph of a gallbladder that contains gallstones. (*b*) A posterior view of a gallbladder that has been removed (cholecystectomy) and cut open to reveal its gallstones (bilary calculi). (A dime has been placed in the photo to show relative size.)

Bile is continuously produced by the liver and drains through the hepatic and common bile ducts to the duodenum. When the small intestine is empty of food, the *sphincter of Oddi* at the end of the common bile duct closes, and bile is forced up to the cystic duct and then to the gallbladder for storage.

Pancreas

The pancreas is a soft, glandular organ that has both exocrine and endocrine functions. The endocrine function is performed by clusters of cells, called the **pancreatic islets,** or **islets of Langerhans,** that secrete the hormones insulin and glucagon into the blood (chapter 19). As an exocrine gland, the pancreas secretes pancreatic juice through the pancreatic duct into the duodenum. Within the lobules of the pancreas are the exocrine secretory units, called **acini** (fig. 18.27). Each acinus consists of a single layer of epithelial cells surrounding a lumen, into which the constituents of pancreatic juice are secreted.

Pancreatic Juice

Pancreatic juice contains water, bicarbonate, and a wide variety of digestive enzymes that are secreted into the duodenum. These enzymes include (1) **amylase,** which digests starch; (2) **trypsin,** which digests protein; and (3) **lipase,** which digests triglycerides. Other pancreatic enzymes are indicated in table 18.6. It should be noted that the complete digestion of food molecules in the small intestine requires the action of both pancreatic enzymes and brush border enzymes.

Most pancreatic enzymes are produced as inactive molecules, or *zymogens,* so that the risk of self-digestion within the pancreas is minimized. The inactive form of trypsin, called trypsinogen, is activated within the small intestine by the catalytic action of the brush border enzyme *enterokinase*. Enterokinase converts trypsinogen to active trypsin. Trypsin, in

Table 18.6 Enzymes Contained in Pancreatic Juice

Enzyme	Zymogen	Activator	Action
Trypsin	Trypsinogen	Enterokinase	Cleaves internal peptide bonds
Chymotrypsin	Chymotrypsinogen	Trypsin	Cleaves internal peptide bonds
Elastase	Proelastase	Trypsin	Cleaves internal peptide bonds
Carboxypeptidase	Procarboxypeptidase	Trypsin	Cleaves last amino acid from carboxyl-terminal end of polypeptide
Phospholipase	Prophospholipase	Trypsin	Cleaves fatty acids from phospholipids such as lecithin
Lipase	None	None	Cleaves fatty acids from glycerol
Amylase	None	None	Digests starch to maltose and short chains of glucose molecules
Cholesterolesterase	None	None	Releases cholesterol from its bonds with other molecules
Ribonuclease	None	None	Cleaves RNA to form short chains
Deoxyribonuclease	None	None	Cleaves DNA to form short chains

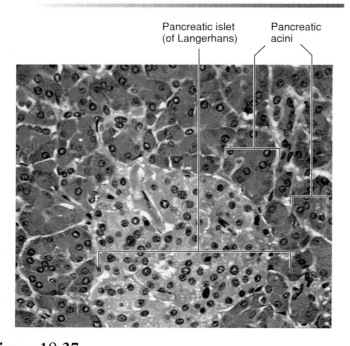

Figure 18.27

A microscopic view of the structure of the pancreas, showing exocrine acini and the pancreatic islet.

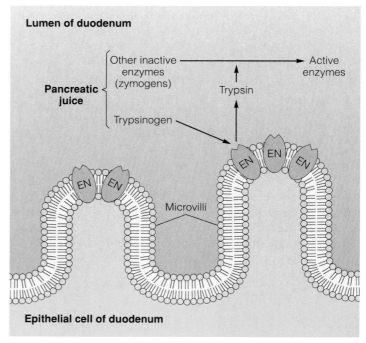

Figure 18.28

The pancreatic protein-digesting enzyme trypsin is secreted in an inactive form known as trypsinogen. This inactive enzyme (zymogen) is activated by a brush border enzyme, enterokinase (*EN*), located in the cell membrane of microvilli. Active trypsin in turn activates other zymogens in pancreatic juice.

Inflammation of the pancreas may result when the various safeguards against self-digestion are insufficient. **Acute pancreatitis** is believed to be caused by the reflux of pancreatic juice and bile from the duodenum into the pancreatic duct. Leakage of trypsin into the blood also occurs, but trypsin is inactive in the blood because of the inhibitory action of two plasma proteins, α_1-antitrypsin and α_2-macroglobulin. Pancreatic amylase may also leak into the blood, but it is not active because its substrate (glycogen) is not present in blood. Pancreatic amylase activity can be measured in vitro, however, and these measurements are commonly performed to assess the health of the pancreas.

turn, activates the other zymogens of pancreatic juice (fig. 18.28) by cleaving off polypeptide sequences that inhibit the activity of these enzymes.

The activation of trypsin, therefore, is the triggering event for the activation of other pancreatic enzymes. Actually, the pancreas does produce small amounts of active trypsin, yet the other enzymes are not activated until pancreatic juice has entered the duodenum. This is because pancreatic juice also contains a small protein called *pancreatic trypsin inhibitor* that attaches to trypsin and inactivates it in the pancreas.

Table 18.7 Physiological Effects of Gastrointestinal Hormones

Secreted by	Hormone	Effects
Stomach	Gastrin	Stimulates parietal cells to secrete HCl
		Stimulates chief cells to secrete pepsinogen
		Maintains structure of gastric mucosa
Small intestine	Secretin	Stimulates water and bicarbonate secretion in pancreatic juice
		Potentiates actions of cholecystokinin on pancreas
Small intestine	Cholecystokinin (CCK)	Stimulates contraction of gallbladder
		Stimulates secretion of pancreatic juice enzymes
		Potentiates action of secretin on pancreas
		Maintains structure of exocrine pancreas (acini)
Small intestine	Gastric inhibitory peptide (GIP)	Inhibits gastric emptying(?)
		Inhibits gastric acid secretion(?)
		Stimulates secretion of insulin from endocrine pancreas (islets of Langerhans)

---

1. Describe the structure of liver lobules and trace the pathways for the flow of blood and bile in the lobules.

2. Describe the composition and function of bile and trace the flow of bile from the liver and gallbladder to the duodenum.

3. Explain how the liver inactivates and excretes compounds such as hormones and drugs.

4. Describe the enterohepatic circulation of bilirubin and urobilinogen.

5. Explain how the liver helps to maintain a constant blood glucose concentration and how the pattern of venous blood flow permits this function.

6. Describe the endocrine and exocrine structures and functions of the pancreas. How is the pancreas protected against self-digestion?

Neural and Endocrine Regulation of the Digestive System

The activities of different regions of the GI tract are coordinated by the actions of the vagus nerve and various hormones. The stomach begins to increase its secretion in anticipation of a meal, and further increases its activities in response to the arrival of chyme. The entry of chyme into the duodenum stimulates the secretion of hormones that promote contractions of the gallbladder, secretion of pancreatic juice, and inhibition of gastric activity.

Neural and endocrine control mechanisms modify the activity of the digestive system. The sight, smell, or taste of food, for example, can stimulate salivary and gastric secretions via activation of the vagus nerve, which helps to "prime" the digestive system in preparation for a meal. Stimulation of the vagus, in this case, originates in the brain and is a conditioned reflex (as Pavlov demonstrated by training dogs to salivate in response to a bell). The vagus nerve is also involved in the reflex control of one part of the digestive system by another—these are "short reflexes," which do not involve the brain.

The GI tract is both an endocrine gland and a target for the action of various hormones. Indeed, the first hormones to be discovered were gastrointestinal hormones. In 1902 two English physiologists, Sir William Bayliss and Ernest Starling, discovered that the duodenum produced a chemical regulator. They named this substance **secretin** and proposed, in 1905, that it was but one of many yet undiscovered chemical regulators produced by the body. Bayliss and Starling coined the term *hormones* for this new class of regulators. In that same year, other investigators discovered that an extract from the stomach antrum (pyloric region) stimulated gastric secretion. The hormone **gastrin** was thus the second hormone to be discovered.

The chemical structures of gastrin, secretin, and the duodenal hormone **cholecystokinin (CCK)** were determined in the 1960s. More recently, a fourth hormone produced by the small intestine, **gastric inhibitory peptide (GIP),** has been added to the list of proven GI tract hormones. The effects of these hormones are summarized in table 18.7.

Regulation of Gastric Function

Gastric motility and secretion are, to some extent, automatic. Waves of contraction that serve to push chyme through the pyloric sphincter, for example, are initiated spontaneously by pacesetter cells in the greater curvature of the stomach. The secretion of HCl and pepsinogen, likewise, can be stimulated in the absence of neural and hormonal influences by the presence of cooked or partially digested protein in the stomach.

Table 18.8	The Cephalic, Gastric, and Intestinal Phases in the Regulation of Gastric Acid Secretion
Phase of Regulation	**Description**
Cephalic Phase	1. Sight, smell, and taste of food cause stimulation of vagus nuclei in brain
	2. Vagus stimulates acid secretion
	a. Direct stimulation of parietal cells (major effect)
	b. Stimulation of gastrin secretion; gastrin stimulates acid secretion (lesser effect)
Gastric Phase	1. Distension of stomach stimulates vagus nerve; vagus stimulates acid secretion
	2. Amino acids and peptides in stomach lumen stimulate acid secretion
	a. Direct stimulation of parietal cells (lesser effect)
	b. Stimulation of gastrin secretion; gastrin stimulates acid secretion (major effect)
	3. Gastrin secretion inhibited when pH of gastric juice falls below 2.5
Intestinal Phase	1. Neural inhibition of gastric emptying and acid secretion
	a. Arrival of chyme in duodenum causes distension, increase in osmotic pressure
	b. These stimuli activate a neural reflex that inhibits gastric activity
	2. In response to fat in chyme, duodenum secretes a hormone that inhibits gastric acid secretion

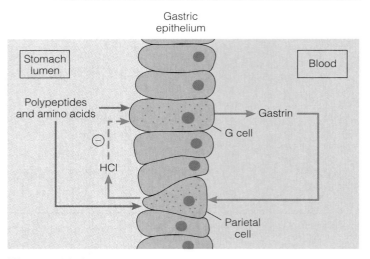

Figure 18.29

The stimulation of gastric acid (HCl) secretion by the presence of proteins in the stomach lumen and by the hormone gastrin. The secretion of gastrin is inhibited by gastric acidity, thus creating a negative feedback loop.

Gastric Phase

The arrival of food into the stomach stimulates the gastric phase of regulation. Gastric secretion is stimulated in response to two factors: (1) distension of the stomach, which is determined by the amount of chyme, and (2) the chemical nature of the chyme. While intact proteins have little stimulatory effect, the presence of short polypeptides and amino acids in the stomach stimulates the G cells to secrete gastrin and the parietal and chief cells to secrete HCl and pepsinogen, respectively. Since gastrin also stimulates HCl and pepsinogen secretion, a *positive feedback mechanism* develops. As more HCl and pepsinogen are secreted, more short polypeptides and amino acids are released from the ingested protein, thus stimulating additional secretion of gastrin and, therefore, additional secretion of HCl and pepsinogen (fig. 18.29). Glucose in the chyme, by contrast, has no effect on gastric secretion, whereas the presence of fat actually inhibits acid secretion.

Secretion of HCl during the gastric phase is also regulated by a *negative feedback mechanism*. As the pH of gastric juice drops, so does the secretion of gastrin—at a pH of 2.5, gastrin secretion is reduced, and at a pH of 1.0 gastrin secretion shuts off entirely. The secretion of HCl, which is largely under the control of gastrin, thus declines accordingly. The presence of proteins and polypeptides in the stomach helps to buffer the acid and thus to prevent a rapid fall in gastric pH; more acid can thus be secreted when proteins are present than when they are absent. The arrival of protein into the stomach thus stimulates acid secretion two ways—by the positive feedback mechanism previously discussed and by inhibition of the negative feedback control of acid secretion. Through these mechanisms, the amount of acid secreted is closely matched to the amount of protein ingested. As the stomach is emptied, the protein buffers exit, the pH thus falls, and the secretion of gastrin and HCl is accordingly inhibited.

The effects of autonomic nerves and hormones are superimposed on this automatic activity. This extrinsic control of gastric function is conveniently divided into three phases: (1) the cephalic phase; (2) the gastric phase; and (3) the intestinal phase. These are summarized in table 18.8.

Cephalic Phase

The cephalic phase of gastric regulation refers to control by the brain via the vagus nerve. As previously discussed, various conditioned stimuli can evoke gastric secretion. This conditioning in humans is, of course, more subtle than that exhibited by Pavlov's dogs in response to a bell. In fact, just talking about appetizing food is sometimes a more potent stimulus for gastric acid secretion than the actual sight and smell of food!

Activation of the vagus nerve can stimulate HCl and pepsinogen secretion by two mechanisms: (1) direct vagus stimulation of the gastric parietal and chief cells and (2) vagus stimulation of gastrin secretion by the G cells, which in turn stimulates the parietal and chief cells to secrete HCl and pepsinogen, respectively. This cephalic phase stimulation of gastric secretion continues into the first 30 minutes of a meal, but gradually declines in importance as the next phase becomes predominant.

The amino acids *tryptophan* and *phenylalanine* have been shown to be the most potent stimulators of gastrin and acid secretion. People who have peptic ulcers must avoid foods that strongly stimulate acid secretion; for example, milk, cola, beer, coffee (including decaffeinated coffee), tea, and wine. Such people should also avoid tryptophan tablets commonly sold in health food stores to promote sleep.

Intestinal Phase

The intestinal phase of gastric regulation refers to the inhibition of gastric activity when chyme enters the small intestine. Investigators in 1886 demonstrated that the addition of olive oil to a meal inhibits gastric emptying, and in 1929 it was shown that the presence of fat inhibits gastric juice secretion. This inhibitory intestinal phase of gastric regulation is due to both a neural reflex originating from the duodenum and to a chemical hormone secreted by the duodenum.

The arrival of chyme into the duodenum increases its osmolality. This stimulus, together with stretch of the duodenum and possibly other stimuli, produces a neural reflex that results in the inhibition of gastric motility and secretion. The presence of fat in the chyme also stimulates the duodenum to secrete a hormone that inhibits gastric function. The general term for such an inhibitory hormone is an **enterogastrone.**

For the past several years, *gastric inhibitory peptide* (GIP) was thought to function as an enterogastrone—hence the name for this hormone. Some researchers, however, now believe that a different polypeptide, known as *somatostatin*, may function in this capacity. Somatostatin is produced by the small intestine (as well as by the brain—see chapter 11), where it appears to serve a number of regulatory roles. In addition, the hormone *cholecystokinin* (CCK), which is secreted by the duodenum in response to the presence of chyme, has been found to inhibit gastric emptying.

It could be that the only physiological role of GIP is stimulation of insulin secretion from the islets of Langerhans in response to the presence of glucose in the intestine. Some scientists therefore propose that the name GIP be retained, but that it serve as an acronym for *glucose-dependent insulinotropic peptide*.

The inhibitory neural and endocrine mechanisms during the intestinal phase prevent the further passage of chyme from the stomach to the duodenum. This gives the duodenum time to process the load of chyme received previously. Since secretion of the enterogastrone is stimulated by fat in the chyme, a breakfast of bacon and eggs takes longer to pass through the stomach—and makes one feel "fuller" for a longer time—than does a breakfast of pancakes and syrup.

Regulation of Intestinal Function

The submucosal and myenteric plexuses within the wall of the intestine contain preganglionic parasympathetic axons, the ganglion cell bodies of postganglionic parasympathetic neurons, postganglionic sympathetic axons, and afferent (sensory) neurons. These plexuses also contain interneurons, as does the CNS. Indeed, some scientists refer to the nervous system within the GI tract as an *enteric brain*. Many of the sensory neurons within the intestinal plexuses send impulses all the way to the CNS, but some sensory neurons synapse with the interneurons in the wall of the intestine. This allows for local reflexes that are controlled within the gastrointestinal tract.

There are several intestinal reflexes controlled both locally and extrinsically. These include: (1) the **gastroilial reflex,** in which increased gastric activity causes increased motility of the ileum and increased movement of chyme through the ileocecal sphincter; (2) the **ileogastric reflex,** in which distension of the ileum causes a decrease in gastric motility; and (3) the **intestino-intestinal reflexes,** in which overdistension of one intestinal segment causes relaxation throughout the rest of the intestine.

Regulation of Pancreatic Juice and Bile Secretion

The arrival of chyme into the duodenum stimulates the intestinal phase of gastric regulation and, at the same time, stimulates reflex secretion of pancreatic juice and bile. The entry of new chyme is thus retarded as the previous load is digested. The secretion of pancreatic juice and bile is stimulated by both neural reflexes initiated in the duodenum and by secretion of the duodenal hormones cholecystokinin and secretin.

Pancreatic Juice

The secretion of pancreatic juice is stimulated by both secretin and CCK. These two hormones, however, are secreted in response to different stimuli and they have different effects on the composition of pancreatic juice. The release of secretin occurs in response to a fall in duodenal pH to below 4.5; this pH fall occurs for only a short time, however, because the acidic chyme is rapidly neutralized by alkaline pancreatic juice. The secretion of CCK occurs in response to the fat content of chyme in the duodenum.

Secretin stimulates the production of bicarbonate by the pancreas. Since bicarbonate neutralizes the acidic chyme and since secretin is released in response to the low pH of chyme, this completes a negative feedback loop in which secretin indirectly inhibits its own secretion. Cholecystokinin, by contrast, stimulates the production of pancreatic enzymes such as trypsin, lipase, and amylase. Secretin and CCK can have different effects on the same cells (the pancreatic acinar cells) because their actions are mediated by different intracellular compounds that act as second messengers. The second messenger of secretin action is cyclic AMP, whereas the second messenger for CCK is Ca^{++} (table 18.9).

Table 18.9 Regulation of Pancreatic Juice and Bile Secretion by the Hormones Secretin and Cholecystokinin (CCK) and by the Neurotransmitter Acetylcholine, Released from Parasympathetic Nerve Endings

Feature	Secretin	CCK	Acetylcholine (Vagus Nerve)
Stimulus for release	Decrease in duodenal pH below 4.5 due to acidity of chyme	Fat and protein in chyme	Sight, smell of food; distension of stomach
Second messenger	Cyclic AMP	Ca^{++}	Ca^{++}
Effect on pancreatic juice	Stimulates water and bicarbonate secretion; potentiates action of CCK	Stimulates enzyme secretion; potentiates action of secretin	Stimulates enzyme secretion
Effect on bile	Stimulates secretion	Potentiates action of secretin; stimulates contraction of gallbladder	Stimulates contraction of gallbladder

Secretion of Bile

The liver secretes bile continuously, but this secretion is greatly augmented following a meal. This increased secretion is due to the release of secretin and CCK from the duodenum. Secretin is the major stimulator of bile secretion by the liver, and CCK enhances this effect. The arrival of chyme in the duodenum also causes the gallbladder to contract and eject bile. Contraction of the gallbladder occurs in response to neural reflexes from the duodenum and in response to stimulation by CCK (but not secretin).

Trophic Effects of Gastrointestinal Hormones

Patients with tumors of the stomach pylorus have excessive acid secretion and hyperplasia (growth) of the gastric mucosa. Surgical removal of the pylorus reduces gastric secretion and prevents growth of the gastric mucosa. Patients with peptic ulcers are sometimes treated by vagotomy—cutting of the vagus nerve. Vagotomy also reduces acid secretion but has no effect on the gastric mucosa. These observations suggest that the hormone gastrin, secreted by the pyloric mucosa, may exert stimulatory, or *trophic*, effects on the gastric mucosa. The structure of the gastric mucosa, in other words, is dependent on the effects of gastrin.

In the same way, the structure of the acinar (exocrine) cells of the pancreas is dependent upon the trophic effects of CCK. Perhaps this explains why the pancreas, as well as the GI tract, atrophies during starvation. Since neural reflexes appear to be capable of regulating digestion, perhaps the primary function of the GI hormones is trophic—that is, maintenance of the structure of their target organs.

---

1. Describe the positive and negative feedback mechanisms that operate during the gastric phase of HCl and pepsinogen secretion.
2. Describe the mechanisms involved in the intestinal phase of gastric regulation and explain why a fatty meal takes longer to leave the stomach than a meal low in fat.
3. Explain the hormonal mechanisms involved in the production and release of pancreatic juice and bile.
4. Describe the composition of the intestinal plexuses and identify some of the short reflexes that regulate intestinal function.

Digestion and Absorption of Carbohydrates, Lipids, and Proteins

Polysaccharides and polypeptides are hydrolyzed into their subunits. These subunits enter the epithelial cells of the intestinal villi and are secreted into blood capillaries. Fat is emulsified by the action of bile salts, hydrolyzed into fatty acids and monoglycerides, and absorbed into the intestinal epithelial cells. Once in the cells, triglycerides are resynthesized and combined with proteins to form particles that are secreted into the lymphatic fluid.

Table 18.10 Characteristics of the Major Digestive Enzymes

Enzyme	Site of Production	Source	Substrate	Optimum pH	Product(s)
Salivary amylase	Mouth	Saliva	Starch	6.7	Maltose
Pepsin	Stomach	Gastric glands	Protein	1.6–2.4	Shorter polypeptides
Pancreatic amylase	Duodenum	Pancreatic juice	Starch	6.7–7.0	Maltose, maltriose, and oligosaccharides
Trypsin, chymotrypsin, carboxypeptidase			Polypeptides	8.0	Amino acids, dipeptides, and tripeptides
Pancreatic lipase			Triglycerides	8.0	Fatty acids and monoglycerides
Maltase		Epithelial membranes	Maltose	5.0–7.0	Glucose
Sucrase			Sucrose	5.0–7.0	Glucose + fructose
Lactase			Lactose	5.8–6.2	Glucose + galactose
Aminopeptidase			Polypeptides	8.0	Amino acids, dipeptides, tripeptides

The caloric (energy) value of food is derived mainly from its content of carbohydrates, lipids, and proteins. In the average American diet, carbohydrates account for approximately 50% of the total calories, protein accounts for 11% to 14%, and lipids make up the balance. These food molecules consist primarily of long combinations of subunits (monomers) that must be digested by hydrolysis reactions into free monomers before absorption can occur. The characteristics of the major digestive enzymes are summarized in table 18.10.

Digestion and Absorption of Carbohydrates

Most carbohydrates are ingested as starch, which is a long polysaccharide of glucose in the form of straight chains with occasional branchings. The most commonly ingested sugars are the disaccharides sucrose (table sugar, consisting of glucose and fructose) and lactose (milk sugar, consisting of glucose and galactose). The digestion of starch begins in the mouth with the action of **salivary amylase,** or **ptyalin.** This enzyme cleaves some of the bonds between adjacent glucose molecules, but most people don't chew their food long enough for sufficient digestion to occur in the mouth. The digestive action of salivary amylase stops when the swallowed bolus enters the stomach because this enzyme is inactivated at the low pH of gastric juice.

The digestion of starch, therefore, occurs mainly in the duodenum as a result of the action of **pancreatic amylase.** This enzyme cleaves the straight chains of starch to produce the disaccharide *maltose* and the trisaccharide *maltriose*. Pancreatic amylase, however, cannot hydrolyze the bond between glucose molecules at the branch points in the starch. As a result, short, branched chains of glucose molecules, called *oligosaccharides,* are released together with maltose and maltriose by the activity of this enzyme (fig. 18.30).

Maltose, maltriose, and oligosaccharides are hydrolyzed to their monosaccharides by brush border enzymes located on the microvilli of the epithelial cells in the small intestine. The brush border enzymes also hydrolyze the disaccharides sucrose and lactose into their component monosaccharides. These monosaccharides are then moved across the epithelial cell

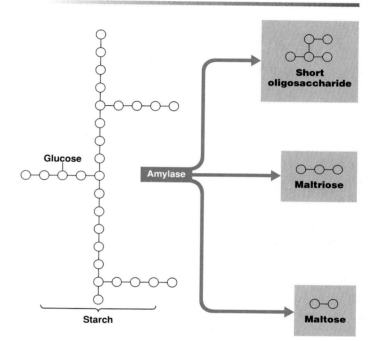

Figure 18.30

Pancreatic amylase digests starch into maltose, maltriose, and short oligosaccharides containing branch points in the chain of glucose molecules.

membrane by secondary active transport, in which the glucose shares a common membrane carrier with Na^+ (chapter 6). Finally, glucose is secreted from the epithelial cells into blood capillaries within the intestinal villi.

Digestion and Absorption of Proteins

Protein digestion begins in the stomach with the action of pepsin. Some amino acids are liberated in the stomach, but the major products of pepsin digestion are short-chain polypeptides. Pepsin digestion helps to produce a more homogenous chyme, but it is not essential for the complete digestion of protein that occurs—even in people with total gastrectomies—in the small intestine.

Most protein digestion occurs in the duodenum and jejunum. The pancreatic juice enzymes **trypsin, chymotrypsin,** and **elastase** cleave peptide bonds within the interior of the polypeptide chains. These enzymes are thus grouped together as *endopeptidases*. Enzymes that remove amino acids from the ends of polypeptide chains, by contrast, are *exopeptidases*. These include the pancreatic juice enzyme **carboxypeptidase,** which removes amino acids from the carboxyl-terminal end of polypeptide chains, and the brush border enzyme **aminopeptidase.** Aminopeptidase cleaves amino acids from the amino-terminal end of polypeptide chains.

As a result of the action of these enzymes, polypeptide chains are digested into free amino acids, dipeptides, and tripeptides. The free amino acids are absorbed by cotransport with Na^+ into the epithelial cells and secreted into blood capillaries. The dipeptides and tripeptides may enter epithelial cells by a different carrier system, but they are then digested within these cells into amino acids, which are secreted into the blood (fig. 18.31).

Newborn babies appear to be capable of absorbing a substantial amount of undigested proteins (hence they can absorb antibodies from their mother's first milk); in adults, however, only the free amino acids enter the portal vein. Foreign food protein, which would be very antigenic, does not normally enter the blood. An interesting exception is the protein toxin that causes botulism, produced by the bacterium *Clostridium botulinum*. This protein is resistant to digestion and is absorbed into the blood.

Digestion and Absorption of Lipids

The salivary glands and stomach of neonates (newborns) produce lipases. In adults, however, there is very little lipid digestion until the lipid globules in chyme arrive in the duodenum. Through mechanisms described in the next section, the arrival of lipids (primarily triglyceride, or fat) in the duodenum serves as a stimulus for the secretion of bile. In a process called **emulsification,** bile salt micelles are secreted into the duodenum and act to break up the fat droplets into tiny *emulsification droplets* of triglycerides. Note that emulsification is not chemical digestion—the bonds joining glycerol and fatty acids are not hydrolyzed by this process.

Digestion of Lipids

The emulsification of fat aids digestion because the smaller and more numerous emulsification droplets present a greater surface area than the unemulsified fat droplets that originally entered the duodenum. Fat digestion occurs at the surface of the droplets through the enzymatic action of **pancreatic lipase,** which is aided in its action by a protein called *colipase* (also secreted by the pancreas) that coats the emulsification droplets and "anchors" the lipase enzyme to them. Through hydrolysis, lipase removes two of the three fatty acids from each triglyceride molecule and thus liberates *free fatty acids* and *monoglycerides* (fig. 18.32). **Phospholipase A** likewise digests phospholipids such as lecithin into fatty acids and lysolecithin (the remainder of the lecithin molecule after two fatty acids are removed).

Free fatty acids, monoglycerides, and lysolecithin are more polar than the undigested lipids and quickly become asso-

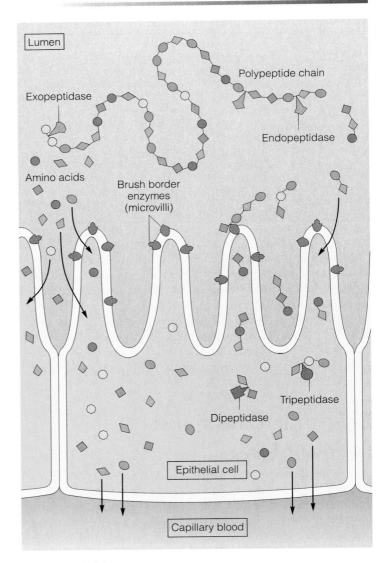

Figure 18.31

Polypeptide chains are digested into free amino acids, dipeptides, and tripeptides by the action of pancreatic juice enzymes and brush border enzymes. The amino acids, dipeptides, and tripeptides enter duodenal epithelial cells. Dipeptides and tripeptides are hydrolyzed into free amino acids within the epithelial cells, and these products are secreted into capillaries that carry them to the hepatic portal vein.

ciated with micelles of bile salts, lecithin, and cholesterol to form "mixed micelles" (fig. 18.33). These micelles then move to the brush border of the intestinal epithelium where absorption occurs.

Absorption of Lipids

Free fatty acids, monoglycerides, and lysolecithin can leave the micelles and pass through the membrane of the microvilli to enter the intestinal epithelial cells. There is also some evidence that the micelles may be transported intact into the epithelial cells and that the lipid digestion products may be removed intracellularly from the micelles. In either event, these products

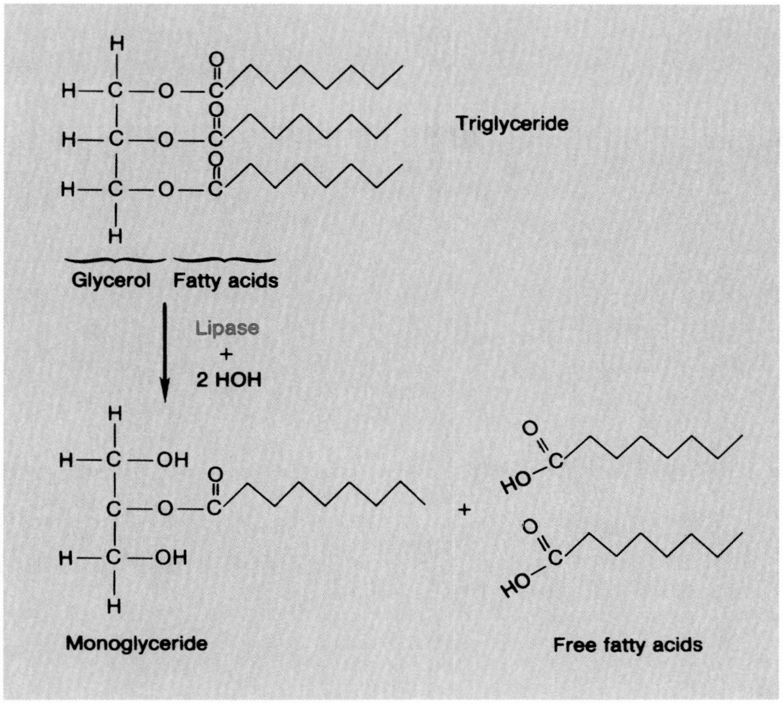

Figure 18.32

Pancreatic lipase digests fat (triglycerides) by cleaving off the first and third fatty acids. This produces free fatty acids and monoglycerides. Sawtooth lines indicate hydrocarbon chains in the fatty acids.

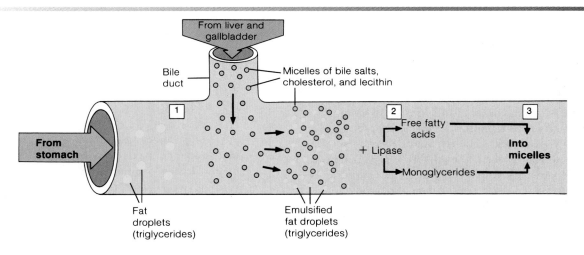

Step 1 Emulsification of fat droplets by bile salts

Step 2 Hydrolysis of triglycerides in emulsified fat droplets into fatty acid and monoglycerides

Step 3 Dissolving of fatty acids and monoglycerides into micelles to produce "mixed micelles"

Figure 18.33

Steps in the digestion of fat (triglycerides) and the entry of fat digestion products (fatty acids and monoglycerides) into micelles of bile salts secreted by the liver into the duodenum.

are used to *resynthesize* triglycerides and phospholipids within the epithelial cells. This process is different from the absorption of amino acids and monosaccharides, which pass through the epithelial cells without being altered.

Triglycerides, phospholipids, and cholesterol are then combined with protein inside the epithelial cells to form small particles called **chylomicrons.** These tiny lipid and protein combinations are secreted into the lymphatic capillaries of the intestinal villi (fig. 18.34). Absorbed lipids thus pass through the lymphatic system, eventually entering the venous blood by way of the thoracic duct (chapter 13). By contrast, amino acids and monosaccharides enter the hepatic portal vein.

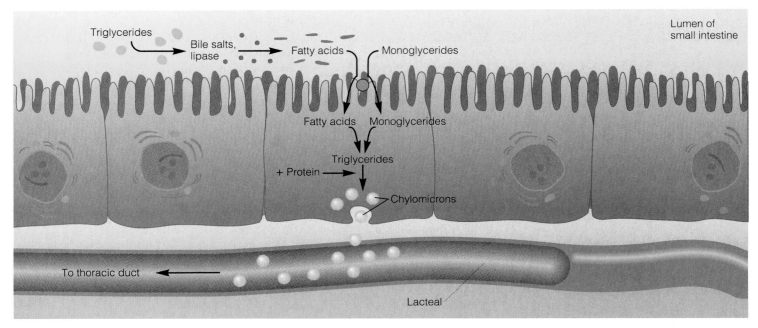

Figure 18.34

Fatty acids and monoglycerides from the micelles within the small intestine are absorbed by epithelial cells and converted intracellularly into triglycerides. These are then combined with protein to form chylomicrons, which enter the lymphatic vessels (lacteals) of the villi. These lymphatic vessels transport the chylomicrons to the thoracic duct, which empties them into the venous blood (of the left subclavian vein).

Transport of Lipids in the Blood

Once the chylomicrons are in the blood, their triglyceride content is removed by the enzyme **lipoprotein lipase,** which is attached to the endothelium of blood vessels. This enzyme hydrolyzes triglycerides and thus provides free fatty acids and glycerol for use by the tissue cells. The remaining *remnant particles*, containing cholesterol, are taken up by the liver. This is a process of endocytosis (chapter 6), which requires membrane receptors for the protein part (or *apoprotein*) of the remnant particle.

Cholesterol and triglycerides produced by the liver are combined with other apoproteins and secreted into the blood as *very-low-density lipoproteins* (*VLDL*), which serve to deliver triglycerides to different organs. Once the triglycerides are removed, the VLDL particles are converted to *low-density lipoproteins* (*LDL*), which transport cholesterol to various organs, including blood vessels. This can contribute to the development of atherosclerosis (chapter 13). Excess cholesterol is returned from these organs to the liver attached to *high-density lipoproteins* (*HDL*). A high ratio of HDL-cholesterol to total cholesterol is believed to offer protection against atherosclerosis (chapter 13). The characteristics of these lipoproteins are summarized in table 18.11.

1. List the enzymes involved in carbohydrate digestion, indicating their origins, sites of action, substrates, and products.

2. List each enzyme involved in protein digestion, indicating its origin and site of action. State whether it is an endopeptidase or exopeptidase. Compare the characteristics of pepsin and trypsin.

3. Describe how bile aids both the digestion and absorption of fats. Explain how the absorption of fat differs from the absorption of amino acids and monosaccharides.

4. Trace the pathway and fate of a molecule of triglyceride and cholesterol in a chylomicron within an intestinal epithelial cell.

5. Cholesterol in the blood may be attached to any of four possible lipoproteins. Distinguish between these proteins in terms of the origin and destination of the cholesterol they carry.

Table 18.11 Characteristics of the Lipid Carrier Proteins (Lipoproteins) Found in Plasma

Lipoprotein Class	Origin	Destination	Major Lipids	Functions
Chylomicrons	Intestine	Many organs	Triglycerides, other lipids	Deliver lipids of dietary origin to body cells
Very-low-density lipoproteins (VLDL)	Liver	Many organs	Triglycerides, cholesterol	Deliver endogenously produced triglycerides to body cells
Low-density lipoproteins (LDL)	Intravascular removal of triglycerides from VLDL	Blood vessels, liver	Cholesterol	Deliver endogenously produced cholesterol to various organs
High-density lipoproteins (HDL)	Liver and intestine	Liver and steroid-hormone-producing glands	Cholesterol	Remove and degrade cholesterol

Summary

Introduction to the Digestive System   p. 540

I. The digestion of food molecules involves the hydrolysis of these molecules into their subunits.
  A. The digestion of food occurs in the lumen of the GI tract and is catalyzed by specific enzymes.
  B. The digestion products are absorbed through the intestinal mucosa and enter the blood or lymph.
II. The layers (tunics) of the GI tract are, from the inside outward, the mucosa, submucosa, muscularis, and serosa.
  A. The mucosa consists of a simple columnar epithelium; a thin layer of connective tissue, called the lamina propria; and a thin layer of smooth muscle, called the muscularis mucosa.
  B. The submucosa is composed of connective tissue; the muscularis consists of layers of smooth muscles; the serosa is connective tissue covered with the visceral peritoneum.
  C. The submucosa contains the submucosal plexus, and the muscularis contains the myenteric plexus of autonomic nerves.

Esophagus and Stomach   p. 543

I. Peristaltic waves of contraction push food through the lower esophageal sphincter into the stomach.

II. The stomach consists of a cardia, fundus, body, and pyloris (antrum); the pyloris terminates with the pyloric sphincter.
  A. The lining of the stomach is thrown into folds, or rugae, and the mucosa is formed into gastric pits and gastric glands.
  B. The parietal cells of the gastric glands secrete HCl; the chief cells secrete pepsinogen.
  C. In the acidic environment of gastric juice, pepsinogen is converted into the active protein-digesting enzyme called pepsin.
  D. Some digestion of protein occurs in the stomach, but the most important function of the stomach is the secretion of intrinsic factor, which is needed for the absorption of vitamin B_{12} in the intestine.

Small Intestine   p. 549

I. Regions of the small intestine include the duodenum, jejunum, and ileum. The common bile duct and pancreatic duct empty into the duodenum.
II. Fingerlike extensions of mucosa called villi project into the lumen, and at the bases of the villi the mucosa forms narrow pouches called the crypts of Lieberkühn.
  A. New epithelial cells are formed in the crypts.

  B. The membrane of intestinal epithelial cells is folded to form microvilli. This brush border of the mucosa increases the surface area.
III. Digestive enzymes, called brush border enzymes, are located in the membranes of the microvilli.
IV. The small intestine exhibits two major types of movements—peristalsis and segmentation.

Large Intestine   p. 552

I. The large intestine is divided into the cecum, colon, rectum, and anal canal.
  A. The appendix is attached to the inferior medial margin of the cecum.
  B. The colon consists of ascending, transverse, descending, and sigmoid portions.
  C. Bulges in the walls of the large intestine are called haustra.
II. Three types of movements occur in the large intestine: peristalsis, haustral churning, and mass movement.
III. The large intestine absorbs water and electrolytes.
  A. Although most of the water that enters the GI tract is absorbed in the small intestine, from 1 to 1.5 L pass to the large intestine each day. The larger intestine absorbs about 90% of this amount.

B. Na^+ is actively absorbed and water follows passively, in a manner analogous to the reabsorption of NaCl and water in the renal tubules.

IV. Defecation occurs when the anal sphincters relax and contraction of other muscles raises the rectal pressure.

Liver, Gallbladder, and Pancreas   *p. 555*

I. The liver, the largest internal organ, is composed of functional units called lobules.

   A. Liver lobules consist of plates of hepatic cells separated by capillary sinusoids.

   B. Blood flows from the periphery of each lobule, where the hepatic artery and portal vein empty, through the sinusoids and out the central vein.

   C. Bile flows within the hepatocyte plates, in canaliculi, to the bile ducts.

   D. Substances excreted in the bile can be returned to the liver in the hepatic portal blood. This is called an enterohepatic circulation.

   E. Bile consists of a pigment called bilirubin, bile salts, cholesterol, and other molecules.

   F. The liver detoxifies the blood by excreting substances in the bile, by phagocytosis, and by chemical inactivation.

   G. The liver modifies the plasma concentrations of proteins, glucose, triglycerides, and ketone bodies.

II. The gallbladder serves to store and concentrate the bile. It releases bile through the cystic duct and common bile duct to the duodenum.

III. The pancreas is both an exocrine and an endocrine gland.

   A. The endocrine portion is known as the islets of Langerhans and secretes the hormones insulin and glucagon.

   B. The exocrine acini of the pancreas produce pancreatic juice, which contains various digestive enzymes and bicarbonate.

Neural and Endocrine Regulation of the Digestive System   *p. 563*

I. The regulation of gastric function occurs in three phases.

   A. In the cephalic phase, the activity of higher brain centers, acting via the vagus nerve, stimulates gastric juice secretion.

   B. In the gastric phase, the secretion of HCl and pepsin is controlled by the gastric contents and by the hormone gastrin, secreted by the gastric mucosa.

   C. In the intestinal phase, the activity of the stomach is inhibited by neural reflexes and hormonal secretion from the duodenum.

II. Intestinal function is regulated, in part by local short reflexes.

   A. The submucosal and myenteric plexuses contain autonomic motor neurons, sensory neurons, and interneurons.

   B. Short reflexes include the intestino-intestinal reflexes, gastroileal reflex, and ileogastric reflex.

III. The secretion of the hormones secretin and cholecystokinin (CCK) regulates pancreatic juice and bile secretion.

   A. Secretin secretion is stimulated by the arrival of acidic chyme into the duodenum.

   B. CCK secretion is stimulated by the presence of fat in the chyme arriving in the duodenum.

   C. Contraction of the gallbladder occurs in response to a neural reflex and to the secretion of CCK by the duodenum.

IV. Gastrointestinal hormones may be needed for the maintenance of the GI tract and accessory digestive organs.

Digestion and Absorption of Carbohydrates, Lipids, and Proteins   *p. 566*

I. The digestion of starch begins in the mouth through the action of salivary amylase.

   A. Pancreatic amylase digests starch into disaccharides and short-chain oligosaccharides.

   B. Complete digestion into monosaccharides is accomplished by brush border enzymes.

II. Protein digestion begins in the stomach by the action of pepsin.

   A. Pancreatic juice contains the protein-digesting enzymes trypsin and chymotrypsin, among others.

   B. The brush border contains digestive enzymes that help to complete the digestion of proteins into amino acids.

   C. Amino acids, like monosaccharides, are absorbed and secreted into capillary blood entering the portal vein.

III. Lipids are digested in the small intestine after being emulsified by bile salts.

   A. Free fatty acids and monoglycerides enter particles called micelles, formed in large part by bile salts, and they are absorbed in this form or as free molecules.

   B. Once inside the mucosal epithelial cells, these subunits are used to resynthesize triglycerides.

   C. Triglycerides in the epithelial cells, together with proteins, form chylomicrons, which are secreted into the central lacteals of the villi.

   D. Chylomicrons are transported by lymph to the thoracic duct and from there enter the blood.

Clinical Investigation

A male college student appears at the student health office complaining of severe but transient pain. He says that this pain often occurs as soon as he has finished a meal. Upon questioning, it is learned that the pain is located inferior to his right scapula, and that it occurs whenever he eats particular foods, such as peanut butter and bacon. The pain does not occur when he eats fish or drinks milk, cola, or alcoholic beverages. The sclera of this student's eyes are markedly yellow. Laboratory tests reveal that he has fatty stools, a prolonged bleeding time, and elevated blood levels of conjugated bilirubin. His blood tests, however, show normal levels of ammonia, urea, free bilirubin, and pancreatic amylase. His white blood cell count is normal, and he does not have fever. What do these observations reveal about the possible cause of his symptoms?

Clues

Read the sections "Gastritis and Peptic Ulcers" and "Bile Production and Secretion." Study the boxed information on jaundice and on gallstones.

Review Activities

Objective Questions

1. Which of the following statements about intrinsic factor is *true?*
   a. It is secreted by the stomach.
   b. It is a polypeptide.
   c. It promotes absorption of vitamin B_{12} in the intestine.
   d. It helps prevent pernicious anemia.
   e. All of the above are true.

2. Intestinal enzymes such as lactase are
   a. secreted by the intestine into the chyme.
   b. produced by the crypts of Lieberkühn.
   c. produced by the pancreas.
   d. attached to the cell membrane of microvilli in the epithelial cells of the mucosa.

3. Which of the following statements about gastric secretion of HCl is *false?*
   a. HCl is secreted by parietal cells.
   b. HCl hydrolyzes peptide bonds.
   c. HCl is needed for the conversion of pepsinogen to pepsin.
   d. HCl is needed for maximum activity of pepsin.

4. Most digestion occurs in
   a. the mouth.
   b. the stomach.
   c. the small intestine.
   d. the large intestine.

5. Which of the following statements about trypsin is *true?*
   a. Trypsin is derived from trypsinogen by the digestive action of pepsin.
   b. Active trypsin is secreted into the pancreatic acini.
   c. Trypsin is produced in the crypts of Lieberkühn.
   d. Trypsinogen is converted to trypsin by the brush border enzyme enterokinase.

6. During the gastric phase, the secretion of HCl and pepsinogen is stimulated by
   a. vagus nerve stimulation that originates in the brain.
   b. polypeptides in the gastric lumen and by gastrin secretion.
   c. secretin and cholecystokinin from the duodenum.
   d. all of the above.

7. The secretion of HCl by the stomach mucosa is inhibited by
   a. neural reflexes from the duodenum.
   b. the secretion of an enterogastrone from the duodenum.
   c. the lowering of gastric pH.
   d. all of the above.

8. The first organ to receive the blood-borne products of digestion is
   a. the liver.
   b. the pancreas.
   c. the heart.
   d. the brain.

9. Which of the following statements about hepatic portal blood is *true?*
   a. It contains absorbed fat.
   b. It contains ingested proteins.
   c. It is mixed with bile in the liver.
   d. It is mixed with blood from the hepatic artery in the liver.

10. Absorption of salt and water is the principal function of which region of the GI tract?
    a. esophagus
    b. stomach
    c. duodenum
    d. jejunum
    e. large intestine

11. Cholecystokinin (CCK) is a hormone that stimulates
    a. bile production.
    b. release of pancreatic enzymes.
    c. contraction of the gallbladder.
    d. both *a* and *b*.
    e. both *b* and *c*.

12. Which of the following statements about vitamin B_{12} is *false?*
    a. Lack of this vitamin can produce pernicious anemia.
    b. Intrinsic factor is needed for absorption of vitamin B_{12}.
    c. Damage to the gastric mucosa may lead to a deficiency in vitamin B_{12}.
    d. Vitamin B_{12} is absorbed primarily in the jejunum.

13. Which of the following statements about starch digestion is *false?*
    a. It begins in the mouth.
    b. It occurs in the stomach.

c. It requires the action of pancreatic amylase.

d. It requires brush border enzymes for completion.

14. Which of the following statements about fat digestion and absorption is *false*?

  a. Emulsification by bile salts increases the rate of fat digestion.

b. Triglycerides are hydrolyzed by the action of pancreatic lipase.

c. Triglycerides are resynthesized from monoglycerides and fatty acids in the intestinal epithelial cells.

d. Triglycerides, as particles called chylomicrons, are absorbed into blood capillaries within the villi.

15. Which of the following statements about contraction of intestinal smooth muscle is *true*?

  a. It occurs automatically.

  b. It is increased by parasympathetic nerve stimulation.

  c. It produces segmentation.

  d. All of the above are true.

Essay Questions

1. Explain how the gastric secretion of HCl and pepsin is regulated during the cephalic, gastric, and intestinal phases.[1]

2. Describe how pancreatic enzymes become activated in the lumen of the intestine. Why are these mechanisms needed?

3. Explain the function of bicarbonate in pancreatic juice. How may peptic ulcers in the duodenum be produced?

4. Describe the mechanisms that are believed to protect the gastric mucosa from self-digestion and state some proposed reasons for the development of a peptic ulcer in the stomach.

5. Explain why the pancreas is considered to be both an exocrine and an endocrine gland. Given this information, predict what effects tying of the pancreatic duct would have on pancreatic structure and function.

6. Explain how jaundice is produced when (a) the person has gallstones, (b) the person has a high rate of red blood cell destruction, and (c) the person has liver disease. In which of these cases would phototherapy for the jaundice be effective? Explain.

7. Describe the steps involved in the digestion and absorption of fat.

8. Distinguish between chylomicrons, very-low-density lipoproteins, low-density lipoproteins, and high-density lipoproteins.

9. Identify the different neurons present in the wall of the intestine and explain how these neurons are involved in "short reflexes."

10. Trace the course of blood flow through the liver and discuss the significance of this pattern in terms of the detoxication of the blood. Describe the enzymes and the reactions involved in this detoxication.

Selected Readings

Achord, J. L. 1995. Alcohol and the liver. *Science and Medicine* 2:16.

Binder, H. J. 1984. The pathophysiology of diarrhea. *Hospital Practice* 19:107.

Bleich, H. L., and E. S. Boro. 1979. Protein digestion and absorption. *New England Journal of Medicine* 300:659.

Bortoff, A. 1972. Digestion. *Annual Review of Physiology* 28:201.

Brasitus, T. A., and M. D. Sitrin. 1990. Intestinal malabsorption syndromes. *Annual Review of Medicine* 41:339.

Buller, H. A., and R. J. Grand. 1991. Lactose intolerance. *Annual Review of Medicine* 41:141.

Carey, M. C., D. M. Small, and C. M. Bliss. 1983. Lipid digestion and absorption. *Annual Review of Physiology* 45:651.

Chou, C. C. 1982. Relationship between intestinal blood flow and motility. *Annual Review of Physiology* 44:29.

Christensen, R. R. 1971. The controls of gastrointestinal movements: Some old and new views. *New England Journal of Medicine* 285:483.

Cohen, S. 1983. Neuromuscular disorders of the gastrointestinal tract. *Hospital Practice* 18:121.

Davenport, H. W. 1982. *Physiology of the Digestive Tract.* 5th ed. Chicago: Year Book Medical Publishers.

DelValle, J., and T. Yamada. 1991. The gut as an endocrine organ. *Annual Review of Medicine* 41:447.

Dockray, G. J. 1979. Comparative biochemistry and physiology of gut hormones. *Annual Review of Physiology* 41:83.

Duane, W. C. 1990. Pathogenesis of gallstones: Implications for management. *Hospital Practice* 24:65.

Feldman, M., and M. E. Burton. 1990. Histamine$_2$ receptor antagonists: Standard therapy for acid-peptic diseases. *New England Journal of Medicine* 323:1672.

Freeman, H. J., and Y. S. Kim. 1978. Digestion and absorption of proteins. *Annual Review of Physiology* 29:99.

Gardner, J. D., and R. T. Jensen. 1986. Receptors and cell activation associated with pancreatic enzyme secretion. *Annual Review of Physiology* 48:103.

Gollan, J. L., and A. B. Knapp. 1985. Bilirubin metabolism and congenital jaundice. *Hospital Practice* 20:83.

Gonzalez-Gallego, J. 1995. New concepts in hepatocellular transport and metabolism of bilirubin. *News in Physiological Sciences* 10:35.

Gray, G. M. 1975. Carbohydrate digestion and absorption: Role of the small intestine. *New England Journal of Medicine* 292:1225.

Grelot, L., and A. D. Miller. 1994. Vomiting—its ins and outs. *News in Physiological Sciences* 9:142.

Grossman, M. I. 1979. Neural and hormonal regulation of gastrointestinal function: An overview. *Annual Review of Physiology* 41:27.

Guengerich, F. P. 1993. Cytochrome P450 enzymes. *American Scientist* 81:440.

[1]Note: This question is answered on page 231 of the Student Study Guide.

Chapter Eighteen

Guth, P. H. 1982. Stomach blood flow and acid secretion. *Annual Review of Physiology* 44:3.

Hersey, S. J., S. H. Norris, and A. J. Gilbert. 1984. Cellular control of pepsinogen secretion. *Annual Review of Physiology* 46:393.

Hoffman, A. F. 1990. Nonsurgical treatment of gallstone disease. *Annual Review of Medicine* 41:401.

Holt, K. M., and J. I. Isenberg. 1985. Peptic ulcer disease: Physiology and pathophysiology. *Hospital Practice* 20:89.

Johnston, D. E., and M. M. Kaplan. 1993. Pathogenesis and treatment of gallstones. *New England Journal of Medicine* 328:412.

Kappas, A., and A. P. Alvarez. June 1975. How the liver metabolizes foreign substances. *Scientific American*.

Keef, K. D. et al. 1993. Enteric inhibitory neural regulation of human colonic circular muscle: Role of nitric oxide. *Gastroenterology* 105:1009.

Lewin, M. J. M. 1992. The somatostatin receptor in the GI tract. *Annual Review of Physiology* 54:455.

Lewis, L. D., and J. A. Williams. 1990. Cholecystokinin: A key integrator of nutrient assimilation. *News in Physiological Sciences* 5:163.

Livingston, E. H., and P. H. Guth. 1992. Peptic ulcer disease. *American Scientist* 80:592.

Makhlouf, G. M. 1990. Neural and hormonal regulation of function in the gut. *Hospital Practice* 24:79.

Makhlouf, G. M., and J. R. Grider. 1993. Nonadrenergic noncholinergic inhibitory transmitters of the gut. *News in Physiological Sciences* 8:195.

McGuigan, J. E. 1978. Gastrointestinal hormones. *Annual Review of Physiology* 29:99.

Moog, F. November 1981. The lining of the small intestine. *Scientific American*.

Potter, G. D. 1990. Intestinal development and regeneration. *Hospital Practice* 24:131.

Rink, T. J. 1994. In search of a satiety factor. *Nature* 372:406.

Salen, G., and S. Shefer. 1983. Bile acid synthesis. *Annual Review of Physiology* 45:679.

Sanders, M. J., and A. H. Soll. 1986. Characterization of receptors regulating secretory function in the fundic mucosa. *Annual Review of Physiology* 48:89.

Schiller, L. R. 1994. Peristalsis. *Science and Medicine* 1:38.

Smith, B. F., and T. Lamont. 1984. The pathogenesis of gallstones. *Hospital Practice* 19:93.

Soll, A., and J. H. Walsh. 1979. Regulation of gastric acid secretion. *Annual Review of Physiology* 41:35.

Ulvnas-Mosberg, K. July 1989. The gastrointestinal tract in growth and reproduction. *Scientific American*.

Waisbren, S. J. et al. 1994. Unusual permeability properties of gastric gland cells. *Nature* 368:332.

Walsh, J. H. 1988. Peptides as regulators of gastric acid secretion. *Annual Review of Physiology* 50:41.

Walsh, J. H., and M. I. Grossman. 1975. Gastrin. *New England Journal of Medicine* 292: first part, p. 1324; second part, p. 1377.

Weisbrodt, N. W. 1981. Patterns of intestinal motility. *Annual Review of Physiology* 43:21.

Williams, J. A. 1984. Regulatory mechanisms in pancreas and salivary acini. *Annual Review of Physiology* 46:361.

Wolfe, M. M., and A. H. Soll. 1988. The physiology of gastric acid secretion. *New England Journal of Medicine* 319:1707.

Life Science Animations

The animations that relate to chapter 18 are #33 Peristalsis, #34 Digestion of Carbohydrates, #35 Digestion of Proteins, and #36 Digestion of Lipids.

Regulation of Metabolism

19

OBJECTIVES

After studying this chapter, you should be able to . . .

1. explain the significance of the basal metabolic rate and state how the metabolic rate is affected by physical activity, temperature, and eating.

2. distinguish between the caloric and anabolic requirements for food and define the terms *essential amino acids* and *essential fatty acids*.

3. distinguish between fat-soluble and water-soluble vitamins and describe some of the functions of different vitamins.

4. define the terms *energy reserves* and *circulating energy substrates* and explain how these sources of energy interact during anabolism and catabolism.

5. describe the regulation of eating and discuss the endocrine control of metabolism in general terms.

6. describe the actions of insulin and glucagon and explain how the secretion of these hormones is regulated.

7. explain how insulin and glucagon regulate metabolism during feeding and fasting.

8. describe the symptoms of insulin-dependent and non-insulin-dependent diabetes mellitus and explain how these conditions are produced.

9. describe the metabolic effects of epinephrine and the glucocorticoids.

10. describe the effects of thyroxine on cell respiration and explain the relationship between thyroxine levels and the basal metabolic rate.

11. describe the symptoms of hypothyroidism and hyperthyroidism and explain how these conditions are produced.

12. describe the metabolic effects of growth hormone and explain why growth hormone and thyroxine are needed for proper body growth.

13. describe the actions of parathyroid hormone, 1,25-dihydroxyvitamin D_3, and calcitonin and explain how the secretion of these hormones is regulated.

14. describe how 1,25-dihydroxyvitamin D_3 is produced and explain why this compound is needed to prevent osteomalacia and rickets.

OUTLINE

Nutritional Requirements

The body's energy requirements must be met by the caloric value of food to prevent catabolism of the body's own fat, carbohydrates, and protein. Additionally, food molecules—particularly the essential amino acids and fatty acids—are needed for replacement of molecules in the body that are continuously degraded. Vitamins and elements do not directly provide energy but instead are required for diverse enzymatic reactions.

Living tissue is maintained by the constant expenditure of energy. This energy is obtained directly from ATP and indirectly from the cell respiration of glucose, fatty acids, ketone bodies, amino acids, and other organic molecules. These molecules are ultimately obtained from food, but they can also be obtained from the glycogen, fat, and protein stored in the body.

The energy value of food is commonly measured in **kilocalories,** which are also called "big calories" and spelled with a capital letter (Calories). One kilocalorie is equal to 1,000 calories; one calorie is defined as the amount of heat required to raise the temperature of one cubic centimeter of water from 14.5° to 15.5°C. As described in chapter 5, the amount of energy released as heat when a quantity of food is combusted in vitro is equal to the amount of energy released within cells through the process of aerobic respiration. This is 4 calories per gram for carbohydrates or proteins, and 9 calories per gram for fat. When this energy is released by cell respiration, some is transferred to the high-energy bonds of ATP and some is lost as heat.

Metabolic Rate and Caloric Requirements

The total rate of body metabolism, or the **metabolic rate,** can be measured by either the amount of heat generated by the body or by the amount of oxygen consumed by the body per minute. This rate is influenced by a variety of factors. For example, the metabolic rate is increased by physical activity and by eating. The increased rate of metabolism that accompanies the assimilation of food can last more than 6 hours after a meal.

Temperature is also an important factor in determining metabolic rate. The reasons for this are twofold: (1) temperature itself is a determinant of the rate of chemical reactions and (2) the hypothalamus contains *temperature control centers,* as well as temperature-sensitive cells that act as sensors for changes in body temperature. In response to deviations from a "set point" for body temperature (chapter 1), the control areas of the hypothalamus can direct physiological responses that help to correct the deviations and maintain a constant body temperature. Changes in body temperature are thus accompanied by physiological responses that influence the total metabolic rate.

Hypothermia (low body temperature)—where the core body temperature is lowered to between 21° and 24°C—is often induced during open heart or brain surgery. Compensatory responses are dampened by the general anesthetic, and the lower body temperature drastically reduces the needs of the tissues for oxygen. Under these conditions, the heart can be stopped, and bleeding is significantly reduced.

The metabolic rate (measured by the rate of oxygen consumption) of an awake, relaxed person 12 to 14 hours after eating and at a comfortable temperature is known as the **basal metabolic rate (BMR).** The BMR is determined primarily by a person's age, sex, and overall body surface area, but it is also strongly influenced by the level of thyroid secretion. A person with hyperthyroidism has an abnormally high BMR, and a person with hypothyroidism has a low BMR. An interesting recent finding is that the BMR may be influenced by genetic inheritance, and that at least some families that are prone to obesity may have a genetically determined low BMR.

In general, however, individual differences in energy requirements are due primarily to differences in physical activity. Daily energy expenditures may range from about 1,300 to 5,000 kilocalories per day. The average values for people not engaged in heavy manual labor but who are active during their leisure time is about 2,900 kilocalories per day for men and 2,100 kilocalories per day for women. People engaged in office work, the professions, sales, and comparable occupations consume up to 5 kilocalories per minute during work. More physically demanding occupations may require energy expenditures of 7.5 to 10 kilocalories per minute.

When the caloric intake is greater than the energy expenditures, excess calories are stored primarily as fat. This is true regardless of the source of the calories—carbohydrates, protein, or fat—because these molecules can be converted to fat by the metabolic pathways described in chapter 5. Appropriate body weights are indicated in table 19.1.

Weight is lost when the caloric value of the food ingested is less than the amount required in cell respiration over a period of time. Weight loss, therefore, can be achieved by dieting alone or in combination with an exercise program to raise the metabolic rate. A summary of the caloric expenditure associated with different forms of exercise is provided in table 19.2.

Anabolic Requirements

In addition to providing the body with energy, food also supplies the raw materials for synthesis reactions—collectively termed **anabolism**—that occur constantly within the cells of the body. Anabolic reactions include those that synthesize DNA and RNA, protein, glycogen, triglycerides, and other polymers. These anabolic reactions must occur constantly to replace those molecules that are hydrolyzed into their component

Table 19.1 Body Weights

Height Feet	Inches	Small Frame	Medium Frame	Large Frame		Height Feet	Inches	Small Frame	Medium Frame	Large Frame
Men 5	2	128–134	131–141	138–150	Women	4	10	102–111	109–121	118–131
5	3	130–136	133–143	140–153		4	11	103–113	111–123	120–134
5	4	132–138	135–145	142–156		5	0	104–115	113–126	122–137
5	5	134–140	137–148	144–160		5	1	106–118	115–129	125–140
5	6	136–142	139–151	146–164		5	2	108–121	118–132	128–143
5	7	138–145	142–154	149–168		5	3	111–124	121–135	131–147
5	8	140–148	145–157	152–172		5	4	114–127	124–138	134–151
5	9	142–151	148–160	155–176		5	5	117–130	127–141	137–155
5	10	144–154	151–163	158–180		5	6	120–133	130–144	140–159
5	11	146–157	154–166	161–184		5	7	123–136	133–147	143–163
6	0	149–160	157–170	164–188		5	8	126–139	136–150	146–167
6	1	152–164	160–174	168–192		5	9	129–142	139–153	149–170
6	2	155–168	164–178	172–197		5	10	132–145	142–156	152–173
6	3	158–172	167–182	176–202		5	11	135–148	145–159	155–176
6	4	162–176	171–187	181–207		6	0	138–151	148–162	158–179

Weights at ages 25–59 based on lowest mortality. Weight in pounds according to frame (in indoor clothing weighing 5 lbs., shoes with 1″ heels).

Weights at ages 25–59 based on lowest mortality. Weight in pounds according to frame (in indoor clothing weighing 3 lbs., shoes with 1″ heels).

Source: Statistical Bulletin, 1983. Courtesy of Metropolitan Life Insurance Company.

Table 19.2 Energy Consumed (in Kilocalories per Minute) by Different Types of Activities

Activity	Weight in Pounds			
	105–115	127–137	160–170	182–192
Bicycling				
10 mph	5.41	6.16	7.33	7.91
Stationary, 10 mph	5.50	6.25	7.41	8.16
Calisthenics	3.91	4.50	7.33	7.91
Dancing				
Aerobic	5.83	6.58	7.83	8.58
Square	5.50	6.25	7.41	8.00
Gardening, Weeding, and Digging	5.08	5.75	6.83	7.50
Jogging				
5.5 mph	8.58	9.75	11.50	12.66
6.5 mph	8.90	10.20	12.00	13.20
8.0 mph	10.40	11.90	14.10	15.50
9.0 mph	12.00	13.80	16.20	17.80
Rowing, Machine				
Easily	3.91	4.50	5.25	5.83
Vigorously	8.58	9.75	11.50	12.66
Skiing				
Downhill	7.75	8.83	10.41	11.50
Cross-country, 5 mph	9.16	10.41	12.25	13.33
Cross-country, 9 mph	13.08	14.83	17.58	19.33
Swimming, Crawl				
20 yards per minute	3.91	4.50	5.25	5.83
40 yards per minute	7.83	8.91	10.50	11.58
55 yards per minute	11.00	12.50	14.75	16.25
Walking				
2 mph	2.40	2.80	3.30	3.60
3 mph	3.90	4.50	5.30	5.80
4 mph	4.50	5.20	6.10	6.80

Regulation of Metabolism

Table 19.3 Recommended Daily Allowances For Vitamins and Minerals[1]

Category	Age (Years) or Condition	Weight[2] (kg)	(lb)	Height[2] (cm)	(in)	Protein (g)	Fat-Soluble Vitamins Vitamin A (μg RE)[3]	Vitamin D (μg)[4]	Vitamin E (mg α-TE)[5]	Vitamin K (μg)
Infants	0.0–0.05	6	13	60	24	13	375	7.5	3	5
	0.5–1	9	20	71	28	14	375	10	4	10
Children	1–3	13	29	90	35	16	400	10	6	15
	4–6	20	44	112	44	24	500	10	7	20
	7–10	28	62	132	52	28	700	10	7	30
Males	11–14	45	99	157	62	45	1,000	10	10	45
	15–18	66	145	176	69	59	1,000	10	10	65
	19–24	72	160	177	70	58	1,000	10	10	70
	25–50	79	174	176	70	63	1,000	5	10	80
	51+	77	170	173	68	63	1,000	5	10	80
Females	11–14	46	101	157	62	46	800	10	8	45
	15–18	55	120	163	64	44	800	10	8	55
	19–24	58	128	164	65	46	800	10	8	60
	25–50	63	138	163	64	50	800	5	8	65
	51+	65	143	160	63	50	800	5	8	65
Pregnant						60	800	10	10	65
Lactating	1st 6 months					65	1,300	10	12	65
	2nd 6 months					62	1,200	10	11	65

Source: Reprinted with permission from Recommended Dietary Allowances, 10th Edition. Copyright 1989 by the National Academy of Sciences. Courtesy of the National Academy Press, Washington, D.C.

[1] The allowances, expressed as average daily intakes over time, are intended to provide for individual variations among most normal persons as they live in the United States under usual environmental stresses. Diets should be based on a variety of common foods in order to provide other nutrients for which human requirements have been less well defined.

[2] Weights and heights of Reference Adults are actual medians for the U.S. population of the designated age, as reported by NHANES II. The use of these figures does not imply that the height-to-weight ratios are ideal.

[3] Retinol equivalents. 1 RE = 1 μg retinol or 6 μg β-carotene.

[4] As cholecalciferol. 10 μg cholecalciferol = 400 w of vitamin D.

[5] α-tocopherol equivalents. 1 mg d-α-tocopherol = 1 α-TE.

Obesity is a risk factor for cardiovascular diseases, diabetes mellitus, gallbladder disease, and some malignancies (particularly endometrial and breast cancer). The distribution of fat in the body is also important; there is a greater risk of cardiovascular disease when fat produces a high waist-to-hip ratio, or an "apple shape," as compared to a "pear shape." Obesity in childhood is due to an increase in both the size and number of adipose cells; weight gain in adulthood is due mainly to an increase in adipose cell size, although the number of adipose cells may also increase in extreme weight gains. When weight is lost, the adipose cells get smaller, but their number remains constant. It is thus important to prevent further increases in weight in all overweight people but particularly so in children.

monomers. These hydrolysis reactions, together with the reactions of cell respiration that ultimately break the monomers down to carbon dioxide and water, are collectively termed **catabolism.**

Acting through changes in hormonal secretion, exercise and fasting increase the catabolism of stored glycogen, fat, and body protein. These molecules are also broken down at a certain rate in a person who is neither exercising nor fasting. Some of the monomers thus formed (amino acids, glucose, and fatty acids) are used immediately to resynthesize body protein, glycogen, and fat. However, some of the glucose derived from stored glycogen, for example, or fatty acids derived from stored triglycerides, are used to provide energy in the process of cell respiration. For this reason, new monomers must be obtained from food to prevent a continual decline in the amount of protein, glycogen, and fat in the body.

The *turnover rate* of a particular molecule is the rate at which it is broken down and resynthesized. For example, the average daily turnover for carbohydrates is 250 g/day. Since some of the glucose in the body is reused to form glycogen, the average daily dietary requirement for carbohydrate is somewhat less than this amount—about 150 g/day. The average daily turnover for protein is 150 g/day, but since many of the amino acids derived from the catabolism of body proteins can be reused in protein synthesis, a person needs only about 35 g/day of protein in the diet. It should be noted that these are average figures and will vary in accordance with individual differences in size, sex, age, genetics, and physical activity. The average daily turnover for fat is about 100 g/day, but very little is required in the diet (other than that which supplies fat-soluble vitamins and essential fatty acids), since fat can be produced from excess carbohydrates.

The minimal amounts of dietary protein and fat required to meet the turnover rate are adequate only if they supply sufficient amounts of the essential amino acids and fatty acids.

Water-Soluble Vitamins							Minerals						
Vitamin C (mg)	Thiamin (mg)	Riboflavin (mg)	Niacin (mg NE)[6]	Vitamin B_6 (mg)	Folate (μg)	Vitamin B_{12} (μg)	Calcium (mg)	Phosphorus (mg)	Magnesium (mg)	Iron (mg)	Zinc (mg)	Iodine (μg)	Selenium (μg)
30	0.3	0.4	5	0.3	25	0.3	400	300	40	6	5	40	10
35	0.4	0.5	6	0.6	35	0.5	600	500	60	10	5	50	15
40	0.7	0.8	9	1.0	50	0.7	800	800	80	10	10	70	20
45	0.9	1.1	12	1.1	75	1.0	800	800	120	10	10	90	20
45	1.0	1.2	13	1.4	100	1.4	800	800	170	10	10	120	30
50	1.3	1.5	17	1.7	150	2.0	1,200	1,200	270	12	15	150	40
60	1.5	1.8	20	2.0	200	2.0	1,200	1,200	400	12	15	150	50
60	1.5	1.7	19	2.0	200	2.0	1,200	1,200	350	10	15	150	70
60	1.5	1.7	19	2.0	200	2.0	800	800	350	10	15	150	70
60	1.2	1.4	15	2.0	200	2.0	800	800	350	10	15	150	70
50	1.1	1.3	15	1.4	150	2.0	1,200	1,200	280	15	12	150	45
60	1.1	1.3	15	1.5	180	2.0	1,200	1,200	300	15	12	150	50
60	1.1	1.3	15	1.6	180	2.0	1,200	1,200	280	15	12	150	55
60	1.1	1.3	15	1.6	180	2.0	800	800	280	15	12	150	55
60	1.0	1.2	13	1.6	180	2.0	800	800	280	10	12	150	55
70	1.5	1.6	17	2.2	400	2.2	1,200	1,200	300	30	15	175	65
95	1.6	1.8	20	2.1	280	2.6	1,200	1,200	355	15	19	200	75
90	1.6	1.7	20	2.1	260	2.6	1,200	1,200	340	15	16	200	75

[6] Niacin equivalents. 1 NE = 1 mg of niacin or 60 mg of dietary tryptophan.

These molecules are termed *essential* because they are necessary for proper protein and fat synthesis but must be obtained in the diet, since the body cannot make them. The nine **essential amino acids** are lysine, methionine, valine, leucine, isoleucine, tryptophane, phenylalanine, threonine, and histidine. The **essential fatty acids** are linoleic acid and linolenic acid.

Vitamins and Elements

Vitamins are small organic molecules that serve as part of coenzymes in metabolic reactions or that have other, specific functions. They must be obtained in the diet because the body either doesn't produce them, or it produces them in insufficient quantities. (Vitamin D is produced in small quantities by the skin, and the B vitamins and vitamin K are produced by intestinal bacteria.) There are two classes of vitamins: fat-soluble and water-soluble. The **fat-soluble vitamins** include vitamins A, D, E, and K. The **water-soluble vitamins** include thiamine (B_1), riboflavin (B_2), niacin (B_3), pyridoxine (B_6), pantothenic acid, biotin, folic acid, vitamin B_{12}, and vitamin C (ascorbic acid). Recommended daily allowances for these vitamins are listed in table 19.3.

Derivatives of water-soluble vitamins serve as coenzymes in the metabolism of carbohydrates, lipids, and proteins. Thiamine, for example, is needed for the activity of the enzyme that converts pyruvic acid to acetyl coenzyme A. Riboflavin and niacin are needed for the production of FAD and NAD, respectively. FAD and NAD serve as coenzymes that transfer hydrogens during cell respiration (chapter 4). Pyridoxine is a cofactor for the enzymes involved in amino acid metabolism. Deficiencies of the water-soluble vitamins can thus have widespread effects in the body (table 19.4).

Beta carotene is a provitamin; it is obtained in the diet and converted within the body cells into vitamin A. This conversion occurs when the β-carotene molecules bind free electrons. β-carotene thus serves as an "electron scavenger," or an *antioxidant*. In this role, it may help protect against atherosclerosis and cancer. Other antioxidants include the tocopherols (vitamin E) and ascorbic acid (vitamin C).

Many fat-soluble vitamins have highly specialized functions. Vitamin K, for example, is required for the production of prothrombin and for clotting factors VII, IX, and X. Vitamin D is converted into a hormone that participates in the regulation of calcium balance. The visual pigments in the rods and cones of the retina are derived from vitamin A. Vitamin A and related compounds, called *retinoids*, also have effects on genetic expression in epithelial cells; these compounds are now used clinically in the treatment of some skin conditions, and researchers are attempting to derive related compounds that may aid the treatment of some cancers.

Elements (minerals) are needed as cofactors for specific enzymes and for a wide variety of other critical functions. Elements that are required daily in relatively large amounts include sodium, potassium, magnesium, calcium, phosphorus, and chlorine (table 19.4). In addition, the following **trace elements** are recognized as essential: iron, zinc, manganese, fluorine, copper, molybdenum, chromium, and selenium. These must be ingested in amounts ranging from 50 μg to 18 mg per day (table 19.5).

Table 19.4 The Major Vitamins

Vitamin	Sources	Action	Deficiency Symptom(s)
A	Yellow vegetables and fruit	Constituent of visual pigment; strengthens epithelial membranes	Night blindness; dry skin
B_1 (Thiamine)	Liver, unrefined cereal grains	Cofactor for enzymes that catalyze decarboxylation	Beriberi; neuritis
B_2 (Riboflavin)	Liver, milk	Part of flavoproteins (such as FAD)	Glossitis; cheilosis
B_6 (Pyridoxine)	Liver, corn, wheat, and yeast	Coenzyme for decarboxylase and transaminase enzymes	Convulsions
B_{12} (Cyanocobalamin)	Liver, meat, eggs, milk	Coenzyme for amino acid metabolism; needed for erythropoiesis	Pernicious anemia
Biotin	Egg yolk, liver, tomatoes	Needed for fatty acid synthesis	Dermatitis; enteritis
C	Citrus fruits, green leafy vegetables	Needed for collagen synthesis in connective tissues	Scurvy
D	Fish liver	Needed for intestinal absorption of calcium and phosphate	Rickets; osteomalacia
E	Milk, eggs, meat, leafy vegetables	Antioxidant	Muscular dystrophy
Folates	Green leafy vegetables	Needed for reactions that transfer one carbon	Sprue: anemia
K	Green leafy vegetables	Promotes reactions needed for function of clotting factors	Hemorrhage; inability to form clot
Niacin	Liver, meat, yeast	Part of NAD and NADP	Pellagra
Pantothenic acid	Liver, eggs, yeast	Part of coenzyme A	Dermatitis; enteritis; adrenal insufficiency

Table 19.5 Estimated Safe and Adequate Daily Dietary Intakes of Selected Vitamins and Minerals[1]

Category	Age (Years)	Vitamins		Trace Elements[2]				
		Biotin (μg)	Pantothenic Acid (mg)	Copper (mg)	Manganese (mg)	Fluoride (mg)	Chromium (μg)	Molybdenum (μg)
Infants	0–0.5	10	2	0.4–0.6	0.3–0.6	0.1–0.5	10–40	15–30
	0.5–1	15	3	0.6–0.7	0.6–1.0	0.2–1.0	20–60	20–40
Children and adolescents	1–3	20	3	0.7–1.0	1.0–1.5	0.5–1.5	20–80	25–50
	4–6	25	3–4	1.0–1.5	1.5–2.0	1.0–2.5	30–120	30–75
	7–10	30	4–5	1.0–2.0	2.0–3.0	1.5–2.5	50–200	50–150
	11+	30–100	4–7	1.5–2.5	2.0–5.0	1.5–2.5	50–200	75–250
Adults		30–100	4–7	1.5–3.0	2.0–5.0	1.5–4.0	50–200	75–250

Source: Reprinted with permission from Recommended Dietary Allowances, 10th Edition. Copyright 1989 by the National Academy of Sciences. Courtesy of the National Academy Press, Washington, D.C.

[1] Because there is less information on which to base allowances, these figures are not given in the main table of RDA and are provided here in the form of ranges of recommended intakes.

[2] Since the toxic levels for many trace elements may be only several times usual intakes, the upper levels for the trace elements given in this table should not be habitually exceeded.

1. Explain how the metabolic rate is influenced by exercise, ambient temperature, and the assimilation of food.
2. Distinguish between the caloric and anabolic requirements of the diet.
3. List the fat-soluble vitamins and describe some of their functions.
4. Explain the roles of vitamins B_1, B_2, and B_3 in energy metabolism.

Regulation of Energy Metabolism

The blood plasma contains circulating glucose, fatty acids, amino acids, and other molecules that can be used by the body tissues for cell respiration. These circulating molecules may be derived from food or from the breakdown of the body's own glycogen, fat, and protein. The building of the body's energy reserves following a meal, and the utilization of these reserves between meals, is regulated by the action of a number of hormones that act to promote either anabolism or catabolism.

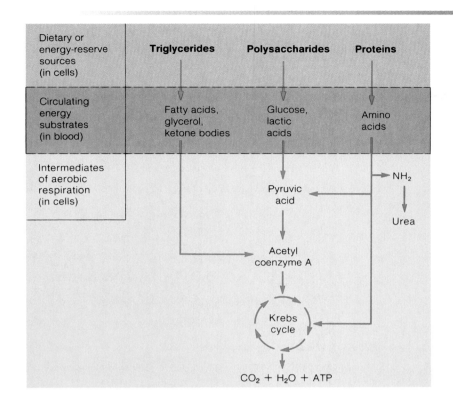

Figure 19.1

A flowchart of energy pathways in the body.

The molecules that can be oxidized for energy by the processes of cell respiration may be derived from the **energy reserves** of glycogen, fat, or protein. Glycogen and fat function primarily as energy reserves; for proteins, by contrast, this represents a secondary, emergency function. Although body protein can provide amino acids for energy, it can do so only through the breakdown of proteins needed for muscle contraction, structural strength, enzymatic activity, and other functions. Alternatively, the molecules used for cell respiration can be derived from the products of digestion that are absorbed through the small intestine. Since these molecules—glucose, fatty acids, amino acids, and others—are carried by the blood to the tissue cells for use in cell respiration, they can be called **circulating energy substrates** (fig. 19.1).

Because of differences in cellular enzyme content, different organs have different *preferred energy sources*. The brain has an almost absolute requirement for blood glucose as its energy source, for example. A fall in the plasma concentration of glucose to below about 50 mg per 100 ml can thus "starve" the brain and have disastrous consequences. Resting skeletal muscles, in contrast, use fatty acids as their preferred energy source. Similarly, ketone bodies (derived from fatty acids), lactic acid, and amino acids can be used to different degrees as energy sources by various organs. The plasma normally contains adequate concentrations of all of these circulating energy substrates to meet the energy needs of the body.

Eating

Ideally, one should eat the kinds and amounts of foods that provide adequate vitamins, minerals, essential amino acids and fatty acids, and calories. Proper caloric intake maintains energy reserves (primarily fat and glycogen) and results in a body weight within an optimum range for health.

There is a tendency for body weight to be stable despite short-term changes in caloric intake. It has been proposed that there may be some mechanism that is sensitive to either body weight or the amount of body fat. Although this mechanism is not known, it is clear that there is a relationship between body fat and endocrine function. The secretion of anterior pituitary hormones is affected in a variety of ways. Obese women, for example, may experience menstrual cycle abnormalities and hirsutism (hairiness), whereas women with little body fat—perhaps as the result of extremely strenuous exercise—may experience amenorrhea (cessation of the menstrual cycle). Abnormalities in growth hormone, ACTH, and prolactin secretion have also been observed in obese people. In experimental animals, prolactin injections given late in the day promote an increase in body fat. In hamsters, a drug that stimulates prolactin secretion was found to promote a loss of body fat.

Eating behavior appears to be at least partially controlled by areas of the hypothalamus. Lesions (destruction) in the ventromedial area of the hypothalamus produce *hyperphagia,* or overeating, and obesity in experimental animals. Lesions of the lateral hypothalamus, by contrast, produce *hypophagia* and weight loss. More recent experiments demonstrate that other brain regions are also involved in the control of eating behavior.

The chemical neurotransmitters that may be involved in neural pathways mediating eating behavior are being investigated. There is evidence, for example, that endorphins may be involved because injections of naloxone (a morphine-blocking drug) suppress overeating in rats. There is also evidence that the neurotransmitters norepinephrine and serotonin may be involved; injections of norepinephrine into the brain cause overeating in rats, whereas injections of serotonin have the opposite effect. Interestingly, the intestinal hormone cholecystokinin (CCK) also appears to function as a neurotransmitter in the brain, and it has been shown that injections of CCK cause experimental animals and humans to stop eating. Drugs that block the interaction of CCK with its membrane receptors, conversely, produce overeating in experimental animals.

 Recent evidence suggests that, in humans, regulatory mechanisms generally match the rate of carbohydrate oxidation (aerobic respiration) to the amount ingested over a 24-hour period. It has therefore been proposed that, in the typical diet, there is little "excess" carbohydrates available to be converted into fat. Obesity, then, may result from eating more fat than can be oxidized day by day. Weight loss (as well as cardiovascular health) might thus be promoted by choosing a high-carbohydrate, low-fat diet. Prolonged exercise of low to moderate intensity also promotes weight loss because, under these conditions, skeletal muscles use fatty acids as their primary source of energy.

Hormonal Regulation of Metabolism

The absorption of energy carriers from the intestine is not continuous; it rises to high levels during a 4-hour period following each meal (the **absorptive state**) and tapers toward zero between meals, after each absorptive state is concluded (the **postabsorptive state**). The postabsorptive state occurs during fasting. Despite this fluctuation, the plasma concentration of glucose and other energy substrates does not remain high during periods of absorption and does not normally fall below a certain level during periods of fasting. During the absorption of digestion products from the intestine, energy substrates are removed from the blood and deposited as energy reserves from which

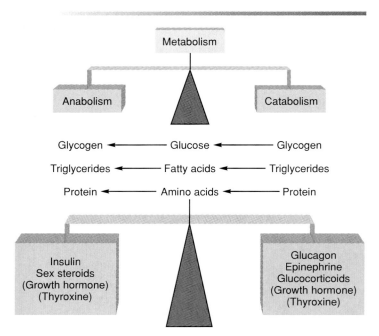

Figure 19.2

The balance of metabolism can be tilted toward anabolism (synthesis of energy reserves) or catabolism (utilization of energy reserves) by the combined actions of various hormones. Growth hormone and thyroxine have both anabolic and catabolic effects.

withdrawals can be made during times of fasting (fig. 19.2). This ensures an adequate plasma concentration of energy substrates to sustain tissue metabolism at all times.

The rate of deposit and withdrawal of energy substrates into and from the energy reserves and the conversion of one type of energy substrate into another are regulated by the actions of hormones. The balance between anabolism and catabolism is determined by the antagonistic effects of insulin, glucagon, growth hormone, thyroxine, and other hormones (fig. 19.2). The specific metabolic effects of these hormones are summarized in table 19.6, and some of their actions are illustrated in figure 19.3.

1. Define the terms *energy reserves* and *circulating energy carriers*. Give examples of each type of molecule.
2. Describe the structures and neurotransmitters that may be involved in the regulation of eating.
3. Which hormones promote an increase in blood glucose? Which promote a decrease? List the hormones that stimulate fat synthesis (lipogenesis) and fat breakdown (lipolysis).

Table 19.6 Endocrine Regulation of Metabolism

Hormone	Blood Glucose	Carbohydrate Metabolism	Protein Metabolism	Lipid Metabolism
Insulin	Decreased	↑ Glycogen formation ↓ Glycogenolysis ↓ Gluconeogenesis	↑ Protein synthesis	↑ Lipogenesis ↓ Lipolysis ↓ Ketogenesis
Glucagon	Increased	↓ Glycogen formation ↑ Glycogenolysis ↑ Gluconeogenesis	No direct effect	↑ Lipolysis ↑ Ketogenesis
Growth hormone	Increased	↑ Glycogenolysis ↑ Gluconeogenesis ↓ Glucose utilization	↑ Protein synthesis	↓ Lipogenesis ↑ Lipolysis ↑ Ketogenesis
Glucocorticoids	Increased	↑ Glycogen formation ↑ Gluconeogenesis	↓ Protein synthesis	↓ Lipogenesis ↑ Lipolysis ↑ Ketogenesis
Epinephrine	Increased	↓ Glycogen formation ↑ Glycogenolysis ↑ Gluconeogenesis	No direct effect	↑ Lipolysis ↑ Ketogenesis
Thyroxine	No effect	↑ Glucose utilization	↑ Protein synthesis	No direct effect

Energy Regulation by the Islets of Langerhans

Insulin secretion is stimulated by a rise in the blood glucose concentration, and insulin promotes the entry of blood glucose into tissue cells. Insulin thus increases the storage of glycogen and fat while causing the blood glucose concentration to fall. Glucagon secretion is stimulated by a fall in blood glucose, and glucagon acts to raise the blood glucose concentration by promoting glycogenolysis in the liver.

Scattered within a "sea" of pancreatic exocrine tissue (the acini) are islands of hormone-secreting cells. These islets of Langerhans (fig. 19.4) contain three distinct cell types that secrete different hormones. The most numerous are the *beta cells*, which secrete the hormone **insulin.** About 60% of each islet consists of beta cells. The *alpha cells* form about 25% of each islet and secrete the hormone **glucagon.** The least numerous cell type, the *delta cells*, produce **somatostatin,** the composition of which is identical to the somatostatin produced by the hypothalamus and the intestine.

All three pancreatic hormones are polypeptides. Insulin consists of two polypeptide chains—one that is twenty-one amino acids long and another that is thirty amino acids long—joined together by disulfide bonds. Glucagon is a twenty-one-amino-acid polypeptide, and somatostatin contains fourteen amino acids. Insulin was the first of these hormones to be discovered (in 1921). The importance of insulin in diabetes mellitus was immediately recognized, and clinical use of insulin in the treatment of this disease began almost immediately after its discovery. The physiological role of glucagon was discovered later, and the importance of glucagon in the development of diabetes has only recently been suspected. The physiological significance of islet-secreted somatostatin is not currently known.

Regulation of Insulin and Glucagon Secretion

Insulin and glucagon secretion is largely regulated by the plasma concentrations of glucose and, to a lesser degree, of amino acids. The alpha and beta cells, therefore, act as both the sensors and effectors in this control system. Since the plasma concentration of glucose and amino acids rises during the absorption of a meal and falls during fasting, the secretion of insulin and glucagon likewise fluctuates between the absorptive and postabsorptive states. These changes in insulin and glucagon secretion, in turn, cause changes in plasma glucose and amino acid concentrations and thus help to maintain homeostasis via negative feedback loops (fig. 19.5).

Effects of Glucose and Amino Acids

During the absorption of a carbohydrate meal, the plasma glucose concentration rises. This rise in plasma glucose (1) stimulates the beta cells to secrete insulin and (2) inhibits the secretion of glucagon from the alpha cells. Insulin acts to stimulate the cellular uptake of plasma glucose. A rise in insulin secretion therefore lowers the plasma glucose concentration. Since glucagon has the antagonistic effect of raising the plasma glucose concentration by stimulating glycogenolysis in the liver, the inhibition of glucagon secretion complements the effect of increased insulin during the absorption of a carbohydrate meal. A rise in insulin and a fall in glucagon secretion thus help to lower the high plasma glucose concentration that occurs during periods of absorption.

Figure 19.3

Different hormones participate both synergistically and antagonistically in the regulation of metabolism.

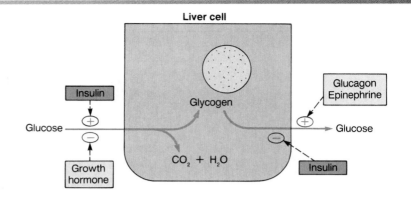

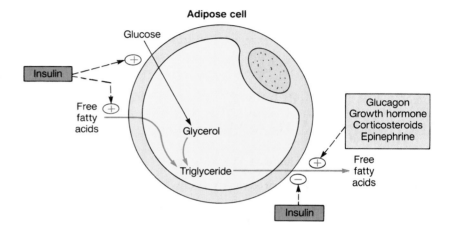

During fasting, the plasma glucose concentration falls. At this time, therefore, (1) insulin secretion decreases and (2) glucagon secretion increases. These changes in hormone secretion prevent the cellular uptake of blood glucose into organs such as the muscles, liver, and adipose tissue and promote the release of glucose from the liver (through the actions of glucagon). A negative feedback loop is therefore completed (fig. 19.5), helping to retard the fall in plasma glucose concentration that occurs during fasting.

The **oral glucose tolerance test** (fig. 19.6) is a measure of the ability of the beta cells to secrete insulin and of the ability of insulin to lower blood glucose. In this procedure, a person drinks a glucose solution and blood samples are taken periodically for plasma glucose measurements. In a normal person, the rise in blood glucose produced by drinking this solution is reversed to normal levels within 2 hours following glucose ingestion.

Insulin secretion is also stimulated by particular amino acids derived from dietary proteins. Meals that are high in protein, therefore, stimulate the secretion of insulin; if the meal is high in protein and low in carbohydrates, glucagon secretion will be stimulated as well. The increased glucagon secretion acts to raise the blood glucose, while the increased insulin promotes the entry of amino acids into tissue cells.

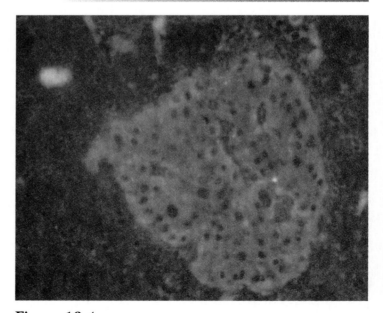

Figure 19.4

A normal pancreatic islet (of Langerhans) visualized with the aid of fluorescently labeled antibodies that stain the cytoplasm green. The dark dots are nuclei.

Chapter Nineteen

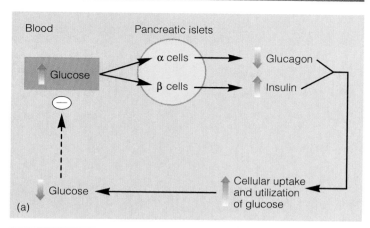

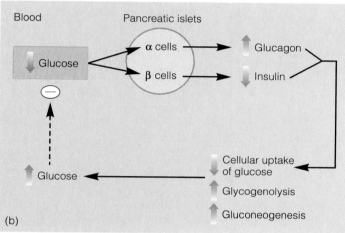

Figure 19.5

The secretion from the β (beta) cells and α (alpha) cells of the pancreatic islets (of Langerhans) is regulated largely by the blood glucose concentration. (*a*) A high blood glucose concentration stimulates insulin and inhibits glucagon secretion. (*b*) A low blood glucose concentration stimulates glucagon and inhibits insulin secretion.

 People with **diabetes mellitus**—due to the inadequate secretion or action of insulin—maintain a state of high plasma glucose concentration (hyperglycemia) during the oral glucose tolerance test (fig. 19.6). People who have **reactive hypoglycemia** (low plasma glucose concentration due to excessive insulin secretion) have lower-than-normal blood glucose concentrations 5 hours following glucose ingestion. These conditions are described in more detail in a later section.

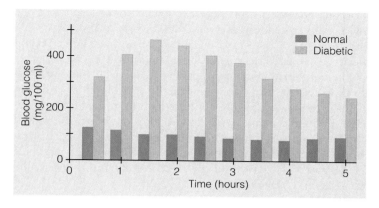

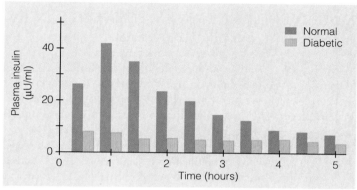

Figure 19.6

Changes in blood glucose and plasma insulin concentrations after the ingestion of 100 grams of glucose in an oral glucose tolerance test. (Insulin is measured in activity units [U].)

Effects of Autonomic Nerves

The islets of Langerhans receive both parasympathetic and sympathetic innervation. The activation of the parasympathetic system during meals stimulates insulin secretion at the same time that gastrointestinal function is stimulated. The activation of the sympathetic system, by contrast, stimulates glucagon secretion and inhibits insulin secretion. The effects of glucagon, together with those of epinephrine, produce a "stress hyperglycemia" when the sympathoadrenal system is activated.

Effects of GIP

Surprisingly, insulin secretion increases more rapidly following glucose ingestion than it does following an intravenous injection of glucose. This is due to the fact that the intestine, in response to glucose ingestion, secretes a hormone that stimulates insulin secretion before the glucose has been absorbed. Insulin secretion thus begins to rise in anticipation of a rise in blood glucose. The intestinal hormone that mediates this effect is believed to be GIP—gastric inhibitory peptide, or, more appropriately in this context, *glucose-dependent insulinotropic peptide* (chapter 18).

The mechanisms that regulate insulin and glucagon secretion and the actions of these hormones normally prevent the plasma glucose concentration from rising above 170 mg per 100 ml after a meal or from falling below about 50 mg per 100 ml between meals. This regulation is important because abnormally high blood glucose can damage tissue cells (as may occur in diabetes mellitus) and abnormally low blood glucose can damage the brain. The later effect results from the fact that glucose enters the brain by facilitated diffusion; when the rate of this diffusion is too low, due to low plasma glucose concentrations, the supply of metabolic energy for the brain may become inadequate. This can result in weakness, dizziness, personality changes, and ultimately in coma and death.

Insulin and Glucagon: Absorptive State

The lowering of plasma glucose by insulin is, in a sense, a side effect of the primary action of this hormone. Insulin is the major hormone that promotes anabolism in the body. During absorption of the products of digestion and the subsequent rise in the plasma concentrations of circulating energy substrates, insulin promotes the cellular uptake of plasma glucose and its incorporation into energy-reserve molecules of glycogen in the liver and muscles, and of triglycerides in adipose cells (fig. 19.7). Quantitatively, skeletal muscles are responsible for most of the insulin-stimulated glucose uptake. Insulin also promotes the cellular uptake of amino acids and their incorporation into proteins. The stores of large energy-reserve molecules are thus increased while the plasma concentrations of glucose and amino acids are decreased.

A nonobese 70-kg (154-lb) man has approximately 10 kg (about 82,500 kcal) of stored fat. Since 250 g of fat can supply the energy requirements for 1 day, this reserve fuel is sufficient for about 40 days. Glycogen is less efficient as an energy reserve, and less is stored in the body; there are about 100 g (400 kcal) of glycogen stored in the liver and 375 to 400 g (1,500 kcal) in skeletal muscles. Insulin promotes the cellular uptake of glucose into the liver and muscles and the conversion of glucose into glucose–6–phosphate. In the liver and muscles, this can be changed into glucose–1–phosphate, which is used as the precursor of glycogen. Once the stores of glycogen have been filled, the continued ingestion of excess calories results in the continued production of fat rather than of glycogen.

Insulin and Glucagon: Postabsorptive State

Glucagon stimulates and insulin suppresses the hydrolysis of liver glycogen, or **glycogenolysis.** Thus during times of fasting, when glucagon secretion is high and insulin secretion is low, liver glycogen is used as a source of additional blood glucose. This results in the liberation of free glucose from glucose–6–phosphate by the action of an enzyme called *glucose–6–phosphatase* (chapter 5). Only the liver has this en-

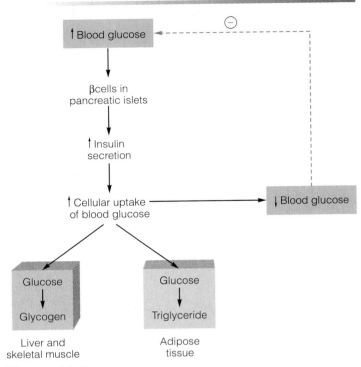

Figure 19.7

A rise in blood glucose concentration stimulates insulin secretion. Insulin promotes a fall in blood glucose by stimulating the cellular uptake of glucose and the conversion of glucose to glycogen and fat.

zyme, and therefore only the liver can use its stored glycogen as a source of additional blood glucose. Since muscles lack glucose 6-phosphatase, the glucose–6–phosphate produced from muscle glycogen can be used for glycolysis only by the muscle cells themselves.

Since there are only about 100 g of stored glycogen in the liver, adequate blood glucose levels could not be maintained for very long during fasting using this source alone. The low levels of insulin secretion during fasting, together with elevated glucagon secretion, however, promote **gluconeogenesis:** the formation of glucose from noncarbohydrate molecules. Low insulin allows the release of amino acids from skeletal muscles, while glucagon and cortisol (discussed later) stimulate the production of enzymes in the liver that convert amino acids to pyruvic acid and pyruvic acid to glucose. During prolonged fasting and exercise, gluconeogenesis in the liver using amino acids from muscles may be the only source of blood glucose.

The secretion of glucose from the liver during fasting compensates for the low blood glucose concentrations and helps to provide the brain with the glucose that it needs. But because insulin secretion is low during fasting, skeletal muscles cannot utilize blood glucose as an energy source. Instead, skeletal muscles—as well as the heart, liver, and kidneys—use free fatty acids as their major source of fuel. This helps to spare glucose for the brain.

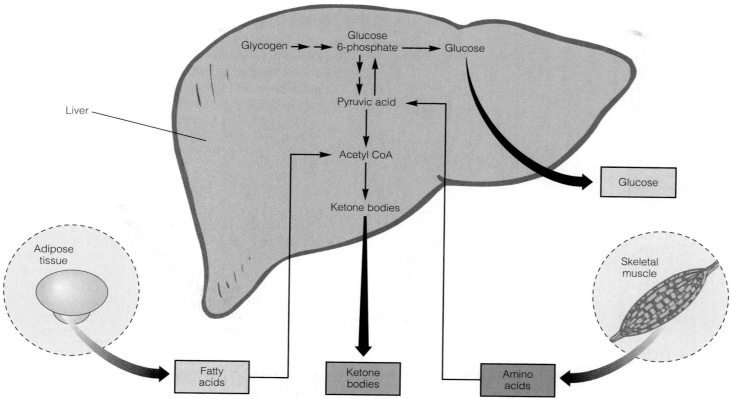

Fasting (↓insulin, ↑glucagon)

Liver

Glycogen → → Glucose 6-phosphate → Glucose

Pyruvic acid

Acetyl CoA

Ketone bodies

Glucose

Adipose tissue

Skeletal muscle

Fatty acids

Ketone bodies

Amino acids

Figure 19.8

Increased glucagon secretion and decreased insulin secretion during fasting favors catabolism. These hormonal changes promote the release of glucose, fatty acids, ketone bodies, and amino acids into the blood. Notice that the liver secretes glucose that is derived both from the breakdown of liver glycogen and from the conversion of amino acids in gluconeogenesis.

The free fatty acids are made available by the action of glucagon. In the presence of low insulin levels, glucagon stimulates in adipose cells an enzyme called *hormone-sensitive lipase*. This enzyme catalyzes the hydrolysis of stored triglycerides and the release of free fatty acids and glycerol into the blood. Glucagon also stimulates enzymes in the liver that convert some of these fatty acids into ketone bodies, which are secreted into the blood (fig. 19.8). Several organs in the body can use ketone bodies, as well as fatty acids, as a source of acetyl CoA in aerobic respiration.

Through the stimulation of **lipolysis** (the breakdown of fat) and **ketogenesis** (the formation of ketone bodies) the high glucagon and low insulin levels that occur during fasting provide circulating energy substrates for use by the muscles, liver, and other organs. Through liver glycogenolysis and gluconeogenesis, these hormonal changes help to provide adequate levels of blood glucose to sustain the metabolism of the brain. The antagonistic action of insulin and glucagon (fig. 19.9) thus promotes appropriate metabolic responses during periods of fasting and periods of absorption.

1. Describe how the secretions of insulin and glucagon change during periods of absorption and periods of fasting. How are these changes in hormone secretion produced?

2. Explain how the synthesis of fat in adipose cells is regulated by insulin. Also, explain how fat metabolism is regulated by insulin and glucagon during periods of absorption and fasting.

3. Define the following terms: *glycogenolysis, gluconeogenesis,* and *ketogenesis.* How do insulin and glucagon affect each of these processes during periods of absorption and fasting?

4. Describe two pathways used by the liver to produce glucose for secretion into the blood. Why can't skeletal muscles secrete glucose into the blood?

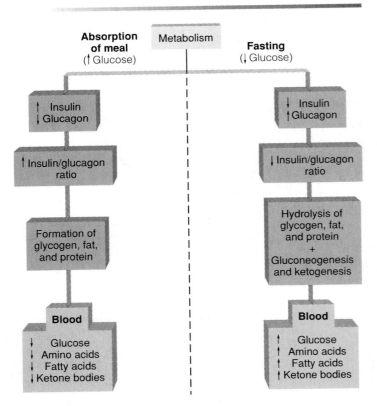

Absorption of meal (↑Glucose) — Metabolism — **Fasting** (↓Glucose)

↑ Insulin
↓ Glucagon

↑ Insulin/glucagon ratio

Formation of glycogen, fat, and protein

Blood
↓ Glucose
↓ Amino acids
↓ Fatty acids
↓ Ketone bodies

↓ Insulin
↑ Glucagon

↓ Insulin/glucagon ratio

Hydrolysis of glycogen, fat, and protein
+
Gluconeogenesis and ketogenesis

Blood
↑ Glucose
↑ Amino acids
↑ Fatty acids
↑ Ketone bodies

Figure 19.9

The inverse relationship between insulin and glucagon secretion during the absorption of a meal and during fasting. Changes in the insulin-to-glucagon ratio tilt metabolism toward anabolism during the absorption of food and toward catabolism during fasting.

Diabetes Mellitus and Hypoglycemia

Inadequate secretion of insulin or defects in the action of insulin, produce metabolic disturbances that are characteristic of the disease diabetes mellitus. A person with type I diabetes requires injections of insulin; a person with type II diabetes can control the condition by other methods. In both types, hyperglycemia and glycosuria result from the deficiency and/or defective action of insulin. A person with reactive hypoglycemia, by contrast, secretes excessive amounts of insulin, and thus experiences hypoglycemia in response to the stimulus of a carbohydrate meal.

Chronic high blood glucose, or hyperglycemia, is the hallmark of the disease **diabetes mellitus.** The name of this disease is derived from the fact that glucose "spills over" into the urine when the blood glucose concentration is too high (*mellitus* is derived from the Latin word meaning "honeyed" or "sweet"). The hyperglycemia of diabetes mellitus results from either the insufficient secretion of insulin by the beta cells of the islets of Langerhans or the inability of secreted insulin to stimulate the cellular uptake of glucose from the blood. Diabetes mellitus, in short, results from the inadequate secretion or action of insulin.

There are two forms of diabetes mellitus. In **insulin-dependent diabetes mellitus (IDDM),** also called *type I diabetes*, the beta cells are progressively destroyed and secrete little or no insulin. This form of the disease accounts for only about 10% of the known cases of diabetes. About 90% of the people who have diabetes have **non-insulin-dependent diabetes mellitus (NIDDM),** also called *type II diabetes*. Type I diabetes was once known as *juvenile-onset diabetes* because this condition is usually diagnosed in people under the age of 20. Type II diabetes has also been called *maturity-onset diabetes*, because it is usually diagnosed in people over the age of 40. Some comparisons of these two forms of diabetes mellitus are shown in table 19.7. It should be noted that only the early stages of IDDM and NIDDM are compared; in some people, NIDDM can grade into and later become IDDM.

Insulin-Dependent Diabetes Mellitus

Insulin-dependent diabetes mellitus results when the beta cells of the islets of Langerhans are progressively destroyed by autoimmune attack. Recent evidence in mice suggests that killer T lymphocytes (chapter 15) may target an enzyme known as glutamate decarboxylase in the beta cells. This autoimmune destruction of the beta cells may be provoked by an environmental agent, such as infection by viruses. In other cases, however, the cause is currently unknown. Removal of the insulin-secreting beta cells causes hyperglycemia and the appearance of glucose in the urine. Without insulin, glucose cannot enter the adipose cells; the rate of fat synthesis thus lags behind the rate of fat breakdown and large amounts of free fatty acids are released from the adipose cells.

In a person with uncontrolled IDDM, many of the fatty acids released from adipose cells are converted into ketone bodies in the liver. This may result in an elevated ketone body concentration in the blood (ketosis), and if the buffer reserve of bicarbonate is neutralized, it may also result in *ketoacidosis*. During this time, the glucose and excess ketone bodies that are excreted in the urine act as osmotic diuretics and cause the excessive excretion of water in the urine. This can produce severe dehydration, which, together with ketoacidosis and associated disturbances in electrolyte balance, may lead to coma and death (fig. 19.10).

In addition to the lack of insulin, people with IDDM have an abnormally high secretion of glucagon from the alpha cells of the islets. Glucagon stimulates glycogenolysis in the liver and

Table 19.7 Comparison of Insulin-Dependent and Non-Insulin-Dependent Diabetes Mellitus

	Insulin-Dependent (Type I)	Non-Insulin-Dependent (Type II)
Usual Age at Onset	Under 20 years	Over 40 years
Development of Symptoms	Rapid	Slow
Percentage of Diabetic Population	About 10%	About 90%
Development of Ketoacidosis	Common	Rare
Association with Obesity	Rare	Common
Beta Cells of Islets (at Onset of Disease)	Destroyed	Not destroyed
Insulin Secretion	Decreased	Normal or increased
Autoantibodies to Islet Cells	Present	Absent
Associated with Particular MHC Antigens*	Yes	Unclear
Treatment	Insulin injections	Diet; oral stimulators of insulin secretion

*Discussed in chapter 15.

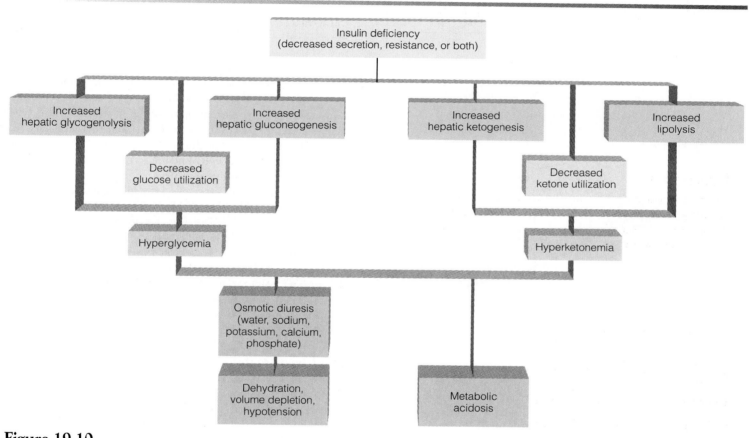

Figure 19.10

The sequence of events by which an insulin deficiency may lead to coma and death.

thus helps to raise the blood glucose concentration. Glucagon also stimulates the production of enzymes in the liver that convert fatty acids into ketone bodies. Some researchers believe that the full symptoms of diabetes result from high glucagon secretion as well as from the absence of insulin. The lack of insulin may be largely responsible for hyperglycemia and for the release of large amounts of fatty acids into the blood. The high glucagon secretion may contribute to the hyperglycemia and be largely responsible for the development of ketoacidosis.

Non-Insulin-Dependent Diabetes Mellitus

The effects produced by insulin, or any hormone, depend on the concentration of that hormone in the blood and on the sensitivity of the target tissue to given amounts of the hormone. Tissue responsiveness to insulin, for example, varies under normal conditions. For reasons that are incompletely understood, exercise increases insulin sensitivity and obesity decreases

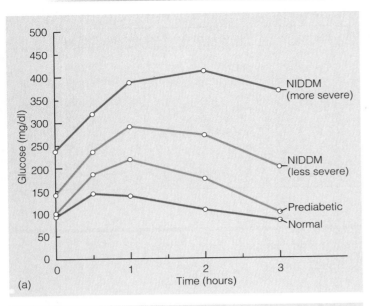

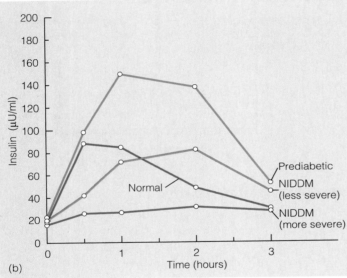

Figure 19.11

The oral glucose tolerance test showing (*a*) blood glucose concentrations and (*b*) insulin values over an interval of 3 hours following ingestion of a glucose solution. Prediabetics often show impaired glucose tolerance without fasting hyperglycemia. (NIDDM = non-insulin-dependent diabetes mellitus.)

Source: Data from Simeon I. Taylor, et al., "Insulin Resistance of Insulin Deficiency: Which is the Primary Cause of NIDDM?" in *Diabetes*, vol. 43, June 1994, p. 735.

insulin sensitivity of the target tissues. The islets of a nondiabetic obese person, therefore, must secrete high amounts of insulin to maintain the blood glucose concentration in the normal range. Conversely, nondiabetic people who are thin and who exercise regularly require lower amounts of insulin to maintain the proper blood glucose concentration.

Non-insulin-dependent diabetes is usually slow to develop, is hereditary, and occurs most often in people who are overweight. Genetic factors are very significant; people at high-

est risk are those who have both parents with NIDDM and those who are members of certain ethnic groups, particularly Mexican-Americans and Pima Indians. Unlike people with IDDM, those who have NIDDM can have normal or even elevated levels of insulin in their blood. Despite this, people with NIDDM have hyperglycemia if untreated. This must mean that, even though the insulin levels may be in the normal range, the amount of insulin secreted is inadequate.

Much evidence has been obtained to show that people with NIDDM have an abnormally low tissue sensitivity to insulin, or an *insulin resistance*. This is true even if the person is not obese, but the problem is compounded by the decreased tissue sensitivity that accompanies obesity, particularly of the "apple-shape" variety in which the adipose cells are enlarged. There is also evidence that the beta cells are not functioning correctly: whatever amount of insulin they secrete is inadequate to the task. People who are prediabetic often have elevated levels of insulin without hypoglycemia, suggesting insulin resistance. People with established NIDDM have both an insulin resistance and insulin deficiency (fig. 19.11).

Since obesity decreases insulin sensitivity, people who are genetically predisposed to insulin resistance may develop symptoms of diabetes when they gain weight. Conversely, this type of diabetes mellitus can usually be controlled by increasing tissue sensitivity to insulin through diet and exercise. If this is not sufficient, oral drugs (generically known as the *sulfonylureas*) are available that increase insulin secretion There is some evidence that overstimulation of the beta cells with sulfonylureas may promote the conversion of NIDDM to IDDM, however, so research into alternative approaches of decreasing insulin resistance is ongoing.

People with type II diabetes do not usually develop ketoacidosis. The hyperglycemia itself, however, can be dangerous on a long-term basis. Diabetes is the leading cause of blindness, kidney failure, and amputation of the lower extremities in the United States. People with diabetes frequently have circulatory problems that increase the tendency to develop gangrene and increase the risk of atherosclerosis. The causes of damage to the retina and lens of the eyes and to blood vessels are not well understood. It is believed, however, that these problems may result from a long-term exposure to high blood glucose, which (1) causes water to leave tissue cells by osmosis, and thus produces dehydration of capillary endothelial cells, and (2) results in glycosylation of tissue proteins.

Hypoglycemia

A person with type I diabetes mellitus depends on insulin injections to prevent hyperglycemia and ketoacidosis. If inadequate insulin is injected, the person may enter a coma as a result of the ketoacidosis, electrolyte imbalance, and dehydration that develop. An overdose of insulin, however, can also produce a coma as a result of the hypoglycemia (abnormally low blood glucose levels) produced. The physical signs and symptoms of diabetic and hypoglycemic coma are sufficiently different (table 19.8) to allow hospital personnel to distinguish between these two types.

Less severe symptoms of hypoglycemia are usually produced by an oversecretion of insulin from the islets of Langerhans after a carbohydrate meal. This **reactive hypoglycemia,** caused

Table 19.8 Comparison of Diabetic and Hypoglycemic Coma

	Diabetic Ketoacidosis	Hypoglycemia
Onset	Hours to days	Minutes
Causes	Insufficient insulin; other diseases	Excess insulin; insufficient food; excessive exercise
Symptoms	Excessive urination and thirst; headache, nausea, and vomiting	Hunger, headache, confusion, stupor
Physical Findings	Deep, labored breathing; acetone odor on breath; blood pressure decreased, pulse weak; skin dry	Pulse, blood pressure, and respiration normal; skin pale and moist
Laboratory Findings	Urine: glucose present, ketone bodies increased Plasma: glucose and ketone bodies increased, bicarbonate decreased	Urine: normal Plasma: glucose concentration low, bicarbonate normal

1. Explain how ketoacidosis and dehydration are produced in a person with type I diabetes mellitus.
2. Describe the causes of hyperglycemia in a person with type II diabetes. How may weight loss help to control this condition?
3. Explain how reactive hypoglycemia is produced and describe the dangers of this condition.

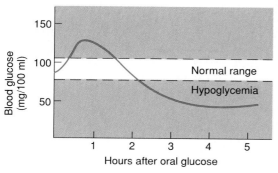

Figure 19.12

An idealized oral glucose tolerance test in a person with reactive hypoglycemia. The blood glucose concentration falls below the normal range within 5 hours of glucose ingestion as a result of excessive insulin secretion.

by an exaggerated response of the beta cells to a rise in blood glucose, is most commonly seen in adults who are genetically predisposed to type II diabetes. For this reason, people with reactive hypoglycemia must limit their intake of carbohydrates and eat small meals at frequent intervals, rather than two or three meals per day.

The symptoms of reactive hypoglycemia include tremor, hunger, weakness, blurred vision, and impaired mental ability. The appearance of some of these symptoms, however, does not necessarily indicate reactive hypoglycemia, and a given level of blood glucose does not always produce these symptoms. To confirm a diagnosis of reactive hypoglycemia, a number of tests must be performed. In the oral glucose tolerance test, for example, reactive hypoglycemia is shown when the initial rise in blood glucose produced by the ingestion of a glucose solution triggers excessive insulin secretion, so that the blood glucose levels fall below normal within 5 hours (fig. 19.12).

Metabolic Regulation by Adrenal Hormones, Thyroxine, and Growth Hormone

Epinephrine, the glucocorticoids, thyroxine, and growth hormone stimulate the catabolism of carbohydrates and lipids. These hormones are thus antagonistic to insulin in their regulation of carbohydrate and lipid metabolism. Thyroxine and growth hormone, however, stimulate protein synthesis and are needed for body growth and proper development of the central nervous system. These hormones thus have an anabolic effect on protein synthesis, which is complementary to that of insulin.

The anabolic effects of insulin are antagonized by glucagon, as previously described, and by the actions of a variety of other hormones. The hormones of the adrenals, thyroid, and anterior pituitary (specifically growth hormone) antagonize the action of insulin on carbohydrate and lipid metabolism. The actions of insulin, thyroxine, and growth hormone, however, can act synergistically in the stimulation of protein synthesis.

Adrenal Hormones

As described in chapter 11, the adrenal gland consists of two different parts with different embryonic origins that function as separate glands. The two parts secrete different hormones and are regulated by different control systems. The **adrenal medulla** secretes catecholamine hormones—epinephrine and lesser amounts of norepinephrine—in response to sympathetic nerve stimulation. The **adrenal cortex** secretes corticosteroid hormones. These are grouped into two functional categories: **mineralocorticoids,** such as aldosterone, which regulate Na^+ and K^+ balance (chapter 17), and **glucocorticoids,** such as hydrocortisone (cortisol), which participate in metabolic regulation.

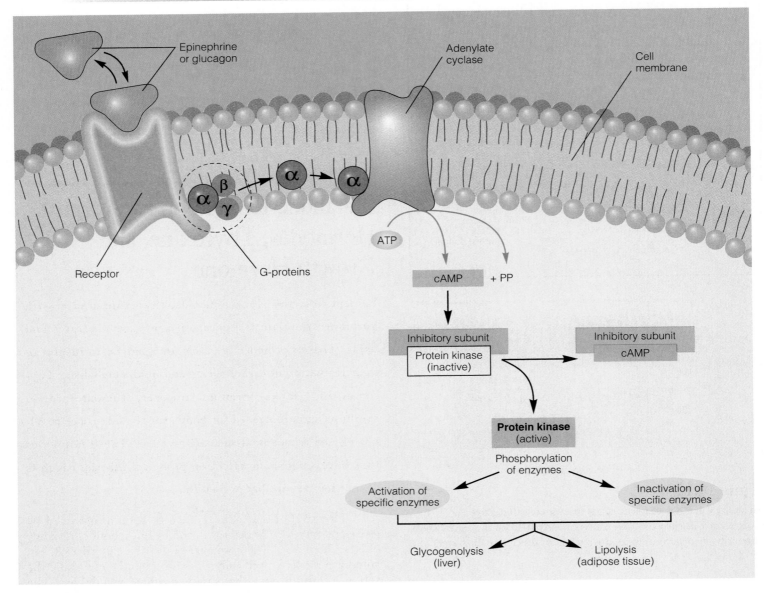

Figure 19.13

Cyclic AMP (cAMP) serves as a second messenger in the actions of epinephrine and glucagon on liver and adipose tissue metabolism. (The mechanisms of hormone action are discussed in more detail in chapter 11.)

Metabolic Effects of Epinephrine

The metabolic effects of epinephrine are similar to those of glucagon. Both stimulate glycogenolysis and the release of glucose from the liver, as well as lipolysis and the release of fatty acids from adipose tissue. These actions occur in response to glucagon during fasting, when low blood glucose stimulates glucagon secretion, and in response to epinephrine during the fight-or-flight reaction to stress. The latter effect provides circulating energy substrates in anticipation of the need for intense physical activity. Glucagon and epinephrine have similar mechanisms of action; the actions of both are mediated by cyclic AMP (fig. 19.13).

Metabolic Effects of Glucocorticoids

Hydrocortisone (cortisol) and other glucocorticoids are secreted by the adrenal cortex in response to ACTH stimulation. The secretion of ACTH from the anterior pituitary occurs as part of the general adaptation syndrome in response to stress (chapter 11). Since prolonged fasting or prolonged exercise certainly qualify as stressors, ACTH—and thus glucocorticoid secretion—is stimulated under these conditions. The increased secretion of glucocorticoids during prolonged fasting or exercise supports the effects of increased glucagon and decreased insulin secretion from the pancreatic islets.

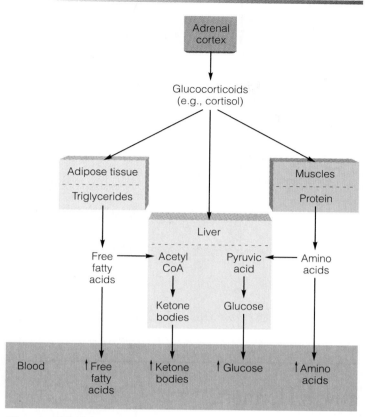

Figure 19.14

The catabolic actions of glucocorticoids help raise the blood concentration of glucose and other energy-carrier molecules.

Like glucagon, hydrocortisone promotes lipolysis and ketogenesis; it also stimulates the synthesis of hepatic enzymes that promote gluconeogenesis. Although hydrocortisone stimulates enzyme (protein) synthesis in the liver, it promotes protein breakdown in the muscles. This latter effect increases the blood levels of amino acids and thus provides the substrates needed by the liver for gluconeogenesis. The release of circulating energy substrates—amino acids, glucose, fatty acids, and ketone bodies—into the blood in response to hydrocortisone (fig. 19.14) helps to compensate for a state of prolonged fasting or exercise. Whether these metabolic responses are beneficial in other stressful states is open to question.

Thyroxine

The thyroid follicles secrete thyroxine, also called tetraiodothyronine (T_4), in response to stimulation by thyroid-stimulating hormone (TSH) from the anterior pituitary. Almost all organs in the body are targets of thyroxine action. Thyroxine itself, however, is not the active form of the hormone within the target cells; thyroxine is a prehormone that must first be converted to triiodothyronine (T_3) within the target cells to be active (chapter 11). Acting via its conversion to T_3, thyroxine (1) regulates the rate of cell respiration and (2) contributes to proper growth and development, particularly during early childhood.

Thyroxine and Cell Respiration

Thyroxine stimulates the rate of cell respiration in almost all cells in the body. This is believed to be due to thyroxine-induced lowering of cellular ATP concentrations. ATP exerts an end-product inhibition (chapter 4) of cell respiration, so that when ATP concentrations increase, the rate of cell respiration decreases. Conversely, a lowering of ATP concentrations, as may occur in response to thyroxine, stimulates cell respiration.

The metabolic rate under resting and carefully defined conditions is known as the *basal metabolic rate (BMR)*, as previously described. Acting through its stimulation of cell respiration, thyroxine acts to "set" the BMR. The BMR can thus be used as an index of thyroid function. Indeed, such measurements were used clinically to evaluate thyroid function prior to the development of direct chemical determinations of T_4 and T_3 in the blood.

The coupling of energy-releasing reactions to energy-requiring reactions is never 100% efficient; a proportion of the energy is always lost as heat. Much of the energy liberated during cell respiration and much of the energy released by the hydrolysis of ATP escapes as heat. Since thyroxine stimulates both ATP consumption and cell respiration, the actions of thyroxine result in the production of metabolic heat.

The heat-producing, or *calorigenic (calor = heat) effects* of thyroxine are required for cold adaptation. This does not mean that people who are cold-adapted have high levels of thyroxine secretion. Rather, thyroxine levels in the normal range coupled with the increased activity of the sympathoadrenal system and other responses previously discussed are responsible for cold adaptation. Thyroxine exerts a permissive effect on the ability of the sympathoadrenal system to increase heat production in response to cold stress.

Thyroxine in Growth and Development

Through its stimulation of cell respiration, thyroxine stimulates the increased consumption of circulating energy substrates such as glucose, fatty acids, and other molecules. These effects, however, are mediated at least in part by the activation of genes; thyroxine thus stimulates both RNA and protein synthesis. As a result of its stimulation of protein synthesis throughout the body, thyroxine is considered to be an anabolic hormone like insulin and growth hormone.

Because of its stimulation of protein synthesis, thyroxine is needed for growth of the skeleton and, most importantly, for the proper development of the central nervous system. Recent evidence has demonstrated the presence of receptor proteins for T_3 in the neurons and astrocytes of the brain. This need for thyroxine is particularly great when the brain is undergoing its greatest rate of development—from the end of the first trimester of prenatal life to 6 months after birth. Hypothyroidism during this time may result in **cretinism** (fig. 19.15). Unlike people with dwarfism, who have normal thyroxine secretion but a low secretion of growth hormone, people with cretinism exhibit severe mental retardation. Treatment with thyroxine soon after birth, particularly before 1 month of age, has been found to completely or almost completely restore development of intelligence as measured by IQ tests administered 5 years later.

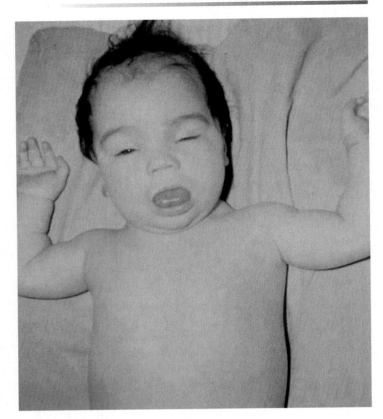

Figure 19.15
Cretinism is a disease of infancy caused by an underactive thyroid gland.

Table 19.9	Comparison of Hypothyroidism and Hyperthyroidism	
	Hypothyroid	**Hyperthyroid**
Growth and Development	Impaired growth	Accelerated growth
Activity and Sleep	Lethargy; increased sleep	Increased activity; decreased sleep
Temperature Tolerance	Intolerance to cold	Intolerance to heat
Skin Characteristics	Coarse, dry skin	Smooth skin
Perspiration	Absent	Excessive
Pulse	Slow	Rapid
Gastrointestinal Symptoms	Constipation; decreased appetite; increased weight	Frequent bowel movements; increased appetite; decreased weight
Reflexes	Slow	Rapid
Psychological Aspects	Depression and apathy	Nervous, "emotional" state
Plasma T_4 Levels	Decreased	Increased

Hypothyroidism and Hyperthyroidism

As might be predicted from the effects of thyroxine, people who are hypothyroid have an abnormally low basal metabolic rate and experience weight gain and lethargy. There is also a decreased ability to adapt to cold stress when there is a thyroxine deficiency. Another symptom of hypothyroidism is **myxedema**—accumulation of mucoproteins and fluid in subcutaneous connective tissues. Hypothyroidism can be produced by a variety of causes, including insufficient thyrotropin-releasing hormone (TRH) secretion from the hypothalamus, insufficient TSH secretion from the pituitary, or insufficient iodine in the diet. Hypothyroidism due to lack of iodine is accompanied by excessive TSH secretion, which stimulates abnormal growth of the thyroid (goiter). This condition can be reversed by iodine supplements.

A goiter can also be produced by another mechanism. In **Graves' disease,** apparently an autoimmune disease, autoantibodies (chapter 15) exert TSH-like effects on the thyroid. Since the production of these autoantibodies is not controlled by negative feedback, the thyroid is stimulated excessively, thus producing the goiter associated with a hyperthyroid state. Hyperthyroidism produces a high BMR accompanied by weight loss, nervousness, irritability, and an intolerance to heat. The symptoms of hypothyroidism and hyperthyroidism are compared in table 19.9.

Growth Hormone

The anterior pituitary secretes *growth hormone*, also called *somatotropic hormone*, in larger amounts than any other of its hormones. As its name implies, growth hormone stimulates growth in children and adolescents. The continued high secretion of growth hormone in adults, particularly under the conditions of fasting and other forms of stress, implies that this hormone can have important metabolic effects even after the growing years have ended.

Regulation of Growth Hormone Secretion

The secretion of growth hormone is inhibited by somatostatin, which is produced by the hypothalamus and secreted into the hypothalamo-hypophyseal portal system (chapter 11). In addition, a recently discovered hypothalamic-releasing hormone stimulates growth hormone secretion. Growth hormone thus appears to be unique among the anterior pituitary hormones in that its secretion is controlled by both a releasing and an inhibiting hormone from the hypothalamus. The secretion of growth hormone follows a circadian ("about a day") pattern, increasing during sleep and decreasing during periods of wakefulness.

Growth hormone secretion is stimulated by an increase in the plasma concentrations of amino acids and by a decrease in the plasma glucose concentration. These events occur during absorption of a high protein meal, when amino acids are absorbed. The secretion of growth hormone is also increased during prolonged fasting when plasma glucose is low and plasma amino acid concentration is raised by the breakdown of muscle protein.

Insulin-like Growth Factors

Insulin-like growth factors (IGF) are polypeptides produced by many tissues that are similar in structure to proinsulin (chapter 2), that have insulin-like effects, and that serve as mediators for some of growth hormone's actions. The term **somatomedins** is often used to refer to two of these factors, designated IGF-1 and IGF-2. The liver produces and secretes IGF-1 in response to growth hormone stimulation, and this secreted IGF-1 then functions as a hormone in its own right, traveling in the blood to the target tissue. A major target is cartilage, where IGF-1 stimulates cell division and growth. IGF-1 also functions as an autocrine regulator (chapter 11), because the chondrocytes (cartilage cells) themselves produce some IGF-1 in response to growth hormone stimulation. The growth-promoting actions of IGF-1 thus directly mediate the effects of growth hormone on cartilage. These actions are supported by IGF-2, which has more insulin-like actions.

Effects of Growth Hormone on Metabolism

The fact that growth hormone secretion is increased during fasting and also during absorption of a protein meal reflects the complex nature of this hormone's action. Growth hormone has both anabolic and catabolic effects; it promotes protein synthesis (anabolism), and in this respect is similar to insulin. It also stimulates the catabolism of fat and release of fatty acids from adipose tissue—effects similar to those of glucagon. A rise in the plasma fatty acid concentration induced by growth hormone results in decreased rates of glycolysis in many organs. This inhibition of glycolysis by fatty acids, perhaps together with a more direct action of growth hormone, results in decreased glucose utilization by the tissues. Growth hormone thus acts to raise the blood glucose concentration.

Growth hormone stimulates the cellular uptake of amino acids and protein synthesis in many organs of the body. These actions are useful when eating a protein-rich meal; amino acids are removed from the blood and used to form proteins, and the plasma concentration of glucose and fatty acids is increased to provide alternate energy sources (fig. 19.16). The anabolic effect of growth hormone on protein synthesis is particularly important during the growing years, when it contributes to increases in bone length and in the mass of many soft tissues.

Effects of Growth Hormone on Body Growth

The stimulatory effects of growth hormone on skeletal growth results from stimulation of mitosis in the epiphyseal discs of cartilage present in the long bones of growing children and adolescents. This action is mediated by the somatomedins, IGF-1 and IGF-2, which stimulate the chondrocytes to divide and secrete more cartilage matrix. Part of this growing cartilage is converted to bone, enabling the bone to grow in length. This skeletal growth stops when the epiphyseal discs are converted to bone after the growth spurt during puberty, despite the fact that growth hormone secretion continues throughout adulthood.

An excessive secretion of growth hormone can produce **gigantism** in children, who may grow up to 8 feet tall, at the same time maintaining normal body proportions. An excessive growth hormone secretion that occurs after the epiphyseal discs

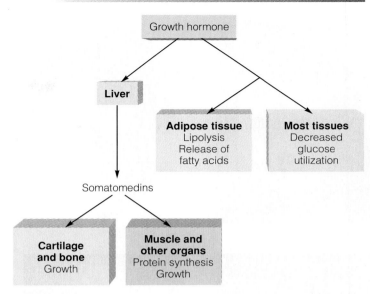

Figure 19.16

The effects of growth hormone. The growth-promoting, or anabolic, effects of growth hormone are mediated indirectly via stimulation of somatomedin production by the liver.

have sealed, however, cannot produce increases in height. The oversecretion of growth hormone in adults results in an elongation of the jaw and deformities in the bones of the face, hands, and feet. This condition, called **acromegaly,** is accompanied by the growth of soft tissues and coarsening of the skin (fig. 19.17). It is interesting that athletes who (illegally) take growth hormone supplements to increase their muscle mass may also experience body changes like those of acromegaly.

An inadequate secretion of growth hormone during the growing years results in **dwarfism.** An interesting variant of this condition is *Laron dwarfism,* in which there is a genetic insensitivity to the effects of growth hormone. This insensitivity is associated with, but may not be caused by, a reduction in the number of growth hormone receptors in the target cells. Genetic engineering has made available recombinant IGF-1, which has recently been approved by the FDA for the medical treatment of Laron dwarfism.

 An adequate diet, particularly with respect to proteins, is required for the production of IGF-1. This helps to explain the common observation that many children are significantly taller than their parents, who may not have had an adequate diet in their youth. Children with protein malnutrition (kwashiorkor) have low growth rates and low levels of IGF-1 in the blood, despite the fact that their growth hormone secretion may be abnormally elevated. When these children are provided with an adequate diet, IGF-1 levels and growth rates increase.

Age 9

Age 33

Age 16

Age 52

Figure 19.17

The progression of acromegaly in one individual. The coarsening of features and disfigurement are evident by age 33 and severe at age 52.

1. Describe the effects of epinephrine and the glucocorticoids on the metabolism of carbohydrates and lipids. What is the significance of these effects as a response to stress?
2. Explain the actions of thyroxine on the basal metabolic rate. Why do people who are hypothyroid have a tendency to gain weight and why are they less resistant to cold stress?
3. Describe the effects of growth hormone on the metabolism of lipids, glucose, and amino acids.
4. Explain how growth hormone stimulates skeletal growth.

Regulation of Calcium and Phosphate Balance

A normal blood Ca^{++} concentration is critically important for contraction of muscles and maintenance of proper membrane permeability. Parathyroid hormone promotes an elevation in blood Ca^{++} by stimulating resorption of the calcium phosphate crystals from bone and renal excretion of phosphate. A derivative of vitamin D produced in the body, 1,25-dihydroxyvitamin D_3, promotes the intestinal absorption of calcium and phosphate.

The calcium and phosphate concentrations of plasma are affected by bone formation and resorption, intestinal absorption, and urinary excretion of these ions. These processes are regulated by parathyroid hormone, 1,25-dihydroxyvitamin D_3, and calcitonin, as summarized in table 19.10.

The skeleton, in addition to providing support for the body, serves as a large store of calcium and phosphate in the form of crystals of *hydroxyapatite*, which have the formula $Ca_{10}(PO_4)_6(OH)_2$. The calcium phosphate in these hydroxyapatite crystals is derived from the blood by the action of bone-forming cells, or **osteoblasts.** The osteoblasts secrete an organic matrix, composed largely of collagen protein, which becomes hardened by deposits of hydroxyapatite. This process is called **bone deposition. Bone resorption** (dissolution of hydroxyapatite), produced by the action of **osteoclasts** (fig. 19.18), results in the return of bone calcium and phosphate to the blood.

The formation and resorption of bone occur constantly at rates determined by the hormonal balance. Body growth during the first two decades of life occurs because bone formation proceeds at a faster rate than bone resorption. By age 50 or 60, the rate of bone resorption often exceeds the rate of bone deposition. The constant activity of osteoblasts and osteoclasts allows bone to be remodeled throughout life. The position of the teeth, for example, can be changed by orthodontic appliances (braces), which cause bone resorption on the pressure-bearing side and bone formation on the opposite side of the alveolar sockets.

Despite the changing rates of bone formation and resorption, the plasma concentrations of calcium and phosphate are maintained by hormonal control of the intestinal absorption and urinary excretion of these ions. These hormonal control mechanisms are very effective in maintaining the plasma calcium and phosphate concentrations within narrow limits. Plasma calcium, for example, is normally maintained at about 2.5 millimolar, or 5 milliequivalents per liter (a milliequivalent equals a millimole times the valence of the ion, in this case, times 2).

Table 19.10 Endocrine Regulation of Calcium and Phosphate Balance

Hormone	Effect on Intestine	Effect on Kidneys	Effect on Bone	Associated Diseases
Parathyroid hormone (PTH)	No direct effect	Stimulates Ca^{++} reabsorption; inhibits PO_4^{-3} reabsorption	Stimulates resorption	Osteitis fibrosa cystica with hypercalcemia due to excess PTH
1,25-dihydroxyvitamin D_3	Stimulates absorption of Ca^{++} and PO_4^{-3}	Stimulates reabsorption of Ca^{++} and PO_4^{-3}	Stimulates resorption	Osteomalacia (adults) and rickets (children) due to deficiency of 1,25-dihydroxyvitamin D_3
Calcitonin	None	Inhibits resorption of Ca^{++} and PO_4^{-3}	Stimulates deposition	None

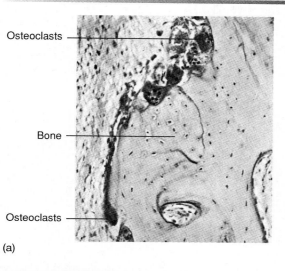

Osteoclasts

Bone

Osteoclasts

(a)

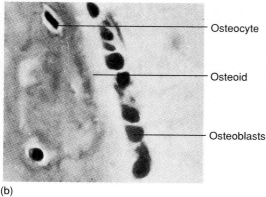

Osteocyte

Osteoid

Osteoblasts

(b)

Figure 19.18

(a) The resorption of bone by osteoclasts and (b) the formation of new bone by osteoblasts. Both resorption and deposition (formation) occur simultaneously throughout the body. Osteoid is the organic matrix of bone prior to calcification.

The maintenance of normal plasma calcium concentrations is important because of the wide variety of effects that calcium has in the body. In addition to its role in bone formation, calcium is needed for excitation-contraction coupling in muscles and as a second messenger in the action of some hormones. It is also needed to maintain proper membrane permeability. An abnormally low plasma calcium concentration increases the permeability of the cell membranes to Na^+ and other ions. Hypocalcemia, therefore, enhances the excitability of nerves and muscles and can result in muscle spasm (tetany).

Parathyroid Hormone

Whenever the plasma concentration of Ca^{++} begins to fall, the parathyroid glands are stimulated to secrete increased amounts of *parathyroid hormone* (*PTH*), which acts to raise the blood Ca^{++} back to normal levels. As might be predicted from this action of PTH, people who have their parathyroid glands removed (as may occur accidentally during surgical removal of the thyroid) will experience hypocalcemia. This can cause severe muscle tetany, for reasons previously discussed, and serves as a dramatic reminder of the importance of PTH.

Parathyroid hormone helps to raise the blood Ca^{++} concentration primarily by stimulating the activity of osteoclasts to resorb bone. In addition, PTH stimulates the kidneys to reabsorb Ca^{++} from the glomerular filtrate while inhibiting the reabsorption of PO_4^{-3}. This raises blood Ca^{++} levels without promoting the deposition of calcium phosphate crystals in bone. Finally, PTH promotes the formation of 1,25-dihydroxyvitamin D_3 (as described in the next section), and so it also helps to raise the blood calcium levels indirectly through the effects of this other hormone.

1,25-Dihydroxyvitamin D_3

The production of **1,25-dihydroxyvitamin D_3** begins in the skin, where vitamin D_3 is produced from its precursor molecule (7-dehydrocholesterol) under the influence of sunlight. When the skin does not make sufficient vitamin D_3 because of insufficient exposure to sunlight, this compound must be ingested in

The most common bone disorder in elderly people is **osteoporosis**. Osteoporosis is characterized by parallel losses of mineral and organic matrix from bone, reducing bone mass (fig. 19.19) and increasing the risk of fractures. Although the causes of osteoporosis are not well understood, age-related bone loss occurs more rapidly in women than men (osteoporosis is almost ten times more common in women after menopause than in men at comparable ages), suggesting that the fall in estrogen secretion at menopause contributes to this condition. Premenopausal women who have a very low percentage of body fat and amenorrhea can also have osteoporosis. It is interesting in this regard that estrogen receptors have recently been identified in osteoblasts, suggesting that estrogen may exert a direct effect on bone production.

Physicians advise teenage girls, who are attaining their maximum bone mass, to eat such calcium-rich foods as milk and other dairy products. This may reduce the progression of osteoporosis when they get older. The National Osteoporosis Foundation recommends that women supplement their diet with 1,200 mg of calcium per day before age 24, and with 1,000 mg of calcium thereafter until menopause. They recommend that postmenopausal women who are not being treated with estrogen supplement their diet with 1,500 mg of calcium per day. Hormone replacement therapy for postmenopausal women is common because it helps reduce osteoporosis and also reduces the risk of ischemic heart disease, which otherwise increases substantially after menopause.

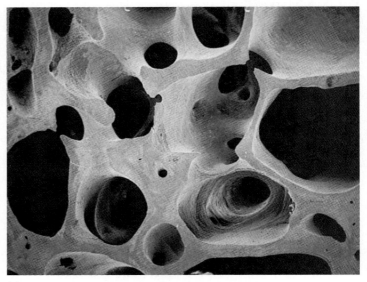

(a)

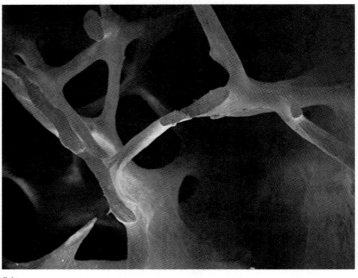

(b)

Figure 19.19

Scanning electron micrographs of bone biopsy specimens from the iliac crest. (*a*) A normal specimen and (*b*) a specimen from a person with osteoporosis.

the diet—that is why it is called a vitamin. Whether this compound is secreted into the blood from the skin or enters the blood after being absorbed from the intestine, vitamin D_3 functions as a *prehormone*; in order to be biologically active, it must be chemically changed (chapter 11).

An enzyme in the liver adds a hydroxyl group (OH) to carbon number 25, which converts vitamin D_3 into 25-hydroxy-vitamin D_3. In order to be active, however, another hydroxyl group must be added to carbon number 1. Hydroxylation of the first carbon is accomplished by an enzyme in the kidneys, which converts the molecule to 1,25-dihydroxyvitamin D_3 (fig. 19.20). The activity of this enzyme in the kidneys is stimulated by parathyroid hormone (fig. 19.21). Increased secretion of PTH, stimulated by low blood Ca^{++}, is thus accompanied by the increased production of 1,25-dihydroxyvitamin D_3.

The hormone 1,25-dihydroxyvitamin D_3 helps to raise the plasma concentrations of calcium and phosphate by stimulating (1) the intestinal absorption of calcium and phosphate, (2) the resorption of bones, and (3) the renal reabsorption of calcium and phosphate so that less is excreted in the urine.

Notice that 1,25-dihydroxyvitamin D_3, but not parathyroid hormone, directly stimulates intestinal absorption of calcium and phosphate and promotes the reabsorption of phosphate in the kidneys. The effect of simultaneously raising the blood concentrations of Ca^{++} and PO_4^{-3} results in the increased tendency of these two ions to precipitate as hydroxyapatite crystals in bone.

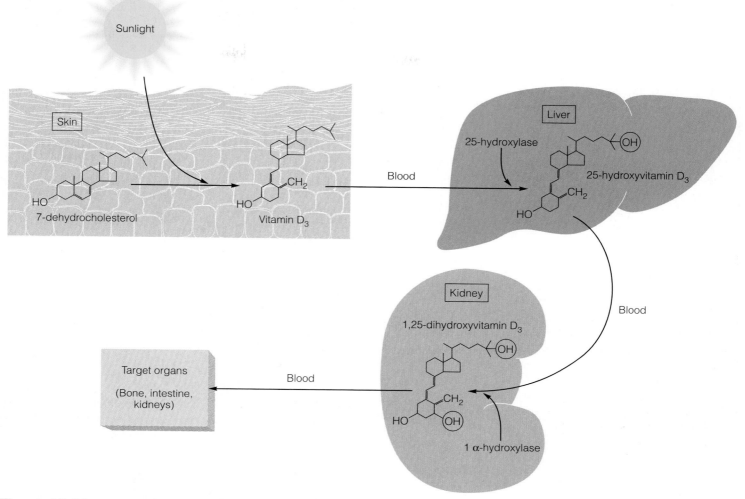

Figure 19.20

The pathway for the production of the hormone 1,25-dihydroxyvitamin D₃. This hormone is produced in the kidneys from the inactive precursor 25-hydroxyvitamin D₃ (formed in the liver). This latter molecule is produced from vitamin D₃ secreted by the skin.

Since 1,25-dihydroxyvitamin D_3 directly stimulates bone resorption, it seems paradoxical that this hormone is needed for proper bone deposition and, in fact, that inadequate amounts of 1,25-dihydroxyvitamin D_3 result in the bone demineralization of osteomalacia and rickets. This apparent paradox may be explained logically by the fact that the primary function of 1,25-dihydroxyvitamin D_3 is stimulation of intestinal Ca^{++} and PO_4^{-3} absorption. When calcium intake is adequate, the major result of 1,25-dihydroxyvitamin D_3 action is the availability of Ca^{++} and PO_4^{-3} in sufficient amounts to promote bone deposition. Only when calcium intake is inadequate does the direct effect of 1,25-dihydroxyvitamin D_3 on bone resorption become significant, acting to ensure proper blood Ca^{++} levels.

 In addition to osteoporosis, a number of other bone disorders are associated with abnormal calcium and phosphate balance. In **osteomalacia** (in adults) and **rickets** (in children), inadequate intake of vitamin D results in inadequate mineralization of the organic matrix of collagen. Excessive secretion of parathyroid hormone results in **osteitis fibrosa cystica**, in which excessive osteoclast activity causes resorption of both the mineral and organic components of bone, which are then replaced by fibrous tissue.

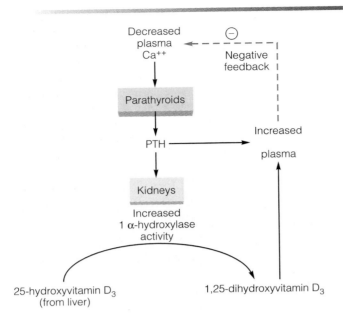

Figure 19.21

A decrease in plasma Ca^{++} directly stimulates the secretion of parathyroid hormone (PTH). The production of 1,25-dihydroxyvitamin D$_3$ also rises when Ca^{++} is low because PTH stimulates the final hydroxylation step in the formation of this compound in the kidneys.

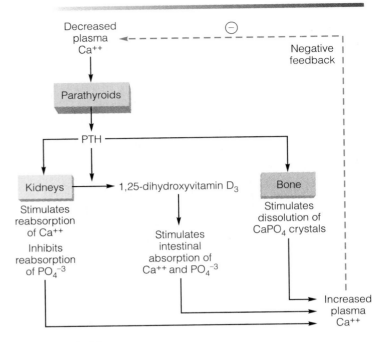

Figure 19.22

The negative feedback loop that returns low blood Ca^{++} concentrations to normal without simultaneously raising blood phosphate levels above normal.

Negative Feedback Control of Calcium and Phosphate Balance

The secretion of parathyroid hormone is controlled by the plasma calcium concentrations. Its secretion is stimulated by low calcium concentrations and inhibited by high calcium concentrations. Since parathyroid hormone stimulates the final hydroxylation step in the formation of 1,25-dihydroxyvitamin D$_3$, a rise in parathyroid hormone results in an increase in production of 1,25-dihydroxyvitamin D$_3$. Low blood calcium can thus be corrected by the effects of increased parathyroid hormone and 1,25-dihydroxyvitamin D$_3$ (fig. 19.22).

It is possible for plasma calcium levels to fall while phosphate levels remain normal. In this case, the increased secretion of parathyroid hormone and the production of 1,25-dihydroxyvitamin D$_3$ that result could abnormally raise phosphate levels while acting to restore normal calcium levels. This is prevented by the inhibition of phosphate reabsorption in the kidneys by parathyroid hormone,

so that more phosphate is excreted in the urine (fig. 19.22). In this way, blood calcium levels can be raised to normal without excessively raising blood phosphate concentrations.

Calcitonin

Experiments in the 1960s revealed that high blood calcium in dogs may be lowered by a hormone secreted from the thyroid gland. This hormone thus has an effect opposite to that of parathyroid hormone and 1,25-dihydroxyvitamin D$_3$. The calcium-lowering hormone, called **calcitonin,** was found to be a thirty-two-amino-acid polypeptide secreted by *parafollicular cells,* or *C cells,* in the thyroid, which are distinct from the follicular cells that secrete thyroxine.

The secretion of calcitonin is stimulated by high plasma calcium levels and acts to lower calcium levels by (1) inhibiting the activity of osteoclasts, thus reducing bone resorption, and (2) stimulating the urinary excretion of calcium and phosphate by inhibiting their reabsorption in the kidneys (fig. 19.23).

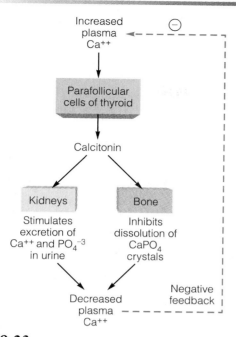

Figure 19.23

Negative feedback control of calcitonin secretion.

Although it is attractive to think that calcium balance is regulated by the effects of antagonistic hormones, the significance of calcitonin in human physiology remains unclear. Patients who have had their thyroid gland surgically removed (as for thyroid cancer) are *not* hypercalcemic, as one would expect them to be if calcitonin were needed to lower blood calcium levels. The ability of very large pharmacological doses of calcitonin to inhibit osteoclast activity and bone resorption, however, is clinically useful in the treatment of *Paget's disease*, in which osteoclast activity causes softening of bone.

---

1. Describe the mechanisms by which the secretion of parathyroid hormone and of calcitonin is regulated.

2. List the steps involved in the formation of 1,25-dihydroxyvitamin D_3, and state how this formation is influenced by parathyroid hormone.

3. Describe the actions of parathyroid hormone, 1,25-dihydroxyvitamin D_3, and calcitonin on the intestine, skeletal system, and kidneys, and explain how these actions affect the blood levels of calcium.

4. Explain the differences in the effects of 1,25-dihydroxyvitamin D_3 on bones according to whether calcium intake is adequate or inadequate.

Summary

Nutritional Requirements   *p. 578*

I. Food provides molecules used in cell respiration for energy.
   A. The metabolic rate is influenced by physical activity, temperature and eating; the basal metabolic rate is measured as the rate of oxygen consumption when such influences are standardized and minimal.
   B. The energy provided in food and the energy consumed by the body are measured in units of kilocalories.
   C. When the caloric intake is greater than the energy expenditure over a period of time, the excess calories are stored primarily as fat.

II. Vitamins and elements serve primarily as cofactors and coenzymes.
   A. Vitamins are divided into those that are fat-soluble (A, D, E, and K) and those that are water-soluble.
   B. Many water-soluble vitamins are needed for the activity of the enzymes involved in cell respiration.

Regulation of Energy Metabolism   *p. 582*

I. The body tissues can use circulating energy substrates, including glucose, fatty acids, ketone bodies, lactic acid, amino acid, and others, for cell respiration.

A. Different organs have different preferred energy sources.
   B. Circulating energy substrates can be obtained from food or from the energy reserves of glycogen, fat, and protein in the body.

II. Eating behavior is regulated, at least in part, by the hypothalamus.
   A. Lesions of the ventromedial area of the hypothalamus produce hyperphagia, whereas lesions of the lateral hypothalamus produce hypophagia.
   B. A variety of neurotransmitters have been implicated in the control of eating behavior; these include the endorphins, norepinephrine, serotonin, and cholecystokinin.

III. The control of energy balance in the body is regulated by the anabolic and catabolic effects of a variety of hormones.

Energy Regulation by the Islets of Langerhans *p. 585*

I. A rise in plasma glucose concentration stimulates insulin and inhibits glucagon secretion.
   A. Amino acids stimulate the secretion of both insulin and glucagon.
   B. Insulin secretion is also stimulated by parasympathetic innervation of the islets and by the action of gastric inhibitory peptide (GIP), secreted by the intestine.
II. During the intestinal absorption of a meal, insulin promotes the uptake of blood glucose into tissue cells.
   A. This lowers the blood glucose concentration and increases the energy reserves of glycogen, fat, and protein.
   B. Skeletal muscles are the major organs that remove blood glucose in response to insulin stimulation.
III. During periods of fasting, insulin secretion decreases and glucagon secretion increases.
   A. Glucagon stimulates glycogenolysis in the liver, gluconeogenesis, lipolysis, and ketogenesis.
   B. These effects help to maintain adequate levels of blood glucose for the brain and provide alternate energy sources for other organs.

Diabetes Mellitus and Hypoglycemia *p. 590*

I. Diabetes mellitus and reactive hypoglycemia represent disorders of the islets of Langerhans.
   A. Insulin-dependent diabetes mellitus occurs when the beta cells are destroyed; the resulting lack of insulin and excessive glucagon secretion produce the symptoms of this disease.
   B. Non-insulin-dependent diabetes mellitus occurs as a result of a relative tissue insensitivity to

insulin and inadequate insulin secretion; this condition is aggravated by obesity.
   C. Reactive hypoglycemia occurs when the islets secrete excessive amounts of insulin in response to a rise in blood glucose concentration.

Metabolic Regulation by Adrenal Hormones, Thyroxine, and Growth Hormone *p. 593*

I. The adrenal hormones involved in energy regulation include epinephrine from the adrenal medulla and glucocorticoids (mainly hydrocortisone) from the adrenal cortex.
   A. The effects of epinephrine are similar to those of glucagon.
   B. Glucocorticoids promote the breakdown of muscle protein and the conversion of amino acids to glucose in the liver.
II. Thyroxine stimulates the rate of cell respiration in almost all cells in the body.
   A. Thyroxine thus sets the basal metabolic rate (BMR), which is the rate at which energy is consumed by the body under resting conditions.
   B. Thyroxine also promotes protein synthesis and is needed for proper body growth and development, particularly of the central nervous system.
III. The secretion of growth hormone is regulated by releasing and inhibiting hormones from the hypothalamus.
   A. The secretion of growth hormone is stimulated by a protein meal and by a fall in glucose, as occurs during fasting.
   B. Growth hormone stimulates catabolism of lipids and inhibits glucose utilization.
   C. Growth hormone also stimulates protein synthesis and thus promotes body growth.
   D. The anabolic effects of growth hormone, including the stimulation of bone growth in childhood, are believed to be produced indirectly via polypeptides called somatomedins.

Regulation of Calcium and Phosphate Balance *p. 598*

I. Bone contains calcium and phosphate in the form of hydroxyapatite crystals. This serves as a reserve supply of calcium and phosphate for the blood.
   A. The formation and resorption of bone are produced by the action of osteoblasts and osteoclasts, respectively.
   B. The plasma concentrations of calcium and phosphate are also affected by absorption from the intestine and by the urinary excretion of these ions.
II. Parathyroid hormone stimulates bone resorption and calcium reabsorption in the kidneys. This hormone thus acts to raise the blood calcium concentration.
   A. The secretion of parathyroid hormone is stimulated by a fall in blood calcium levels.
   B. Parathyroid hormone also inhibits reabsorption of phosphate in the kidneys, so that more phosphate is excreted in the urine.
III. 1,25-dihydroxyvitamin D_3 is derived from vitamin D by hydroxylation reactions in the liver and kidneys.
   A. The last hydroxylation step is stimulated by parathyroid hormone.
   B. 1,25-dihydroxyvitamin D_3 stimulates the intestinal absorption of calcium and phosphate, resorption of bone, and renal reabsorption of phosphate.
IV. A rise in parathyroid hormone, accompanied by the increased production of 1,25-dihydroxyvitamin D_3, helps to maintain proper blood levels of calcium and phosphate in response to a fall in calcium levels.
V. Calcitonin is secreted by the parafollicular cells of the thyroid gland.
   A. Calcitonin secretion is stimulated by a rise in blood calcium levels.
   B. Calcitonin, at least at pharmacological levels, acts to lower blood calcium by inhibiting bone resorption and stimulating the urinary excretion of calcium and phosphate.

Clinical Investigation

A middle-aged woman goes to her physician complaining of nausea, headaches, and continuous thirst. Further questioning reveals that she has frequent urination and that she has experienced a recent and rapid weight loss. In addition, both her mother and uncle are diabetics. She provides a sample of urine, which does not give evidence of glycosuria or ketonuria. She is told to return the next day to provide a fasting blood sample. When this is analyzed, a blood glucose concentration of 150 mg/dl is measured. An oral glucose tolerance test is subsequently performed, and a blood glucose concentration of 220 mg/dl is measured 2 hours following the ingestion of the glucose solution. The physician places this patient on a weight-reduction program and advises her to begin a mild but regular exercise regimen. He mentions that, if this program is not effective in relieving her symptoms, he will prescribe sulfonylureas. What diagnosis did this physician make? Why did he make this diagnosis and subsequent recommendations?

Clues

Read about the oral glucose tolerance test in the section "Regulation of Insulin and Glucagon Secretion." Study the section on non-insulin-dependent diabetes mellitus.

Review Activities

Objective Questions

Match the following:

1. absorption of carbohydrate meal
2. fasting

   a. rise in insulin; rise in glucagon
   b. fall in insulin; rise in glucagon
   c. rise in insulin; fall in glucagon
   d. fall in insulin; fall in glucagon

Match the following:

3. growth hormone
4. thyroxine
5. hydrocortisone

   a. increased protein synthesis; increased cell respiration
   b. protein catabolism in muscles; gluconeogenesis in liver
   c. protein synthesis in muscles; decreased glucose utilization
   d. fall in blood glucose; increased fat synthesis

6. A lowering of blood glucose concentration promotes
   a. decreased lipogenesis.
   b. increased lipolysis.
   c. increased glycogenolysis.
   d. all of the above.
7. Glucose can be secreted into the blood by
   a. the liver.
   b. the muscles.
   c. the liver and muscles.
   d. the liver, muscles, and brain.
8. The basal metabolic rate is determined primarily by
   a. hydrocortisone.
   b. insulin.
   c. growth hormone.
   d. thyroxine.
9. Somatomedins are required for the anabolic effects of
   a. hydrocortisone.
   b. insulin.
   c. growth hormone.
   d. thyroxine.
10. The increased intestinal absorption of calcium is stimulated directly by
    a. parathyroid hormone.
    b. 1,25-dihydroxyvitamin D_3.
    c. calcitonin.
    d. all of the above.
11. A rise in blood calcium levels directly stimulates
    a. parathyroid hormone secretion.
    b. calcitonin secretion.
    c. 1,25-dihydroxyvitamin D_3 formation.
    d. all of the above.

12. At rest, about 12% of the total calories consumed are used for
    a. protein synthesis.
    b. cell transport.
    c. the Na^+/K^+ pumps.
    d. DNA replication.
13. Which of the following hormones stimulates anabolism of proteins and catabolism of fat?
    a. growth hormone
    b. thyroxine
    c. insulin
    d. glucagon
    e. epinephrine
14. If a person eats 600 kilocalories of protein in a meal, which of the following statements will be *false*?
    a. Insulin secretion will be increased.
    b. The metabolic rate will be increased over basal conditions.
    c. The tissue cells will use some of the amino acids for resynthesis of body proteins.
    d. The tissue cells will obtain 600 kilocalories worth of energy.
    e. Body-heat production and oxygen consumption will be increased over basal conditions.
15. Ketoacidosis in untreated diabetes mellitus is due to
    a. excessive fluid loss.
    b. hypoventilation.
    c. excessive eating and obesity.
    d. excessive fat catabolism.

Essay Questions

1. Compare the metabolic effects of fasting to the state of uncontrolled insulin-dependent diabetes mellitus. Explain the hormonal similarities of these conditions.[1]

2. Glucocorticoids stimulate the breakdown of protein in muscles but the synthesis of protein in the liver. Explain the significance of these different effects.

3. Describe how thyroxine affects cell respiration. Why does a person who is hypothyroid have a tendency to gain weight and less tolerance for cold?

4. Compare and contrast the metabolic effects of thyroxine and growth hormone.

5. Why is vitamin D considered to be both a vitamin and a prehormone?

Explain why people with osteoporosis might be helped by taking controlled amounts of vitamin D.

6. Explain what is meant by the term *insulin resistance*. What is the relationship between insulin resistance, obesity, exercise, and non-insulin-dependent diabetes mellitus?

7. Describe the chemical nature and origin of the somatomedins and explain the physiological significance of these growth factors.

8. Explain how the secretion of insulin and glucagon are influenced by fasting; a meal that is high in carbohydrate and low in protein; and a meal that is high in protein and high in carbohydrate. Also,

explain how the changes in insulin and glucagon secretion under these conditions function to maintain homeostasis.

9. Using a cause-and-effect sequence, explain how an inadequate intake of dietary calcium or vitamin D can cause bone resorption. Also, describe the cause-and-effect sequence whereby an adequate intake of calcium and vitamin D may promote bone deposition.

10. Describe the conditions of gigantism, acromegaly, Laron dwarfism, and kwashiorkor and explain how these conditions relate to blood levels of growth hormone and IGF-1.

Selected Readings

Atkinson, M. A., and N. K. Maclaren. July 1990. What causes diabetes? *Scientific American*.

Austin, L. A., and H. Heath III. 1981. Calcitonin: Physiology and pathophysiology. *New England Journal of Medicine* 304:269.

Bikle, D. D. 1995. A bright future for the sunshine hormone. *Science and Medicine* 2:58.

Bray, G. A. 1991. Weight homeostasis. *Annual Review of Medicine* 42:205.

Brent, G. A. 1994. The molecular basis of thyroid hormone action. *New England Journal of Medicine* 331:847.

Brownlee, M. 1991. Glycosylation products as toxic mediators of diabetic complications. *Annual Review of Medicine* 42:159.

Cahill, G. F., and H. O. McDevitt. 1981. Insulin-dependent diabetes mellitus: The initial lesion. *New England Journal of Medicine* 304:454.

Cheng, K., and J. Larner. 1985. Intracellular mediators of insulin action. *Annual Review of Physiology* 47:405.

Cohick, W. S., and D. R. Clemmons. 1993. The insulin-like growth factors. *Annual Review of Physiology* 55:131.

DeLuca, H. F. 1980. The vitamin D hormonal system: Implications for bone disease. *Hospital Practice* 15:57.

Dineen, S. et al. 1992. Carbohydrate metabolism in non-insulin-dependent diabetes mellitus. *New England Journal of Medicine* 327:707.

Dussault, H., and J. Ruel. 1987. Thyroid hormones and brain development. *Annual Review of Physiology* 49:321.

Eisenbarth, G. S. 1986. Type I diabetes mellitus: A chronic autoimmune disease. *New England Journal of Medicine* 314:1360.

Gardner, L. I. July 1972. Deprivation dwarfism. *Scientific American*.

Goodman, D. S. 1984. Vitamin A and retinoids in health and disease. *New England Journal of Medicine* 310:1023.

Habener, J. F., and J. E. Mahaffey. 1978. Osteomalacia and disorders of vitamin D metabolism. *Annual Review of Medicine* 29:327.

Hahn, T. J. 1986. Physiology of bone: Mechanisms of osteopenic disorders. *Hospital Practice* 21:73.

Haussler, M. R., and T. A. McCain. 1977. Basic and clinical concepts related to vitamin D metabolism and action. *New England Journal of Medicine* 297: first part, p. 974; second part, p. 1041.

Hirsch, J. 1984. Hypothalamic control of appetite. *Hospital Practice* 19:131.

Isaksson, O. G. P., S. Edén, and J. O. Jansson. 1985. Mode of action of pituitary growth hormone on target cells. *Annual Review of Physiology* 47:483.

Jequier, E. 1993. Body weight regulation in humans: The importance of nutrient balance. *News in Physiological Sciences* 8: 273.

Kitabchi, A. E., and R. C. Goodman. 1987. Hypoglycemia: Pathophysiology and diagnosis. *Hospital Practice* 22:45.

Levine, M. 1986. New concepts in the biology and biochemistry of ascorbic acid. *New England Journal of Medicine* 314:892.

Liebel, R. L., M. Rosenbaum, and J. Hirsch. 1995. Changes in energy expenditure resulting from altered body weight. *New England Journal of Medicine* 332:621.

Marcus, R. 1989. Understanding and preventing osteoporosis. *Hospital Practice* 24:189.

Martin, R. J. et al. 1991. The regulation of body weight. *American Scientist* 79:528.

Miller, D. E., and J. S. Flier. 1991. Insulin resistance—mechanisms, syndromes, and implications. *New England Journal of Medicine* 325:938.

Mitlak, B. H., and S. R. Nussbaum. 1993. Diagnosis and treatment of osteoporosis. *Annual Review of Medicine* 44:265.

Mueckler, M. M. 1995. Glucose transport and glucose homeostasis: new insight from transgenic mice. *News in Physiological Sciences* 10:22.

Nadel, E. R., and S. R. Bussolari. July–August 1988. The Daedalus project: Physiological problems and solutions. *American Scientist*, p. 351.

---

[1] *Note:* This question is answered on page 248 of the Student Study Guide.

Nathan, D. M. 1993. Long-term complications of diabetes mellitus. *New England Journal of Medicine* 328:1676.

Nestler, J. E. 1994. Assessment of insulin resistance. *Science and Medicine* 1:58.

Notkins, A. L. November 1979. The cause of diabetes. *Scientific American.*

Oppenheimer, J. H. 1979. Thyroid hormone action at the cellular level. *Science* 203:971.

Orci, L., J. D. Vassalli, and A. Perrelet. September 1988. The insulin factory. *Scientific American.*

Phillips, L. S., and R. Vassilopoulou-Sellin. 1980. Somatomedins. *New England Journal of Medicine* 302: first part, p. 371; second part, p. 438.

Raisz, L. G. 1988. Local and systemic factors in the pathogenesis of osteoporosis. *New England Journal of Medicine* 318:818.

Raisz, L. G., and B. E. Kream. 1981. Hormonal control of skeletal growth. *Annual Review of Physiology* 43:225.

Reichel, H., H. P. Koeffler, and A. W. Norman. 1989. The role of the vitamin D endocrine system in health and disease. *New England Journal of Medicine* 320:980.

Riggs, B. L., and L. J. Melton III. 1992. The prevention and treatment of osteoporosis. *New England Journal of Medicine* 327:620.

Siperstein, M. D. 1985. Type II diabetes: Some problems in diagnosis and treatment. *Hospital Practice* 20:55.

Sterling, S. 1979. Thyroid hormone action at the cellular level. *New England Journal of Medicine* 300: first part, p. 117; second part, p. 173.

Strauss, R. H., and C. E. Yesalis. 1991. Anabolic steroids in the athlete. *Annual Review of Medicine* 42:449.

Suter, P. M. et al. 1992. The effect of ethanol on fat storage in healthy subjects. *New England Journal of Medicine* 326:983.

Taylor, S. I. et al. 1994. Insulin resistance or insulin deficiency: Which is the primary cause of NIDDM? *Diabetes* 43:745.

Tepperman, J. 1980. *Metabolic and Endocrine Physiology.* 4th ed. Chicago: Year Book Medical Publishers.

Unger, R. H., R. E. Dobbs, and L. Orci. 1979. Insulin, glucagon, and somatostatin secretion in the regulation of metabolism. *Annual Review of Physiology* 40:307.

Unger, R. H., and L. Orci. 1981. Glucagon and the A cell: Physiology and pathophysiology. *New England Journal of Medicine* 304: first part, p. 1518; second part, p. 1575.

Van Wyk, J., and L. E. Underwood. 1978. Growth hormone, somatomedins, and growth failure. *Hospital Practice* 13:57.

Verhaeghe, J., and R. Bouillon. 1994. Action of IGFs on bone. *News in Physiological Sciences* 9:20.

Wasserman, D. H. 1995. Regulation of glucose fluxes during exercise in the postabsorptive state. *Annual Review of Physiology* 57:191.

Explorations CD-ROM

The modules accompanying chapter 19 are #7 Diet and Weight Loss and #11 Hormone Action.

Reproduction

OBJECTIVES

After studying this chapter, you should be able to . . .

1. describe how the chromosomal content determines the sex of an embryo and how this relates to the development of testes or ovaries.

2. explain how the development of accessory sex organs and external genitalia is affected by the presence or absence of testes in the embryo.

3. describe the hormonal changes that occur during puberty, the mechanisms that may control the onset of puberty, and the secondary sex characteristics that develop during puberty.

4. explain how the secretions of pituitary gonadotropic hormones (FSH and LH) are regulated in the male and describe the actions of FSH and LH on the testis.

5. describe the structure of the testis and the interaction between the interstitial Leydig cells and seminiferous tubules.

6. describe the stages of spermatogenesis and the roles of Sertoli cells in spermatogenesis.

7. explain the hormonal control of spermatogenesis and describe the effects of androgens on the male accessory sex organs.

8. describe the composition of semen, explain the physiology of erection and ejaculation, and discuss the various factors that affect male fertility.

9. describe oogenesis and the stages of follicle development through ovulation and the formation of a corpus luteum.

10. explain the hormonal interactions involved in the control of ovulation.

11. describe the changes in the secretion of ovarian sex steroids during a nonfertile cycle and explain the function and fate of the corpus luteum.

12. explain how the secretion of FSH and LH is controlled through negative and positive feedback mechanisms during a menstrual cycle.

13. explain how contraceptive pills prevent ovulation.

14. describe the cyclic changes that occur in the endometrium and the hormonal mechanisms that cause these changes.

15. describe the acrosomal reaction and the events that occur at fertilization, blastocyst formation, and implantation.

16. explain how menstruation and further ovulation are normally prevented during pregnancy.

17. describe the structure and functions of the placenta.

18. list the hormones secreted by the placenta and describe their actions.

19. discuss the factors that stimulate uterine contractions during labor and parturition and explain how the onset of labor may be regulated.

20. describe the hormonal requirements for development of the mammary glands during pregnancy and explain how lactation is prevented during pregnancy.

21. describe the milk-ejection reflex.

OUTLINE

Sexual Reproduction

Early embryonic gonads can become either testes or ovaries. A particular gene on the Y chromosome induces the embryonic gonads to become testes. Females lack a Y chromosome, and the absence of this gene causes the development of ovaries. The embryonic testes secrete testosterone, which induces the development of male accessory sex organs and external genitalia. The absence of testes (rather than the presence of ovaries) in a female embryo causes the development of the female accessory sex organs.

"A chicken is an egg's way of making another egg." Phrased in more modern terms, genes are "selfish." Genes, according to this view, do not exist in order to make a well-functioning chicken (or other organism). The organism, rather, exists and functions so that the genes can survive beyond the mortal life of individual members of a species. Whether or not one accepts this rather cynical view, it is clear that reproduction is one of life's essential functions. The incredible complexity of structure and function in living organisms could not be produced in successive generations by chance; mechanisms must exist to transmit the blueprint (genetic code) from one generation to the next. Sexual reproduction, in which genes from two individuals are combined in random and novel ways with each new generation, offers the further advantage of introducing great variability into a population. This variability of genetic constitution helps to ensure that some members of a population will survive changes in the environment over evolutionary time.

In sexual reproduction, **germ cells,** or **gametes** (sperm and ova), are formed within the *gonads* (testes and ovaries) by a process of reduction division, or *meiosis* (chapter 3). During this type of cell division, the normal number of chromosomes in most human cells—forty-six—is halved, so that each gamete receives twenty-three chromosomes. Fusion of a sperm cell and ovum (egg cell) in the act of **fertilization** results in restoration of the original chromosome number of forty-six in the **zygote,** or fertilized egg. Growth of the zygote into an adult member of the next generation occurs by means of mitotic cell divisions, as described in chapter 3. When this individual reaches puberty, mature sperm or ova will be formed by meiosis within the gonads so that the life cycle can be continued (fig. 20.1).

Sex Determination

Each zygote inherits twenty-three chromosomes from its mother and twenty-three chromosomes from its father. This does not produce forty-six different chromosomes, but rather twenty-three pairs of *homologous chromosomes*. The members of a homologous pair, with the important exception of the sex chromosomes, look like each other and contain similar genes (such as those coding for eye color, height, and so on). These homologous pairs of chromosomes can be photographed and

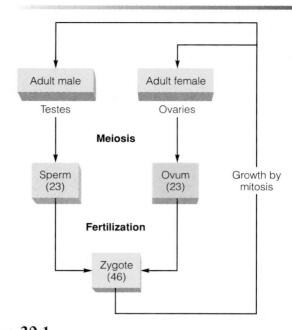

Figure 20.1

The human life cycle. Numbers in parentheses indicate the haploid state (twenty-three chromosomes) or diploid state (forty-six chromosomes).

numbered (as shown in fig. 20.2). Each cell that contains forty-six chromosomes (that is *diploid*) has two number 1 chromosomes, two number 2 chromosomes, and so on through pair number 22. The first twenty-two pairs of chromosomes are called **autosomal chromosomes.**

The twenty-third pair of chromosomes are the **sex chromosomes.** In a female, these consist of two X chromosomes, whereas in a male there is one X chromosome and one Y chromosome. The X and Y chromosomes look different and contain different genes. This is the exceptional pair of homologous chromosomes mentioned earlier.

When a diploid cell (with forty-six chromosomes) undergoes meiotic division, its daughter cells receive only one chromosome from each homologous pair of chromosomes. The gametes are therefore said to be *haploid* (they contain only half the number of chromosomes in the diploid parent cell). Each sperm cell, for example, will receive only one chromosome of homologous pair number 5—either the one originally contributed by the organism's mother, or the one originally contributed by the father (modified by the effects of crossing-over, as discussed in chapter 3). Which of the two chromosomes—maternal or paternal—ends up in a given sperm cell is completely random. This is also true for the sex chromosomes, so that approximately half of the sperm produced will contain an X and approximately half will contain a Y chromosome.

The egg cells (ova) in a woman's ovary will receive a similar random assortment of maternal and paternal chromosomes. Since the body cells of females have two X chromosomes, however, all of the ova will normally contain one X chromosome. Because all ova contain one X chromosome, whereas some sperm are X bearing and others are Y bearing, *the chromosomal*

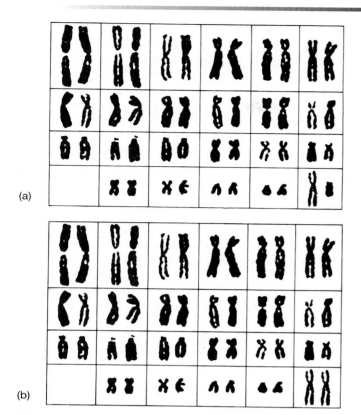

(a)

(b)

Figure 20.2

Homologous pairs of chromosomes obtained from a human diploid cell. The first twenty-two pairs of chromosomes are called the autosomal chromosomes. The sex chromosomes are (a) XY for a male and (b) XX for a female.

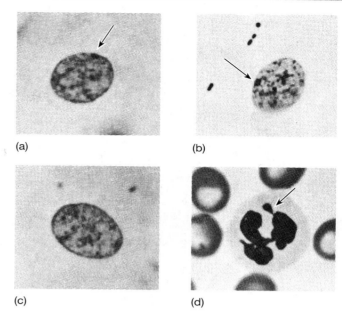

(a)

(b)

(c)

(d)

Figure 20.3

The nuclei of cheek cells obtained from females (a, b) have Barr bodies (arrows). These are formed from one of the X chromosomes, which is inactive. No Barr body is present in the cell obtained from a male (c) because males have only one X chromosome, which remains active. Some neutrophils obtained from females (d) have a "drumsticklike" appendage (arrow) that is not found in the white blood cells of males.

sex of the zygote is determined by the fertilizing sperm cell. If a Y-bearing sperm cell fertilizes the ovum, the zygote will be XY and male; if an X-bearing sperm cell fertilizes the ovum, the zygote will be XX and female.

Although each diploid cell in a woman's body inherits two X chromosomes, it appears that only one of each pair of X chromosomes remains active. The other X chromosome forms a clump of inactive "heterochromatin," which can often be seen as a dark spot, called a *Barr body,* at the edge of the nucleus of cheek cells (fig. 20.3). This provides a convenient test for chromosomal sex in cases where it is suspected that the chromosomal sex may differ from the apparent ("phenotypic") sex of the individual. Also, some of the nuclei in polymorphonuclear leukocytes of females have a "drumstick" appendage not seen in neutrophils from males.

Formation of Testes and Ovaries

The gonads of males and females are similar in appearance for the first forty or so days of development following conception. During this time, cells that will give rise to sperm (called *spermatogonia*) and cells that will give rise to ova (called *oogonia*) migrate from the yolk sac to the developing embryonic gonads. At this stage, the embryonic structures have the potential to become either testes or ovaries. The hypothetical substance that promotes their conversion to testes has been called the **testis-determining factor (TDF).**

Although it has long been recognized that male sex is determined by the presence of a Y chromosome and female sex by the absence of the Y chromosome, the genes involved have only recently been localized. In rare male babies with XX genotypes, scientists have discovered that one of the X chromosomes contains a segment of the Y chromosome—the result of an error that occurred during the meiotic cell division that formed the sperm. Similarly, rare female babies with XY genotypes were found to be missing the same portion of the Y chromosome erroneously inserted into the X chromosome of XX males.

Through these and other observations, it has been shown that the gene for the testis-determining factor is located on the short arm of the Y chromosome (fig. 20.4). Evidence suggests that it may be a particular gene known as *SRY* (for "sex-determining region of the Y"). This gene is found in the Y chromosome of all mammals and is highly conserved, meaning that it shows little variation in structure over evolutionary time.

The structures that will eventually produce sperm within the testes, the **seminiferous tubules,** appear very early in embryonic development—between 43 and 50 days following conception. Although spermatogenesis begins during embryonic life, it is arrested until the onset of puberty (this will be described in a later section). The tubules contain two major cell types: germinal and nongerminal. The **germinal cells** are those

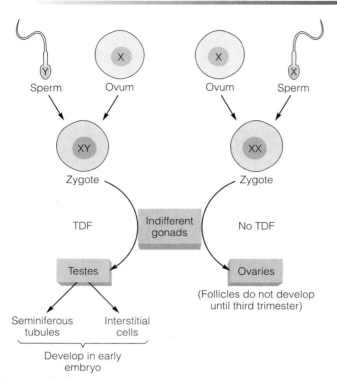

Figure 20.4

The formation of the chromosomal sex of the embryo and the development of the gonads. The very early embryo has "indifferent gonads" that can develop into either testes or ovaries. The testis-determining factor (TDF) is a gene located on the Y chromosome. In the absence of TDF, ovaries will develop.

Notice that it is normally the presence or absence of the Y chromosome that determines whether the embryo will have testes or ovaries. This point is well illustrated by two genetic abnormalities. In **Klinefelter's syndrome** the affected person has forty-seven instead of forty-six chromosomes because of the presence of an extra X chromosome. This person, with an XXY genotype, will develop testes despite the presence of two X chromosomes. Patients with **Turner's syndrome,** who have the genotype XO (and therefore have only forty-five chromosomes), have poorly developed ("streak") gonads.

that, through meiosis and subsequent specialization, will eventually become sperm. The nongerminal cells are called **Sertoli cells,** but are also known as nurse cells or sustentacular cells. The Sertoli cells appear at about day 42. At about day 65, the **Leydig cells** appear in the embryonic testes. The Leydig cells are located in the **interstitial tissue,** located outside of the seminiferous tubules between adjacent convolutions of the tubules. The interstitial Leydig cells constitute the endocrine tissue of the testes. In contrast to the rapid development of the testes, the functional units of the ovaries—called the **ovarian follicles**—do not appear until the second trimester of pregnancy (at about day 105).

The early appearing Leydig cells in the embryonic testes secrete large amounts of male sex hormones, or *androgens* (*andro* = man; *gen* = forming). The major androgen secreted by these cells is **testosterone.** Testosterone secretion begins as early as 8 weeks after conception, reaches a peak at 12 to 14 weeks, and thereafter declines to very low levels by the end of the second trimester (at about 21 weeks). Testosterone secretion during embryonic development in the male serves a very important function (described in the next section); similarly high levels of testosterone will not appear again in the life of the individual until the time of puberty.

As the testes develop, they move within the abdominal cavity and gradually descend into the *scrotum*. Descent of the testes is sometimes not complete until shortly after birth. The temperature of the scrotum is maintained at about 35°C—about 3°C below normal body temperature. This cooler temperature is needed for spermatogenesis. The fact that spermatogenesis does not occur in males with undescended testes—a condition called *cryptorchidism* (*crypt* = hidden; *orchid* = testes) illustrates this requirement.

The cremaster muscle is a strand of skeletal muscle associated with each spermatic cord. In cold weather, these muscles contract and elevate the testes, bringing them closer to the warmth of the trunk. The **cremasteric reflex** produces the same effect when the inside of a man's thigh is stroked. In a baby, however, this stimulation can cause the testes to be drawn up through the inguinal canal into the body cavity. The testes can also be drawn up into the body cavity voluntarily by trained Sumo wrestlers.

Development of Accessory Sex Organs and External Genitalia

In addition to testes and ovaries, various internal accessory sex organs are needed for reproductive function. Most of these are derived from two systems of embryonic ducts. Male accessory organs are derived from the **wolffian (mesonephric) ducts,** and female accessory organs are derived from the **müllerian (paramesonephric) ducts** (fig. 20.5). Interestingly, both male and female embryos between day 25 and day 50 have both duct systems and, therefore, have the potential to form the accessory organs characteristic of either sex.

Experimental removal of the testes (castration) from male embryonic animals results in regression of the wolffian ducts and development of the müllerian ducts into female accessory organs: the **uterus** and **uterine (fallopian) tubes.** Female accessory sex organs, therefore, develop as a result of the absence of testes rather than as a result of the presence of ovaries.

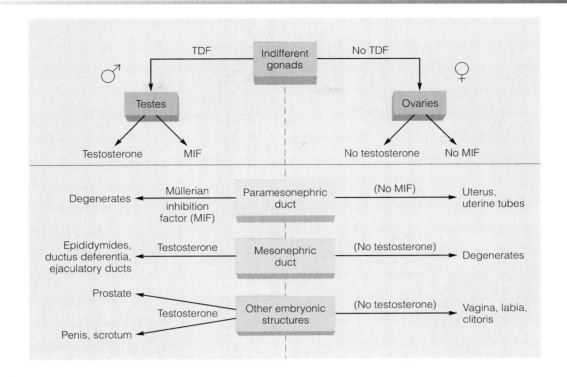

Figure 20.5

The embryonic development of male and female accessory sex organs and external genitalia. In the presence of testosterone and müllerian inhibition factor (MIF) secreted by the testes, male structures develop. In the absence of these secretions, female structures develop.

In a male, the Sertoli cells of the seminiferous tubules secrete a polypeptide called *müllerian inhibition factor* (MIF), which causes regression of the müllerian ducts beginning at about day 60. The secretion of testosterone by the Leydig cells of the testes subsequently causes growth and development of the wolffian ducts into male *accessory sex organs*: the **epididymis, vas deferens, seminal vesicles,** and **ejaculatory duct.** The structure and function of the accessory sex organs will be described in later sections.

The external genitalia of males and females are essentially identical during the first 6 weeks of development, sharing in common a *urogenital sinus, genital tubercle, urethral folds,* and a pair of *labioscrotal swellings*. The secretions of the testes masculinize these structures to form the **penis** and spongy (penile) urethra, **prostate,** and **scrotum.** The genital tubercle that forms the penis in a male will, in the absence of secreted testosterone, become the **clitoris** in a female. The penis and clitoris are thus said to be *homologous structures*. Similarly, the labioscrotal swellings form the scrotum in a male or the **labia majora** in a female; these structures are therefore also homologous (fig. 20.6).

Masculinization of the embryonic structures described occurs as a result of testosterone, secreted by the embryonic testes. Testosterone itself, however, is not the active agent within all of the target organs. Once inside particular target cells, testosterone is converted by the enzyme *5α-reductase* into the active hormone known as **dihydrotestosterone (DHT)** (fig. 20.7). DHT is needed for the development and maintenance of the penis, spongy urethra, scrotum, and prostate. Evidence suggests that testosterone itself directly stimulates the wolffian duct derivatives—epididymis, vas deferens, ejaculatory duct, and seminal vesicles.

In summary, the genetic sex is determined by whether a Y-bearing or an X-bearing sperm cell fertilizes the ovum; the presence or absence of a Y chromosome in turn determines whether the gonads of the embryo will be testes or ovaries; the presence or absence of testes, finally, determines whether the accessory sex organs and external genitalia will be male or female (table 20.1). This regulatory pattern of sex determination makes sense in light of the fact that both male and female embryos develop within an environment high in estrogen, which is secreted by the mother's ovaries and the placenta. If the secretions of the ovaries determined the sex, all embryos would be female.

Disorders of Embryonic Sexual Development

Hermaphroditism is a condition in which both ovarian and testicular tissue is present in the body. About 34% of hermaphrodites have an ovary on one side and a testis on the other. About 20% have ovotestes—part testis and part ovary—on both sides. The remaining 46% have an ovotestis on one side and an ovary or testis on the other. This condition is extremely rare and appears to be caused by the fact that some embryonic cells receive the short arm of the Y chromosome, with its testis-determining factor, whereas others do not. More common (though still rare) disorders of sex determination involve individuals with either

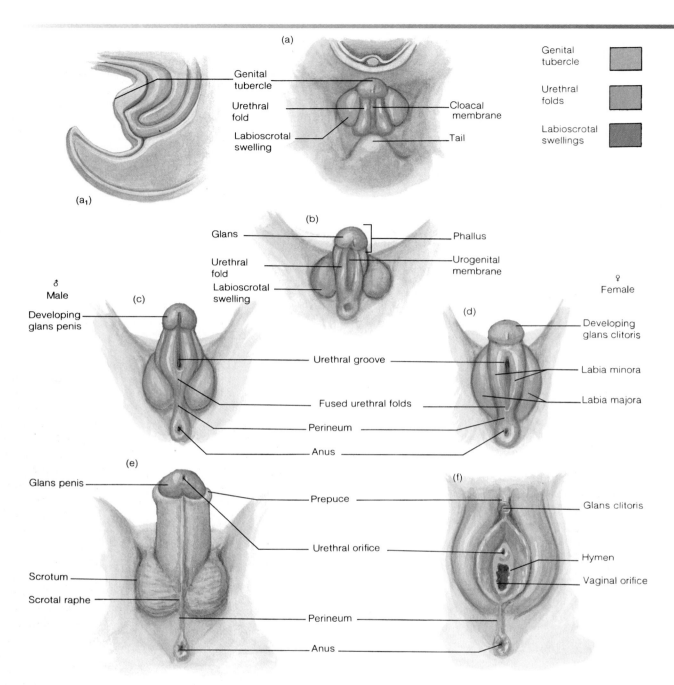

Figure 20.6

Differentiation of the external genitalia in the male and female. (*a, a₁* [sagittal view]) At 6 weeks, the genital tubercle, urethral fold, and labioscrotal swelling have differentiated from the genital tubercle. (*b*) At 8 weeks, a distinct phallus is present during the indifferent stage. By week 12, the genitalia have become distinctly male (*c*) or female (*d*), being derived from homologous structures. (*e, f*) At 16 weeks, the genitalia are formed.

testes or ovaries, but not both, who have accessory sex organs and external genitalia that are incompletely developed or that are inappropriate for their chromosomal sex. These individuals are called *pseudohermaphrodites* (*pseudo* = false).

The most common cause of female pseudohermaphroditism is *congenital adrenal hyperplasia*. This condition, which is inherited as a recessive trait, is caused by the excessive secretion of androgens from the adrenal cortex. A female with this condition would have müllerian duct derivatives (uterus and

fallopian tubes) because the adrenal doesn't secrete müllerian inhibition factor, but she would have partially masculinized external genitalia and wolffian duct derivatives.

An interesting cause of male pseudohermaphroditism is *testicular feminization syndrome*. Individuals with this condition have normally functioning testes but lack receptors for testosterone. Thus, although large amounts of testosterone are secreted, the embryonic tissues cannot respond to this hormone. Female genitalia therefore develop, but the vagina ends blindly (a uterus and

Table 20.1 A Developmental Timetable for the Reproductive System

Approximate Time after Fertilization			Developmental Changes	
Days	Trimester	Indifferent	Male	Female
19	First	Germ cells migrate from yolk sac.		
25–30		Wolffian ducts begin development.		
44–48		Müllerian ducts begin development.		
50–52		Urogenital sinus and tubercle develop.		
53–60			Tubules and Sertoli cells appear. Müllerian ducts begin to regress.	
60–75			Leydig cells appear and begin testosterone production. Wolffian ducts grow.	Formation of vagina begins. Regression of wolffian ducts begins.
105	Second			Development of ovarian follicles begins.
120				Uterus is formed.
160–260	Third		Testes descend into scrotum. Growth of external genitalia occurs.	Formation of vagina complete.

Source: Reproduced, with permission, from the *Annual Review of Physiology*, Volume 40, p. 279. Copyright © 1978 by Annual Reviews, Inc.

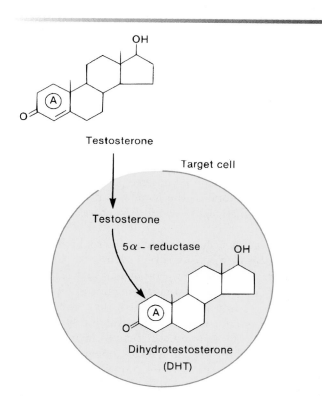

Figure 20.7

The conversion of testosterone, secreted by the interstitial (Leydig) cells of the testes, into dihydrotestosterone (DHT) within the target cells. This reaction involves the addition of a hydrogen (and the removal of the double carbon bond) in the first (A) ring of the steroid.

fallopian tubes do not develop because of the secretion of müllerian inhibition factor). Male accessory sex organs likewise cannot develop because the wolffian ducts lack testosterone receptors. A child with this condition appears externally to be a normal prepubertal girl, but she has testes in her body cavity and no accessory sex organs. These testes secrete an exceedingly large amount of testosterone at puberty because of the absence of negative feedback inhibition. This abnormally large amount of testosterone is converted by the liver and adrenal cortex into estrogens. As a result, the person with testicular feminization syndrome develops into a female with well-developed breasts who never menstruates (and who, of course, can never become pregnant).

Some male pseudohermaphrodites have normally functioning testes and normal testosterone receptors, but genetically lack the ability to produce the enzyme 5α-reductase. Individuals with *5α-reductase deficiency* have normal epididymides, vasa deferentia, seminal vesicles, and ejaculatory ducts because the development of these structures is stimulated directly by testosterone. The external genitalia are poorly developed and more female in appearance, however, because DHT, which cannot be produced from testosterone in the absence of 5α-reductase, is required for the development of male external genitalia.

1. Define the terms *diploid* and *haploid* and explain how the chromosomal sex of an individual is determined.

2. Explain how the chromosomal sex determines whether testes or ovaries will be formed.

3. List the male and female accessory sex organs and explain how the development of one or the other set of organs is determined.

4. Describe the abnormalities characteristic of testicular feminization syndrome and of 5α-reductase deficiency and explain how these abnormalities are produced.

Endocrine Regulation of Reproduction

The functions of the testes and ovaries are regulated by gonadotropic hormones secreted by the anterior pituitary. The gonadotropic hormones stimulate the gonads to secrete their sex steroid hormones, and these steroid hormones, in turn, have an inhibitory effect on the secretion of the gonadotropic hormones. This interaction between the anterior pituitary and the gonads forms a negative feedback loop.

The embryonic testes during the first trimester of pregnancy are active endocrine glands, secreting the high amounts of testosterone needed to masculinize the male embryo's external genitalia and accessory sex organs. Ovaries, by contrast, do not mature until the third trimester of pregnancy. During the second trimester of pregnancy, testosterone secretion in the male declines, so that the gonads of both sexes are relatively inactive at the time of birth.

Before puberty, there are equal blood concentrations of *sex steroids*—androgens and estrogens—in both males and females. Apparently, this is not due to deficiencies in the ability of the gonads to produce these hormones, but rather to lack of sufficient stimulation. During *puberty*, the gonads secrete increased amounts of sex steroid hormones as a result of increased stimulation by **gonadotropic hormones** from the anterior pituitary.

Interactions between the Hypothalamus, Pituitary Gland, and Gonads

The anterior pituitary produces and secretes two gonadotropic hormones—**FSH (follicle-stimulating hormone)** and **LH (luteinizing hormone).** Although these two hormones are named according to their actions in the female, the same hormones are secreted by the male's pituitary gland. The gonadotropic hormones of both sexes have three primary effects on the gonads: (1) stimulation of spermatogenesis or oogenesis (formation of sperm or ova); (2) stimulation of gonadal hormone secretion; and (3) maintenance of the structure of the gonads (the gonads atrophy if the pituitary is removed).

The secretion of both LH and FSH from the anterior pituitary is stimulated by a hormone produced by the hypothalamus and secreted into the hypothalamo-hypophyseal portal vessels (see chapter 11). This releasing hormone is sometimes called LHRH (*luteinizing hormone–releasing hormone*). Since attempts to find a separate FSH-releasing hormone have thus far failed, and since LHRH stimulates FSH as well as LH secretion, LHRH is often referred to as **gonadotropin-releasing hormone (GnRH).**

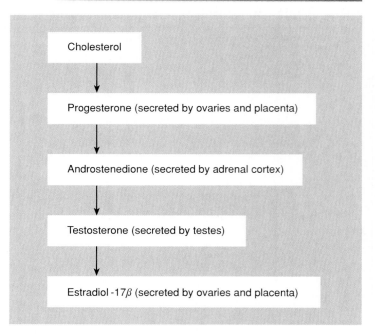

Figure 20.8

A simplified biosynthetic pathway for the steroid hormones. Sources of sex hormones in the blood are also indicated.

If a male or female animal is castrated (has its gonads surgically removed), the secretion of FSH and LH increases to much higher levels than those measured in the intact animal. This demonstrates that the gonads secrete products that exert a **negative feedback inhibition** on gonadotropin secretion. This negative feedback is exerted in large part by sex steroids: estrogen and progesterone in the female, and testosterone in the male. A biosynthetic pathway for these steroids is shown in figure 20.8.

The negative feedback effects of steroid hormones are believed to occur by means of two mechanisms: (1) inhibition of GnRH secretion from the hypothalamus and (2) inhibition of the pituitary's response to a given amount of GnRH. In addition to steroid hormones, the testes and ovaries secrete a polypeptide hormone called **inhibin.** Inhibin is secreted by the Sertoli cells of the seminiferous tubules in males and by the granulosa cells of the ovarian follicles in females. This hormone specifically inhibits the anterior pituitary's secretion of FSH, without affecting the secretion of LH. Inhibin will be discussed more fully in later sections.

Figure 20.9 illustrates the process of gonadal regulation. Although hypothalamus-pituitary-gonad interactions are similar in males and females, there are important differences. Secretion of gonadotropins and sex steroids is more or less constant in adult males. Secretion of gonadotropins and sex steroids in adult females, by contrast, shows cyclic variations (during the menstrual cycle). Also, during one phase of the female cycle, estrogen exerts a positive feedback effect on LH secretion. This will be discussed in a later section.

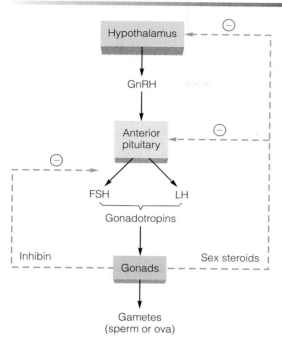

Figure 20.9

Interactions between the hypothalamus, anterior pituitary, and gonads.
Sex steroids secreted by the gonads have a negative feedback effect on the
secretion of GnRH (gonadotropin-releasing hormone) and on the
secretion of gonadotropins. The gonads may also secrete a polypeptide
hormone called inhibin that functions in the negative feedback control
of FSH secretion.

Studies have shown that secretion of GnRH from the hypothalamus is *pulsatile* rather than continuous, and thus the secretion of FSH and LH follows this pulsatile pattern. Pulsatile patterns of secretion are needed to prevent desensitization and down-regulation of the target glands (discussed in chapter 11). It appears that the frequency of the pulses of secretion, as well as their amplitude (how much hormone is secreted per pulse), affects the target gland's response to the hormone.

The Onset of Puberty

Secretion of FSH and LH is high in the newborn, but falls to very low levels a few weeks after birth. Gonadotropin secretion remains low until the beginning of puberty, which is marked by rising levels of FSH followed by LH secretion. Experimental evidence suggests that this rise in gonadotropin secretion is a result of two processes: (1) maturational changes in the brain that result in increased GnRH secretion by the hypothalamus and (2) decreased sensitivity of gonadotropin secretion to the negative feedback effects of sex steroid hormones.

The maturation of the hypothalamus or other regions of the brain that leads to increased GnRH secretion at the time of puberty appears to be programmed—children without gonads show increased FSH secretion at the normal time. Also during this period of time, a given amount of sex steroids has less of a suppressive effect on gonadotropin secretion than the same dose

If a powerful synthetic analogue of GnRH (such as *nafarelin*) is administered, the anterior pituitary first increases and then decreases its secretion of FSH and LH. This decrease is contrary to the normal stimulatory action of GnRH, and is due to a desensitization of the anterior pituitary evoked by continuous exposure to GnRH. The decrease in LH causes a fall in testosterone secretion from the testes, or of estradiol secretion from the ovaries. The decreased testosterone secretion is useful in the treatment of men who have **benign prostatic hypertrophy.** In this condition, common in older men, testosterone supports abnormal growth of the prostate. The fall in estradiol secretion in women given synthetic GnRH analogues can be useful in the treatment of **endometriosis.** In this condition, ectopic endometrial tissue from the uterus (dependent on estradiol for growth) is found growing outside the uterus—for example, on the ovaries or on the peritoneum. These treatments, which illustrate the reasons why GnRH and the gonadotropins are normally secreted in a pulsatile fashion, are particularly beneficial clinically because they are reversible.

would have if administered prior to puberty. This suggests that the sensitivity of the hypothalamus and the pituitary to negative feedback effects decreases at puberty, which would also help to account for rising gonadotropin secretion at this time.

During late puberty there is a pulsatile secretion of gonadotropins—FSH and LH secretion increase during periods of sleep and decrease during periods of wakefulness. These pulses of increased gonadotropin secretion during puberty stimulate a rise in sex steroid secretion from the gonads. Increased secretion of testosterone from the testes and of **estradiol-17β** (estradiol is the major *estrogen,* or female sex steroid) from the ovaries during puberty, in turn, produces changes in body appearance characteristic of the two sexes. Such **secondary sex characteristics** (tables 20.2 and 20.3) are the physical manifestations of the hormonal changes occurring during puberty. These changes are accompanied by a growth spurt, which begins at an earlier age in girls than in boys (fig. 20.10).

The age at which puberty begins is related to the amount of body fat and level of physical activity of the child. The average age of *menarche*—the first menstrual flow—is later (age 15) in girls who are very active physically than in the general population (age 12.6). This appears to be due to a requirement for a minimum percentage of body fat for menstruation to begin, and may represent a mechanism favored by natural selection to ensure the ability to successfully complete a pregnancy and nurse the baby. Later in life, women who are very lean and physically active may have irregular cycles and *amenorrhea* (cessation of menstruation). This may also be related to the percentage of body fat. In addition, there is evidence that physical exercise may, through activation of neural pathways involving endorphin neurotransmitters (chapter 7), act to inhibit GnRH and gonadotropin secretion.

Table 20.2 Development of Secondary Sex Characteristics and Other Changes That Occur During Puberty in Girls

Characteristic	Age of First Appearance	Hormonal Stimulation
Appearance of breast bud	8–13	Estrogen, progesterone, growth hormone, thyroxine, insulin, cortisol
Pubic hair	8–14	Adrenal androgens
Menarche (first menstrual flow)	10–16	Estrogen and progesterone
Axillary (underarm) hair	About 2 years after the appearance of pubic hair	Adrenal androgens
Eccrine sweat glands and sebaceous glands; acne (from blocked sebaceous glands)	About the same time as axillary hair growth	Adrenal androgens

Table 20.3 Development of Secondary Sex Characteristics and Other Changes That Occur During Puberty in Boys

Characteristic	Age of First Appearance	Hormonal Stimulation
Growth of testes	10–14	Testosterone, FSH, growth hormone
Pubic hair	10–15	Testosterone
Body growth	11–16	Testosterone, growth hormone
Growth of penis	11–15	Testosterone
Growth of larynx (voice lowers)	Same time as growth of penis	Testosterone
Facial and axillary (underarm) hair	About 2 years after the appearance of pubic hair	Testosterone
Eccrine sweat glands and sebaceous glands; acne (from blocked sebaceous glands)	About the same time as facial and axillary hair growth	Testosterone

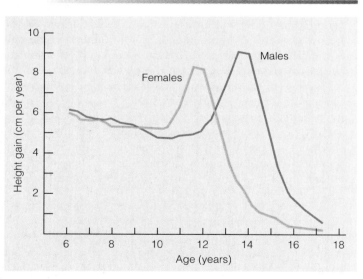

Figure 20.10

Growth in height of females and males as a function of age. Notice that the growth spurt during puberty occurs at an earlier age in females than in males.

Pineal Gland

The role of the pineal gland in human physiology is poorly understood. It is known that the pineal, a gland located deep within the brain, secretes the hormone **melatonin** as a derivative of the amino acid tryptophane (fig. 20.11) and that production of this hormone is influenced by light-dark cycles.

The pineal glands of some vertebrates have photoreceptors that are directly sensitive to environmental light. Although no such photoreceptors are present in the pineal glands of mammals, the secretion of melatonin has been shown to increase at night and decrease during daylight. The inhibitory effect of light on melatonin secretion in mammals is indirect. Pineal secretion is stimulated by postganglionic sympathetic neurons that originate in the superior cervical ganglion; activity of these neurons, in turn, is inhibited by nerve tracts that are activated by light striking the retina.

There is abundant experimental evidence that, in rats and some other vertebrates, melatonin can inhibit gonadotropin secretion and thus have an "antigonad" effect. In other experimental animals (such as sheep), however, melatonin can stimulate the reproductive system. That melatonin plays a role in the regulation of human reproduction has long been suspected but, because of conflicting and inconclusive data, has not as yet been established.

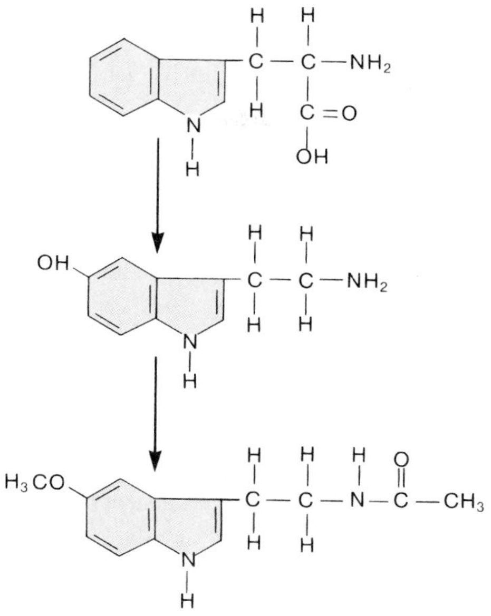

Figure 20.11

Tryptophane
(an amino acid)

Serotonin
(a biogenic amine)

Melatonin
(a pineal gland hormone)

A simplified biosynthetic pathway for the pineal gland hormone melatonin.

Receptors for melatonin have been localized to the **suprachiasmatic nucleus** of the hypothalamus. This area of the brain is believed to be the site of the biological clock, which entrains (synchronizes) various processes of the body to a *circadian rhythm* (one that repeats every 24 hours). The rhythms of melatonin secretion may be entrained to cycles of light and dark through the action of the suprachiasmatic nucleus, and melatonin, in turn, may have some effect on this area of the brain. It is interesting in this regard that melatonin has been used clinically to relieve the symptoms of jet lag and to help reestablish the disturbed circadian rhythms of people who are blind.

1. Using a flow diagram, show the negative feedback control that the gonads exert on GnRH and gonadotropin secretion. Explain the effects of castration on FSH and LH secretion and the effects of removal of the pituitary on the structure of the gonads and accessory sex organs.

2. Explain the significance of the pulsatile secretion of GnRH and the gonadotropic hormones.

3. Describe the two mechanisms that have been proposed to explain the rise in sex steroid secretion that occurs at puberty. Explain the possible effects of body fat and intense exercise on the timing of puberty.

4. Describe the effect of light on the pineal secretion of melatonin and discuss the possible functions of this hormone.

Male Reproductive System

The Leydig cells in the interstitial tissue of the testes are stimulated by LH to secrete testosterone, a potent androgen that acts to maintain the structure and function of the male accessory sex organs and to promote the development of male secondary sex characteristics. The Sertoli cells in the seminiferous tubules of the testes are stimulated by FSH. Spermatogenesis requires the cooperative actions of both FSH and testosterone.

The testes consist of two parts, or "compartments"—the seminiferous tubules, where spermatogenesis occurs, and the interstitial tissue, which contains androgen-secreting **Leydig cells** (fig. 20.12). The seminiferous tubules account for about 90% of the weight of an adult testis (which averages 20 g). The interstitial tissue is a thin web of connective tissue (containing Leydig cells) between convolutions of the tubules.

There is a strict compartmentation in the testes with regard to gonadotropin action. Cellular receptor proteins for FSH are located exclusively in the seminiferous tubules, where they are confined to the **Sertoli cells** (discussed in a later section). LH receptor proteins are located exclusively in the interstitial Leydig cells. Secretion of testosterone by the Leydig cells is stimulated by LH but not by FSH. Spermatogenesis in the tubules is stimulated by FSH. The apparent simplicity of this compartmentation, however, is an illusion because the two compartments can interact with each other in complex ways.

Control of Gonadotropin Secretion

Castration of a male animal results in an immediate rise in FSH and LH secretion. This demonstrates that hormones secreted by the testes exert negative feedback inhibition of gonadotropin secretion. If testosterone is injected into the castrated animal, the secretion of LH can be returned to the previous (precastration) levels. This provides a classical example of negative feedback—LH stimulates testosterone secretion by the Leydig cells, and testosterone inhibits pituitary secretion of LH (fig. 20.13).

The amount of testosterone that is sufficient to suppress LH, however, is not sufficient to suppress the postcastration rise in FSH secretion in most experimental animals. In rams and bulls, a water-soluble (and, therefore, peptide rather than steroid) product of the seminiferous tubules specifically suppresses FSH secretion. This hormone, produced by the Sertoli cells, is called **inhibin.** There is now good evidence that the seminiferous tubules of the human testes also produce inhibin.

Testosterone Derivatives in the Brain

The brain contains testosterone receptors and is a target organ for this hormone. The effects of testosterone on the brain, such as the suppression of LH secretion, are not mediated directly by testosterone, however, but rather by its derivatives that are produced within the brain cells. Testosterone may be converted by

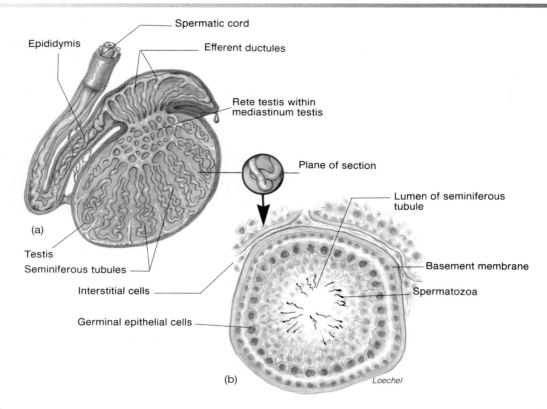

Figure 20.12

A diagrammatic representation of seminiferous tubules. (*a*) A sagittal section of a testis and (*b*) a transverse section of a seminiferous tubule.

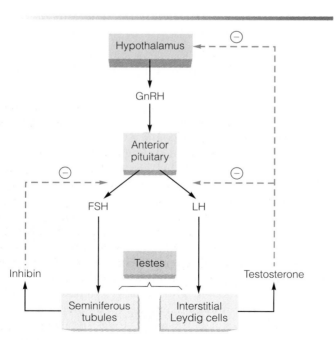

Figure 20.13

Negative feedback relationships between the anterior pituitary and testes. The seminiferous tubules are the targets of FSH action; the interstitial Leydig cells are targets of LH action. Testosterone secreted by the Leydig cells inhibits LH secretion; inhibin secreted by the tubules may inhibit FSH secretion.

the enzyme 5α-reductase to dihydrotestosterone (DHT), as previously described. The DHT, in turn, can be changed by other enzymes into other 5α-reduced androgens—abbreviated 3α-diol and 3β-diol (fig. 20.14). Alternatively, testosterone is also converted within the brain to estradiol-17β. Although usually regarded as a female sex steroid, estradiol is therefore an active compound in normal male physiology! Estradiol is formed from testosterone by an enzyme called *aromatase*, in a reaction known as *aromatization* (this term refers to the presence of an aromatic carbon ring—see chapter 2—not to an odor). The estradiol formed from testosterone in the brain is believed to be required for the negative feedback effects of testosterone on LH secretion.

Testosterone Secretion and Age

The negative feedback effects of testosterone and inhibin help to maintain a relatively constant (that is, noncyclic) secretion of gonadotropins in males, resulting in relatively constant levels of androgen secretion from the testes. This contrasts with the cyclic secretion of gonadotropins and ovarian steroids in females. Women experience an abrupt cessation in sex steroid secretion during menopause. By contrast, the secretion of androgens declines only gradually and to varying degrees in men over 50 years of age. The causes of this age-related change in testicular function are not currently known. The decline in testosterone secretion cannot be due to decreasing gonadotropin secretion, since gonadotropin levels in the blood

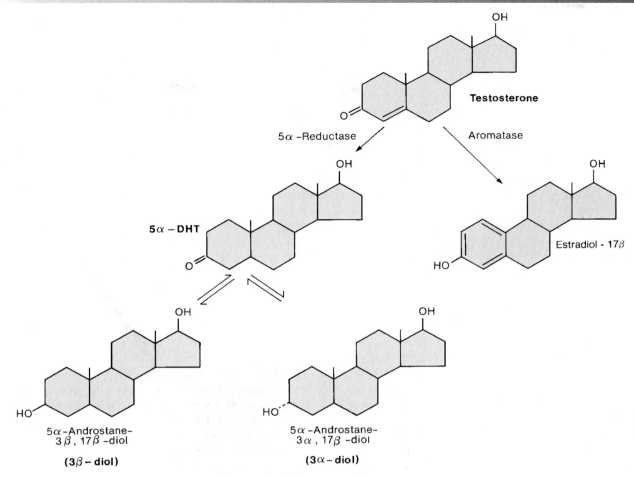

Figure 20.14

Testosterone secreted by the Leydig cells of the testes can be converted into active metabolites in the brain and other target organs. These active metabolites include DHT and other 5α-reduced androgens and estradiol.

are, in fact, elevated (due to less negative feedback) at the time that testosterone levels are declining.

Endocrine Functions of the Testes

Testosterone is by far the major androgen secreted by the adult testis. This hormone and its derivatives (the 5α-reduced androgens) are responsible for initiation and maintenance of the body changes associated with puberty in males. Androgens are sometimes called *anabolic steroids* because they stimulate the growth of muscles and other structures (table 20.4). Increased testosterone secretion during puberty is also required for growth of the accessory sex organs—primarily the seminal vesicles and prostate. Removal of androgens by castration results in atrophy of these organs.

Androgens stimulate growth of the larynx (causing lowering of the voice), increased hemoglobin synthesis (males have higher hemoglobin levels than females), and bone growth. The effect of androgens on bone growth is self-limiting, however, because androgens ultimately cause re-

placement of cartilage by bone in the epiphyseal discs, thus "sealing" the discs and preventing further lengthening of the bones (as described in chapter 18).

Although androgens are by far the major secretory product of the testes, the testes do produce and secrete small amounts of estradiol. There is evidence that both the Sertoli cells of the tubules and the Leydig cells can produce estradiol, although estradiol receptors in the testes appear to be located only in the Leydig cells. Experiments suggest that, when LH is present in high amounts and not secreted in a pulsatile fashion, the desensitization and down-regulation of Leydig cell function that results may be partly mediated by estradiol.

The two compartments of the testes interact with each other in an autocrine fashion (fig. 20.15). Autocrine regulation, as described in chapter 11, refers to chemical regulation that occurs within an organ. Testosterone from the Leydig cells is metabolized by the tubules into other active androgens and is required for spermatogenesis (as described in the next section). The tubules also secrete products that might influence Leydig cell function. Such interactions are suggested by evidence that, in the

Table 20.4	Actions of Androgens in the Male
Category	**Action**
Sex Determination	Growth and development of wolffian ducts into epididymis, vas deferens, seminal vesicles, and ejaculatory ducts
	Development of urogenital sinus and tubercle into prostate
	Development of male external genitalia (penis and scrotum)
Spermatogenesis	At puberty: Completion of meiotic division and early maturation of spermatids
	After puberty: Maintenance of spermatogenesis
Secondary Sex Characteristics	Growth and maintenance of accessory sex organs
	Growth of penis
	Growth of facial and axillary hair
	Body growth
Anabolic Effects	Protein synthesis and muscle growth
	Growth of bones
	Growth of other organs (including larynx)
	Erythropoiesis (red blood cell formation)

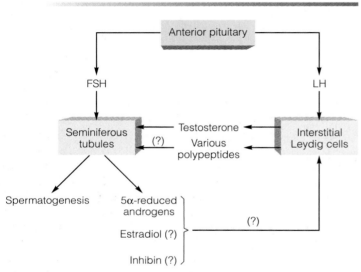

Figure 20.15

Interactions between the two compartments of the testes. Testosterone secreted by the interstitial Leydig cells stimulates spermatogenesis in the tubules. Leydig cells may also secrete ACTH, MSH, and β-endorphin. Secretion of inhibin by the tubules may affect the sensitivity of the interstitial cells to LH stimulation.

pubertal male rat, exposure to FSH augments the responsiveness of the Leydig cells to LH. Since FSH can directly stimulate only the Sertoli cells of the tubules, the FSH-induced enhancement of LH responsiveness must be mediated by products secreted from the Sertoli cells.

Inhibin secreted by the Sertoli cells in response to FSH can facilitate the Leydig cells' response to LH, as measured by the amount of testosterone secreted. Further, it has been shown that the Leydig cells are capable of producing a family of polypeptides previously associated only with the pituitary gland—ACTH, MSH, and β-endorphin. Experiments suggest that ACTH and MSH can stimulate Sertoli cell function,

whereas β-endorphin can inhibit Sertoli function. The physiological significance of these fascinating autocrine interactions between the two compartments of the testes remains to be demonstrated.

Spermatogenesis

The germ cells that migrate from the yolk sac to the testes during early embryonic development become "stem cells" called **spermatogonia** within the outer region of the seminiferous tubules. Spermatogonia are diploid cells (with forty-six chromosomes) that ultimately give rise to mature haploid gametes by a process of cell division called meiosis.

Meiosis, or reduction division, occurs in two parts. In the first part of this process, the DNA duplicates, and homologous chromosomes are separated into two daughter cells. Since each daughter cell contains only one of each homologous pair of chromosomes, the cells formed at the end of this first meiotic division contain twenty-three chromosomes each and are haploid. Each of the twenty-three chromosomes at this stage, however, consists of two strands (called chromatids) of identical DNA. During the second meiotic division, these duplicate chromatids are separated into daughter cells. Meiosis of one diploid spermatogonia cell therefore produces four haploid cells.

Actually, only about 1,000 to 2,000 stem cells migrate from the yolk sac into the embryonic testes. In order to produce many millions of sperm throughout adult life, these spermatogonia cells duplicate themselves by mitotic division, and only one of the two cells—now called a **primary spermatocyte**—undergoes meiotic division (fig. 20.16). In this way, spermatogenesis can occur continuously without exhausting the number of spermatogonia.

When a diploid primary spermatocyte completes the first meiotic division (at telophase I), the two haploid cells thus produced are called **secondary spermatocytes.** At the end of the second meiotic division, each of the two secondary spermatocytes produces two haploid **spermatids.** One primary spermatocyte therefore produces four spermatids.

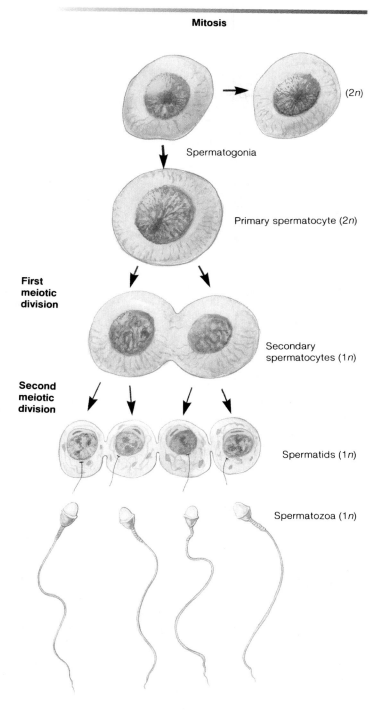

Mitosis

Spermatogonia

Primary spermatocyte (2n)

First meiotic division

Secondary spermatocytes (1n)

Second meiotic division

Spermatids (1n)

Spermatozoa (1n)

(2n)

Figure 20.16

Spermatogonia undergo mitotic division to replace themselves and produce a daughter cell that will undergo meiotic division. This cell is called a primary spermatocyte. Upon completion of the first meiotic division, the daughter cells are called secondary spermatocytes. Each of these completes a second meiotic division to form spermatids. Notice that the four spermatids produced by the meiosis of a primary spermatocyte are interconnected. Each spermatid forms a mature spermatozoon.

The sequence of events in spermatogenesis is reflected in the cellular arrangement of the wall of the seminiferous tubule. The spermatogonia and primary spermatocytes are located toward the outer side of the tubule, whereas spermatids and mature spermatozoa are located on the side of the tubule facing the lumen.

At the end of the second meiotic division, the four spermatids produced by meiosis of one primary spermatocyte are interconnected with each other—their cytoplasm does not completely pinch off at the end of each division. Development of these interconnected spermatids into separate, mature **spermatozoa** (singular, *spermatozoon*)—a process called **spermiogenesis**—requires the participation of the Sertoli cells (fig. 20.17).

Sertoli Cells

The Sertoli cells are the only nongerminal cell type in the tubules. They form a continuous layer, connected by tight junctions, around the circumference of each tubule. In this way, the Sertoli cells constitute a **blood-testis barrier;** molecules from the blood must pass through the cytoplasm of the Sertoli cells before entering germinal cells. Similarly, this barrier normally prevents the immune system from becoming sensitized to antigens in the developing sperm, and thus prevents autoimmune destruction of the sperm. The cytoplasm of the Sertoli cells extends through the width of the germinal epithelium and envelops the developing germ cells, so that it is often difficult to tell where the cytoplasm of the Sertoli cells and that of germ cells is separated.

In the process of spermiogenesis (conversion of spermatids to spermatozoa), most of the spermatid cytoplasm is eliminated. This occurs through phagocytosis by Sertoli cells of the "residual bodies" of cytoplasm from the spermatids (fig. 20.18). Phagocytosis of residual bodies may transmit informational molecules from germ cells to Sertoli cells. The Sertoli cells, in turn, may provide molecules needed by the germ cells. It is known, for example, that the X chromosome of germ cells is inactive during meiosis. Since this chromosome contains genes needed to produce many essential molecules, it is believed that these molecules are provided by the Sertoli cells during this time.

Sertoli cells produce a protein called **androgen binding protein (ABP)** into the lumen of the seminiferous tubules. This protein, as its name implies, binds to testosterone and thereby concentrates it within the tubules. The importance of Sertoli cells in tubular function is further evidenced by the fact that FSH receptors are confined to the Sertoli cells. Any effect of FSH on the tubules, therefore, must be mediated through the action of Sertoli cells. These include the FSH-induced stimulation of spermiogenesis and the autocrine interactions between Sertoli cells and Leydig cells that have been previously described.

Hormonal Control of Spermatogenesis

The very beginning of spermatogenesis—formation of primary spermatocytes and entry into early prophase I—is apparently somewhat independent of hormonal control and, in fact, starts during embryonic development. Spermatogenesis is arrested at this stage, however, until puberty, when testosterone secretion

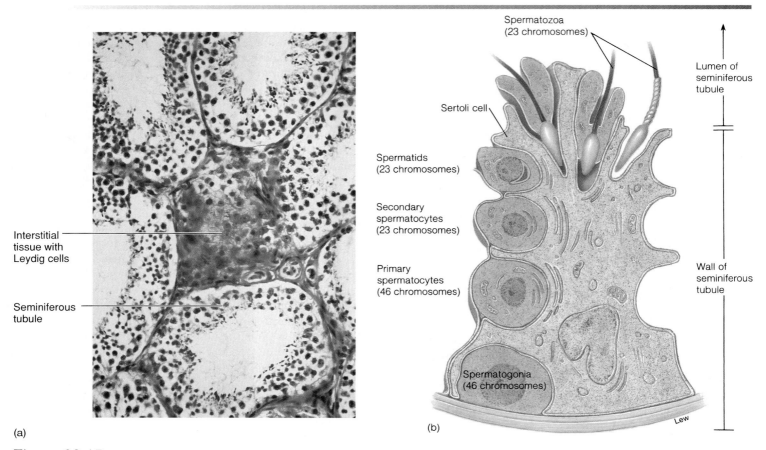

Spermatozoa
(23 chromosomes)

Lumen of
seminiferous
tubule

Sertoli cell

Spermatids
(23 chromosomes)

Secondary
spermatocytes
(23 chromosomes)

Primary
spermatocytes
(46 chromosomes)

Wall of
seminiferous
tubule

Spermatogonia
(46 chromosomes)

Interstitial
tissue with
Leydig cells

Seminiferous
tubule

(a)

(b)

Lew

Figure 20.17

Seminiferous tubules. (*a*) A cross section with surrounding interstitial tissue. (*b*) The stages of spermatogenesis within the germinal epithelium of a seminiferous tubule in which the relationship between Sertoli cells and developing spermatozoa is shown.

rises. Testosterone is required for completion of meiotic division and for the early stages of spermatid maturation. This effect is probably not produced by testosterone directly, but rather by some of the molecules derived from testosterone in the tubules.

The later stages of spermatid maturation during puberty appear to require stimulation by FSH (fig. 20.19). This FSH effect is mediated by the Sertoli cells, as previously described. During puberty, therefore, both FSH and androgens are needed for the initiation of spermatogenesis.

Experiments in rats and evidence in humans reveal that spermatogenesis within the adult testis can be maintained by androgens alone, in the absence of FSH. It appears, in other words, that FSH is needed to initiate spermatogenesis at puberty, but that it may no longer be required for this function once spermatogenesis has begun.

 Men who have had a *hypophysectomy* (surgical removal of the pituitary) experience a cessation of spermatogenesis. Spermatogenesis is restored in these patients by injections of FSH and a hormone derived from the placenta called *hCG* (*human chorionic gonadotropin*), which has the same biological activity as LH. In this case, hCG acts like LH to stimulate the Leydig cells to secrete testosterone, which is needed along with FSH to initiate sperm production. After spermatogenesis has been restored, it can be maintained in hypophysectomized patients with hCG injections alone, demonstrating that testosterone (secreted in response to hCG stimulation) can maintain spermatogenesis by itself.

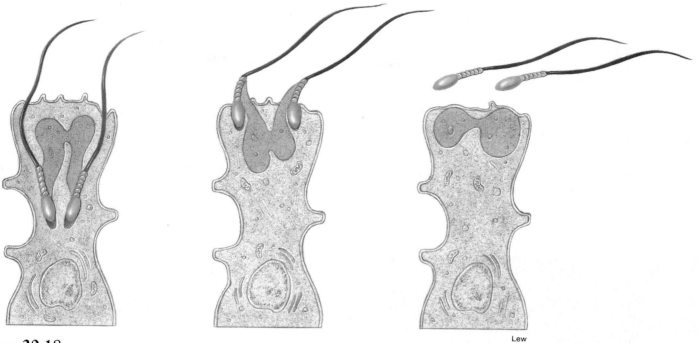

Lew

Figure 20.18

The processing of spermatids into spermatozoa (spermiogenesis). As the spermatids develop into spermatozoa, most of their cytoplasm is pinched off as residual bodies and ingested by the surrounding Sertoli cell cytoplasm.

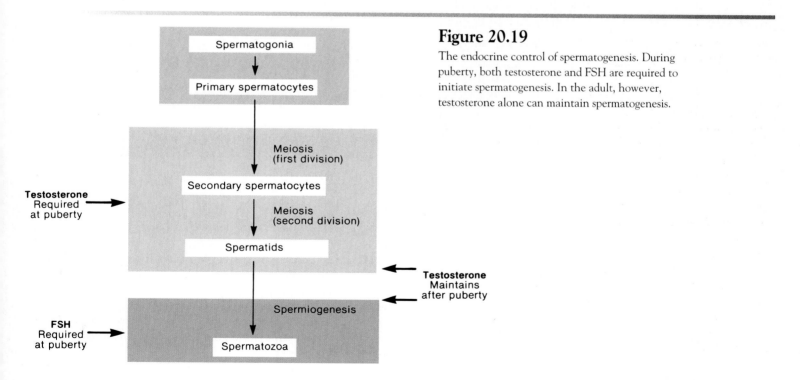

Figure 20.19

The endocrine control of spermatogenesis. During puberty, both testosterone and FSH are required to initiate spermatogenesis. In the adult, however, testosterone alone can maintain spermatogenesis.

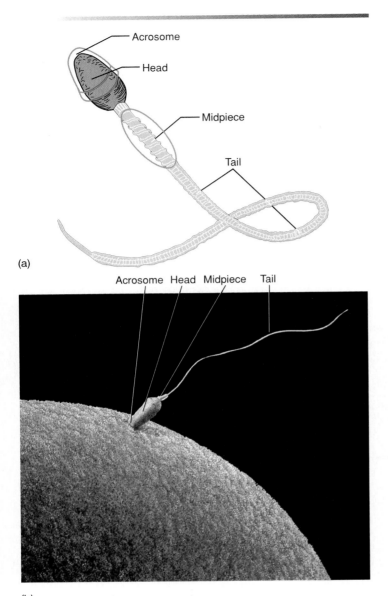

(a)

(b)

Figure 20.20

A human spermatozoon. (*a*) A diagrammatic representation and (*b*) a scanning electron micrograph in which it is seen in contact with an ovum.

At the conclusion of spermiogenesis, spermatozoa are released into the lumen of the seminiferous tubules. The spermatozoa consist of an oval-shaped *head* (with DNA inside), a *midpiece*, and a *tail* (fig. 20.20). Although the tail will ultimately be capable of flagellar movement, the sperm at this stage are nonmotile. Motility and other maturational changes occur outside the testes in the epididymis.

Male Accessory Sex Organs

The seminiferous tubules are connected at both ends to the *rete testis* (see fig. 20.12). Spermatozoa and tubular secretions are moved to this area of the testis and are drained via the *efferent ductules* into the **epididymis.** The epididymis is a single-coiled tube, 4 to 5 meters

long if stretched out, that receives the tubular products. Spermatozoa enter at the "head" of the epididymis and are drained from its "tail" by a single tube, the **vas,** or **ductus, deferens.**

Spermatozoa that enter the head of the epididymis are nonmotile. During their passage through the epididymis, they gain motility and undergo other maturational changes. When they leave the epididymis, the sperm are more resistant to changes in pH and temperature, they are motile, and they can become capable of fertilizing an ovum once they spend some time in the female reproductive tract. Sperm obtained from the seminiferous tubules, by contrast, cannot fertilize an ovum. The epididymis serves as a site for sperm maturation and for the storage of sperm between ejaculations.

The vas deferens carries sperm from the epididymis out of the scrotum and into the body cavity. In its passage, the vas deferens obtains fluid secretions of the **seminal vesicles** and **prostate.** This fluid, now called *semen,* is carried by the ejaculatory duct to the *urethra* (fig. 20.21).

The seminal vesicles and prostate are androgen-dependent accessory sex organs—they will atrophy if androgen is withdrawn by castration. The seminal vesicles secrete fluid containing fructose, which serves as an energy source for the spermatozoa. This fluid secretion accounts for about 60% of the volume of the semen. The prostate also contributes fluid to the semen; this fluid contains citric acid, calcium, and coagulation proteins. Clotting proteins cause the semen to coagulate after ejaculation, but the hydrolytic action of fibrinolysin later causes the coagulated semen to again assume a more liquid form, thereby freeing the sperm. The prostate releases the enzyme *acid phosphatase* into the blood, which is often measured clinically to assess prostate function. An immunoassay for *prostate-specific antigen* (*PSA*) is a more specific laboratory test for prostate disorders, including prostate cancer.

Erection, Emission, and Ejaculation

Erection, accompanied by increases in the length and width of the penis, is achieved as a result of blood flow into the "erectile tissues" of the penis. These erectile tissues include two paired structures—the *corpora cavernosa*—located on the dorsal side of the penis, and one unpaired *corpus spongiosum* on the ventral side (fig. 20.22). The urethra runs through the center of the corpus spongiosum. The erectile tissue forms columns extending the length of the penis, although the corpora cavernosa do not extend all the way to the tip.

Erection is achieved by parasympathetic nerve-induced vasodilation of arterioles that allows blood to flow into the corpora cavernosa of the penis. The neurotransmitter that mediates this increased blood flow and causes penile erection is now believed to be nitric oxide. As the erectile tissues become engorged with blood and the penis becomes turgid, venous

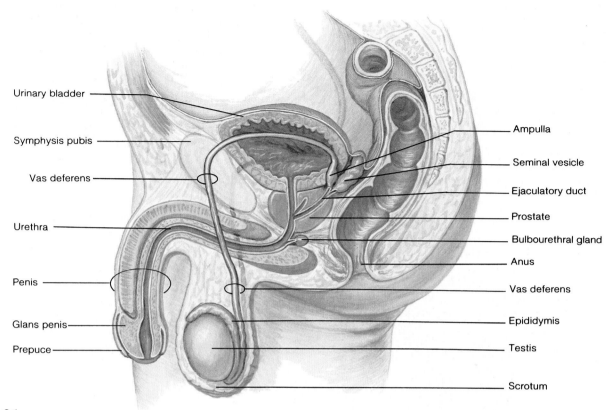

Figure 20.21

Organs of the male reproductive system in sagittal view.

outflow of blood is partially occluded, thus aiding erection. The term *emission* refers to the movement of semen into the urethra, and *ejaculation* refers to the forcible expulsion of semen from the urethra out of the penis. Emission and ejaculation are stimulated by sympathetic nerves, which cause peristaltic contractions of the tubular system; contractions of the seminal vesicles and prostate; and contractions of muscles at the base of the penis. Sexual function in the male thus requires the synergistic action (rather than antagonistic action) of the parasympathetic and sympathetic systems.

Erection is controlled by two portions of the central nervous system—the hypothalamus in the brain and the sacral portion of the spinal cord. Conscious sexual thoughts originating in the cerebral cortex act via the hypothalamus to control the sacral region, which in turn increases parasympathetic nerve activity to promote vasodilation and erection in the penis. Conscious thought is not required for erection, however, because sensory stimulation of the penis can more directly activate the sacral region of the spinal cord and cause an erection.

Male Fertility

The approximate volume of semen for each ejaculation is 1.5 to 5.0 ml. The bulk of this fluid (45% to 80%) is produced by the seminal vesicles, and 15% to 30% is contributed by the prostate. The sperm content in human males ranges between 40 and 250 million per milliliter in the ejaculated semen. Normal human semen values are summarized in table 20.5.

A sperm concentration below about 20 million per milliliter is termed *oligospermia* (*oligo* = few), and is associated with decreased fertility. A total sperm count below about 50 million per ejaculation is clinically significant in male infertility. In addition to low sperm counts as a cause of infertility, some men and women have antibodies against sperm antigens (this is very common in men with vasectomies). Such antibodies do not appear to affect health, but do reduce fertility.

Vasectomy (fig. 20.23) is commonly performed as a contraceptive method. In this procedure, each vas deferens is cut and tied or, in some cases, a valve or similar device is

Table 20.5 Semen Analysis

Characteristic	Reference Value
Volume of ejaculate	1.5–5.0 ml
Sperm count	40–250 million/ml
Sperm motility	
Percentage of motile forms:	
1 hour after ejaculation	70% or more
3 hours after ejaculation	60% or more
Leukocyte count	0–2,000/ml
pH	7.2–7.8
Fructose concentration	150–600 mg/100 ml

Source: Modified from L. Glasser, "Seminal Fluid and Subfertility," *Diagnostic Medicine,* July/August 1981, p. 28. Used by permission.

inserted. This procedure interferes with sperm transport but does not directly affect the secretion of androgens from Leydig cells in the interstitial tissue. Since spermatogenesis continues, the sperm produced cannot be drained from the testes and instead accumulate in "crypts" that form in the seminiferous tubules, epididymis, and vas deferens. These crypts present sites of inflammatory reactions in which spermatozoa are phagocytosed and destroyed by the immune system. It is thus not surprising that approximately 70% of men with vasectomies develop antisperm antibodies. These antibodies do not appear to cause autoimmune damage to the testes, but they do significantly diminish the possibility of reversing the procedure and restoring fertility.

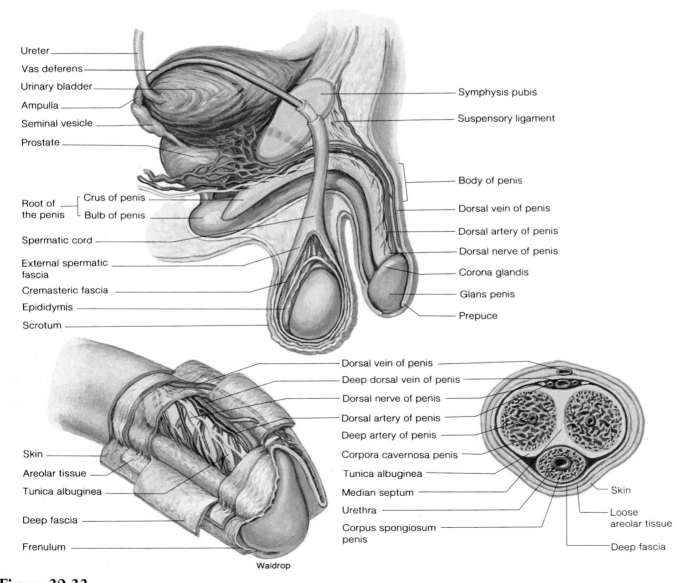

Figure 20.22

The structure of the penis, showing the attachment, blood and nerve supply, and arrangement of the erectile tissue.

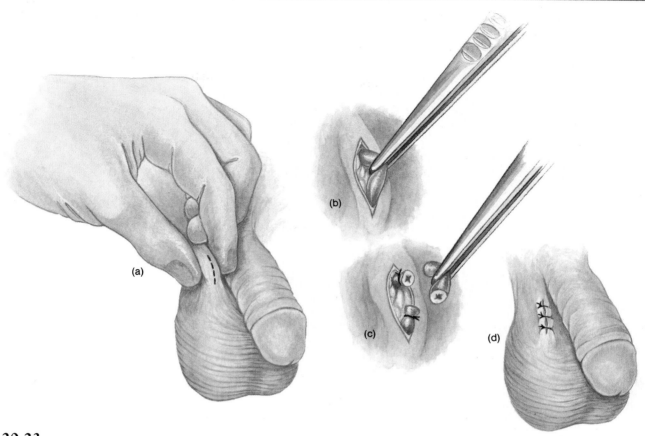

Figure 20.23

A simplified illustration of a vasectomy, in which a segment of the vas deferens is removed through an incision in the scrotum.

1. Describe the effects of castration on FSH and LH secretion in the male. Explain the experimental evidence suggesting that the testes produce a polypeptide that specifically inhibits FSH secretion.

2. Describe the two compartments of the testes with respect to (a) structure, (b) function, and (c) response to gonadotropin stimulation. Describe two ways in which these compartments interact.

3. Using a diagram, describe the stages of spermatogenesis. Explain why spermatogenesis can continue throughout life without using up all of the spermatogonia.

4. Describe the structure and proposed functions of the Sertoli cells in the seminiferous tubules.

5. Explain how FSH and androgens synergize to stimulate sperm production at puberty. Describe the hormonal requirements for spermatogenesis after puberty.

Female Reproductive System

The ovaries contain a large number of follicles, each of which encloses an ovum. Some of these follicles mature during the ovarian cycle, and the ova they contain progress to the secondary oocyte stage of meiosis. At ovulation, the largest follicle breaks open to extrude a secondary oocyte from the ovary. The empty follicle then becomes a corpus luteum, which ultimately degenerates at the end of a nonfertile cycle.

The two ovaries (fig. 20.24) are located within the body cavity, suspended by means of ligaments from the pelvic girdle. Extensions, or fimbriae, of the **uterine**, or **fallopian, tubes** partially cover each ovary. Ova that are released from the ovary—in a process called *ovulation*—are normally drawn into the

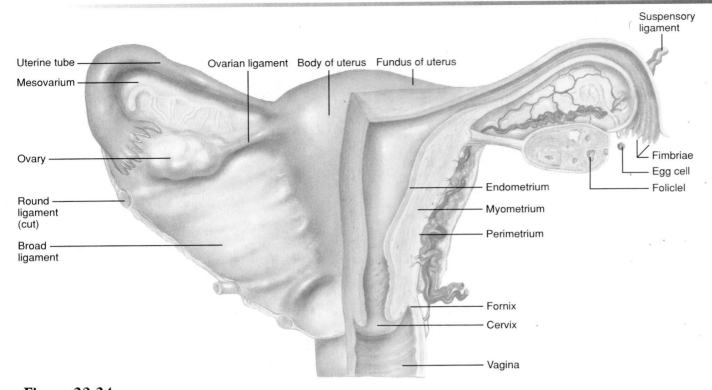

Figure 20.24

The organs of the female reproductive system and the supporting ligaments seen in posterior view.

fallopian tubes by the action of the ciliated epithelial lining of the tubes. The lumen of each fallopian tube is continuous with the **uterus** (or womb), a pear-shaped muscular organ also suspended within the pelvic girdle by ligaments.

The uterus consists of three layers. The outer layer of connective tissue is the *perimetrium*, the middle layer of smooth muscle is the *myometrium*, and the inner epithelial layer is the **endometrium.** The endometrium is a stratified, squamous, nonkeratinized epithelium (see chapter 1) that consists of a deeper *stratum basale* and a more superficial *stratum functionale*. The stratum functionale cyclically grows thicker as a result of estrogen and progesterone stimulation and is shed, with bleeding, at menstruation. The structure and cyclic changes of the endometrium will be described later as part of the menstrual cycle.

The uterus narrows to form the *cervix* (= neck), which opens to the **vagina.** The only physical barrier between the vagina and uterus is a plug of *cervical mucus*. These structures—the vagina, uterus, and fallopian tubes—constitute the accessory sex organs of the female (fig. 20.25). Like the accessory sex organs of the male, the female reproductive tract is affected by gonadal steroid hormones. Cyclic changes in ovarian secretion, as will be described in the next section, cause cyclic changes in the epithelial lining of the tract.

The vaginal opening is located immediately posterior to the opening of the urethra. Both openings are covered by inner **labia minora** and outer **labia majora** (fig. 20.26). The **clitoris** is located at the anterior margin of the labia minora.

Ovarian Cycle

The germ cells that migrate into the ovaries during early embryonic development multiply, so that by about 5 months of gestation (prenatal life) the ovaries contain approximately 6 million to 7 million oogonia. The production of new oogonia stops at this point and never resumes again. The oogonia begin meiosis toward the end of gestation, at which time they are called **primary oocytes** (fig. 20.27a). Like spermatogenesis in the prenatal male, oogenesis is arrested at prophase I of the first meiotic division. The primary oocytes are thus still diploid. The number of primary oocytes decreases throughout a woman's life. The ovaries of a newborn girl contain about 2 million oocytes—all she will ever have. By the time she reaches puberty, this number has been reduced to 300,000 to 400,000. Oogenesis ceases entirely at menopause (the time menstruation stops).

Primary oocytes that are not stimulated to complete the first meiotic division are contained within tiny follicles,

Chapter Twenty

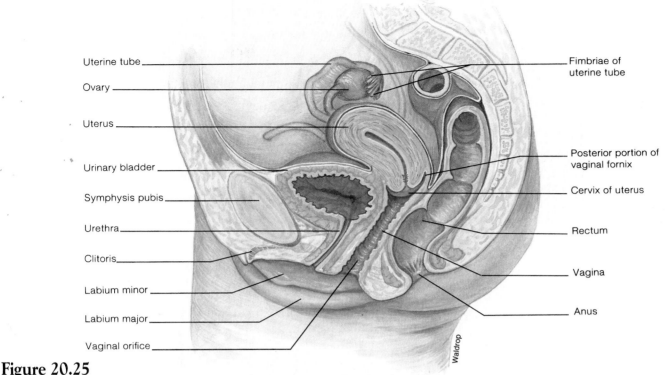

Figure 20.25

The organs of the female reproductive system seen in sagittal section.

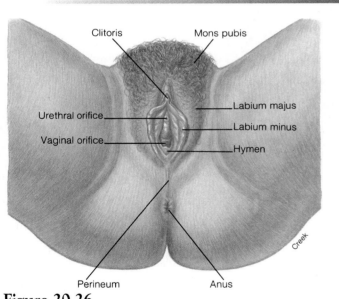

Figure 20.26

The external female genitalia.

called **primary follicles.** Immature primary follicles consist of only a single layer of *granulosa cells.* In response to FSH stimulation, some of these oocytes and follicles get larger, and the follicular cells divide to produce numerous layers of granulosa cells that surround the oocyte and fill the follicle. Some primary follicles will be stimulated to grow still bigger and develop a number of fluid-filled cavities called *vesicles;* at this point, they are called **secondary follicles** (fig. 20.27*a*). Continued growth of one of these follicles will be accompanied by the fusion of its vesicles to form a single, fluid-filled cavity called an *antrum.* At this stage the follicle is known as a **graafian follicle** (fig. 20.27*b*).

As the follicle develops, the primary oocyte completes its first meiotic division. This does not form two complete cells, however, because only one cell—the **secondary oocyte**—gets all the cytoplasm. The other cell formed at this time becomes a small *polar body* (fig. 20.28) that eventually fragments and disappears. This unequal division of cytoplasm ensures that the ovum will be large enough to become a viable embryo should fertilization later occur. The secondary oocyte then begins the

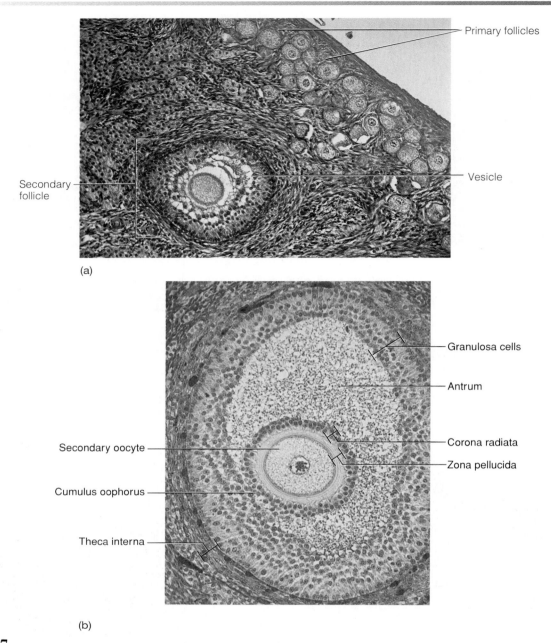

(a)

Primary follicles

Secondary follicle

Vesicle

(b)

Granulosa cells

Antrum

Secondary oocyte

Corona radiata

Zona pellucida

Cumulus oophorus

Theca interna

Figure 20.27

Photomicrographs (*a*) of primary follicles and (*b*) of a graafian follicle.

second meiotic division, but meiosis is arrested at metaphase II. The second meiotic division is completed only by an ovum that has been fertilized.

The secondary oocyte, arrested at metaphase II, is contained within a graafian follicle. The granulosa cells of this follicle form a ring around the oocyte and form a mound that supports the oocyte. This mound is called the *cumulus oophorus*. The ring of granulosa cells surrounding the oocyte is the *corona radiata*. Between the oocyte and the corona radiata is a thin gel-like layer of proteins and polysaccharides called the *zona pellucida* (fig. 20.27*b*).

Under the stimulation of FSH from the anterior pituitary, the granulosa cells of the ovarian follicles secrete increasing amounts of estrogen as the follicles grow. Interestingly, the granulosa cells produce estrogen from its precursor testosterone, which is supplied by cells of the *theca interna*, the layer immediately outside the follicle (fig. 20.27*b*).

Ovulation

Usually, by about 10 to 14 days after the first day of menstruation, only one follicle has continued its growth to become a fully mature graafian follicle (fig. 20.29). Other secondary follicles during that cycle regress and become *atretic*, a term that means "without an opening" in reference to their failure to

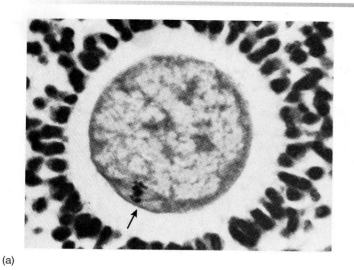

(a)

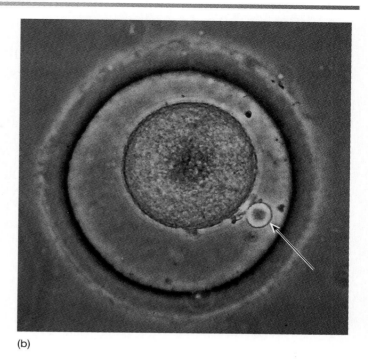

(b)

Figure 20.28

(*a*) A primary oocyte at metaphase I of meiosis. Notice the alignment of chromosomes (*arrow*). (*b*) A human secondary oocyte formed at the end of the first meiotic division and the first polar body (*arrow*).

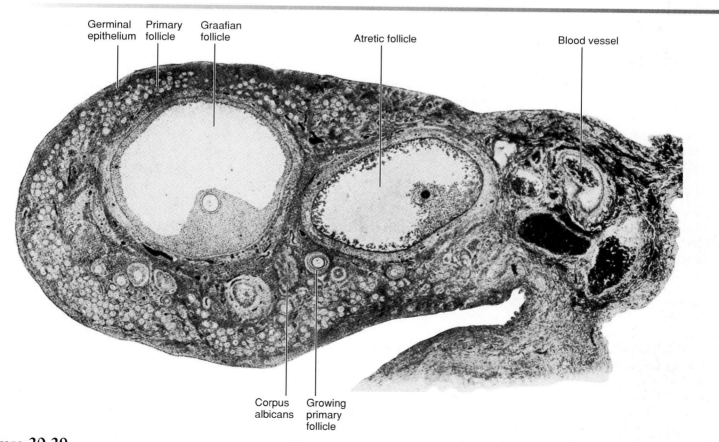

Germinal epithelium Primary follicle Graafian follicle Atretic follicle Blood vessel

Corpus albicans Growing primary follicle

Figure 20.29

An ovary containing follicles at different stages of development.

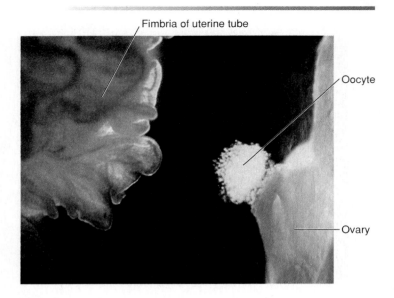

Figure 20.30

Ovulation from a human ovary.

rupture. The graafian follicle is so large that it forms a bulge on the surface of the ovary. Under proper hormonal stimulation, this follicle will rupture—much like the popping of a blister—and extrude its oocyte into the uterine tube in the process of **ovulation** (fig. 20.30).

The released cell is a secondary oocyte, surrounded by the zona pellucida and corona radiata. If it is not fertilized, it will degenerate in a couple of days. If a sperm passes through the corona radiata and zona pellucida and enters the cytoplasm of the secondary oocyte, the oocyte will then complete the second meiotic division. In this process, the cytoplasm is again not divided equally; most remains in the zygote (fertilized egg), leaving another polar body which, like the first, degenerates (fig. 20.31).

Changes continue in the ovary following ovulation. The empty follicle, under the influence of luteinizing hormone from the anterior pituitary, undergoes structural and biochemical changes to become a **corpus luteum** (= yellow body). Unlike the ovarian follicles, which secrete only estrogen, the corpus luteum secretes two sex steroid hormones: estrogen and progesterone. Toward the end of a nonfertile cycle, the corpus luteum regresses to become a nonfunctional *corpus albicans*. These cyclic changes in the ovary are summarized in figure 20.32.

Pituitary-Ovarian Axis

The term *pituitary-ovarian axis* refers to the hormonal interactions between the anterior pituitary and the ovaries. The anterior pituitary secretes two gonadotropic hormones—follicle-stimulating hormone (FSH) and luteinizing hormone (LH)—that promote cyclic changes in the structure and function of the ovaries. The secretion of both gonadotropic hormones, as previously discussed, is controlled by a single releasing hormone from the

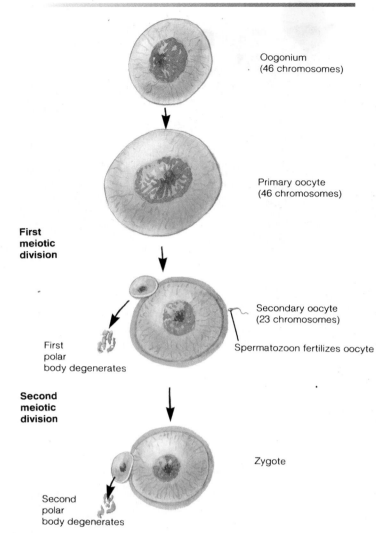

Figure 20.31

A diagram showing the process of oogenesis. During meiosis, each primary oocyte produces a single haploid gamete. If the secondary oocyte is fertilized, it will form a secondary polar body and become a zygote.

hypothalamus—called gonadotropin-releasing hormone (GnRH)—and by feedback effects from hormones secreted by the ovaries. The nature of these interactions will be described in detail in the next section.

Since one releasing hormone can stimulate the secretion of both FSH and LH, one might expect always to see parallel changes in the secretion of these gonadotropins. This, however, is not the case. During an early phase of the menstrual cycle FSH secretion is slightly greater than LH secretion, and just prior to ovulation LH secretion greatly exceeds FSH secretion. These differences are believed to be a result of the feedback effects of ovarian sex steroids, which can change the amount of GnRH secreted, the pulse frequency of GnRH secretion, and the ability of the anterior pituitary cells to secrete FSH and/or LH. These complex interactions result in a pattern of hormone secretion that regulates the phases of the menstrual cycle.

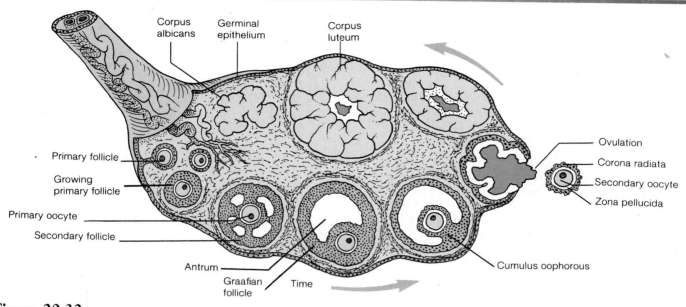

Figure 20.32

A diagram of an ovary showing the various stages of ovum and follicle development.

Menstrual Cycle

Cyclic changes in the secretion of gonadotropic hormones from the anterior pituitary cause the ovarian changes during a monthly cycle. The ovarian cycle is accompanied by cyclic changes in the secretion of estradiol and progesterone, which interact with the hypothalamus and pituitary to regulate gonadotropin secretion. The cyclic changes in ovarian hormone secretion also cause changes in the endometrium of the uterus during a menstrual cycle.

Humans, apes, and Old-World monkeys have cycles of ovarian activity that repeat at approximately one-month intervals; hence the name menstrual cycle (*menstru* = monthly). The term *menstruation* is used to indicate the periodic shedding of the stratum functionale of the endometrium, which becomes thickened prior to menstruation under stimulation by ovarian steroid hormones. In primates (other than New-World monkeys) this shedding of the endometrium is accompanied by bleeding. There is no bleeding, in contrast, when most other mammals shed their endometrium; their cycles, therefore, are not called menstrual cycles.

In human females and other primates that have menstrual cycles, coitus (sexual intercourse) may be permitted at any time of the cycle. Nonprimate female mammals, by contrast, are sexually receptive (in "heat" or "estrus") only at a particular time in their cycles, shortly before or after ovulation. These animals are therefore said to have *estrous cycles*. Bleeding occurs in some animals (such as dogs and cats) that have estrous cycles shortly before they permit coitus. This bleeding is a result of high estrogen secretion and is not associated with shedding of the endometrium. The bleeding that accompanies menstruation, in contrast, is caused by a fall in estrogen and progesterone secretion.

Phases of the Menstrual Cycle: Cyclic Changes in the Ovaries

The average menstrual cycle has a duration of about 28 days. Since it is a cycle, there is no beginning or end, and the changes that occur are generally gradual. It is convenient, however, to call the first day of menstruation "day one" of the cycle, because menstrual blood flow is the most apparent of the changes that occur. It is also convenient to divide the cycle into phases based on changes that occur in the ovary and in the endometrium. The ovaries are in the **follicular phase** starting on the first day of menstruation and ending on the day of ovulation. After ovulation, the ovaries are in the **luteal phase** until the first day of menstruation. The cyclic changes that occur in the endometrium are called the *menstrual, proliferative,* and *secretory phases* and will be discussed separately. Note that the time frames used for discussion of these phases are only averages and individual cycles may vary greatly from these averages.

Follicular Phase

Menstruation lasts from day 1 to day 4 or 5 of the average cycle. During this time, the secretions of ovarian steroid hormones are at their lowest ebb, and the ovaries contain only primordial and primary follicles. During the *follicular phase of the ovaries,* which lasts from day 1 to about day 13 of the cycle (this is highly variable), some of the primary follicles grow, develop vesicles, and become secondary follicles. Toward the end of the follicular phase, one follicle in one ovary reaches maturity and becomes a graafian follicle. As follicles grow, the granulosa cells secrete an increasing amount of **estradiol** (the principal estrogen), which reaches its highest concentration in the blood at about day 12 of the cycle, 2 days before ovulation.

The growth of the follicles and the secretion of estradiol are stimulated by, and dependent upon, FSH secreted from the anterior pituitary. The amount of FSH secreted during the early follicular phase is believed to be slightly greater than the amount secreted in the late follicular phase, although this can vary from cycle to cycle (a measure of variance is shown by vertical bars in fig. 20.33). FSH stimulates the production of FSH receptors in the granulosa cells, so that the follicles become increasingly sensitive to a given amount of FSH. This increased sensitivity is augmented by estradiol, which also stimulates the production of new FSH receptors in the follicles. As a result, the stimulatory effect of FSH on the follicles increases despite the fact that FSH levels in the blood do not increase throughout the follicular phase. Toward the end of the follicular phase, FSH and estradiol also stimulate the production of LH receptors in the graafian follicle. This prepares the graafian follicle for the next major event in the cycle.

The rapid rise in estradiol secretion from the granulosa cells during the follicular phase acts on the hypothalamus to increase the frequency of GnRH pulses. In addition, estradiol augments the ability of the pituitary to respond to GnRH with an increase in LH secretion. As a result of this stimulatory, or **positive feedback,** effect of estradiol on the pituitary, there is an increase in LH secretion in the late follicular phase that culminates in an **LH surge** (fig. 20.33).

The LH surge begins about 24 hours before ovulation and reaches its peak about 16 hours before ovulation. It is this surge that acts to trigger ovulation. Since GnRH stimulates the anterior pituitary to secrete both FSH and LH, there is a simultaneous, though smaller, surge in FSH secretion. Some investigators believe that this midcycle peak in FSH acts as a stimulus for the development of new follicles for the next month's cycle.

Ovulation

Under the influence of FSH stimulation, the graafian follicle grows so large that it becomes a thin-walled "blister" on the surface of the ovary. The growth of the follicle is accompanied by a rapid rate of increase in estradiol secretion. This rapid increase in estradiol, in turn, triggers the LH surge at about day 13. Finally, the surge in LH secretion causes the wall of the graafian follicle to rupture at about day 14 (fig. 20.34 *top*). In ovulation, a secondary oocyte, arrested at

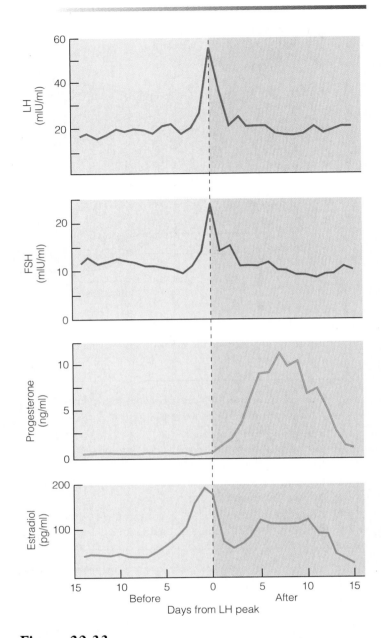

Figure 20.33

Sample values for LH, FSH, progesterone, and estradiol during the menstrual cycle. The midcycle peak of LH is used as a reference day. (IU = international unit.)

metaphase II of meiosis, is released into a uterine tube. The ovulated oocyte is still surrounded by a zona pellucida and corona radiata as it begins its journey to the uterus.

Ovulation occurs, therefore, as a result of the sequential effects of FSH and LH on the ovarian follicles. By means of the positive feedback effect of estradiol on LH secretion, the follicle, in a sense, sets the time for its own ovulation. This is because ovulation is triggered by an LH surge, and the LH surge is triggered by increased estradiol secretion that occurs while the follicle grows. In this way, the graafian follicle is not normally ovulated until it has reached the proper size and degree of maturation.

Chapter Twenty

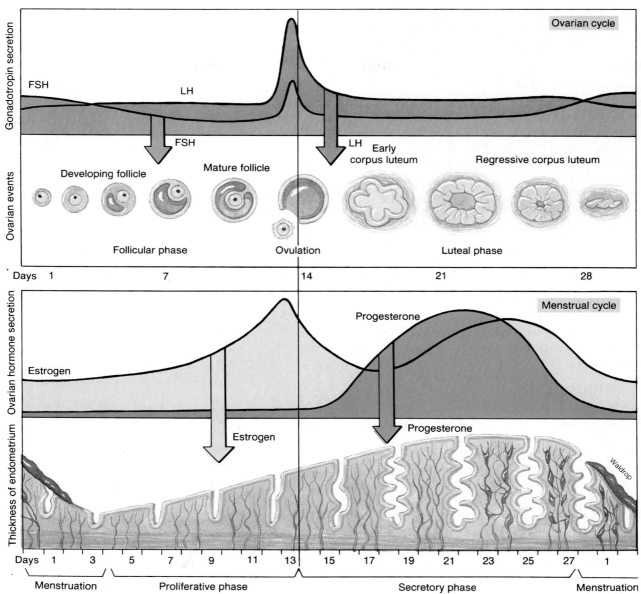

Figure 20.34

The cycle of ovulation and menstruation.

Luteal Phase

After ovulation, the empty follicle is stimulated by LH to become a new structure—the corpus luteum (fig. 20.35). This change in structure is accompanied by a change in function. Whereas the developing follicles secrete only estradiol, the corpus luteum secretes both estradiol and **progesterone.** Progesterone levels in the blood are negligible before ovulation but rise rapidly to reach a peak during the luteal phase, approximately 1 week after ovulation (see figs. 20.33 and 20.34).

The high levels of progesterone combined with estradiol during the luteal phase exert a **negative feedback inhibition** of FSH and LH secretion. This serves to retard development of new follicles, so that further ovulation does not normally occur during that cycle. In this way, multiple ovulations (and possible pregnancies) on succeeding days of the cycle are prevented.

High levels of estrogen and progesterone during the non-fertile cycle do not persist for very long, however, and new follicles do start to develop toward the end of one cycle, in preparation for the next cycle. Estrogen and progesterone levels fall during the late luteal phase (starting about day 22) because the corpus luteum regresses and stops functioning. In lower mammals, the decline in corpus luteum function is caused by a hormone secreted by the uterus called *luteolysin.* There is evidence that the luteolysin in humans may be prostaglandin $F_{2\alpha}$ (chapters 2 and 11), but the mechanisms of corpus luteum regression in humans is still incompletely understood. Luteolysis (breakdown of the corpus luteum) can be prevented by high levels of LH, but LH levels remain low during the luteal phase as a result of negative feedback inhibition by ovarian steroids. In a sense, therefore, the corpus causes its own demise.

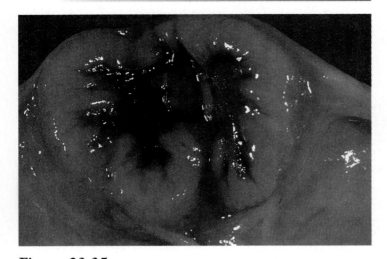

Figure 20.35

A corpus luteum in a human ovary.

With the declining function of the corpus luteum, estrogen and progesterone fall to very low levels by day 28 of the cycle. The withdrawal of ovarian steroids causes menstruation and permits a new cycle of follicle development to progress.

Cyclic Changes in the Endometrium

In addition to a description of the female cycle in terms of the phases of ovarian function, the cycle can also be described in terms of the changes that occur in the endometrium. Three phases can be identified on this basis: (1) the proliferative phase; (2) the secretory phase; and (3) the menstrual phase (fig. 20.34 *bottom*).

The **proliferative phase** of the endometrium occurs while the ovary is in its follicular phase. The increasing amounts of estradiol secreted by the developing follicles stimulate growth (proliferation) of the stratum functionale of the endometrium. In humans and other primates, spiral arteries develop in the endometrium during this phase. Estradiol may also stimulate the production of receptor proteins for progesterone at this time, in preparation for the next phase of the cycle.

The **secretory phase** of the endometrium occurs when the ovary is in its luteal phase. In this phase, increased progesterone secretion stimulates the development of uterine glands. As a result of the combined actions of estradiol and progesterone, the endometrium becomes thick, vascular, and "spongy" in appearance, and the uterine glands become engorged with glycogen during the phase following ovulation. The endometrium is therefore well prepared to accept and nourish an embryo should fertilization occur.

The **menstrual phase** occurs as a result of the fall in ovarian hormone secretion during the late luteal phase. Necrosis (cellular death) and sloughing of the stratum functionale of the endometrium may be produced by constriction of the spiral arteries. The spiral arteries appear to be responsible for bleeding during menstruation, since lower animals that lack spiral arteries do

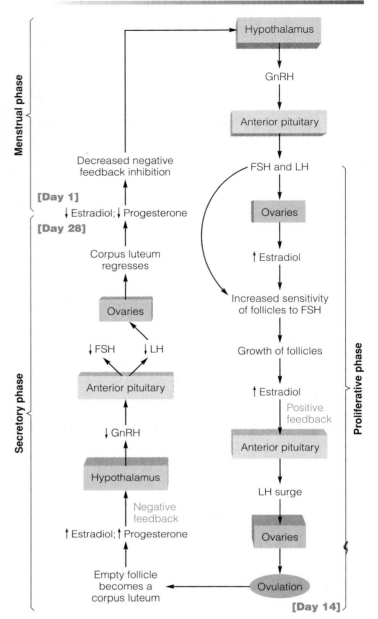

Figure 20.36

The sequence of events in the endocrine control of the ovarian cycle in context of the phases of the endometrium during the menstrual cycle.

not bleed when they shed their endometrium. The phases of the menstrual cycle are summarized in figure 20.36 and in table 20.6.

The cyclic changes in ovarian secretion cause other cyclic changes in the female reproductive tract. High levels of estradiol secretion, for example, cause cornification of the vaginal epithelium (the upper cells die and become filled with keratin). High levels of estradiol also cause the production of a thin, watery cervical mucus that can easily be penetrated by spermatozoa. During the luteal phase of the cycle, the high levels of progesterone cause the cervical mucus to become thick and sticky after ovulation has occurred.

Table 20.6 Phases of the Menstrual Cycle

Phase of Cycle		Hormonal Changes		Tissue Changes	
Ovarian	*Endometrial*	*Pituitary*	*Ovary*	*Ovarian*	*Endometrial*
Follicular (days 1–4)	Menstrual	FSH and LH secretion low	Estradiol and progesterone remain low	Primary follicles grow	Outer two-thirds of endometrium is shed with accompanying bleeding
Follicular (days 5–13)	Proliferative	FSH slightly higher than LH secretion in early follicular phase	Estradiol secretion rises (due to FSH stimulation of follicles)	Follicles grow; graafian follicle develops (due to FSH stimulation)	Mitotic division increases thickness of endometrium; spiral arteries develop (due to estradiol stimulation)
Ovulatory (day 14)	Proliferative	LH surge (and increased FSH) stimulated by positive feedback from estradiol	Fall in estradiol secretion	Graafian follicle is ruptured and secondary oocyte is extruded into fallopian tube	No change
Luteal (days 15–28)	Secretory	LH and FSH decrease (due to negative feedback of steroids)	Progesterone and estrogen secretion increase, then fall	Development of corpus luteum (due to LH stimulation); regression of corpus luteum	Glandular development in endometrium (due to progesterone stimulation)

Abnormal menstruations are among the most common disorders of the female reproductive system. The term *amenorrhea* refers to the absence of menstruation. *Dysmenorrhea* is painful or difficult menstruation accompanied by severe menstrual cramps. *Menorrhagia* is excessive bleeding during the menstrual period, and *metrorrhagia* refers to spotting between menstrual periods.

Cyclic changes in ovarian hormone secretion also cause cyclic changes in *basal body temperature*. In the **rhythm method** of birth control, a woman measures her oral basal body temperature upon waking to determine when ovulation has occurred. On the day of the LH peak, when estradiol secretion begins to decline, there is a slight drop in basal body temperature. Starting about 1 day after the LH peak, the basal body temperature sharply rises as a result of progesterone secretion and remains elevated throughout the luteal phase of the cycle (fig. 20.37). The day of ovulation can be accurately determined by this method, making the method useful in increasing fertility if conception is desired. Since the day of the cycle in which ovulation occurs is quite variable in many women, however, the rhythm method is not very reliable for contraception by predicting when the next ovulation will occur. The contraceptive pill is a statistically more effective means of birth control.

Contraceptive Pill

About 10 million women in the United States and 60 million women worldwide are currently using **oral contraceptives.** These contraceptives usually consist of a synthetic estrogen combined with a synthetic progesterone in the form of pills that are taken once each day for 3 weeks after the last day of a menstrual period. This procedure causes an immediate increase in blood levels of ovarian steroids (from the pill), which is maintained for the normal duration of a monthly cycle. As a result of *negative feedback inhibition* of gonadotropin secretion, *ovulation never occurs*. The entire cycle is like a false luteal phase, with high levels of progesterone and estrogen and low levels of gonadotropins.

Since the contraceptive pills contain ovarian steroid hormones, the endometrium proliferates and becomes secretory just as it does during a normal cycle. In order to prevent an abnormal growth of the endometrium, women stop taking the steroid pills after 3 weeks (placebo pills are taken during the fourth week). This causes estrogen and progesterone levels to fall, permitting menstruation to occur.

The side effects of earlier versions of the birth control pill have been reduced through a decrease in the content of estrogen and through the use of newer generations of progestogens (analogues of progesterone). The newer contraceptive pills are very effective and have a number of beneficial side effects, including a reduced risk for endometrial and ovarian cancer, re-

duced risk for cardiovascular disease, and a reduction in osteoporosis. However, there may be an increased risk for breast cancer, and possibly cervical cancer, with oral contraceptives. The current consensus is that the health benefits of oral contraceptives generally outweigh the risks.

Newer systems for delivery of contraceptive steroids are designed so that the steroids are not taken orally, and as a result do not have to pass through the liver before entering the general circulation. (All drugs taken orally pass from the hepatic portal vein to the liver before they are delivered to any other organ—see chapter 17.) This permits lower doses of hormones to be effective. Such newer systems include a subcutaneous implant (Norplant), which need only be replaced after 5 years, and vaginal rings, which can be worn for three weeks. The long-term safety of these newer methods has not yet been established.

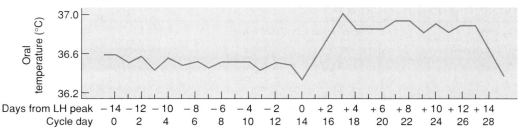

Figure 20.37

Changes in basal body temperature during the menstrual cycle.

Menopause

The term *menopause* means literally "pause in the menses" and refers to the cessation of ovarian activity that occurs at about the age of 50. During the postmenopausal years, which account for about a third of a woman's life span, the ovaries are depleted of follicles and cease secreting estradiol. This fall in estradiol is due to changes in the ovaries, not in the pituitary; indeed, FSH and LH secretion by the pituitary is elevated due to the absence of negative feedback inhibition from estradiol. Like prepubertal boys and girls, the only estrogen found in the blood of postmenopausal women is that formed by aromatization of the weak androgen androstenedione, secreted by the adrenal cortex and converted in the adipose tissue into a weak estrogen called estrone.

It is the withdrawal of estradiol secretion from the ovaries that is most responsible for the many symptoms of menopause. These include vasomotor disturbances and urogenital atrophy. Vasomotor disturbances produce the "hot flashes" of menopause, where a fall in core body temperature is followed by feelings of heat and profuse perspiration. Atrophy of the urethra, vaginal wall, and vaginal glands occurs, with loss of lubrication. There is also increased risk of atherosclerotic cardiovascular disease and increased progression of osteoporosis (see chapter 18). These changes can be reversed, to a significant degree, by estrogen treatments.

1. Describe the changes that occur in the ovary and endometrium during the follicular phase and explain how these changes are regulated by hormones.
2. Describe the hormonal regulation of ovulation.
3. Describe the formation, function, and fate of the corpus luteum. Also describe the changes that occur in the endometrium during the luteal phase.
4. Explain the significance of negative feedback inhibition during the luteal phase and describe the hormonal control of menstruation.

Fertilization, Pregnancy, and Parturition

Once fertilization has occurred, the secondary oocyte completes meiotic division and then undergoes mitosis to form first a ball of cells and then an early embryonic structure called a blastocyst. Cells of the blastocyst secrete a hormone called human chorionic gonadotropin that maintains the mother's corpus luteum and its production of estradiol and progesterone. This prevents menstruation, so that the embryo can implant into the endometrium, develop, and form a placenta. Birth is dependent upon strong contractions of the uterus, which are stimulated by oxytocin from the posterior pituitary.

During the act of sexual intercourse, the male ejaculates an average of 300 million sperm into the vagina of the female. This tremendous number is needed because of the high sperm fatality—only about 100 survive to enter each fallopian tube. During their passage through the female reproductive tract, the sperm gain the ability to fertilize an ovum. This process is called **capacitation.** Although the changes that occur in capacitation are incompletely understood, experiments have shown that freshly ejaculated sperm are infertile; they must be present in the female tract for at least 7 hours before they can fertilize an ovum.

A woman usually ovulates only one ovum a month, for a total of less than 450 ova during her reproductive years. Each ovulation releases a secondary oocyte arrested at metaphase of the second meiotic division. The secondary oocyte, as previously described, enters the fallopian tube surrounded by its zona pellucida (a thin transparent layer of protein and polysaccharides) and corona radiata of granulosa cells (fig. 20.38).

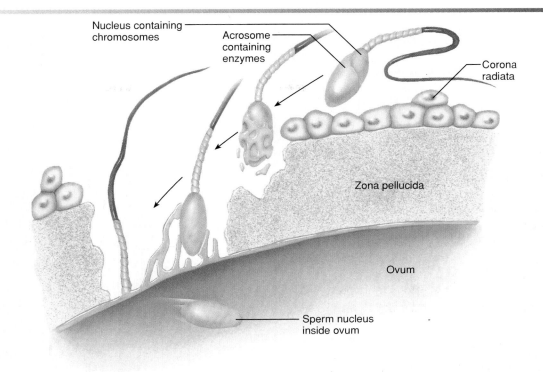

Figure 20.38

A diagram showing the process of fertilization. As the head of the sperm cell encounters the gelatinous corona radiata of the ovum, which has progressed in meiotic development to the secondary oocyte stage, the acrosomal vesicle ruptures and the sperm cell digests a path for itself by the action of the enzymes released from the acrosome. When the cell membrane of the sperm cell contacts the cell membrane of the ovum, they become continuous, and the sperm cell nucleus and other contents move into the cytoplasm of the ovum.

Fertilization normally occurs in the fallopian tubes. The head of each spermatozoan is capped by an organelle called an *acrosome* (fig. 20.39). The acrosome contains a trypsinlike protein-digesting enzyme and hyaluronidase (which digests hyaluronic acid, an important constituent of connective tissues). When sperm cell meets ovum, an **acrosomal reaction** exposes the acrosome's digestive enzymes and allows the sperm cell to penetrate the corona radiata and the zona pellucida. The acrosomal enzymes are not released in this process; rather, the sperm cell tunnels its way through these barriers by digestion reactions that are restricted to its acrosomal cap.

As the first sperm cell tunnels its way through the zona pellucida, a chemical change in the zona occurs that prevents other sperm from entering. Only a single sperm cell, therefore, is allowed to fertilize the ovum. As fertilization occurs, the secondary oocyte is stimulated to complete its second meiotic division (fig. 20.40). Like the first meiotic division, the second produces one cell that contains all of the cytoplasm—the mature ovum—and one polar body. The second polar body, like the first, ultimately fragments and disintegrates.

At fertilization, the sperm cell enters the cytoplasm of the much larger egg cell. Within 12 hours, the nuclear membrane in the ovum disappears, and the haploid number of chromosomes (twenty-three) in the ovum is joined by the haploid number of chromosomes from the sperm cell. A fertilized egg, or **zygote,** containing the diploid number of chromosomes (forty-six) is thus formed (fig. 20.40).

It should be noted that the sperm cell contributes more than the paternal set of chromosomes to the zygote. Recent evidence demonstrates that the centrosome of the human zygote is derived from the sperm cell and not from the oocyte. As described in chapter 3, the centrosome is needed for the organization of microtubules into a spindle apparatus, so that duplicated chromosomes can be separated during mitosis. Without a centrosome to form the spindle apparatus, cell division (and hence embryonic development) cannot proceed.

A secondary oocyte that has been ovulated but not fertilized does not complete its second meiotic division, but instead disintegrates 12 to 24 hours after ovulation. Fertilization therefore cannot occur if intercourse takes place later than 1 day following ovulation. Sperm, by contrast, can survive up to 3 days in the female reproductive tract. Fertilization therefore can occur if intercourse takes place within a 3-day period prior to the day of ovulation.

The process of **in vitro fertilization** is sometimes used to produce pregnancies in women with absent or damaged fallopian tubes or in women who are infertile for a variety of other reasons. An ovum may be collected by aspiration following ovulation (as estimated by waiting 36 to 38 hours after the LH surge). Alternatively, a woman may be treated with powerful FSH-like hormones, which cause the development of multiple follicles, and ova may be collected by aspiration guided by ultrasound and laparoscopy. The donor's sperm are treated with procedures that duplicate normal capacitation. The ova may be placed in a petri dish for 2 to 3 days with sperm collected from the donor, or newer techniques may be used to promote fertilization. These newer techniques include ICSI (for intracytoplasmic injection), which is the microinjection of sperm through the zona pellucida directly into the ovum (fig. 20.41). A number of embryos may be produced at the same time, and the surplus stored frozen in liquid nitrogen for later use. The embryos are usually transferred, three or more at a time, to the woman's uterus at their four-cell stage, 48 to 72 hours after fertilization. In some cases, the embryos may be transferred to the end of the fallopian tube. The chance of a successful implantation is low, and the procedure is expensive. The long-term safety of fertility drugs has also been questioned.

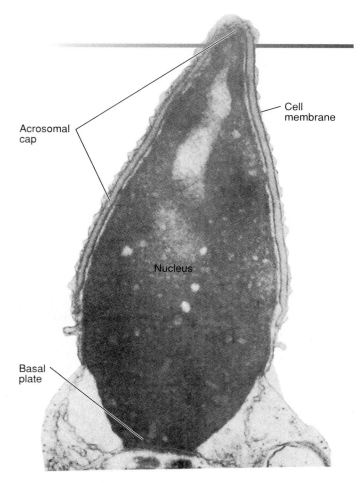

Figure 20.39

An electron micrograph showing the head of a human sperm cell with its nucleus and acrosomal cap.

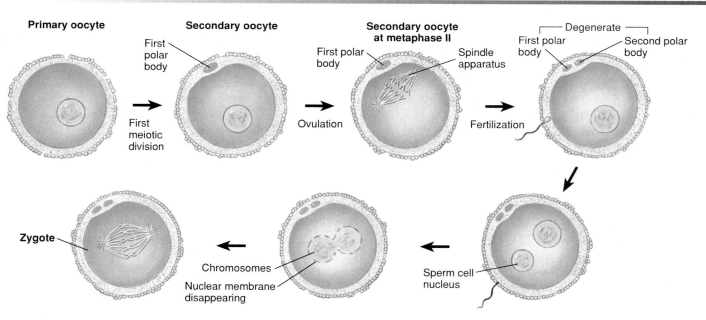

Figure 20.40

A secondary oocyte, arrested at metaphase II of meiosis, is released at ovulation. If this cell is fertilized, it will complete its second meiotic division and produce a second polar body. The chromosomes of the two gametes are joined in the zygote.

Chapter Twenty

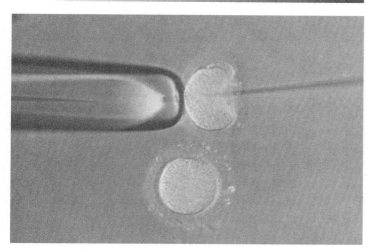

Figure 20.41

A needle (the shadow on the right) is used to inject a single spermatozoon into a human oocyte in vitro.

Cleavage and Formation of a Blastocyst

At about 30 to 36 hours after fertilization, the zygote divides by mitosis—a process called *cleavage*—into two smaller cells. The rate of cleavage is thereafter accelerated. A second cleavage, which occurs about 40 hours after fertilization, produces four cells. About 50 to 60 hours after fertilization, a third cleavage produces a ball of eight cells called a **morula** (= mulberry). This very early embryo enters the uterus three days after ovulation has occurred (fig. 20.42).

Cleavage continues so that a morula consisting of thirty-two to sixty-four cells is produced by the fourth day after fertilization. The embryo remains unattached to the uterine wall for the next 2 days, during which time it undergoes changes that convert it into a hollow structure called a **blastocyst** (fig. 20.43). The blastocyst consists of two parts: (1) an *inner cell mass*, which will become the fetus, and (2) a surrounding *chorion*, which will become part of the placenta. The cells that form the chorion are called *trophoblast cells*.

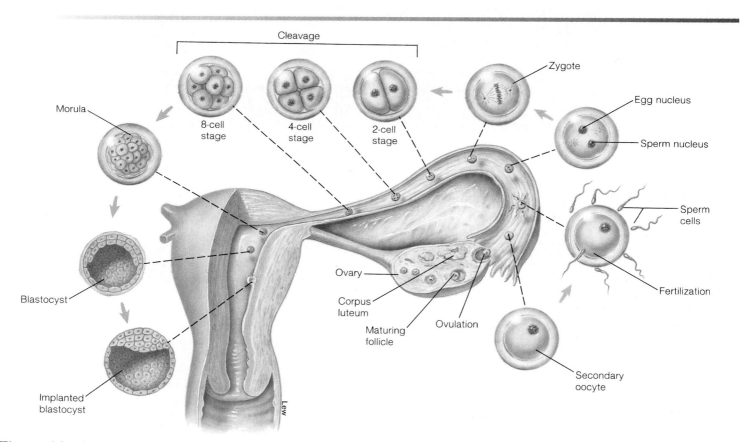

Figure 20.42

A diagram showing the ovarian cycle, fertilization, and the morphogenic events of the first week. Implantation of the blastocyst begins between the fifth and seventh day and is generally complete by the tenth day.

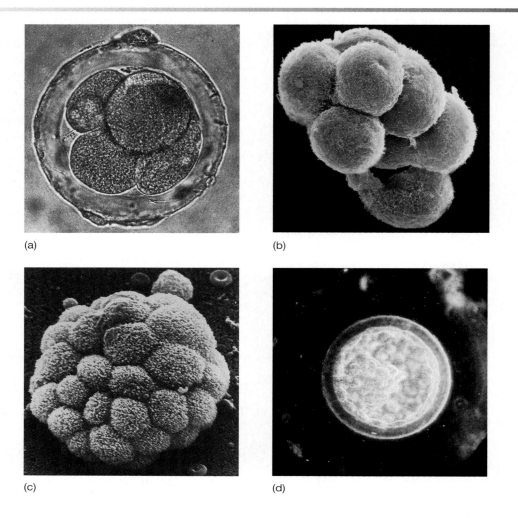

(a)

(b)

(c)

(d)

Figure 20.43

Stages of pre-embryonic development of a human ovum fertilized in a laboratory (in vitro) as seen in scanning electron micrographs. (*a*) The 4-cell stage, (*b*) cleavage at the 16-cell stage, (*c*) a morula, and (*d*) a blastocyst.

Progesterone, secreted from the woman's corpus luteum, is required for the endometrium to support the implanted embryo and maintain the pregnancy. A new drug, developed in France, promotes abortion by blocking the actions of progesterone. This drug—*RU 486*—competes with progesterone for cytoplasmic receptor proteins in the endometrium. When the RU 486 binds to progesterone receptors, the receptors are unable to translocate to the nucleus and activate the genes that are normally activated by progesterone. This results in the breakup of the endometrium and consequent loss of the embryo. Sometimes called the "abortion pill," RU 486 has generated bitter controversy in the United States.

On the sixth day following fertilization, the blastocyst attaches to the uterine wall, with the side containing the inner cell mass against the endometrium. The trophoblast cells produce enzymes that allow the blastocyst to "eat its way" into the thick endometrium. This begins the process of **implantation,** or **nidation,** and by the seventh to tenth day the blastocyst is completely buried in the endometrium (fig. 20.44).

Implantation and Formation of a Placenta

If fertilization does not take place, the corpus luteum begins to decrease its secretion of steroids about 10 days after ovulation. This withdrawal of steroids, as previously described, causes necrosis and sloughing of the endometrium following day 28 of the cycle. If fertilization and implantation have occurred, however, these events must obviously be prevented to maintain the pregnancy.

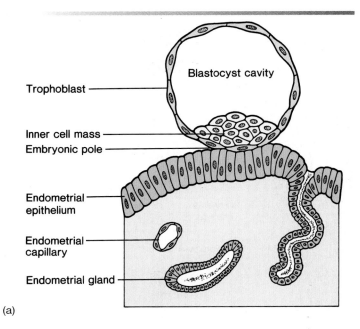

Trophoblast

Blastocyst cavity

Inner cell mass

Embryonic pole

Endometrial epithelium

Endometrial capillary

Endometrial gland

(a)

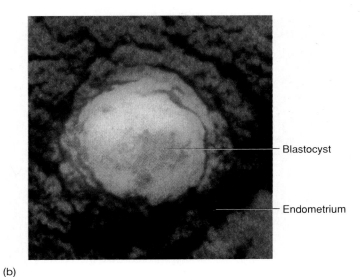

Blastocyst

Endometrium

(b)

Figure 20.44

(a) A diagram showing the blastocyst attached to the endometrium on about the sixth day. (b) A scanning electron micrograph showing the surface of the endometrium and implantation on day 12 following fertilization.

Chorionic Gonadotropin

The blastocyst saves itself from being eliminated with the endometrium by secreting a hormone that indirectly prevents menstruation. Even before the sixth day when implantation occurs, the trophoblast cells of the chorion secrete **chorionic gonadotropin,** or hCG (the *h* stands for *human*). This hormone is identical to LH in its effects and therefore is able to maintain the corpus luteum past the time when it would otherwise regress. The secretion of estradiol and progesterone is thus maintained and menstruation is normally prevented.

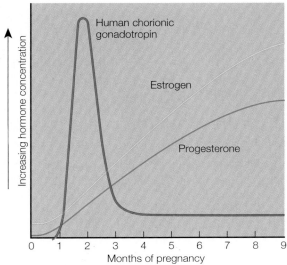

Figure 20.45

Human chorionic gonadotropin (hCG) is secreted by trophoblast cells during the first trimester of pregnancy. This hormone maintains the mother's corpus luteum for the first 5 ½ weeks. After that time, the placenta becomes the major sex-hormone-producing gland, secreting increasing amounts of estrogen and progesterone throughout pregnancy.

All **pregnancy tests** assay for the presence of hCG in blood or urine because this hormone is secreted by the blastocyst but not by the mother's endocrine glands. Modern pregnancy tests detect the presence of hCG by the use of antibodies against hCG or by use of cellular receptor proteins for hCG. Extremely sensitive techniques utilizing monoclonal antibodies against a subunit of hCG and radioimmunoassay (chapter 15) make it possible for pregnancy to be detected in a clinical laboratory as early as 7 to 10 days after conception.

The secretion of hCG declines by the tenth week of pregnancy (fig. 20.45). Actually, this hormone is required for only the first 5 to 6 weeks of pregnancy because the placenta itself becomes an active steroid hormone-secreting gland. At the fifth to sixth week, the mother's corpus luteum begins to regress (even in the presence of hCG), but by this time the placenta is secreting more than sufficient amounts of steroids to maintain the endometrium and prevent menstruation.

Chorionic Membranes

Between days 7 and 12, as the blastocyst becomes completely embedded in the endometrium, the chorion becomes a two-cell-thick structure, consisting of an inner *cytotrophoblast* layer and an outer *syncytiotrophoblast* layer. The inner cell mass (which will become the fetus), meanwhile, also develops two cell layers. These are the *ectoderm* (which will form such organs as the nervous system and skin) and the *endoderm* (which will

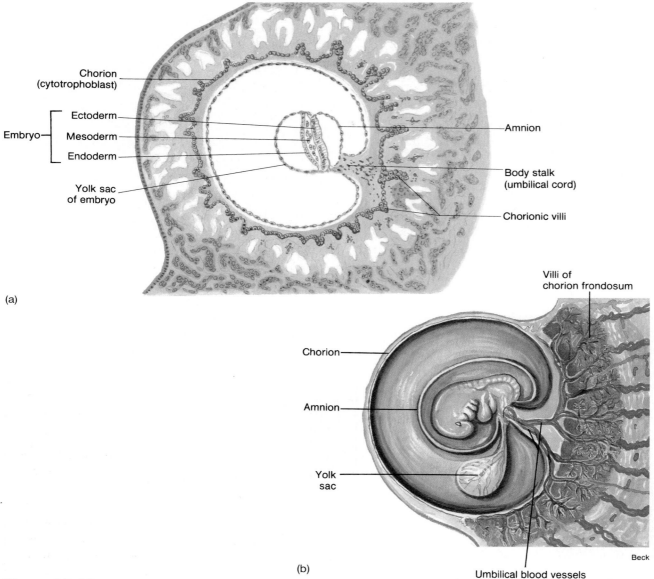

(a)

Chorion
(cytotrophoblast)

Ectoderm

Embryo — Mesoderm

Endoderm

Yolk sac
of embryo

Amnion

Body stalk
(umbilical cord)

Chorionic villi

Villi of
chorion frondosum

Chorion

Amnion

Yolk
sac

Umbilical blood vessels

Beck

(b)

Figure 20.46

After the syncytiotrophoblast has created blood-filled cavities in the endometrium, these cavities are invaded by extensions of the cytotrophoblast (a). These extensions, or villi, branch extensively to produce the chorion frondosum (b). The developing embryo is surrounded by a membrane called the amnion.

eventually form the gut and its derivatives). A third, middle embryonic layer—the *mesoderm*—is not yet seen at this stage. The embryo at this stage is a two-layer-thick disc separated from the cytotrophoblast of the chorion by an *amniotic cavity.*

As the syncytiotrophoblast invades the endometrium, it secretes protein-digesting enzymes that create many small, blood-filled cavities in the maternal tissue. The cytotrophoblast then forms projections, or *villi* (fig. 20.46), that grow into these pools of venous blood, producing a leafy-appearing structure called the *chorion frondosum (frond* = leaf). This occurs only on the side of the chorion that faces the uterine wall. As the embryonic structures grow, the other side of the chorion bulges into the cavity of the uterus, loses its villi, and takes on a smooth appearance.

Formation of the Placenta and Amniotic Sac

As the blastocyst is implanted in the endometrium and the chorion develops, the cells of the endometrium also undergo changes. These changes, including cellular growth and the accumulation of glycogen, are collectively called the **decidual reaction.** The maternal tissue in contact with the chorion frondosum is called the *decidua basalis.* These two structures—chorion frondosum (fetal tissue) and decidua basalis (maternal tissue)—together form the functional unit known as the **placenta.**

The human placenta is a disc-shaped structure that is continuous at its outer surface with the smooth part of the chorion, which bulges into the uterine cavity. Immediately beneath the chorionic membrane is the amnion, which has grown

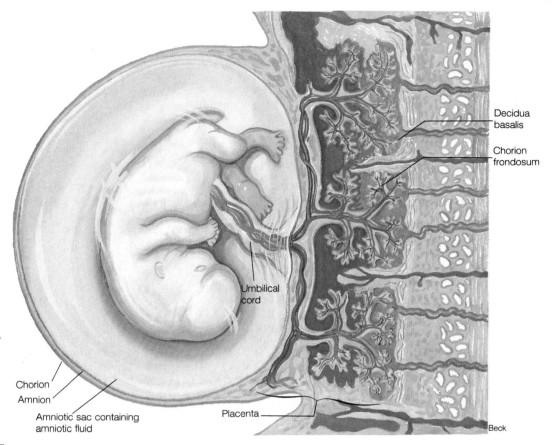

Decidua
basalis

Chorion
frondosum

Umbilical
cord

Chorion
Amnion
Amniotic sac containing
amniotic fluid

Placenta

Beck

Figure 20.47

Blood from the fetus is carried to and from the chorion frondosum by umbilical arteries and veins. The maternal tissue between the chorionic villi is known as the decidua basalis; this tissue, together with the villi, forms the functioning placenta. The space between chorion and amnion is obliterated, and the fetus lies within the fluid-filled amniotic sac.

to envelop the entire fetus (fig. 20.47). The fetus, together with its umbilical cord, is therefore located within the fluid-filled *amniotic sac.*

Amniotic fluid is formed initially as an isotonic secretion. Later, the volume is increased and the concentration changed by urine from the fetus. Amniotic fluid also contains cells that are sloughed off from the fetus, placenta, and amniotic sac. Since all of these cells are derived from the same fertilized ovum, all have the same genetic composition. Many genetic abnormalities can be detected by aspiration of this fluid and examination of the cells thus obtained. This procedure is called **amniocentesis** (fig. 20.48).

Amniocentesis is usually performed after the sixteenth week of pregnancy, when the amniotic sac contains 175 to 225 ml of fluid. Genetic diseases such as Down syndrome (characterized by three instead of two chromosomes number 21) can be detected by examining chromosomes; diseases such as Tay–Sachs disease, in which there is a defective enzyme involved in the formation of myelin sheaths, can be detected by biochemical techniques.

The amniotic fluid that is withdrawn contains fetal cells at a concentration that is too low to permit direct determination of genetic or chromosomal disorders. These cells must therefore be cultured in vitro for 10 to 14 days before they are present in sufficient numbers for the laboratory tests required. A newer method, called **chorionic villus biopsy,** is now available to detect genetic disorders earlier than permitted by amniocentesis. In chorionic villus biopsy, a catheter is inserted through the cervix to the chorion, and a sample of a chorionic villus is obtained by suction or cutting. Genetic tests can be performed directly on the villus sample, since this sample contains much larger numbers of fetal cells than does a sample of amniotic fluid. Chorionic villus biopsy can provide genetic information at 12 weeks' gestation. Amniocentesis, by contrast, cannot provide such information before about 20 weeks.

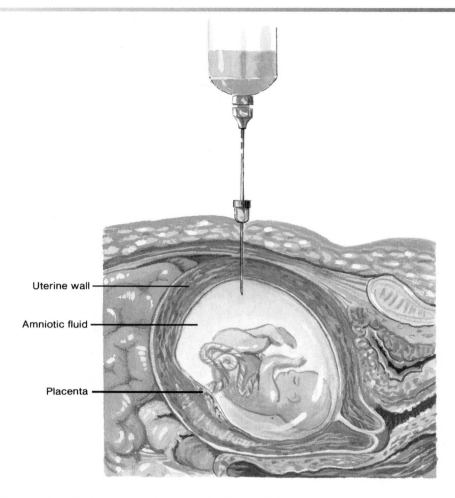

Figure 20.48

Amniocentesis. In this procedure, amniotic fluid, together with suspended cells, is withdrawn for examination. Various genetic diseases can be detected prenatally by this means.

Labels in figure: Uterine wall, Amniotic fluid, Placenta

Major structural abnormalities, which may not be predictable from genetic analysis, can often be detected by *ultrasound*. Sound-wave vibrations are reflected from the interface of tissues with different densities—such as the interface between the fetus and amniotic fluid—and used to produce an image. This technique is so sensitive that it can be used to detect a fetal heartbeat several weeks before it can be heard using a stethoscope.

Exchange of Molecules across the Placenta

The *umbilical arteries* deliver fetal blood to vessels within the villi of the chorion frondosum of the placenta. This blood circulates within the villi and returns to the fetus via the *umbilical vein*. Maternal blood is delivered to and drained from the cavities within the decidua basalis that are located between the chorionic villi (fig. 20.49). In this way, maternal and fetal blood are brought close together but never mix within the placenta.

The placenta serves as a site for the exchange of gases and other molecules between the maternal and fetal blood. Oxygen diffuses from mother to fetus, and carbon dioxide diffuses in the opposite direction. Nutrient molecules and waste products likewise pass between maternal and fetal blood; the placenta is, after all, the only link between the fetus and the outside world.

The placenta is not merely a passive conduit for exchange between maternal and fetal blood, however. It has a very high metabolic rate, utilizing about a third of all the oxygen and glucose supplied by the maternal blood. The rate of protein synthesis is, in fact, higher in the placenta than it is in the liver. Like the liver, the placenta produces a great variety of enzymes capable of converting hormones and exogenous drugs into less active molecules. In this way, potentially dangerous molecules in the maternal blood are often prevented from harming the fetus.

Endocrine Functions of the Placenta

The placenta secretes both steroid hormones and protein hormones that have actions similar to those of some anterior pituitary hormones. The protein hormones include **chorionic gonadotropin (hCG)** and **chorionic somatomammotropin (hCS)** (table 20.7). Chorionic gonadotropin has LH-like

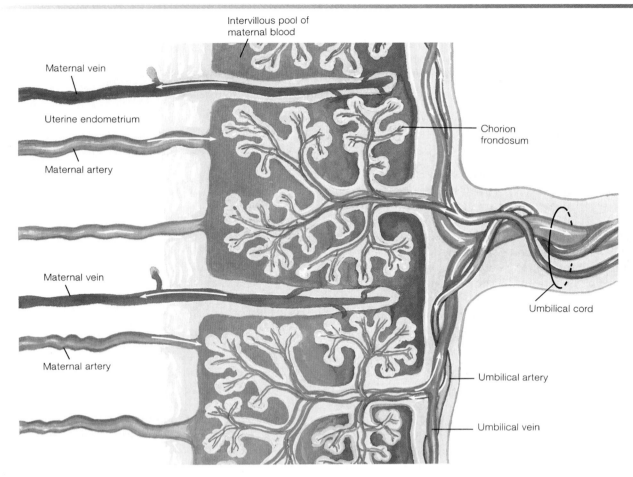

Figure 20.49

The circulation of blood within the placenta. Maternal blood is delivered to and drained from the spaces between the chorionic villi. Fetal blood is brought to blood vessels within the villi by branches of the umbilical artery and is drained by branches of the umbilical vein.

Table 20.7 Hormones Secreted by the Placenta

Hormones	Effects
Pituitary-like Hormones	
Chorionic gonadotropin (hCG)	Similar to LH; maintains mother's corpus luteum for first 5 ½ weeks of pregnancy; may be involved in suppressing immunological rejection of embryo; also exhibits TSH-like activity
Chorionic somatomammotropin (hCS)	Similar to prolactin and growth hormone; in the mother, hCS acts to promote increased fat breakdown and fatty acid release from adipose tissue and to promote the sparing of glucose use by maternal tissues ("diabetic-like" effects)
Sex Steroids	
Progesterone	Helps maintain endometrium during pregnancy; helps suppress gonadotropin secretion; stimulates development of alveolar tissue in mammary glands
Estrogens	Help maintain endometrium during pregnancy; help suppress gonadotropin secretion; help stimulate mammary gland development; inhibit prolactin secretion; promote uterine sensitivity to oxytocin; stimulate duct development in mammary glands

effects, as previously described; it also has thyroid-stimulating ability, like pituitary TSH. Chorionic somatomammotropin likewise has actions that are similar to two pituitary hormones: growth hormone and prolactin. The placental hormones hCG and hCS thus duplicate the actions of four anterior pituitary hormones.

Pituitary-like Hormones from the Placenta

The importance of chorionic gonadotropin in maintaining the mother's corpus luteum for the first 5 ½ weeks of pregnancy has been previously discussed. There is also some evidence that hCG may in some way help to prevent immunological rejection of the implanting embryo. Chorionic somatomammotropin acts together (synergizes) with growth hormone from the mother's pituitary to produce a diabetic-like effect in the pregnant woman. The effects of these two hormones stimulate (1) lipolysis and increased plasma fatty acid concentration; (2) decreased maternal utilization of glucose and, therefore, increased blood glucose concentrations; and (3) polyuria (excretion of large volumes of urine), thereby producing a degree of dehydration and thirst. This diabetic-like effect in the mother helps to spare glucose for the placenta and fetus that, like the brain, use glucose as their primary energy source.

Steroid Hormones from the Placenta

After the first 5½ weeks of pregnancy, when the corpus luteum regresses, the placenta becomes the major sex-steroid-producing gland. The blood concentration of estrogens, as a result of placental secretion, rises to levels more than 100 times greater than those existing at the beginning of pregnancy. The placenta also secretes large amounts of progesterone, changing the estrogen/progesterone ratio in the blood from 100:1 at the beginning of pregnancy to a ratio of close to 1:1 toward full-term.

The placenta, however, is an "incomplete endocrine gland" because it cannot produce estrogen and progesterone without the aid of precursors supplied to it by both the mother and the fetus. The placenta, for example, cannot produce cholesterol from acetate and so must be supplied with cholesterol from the mother's circulation. Cholesterol, which is a steroid containing twenty-seven carbons, can then be converted by enzymes in the placenta into steroids that contain twenty-one carbons—such as progesterone. The placenta, however, lacks the enzymes needed to convert progesterone into androgens (which have nineteen carbons). Therefore, androgens produced by the fetus are needed as substrates for the placenta to convert into estrogens (fig. 20.50), which have eighteen carbons.

In order for the placenta to produce estrogens, it therefore needs to cooperate with steroid-producing tissues in the fetus. Fetus and placenta thus form a single functioning system in terms of steroid hormone production. This system has been called the **fetal-placental unit** (fig. 20.50).

The ability of the placenta to convert androgens into estrogen helps to protect the female embryo from becoming masculinized by the androgens secreted from the mother's adrenal glands. In addition to producing estradiol, the placenta secretes large amounts of a weak estrogen called **estriol.** The production of estriol increases tenfold during pregnancy, so that by the third trimester estriol accounts for about 90% of the estrogens excreted in the mother's urine. Since almost all of this estriol comes from the placenta (rather than from maternal tissues), measurements of urinary estriol can be used clinically to assess the health of the placenta.

Labor and Parturition

Powerful contractions of the uterus in *labor* are needed for childbirth (*parturition*). These uterine contractions are known to be stimulated by two agents: (1) **oxytocin,** a polypeptide hormone produced in the hypothalamus and secreted by the posterior pituitary, and (2) **prostaglandins,** a class of cyclic fatty acids with autocrine function produced within the uterus itself. Labor

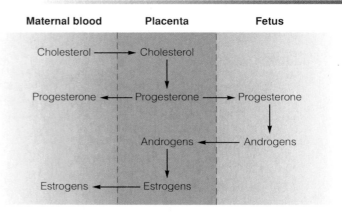

Figure 20.50

The secretion of progesterone and estrogen from the placenta requires a supply of cholesterol from the mother's blood and the cooperation of fetal enzymes that convert progesterone to androgens.

can indeed be induced artificially by injections of oxytocin or by insertion of prostaglandins into the vagina as a suppository.

Although the mechanisms operating during labor are well understood, the factors that lead to the initiation of parturition in humans remain unclear. In some lower mammals, the concentrations of estrogen and progesterone drop prior to parturition; this does not appear to be true in humans. In sheep and goats, labor seems to be initiated by corticosteroids from the fetal adrenal cortex that stimulate production of prostaglandins in the uterus. The uterine contractions that result stimulate increased secretion of oxytocin. The timing of parturition in humans, however, does not appear to be similarly dependent on the fetal adrenal cortex.

It appears that both oxytocin and prostaglandins are required in humans; oxytocin-induced contractions do not lead to dilation of the cervix and progressive labor in the absence of prostaglandins. Indeed, both alcohol (which inhibits oxytocin secretion) and indomethacin (which inhibits prostaglandin production) can be used to prevent preterm births. A mechanism has recently been proposed to explain how oxytocin and prostaglandins might cooperate to initiate labor in humans.

It has been demonstrated that the concentration of oxytocin receptors in the myometrium (smooth muscle layer) of the uterus increases dramatically during gestation. Although the blood concentrations of oxytocin remain constant, the sensitivity of the uterus to oxytocin increases as a result of the increased receptors. The threshold amount of oxytocin required to induce uterine contractions in a woman at term is approximately one-hundredth the amount required in a nonpregnant woman. Increased oxytocin receptors and sensitivity may result from the effects of the rising levels of estrogens that occur during gestation.

In addition to increased oxytocin receptors in the myometrium, evidence of increased oxytocin receptors in the nonmuscular decidua has also been found. It has been proposed that oxytocin may stimulate prostaglandin production in the decidua. The oxytocin secretion from the mother's posterior pituitary does not increase during labor, however. Some researchers

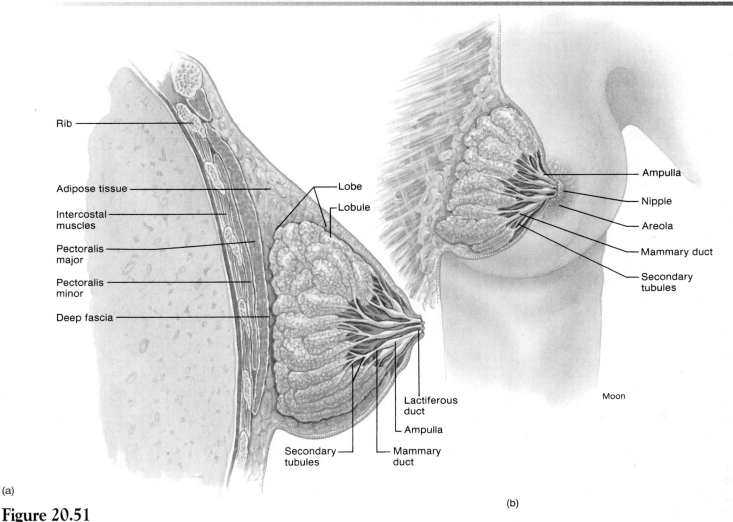

Rib

Adipose tissue

Intercostal muscles

Pectoralis major

Pectoralis minor

Deep fascia

Lobe

Lobule

Lactiferous duct

Ampulla

Secondary tubules

Mammary duct

Ampulla

Nipple

Areola

Mammary duct

Secondary tubules

Moon

(a)

(b)

Figure 20.51

The structure of the breast and mammary glands: (*a*) a sagittal section and (*b*) an anterior view partially sectioned.

have proposed that oxytocin may be secreted by the fetus, and there is evidence that, at least in the rat, the uterus may produce oxytocin as an autocrine regulator. Prostaglandins diffusing into the myometrium from the decidua may then act together with the increased sensitivity of the myometrium to oxytocin and stimulate the onset of labor. These events are summarized in table 20.8. Additionally, there is now evidence that the rat uterus produces nitric oxide, but that this production is decreased during labor. Since nitric oxide causes the relaxation of uterine smooth muscle, this mechanism may also contribute to labor. More research is required to understand the physiology of labor and parturition in humans.

Lactation

Each mammary gland is composed of fifteen to twenty *lobes*, divided by adipose tissue. Each lobe has its own drainage pathway to the outside. The amount of adipose tissue determines the size and shape of the breast but has nothing to do with the ability of a woman to nurse. Each lobe is subdivided into *lobules*, which contain the glandular **alveoli** (fig. 20.51) that secrete the milk of a

Table 20.8	Possible Sequence of Events Leading to the Onset of Labor in Humans
Step	**Event**
1	High estrogen secretion from the placenta stimulates production of oxytocin receptors in the uterus
2	Uterine muscle (myometrium) becomes increasingly sensitive to effects of oxytocin during pregnancy
3	Oxytocin may stimulate production of prostaglandins in the uterus
4	Prostaglandins may stimulate uterine contractions
5	Contractions of the uterus stimulate oxytocin secretion from the posterior pituitary
6	Increased oxytocin secretion stimulates increased uterine contractions, creating a positive feedback loop and resulting in labor

lactating female. The clustered alveoli secrete milk into a series of *secondary tubules*. These tubules converge to form a series of *mammary ducts*, which in turn converge to form a **lactiferous duct** that drains at the tip of the nipple. The lumen of each lactiferous duct

Table 20.9 Hormonal Factors Affecting Lactation

Hormones	Major Source	Effects
Insulin, cortisol, thyroid hormones	Pancreas, adrenal cortex, and thyroid	Permissive effects—adequate amounts must be present for other hormones to exert their effects on mammary glands
Estrogen and progesterone	Placenta	Growth and development of secretory units (alveoli) and ducts in mammary glands
Prolactin	Anterior pituitary	Production of milk proteins, including casein and lactalbumin
Oxytocin	Posterior pituitary	Stimulation of milk ejection

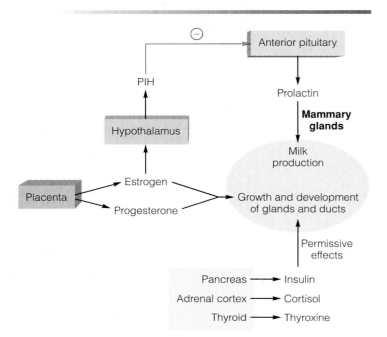

Figure 20.52

The hormonal control of mammary gland development during pregnancy and lactation. Notice that milk production is prevented during pregnancy by estrogen inhibition of prolactin secretion. This inhibition is accomplished by the stimulation of PIH (prolactin-inhibiting hormone) secretion from the hypothalamus.

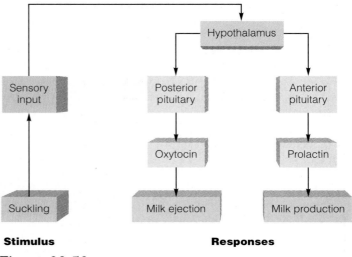

Figure 20.53

Lactation occurs in two stages: milk production (stimulated by prolactin) and milk ejection (stimulated by oxytocin). The stimulus of sucking triggers a neuroendocrine reflex that results in increased secretion of oxytocin and prolactin.

expands just beneath the surface of the nipple to form an *ampulla*, where milk accumulates during nursing.

The changes that occur in the mammary glands during pregnancy and the regulation of lactation provide excellent examples of hormonal interactions and neuroendocrine regulation (table 20.9). Growth and development of the mammary glands during pregnancy requires the permissive actions of insulin, cortisol, and thyroid hormones; in the presence of adequate amounts of these hormones, high levels of progesterone stimulate the development of the mammary alveoli and estrogen stimulates proliferation of the tubules and ducts (fig. 20.52).

The production of milk proteins, including casein and lactalbumin, is stimulated after parturition by **prolactin,** a hormone secreted by the anterior pituitary. The secretion of prolactin is controlled primarily by *prolactin-inhibiting hormone* (PIH), which is believed to be dopamine, produced by the hy-

pothalamus and secreted into the portal blood vessels. The secretion of PIH is stimulated by high levels of estrogen. In addition, high levels of estrogen act directly on the mammary glands to block their stimulation by prolactin and hCS. During pregnancy, consequently, the high levels of estrogen prepare the breasts for lactation but prevent prolactin secretion and action.

After parturition, when the placenta is eliminated, declining levels of estrogen are accompanied by an increase in the secretion of prolactin. Milk production is therefore stimulated. If a woman does not wish to breast-feed her baby, she may take oral estrogens to inhibit prolactin secretion. A different drug commonly given in these circumstances, and in other conditions in which it is desirable to inhibit prolactin secretion, is *bromocriptine.* This drug binds to dopamine receptors, and thus promotes the action of dopamine. The fact that this action inhibits prolactin secretion offers additional evidence that dopamine functions as the prolactin-inhibiting hormone (PIH).

The act of nursing helps to maintain high levels of prolactin secretion via a *neuroendocrine reflex* (fig. 20.53). Sensory endings in the breast, activated by the stimulus of suckling, relay impulses to the hypothalamus and inhibit the secretion of

PIH. There is also indirect evidence that the stimulus of suckling may cause the secretion of a *prolactin-releasing hormone,* but this is controversial. Suckling thus results in the reflex secretion of high levels of prolactin, which promotes the secretion of milk from the alveoli into the ducts. In order for the baby to get the milk, however, the action of another hormone is needed.

The stimulus of suckling also results in the reflex secretion of oxytocin from the posterior pituitary. This hormone is produced in the hypothalamus and stored in the posterior pituitary; its secretion results in the **milk-ejection reflex,** or milk letdown. Oxytocin stimulates contraction of the lactiferous ducts, as well as of the uterus (this is why women who breast-feed regain uterine muscle tone faster than women who do not).

Breast-feeding, acting through reflex inhibition of GnRH secretion, can inhibit the secretion of gonadotropins from the mother's anterior pituitary and thus inhibit ovulation. Breast-feeding is thus a natural contraceptive mechanism that helps to space births. This mechanism appears to be most effective in women with limited caloric intake and those who breast-feed their babies at frequent intervals throughout the day and night. In the traditional societies of the less industrialized nations, therefore, breast-feeding is an effective contraceptive. Breast-feeding has much less of a contraceptive effect in women who are well nourished and who breast-feed their babies at more widely spaced intervals.

Milk let-down can become a conditioned reflex made in response to visual or auditory cues; the crying of a baby can elicit oxytocin secretion and the milk-ejection reflex. On the other hand, this reflex can be suppressed by the adrenergic effects produced in the fight-or-flight reaction. Thus, if a woman becomes nervous and anxious while breast-feeding, her milk will be produced but it will not flow (there will be no milk let-down). This can cause increased pressure, intensifying her anxiety and frustration and further inhibiting the milk-ejection reflex. It is therefore important for mothers to nurse their babies in a quiet and calm environment.

1. Describe the changes that occur in the sperm cell and ovum during fertilization.
2. Identify the source of hCG and explain why this hormone is needed to maintain pregnancy for the first 10 weeks.
3. List the fetal and maternal components of the placenta and describe the circulation in these two components. Explain how fetal and maternal gas exchange occurs.
4. List the protein hormones and sex steroids secreted by the placenta and describe their functions.
5. Identify the two factors that stimulate uterine contraction during labor and describe the proposed mechanisms that may initiate labor in humans.
6. Describe the hormonal interactions required for breast development during pregnancy and for lactation after delivery.

Concluding Remarks

It may seem strange to end a textbook on physiology with the topics of pregnancy and parturition. This is done in part for practical reasons; these topics are complex, and to understand them requires a grounding in subjects covered earlier. Also, it seems appropriate to end at the beginning, at the start of a new life. Although generations of researchers have accumulated an impressive body of knowledge, the study of physiology is still young and rapidly growing. I hope that this introductory textbook will serve students' immediate practical needs as a resource for understanding current applications, and that it will provide a good foundation for a lifetime of further study.

Summary

Sexual Reproduction p. 610

I. Sperm that bear X chromosomes produce XX zygotes when they fertilize an ovum; sperm that bear Y chromosomes produce XY zygotes.
  A. Embryos that have the XY genotype develop testes; those without a Y chromosome produce ovaries.

B. The testes of a male embryo secrete testosterone and müllerian inhibition factor. MIF causes degeneration of female accessory sex organs and testosterone promotes the formation of male accessory sex organs.

II. The male accessory sex organs are the epididymis, vas deferens, seminal vesicles, prostate, and ejaculatory duct.
  A. The female accessory sex organs are the uterus and uterine (fallopian) tubes. They develop when testosterone and müllerian inhibition factor are absent.

B. Testosterone indirectly (acting via conversion to dihydrotestosterone) promotes the formation of male external genitalia; female genitalia are formed when testosterone is absent.

III. There are a variety of disorders of embryonic sexual development that can be understood in terms of the normal physiology of the developmental processes.

Endocrine Regulation of Reproduction   p. 616

I. The gonads are stimulated by two anterior pituitary hormones: FSH (follicle-stimulating hormone) and LH (luteinizing hormone).

  A. The secretion of FSH and LH is stimulated by gonadotropin-releasing hormone (GnRH), which is secreted by the hypothalamus.

  B. The secretion of FSH and LH is also under the control of the gonads by means of negative feedback inhibition by gonadal steroid hormones and by a peptide called inhibin.

II. The rise in FSH and LH secretion that occurs at puberty may be due to maturational changes in the brain and to decreased sensitivity of the hypothalamus and pituitary to the negative feedback effects of sex steroid hormones.

III. The pineal gland secretes melatonin. This hormone has an inhibitory effect on gonadal function in some species of mammals, but its role in human physiology is presently controversial.

Male Reproductive System   p. 619

I. In the male, the pituitary secretion of LH is controlled by negative feedback from testosterone, whereas the secretion of FSH is controlled by the secretion of inhibin from the testes.

  A. The negative feedback effect of testosterone is actually produced by the conversion of testosterone to 5α-reduced androgens and to estradiol.

  B. The secretion of testosterone is relatively constant rather than cyclic, and it does not decline sharply at a particular age.

II. Testosterone promotes the growth of soft tissue and bones before the epiphyseal discs have sealed; thus, testosterone and related androgens are anabolic steroids.

  A. Testosterone is secreted by the interstitial Leydig cells under stimulation by LH.

  B. LH receptor proteins are located in the interstitial tissue; FSH receptors are located in the Sertoli cells within the seminiferous tubules.

  C. The Leydig cells of the interstitial compartment and the Sertoli cells of the tubular compartment of the testes secrete autocrine regulatory molecules that allow the two compartments to interact.

III. Diploid spermatogonia in the seminiferous tubules undergo meiotic cell division to produce haploid sperm.

  A. At the end of meiosis, four spermatids are formed. They change into spermatozoa by a maturational process called spermiogenesis.

  B. Sertoli cells in the seminiferous tubules are required for spermatogenesis.

  C. At puberty, testosterone is required for the completion of meiosis, and FSH is required for spermiogenesis.

IV. Spermatozoa in the seminiferous tubules are conducted to the epididymis and drained from the epididymis into the vas deferens. The prostate and seminal vesicles add fluid to the semen.

V. Penile erection is produced by parasympathetic-induced vasodilation. Ejaculation is produced by sympathetic nerve stimulation of peristaltic contraction of the male accessory sex organs.

Female Reproductive System   p. 629

I. Primordial follicles in the ovary contain primary oocytes that have become arrested at prophase of the first meiotic division. Their number is maximal at birth and declines thereafter.

  A. A small number of oocytes in each cycle are stimulated to complete their first meiotic division and become secondary oocytes.

  B. At the completion of the first meiotic division, the secondary oocyte is the only complete cell formed. The other product of this division is a tiny polar body, which disintegrates.

II. One of the secondary follicles grows very large, becomes a graafian follicle, and is ovulated.

  A. Upon ovulation, the secondary oocyte is extruded from the ovary; it does not complete the second meiotic division unless it becomes fertilized.

  B. After ovulation, the empty follicle becomes a new endocrine gland called a corpus luteum.

  C. Whereas the ovarian follicles secrete only estradiol, the corpus luteum secretes both estradiol and progesterone.

III. The hypothalamus secretes GnRH in a pulsatile fashion, causing pulsatile secretion of gonadotropins. This is needed to prevent desensitization and down-regulation of the target glands.

Menstrual Cycle   p. 635

I. During the follicular phase of the cycle, the ovarian follicles are stimulated by FSH from the anterior pituitary.

  A. Under FSH stimulation, the follicles grow, mature, and secrete increasing concentrations of estradiol.

B. At about day 13, the rapid rise in estradiol secretion stimulates a surge of LH secretion from the anterior pituitary. This represents positive feedback.

C. The LH surge stimulates ovulation at about day 14.

D. After ovulation, the empty follicle is stimulated by LH to become a corpus luteum, at which point the ovary is in a luteal phase.

E. The secretion of progesterone and estradiol rises during the first part of the luteal phase and exerts negative feedback inhibition of FSH and LH secretion.

F. Without continued stimulation by LH, the corpus luteum regresses at the end of the luteal phase, and the secretion of estradiol and progesterone declines. This decline results in menstruation and the beginning of a new cycle.

II. The rising estradiol concentration during the follicular phase produces the proliferative phase of the endometrium. The secretion of progesterone during the luteal phase produces the secretory phase of the endometrium.

III. Oral contraceptive pills usually contain combinations of estrogen and progesterone that produce negative feedback inhibition of FSH and LH secretion.

Fertilization, Pregnancy, and Parturition   p. 640

I. The sperm undergoes an acrosomal reaction, which allows it to penetrate the corona radiata and zona pellucida.

A. Upon fertilization, the secondary oocyte completes meiotic division and produces a second polar body, which degenerates.

B. A diploid zygote is formed, which undergoes cleavage to form a morula and then a blastocyst. Implantation of the blastocyst in the endometrium begins between the fifth and seventh day.

II. The trophoblast cells of the blastocyst secrete human chorionic gonadotropin (hCG), which functions in the manner of LH and maintains the mother's corpus luteum for the first 10 weeks of pregnancy.

A. The trophoblast cells become the fetal contribution to the placenta. The placenta is also formed from adjacent maternal tissue in the endometrium.

B. Oxygen, nutrients, and wastes are exchanged by diffusion between the fetal and maternal blood.

III. The placenta secretes chorionic somatomammotropin (hCS), chorionic gonadotropin (hCG), and steroid hormones.

A. hCS is similar in its action to prolactin and growth hormone; hCG is similar to LH and TSH.

B. The major steroid hormone secreted by the placenta is estriol. The placenta and fetal glands cooperate in the production of steroid hormones.

IV. Contraction of the uterus in labor is stimulated by oxytocin from the posterior pituitary and by prostaglandins, produced within the uterus.

V. The high levels of estrogen during pregnancy, acting synergistically with other hormones, stimulate growth and development of the mammary glands.

A. Prolactin (and the prolactin-like effects of hCS) can stimulate the production of milk proteins. Prolactin secretion and action, however, are blocked during pregnancy by the high levels of estrogen secreted by the placenta.

B. After delivery, when the estrogen levels fall, prolactin stimulates milk production.

C. The milk-ejection reflex is a neuroendocrine reflex. The stimulus of suckling causes reflex secretion of oxytocin. This stimulates contractions of the lactiferous ducts and the ejection of milk from the nipple.

Clinical Investigation

A first-year college student goes to the student health center complaining of amenorrhea of several months' duration. She states that she does not have a history of dysmenorrhea, menorrhagia, or metrorrhagia. The young woman's body weight is significantly below average for her height, but she otherwise exhibits normal development of secondary sex characteristics. She had been diagnosed a few years earlier as hypothyroid, and thyroxine pills were prescribed for this condition. She states that she does aerobic exercises regularly, and indeed leads an aerobics class at a local gym. A pregnancy test is performed and is negative. Blood tests are ordered for thyroxine, prolactin, androgens, estradiol, and gonadotropins. These all prove to be normal. What is the most likely diagnosis of this student's condition, and what should she do about it?

Clue

See the boxed definitions of the terms used to describe abnormal menstrual periods.

Review Activities

Objective Questions

Match the following:

1. menstrual phase
2. follicular phase
3. luteal phase
4. ovulation

   a. high estrogen and progesterone; low FSH and LH

   b. low estrogen and progesterone

   c. LH surge

   d. increasing estrogen; low LH and low progesterone

5. A person with the genotype XO has
   a. ovaries.
   b. testes.
   c. both ovaries and testes.
   d. neither ovaries nor testes.

6. An embryo with the genotype XX develops female accessory sex organs because of
   a. androgens.
   b. estrogens.
   c. absence of androgens.
   d. absence of estrogens.

7. In the male,
   a. FSH is not secreted by the pituitary.
   b. FSH receptors are located in the Leydig cells.
   c. FSH receptors are located in the spermatogonia.
   d. FSH receptors are located in the Sertoli cells.

8. The secretion of FSH in a male is inhibited by negative feedback effects of
   a. inhibin secreted from the tubules.
   b. inhibin secreted from the Leydig cells.
   c. testosterone secreted from the tubules.
   d. testosterone secreted from the Leydig cells.

9. Which of the following statements is *true*?
   a. Sperm are not motile until they pass through the epididymis.
   b. Sperm require capacitation in the female reproductive tract before they can fertilize an ovum.
   c. An ovum does not complete meiotic division until it has been fertilized.
   d. All of the above are true.

10. The corpus luteum is maintained for the first 10 weeks of pregnancy by
    a. hCG.
    b. LH.
    c. estrogen.
    d. progesterone.

11. Fertilization normally occurs in
    a. the ovaries.
    b. the fallopian tubes.
    c. the uterus.
    d. the vagina.

12. The placenta is formed from
    a. the fetal chorion frondosum.
    b. the maternal decidua basalis.
    c. both a and b.
    d. neither a nor b.

13. Uterine contractions are stimulated by
    a. oxytocin.
    b. prostaglandins.
    c. prolactin.
    d. both a and b.
    e. both b and c.

14. Contraction of the mammary glands and ducts during the milk-ejection reflex is stimulated by
    a. prolactin.
    b. oxytocin.
    c. estrogen.
    d. progesterone.

15. If GnRH were secreted in large amounts and at a constant rate rather than in a pulsatile fashion, which of the following statements would be *true*?
    a. LH secretion will increase at first and then decrease.
    b. LH secretion will increase indefinitely.
    c. Testosterone secretion in a male will be continuously high.
    d. Estradiol secretion in a woman will be continuously high.

Essay Questions

1. Identify the conversion products of testosterone and describe their functions in the brain, prostate, and seminiferous tubules.[1]

2. Explain why a testis is said to be composed of two separate compartments. Describe the interactions that may occur between these compartments.

3. Describe the roles of the Sertoli cells in the testes.

4. Describe the steps of spermatogenesis and explain its hormonal control.

5. Explain the hormonal interactions that control ovulation and cause it to occur at the proper time.

6. Compare menstrual bleeding and bleeding that occurs during the estrous cycle of a dog in terms of hormonal control mechanisms and the ovarian cycle.

7. "The [contraceptive] pill tricks the brain into thinking you're pregnant." Interpret this popularized explanation in terms of physiological mechanisms.

8. Why does menstruation normally occur? Under what conditions does menstruation not occur? Explain.

9. Explain the proposed mechanisms whereby the act of a mother nursing her baby results in lactation. By what mechanisms might the sound of a baby crying elicit the milk-ejection reflex?

10. Describe the steps of oogenesis when fertilization occurs and when it does not occur. Why are polar bodies produced?

---

[1] *Note:* This question is answered on page 267 of the Student Study Guide.

11. Identify the hormones secreted by the placenta. Why is the placenta considered an "incomplete endocrine gland"?
12. Describe the endocrine changes that occur at menopause and discuss the consequences of these changes. What are the benefits and risks of hormone replacement therapy?
13. Explain the sequence of events by which the male accessory sex organs and external genitalia are produced. What occurs when a male embryo lacks receptor proteins for testosterone? What occurs when a male embryo lacks the enzyme 5α-reductase?

Selected Readings

Baird, D. T., and A. F. Glasier. 1993. Hormonal contraception. *New England Journal of Medicine* 328:1543.

Bardin, C. W. 1979. The neuroendocrinology of male reproduction. *Hospital Practice* 14:65.

Bartke, A., A. A. Hafiez, F. J. Bex, and S. Dalterio. 1978. Hormonal interaction in the regulation of androgen secretion. *Biology of Reproduction* 18:44.

Beaconsfield, P., G. Birdwood, and R. Beaconsfield. July 1980. The placenta. *Scientific American*.

Belchetz, P. E. 1994. Hormonal treatment of postmenopausal women. *New England Journal of Medicine* 330:1062.

Ben-Jonathon, N., J. F. Hyde, and I. Murai. 1988. Suckling-induced rise in prolactin: Mediation by a prolactin-releasing factor from the posterior pituitary. *News in Physiological Sciences* 3:172.

Boyar, R. M. 1978. Control of the onset of puberty. *Annual Review of Medicine* 31:329.

Brann, D. W. 1993. Progesterone: The forgotten hormone? *Perspectives in Biology and Medicine* 36:642.

Bredt, D. S., and S. H. Snyder. 1994. Nitric oxide: A physiologic messenger molecule. *Annual Review of Biochemistry* 63:175.

Chervenak, F. A., G. Isaacson, and M. J. Mahoney. 1986. Advances in the diagnosis of fetal defects. *New England Journal of Medicine* 315:305.

Conn, P. M., and W. F. Crowley Jr. 1991. Gonadotropin releasing hormone and its analogues. *New England Journal of Medicine* 324:93.

Cross, J. C., Z. Werb, and S. J. Fisher. 1994. Implantation and the placenta: Key pieces of the development puzzle. *Science* 266:1508.

D'Alton, M. E., and A. H. DeCherney. 1993. Prenatal diagnosis. *New England Journal of Medicine* 328:114.

Dufau, M. L. 1988. Endocrine regulation and communicating functions of the Leydig cell. *Annual Review of Physiology* 50:483.

Fink, G. 1979. Feedback action of target hormones on hypothalamus and pituitary with special reference to gonadal steroids. *Annual Review of Physiology* 41:571.

Frantz, A. G. 1979. Prolactin. *New England Journal of Medicine* 298:112.

Frisch, R. E. March 1988. Fatness and fertility. *Scientific American*.

Garnick, M. B. April 1994. The dilemmas of prostate cancer. *Scientific American*.

Goldzeiher, J. W. 1993. The history of steroidal contraceptive development: The estrogens. *Perspectives in Biology and Medicine* 36:363.

Grobstein, C. March 1979. External human fertilization. *Scientific American*.

Grumbach, M. M. 1979. The neuroendocrinology of puberty. *Hospital Practice* 14:65.

Gustafson, M. L., and P. K. Donahoe. 1994. Male sex determination: Current concepts of male sexual differentiation. *Annual Review of Medicine* 45:505.

Hagg, C. M. et al. 1994. Molecular basis of mammalian sexual determination: Activation of Müllerian inhibiting substance gene expression by SRY. *Science* 266:1494.

Hoberman, J. M., and C. E. Yesalis. February 1995. The history of synthetic testosterone. *Scientific American*.

Jackson, L. G. 1985. First-trimester diagnosis of fetal genetic disorders. *Hospital Practice* 20:39.

Keys, P. L., and M. C. Wiltbank. 1988. Endocrine regulation of the corpus luteum. *Annual Review of Physiology* 50:465.

Lagerkrantz, H., and T. A. Slotkin. April 1986. The "stress" of being born. *Scientific American*.

Leong, D. A., L. S. Frawley, and J. D. Neill. 1983. Neuroendocrine control of prolactin secretion. *Annual Review of Physiology* 45:109.

Lipsett, M. B. 1980. Physiology and pathology of the Leydig cell. *New England Journal of Medicine* 303:682.

Mader, S. S. 1992. *Human Reproductive Biology.* Dubuque: Wm. C. Brown.

Masters, W. H. 1986. Sex and aging—expectations and reality. *Hospital Practice* 21:175.

Means, A. R. et al. 1980. Regulation of the testis Sertoli cell by follicle-stimulating hormone. *Annual Review of Physiology* 42:59.

Nabulsi, A. A. et al. 1993. Association of hormone replacement therapy with various cardiovascular risk factors in postmenopausal women. *New England Journal of Medicine* 328:1069.

Naftolin, F. March 1981. Understanding the basis of sex differences. *Science* 211 (No. 4488):1263.

Neumann, P. J. et al. 1994. The cost of a successful delivery with in vitro fertilization. *New England Journal of Medicine* 331:239.

Odell, W. D., and D. L. Moyer. 1971. *Physiology of Reproduction.* St. Louis: Mosby.

Ojeda, S. R. 1991. The mystery of mammalian puberty: How much more do we know? *Perspectives in Biology and Medicine* 34:365.

Perone, N. 1993. The history of steroidal contraceptive development: the progestins. *Perspectives in Biology and Medicine* 36:347.

Reiter, E. O., and M. M. Grumbach. 1982. Neuroendocrine control mechanisms and the onset of puberty. *Annual Review of Physiology* 44:595.

Richards, J. S., and L. Hedin. 1988. Molecular aspects of hormone action in ovarian follicular development, ovulation, and luteinization. *Annual Review of Physiology* 50:441.

Segal, S. J. September 1974. The physiology of human reproduction. *Scientific American*.

Seibel, M. M. 1988. A new era in reproductive technology. *New England Journal of Medicine* 318:828.

Simerly, C. et al. 1995. The paternal inheritance of the centrosome, the cell's microtubule organizing center, in humans, and the implications for infertility. *Nature and Medicine* 1:47.

Simpson, E. R., and P. C. MacDonald. 1981. Endocrine physiology of the placenta. *Annual Review of Physiology* 43:163.

Tamarkin, L., C. J. Baird, and O. F. X. Almeida. 1985. Melatonin: A coordinating signal for mammalian reproduction? *Science* 227:714.

Ulmann, A., G. Teutsch, and D. Philibert. June 1990. RU 486. *Scientific American*.

Warkentin, D. L. April 1984. From A to hCG in pregnancy testing. *Diagnostic Medicine*, p. 35.

Wilson, J. D. 1978. Sexual differentiation. *Annual Review of Physiology* 40:279.

Wilson, J. D., F. W. George, and J. E. Griffin. March 1981. The hormonal control of sexual development. *Science* 211 (No. 4488):1278.

Winston, R. M. L., and A. H. Handyside. 1993. New challenges in human in vitro fertilization. *Science* 260:932.

Woodruff, T. K. and J. P. Mather. 1995. Inhibin, activin, and the female reproductive axis. Annual Review of Physiology 57:219.

Yallampalli, C. et al. 1993. Nitric oxide inhibits uterine contractility during pregnancy but not during delivery. *Endocrinology* 133:1899.

Yen, S. S. C. 1977. Regulation of the hypothalamic-pituitary-ovarian axis in women. *Journal of Reproduction and Fertility* 51:181.

Yen, S. S. C. 1979. Neuroendocrine regulation of the menstrual cycle. *Hospital Practice* 14:83.

Life Science Animations

▣ The animations that relate to chapter 20 are #19 Spermatogenesis, #20 Oogenesis, and #21 Human Embryonic Development.

Appendix A

Exercise Physiology: Summary and Text References

Metabolism

1. Anaerobic respiration produces lactic acid in exercising skeletal muscles when the ratio of oxygen supply to oxygen need falls below a critical level. *Chapter 5, p. 104.*
2. Lactic acid produced by skeletal muscles is converted to glucose in the liver and recycled to the muscles after exercise; this two-way traffic between skeletal muscles and the liver is called the Cori cycle. *Chapter 5, p. 105.*
3. Glycogenolysis in exercising muscles provides glucose for cell respiration. *Chapter 5, p. 111.*
4. Resting skeletal muscles use fatty acids as their major energy source. *Chapter 5, pp. 114 and 117.*
5. Exercise increases total caloric expenditures, increasing the metabolic rate. *Chapter 19, p. 578.*
6. Acting through secretion of glucocorticoids, the stress of physical exercise may promote the catabolism of muscle proteins and the conversion of amino acids to glucose. *Chapter 19, p. 594.*
7. The maximal oxygen uptake can be increased by endurance training. *Chapter 12, p. 333.*
8. Phosphocreatine provides high-energy phosphate for rapid production of ATP. *Chapter 12, p. 331.*
9. Aerobic respiration usually predominates until the exercise level requires more than 60% of the maximal oxygen uptake. *Chapter 12, p. 333.*

Skeletal Muscles

1. Muscle fibers that are fast-twitch have different adaptations than those that are slow-twitch. *Chapter 12, p. 332.*
2. During exercise, smaller motor units with slow-twitch fibers are used at lower levels of effort than larger motor units containing fast-twitch fibers. *Chapter 12, p. 332.*
3. Muscle fatigue during exercise may be produced by a fall in cytoplasmic pH due to lactic acid accumulation; endurance training reduces the amount of lactic acid produced at a given level of exercise. *Chapter 12, p. 333.*
4. As muscles adapt to endurance training, their triglyceride content increases, and they rely more on fat metabolism for energy. *Chapter 12, p. 334.*
5. Skeletal muscles show an increase in their number of mitochondria and aerobic respiratory enzymes as a result of endurance training. *Chapter 12, p. 333.*
6. Exercise against a high resistance, as in weight training, causes muscle hypertrophy as a result of the growth of individual muscle fibers. *Chapter 12, p. 334.*
7. Prolonged exercise of low-to-moderate intensity causes skeletal muscles to oxidize fatty acids as their primary source of energy, helping to reduce the amount of stored fat in the body. *Chapter 19, p. 584.*

Fluid and Electrolyte Balance

1. Dehydration increases plasma osmolality, causing thirst and the secretion of antidiuretic hormone (ADH). *Chapter 6, p. 130; chapter 14, p. 395.*
2. Contraction of skeletal muscles massages veins, thereby decreasing the blood contained in the venous system as the return of blood to the heart is improved. *Chapter 13, p. 371; chapter 14, p. 404.*
3. Fluid in skeletal muscle capillaries is filtered out, due to the blood pressure, and returned to the vascular system, due to the plasma oncotic pressure. *Chapter 14, p. 393.*
4. The loss of Na^+ in sweat during exercise may interfere with the ability to restore blood volume simply by drinking water until the sense of thirst is satisfied. *Chapter 14, p. 395.*

5. Only about a week's worth of endurance training is needed to achieve a significant increase in the blood volume. *Chapter 14, p. 404.*

Cardiovascular System

1. Venous return to the heart increases during exercise due to action of the skeletal muscle pumps and to movements of the diaphragm. *Chapter 13, p. 371; chapter 14, p. 404.*
2. Exercise increases the ratio of HDL to LDL cholesterol; this is believed to protect against atherosclerosis. *Chapter 13, p. 373.*
3. During exercise, the cardiac rate first increases as a result of less parasympathetic nerve inhibition, and then because of sympathetic nerve stimulation of the SA node. *Chapter 14, p. 388.*
4. Sympathetic nerves produce vasoconstriction in the viscera and skin through alpha-adrenergic stimulation and vasodilation in skeletal muscles through cholinergic and beta-adrenergic stimulation. *Chapter 14, p. 403.*
5. The coronary blood flow increases during exercise due to intrinsic metabolic vasodilation. *Chapter 14, p. 402.*
6. Blood flow through exercising skeletal muscles may increase to as much as 85% of the total blood flow in the body; this is largely due to intrinsic metabolic vasodilation. *Chapter 14, p. 404.*
7. The cardiac output may increase fivefold or more during strenuous activity; this is due to increases in the cardiac rate and stroke volume. *Chapter 14, p. 404.*

8. Endurance training is often associated with a low resting cardiac rate in conjunction with an increased blood volume and stroke volume. *Chapter 14, p. 404.*
9. People who lift weights should avoid the Valsalva maneuver because of its effect on the cardiac output and blood pressure. *Chapter 14, p. 410.*
10. Isometric contraction of skeletal muscles reduces the blood flow, causing fatigue to occur more quickly than when rhythmic isotonic contractions are performed. *Chapter 14, p. 403.*
11. During exercise, there is an increase in arterial blood pressure and a marked increase in the pulse pressure. *Chapter 14, p. 415.*

Respiratory System

1. There is an increase in total minute volume during exercise; this is generally hyperpnea rather than hyperventilation. *Chapter 16, p. 495.*
2. The increased tidal volume during exercise brings a lower proportion of air from the anatomical dead space into the alveoli, and a higher proportion of fresh air. *Chapter 16, p. 472.*
3. Divers should not ascend too rapidly because of the dangers associated with decompression sickness. *Chapter 16, p. 480.*
4. During exercise, the P_{O_2} of tissue capillaries and venous blood is decreased, so that more oxygen is unloaded. *Chapter 16, p. 488.*
5. The lower pH and increased temperature of exercising muscles causes a decrease in the affinity of hemoglobin for oxygen; more oxygen is therefore unloaded to the tissues. *Chapter 16, p. 489.*

6. Neurogenic and humoral theories have been proposed to explain the hyperpnea of exercise. *Chapter 16, p. 495.*
7. After exercise has ceased, there is still an increased total minute volume over resting levels; the extra oxygen required to oxidize lactic acid and supply the warmed muscles is called the oxygen debt. *Chapter 12, p. 331.*
8. Endurance-trained athletes, particularly those who are world-class, have extremely high maximal oxygen uptakes. *Chapter 12, p. 331.*
9. The anaerobic threshold—the maximum rate of oxygen consumption that can be attained before blood lactic acid levels rise—is increased by endurance training. *Chapter 16, p. 496.*

Endocrine and Reproductive Systems

1. Anabolic steroids, taken by some athletes, cause inhibition of the pituitary-testicular axis and consequent atrophy of the testes, abnormal growth of the breasts, and other ill effects. *Chapter 11, p. 276.*
2. Girls who are physically active and who have a low percentage of body fat reach menarche at an age later than average. *Chapter 20, p. 617.*
3. Intense physical activity and low body fat in women may produce a secondary amenorrhea. *Chapter 20, p. 617.*
4. Athletes who take growth hormone may experience changes similar to those in acromegaly. *Chapter 19, p. 597.*

Appendix B

Solutions to Clinical Investigations

Chapter 2

Since our enzymes can recognize only L-amino acids and D-sugars, the opposite stereoisomers that the student was eating could not be used by his body. He was weak because he was literally starving. The ketonuria also may have contributed to his malaise. Since he was starving, his stored fat was being rapidly hydrolyzed into glycerol and fatty acids for use as energy sources. The excessive release of fatty acids from his adipose tissue resulted in the excessive production of ketone bodies by his liver; hence, his ketonuria.

Chapter 3

The substance abuse could have resulted in the development of an extensive smooth endoplasmic reticulum, which contains many of the enzymes required to metabolize drugs. Liver disease could have been caused by the substance abuse, but there is an alternative explanation. The low amount of the enzyme that breaks down glycogen signals the presence of glycogen storage disease, a genetic condition in which a key lysosomal enzyme is lacking. This enzymatic evidence is supported by the observations of large amounts of glycogen granules and the lack of partially digested glycogen granules within secondary lysosomes. (In reality, such a genetic condition would more likely be diagnosed in early childhood.)

Chapter 4

The high blood concentrations of the MB isoenzyme form of creatine phosphokinase (CPK) following severe chest pains suggest that this man experienced a myocardial infarction ("heart attack"—see chapter 13). His difficulty in urination, together with his high blood levels of acid phosphatase, suggest prostate disease. (The relationship between the prostate gland and the urinary system is described in chapter 20.) Further tests—including one for *prostate-specific antigen (PSA)*—can be performed to confirm this diagnosis.

Chapter 5

The student's great fatigue following workouts is partially related to the depletion of her glycogen reserves and extensive utilization of anaerobic respiration (with consequent production of lactic acid) for energy. Production of large amounts of lactic acid during exercise causes her need for extra oxygen to metabolize the lactic acid following exercise (the oxygen debt)—hence, her gasping and panting. Eating more carbohydrates would help this student to maintain the glycogen stores in her liver and muscles, and training more gradually could increase the ability of her muscles to obtain more of their energy through aerobic respiration, so that she would experience less pain and fatigue.

The pain in her arms and shoulders is probably the result of lactic acid production by the exercised skeletal muscles. However, the intense pain in her left pectoral region could be angina pectoris, caused by anaerobic respiration of the heart. If this is the case, it would indicate that the heart became ischemic due to a blood flow that was inadequate for the demands placed upon it. Blood tests for particular enzymes released by damaged heart muscle (chapter 4) and an electrocardiogram (ECG) should be performed.

Chapter 6

The student's hyperglycemia caused his renal carrier proteins to become saturated, resulting in glycosuria (glucose in the urine). The elimination of glucose in the urine and its consequent osmotic effects caused the urinary excretion of an excessive amount of water, resulting in dehydration. This raised the plasma osmolality, stimulating the thirst center in the hypothalamus. (Hyperglycemia and excessive thirst and urination are cardinal signs of diabetes mellitus.) Further, the loss of plasma water (increased plasma osmolality) caused the concentration of plasma solutes, including K^+, to be raised. The resulting hyperkalemia affected the membrane potential of myocardial cells in the heart, producing electrical abnormalities that were revealed in the student's electrocardiogram.

Chapter 7

The muscular paralysis and difficulty in breathing (due to paralysis of the diaphragm) could have been caused by saxitoxin poisoning from the shellfish, if they had been gathered during a red tide. A positive chemical analysis of the student's blood and of the shellfish for saxitoxin would confirm this diagnosis. The monoamine oxidase (MAO) inhibitor was probably prescribed to treat her depression. It turns out, however, that there are significant drug-food interactions with MAO inhibitors—in fact, shellfish is specifically contraindicated! Other drugs are now available to treat depression that have fewer side effects.

Chapter 8

The patient evidently suffered a cerebrovascular accident (CVA), otherwise known as a "stroke." The obstruction of blood flow in a cerebral artery damaged part of the precentral gyrus (motor cortex) in the left hemisphere. Since most corticospinal tracts decussate in the pyramids, this caused paralysis on the right side of his body. The corticospinal tracts themselves, however, were not damaged, so the patient did not have a Babinski sign. Likewise, spinal nerves were undamaged, so his knee-jerk reflex was intact. The damage to the left cerebral hemisphere apparently included damage to Broca's area, producing a characteristic aphasia that accompanied the paralysis of the right side of his body.

Chapter 9

Because of her final examinations, the student had been under prolonged stress, which overstimulated her sympathoadrenal system. The increased sympathoadrenal activity could account for her rapid pulse (due to increased heart rate) and her hypertension (due to increased heart rate and vasoconstriction). Her headache was probably due to the fact that her pupils were dilated, thus admitting excessive amounts of light. Since she had been preparing drugs for a laboratory exercise on autonomic control, she may have been exposed to atropine, which would have caused dilation of her pupils. This possibility is strengthened by the fact that she felt her mouth to be excessively dry.

Chapter 10

The woman apparently has myopia (with or without astigmatism), because her glasses allow her to see distant objects that would otherwise be blurry. In the past year or so, she also seems to have developed presbyopia, which would be expected at her age. Her inability to focus on small print held close to her eyes while wearing her glasses indicates a loss of ability to accommodate. Bifocals contain two different lenses—one to help focus on distant objects, and one for close vision—and would help compensate for this woman's visual impairments. The fact that her intraocular pressure is normal indicates that she does not have glaucoma.

Chapter 11

The hyperglycemia cannot be attributed to diabetes mellitus because insulin activity is normal. The symptoms might be due to hyperthyroidism or to excessive catecholamine action (as in pheochromocytoma), but these possibilities are ruled out by the blood tests. The high blood levels of corticosteroids are not the result of ingestion of these compounds as drugs, and so might be due to their hypersecretion from the adrenal cortex. The patient might thus have Cushing's syndrome, in which case an adrenal tumor could be responsible for the hypersecretion of corticosteroids and, as a result of negative feedback inhibition, a decrease in blood ACTH levels. Excessive corticosteroid levels cause the mobilization of glucose from the liver, thus increasing the blood glucose to hyperglycemic levels.

Chapter 12

Since this woman has a high maximal oxygen uptake, she should have good endurance with little fatigue and pain during exercise. The fact that her muscles are not large but have good tone supports her statement that she frequently engages in endurance-type exercise. The normal concentration of creatine phosphokinase suggests that her skeletal muscles and heart may not be damaged, but further tests should be done to confirm this, particularly since she has a history of hypertension. The fatigue and muscle pain might simply be due to excessive workouts, but

the high blood Ca^{++} concentration suggests another possibility. The high blood Ca^{++} could be responsible for her excessively high muscle tone; this inability of her muscles to relax might, in fact, be responsible for the pain and fatigue. This person, therefore, should undergo an endocrinological workup (for parathyroid hormone, for example) to determine the cause of her high blood Ca^{++} levels.

Chapter 13

The heart murmur is due to the ventricular septal defect and mitral stenosis, which were probably congenital. These conditions could reduce the amount of blood pumped by the left ventricle through the systemic arteries, and thus weaken the pulse. The reduced blood flow and consequent reduced oxygen delivery to the tissues could be the cause of the girl's chronic fatigue. The lowered volume of blood pumped by the left ventricle could cause a reflex increase in the heart rate, as detected by her rapid pulse and the ECG tracing showing sinus tachycardia. The girl's high blood cholesterol and LDL/HDL ratio is probably unrelated to her symptoms. This condition could be dangerous, however, as it increases her risk of atherosclerosis. This girl should therefore be placed on a special diet, and perhaps medication, to lower her blood cholesterol.

Chapter 14

The man was suffering from dehydration, which lowered his blood volume and thus lowered his blood pressure. This stimulated the baroreceptor reflex, resulting in intense activation of sympathetic nerves. Sympathetic nerve activation caused vasoconstriction in cutaneous vessels—hence the cold skin—and an increase in cardiac rate (hence the high pulse rate). The intravenous albumin solution was given in the hospital in order to increase his blood volume and pressure. His urine output was low as a result of (1) sympathetic nerve-induced vasoconstriction of arterioles in the kidneys, which decreased blood flow to the kidneys; (2) water reabsorption in response to high ADH secretion, which resulted from stimulation of osmoreceptors in the hypothalamus; and (3) water and salt retention in response to aldosterone secretion, which was stimulated by activation of the renin-angiotensin system. The absence of sodium in his urine resulted from the high aldosterone secretion.

Chapter 15

While crawling through the underbrush, the little girl may have been exposed to poison oak, causing a contact dermatitis. Since this is a delayed hypersensitivity response mediated by T cells, antihistamines would not have alleviated the symptoms. Cortisone helped, however, due to its immunosuppressive effect. The first bee sting did not have much of an effect, but served to sensitize the girl (through the development of B cell clones) to the second bee sting. The second sting resulted in an immediate hypersensitivity response (mediated by IgE), which caused the release of histamine. This allergic reaction could thus be treated effectively with antihistamines.

Chapter 16

The puncture wound must have admitted air into the pleural cavity, raising the intrapleural pressure and causing a pneumothorax of the right lung. Since the left lung is located in a separate pleural compartment, it was unaffected by the wound. As a result of the collapse of his right lung, the patient was hypoventilating. This caused retention of CO_2, thus raising his arterial P_{CO_2} and resulting in respiratory acidosis (as indicated by an arterial pH lower than 7.35). Upon recovery, analysis of his arterial blood revealed that he was breathing adequately but that he had a carboxyhemoglobin saturation of 20%. This very high level is probably due to a combination of smoking and driving in heavily congested areas, with much automobile exhaust. The high carboxyhemoglobin would reduce oxygen transport, thus aggravating any problems he might have with his cardiovascular or pulmonary system.

The significantly low $FEV_{1.0}$ indicates that he has an obstructive pulmonary problem, possibly caused by smoking and the inhalation of polluted air. The low $FEV_{1.0}$ could simply indicate bronchoconstriction, but the fact that his vital capacity was a little low suggests that he may have early stage lung damage, possibly emphysema. He should be strongly advised to quit smoking, and further pulmonary tests should be administered at regular intervals.

Chapter 17

The location of the pain and the discoloration of the urine are indicative of a renal disorder. The hematuria (blood in the urine) was responsible for the discolored urine, and the presence of casts with associated red blood cells indicated glomerulonephritis. The elevated blood creatinine concentration indicated a reduction in the glomerular filtration rate (GFR) as a result of the glomerulonephritis, and this reduced GFR could have been responsible for the fluid retention and observed edema. The presence of only trace amounts of protein in the urine, however, was encouraging, and could be explained by the boy's running activity (proteinuria in this case would have been an ominous sign). The streptococcus infection, acting via an autoimmune reaction, was probably responsible for the glomerulonephritis. This was confirmed by the fact that the symptoms of glomerulonephritis disappeared after treatment with an antibiotic. Hydrochlorothiazide is a diuretic that helped to alleviate the edema by (1) promoting the excretion of larger amounts of urine and (2) shifting of edematous fluid from the interstitial to the vascular compartment.

Chapter 18

The student does not appear to have a peptic ulcer, because the food and drinks that stimulate gastric acid secretion do not cause pain. The lack of fever and the normal white blood cell count suggest that the inflammation associated with appendicitis is absent. The yellowing of the sclera indicates jaundice, and this symptom—together with the prolonged clotting time—could be caused by liver disease. However, liver disease would elevate the blood levels of free bilirubin, which were found to be normal. The normal levels of urea and ammonia in the blood likewise suggest normal liver function. Similarly, the normal pancreatic amylase levels suggest that the pancreas is not affected.

This student's symptoms are most likely due to the presence of gallstones. Gallstones could obstruct the normal flow of bile, and thus prevent normal fat digestion. This would explain the fatty stools. The resulting loss of dietary fat could cause a deficiency in vitamin K, which is a fat-soluble vitamin required for the production of a number of clotting factors (chapter 13)—hence, the prolonged clotting time. The pain would be provoked by oily or fatty foods (peanut butter and bacon), which trigger a reflex contraction of the gallbladder once the fat arrives in the duodenum. Contraction of the gallbladder against an obstructed cystic duct or common bile duct often produces a severe referred pain below the right scapula.

Chapter 19

The woman's frequent urinations (polyuria) probably are causing her thirst and other symptoms. The weight loss associated with these symptoms and the fact that her mother and uncle were diabetics suggested that this woman might have diabetes mellitus. Indeed, polyuria, polyphagia (frequent eating), and polydipsia (frequent drinking)—the "three P's"—are cardinal symptoms of diabetes mellitus. The fasting hyperglycemia (blood glucose concentration of 150 mg/dl) confirmed the diagnosis of diabetes mellitus. This abnormally high fasting blood glucose is too low to result in glycosuria. She could have glycosuria after meals, however, which would be responsible for her polyuria. The oral glucose tolerance test further confirmed the diagnosis of diabetes mellitus, and the observations that this condition appeared to have begun in middle age and that it was not accompanied by ketosis and ketonuria suggested that it was non-insulin-dependent diabetes mellitus. This being the case, she could increase her tissue sensitivity to insulin by diet and exercise. If this failed, she could probably control her symptoms with sulfonylureas, which increase insulin secretion and also increase the tissue sensitivity to the effects of insulin.

Chapter 20

The young woman's lack of menstruation was not accompanied by pain, and she did not have a history of spotting or excessive menstrual bleeding. The fact that she had menstruated prior to her amenorrhea ruled out the possibility of primary amenorrhea. Her secondary amenorrhea could have been the result of pregnancy, but this was ruled out by the negative pregnancy test. The amenorrhea could have been caused by her hypothyroidism, but she stated that she took her thyroid pills regularly, and her blood test demonstrated normal thyroxine levels. Blood levels of estradiol and the gonadotropins were normal, suggesting that the amenorrhea was not secondary to a pituitary or ovarian neoplasm or other problem.

This student most likely has a secondary amenorrhea that is due to emotional stress, low body weight, and/or her strenuous exercise program. She should take steps to alleviate these conditions if she wants to resume her normal menstrual periods. If she refuses to gain weight and reduce her level of physical activity, her physician might recommend the use of oral contraceptives to help regulate her cycles.

Appendix C

Answers to Objective Questions

Chapter 1
1.	d	5.	d	9.	a
2.	d	6.	c	10.	c
3.	b	7.	b	11.	c
4.	b	8.	b		

Chapter 2
1.	c	5.	c	9.	d
2.	b	6.	b	10.	b
3.	a	7.	c	11.	d
4.	d	8.	d	12.	b

Chapter 3
1.	d	6.	b	11.	d
2.	b	7.	a	12.	e
3.	a	8.	c	13.	b
4.	c	9.	a	14.	a
5.	d	10.	b		

Chapter 4
1.	b	5.	d	8.	d
2.	d	6.	e	9.	d
3.	d	7.	e	10.	d
4.	a				

Chapter 5
1.	b	6.	c	10.	d
2.	a	7.	a	11.	b
3.	c	8.	c	12.	d
4.	e	9.	a	13.	b
5.	d				

Chapter 6
1.	c	5.	b	9.	b
2.	b	6.	d	10.	d
3.	a	7.	a	11.	b
4.	c	8.	a	12.	b

Chapter 7
1.	c	7.	d	13.	d
2.	d	8.	a	14.	b
3.	a	9.	c	15.	a
4.	a	10.	c	16.	c
5.	c	11.	b	17.	a
6.	d	12.	d	18.	e

Chapter 8
1.	d	6.	e	11.	a
2.	b	7.	c	12.	b
3.	e	8.	d	13.	d
4.	a	9.	b	14.	a
5.	b	10.	c		

Chapter 9
1.	c	5.	c	9.	c
2.	c	6.	b	10.	c
3.	c	7.	b	11.	b
4.	a	8.	e	12.	c

Chapter 10
1.	d	7.	c	13.	b
2.	a	8.	c	14.	c
3.	c	9.	d	15.	b
4.	d	10.	b	16.	c
5.	c	11.	d	17.	c
6.	a	12.	b	18.	b

Chapter 11
1.	d	6.	a	11.	d
2.	d	7.	b	12.	c
3.	e	8.	e	13.	b
4.	e	9.	d	14.	d
5.	d	10.	a	15.	c
				16.	b

Chapter 12
1.	b	6.	b	11.	d
2.	d	7.	a	12.	a
3.	c	8.	c	13.	e
4.	b	9.	b	14.	c
5.	e	10.	b	15.	b

Chapter 13
1.	c	7.	c	13.	d
2.	b	8.	a	14.	c
3.	e	9.	d	15.	c
4.	a	10.	b	16.	b
5.	b	11.	c	17.	d
6.	c	12.	d	18.	c

Chapter 14

1.	a	8.	c	15.	c
2.	d	9.	d	16.	e
3.	c	10.	b	17.	b
4.	e	11.	c	18.	d
5.	b	12.	d	19.	d
6.	c	13.	b	20.	c
7.	a	14.	d		

Chapter 15

1.	c	7.	d	13.	d
2.	b	8.	b	14.	c
3.	d	9.	e	15.	b
4.	a	10.	a	16.	c
5.	c	11.	d	17.	a
6.	d	12.	a	18.	d

Chapter 16

1.	c	7.	b	13.	a
2.	d	8.	a	14.	d
3.	c	9.	e	15.	b
4.	a	10.	c	16.	a
5.	c	11.	a	17.	c
6.	c	12.	c		

Chapter 17

1.	b	7.	d	12.	c
2.	a	8.	c	13.	a
3.	c	9.	d	14.	d
4.	b	10.	a	15.	b
5.	e	11.	e	16.	e
6.	d				

Chapter 18

1.	e	6.	b	11.	e
2.	d	7.	d	12.	d
3.	b	8.	a	13.	b
4.	c	9.	d	14.	d
5.	d	10.	e	15.	d

Chapter 19

1.	c	6.	d	11.	b
2.	b	7.	a	12.	c
3.	c	8.	d	13.	a
4.	a	9.	c	14.	d
5.	b	10.	b	15.	d

Chapter 20

1.	b	6.	c	11.	b
2.	d	7.	d	12.	c
3.	a	8.	a	13.	d
4.	c	9.	d	14.	b
5.	a	10.	a	15.	a

Glossary

Keys to Pronunciation

Most of the words in this glossary are followed by a phonetic spelling that serves as a guide to pronunciation. The phonetic spellings reflect standard scientific usage and can easily be interpreted following a few basic rules.

1. Any unmarked vowel that ends a syllable or that stands alone as a syllable has the long sound. For example, *ba, ma,* and *na* rhyme with *fay; be, de,* and *we* rhyme with *fee; bi, di,* and *pi* rhyme with *sigh; bo, do,* and *mo* rhyme with *go.* Any unmarked vowel that is followed by a consonant has the short sound (for example, the vowel sounds in *hat, met, pit, not,* and *but*).

2. If a long vowel appears in the middle of a syllable (followed by a consonant), it is marked with a macron (‾). Similarly, if a vowel stands alone or ends a syllable but should have short sound, it is marked with a breve (˘).

3. Syllables that are emphasized are indicated by stress marks. A single stress mark (′) indicates the primary emphasis; a secondary emphasis is indicated by a double stress mark (″).

A

a-, an- (Gk.) Not, without, lacking.

ab- (L.) Off, away from.

abdomen (*ab′dŏ-men, ab-do′men*) The portion of the trunk between the diaphragm and pelvis.

abductor (*ab-duk′tor*) A muscle that moves the skeleton away from the midline of the body or away from the axial line of a limb.

ABO system The most common system of classification for red blood cell antigens. On the basis of antigens on the red blood cell surface, individuals can be type A, type B, type AB, or type O.

absorption (*ab-sorp′shun*) The transport of molecules across epithelial membranes into the body fluids.

accommodation (*ă-kom″ŏ-da′shun*) Adjustment; specifically, the process whereby the focal length of the eye is changed by automatic adjustment of the curvature of the lens to bring images of objects from various distances into focus on the retina.

acetyl (*as′ĕ-tl, ă-set′l*) **CoA** Acetyl coenzyme A. An intermediate molecule in aerobic cell respiration that, together with oxaloacetic acid, begins the Krebs cycle. Acetyl CoA is also an intermediate in the synthesis of fatty acids.

acetylcholine (*ă-set″l-ko′lēn*) **(ACh)** An acetic acid ester of choline—a substance that functions as a neurotransmitter chemical in somatic motor nerve and parasympathetic nerve fibers.

acetylcholinesterase (*ă-set″l-ko″lĭ-nes′tĕ-rās*) An enzyme in the membrane of postsynaptic cells that catalyzes the conversion of ACh into choline and acetic acid. This enzymatic reaction inactivates the neurotransmitter.

acidosis (*as″ĭ-do′sis*) An abnormal increase in the H⁺ concentration of the blood that lowers arterial pH below 7.35.

acromegaly (*ak″ro-meg′ă-le*) A condition caused by the hypersecretion of growth hormone from the pituitary after maturity and characterized by enlargement of the extremities, such as the nose, jaws, fingers, and toes.

ACTH Adrenocorticotropic (*ă-dre″no-kor″tĭ-ko-trop′ik*) hormone. A hormone secreted by the anterior pituitary that stimulates the adrenal cortex.

actin (*ak′tin*) A structural protein of muscle that, along with myosin, is responsible for muscle contraction.

action potential An all-or-none electrical event in an axon or muscle fiber in which the polarity of the membrane potential is rapidly reversed and reestablished.

active immunity Immunity involving sensitization, in which antibody production is stimulated by prior exposure to an antigen.

active transport The movement of molecules or ions across the cell membranes of epithelial cells by membrane carriers. An expenditure of cellular energy (ATP) is required.

ad- (L.) Toward, next to.

adductor (*ă-duk′tor*) A muscle that moves the skeleton toward the midline of the body or toward the axial plane of a limb.

adenohypophysis (*ad″n-o-hi-pof′ĭ-sis*) The anterior, glandular lobe of the pituitary gland that secretes FSH (follicle-stimulating hormone), LH (luteinizing hormone), ACTH (adrenocorticotropic hormone), TSH (thyroid-stimulating hormone), GH (growth hormone), and prolactin. Secretions of the anterior pituitary are controlled by hormones secreted by the hypothalamus.

adenylate cyclase (*ă-den′l-it si′klāse*) An enzyme found in cell membranes that catalyzes the conversion of ATP to cyclic AMP and pyrophosphate (PP_1). This enzyme is activated by an interaction between a specific hormone and its membrane receptor protein.

ADH Antidiuretic (*an″te-di″yŭ-ret′ik*) hormone, also known as *vasopressin*. A hormone produced by the hypothalamus and secreted by the posterior pituitary. It acts on the kidneys to promote water reabsorption, thus decreasing the urine volume.

adipose (*ad′ĭ-pōs*) **tissue** Fatty tissue. A type of connective tissue consisting of fat cells in a loose connective tissue matrix.

ADP Adenosine diphosphate (ă-den'ŏ-sēn di-fos'făt). A molecule that, together with inorganic phosphate, is used to make ATP (adenosine triphosphate).

adrenal cortex (ă-dre'nal kor'teks) The outer part of the adrenal gland. Derived from embryonic mesoderm, the adrenal cortex secretes corticosteroid hormones (such as aldosterone and hydrocortisone).

adrenal medulla (mĕ-dul'ă) The inner part of the adrenal gland. Derived from embryonic postganglionic sympathetic neurons, the adrenal medulla secretes catecholamine hormones—epinephrine and (to a lesser degree) norepinephrine.

adrenergic (ad"rĕ-ner'jik) Denoting the actions of epinephrine, norepinephrine, or other molecules with similar activity (as in *adrenergic receptor* and *adrenergic stimulation*).

aerobic (ă-ro'bik) **capacity** The ability of an organ to utilize oxygen and respire aerobically to meet its energy needs.

afferent (af'er-ent) Conveying or transmitting inward, toward a center. Afferent neurons, for example, conduct impulses toward the central nervous system; afferent arterioles carry blood toward the glomerulus.

agglutinate (ă-gloot'n-āt) A clumping of cells (usually erythrocytes) formed as a result of specific chemical interaction between surface antigens and antibodies.

agranular leukocytes (ă-gran'yŭ-lar loo'kŏ-sītz) White blood cells (leukocytes that do not contain cytoplasmic granules; specifically, lymphocytes and monocytes.

albumin (al-byoo'min) A water-soluble protein produced in the liver that is the major component of the plasma proteins.

aldosterone (al-dos'ter-ōn) The principal corticosteroid hormone involved in the regulation of electrolyte balance (mineralocorticoid).

alkalosis (al"kă-lo'sis) An abnormally high alkalinity of the blood and body fluids (blood pH > 7.45).

allergen (al'er-jen) An antigen that evokes an allergic response rather than a normal immune response.

allergy (al'er-je) A state of hypersensitivity caused by exposure to allergens. It results in the liberation of histamine and other molecules with histamine-like effects.

all-or-none law The statement that a given response will be produced to its maximum extent in response to any stimulus equal to or greater than a threshold value. Action potentials obey an all-or-none law.

allosteric (al"o-ster'ik) Denoting the alteration of an enzyme's activity by its combination with a regulator molecule. Allosteric inhibition by an end product represents negative feedback control of an enzyme's activity.

alpha motoneuron (al'fă mo"tŏ-noor'on) The type of somatic motor neuron that stimulates extrafusal skeletal muscle fibers.

alveoli (al-ve'ŏ-li) sing. alveolus Small, saclike dilations (as in *lung alveoli*).

amniocentesis (am"ne-o-sen-te'sis) A procedure to obtain amniotic fluid and fetal cells in this fluid through transabdominal perforation of the uterus.

amnion (am'ne-on) A developmental membrane surrounding the fetus that contains amniotic fluid; commonly called the "bag of waters."

amphoteric (am-fo-ter'ik) Having both acidic and basic characteristics; used to denote a molecule that can be positively or negatively charged, depending on the pH of its environment.

an- (Gk.) Without, not.

anabolic steroids (an"ă-bol'ik ster'oidz) Steroids with androgen-like stimulatory effects on protein synthesis.

anabolism (ă-nab'ŏ-liz"em) Chemical reactions within cells that result in the production of larger molecules from smaller ones; specifically, the synthesis of protein, glycogen, and fat.

anaerobic respiration (an-ă-ro'bik res"pǐ-ra'shun) A form of cell respiration involving the conversion of glucose to lactic acid in which energy is obtained without the use of molecular oxygen.

anaerobic threshold The maximum rate of oxygen consumption that can be attained before a significant amount of lactic acid is produced by the exercising skeletal muscles through anaerobic respiration. This generally occurs when about 60% of the person's total maximal oxygen uptake has been reached.

anaphylaxis (an"ă-fǐ-lak'sis) An unusually severe allergic reaction that can result in cardiovascular shock and death.

androgen (an'dro-jen) A steroid hormone that controls the development and maintenance of masculine characteristics; primarily testosterone secreted by the testes, although weaker androgens are secreted by the adrenal cortex.

anemia (ă-ne'me-ă) An abnormal reduction in the red blood cell count, hemoglobin concentration, or hematocrit, or any combination of these measurements. This condition is associated with a decreased ability of the blood to carry oxygen.

angina pectoris (an-ji'nă pek'tŏ-ris) A thoracic pain, often referred to the left pectoral and arm area, caused by myocardial ischemia.

angiotensin II (an"je-o-ten'sin) An eight-amino-acid polypeptide formed from angiotensin I (a ten-amino-acid precursor), which in turn is formed from the cleavage of a protein (angiotensinogen) by the action of renin, an enzyme secreted by the kidneys. Angiotensin II is a powerful vasoconstrictor and a stimulator of aldosterone secretion from the adrenal cortex.

anion (an'i-on) An ion that is negatively charged, such as chloride, bicarbonate, or phosphate.

antagonistic effects Actions of regulators such as hormones or nerves that counteract the effects of other regulators. The actions of sympathetic and parasympathetic neurons on the heart, for example, are antagonistic.

anterior (an-tir'e-or) At or toward the front of an organism, organ, or part; the ventral surface.

anterior pituitary (pǐ-too'ǐ-ter-e) See adenohypophysis.

antibodies (an'tǐ-bod"ēz) Immunoglobulin proteins secreted by B lymphocytes that have transformed into plasma cells. Antibodies are responsible for humoral immunity. Their synthesis is induced by specific antigens, and they combine with these specific antigens but not with unrelated antigens.

anticoagulant (an"te-ko-ag'yŭ-lant) A substance that inhibits blood clotting.

anticodon (an"te-ko'don) A base triplet provided by three nucleotides within a loop of transfer RNA that is complementary in its base pairing properties to a triplet (the codon in mRNA). The matching of codon to anticodon provides the mechanism for translation of the genetic code into a specific sequence of amino acids.

antigen (an'tǐ-jen) A molecule able to induce the production of antibodies and to react in a specific manner with antibodies.

antigenic (an-tǐ-jen'ik) **determinant site** The region of an antigen molecule that specifically reacts with particular antibodies. A large antigen molecule may have a number of such sites.

antiserum (an'tǐ-sir"um) A serum containing antibodies that are specific for one or more antigens.

aphasia (ă-fa'zhă) Defects in speech, writing, or in the comprehension of written or spoken language caused by brain damage or disease. Broca's area, Wernicke's area, the arcuate fasciculus, or the angular gyrus may be involved.

apnea (ap'ne-ă) The temporary cessation of breathing.

apneustic (ap-noo'stik) **center** A collection of neurons in the brain stem that participates in the rhythmic control of breathing.

apoptosis (ap"ŏ-to'sis) Cellular death in which the cells show characteristic histological changes. It occurs as part of programmed cell death and other events in which cell death is a physiological response.

aqueous humor (a'kwe-us) A fluid produced by the ciliary body that fills the anterior and posterior chambers of the eye.

arteriosclerosis (ar-tir"e-o-sklĕ-ro'sis) Any of a group of diseases characterized by thickening and hardening of the artery wall and narrowing of its lumen.

arteriovenous anastomosis (ar-tir"e-o-ve'nus ă-nas"tŏ-mo'sis) Direct connection between an artery and a vein that bypasses the capillary bed.

artery (ar'tĕ-re) A vessel that carries blood away from the heart.

astigmatism (ă-stig'mă-tiz"em) Unequal curvature of the refractive surfaces of the eye (cornea and/or lens), so that light that enters the eye along certain meridians does not focus on the retina.

atherosclerosis (ath"ĕ-ro-sklĕ'ro'sis) A common type of arteriosclerosis found in medium and large arteries in which raised areas, or plaques, within the tunica intima are formed from smooth muscle cells, cholesterol, and other lipids. These plaques occlude arteries and serve as sites for the formation of thrombi.

atomic number A whole number representing the number of positively charged protons in the nucleus of an atom.

atopic dermatitis (ă-top'ik der"mă-ti'tis) An allergic skin reaction to agents such as poison ivy and poison oak; a type of delayed hypersensitivity.

ATP Adenosine triphosphate (ă-den'ŏ-sēn tri-fos'făt). The universal energy carrier of the cell.

atretic (ă-tret'ik) Without an opening. Atretic ovarian follicles are those that fail to ovulate.

atrial natriuretic (a'tre-al na"trǐ-yoo-ret'ik) **factor** A chemical secreted by the atria that acts as a natriuretic hormone (a hormone that promotes the urinary excretion of sodium).

atrioventricular node (a"tre-o-ven-trik'yŭ-lar nōd) A specialized mass of conducting tissue located in the right atrium near the junction of the interventricular septum. It transmits the impulse into the bundle of His; also called the *AV node*.

atrioventricular valves One-way valves located between the atria and ventricles. The AV valve on the right side of the heart is the tricuspid, and the AV valve on the left side is the bicuspid or mitral valve.

atrophy (at'rŏ-fe) A gradual wasting away, or decrease in mass and size of an organ; the opposite of hypertrophy.

atropine (at'rŏ-pēn) An alkaloid drug, obtained from a plant of the species *Belladonna*, that acts as an anticholinergic agent. It is used medically to inhibit parasympathetic nerve effects, dilate the pupils of the eye, increase the heart rate, and inhibit intestinal movements.

auto- (Gk.) Self, same.

autoantibody (aw"to-an'tǐ-bod"e) An antibody that is formed in response to and that reacts with molecules that are part of one's own body.

autocrine (*aw'tŏ-krin*) A type of regulation in which one part of an organ releases chemicals that help regulate another part of the same organ. Prostaglandins, for example, are autocrine regulators.

autonomic (*aw"tŏ-nom'ik*) **nervous system** The part of the nervous system that involves control of smooth muscle, cardiac muscle, and glands. The autonomic nervous system is subdivided into the sympathetic and parasympathetic divisions.

autoregulation (*aw"to-reg'yŭ-la'shun*) The ability of an organ to intrinsically modify the degree of constriction or dilation of its small arteries and arterioles and to thus regulate the rate of its own blood flow. Autoregulation may occur through myogenic or metabolic mechanisms.

autosomal chromosomes (*aw"to-so'mal kro'mŏ-sōmz*) The paired chromosomes; those other than the sex chromosomes.

axon (*ak'son*) The process of a nerve cell that conducts impulses away from the cell body.

axonal (*ak'sŏ-nal, ak-son'al*) **transport** The transport of materials through the axon of a neuron. This usually occurs from the cell body to the end of the axon, but retrograde transport in the opposite direction can also occur.

B

baroreceptors (*bar"o-re-sep'torz*) Receptors for arterial blood pressure located in the aortic arch and the carotid sinuses.

Barr body A microscopic structure in the cell nucleus produced from an inactive X chromosome in females.

basal ganglia (*ba'sal gang'gle-ă*) Gray matter, or nuclei, within the cerebral hemispheres, forming the corpus striatum, amygdaloid nucleus, and claustrum.

basal metabolic (*ba'sal met"ă-bol'ik*) **rate** (**BMR**) The rate of metabolism (expressed as oxygen consumption or heat production) under resting or basal conditions 14 to 18 hours after eating.

basophil (*ba'sŏ-fil*) The rarest type of leukocyte; a granular leukocyte with an affinity for blue stain in the standard staining procedure.

B cell lymphocytes Lymphocytes that can be transformed by antigens into plasma cells that secrete antibodies (and are thus responsible for humoral immunity). The B stands for *bursa equivalent*, which is believed to be the bone marrow.

benign (*bĭ-nīn'*) Not malignant or life threatening.

bi- (L.) Two, twice.

bile (*bīl*) Fluid produced by the liver and stored in the gallbladder that contains bile salts, bile pigments, cholesterol, and other molecules. The bile is secreted into the small intestine.

bile salts Salts of derivatives of cholesterol in bile that are polar on one end and nonpolar on the other end of the molecule. Bile salts have detergent or surfactant effects and act to emulsify fat in the lumen of the small intestine.

bilirubin (*bil"ĭ-roo'bin*) Bile pigment derived from the breakdown of the heme portion of hemoglobin.

blastocyst (*blas'tŏ-sist*) The stage of early embryonic development that consists of an embroblast, which will become the embryo, and the trophoblast, which will become the chorionic membrane. This is the form of the embryo that implants into the endometrium of the uterus beginning at about the fifth day following fertilization.

blood-brain barrier The structures and cells that selectively prevent particular molecules in the plasma from entering the central nervous system.

Bohr effect The effect of blood pH on the dissociation of oxyhemoglobin. Dissociation is promoted by a decrease in the pH.

Boyle's law The statement that the pressure of a given quantity of a gas is inversely proportional to its volume.

bradycardia (*brad"ĭ-kar'de-ă*) A slow cardiac rate; less than sixty beats per minute.

bradykinin (*brad"ĭ-ki'nin*) A short polypeptide that stimulates vasodilation and other cardiovascular changes.

bronchiole (*brong'ke-ōl*) The smallest of the air passages in the lungs, containing smooth muscle and cuboidal epithelial cells.

brown fat A type of fat most abundant at birth that provides a unique source of heat energy for infants, protecting them against hypothermia.

brush border enzymes Digestive enzymes that are located in the cell membrane of the microvilli of intestinal epithelial cells.

buffer A molecule that serves to prevent large changes in pH by either combining with H^+ or by releasing H^+ into solution.

bundle of His (*hiss*) A band of rapidly conducting cardiac fibers originating in the AV node and extending down the atrioventricular septum to the apex of the heart. This tissue conducts action potentials from the atria into the ventricles.

C

cable properties The ability of neurons to conduct an electrical current. This occurs, for example, between nodes of Ranvier, where action potentials are produced in a myelinated fiber.

calcitonin (*kal"sĭ-to'nin*) Also called *thyrocalcitonin*. A polypeptide hormone produced by the parafollicular cells of the thyroid and secreted in response to hypercalcemia. It acts to lower blood calcium and phosphate concentrations and may serve as an antagonist of parathyroid hormone.

calmodulin (*kal"mod'yŭ-lin*) A receptor protein for Ca^{++} located within the cytoplasm of target cells. It appears to mediate the effects of this ion on cellular activities.

calorie (*kal'ŏ-re*) A unit of heat equal to the amount of heat needed to raise the temperature of one gram of water by 1°C.

cAMP Cyclic adenosine monophosphate (*ă-den'ŏ-sēn mon"o-fos'fāt*). A second messenger in the action of many hormones, such as catecholamine, polypeptide, and glycoprotein hormones. It serves to mediate the effects of these hormones on their target cells.

cancer A tumor characterized by abnormally rapid cell division and the loss of specialized tissue characteristics. This term usually refers to malignant tumors.

capacitation (*kă-pas"ĭ-ta'shun*) Changes that occur within spermatozoa in the female reproductive tract that enable the sperm to fertilize ova; sperm that have not been capacitated in the female tract cannot fertilize ova.

capillary (*kap'ĭ-lar"e*) The smallest vessel in the vascular system. Capillary walls are only one cell thick, and all exchanges of molecules between the blood and tissue fluid occur across the capillary wall.

carbohydrate (*kar"bo-hi'drāt*) An organic molecule containing carbon, hydrogen, and oxygen in a ratio of 1:2:1. The carbohydrate class of molecules is subdivided into monosaccharides, disaccharides, and polysaccharides.

carbonic anhydrase (*kar-bon'ik an-hi'drās*) An enzyme that catalyzes the formation or breakdown of carbonic acid. When carbon dioxide concentrations are relatively high, this enzyme catalyzes the formation of carbonic acid from CO_2 and H_2O. When carbon dioxide concentrations are low, the breakdown of carbonic acid to CO_2 and H_2O is catalyzed. These reactions aid the transport of carbon dioxide from tissues to alveolar air.

carboxyhemoglobin (*kar-bok"se-he"mŏ-glo'bin*) An abnormal form of hemoglobin in which the heme is bonded to carbon monoxide.

cardiac (*kar'de-ak*) **muscle** Muscle of the heart, consisting of striated muscle cells. These cells are interconnected into a mass called the myocardium.

cardiac output The volume of blood pumped by either the right or the left ventricle each minute.

cardiogenic (*kar"de-o-jen'ik*) **shock** Shock that results from low cardiac output in heart disease.

carrier-mediated transport The transport of molecules or ions across a cell membrane by means of specific protein carriers. It includes both facilitated diffusion and active transport.

cast An accumulation of proteins molded from the kidney tubules that appear in urine sediment.

catabolism (*kă-tab'ŏ-liz-em*) Chemical reactions in a cell whereby larger, more complex molecules are converted into smaller molecules.

catalyst (*kat'ă-list*) A substance that increases the rate of a chemical reaction without changing the nature of the reaction or being changed by the reaction.

catecholamine (*kat"ĕ-kol'ă-mēn*) Any one of a group of molecules including epinephrine, norepinephrine, and L-dopa. The effects of catecholamines are similar to those produced by activation of the sympathetic nervous system.

cations (*kat'i-ons*) Positively charged ions, such as sodium, potassium, calcium, and magnesium.

cell-mediated immunity Immunological defense provided by T cell lymphocytes that come into close proximity with their victim cells (as opposed to humoral immunity provided by the secretion of antibodies by plasma cells).

cellular respiration (*sel'yŭ-lar res"pĭ-ra'shun*) The energy-releasing metabolic pathways in a cell that oxidize organic molecules such as glucose and fatty acids.

centri- (L.) Center.

centriole (*sen'trĭ-ōl*) The cell organelle that forms the spindle apparatus during cell division.

centromere (*sen'trŏ-mēr*) The central region of a chromosome to which the chromosomal arms are attached.

cerebellum (*ser"ĕ-bel'um*) A part of the metencephalon of the brain that serves as a major center of control in the extrapyramidal motor system.

cerebral lateralization (*ser"ĕ-bral lat"er-al-ĭ-za'shun*) The specialization of function of each cerebral hemisphere. Language ability, for example, is lateralized to the left hemisphere in most people.

chemiosmotic (*kem"e-o-os-mo'tik*) **theory** The theory that oxidative phosphorylation within mitochondria is driven by the development of a H^+ gradient across the inner mitochondrial membrane.

chemoreceptor (ke"mo-re-sep'tor) A neural receptor that is sensitive to chemical changes in blood and other body fluids.

chemotaxis (ke"mo-tak'sis) The movement of an organism or a cell, such as a leukocyte, toward a chemical stimulus.

Cheyne–Stokes (chān'stōks) **respiration** Breathing characterized by rhythmic waxing and waning of the depth of respiration, with regularly occurring periods of apnea (failure to breathe).

chloride (klor'īd) **shift** The diffusion of Cl⁻ into red blood cells as HCO_3^- diffuses out of the cells. This occurs in tissue capillaries due to the production of carbonic acid from carbon dioxide.

cholecystokinin (ko"lĭ-sis"to-ki'nin) **(CCK)** A hormone secreted by the duodenum that acts to stimulate contraction of the gallbladder and to promote the secretion of pancreatic juice.

cholesterol (kŏ-les'ter-ol) A twenty-seven-carbon steroid that serves as the precursor of steroid hormones.

cholinergic (ko"lĭ-ner'jik) Denoting nerve endings that liberate acetylcholine as a neurotransmitter, such as those of the parasympathetic system.

chondrocyte (kon'dro-sīt) A cartilage-forming cell.

chorea (kŏ-re'ă) The occurrence of a wide variety of rapid, complex, jerky movements that appear to be well coordinated but that are performed involuntarily.

chromatids (kro'mă-tidz) Duplicated chromosomes, joined together at the centromere, that separate during cell division.

chromatin (kro'mă-tin) Threadlike structures in the cell nucleus consisting primarily of DNA and protein. They represent the extended form of chromosomes during interphase.

chromosome (kro'mŏ-sōm) A structure in the cell nucleus, containing DNA and associated proteins, as well as RNA, that is made according to the genetic instructions in the DNA. Chromosomes are in a compact form during cell division and thus become visible as discrete structures in the light microscope at this time.

chylomicron (ki"lo-mi'kron) A particle of lipids and protein secreted by the intestinal epithelial cells into the lymph and transported by the lymphatic system to the blood.

chyme (kīm) A mixture of partially digested food and digestive juices that passes from the pylorus of the stomach into the duodenum.

cilia (sil'e-ă) sing. *cilium* Tiny hairlike processes extending from the cell surface that beat in a coordinated fashion.

circadian (ser"kă-de'an) **rhythms** Physiological changes that repeat at about a 24-hour period. They are often synchronized to changes in the external environment, such as the day-night cycles.

cirrhosis (sĭ-ro'sis) Liver disease characterized by the loss of normal microscopic structure, which is replaced by fibrosis and nodular regeneration.

clonal (klōn'al) **selection theory** The theory in immunology that active immunity is produced by the development of clones of lymphocytes able to respond to a particular antigen.

clone (klōn) **1.** A group of cells derived from a single parent cell by mitotic cell division; since reproduction is asexual, the descendants of the parent cell are genetically identical. **2.** A term used to refer to cells as separate individuals (as in white blood cells) rather than as part of a growing organ.

CNS Central nervous system. That part of the nervous system consisting of the brain and spinal cord.

cochlea (kok'le-ă) The organ of hearing in the inner ear where nerve impulses are generated in response to sound waves.

codon (ko'don) The sequence of three nucleotide bases in mRNA that specifies a given amino acid and determines the position of that amino acid in a polypeptide chain through complementary base pairing with an anticodon in transfer RNA.

coenzyme (ko-en'zīm) An organic molecule, usually derived from a water-soluble vitamin, that combines with and activates specific enzyme proteins.

cofactor (ko'fak-tor) A substance needed for the catalytic action of an enzyme; generally used in reference to inorganic ions such as Ca⁺⁺ and Mg⁺⁺.

colloid osmotic (kol'oid oz-mot'ik) **pressure** Osmotic pressure exerted by plasma proteins that are present as a colloidal suspension; also called *oncotic pressure*.

com-, con- (L.) With, together.

compliance (kom-pli'ans) A measure of the ease with which a structure such as the lung expands under pressure; a measure of the change in volume as a function of pressure changes.

conducting zone The structures and airways that transmit inspired air into the respiratory zone of the lungs, where gas exchange occurs. The conducting zone includes such structures as the trachea, bronchi, and larger bronchioles.

cone Photoreceptor in the retina of the eye that provides color vision and high visual acuity.

congestive (kon-jes'tiv) **heart failure** The inability of the heart to deliver an adequate blood flow due to heart disease or hypertension. It is associated with breathlessness, salt and water retention, and edema.

conjunctivitis (kon-jungk"tĭ-vi'tis) Inflammation of the conjunctiva of the eye, which is sometimes called "pink eye."

connective tissue One of the four primary tissues, characterized by an abundance of extracellular material.

Conn's syndrome Primary hyperaldosteronism in which excessive secretion of aldosterone produces electrolyte imbalances.

contralateral (kon"tră-lat'er-al) Taking place or originating in a corresponding part on the opposite side of the body.

cornea (kor'ne-ă) The transparent structure forming the anterior part of the connective tissue covering of the eye.

corpora quadrigemina (kor'por-ă kwad"rĭ-jem'ĭ-na) A region of the mesencephalon consisting of the superior and inferior colliculi. The superior colliculi are centers for the control of visual reflexes, whereas the inferior colliculi are centers for the control of auditory reflexes.

corpus callosum (kor'pus kă-lo'sum) A large transverse tract of nerve fibers connecting the cerebral hemispheres.

cortex (kor'teks) **1.** The outer layer of an internal organ or body structure, as of the kidney or adrenal gland. **2.** The convoluted layer of gray matter that covers the surface of the cerebral hemispheres.

corticosteroid (kor"tĭ-ko-ster'oid) Any of a class of steroid hormones of the adrenal cortex, consisting of glucocorticoids (such as hydrocortisone) and mineralocorticoids (such as aldosterone).

cotransport Also called *coupled transport* or *secondary active transport*. Carrier-mediated transport in which a single carrier transports an ion (e.g., Na⁺) down its concentration gradient while transporting a specific molecule (e.g., glucose) against its concentration gradient. The hydrolysis of ATP is indirectly required for cotransport because it is needed to maintain the steep concentration gradient of the ion.

countercurrent exchange The process that occurs in the vasa recta of the renal medulla in which blood flows in U-shaped loops. This allows sodium chloride to be trapped in the interstitial fluid while water is carried away from the kidneys.

countercurrent multiplier system The interaction that occurs between the descending limb and the ascending limb of the loop of Henle in the kidney. This interaction results in the multiplication of the solute concentration in the interstitial fluid of the renal medulla.

creatine phosphate (kre'ă-tin fos'fāt) An organic phosphate molecule in muscle cells that serves as a source of high-energy phosphate for the synthesis of ATP; also called *phosphocreatine*.

crenation (krĭ-na'shun) A notched or scalloped appearance of the red blood cell membrane caused by the osmotic loss of water from these cells.

cretinism (krēt'n-iz"em) A condition caused by insufficient thyroid secretion during prenatal development or the years of early childhood. It results in stunted growth and inadequate mental development.

crypt- (Gk.) Hidden, concealed.

cryptorchidism (krip-tor'kĭ-diz"em) A developmental defect in which the testes fail to descend into the scrotum, and instead remain in the body cavity.

curare (koo-ră're) A chemical derived from plant sources that causes flaccid paralysis by blocking ACh receptor proteins in muscle cell membranes.

Cushing's syndrome Symptoms caused by hypersecretion of adrenal steroid hormones as a result of tumors of the adrenal cortex or ACTH-secreting tumors of the anterior pituitary.

cyanosis (si'ă-no"sis) A bluish discoloration of the skin or mucous membranes due to excessive concentration of deoxyhemoglobin; indicative of inadequate oxygen concentration in the blood.

cyto- (Gk.) Cell.

cytochrome (si'tŏ-krōm) A pigment in mitochondria that transports electrons in the process of aerobic respiration.

cytochrome P450 enzymes Enzymes of a particular kind, not related to the mitochondrial cytochromes, that metabolize a broad spectrum of biological molecules, such as steroid hormones, and toxic drugs. They are prominent in the liver, where they help in detoxication of the blood.

cytokinesis (si"to-kĭ-ne'sis) The division of the cytoplasm that occurs in mitosis and meiosis when a parent cell divides to produce two daughter cells.

cytoplasm (si'tŏ-plaz"em) The semifluid part of the cell between the cell membrane and the nucleus, exclusive of membrane-bound organelles. It contains many enzymes and structural proteins.

cytoskeleton (si"to-skel'ĕ-ton) A latticework of structural proteins in the cytoplasm arranged in the form of microfilaments and microtubules.

D

Dalton's law The statement that the total pressure of a gas mixture is equal to the sum that each individual gas in the mixture would exert independently. The part contributed by each gas is known as the partial pressure of the gas.

dark adaptation The ability of the eyes to increase their sensitivity to low light levels over a period of time. Part of this adaptation involves increased amounts of visual pigment in the photoreceptors.

dark current The steady inward diffusion of Na$^+$ into the rods and cones when the photoreceptors are in the dark. Stimulation by light causes this dark current to be blocked, and thus hyperpolarizes the photoreceptors.

delayed hypersensitivity An allergic response in which the onset of symptoms may not occur until 2 or 3 days after exposure to an antigen. Produced by T cells, it is a type of cell-mediated immunity.

dendrite (*den'drīt*) A relatively short, highly branched neural process that carries electrical activity to the cell body.

denervation (*de″ner-va'shun*) **hypersensitivity** The increased sensitivity of smooth muscles to neural stimulation after their innervation has been blocked or removed for a period of time.

dentin (*den'tin*) One of the hard tissues of the teeth. It covers the pulp cavity and is itself covered on its exposed surface by enamel and on its root surface by cementum.

deoxyhemoglobin (*de-ok″se-he″mŏ-glo'bin*) The form of hemoglobin in which the heme groups are in the normal reduced form but are not bonded to a gas. Deoxyhemoglobin is produced when oxyhemoglobin releases oxygen.

depolarization (*de-po″lar-ĭ-za'shun*) The loss of membrane polarity in which the inside of the cell membrane becomes less negative in comparison to the outside of the membrane. The term is also used to indicate the reversal of membrane polarity that occurs during the production of action potentials in nerve and muscle cells.

deposition (*dep-ŏ-zish'on*), **bone** The formation of the extracellular matrix of bone by osteoblasts. This process includes secretion of collagen and precipitation of calcium phosphate in the form of hydroxyapatite crystals.

detoxication (*de-tok″sĭ-ka'shun*) The removal of the toxic properties of molecules. This occurs through chemical transformation of the molecules and takes place, to a large degree, in the liver.

diabetes insipidus (*di″ă-be'tēz in-sip'ĭ-dus*) A condition in which inadequate amounts of antidiuretic hormone (ADH) are secreted by the posterior pituitary. It results in inadequate reabsorption of water by the kidney tubules, and thus in the excretion of a large volume of dilute urine.

diabetes mellitus (*mĕ-li'tus*) The appearance of glucose in the urine due to the presence of high plasma glucose concentrations, even in the fasting state. This disease is caused by either a lack of sufficient insulin secretion or by inadequate responsiveness of the target tissues to the effects of insulin.

dialysis (*di-al'ĭ-sis*) A method of removing unwanted elements from the blood by selective diffusion through a semipermeable membrane.

diapedesis (*di″ă-pĕ-de'sis*) The migration of white blood cells through the endothelial walls of blood capillaries into the surrounding connective tissues.

diarrhea (*di″ă-re'ă*) Abnormal frequency of defecation accompanied by abnormal liquidity of the feces.

diastole (*di-as'tŏ-le*) The phase of relaxation in which the heart fills with blood. Unless accompanied by the modifier *atrial*, diastole refers to the resting phase of the ventricles.

diastolic (*di″ă-stol-ik*) **blood pressure** The minimum pressure in the arteries that is produced during the phase of diastole of the heart. It is indicated by the last sound of Korotkoff when taking a blood pressure measurement.

diffusion (*dĭ-fyoo'zhun*) The net movement of molecules or ions from regions of higher to regions of lower concentration.

digestion The process of converting food into molecules that can be absorbed through the intestine into the blood.

1,25-dihydroxyvitamin (*di″hi-drok″se-vi'tă-min*) **D₃** The active form of vitamin D produced within the body by hydroxylation reactions in the liver and kidneys of vitamin D formed by the skin. This is a hormone that promotes the intestinal absorption of Ca^{++}.

diploid (*dip'loid*) Denoting cells having two of each chromosome or twice the number of chromosomes that are present in sperm or ova.

disaccharide (*di-sak'ă-rīd*) Any of a class of double sugars; carbohydrates that yield two simple sugars, or monosaccharides, upon hydrolysis.

diuretic (*di″yŭ-ret'ik*) A substance that increases the rate of urine production, thereby lowering the blood volume.

DNA Deoxyribonucleic (*de-ok″se-ri″bo-noo-kle'ik*) acid. A nucleic acid composed of nucleotide bases and deoxyribose sugar that contains the genetic code.

dopa (*do'pă*) Dihydroxyphenylalanine (*di″hi-drok″se-fen″al-ă-lă-nīn*). An amino acid formed in the liver from tyrosine and converted to dopamine in the brain. L-dopa is used in the treatment of Parkinson's disease to stimulate dopamine production.

dopamine (*do'pă-mīn*) A type of neurotransmitter in the central nervous system; also is the precursor of norepinephrine, another neurotransmitter molecule.

2,3-DPG 2,3-diphosphoglyceric (*di-fos'fo-glis-er″ik*) acid. A product of red blood cells, 2,3-DPG bonds with the protein component of hemoglobin and increases the ability of oxyhemoglobin to dissociate and release its oxygen.

ductus arteriosus (*duk'tus ar-tir″e-o'sus*) A fetal blood vessel connecting the pulmonary artery directly to the aorta.

dwarfism A condition in which a person is undersized due to inadequate secretion of growth hormone.

dyspnea (*disp-ne'ă*) Subjective difficulty in breathing.

dystrophin (*dis-trof'in*) A protein associated with the sarcolemma of skeletal muscle cells that is produced by the defective gene of people with Duchenne's muscular dystrophy.

E

ECG Electrocardiogram (*ĕ-lek″tro-kar'de-ŏ-gram*) (also abbreviated EKG). A recording of electrical currents produced by the heart.

E. coli (*e ko'li*) A species of bacteria normally found in the human intestine; full name is *Escherichia* (*esh″ĭ-rik'e-ă*) *coli.*

ecto- (Gk.) Outside, outer.

-ectomy (Gk.) Surgical removal of a structure.

ectopic (*ek-top'ik*) Foreign, out of place.

ectopic focus An area of the heart other than the SA node that assumes pacemaker activity.

ectopic pregnancy Embryonic development that occurs anywhere other than in the uterus (as in the fallopian tubes or body cavity).

edema (*ĕ-de'mă*) Swelling due to an increase in tissue fluid.

EEG Electroencephalogram (*ĕ-lek″tro-en-sef'ă-lŏ-gram*) A recording of the electrical activity of the brain from electrodes placed on the scalp.

effector (*ĕ-fek'tor*) **organs** A collective term for muscles and glands that are activated by motor neurons.

efferent (*ef'er-ent*) Conveying or transporting something away from a central location. Efferent nerve fibers conduct impulses away from the central nervous system, for example, and efferent arterioles transport blood away from the glomerulus.

elasticity (*ĕ″las-tis'ĭ-te*) The tendency of a structure to recoil to its initial dimensions after being distended (stretched).

electrolyte (*ĕ-lek'tro-līt*) An ion or molecule that is able to ionize and thus carry an electric current. The most common electrolytes in the plasma are Na$^+$, HCO$_3^-$, and K$^+$.

electrophoresis (*ĕ-lek″tro-fŏ-re'sis*) A biochemical technique in which different molecules can be separated and identified by their rate of movement in an electric field.

element, chemical A substance that cannot be broken down by chemical means into simpler compounds. An element is composed of atoms that all have the same atomic number. An element can, however, include different forms of a given atom (isotopes) that have different numbers of neutrons, and thus different atomic weights.

elephantiasis (*el″ĕ-fan-ti'ă-sis*) A disease caused by infection with a nematode worm in which the larvae block lymphatic drainage and produce edema; the lower areas of the body can become enormously swollen as a result.

EMG Electromyogram (*ĕ-lek″tro-mi'ŏ-gram*). An electrical recording of the activity of skeletal muscles through the use of surface electrodes.

emmetropia (*em″ĭ-tro'pe-ă*) A condition of normal vision in which the image of objects is focused on the retina, as opposed to nearsightedness (myopia) or farsightedness (hypermetropia).

emphysema (*em″fĭ-se'mă, em″fĭ-ze'mă*) A lung disease in which alveoli are destroyed and the remaining alveoli become larger. It results in decreased vital capacity and increased airway resistance.

emulsification (*ĭ-mul″sĭ-fĭ-ka'shun*) The process of producing an emulsion or fine suspension; in the small intestine, fat globules are emulsified by the detergent action of bile.

end-diastolic (*di"ă-stol'ik*) **volume** The volume of blood in each ventricle at the end of diastole, immediately before the ventricles contract at systole.

endergonic (*en"der-gon'ik*) Denoting a chemical reaction that requires the input of energy from an external source in order to proceed.

endo- (Gk.) Within, inner.

endocrine (*en'dŏ-krin*) **glands** Glands that secrete hormones into the circulation rather than into a duct; also called *ductless glands*.

endocytosis (*en"do-si-to'sis*) The cellular uptake of particles that are too large to cross the cell membrane. This occurs by invagination of the cell membrane until a membrane-enclosed vesicle is pinched off within the cytoplasm.

endoderm (*en'dŏ-derm*) The innermost of the three primary germ layers of an embryo. It gives rise to the digestive tract and associated structures, respiratory tract, bladder, and urethra.

endogenous (*en-doj'ĕ-nus*) Denoting a product or process arising from within the body (as opposed to exogenous products or influences, which arise from external sources).

endolymph (*en'dŏ-limf*) The fluid contained within the membranous labyrinth of the inner ear.

endometrium (*en"do-me'tre-um*) The mucous membrane of the uterus, the thickness and structure of which vary with the phase of the menstrual cycle.

endoplasmic reticulum (*en-do-plaz'mik rĭ-tik'yŭ-lum*) An extensive system of membrane-enclosed cavities within the cytoplasm of the cell. Those with ribosomes on their surface are called rough endoplasmic reticulum and participate in protein synthesis.

endorphin (*en-dor'fin*) Any of a group of endogenous opioid molecules that may act as a natural analgesic.

endothelin (*en"do-the'lin*) A polypeptide secreted by the endothelium of a blood vessel that serves as a paracrine regulator, promoting contraction of the smooth muscle and constriction of the vessel.

endothelium (*en"do-the'le-um*) The simple squamous epithelium that lines blood vessels and the heart.

endotoxin (*en"do-tok'sin*) A toxin found within certain types of bacteria that is able to stimulate the release of endogenous pyrogen and produce a fever.

end-plate potential The graded depolarization produced by ACh at the neuromuscular junction. This is equivalent to the excitatory postsynaptic potential produced at neuron-neuron synapses.

end-product inhibition The inhibition of enzymatic steps of a metabolic pathway by products formed at the end of that pathway.

enkephalin (*en-kef'ă-lin*) Either of two short polypeptides, containing five amino acids, that have analgesic effects. The two known enkephalins (which differ in only one amino acid) are endorphins, and may function as neurotransmitters in the brain.

enteric (*en-ter'ik*) A term referring to the intestine.

enterohepatic (*en"ter-o-hĕ-pat'ik*) **circulation** The recirculation of a compound between the liver and small intestine. The compound is present in the bile secreted by the liver into the small intestine, and then is reabsorbed and returned to the liver via the hepatic portal vein.

entropy (*en'trŏ-pe*) The energy of a system that is not available to perform work; a measure of the degree of disorder in a system, entropy increases whenever energy is transformed.

enzyme (*en'zīm*) A protein catalyst that increases the rate of specific chemical reactions.

epi- (Gk.) Upon, over, outer.

epidermis (*ep"ĭ-der'mis*) The stratified squamous epithelium of the skin, the outer layer of which is dead and filled with keratin.

epididymis (*ep"ĭ-did'ĭ-mis*) A tubelike structure outside the testes. Sperm pass from the seminiferous tubules into the head of the epididymis and then pass from the tail of the epididymis to the vas deferens. The sperm mature, becoming motile, as they pass through the epididymis.

epinephrine (*ep"ĭ-nef'rin*) A catecholamine hormone secreted by the adrenal medulla in response to sympathetic nerve stimulation that acts together with norepinephrine released from sympathetic nerve endings to prepare the organism for "fight or flight"; also known as *adrenaline*.

epithelium (*ep"ĭ-the'le-um*) One of the four primary tissue types; the type of tissue that covers and lines the body surfaces and forms exocrine and endocrine glands.

EPSP Excitatory postsynaptic (*pōst"sĭ-nap'tik*) potential. A graded depolarization of a postsynaptic membrane in response to stimulation by a neurotransmitter chemical. EPSPs can be summated, but can be transmitted only over short distances; they can stimulate the production of action potentials when a threshold level of depolarization is attained.

equilibrium (*e"kwĭ-lib're-um*) **potential** The hypothetical membrane potential that would be created if only one ion were able to diffuse across a membrane and reach a stable, or equilibrium, state. In this stable state, the concentrations of the ion would remain constant inside and outside the membrane, and the membrane potential would be equal to a particular value.

erythroblastosis fetalis (*ĕ-rith"ro-blas-to'sis fĭ-tal'is*) Hemolytic anemia in an Rh-positive newborn caused by maternal antibodies against the Rh factor that have crossed the placenta.

erythrocyte (*ĕ-rith'rŏ-sīt*) A red blood cell. Erythrocytes are the formed elements of blood that contain hemoglobin and transport oxygen.

erythropoietin (*ĕ-rith"ro-poi'ĕ-tin*) A hormone secreted by the kidneys that stimulates the bone marrow to produce red blood cells.

essential amino acids The eight amino acids in adults or nine amino acids in children that cannot be made by the human body; therefore, they must be obtained in the diet.

estradiol (*es"tră-di'ol*) The major estrogen (female sex steroid hormone) secreted by the ovaries.

estrus (*es'trus*) **cycle** Cyclic changes in the structure and function of the ovaries and female reproductive tract, accompanied by periods of "heat" (estrus), or sexual receptivity; the lower mammalian equivalent of the menstrual cycle, but differing from the menstrual cycle in that the endometrium is not shed with accompanying bleeding.

ex- (L.) Out, off, from.

excitation-contraction coupling The means by which electrical excitation of a muscle results in muscle contraction. This coupling is achieved by Ca++, which enters the muscle cell cytoplasm in response to electrical excitation and which stimulates the events culminating in contraction.

exergonic (*ek"ser-gon'ik*) Denoting chemical reactions that liberate energy.

exo- (Gk.) Outside or outward.

exocrine (*ek'sŏ-krin*) **gland** A gland that discharges its secretion through a duct to the outside of an epithelial membrane.

exocytosis (*ek"so-si-to'sis*) The process of cellular secretion in which the secretory products are contained within a membrane-enclosed vesicle. The vesicle fuses with the cell membrane so that the lumen of the vesicle is open to the extracellular environment.

exon (*ek'son*) A nucleotide sequence in DNA that codes for the production of messenger RNA.

extensor (*ek-sten'sor*) A muscle that upon contraction increases the angle of a joint.

exteroceptor (*ek"stĕ-ro-sep'tor*) A sensory receptor that is sensitive to changes in the external environment (as opposed to an interoceptor).

extra- (L.) Outside, beyond.

extrafusal (*ek"stră-fyooz'al*) **fibers** The ordinary muscle fibers within a skeletal muscle; not found within the muscle spindles.

extraocular (*ek"stră-ok'yŭ-lar*) **muscles** The muscles that insert into the sclera of the eye. They act to change the position of the eye in its orbit (as opposed to the intraocular muscles such as those of the iris and ciliary body within the eye).

extrapyramidal (*ek"stră-pĭ-ram'ĭ-dl*) **tracts** Neural pathways situated outside or independent of pyramidal tracts. The major extrapyramidal tract is the reticulospinal tract, which originates in the reticular formation of the brain stem and receives excitatory and inhibitory input from both the cerebrum and the cerebellum. The extrapyramidal tracts are thus influenced by activity in the brain involving many synapses, and appear to be required for fine control of voluntary movements.

F

facilitated (*fă-sil'ĭ-ta"tid*) **diffusion** The carrier-mediated transport of molecules through the cell membrane along the direction of their concentration gradients. It does not require the expenditure of metabolic energy.

FAD Flavin adenine dinucleotide (*fla'vin ad'n-ēn di-noo'kle-ŏ-tīd*). A coenzyme derived from riboflavin that participates in electron transport within the mitochondria.

feces (*fe'sēz*) The excrement discharged from the large intestine.

fertilization (*fer'tĭ-lĭ-za"shun*) The fusion of an ovum and spermatozoon.

fiber, muscle A skeletal muscle cell.

fiber, nerve An axon of a motor neuron or the dendrite of a pseudounipolar sensory neuron in the PNS.

fibrillation (*fib"rĭ-la'shun*) A condition of cardiac muscle characterized electrically by random and continuously changing patterns of electrical activity and resulting in the inability of the myocardium to contract as a unit and pump blood. It can be fatal if it occurs in the ventricles.

fibrin (*fi'brin*) The insoluble protein formed from fibrinogen by the enzymatic action of thrombin during the process of blood clot formation.

fibrinogen (*fi-brin'ŏ-jen*) A soluble plasma protein that serves as the precursor of fibrin; also called *factor I*.

flaccid paralysis (*flak'sid pă-ral'ĭ-sis*) The inability to contract muscles, resulting in a loss of muscle tone. This may be due to damage to lower motor neurons or to factors that block neuromuscular transmission.

flagellum (*flă-jel'um*) A whiplike structure that provides motility for sperm.

flare-and-wheal reaction A cutaneous reaction to skin injury or the administration of antigens produced by release of histamine and related molecules and characterized by local edema and a red flare.

flavoprotein (*fla"vo-pro'te-in*) A conjugated protein containing a flavin pigment that is involved in electron transport within the mitochondria.

flexor (*flek'sor*) A muscle that decreases the angle of a joint when it contracts.

follicle (*fol'ĭ-k'l*) A microscopic, hollow structure within an organ. Follicles are the functional units of the thyroid gland and of the ovary.

foramen ovale (*fŏ-ra'men o-val'e*) An opening normally present in the atrial septum of the fetal heart that allows direct communication between the right and left atria.

fovea centralis (*fo've-ă sen-tra'lis*) A tiny pit in the macula lutea of the retina that contains slim, elongated cones. It provides the highest visual acuity (clearest vision).

Frank–Starling law of the heart The statement describing the relationship between end-diastolic volume and stroke volume of the heart. A greater amount of blood in a ventricle prior to contraction results in greater stretch of the myocardium, and by this means produces a contraction of greater strength.

FSH Follicle-stimulating hormone. One of the two gonadotropic hormones secreted from the anterior pituitary. In females, FSH stimulates the development of the ovarian follicles; in males, it stimulates the production of sperm in the seminiferous tubules.

G

GABA Gamma-aminobutyric (*gam"ă-ă-me"no-byoo-tir'ik*) acid. An amino acid believed to function as an inhibitory neurotransmitter in the central nervous system.

gamete (*gam'ēt*) A collective term for haploid germ cells: sperm and ova.

gamma motoneuron (*gam'ă mo"tŏ-noor'on*) The type of somatic motor neuron that stimulates intrafusal fibers within the muscle spindles.

ganglion (*gang'gle-on*) A grouping of nerve cell bodies located outside the brain and spinal cord.

gap junctions Specialized regions of fusion between the cell membranes of two adjacent cells that permit the diffusion of ions and small molecules from one cell to the next. These regions serve as electrical synapses in certain areas, such as in cardiac muscle.

gas exchange The diffusion of oxygen and carbon dioxide down their concentration gradients that occurs between pulmonary capillaries and alveoli, and between systemic capillaries and the surrounding tissue cells.

gastric (*gas'trik*) **intrinsic factor** A glycoprotein secreted by the stomach that is needed for the absorption of vitamin B_{12}.

gastric juice The secretions of the gastric mucosa. Gastric juice contains water, hydrochloric acid, and pepsinogen as the major components.

gastrin (*gas'trin*) A hormone secreted by the stomach that stimulates the gastric secretion of hydrochloric acid and pepsin.

gastroileal (*gas'tro-il"e-al*) **reflex** The reflex in which increased gastric activity causes increased motility of the ileum and increased movement of chyme through the ileocecal sphincter.

gates A term used to describe structures within the cell membrane that regulate the passage of ions through membrane channels. Gates may be chemically regulated (by neurotransmitters) or voltage regulated (in which case they open in response to a threshold level of depolarization).

gen- (Gk.) Producing.

generator (*jen'ĕ-ra"tor*) **potential** The graded depolarization produced by stimulation of a sensory receptor that results in the production of action potentials by a sensory neuron; also called the *receptor potential*.

genetic (*jĕ-net'ik*) **recombination** The formation of new combinations of genes, as by crossing-over between homologous chromosomes.

genetic transcription The process by which RNA is produced with a sequence of nucleotide bases that is complementary to a region of DNA.

genetic translation The process by which proteins are produced with amino acid sequences specified by the sequence of codons in messenger RNA.

gigantism (*ji-gan'tiz"em*) Abnormal body growth due to the excessive secretion of growth hormone.

glomerular (*glo-mer'yŭ-lar*) **filtration rate (GFR)** The volume of blood plasma filtered out of the glomeruli of both kidneys each minute. The GFR is measured by the renal plasma clearance of inulin.

glomerular ultrafiltrate Fluid filtered through the glomerular capillaries into glomerular (Bowman's) capsule of the kidney tubules.

glomeruli (*glo-mer'yŭ-li*) The tufts of capillaries in the kidneys that filter fluid into the kidney tubules.

glomerulonephritis (*glo-mer"yŭ-lo-nĕ-fri'tis*) Inflammation of the renal glomeruli; associated with fluid retention, edema, hypertension, and the appearance of protein in the urine.

glucagon (*gloo'că-gon*) A polypeptide hormone secreted by the alpha cells of the islets of Langerhans in the pancreas that acts to promote glycogenolysis and raise the blood glucose levels.

glucocorticoid (*gloo"ko-kor'tĭ-koid*) Any of a class of steroid hormones secreted by the adrenal cortex (corticosteroids) that affects the metabolism of glucose, protein, and fat. These hormones also have anti-inflammatory and immunosuppressive effects. The major glucocorticoid in humans is hydrocortisone (cortisol).

gluconeogenesis (*gloo"ko-ne"ŏ-jen'ĭ-sis*) The formation of glucose from noncarbohydrate molecules, such as amino acids and lactic acid.

glutamate (*gloo'tă-māt*) The ionized form of glutamic acid, an amino acid that serves as the major excitatory neurotransmitter of the CNS. *Glutamate* and *glutamic acid* are terms that can be used interchangeably.

glycogen (*gli'kŏ-jen*) A polysaccharide of glucose—also called *animal starch*—produced primarily in the liver and skeletal muscles. Similar to plant starch in composition, glycogen contains more highly branched chains of glucose subunits than does plant starch.

glycogenesis (*gli"kŏ-jen'ĭ-sis*) The formation of glycogen from glucose.

glycogenolysis (*gli"ko-jĕ-nol'ĭ-sis*) The hydrolysis of glycogen to glucose 1-phosphate, which can be converted to glucose 6-phosphate, which then may be oxidized via glycolysis or (in the liver) converted to free glucose.

glycolysis (*gli'kol'ĭ-sis*) The metabolic pathway that converts glucose to pyruvic acid; the final products are two molecules of pyruvic acid and two molecules of reduced NAD, with a net gain of two ATP molecules. In anaerobic respiration, the reduced NAD is oxidized by the conversion of pyruvic acid to lactic acid. In aerobic respiration, pyruvic acid enters the Krebs cycle in mitochondria, and reduced NAD is ultimately oxidized by oxygen to yield water.

glycosuria (*gli"kŏ-soor'e-ă*) The excretion of an abnormal amount of glucose in the urine (urine normally contains only trace amounts of glucose).

Golgi (*gol'je*) **apparatus** A network of stacked, flattened membranous sacs within the cytoplasm of cells. Its major function is to concentrate and package proteins within vesicles that bud off from it.

Golgi tendon organ A tension receptor in the tendons of muscles that becomes activated by the pull exerted by a muscle on its tendons.

gonad (*go'nad*) A collective term for testes and ovaries.

gonadotropic (*go-nad"ŏ-tro'pik*) **hormones** Hormones of the anterior pituitary that stimulate gonadal function—the formation of gametes and secretion of sex steroids. The two gonadotropins are FSH (follicle-stimulating hormone) and LH (luteinizing hormone), which are essentially the same in males and females.

graafian (*graf'e-an*) **follicle** A mature ovarian follicle, containing a single fluid-filled cavity, with the ovum located toward one side of the follicle and perched on top of a hill of granulosa cells.

granular leukocytes (*loo'kŏ-sītz*) Leukocytes with granules in the cytoplasm. On the basis of the staining properties of the granules, these cells are of three types: neutrophils, eosinophils, and basophils.

Graves' disease A hyperthyroid condition believed to be caused by excessive stimulation of the thyroid gland by autoantibodies. It is associated with exophthalmos (bulging eyes), high pulse rate, high metabolic rate, and other symptoms of hyperthyroidism.

gray matter The part of the central nervous system that contains neuron cell bodies and dendrites, but few myelinated axons. It forms the cortex of the cerebrum, cerebral nuclei, and the central region of the spinal cord.

growth hormone A hormone secreted by the anterior pituitary that stimulates growth of the skeleton and soft tissues during the growing years and that influences the metabolism of protein, carbohydrate, and fat throughout life.

gyrus (*ji'rus*) A fold or convolution in the cerebrum.

H

haploid (*hap'loid*) A cell that has one of each chromosome type and therefore half the number of chromosomes present in most other body cells. Only the gametes (sperm and ova) are haploid.

hapten (*hap'ten*) A small molecule that is not antigenic by itself, but which—when combined with proteins—becomes antigenic and thus capable of stimulating the production of specific antibodies.

haversian (*hă-ver'shan*) **system** A haversian canal and its concentrically arranged layers, or lamellae, of bone. It constitutes the basic structural unit of compact bone.

hay fever A seasonal type of allergic rhinitis caused by pollen. It is characterized by itching and tearing of the eyes, swelling of the nasal mucosa, attacks of sneezing, and often by asthma.

hCG Human chorionic gonadotropin (*kor'e-on-ik go-nad'ŏ-tro'pin*). A hormone secreted by the embryo that has LH-like actions and that is required for maintenance of the mother's corpus luteum for the first 10 weeks of pregnancy.

heart murmur An abnormal heart sound caused by an abnormal flow of blood in the heart due to structural defects, usually of the valves or septum.

heart sounds The sounds produced by closing of the AV valves of the heart during systole (the first sound) and by closing of the semilunar valves of the aorta and pulmonary trunk during diastole (the second sound).

helper T cells A subpopulation of T cells (lymphocytes) that help stimulate antibody production of B lymphocytes by antigens.

hematocrit (*he-mat'ŏ-krit*) The ratio of packed red blood cells to total blood volume in a centrifuged sample of blood, expressed as a percentage.

heme (*hēm*) The iron-containing red pigment that, together with the protein globin, forms hemoglobin.

hemoglobin (*he'mŏ-glo"bin*) The combination of heme pigment and protein within red blood cells that acts to transport oxygen and (to a lesser degree) carbon dioxide. Hemoglobin also serves as a weak buffer within red blood cells.

Henderson–Hasselbalch (*hen'der-son-has'el-balk*) **equation** A formula used to determine the blood pH produced by a given ratio of bicarbonate to carbon dioxide concentrations.

Henry's law The statement that the concentration of gas dissolved in a fluid is directly proportional to the partial pressure of that gas.

heparin (*hep'ar-in*) A mucopolysaccharide found in many tissues, but in greatest abundance in the lungs and liver, that is used medically as an anticoagulant.

hepatic (*hĕ-pat'ik*) Pertaining to the liver.

hepatitis (*hep"ă-ti'tis*) Inflammation of the liver.

Hering–Breuer reflex A reflex in which distension of the lungs stimulates stretch receptors, which in turn act to inhibit further distension of the lungs.

hermaphrodite (*her-maf'rŏ-dīt*) An organism with both testicular and ovarian tissue.

hetero- (Gk.) Different, other.

heterochromatin (*het"ĕ-ro-kro'mă-tin*) A condensed, inactive form of chromatin.

hiatal hernia (*hi-a'tal her'ne-ă*) A protrusion of an abdominal structure through the esophageal hiatus of the diaphragm into the thoracic cavity.

high-density lipoproteins (*lip"o-pro'te-inz*) **(HDLs)** Combinations of lipids and proteins that migrate rapidly to the bottom of a test tube during centrifugation. HDLs are carrier proteins that are believed to transport cholesterol away from blood vessels to the liver, and thus to offer some protection from atherosclerosis.

higher motor neurons Neurons in the brain that, as part of the pyramidal or extrapyramidal system, influence the activity of the lower motor neurons in the spinal cord.

histamine (*his'tă-mēn*) A compound secreted by tissue mast cells and other connective tissue cells that stimulates vasodilation and increases capillary permeability. It is responsible for many of the symptoms of inflammation and allergy.

histocompatibility (*his"to-kom-pat"ĭ-bil'ĭ-te*) **antigens** A group of cell-surface antigens found on all cells of the body except mature red blood cells. These are important for the function of T lymphocytes, and the greater their variance, the greater will be the chance of transplant rejection.

histone (*his'tōn*) A basic protein associated with DNA that is believed to repress genetic expression.

homeo (Gk.) Same.

homeostasis (*ho"me-o-sta'sis*) The dynamic constancy of the internal environment, the maintenance of which is the principal function of physiological regulatory mechanisms. The concept of homeostasis provides a framework for understanding most physiological processes.

homologous (*hŏ-mol'ŏ-gus*) **chromosomes** The matching pairs of chromosomes in a diploid cell.

hormone (*hor'mōn*) A regulatory chemical produced in an endocrine gland that is secreted into the blood and carried to target cells that respond to the hormone by an alteration in their metabolism.

humoral immunity (*hyoo'mor-al ĭ-myoo'nĭ-te*) The form of acquired immunity in which antibody molecules are secreted in response to antigenic stimulation (as opposed to cell-mediated immunity).

hyaline (*hi'ă-lĭn*) **membrane disease** A disease affecting premature infants who lack pulmonary surfactant that is characterized by collapse of the alveoli (atelectasis) and pulmonary edema; also called *respiratory distress syndrome*.

hydrocortisone (*hi"drŏ-kor'tĭ-sōn*) The principal corticosteroid hormone secreted by the adrenal cortex, with glucocorticoid action; also called *cortisol*.

hydrophilic (*hi"drŏ-fil'ik*) Denoting a substance that readily absorbs water; literally, "water loving."

hydrophobic (*hi"drŏ-fo'bik*) Denoting a substance that repels, and that is repelled by, water; literally, "water fearing."

hyper- (Gk.) Over, above, excessive.

hyperbaric (*hi"per-bar'ik*) **oxygen** Oxygen gas present at greater than atmospheric pressure.

hypercapnia (*hi"per-kap'ne-ă*) Excessive concentration of carbon dioxide in the blood.

hyperemia (*hi"per-e'me-ă*) Excessive blood flow to a part of the body.

hyperglycemia (*hi"per-gli-se'me-ă*) An abnormally increased concentration of glucose in the blood.

hyperkalemia (*hi"per-kă-le'me-ă*) An abnormally high concentration of potassium in the blood.

hyperopia (*hi"per-o'pe-ă*) A refractive disorder in which rays of light are brought to a focus behind the retina as a result of the eyeball being too short; also called *farsightedness*.

hyperplasia (*hi"per-pla'zha*) An increase in organ size due to an increase in cell numbers as a result of mitotic cell division.

hyperpnea (*hi"perp'ne-ă*) Increased total minute volume during exercise. Unlike hyperventilation, the arterial blood carbon dioxide values are not changed during hyperpnea because the increased ventilation is matched to an increased metabolic rate.

hyperpolarization (*hi"per-po"lar-ĭ-za'shun*) An increase in the negativity of the inside of a cell membrane with respect to the resting membrane potential.

hypersensitivity (*hi"per-sen"sĭ-tiv'ĭ-te*) Another name for *allergy;* an abnormal immune response that may be immediate (due to antibodies of the IgE class) or delayed (due to cell-mediated immunity).

hypertension (*hi"per-ten'shun*) High blood pressure. Classified as either primary, or essential, hypertension of unknown cause or secondary hypertension that develops as a result of other, known disease processes.

hypertonic (*hi"per-ton'ik*) A solution with a greater solute concentration and thus a greater osmotic pressure than plasma.

hypertrophy (*hi-per'trŏ-fe*) Growth of an organ due to an increase in the size of its cells.

hyperventilation (*hi-per-ven"tĭ-la'shun*) A high rate and depth of breathing that results in a decrease in the blood carbon dioxide concentration to below normal.

hypo- (Gk.) Under, below, less.

hypodermis (*hi"pŏ-der'mis*) A layer of fat beneath the dermis of the skin.

hypotension (*hi"pŏ-ten'shun*) Abnormally low blood pressure.

hypothalamic (*hi"po-thă-lam'ik*) **hormones** Hormones produced by the hypothalamus. These include antidiuretic hormone and oxytocin, which are secreted by the posterior pituitary, and both releasing and inhibiting hormones that regulate the secretion of the anterior pituitary.

hypothalamo-hypophyseal (*hi"po-thă-lam'o-hi"pof-ĭ-se'al*) **portal system** A vascular system that transports releasing and inhibiting hormones from the hypothalamus to the anterior pituitary.

hypothalamo-hypophyseal tract The tract of nerve fibers (axons) that transports antidiuretic hormone and oxytocin from the hypothalamus to the posterior pituitary.

hypothalamus (*hi"po-thal'ă-mus*) An area of the brain that lies below the thalamus and above the pituitary gland. The hypothalamus regulates the pituitary gland and contributes to the regulation of the autonomic nervous system, among its many other functions.

hypothermia (*hi"pŏ-ther'me-ah*) A low body temperature. This is a dangerous condition that is defended against by shivering and other physiological mechanisms that generate body heat.

hypovolemic (*hi"po-vo-le'mik*) **shock** A rapid fall in blood pressure as a result of diminished blood volume.

hypoxemia (*hi"pok-se'me-ă*) A low oxygen concentration of the arterial blood.

I

ileogastric (*il″e-o-gas′trik*) **reflex** The reflex in which distension of the ileum causes decreased gastric motility.

immediate hypersensitivity Hypersensitivity (allergy) that is mediated by antibodies of the IgE class and that results in the release of histamine and related compounds from tissue cells.

immunization (*im″yŭ-nĭ-za′shun*) The process of increasing one's resistance to pathogens. In active immunity a person is injected with antigens that stimulate the development of clones of specific B or T lymphocytes; in passive immunity a person is injected with antibodies made by another organism.

immunoassay (*im″yŭ-no-as′a*) Any of a number of laboratory or clinical techniques that employ specific bonding between an antigen and its homologous antibody in order to identify and quantify a substance in a sample.

immunoglobulins (*im″yŭ-no-glob′yŭ-linz*) Subclasses of the gamma globulin fraction of plasma proteins that have antibody functions, providing humoral immunity.

immunosurveillance (*im″yŭ-no-ser-va′lens*) The function of the immune system to recognize and attack malignant cells that produce antigens not recognized as "self." This function is believed to be cell mediated rather than humoral.

implantation (*im″plan-ta′shun*) The process by which a blastocyst attaches itself to and penetrates into the endometrium of the uterus.

infarct (*in′farkt*) An area of necrotic (dead) tissue as a result of inadequate blood flow (ischemia).

inhibin (*in-hib′in*) Believed to be a water-soluble hormone secreted by the seminiferous tubules of the testes that specifically exerts negative feedback inhibition of FSH secretion from the anterior pituitary.

inositol triphosphate (*ĭ-no′sĭ-tol tri-fos′fāt*) A second messenger in hormone action that is produced by the cell membrane of a target cell in response to the action of a hormone. This compound is believed to stimulate the release of Ca⁺⁺ from the endoplasmic reticulum of the cell.

insulin (*in′sŭ-lin*) A polypeptide hormone secreted by the beta cells of the islets of Langerhans in the pancreas that promotes the anabolism of carbohydrates, fat, and protein. Insulin acts to promote the cellular uptake of blood glucose and, therefore, to lower the blood glucose concentration; insulin deficiency produces hyperglycemia and diabetes mellitus.

inter- (L.) Between, among.

interferons (*in″ter-fēr′onz*) Small proteins that inhibit the multiplication of viruses inside host cells and that also have antitumor properties.

interleukin-2 (*in″ter-loo′kin-2*) A lymphokine secreted by T lymphocytes that stimulates the proliferation of both B and T lymphocytes.

interneurons (*in″ter-noor′onz*) Those neurons within the central nervous system that do not extend into the peripheral nervous system; they are interposed between sensory (afferent) and motor (efferent) neurons; also called *association neurons*.

interoceptors (*in″ter-o-sep′torz*) Sensory receptors that respond to changes in the internal environment (as opposed to exteroceptors).

interphase The interval between successive cell divisions, during which time the chromosomes are in an extended state and are active in directing RNA synthesis.

intestino-intestinal (*in-tes′tĭ-no-in-tes′tĭ-nal*) **reflex** The reflex in which overdistension to one region of the intestine causes relaxation throughout the rest of the intestine.

intra- (L.) Within, inside.

intrafusal (*in″tră-fyoo′sal*) **fibers** Modified muscle fibers that are encapsulated to form muscle spindle organs, which are muscle stretch receptors.

intrapleural (*in″tră-ploor′al*) **space** An actual or potential space between the visceral pleural membrane covering the lungs and the somatic pleural membrane lining the thoracic wall.

intrapulmonary (*in″tră-pul′mŏ-nar″e*) **space** The space within the air sacs and airways of the lungs.

intron (*in′tron*) A noncoding nucleotide sequence in DNA that interrupts the coding regions (exons) for mRNA.

inulin (*in′yŭ-lin*) A polysaccharide of fructose, produced by certain plants, that is filtered by the human kidneys but neither reabsorbed nor secreted. The clearance rate of injected inulin is thus used to measure the glomerular filtration rate.

in vitro (*in ve′tro*) Occurring outside the body, in a test tube or other artificial environment.

in vivo (*in ve′vo*) Occurring within the body.

ion (*i′on*) An atom or a group of atoms that has either lost or gained electrons and thus has a net positive or a net negative charge.

ionization (*i″on-ĭ-za′shun*) The dissociation of a solute to form ions.

ipsilateral (*ip″sĭ-lat′er-al*) On the same side (as opposed to contralateral).

IPSP Inhibitory postsynaptic potential. A hyperpolarization of the postsynaptic membrane in response to a particular neurotransmitter chemical, which makes it more difficult for the postsynaptic cell to attain a threshold level of depolarization required to produce action potentials. IPSPs are responsible for postsynaptic inhibition.

ischemia (*ĭ-ske′me-ă*) A rate of blood flow to an organ that is inadequate to supply sufficient oxygen and maintain aerobic respiration in that organ.

islets of Langerhans (*i′letz of lang′er-hanz*) Encapsulated groupings of endocrine cells within the exocrine tissue of the pancreas, including alpha cells that secrete glucagon and beta cells that secrete insulin; also called *pancreatic islets*.

iso- (Gk.) Equal, same.

isoenzymes (*i″so-en′zīmz*) Enzymes, usually produced by different organs, that catalyze the same reaction but that differ from each other in amino acid composition.

isometric (*i″sŏ-met′rik*) **contraction** Muscle contraction in which there is no appreciable shortening of the muscle.

isotonic (*i″sŏ-ton′ik*) **contraction** Muscle contraction in which the muscle shortens in length and maintains approximately the same amount of tension throughout the shortening process.

isotonic solution A solution having the same total solute concentration, osmolality, and osmotic pressure as the solution with which it is compared; a solution with the same solute concentration and osmotic pressure as plasma.

J

jaundice (*jawn′dis*) A condition characterized by high blood bilirubin levels and staining of the tissues with bilirubin, which imparts a yellow color to the skin and mucous membranes.

junctional (*jungk′shun-al*) **complexes** Structures that join adjacent epithelial cells together, including the zonula occludens, zonula adherens, and macula adherens (desmosome).

juxta- (L.) Near to, next to.

juxtaglomerular (*juk″stă-glo-mer′yŭ-lar*) **apparatus** A renal structure in which regions of the nephron tubule and afferent arteriole are in contact with each other. Cells in the afferent arteriole of the juxtaglomerular apparatus secrete the enzyme renin into the blood, which activates the renin-angiotensin system.

K

keratin (*ker′ă-tin*) A protein that forms the principal component of the outer layer of the epidermis and of hair and nails.

ketoacidosis (*ke″to-ă-sĭ-do′sis*) A type of metabolic acidosis resulting from the excessive production of ketone bodies, as in diabetes mellitus.

ketogenesis (*ke″to-jen′ĭ-sis*) The production of ketone bodies.

ketone (*ke′tŏn*) **bodies** The substances derived from fatty acids via acetyl coenzyme A in the liver; namely, acetone, acetoacetic acid, and β-hydroxybutyric acid. Ketone bodies are oxidized by skeletal muscles for energy.

ketosis (*ke-to′sis*) An abnormal elevation in the blood concentration of ketone bodies that does not necessarily produce acidosis.

kilocalorie (*kil′ŏ-kal″ŏ-re*) A unit of measurement equal to 1,000 calories, which are units of heat. (A kilocalorie is the amount of heat required to raise the temperature of 1 kilogram of water 1°C.) In nutrition, the kilocalorie is called a big calorie (Calorie).

Klinefelter's (*klīn′fel-terz*) **syndrome** The syndrome produced in a male by the presence of an extra X chromosome (genotype XXY).

Krebs (*krebz*) **cycle** A cyclic metabolic pathway in the matrix of mitochondria by which the acetic acid part of acetyl CoA is oxidized and substrates provided for reactions that are coupled to the formation of ATP.

Kupffer (*koop′fer*) **cells** Phagocytic cells that line the sinusoids of the liver and are part of the reticuloendothelial system.

L

lactose (*lak′tōs*) Milk sugar; a disaccharide of glucose and galactose.

lactose intolerance The inability of many adults to digest lactose because of an enzyme, lactase, deficiency.

LaPlace, law of The statement that the pressure within an alveolus is directly proportional to its surface tension and inversely proportional to its radius.

larynx (*lar′ingks*) A structure consisting of epithelial tissue, muscle, and cartilage that serves as a sphincter guarding the entrance of the trachea. It is the organ responsible for voice production.

lateral inhibition The sharpening of perception that occurs in the neural processing of sensory input. Input from those receptors that are most greatly stimulated is enhanced, while input from other receptors is reduced. This results, for example, in improved pitch discrimination in hearing.

lesion (*le'zhun*) A wounded or damaged area.

leukocyte (*loo'kŏ-sīt*) A white blood cell.

Leydig (*li'dig*) **cells** The interstitial cells of the testes that serve an endocrine function by secreting testosterone and other androgenic hormones.

ligament (*lig'ă-ment*) A tough cord or fibrous band of dense regular connective tissue that contains many parallel arrangements of collagen fibers. It connects bones or cartilages and serves to strengthen joints.

ligand (*li'gand, lig'and*) A smaller molecule that chemically binds to a larger molecule, which is usually a protein. Oxygen, for example, is the ligand for the heme in hemoglobin, and hormones or neurotransmitters can be the ligands for specific membrane proteins.

limbic (*lim'bik*) **system** A group of brain structures, including the hippocampus, cingulate gyrus, dentate gyrus, and amygdala. The limbic system appears to be important in memory, the control of autonomic function, and some aspects of emotion and behavior.

lipid (*lip'id*) An organic molecule that is nonpolar and thus insoluble in water. Lipids include triglycerides, steroids, and phospholipids.

lipogenesis (*lip'ŏ-jen'ě-sis*) The formation of fat or triglycerides.

lipolysis (*li-pol'ĭ-sis*) The hydrolysis of triglycerides into free fatty acids and glycerol.

long-term potentiation (*pŏ-ten"she-a'shun*) The improved ability of a presynaptic neuron that has been stimulated at high frequency to subsequently stimulate a postsynaptic neuron over a period of weeks or even months. This may represent a mechanism of neural learning.

low-density lipoproteins (*lip"o-pro'te-inz*) **(LDLs)** Plasma proteins that transport triglycerides and cholesterol to the arteries and that are believed to contribute to arteriosclerosis.

lower motor neuron The motor neuron that has its cell body in the gray matter of the spinal cord and that contributes axons to peripheral nerves. This neuron innervates muscles and glands.

lumen (*loo'men*) The cavity of a tube or hollow organ.

lung surfactant (*sur-fak'tant*) A mixture of lipoproteins (containing phospholipids) secreted by type II alveolar cells into the alveoli of the lungs. It lowers surface tension and prevents collapse of the lungs, as occurs in hyaline membrane disease when surfactant is absent.

luteinizing (*loo'te-ĭ-ni"zing*) **hormone (LH)** A gonadotropic hormone secreted by the anterior pituitary that, in a female, stimulates ovulation and the development of a corpus luteum. In a male, LH stimulates the Leydig cells to secrete androgens.

lymph (*limf*) The fluid in lymphatic vessels that is derived from tissue fluid.

lymphatic (*lim-fat'ik*) **system** The lymphatic vessels and lymph nodes.

lymphocyte (*lim'fŏ-sīt*) A type of mononuclear leukocyte; the cell responsible for humoral and cell-mediated immunity.

lymphokine (*lim'fŏ-kīn*) Any of a group of chemicals released from T cells that contribute to cell-mediated immunity.

-lysis (Gk.) Breakage, disintegration.

lysosome (*li'sŏ-sōm*) An organelle containing digestive enzymes that is responsible for intracellular digestion.

M

macro- (Gk.) Large.

macromolecule (*mak"rŏ-mol'ĭ-kyool*) A large molecule; a term commonly used to refer to protein, RNA, and DNA.

macrophage (*mak'rŏ-fāj*) A large phagocytic cell in connective tissue that contributes to both specific and nonspecific immunity.

macula densa (*mak'yŭ-lă den'să*) The region of the distal tubule of the renal nephron in contact with the afferent arteriole. This region functions as a sensory receptor for the amount of sodium excreted in the urine and acts to inhibit the secretion of renin from the juxtaglomerular apparatus.

macula lutea (*loo'te-ă*) A yellowish depression in the retina of the eye that contains the fovea centralis, the area of keenest vision.

malignant A structure or process that is life threatening. Of a tumor, tending to metastasize.

mast cell A type of connective tissue cell that produces and secretes histamine and heparin.

maximal oxygen uptake The maximum amount of oxygen that can be consumed by the body per unit time during heavy exercise.

mean arterial pressure An adjusted average of the systolic and diastolic blood pressures. It averages about 100 mmHg in the systemic circulation and 10 mmHg in the pulmonary circulation.

mechanoreceptor (*mek"ă-no-re-sep'tor*) A sensory receptor that is stimulated by mechanical means. Mechanoreceptors include stretch receptors, hair cells in the inner ear, and pressure receptors.

medulla oblongata (*mě-dul'ă ob"long-gă'tă*) A part of the brain stem that contains neural centers for the control of breathing and for regulation of the cardiovascular system via autonomic nerves.

mega- (Gk.) Large, great.

megakaryocyte (*meg'ă-kar'e-o-sīt*) A bone marrow cell that gives rise to blood platelets.

meiosis (*mi-o'sis*) A type of cell division in which a diploid parent cell gives rise to haploid daughter cells. It occurs in the process of gamete production in the gonads.

melanin (*mel'ă-nin*) A dark pigment found in the skin, hair, choroid layer of the eye, and substantia nigra of the brain; it may also be present in certain tumors (melanomas).

melatonin (*mel'ă-to'nin*) A hormone secreted by the pineal gland that produces lightening of the skin in lower animals and that may contribute to the regulation of gonadal function in mammals. Secretion follows a circadian rhythm and peaks at night.

membrane potential The potential difference or voltage that exists between the inner and outer sides of a cell membrane. It exists in all cells, but is capable of being changed by excitable cells (neurons and muscle cells).

membranous labyrinth (*mem'bră-nus lab'ĭ-rinth*) A system of communicating sacs and ducts within the bony labyrinth of the inner ear.

menarche (*mě-nar'ke*) The first menstrual discharge, normally occurring during puberty.

Ménière's (*mān-yarz'*) **disease** Deafness, tinnitus, and vertigo resulting from a disease of the labyrinth.

menopause (*men'ŏ-pawz*) The cessation of menstruation, usually occurring at about age 50.

menstrual (*men'stroo-al*) **cycle** The cyclic changes in the ovaries and endometrium of the uterus that lasts about a month. It is accompanied by shedding of the endometrium, with bleeding, and occurs only in humans and the higher primates.

menstruation (*men"stroo-a'shun*) Shedding of the outer two-thirds of the endometrium with accompanying bleeding as a result of a lowering of estrogen secretion by the ovaries at the end of the monthly cycle. The first day of menstruation is taken as day 1 of the menstrual cycle.

meso- (Gk.) Middle.

mesoderm (*mes'ŏ-derm*) The middle embryonic tissue layer that gives rise to connective tissue (including blood, bone, and cartilage); blood vessels; muscles; the adrenal cortex; and other organs.

messenger RNA (mRNA) A type of RNA that contains a base sequence complementary to a part of the DNA that specifies the synthesis of a particular protein.

meta- (Gk.) Change.

metabolic acidosis (*as"ĭ-do'sis*) and **alkalosis** (*al"kă-lo'sis*) Abnormal changes in arterial blood pH due to changes in nonvolatile acid concentration (for example, changes in lactic acid or ketone body concentrations) or to changes in blood bicarbonate concentration.

metabolism (*mě-tab'ŏ-liz-em*) All of the chemical reactions in the body; it includes those that result in energy storage (anabolism) and those that result in the liberation of energy (catabolism).

metastasis (*mě-tas'tă-sis*) A process whereby cells of a malignant tumor separate from the tumor, travel to a different site, and divide to produce a new tumor.

methemoglobin (*met-he'mŏ-glo"bin*) The abnormal form of hemoglobin in which the iron atoms in heme are oxidized to the ferrous form. Methemoglobin is incapable of bonding with oxygen.

micelle (*mi-sel'*) A colloidal particle formed by the aggregation of many molecules.

micro- (L.) Small; also, one-millionth.

microvilli (*mi"kro-vil'i*) Tiny fingerlike projections of a cell membrane. They occur on the apical (lumenal) surface of the cells of the small intestine and in the renal tubules.

micturition (*mik"tŭ-rish'un*) Urination.

milliequivalent (*mil"ĭ-e-kwiv'ă-lent*) The millimolar concentration of an ion multiplied by its number of charges.

mineralocorticoid (*min"er-al-o-kor'tĭ-koid*) Any of a class of steroid hormones of the adrenal cortex (corticosteroids) that regulate electrolyte balance.

mitosis (*mi-to'sis*) Cell division in which the two daughter cells receive the same number of chromosomes as the parent cell (both daughters and parent are diploid).

molal (*mo'lal*) Pertaining to the number of moles of solute per kilogram of solvent.

molar (*mo'lar*) Pertaining to the number of moles of solute per liter of solution.

mole (*mōl*) The number of grams of a chemical that is equal to its formula weight (atomic weight for an element or molecular weight for a compound).

mono- (Gk.) One, single.

monoamine (*mon"o-am'ēn*) Any of a class of neurotransmitter molecules that includes serotonin, dopamine, and norepinephrine.

monoamine oxidase (*mon″o-am′ēn ok′sĭ-dās*) **(MAO)** An enzyme that degrades monoamine neurotransmitters within presynaptic axon endings. Drugs that inhibit the action of this enzyme thus potentiate the pathways that use monoamines as neurotransmitters.

monoclonal (*mon′ŏ-klōn′al*) **antibodies** Identical antibodies derived from a clone of genetically identical plasma cells.

monocyte (*mon′o-sīt*) A mononuclear, nongranular leukocyte that is phagocytic and that can be transformed into a macrophage.

monomer (*mon′ŏ-mer*) A single molecular unit of a longer, more complex molecule. Monomers are joined together to form dimers, trimers, and polymers; the hydrolysis of polymers eventually yields separate monomers.

mononuclear leukocyte (*mon″o-noo′kle-ar loo′kŏ-sīt*) Any of a category of white blood cells that includes the lymphocytes and monocytes.

mononuclear phagocyte (*fag′ŏ-sīt*) **system** A term used to describe monocytes and tissue macrophages.

monosaccharide (*mon″ŏ-sak′ă-rīd*) The monomer of the more complex carbohydrates. Examples of monomers include glucose, fructose, and galactose. Also called a *simple sugar*.

-morph, morpho- (Gk.) Form, shape.

motile (*mo′til*) Capable of self-propelled movement.

motor cortex (*kor′teks*) The precentral gyrus of the frontal lobe of the cerebrum. Axons from this area form the descending pyramidal motor tracts.

motor neuron (*noor′on*) An efferent neuron that conducts action potentials away from the central nervous system and innervates effector organs (muscles and glands). It forms the ventral roots of spinal nerves.

motor unit A lower motor neuron and all of the skeletal muscle fibers stimulated by branches of its axon. Larger motor units (more muscle fibers per neuron) produce more power when the unit is activated, but smaller motor units afford a finer degree of neural control over muscle contraction.

mucous (*myoo′kus*) **membrane** The layers of visceral organs that include the lining epithelium, submucosal connective tissue, and (in some cases) a thin layer of smooth muscle (the muscularis mucosa).

muscarinic receptors (*mus″kă-rin′ik re-sep′torz*) Receptors for acetylcholine that are stimulated by postganglionic parasympathetic neurons. Their name is derived from the fact that they are also stimulated by the chemical muscarine, derived from a mushroom.

muscle spindle (*mus′el spin′d′l*) A sensory organ within skeletal muscle that is composed of intrafusal fibers. It is sensitive to muscle stretch and provides a length detector within muscles.

myelin (*mi′ĕ-lin*) **sheath** A sheath surrounding axons formed by successive wrappings of a neuroglial cell membrane. Myelin sheaths are formed by Schwann cells in the peripheral nervous system and by oligodendrocytes within the central nervous system.

myocardial infarction (*mi′ŏ-kar′de-al in-fark′shun*) An area of necrotic tissue in the myocardium that is filled in by scar (connective) tissue.

myofibril (*mi′ŏ-fi′bril*) A subunit of striated muscle fiber that consists of successive sarcomeres. Myofibrils run parallel to the long axis of the muscle fiber, and the pattern of their filaments provides the striations characteristic of striated muscle cells.

myogenic (*mi′ŏ-jen′ik*) Originating within muscle cells; this term is used to describe self-excitation by cardiac and smooth muscle cells.

myoglobin (*mi′ŏ-glo′bin*) A molecule composed of globin protein and heme pigment. It is related to hemoglobin but contains only one subunit (instead of the four in hemoglobin) and is found in striated muscles, where it serves to store oxygen in skeletal and cardiac muscle cells.

myoneural (*mi′ŏ-noor′al*) **junction** A synapse between a motor neuron and the muscle cell that it innervates; also called the *neuromuscular junction*.

myopia (*mi-o′pe-ă*) A condition of the eyes in which light is brought to a focus in front of the retina due to the eye being too long; also called *nearsightedness*.

myosin (*mi′ŏ-sin*) The protein that forms the A bands of striated muscle cells. Together with the protein actin, myosin provides the basis for muscle contraction.

myxedema (*mik″sĭ-de′mă*) A type of edema associated with hypothyroidism. It is characterized by accumulation of mucoproteins in tissue fluid.

N

NAD Nicotinamide adenine dinucleotide (*nik″ŏ-tin′ă-mīd ad′n-ēn di-noo′kle-ŏ-tīd*). A coenzyme derived from niacin that functions to transport electrons in oxidation-reduction reactions. It helps to transport electrons to the electron-transport chain within mitochondria.

naloxone (*nal′ok-sōn, nă-lok′-sōn*) A drug that antagonizes the effects of morphine and endorphins.

natriuretic (*na″trī-yoo-ret′ik*) **hormone** A hormone that increases the urinary excretion of sodium. This hormone is currently believed to be secreted by the atria of the heart.

necrosis (*nĕ-kro′sis*) Cellular death within tissues and organs as a result of pathological conditions. This differs histologically from the physiological cell death of apoptosis.

negative feedback A mechanism of response that serves to maintain a state of internal constancy, or homeostasis. Effectors are activated by changes in the internal environment, and the actions of the effectors serve to counteract these changes and maintain a state of balance.

neoplasm (*ne′ŏ-plazm*) A new, abnormal growth of tissue, as in a tumor.

nephron (*nef′ron*) The functional unit of the kidneys, consisting of a system of renal tubules and a vascular component that includes capillaries of the glomerulus and the peritubular capillaries.

Nernst equation The equation used to calculate the equilibrium membrane potential for given ions when the concentrations of those ions on each side of the membrane are known.

nerve A collection of motor axons and sensory dendrites in the peripheral nervous system.

neurilemma (*noor″ĭ-lem′ă*) The sheath of Schwann and its surrounding basement membrane that encircles nerve fibers in the peripheral nervous system.

neuroglia (*noo-rog′le-ă*) The support cells of the nervous system that aid the functions of neurons. In addition to providing support, the neuroglial, or glial, cells participate in the metabolic and bioelectrical processes of the nervous system.

neurohypophysis (*noor″o-hi-pof′ĭ-sis*) The posterior part of the pituitary gland that is derived from the brain. It secretes vasopression (ADH) and oxytocin, both of which are produced in the hypothalamus.

neuron (*noor′on*) A nerve cell, consisting of a cell body that contains the nucleus, short branching processes, called dendrites, that carry electrical charges to the cell body, and a single fiber, or axon, that conducts nerve impulses away from the cell body.

neurotransmitter (*noor″o-trans′mit-er*) A chemical contained in synaptic vesicles in nerve endings that is released into the synaptic cleft, where it stimulates the production of either excitatory or inhibitory postsynaptic potentials.

neurotrophin (*noor″ŏ-trof′in*) Any of a family of autocrine regulators secreted by neurons and neuroglial cells that promote axon growth and other effects. Nerve growth factor is an example.

neutron (*noo′tron*) An electrically neutral particle that exists together with positively charged protons in the nucleus of atoms.

nexus (*nek′sus*) A bond between members of a group; the type of intercellular connection found in single-unit smooth muscles.

niacin (*ni′ă-sin*) A water-soluble B vitamin needed for the formation of NAD, which is a coenzyme that participates in the transfer of hydrogen atoms in many of the reactions of cell respiration.

nicotinic receptors (*nik″ŏ-tin′ik re-sep′torz*) Receptors for acetylcholine located in the autonomic ganglia and in neuromuscular junctions. Their name is derived from the fact that they can also be stimulated by nicotine, derived from the tobacco plant.

nidation (*ni-da′shun*) Implantation of the blastocyst into the endometrium of the uterus.

Nissl (*nis′l*) **bodies** Granular-appearing structures in the cell bodies of neurons that have an affinity for basic stain; they correspond to ribonucleoprotein; also called *chromatophilic substances*.

Nitric oxide A gas that functions as a neurotransmitter in both the CNS and in peripheral autonomic neurons and as an autocrine and paracrine regulator in many organs. It promotes vasodilation, intestinal relaxation, penile erection, and aids long-term potentiation in the brain.

nociceptor (*no″sĭ-sep′tor*) A receptor for pain that is stimulated by tissue damage.

nodes of Ranvier (*ran′ve-a*) Gaps in the myelin sheath of myelinated axons, located approximately 1 mm apart. Action potentials are produced only at the nodes of Ranvier in myelinated axons.

norepinephrine (*nor″ep-ĭ-nef′rin*) A catecholamine released as a neurotransmitter from postganglionic sympathetic nerve endings and as a hormone (together with epinephrine) by the adrenal medulla.

nucleolus (*noo-kle′ŏ-lus*) A dark-staining area within a cell nucleus; the site where ribosomal RNA is produced.

nucleoplasm (*noo′kle-ŏ-plaz″em*) The protoplasm of a nucleus.

nucleosome (*noo'kle-ŏ-sōm*) A complex of DNA and histone proteins that is believed to constitute an inactive form of DNA. In the electron microscope, the histones look like beads threaded on a string of chromatin.

nucleotide (*noo'kle-ŏ-tīd*) The subunit of DNA and RNA macromolecules. Each nucleotide is composed of a nitrogenous base (adenine, guanine, cytosine, and thymine or uracil); a sugar (deoxyribose or ribose); and a phosphate group.

nucleus (*noo'kle-us*), **brain** An aggregation of neuron cell bodies within the brain. Nuclei within the brain are surrounded by white matter and are deep to the cerebral cortex.

nucleus, cell The organelle, surrounded by a double saclike membrane called the nuclear envelope, that contains the DNA and genetic information of the cell.

nystagmus (*nĭ-stag'mus*) Involuntary oscillatory movements of the eye.

O

obese (*o-bēs*) Excessively fat.

oligo- (Gk.) Few, small.

oligodendrocyte (*ol''ĭ-go-den'drŏ-sīt*) A type of neuroglial cell that forms myelin sheaths around axons in the central nervous system.

oncogene (*on'kŏ-jēn*) A gene that contributes to cancer. Oncogenes are believed to be abnormal forms of genes that participate in normal cellular regulation.

oncology (*on-kol'ŏ-je*) The study of tumors.

oncotic (*on-kot'ik*) **pressure** The colloid osmotic pressure of solutions produced by proteins. In plasma, it serves to counterbalance the outward filtration of fluid from capillaries due to hydrostatic pressure.

oo- (Gk.) Pertaining to an egg.

oocyte (*o'ŏ-sīt*) An immature egg cell (ovum). A primary oocyte has not yet completed the first meiotic division; a secondary oocyte has begun the second meiotic division. A secondary oocyte, arrested at metaphase II, is ovulated.

oogenesis (*o''ŏ-jen'ĕ-sis*) The formation of ova in the ovaries.

opsonization (*op''sŏ-nĭ-za'shun*) The process by which antibodies enhance the ability of phagocytic cells to attack bacteria.

optic (*op'tik*) **disc** The area of the retina where axons from ganglion cells gather to form the optic nerve and where blood vessels enter and leave the eye. It corresponds to the blind spot in the visual field due to the absence of photoreceptors.

organ A structure in the body composed of two or more primary tissues that performs a specific function.

organelle (*or''gă-nel*) A structure within cells that performs specialized tasks. The term includes mitochondria, Golgi apparatus, endoplasmic reticulum, nuclei, and lysosomes; it is also used for some structures not enclosed by a membrane, such as ribosomes and centrioles.

organ of Corti The structure within the cochlea that is the functional unit of hearing. It consists of hair cells and supporting cells on the basilar membrane that help transduce sound waves into nerve impulses.

osmolality (*oz''mŏ-lal'ĭ-te*) A measure of the total concentration of a solution; the number of moles of solute per kilogram of solvent.

osmoreceptor (*oz''mŏ-re-cep'tor*) A sensory neuron that responds to changes in the osmotic pressure of the surrounding fluid.

osmosis (*oz-mo'sis*) The passage of solvent (water) from a more dilute to a more concentrated solution through a membrane that is more permeable to water than to the solute.

osmotic (*oz-mot'ik*) **pressure** A measure of the tendency for a solution to gain water by osmosis when separated by a membrane from pure water. Directly related to the osmolality of the solution, it is the pressure required to just prevent osmosis.

osteo- (Gk.) Pertaining to bone.

osteoblast (*os'te-ŏ-blast'*) A bone-forming cell.

osteoclast (*os'te-ŏ-klast'*) A cell that resorbs bone by promoting the dissolution of calcium phosphate crystals.

osteocyte (*os'te-ŏ-sīt*) A mature bone cell that has become entrapped within a matrix of bone. This cell remains alive due to nourishment supplied by canaliculi within the extracellular material of bone.

osteomalacia (*os''te-o-mă-la'shă*) Softening of bones due to a deficiency of vitamin D and calcium.

osteoporosis (*os''te-o-pŏ-ro'sis*) Demineralization of bone, seen most commonly in postmenopausal women and patients who are inactive or paralyzed. It may be accompanied by pain, loss of stature, and other deformities and fractures.

ovary (*o'vă-re*) The gonad of a female that produces ova and secretes female sex steroids.

ovi- (L.) Pertaining to an egg.

oviduct (*o'vĭ-dukt*) The part of the female reproductive tract that transports ova from the ovaries to the uterus. Also called the *uterine*, or *fallopian*, *tube*.

ovulation (*ov-yŭ-la'shun*) The extrusion of a secondary oocyte from the ovary.

oxidation-reduction (*ok''sĭ-da'shun-re-duk'shun*) The transfer of electrons or hydrogen atoms from one atom or molecule to another. The atom or molecule that loses the electrons or hydrogens is oxidized; the atom or molecule that gains the electrons or hydrogens is reduced.

oxidative phosphorylation (*ok''sĭ-da'tiv fos''for-ĭ-la'shun*) The formation of ATP by using energy derived from electron transport to oxygen. It occurs in the mitochondria.

oxidizing (*ok'sĭ-dizing*) **agent** An atom that accepts electrons in an oxidation-reduction reaction.

oxygen (*ok'sĭ-jen*) **debt** The extra amount of oxygen required by the body after exercise to metabolize lactic acid and to supply the higher metabolic rate of muscles warmed during exercise.

oxyhemoglobin (*ok''se-he''mŏ-glo'bin*) A compound formed by the bonding of molecular oxygen to hemoglobin.

oxyhemoglobin saturation The ratio, expressed as a percentage, of the amount of oxyhemoglobin compared to the total amount of hemoglobin in blood.

oxytocin (*ok''sĭ-to'sin*) One of the two hormones produced in the hypothalamus and secreted by the posterior pituitary (the other being vasopressin). Oxytocin stimulates the contraction of uterine smooth muscles and promotes milk ejection in females.

P

pacemaker A group of cells that has the fastest spontaneous rate of depolarization and contraction in a mass of electrically coupled cells; in the heart, this is the sinoatrial, or SA, node.

pacesetter potentials Changes in membrane potential produced spontaneously by pacemaker cells of single-unit smooth muscles.

pacinian corpuscle (*pă-sin'e-an kor'pus'l*) A cutaneous sensory receptor sensitive to pressure. It is characterized by an onionlike layering of cells around a central sensory dendrite.

PAH Para-aminohippuric (*par'ă-ă-me'no-hi-pyoor''ik*) **acid.** A substance used to measure total renal plasma flow because its clearance rate is equal to the total rate of plasma flow to the kidneys; PAH is filtered and secreted but not reabsorbed by the renal nephrons.

pancreatic (*pan''kre-at'ik*) **islets** *See* islets of Langerhans.

pancreatic juice The secretions of the pancreas that are transported by the pancreatic duct to the duodenum. Pancreatic juice contains bicarbonate and the digestive enzymes trypsin, lipase, and amylase.

paracrine (*par'ă-krin*) A regulatory molecule that acts on a different tissue of the same organ than the tissue that produced it. For example, the endothelium of blood vessels secretes a number of paracrine regulators that act on the smooth muscle layer of the vessels to cause vasoconstriction or vasodilation.

parasympathetic (*par''ă-sim''pă-thet'ik*) Pertaining to the craniosacral division of the autonomic nervous system.

parathyroid (*par''ă-thi'roid*) **hormone** (**PTH**) A polypeptide hormone secreted by the parathyroid glands, PTH acts to raise the blood Ca^{++} levels primarily by stimulating resorption of bone.

Parkinson's disease A tremor of the resting muscles and other symptoms caused by inadequate dopamine-producing neurons in the basal ganglia of the cerebrum. Also called *paralysis agitans*.

parturition (*par''tyoo-rish'un*) The process of giving birth; childbirth.

passive immunity Specific immunity granted by the administration of antibodies made by another organism.

Pasteur effect A decrease in the rate of glucose utilization and lactic acid production in tissues or organisms by their exposure to oxygen.

pathogen (*path'ŏ-jen*) Any disease-producing microorganism or substance.

pepsin (*pep'sin*) The protein-digesting enzyme secreted in gastric juice.

peptic ulcer (*pep'tik ul'ser*) An injury to the mucosa of the esophagus, stomach, or small intestine caused by acidic gastric juice.

perfusion (*per'fyoo''zhun*) The flow of blood through an organ.

peri- (Gk.) Around, surrounding.

perilymph (*per'ĭ-limf*) The fluid between the membranous and bony labyrinths of the inner ear.

perimysium (*per''ĭ-mis'e-um*) The connective tissue surrounding a fascicle of skeletal muscle fibers.

periosteum (*per''e-os'te-um*) Connective tissue covering bones. It contains osteoblasts and is therefore capable of forming new bone.

peripheral resistance The resistance to blood flow through the arterial system. Peripheral resistance is largely a function of the radius of small arteries and arterioles. The resistance to blood flow is proportional to the fourth power of the radius of the vessel.

peristalsis (*per″ĭ-stal′sis*) Waves of smooth muscle contraction in smooth muscles of the tubular digestive tract that involve circular and longitudinal muscle fibers at successive locations along the tract. It serves to propel the contents of the tract in one direction.

permissive effect The phenomenon in which the presence of one hormone "permits" the full exertion of the effects of another hormone. This may be due to promotion of the synthesis of the active form of the second hormone, or it may be due to an increase in the sensitivity of the target tissue to the effects of the second hormone.

pH The symbol (short for *p*otential of *h*ydrogen) used to describe the hydrogen ion (H^+) concentration of a solution. The pH scale in common use ranges from 0 to 14. Solutions with a pH of 7 are neutral; those with a pH lower than 7 are acidic; and those with a higher pH are basic.

phagocytosis (*fag″ŏ-si-to′sis*) Cellular eating; the ability of some cells (such as white blood cells) to engulf large particles (such as bacteria) and digest these particles by merging the food vacuole in which they are contained with a lysosome containing digestive enzymes.

phenylalanine (*fen″il-al′ă-nēn*) An amino acid that also serves as the precursor of L-dopa, dopamine, norepinephrine, and epinephrine.

phenylketonuria (*fen″il-kēt″on-our′e-ă*)(**PKU**) An inborn error of metabolism that results in the inability to convert the amino acid phenylalanine into tyrosine. This defect can cause central nervous system damage if the child is not placed on a diet low in phenylalanine.

phonocardiogram (*fo″nŏ-kar′de-ŏ-gram*) A visual display of the heart sounds.

phosphodiesterase (*fos″fo-di-es′ter-ās*) An enzyme that cleaves cyclic AMP into inactive products, thus inhibiting the action of cyclic AMP as a second messenger.

phospholipid (*fos″fo-lip′id*) A lipid containing a phosphate group. These molecules (such as lecithin) are polar on one end and nonpolar on the other end. Phospholipids make up a large part of the cell membrane and function in the lung alveoli as surfactants.

phosphorylation (*fos″for-ĭ-la′shun*) The addition of an inorganic phosphate group to an organic molecule. Examples include the addition of a phosphate group to ADP to make ATP, and the addition of a phosphate group to specific proteins as a result of the action of protein kinase enzymes.

photoreceptors (*fo″to-re-sep′torz*) Sensory cells (rods and cones) that respond electrically to light; they are located in the retina of the eyes.

pia mater (*pi′ă ma′ter*) The innermost of the connective tissue meninges that envelops the brain and spinal cord.

pineal (*pin′e-al*) **gland** A gland within the brain that secretes the hormone melatonin. It is affected by sensory input from the photoreceptors of the eyes.

pinocytosis (*pin″ŏ-si-to′sis*) Cell drinking; invagination of the cell membrane to form narrow channels that pinch off into vacuoles. This permits cellular intake of extracellular fluid and dissolved molecules.

pituitary (*pĭ-too′ĭ-ter-e*) **gland** Also called the *hypophysis*. A small endocrine gland joined to the hypothalamus at the base of the brain. The pituitary gland is functionally divided into anterior and posterior portions. The anterior pituitary secretes ACTH, TSH, FSH, LH, growth hormone, and prolactin. The posterior pituitary secretes oxytocin and antidiuretic hormone (ADH).

plasma (*plaz′mă*) The fluid portion of the blood. Unlike serum (which lacks fibrinogen), plasma is capable of forming insoluble fibrin threads when in contact with test tubes.

plasma cells Cells derived from B lymphocytes that produce and secrete large amounts of antibodies. They are responsible for humoral immunity.

plasmalemma (*plaz″mă-lem′ă*) The cell membrane; an alternate term for the semipermeable membrane that encloses the cytoplasm of a cell.

platelet (*plāt′let*) A disc-shaped structure, 2 to 4 micrometers in diameter, that is derived from bone marrow cells called megakaryocytes. Platelets circulate in the blood and participate (together with fibrin) in forming blood clots.

pluripotent (*ploo-rip′ŏ-tent*) A term used to describe the ability of early embryonic cells to specialize in a number of ways to produce tissues characteristic of different organs.

pneumotaxic (*noo″mŏ-tak′sik*) **center** A neural center in the pons that rhythmically inhibits inspiration in a manner independent of sensory input.

pneumothorax (*noo″mo-thor′aks*) An abnormal condition in which air enters the intrapleural space, either through an open chest wound or from a tear in the lungs. This can lead to the collapse of a lung.

PNS The peripheral nervous system, including nerves and ganglia.

-pod, -podium (Gk.) Foot, leg, extension.

Poiseuille's (*pwă-zŭ′yez*) **law** The statement that the rate of blood flow through a vessel is directly proportional to the pressure difference between the two ends of the vessel and inversely proportional to the length of the vessel, the viscosity of the blood, and the fourth power of the radius of the vessel.

polar body A small daughter cell formed by meiosis that degenerates in the process of oocyte production.

polar molecule A molecule in which the shared electrons are not evenly distributed, so that one side of the molecule is negatively (or positively) charged in comparison with the other side. Polar molecules are soluble in polar solvents such as water.

poly- (Gk.) Many.

polycythemia (*pol″e-si-the′me-ă*) An abnormally high red blood cell count.

polydipsia (*pol″e-dip′se-ă*) Excessive thirst.

polymer (*pol′ĭ-mer*) A large molecule formed by the combination of smaller subunits, or monomers.

polymorphonuclear (*pol″e-mor″fŏ-noo′kle-ar*) **leukocyte** A granular leukocyte containing a nucleus with a number of lobes connected by thin, cytoplasmic strands. This term includes neutrophils, eosinophils, and basophils.

polypeptide (*pol″e-pep′tīd*) A chain of amino acids connected by covalent bonds called peptide bonds. A very large polypeptide is called a protein.

polyphagia (*pol′e-fa″je-ă*) Excessive eating.

polysaccharide (*pol″e-sak′ă-rīd*) A carbohydrate formed by covalent bonding of numerous monosaccharides. Examples include glycogen and starch.

polyuria (*pol″e-yoor′e-ă*) Excretion of an excessively large volume of urine in a given period.

portal (*por′tal*) **system** A system of vessels consisting of two capillary beds in series, where blood from the first is drained by veins into a second capillary bed, which in turn is drained by veins that return blood to the heart. The two major portal systems in the body are the hepatic portal system and the hypothalamo-hypophyseal portal system.

positive feedback A mechanism of response that results in the amplification of an initial change. Positive feedback results in avalanche-like effects, as occur in the formation of a blood clot or in the production of the LH surge by the stimulatory effect of estrogen.

posterior (*pos-tēr′e-or*) At or toward the back of an organism, organ, or part; the dorsal surface.

posterior pituitary *See* neurohypophysis.

postsynaptic (*pōst″sĭ-nap′tik*) **inhibition** The inhibition of a postsynaptic neuron by axon endings that release a neurotransmitter that induces hyperpolarization (inhibitory postsynaptic potentials).

post-tetanic potentiation (*pōst″tĕ-tan′ik pŏ-ten″she-ă′shun*) The improved ability of a presynaptic neuron that has been stimulated at high frequency to release neurotransmitter when stimulated again at a later time.

potential (*pŏ-ten′shal*) **difference** In biology, the difference in charge between two solutions separated by a membrane. The potential difference is measured in voltage.

prehormone (*pre-hor′mōn*) An inactive form of a hormone that is secreted by an endocrine gland. The prehormone is converted within its target cells to the active form of the hormone.

presynaptic (*pre″sĭ-nap′tik*) **inhibition** Neural inhibition in which axoaxonic synapses inhibit the release of neurotransmitter chemicals from the presynaptic axon.

pro- (Gk.) Before, in front of, forward.

process (*pros′es, pro′ses*), **cell** Any thin cytoplasmic extension of a cell, such as the dendrites and axon of a neuron.

progesterone (*pro-jes′tĕ-rōn*) A steroid hormone secreted by the corpus luteum of the ovaries and by the placenta. Secretion of progesterone during the luteal phase of the menstrual cycle promotes the final maturation of the endometrium.

prohormone (*pro-hor′mōn*) The precursor of a polypeptide hormone that is larger and less active than the hormone. The prohormone is produced within the cells of an endocrine gland and is normally converted into the shorter, active hormone prior to secretion.

prolactin (*pro-lak′tin*) A hormone secreted by the anterior pituitary that stimulates lactation (acting together with other hormones) in the postpartum female. It may also participate (along with the gonadotropins) in regulating gonadal function in some mammals.

prophylaxis (*pro″fĭ-lak′sis*) Prevention or protection.

proprioceptor (*pro″pre-o-sep′tor*) A sensory receptor that provides information about body position and movement. Examples include receptors in muscles, tendons, and joints and in the semicircular canals of the inner ear.

prostaglandin (*pros″tă-glan′din*) Any of a family of fatty acids containing a cyclic ring that serve numerous autocrine regulatory functions.

protein (*pro′te-in*) The class of organic molecules composed of large polypeptides, in which over a hundred amino acids are bonded together by peptide bonds.

protein kinase (*ki′nās*) The enzyme activated by cyclic AMP that catalyzes the phosphorylation of specific proteins (enzymes). Such phosphorylation may activate or inactivate enzymes.

proto- (Gk.) First, original.

proton (*pro′ton*) A unit of positive charge in the nucleus of atoms.

protoplasm (*pro′tŏ-plaz″em*) A general term for the colloidal complex of protein that includes cytoplasm and nucleoplasm.

pseudo- (Gk.) False.

pseudohermaphrodite (*soo″dŏ-her-maf′rŏ-dīt*) An individual who has the gonads of one sex only, but some of the body features of the opposite sex. (A true hermaphrodite has both ovarian and testicular tissue.)

pseudopod (*soo′dŏ-pod*) A footlike extension of the cytoplasm that enables some cells (with amoeboid motion) to move across a substrate. Pseudopods also are used to surround food particles in the process of phagocytosis.

ptyalin (*ti′ă-lin*) An enzyme in saliva that catalyzes the hydrolysis of starch into smaller molecules; also called *salivary amylase*.

puberty (*pyoo′ber-te*) The period in an individual's life span when secondary sexual characteristics and fertility develop.

pulmonary (*pul′mŏ-ner″e*) **circulation** The part of the vascular system that includes the pulmonary arteries and pulmonary veins. It transports blood from the right ventricle of the heart, through the lungs, and back to the left atrium of the heart.

pupil The opening at the center of the iris of the eye.

Purkinje (*pur-kin′je*) **fibers** Specialized conducting tissue in the ventricles of the heart that carry impulses from the bundle of His to the myocardium of the ventricles.

pyramidal (*pĭ-ram′ĭ-dal*) **tracts** Motor tracts that descend without synaptic interruption from the cerebrum to the spinal cord, where they synapse either directly or indirectly (via spinal interneurons) with the lower motor neurons of the spinal cord; also called *cortiscospinal tracts*.

pyrogen (*pi′rŏ-jen*) A fever-producing substance.

Q

QRS complex The principal deflection of an electrocardiogram, produced by depolarization of the ventricles.

R

reabsorption (*re″ab-sorp′shun*) The transport of a substance from the lumen of the renal nephron into the peritubular capillaries.

receptive field An area of the body that, when stimulated by a sensory stimulus, activates a particular sensory receptor.

reciprocal innervation (*rĭ-sip′rŏ-kal in″er-va′shun*) The process whereby the motor neurons to an antagonistic muscle are inhibited when the motor neurons to an agonist muscle are stimulated. In this way, for example, the extensor muscle of the elbow joint is inhibited when the flexor muscles of this joint are stimulated to contract.

recruitment (*rĭ-kroot′ment*) In terms of muscle contraction, the successive stimulation of more and larger motor units in order to produce increasing strengths of muscle contraction.

reduced hemoglobin Hemoglobin with iron in the reduced ferrous state. It is able to bond with oxygen but is not combined with oxygen. Also called *deoxyhemoglobin*.

reducing agent An electron donor in a coupled oxidation-reduction reaction.

refraction (*rĭ-frak′shun*) The bending of light rays when light passes from a medium of one density to a medium of another density. Refraction of light by the cornea and lens acts to focus the image on the retina of the eye.

refractory (*rĭ-frak′tŏ-re*) **period** The period of time during which a region of axon or muscle cell membrane cannot be stimulated to produce an action potential (absolute refractory period), or can be stimulated only by a very strong stimulus (relative refractory period).

releasing hormones Polypeptide hormones secreted by neurons in the hypothalamus that travel in the hypothalamo-hypophyseal portal system to the anterior pituitary and stimulate the anterior pituitary to secrete specific hormones.

REM sleep The stage of sleep in which dreaming occurs. It is associated with rapid eye movements (REMs). REM sleep occurs three to four times each night and lasts from a few minutes to over an hour.

renal (*re′nal*) Pertaining to the kidneys.

renal plasma clearance The milliliters of plasma that are cleared of a particular solute each minute by the excretion of that solute in the urine. If there is no reabsorption or secretion of that solute by the nephron tubules, the renal plasma clearance is equal to the glomerular filtration rate.

renal pyramid (*pĭ′ră-mid*) One of a number of cone-shaped tissue masses that compose the renal medulla.

renin (*re′nin*) An enzyme secreted into the blood by the juxtaglomerular apparatus of the kidneys. Renin catalyzes the conversion of angiotensinogen into angiotensin II.

rennin (*ren′in*) A digestive enzyme secreted in the gastric juice of infants that catalyzes the digestion of the milk protein casein.

repolarization (*re-po″lar-ĭ-za′shun*) The reestablishment of the resting membrane potential after depolarization has occurred.

resorption (*re-sorp′shun*), **bone** The dissolution of the calcium phosphate crystals of bone by the action of osteoclasts.

respiratory acidosis (*rĭ-spīr′ă-tor-e as″ĭ-do′sis*) A lowering of the blood pH to below 7.35 due to the accumulation of CO_2 as a result of hypoventilation.

respiratory alkalosis (*al″kă-lo′sis*) A rise in blood pH to above 7.45 due to the excessive elimination of blood CO_2 as a result of hyperventilation.

respiratory distress syndrome A lung disease of the newborn, most frequently occurring in premature infants, that is caused by abnormally high alveolar surface tension as a result of a deficiency in lung surfactant; also called *hyaline membrane disease*.

respiratory zone The region of the lungs including the respiratory bronchioles, in which individual alveoli are found, and the terminal alveoli. Gas exchange between the inspired air and pulmonary blood occurs in the respiratory zone of the lungs.

resting potential The potential difference across a cell membrane when the cell is in an unstimulated state. The resting potential is always negatively charged on the inside of the membrane compared to the outside.

reticular (*rĭ-tik′yu-lar*) **activating system (RAS)** A complex network of nuclei and fiber tracts within the brain stem that produces nonspecific arousal of the cerebrum to incoming sensory information. The RAS thus maintains a state of alert consciousness and must be depressed during sleep.

retina (*ret′-n-ă*) The layer of the eye that contains neurons and photoreceptors (rods and cones).

rhodopsin (*ro-dop′sin*) Visual purple. A pigment in rod cells that undergoes a photochemical dissociation in response to light and, in so doing, stimulates electrical activity in the photoreceptors.

riboflavin (*rĭ′bo-fla″vin*) Vitamin B_2. Riboflavin is a water-soluble vitamin that is used to form the coenzyme FAD, which participates in the transfer of hydrogen atoms.

ribosome (*ri′bo-sōm*) A cytoplasmic organelle composed of protein and ribosomal RNA that is responsible for the translation of messenger RNA and protein synthesis.

rickets (*rik′ets*) A condition caused by a deficiency of vitamin D and associated with interference of the normal ossification of bone.

rigor mortis (*rig′or mor′tis*) The stiffening of a dead body due to the depletion of ATP and the production of rigor complexes between actin and myosin in muscles.

RNA Ribonucleic (*ri″bo-noo-kle′ik*) acid. A nucleic acid consisting of the nitrogenous bases adenine, guanine, cytosine, and uracil; the sugar ribose; and phosphate groups. There are three types of RNA found in cytoplasm: messenger RNA (mRNA), transfer RNA (tRNA), and ribosomal RNA (rRNA).

rods One of the two categories of photoreceptors (the other being cones) in the retina of the eye. Rods are responsible for black-and-white vision under low illumination.

S

saccadic (*să-kad'ik*) **eye movements** Very rapid eye movements that occur constantly and that change the focus on the retina from one point to another.

saltatory (*sal'tă-tor-e*) **conduction** The rapid passage of action potentials from one node of Ranvier to another in myelinated axons.

sarcolemma (*sar″cŏ-lem'ă*) The cell membrane of striated muscle cells.

sarcomere (*sar'kŏ-mēr*) The structural subunit of a myofibril in a striated muscle, equal to the distance between two successive Z lines.

sarcoplasm (*sar'kŏ-plaz″em*) The cytoplasm of striated muscle cells.

sarcoplasmic reticulum (*sar″kŏ-plaz′mik rĭ-tik'yŭ-lum*) The smooth or agranular endoplasmic reticulum of striated muscle cells. It surrounds each myofibril and serves to store Ca^{++} when the muscle is at rest.

Schwann (*shvan*) **cell** A neuroglial cell of the peripheral nervous system that forms sheaths around peripheral nerve fibers. Schwann cells also direct regeneration of peripheral nerve fibers to their target cells.

sclera (*skler'ă*) The tough white outer coat of the eyeball that is continuous anteriorly with the clear cornea.

second messenger A molecule or ion whose concentration within a target cell is increased by the action of a regulator compound (e.g., hormone or neurotransmitter) and which stimulates the metabolism of that target cell in a way characteristic of the actions of that regulator molecule—that is, in a way that mediates the intracellular effects of that regulatory compound.

secretin (*sĕ-kre'tin*) A polypeptide hormone secreted by the small intestine in response to acidity of the intestinal lumen. Along with cholecystokinin, secretin stimulates the secretion of pancreatic juice into the small intestine.

secretion (*sĕ-kre'shun*), **renal** The transport of a substance from the blood through the wall of the nephron tubule into the urine.

semen (*se'men*) The fluid ejaculated by a male, containing sperm and the secretions of the prostate and seminal vesicles.

semicircular canals Three canals of the bony labyrinth that contain endolymph, which is continuous with the endolymph of the membranous labyrinth of the cochlea. The semicircular canals provide a sense of equilibrium.

semilunar (*sem″e-loo'nar*) **valves** The valve flaps of the aorta and pulmonary artery at their juncture with the ventricles.

seminal vesicles (*sem'ĭ-nal ves'ĭ-k'lz*) The paired organs located on the posterior border of the urinary bladder that empty their contents into the vas deferens and thus contribute to the semen.

seminiferous tubules (*sem″ĭ-nif'er-us too'byoolz*) The tubules within the testes that produce spermatozoa by meiotic division of their germinal epithelium.

semipermeable (*sem″e-per′me-ă-b'l*) **membrane** A membrane with pores of a size that permits the passage of solvent and some solute molecules but restricts the passage of other solute molecules.

sensory neuron (*noor'on*) An afferent neuron that conducts impulses from peripheral sensory organs into the central nervous system.

serosa (*sĭ-ro'să*) An outer epithelial membrane that covers the surface of a visceral organ.

Sertoli (*ser-to'le*) **cells** Nongerminal supporting cells in the seminiferous tubules. Sertoli cells envelop spermatids and appear to participate in the transformation of spermatids into spermatozoa.

serum (*ser'um*) The fluid squeezed out of a clot as it retracts; the supernatant when a sample of blood clots in a test tube and is centrifuged. Serum is plasma from which fibrinogen and other clotting proteins have been removed as a result of clotting.

sex chromosomes The X and Y chromosomes. The unequal pairs of chromosomes involved in sex determination (which is due to the presence or absence of a Y chromosome). Females lack a Y chromosome and normally have the genotype XX; males have a Y chromosome and normally have the genotype XY.

shock As it relates to the cardiovascular system, a rapid, uncontrolled fall in blood pressure, which in some cases becomes irreversible and leads to death.

sickle-cell anemia A hereditary, autosomal recessive trait that occurs primarily in people of African ancestry, in which it evolved apparently as a protection (in the carrier state) against malaria. In the homozygous state, hemoglobin S is made instead of hemoglobin A; this leads to the characteristic sickling of red blood cells, hemolytic anemia, and organ damage.

sinoatrial (*si″no-a'tre-al*) **node** A mass of specialized cardiac tissue in the wall of the right atrium that initiates the cardiac cycle; the SA node; also called the *pacemaker*.

sinus (*si'nus*) A cavity.

sinusoid (*si'nŭ-soid*) A modified capillary with a relatively large diameter that connects the arterioles and venules in the liver, bone marrow, lymphoid tissues, and some endocrine organs. In the liver, sinusoids are partially lined by phagocytic cells of the reticuloendothelial system.

skeletal muscle pump A term used with reference to the effect of skeletal muscle contraction on the flow of blood in veins. As the muscles contract, they squeeze the veins, and in this way help move the blood toward the heart.

sleep apnea A temporary cessation of breathing during sleep, usually lasting for several seconds.

sliding filament theory The theory that the thick and thin filaments of a myofibril slide past each other, while maintaining their initial length, during muscle contraction.

smooth muscle A specialized type of nonstriated muscle tissue composed of fusiform, single-nucleated fibers. It contracts in an involuntary, rhythmic fashion in the walls of visceral organs.

sodium/potassium (*so'de-um/pŏ-tas'e-um*) **pump** An active transport carrier, with ATPase enzymatic activity that acts to accumulate K^+ within cells and extrude Na^+ from cells, thus maintaining gradients for these ions across the cell membrane.

soma-, somato-, -some (Gk.) Body, unit.

somatesthetic (*so″mat-es-thet'ik*) **sensations** Sensations arising from cutaneous, muscle, tendon, and joint receptors. These sensations project to the postcentral gyrus of the cerebral cortex.

somatic (*so-mat'ik*) **motor neuron** A motor neuron in the spinal cord that innervates skeletal muscles. Somatic motor neurons are divided into alpha and gamma motoneurons.

somatomammotropic (*so'mă-tŏ-mam″ŏ-trop'ik*) **hormone** A hormone secreted by the placenta that has actions similar to the pituitary hormones growth hormone and prolactin.

somatomedin (*so″mă-tŏ-med'n*) Any of a group of small polypeptides that are believed to be produced in the liver in response to growth hormone stimulation and to mediate the actions of growth hormone on the skeleton and other tissues.

somatostatin (*so″mat-ŏ-stāt'n*) A polypeptide produced in the hypothalamus that acts to inhibit the secretion of growth hormone from the anterior pituitary. Somatostatin is also produced in the islets of Langerhans of the pancreas, but its function there has not been established.

somatotropic (*sŏ″mat'ă-trop'ik*) **hormone** Growth hormone. An anabolic hormone secreted by the anterior pituitary that stimulates skeletal growth and protein synthesis in many organs.

sounds of Korotkoff (*kŏ-rot'kof*) The sounds heard when blood pressure measurements are taken. These sounds are produced by the turbulent flow of blood through an artery that has been partially constricted by a pressure cuff.

spastic paralysis (*spas'tik pă-ral'ĭ-sis*) Paralysis in which the muscles have such a high tone that they remain in a state of contracture. This may be caused by inability to degrade ACh released at the neuromuscular junction (as caused by certain drugs) or by damage to the spinal cord.

spermatid (*sper'mă-tid*) Any of the four haploid cells formed by meiosis in the seminiferous tubules that mature to become spermatozoa without further division.

spermatocyte (*sper-mat'ŏ-sīt*) A diploid cell of the seminiferous tubules in the testes that divides by meiosis to produce spermatids.

spermatogenesis (*sper-mat″ŏ-jen'ĭ-sis*) The formation of spermatozoa, including meiosis and maturational processes in the seminiferous tubules.

spermatozoon (*sper-mat″ŏ-zo'on*) pl. *spermatozoa* or, loosely, *sperm*. A mature sperm cell, formed from a spermatid.

spermiogenesis (*sper″me-ŏ-jen'ĕ-sis*) The maturational changes that transform spermatids into spermatozoa.

sphygmo- (Gk.) The pulse.

sphygmomanometer (*sfig″mo-mă-nom'ĭ-ter*) A manometer (pressure transducer) used to measure the blood pressure.

spindle fibers Filaments that extend from the poles of a cell to its equator and attach to chromosomes during the metaphase stage of cell division. Contraction of the spindle fibers pulls the chromosomes to opposite poles of the cell.

spironolactone (*spi-rŏ″no-lak'tōn*) A diuretic drug that acts as an aldosterone antagonist.

Starling forces The hydrostatic pressures and the colloid osmotic pressures of the blood and tissue fluid. The balance of the Starling forces determines the net movement of fluid out of or into blood capillaries.

steroid (*ster'oid*) A lipid, derived from cholesterol, that has three six-sided carbon rings and one five-sided carbon ring. These form the steroid hormones of the adrenal cortex and gonads.

stretch reflex The monosynaptic reflex whereby stretching a muscle results in a reflex contraction. The knee-jerk reflex is an example of a stretch reflex.

striated (*stri'āt-ed*) **muscle** Skeletal and cardiac muscle, the cells of which exhibit cross banding, or striations, due to the arrangement of thin and thick filaments into sarcomeres.

stroke volume The amount of blood ejected from each ventricle at each heartbeat.

sub- (L.) Under, below.

substrate (*sub'strāt*) In enzymatic reactions, the molecules that combine with the active sites of an enzyme and are converted to products by catalysis of the enzyme.

sulcus (*sul'kus*) A groove or furrow; a depression in the cerebrum that separates the folds, or gyri, of the cerebral cortex.

summation (*su-ma'shun*) In neural physiology, the additive effects of graded synaptic potentials. In muscle physiology, the additive effects of contractions of different muscle fibers.

super-, supra- (L.) Above, over.

suppressor T cells A subpopulation of T lymphocytes that acts to inhibit the production of antibodies against specific antigens by B lymphocytes.

surfactant (*sur-fak'tant*) In the lungs, a mixture of phospholipids and proteins produced by alveolar cells that reduces the surface tension of the alveoli and contributes to the lungs' elastic properties.

sym-, syn- (Gk.) With, together.

synapse (*sin'aps*) The junction across which a nerve impulse is transmitted from an axon terminal to a neuron, a muscle cell, or a gland cell either directly or indirectly (via the release of chemical neurotransmitters).

synapsin (*sĭ-nap'sin*) A protein within the membrane of the synaptic vesicles of axons. When activated by the arrival of action potentials, synapsins aid the fusion of the synaptic vesicles with the cell membrane so that the vesicles may undergo exocytosis and release their content of neurotransmitters.

synaptic plasticity (*sĭ-nap'tik plas-tis'ĭ-te*) The ability of synapses to change at a cellular or molecular level. At a cellular level, plasticity refers to the ability to form new synaptic associations. At a molecular level, plasticity refers to the ability of a presynaptic axon to release more than one type of neurotransmitter.

syncytium (*sin-sish'e-um*) The merging of cells in a tissue into a single functional unit. Because the atria and ventricles of the heart have gap junctions between their cells, these myocardia behave as syncytia.

synergistic (*sin"er-jis'tik*) Pertaining to regulatory processes or molecules (such as hormones) that have complementary or additive effects.

systemic (*sis-tem'ik*) **circulation** The circulation that carries oxygenated blood from the left ventricle via arteries to the tissue cells and that carries blood depleted in oxygen via veins to the right atrium; the general circulation, as compared to the pulmonary circulation.

systole (*sis'tŏ-le*) The phase of contraction in the cardiac cycle. Used alone, this term refers to contraction of the ventricles; the term *atrial systole* refers to contraction of the atria.

T

tachycardia (*tak"ĭ-kar'de-ă*) An excessively rapid heart rate, usually applied to rates in excess of 100 beats per minute (in contrast to bradycardia, in which the heart rate is very slow—below 60 beats per minute).

target organ The organ that is specifically affected by the action of a hormone or other regulatory process.

T cell A type of lymphocyte that provides cell-mediated immunity (in contrast to B lymphocytes, which provide humoral immunity through the secretion of antibodies). There are three subpopulations of T cells: cytotoxic (killer), helper, and suppressor.

telo- (Gk.) An end, complete, final.

telophase (*tel'ŏ-fāz*) The last step of mitosis and the last step of the second division of meiosis.

tendon (*ten'dun*) The dense regular connective tissue that attaches a muscle to the bones of its origin and insertion.

testes (*tes'tēz*) Sing. *testis*. Male gonads. Testes are also known as *testicles*.

testis-determining factor The product of a gene located on the short arm of the Y chromosome that causes the indeterminate embryonic gonads to develop into testes.

testosterone (*tes-tos'tĕ-rōn*) The major androgenic steroid secreted by the Leydig cells of the testes after puberty.

tetanus (*tet'n-us*) In physiology, a term used to denote a smooth, sustained contraction of a muscle, as opposed to muscle twitching.

tetraiodothyronine (*tet"răi"ŏ-dŏ-thi'ro-nēn*) (T_4) A hormone containing four iodine atoms; also known as *thyroxine*.

thalassemia (*thal"ă-se'me-ă*) Any of a group of hemolytic anemias caused by the hereditary inability to produce either the alpha or beta chain of hemoglobin. It is found primarily among Mediterranean people.

theophylline (*the-of'ĭ-lin*) A drug found in some tea leaves that promotes dilation of the bronchioles by increasing the intracellular concentration of cyclic AMP (cAMP) in the smooth muscle cells. This effect is due to inhibition of the enzyme phosphodiesterase, which breaks down cAMP.

thermiogenesis (*ther-me-o-jen'ĭ-sis*) The production of heat by the body through mechanisms such as increased metabolic rate.

thorax (*thor'aks*) The part of the body cavity above the diaphragm; the chest.

threshold The minimum stimulus that just produces a response.

thrombin (*throm'bin*) A protein formed in blood plasma during clotting that enzymatically converts the soluble protein fibrinogen into insoluble fibrin.

thrombocyte (*throm'bŏ-sīt*) A blood platelet; a disc-shaped structure in blood that participates in clot formation.

thrombosis (*throm-bo'sis*) The development or presence of a thrombus.

thrombus (*throm'bus*) A blood clot produced by the formation of fibrin threads around a platelet plug.

thymus (*thi'mus*) A lymphoid organ located in the superior portion of the anterior mediastinum. It processes T lymphocytes and secretes hormones that regulate the immune system.

thyroglobulin (*thi-ro-glob'yŭ-lin*) An iodine-containing protein in the colloid of the thyroid follicles that serves as a precursor for the thyroid hormones.

thyroxine (*thi-rok'sin*) Also called *tetraiodothyronine*, or T_4. The major hormone secreted by the thyroid gland. It regulates the basal metabolic rate and stimulates protein synthesis in many organs. A deficiency of this hormone in early childhood produces cretinism.

tinnitus (*tĭ-ni'tus*) The spontaneous sensation of a ringing sound or other noise without sound stimuli.

tolerance, immunological The ability of the immune system to distinguish self from nonself and resist attacking those antigens that are part of one's own tissues.

total minute volume The product of tidal volume (ml per breath) and ventilation rate (breaths per minute).

toxin (*tok'sin*) A poison.

toxoid (*tok'soid*) A modified bacterial endotoxin that has lost toxicity but still has the ability to act as an antigen and stimulate antibody production.

tracts A collection of axons within the central nervous system that forms the white matter of the CNS.

trans- (L.) Across, through.

transamination (*trans"am-ĭ-na'shun*) The transfer of an amino group from an amino acid to an alpha-keto acid, forming a new keto acid and a new amino acid without the appearance of free ammonia.

transcription (*tran-skrip'shun*), **genetic** The process by which messenger RNA is synthesized from a DNA template resulting in the transfer of genetic information from the DNA molecule to the mRNA.

translation (*trans-la'shun*), **genetic** The process by which messenger RNA directs the amino acid sequence of a growing polypeptide during protein synthesis.

transplantation (*trans"plan-ta'shun*) The grafting of tissue from one part of the body to another part, or from a donor to a recipient.

transpulmonary (*trans"pul'mŏ-ner"-e*) **pressure** The pressure difference across the wall of the lung, equal to the difference between intrapulmonary pressure and intrapleural pressure.

triiodothyronine (*tri"i-ŏ"dŏ-thi'ro-nēn*) (T_3) A hormone secreted in small amounts by the thyroid; the active hormone in target cells formed from thyroxine.

tropomyosin (*tro"pŏ-mi'ŏ-sin*) A filamentous protein that attaches to actin in the thin filaments and that acts, together with another protein called troponin, to inhibit and regulate the attachment of myosin cross bridges to actin.

troponin (*tro'pŏ-nin*) A protein found in the thin filaments of the sarcomeres of skeletal muscle. A subunit of troponin binds to Ca++, and as a result, causes tropomyosin to change position in the thin filament.

trypsin (*trip'sin*) A protein-digesting enzyme in pancreatic juice that is released into the small intestine.

tryptophan (*trip'tŏ-fān*) An amino acid that also serves as the precursor for the neurotransmitter molecule serotonin.

TSH Thyroid-stimulating hormone, also called thyrotropin (*thi"rŏ-tro'pin*). A hormone secreted by the anterior pituitary that stimulates the thyroid gland.

tubuloglomerular (*too″byŭ-lo-glo-mer′yŭ-lar*) **feedback** A control mechanism whereby an increased flow of fluid though the nephron tubules causes a reflex reduction in the glomerular filtration rate.

turgid (*tur′jid*) Swollen and congested.

twitch A rapid contraction and relaxation of a muscle fiber or a group of muscle fibers.

tympanic (*tim-pan′ik*) **membrane** The eardrum; a membrane separating the external from the middle ear that transduces sound waves into movements of the middle-ear ossicles.

U

universal donor A person with blood type O, who is able to donate blood to people with other blood types in emergency blood transfusions.

universal recipient A person with blood type AB, who can receive blood of any type in emergency transfusions.

urea (*yoo-re′ă*) The chief nitrogenous waste product of protein catabolism in the urine, formed in the liver from amino acids.

uremia (*yoo-re′me-ă*) The retention of urea and other products of protein catabolism as a result of inadequate kidney function.

urobilinogen (*yoo″rŏ-bi-lin′ŏ-jen*) A compound formed from bilirubin in the intestine. Some is excreted in the feces and some is absorbed and enters the enterohepatic circulation, where it may be excreted either in the bile or in the urine.

V

vaccination (*vak″sĭ-na′shun*) The clinical induction of active immunity by introducing antigens into the body, so that the immune system becomes sensitized to them. The immune system will mount a secondary response to those antigens upon subsequent exposures.

vagina (*vă-ji′nă*) The tubular organ in the female leading from the external opening of the vulva to the cervix of the uterus.

vagus (*va′gus*) **nerve** The tenth cranial nerve, composed of sensory dendrites from visceral organs and preganglionic parasympathetic nerve fibers. The vagus is the major parasympathetic nerve in the body.

Valsalva (*val-sal′vă*) **maneuver** Exhalation against a closed glottis, so that intrathoracic pressure rises to the point that the veins returning blood to the heart are partially constricted. This produces circulatory and blood pressure changes that could be dangerous.

vasa-, vaso- (L.) Pertaining to blood vessels.

vasa vasora (*va′să va-sor′ă*) Blood vessels that supply blood to the walls of large blood vessels.

vasectomy (*vă-sek′tŏ-me, va-zek′tŏ-me*) Surgical removal of a portion of the vas (ductus) deferens to induce infertility.

vasoconstriction (*va″zo-kon-strik′shun*) A narrowing of the lumen of blood vessels as a result of contraction of the smooth muscles in their walls.

vasodilation (*va″zo-di-la′shun*) A widening of the lumen of blood vessels as a result of relaxation of the smooth muscles in their walls.

vasopressin (*va″zo-pres′in*) Another name for antidiuretic hormone (ADH), secreted by the posterior pituitary. The name *vasopressin* is derived from the fact that this hormone can stimulate constriction of blood vessels.

vein A blood vessel that returns blood to the heart.

ventilation (*ven″tĭ-la′shun*) Breathing; the process of moving air into and out of the lungs.

vertigo (*ver′tĭ-go*) A feeling of movement or loss of equilibrium.

vestibular (*vĕ-stib′yŭ-lar*) **apparatus** The parts of the inner ear, including the semicircular canals, utricle, and saccule, which function to provide a sense of equilibrium.

villi (*vil′i*) Fingerlike folds of the mucosa of the small intestine.

virulent (*vir′yŭ-lent*) Pathogenic, or able to cause disease.

vital capacity The maximum amount of air that can be forcibly expired after a maximal inspiration.

vitamin (*vi′tă-min*) Any of various unrelated organic molecules present in foods that are required in small amounts for normal metabolic function of the body. Vitamins are classified as water-soluble or fat-soluble.

W

white matter The portion of the central nervous system composed primarily of myelinated fiber tracts. This forms the region deep to the cerebral cortex in the brain and the outer portion of the spinal cord.

Z

zygote (*zi′gōt*) A fertilized ovum.

zymogen (*zi′mŏ-jen*) An inactive enzyme that becomes active when part of its structure is removed by another enzyme or by some other means.

Credits

Illustrators

Ernest Beck
Figures 13.13, 20.46b, 20.47, 20.49

Samuel Collins
Figures 15.1, 20.23

Chris Creek
Figures 1.15, 1.20, 10.5, 11.1a, 12.7, 13.23, 13.26, 14.27, 16.1, 16.14, 16.15, 17.1, 17.4a, 17.5, 20.26

Fineline
Figure 20.45

Rob Gordon
Figures 8.5, 8.6, 8.15a, 8.17

Rob Gordon/Tom Waldrop
Figure 10.19

Rolin Graphics
Figures 2.14, 11.8

Illustrious, Inc.
Figures 1.1, 1.2, 1.3, 1.4, 1.6, 2.1, 2.2, 2.3, 2.4, 2.5, 2.6, 2.8, 2.19, 2.20, 2.26, 3.16, 3.19, 3.20, 3.22, 3.23, 3.24, 3.25, 3.26, 3.28, 4.1, 4.2, 4.5, 4.7, 4.8, 4.9, 4.10, 4.11, 4.12, 4.13, 5.1, 5.2, 5.3, 5.5, 5.8, 5.10, 5.12, 5.13, 5.14, 5.16, 5.17, 5.18, 6.4, 6.5, 6.6, 6.7, 6.8, 6.9, 6.11, 6.15, 6.16, 6.17, 6.21, 7.13, 7.16, 7.17, 7.27, 8.21, 9.8, 10.17, 10.21, 10.30b, 10.45, 10.46, 11.4, 11.5, 11.13, 11.14, 11.17, 11.22, 11.15, 13.6, 14.3, 14.5, 14.7, 14.9, 14.10, 14.18, 14.21, 14.25, 15.2, 15.3, 15.4, 15.5, 15.9, 15.11, 15.13, 15.14, 15.17, 15.18b, 15.19, 15.20, 15.21, 15.24, 16.21, 16.24, 16.25, 16.28, 16.29, 16.37, 16.38, 17.11, 17.23, 17.25, 18.28, 18.29, 18.30, 18.31, 19.2, 19.7, 19.9, 19.10, 19.14, 19.16, 19.20, 19.21, 19.22, 19.23, 20.1, 20.4, 20.5, 20.9, 20.13, 20.15, 20.36, 20.50, 20.52, 20.53

J&R Art Services
Figure 14.1

Ruth Krabach
Figures 3.11b-c, 8.1, 8.18, 10.9, 13.34

Rictor Lew
Figures 1.21, 3.17, 7.1, 7.2, 7.5, 7.7, 7.8, 7.18, 7.22, 7.25, 8.4, 8.7, 8.12, 8.13, 8.14, 8.16, 8.19, 8.20, 8.22, 9.1, 9.2, 9.4, 9.5, 9.6, 10.1, 10.12, 10.13, 10.20, 10.23, 10.24, 10.27, 10.28, 10.29, 10.33a, 10.34, 10.40, 10.41, 10.43, 11.10, 11.12, 11.15, 11.26, 11.28, 12.19, 12.20, 13.1, 13.3, 13.7, 13.8, 13.12, 13.19a, 13.24, 13.28b, 13.33, 14.26, 16.5, 16.8, 17.8, 17.13, 17.19, 17.24, 18.10, 18.11, 18.15, 18.17, 20.17b, 20.18, 20.24, 20.38, 20.40, 20.42

Bill Loechel
Figures 17.2, 17.6, 18.3, 18.5, 18.8, 18.21

Steve Moon
Figures 10.10b, 10.35, 11.24, 16.27, 20.51

Diane Nelson
Figures 11.18a, 13.32, 14.16

Mark Nero
Figures 1.5, 2.12, 6.2, 6.3, 6.13, 6.14, 7.3, 7.10, 7.11, 7.21, 7.23, 7.24, 7.28, 10.4, 10.7, 11.6, 11.7, 11.29, 12.12, 12.13, 12.14, 12.15, 13.17, 13.29, 14.4, 14.17, 14.22, 14.29, 15.15, 16.4, 16.17, 16.18, 16.23, 16.30, 16.40, 17.10, 17.14, 17.16, 17.20, 17.21, 17.22, 17.27, 17.28, 18.34, 19.5, 19.6, 19.8, 19.11, 19.13, 20.10, 20.33

Felecia Paras
Figures 1.11, 1.13, 1.14, 3.8, 3.27b, 3.30, 3.31a-e, 3.32b, 3.34, 3.35, 7.4, 7.31, 9.3, 9.7, 10.8, 10.10a, 10.15, 10.16, 14.20, 16.12, 16.32a, 18.4a, 18.9, 20.16, 20.31

Precision Graphics
Figures 17.18, 20.32

Mike Schenk
Figures 10.31, 11.9, 12.11b

Tom Sims/Mike Schenk
Figures 10.42, 14.24, 16.26, 16.31, 20.44a

Waldrop
Figures 11.11, 12.2, 12.8, 12.16, 12.18, 16.22, 18.13, 20.6, 20.21, 20.25, 20.34

Walters & Assoc
Figure 10.14b-d

Line Art

Chapter 1
Figure 1.15: From Kent M. Van De Graaff, *Human Anatomy*, 4th ed. Copyright © 1995 Wm. C. Brown Communications, Inc., Dubuque, Iowa. Reprinted by permission of Times Mirror Higher Education Group, Inc., Dubuque, Iowa. All Rights Reserved.

Chapter 3
Figure 3.10b: From Leland G. Johnson, *Biology*, 2d ed. Copyright © 1987 Wm. C. Brown Communications, Dubuque, Iowa. Reprinted by permission of Times Mirror Higher Education Group, Inc., Dubuque, Iowa. All Rights Reserved; **Figure 3.23:** From Bruce Alberts, et al., *Molecular Biology of the Cell*, 2d ed., figure 5.7, page 206. Copyright © 1989 Garland Publishing Inc., New York, NY. Reprinted by permission; **Figure 3.29:** From Kent M. Van De Graaff, *Human Anatomy*, 4th ed. Copyright © 1995 Wm. C. Brown Communications, Inc., Dubuque, Iowa. Reprinted by permission of Times Mirror Higher Education Group, Inc., Dubuque, Iowa. All Rights Reserved.

Chapter 7
Figure 7.19b: From Leland G. Johnson, *Biology*, 2d ed. Copyright © 1987 Wm. C. Brown Communications, Dubuque, Iowa. Reprinted by permission of Times Mirror Higher Education Group, Inc., Dubuque, Iowa. All Rights Reserved.

Chapter 8
Figure 8.10: From Kent M. Van De Graaff, *Human Anatomy*, 4th ed. Copyright © 1995 Wm. C. Brown Communications, Inc., Dubuque, Iowa. Reprinted by permission of Times Mirror Higher Education Group, Inc., Dubuque, Iowa. All Rights Reserved.

Chapter 10
Table 10.4, Figure 10.36: From Kent M. Van De Graaff, *Human Anatomy*, 4th ed. Copyright © 1995 Wm. C. Brown Communications, Inc., Dubuque, Iowa. Reprinted by permission of Times Mirror Higher Education Group, Inc., Dubuque, Iowa. All Rights Reserved; **Figures 10.18, 10.37a (left), 10.37a (right):** From John W. Hole, Jr., *Human Anatomy and Physiology*, 6th ed. Copyright © 1993 Wm. C. Brown Communications, Dubuque, Iowa. Reprinted by permission of Times Mirror Higher Education Group, Inc., Dubuque, Iowa. All Rights Reserved.

Chapter 11
Figures 11.1a, 11.9: From Kent M. Van De Graaff, *Human Anatomy*, 4th ed. Copyright © 1995 Wm. C. Brown Communications, Inc., Dubuque, Iowa. Reprinted by permission of Times Mirror Higher Education Group, Inc., Dubuque, Iowa. All Rights Reserved; **Figure 11.18a:** From Eldon D. Enger, et al., *Concepts in Biology*, 6th ed. Copyright © 1991 Wm. C. Brown Communications, Inc., Dubuque, Iowa. Reprinted by permission of Times Mirror Higher Education Group, Inc., Dubuque, Iowa. All Rights Reserved; **Figure 11.24:** From John W. Hole, Jr., *Human Anatomy and Physiology*, 6th ed. Copyright © 1993 Wm. C. Brown Communications, Dubuque, Iowa. Reprinted by permission of Times Mirror Higher Education Group, Inc., Dubuque, Iowa. All Rights Reserved.

Chapter 12

Figure 12.2: From John W. Hole, Jr., *Human Anatomy and Physiology*, 6th ed. Copyright © 1993 Wm. C. Brown Communications, Dubuque, Iowa. Reprinted by permission of Times Mirror Higher Education Group, Inc., Dubuque, Iowa. All Rights Reserved.

Chapter 13

Figure 13.1: From John W. Hole, Jr., *Human Anatomy and Physiology*, 6th ed. Copyright © 1993 Wm. C. Brown Communications, Dubuque, Iowa. Reprinted by permission of Times Mirror Higher Education Group, Inc., Dubuque, Iowa. All Rights Reserved; **Figure 13.6:** Adapted from A. Marchand, "Case of the Month, Circulating Anticoagulants: Chasing the Diagnosis" in Diagnostic Medicine, June 1983, page 14. Copyright © 1984 Medical Economics Company, Inc., Oradell, NJ. Used with permission; **Figure 13.32:** From Kent M. Van De Graaff, *Human Anatomy*, 3d ed. Copyright © 1992 Wm. C. Brown Communications, Inc., Dubuque, Iowa. Reprinted by permission of Times Mirror Higher Education Group, Inc., Dubuque, Iowa. All Rights Reserved.

Chapter 14

Figure 14.17: Adapted P. Astrand and K. Rodahl, *Textbook of Work Physiology*, 3d ed. Copyright (c) 1986 McGraw-Hill, Inc., New York, NY. Used by permission of the authors.

Chapter 16

Figures 16.1, 16.14, 16.15: From Kent M. Van De Graaff, *Human Anatomy*, 4th ed. Copyright © 1995 Wm. C. Brown Communications, Inc., Dubuque, Iowa. Reprinted by permission of Times Mirror Higher Education Group, Inc., Dubuque, Iowa. All Rights Reserved.

Chapter 17

Figures 17.2, 17.6: From John W. Hole, Jr., *Human Anatomy and Physiology*, 6th ed. Copyright © 1993 Wm. C. Brown Communications, Dubuque, Iowa. Reprinted by permission of Times Mirror Higher Education Group, Inc., Dubuque, Iowa. All Rights Reserved.

Chapter 18

Figure 18.8: From John W. Hole, Jr., *Human Anatomy and Physiology*, 6th ed. Copyright © 1993 Wm. C. Brown Communications, Dubuque, Iowa. Reprinted by permission of Times Mirror Higher Education Group, Inc., Dubuque, Iowa. All Rights Reserved.

Chapter 20

Figure 20.10: From F. K. Shuttleworth, *Monograph of The Society for Research in Child Development, Inc.*, Chicago, IL. Copyright © 1939 *The Society for Research in Child Development, Inc.* Reprinted by permission; **Figures 20.20a, 20.21, 20.25, 20.34, 20.45:** From John W. Hole, Jr., *Human Anatomy and Physiology*, 6th ed. Copyright © 1993 Wm. C. Brown Communications, Dubuque, Iowa. Reprinted by permission of Times Mirror Higher Education Group, Inc., Dubuque, Iowa. All Rights Reserved; **Figures 20.26, 20.32:** From Kent M. Van De Graaff, *Human Anatomy*, 4th ed. Copyright © 1995 Wm. C. Brown Communications, Inc., Dubuque, Iowa. Reprinted by permission of Times Mirror Higher Education Group, Inc., Dubuque, Iowa. All Rights Reserved.

Photos

Chapter 1

1.7, 1.8, 1.9, 1.10, 1.12, 1.16, 1.17, 1.18b: © Edwin Reschke; **1.19:** © SIU Biomedical Communications/Journalism Services, Inc.

Chapter 2

2.27: Edwin Reschke

Chapter 3

3.3a,b: Courtesy Kwang W. Jeon; **3.4 (All):** M. M. Perry & A. B. Gilbert, *Journal of Cell Science* 39:257–272, 1979.

Courtesy The Company of Biologists Ltd.; **3.5a:** Courtesy Keith R. Porter.; **3.5b:** Courtesy Richard Chao; **3.6a:** Courtesy Dr. Carolyn Chambers; **3.6b:** From R. G. Kessel and R. H. Kardon: *Tissues and Organs: A Text-Atlas of Scanning Electron Microscopy*, W. H. Freeman and Co. © 1979; **3.7:** © K. G. Murti/Visuals Unlimited; **3.9:** Courtesy Richard Chao; **3.10a, 3.11a:** Courtesy Keith R. Porter; **3.12:** Courtesy Richard Chao; **3.13:** Courtesy E. G. Pollack; **3.21:** Courtesy Dr. Alexander Rich; **3.27a:** Courtesy Daniel S. Friend; **3.31a, 3.31b, 3.31c, 3.31d, 3.31e:** © Edwin Reschke; **3.32a:** © David M. Phillips/Visuals Unlimited; **3.33:** CNRI/Science Photo Library/Photo Researchers, Inc.

Chapter 5

5.9: Fernandes-Moran V. M. D. P.H.D.

Chapter 6

6.10: Courtesy Richard Chao

Chapter 7

7.6: H. Webster, from Hubbard, John, *The Vertebrate Peripheral Nervous System*, © Plenum Press, 1974; **7.9:** © Andreas Karschin, Heinz Wassle, and Jutta Schnitzer/Scientific America; **7.14:** Bell et al., *Textbook of Physiology and Biochemistry*, 10th ed. © Churchill Livingstone, Edinburgh; **7.19a:** Gilula, Reverand Steimbach, Nature 335:262–265 © Macmillan Journals Limited; **7.20:** © John Heuser, Washington University, School of Medicine, St. Louis, MO

Chapter 8

8.8a: Dr. R. Llinas/New York University Medical Center; **8.9:** From W. T. Carpenter and R. W. Buchanan, "Schizophrenia" *New England Journal of Medicine*, 330:685, 1994; **8.15b:** Times Mirror Higher Education Group/Karl Rubin, photographer

Chapter 10

10.11: © 1993 Discover Magazine, reprinted with permission; **10.14a:** Courtesy Dean E. Hillman; **10.22:** From *Tissues and Organs: A Text Atlas of Scanning Electron Microscopy* by Richard G. Kessel & Randy H. Kardon. © 1979 by W. H. Freeman & Company. Reprinted with permission; **10.30a:** © Thomas Sims; **10.33b:** P. N. Farnsworth/University of Medicine & Dentistry, New Jersey Medical School; **10.37b:** © Frank Werblin; **10.44:** D. Hubel, American Scientist 67 1979 534

Chapter 11

11.1b: © Edwin Reschke; **11.18b:** © SIU/Nawrocki Stock Photo, Inc.; **11.19:** © Martin M. Rotker; **11.21, 11.23:** © Lester V. Bergman & Associates, Inc.; **11.27:** © Edwin Reschke

Chapter 12

12.3: © Edwin Reschke; **12.4a:** Courtesy International Bio-Medical, Inc.; **12.4b:** Stuart Fox; **12.5a, 12.5b:** Kent M. Van De Graaff; **12.6b:** © John D. Cunningham/Visuals Unlimited; **12.9, 12.10a, 12.10b:** © Dr. H. E. Huxley; **12.10c:** From R. G. Kessel and R. H. Kardon: *Tissues and Organs: A Text-Atlas of Scanning Electron Microscopy*, W. H. Freeman and Company © 1979; **12.11a:** © Dr. H. E. Huxley; **12.25:** Hans Hoppler, *Respiratory Physiology* 44:94 (1981); **12.27:** © Edwin Reschke; **12.28:** Avril V. Somlyo, University of Pennsylvania, Pennsylvania Muscle, Institute

Chapter 13

13.2: Reginald J. Poole/Polaroid International Instant Photomicrography Competition; **13.4:** Stuart Fox; **13.5:** © David Phillips/Visuals Unlimited; **13.9:** © Igaku-Shoin, Ltd.; **13.25:** Courtesy Don W. Fawcett; **13.28a:** © Biophoto Assoc./Science Source/Photo Researchers, Inc.; **13.30a, 13.30b:** © Richard Manard

Chapter 14

14.8: From E. K. Markell and M. Vogue, *Medical Parasitology*, 5th ed, W. B. Saunders; **14.15a, 14.15b:** Donald S. Bain from Hurst et al: The Heart, 5/E © McGraw Hill Book 6, 1982; **14.19a–d:** © Niels A. Lassen, Copenhagen, Denmark

Chapter 15

15.6: © Times Mirror Higher Education Group; **15.7a:** Courtesy Arthur J. Olson, Ph.D., The Scripps Research Institute, Molecular Graphics Laboratory; **15.8a, 15.8b:** From Dr. A. G. Amit, "Three Dimensional Structure of an Antigen Antibody Complex at 2.8 Resolution," *Science* 233:747–753, 15 August 1986, 2 figures © 1986 by AAAS; **15.18a:** From Alan S. Rosenthal, *New England Journal of Medicine* 303:1153, 1980; **15.22a, 15.22b:** © Dr. Andrejs Liepins; **15.23a, 15.23b:** Courtesy of Dr. Noel Rose; **15.25:** SIU Biomed Comm./Custom Medical Stock; **15.26a:** From R. G. Kessel and C. Y. Shih, *Scanning Electron Microscopy in Biology*, © 1976, Springer-Verlag; **15.26b:** © Dr. Jeremy Burgess/SPL/Photo Researchers, Inc.

Chapter 16

16.2: Murray, John F.: The Normal Lung, 2/E. © W. B. Saunders Company 1986; **16.3a, 16.3b:** American Lung Association; **16.5b:** © John Watney Photo Library; **16.6:** © CNRI/Phototake; **16.7:** American Lung Association; **16.9a, 16.9b, 16.10:** Edward C. Vasquez R. T. C.R.T./Dept. of Radiologic Technology, Los Angeles City College; **16.13a, 16.13b:** From J. H. Comroe, Jr., *Physiology of Respiration*, © 1974, Yearbook Medical Publishers, Inc., Chicago; **16.16:** Courtesy of Warren E. Collins, Inc., Braintree, MA; **16.19a:** © 1988, R. Calentine/Visuals Unlimited; **16.19b:** © M. Moore/Visuals Unlimited; **16.35a, 16.35b, 16.35c:** McCurdy, P. R., Sickle-Cell Disease. © Medcom, Inc. 1973

Chapter 17

17.3: © SPL/Photo Researchers, Inc.; **17.4b:** © Biophoto Assoc./Photo Researchers; **17.7:** © F. Spinnelli - D. W. Fawcett/Visuals Unlimited; **17.9:** © Daniel Friend from William Bloom and Don Fawcett, *Textbook of Histology*, 10th, ed, W. B. Saunders, Co.

Chapter 18

18.4b, 18.6: Courtesy of Utah Valley Regional Medical Center, Dept. of Radiation; **18.7:** © Edwin Reschke; **18.12:** © Manfred Kage/Peter Arnold, Inc.; **18.14a, 18.14b:** Keith R. Porter/Alpers D. H. and Seetharan D. *New England Journal of Medicine*, 296/1977–1047; **18.18:** From W. A. Sodeman and T. M. Watson, *Pathologic Physiology*, 6th ed., W. B. Saunders Co., 1969; **18.19:** After Sobotta, from Bloom, W. and Fawcett, D. W.: *A Textbook of Histology*, 10/E © W. B. Saunders Company, 1975; **18.20a:** From *Tissues and Organs: A Text Atlas of Scanning Electron Microscopy*, by Richard G. Kessel & Randy H. Kardon © 1979 by W. H. Freeman & Company. Reprinted with permission; **18.26a:** © Carroll Weiss/Camera M. D. Studios; **18.26b:** © Dr. Sheril D. Burton; **18.27:** © Edwin Reschke

Chapter 19

19.4: Courtesy Dr. Mark Atkinson; **19.15:** © Lester V. Bergman and Associates; **19.17a, 19.17b, 19.17c, 19.17d:** *American Journal of Medicine* 20 (1956) 133; **19.18a, 19.18b:** Bhasker, S. N., ed: Orban's Oral *Histology and Embryology*, 9/E. © C. V. Mosby Company, 1980; **19.19a, 19.19b:** From Raisz, L. G., Dempster, D. W. et al, 1986. *New England Journal of Medicine*. 318 (13): 818

Chapter 20

20.2: March of Dimes; **20.3a, 20.3b, 20.3c, 20.3d:** Williams, R. H.; *A Textbook of Endocrinology*, 6/E © W. B. Saunders Company, 1981; **20.17a:** © Biophoto Associates/Photo Reseachers, Inc.; **20.20b:** © Francis Leroy/SPL/Photo Researchers, Inc.; **20.27a, 20.27b:** © Edwin Reschke; **20.28a, 20.28b:** From R. J. Blandau A *Textbook of Histology*, 10th ed. 1975 W. B. Saunders, Co.; **20.29:** Bloom/Fawcett, *Histology*, 10/E pg. 859, W. B. Saunders, Co; **20.30:** © Dr. Landrum Shettles; **20.35:** © Martin M. Rotker; **20.39:** Lucian Zamboni, from Greep, Roy and Weiss, Leon: *Histology*, 3/E. © McGraw-Hill Book Company, 1973; **20.41:** Dr. David Hill/Center for Reproductive Medicine; **20.43a:** © Petit Format/Nestle/Science Source/Photo Researchers, Inc.; **20.43b:** © Motta and Blerkom/SPL/Photo Researchers, Inc.; **20.43c:** © Petit Format/Nestle/Science Source/Photo Researchers, Inc.; **20.43d:** © Biophoto Assoc./Photo Researchers, Inc.; **20.44b:** © Dr. Landrum B. Shettles

Index

B

Cortical neurons
    classification by stimulus requirements, 263
    complex, 264
    hypercomplex, 264
    simple, 263, 263f–264f
Corticoids, 289
Corticospinal tracts, 197–198, 198f
Corticosteroids, 37, 273, 289. *See also* Steroid hormones
    types of, 289
Corticotropin-releasing hormone, 285, 287t
Cortisol (hydrocortisone), 37f, 289
Cortisone, effects, on immune system, 445
Cotransport, 133, 133f–134f
Coumarin(s), 353, 353t
Countercurrent exchange, 518, 518f
Countercurrent multiplication, 517–518
Countercurrent multiplier system, 516–518, 517f
Coupled reactions, 93f
    adenosine triphosphate, 92
    oxidation-reduction, 92–94
Covalent bonds, 25f, 25–26
    double, 29, 30f
    single, 29, 30f
Cranial nerve(s), 144, 199–200
    abducens (VI) nerve, 194, 200t
    accessory (XI) nerve, 195, 200t
    facial (VII) nerve, 194, 200t, 234
    glossopharyngeal (IX) nerve, 195, 200t, 234
    hypoglossal (XII) nerve, 195, 200t
    oculomotor (III) nerve, 200t
    olfactory (I) nerve, 200t
    optic (II) nerve, 200t
    trigeminal (V) nerve, 194, 200t
    trochlear (IV) nerve, 200t
    vagus (X) nerve, 195, 200t
        parasympathetic fibers in, 213, 214f
        transmission of sensory input to medulla, effects of, 222t
    vestibulocochlear (VIII) nerve, 194–195, 200t
        in neural pathway for hearing, 247
        and vestibular apparatus, 239
Craniosacral division. *See* Parasympathetic nervous system
Creatine kinase, 331
    diseases associated with abnormal plasma concentrations of, 86t
Creatine phosphokinase, 86
    diseases associated with abnormal plasma concentrations of, 86t
Creatinine, 523
Cremasteric reflex, 612
Crenation, 129
Cretinism, 294, 595, 596f
Cross bridge cycles, 316–318, 319f
Cross bridges, 316
Crossed-extensor reflex, 327–329, 329f
Crossing over, 76, 76f
Cryptochidism, 612
Cuboidal epithelial cells, 11
Cumulus oophorus, 632
Cupula, 239, 241f
Curare, 164, 164t, 219
Cushing's syndrome, 290
Cutaneous effectors, 212
Cutaneous receptors, 228, 231t, 231f
Cutaneous sensations, 231–233
    neural pathways for, 232
Cyanide, effects of, 110
Cyclic adenosine monophosphate, 166
    as second messenger
        in actions of epinephrine and glucagon, 594f
        hormonal mechanism of action with, 279–280, 280f
        hormones using, 279t, 279–280
        sequence of events involving, 279t
Cyclic compounds, 30f
Cyclic guanosine monophosphate, 279–280
Cyclin(s), 69–70
Cyclin D, 69
Cyclo-oxygenase, 300
Cysteine, 39f
Cystic duct, 560
Cystic fibrosis, 125
Cystic fibrosis transmembrane conductance regulator, 125
Cytochrome(s), 107–108
Cytochrome 450 enzymes, 559
Cytokines, 299, 441
    regulation of immune system, 442t
Cytokinesis, 71

Cytoplasm, 54–55
    definition of, 48
    division of, 71
    function of, 48t
    structure of, 48t
Cytoplasmic receptor proteins, 278
Cytosine, 59
Cytoskeleton, 54f, 54–55
Cytotrophoblast, 645

D

Dalton's law, 475
    and ventilation and gas exchange, 477t
Decibels (dB), 242
Decidual reaction, 646
Decompression sickness, 480
Decussation, 187
Dedifferentiation, 447
Defecation, 554–555
Defecation reflex, 554
Defense mechanisms, 426–430
Dehydration synthesis, 33, 34f
Dehydrogenases, 85
Delta waves, 186, 187f
Dendrites, 10, 145, 145f
Dendrodendritic synapses, 157, 158f
Denervation hypersensitivity, 209
Dense fibrous connective tissue, 13, 15t
Dense irregular connective tissue, 13, 15f
Dense regular connective tissue, 13, 15f
Dentin, 15, 17f
Dentinal tubules, 15
Deoxyhemoglobin, 487
Deoxyribonucleic acid, 57–59
    nucleotides and sugars in, versus RNA, 60f
    replication, 68, 69f
    structure of, 59, 60f
    synthesis, 68–76
Deoxyribose, 59
Depolarization, 152, 153f
Deprenyl, 166
Dermatitis, contact, 453
Dermis, 13f, 16
Descending tracts, 197t, 197–199
Desensitization, 276–277
Deuteranopia, 259
Deuterium, 24
Diabetes insipidus, 521
Diabetes mellitus, 131, 297, 526, 587, 590
    insulin-dependent (type I), 443, 450, 590–591, 591t
    ketone bodies in, 114
    non-insulin-dependent (type II), 590, 591t, 591–592
Diabetic coma, 593t
Diacylglycerol, 281
Dialysis, 124, 533
Diapedesis, 346, 427, 427f
Diaphragm, 371, 464
Diarrhea, mechanisms of, 555
Diastole, definition of, 358
Dichromats, 259
Diencephalon, 180, 181f, 192–193
Diet
    and cholesterol, 373
    fats in, 35, 373
Diffusion, 124–125, 125f
    rate of, 126
        factors affecting, 126
    simple, 124
    through cell membrane, 125
Digestion, 541
    hydrolysis reactions in, 540, 540f
    in stomach, 548
Digestive system, 19t, 538–571, 541f
    functions of, 540–541
    prostaglandin regulation in, 301
    regions of, 541
    regulation of, 563–566
    structures of, 541
Digitalis, 418
Dihydrotestosterone, 613, 615f
Dihydroxyphenylalanine, 90

1,25-Dihydroxyvitamin D$_3$, 599–601
    actions of, 600–601
    production of, 599–600, 601f, 602, 602f
Diiodotyrosine, 292
2,3-Diphosphoglyceric acid, effect on oxygen transport, 489–490
Disaccharides, 32
Distal convoluted tubule, 508
    transport properties of, 519t
Disulfide bonds, 41
Disynaptic reflex, 327, 328f
Diuretics
    and loss of potassium, 529, 533
    mechanisms of, 532t, 532–533
DNA. *See* Deoxyribonucleic acid
DNA polymerase, 68
Dominant retinitis pigmentosa, 256
Dopamine, 151, 166
    as neurotransmitter, 166–167
Dopaminergic neurons, 166
Dormitory effect, 288
Dorsal horns, of spinal cord gray matter, 196
Dorsal root, 200–201
Dorsal root ganglion, 201
Double helix, 59, 60f
Downregulation, 276–277
Drug trials, 5
Dual innervation
    antagonistic effects, 220–221
    complementary and cooperative effects, 221
    organs with, 220–221
    organs without, 221
Duchenne muscular dystrophy, 310
Duct of Hensen, 240
Ducts, 12
Ductus arteriosus, 361, 361f
Ductus deferens, 626
Duodenum, 549, 550f
    histology of, 550f
Dust mites, 452, 453f
Dwarfism, 597
Dynamic constancy, 6
Dynorphin, 170
Dysmenorrhea, 639
Dyspnea, 473
    definition of, 473t

E

Ear(s), 241–247, 242f
    inner, 244
        labyrinths of, 238f
    middle, 242–243, 243f
    outer, 242, 242f
Eardrum. *See* Tympanic membrane
Eating, 583–584
    effects, on metabolic rate, 578
    regulation of, 583–584
Eccrine sweat glands, 12
Ectoderm, 180, 181f, 645
Ectopic pacemaker(s), 362, 375
Edema, 127
    causes of, 394, 394t
    definition of, 394
Effectors, 6
    antagonistic, 6–7
Efferent, definition of, 324
Efferent arteriole, 508, 509f
Efferent neurons. *See* Motor neuron(s)
Ejaculation
    definition of, 627
    physiology of, 627
Ejaculatory duct, 613
Ejection fraction, 389
Ejection phase, of cardiac cycle, 358
Elastase, in digestion of proteins, 568
Elastic connective tissue, 15t
Elasticity
    definition of, 467
    and ventilation, 467
Electrical defibrillation, 375

Nitric oxide, 170
    formation of, 443
    medical uses of, 170
    as neurotransmitter, 170, 219–220
    as paracrine regulator, 299
        of blood flow, 400
Nitric oxide synthetase, 443
Nitrogen, partial pressure, in inspired versus alveolar air, 476t
Nitrogen atom, 24t
Nitrogen balance, 114
Nitrogen narcosis, 480
Nitrogenous bases, 58, 59f
Nitroglycerin, 105
Nocireceptors, 228
Nodes of Ranvier, 148, 156
Noncatalyzed reaction, 85f
Nonkeratinized membrane, 12
Nonpolar molecules, 25–26
Nonself antigens, 445
Nonshivering thermogenesis, 221
Nonsteroidal antiinflammatory drugs, 301
Nonvolatile acids, 354
Noradrenergic noncholinergic fibers, 219
Norepinephrine, 167, 212
    effects of, 290, 291t
    as neurotransmitter, 167–168, 168f
    structure of, 218f
    in synaptic transmission, 216
Norplant, 639
Nuclear bag fibers, 324
Nuclear chain fibers, 324
Nuclear envelope, 57, 58f
Nuclear membrane
    function of, 48t
    structure of, 48t
Nuclear pore(s), 57, 58f
Nuclear pore complexes, 57
Nucleic acids, 58–60
Nucleolus, 58, 58f
    function of, 48t
    structure of, 48t
Nucleosomes, 59, 60f
Nucleotides, 57–58
    structure of, 58f
Nucleus/nuclei, 24, 58f
    cerebral, 182
    definition of, 48
    nervous, 144
        definition of, 145t, 186
Nutritional requirements, 578–581, 597
    anabolic, 578–581
Nystagmus, 240

Oocyte(s)
    primary, 630, 633f
    secondary, 631, 633f, 642f
Oogenesis, 630, 634, 634f
Oogonia, 611
Ophthalmia, sympathetic, 449
Opioids, 169
    endogenous, 169–170
Opsin, 256
Opsonization, 432
Optic chiasma, 260
Optic disc, 251, 252f
Optic (II) nerve, 200t
Optic radiation, 261f, 263
Optic tectum, 261
Oral cavity, 541, 541f
Oral contraceptives, 276
    actions of, 639
Oral glucose tolerance test, 586, 587f
    in diabetic and prediabetic individuals, 592f
    in hypoglycemic individual, 593f
Oral rehydration therapy, 134
Orbital, 24
Organelles, 49f, 54–57
    definition of, 48
Organic acids, 30, 31f
Organic molecules, 29–31, 31f
Organ of Corti, 244–247, 246f
Organs, 9, 16–17
Organ systems, 19t
Origin, 308
Osmolality, 128, 128f
    measurement of, 128
Osmoreceptors, 130, 395
Osmosis, 33, 126f, 126–129, 127f
    definition of, 124
    effect of ionization on, 128, 129f
    in loop of Henle, 517
Osmotically active solutes, 127
Osmotic pressure, 126–127, 127f
Ossicles, middle-ear, 242
Osteitis fibrosa cystica, 601
Osteoblasts, 15, 598
Osteoclasts, 598, 599f
Osteocytes, 15
Osteomalacia, 601
Osteoporosis, 600, 600f
Otitis media, 243
Otolith membrane, 238
Otolith organ(s), 237, 237f, 240f
Otosclerosis, 243
Oval window, 242
Ovarian cycle, 630–632, 635–638, 637f, 643f
    endocrine control of, 638f
Ovarian follicles, 298, 612
Ovary(ies), 13, 629, 630f
    development of, 611–612, 612f
    endocrine function of, 273t
    follicular phase, 635–636
    hormones of, 298
    luteal phase, 635, 637–638
    pituitary's influence on, 634
    structure of, 635f
Ovulation, 287, 629, 632–634, 634f, 636
    prevention of, 639
    regulation of, hormonal, 634
Ovum, 298
    fertilized, development of, 644f
Oxidation, 92, 94
β-Oxidation, 113, 113f
Oxidation-reduction, 95f
Oxidative deamination, 115f, 115–116
Oxidative phosphorylation, 107–110, 108f–109f
Oxidizing agent, 94
Oxygen
    in electron transport, 110
    hyperbaric, 480
    partial pressure
        altitude's effects on, 475–476, 476t
        calculation of, 475–477
        effect on loading and unloading reactions, 487
        in inspired versus alveolar air, 476t
        as result of gas exchange, 179f
        significance of, 478
    unloading percentage for, derivation of, 487–488

Oxygen atom, 24t
Oxygen debt, 117, 331
Oxygen electrode, blood gas measurement with, 476–477, 478f
Oxygen toxicity, 480
Oxygen transport
    2,3-diphosphoglyceric acid and, 489–490
    factors affecting, 488–489, 489t
Oxygen utilization, 460
Oxyhemoglobin, 486–487
Oxyhemoglobin dissociation curve, 487–488
    position of, factors affecting, 490t
Oxytocin, 193, 273t, 284, 650
    action of, 650
    and breastfeeding, 653
    origin of, 285

P

p53, 69–70
Pacemaker(s), 335
    artificial, 377
    ectopic, 362, 375
    sinoatrial node, 361–362
Pacemaker potential(s)
    rhythm set by, 389f
    in sinoatrial node, 361–362, 362f
Pacesetter potentials, 552, 552f
Pacinian corpuscle(s), 17, 19f, 229f, 230–232
Paget's disease, 603
Pain, blockers, 170
Pancreas, 12, 295–298, 296f, 541, 541f, 561–562, 562f
    endocrine function of, 561
    as endocrine gland, 296
    exocrine function of, 561
    as exocrine gland, 296
    islets. See Islets of Langerhans
Pancreatic amylase
    characteristics of, 567t
    in digestion of carbohydrates, 567, 567f
Pancreatic juice, 12–13, 561–562
    regulation of, 565, 566t
Pancreatic lipase
    characteristics of, 567t
    in digestion of lipids, 568, 569f
Pancreatitis, acute, 562
Pantothenic acid, safe and adequate daily intake, 582t
Papez circuit, 189
Papillary muscles, 357
Para-aminohippuric acid, clearance of, 524, 525f
Paracrine regulation, 299–301
Paracrine regulators, 299, 299t
Paradoxical cold, 229
Parafollicular cells, of thyroid, 292
Paralysis agitans, 330
Paraplegia, 330
Parasternal intercostal muscles, 470, 470f
Parasympathetic nervous system, 209, 212–214, 215f, 215t
    adrenergic effects, 220t
    cholinergic effects, 220t
    and effector organs, 217t
    functions of, 213–214, 216
    structure of, 212–213
        versus sympathetic, 216t
Parathyroid glands, 295, 295f
    endocrine function of, 273t
    location of, 295
Parathyroid hormone, 273t
    actions of, 295, 296f, 599, 602f
    regulation of, 296f
    secretion of, 599
Paraventricular nucleus/nuclei, of hypothalamus, 193, 194f, 285
Paravertebral ganglia, 210
    sympathetic chain of, 210, 210f
Parietal cells, 545
Parietal lobe, 183, 184f
    functions of, 184t
Parietal pleural membranes, 464
Parkinson's disease, 151, 166–167, 187, 330
    resting tremor in, 195

T

W

Water, 24, 26, 26f
    in body, distribution of, 392, 392f
    obligatory loss, 514
    partial pressure of, in inspired versus alveolar air, 476t
    properties of, 28
    reabsorption of, 514–521
        in proximal tubule, 514–516, 515f
    transport of, 514
Weight loss, 578, 584
Wernicke's aphasia, 188
Wernicke's area, 188–189, 190f
White matter, 148, 182
Wolffian (mesonephric) ducts, 612

Z

Z line(s), 310, 315
Zollinger-Ellison syndrome, 548
Zona pellucida, 632
Zonular fibers, 251
    in accommodation, 253, 253f
Zygote, 610, 642f
Zymogens, 561